S0-CAH-287

THE ALKALOIDS

Studies in Organic Chemistry

Executive Editor

Paul G. Gassman

Professor and Chairman
Department of Chemistry
University of Minnesota
Minneapolis, Minnesota

Volume 1 Organoboranes in Organic Synthesis
Gordon M. L. Cragg

Volume 2 Molecular Orbital Methods in Organic Chemistry HMO and PMO: An Introduction
William B. Smith

Volume 3 Highlights of Organic Chemistry
William J. le Noble

Volume 4 Physical Organic Chemistry: The Fundamental Concepts
Calvin D. Ritchie

Volume 5 Adamantane: The Chemistry of Diamond Molecules
Raymond C. Fort, Jr.

Volume 6 Principles of Organic Stereochemistry
Bernard Testa

Volume 7 The Alkaloids: The Fundamental Chemistry–A Biogenetic Approach
David R. Dalton

Other Volumes in Preparation

THE ALKALOIDS

The Fundamental Chemistry

A Biogenetic Approach

DAVID R. DALTON

Temple University
Philadelphia, Pennsylvania

Marcel Dekker, Inc. New York and Basel

COPYRIGHT © 1979 BY MARCEL DEKKER, INC. ALL RIGHTS RESERVED

Neither this book nor any part may be reproduced or transmitted in any form or by any means, electronic or mechanical, including photocopying, microfilming, and recording, or by any information storage and retrieval system, without permission in writing from the publisher.

MARCEL DEKKER, INC.
270 Madison Avenue, New York, New York 10016

Current printing (last digit):
10 9 8 7 6 5 4 3 2 1

PRINTED IN THE UNITED STATES OF AMERICA

Library of Congress Cataloging in Publication Data

Dalton, David R
The alkaloids.

(Studies in organic chemistry ; v. 7)
Includes bibliographical references and indexes.
1. Alkaloids. I. Title
QD421.D16 547'.72 79-4538
ISBN 0-8247-6788-8

QD421
D16
1979
CHEM

To my family

ותוצא הארץ דשא עשב מזריע זרע למינהו ועץ עשה־פרי אשר זרעו־בו למינהו וירא אלהים כי־טוב: ויהי־ערב ויהי־בקר יום שלישי:

בראשית

And the earth brought forth grass, herb yielding seed after its kind and tree bearing fruit, wherein is the seed thereof, after its kind; and God saw that it was good. And there was evening and there was morning, a third day.

Genesis

ACKNOWLEDGMENTS

The development of this book has been aided in no small measure by several graduate classes at Temple University who worked with various stages of manuscript copy as text. I am grateful to those students for helping improve the work. Additionally, valuable in-depth critique on different portions of the text during its development has been received from my wife, Cecile, and my friends and colleagues, Angeline Cardis, Michael Cava, Edward Leete, Anita Lewin, Ian Scott, Ian Spenser, John Tierney, and Kathleen Wert. The final stages of manuscript preparation were aided by Edgar Howard, Jr. who provided help with the indexes, and the sterling efforts of Gloria Goss who typed the entire work in an incredibly short time.

Despite our combined efforts, the careful reader will certainly find some mistakes: these are mine alone.

CONTENTS

PROLOGUE

A large volume of experimental in vivo, or biosynthetic work has been carried out on secondary plant metabolites, a group of vaguely defined compounds.* For the most part, the work on the secondary metabolites has sought to link the more or less well-understood pathways and relationships of the primary metabolism of amino acids, proteins, carbohydrates, etc., with the often elaborate molecular framework of their secondary derivatives.

In many cases, the biosynthetic work on secondary metabolites has followed in the footsteps of speculations, based upon visual dissection of these molecules into smaller fragments, which were, on rational chemical grounds, derivable from the intermediates of primary metabolism. These biogenetic hypotheses, although occasionally misleading, nevertheless provided insight into structural relationships present in the compounds analyzed, and their successes have promoted additional experimental work, as well as further speculations.

The current level of biogenetic perception regarding the overall production of many different groups of secondary metabolites is quite high. Indeed, in a few cases, the intimate details of the actual biosynthetic parallel to the biogenetic speculation have been made clear. Thus, while a broad brush of hypothesis may be used in generalizing concepts which formulate entire families of compounds, from time to time, the fine detail of experimental reality necessary for clarity can be painted in.

Most of the reactions utilized to build the primary metabolites and then combine them to form the secondary metabolites are the fundamental reactions of organic chemistry, and although it is true that they are mediated by enzymes, they are nevertheless reactions with which the organic chemist is familiar. Indeed, herein lies the major strength of the biogenetic approach to the rationalization of structure: using a very few building blocks known to be present, and some simple reactions to cement them, different structures can be created, from the very simple to the most complex. The elaborate structural edifices commonplace among, for example, secondary plant metabolites called alkaloids, with which this book deals, are thus seen as the logical sum of their components rather than a haphazard accumulation of bonds and atoms.

Some of the biogenetic relationships and events depicted, which will be used to build alkaloids and other required materials, may not be directly analogous or, perhaps, entirely relevant to events which actually occur. Nevertheless, such "speculative" relationships are

---

*The problem in the definition arises in that these chemical entities which are present in the living systems, and which are often accumulated in quantity, are generally presumed as not being required, in any obvious way, for life. As we learn more, the definition may change. These materials, which are usually species-specific, occasionally include some required (primary) metabolites (which can also accumulate for further elaboration), and some of the biosynthetic work on secondary metabolites therefore overlaps that of these primary metabolites.

rooted in biosynthetic fact and, aside from having provided direction to some laboratory work, help in understanding the interweaving of a few precursors into the molecular fabric.

The approach that will be used throughout this book will be to discuss the families of these bases from the point of view of biogenesis. We shall start by examining the reactions which can be used to create and combine some necessary amino acids and other simple plant constituents* and then proceed to the products of primary metabolism. This will be followed by—and the order is one of personal preference—examination of some of the families of alkaloids and an exposition detailing how, using, for the most part, the same few reactions already discussed, the fragments can be combined and modified to produce the plant bases.

Within each of the families so created, a few typical members will be considered, some of the chemistry which originally led to elucidation of the structures of the examples chosen will be discussed, rearrangements—often first discovered in the very systems shown and then applied to simpler cases—will be elaborated upon, and occasionally, spectroscopic results which were utilized to confirm or lead in the struggle to define the structure will be brought out. Finally, for the specific compounds chosen as typical, a synthesis, where it is available, is provided so that some feeling for the labor involved in laboratory preparation is gained and a variety of useful chemical transformations illustrated.

It is clear, I hope, that a work of this size, which is largely based upon a one-semester series of lectures, cannot be encyclopedic, and many interesting facets of alkaloid chemistry and biosynthesis have necessarily been left out. However, it is my intent that the material provided whet the appetite for further exploration and thought on the part of the reader. Thus, to those expert in the art, whose areas of interest have been mentioned only briefly, oversimplified, or ignored, I offer apologies, and hope they will recognize that my own interests and tastes have been given precedence and that this, and this alone, accounts for the choices made.

David R. Dalton

---

*Many related and very important plant constituents will be ignored or mentioned only briefly in passing because they are not obviously relevant to the alkaloids to be discussed.

THE ALKALOIDS

Part 1

AN INTRODUCTION

1

THE REACTIONS

> The time has come, the walrus said,
> to talk of many things
> Of shoes and ships and sealing wax,
> of cabbages and kings . . . .
>
> Lewis Carroll, The Walrus and the Carpenter

The reactions to be discussed here are just those few types required for explication of the amino acids and their elaboration, in a biogenetic sense, into alkaloids. While each reaction with each individual substrate is, of course, different, and while each reaction is mediated by its own enzyme or enzymes and appropriate cofactor or cofactors, there is a similarity between reactions and between in vivo and in vitro experiences. The reactions we will need are broadly grouped into: (a) oxidation-reduction; (b) carbon-carbon bond forming and breaking; and (c) carbon-heteroatom bond forming and breaking.

I. OXIDATION AND REDUCTION

The first two readily enzyme-mediated reactions with which we shall concern ourselves are oxidation and reduction. The variety of oxidations (and reductions) known to organic chemists is legion but those with which we are concerned rather limited. Typically, these will be the oxidative conversion of an alcohol to an aldehyde (or ketone) and an aldehyde to a carboxylic acid (and the reductive reverse; Equation 1.1). We readily appreciate that these reactions

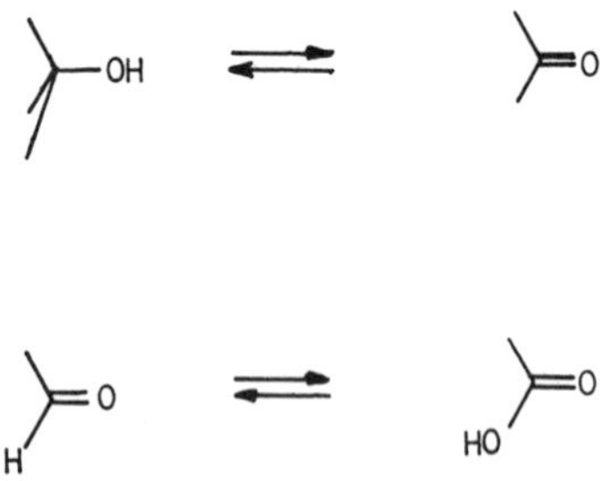

(Eq. 1.1)

are the two edges of the same sword since, for example, in the Meerwein-Ponndorf-Oppenauer-Verley oxidation-reduction system with, e.g., aluminum isopropoxide, it is only a change in solvent which permits the reaction of the substrate to go either to the oxidized or to the reduced side.

In the same way, a change from the reduced to the oxidized form of the enzyme, for example, the reduced and oxidized forms of nicotinamide adenine dinucleotide (see below, NADH and NAD^+, respectively), which serves to add or remove hydrogen, accomplishes the same result, and the reaction is predictable on rational, mechanistic grounds. Therefore, introduction of an oxygen in any oxidation state permits utilization of all its corresponding oxidized and reduced forms within structural limits imposed by the system itself, while the function of the enzyme is to catalyze the specific reaction, usually with fixed stereochemistry, relative to others which might equally well have occurred. In this regard, if a portion of the enzyme (i.e., the active site) is known to participate, we may invoke it and write the corresponding intermediate.

Now, in addition to the oxidation-reduction relationship of alcohols, aldehydes (ketones), and carboxylic acids, there are other redox systems which are common . Among these, we recognize (a) the conversion of amines to imines and nitriles (and the reverse; Equation 1.2)

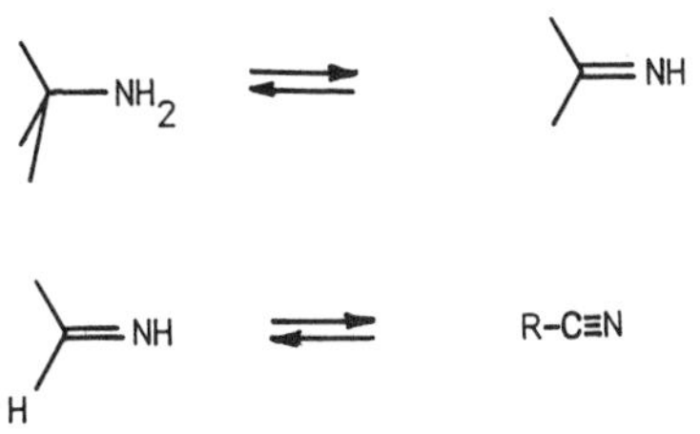

NH_2 ⇌ NH

NH ⇌ R-C≡N
H

(Eq. 1.2)

as being the nitrogen equivalent of the alcohol, aldehyde (ketone), carboxylic and oxidation-reduction system; (b) the epoxidation (oxygenation) of carbon-carbon double bonds (Equation 1.3); (c) the oxidation of phenols to phenoxy radicals (Equation 1.4), in which case the resonance structures will give a clear picture as to which sites on the aromatic ring might be later utilized for carbon-carbon bond formation; and (d) the direct insertion of an oxygen into an activated (e.g., allylic, etc.) position (Equation 1.5).

(Eq. 1.3)

OH ⇌ O• ↔ O ↔ O

(Eq. 1.4)

OH

(Eq. 1.5)

Finally, an uncommon organic oxidation reaction must be considered. This reaction is the introduction of a hydroxyl group onto an unoxidized aromatic ring (see Section III.C; [1, 2]). The reaction is known to be enzyme mediated and is assumed to proceed via an epoxide intermediate in which the hydroxyl replaces a hydrogen which migrates to an adjacent position (Equation 1.6). Subsequent hydroxyl group introduction although formally an oxidation may be considered either oxidative or by substitution, depending upon whether the hydrogen lost leaves as a proton or as a hydride ion.

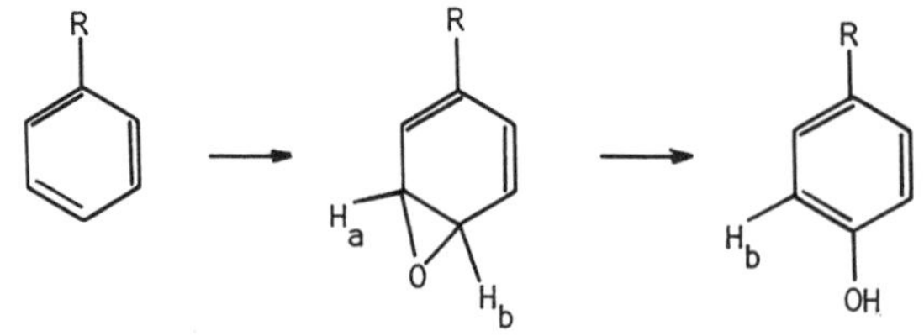

(Eq. 1.6)

II. CARBON-CARBON BOND FORMING AND BREAKING

A. Spontaneous

As is well recognized in certain activated molecules, when the pH and temperature are properly adjusted within broad but reasonable limits, the reactions herein called "spontaneous" will occur. Since these reactions proceed very readily in vitro we accept that the same reactions could reasonably occur in vivo, with, of course, enzyme mediation to further lower whatever barrier might be present and direct us to the right product [3]. In this group we place the aldol and retroaldol reactions (Equation 1.7),* irreversible decarboxylation of β-keto acids (Equation 1.8), and certain electrocyclic reactions (Equation 1.9).

(Eq. 1.7)

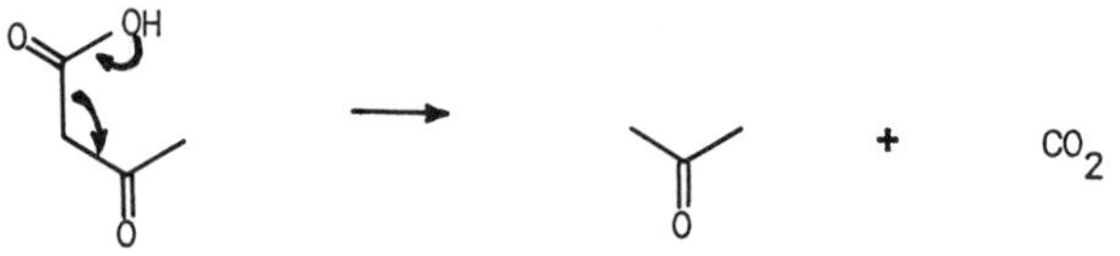

(Eq. 1.8)

(Eq. 1.9)

*The generalized "aldol" reaction has been masked by organic chemists behind a variety of names which are completely substrate and condition dependent. Here, regardless of the substitution at a carbonyl group, all reactions involving carbanions α to the carbonyl attacking a second carbonyl species, again regardless of the substitution and whether dehydration ultimately occurs, will be called "aldol" reactions, and the reverse, again, regardless of the substitution pattern, "retroaldol" reactions. The pedant will, not without some cause, object that these carbon-carbon bond-forming reactions should retain the distinguishing names which allow others to immediately recognize different substitution patterns and conditions of reaction, but there is a certain similarity of pathways which, in the author's opinion, given the sophistication expected here, is overriding.

B. Nonspontaneous

In this category are placed only two carbon-carbon bond-forming reactions. Here, having noted that the in vitro reaction is more difficult to carry out than in the cases called "spontaneous," we assume the parallel true in vivo. Clearly, enzyme mediation is required in both spontaneous and "nonspontaneous" systems, but as we shall see later, such an artificial construct will prove useful. These reactions are (a) alkylation at an activated position (Equation 1.10) and (b) phenyl-phenyl coupling to biphenyl derivatives (Equation 1.11). Finally,

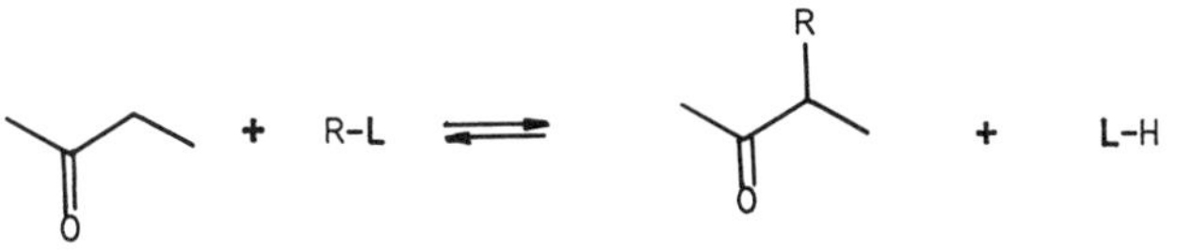

+ R-L ⇌ + L-H (Eq. 1.10)

L = leaving group

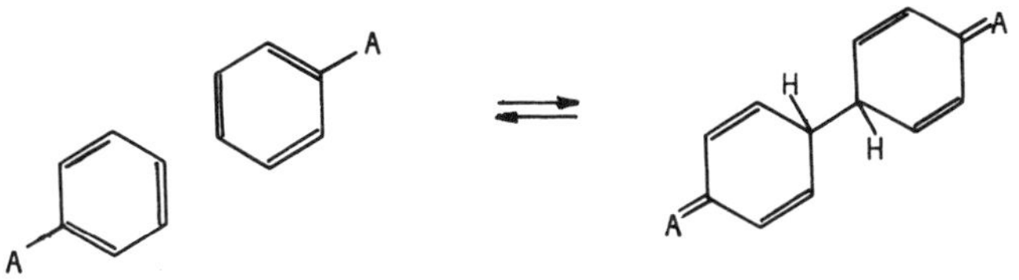

(Eq. 1.11)

A = activating group

both spontaneous and nonspontaneous reactions shown here often occur in a vinylogous sense. That is, a carbon-carbon double bond may be interposed between the seat of the reaction and the leaving group or activating site. Thus, the vinylogous analog of the aldol reaction is shown (Equation 1.12) as well as the vinylogous analog of the alkylation reaction (Equation 1.13).

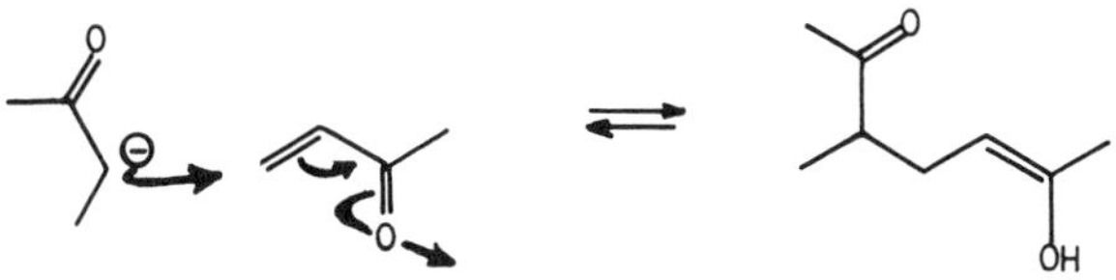

(Eq. 1.12)

(Eq. 1.13)

III. CARBON-HETEROATOM BOND FORMING AND BREAKING

A. Acylation, Alkylation, and Arylation

These reactions are formally analogous to the reactions already shown for carbon-carbon bond formation and breaking but may be even more facile. This broad category, therefore, includes alkylation (acylation, arylation) on oxygen (Equation 1.14) and nitrogen (Equation

—OH + R-L ⇌ —OR + L-H (Eq. 1.14)

L = leaving group

1.15), as well as the special cases of O—N acyl migration (Equation 1.16) and lactone (Equation 1.17) and lactam (Equation 1.18) formation. For the last two reactions (i.e., lactone and lactam formation) we shall, in the case of medium ring compounds, consider the open and closed forms equivalent if the geometry of the particular system permits it.

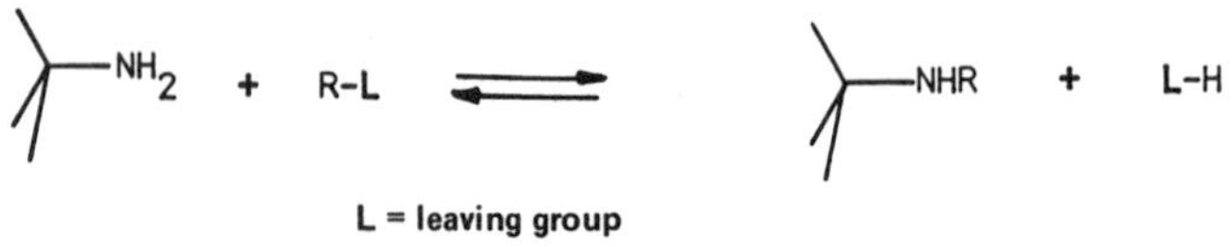

$$\text{—NH}_2 + \text{R-L} \rightleftharpoons \text{—NHR} + \text{L-H}$$ (Eq. 1.15)

L = leaving group

(Eq. 1.16)

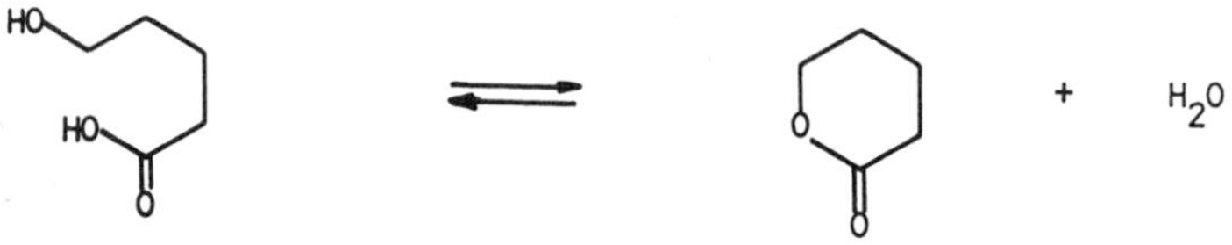

(Eq. 1.17)

(Eq. 1.18)

B. Addition-Elimination Reactions

In the systems we will be considering, most addition and elimination reactions are acknowledged to be, at least in principle, microscopically reversible, and will involve only gain or loss of water or ammonia. Nevertheless, the proclivity any system may possess to undergo either addition or elimination in an overall syn (or cis or Z or suprafacial) or anti (trans or E or antarafacial) sense in solution may not necessarily apply in the enzyme-mediated reaction. Second, it is well appreciated by organic chemists that simple alcohols and amines do not readily undergo elimination but they must be converted to labile derivatives (i.e., esters of strong acids, quaternary amines, etc.) to raise the ground (and/or lower the transition) state for the elimination reaction. Thus, phosphorylation of alcohols (to generate esters of phosphoric acid) and quaternization of amines, often simply through protonation, will be necessary.* These reactions are shown in Equations 1.19 and 1.20.†

---

*Elimination in ammonium ions (e.g., Equation 1.20) does not, of course, generally occur in solution to any appreciable extent.

†Throughout this text, the phosphate ester will be denoted by the symbol (P) attached to the oxygen to which it is esterified. The advantage (showing that it is indeed a good leaving group, in part through electron delocalization within the anion, to drawing it out is overshadowed by the additional effort required and the minor strain it places on the memory of the reader.

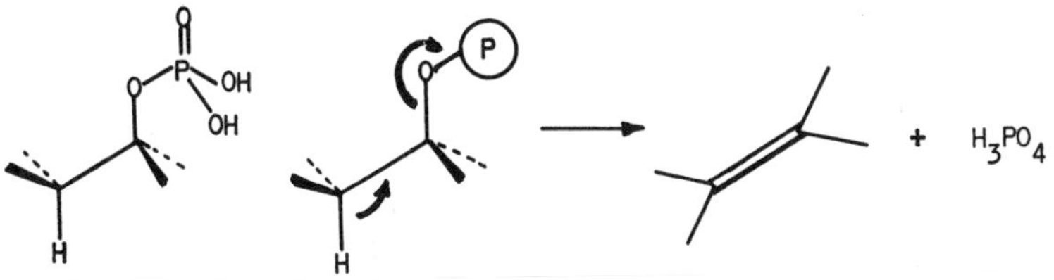

(Eq. 1.19)

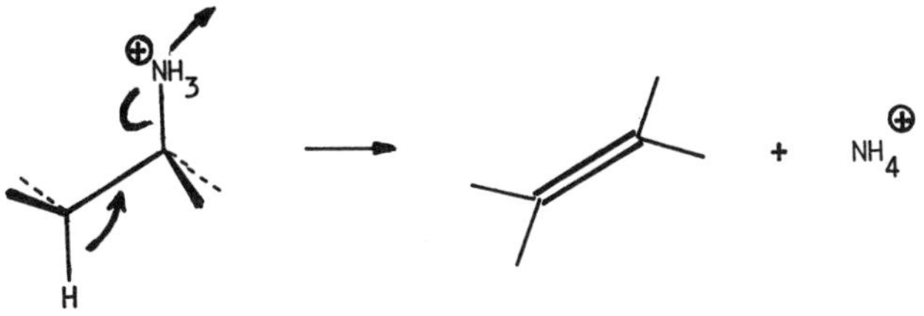

(Eq. 1.20)

C. Substitution

For the most part, the substitution reactions with which we shall deal are (a) the replacement of oxygen by nitrogen and vice versa (Equation 1.21) and (b) the replacement of hydrogen, particularly on an aromatic nucleus, by oxygen (Equation 1.22). In the first reaction, it will

(Eq. 1.21)

(Eq. 1.22)

often be found that a simple (i.e., S_N1 or S_N2) displacement which might typically be visualized does not apply. Instead, substitution will involve initial oxidation of, for example, an amine to an imine, followed by addition of water to the imine and elimination of ammonia from the alcohol-amine thus formed to an aldehyde or ketone. Reduction to the alcohol follows (Scheme 1.1). In the second reaction, the oxygen which is substituted for a hydrogen is invariably introduced para to an alkyl substituent or, if that position is already occupied by an oxygen, ortho the oxygen that is already present.*

*It can be argued that this is really an oxidation reaction since oxygen is being substituted for hydrogen, but it is not clear whether the hydrogen being replaced is lost as a hydride or a proton. There is some evidence that the hydrogen being replaced is capable of migration to an adjacent position, and indeed, this reaction might proceed either via the oxygen already present attacking that position, forming an epoxide, which then undergoes ring opening to the catechol [1] or by way of some other oxygen-arene complex [2].

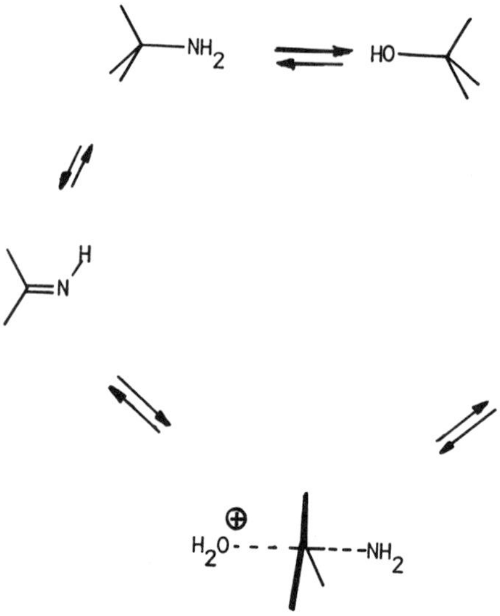

NH_2 HO H O H_2O NH_2 N H

SCHEME 1.1 The conversion of an amine to an alcohol. The reaction is one of net substitution.

D. Electrocyclic Reactions

As was the case with carbon-carbon forming reactions, certain electrocyclic processes can be envisioned where heteroatoms are involved. Whether these reactions occur as indicated (Equation 1.23) without involvement of an enzyme surface holding the fragments in the appropriate positions is doubtful; however, the classification of this form of reaction as being biogenetically useful and chemically viable does not depend, as previously noted, on enzyme mediation.

O O (Eq. 1.23)

2

THE REACTANTS

To everything there is a season,
and a time to every purpose under heaven. . . .

Ecclesiastes 3:1

In this chapter we shall briefly consider from a chemical point of view some of the processes required to convert carbon dioxide into the fragments necessary for alkaloid elaboration. The intermediates between carbon dioxide and the alkaloids will be looked upon as reasonable products derived from each other as part of ongoing cycles through simple chemical transformations. To do this, mediation by enzyme prosthetic groups, where known, will be invoked in a rational, chemical sense. Of course, the pathways thus created may actually be far from correct or, indeed, simply wrong, but their service as a rationale for what has occurred ameliorates the anguish of the error.

I. THE CARBON BACKBONE

A. The Origin of 3-Phosphoglyceric Acid

In the beginning there was light and the photosynthetic processes provided the building blocks necessary for plant metabolism (Equation 2.1). We will not concern ourselves here with how the energy transfers from the sunlight to the molecules that power the organic chemistry performed by the plant; rather, we will start with the absorption of carbon dioxide.*

$$nCO_2 + nH_2O + \text{light energy } (h\nu) = nO_2 + n(112 \text{ kcal}) + (H_2CO)_n \qquad \text{(Eq. 2.1)}$$

---

*The reader will realize that each of the steps utilized in reactions given below is known to be associated with a specific enzyme or group of enzymes. In general, no reference will be made to the specific enzyme (even where it is known) unless details of the involvement of the prosthetic group, where reasonable on the grounds of mechanistic organic chemistry, can be invoked to explain what has occurred. The General Reading section found at the end of this part will direct the interested reader to more complete information on the material presented.

A word of caution is in order here for those unfamiliar with cycles. By the nature of a "cycle" one cannot put several pieces together, de novo, to begin, for then you would have a beginning, which a cycle cannot have. Instead, the system must be entered at a convenient place, tangentially, by combining a fragment with one already present in the cycle and recognizing that the fragment added will become part of the cycle and another fragment withdrawn

Carbon dioxide combines with ribulose-1,5-diphosphate (2.1) in an aldol-type process, and apparently, the resultant unstable product undergoes a retroaldol reaction to two equivalents of 3-phosphoglyceric acid (2.2; Scheme 2.1). Then, in the second step of the cycle,

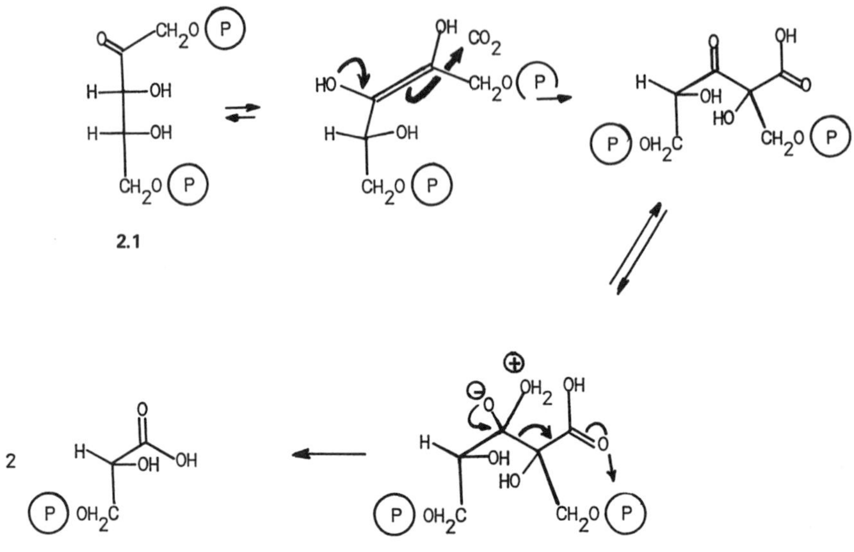

SCHEME 2.1 The incorporation of CO_2 into a carbon cycle, the reaction of CO_2 with ribulose-1,5-diphosphate, and subsequent fragmentation of the adduct to 3-phosphoglyceric acid.

the 3-phosphoglyceric acid (2.2), very probably after being phosphorylated a second time to 1,3-diphosphoglyceric acid (2.3) for activation, is reduced to 3-phosphoglyceraldehyde (2.4), Equation 2.2. From this point in the cycle, we start to build again toward ribulose-1,5-diphosphate (2.1).

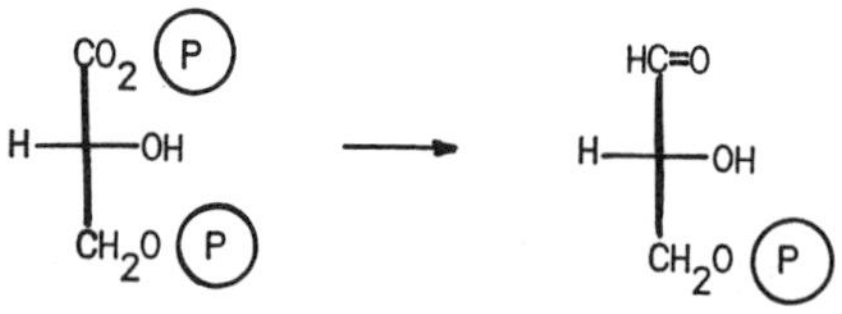

(Eq. 2.2)

It will be noted that we have generated a ubiquitous three-carbon fragment (3-phosphoglyceraldehyde, 2.4) for use later. We can recognize that in the generalized α-hydroxycarbonyl system, either of the carbons involved may be in the oxidized or reduced form through

---

so that the cycle remains balanced. Turning the cycle is work and costs energy, ultimately derived in this case from the absorption of sunlight, and each fragment present must be explained in terms of its fellows in the cycle.

the common enol (Equation 2.3). Thus, it comes as no surprise that 3-phosphoglyceraldehyde (2.4) will equilibrate with dihydroxyacetone monophosphate (2.5; Equation 2.4). While this

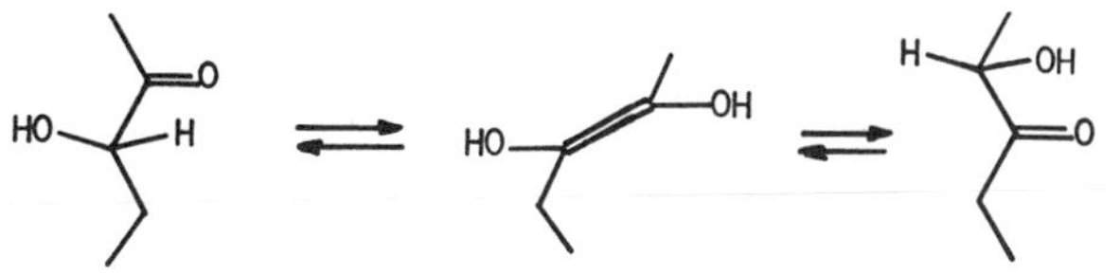

(Eq. 2.3)

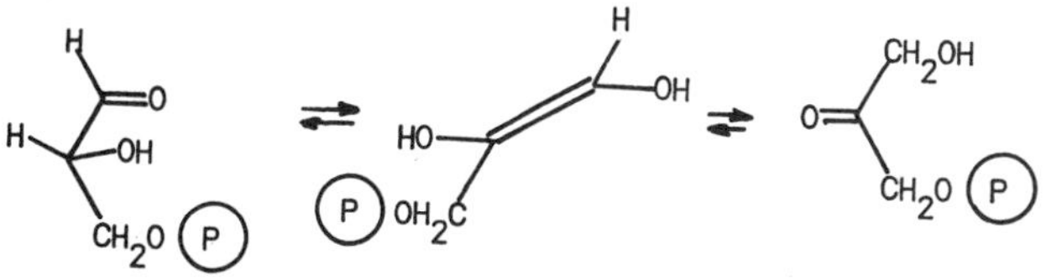

(Eq. 2.4)

2.4 2.5

process is visualized as occurring in vitro without specificity as to the face of the double bond, the two possible orientations are clearly different and are readily distinguished by enzyme mediation [3].

Now, if the same (common) enol of 3-phosphoglyceraldehyde (2.4; or of dihydroxyacetone monophosphate, 2.5) undergoes aldol condensation with 3-phosphoglyceraldehyde (2.4), fructose-1, 6-diphosphate (2.6) is generated (Equation 2.5). Subsequent hydrolysis of the phosphate

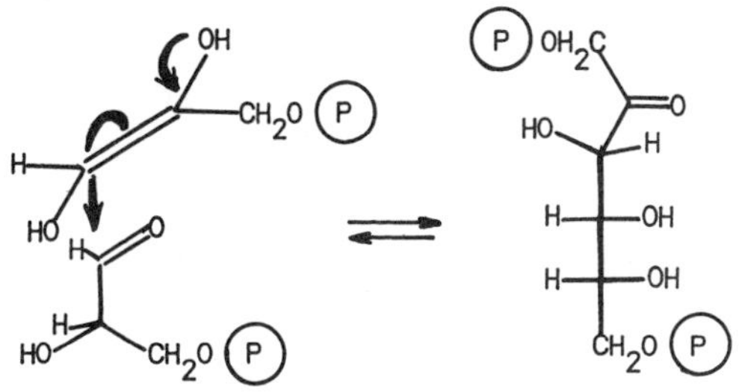

(Eq. 2.5)

2.4 2.6

at C_1 of fructose-1, 6-diphosphate (2.6) (the normal hydrolysis of the ester of a strong acid) results in the formation of fructose-6-phosphate (2.7), which on enolization toward C_1 generates a double bond isomer (2.8) which can also be derived from glucose-6-phosphate (2.9). Reversion of that enol to its carbonyl form results in the formation of glucose-6-phosphate (2.9) itself (Equation 2.6). There is some evidence that enzyme participation, in the form of

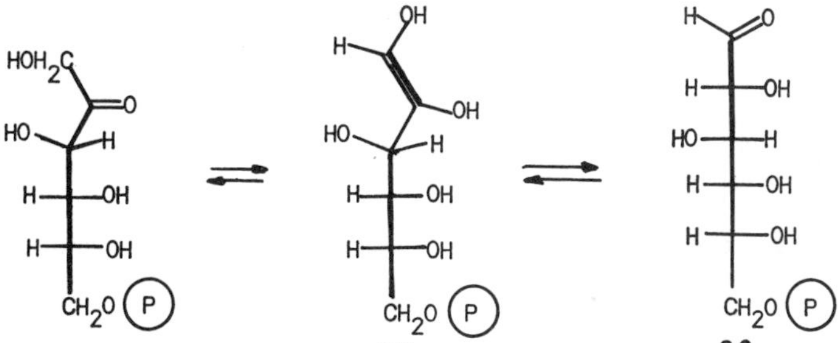

(Eq. 2.6)

2.7 2.8 2.9

of the aldimine, is involved in this process. It should be noted that both fructose-6-phosphate (2.7) and glucose-6-phosphate (2.9) are, in fact, reduced (and monodephosphorylated) relatives of the original six-carbon compound formed from ribulose-1,5-diphosphate (2.1) and carbon dioxide.

The regeneration of ribulose-1,5-diphosphate (2.1) is accomplished through the use of an enzyme whose mode of action has been investigated. The enzyme is "transketolase," and the prosthetic group for this enzyme is thiamine pyrophosphate (2.10; Figure 2.1). Thiamine

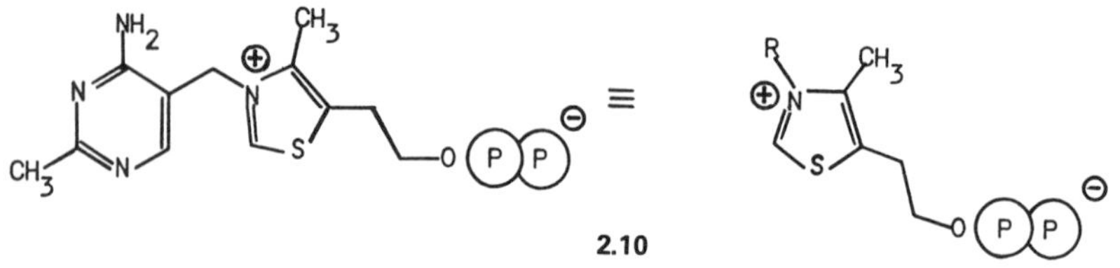

FIGURE 2.1 Thiamine pyrophosphate and its abbreviation.

pyrophosphate (2.10) acts by addition to the carbon of the carbonyl,* followed by a retroaldol reaction which generates, in the case of fructose-6-phosphate (2.7), erythrose-4-phosphate (2.11) and a two-carbon fragment attached to the thiamine pyrophosphate (2.12). This

---

*It can be argued that this addition occurs by one of two possible pathways. In one, the thiamine pyrophosphate (2.10) attacks the carbonyl compound directly and subsequently loses a proton (Equation 2.7), while in the other, proton loss occurs prior to addition, to generate a carbanion which then attacks the carbonyl function (Equation 2.8). Justification for the

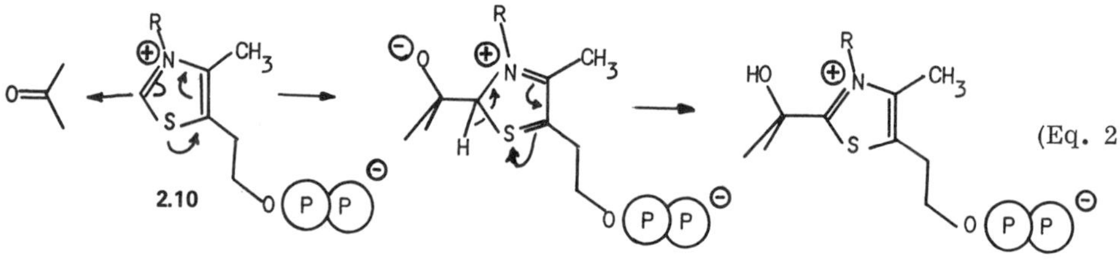

(Eq. 2.7)

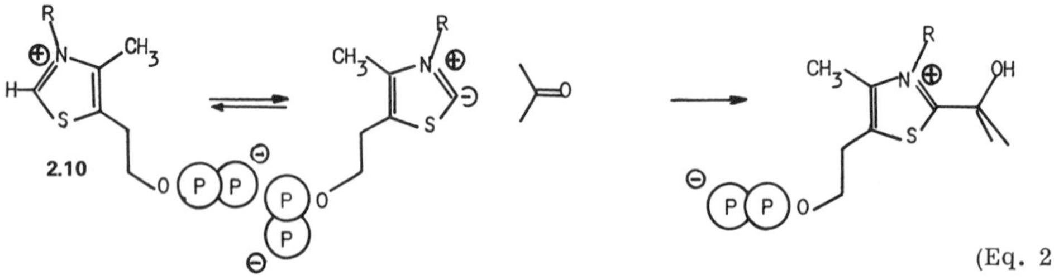

(Eq. 2.8)

latter is based upon solution chemistry in which such Zwitterions or nitrogen ylids have presumably been generated by decarboxylation reactions and metal-catalyzed deprotonation processes [4, 5, 6]. It is clear, however, that no enough is yet known about the enzyme-mediated process to decide which of these two paths is correct. However, for the sake of simplicity, and because it does not affect the outcome, the Zwitterionic form will be used hereafter.

two-carbon fragment is called "active glycolaldehyde." Then, much in the same way as a tertiary amine catalyzes amide formation, the activated acyl function is transferred from the thiamine pyrophosphate (2.10) to a new aldehyde.

If this new aldehyde is 3-phosphoglyceraldehyde (2.4), the addition of the "active glycoaldehyde" (2.12) results in the generation of xylulose-5-phosphate (2.13), which can then be isomerized via the common enol to ribulose-5-phosphate (2.14) and the latter phosphorylated to ribulose-1,5-diphosphate (2.1). These reactions are shown in Scheme 2.2 and may be summarized as follows: carbon dioxide + ribulose-1,5-diphosphate (2.1) yields two 1,3-diphosphoglyceric acid (2.3) fragments (1 + 5 = 3 + 3). Reduction of the latter (2.3) provides two equivalents of 3-phosphoglyceraldehyde (2.4) which generate one fructose-1,6-diphosphate (2.6; 3 + 3 = 6). Fructose-6-phosphate (2.7) is cleaved to active glycoaldehyde (2.12) and erythrose-4-phosphate (2.11; 6 = 2 + 4). Then, to regenerate ribulose-1,5-diphosphate (2.1), the active glycolaldehyde (2.12) adds to a 3-phosphoglyceraldehyde (2.4; 2 + 3 = 5). In the balance, the cycle has generated an equivalent of erythrose-4-phosphate (2.11) and consumed an equivalent of 3-phosphoglyceraldehyde (2.4) that had not been made. So, we are not finished unless the erythrose-4-phosphate (2.11) can be converted into 3-phosphoglyceraldehyde (2.4) to balance the cycle, i.e., we cannot use up that which has not been produced.

The erythrose-4-phosphate (2.11), in addition to being available for use in further syntheses (see below), can in fact be used to made more ribulose-1,5-diphosphate (2.1). Thus, aldol condensation of erythrose-4-phosphate (2.11) with the common enol derived from either dihydroxyacetone monophosphate (2.5) or glyceraldehyde monophosphate (3-phosphoglyceraldehyde, 2.4; Equation 2.4) will generate a seven-carbon sugar, sedoheptulose-1,7-diphosphate (2.15; Equation 2.9). Now, operation of thiamine pyrophosphate (2.10; the transketolase enzyme prosthetic group) on the sedoheptulose-7-phosphate [2.16; obtained by dephosphorylation

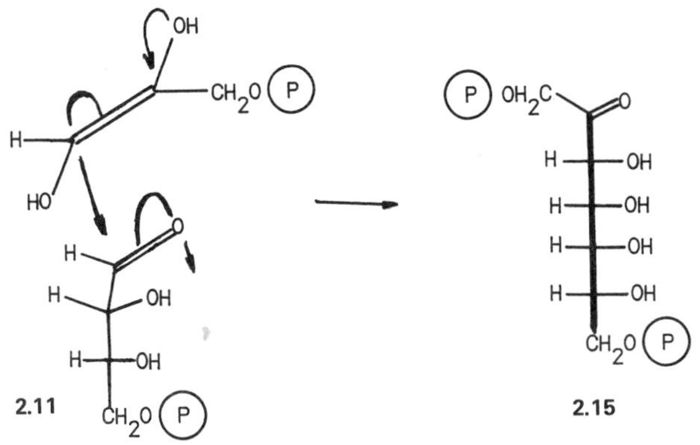

2.11 → 2.15 (Eq. 2.9)

(saponification, hydrolysis, etc.) of sedoheptulose-1,7-diphosphate, 2.15)] generates ribose-5-phosphate (2.17) and, once again, "active glycolaldehyde" (2.12). Ribose-5-phosphate (2.17), through the common enol, is convertible to ribulose-5-phosphate (2.14), and phosphorylation of the latter results in generation of ribulose-1,5-diphosphate (2.1). This is the second equivalent of ribulose-1,5-diphosphate (2.1) we have formed in this cycle, this time at the expense of erythrose-4-phosphate (2.11), but now we have an active glycolaldehyde (2.12) in excess and we have used up another 3-phosphoglyceraldehyde (2.4).

Now, if the active glycolaldehyde (2.12) also formed in the above reaction is allowed to react with 3-phosphoglyceraldehyde (2.4), xylulose-5-phosphate (2.13) is formed, and this five-carbon sugar is convertible, as noted before, through the common enol equilibration

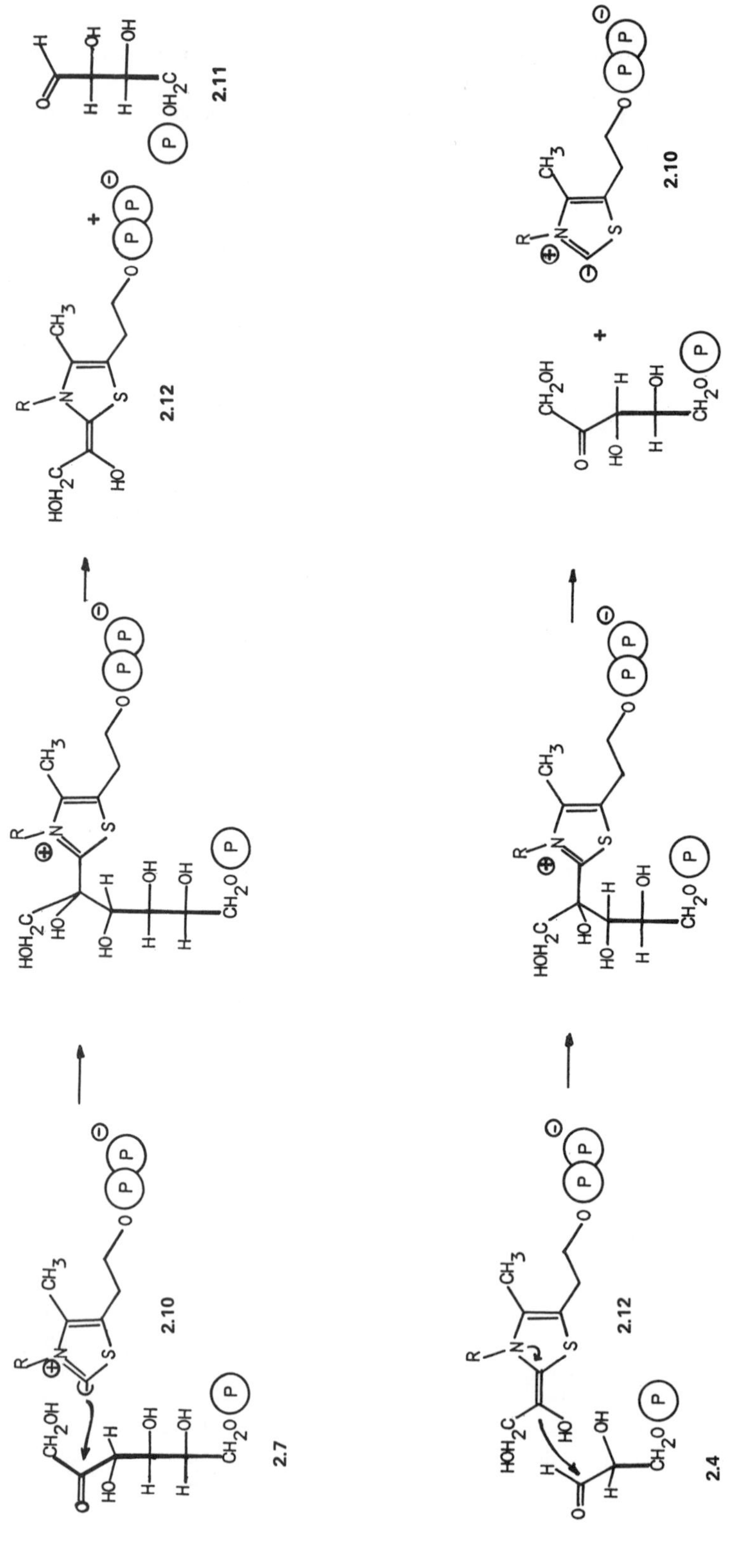

2.7 2.10 2.12 2.11 2.4 2.12 2.13 2.10

SCHEME 2.2 A possible scheme for the operation of thiamine pyrophosphate of the transketolase enzyme in the formation of xylulose-5-phosphate from fructose-6-phosphate.

to ribulose-5-phosphate (2.14) and thence to ribulose-1,5-diphosphate (2.1).* The latter, of course, was shown earlier to yield two equivalents of phosphoglyceraldehyde by fragmentation and reduction (Scheme 2.1).

The reactions involved above are shown in Scheme 2.3, and the completed Calvin-Bassham cycle is shown in an abbreviated form in Scheme 2.4 [8] and is summarized as follows:

1. If three molecules of CO_2 enter the cycle, three ribulose-1,5-diphosphates (2.1) are needed and three unstable C_6 adducts are formed ($3CO_2 + 3C_5 = 3C_6$).
2. The three C_6 molecules will break down into six 3-phosphoglyceric acid (2.2) molecules which are reduced to six 3-phosphoglyceraldehydes (2.4; $3C_6 = 6C_3$).
3. Of the six C_3 molecules (i.e., six 3-phosphoglyceraldehydes, 2.4), five will be used to complete the cycle once and one will be gained as a result of turning the cycle once ($5C_3 + 1C_3 = 6C_3$). The five 3-phosphoglyceraldehydes (2.4; the first of the six having been stored) are used as follows:
   a. Two and three undergo an aldol condensation to fructose-1,6-diphosphate (2.6; $C_3 + C_3 = C_6$).
   b. Four and fructose-1,6-diphosphate (2.6) react (transketolase) to erythrose-4-phosphate (2.11) and xylulose-5-phosphate (2.13): xylulose-5-phosphate (2.13) isomerizes to ribulose-5-phosphate (2.14; $C_3 + C_6 = C_4 + C_5$).
   c. Five and erythrose-4-phosphate (2.11) undergo an aldol condensation to sedoheptulose-7-phosphate (2.16; $C_3 + C_4 = C_7$).
   d. Six and sedoheptulose-7-phosphate (2.16) react (transketolase) to ribose-5-phosphate (2.17) and xylulose-5-phosphate (2.13; $C_3 + C_7 = C_5 + C_5$); ribose-5-phosphate (2.17) isomerizes to ribulose-5-phosphate (2.14); xylulose-5-phosphate (2.13) isomerizes to ribulose-5-phosphate (2.14).
4. Additionally, it should be noted that:
   a. 3-Phosphoglyceraldehyde (2.14) is interchangeable with dihydroxyacetone monophosphate (2.5) via the common enol.
   b. Fructose-6-phosphate (2.17) is interchangeable with glucose-6-phosphate (2.9) via the common enol.
   c. The aldol condensation which generates fructose-1,6-diphosphate (2.6) from 3-phosphoglyceraldehyde (2.4) and its enol is reversible. Therefore, glucose (2.18), which can be stored, can be broken down to 3-phosphoglyceraldehyde (2.4; Scheme 2.5).

---

*It is interesting to note that active glycolaldehyde (2.12) derived from fructose-6-phosphate (2.7), when added to 3-phosphoglyceraldehyde (2.4), generates xylulose-5-phosphate (2.13; Scheme 2.2), and addition of the same three-carbon fragment, now derived from sedoheptulose-7-phosphate (2.16) to 3-phosphoglyceraldehyde (2.4), generates the same hexose, not the isomer, ribulose-5-phosphate (2.14). From this it can be argued, for example, that the addition to the prochiral carbonyl [7] may be mediated by the same enzyme in the two cases, or that the formation and utilization of active glycolaldehyde (2.12) is not substrate dependent, but the site available for complexation of the 3-phosphoglyceraldehyde (2.4) prior to reaction is either, in an absolute sense, to the right (re) [7] or to the left (si) [7] of the adduct of thiamine pyrophosphate (2.10) [3].

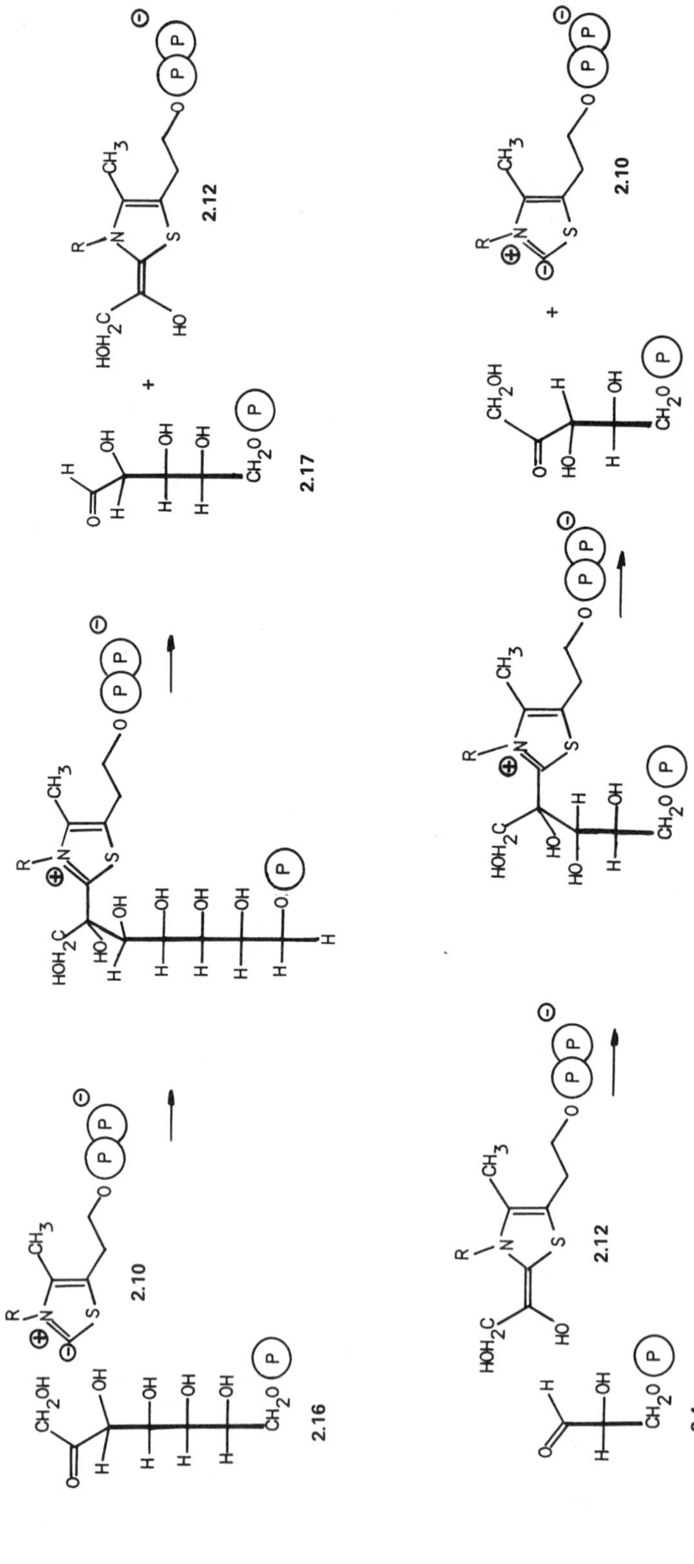

SCHEME 2.3 The generation of ribose-5-phosphate and xylulose-5-phosphate from sedoheptulose-7-phosphate and thiamine pyrophosphate.

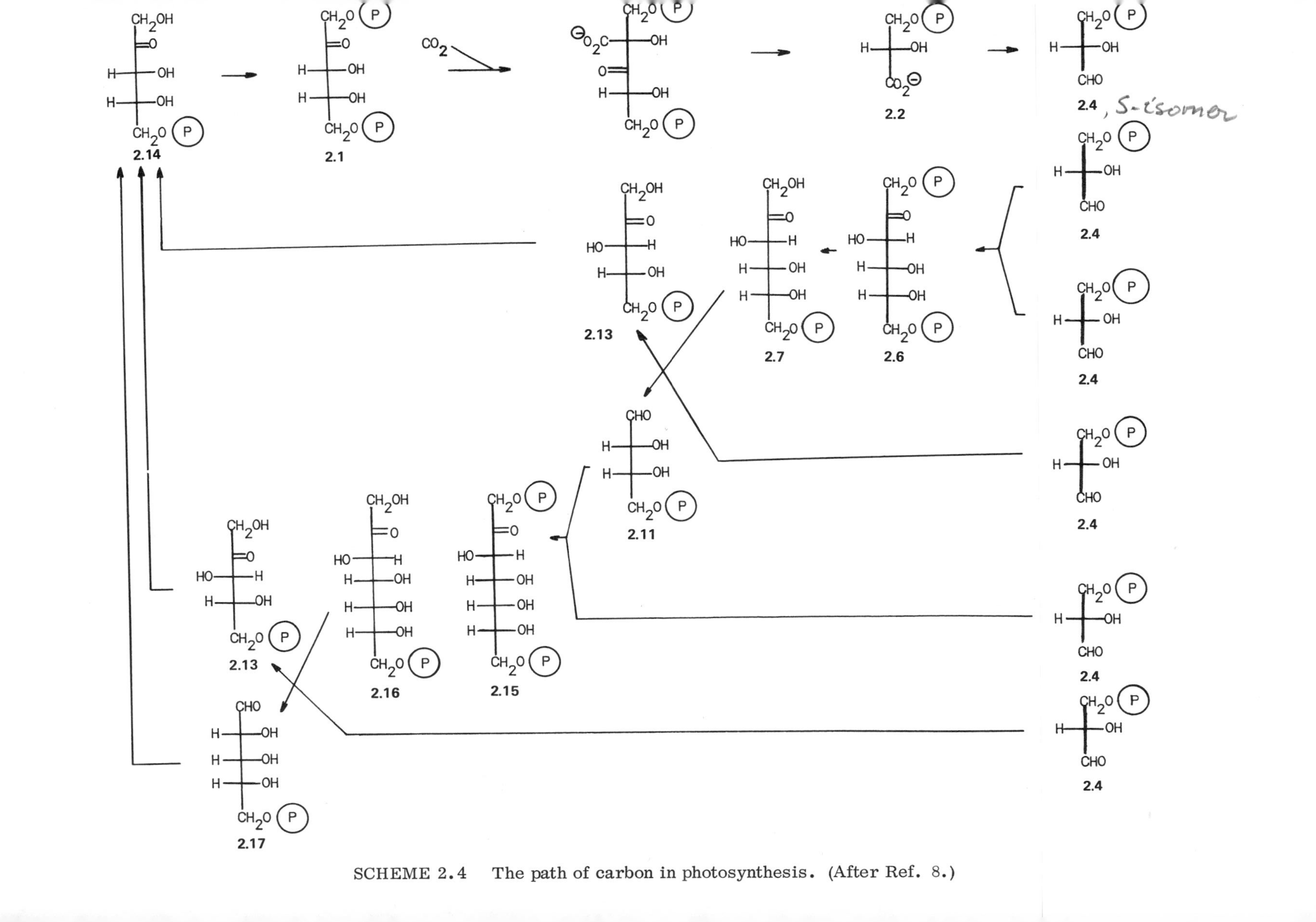

SCHEME 2.4 The path of carbon in photosynthesis. (After Ref. 8.)

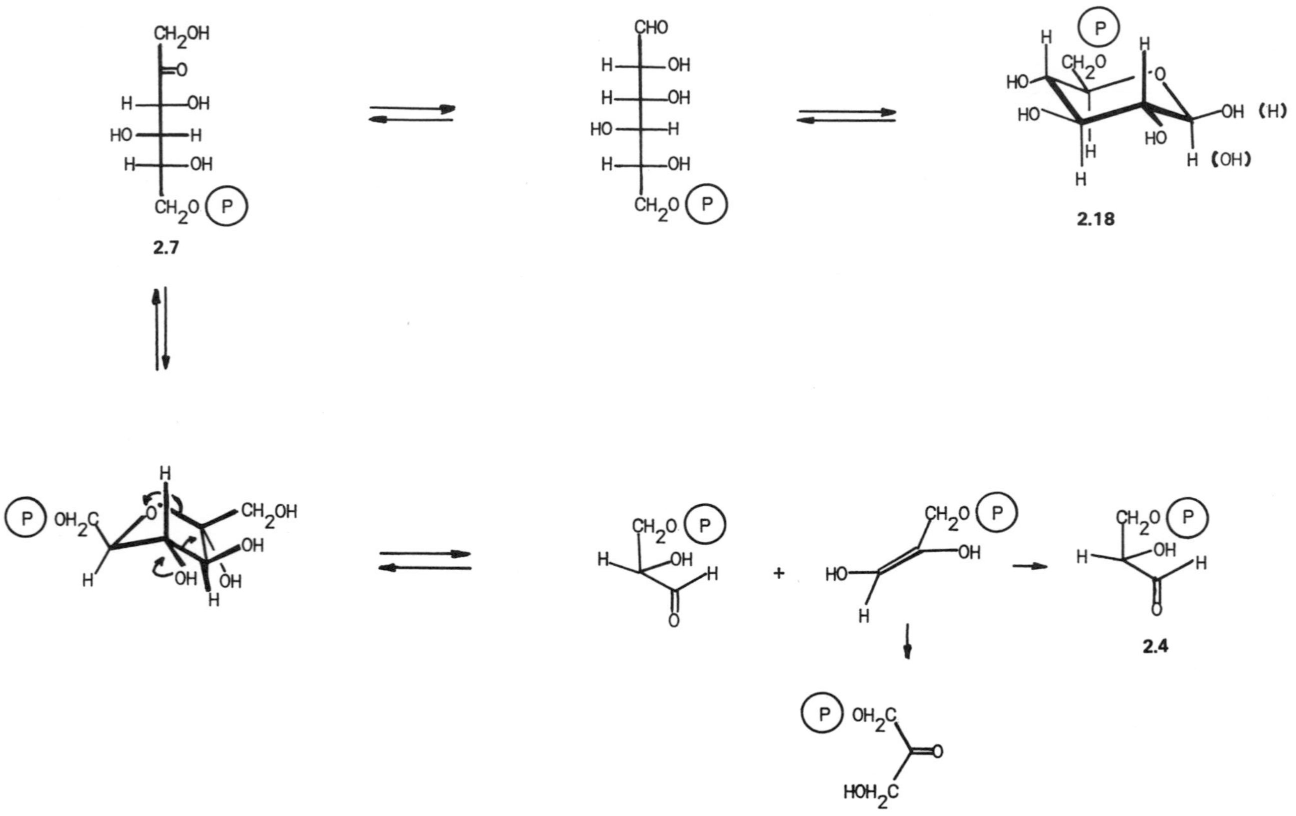

SCHEME 2.5 Glycolysis of glucose to 3-phosphoglyceraldehyde and dihydroxyacetone monophosphate.

d. Erythrose-4-phosphate (2.11) is interchangeable with threose-4-phosphate (2.19) via the common enol.

The result of the operation of the Calvin-Bassham cycle for the absorption of CO_2, which we must utilize for elaboration of the plant bases, is that there has been an overall gain of one C_3 unit for every three CO_2 molecules absorbed. We shall now see what this one C_3 unit, in combination with its fellows, can do.

B. The Use of 3-Phosphoglyceric Acid

3-Phosphoglyceric acid (2.3), either directly from fragmentation of the original unstable adduct from carbon dioxide and ribulose-1,5-diphosphate (2.1), or by reoxidation of 3-phosphoglyceraldehyde (2.4), is isomerized to 2-phosphoglyceric acid (2.20), and the latter undergoes elimination of water to yield the corresponding α,β-unsaturated carboxylic acid, phosphoenolpyruvic acid (2.21). Hydrolysis of this enol ester generates pyruvic acid (2.22). These reactions are illustrated in Scheme 2.6.

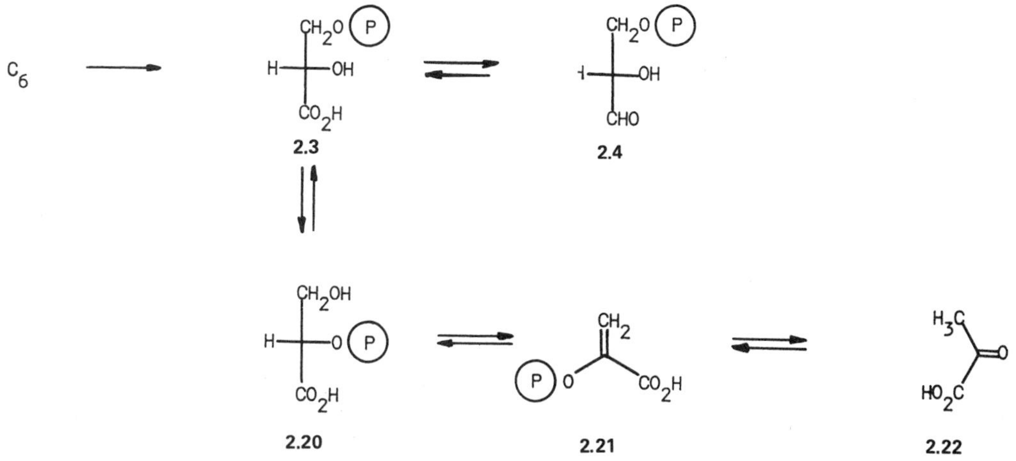

SCHEME 2.6 The conversion of 3-phosphoglyceric acid to pyruvic acid.

Our major concern with pyruvic acid (2.22; aside from the fact that it serves as a precursor to amino acids; see below) is that, by decarboxylation, it becomes the ubiquitous two-carbon fragment "acetate" which is held ready for use as a building block as the thio ester of coenzyme A (2.23; Figure 2.2). Conversion of pyruvic acid (2.22) to acetyl coenzyme A

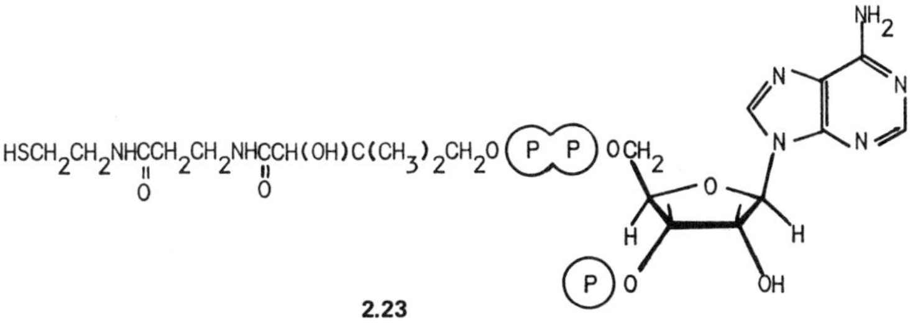

FIGURE 2.2 Coenzyme A (CoA-SH; 2.23).

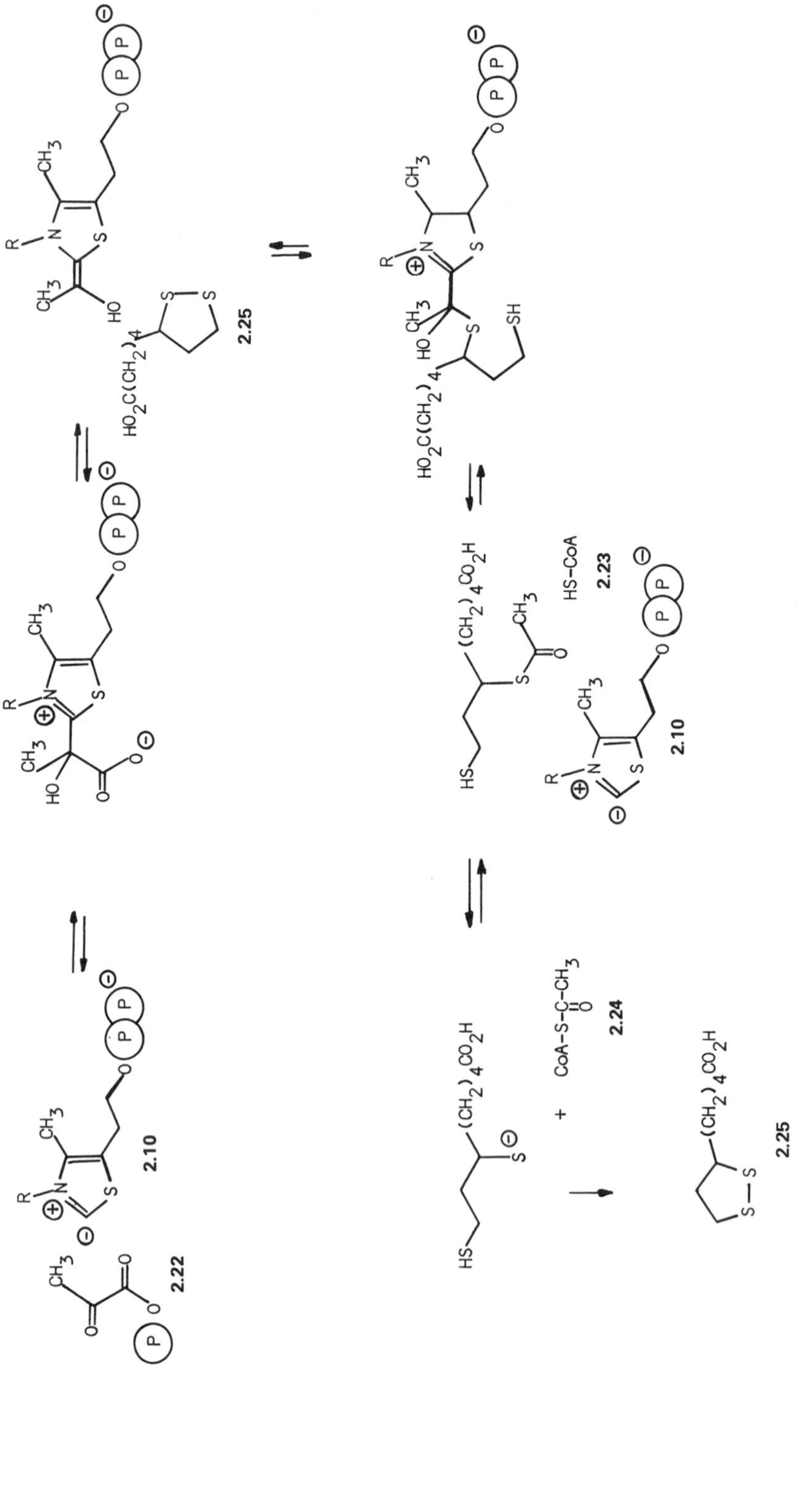

SCHEME 2.7 The formation of acetyl coenzyme A.

(Scheme 2.7) again involves the use of thiamine pyrophosphate (2.10). The prosthetic group (2.10) acts the same way it did in the formation of active glycolaldehyde (2.12), but, in this case, instead of forming that two-carbon fragment, a different two-carbon fragment, in a lower oxidation state, is generated.

Thus, the anion (see footnote, p. 12 and Equations 2.7 and 2.8) of thiamine pyrophosphate (2.10) attacks the pyruvic acid (2.22), and the resultant adduct undergoes decarboxylation. The decarboxylation product, an activated two-carbon fragment, is now transferred to the coenzyme A (2.23) throgh the intermediacy of a compound formed with lipoic acid (2.25; necessary because a net oxidation must occur, and the facile reoxidation of this dithiol product back to the corresponding disulfide serves this purpose).

As a result of the attachment of the acyl group to coenzyme A (2.23), a difunctionally active moiety is generated. The acyl methyl group readily loses a proton to produce a carbanion (Equation 2.10), and the carbonyl carbon of the acyl ligand, bearing a good leaving

$$\underset{\mathbf{2.24}}{CoA\text{-}S\text{-}C(=O)\text{-}CH_3} \rightleftharpoons CoA\text{-}S\text{-}C(=O)\text{-}CH_2^{\ominus} \qquad \text{(Eq. 2.10)}$$

group, is available for acylation reactions (Equation 2.11). Indeed, using these two reactions, acetyl groups can be combined (Equation 2.12) and larger molecules built.

$$\underset{\mathbf{2.24}}{CH_3C(=O)\text{-}S\text{-}CoA} + R^{\ominus} \rightleftharpoons {}^{\ominus}O\text{-}C(CH_3)(R)\text{-}S\text{-}CoA \rightleftharpoons CH_3C(=O)R + CoA\text{-}S^{\ominus} \qquad \text{(Eq. 2.11)}$$

$$\underset{\mathbf{2.24}}{CH_3C(=O)\text{-}S\text{-}CoA} + {}^{\ominus}CH_2C(=O)\text{-}S\text{-}CoA \rightleftharpoons {}^{\ominus}O\text{-}C(CH_3)(S\text{-}CoA)\text{-}CH_2C(=O)\text{-}S\text{-}CoA \rightleftharpoons H_3C\text{-}C(=O)\text{-}CH_2\text{-}C(=O)\text{-}S\text{-}CoA + CoA\text{-}S^{\ominus} \qquad \text{(Eq. 2.12)}$$

Finally, acetyl coenzyme A (2.24) serves as a convenient point from which to enter the citric acid cycle (Scheme 2.8) for the complete oxidation of pyruvic acid (2.22). The entry into this already turning cycle from acetyl coenzyme A (2.24), which may be considered a "pregenitor" of the cycle, will serve to generate a number of useful small molecules, e.g., oxalacetate (2.26), citrate (2.27), cis-aconitate (2.28), isocitrate (2.29), oxalosuccinate (2.30), α-ketoglutarate (2.31), succinate (2.32), fumarate (2.33), and malate (2.34). As an aside, which will nevertheless prove useful, the stereochemistry at each step in this cycle has been defined and is as shown in Scheme 2.8. Thus, in the formation of citrate (2.27) from acetyl coenzyme A (2.24) and oxalacetate (2.26), the carbon introduced (i.e., the acetyl) becomes the pro-S carbon in citrate (2.27). In the dehydration of citrate (2.27) to cis-aconitate (2.28), water is eliminated from the subunit that was originally the oxalacetate moiety (2.26) in a trans (antarafacial) manner from the conformer in which the carboxyl on the oxalacetate-derived end is anti to the carboxymethyl group derived from the introduced acetate. In the hydration of cis-aconitate (2.28) to erythro-2R,3S-isocitrate (2.29), the hydroxyl adds to the position from which the hydrogen was lost in the preceding step and vice versa. This is a trans (antarafacial) addition. Oxalosuccinate (2.30) undergoes decarboxylation to α-ketoglutarate (2.31) by loss of the carboxyl that is replaced with retention by a hydrogen from exogenous water, and this hydrogen is now the pro-S hydrogen in succinate (2.32). The formation of fumarate (2.33) from succinate (2.32) involves the loss of one

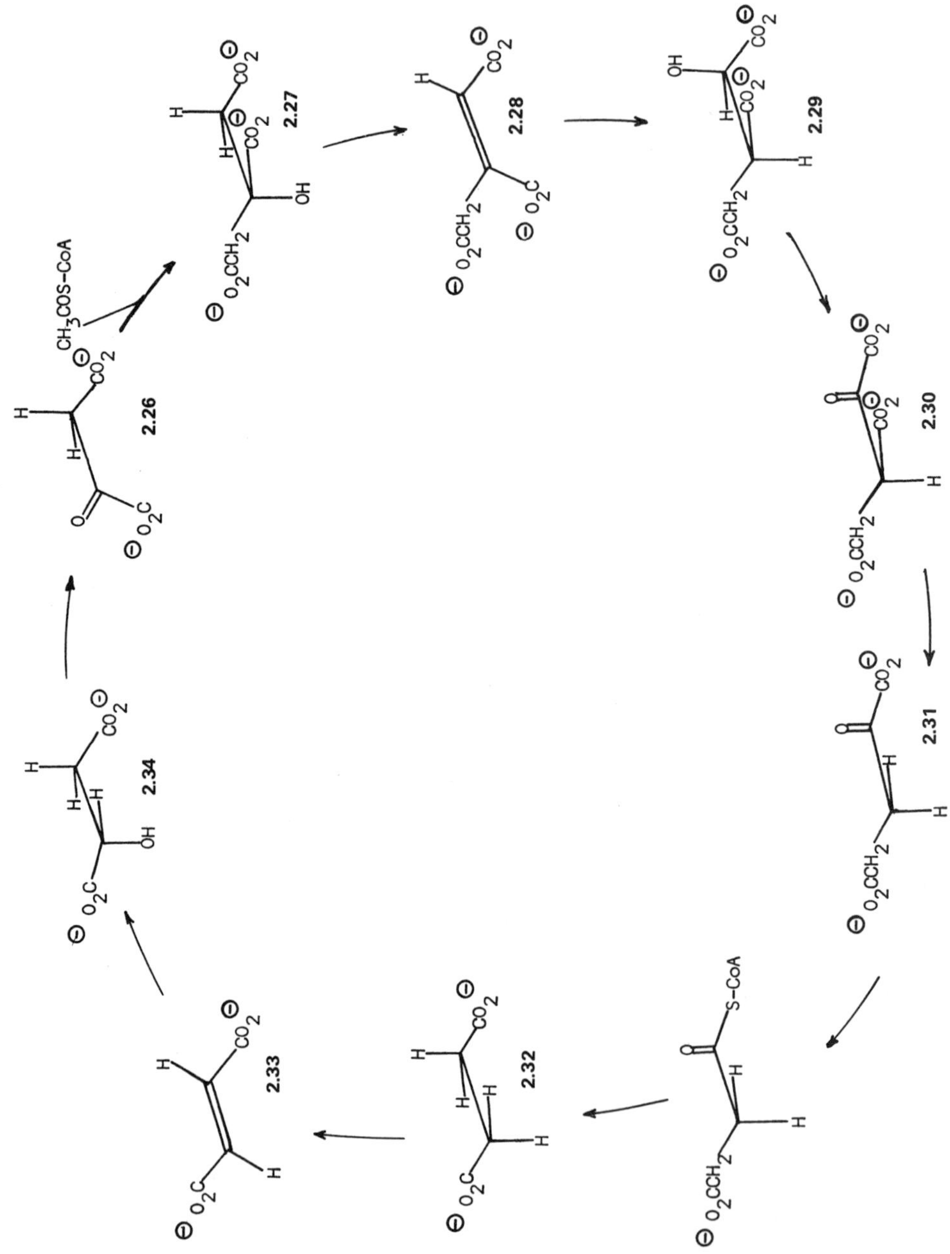

SCHEME 2.8 The citric acid cycle. (After Ref. 9.)

pro-R and one pro-S hydrogen, and finally, in the hydration of fumarate (2.33) to malate (2.34), the addition of water occurs stereospecifically to yield only S-malate while the pro-R hydrogen is derived from exogenous sources and the pro-S hydrogen from the fumarate (2.33) substrate. Additionally, it is important to note that, in plants, erythro-2R,3S-isocitrate (2.29) can be converted directly to succinate (2.32) and glyoxylate (2.35; Equation 2.13).

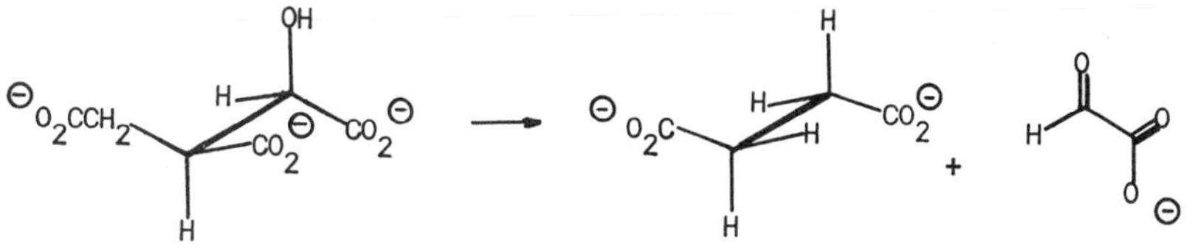

(Eq. 2.13)

2.29 2.32 2.35

Now, in addition to the conversion of phosphoenolpyruvate (2.21) into pyruvic acid (2.22) and thence to acetyl coenzyme A (2.24), phosphoenolpyruvic acid (2.21) is utilized in the formation of shikimic acid (2.36).

It has long been recognized [10] that the hydrolysis of phosphate esters is pH and ligand dependent. Thus, depending upon the particular circumstances, either phosphorous-oxygen or carbon-oxygen bond breaking might occur. For this reason, it comes as no surprise that phosphoenolpyruvic acid (2.21) can undergo hydrolysis by two different pathways (Equations 2.14 and 2.15).* Thus, using phosphorous-oxygen bond cleavage, the condensation of phosphoenolpyruvic acid (2.21) with erythrose-4-phosphate (2.11; from the Calvan-Bassham cycle, p. 15 and Scheme 2.2) generates a new seven-carbon sugar, namely, D-arabinoheptulasonic acid-7-phosphate (2.37).

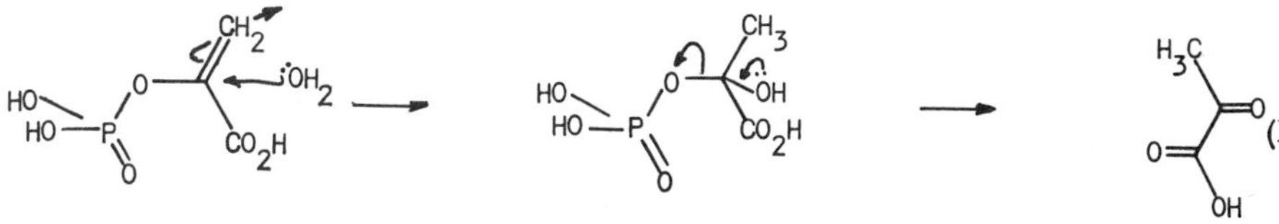

(Eq. 2.14)

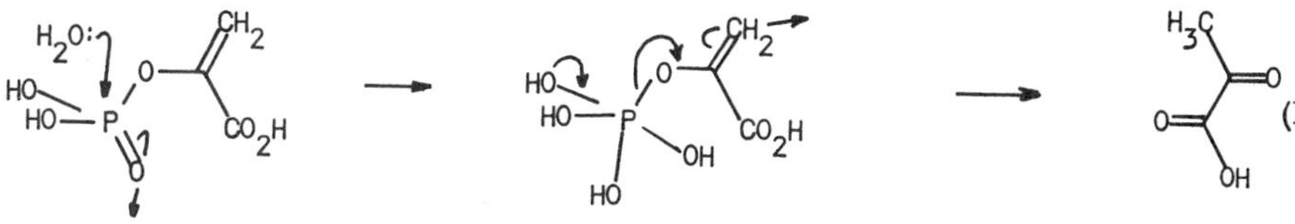

(Eq. 2.15)

From here, elimination of phosphoric acid across C_6 and C_7 forms an olefin which leads directly to 5-dehydroquinic acid (2.38) on cyclization. Elimination of water from 5-dehydroquinic acid (2.38) results, in turn, in the formation of 5-dehydroshikimic acid (2.39), which on reduction generates shikimic acid (2.36) itself. The elimination of water from the dehydroquinic acid (2.38) removes the pro-R hydrogen (a cis or suprafacial, elimination), and this presumably occurs via an enamine [11] formed with stereospecific enolization [12]. The reactions mentioned above are shown in Scheme 2.9.

From this point, if the axial hydroxyl of shikimic acid (2.36) is phosphorylated and one of the free hydroxyl groups of the resultant ester is exchanged for phosphate of phosphoenolpyruvate (2.21), trans (antarafacial) conjugate elimination of the pro-R hydrogen and axial

*In both illustrated pathways, various proton shifts, etc., have been omitted for simplicity.

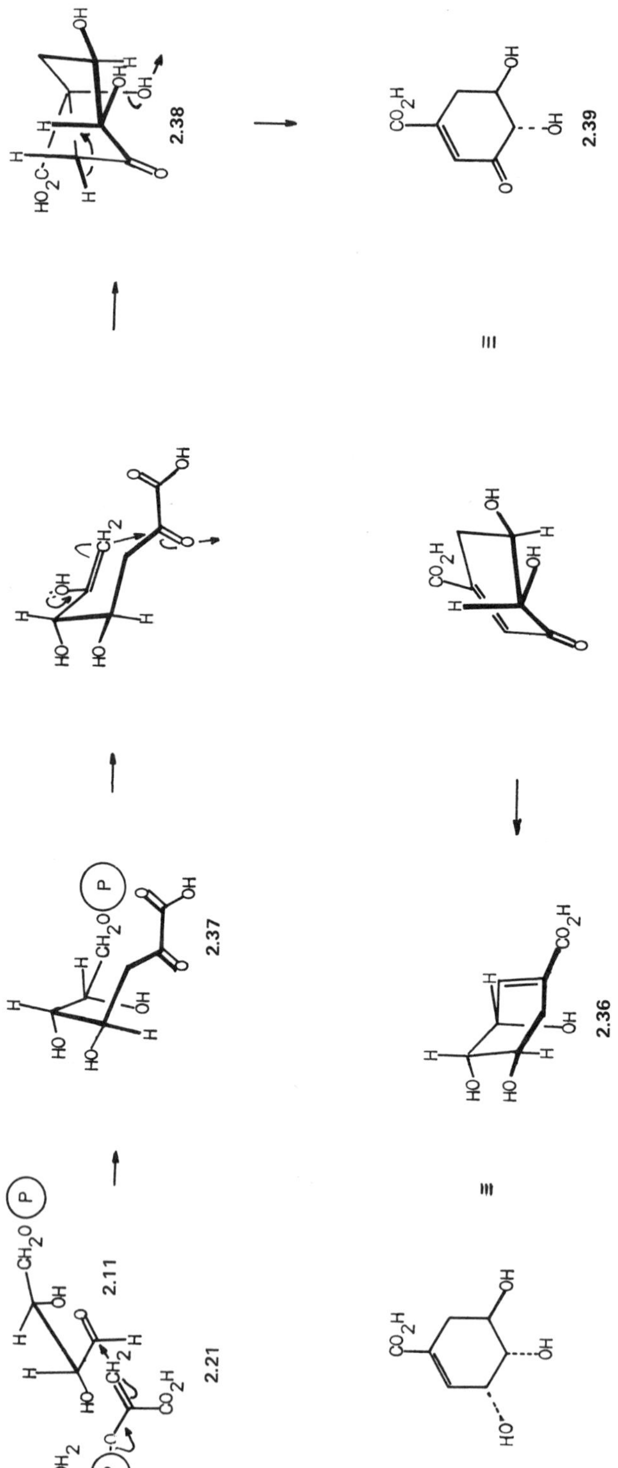

SCHEME 2.9 A pathway for the formation of shikimic acid.

phosphate generates the known chorismic acid (2.40).* If chorismic acid (2.40) is permitted to undergo an allowed ${}_{\pi}2_{s} + {}_{\pi}2_{s} + {}_{\sigma}2_{s}$ reorganization (1,5-sigamatropic rearrangement), prephenic acid (2.41) is formed, and decarboxylation coupled with elimination of water leads to phenylpyruvic acid (2.42). These reactions are shown in Scheme 2.10. Finally, along this line, p-hydroxyphenylpyruvic acid (2.43) can be generated (Equation 2.16) by oxidation of

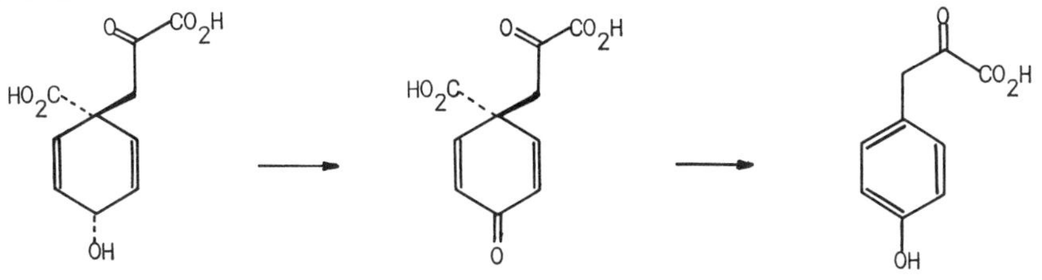

2.41 (Eq. 2.16) 2.43

prephenic acid (2.41) to the corresponding ketone and decarboxylation. Thus, p-hydroxyphenylpyruvic acid (2.43) is not formed, in plants, by hydroxylation of phenylpyruvic acid (2.42), although, as we shall see, late-stage hydroxylation (oxidation) can occur. This late-stage process is thought to occur after further transformations of the p-hydroxyphenylpyruvic acid (2.43).

In addition to its participation in the tricarboxylic (or citric) acid cycle, acetyl coenzyme A (2.24) is also implicated in two other reaction sequences of interest to us. One of these, already alluded to (p. 21), is the formation of polyketide chains which can ultimately, through a series of reductions, dehydrations, and further reductions, lead to long-chain carboxylic acids. For these acids, the methyl terminus is derived from acetyl coenzyme A (2.24) while the intermediate carbon atoms appear to come from malonyl coenzyme A (2.44). Now, malonyl coenzyme A (2.44) is generated by nucleophilic addition of the anion of acetyl coenzyme A (2.24) onto carbon dioxide. As is well appreciated, the anion of malonyl coenzyme A (2.44) corresponding to that presumed for acetyl coenzyme A (2.24) would be even easier to generate (compare malonyl and acetyl esters) although there is some question as to whether the decarboxylation of malonyl coenzyme A (2.44) is concomitant with, or subsequent to, condensation.

Further, while the actual carboxylation reaction could be written as the attack of the anion of acetyl coenzyme A (2.24) on carbon dioxide, it is known that biotin (2.45) is the prosthetic group of the enzyme that catalyzes the reaction. The formation of malonyl coenzyme A (2.44) can then be formulated as shown in Scheme 2.11. Finally, in this regard, it has been shown that the actual condensation reactions occur through mediation not of acetyl coenzyme A (2.24) but rather another enzyme (a "synthetase") which also bears sulfhydryl groups and thus exchanges with coenzyme A (2.23). Presumably, there are several sulfhydryl groups properly positioned on the synthetase to promote the reaction by holding the fragments in juxtaposition. The overall scheme for the production of fatty acids can then be shown in an abbreviated fashion (Scheme 2.12).

---

*Several mechanistic possibilities for the exchange reaction exist. It does not appear to be known whether the oxygen originally present in shikimic acid (2.36) is retained in chorismic acid (2.40) or whether the oxygen is the one from phosphoenolpyruvate (2.21). Thus, processes involving, for example, addition-elimination, direct displacement, or hydrolysis to pyruvate (2.22) followed by hemiketal formation and subsequent elimination of water are all, a priori, possible.

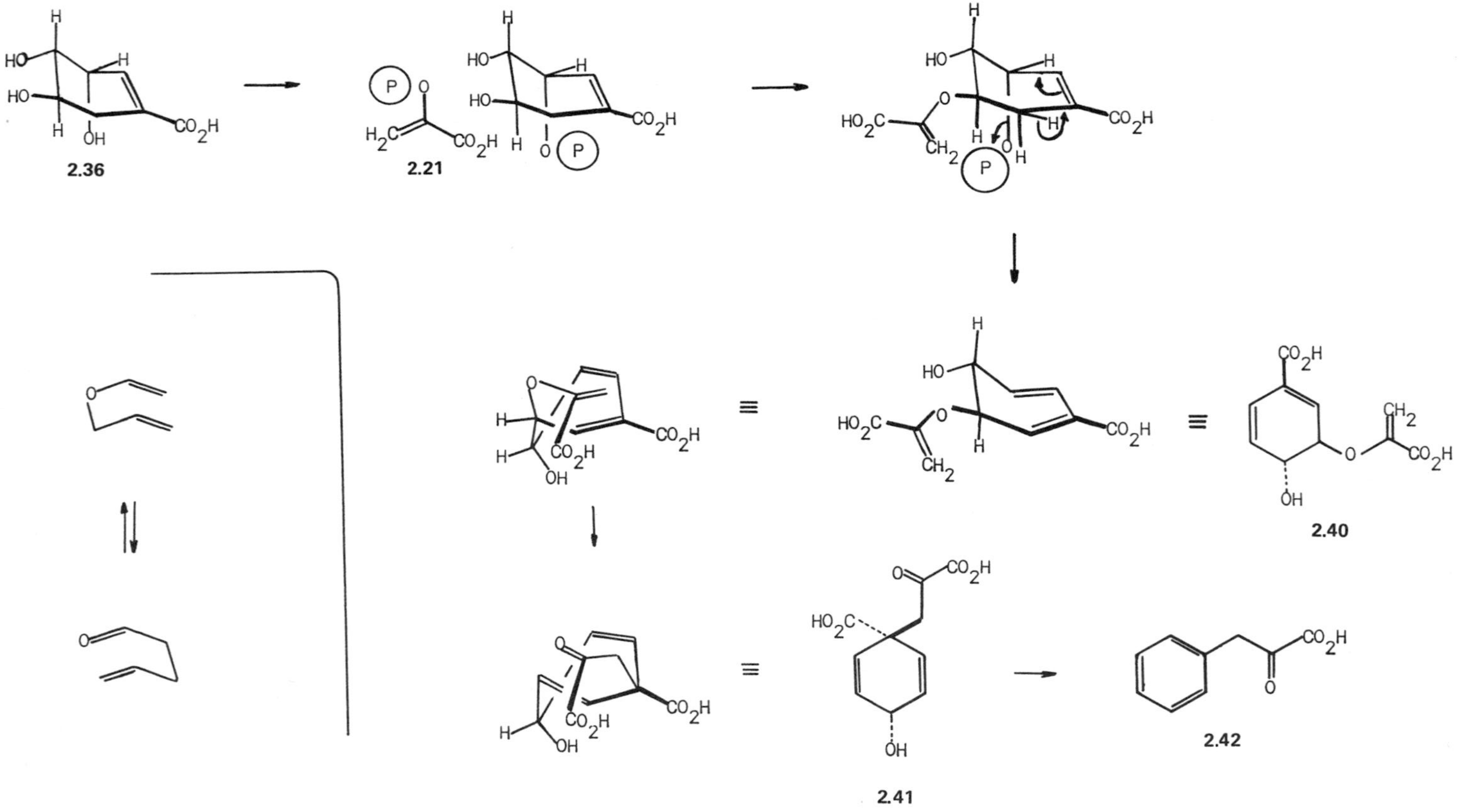

SCHEME 2.10 A possible pathway for the formation of phenylpyruvic acid from shikimic acid.

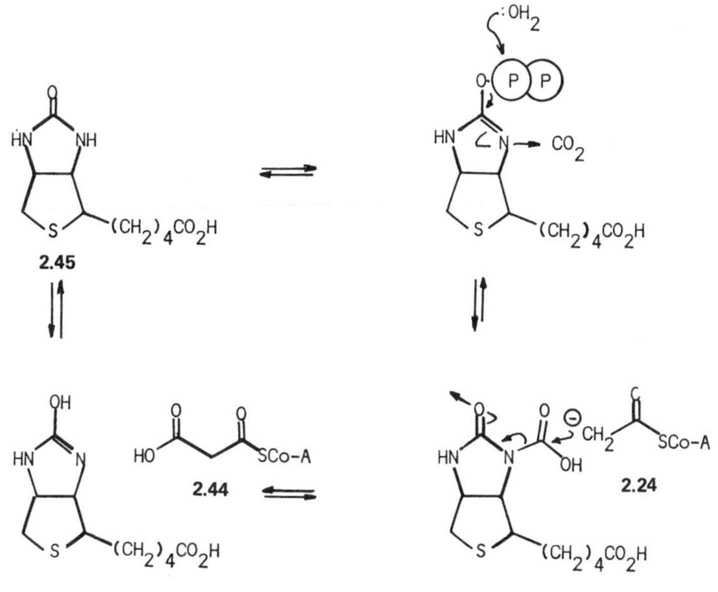

SCHEME 2.11 A potential pathway for the formation of malonyl coenzyme A from acetyl coenzyme A and carbon dioxide using biotin.

The third process utilizing acetyl coenzyme A (2.24) will have wide applicability to the biogenesis of some alkaloids to be considered later. This use of acetyl coenzyme A is the formation of isopentenyl pyrophosphate (2.46) and its double bond isomer, 3,3-dimethylallyl pyrophosphate (2.47). Briefly, three units of acetyl coenzyme A (2.24) combine, and this is followed by a reduction to generate mevalonic acid (2.48; or mevalonolactone, 2.48a), which via the corresponding pyrophosphate is converted to isopentenyl pyrophosphate (2.46; Scheme 2.13). The details of the elaboration of isopentenyl pyrophosphate (2.46) into much larger systems have been thoroughly examined, and when appropriate, we shall consider them. For the moment, it is sufficient to note that if isopentenyl pyrophosphate (2.46) is allowed to condense with 3,3-dimethylallyl pyrophosphate (2.47), a ten-carbon compound, geranyl pyrophosphate (2.49), results (Equation 2.17). While this process serves as the basis for biosynthesis of terpenes, sesquiterpenes, diterpenes, etc., and finds application in the genesis of alkaloids related to these terpenoids, we also see that a second important group of compounds can be generated from geranyl pyrophosphate (2.49). The compounds belonging to this group are called iridoids.

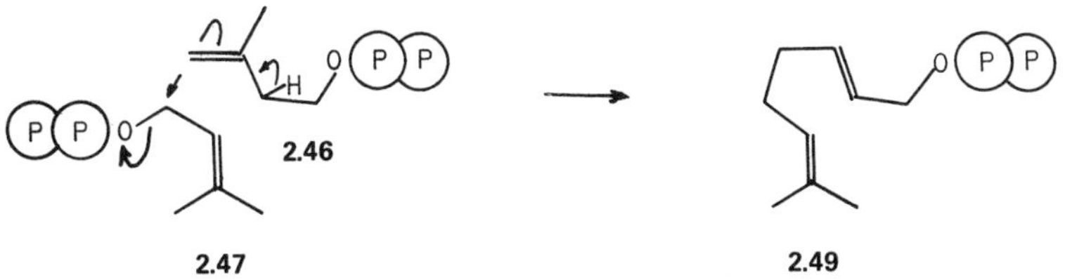

(Eq. 2.17)

EQUATION 2.17 A pathway for the formation of geranyl pyrophosphate from isopentenyl pyrophosphate and 3,3-dimethylallyl pyrophosphate.

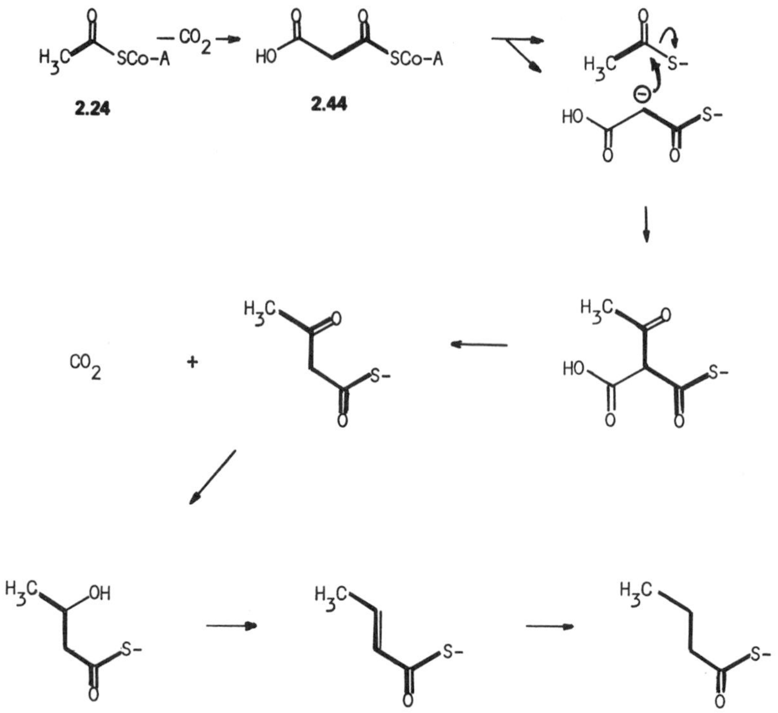

SCHEME 2.12 An abbreviated pathway for the formation of fatty acids from acetyl coenzyme A and carbon dioxide.

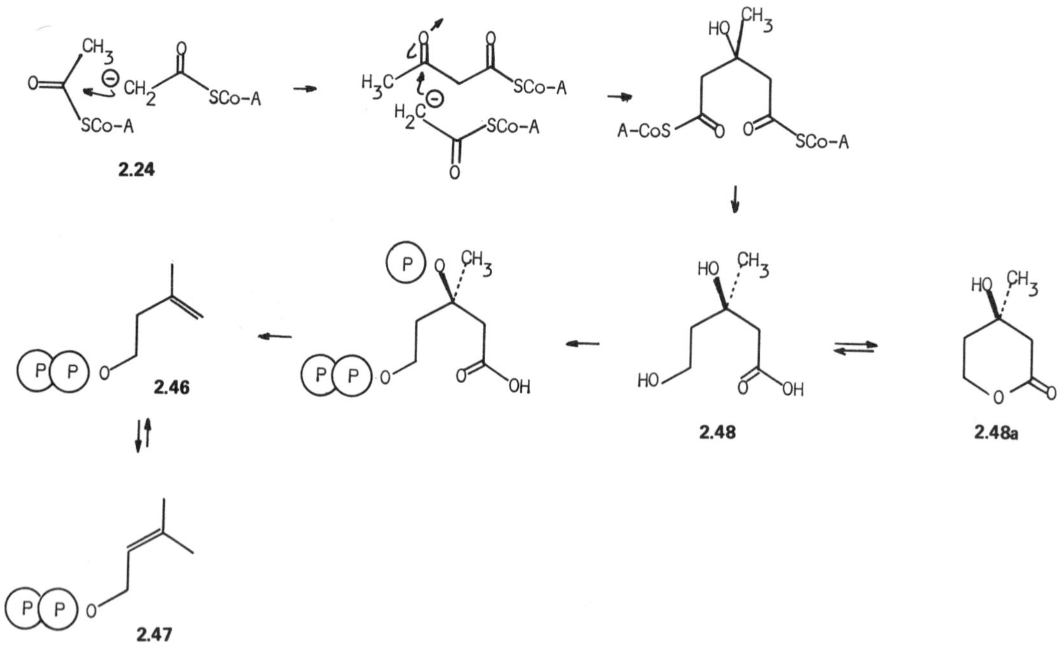

SCHEME 2.13 A pathway for the generation of isopentenyl pyrophosphate from acetyl coenzyme A via mevalonic acid.

Iridoids are highly oxygenated, mevalonate-derived (2.48), ubiquitous compounds almost always found attached to sugars. Our interest here revolves about a typical member of the class, i.e., loganin (2.50; Figure 2.3), because it has been widely implicated in the biosynthesis of indole and isoquinoline alkaloids. Labeling experiments (with both ^3H and ^{14}C)

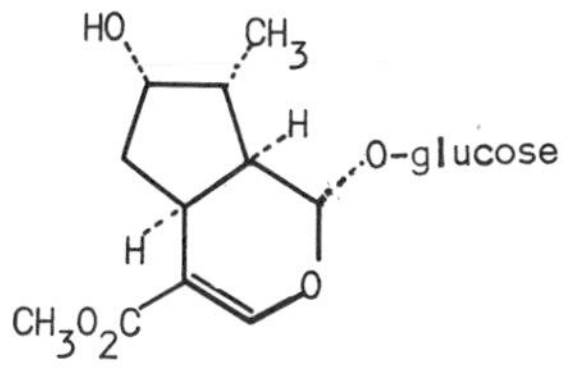

2.50

FIGURE 2.3 Loganin (2.50).

have shown that there is a direct path from geranyl pyrophosphate (2.49) to loganin (2.50), and although the details of the process are presently unknown, certain transformations must occur and a reasonable series of steps can be postulated. For now (and much of this material will be considered again later), Scheme 2.14 sets forth some ideas about what must occur and what is presumed to occur. Thus, it is likely that the first step in the conversion of geranyl pyrophosphate (2.49) to loganin (2.50) is oxidative and involves epoxidation of a double bond (similar to what is known for the cyclization of squalene). This is followed by cyclization to the required cyclopentane derivative. Subsequent dehydration and allylic oxidations serve to generate the loganin precursor (2.52), which forms an acetal with glucose to yield loganin (2.50) itself.

Alternate postulates have been made. First, both geraniol (2.53) and nerol (2.54) are incorporated into loganin (2.50), the former somewhat better than the latter. Second, since oxidation of the hydroxyl group to the corresponding aldehyde in both geraniol (2.53) and nerol (2.54) and introduction of an oxygen at C_{10} are, among other transformations, required, 10-hydroxygeraniol (2.55) and 10-hydroxynerol (2.56) have been considered as precursors. Indeed [13-15], in _Catharanthus roseus_, administration of [9-^{14}C]-10-hydroxygeraniol (2.55). and [9-^{14}C]-10-hydroxynerol (2.56) leads to specific ^{14}C incorporations (of 40-50%) into isolated loganin (2.50), showing that the terminal oxygen-bearing positions had randomized, and indicating that the immediate precursor to loganin (2.50) was probably deoxyloganin (2.57). Scheme 2.15 contains the material presented above. Finally, for use in alkaloid elaboration, loganin (2.50) is transformed into secologanin (2.58). This conversion involves oxidative cleavage of the five-membered ring (Scheme 2.16) and, possibly, direct hydride loss may be involved, although a different mechanism, which involves initial oxygenation at the methyl group, has also been proposed [16].

The final beam required for the carbon framework is the smallest, but perhaps the most ubiquitous. Throughout all schemes we will subsequently examine, compounds bearing methyl groups on oxygen and nitrogen (and, on occasion, on carbon, but not derived from the biosynthetic pathways we have already examined) will be found. The principal source of methyl groups which become attached to nucleophilic sites is S-adenosylmethionine (2.59): a sulfonium salt (Figure 2.4). As is characteristic of such methyl sulfonium salts, the methyl group readily acts as an alkylating agent. Thus, attack by any electron-rich species will permit transfer of the methyl group. This attack on the methyl can be by $-OH$, $-NH_2$, or, possibly, by the carbon α to a carbonyl group. In the first case, O-methyl ethers will be formed; in the second case, N-methyl amines; and in the third case, for example, C-methyl derivatives which do not fit the "isoprene" or "acetate" rules because of an _additional_ methyl.

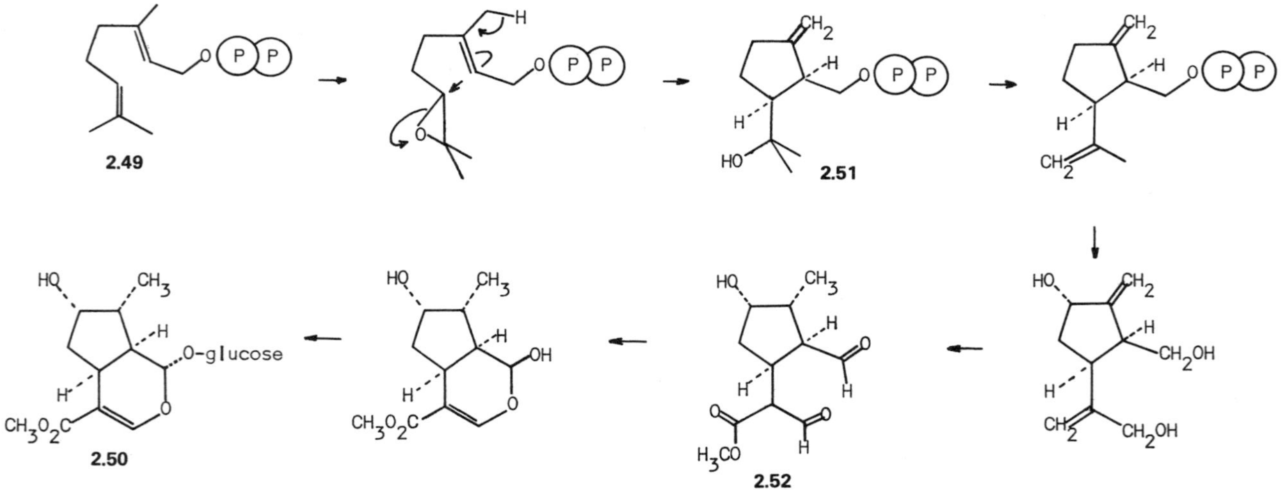

SCHEME 2.14 A possible pathway for the generation of loganin from geranyl pyrophosphate.

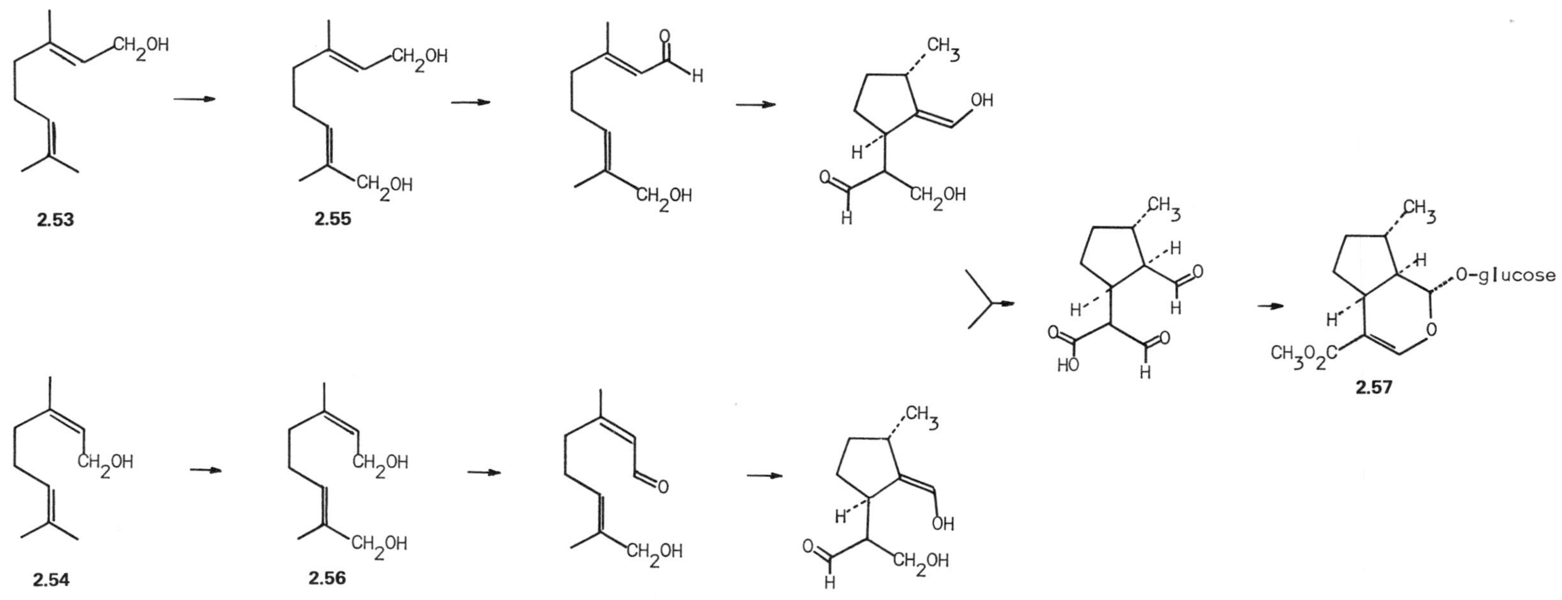

SCHEME 2.15 Pathways for the formation of deoxyloganin, a likely immediate precursor to loganin.

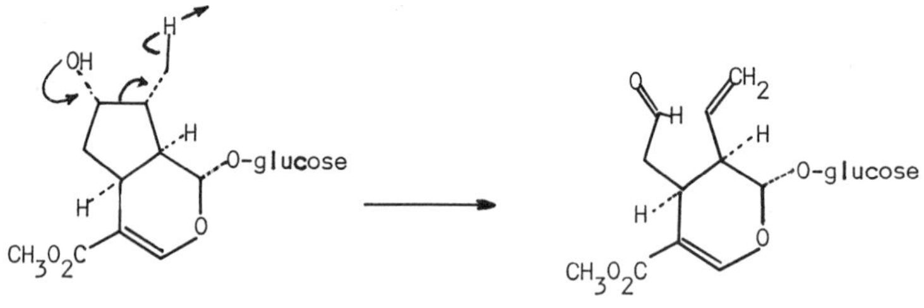

2.50 2.58

SCHEME 2.16 A pathway from loganin to secologanin.

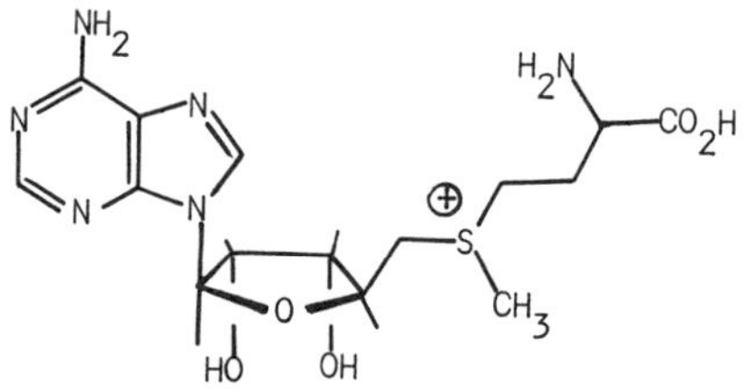

2.59

FIGURE 2.4 S-Adenosylmethionine, the source of C_1 methyl groups attached to oxygen and nitrogen.

In the particular cases we shall be examining, as already pointed out, C-methylation will be rare. Now, the methyl group in methionine (2.60) itself is usually derived from a tetrahydrofolic acid (2.61) (Figure 2.5) which originally was formylated by, e.g., formic acid, formaldehyde, etc. Thus, in tetrahydrofolic acid (2.61), formylation at nitrogen generates N^{10}-formyltetrahydrofolic acid (N^5 formylation is also important) which can be reduced through the corresponding alcohol and imine to the N-methyl derivative. It is this methyl group which is transferred to methionine (2.60).

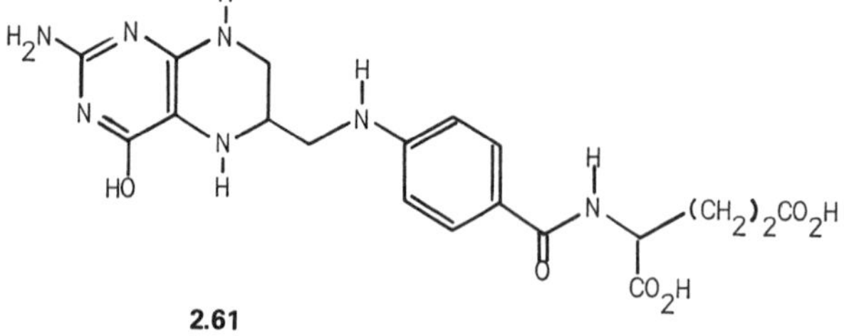

2.61

FIGURE 2.5 Tetrahydrofolic acid. Initial formylation occurs at N^{10}, and subsequent reduction to methyl occurs. This methyl is transferred to sulfur and stored as methionine.

II. THE AMINO ACIDS

Except for the very few families of alkaloids where nitrogen appears to be incorporated directly into an already elaborated skeleton (e.g., diterpene alkaloids, steroid alkaloids, systems derived directly from a linear combination of acetate units, etc.), only a few amino acids are necessary to account for the many diverse alkaloids. In fact, except for a few amino acids which are required for the preparation of others, we need to consider only nine such compounds, namely, glutamic acid (2.62), ornithine (2.63), lysine (2.64), phenylalanine (2.65), tyrosine (2.66), tryptophan (2.67), aspartic acid (2.68), methionine (2.60; for methylation), and histidine (2.69), to account for the majority of plant bases. However, before considering how these amino acids, the carbon compounds we have already generated (most of which will be used to prepare the amino acids themselves), and the few reactions we will be allowed to use, can all be combined to produce the alkaloids, we must briefly digress into the biosynthesis of the amino acids themselves.

The ultimate source of nitrogen in the amino acids is atmospheric nitrogen and the oxides of nitrogen found in the soil, both of which must be reduced to ammonia. The reduction of nitrogen by, for example, nitrogen-fixing bacteria, is an energy-intensive process (i.e., costly) and is poorly understood. The cost for reducing nitrogen to diimide which, via hydrazine, yields ammonia (none of the intermediates being detected) is estimated at about 50 equivalents of pyruvic acid per equivalent of ammonia produced [17]. In some systems, the required electron transfer is apparently accomplished by the enzyme ferrodoxin, but its mode of action is not yet well understood. On the other hand, reduction of nitrate to nitrite and nitrite to ammonia has been linked to a pyridine nucleotide which appears to be better understood and involves an extensive series of one-electron and hydrogen transfers.

Now, once the ammonia has been formed, it must be introduced into the organic molecules. Primary* formation of glutamic acid will be considered first.

A. Glutamic Acid

L-Glutamic acid (2.62) is formed by the enzyme-catalyzed introduction of ammonia into α-ketoglutaric acid (2.31; citric acid cycle, p. 21 and Scheme 2.8). The reversible formation of the imine with loss of water can be immediately appreciated in the reaction of ammonia with the α-ketoglutaric acid (2.31) substrate, and the reduction of the imine to the amine is enzymatically controlled, as would be expected. The details of this are not yet known, but the reaction involves the conversion of reduced nicotinamide adenine dinucleotide (NADH; 2.70) to the corresponding oxidized form (NAD^+; 2.71; Figure 2.6). The overall reaction is outlined in Scheme 2.17, where a proposed pathway involving NADH-NAD^+ (i.e., 2.70-2.71) is shown.

Other amino acids we require can be generated by the transamination of the corresponding ketoacid with glutamic acid (2.62). In the transamination reaction, the amino group is removed from glutamic acid (2.62) and added to the keto carbonyl of the α-keto acid. For this reaction, the involvement of pyridoxyl (2.72) and (perhaps) a metal appears to be required. This prosthetic group (pyridoxyl, 2.72) is also involved in the decarboxylation of amino acids, as we shall see later. The use of pyridoxal (2.72) in the transamination process is indicated is Scheme 2.18. The reasonable pathway shown involves formation of an imine (2.73) by condensation with the aldehyde carbon of pyridoxal (2.72) and with the amino group of glutamic acid (2.62). Coordination of this imine (2.73) with a suitable metal is then followed by proton transfer from carbon to nitrogen, and the formation of a new imine (2.74),

---

*The term "primary" is specifically meant to exclude the transamination reaction which converts one amino acid into another. This reaction will be considered later.

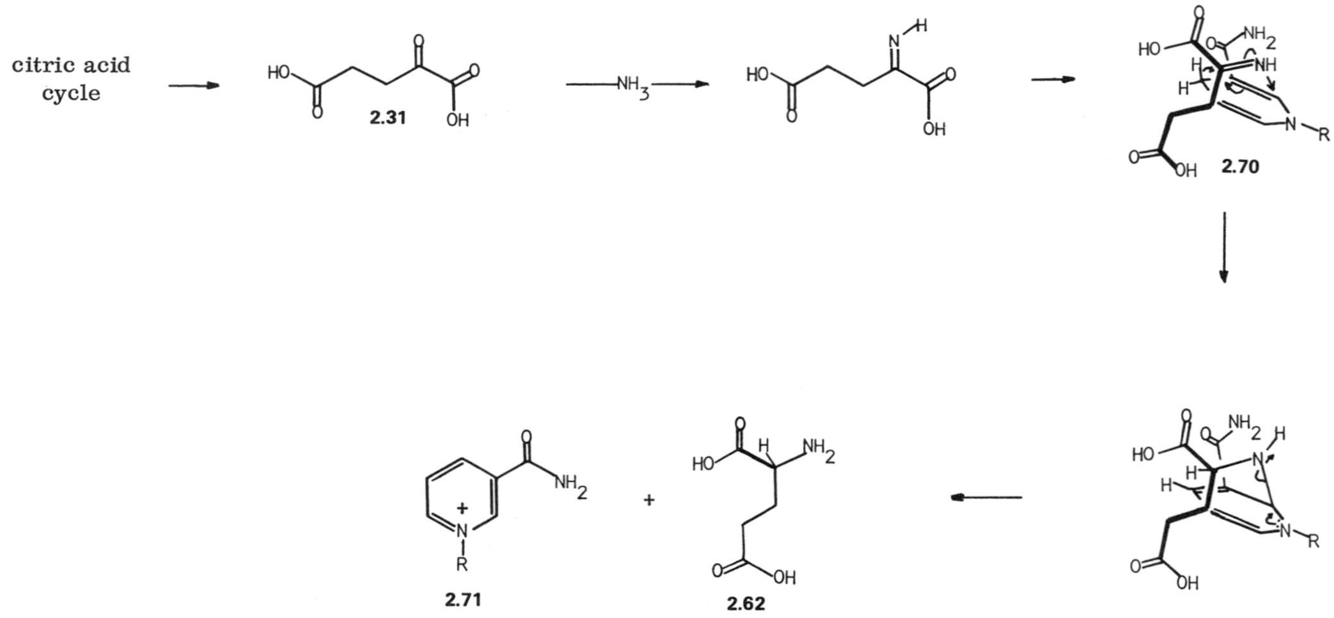

SCHEME 2.17 A projected pathway for the formation of glutamic acid from ammonia and α-ketoglutaric acid involving imine reduction and oxidation of reduced nicotinamide adenine dinucleotide (NADH) to the oxidized form (NAD^+).

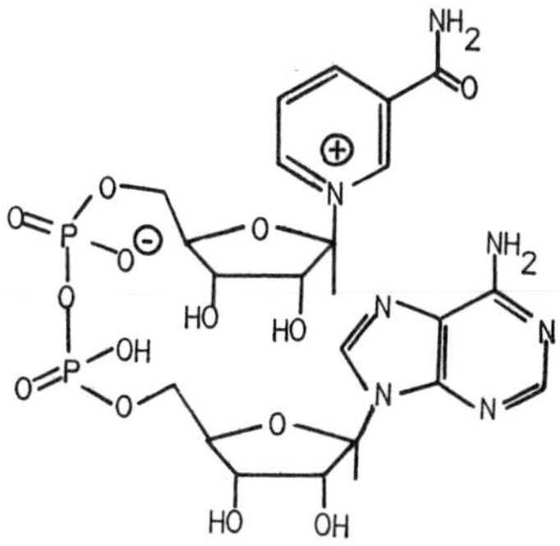

FIGURE 2.6 (2.70) Reduced nicotinamide adenine dinucleotide (NADH). (2.71) Nicotinamide adenine dinucleotide (NAD^+).

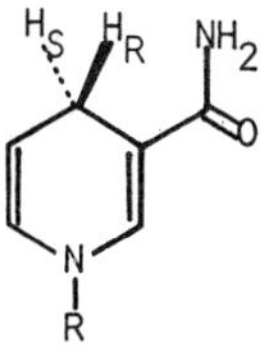

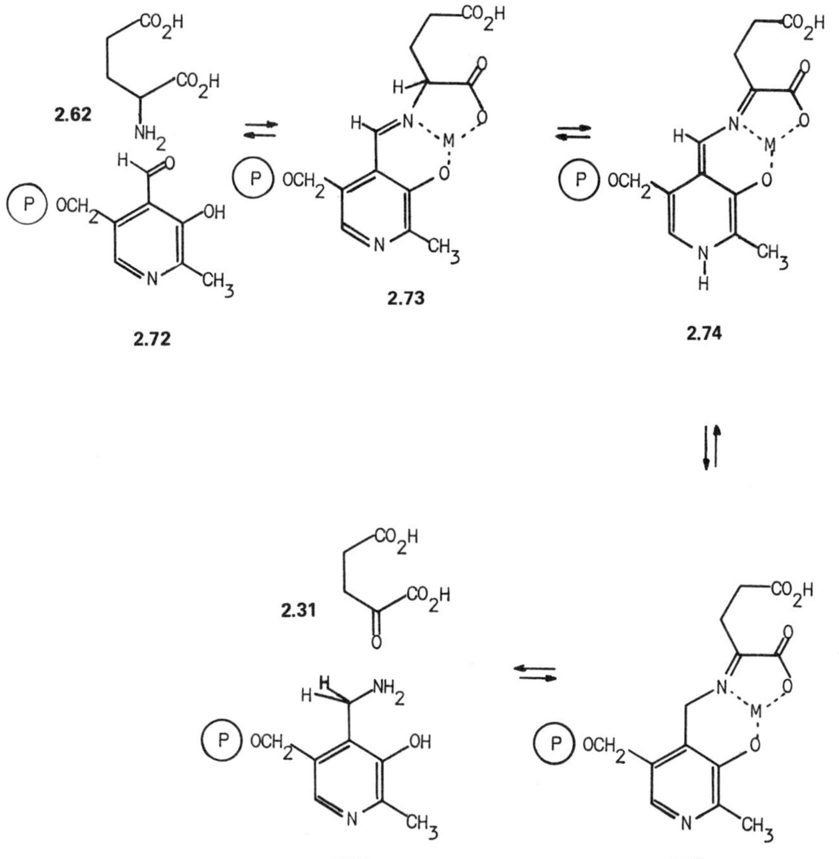

SCHEME 2.18 Transamination. The involvement of pyridoxal in the removal of the amino group from glutamic acid. Reversal of the scheme, substituting any other α-ketoacid for α-ketoglutaric acid, serves to generate the corresponding amino acid.

which when attacked by water and hydrolyzed yields α-ketoglutaric acid (2.31) and pyridoxamine (2.75). The amino nitrogen of pyridoxamine (2.75) can now be transferred to a new keto group of a different α-keto acid by reversing the process shown in the scheme and losing water.

B. Ornithine

Ornithine (2.63), an amino acid not found in proteins but important in the synthesis of urea and a number of alkaloids containing tetrahydropyrrole rings, may be directly derived from glutamic acid (2.62) via reduction of the latter to glutamic-δ-semialdehyde (2.76) and subsequent transamination (Equation 2.18).

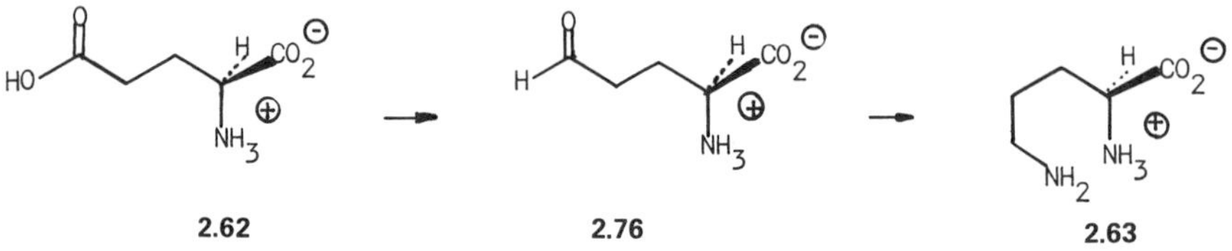

2.62 → 2.76 → 2.63 (Eq. 2.18)

C. Lysine

Just as ornithine (2.63) will be implicated in the formation of tetrahydropyrrole rings, lysine (2.64), which contains one more methylene unit than ornithine (2.63), will be involved in piperidine ring formation. However, while ornithine (2.63) may be generated from glutamic acid (2.62) directly (Equation 2.18), a more circuitous route is required for lysine (2.64). Thus, in what amounts to a homologous analog of the citric acid cycle, the aldol condensation of α-ketoglutaric acid (2.31) with acetyl coenzyme A (2.24) generates homocitric acid (2.77), which on dehydration and rehydration provides homoisocitric acid (2.79) via homoaconitic acid (2.78). Oxidation of homoisocitric acid (2.79) leads, in turn, to oxaloglutaric acid (2.80) and, by decarboxylation of this β-keto acid, to α-ketoadipic acid (2.81). Then, transamination (from glutamic acid, 2.62) generates α-aminoadipic acid (2.82), which on reduction forms the terminal semialdehyde, α-aminoadipic-ϵ-semialdehyde (2.83). This reaction is the homolog of semialdehyde formation in the case of ornithine (2.63) formation from glutamic acid (2.62), and transamination here leads to lysine (2.64). These reactions, which are only one possible route to lysine (2.64), are shown in Scheme 2.19.

D. Phenylalanine and Tyrosine

Both phenylalanine (2.65) and tyrosine (2.66) are formed in the same fashion, i.e., transamination from glutamic acid (2.62; forming, in the process, α-ketoglutarate, 2.31) to phenylpyruvic acid (2.42) or p-hydroxyphenylpyruvic acid (2.43) respectively (Equation 2.19)

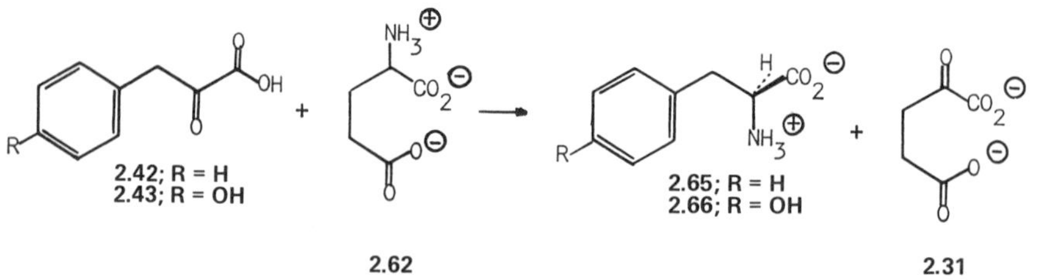

2.42; R = H
2.43; R = OH
+ 2.62 → 2.65; R = H
2.66; R = OH
+ 2.31 (Eq. 2.19)

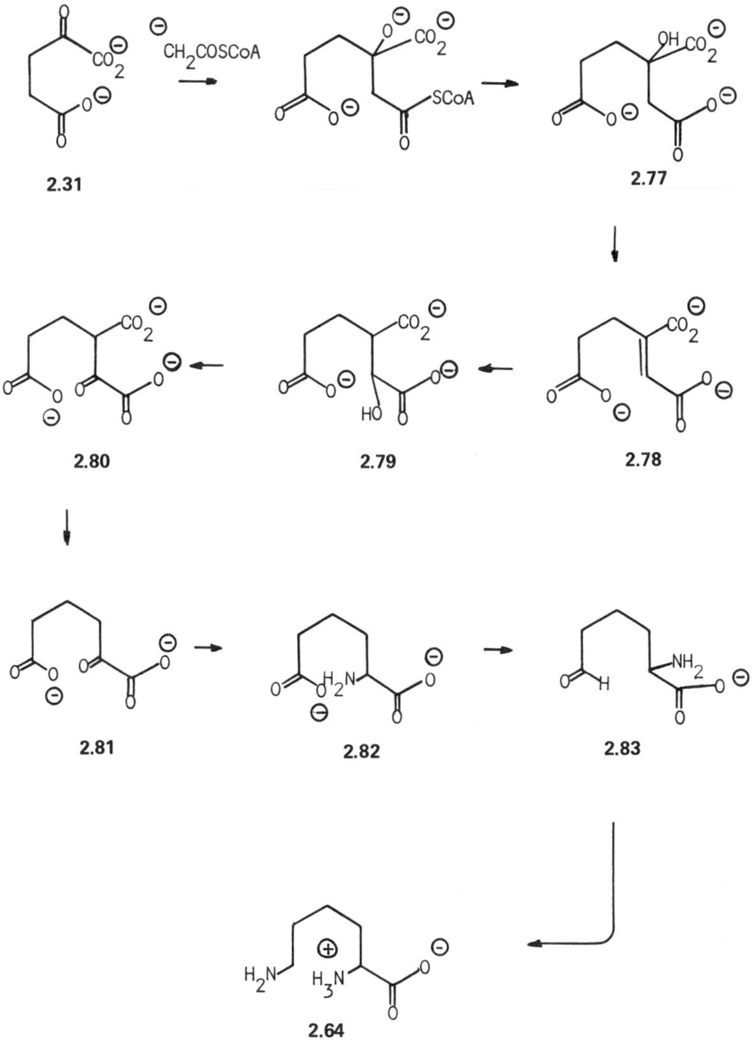
$CH_2COSCoA$
2.31
2.77
2.78
2.79
2.80
2.81
2.82
2.83
2.64

SCHEME 2.19 A route to lysine from α-ketoglutaric acid.

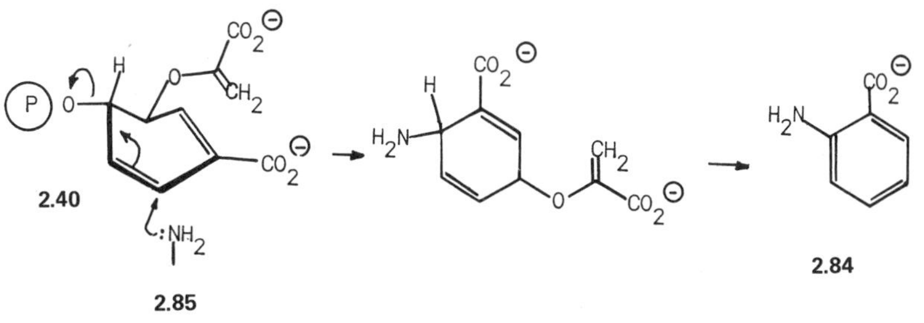
2.40
2.85
2.84

SCHEME 2.20 A potential pathway for the formation of anthranilic acid from chorismic acid.

E. Tryptophan

Tryptophan (2.67), at least in yeasts and Escherichia coli (a bacterium), is known to be formed from anthranilic acid (2.84). The phenylpyruvic acid (2.42) precursor, chorismic acid (2.40), is known to yield anthranilic acid (2.84) and to involve glutamine (2.85) in the process. Not much more information is currently available on this reaction, but the process shown in Scheme 2.20 is plausible. Here, the source of ammonia is shown as glutamine (2.85; generated by aminolysis or transamination of glutamate, 2.62; Equation 2.20).

2.62 → 2.85 (Eq. 2.20)

For tryptophan (2.67), then, anthranilic acid (2.84) reacts with 5-phosphoribosyl-1-pyrophosphate (2.86) at the hemiacetal carbon (C_1) to generate N-(5'-phosphoribosyl)anthranilic acid (2.87). Since this step is reported to proceed with retention of configuration, it may well go through the open form of the sugar (i.e., the aldehyde) which is in equilibrium, at least in principle, with the hemiacetal, rather than through a displacement reaction, which should invert the configuration at the seat of the reaction (Scheme 2.21). Then, ring opening of N-(5'-phosphoribosyl)anthranilic acid (2.87) to the keto form of enol-1-(o-carboxyphenylamine)-1-deoxyribulose-5-phosphate (2.88), followed by electrophilic aromatic subsitution at the carbon bearing the carboxylate (ipso) and rearomatization with loss of carbon dioxide and water, yields indole-3-glycerol phosphate (2.89). These reactions are shown in Scheme 2.21.

Finally, the glycerol phosphate (2.90) is lost as 3-phosphoglyceraldehyde (2.4) and is replaced by serine (2.91; which has arisen via transamination and hydrolysis of 2-phosphohydroxypyruvate, 2.92, obtained from oxidation of 3-phosphoglyceric acid, 2.2). A plausible pathway for these last steps in the formation of tryptophan (2.67) from anthranilic acid (2.84) is shown in Scheme 2.22.

F. Methionine

Methionine (2.60), the amino acid required for methylation of oxygen, nitrogen, and (occasionally) carbon, is generated by methylation of homocysteine (2.93). Homocysteine (2.93) and pyruvate (2.22) result from the decomposition of cystathionine (2.94), which in turn is generated by displacement of succinate (2.32) from O-succinylhomoserine (2.95) by cysteine (2.96). O-Succinylhomoserine (2.95) arises from the reaction of succinyl coenzyme A (2.97; citric acid cycle, p. 21) with homoserine (2.98), which is generated by reduction of aspartate (2.99).* Cysteine (2.96) is formed from the reaction of hydrogen sulfide on serine (2.91). The entire process described above is shown in Scheme 2.23.

G. Histidine

A number of imidazole ring containing alkaloids have been reported but will not be considered in this volume beyond this introduction. It is not yet known, in detail, whether the imidazole ring in the alkaloids in which it occurs is in fact derived specifically from histidine (2.69) or from purine precursors to histidine (2.69). Therefore, the scheme shown (Scheme 2.24)

*Aspartate (2.99), which will not concern us much after this, is formed either by direct amination of fumarate (2.33) or transamination from glutamate (2.62) to oxaloacetic acid (2.26).

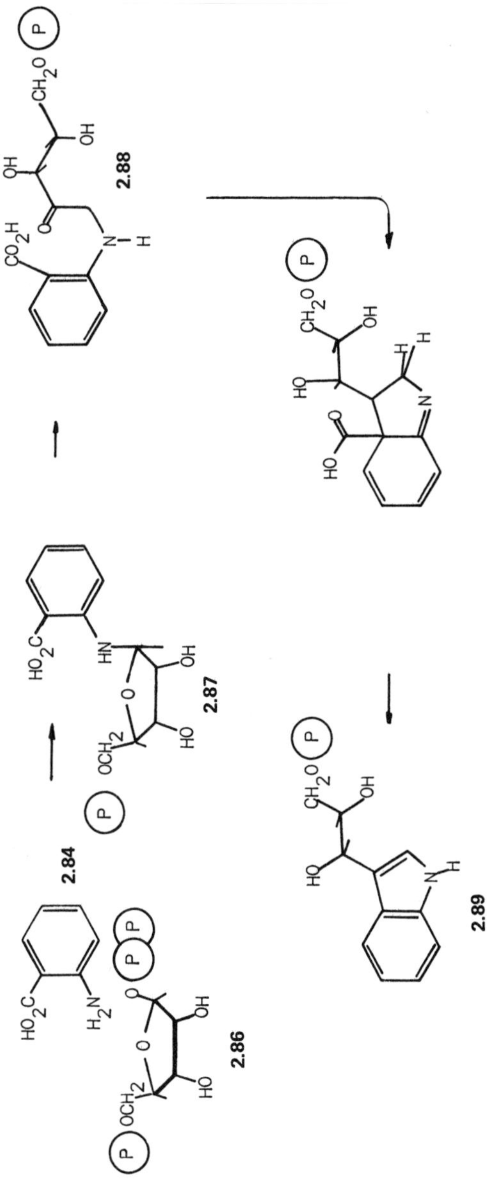

SCHEME 2.21 A pathway to indole-3-glycerol phosphate from anthranilic acid.

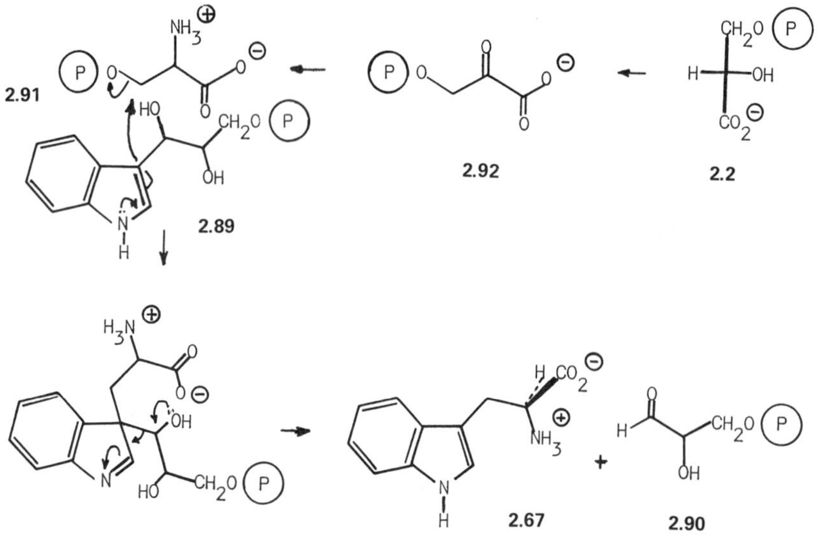

2.91 2.92 2.2

2.89

2.67 2.90

SCHEME 2.22 A pathway for the formation of tryptophan from indole-3-glycerol phosphate and serine.

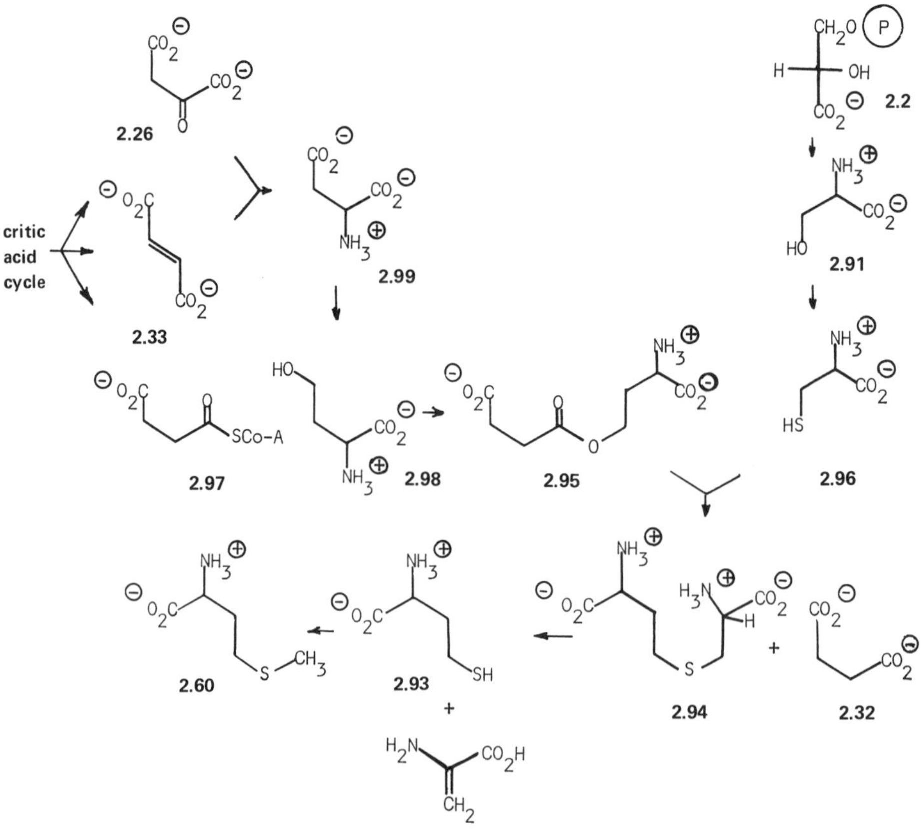

2.26 2.2

critic acid cycle

2.33 2.99 2.91

2.97 2.98 2.95 2.96

2.60 2.93 2.94 2.32

SCHEME 2.23 The formation of methionine.

accounts first for the purine precursor (adenosine monophosphate, AMP, 2.100) and then for histidine (2.69).

As indicated in Scheme 2.24, phosphorylation of ribose-5-phosphate (2.17) generates ribosylpyrophosphate-5'-phosphate (2.86), which as we have just seen is intimately involved in the biosynthesis of tryptophan (2.67; Scheme 2.21). Then, if (as in the biosynthesis of anthranilic acid, 2.84) glutamine (2.85) serves to donate an amino group and replace, in that process, the pyrophosphate residue on the hemiacetal hydroxyl of the ribosylpyrophosphate-5'-phosphate (2.86), 5-phosphoribosylamine (2.101) is produced. Conversion of this amine to an amide of glycine (i.e., 2.102),* followed by a second formylation at the nitrogen terminus from an N-formyltetrahydrofolic acid derivative, produces a formylglycinamide ribotide (2.104). Then, exchange of the amide oxygen for nitrogen (derived from glutamine, 2.85) and cyclization yield aminoimidazole ribotide (2.105) directly. The aminoimidazole ribotide (2.105) is then converted into a purine derivative as follows.

Direct carbon dioxide (or, perhaps, biotin-assisted, 2.45) addition to (2.105) yields 5-amino-4-imidazole carboxylic acid ribotide (2.106). The latter is then converted to the corresponding amide, 5-amino-4-imidazolcarboxamide ribotide (2.107), utilizing aspartic acid (2.68) as the nitrogen source. Elimination to fumaric acid (2.33) leaves the nitrogen attached to the carbonyl carbon. Then, another formylation, followed by cyclization, leads to inosinic acid (2.108). Transamination of inosinic acid (2.108) with aspartate (2.68; the latter going again to fumarate, 2.33) leads to adenosine monophosphate (AMP; 2.100), which as noted above might directly provide imidazole rings to the appropriate alkaloids.

Finally, adenosine triphosphate (2.109), generated by phosphorylation of the monophosphate (2.100), condenses with ribosylpyrophosphate-5'-phosphate (2.86) to form N-1-(5'-phosphoribosyl)adenosine triphosphate (2.110), which in the presence of ammonium ions (presumably glutamine, 2.85, could again serve as the source of nitrogen) fragments to imidazole glycerol phosphate (2.111) and 5-amino-1-ribosyl-4-imidazolcarboxamide-5'-phosphate (2.112). The imidazole glycerol phosphate (2.111), on loss of water, generates an enol (2.113), and the latter, on ketonization and transamination, leads to histidinol phosphate (2.114). Hydrolysis and oxidation of histidinol phosphate (2.114) yield histidine (2.69). Scheme 2.24 summarizes the reactions indicated above.

III. THE AMINES

The final set of intermediates we need to consider for biogenetic elaboration of the alkaloids are amines derived from some amino acids already discussed. Although the compounds mentioned here will be dealt with again later when each of the amino acids we will be utilizing is discussed, the generality of the decarboxylation of amino acids to amines requires emphasis. Some of the amines generated by the decarboxylation of amino acids (which is pyridoxal-mediated, 2.72) have, as we shall see, been used as presumed biosynthetic precursors in actual incorporation experiments with excellent results. Also, as we shall see, it is not necessarily true that incorporation of the amine requires that it be an obligatory precursor. That is, an aberrant biosynthetic pathway may exist which utilizes amine when it is present but does not require early-stage decarboxylation of the amino acid. Regardless, the pathway to these decarboxylated species is presumably available, and some amines have been detected in plants which elaborate alkaloids. Scheme 2.25 sets forth a pathway to the amines, and some of them, with the corresponding amino acid from which they are generated, are presented in Table 2.1.

---

*Glycine (2.103) is derived from serine (2.91) by loss of a carbon to tetrahydrofolate (2.61) to generate N-formyltetrahydrofolate, which, it will be remembered, is required for the synthesis of methionine (2.60).

2.17 2.86 2.101 2.102 2.103 2.91 2.104 2.105 2.106 2.68 2.33 2.107 2.108 2.100 2.110 2.86

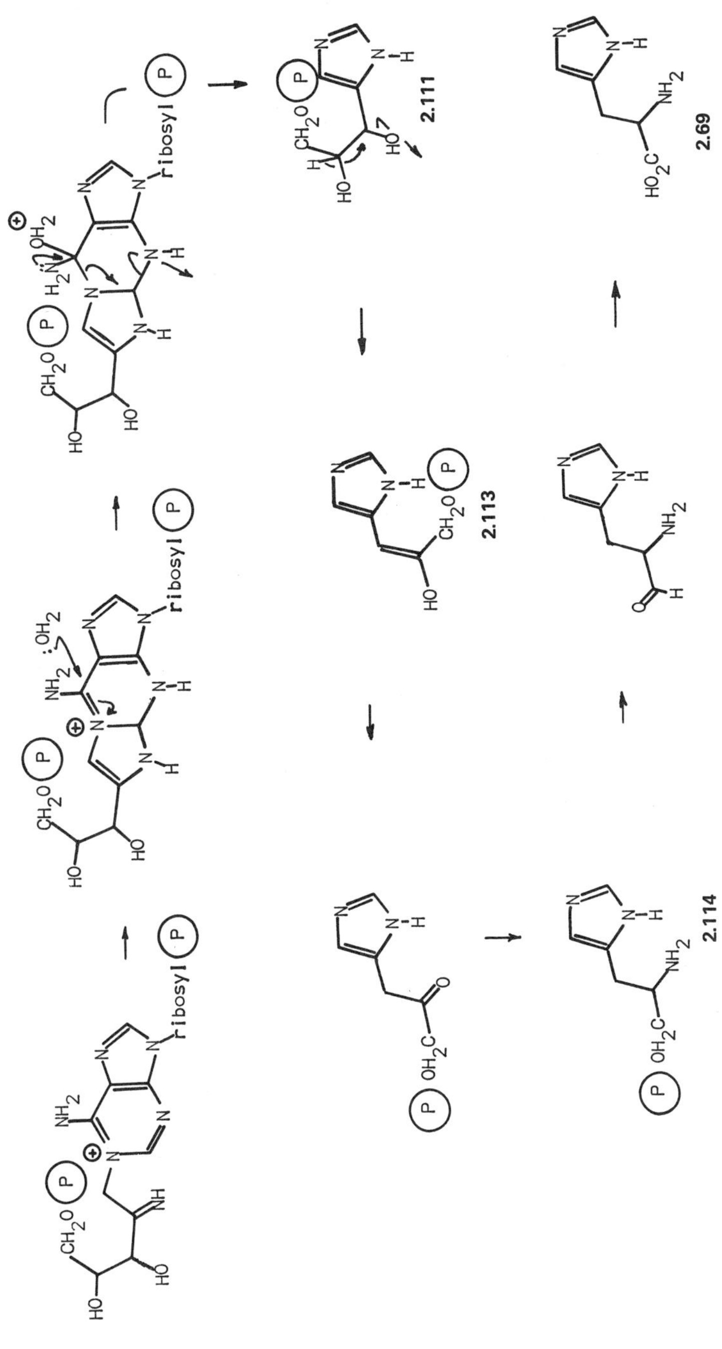

SCHEME 2.24 A pathway for the formation of histidine.

TABLE 2.1

Some Amines Present in Some Alkaloid-bearing Plants and the Amino Acids from Which They Are Derived

Amino Acid	Amine
Ornithine (2.63)	Putrescine (2.114)
$^{\ominus}O_2CCHCH_2CH_2CH_2NH_2$ $NH_3^{\oplus}$	$H_2NCH_2CH_2CH_2CH_2NH_2$
Lysine (2.64)	Cadaverine (2.115)
$^{\ominus}O_2CCHCH_2CH_2CH_2CH_2NH_2$ $NH_3^{\oplus}$	$H_2NCH_2CH_2CH_2CH_2CH_2NH_2$
Phenylalanine (2.65)	β-Phenethylamine (2.116)
$C_6H_5CH_2CHCO_2^{\ominus}$ $NH_3^{\oplus}$	$C_6H_5CH_2CH_2NH_2$
Tyrosine (2.66)	Tyramine (2.117)
p-$HOC_6H_4CH_2CHCO_2^{\ominus}$ $NH_3^{\oplus}$	p-$HOC_6H_4CH_2CH_2NH_2$
Tryptophan (2.67)	Tryptamine (2.118)
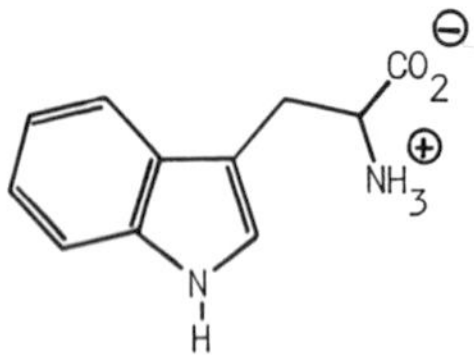	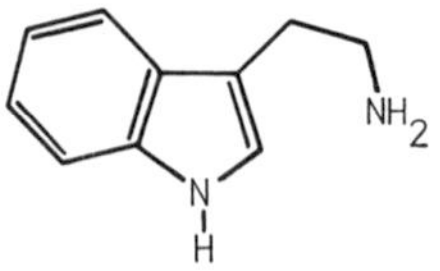

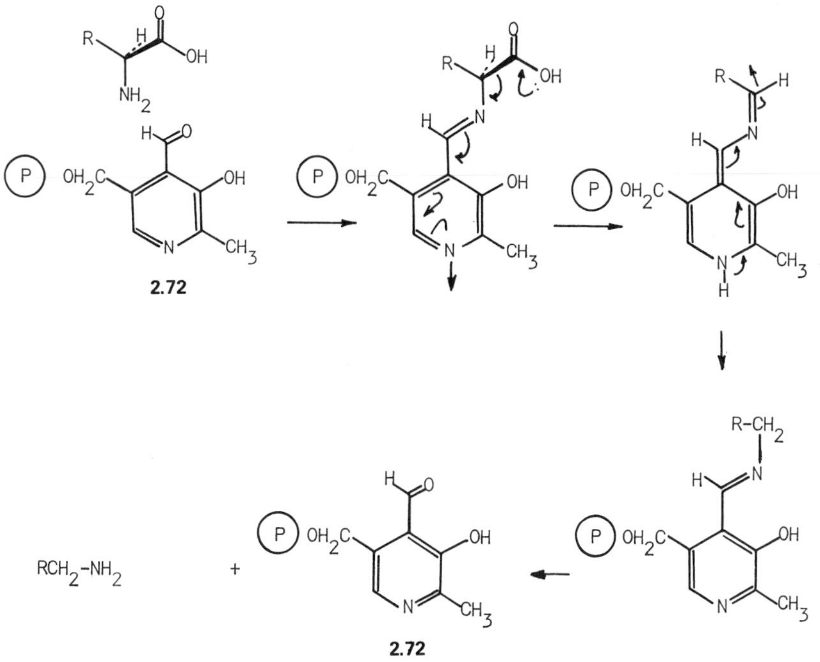

SCHEME 2.25 A generalized pathway for the decarboxylation of amino acids to amines.

GENERAL READING FOR PART 1

P. Bernfeld, ed., Biogenesis of Natural Compounds, 2nd ed. Pergamon, New York, 1967.
A. L. Lehninger, Biochemistry, 2nd ed. Worth, New York, 1975.
H. R. Mahler and E. H. Cordes, Biological Chemistry. Harper and Row, New York, 1966.

Part 2

ALKALOIDS DERIVED FROM ORNITHINE (EXCLUDING THE TOBACCO ALKALOIDS)

Ornithine (3.1) is known to be incorporated, intact and in a nonsymmetric fashion, into a variety of plant bases. These include the simple pyrrolidines hygrine (3.2) and cyscohygrine (3.3), the tropane (7-azabicyclo[3.2.1]heptane) alkaloids of which cocaine (3.4) is an example, and the pyrrolizidine (1-azabicyclo[3.3.0]heptane) alkaloids typified by retronecine (3.5). In addition, ornithine (3.1) may be considered, on reasonable biogenetic grounds, to be incorporated into some of the Securingea alkaloids, e.g., norsecurinine (3.6), as well as into the *Elaeocarpus* alkaloids such as elaeocarpine (3.7).

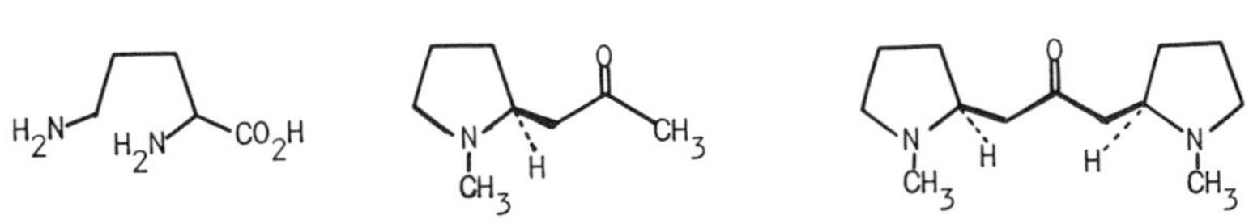

H_2N H_2N CO_2H

3.1

N CH_3 H CH_3

3.2

N CH_3 H H N CH_3

3.3

3.5

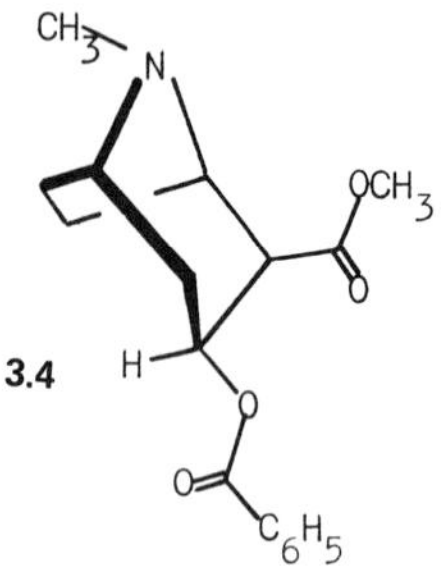

3.4

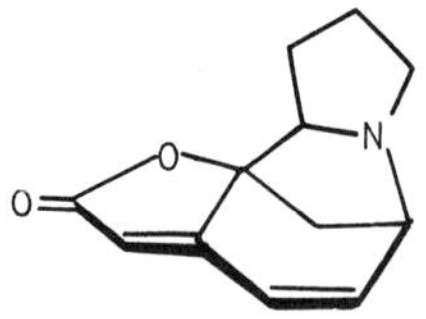

3.6

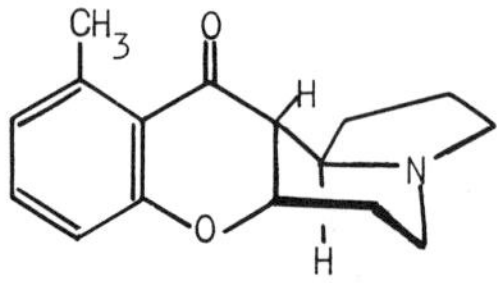

3.7

3

BIOSYNTHESIS OF PYRROLIDINE, TROPANE, AND PYRROLIZIDINE ALKALOIDS

> Why is it that we entertain the belief that for every purpose odd numbers are the most effectual?
>
> Pliny the Elder, Natural History, Book XXVIII, Section 23

I. BIOSYNTHESIS OF HYGRINE, CUSCOHYGRINE, AND THE TROPANE ALKALOIDS

Classical hypotheses suggested that ornithine (3.1) underwent decarboxylation to putrescine (3.8), and conversion of this to 4-aminobutanal (3.9) was followed by cyclization to the Δ^1-pyrroline (3.10). Since 2-^{14}C-labeled ornithine (3.1) was incorporated specifically (i.e., in a nonsymmetrical fashion) in many compounds containing such pyrrolidine nuclei, and since nitrogen bore a methyl group in most of these, it was believed that N-methyl-Δ^1-pyrrolinium salts (3.11) could not undergo 1,2- to 1,5- double bond isomerism readily. Now, however, it is fairly well established that although putrescine (3.8) can serve as a precursor, it is not an obligatory one. Thus, labeling experiments demonstrate that the pathway from ornithine (3.1) to hygrine (3.2) requires the terminal (δ) nitrogen in ornithine (3.1) be utilized in ring formation with loss of the α-amino function [that is, α-^{15}N-ornithine (3.1) is incorporated with loss of ^{15}N while δ-^{15}N-ornithine (3.1) retains activity, as does N-[δ-^{14}C-methyl]-δ-^{15}N-ornithine (3.1)].

It is, therefore, not unreasonable to postulate (based upon a published suggestion; [18]) that it is the coordination of the α-nitrogen with pyridoxal (3.12) required for deamination and decarboxylation which begins the sequence,* and ultimately, it is this coordinated nitrogen

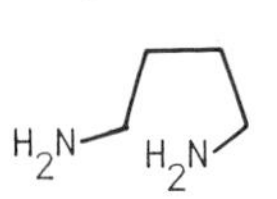

3.8 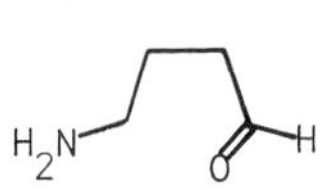 3.9 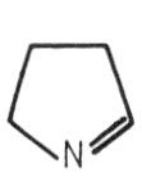 3.10 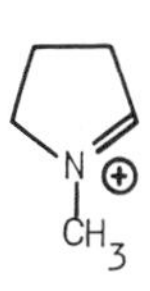 3.11

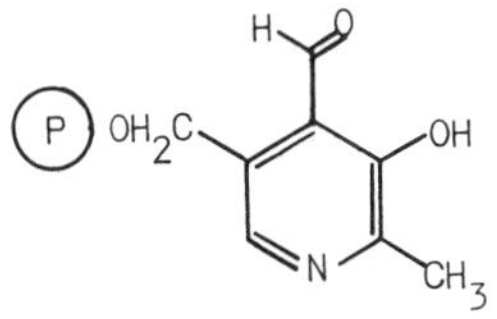

3.12 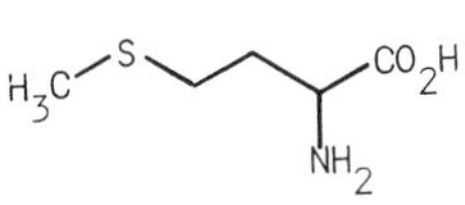 3.13

*Methylation at the terminal nitrogen by methionine (3.13) may be early or late stage.

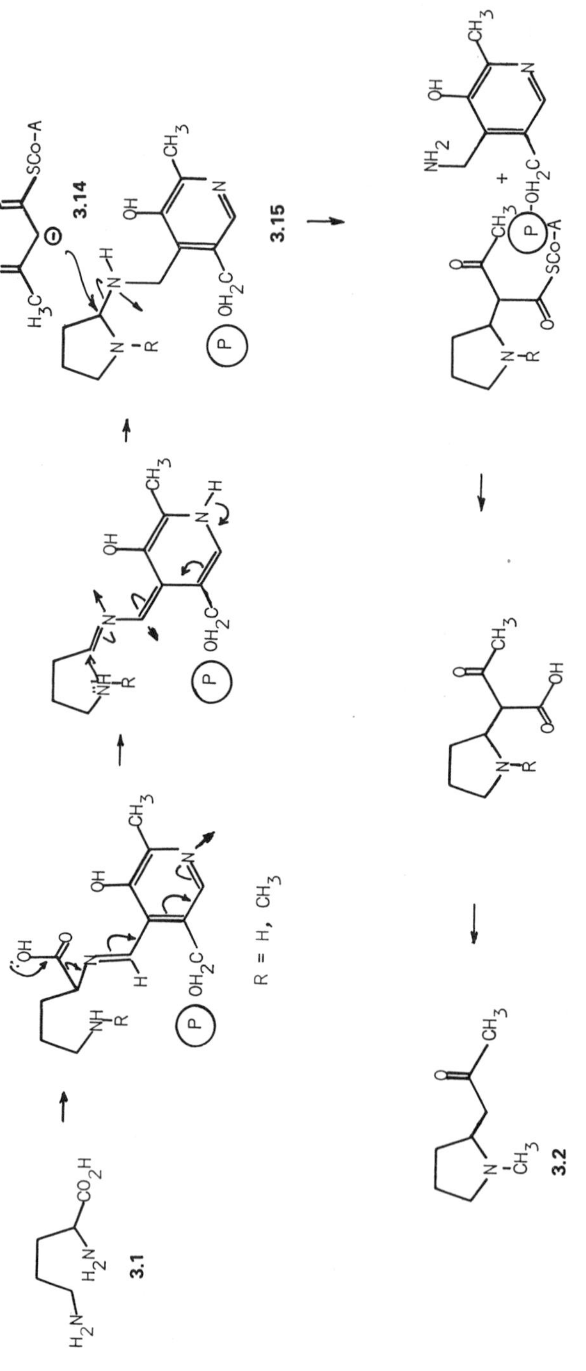

SCHEME 3.1 A pathway for the biosynthesis of hygrine.

which, as expected, is lost on cyclization. Attack on the still coordinated cyclic species by a four-carbon, acetate-derived unit (i.e., 3.14; in concert with acetate-labeling experiments) results in displacement of the pyridoxamine (3.15) leaving group and generation of a β-keto-acid (3.16), which leads on decarboxylation to hygrine (3.2; Scheme 3.1). Reaction of hygrine (3.2) itself with a second ornithine (3.1) coordinated with pyridoxal (3.12; possibly after conversion of hygrine, 3.2, to a malonyl derivative, i.e., 3.17, by direct carboxylation of the terminus, for activation) would then lead to cuscohygrine (3.3). Alternatively, oxidation of either hygrine (3.2) or the hypothetical activated hygrine (3.17) to the pyrroline (3.18) followed by cyclization and reduction leads either to tropine (3.19) or the carboxyl derivative (3.20), respectively, which could conveniently serve as a precursor for cocaine (3.4; Scheme 3.2). In this regard, suitably labeled tropine (3.19) has been examined as a precursor for

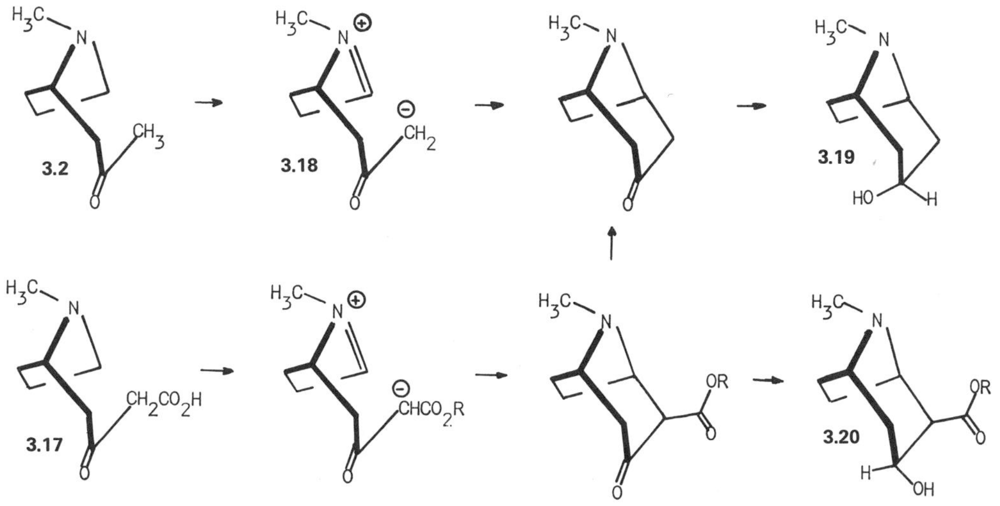

SCHEME 3.2 A pathway for the generation of the tropane alkaloids.

the other tropane alkaloids, and, indeed, it was efficiently converted to meteloidine (3.21), hyoscyamine (3.22), and hyoscine (3.23; Scheme 3.3).* The details of the late-stage oxygenations in evidence here are unknown, but since oxygen is introduced onto the exo face of the bicyclic system, either oxygen addition to an appropriate alkene (which has itself not yet been found) or N-oxide formation, followed by suitable one-electron transfer processes and hydrogen and hydroxyl migrations, might be responsible.

The nonsymmetrical incorporation of ornithine (3.1) into hyoscyamine (3.22) was demonstrated in an elegant fashion. Hyoscyamine (3.22), isolated from feeding experiments of (±)-[2-^{14}C]ornithine (3.1) to Datura stramonium L. is chiral. When the base is induced to undergo elimination by treatment with sulfuric acid, a mixture of enantiomeric alkenes is obtained. Hofmann elimination on the exhaustively methylated alkenes then produces a mixture of enantiomeric α-methyl tropidines (3.24 and 3.25, respectively). Since the Hofmann reaction is equally likely from either enantiomer, nonsymmetrical incorporation of (±)-[2-^{14}C]ornithine (3.1) must generate one α-methyltropidine enantiomer bearing nitrogen at labeled (i.e., ^{14}C-labeled) carbon and its mirror image bearing nitrogen at unlabeled carbon.

*The origin of the carboxylic acid portion of these esters will be considered in the next section.

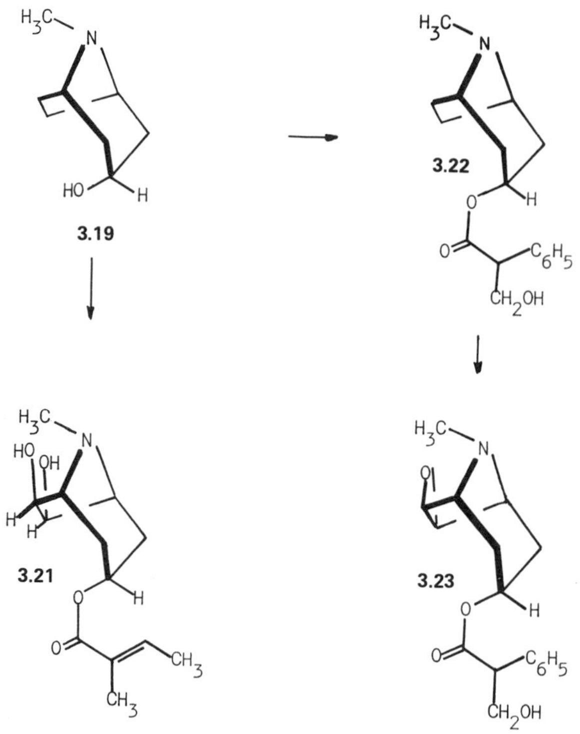

SCHEME 3.3 Some tropane alkaloids derived (labeled feeding experiments) from tropine.

Resolution of the enantiomers 3.24 and 3.25, followed by isomerization to the corresponding enamines, hydrolysis, and hydrogenation, yields a pair of cycloheptanones (3.26 and 3.27), only one of which can bear labeled carbon at the carbonyl. Thus, excision of the carbonyl carbon by reaction of the ketones 3.26 and 3.27 with p-chlorophenylmagnesium bromide and oxidation to the corresponding benzoic acids provide unambiguous proof that (+)-S-[2-^{14}C]ornithine (3.1) is incorporated into hyoscyamine (3.22) in such a fashion as to place the label exclusively at C_1 of hyoscyamine (3.22). The degradation discussed above is shown in Scheme 3.4.

II. THE ACIDS OF THE TROPANE ALKALOIDS

As already noted, all tropane alkaloids, regardless of other substitution, bear a hydroxyl (α or β)* at C_3 as required by genesis of that portion of the molecule from acetate. Except for the parent bases, tropine (3.19) and ψ-tropine (3.28), the remainder of the tropane alkaloids occur as esters of carboxylic acids. These acids are either aliphatic [e.g., tiglic (E-2-methyl-2-butenoic acid), 3.29 and 3-methylbutenoic, 3.30] C_5 amino acid derived or aromatic [e.g., benzoic, tropic (3-hydroxy-2-phenylpropanoic acid), 3.31, and cinnamic] amino acid

---

*Because the six-membered ring can flex, axial and equatorial positions may interchange. Thus, in concert with accepted usage, if the hydroxyl is cis (Z) to the nitrogen bridge, it will be β, while if trans (E) it will be α.

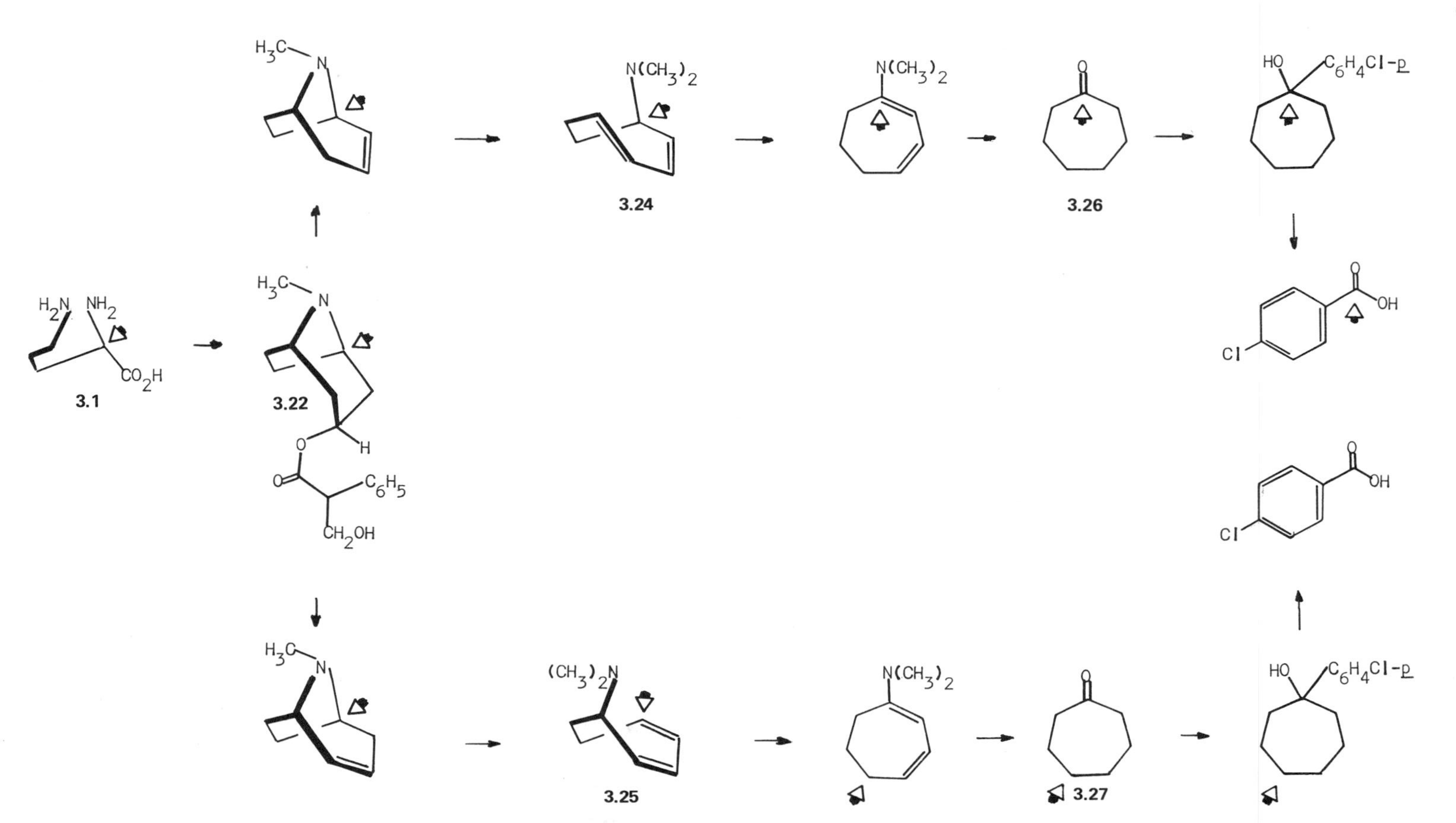

SCHEME 3.4 Degradation of hyoscamine demonstrating that ornithine is incorporated in a nonsymmetrical fashion.

derived. Hydrolysis of the naturally occurring esters yields the various tropines, and the carboxylic acids are generally isolated by acidification with aqueous mineral acid.

Tiglic acid (3.29) from, e.g., meteloidine (3.21) is derived from isoleucine (3.32; Scheme 3.5), and 3-methylbutyric acid (3.30) from, e.g., valeroidine (3.33) is derived from leucine (3.34; Scheme 3.6). Benzoic acid from, e.g., tropacocaine (3.35), tropic acid (3.31) from, e.g., hyoscyamine (3.22), and hyoscine (3.23) and phenyllactic acid (3.36) from, e.g., littorine (3.37), are derived from phenylalanine (3.38). The formation of benzoic acid from phenylalanine (3.38) is straightforward (Scheme 3.7) as is the reduction of phenylpyruvic acid (3.39) to phenyllactic acid (3.36). However, as indicated in Scheme 3.8, the pathway from phenylalanine (3.38) to tropic acid (3.31) is not as clear. Indeed, although a good deal of information is currently available, the transformations indicated are speculative.

H_3C N HO H

3.19

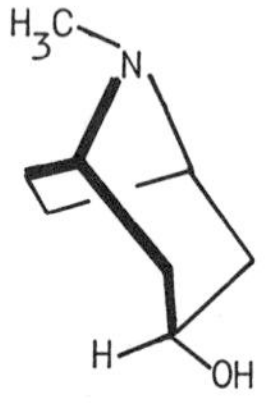

H_3C N H OH

3.28

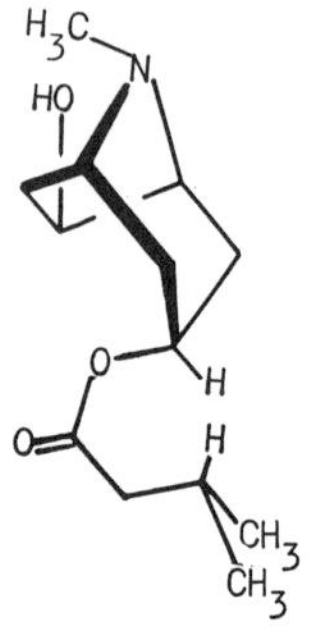

H_3C N HO O H O H CH_3 CH_3

3.33

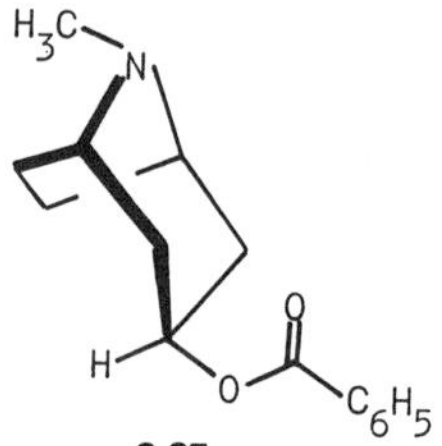

H_3C N H O O C_6H_5

3.35

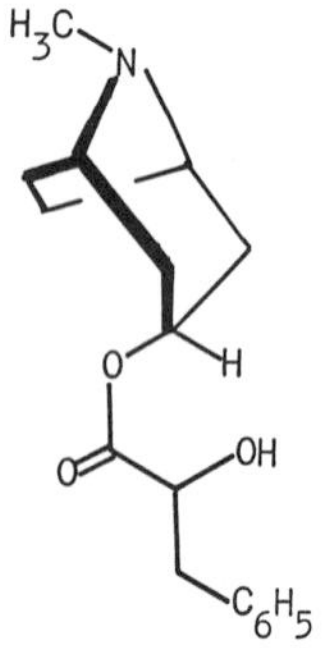

H_3C N O H O OH C_6H_5

3.37

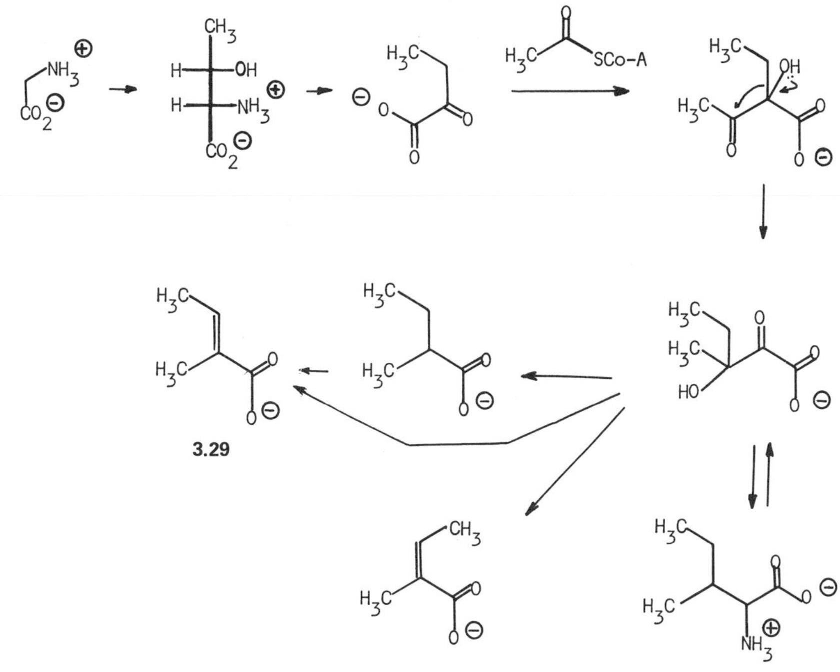

3.29

3.40

3.32

SCHEME 3.5 A pathway for the formation of tiglic acid (and angelic acid) from isoleucine and the formation of isoleucine from glycine.

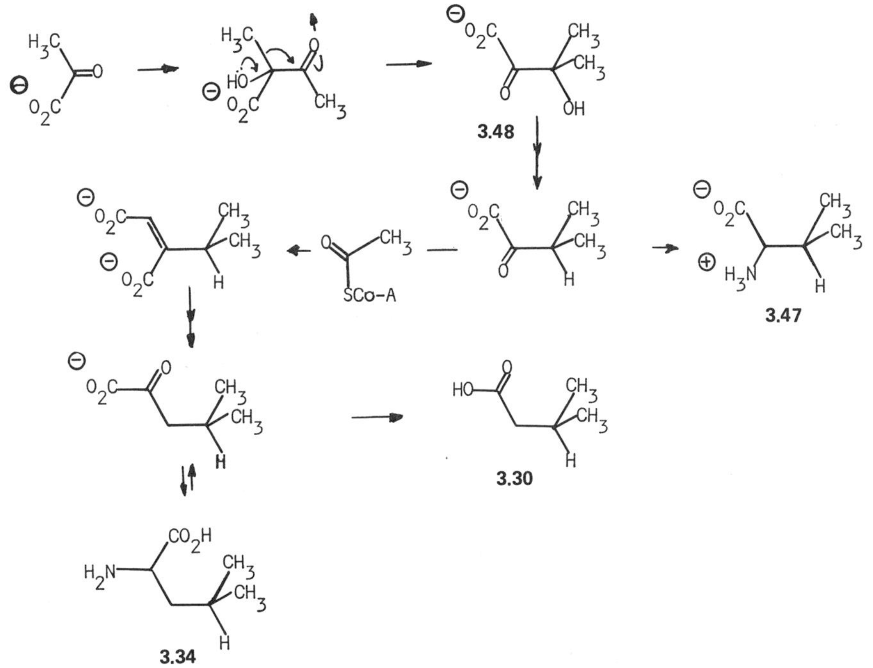

3.48

3.47

3.30

3.34

SCHEME 3.6 The formation of leucine, valine, and 3-methylbutyric acid from, presumably, pyruvate.

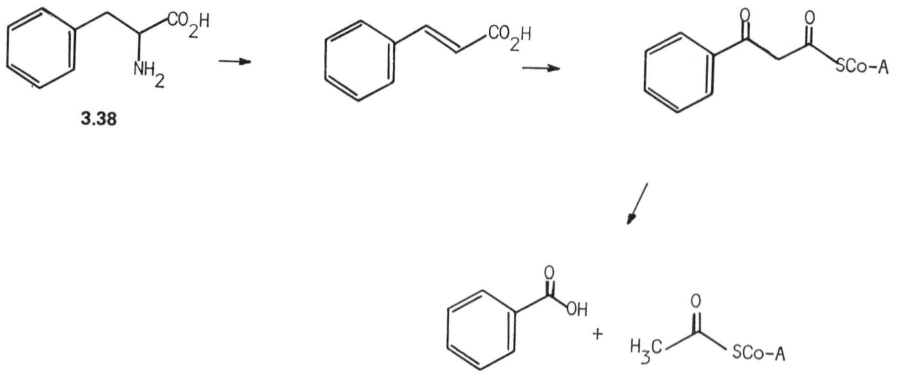

SCHEME 3.7 A pathway for the formation of benzoic acid from phenylalanine.

Equation 3.1 contains, in an abbreviated form, the results of a series of labeling studies which Scheme 3.8 attempts to rationalize. While other precursors which may be derived from phenylalanine (3.38) might be considered possible (because rational, mechanistic pictures can

Equation 3.1 A summary of the labeling experiments carried out with phenylalanine to determine the pathway to tropic acid.

be drawn), many of them have been examined and discarded. The suggestion [19] that cinnamic acid might be a precursor was particularly attractive; however, no incorporation was found when it was fed.*

Finally, although the details are not clear, it appears that feeding [2-^{14}C]-2,3-[$^{3}H_3$]-phenylalanine (3.38) resulted in the retention of only about one-third of the ^{3}H when the phenylalanine (3.38) was converted to tropic acid (3.27). This is consistent with retention of only one of the two hydrogens at C_3 of phenylalanine (3.38) and is also consistent with the observation that both phenylpyruvic (3.39) and phenyllactic (3.36) acids also serve as precursors to tropic acid (3.31). Therefore, as outlined in Scheme 3.8, it is suggested that a pyridoxal-complexed (3.12) imine, which can be wirtten for phenylalanine (3.38) and can also be readily derived from phenylpyruvic acid (3.39) and phenyllactic acid (3.36), may be involved.

In such a pathway, instead of the normal decarboxylation occurring, proton loss from the benzylic carbon generates an enamine. Now, although α,β-unsaturated carboxylic acids are normally coplanar, here, because of the high nucleophilicity of the enamine anion, it is

*At this writing, the details of this experimental work have not been made available, and, indeed, lack of incorporation may have a variety of causes, not the least of which is that the material is not a precursor.

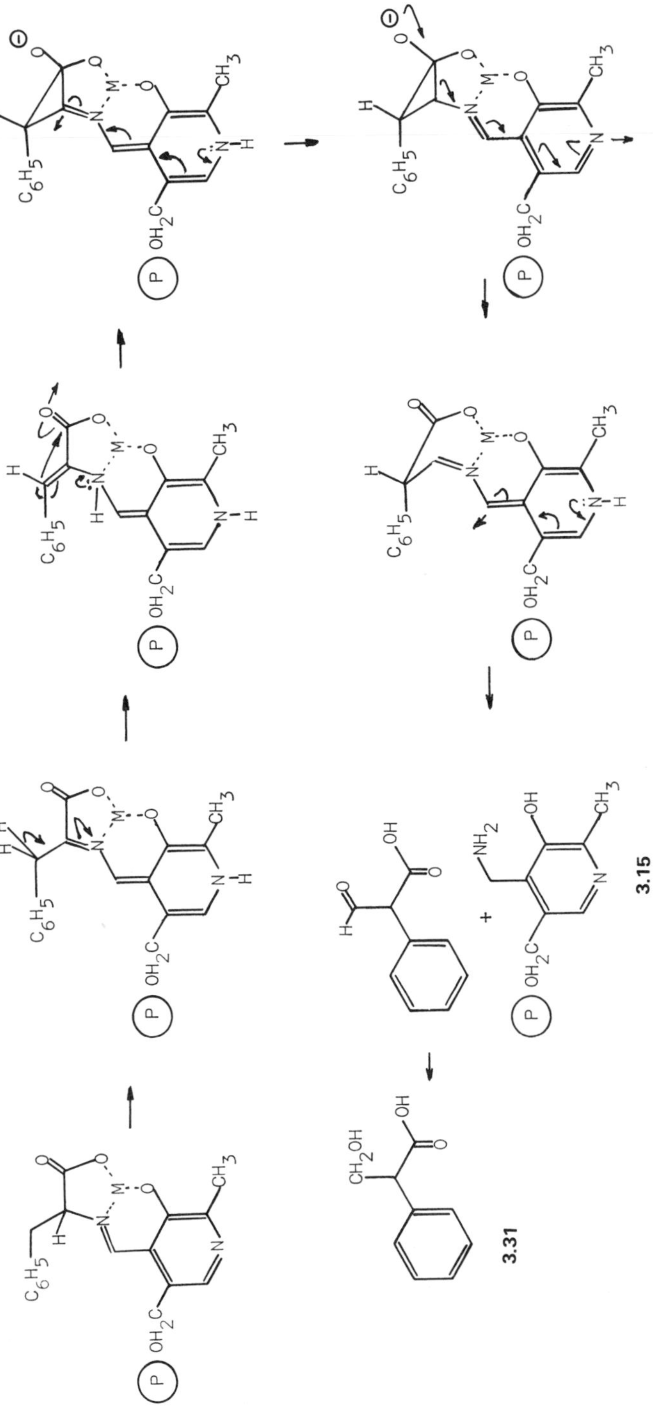

SCHEME 3.8 A possible pathway from phenylalanine to tropic acid which accounts for the experimental observations regarding labeling patterns.

possible that attack on the unusually electrophilic metal-coordinated carbonyl can occur.* The resulting intermediate, similar to that found in a Favorskii rearrangement, can proceed to generate an imine, which on hydrolysis forms an aldehyde whose reduction leads to tropic acid (3.31), or the imine can lead to alcohol via an alcoholamine and direct reduction.

III. BIOSYNTHESIS OF THE PYRROLIZIDINE ALKALOIDS

The bicyclic pyrrolizidine alkaloids, of which well over 100 are known, occur largely as esters of amino alcohols (called necines) and carboxylic acids (called necic acids). Most of these bases fall into four broad categories (which specifically exclude the unesterified amino alcohols themselves, a fifth category, since they rarely occur in nature). These classes are (a) simple monoesters of the necines; (b) diesters of the necines in which one of the acids is angelic acid (Z-2-methyl-2-butanoic acid; 3.40); (c) cyclic diesters of the necines in which there is a single dicarboxylic acid which has esterified both hydroxyl functions; and (d) N-oxides of the above bases. Typical members of these groups include heliosupine (3.41), which is composed of the necine heliotridine (3.42), and the necic acids angelic (3.40) and echimidinic (3.43); senecionine (3.44), made up of retronecine (3.5) and senecic acid (3.45); and 9-angelylretronecine-N-oxide (3.46), which is the N-oxide of the angelic acid (3.40) ester of retronecine (3.5; [20]).

Alkaline hydrolysis suffices to separate the necines from the necic acids. The latter, when they do not undergo spontaneous decomposition, are liberated from the basic solution on acidification. Similarly, hydrogenolysis of the acid attached to the allylic alcohol (with concomitant loss of the hydroxyl function to the carboxylic acid) permits separation of the carboxylic acid attached to the secondary alcohol (at C_7, which is not lost) from the acid attached to the primary, and allylic, alcohol (which is lost).

IV. THE NECINES

Information resulting from a study of the feeding of labeled ornithine to, for example, Senecio istideus, shows that two ornithine-derived (3.1) fragments have been incorporated. Scheme 3.9 attempts to take this observation into account by generating the same hypothetical precursor as seen earlier in the pathway for the formation of pyrrolidine and tropane alkaloids (Scheme 3.1) and permitting it to react with another identical fragment. The nonsymmetrical intermediate so formed, on rearrangement, generates an aldehyde which can serve as precursor to the necines.

Data based upon feeding experiments (acetate) and contradictory to those shown in Scheme 3.9 have been reported. This feeding work implies a symmetrical intermediate is present and would then suggest that ornithine (3.1) may be converted to putrescine (3.8) prior to further elaboration. Regardless of the details, however, it is sufficient for our purpose to indicate that there is every indication that, in fact, ornithine is a precursor of the Senecio alkaloids and may be incorporated in a symmetrical or a nonsymmetrical fashion.

---

*It is very important to recognize that the material presented in this scheme, in contradistinction to much of that presented previously, is highly speculative and without precedent but, nevertheless, consonant with the known experimental details.

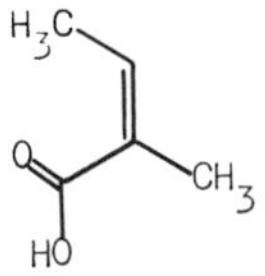

H_3C CH_3 HO O

3.40

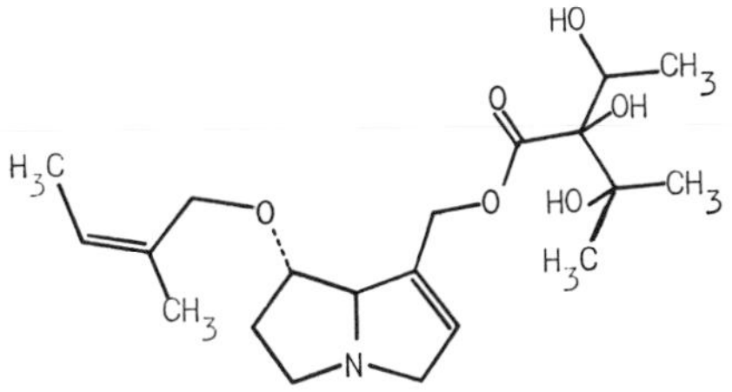

H_3C O CH_3 N O O HO CH_3 OH HO CH_3 H_3C

3.41

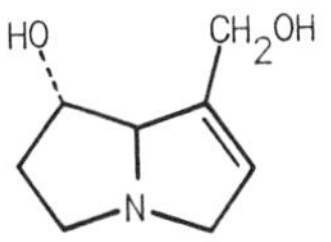

HO CH_2OH N

3.42

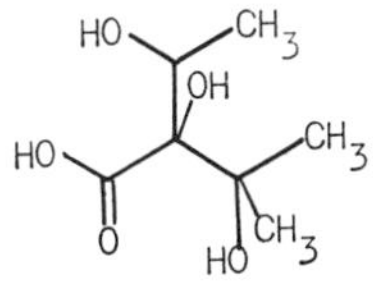

HO CH_3 OH HO CH_3 O HO CH_3

3.43

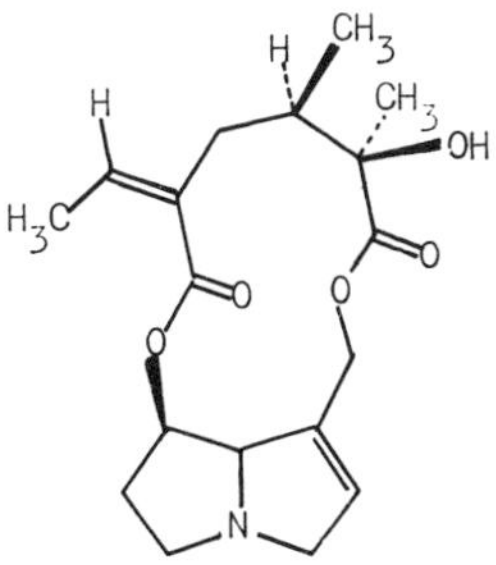

H H_3C H CH_3 CH_3 OH O O O O N

3.44

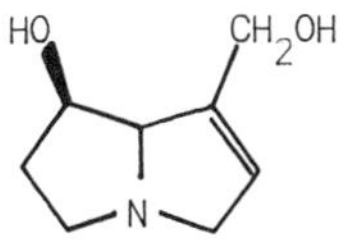

HO CH_2OH N

3.5

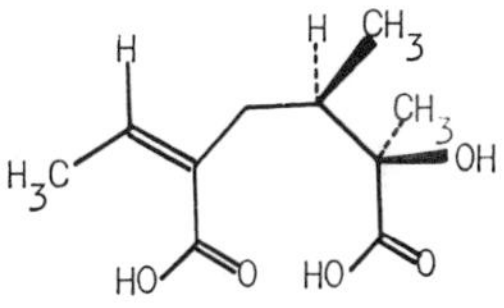

H H_3C H CH_3 CH_3 OH HO O HO O

3.45

H_3C HO O O CH_3 N O

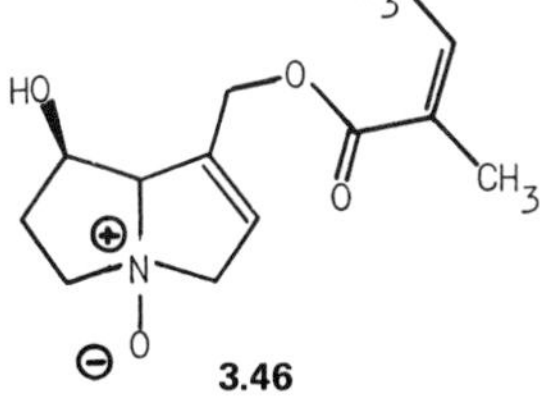

3.46

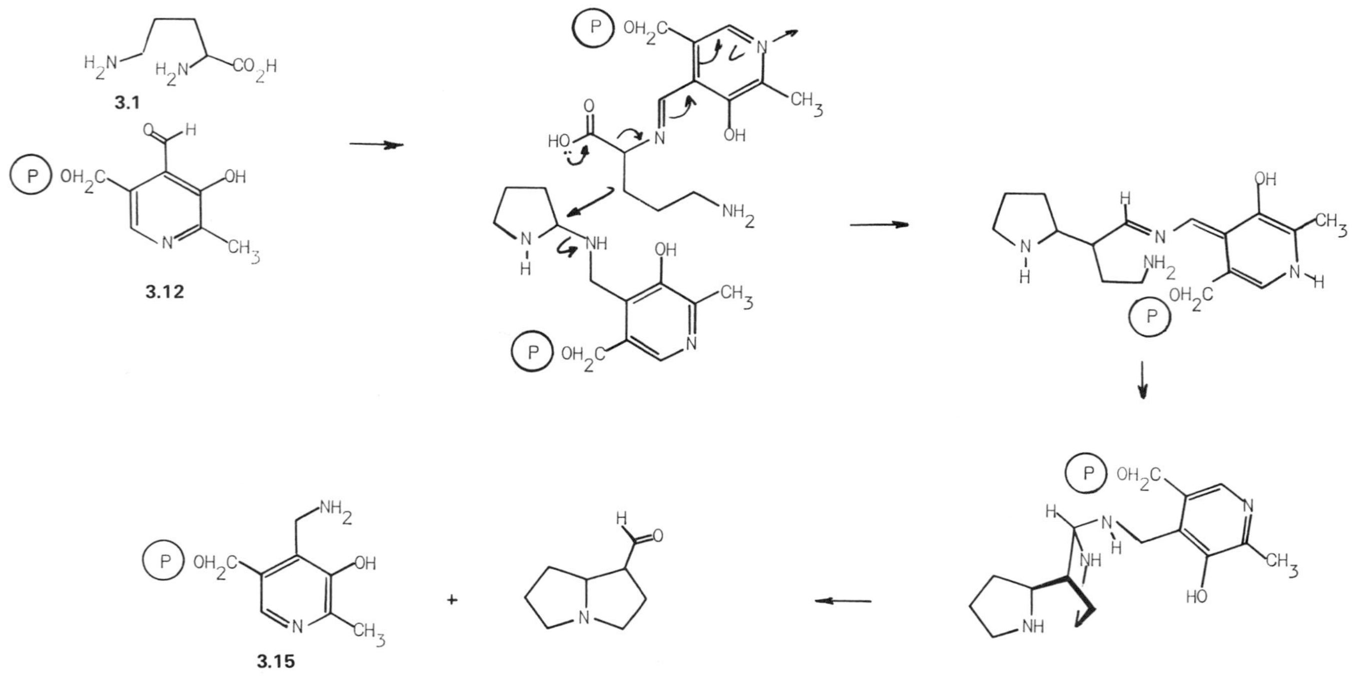

SCHEME 3.9 A possible pathway for the formation of the fundamental necine skeleton

V. THE NECIC ACIDS

The acids with which the necines are esterified are generally not found elsewhere in alkaloids. For some time, these acids, which include the C_5 monocarboxylic acid, angelic acid (3.40), the C_7 monocarboxylic acid echimidinic acid (3.43), and a number of C_{10} dicarboxylic acids such as senecic acid (3.45), were thought to be acetate (mevalonate) derived. It is now clear that, at least for angelic (3.40), echimidinic (3.43), and senecic (3.45) acids (and by analogy for the others), that they are derived from simple amino acids. Thus, angelic acid (3.40), like its isomer tiglic acid (3.29), is derived from isoleucine (3.32; Scheme 3.5), while echimidinic acid (3.43) comes from valine (3.47; Scheme 3.6) by addition, at the α-carbon, of an activated (thiamine pyrophosphate) two-carbon fragment. In any event, β-hydroxy-α-ketoisovaleric acid (3.48; Scheme 3.6) is the intermediate to which the acetate-derived fragment is presumably added (Scheme 3.10). In the case of the C_{10} acid, senecic acid (3.45), two C_5 fragments which come from isoleucine (3.32) are joined together. Scheme 3.11 presents a plausible pathway for the generation of senecic acid (3.45).

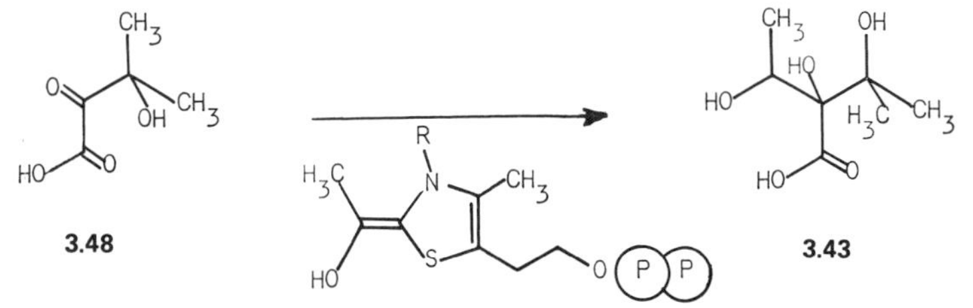

3.48 → 3.43

SCHEME 3.10 A suggested pathway for the formation of echimidinic acid.

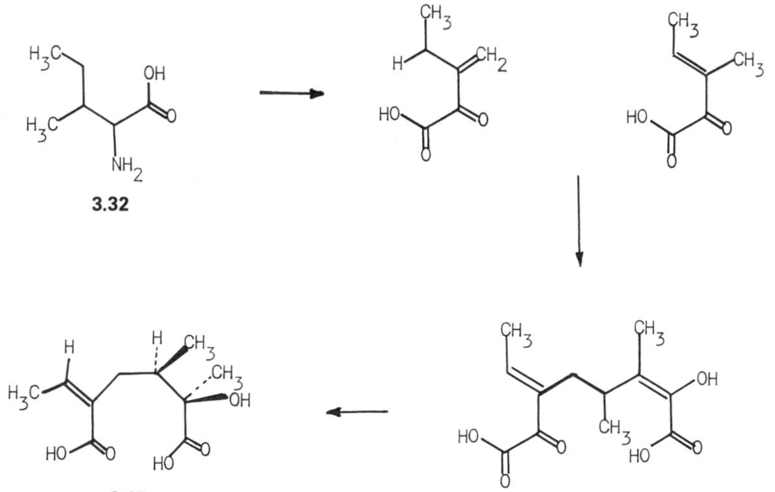

3.32 → … → 3.45

SCHEME 3.11 A possible pathway for the formation of senecic acid.

4

CHEMISTRY OF THE PYRROLIDINE ALKALOIDS

> So very difficult a matter is it to trace and find
> out the truth of anything by history. . . .
>
> Plutarch, Lives of the Ancient Greeks

Along with cocaine (4.1), the Peruvian coca shrub (Erythroxylon truxillence Rusby, not Theobroma cacao, the source of cocoa) contains hygrine (4.2) and cuscohygrine (4.3), the only members of the pyrrolidine alkaloids to be discussed here. The oily base hygrine (4.2; $C_8H_{15}NO$), which distills at 193-195°C without decomposition, was first isolated in an impure form by Wohler and Lossen in 1862 and later (1889) in a more highly purified state by Liebermann.

Because hygrine (4.2) readily forms an oxime, it was recognized as containing a carbonyl group. Chromic acid, in sulfuric acid, oxidizes hygrine (4.2) to hygrinic acid (4.4; $C_6H_{11}NO_2$), with a melting point (mp) of 164°C, different from any of the three piperidine carboxylic acids and convertible, on dry distillation, to N-methylpyrrolidine (4.5). This ready decarboxylation led to the inference that hygrinic acid (4.4) was an α-amino acid. The position of the carbonyl group in the side chain was established by synthesis. Thus, the product from the reaction of N-methylpyrrole (4.6) with diazoacetone, followed by catalytic hydrogenation yields hygrine on oxidation (4.2, and its mirror image; Scheme 4.1). Resolution of the synthetic hygrine (4.2) with tartaric acid yields (-)-hygrine (4.2) $[\alpha]_D^{18}$, -1.8° (water), which undergoes rapid racemization, thus accounting for the difficulties originally experienced in determining whether the naturally occurring hygrine (4.2) was optically active

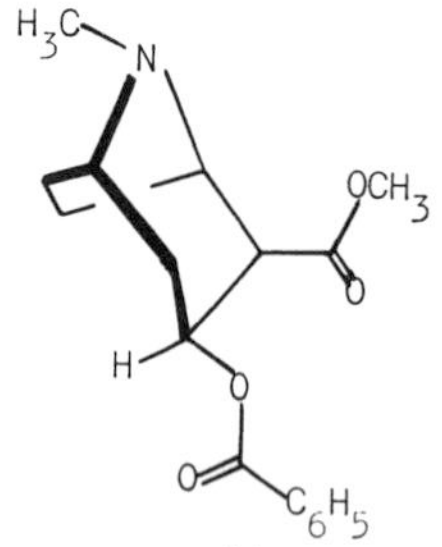

4.1

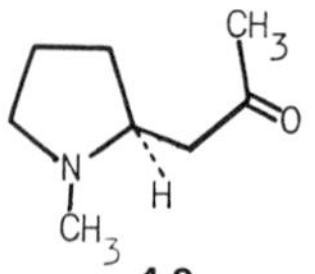

4.2

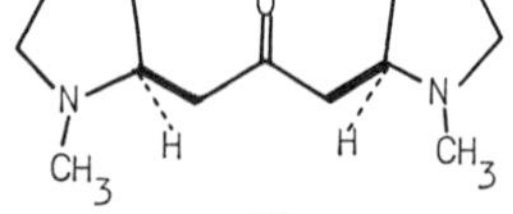

4.3

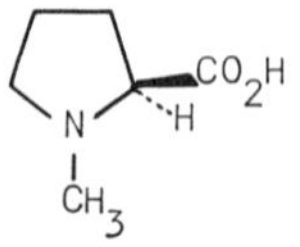

4.4

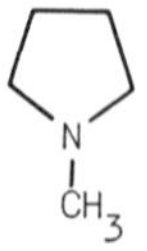

4.5

or whether the small negative rotation observed was due to impurities. (-)-Hygrinic acid (4.4), obtained by chromic and sulfuric acid oxidation of hygrine (4.2), was related to L-(-)-proline (4.7) since both yielded (-)-stachydrine (4.8; obtained from alfalfa) on methylation. [L-(-)-Proline (4.7) has been related to L-(+)-glutamic acid.] From the higher boiling residue of the distillate, Leibermann isolated an oil, with a boiling point (bp) of 152°C/14 mm ($C_{13}H_{24}NO_2$), which forms a very hygroscopic dihydrochloride. This new base, cuscohygrine (4.3), also gives hygrinic acid (4.4) on oxidation and two (α and β) dihydrocuscohygrines (4.9) on reduction with sodium in ethanol, either or both of which can be reoxidized with chromic acid to cuscohygrine (4.3). Successive Hofmann degradations and hydrogenations carried out

4.9

on the mixture of dihydrocuscohygrines (4.9) yield a mixture of n-undecane and n-undecan-6-ol, confirming Liebermann's assignment of the structure of cuscohygrine (4.3). Cuscohygrine (4.3) has been synthesized from N-methyl-2-hydroxypyrrolidine (4.10) and acetonedicarboxylic acid. Hygrine (4.2) accompanies cyscohygrine (4.3) in the synthesis. Naturally occurring cuscohygrine (4.3) is optically inactive and must, of course, be meso rather than racemic, since two (α and β) epimeric alcohols are formed on reduction (see above). A racemate can only afford a single racemic alcohol.

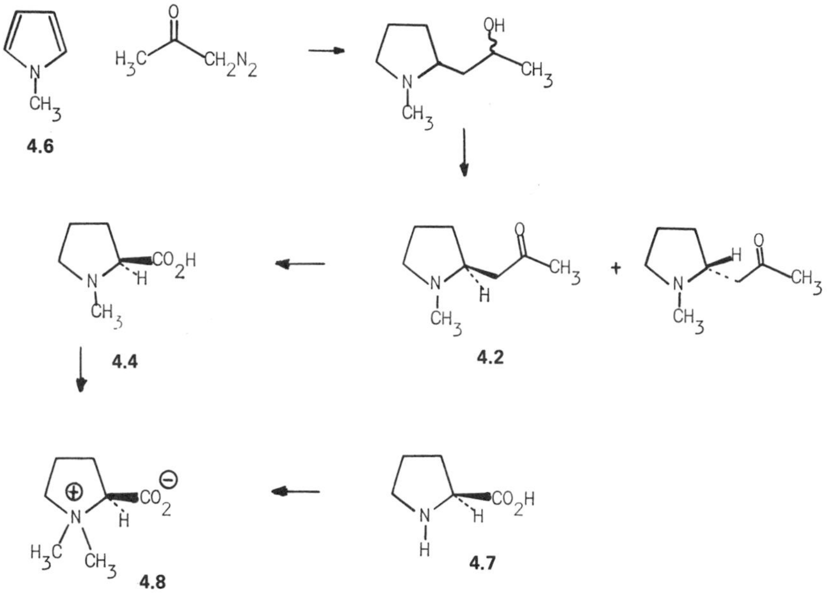

SCHEME 4.1 A synthesis of hygrine and the relationship of (-)-hygrine to L-(-)-proline.

SELECTED READING

L. Marion, in The Alkaloids, Vol. 1, R. H. F. Manske (ed.). Academic, New York, 1950, pp. 91 ff.

L. Marion, in The Alkaloids, Vol. 6, R. H. F. Manske (ed.). Academic, New York, 1960, pp. 31 ff.

5

CHEMISTRY OF THE TROPANE ALKALOIDS

Belladonna, n. In Italian, a beautiful lady; in English, a deadly poison. A striking example of the essential identity of the two tongues.

Ambrose Bierce, The Devil's Dictionary

The tropane alkaloids to be discussed here are found in two different families, namely, Erythroxylaceae and Solanaceae. Although the same compounds also occur in a few other families, and additional compounds occur in these two, the compounds with which we shall deal are typical. The dried leaves of Erythroxylon coca Lam. have been chewed for many years by the natives of Peru, where it grows wild. The leaves have long been recognized as containing a central nervous system (CNS) stimulant. The isolation of cocaine, which numbs the tongue and lips and proved to be a good local and surface anesthetic, led to excitement in the medical community. Since cocaine was not originally recognized as causing true addiction when taken internally or by injection, the creation of plantations in Bolivia, Brazil, and Java where, particularly in the latter case, a higher concentration of alkaloids per kilogram of dried leaf is typical, seemed justified. Related tropane alkaloids have been isolated from species of the family Solanaceae. Members of this family, which includes henbane (Hyoscyamus niger L.), the deadly nightshade (Atropa belladonna L.), and the thorn apple (Datura stramonium L.), have a history of inducing hallucinations, severe illness, or even death. In specified small doses, even crude plant extracts prepared from members of the Solanaceae family have been used to relax smooth muscle, thus easing intestinal and bronchial spasms, to dilate the pupil of the eye, and as a specific antagonist of morphine.

I. (±)-HYOSCYAMINE (ATROPINE)

In plants of the Solanaceae family, the most ubiquitous alkaloid is (-)-hyoscyamine (5.1), which has been known since 1833. Heating (-)-hyoscyamine (5.1) under vacuum or boiling it in chloroform is sufficient to cause its racemization. The racemic modification is called "atropine," and although it too has been reported as occurring naturally, the ease of racemization of (-)-hyoscyamine (5.1) makes such reports questionable.

(-)-Hyoscyamine (5.1), and thus atropine, correspond to $C_{17}H_{23}NO_3$, mp 108-111°C, atropine 116-117°C. Hydrolysis with aqueous acid or base generates an optically inactive base (called "tropine," 5.2), $C_8H_{15}NO$, and a racemic acid (called "tropic" acid, 5.3),* $C_9H_{10}O$, which, depending upon the conditions, may undergo dehydration to optically inactive atropic acid (5.4; Equation 5.1). Very gentle (neutral) hydrolysis of (-)-hyoscyamine (5.1)

---

*Shown as S-(-)-tropic acid (5.3).

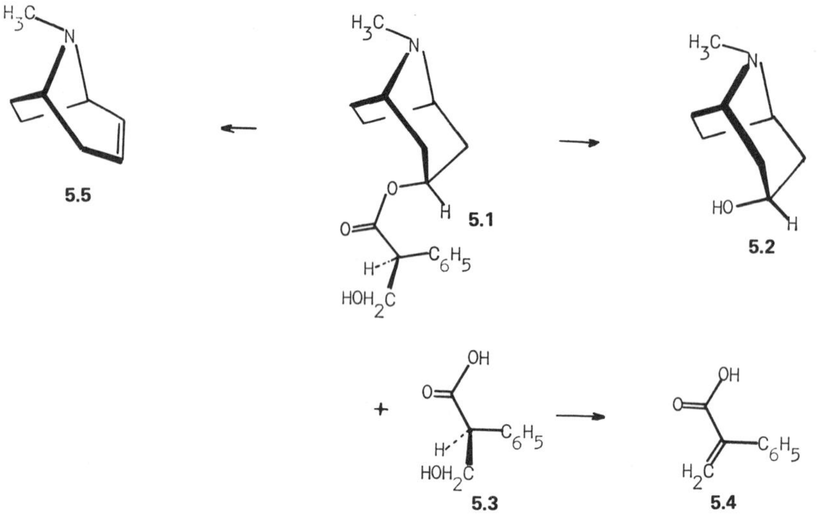

5.5 5.1 5.2 (Eq. 5.1)

5.3 5.4

generates tropine (5.2) and (S)-(-)-tropic acid (5.3). The use of hydrochloric or hydrobromic acids in the hydrolysis causes formation of a halotropine, in which a halogen corresponding to that in the acid has replaced a hydroxyl function. The use of sulfuric acid for the hydrolysis causes dehydration and leads to an unsaturated base ($C_8H_{13}N$), called tropidine (5.5; Equation 5.1). Attempted bromination of tropidine (5.5) with a large excess of bromine leads on heating to 3,5-dibromopyridine, 1,2-dibromoethane, and methyl bromide (Equation 5.2).

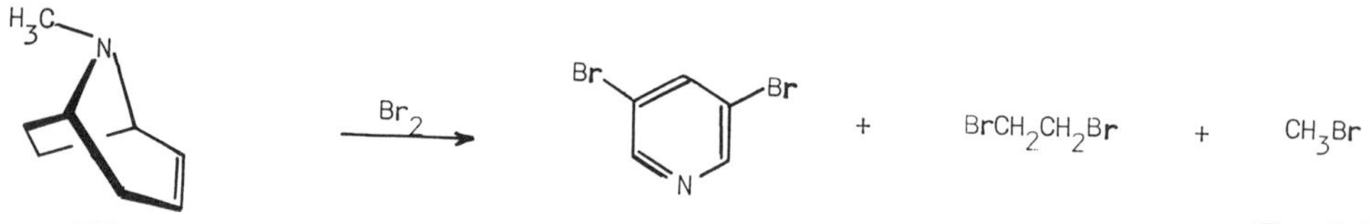

5.5 $\xrightarrow{Br_2}$ 3,5-dibromopyridine + $BrCH_2CH_2Br$ + CH_3Br (Eq. 5.2)

Tropine (5.2) reacts with one equivalent of methyl iodide to form a quaternary base, which when digested with moist silver oxide and heated (Hofmann degradation) decomposes to α-methyltropidine (5.6; Equation 5.3). Heating α-methyltropidine (5.6) ($C_9H_{15}N$) to about

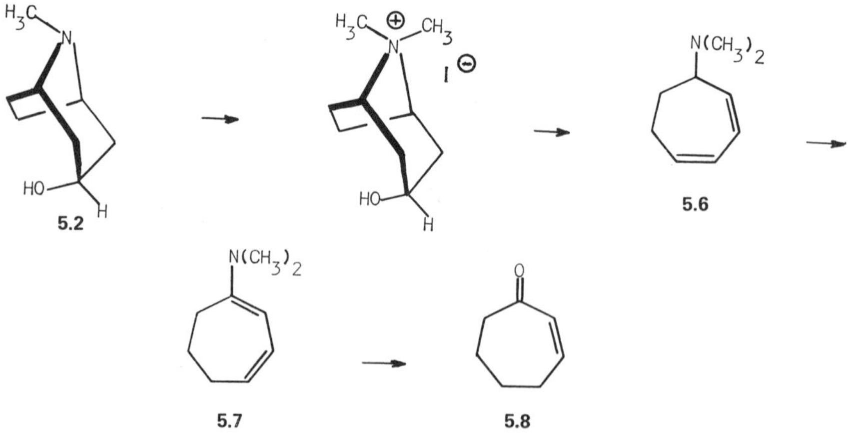

5.2 → → 5.6 → (Eq. 5.3)

5.7 → 5.8

150°C causes it to isomerize to β-methyltropidine (5.7), which in contradistinction to its isomer α-methyltropidine (5.6) readily undergoes hydrolysis to dimethylamine and tropiline (5.8), $C_7H_{10}O$, a carbonyl-containing compound (Equation 5.3). When α-methyltropidine (5.6) is again methylated with methyl iodide and allowed to undergo a Hofmann degradation, trimethylamine and the hydrocarbon tropilidine (5.9; C_7H_8) are isolated (Equation 5.4).

$N(CH_3)_2$

5.6 → 5.9 (Eq. 5.4)

Careful oxidation of tropine (5.2) with chromic acid leads to a ketone, tropinone (5.10a), which forms a dibenzylidine derivative (Equation 5.5). Of the possible structures (5.10a, 5.10b, and 5.10c) which might be written for tropinone, 5.10b and 5.10c can be rejected because they should be capable of resolution (which tropinone, 5.10a, is not) and they cannot yield tropilidine (5.9) from the Hofmann degradation.

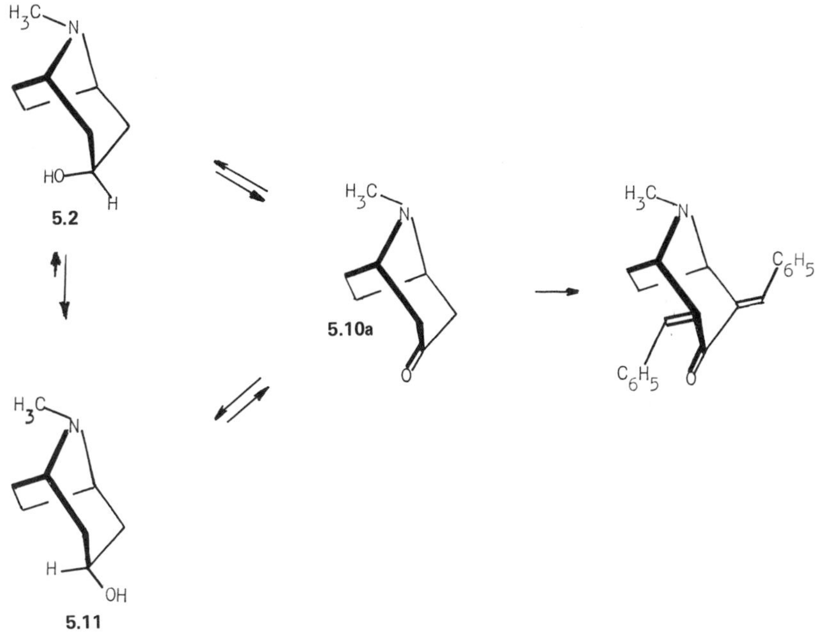

H_3C N HO H 5.2

H_3C N 5.10a O

H_3C N C_6H_5 C_6H_5 O (Eq. 5.5)

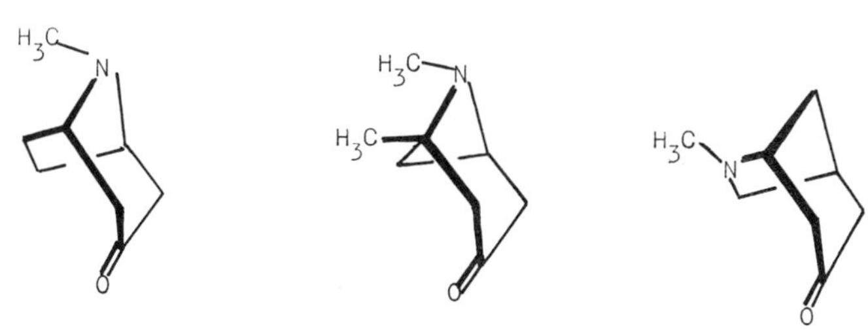

H_3C N H OH 5.11

H_3C N O 5.10a

H_3C N H_3C O 5.10b

H_3C N O 5.10c

Although it was clear that the relationship between tropine (5.2) and tropinone (5.10a) was that of a secondary alcohol to its corresponding ketone, reduction of tropinone (5.10a) with sodium in alcohol does not yield tropine (5.2), i.e., ψ-tropine (5.11) forms. Both tropine (5.2) and ψ-tropine (5.11) can be oxidized to tropinone (5.10a), but at different rates, and indeed, reduction of tropinone (5.10a) with zinc and hydroiodic acid yields a mixture of tropine (5.2) and ψ-tropine (5.11). Finally, ψ-tropine (5.11) was recognized as being the "thermodynamic" isomer since refluxing tropine (5.2) with sodium amylate (in amyl alcohol) converts tropine (5.2) into ψ-tropine (5.11) while the reverse conversion can only be accomplished by oxidation and reduction under nonequilibrating conditions, allowing the "kinetic" isomer (5.2) to be trapped.

The more subtle question of which isomer was tropine (5.2) and which was ψ-tropine (5.11) was solved by converting tropine (5.2) and ψ-tropine (5.11) into their corresponding N-normethylacetamides, 5.12 and 5.13, respectively. This conversion was accomplished (one of a number of ways) by preparation of the N-oxides of the bases themselves with hydrogen peroxide and treatment of the N-oxides with acetic anhydride. Here, stereoelectronic factors prevent other elimination reactions from occurring, and selective hydrolysis of the amide-acetates generates the amides 5.12 and 5.13 (Equation 5.6). Of these two, only 5.13, in concert with experimental observation, can undergo N ⟶ O acyl migration (Equations 5.7 and 5.8). Therefore, in ψ-tropine (5.11), the hydroxyl is equatorial (chair conformation) or β (cis or Z to the nitrogen bridge) while the less stable isomer, tropine (5.2), possesses the hydroxyl axial (chair conformation) or α (trans or E to the nitrogen bridge). Finally, the structure of tropanone (5.10a; and hence tropine, 5.2, and ψ-tropine, 5.11) was established by synthesis. While several syntheses have been reported, that of Robinson served not only to prove the structure but also to give impetus to biogenetic speculation. Thus, at 25°C in aqueous solution at pH 5, after 72 hr, succinic dialdehyde (obtained from pyrrole), methylamine, and the calcium salt of acetonedicarboxylic acid combine to form tropanone (5.10a) in 89% yield (Equation 5.9). As we shall later see, this same general <u>Robinson-Mannich</u> base synthesis has been used time and gain to generate members of this and related families of alkaloids.

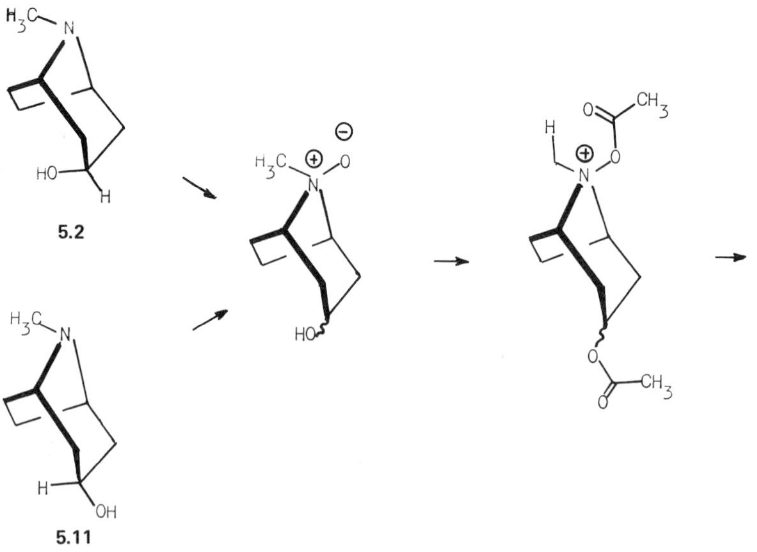

5.2

5.11

(Eq. 5.6)

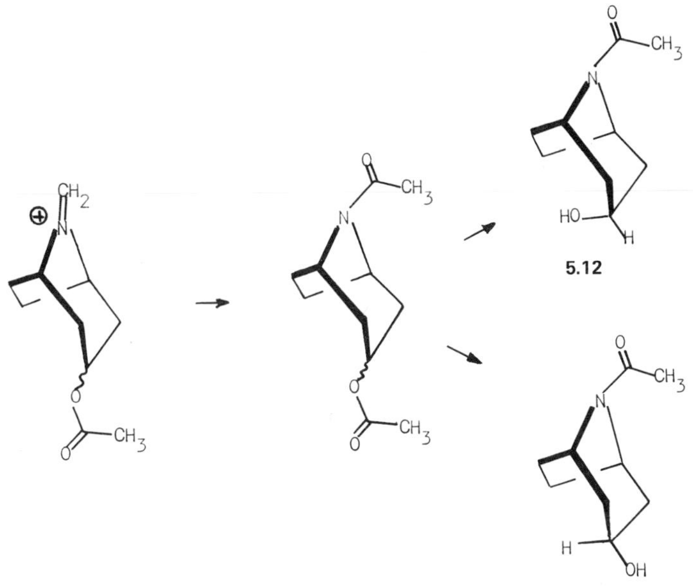

5.12

(Eq. 5.6)
(continued)

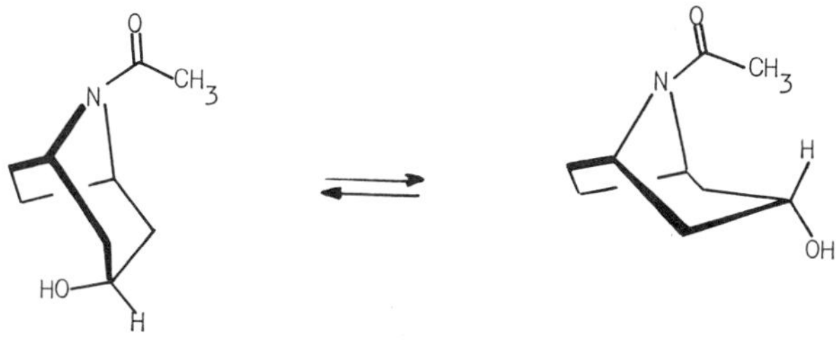

5.13

5.12

(Eq. 5.7)

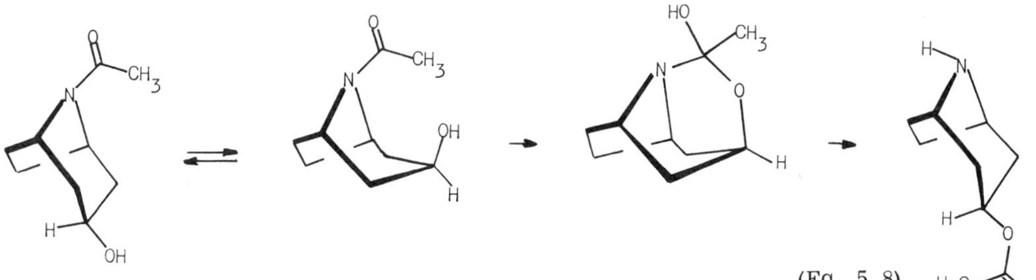

5.13

(Eq. 5.8)

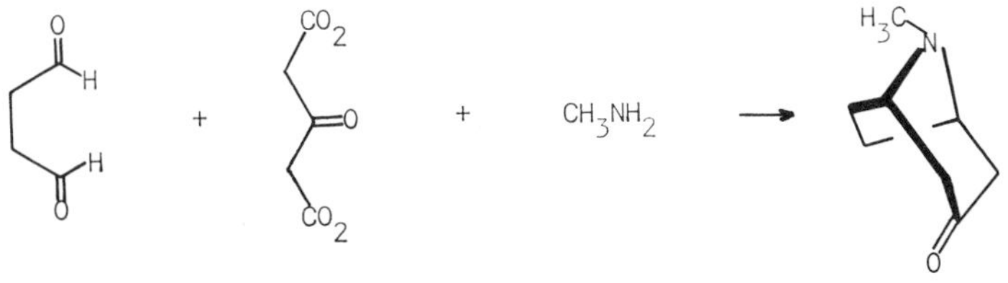

(Eq. 5.9)

5.10a

II. TROPIC ACID

As already pointed out, S-(-)-tropic acid (5.3) is the carboxylic acid with which tropine (5.2) is esterified. Tropic acid (5.3) readily undergoes racemization, and it dehydrates to atropic acid (5.4) under the most mild conditions. Oxidation of atropic acid (5.4) generates benzoic acid, and since atropic acid (5.4) absorbs 1 equivalent of bromine, only three structures for the C_9 carboxylic acid can be written, i.e., 5.14, 5.15, and 5.4.

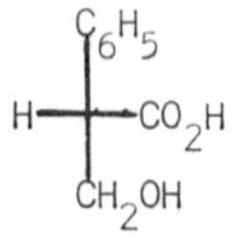

5.3 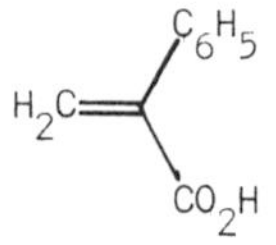 5.4 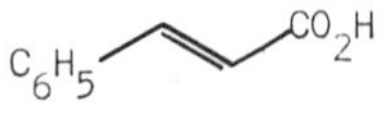 5.14 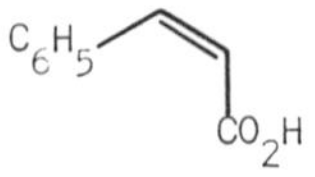 5.15

Structure 5.4 was confirmed by synthesis. Thus, acetophenone cyanohydrin, on hydrolysis, generates atrolactic acid (5.16), the rapid distillation of which (at 10-15 mm) causes dehydration to atropic acid (5.4; Equation 5.10). Now, since atrolactic acid (5.16) is not identical with tropic acid (5.3) but both are capable of resolution and both, on dehydration, lead

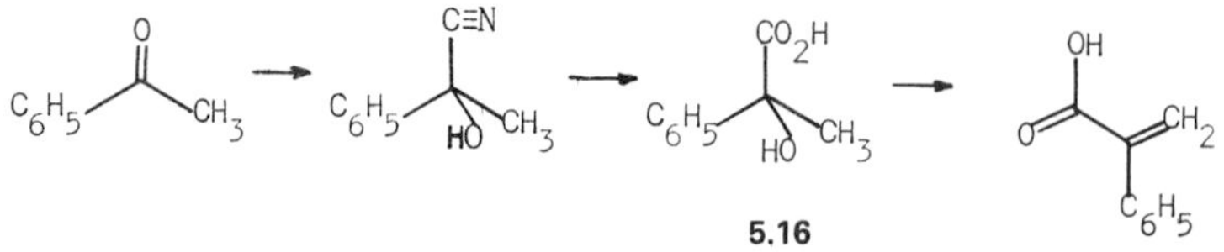

5.16 5.4 (Eq. 5.10)

to atropic acid (5.4), it was assumed that 5.3 accurately represented tropic acid (5.3). That this is indeed the case was confirmed when various phenylacetic acid esters were permitted to condense with formic acid esters to yield α-formylphenylacetic acid esters which were capable of reduction (aluminum amalgam) to racemic tropic acid esters (Equation 5.11).

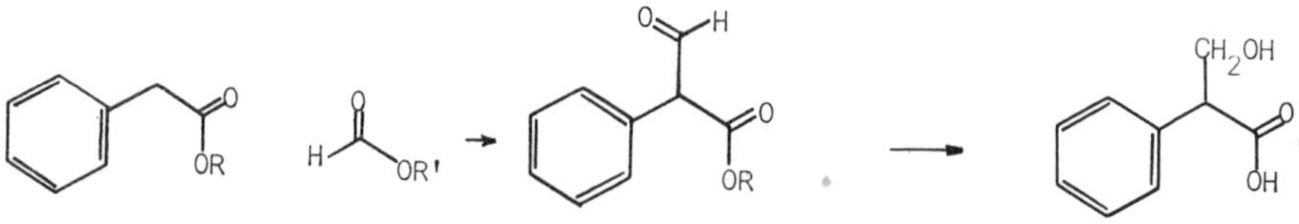

5.3 (Eq. 5.11)

The enantiomers of tropic acid (5.3, and its mirror image) were resolved, e.g., with quinine, and S-(-)-tropic acid affords an acetyl derivative, which after conversion to the corresponding acid chloride yields (-)-acetylhyoscyamine on reaction with tropine (5.2). Gentle hydrolytic removal of the acetyl group affords (-)-hyoscyamine (5.1). The absolute configuration of S-(-)-tropic acid (5.3) has been established by relating it to S-(+)-alanine (5.17; Scheme 5.1). Thus, (-)-β-chlorohydratropic acid (5.18) yields (-)-tropic acid (5.3) on hydrolysis and (-)-methylphenylacetic acid (5.19) on catalytic hydrogenolysis. Neither of these reactions should cause inversion.

Now, (+)-methylphenylacetic acid (5.20) has been converted, via the Curtius degradation, to (-)-α-phenylethylamine (5.21) and thence to S-(+)-alanine (5.17). Since the Curtius degradation

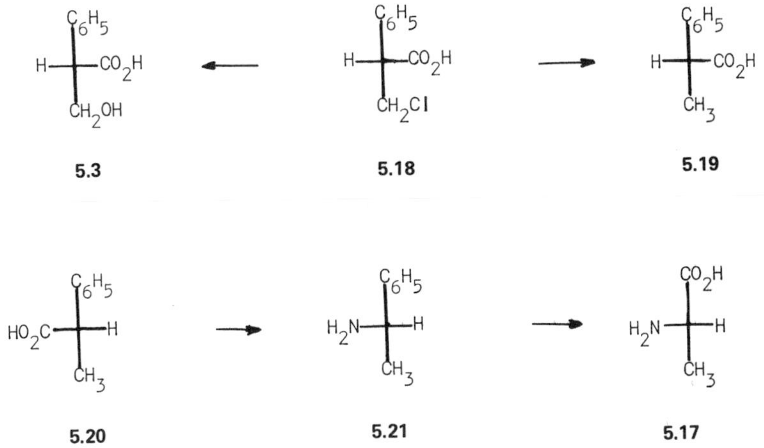

5.3 5.18 5.19

5.20 5.21 5.17

SCHEME 5.1 Proof of the absolute configuration of S-(-)-tropic acid.

is known to proceed with retention of configuration at the seat of migration, and the oxidation to 5.17 does not involve the chiral carbon and therefore also must proceed with retention of configuration, it follows that the configuration of S-(-)-tropic acid (5.3) must be the same as S-(+)-alanine (i.e., since 5.17 is S, 5.20, 5.21, and hence 5.3, must also be S). Thus, S-(-)-hyoscyamine (5.1) must be S-(-)-tropyltropan-3α-ol (5.1).

III. (-)-HYOSCINE (SCOPOLAMINE)

(-)-Hyoscine (5.22) was initially obtained in 1881 from the mother liquors which remained after the isolation of (-)-hysocyamine (5.1) from Hyoscyamus muticus L. (-)-Scopolamine (5.22) was isolated later (1892) from Scopolia atropoides Bercht and Presl, and it was shown

5.1

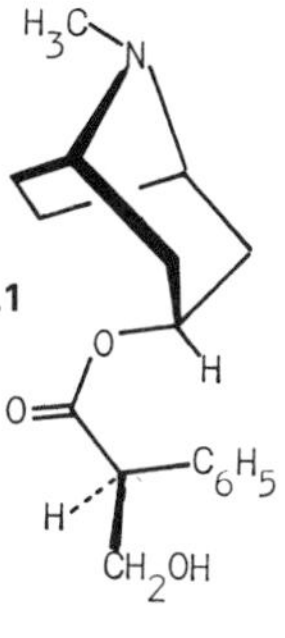

to be identical to (-)-hyoscine (5.22) by direct comparison of the two bases and their hydrobromides. (-)-Hyoscine (scopolamine; 5.22) corresponds to $C_{17}H_{21}NO_4$ and contains hydroxyl (formation of a monoacetate) and tertiary nitrogen (formation of a quaternary base with one equivalent of methyl iodide) groups.

Hydrolysis of (-)-hyoscine (5.22) with dilute base or acid leads to tropic acid (5.3) and a new base, oscine (scopoline) $C_8H_{13}NO_2$ (5.23; Equation 5.12). The tropic acid (5.3) could be obtained in either racemic or optically active S-(-) modifications, depending upon the conditions of the hydrolysis; atropic acid (5.4) could also occasionally be obtained if the hydrolysis was particularly vigorous. Oscine (5.23), although optically inactive as isolated, formed a benzoate derivative capable of resolution and was, therefore, a racemic compound.

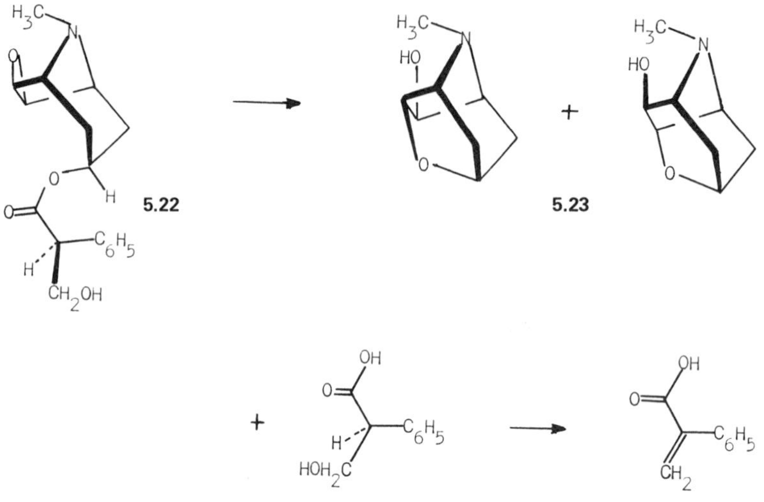

5.22 5.23

5.3 5.4

(Eq. 5.12)

The cooccurrence of (-)-hyoscine (5.22) and (-)-hyoscyamine (5.1), the hydrolysis of both of these alkaloids to (-)-tropic acid (5.3), and the similarity of the two alcohols, tropine (5.2; $C_8H_{15}NO$) and oscine (5.23; $C_8H_{14}NO_2$) were presumptive evidence that oscine (5.23) and tropine (5.2) were related. However, there was a major difference. As we have already seen, (-)-hyoscyamine (5.1) can be regenerated from tropine (5.2) and S-(-)-tropic acid (5.3), but hyoscine (5.22) cannot be obtained from the reaction of oscine (5.23) and S-(-)-tropic acid (5.3). Thus, it was recognized quite early that the product of the latter reaction—although not well characterized—was not (-)-hyoscine (5.22) and that, therefore, some deep-seated change, along with hydrolysis, must have occurred on base (or acid) treatment. This conclusion was substantiated when it was found that buffered enzymatic hydrolysis of (-)-hyoscine (5.22) yielded S-(-)-tropic acid (5.3) and a new base, isomeric with oscine (5.23), called scopine (5.24; Equation 5.13). Scopine (5.24), which was optically inactive and could not be

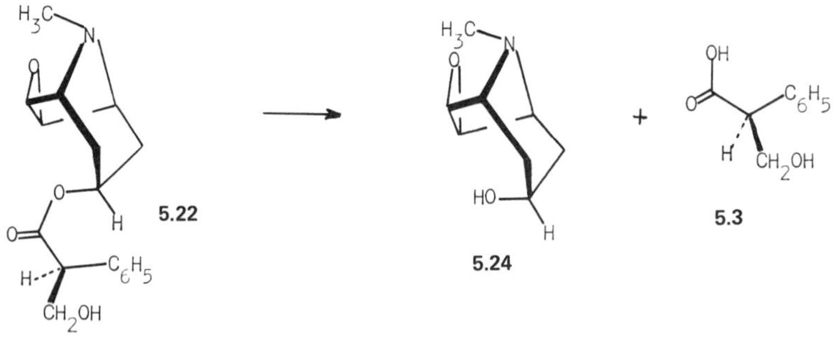

5.22 5.24 5.3

(Eq. 5.13)

resolved, was readily converted, by acid or base, to oscine (5.23) and yielded (±)-hyoscine (5.22, and its mirror image), along with other products, on reaction with (±)-acetyltropoyl chloride, followed by gentle hydrolysis.

Since oscine (5.23), $C_8H_{13}NO_2$, is two hydrogens less and one oxygen more than tropine (5.2), $C_8H_{15}NO$, it was originally thought that oscine (5.23) possessed a carbonyl group; however, no carbonyl derivatives could be formed. Then, since hydrogen bromide adds to oscine (5.23) to form a hydrobromide which (a) forms a diacetate and (b) reverts to oscine (5.23) by heating with hydrogen chloride at 100°C, it was concluded that a cyclic ether bridge must be present. The location of the oxygen in scopine (5.24) as an epoxide, symmetrically placed (in concert with its lack of chirality), opening intramolecularly upon attack by the hydroxyl to yield racemic oscine (5.23; Equation 5.14), was ultimately confirmed by synthesis.

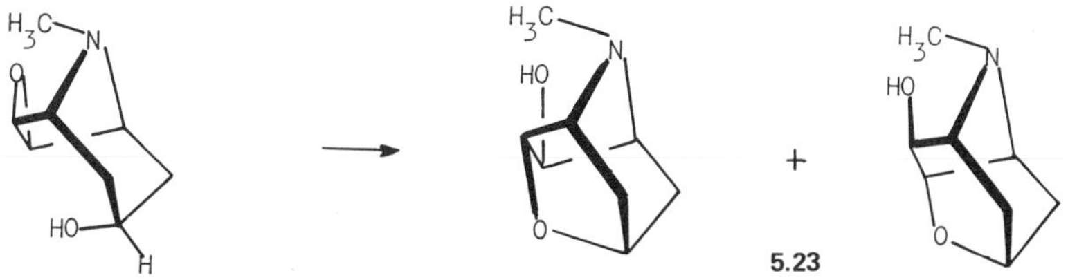

5.24 → 5.23 (Eq. 5.14)

Thus, as shown in Scheme 5.2, 2,5-diethoxy-2,5-dihydropyran (5.25; made by anodic oxidation of furan in ethanol) was converted to its bromohydrin, which on basic hydrogenolysis yields 2,5-diethoxy-3-hydroxytetrahydropyran (5.26). Mild acid-catalyzed hydrolysis of 5.26 leads to malic dialdehyde (5.27) which undergoes Robinson-Mannich condensation to (±)-6-hydroxytropinone (5.28, and its mirror image). Treatment of the Robinson-Mannich product, the hydroxyketone 5.28, with phenylisocyanate leads to a urethane capable of catalytic reduction to a tropan-3α, 6β-diol monophenylurethane 5.29. Acylation of 5.29, followed by distillation, results in cleavage of the phenylcarbamyl moiety to phenylisocyanate and 3α-acetoxy-6β-hydroxytropane (5.30).*

*There appears to have been little speculation on the facile thermal loss of phenylisocyanate. Given the proximate relationship of the 6β oxygen and the tertiary nitrogen, a cyclic process can be envisioned.

Conclusive evidence that the hydroxyl at C_6 is β was obtained by hydrolysis of 5.30 to the corresponding diol (5.31) and treatment of the latter with ethyliodoacetate. This results in the formation of N_b-carboxymethyl-3α, 6′β-dihydroxytropinium iodide (5.32). This cyclic compound could form if, and only if, the hydroxyl at C_6 were cis (Z or β) to the nitrogen bridge. In a similar way, the hydroxyl at C_3 must be α (trans or E) to the nitrogen since dehydration ($POCl_3$) of 5.31 yields (±)-tropene oxide (5.33) which leads to an acetoxybromotropane (presumably 5.34) on acetobromolysis (Equation 5.15). Dehydrohalogenation and reduction provide acetyltropan-3α-ol (5.35), identical to that produced from hyoscyamine (5.1).

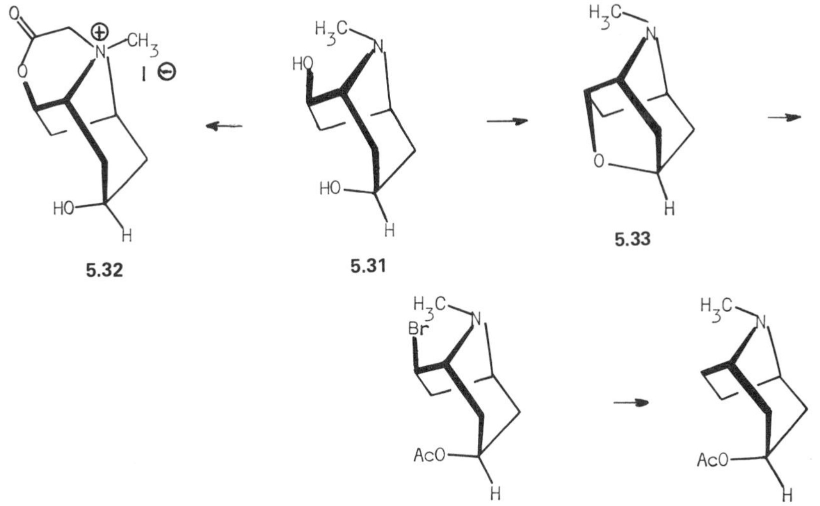

5.32 ← 5.31 → 5.33 → 5.34 → 5.35 (Eq. 5.15)

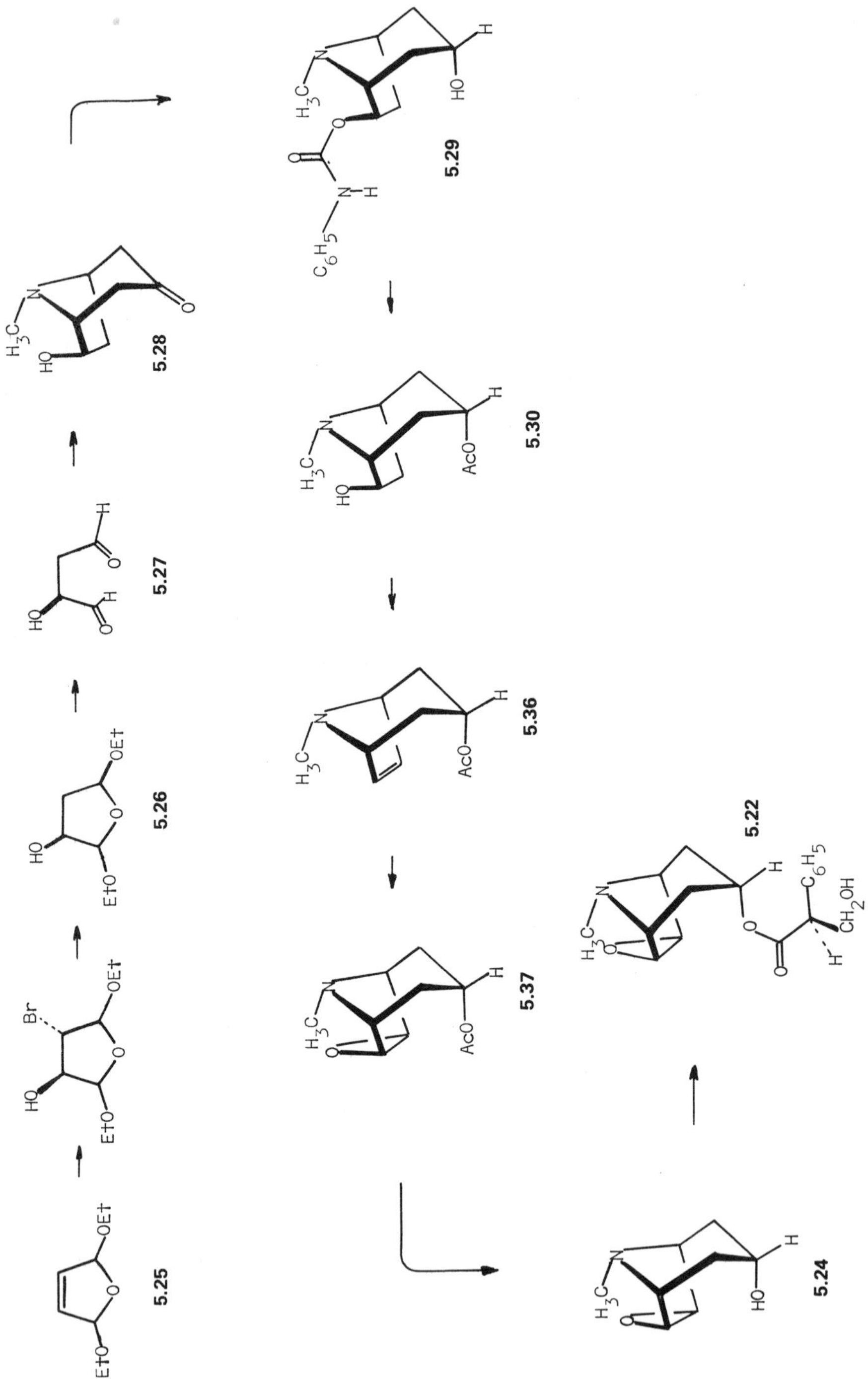

SCHEME 5.2 A total synthesis of hyoscine.

After conversion of 5.30 into its corresponding tosyl ester with p-toluenesulfonyl chloride, elimination of toluenesulfonic acid could be induced with triethylamine to generate acetyl-6-tropene-3α-ol (5.36), which in turn leads to acetylscopine (5.37) on epoxidation of its trifluoroacetate salt with trifluoroperacetic acid (conditions under which N-oxide formation is avoided). Gentle basic hydrolysis (1 N sodium hydroxide in acetone) of acetylscopine (5.37) leads to scopine (5.24), which as previously noted generates (±)-hyoscine on treatment with (±)-acetyltropoyl chloride followed by hydrolysis with dilute hydrochloric acid. Resolution of (±)-hyoscine (5.22) with tartaric acid leads to (-)-hyoscine (5.22), identical in all respects to the naturally occurring material.

IV. (-)-VALEROIDINE

The various parts (leaves, twigs, and berries) of the Australian plant Duboisia myoporoidies contain a number of tropane alkaloids; included among these is valeroidine (5.38). Although 5.38 is levorotatory, its hydrobromide, the form in which it is usually isolated, is dextrorotatory. Hydrolysis of (-)-valeroidine (5.38) generates isovaleric acid (5.39) and a dihydroxytropane ($C_8H_{15}NO_2$; Equation 5.16), found to be identical to the tropan-3α-6β-diol (5.31) which had been prepared during the synthesis of hyoscine (5.22).

Thus, the structure of valeroidine was confirmed by its synthesis from the same phenylisocyanate derivative (5.29) already used as an intermediate in the preparation of hyoscine (5.22). When 3α-hydroxy-6β-phenylcarbamyloxytropane (5.29) is resolved with tartaric acid and each of the enantiomers individually esterified with isovaleric acid (5.39), the reaction being carried out with the acid chloride of 5.39, (+)- and (-)-3α-isovaleryl-6β-phenylcarbamloxytropane (5.40 and its mirror image) are obtained. Vacuum distillation of the enantiomeric esters results in loss of phenylisocyanate from each, and (+)- and (-)-valeroidine (5.38) are obtained (see footnote, p. 73; Equation 5.17).

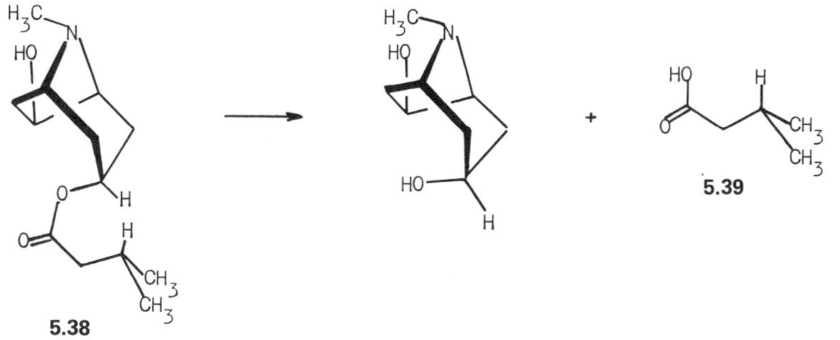

5.38 → + 5.39 (Eq. 5.16)

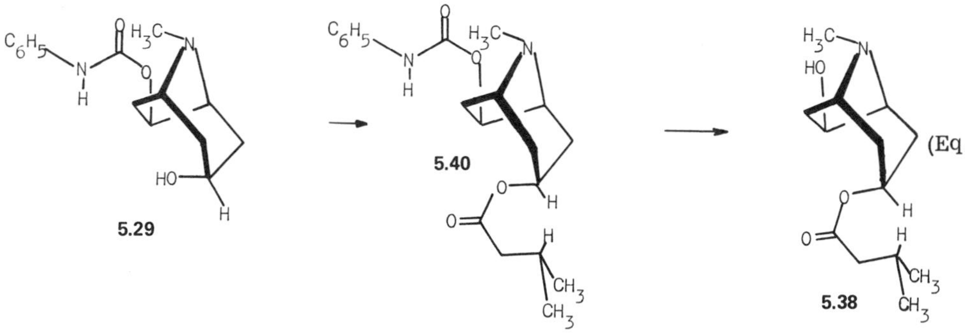

5.29 → 5.40 → 5.38 (Eq. 5.17)

One of the more unusual aspects of the chemistry of (-)-valeroidine (5.38) is its facile demethylation on treatment with thionyl chloride, a reaction examined in a vain attempt to replace the unesterified 6β-hydroxyl with chlorine. Norvaleroidine (5.41) can be isolated in good yield from this reaction, and methylation of 5.41 regenerates valeroidine (5.38). Although it is now known that acid chlorides in general [21] may be used to bring about N-demethylation of tropanes with concimitant amide formation, the ease with which this reaction occurs in the case of valeroidine (5.38) is unusual. Therefore, there is a temptation to invoke a cyclic intermediate in this case, in which, for steric reasons, backside displacement of the 6β-oxygen is impossible and thus displacement at the methyl occurs instead (Equation 5.18).

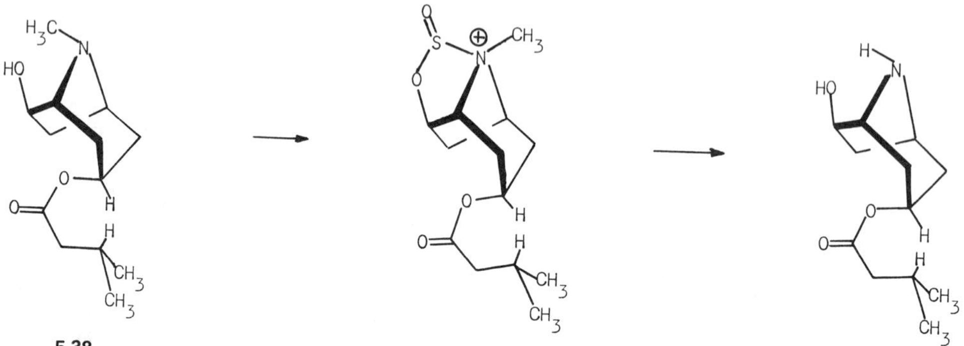

5.38 5.41

(Eq. 5.18)

V. TROPACOCAINE

Tropacocaine (5.42) was isolated from Javanese coca leaves and, for some time, being readily available and easily synthesized, was widely used as a local and spinal anesthetic. Since gentle hydrolysis of 5.42 results in the formation of ψ-tropine (5.11) and benzoic acid, tropacocaine (5.42) was immediately recognized as benzoyl-ψ-tropine (Equation 5.19). In addition to its early pharmaceutical use, tropacocaine (5.42) is interesting because it possesses the less common ψ-tropine (5.11) structure and because it was among the first (1896) alkaloids synthesized [22].

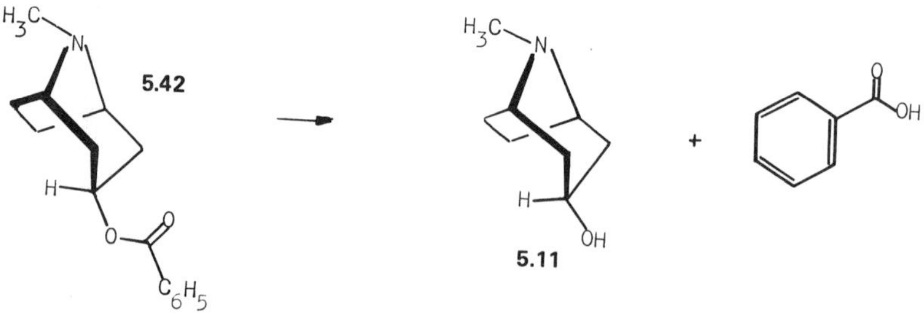

5.42 5.11

(Eq. 5.19)

VI. METELOIDINE

Meteloidine (5.43), $C_{13}H_{21}NO_4$, was isolated along with hyoscyamine (5.1) and hyoscine (5.22) from Datura meteloides and is unusual among tropane alkaloids in that it shows no marked physiologic activity. Since meteloidine (5.43) was incapable of resolution, yielded an alkamine (teloidine, 5.44, $C_4H_{15}NO_3$) and tiglic acid (5.45), $C_5H_8O_2$, on hydrolysis with barium hydroxide (Equation 5.20), and was found along with hyoscyamine (5.1) and hyoscine (5.22), it was assumed that teloidine (5.44) was one of the stereoisomers of 3, 6, 7-trihydroxytropane. This assumption was verified when meso-tartaric dialdehyde, methylamine, and

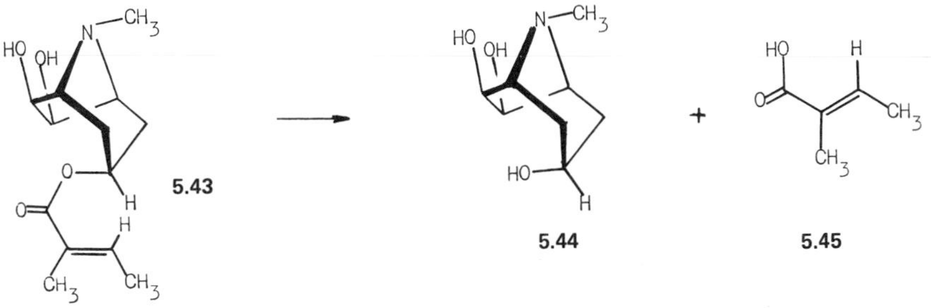

5.43 → 5.44 + 5.45 (Eq. 5.20)

acetonedicarboxylic acid were successfully condensed in a Robinson-Mannich reaction under "physiologic" conditions to yield teloidinone (5.46) which was, in turn, catalytically reduced to teloidine (5.44) and ψ-teloidine (5.47; Equation 5.21). Since the hydroxyl groups in meso-tartaric dialdehyde must necessarily be cis, teloidine (5.44) must have the hydroxyls at C_6

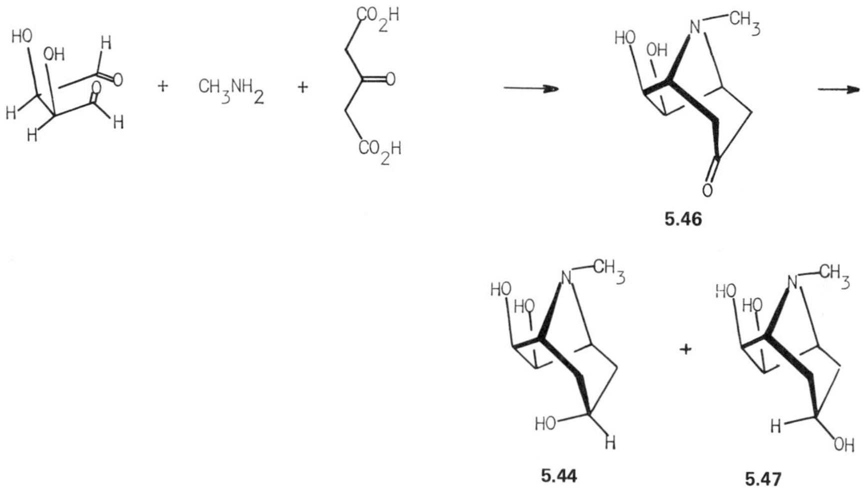

CH_3NH_2 → 5.46 → 5.44 + 5.47 (Eq. 5.21)

and C_7 cis, and the formation of a lactone (5.48) with ethyl iodoacetate demonstrates that they must be β (cis or Z to the nitrogen; Equation 5.22). Further, since ψ-N-normethylteloidine-6, 7-acetonide (5.49) yields a tetrahydrooxazine (5.50) with p-nitrobenzaldehyde (Equation 5.23)

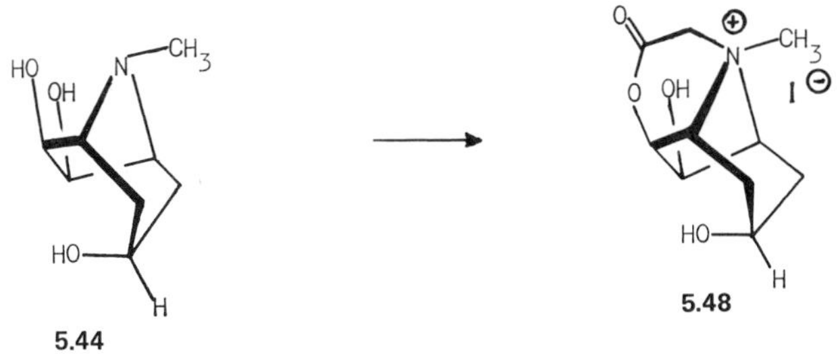

5.44 → 5.48 (Eq. 5.22)

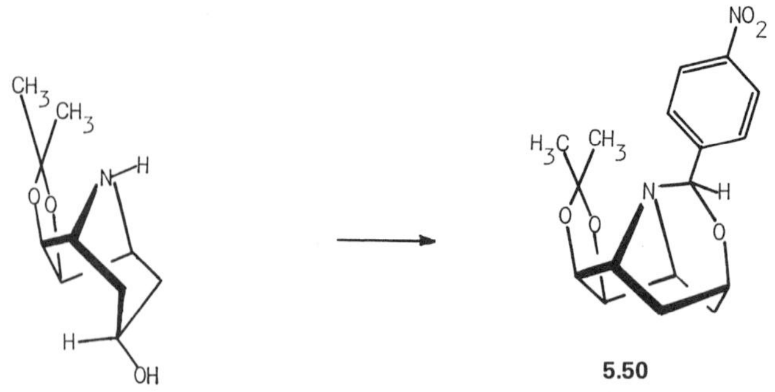

5.49 5.50 (Eq. 5.23)

which N-normethylteloidine-6,7-acetonide (5.51) failed to do, the hydroxyl at C_3 in teloidine (5.44) must be α (trans or E to the nitrogen), and hence teloidine (5.44) is tropan-3α,6β,7β-triol and meteloidine (5.43) is its tiglic acid (5.45) ester at C_3.

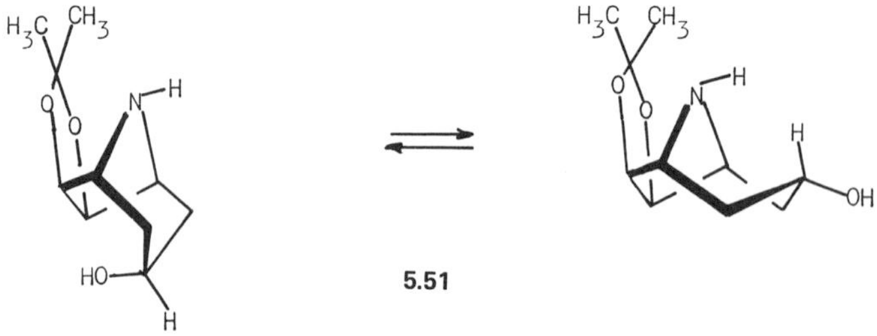

5.51

VII. (-)-LITTORINE

Along with the ubiquitous (-)-hyoscyamine (5.1) and meteloidine (5.43), the reputedly poisonous, but widespread, Australian shrub Anthocercis littorea Labill. contains R-(+)-3α-(2-hydroxy-4-phenylpropionyloxy)tropane, "(-)-littorine" (5.52). In the mass spectrum, (-)-littorine (5.52), which corresponds to $C_{17}H_{23}NO_3$, possesses a molecular ion m/e 289, which can be increased to m/e 290 by treatment with deuterium oxide. This was taken to indicate that one exchangeable hydrogen was present as a hydroxyl. The proton magnetic resonance (pmr) spectrum of (-)-littorine (5.52) closely resembles (-)-hyoscyamine (5.1) for eighteen of the twenty-three protons. In the examination of the remaining protons, there is detected an ABX system which is assigned to an α-hydroxy-β-phenylpropionyl group in which the benzylic protons adjacent to the asymmetric carbon are magnetically nonequivalent. Hydrolysis (base) of (-)-littorine (5.52) yields tropine (5.2) and optically pure (+)-2-hydroxy-3-phenylpropionic acid [R-(+)-phenyllactic acid (5.53; Equation 5.24)] and since the absolute configuration of (-)-phenyllactic (5.53) acid had been established as S, it follows that (-)-littorine must be the tropine ester of R-(+)-phenyllactic acid (5.53); this was confirmed by synthesis. Finally, as is occasionally the case, (±)-littorine (5.52) had been synthesized about 60 years (!) before it was isolated in a search for useful medicinals. It had, in fact, been found to be a potent mydriatic.

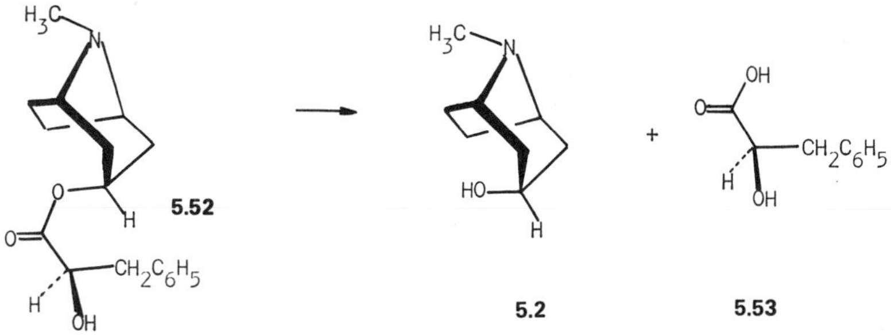

5.52 → 5.2 + 5.53 (Eq. 5.24)

VIII. (-)-COCAINE

(-)-Cocaine (5.54) was isolated in 1862 from the leaves of Erythroxylon coca Lam., and although it is still occasionally used as a topical anesthetic, it can induce drug dependence and has largely been replaced by analogous synthetic materials or other compounds.

(-)-Cocaine (5.54) is a crystalline material, mp 90°C, corresponding to $C_{17}H_{21}NO_4$, which forms a quaternary base on addition of 1 equivalent of methyl iodide. Treatment of (-)-cocaine (5.54) with cyanogen bromide generates methyl bromide and cyanonorcocaine (5.55), indicating that the tertiary nitrogen bears a methyl group (Equation 5.25). On acid

5.54 → 5.55 (Eq. 5.25)

hydrolysis, (-)-cocaine (5.54) yields methanol, benzoic acid, and an alkamine called "(-)-ecgonine" (5.56; Equation 5.26). Mild hydrolysis (neutral, boiling water) of (-)-cocaine (5.54), on the other hand, yields methanol and O-benzyl-(-)-ecgonine (5.57; Equation 5.27).

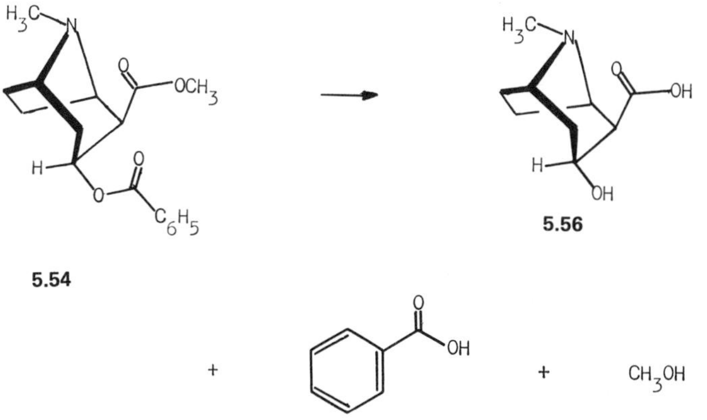

5.54 → 5.56 + benzoic acid + CH_3OH (Eq. 5.26)

5.54 → 5.57 (Eq. 5.27)

Chromic acid oxidation of (-)-ecgonine (5.56) yields, in turn, ecgoninone (5.58), tropinone (5.10a), tropinic acid (5.59), and ecgoninic acid (5.60; Equation 5.28). The formation of tropinone (5.10a) defines the position of the hydroxyl, the ring system, and the nature of ecgoninone (5.58; i.e., a β-ketoacid which has decarboxylated to 5.10a).

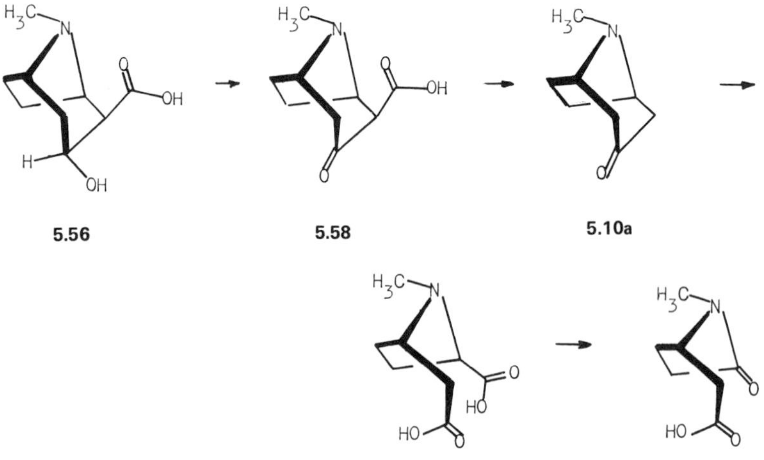

5.56 → 5.58 → 5.10a → 5.59 → 5.60 (Eq. 5.28)

As we have seen in other tropane alkaloids, the ready syntheses of these compounds via the Robinson-Mannich reaction is often used to help establish structure. Cocaine (5.54) was not an exception. Thus, Willstatter, in 1923, reported that succinic dialdehyde, methylamine, and monomethyl acetonedicarboxylate (Equation 5.29) reacted to form ecgoninone methyl ester, which could then be reduced to ecgonine methyl ester (5.61) and ψ-ecgonine methyl ester (5.62). Benzoylation of the former leads to (±)-cocaine (5.54 and its mirror image), while the latter forms (±)-ψ-cocaine (5.63 and its mirror image).

To define the structure of (-)-cocaine (5.54) further, two of the following three relationships and the absolute configuration must be known. The relationships are the relative positions of (a) the nitrogen bridge and the hydroxyl at C_3; (b) the nitrogen bridge and the carboxylic acid at C_2; and (c) the carboxylic acid at C_2 and the hydroxyl at C_3. Now, alkaline permanganate oxidation of (-)-ecgonine (5.56) causes oxidation of the N-methyl (!) to generate formic acid and (-)-norecgonine (5.64; Equation 5.30). Similarly, (+)-ψ-ecgonine (5.65) from 5.62 leads to (+)-nor-ψ-ecgonine (5.66; Equation 5.31). Then, acetyl-nor-ψ-ecgonine ethyl ester (5.67), prepared from 5.66, readily undergoes N ⟶ O acyl migration, indicating that the hydroxyl here, and thus, by implication, in (+)-ψ-ecgonine (5.65), is β (cis or Z to the nitrogen; Equation 5.32). On the other hand, it was already known that O-benzoylnorecgonine, obtained on alkaline permanganate oxidation of O-benzoyl-(-)-ecgonine (5.57; see above) underwent O ⟶ N benzoyl migration. Therefore, in (-)-ecgonine (5.56) the hydroxyl must also be β!

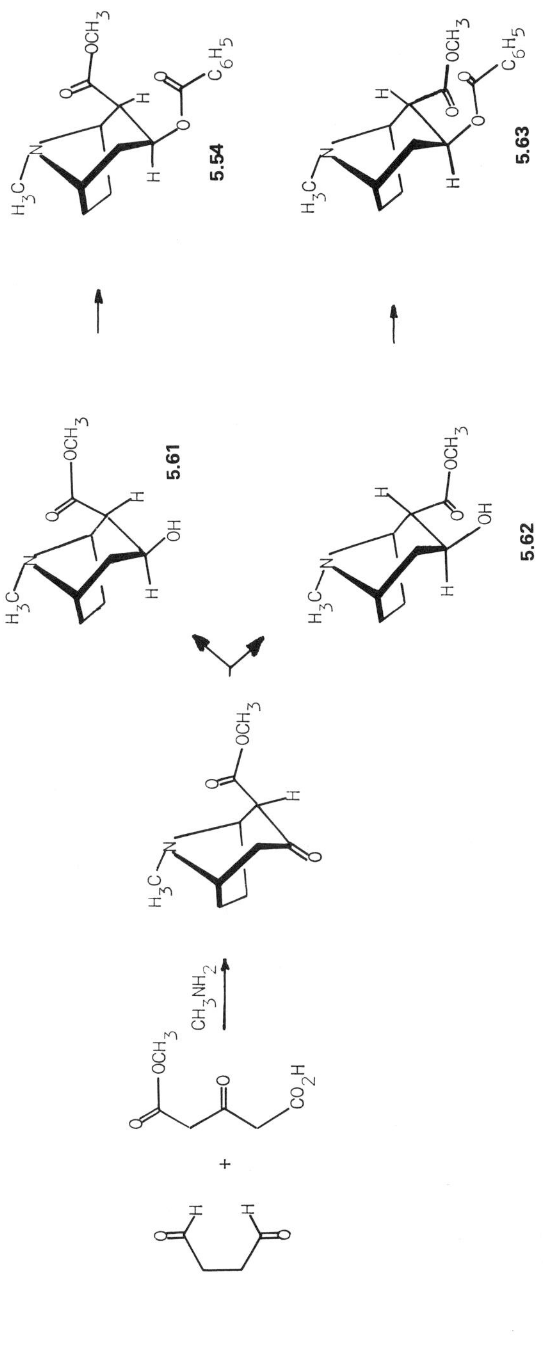

(Eq. 5.29)

5.56 → 5.64 (Eq. 5.30)

5.65 → 5.66 (Eq. 5.31)

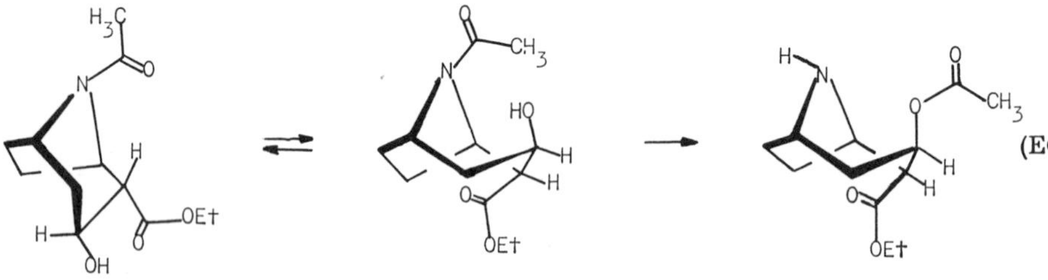

5.67 (Eq. 5.32)

Thus, on the basis of the above experimental information, (-)-ecgonine (5.56) and (+)-ψ-ecgonine (5.65) must differ from one another in the geometric relationship between the carboxylic acid group and the hydroxyl (or nitrogen). This relationship was established by partial hydrolysis of N-cyanonorcocaine (5.55) to the corresponding amido alcohol, which on treatment with sodium methoxide at -15°C, smoothly undergoes cyclization to the lactam of N-carbamylnorecgonine (5.68; Equation 5.33). This means that the carboxylic acid group must be β (cis or Z to the nitrogen) in (-)-cocaine (5.54) and (-)-ecgonine (5.56) and, by difference, α (trans or E to the nitrogen) in (+)-ψ-ecgonine (5.65).

In addition to this evidence, it was known, as shown in Scheme 5.3, that when the Curtius degradation is carried out on O-benzoylecgonine (5.57) and O-benzoyl-ψ-ecgonine (5.69),

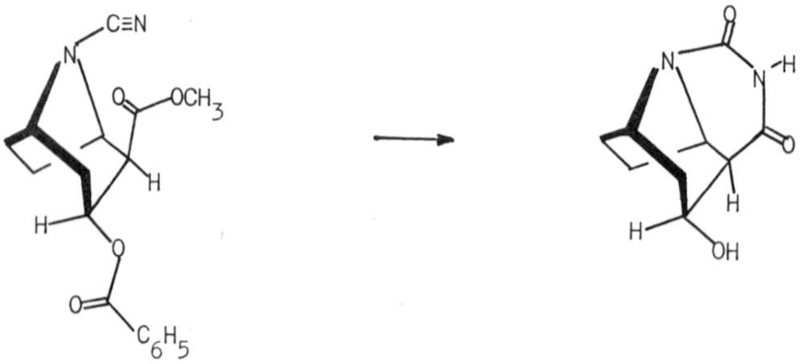

5.55 → 5.68 (Eq. 5.33)

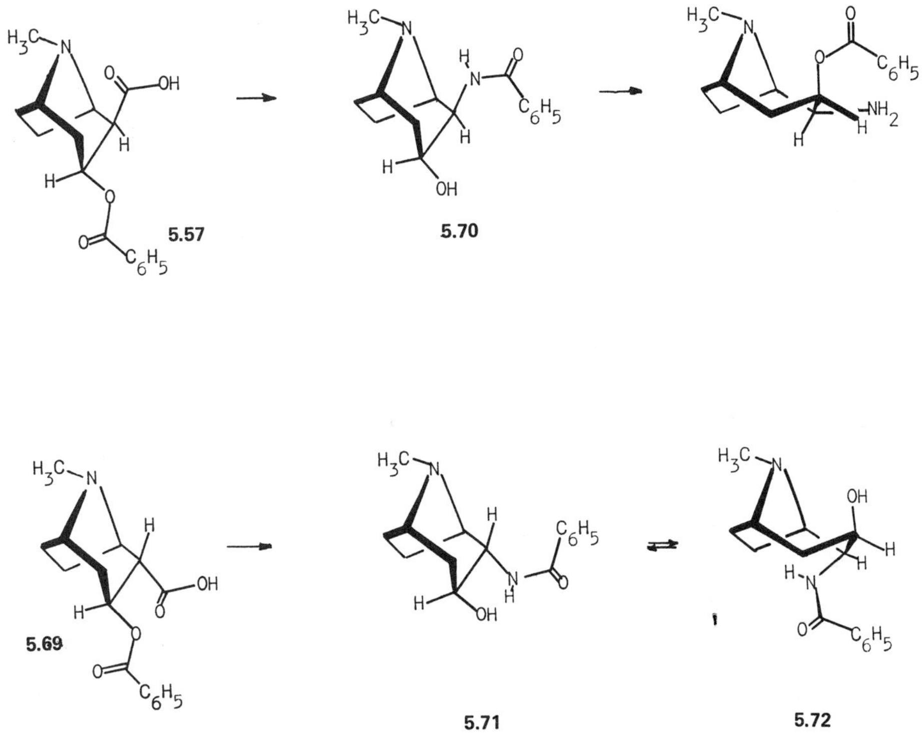

H_3C—N, O, OH, H, H, O, O, C_6H_5

5.57

H_3C—N, H, N, O, C_6H_5, H, H, OH

5.70

H_3C—N, O, O, C_6H_5, NH_2, H, H

H_3C—N, H, OH, O, H, O, O, C_6H_5

5.69

H_3C—N, H, C_6H_5, N, H, O, H, OH

H_3C—N, OH, H, H, H-N, O, C_6H_5

5.71

5.72

SCHEME 5.3 The Curtius degradation of O-benzoylecgonine and O-benzoyl-ψ-ecgonine.

epimeric 2-benzamido-3β-tropanols (5.70 and 5.71, respectively) are obtained. Only the benzamidotropanol (5.70) derived from O-benzoylecgonine (5.57) undergoes rearrangement to the corresponding O-benzoyl derivative. This evidence was taken to mean that the amino and hydroxyl functions are cis (and both therefore β) in (-)-cocaine (5.54) and trans (carboxylate α, hydroxyl β) in (+)-ψ-cocaine (5.63). It is clearly necessary to assume, as is usually mentioned, that the Curtius degradation proceeds with retention, for which there is precedent, and, as is usually not mentioned, that the benzamidotropanol (5.71) derived from O-benzoyl-ψ-ecgonine (5.69) exists as the diaxial conformer (5.72) since the angle subtended between nitrogen and oxygen, were the groups diequatorial trans, would be the same as that in the axial-equatorial cis isomer and the migration of the benzoyl group as facile.

Finally, as shown in Scheme 5.4, (-)-cocaine (5.54) has been correlated with S-(+)-glutamic acid (5.73). The same ecgoninic acid epimer (5.60) obtained from (-)-cocaine (5.54) was prepared from S-(+)-glutamic acid (5.73) via pyroglutamic acid (5.74) and its corresponding nitrile (5.75), thus allowing us to write (-)-2R-methoxycarboxyl-3S-benzoxyloxytropane (5.54) for (-)-cocaine (5.54).

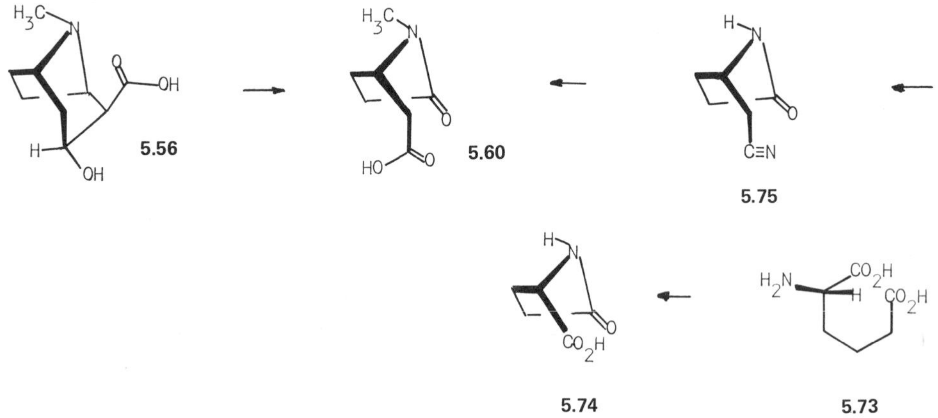

5.56 5.60 5.75

5.74 5.73

SELECTED READING

G. Fodor, in The Alkaloids, Vol. 6, R. H. F. Manske (ed.). Academic, New York, 1960, pp. 145 ff.

G. Fodor, in The Alkaloids, Vol. 9, R. H. F. Manske (ed.). Academic, New York, 1967, pp. 269 ff.

G. Fodor, in The Alkaloids, Vol. 17, R. H. F. Manske (ed.). Academic, New York, 1977, pp. 352 ff.

H. L. Holmes, in The Alkaloids, Vol. 1, R. H. F. Manske (ed.). Academic, New York, 1950, pp. 271 ff.

6

CHEMISTRY OF THE PYRROLIZIDINE ALKALOIDS

They are drunken, but not with wine; they stagger,
but not with strong drink.

Isaiah, 29:9

The pyrrolizidine alkaloids constitute a large (well over 100 are known), cosmopolitan group of compounds which are found in different species of plants within a genus (Senecio, astors, and ragworts), and related members of which are found in different genera (Heliotropium, Trachelanthus, and Trichodesma) within different botanic families (Compositae, Boraginaceae, and Leguminosae). Most of the alkaloids are toxic, affecting the liver, and their ingestion is manifested in animals with the onset of symptoms associated with names such as "horse staggers" or "walking disease." As has already been pointed out (p. 58), the alkaloids are esters of amino alcohols, called "necines" and carboxylic acids, called "necic acids." Only three members of this class, which are nevertheless quite typical, will be considered here. These compounds are heliosupine, (6.1), senecionine (6.2), and 9-angelylretronecine-N-oxide (6.3).

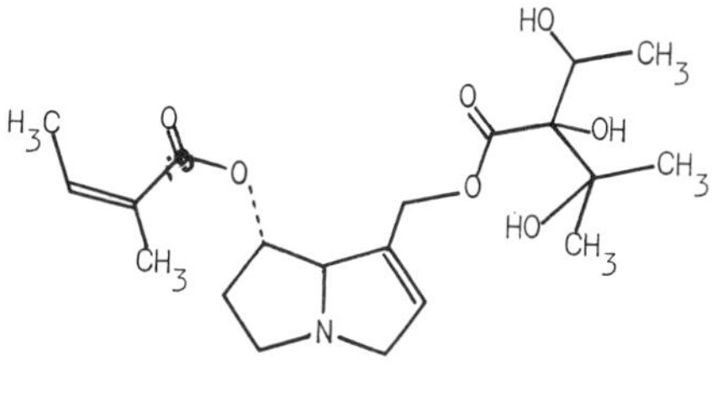

6.1 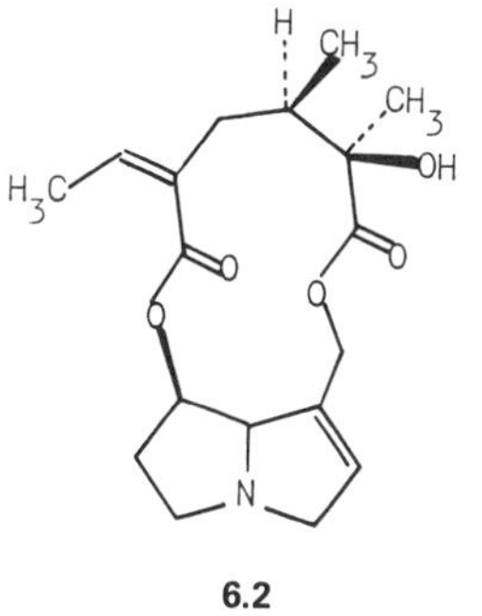 6.2 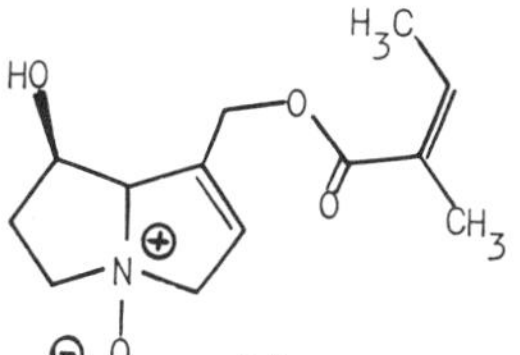 6.3

I. HELIOSUPINE

The alkaloid heliosupine (6.1) is an optically active material ($[\alpha]_D^{20}$ -4.3°, ethanol) which also occurs naturally as its corresponding N-oxide and which has not yet been obtained in a crystalline form. However, it corresponds to $C_{20}H_{31}NO_7$, and on hydrolysis it yields an optically active base, heliotridine (6.4), an acid, angelic acid (6.5), and acetone as the only reported products (Equation 6.1). Hydrogenolysis (platinum oxide, H_2), on the other hand, affords 7-(2'-methylbutyroxy)heliotridane (6.6) and an optically active ($[\alpha]_D^{20}$ +17.5°, ethanol) glassy carboxylic acid, called echimidinic acid (6.7; Equation 6.2).

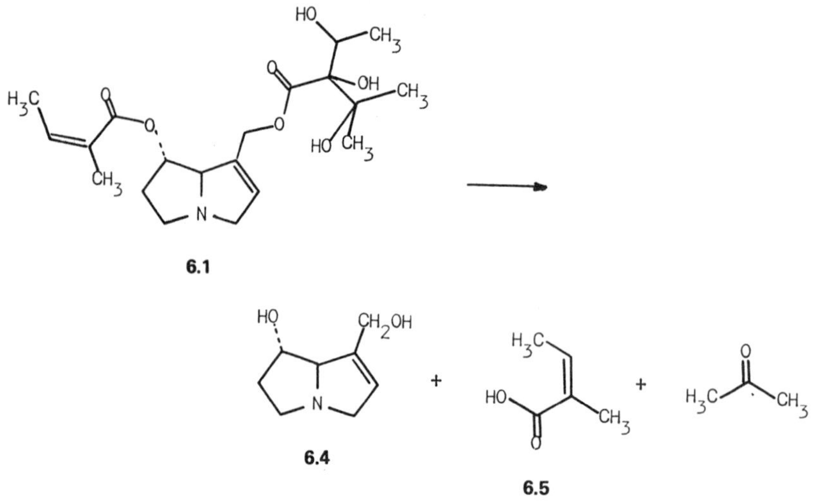

(Eq. 6.1)

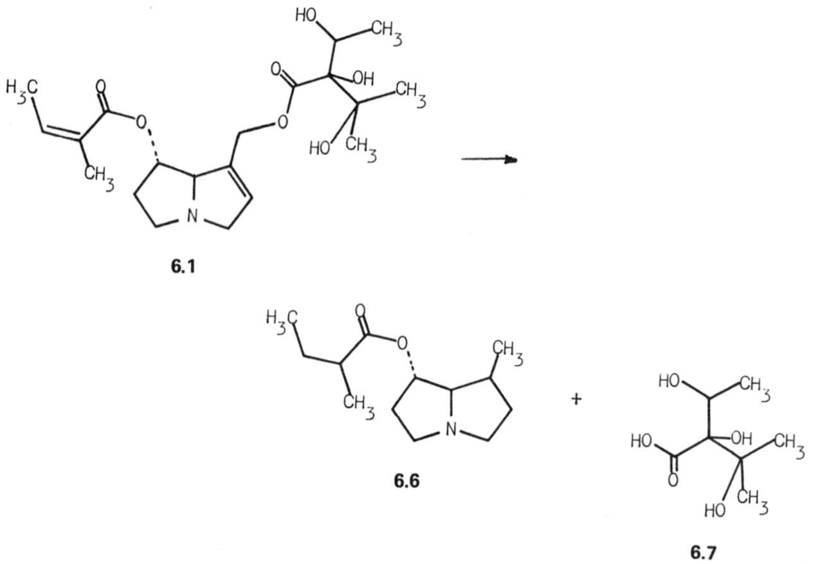

(Eq. 6.2)

A. The Necine of Heliosupine: Heliotridine

Heliotridine (6.4), which is quite common as an esterifying alcohol in the pyrrolizidine bases, is optically active, corresponds to $C_8H_{13}NO_2$, readily forms salts with strong acids, and although it is a tertiary amine (it yields a quaternary salt with one equivalent of methyl iodide) it nevertheless possesses two active hydrogens (Zerewitinoff). Since heliotridine (6.4) forms no carbonyl derivatives but does form a diacetate, the oxygens were quickly recognized as being present as alcohols, accounting, at the same time, for the two active hydrogens.

Catalytic hydrogenation of heliotridine (6.4) on platinum oxide catalyst, yields a new compound, $C_8H_{15}NO$, called oxyheliotridane (6.8; Equation 6.3). The same compound can be obtained by hydrogenolysis of dibenzoyl heliotridine followed by saponification. Thus, in short order it was established that heliotridine (6.4) possesses a tertiary amino group, two esterifiable hydroxyl groups (presumably alcohols), one of which is removed during hydrogenation, and a carbon-carbon double bond (as the final site of reducible unsaturation). When

HO CH$_2$OH → HO CH$_3$ (Eq. 6.3)

6.4 6.8

oxyheliotridane (6.8) is treated with sulfuric acid at 170-175°C, an unsaturated base, $C_8H_{13}N$, heliotridene (6.9), is formed. Heliotridene (6.9) absorbs 1 equivalent of hydrogen on catalytic hydrogenation to yield heliotridane ($C_8H_{15}N$, 6.10; Equation 6.4).

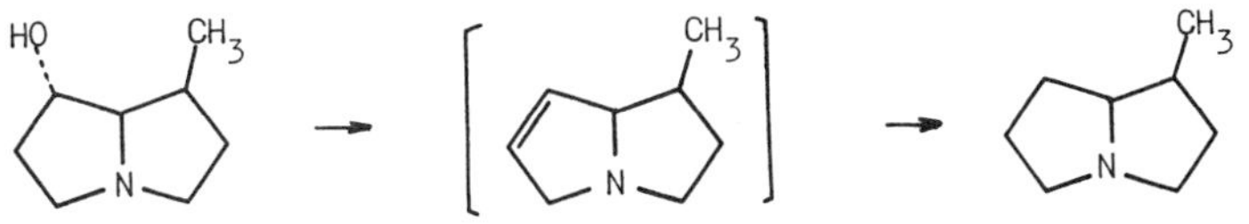

HO CH$_3$ → [CH$_3$] → CH$_3$ (Eq. 6.4)

6.8 6.9 6.10

Exhaustive methylation of heliotridane (6.10) followed by elimination and reduction yields 1,3-dimethyl-2-n-propylpyrrolidine (6.11), which in turn provides 4-dimethylamino-3-methylheptane (6.12) on subsequent methylation, elimination, and reduction, thus establishing the structure of heliotridane (6.10) as 1-methylpyrrolizidine (Equation 6.5). (It is

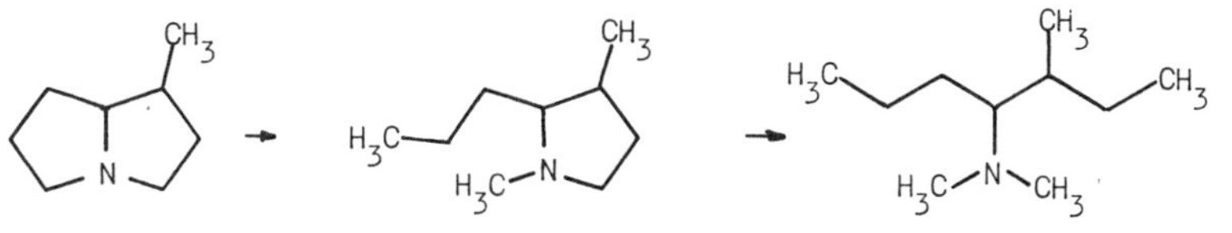

CH$_3$ → H$_3$C CH$_3$ H$_3$C–N → CH$_3$ H$_3$C CH$_3$ H$_3$C–N–CH$_3$ (Eq. 6.5)

6.10 6.11

worthwhile noting that the pyrrolizidine nucleus was first found here, in nature, rather than prepared, de novo, in the laboratory). Heliotridane (6.10) has also been obtained from retronecine (6.13), a stereoisomer of heliotridine (6.4), by a reaction sequence identical to that used for the latter. Since retronecine (6.13) was more readily available, subsequent reactions were carried out on it. Thus, retronecanol (6.14), the epimer at C_7 of oxyheliotridane (6.8), when treated with potassium t-butoxide in the presence of cyclohexanone, yields a ketone, retronecanone (6.15),* demonstrating the presence of the secondary hydroxyl (Equation 6.6).

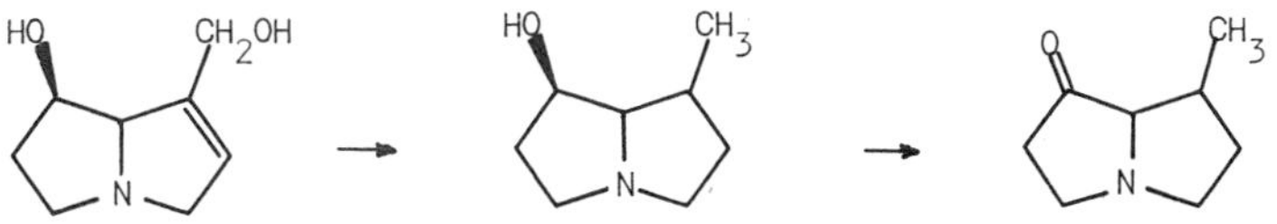

HO CH$_2$OH → HO CH$_3$ → O CH$_3$ (Eq. 6.6)

6.13 6.14 6.15

*Retronecanone (6.15) was also obtained at a later date from oxyheliotridane (6.8), as expected; presumably it was the relative paucity of material which initially prevented this reaction from being run on 6.8.

Reduction of retronecine (6.13) with Raney nickel (instead of platinum oxide) and hydrogen, yields a necine, already found in nature, called platynecine (6.16), which undergoes loss of water on treatment with a variety of dehydrating agents to form anhydroplatynecine, $C_8H_{13}NO$ (6.17), readily identified as an ether (Equation 6.7). Since platynecine (6.16) did

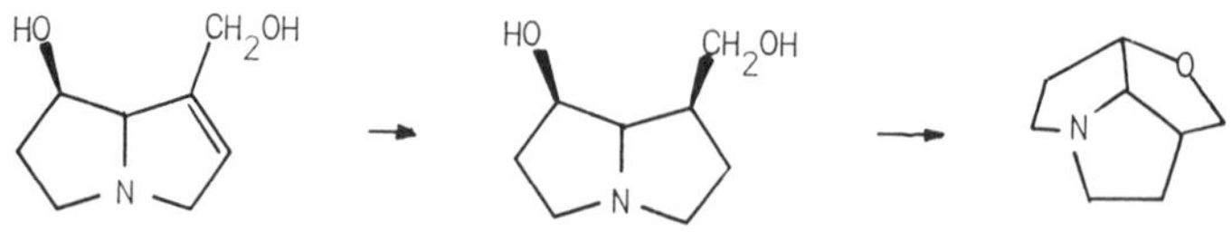

6.13 6.16 6.17 (Eq. 6.7)

not undergo hydrogenolysis, it was assumed that the hydroxyl in retronecine (6.13; and thus in heliotridine, 6.4) that was lost on hydrogenation was allylic and that it had undergone hydrogenolysis. That this hydroxyl was primary was demonstrated by converting platynecine (6.16) to its corresponding monobenzoyl derivative (6.18), the primary alcohol reacting more readily than the secondary, and thence, with thionyl chloride to the chloride (6.19) which undergoes reductive dehalogenation (Raney nickel), hydrolysis, and oxidation to a carboxylic acid, $C_8H_{13}NO_2$, possessing all the carbon atoms with which the reaction sequence was begun (Equation 6.8).

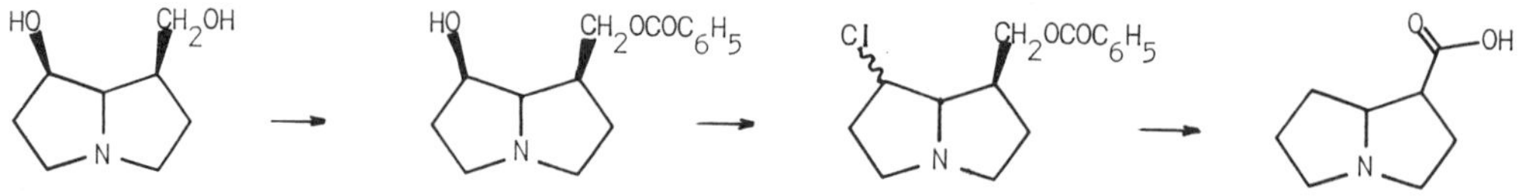

6.16 6.18 6.19 (Eq. 6.8)

The position of the double bond in heliotridine (6.4) and retronecine (6.13) was established as 1,2- rather than 1,8- by (a) the fact that neither of these molecules exhibits the properties expected for an enamine and (b) a sequence of reactions in which desoxyretronecine (6.20)* is converted by thionyl chloride into chloroisoheliotridene (6.21) and thence, reductively, to isoheliotridene (6.22) and heliotridane (6.10; Equation 6.9). Ozonolysis of isoheliotridene (6.22) yields 2-acetyl-1-pyrrolidine acetic acid (6.23; Equation 6.10).

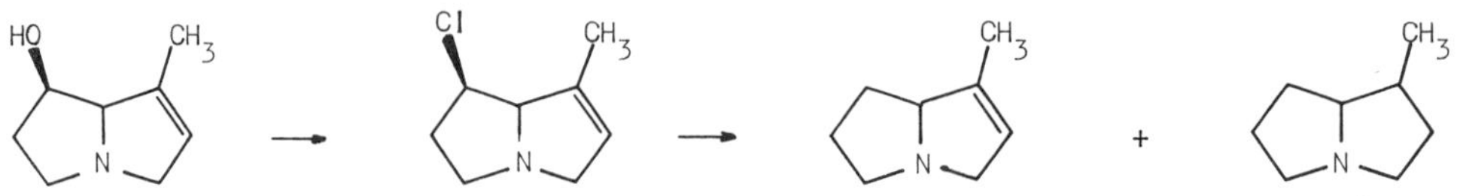

6.20 6.21 6.22 6.10 (Eq. 6.9)

Finally, the position of the secondary hydroxyl at C_7 was established by synthesis of (-)-retronecanone (6.15; Scheme 6.1; [23]). Thus, N-benzoyl-4-methylpiperidine (6.24) was

*Desoxyretronecine (6.20) was obtained by partial reduction of esters of retronecine in which hydrogenolysis but not reduction of the double bond occurs.

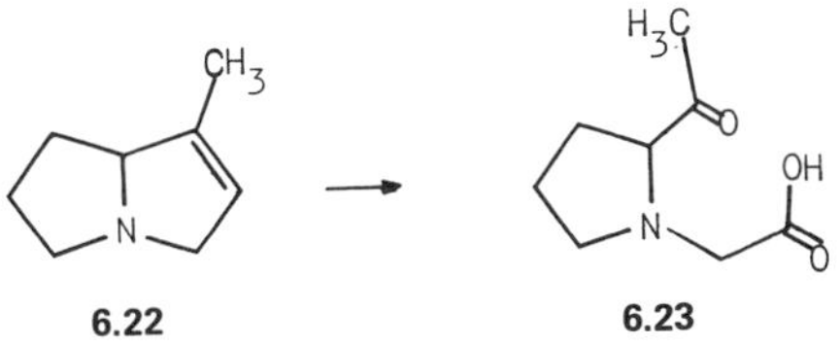

(Eq. 6.10)

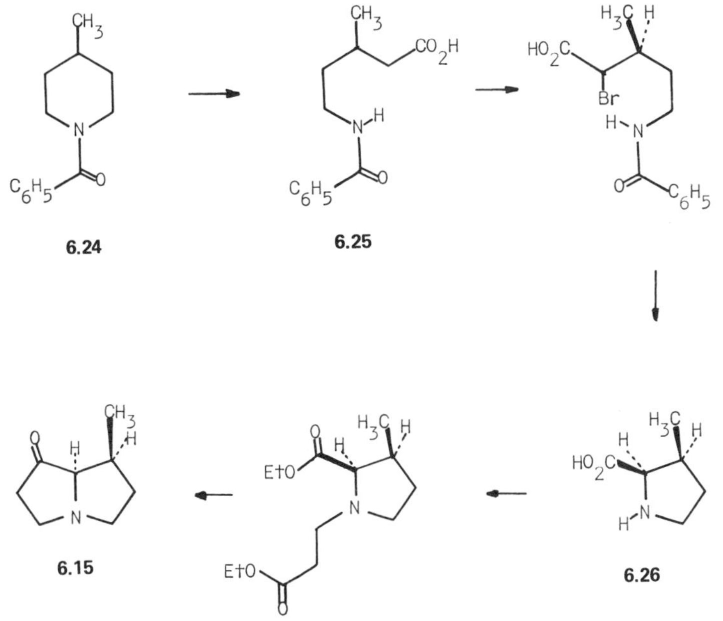

SCHEME 6.1 The synthesis of (-)-retronecanone. (After Reference 23.)

oxidized in base to the amino acid 6.26. (The structure of the resolved amidocarboxylic acid was correlated with S-(-)-3-methyl-5-aminovaleric acid.) Esterification, condensation with ethyl acrylate, Dieckmann cyclization, and hydrolysis yield optically active (-)-retronecanone (6.15).

Now, the stereochemistry of platynecine (6.16) was decided by the formation of anhydroplatynecine (6.17), since the latter can form if, and only if, the hydrogens at all three chiral centers are cis (and the reaction is assumed to involve displacement of the esterified primary hydroxyl by the secondary one). Thus, in retronecine (6.13), the hydrogens at C_7 and C_8 must be cis, and in heliotridine (6.4), trans to each other.

B. The Necic Acids of Heliosupine: Angelic and Echimidinic Acids

Angelic acid (6.5), $C_5H_8O_2$, mp 45-46°C was easily recognized as Z-2-methyl-2-butanoic acid (Z-α,β-dimethylacrylic acid; 6.5) by comparison to synthetic material and also by its reduction to α-methylbutanoic acid. The synthetic material is obtained by the dehydration of α-hydroxy-α-methylbutyric acid, which in turn is prepared by hydrolysis of the cyanohydrin obtained from ethyl methyl ketone. Since angelic acid (6.5) is readily isomerized to tiglic acid (6.28) on treatment with concentrated sulfuric acid or on heating with dilute sodium

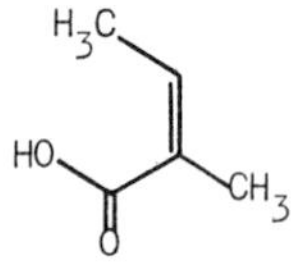

6.5

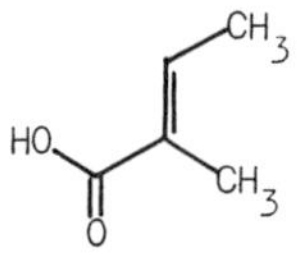

6.28

hydroxide, care is necessary during its isolation and the isolation of the alkaloid to insure isomerization is avoided.*

Echimidinic acid (6.27) is a noncrystalline, glassy compound, $C_7H_{14}O_5$, $[\alpha]_D^{20}$ +17.5° (ethanol), which forms a crystalline brucine salt, yields acetone, acetaldehyde, and oxalic

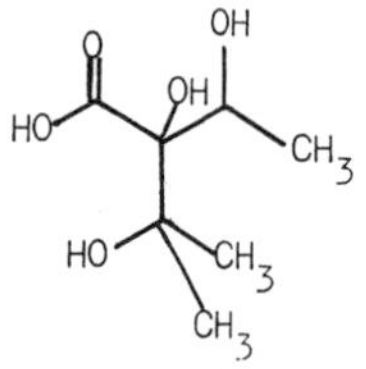

6.27

acid on treatment with periodate (2.0 equivalents are consumed) and, on this basis (and in concert with its spectral data), is thus formulated as 2-methyl-2,3,4-trihydroxypentane-3-carboxylic acid (6.27). Its absolute configuration is unknown.

II. SENECIONINE

Senecionine (6.2) has been isolated from a number of Senecio from which it can often be obtained by ethanol extraction. The alkaloid, mp 236°C (dec.), which analyzes for $C_{18}H_{25}NO_5$, is optically active, $[\alpha]_D^{25}$ -55.1° (chloroform), and forms a methiodide with 1 equivalent of methyl iodide. On refluxing with aqueous barium hydroxide, senecionine (6.2) undergoes hydrolysis to a necine, retronecine (6.13), and a dibasic necic acid called "senecic acid" (6.29). If the hydrolysis is carried out under more vigorous conditions, a lactone, senecic acid lactone (6.30), rather than the acid (6.29) itself is obtained along with retronecine (6.13 Scheme 6.2).

A. The Necine of Senecionine: Retronecine

The structure of retronicine (6.13) was established by degradation, as indicated in the previous section, as the C_7 epimer of heliotridine (6.4). Subsequently, the structure was confirmed by synthesis (Scheme 6.3). Thus, N-carboethoxy-3-aminopropionate (6.31) is allowed to add (sodium metal) to diethyl fumarate (6.32) and the product cyclized to the pyrrolidine 6.33. Hydrolysis, decarboxylation, and reesterification to 6.34 is followed by reduction to the lactone 6.35 which can be induced to undergo decarboxylation to 3-hydroxypyrrolidine-2-acetic acid lactone (6.36). Treatment of the latter with ethyl bromoacetate yields 6.37,

---

*The interesting question of which isomer is which had early been solved by bromination and decarboxylation experiments and later confirmed by x-ray analysis.

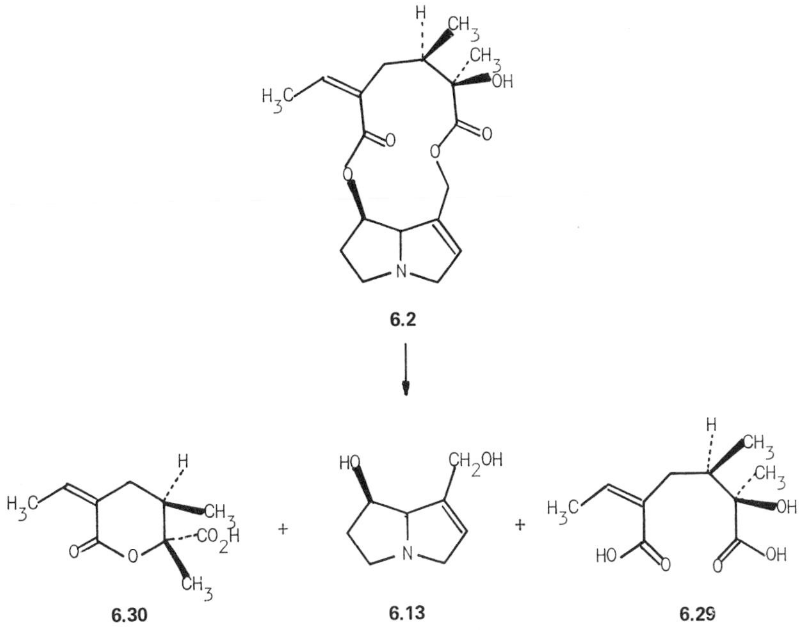

6.2

6.30 + 6.13 + 6.29

SCHEME 6.2 The products of hydrolysis of senecionine.

which then undergoes cyclization and reduction to ethyl 2,7-dihydroxypyrrolizidine-1-carboxylate (6.38). Hydrolysis and dehydration, followed by esterification and reduction yield (±)-retronecine (6.13, and its mirror image; [24]).

B. The Necic Acid of Senecionine: Senecic Acid

The structure of senecic acid (6.29), $C_{10}H_{14}O_4$, mp 156°C, was established by a series of degradative experiments as follows:

1. A Kuhn-Roth determination yields 2.9 equivalents of acetic acid, indicating that at least three methyl groups must be present.
2. Since a lactone is readily formed, the hydroxyl group is most probably δ to the carbonyl.
3. Ozonization in ethyl acetate yields two products, acetaldehyde and a second material which is not isolated but is further oxidized with lead tetraacetate. The second oxidation yields carbon dioxide and a methyl ketone which generates bromoform and methyl succinic acid (6.39) under conditions of the haloform reaction (Equation 6.11).

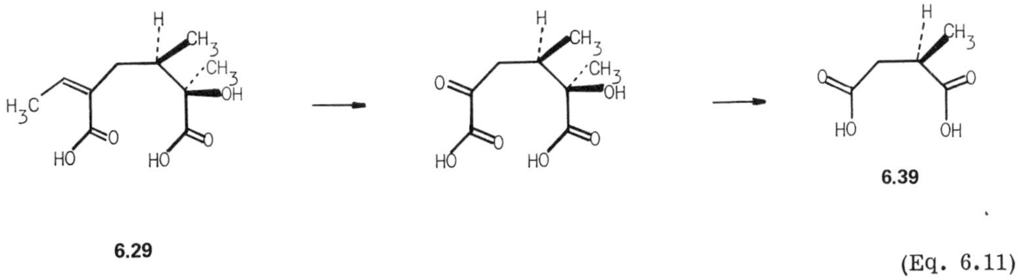

6.29 → → 6.39

(Eq. 6.11)

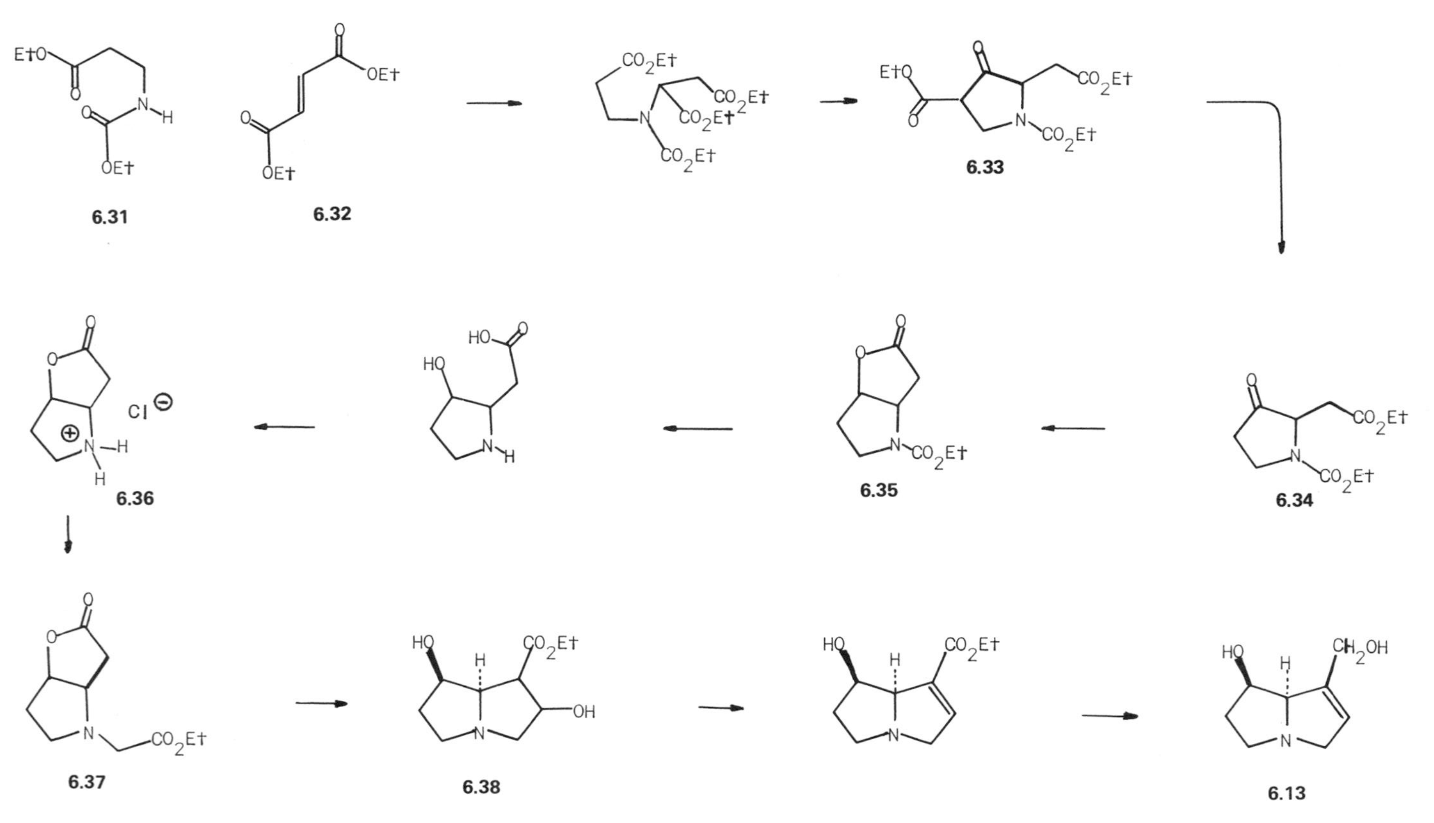

SCHEME 6.3 The synthesis of retronecine. (After Reference 24.)

4. Dihydrosenecic acid (6.40) was subjected to cleavage with lead tetraacetate to yield carbon dioxide and a methyl ketonic acid, which on hypobromite oxidation and treatment with ammonium hydroxide yields cis-α-ethyl-γ-methylglutarimide (6.41), identical to an authentic sample. This series of reactions also, therefore, establishes the absolute configuration of the methyl group as R (Equation 6.12). Finally, senecic

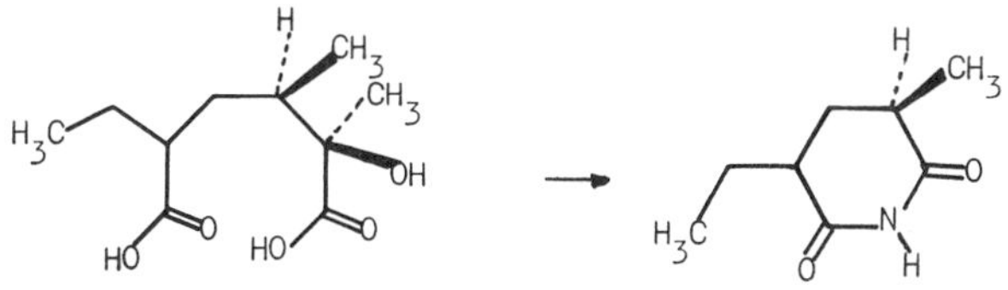

(Eq. 6.12)

6.40 6.41

acid (6.29) was synthesized (Scheme 6.4) by allowing 3-acetoxy-1-butyne (6.42) to react with nickel carbonyl to yield 2-methylene-3-acetoxybutanoate (6.43) and permitting 6.43 to condense with ethyl acetopropanoate. When the resultant adduct is hydrolyzed and decarboxylated,

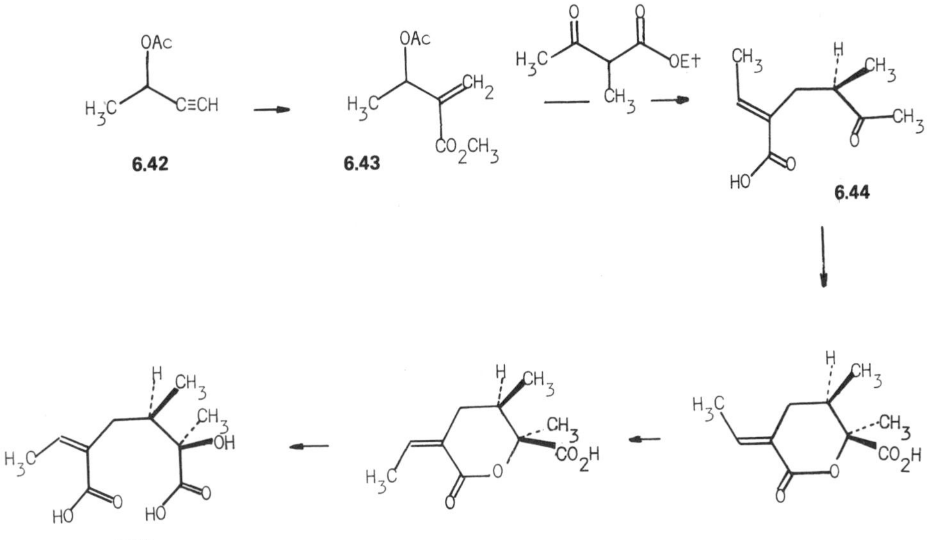

6.42 6.43 6.44

6.29

SCHEME 6.4 The synthesis of senecic acid. (After Reference 25.)

cis- and trans-(±)-5-methylhept-2-en-6-one-3-carboxylic acid (6.44, and its mirror image) are obtained. The (±)-trans acid (6.44) is converted to the cyanohydrin by reaction at the carbonyl and the product hydrolyzed to a lactone which is resolved and which yields a cis lactone on photolysis. The latter can be hydrolyzed to 2R, 3R-senecic acid (6.29), identical with the naturally occurring material [25].

That senecionine (6.2) is the cyclic diester of one molecule of retronecine (6.13) and one of senecic acid (6.29) was established by noting that a tetrahydrosenecionine (an amino acid; 6.45; $C_{18}H_{29}NO_5$) which yields retronecanol (6.14) and senecic acid (6.29) on hydrolysis, can be obtained on Raney nickel-catalyzed reduction of senecionine (6.2) (Equation 6.13).

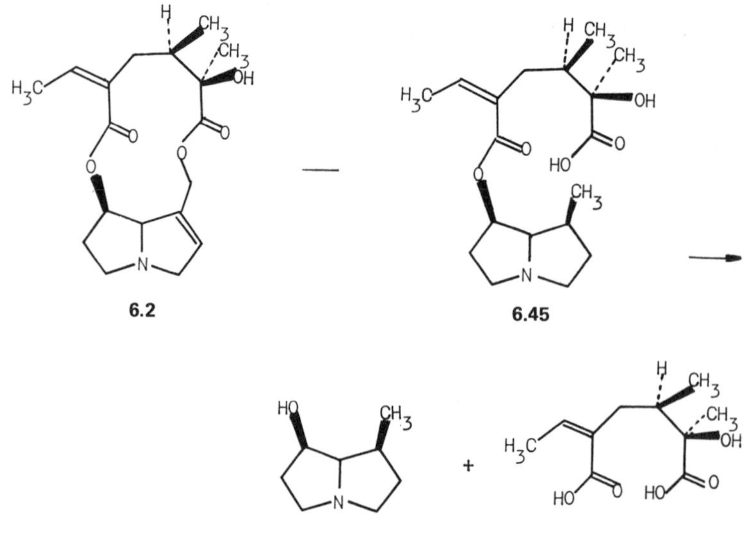

6.2 6.45 (Eq. 6.13)

6.14 6.29

With 3 equivalents of hydrogen (platinum oxide catalyst) under more forcing conditions, an amino acid (6.46) is isolated which yields retronecanol (6.14) and dihydrosenecic acid (6.40) on hydrolysis (Equation 6.14). Thus, it was concluded on the basis of similarity with other

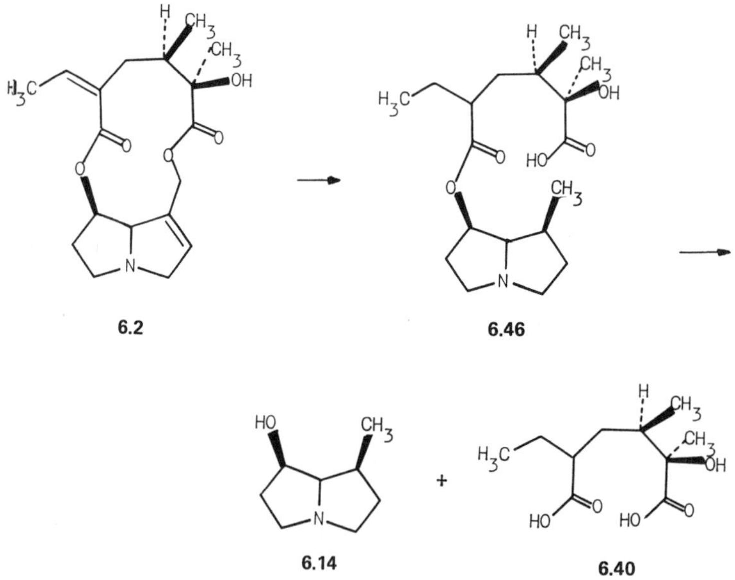

6.2 6.46 (Eq. 6.14)

6.14 6.40

Senecio alkaloids, that hydrogenolysis had occurred in the first case and that senecionine (6.2) must be a diester. The direction of addition was inferred from oxidation reactions carried out on the tetrahydroester 6.45.

III. 9-ANGELYLRETRONECINE N-OXIDE

9-Angelylretronecine N-oxide (6.3), $C_{13}H_{19}NO_4$, mp 153-154°C, $[\alpha]_D^{20}$ +30° (chloroform) was isolated from the bark of a large rain forest tree of the family Celastraceae (Behsa archboldiana Merr. and Perry) found in New Guinea. Since the N-oxide would normally have been discarded with the nonbasic fractions, care was required to insure it was not, and it was isolated on chromatography. The pmr spectrum of 6.3 contains a quartet at δ 6.11 (1H); a singlet, δ 1.91 (3H), and a doublet (fine splitting noted) at δ 1.96 (3H), along with the typical 1,2-dehydropyrrolizidine N-oxide nucleus. The seven protons for which data are given are assigned to the angelyl group (6.47). Although observation of a molecular ion in the mass

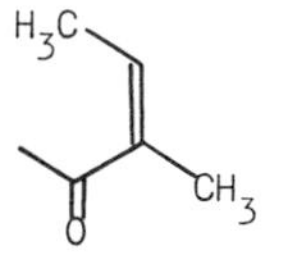

6.47

spectrum of pyrrolizidine N-oxides is unusual, 9-angelylretronecine N-oxide (6.3) possesses such a peak at m/e 253 (22% of base).

When the alkaloid is hydrolyzed with barium hydroxide in aqueous solution, retronecine N-oxide is obtained along with angelic acid (6.5). After reduction with zinc in sulfuric acid, a tertiary base is obtained which undergoes hydrogenolysis to retronecanol (6.14) and 2-methylbutyric acid (Equation 6.15). Thus, the angelic acid ester is at C_9 and the alkaloid

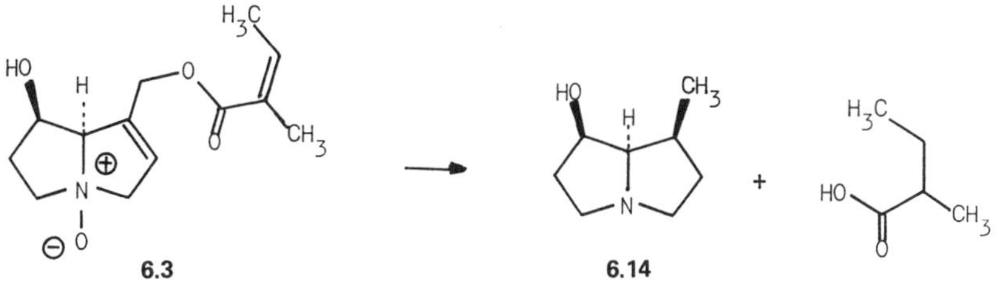

6.3 → 6.14 + (Eq. 6.15)

must be 9-angelylretronecine-N-oxide (6.3). This was confirmed by a synthesis [20] in which 1-chloromethyl-1,2-dehydro-7β-hydroxy-8α-pyrrolizidine (6.48), which could be prepared from retronecine (6.13) itself, and sodium angelate are allowed to react to form 9-angelylretronecine (6.49), which then reacts with 30% hydrogen peroxide for 5 days at room temperature to yield the corresponding N-oxide, identical in all respects to 6.3 (Equation 6.16).

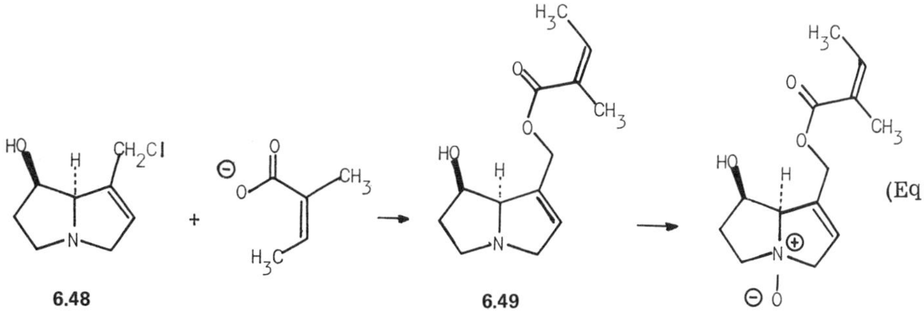

6.48 + → 6.49 → 6.3 (Eq. 6.16)

SELECTED READING

N. J. Leonard, in The Alkaloids, Vol. 1, R. H. F. Manske (ed.). Academic, New York, 1950, pp. 107 ff.

N. J. Leonard, in The Alkaloids, Vol. 6, R. H. F. Manske (ed.). Academic, New York, 1960, pp. 35 ff.

F. L. Warren, in The Alkaloids, Vol. 12, R. H. F. Manske (ed.). Academic, New York, 1970, pp. 246 ff.

Part 3

ALKALOIDS DERIVED FROM LYSINE (EXCLUDING THE TOBACCO ALKALOIDS)

Lysine (7.1), which is known (p. 36) to be derived from acetate and α-ketoglutarate (7.2; the α-aminoadipic pathway), can also be derived from pyruvate (7.3) and aspartic acid (7.4; the

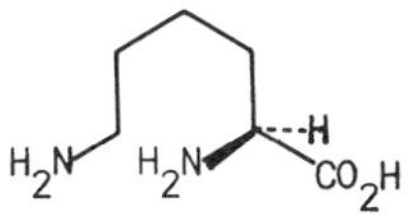

H_2N H_2N H CO_2H

7.1

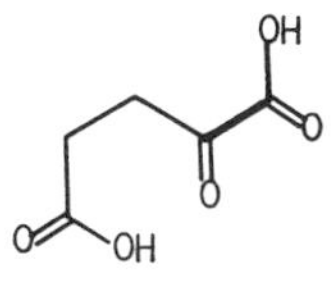

OH O O O OH

7.2

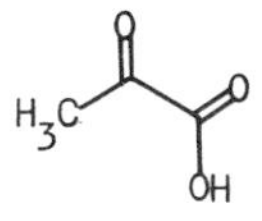

H_3C O O OH

7.3

diaminopimelic pathway). Although nonsymmetrical incorporation of lysine (7.1) can be demonstrated in the alkaloids derived therefrom, it had long been a fundamental assumption of classic biogenetic hypothesis that an "equivalent" of lysine, e.g., cadaverine (7.5), 5-aminopentanal (7.6), or glutaric dialdehyde (7.7), lay on the pathway to a Δ^1-piperideine (7.8) which was the penultimate alkaloid pregenitor.

There are many alkaloids which are derived from lysine (7.1). In this portion of the book, we shall consider a few representative examples and limit our discussion to their syntheses and reactions. Specifically, we shall look at the pomegranate alkaloids pelletierine (7.9) and ψ-pelletierine (7.10), the Sedum alkaloids sedamine (7.11) and sedridine (7.12), the Lobelia alkaloids lobeline (7.13) and lobinaline (7.14), the Lupine alkaloids lupinine (7.15), sparteine (7.16), and matrine (7.17), and the more complex Lycopodium alkaloids, lycopodine (7.18) and annotinine (7.19).

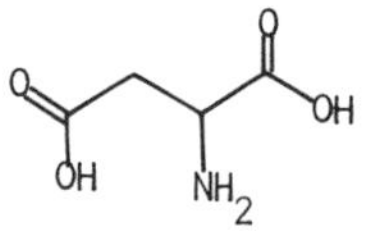

OH NH2 OH O O

7.4

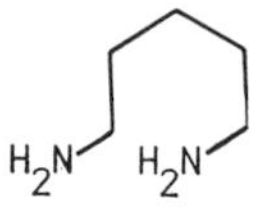

H2N H2N

7.5

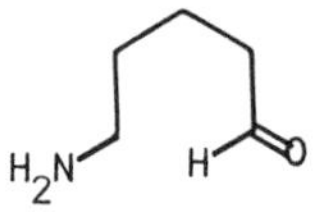

H2N H O

7.6

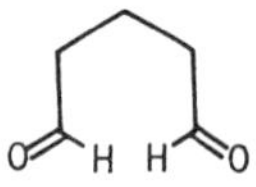

O H H O

7.7

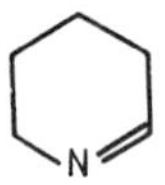

N

7.8

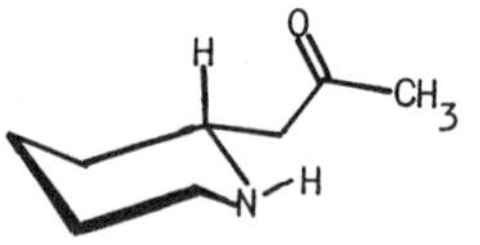

H O CH3 N H

7.9

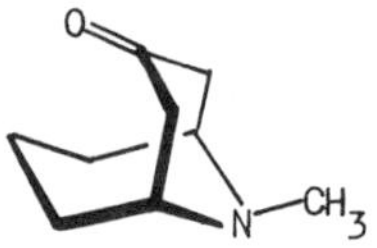

O N CH3

7.10

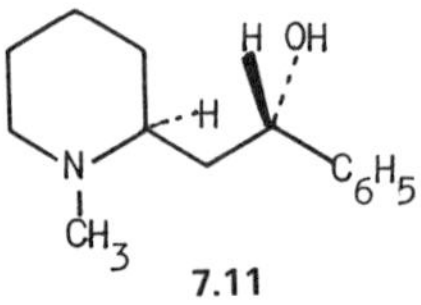

N CH3 H H OH C6H5

7.11

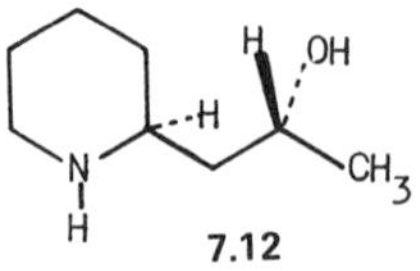

N H H H OH CH3

7.12

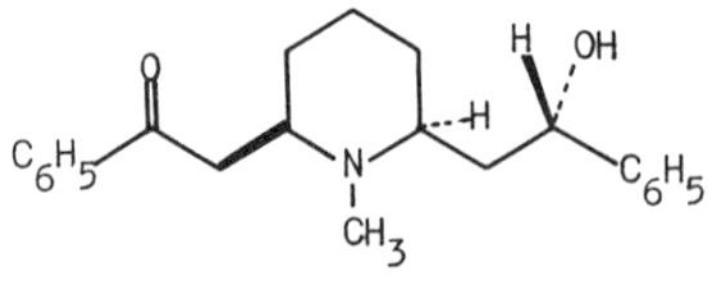

C6H5 O N CH3 H H OH C6H5

7.13

H C6H5 N N H CH3 C6H5

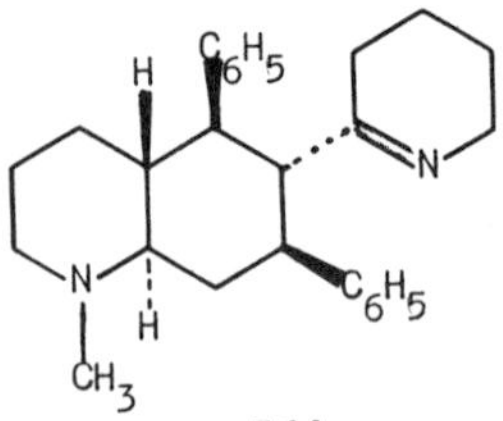

7.14

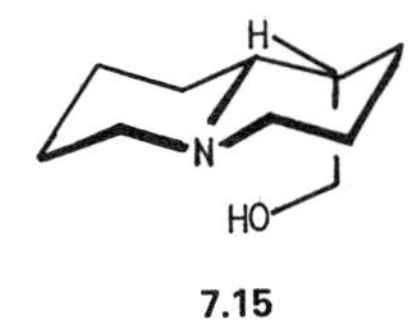

7.15

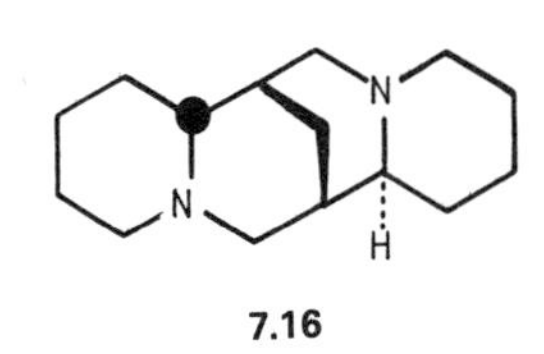

7.16

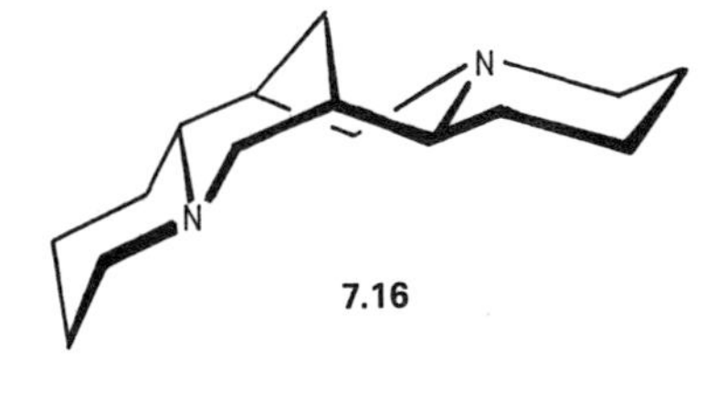

7.16

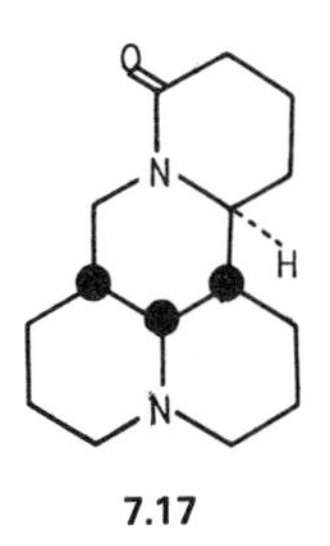

7.17

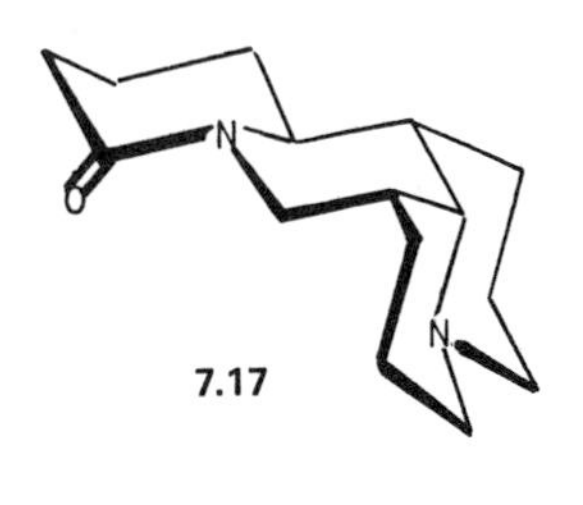

7.17

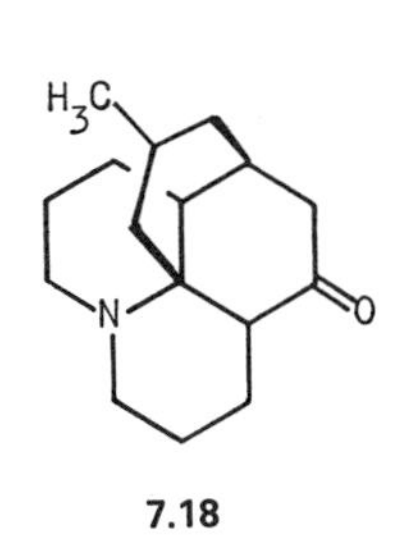

7.18

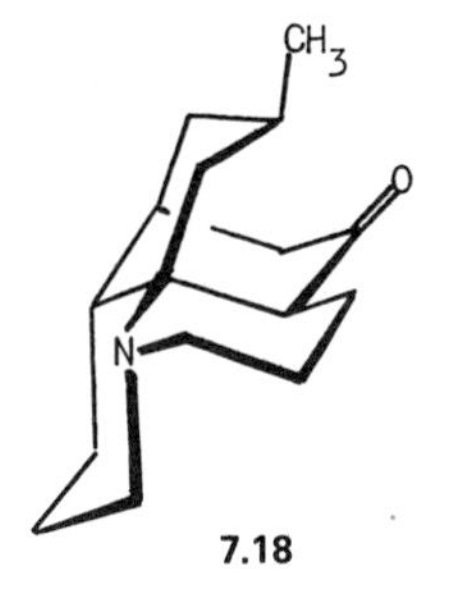

7.18

7.19

7.19

7

BIOSYNTHESIS OF THE POMEGRANATE, *SEDUM*, *LOBELIA*, *LUPIN*, AND *LYCOPODIUM* ALKALOIDS

> Go, sir, gallop, and don't forget that the world was made in six days. You can ask me for anything you like, except time.
>
> Napoleon Bonaparte, The Corsican (1803)

In a vein similar to that already set forth for alkaloids derived from ornithine (see Part 2), it was long considered axiomatic that lysine (7.1) underwent ready decarboxylation to cadaverine (7.5) which was convertible, oxidatively, in several steps, to a Δ^1-piperideine (7.8). Since the Δ^1-piperideine (7.8) system was recognized as being susceptible to nucleophilic addition reactions at the unsaturated carbon, suitable side chains could be added to elaborate the alkaloids. In this scheme of things, the amino acid, pipecolic acid (7.20), was usually considered a final product rather than a potential intermediate; this view has been modified.

Feeding experiments with a variety of labeled precursors in the plants elaborating the alkaloids discussed here could not, until recently, be interpreted in a single, reasonable pathway. Thus, it had been shown that cadaverine (7.5), despite its specific nonrandom incorporation into, e.g., ψ-pelletierine (7.10), isolated from *Puncia granatum* L., could not be an obligatory intermediate between lysine (7.1) and the alkaloids (7.10) since it is symmetrical and randomization between C_2 and C_6 of lysine (7.1) does not occur. Similarly, ϵ-amino-α-ketocaproic acid (7.21) could be excluded be excluded because the hydrogen at C_2 of lysine (7.1) is retained and the α-amino-δ-semialdehyde (7.22) excluded because the

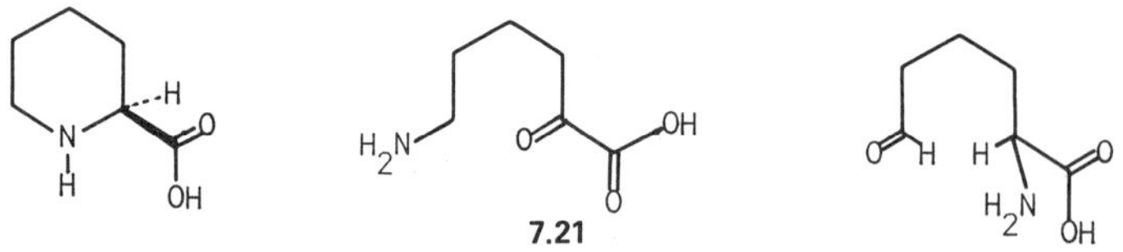

7.20 7.21 7.22

ϵ-amino, but not the α-amino nitrogen, and the hydrogen atoms at C_6 are retained. Although, even with these restrictions, several possible pathways could a priori be suggested, only one appears, at present, consonant with all available data.

This pathway (Scheme 7.1) is similar to that shown on p. 50 [18] but requires stereospecific decarboxylation and reprotonation in order that the hydrogen at C_2 of L-lysine (7.1) be retained and the entrance of cadaverine (7.5) can occur. In addition, cadaverine (7.5) is required to be in equilibrium between "free" and a "pyridoxyl-bound" state. The stereochemistry of decarboxylation is in concert with labeling experiments, but the actual formation of 5-aminobutanal is not required, and a pyridoxyl-bound intermediate may lead directly

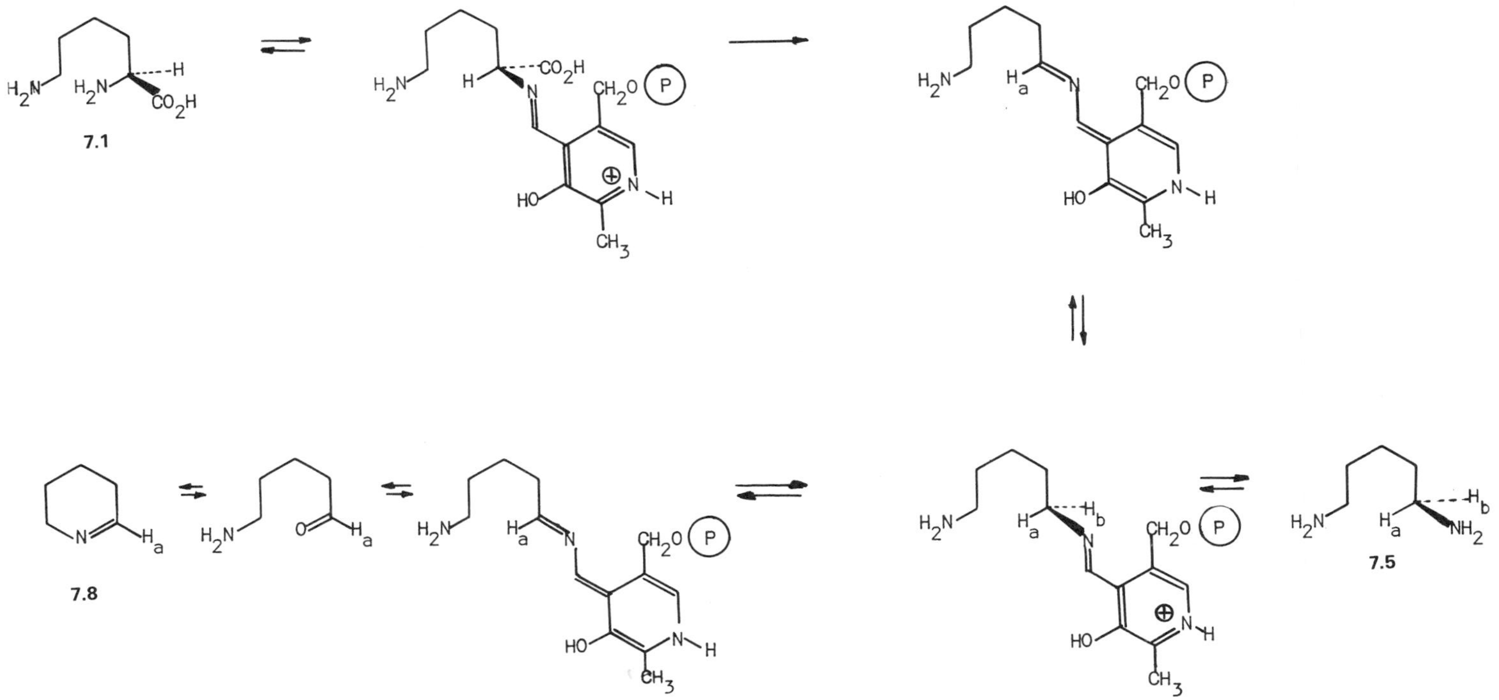

SCHEME 7.1 A pathway for the production of Δ^1-piperideine consistent with all available evidence. (After Reference 18.)

to the Δ^1-piperideine (7.8). These studies have been confirmed in striking detail [26] by the appropriate ^{3}H-labeling experiments which have also shown, among other things, that it is D-lysine (optical antipode of 7.1) which leads to L-pipecolic acid (7.20) through loss of the α-amino group and cyclization. Reopening of the L-pipecolic acid (7.20) now leads, by aminolysis at C_6, to L-lysine (7.1), which can be utilized for alkaloid synthesis.

I. BIOSYNTHESIS OF PELLETIERINE, ψ-PELLETIERINE, SEDAMINE, SEDRIDINE, LOBELINE, AND LOBINALINE

Pelletierine (7.9; the enantiomer of isopelletierine, 7.9) and ψ-pelletierine (7.10) might well be formed in a fashion analogous to that already shown for hygrine (Scheme 3.1) and the tropane alkaloids (Scheme 3.2), respectively. Thus, either the Δ^1-piperideine (7.8) or an appropriately enzyme-bound analog, on reaction with acetoacetate (7.23) or its equivalent, will generate an intermediate suitable for decarboxylation to pelletierine (7.9; Scheme 7.2).

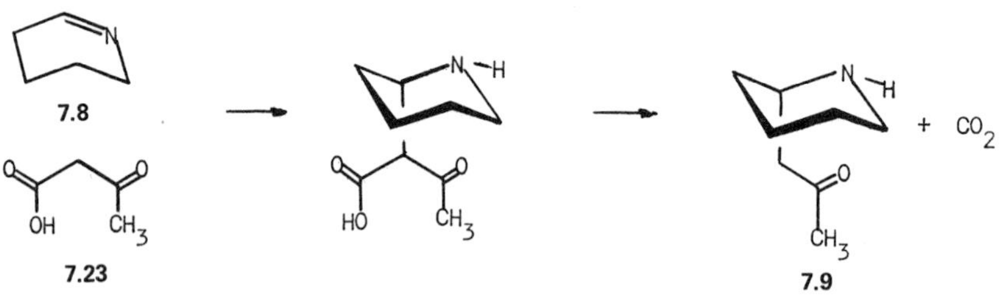

7.8 7.23 → → 7.9 + CO_2

SCHEME 7.2 A possible pathway for the formation of (-)-pelletierine from a Δ^1-piperideine and acetoacetate.

(-)-Pelletierine (7.9), which has the 2R configuration, would result from attack of the bottom (si face); attack on the top (re face) then generates (2S)-(+)-pelletierine (antipode of 7.9). Pelletierine (7.9), or a suitable derivative, after N-methylation and oxidation, can then undergo cyclization to ψ-pelletierine (7.10; Scheme 7.3).

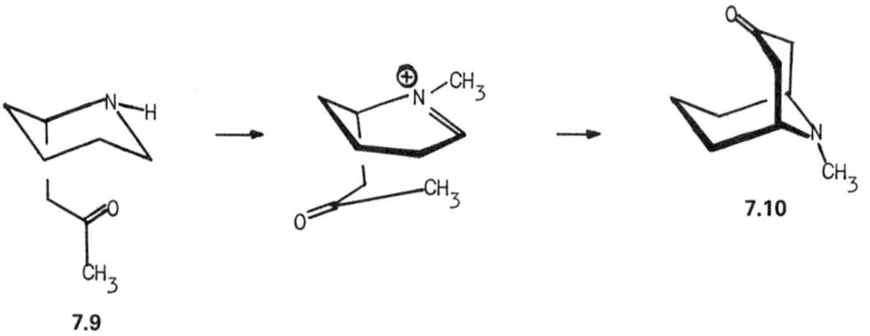

7.9 → → 7.10

SCHEME 7.3 A possible pathway for the formation of ψ-pelletierine from pelletierine.

In the case of sedamine (7.11), the same Δ^1-piperideine (7.8) reacts with a C_6-C_3 fragment derived from cinnamic acid (7.25), which in turn is generated from phenylalanine (7.24). Presumably, cinnamic acid (7.25) first undergoes hydration and oxidation to a benzoylacetate (7.26) which then adds (as did acetoacetate, 7.23) and undergoes decarboxylation and reduction

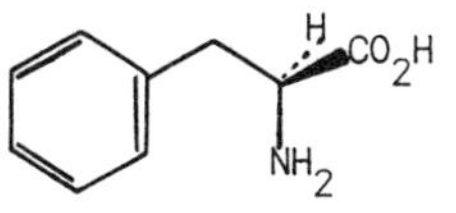

7.24 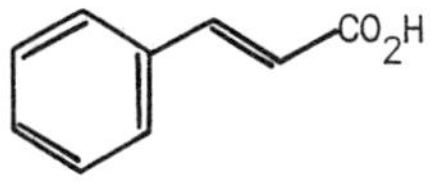 7.25 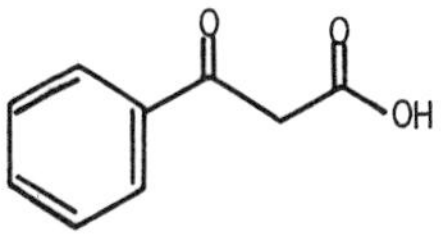 7.26

to sedamine (7.11; Scheme 7.4). Interestingly, (-)-sedamine (7.11) has been shown to have the 2S, 8S configuration, implying that, in this case, attack of the benzoylacetate (7.26) on the Δ^1-piperideine (7.8) is from the re face. If the re face of the Δ^1-piperideine (7.8) is attacked

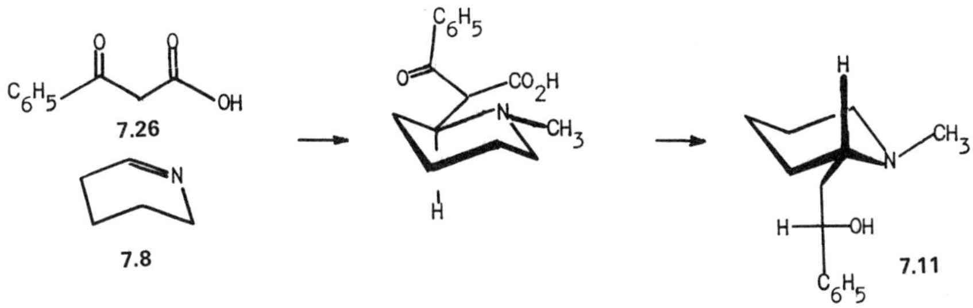

7.26 7.8 7.11

SCHEME 7.4 A pathway for the formation of sedamine from a Δ^1-piperideine and benzoylacetate.

by acetoacetate (7.23) or its equivalent and the carbonyl reduced (after decarboxylation), the 2S, 8S-dihydropelletierine so formed is, in fact, (+)-sedridine (7.12), the oxidation of which results in formation of (+)-pelletierine (optical antipode of 7.9).

The stereochemistry of lobeline (7.13) at the chiral centers also present in sedamine (7.11) is identical to that found in sedamine (7.11). Presumably, therefore the pathway (Scheme 7.4) already shown for sedamine (7.11) is again applicable, and the second alkylation by the C_6-C_3 fragment occurs on the previously unsubstituted side of sedamine (7.11), or its equivalent, after an oxidation to the imine.

Early feeding experiments from which the alkaloid lobinaline (7.14) had been isolated showed that C_6-labeled lysine (7.1) was incorporated into the decahydroquinoline system,

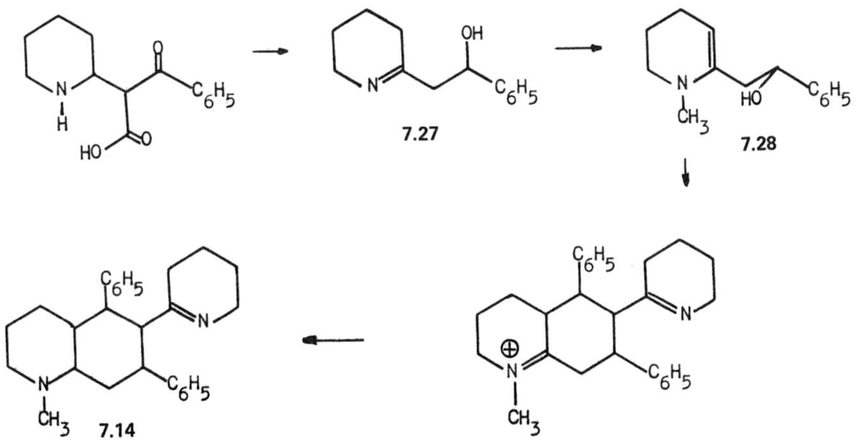

7.27 7.28 7.14

SCHEME 7.5 A possible pathway for the production of lobinaline (7.14) by the dimerization of sedamine-like (7.11) intermediates.

but that only 25% of the label was found at C_2. This argued for symmetrical incorporation of lysine (7.1), which although capable of being explained by Scheme 7.1 is not in concert with the earlier (see above) reports of nonrandom incorporation of lysine (7.1) into the other alkaloids discussed. Repetition of this work [27], which included more complete degradation of the alkaloid (7.14), showed that, as was the case with, e.g., sedamine (7.11), the incorporation was not symmetrical. Thus, there is apparently a dimerization of the sedamine-like (7.11) hypothetical intermediates 7.27 and 7.28 (Scheme 7.5) to generate lobinaline (7.14) directly.

II. BIOSYNTHESIS OF LUPININE, SPARTEINE, AND MATRINE

In concert with feeding experiments, lupinine (7.15) may be considered to arise from 2 equivalents of lysine (7.1), via cadaverine (7.5), since incorporation of lysine (7.1) occurs in a symmetrical fashion. In addition, in lupin seedlings, diamine oxidase activity has been demonstrated. While the presence of such activity is exciting in that a potential pathway from cadaverine (7.5) to 5-aminopentanal (7.6) is actually available, it is not obligatory that it be used, and indeed, the presence of cadaverine (7.5) in lupin has not been demonstrated. Scheme 7.6 presents the use of such a potential pathway in the biosynthesis of lupinine (7.15). Here,

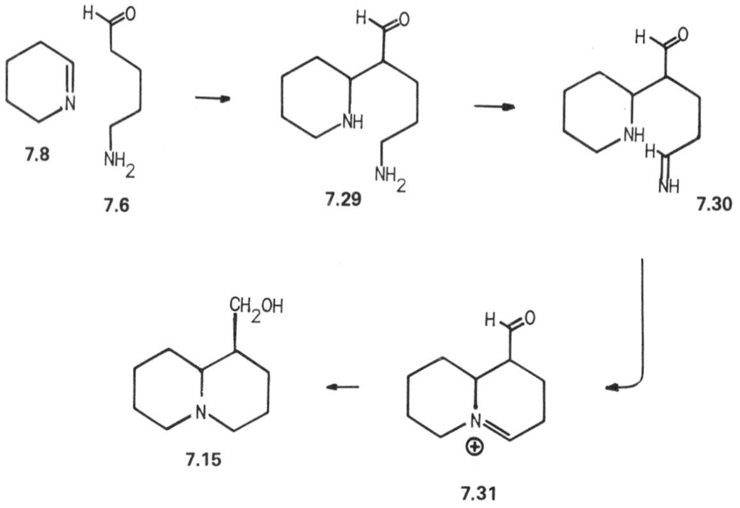

SCHEME 7.6 A pathway to lupinine from 5-aminopentanal and a Δ^1-piperideine.

5-aminopentanal (7.6), or its equivalent, is permitted to condense with the Δ^1-piperideine (7.8) to the intermediate diamine (7.30). Cyclization and elimination lead to the immonium salt (7.31) which is reduced to lupinine (7.15). Alternatively, instead of reduction, 7.31 could (Scheme 7.7) condense with another equivalent of 5-aminopentanal (7.6) to yield the imine (7.32), which on partial reduction and cyclization leads to 7.33. Now, if 7.33 isomerizes to the enamine 7.34 and cyclization results, the ready reduction to sparteine (7.16) would be expected. Finally (Scheme 7.8), 7.33 could isomerize instead to 7.35, the corresponding enamine of which (7.36) leads to 7.37, a potential precursor to matrine (7.17). Thus, 7.37, undergoing isomerization to 7.38, hydration to 7.39, and oxidation, leads to

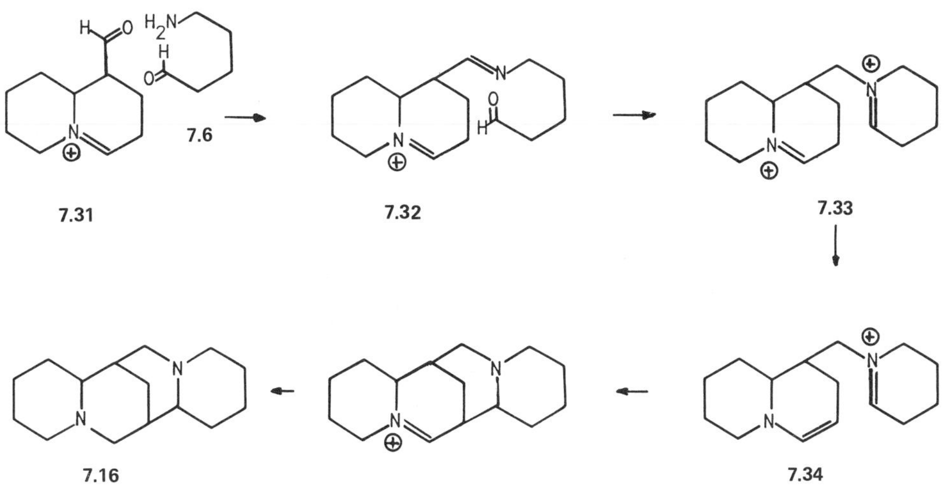

7.6

7.31 7.32 7.33

7.16 7.34

SCHEME 7.7 A pathway for the formation of sparteine by condensation of a lupinine-like precursor with 5-aminopentanal.

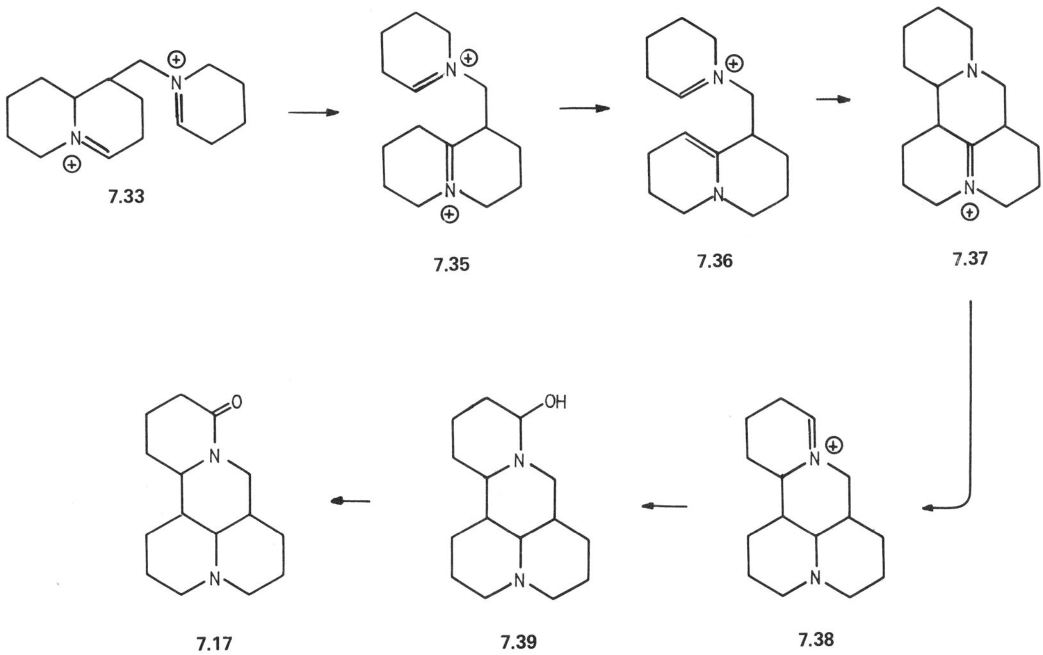

7.33 7.35 7.36 7.37

7.17 7.39 7.38

SCHEME 7.8 A possible pathway for the formation of matrine from an intermediate which might also lead to sparteine.

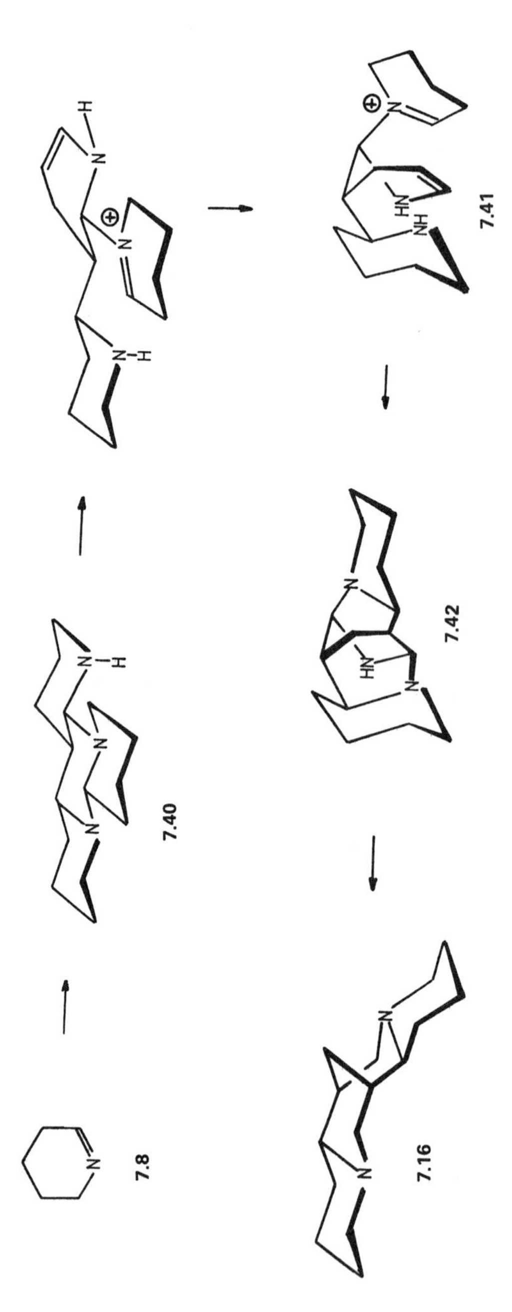

SCHEME 7.9 A possible pathway to sparteine via a Δ^1-piperideine trimer. (After Reference 29.)

matrine (7.17). The merit of these particular hypotheses is sustained by the ready self-condensation reaction of tetrahydropyridines which leads to dimers and trimers similar to these products at pHs between 6 and 10. They are, of course, simply aldol condensations [28].

Although the gross structure of the alkaloids can be accounted for in this way, there is still the problem of stereochemistry. An alternative hypothesis has recently been suggested [29] which addresses this problem. Thus, as shown in Scheme 7.9, if the Δ^1-piperideine (7.8) is allowed to trimerize, the isotripiperidine (7.40), shown in the favored all-trans configuration (one enantiomer), can be readily obtained. Oxidative ring opening to the immonium ion 7.41, followed by recyclization to 7.42, then leads on excision of nitrogen and reduction to sparteine (7.16), with the stereochemistry fixed as (-)-6R:7S:9S:11S or its enantiomer.

III. BIOSYNTHESIS OF LYCOPODINE AND ANNOTININE

Although the structures of these two Lycopodium alkaloids (and others in the family) can be accounted for on the basis of suitably folded polyketide chains, it is now accepted that two pelletierine-like (7.9) fragments combine to yield these C_{16} bases. However, feeding labeled pelletierine (7.9; shown independently to be present in the plant to which it was fed) provides the information that only one-half the molecule is derived from this intact precursor. Since both halves can be derived from lysine (7.1) and cadaverine (7.5), this may mean that either pelletierine (7.9) is not an obligatory precursor or that addition of the three-carbon side chain to pelletierine (7.9) occurs before or after reaction of pelletierine (7.9) with a Δ^1-piperideine (7.8). However, since pelletierine (7.9) has been reported to be an obligatory precursor, it becomes necessary to postulate the formation of lycopodine from one pelletierine only, as in Scheme 7.10.

Thus, if pelletierine (7.9) condenses again with acetoacetate (7.23), and the product is permitted to undergo decarboxylation and oxidation, 7.44 is formed. If 7.44 now condenses with the Δ^1-piperideine (7.8), and oxidation of the resultant amine to an imine occurs, 7.45 results. Two cyclization reactions are now possible: one to the carbonyl carbon and one to the imine carbon, forming, in that sequence, all four required rings and generating 7.46 (related to another Lycopodium alkaloid, α-obscurine, 7.47). Reduction of 7.46, hydrolysis with ring opening, and recyclization followed by another reduction, yield lycopodine (7.18). If, however, instead of reduction to lycopodine (7.18), the presumed intermediate 7.46 is converted to 7.48 (a number of steps, including hydration, ring opening, and recyclization, double-bond isomerization, and reduction of the carbonyl are necessary), intramolecular ketal formation results in generation of another Lycopodium alkaloid (acrofoline, 7.49). Oxidative rearrangement [30] of acrofoline (7.49) to 7.50, followed by epoxidation, generates annotinine (7.19; Scheme 7.11).

(Scheme 7.10 appears on p. 108 and Scheme 7.11 appears on p. 109.)

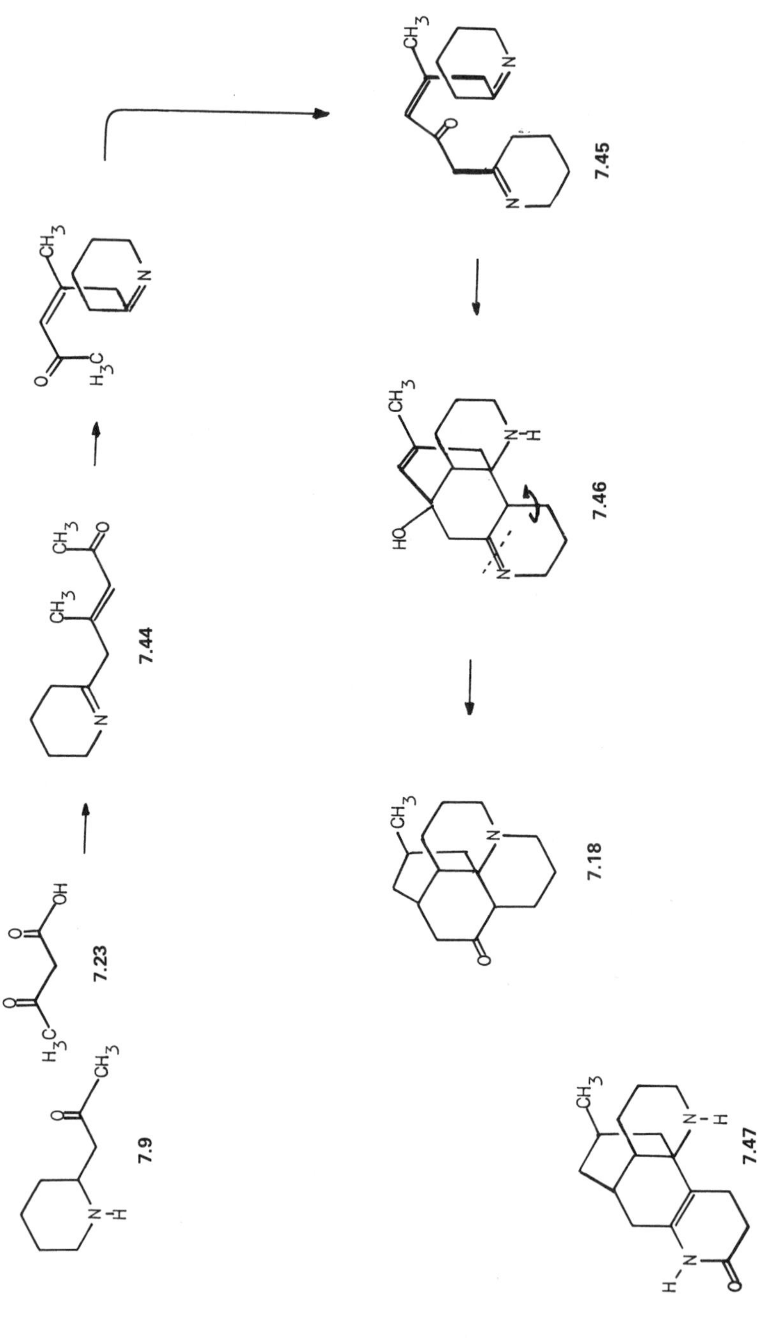

SCHEME 7.10 A pathway for the formation of lycopodine from 1 equivalent of pelletierine, acetoacetate, and Δ^1-piperideine.

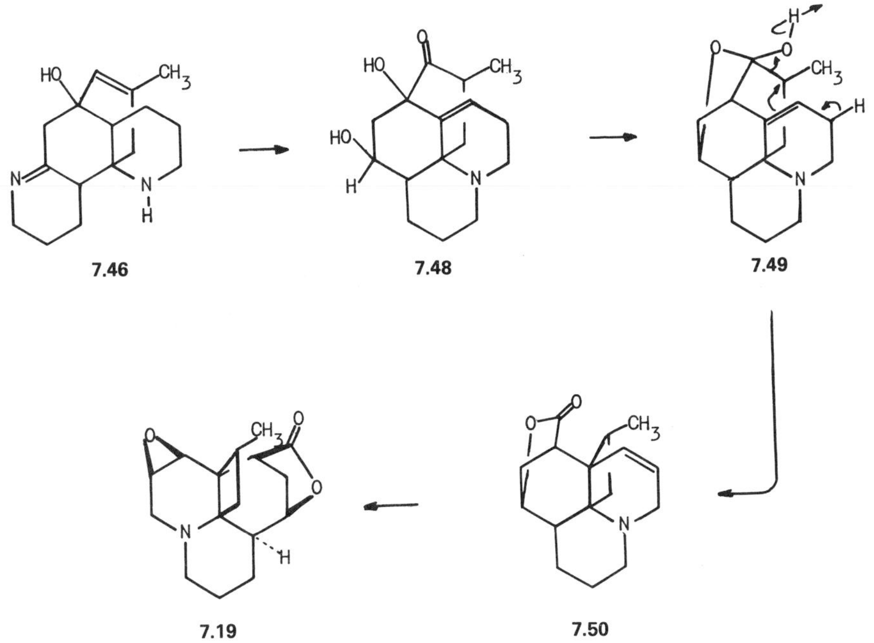

7.46 7.48 7.49

7.19 7.50

SCHEME 7.11 A pathway for the formation of annotinine from a potential intermediate on the pathway to the formation of lycopodine.

8

CHEMISTRY OF THE POMEGRANATE ALKALOIDS

> You have deliberately tasted two worms and you can leave
> Oxford by the town drain.
>
> W. A. Spooner (ca. 1900)

Pelletierine tannate, long used as an anthelmintic for intestinal pinworms, is a mixture of the tannates of all the alkaloids in pomegranate tree bark (Puncia granatum L.). The proportion of alkaloids is somewhat variable, but alkali treatment of the mixture yields the free bases. The alkaloids themselves were first isolated by Tanret in 1878, and they are among the few named after an individual (P. J. Pelletier) rather than the plant from which they were obtained. We shall consider two of the alkaloids found, namely, pelletierine and ψ- pelletierine.

I. PELLETIERINE

Pelletierine (8.1; b_{10} 86°C, HCl salt mp 143-144°C), and an "isomer," isopelletierine (8.2; b_{21} 106°C, HCl salt mp 164°C) are both basic liquids. Wolff-Kishner reduction of N-methylisopelletierine (8.3) yields N-methylconiine (8.4; Equation 8.1), and chromic acid oxidation of N-methylisopelletierine (8.3) gives N-methylpipecolic acid (8.5; Equation 8.2).

8.1

8.2

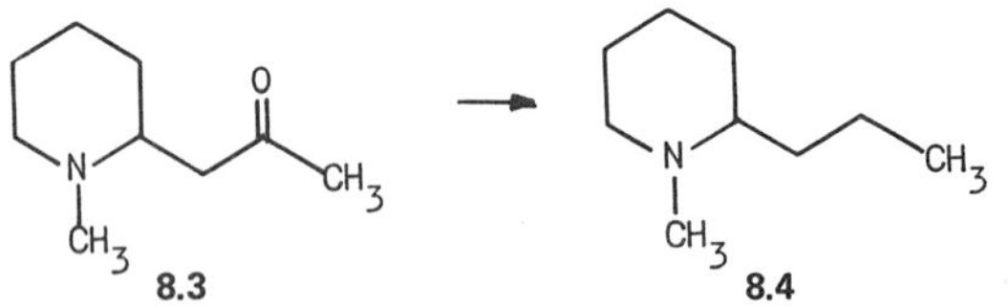

8.3 → 8.4 (Eq. 8.1)

The position of the carbonyl group in the side chain of isopelletierine (8.2) was established by synthesis. Thus, dehydrohalogenation of N-chloropiperidine (8.6) yields 2,3,4,5-tetrahydropyridine (Δ^1-piperideine, 8.7), and condensation of 8.6 with acetoacetic acid (8.8)

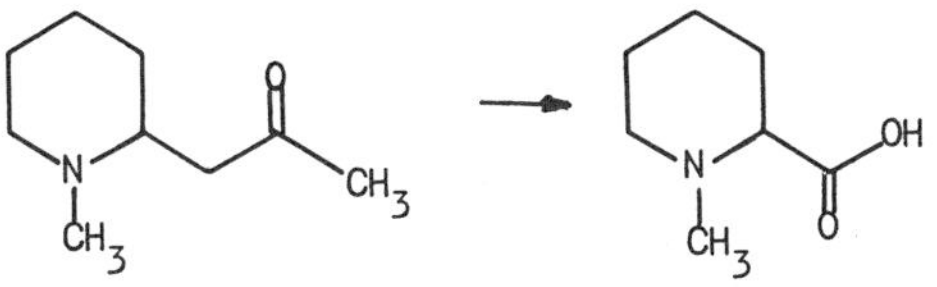

8.5 (Eq. 8.2)

to directly yield the product (8.2) can be effected at low or high pH (Equation 8.3). (The Δ^1-piperideine, 8.7, readily dimerizes and trimerizes at pHs between 5 and 11 but does so only slowly beyond either extreme [28].)

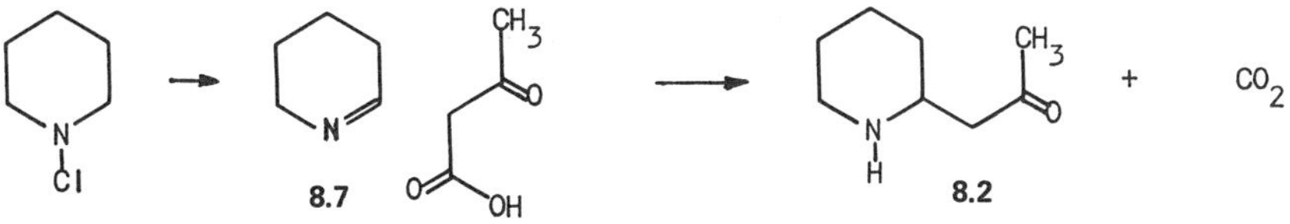

8.6 8.7 8.8 8.2 + CO_2 (Eq. 8.3)

For many years, pelletierine (8.1) was thought to be 3-(2-piperidyl)propionaldehyde (8.9) since the oxime of pelletierine (8.1) was reported to yield a nitrile ($C_8H_{14}N_2$; 8.10) by phosphorus pentachloride dehydration and Wolff-Kishner reduction also yielded coniine (8.11; Equation 8.4). However, the alkaloid was much more stable than the aldehyde (8.9), which

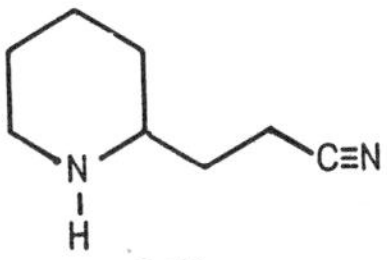

8.10

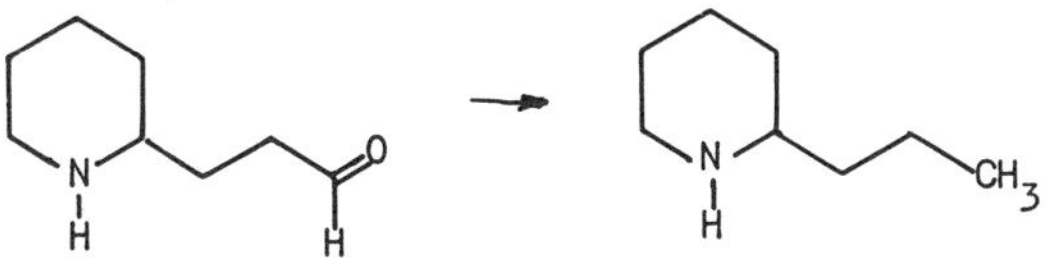

8.9 8.11 (Eq. 8.4)

resisted synthesis because of its liability, and reinvestigation of the nitrile 8.10 showed it to be the amidine (8.12), formed by dehydration of the Beckmann rearrangement product of isopelletierine oxime (8.13; Equation 8.5). If the isolation of isopelletierine (8.2) is carried out very carefully, the R-(-)-form is obtained. This material is pelletierine (8.1). It is amusing to speculate that the act of isolation of the alkaloid fraction had, perhaps, reacemized some of the pelletierine (8.1), and since the racemate and the pure enantiomer had slightly different physical properties, they were mistaken for different compounds. Indeed, isopelletierine (8.2) may be an artifact of isolation.*

*Philosophically, for whatever its merit, the reader should be aware that many (we will encounter more) of the isolated alkaloids may be artifacts of isolation: rearrangements, epimerization, etc., having occurred during the physical wrenching out of the bases from their cellular matrix and/or during their separation and purification. This knowledge does not appear to have detracted much from the spirit of the investigators, and indeed the challenge has mothered new and more gentle isolation techniques.

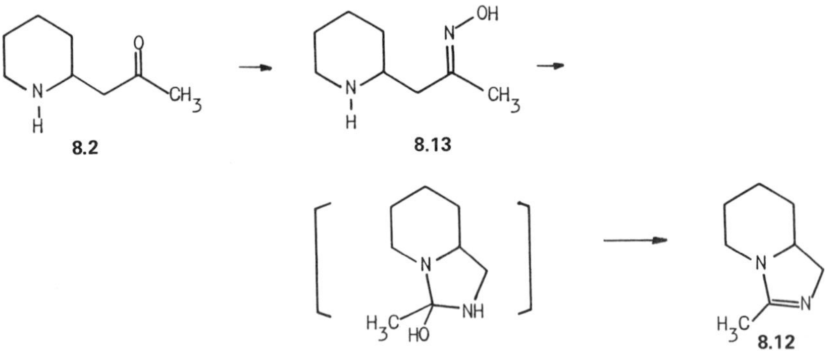

8.2 8.13

(Eq. 8.5)

8.12

II. ψ-PELLETIERINE

ψ-Pelletierine (8.14), $C_9H_{15}NO$, was recognized as a tertiary amine, which because it formed a dibenzylidene derivative, possessed a carbonyl flanked by two methylene units (i.e., 8.15). Electrolytic reduction of the keto group provides the deoxygenated base (8.16) which

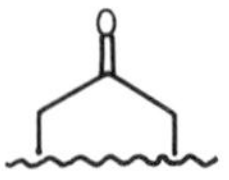

8.15

was degraded via a series of exhaustive methylations and Hofmann eliminations to 1,5-cyclooctadiene (8.17) by Willstatter (Equation 8.6). The gross structure of the 1,5-cyclooctadiene (8.17) was established by its oxidation to succinic acid (8.18). It was later shown that this

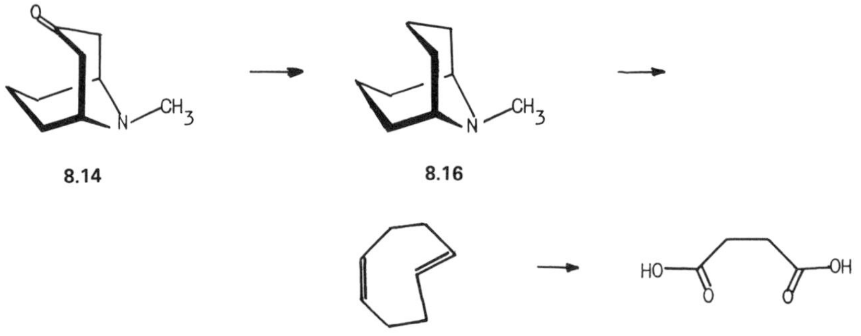

8.14 8.16

(Eq. 8.6)

8.17 8.18

diene (8.17) is not identical to the cyclooctadiene isomer prepared by the photochemical dimerization of butadiene (Z,Z) and that the Willstatter diene is the Z,E isomer. Indeed, if the amine resulting from the first Hofmann elimination (i.e., 8.19) is resolved, the Z,E-1,5-cyclooctadiene (8.17) isolated is chiral.

Willstattër also reduced the carbonyl of ψ-pelletierine (8.14) and dehydrated the resulting alcohol to an alkene which he then carried through the series of Hofmann sequences to 1,3,5-cyclooctatriene (8.20). Bromination, substitution of dimethyl amine for bromine, and a final Hofmann elimination culminated (8 years of work) in the generation of cyclooctatetraene

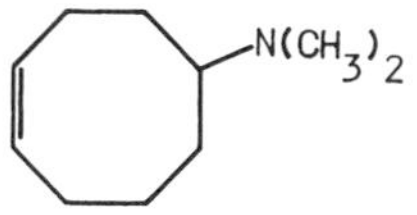

$N(CH_3)_2$

8.19

(8.21; Equation 8.7) which proved to be like a typical olefin rather than like benzene. Although his characterization of this cyclic unsaturated alkene was challenged, the production of cyclooctatetraene (8.21) in ton quantities by catalytic tetramerization of acetylene has given

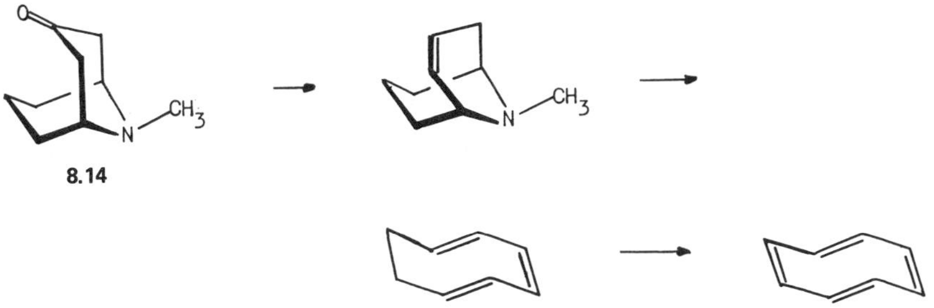

CH_3 N 8.14 → CH_3 N →

(Eq. 8.7)

8.20 → 8.21

those who followed him enough material to vindicate his work. The synthesis of ψ-pelletierine (8.14) is a classic application of the Mannich-type condensation between glutaric dialdehyde, methylamine, and acetone dicarboxylic acid at room temperature (Equation 8.8; [31]).

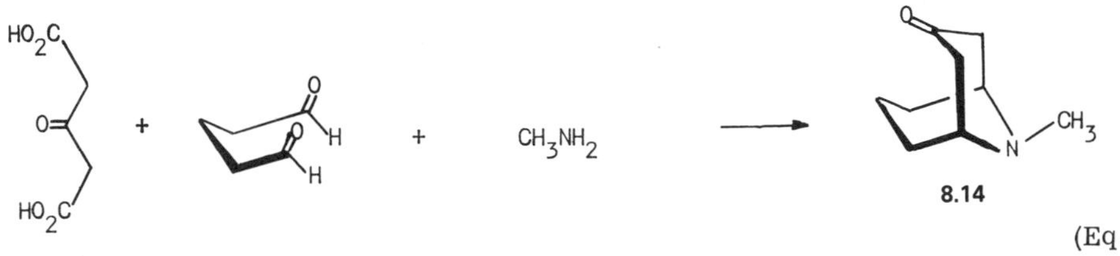

HO_2C O= HO_2C + O H O H + CH_3NH_2 → O N CH_3 8.14

(Eq. 8.8)

SELECTED READING

W. A. Ayer and T. E. Habgood, in The Alkaloids, Vol. 11, R. H. F. Manske (ed.). Academic, New York, 1968, pp. 461 ff.

L. Marion, in The Alkaloids, Vol. 1, R. H. F. Manske (ed.). Academic, New York, 1950, p. 176.

9

CHEMISTRY OF THE SEDUM ALKALOIDS

From the latin sedo; to sit, referring to the manner in which some species attach themselves to stones and walls.

B. J. Healey, A Gardener's Guide to Plant Names

There are some 600 annual, biennial, or perennial succulents belonging to the Sedum genus of the family Crassulaceae, many of which are characterized by the ability to grow where little else can. A number of alkaloid-producing plants have been investigated and diverse products found. We will consider only two typical members, sedamine and sedridine.

I. SEDAMINE

Sedamine (9.1), $C_{14}H_{21}NO$, is a basic levorotatory secondary alcohol which occurs, along with an epimer (allosedamine, 9.2) in Lobelia as well as in Sedum. Both sedamine (9.1) and allosedamine (9.2) are oxidized to the same ketone (9.3) and, under more vigorous conditions, to N-methylpipecolic acid (9.4), $C_7H_{13}O_2$, and benzoic acid (9.5), $C_7H_6O_2$, thus accounting for all the carbon atoms (Equation 9.1). Furthermore, base-catalyzed condensation of

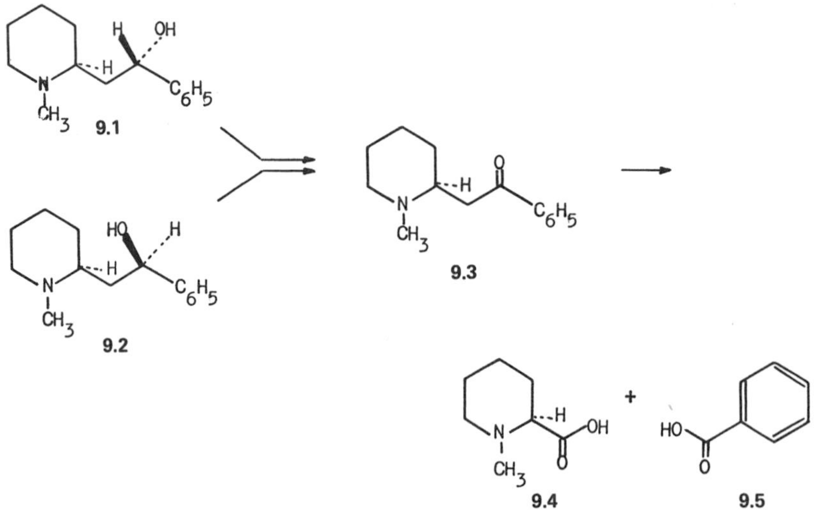

9.1 9.2 9.3 9.4 9.5

(Eq. 9.1)

α-picoline methiodide (9.6) with benzaldehyde (9.7) gives a salt (9.8) which can be reduced to a mixture of racemic sedamine (9.1) and allosedamine (9.2; and their mirror images; Equation 9.2). Finally, Hofmann degradation of allosedamine (9.2) followed by oxidation yields

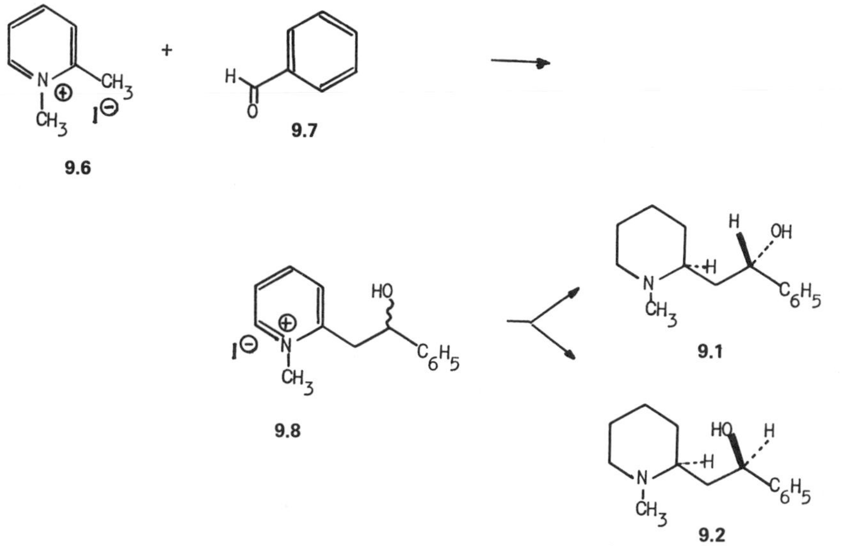

(Eq. 9.2)

S-(-)-β-hydroxy-β-phenylpropanoic acid (9.9), thus establishing the epimer as sedamine (9.1; Equation 9.3).

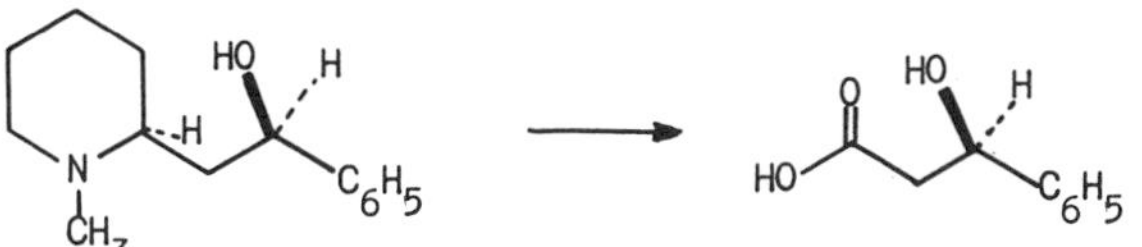

(Eq. 9.3)

II. SEDRIDINE

Sedridine (9.10), $C_8H_{17}NO$, a major alkaloid of the biting stonecrop (Sedum acre L.), is dextrorotatory, has both secondary amino and hydroxyl groups, and on careful oxidation yields (+)-isopelletierine (9.11). Additionally, a series of von Braun degradations on the O,N-dibenzoate (9.12) followed by hydrogenolysis of the dibromide product (9.13) yields S-(+)-2-octanol benzoate (9.14), thus establishing the absolute stereochemistry at that chiral center. These reactions are shown in Equation 9.4.

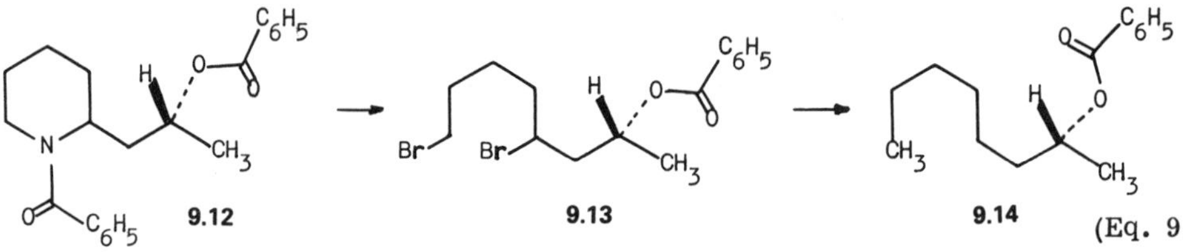

9.12 9.13 9.14 (Eq. 9.4)

SELECTED READING

W. A. Ayer and T. E. Habgood, in The Alkaloids, Vol. 11, R. H. F. Manske (ed.). Academic, New York, 1968, pp. 462 ff.

L. Marion, in The Alkaloids, Vol. 6, R. H. F. Manske (ed.). Academic, New York, 1960, pp. 136 ff.

10

CHEMISTRY OF THE LOBELIA ALKALOIDS

For thy sake, tobacco, I would do anything but die.

C. Lamb, A Farewell to Tobacco

Indian tobacco (*Lobelia inflata* L.) is still used in antismoking devices. This alkaloid-rich herb is a potent emetic which has found some use in asthma treatment and contains, among other bases, lobeline and lobenaline, the only alkaloids we will consider here.

I. LOBELINE

The major alkaloid present in *Lobelia inflata* L. is lobeline (10.1), an optically active tertiary amine which readily undergoes mutarotation and which contains both keto and secondary hydroxyl groups. Careful oxidation of lobeline (10.1) yields a dione (10.2; Equation 10.1)

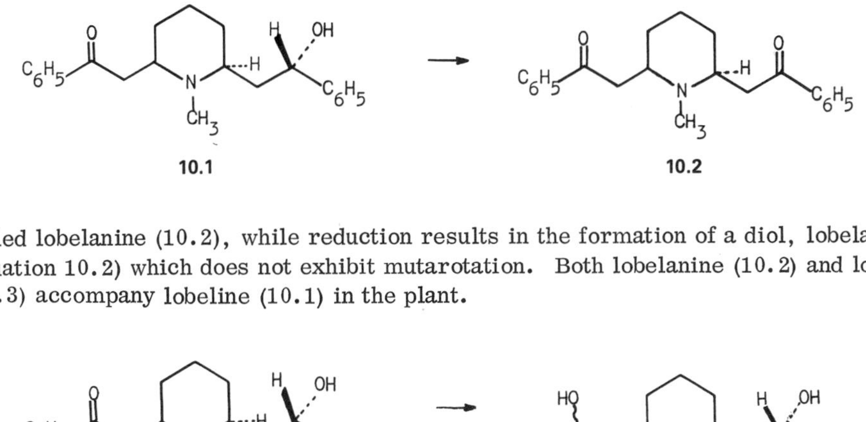

(Eq. 10.1)

10.1 **10.2**

called lobelanine (10.2), while reduction results in the formation of a diol, lobelanidine (10.3; Equation 10.2) which does not exhibit mutarotation. Both lobelanine (10.2) and lobelanidine (10.3) accompany lobeline (10.1) in the plant.

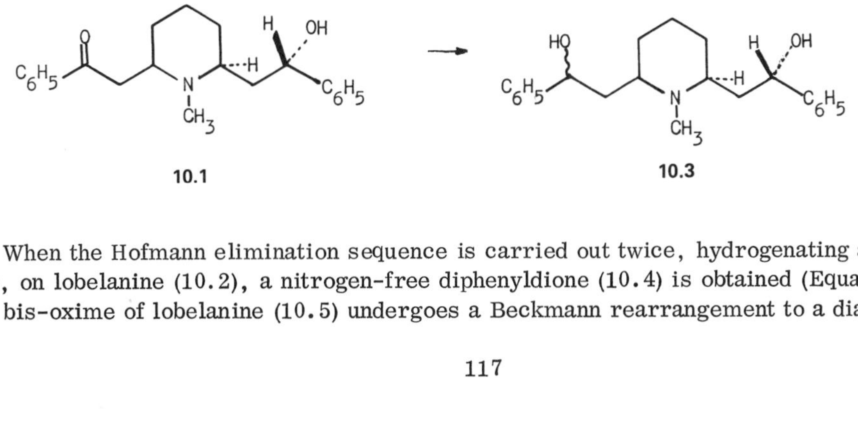

(Eq. 10.2)

10.1 **10.3**

When the Hofmann elimination sequence is carried out twice, hydrogenating after each step, on lobelanine (10.2), a nitrogen-free diphenyldione (10.4) is obtained (Equation 10.3). The bis-oxime of lobelanine (10.5) undergoes a Beckmann rearrangement to a diamide which

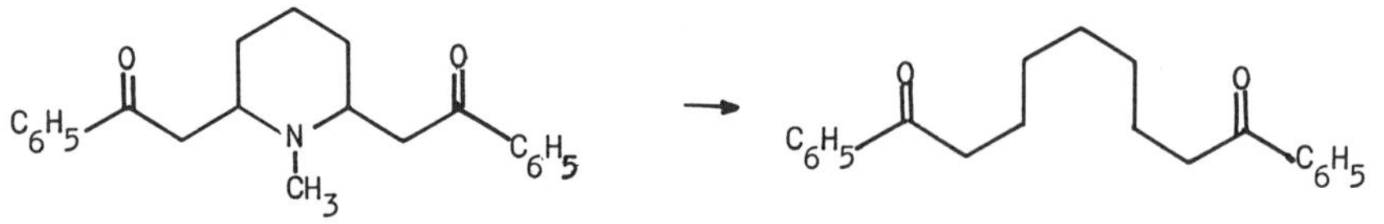

10.2 → 10.4 (Eq. 10.3)

yields aniline (10.6) and the N-methylpiperidine dicarboxylic acid (10.7) on hydrolysis (Equation 10.4). The thus deduced structure of the dione lobelanine (10.2) was confirmed by synthesis via a Mannich condensation between glutaraldehyde (10.8), methylamine (10.9), and

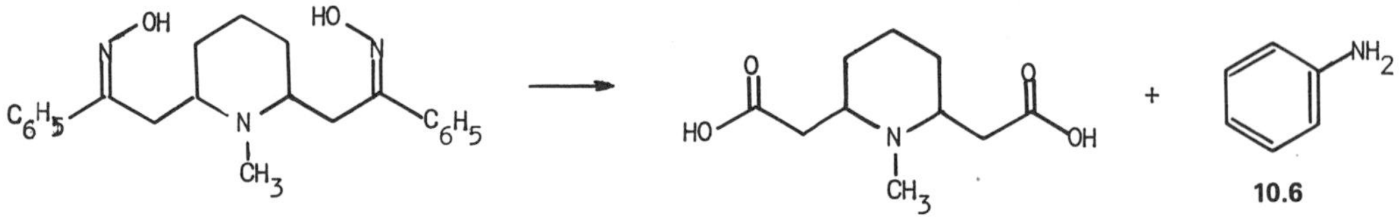

10.5 → 10.7 + 10.6 (Eq. 10.4)

benzoyl acetic acid (10.10) at pH 4 (Equation 10.5), and lobeline (10.1) is the corresponding monool.

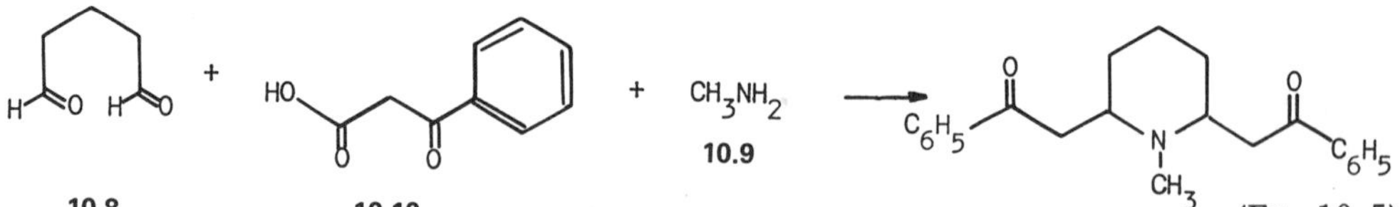

10.8 + 10.10 + CH_3NH_2 10.9 → (Eq. 10.5)

II. LOBINALINE

Lobinaline (10.11), which was first isolated by Manske in 1938, occurs as the major alkaloid in Lobelia cardinalis L. Lobinaline (10.11), by mass spectral and combustion analysis corresponds to $C_{27}H_{34}N_2$, mp 108°C, $[\alpha]_D^{20}$ +38° (chloroform). Initial nuclear magnetic resonance (nmr) measurements provided evidence for two monosubstituted benzene rings and an N-methyl group. Although no N—H functionality is present, lobinaline (10.11) undergoes acetylation to yield an N-acetyl derivative (10.12). Presumably, either the imine first isomerizes to the enamine (10.13) which then yields the N-acetyl derivative (10.12), or isomerization after complexation occurs. (N-acetyllobinaline does have an olefinic proton signal; Equation 10.6).

When N-acetyllobinaline (10.12) is treated with cyanogen bromide under conditions of the von Braun reaction, methyl bromide is evolved, and the resulting cyanamide undergoes hydrolysis to desmethyllobinaline (10.14), dehydrogenation of which yields two compounds, 10.15 and 10.16 (Equation 10.7). The first, 10.15, was degraded by permanganate oxidation to yield picolinic acid (10.17), quinolinic acid (10.18), and benzoic acid (10.19; Equation 10.8). The second, 10.16, was found to be readily synthesized. The synthesis of 10.16 was accomplished by condensing 2-phenylacetylpyridine (10.20) with benzalacetone (10.21) to form 10.22. Dehydration of 10.22 gives the ketone 10.23 (Equation 10.9). Then, conversion of 10.23 to the corresponding hydrazine derivative, dehydrogenation and hydrogenolysis to the amine 10.24, and use of the Skraup cyclization reaction to 10.16, provide a product identical to the material from natural sources (Equation 10.10).

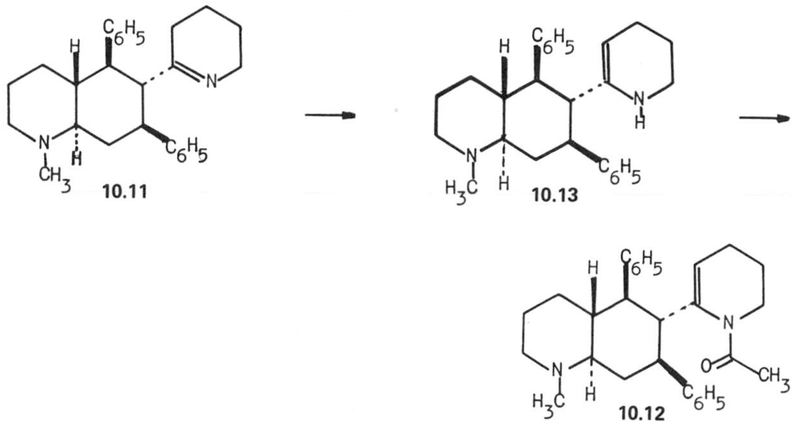

10.11 → 10.13 → 10.12

(Eq. 10.6)

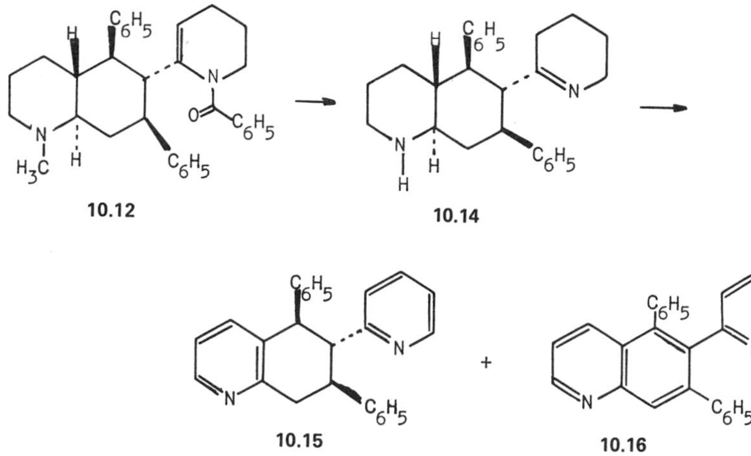

10.12 → 10.14 → 10.15 + 10.16

(Eq. 10.7)

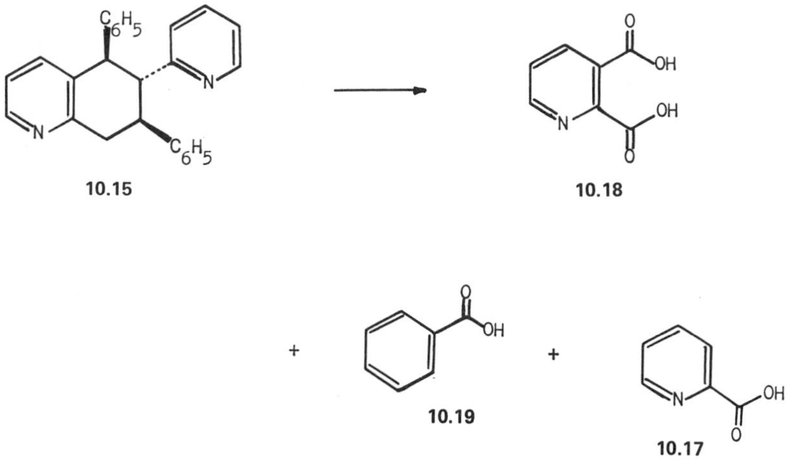

10.15 → 10.18 + 10.19 + 10.17

(Eq. 10.8)

10.20 10.21 → [10.22] → 10.23 (Eq. 10.9)

10.23 → [ ] → 10.24 + $HOCH(CH_2OH)_2$ → 10.16 (Eq. 10.10)

On the other hand, catalytic reduction of the mixture of condensation products (10.23) yields racemic 10.25 (whose structure was assigned on the basis of its pmr spectrum [32]). Thus, the four methylene protons flanking the carbonyl appear as a broad multiplet centered at δ 2.81 ppm, while the C_4 proton appears as a multiplet centered at δ 3.31 ppm. Additionally, the protons at C_3 and C_5 appear as another multiplet (25 Hz wide) centered at δ 3.5–4.0 ppm. The essentially equivalent environments of the protons at C_3 and C_5 (as demonstrated by the closeness of their chemical shifts) and the shielded (by phenyl) proton at C_4 argue for the structure in which all substituents are trans and equatorial. Thus, conversion of 10.25 into its pyrrolidine enamine, cyanoethylation, and hydrolysis generate the ketonitrile (10.26), which on hydrogenation in the presence of a large excess of anhydrous ammonia affords, after methylation, 10.27 (Equation 10.11). This racemic trans-fused tertiary amine (10.27) is identical to a selenium dioxide oxidation product obtained from lobinaline (10.11) itself, and thus the relative stereochemistry at each center has been established in lobinaline (10.11).

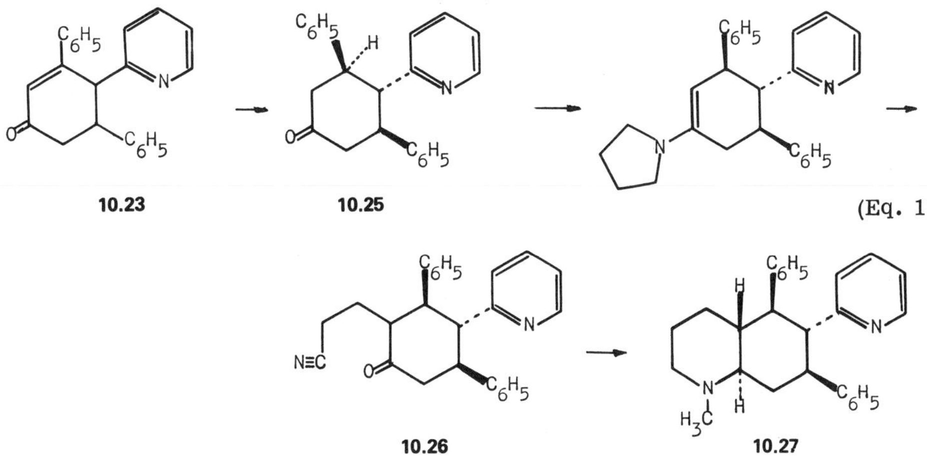

10.23 10.25

(Eq. 10.11)

10.26 10.27

SELECTED READINGS

W. A. Ayer and T. E. Habgood, in The Alkaloids, Vol. 11, R. H. F. Manske (ed.). Academic, New York, 1968, pp. 462 ff.

L. Marion, in The Alkaloids, Vol. 1, R. H. F. Manske (ed.). Academic, New York, 1950, pp. 189 ff.

L. Marion, in The Alkaloids, Vol. 6, R. H. F. Manske (ed.). Academic, New York, 1960, pp. 126 ff.

11

CHEMISTRY OF THE LUPIN ALKALOIDS

'Tis a hundred to one if a man fling two sixes.

A. Cowley (1681-1767)

The common feature of this group of alkaloids occurring in the family Leguminosae is the quinolizidine ring system (11.1). As in the case of the Senecio alkaloids (p. 85), the N-oxides of a number of these bases have been reported and the list of alkaloids present with quinolizidine -derived and other skeletons is quite impressive. In part, this appears to be due to combinations available for oligomerization of a Δ^1-piperideine (11.2) unit, but in the main, it results from the occurrence of numerous stereoisomers of each alkaloid. We shall consider only three members of the group, i.e., lupinine, sparteine, and matrine.

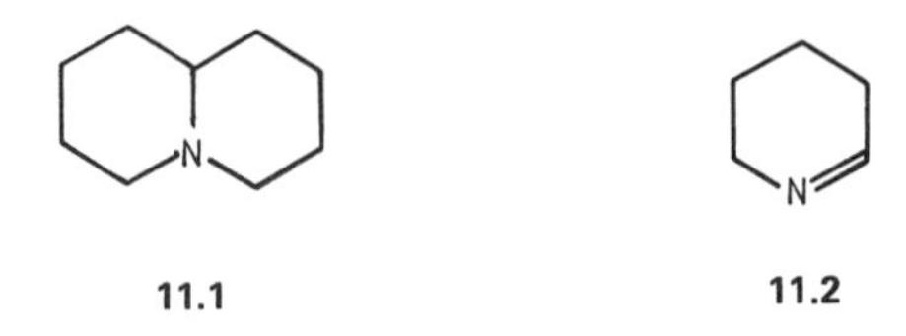

11.1 11.2

I. LUPININE

The simplest member of the family of alkaloids to ne discussed here is lupinine (11.3), $C_{10}H_{19}NO$, mp 70-71°C, $[\alpha]_D^{20}$ -20.9° (water), isolated from the seeds of Lupinus luteus L. In this basic material, the oxygen was found present as a primary hydroxyl, oxidizable to an

N

HO

11.3

aldehyde, and eventually to a carboxylic acid, while the nitrogen was found to be tertiary and common to two rings since its removal by exhaustive methylation and hydrogenation after each elimination required 3 equivalents of methyl iodide.

Since the product of the Hofmann eliminations could be dehydrated (via the corresponding bromide) and the resulting alkene degraded by ozonolysis to formaldehyde and 4-nonanone, the structure shown (stereochemistry omitted) was proposed for lupinine (11.3). The relative configuration of the hydrogen at C_1 to that at C_{10} has been established a number of times in different ways as being cis (Z). Thus, when the methyl ester of 2-pyridylacetate (11.4) is allowed to react with diethyl ethoxymethylenemalonate (11.5), the product, after hydrolysis and decarboxylation, is 1-carbomethoxy-4-quinolizone (11.6; Equation 11.1).

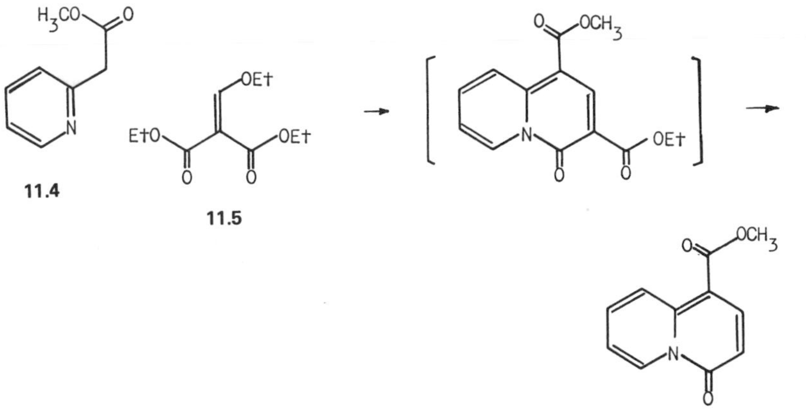

(Eq. 11.1)

Hydrogenation of the quinolizone (11.6), carried out in acid with platinum catalysis, gives 1-carbomethoxy-6,7,8,9-tetrahydro-4-quinolizone (11.7), and further reduction in neutral solution provides 1-carbomethoxy-4-quinolizidone (11.8; Equation 11.2), for which

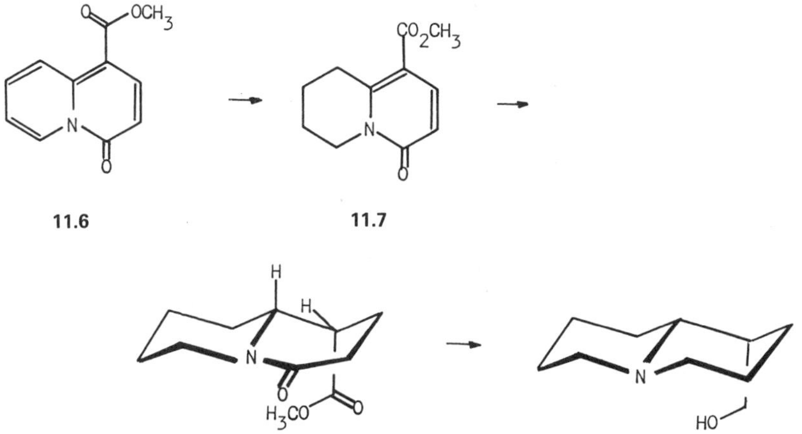

(Eq. 11.2)

only one of the optical isomers is shown but which, nevertheless, must have the hydrogen at the ring juncture (C_{10}) and the carbomethoxy junction (C_1) cis (or Z).

Now, despite the fact that base (e.g., sodium methoxide) is known to isomerize methyl (-)-lupinate (11.9) to its epimer, and the latter is probably the thermodynamic isomer, it is reported [33] that reduction of the lactam in the quinolizidone (11.8) by hydrogen with platinum catalysis in concentrated hydrochloric acid, followed by lithium aluminum hydride reduction of the carbomethoxy group, yields racemic lupinine (11.3, and its mirror image; Equation 11.2). The absolute configuration of lupinine (11.3) rests upon the degradation of (-)-lupinine to R-(-)-4-methylnonane (11.10), $[\alpha]_D^{20}$ -1.3°, whose configuration has been established [34].

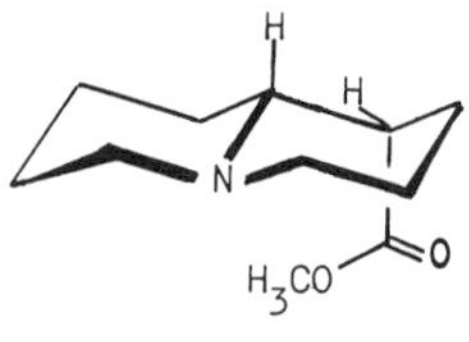

11.9

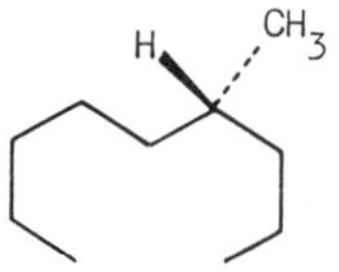

11.10

II. SPARTEINE

Sparteine (11.11), $C_{15}H_{26}N_2$, has been found in nature in both dextrorotatory and levorotatory forms and also as a mixture (see footnote, p. 111). (-)-Sparteine (11.11) was first

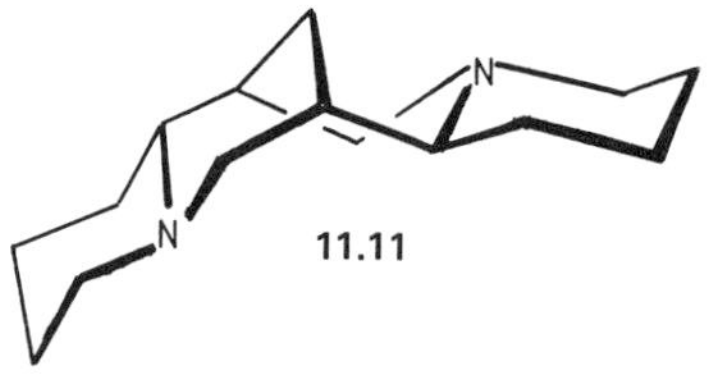

11.11

isolated in 1851, along with several other bases, including one which ultimately provided the structural key to sparteine, i.e., anagyrine (11.12).

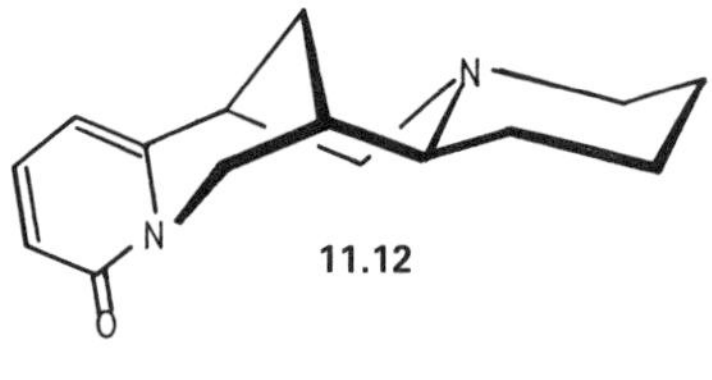

11.12

Electrolytic reduction of anagyrine (11.12) in sulfuric acid at a lead cathode to a "hexahydrodeoxyanagyrine" gave a material, the derivatives of which were identical to those of (+)-sparteine (antipode of 11.11). The structure of anagyrine (11.12) had been decided on the basis of degradative evidence which included the following salient observations:

1. Anagyrine (11.12), which is basic, gives a red color with ferric chloride and can be brominated to yield a dibromide. The dibromide is presumably the product of a substitution rather than an addition reaction, since it can be reconverted to anagyrine (11.12) by treatment with zinc in acetic acid but is stable to alcoholic potassium hydroxide. Additionally, anagyrine (11.12) can be oxidized with barium permanganate to a crystalline product, $C_{15}H_{18}N_8O_2$, mp 201-202°C, called "anagyramide" (11.13; Equation 11.3).
2. Anagyramide (11.13) is very stable to acid and base but when treated with hydrogen iodide and red phosphorous, it yields a base, $C_{14}H_{20}N_2O$ (11.14) and carbon dioxide (Equation 11.4). This implies that hydrolysis of the amide is accompanied by decarboxylation, a well-known property of pyridine-2-acetic acids. In addition, anagyramide (11.13) on ozonolysis loses the elements C_4H_2 and yields a new lactam (Equation 11.5), $C_{11}H_{16}N_2O_2$, mp 258°C (11.15). This, in conjunction with the color

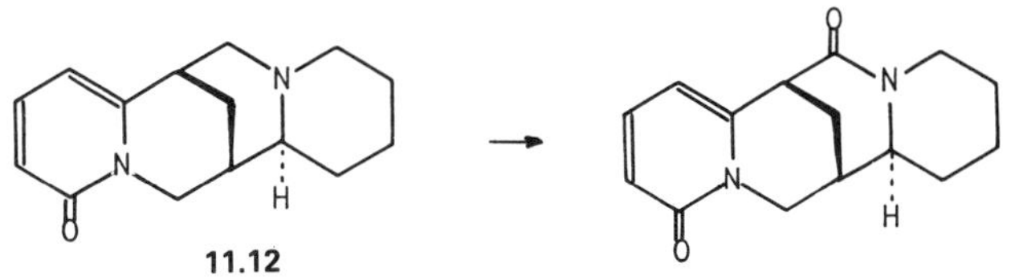

11.12 → 11.13 (Eq. 11.3)

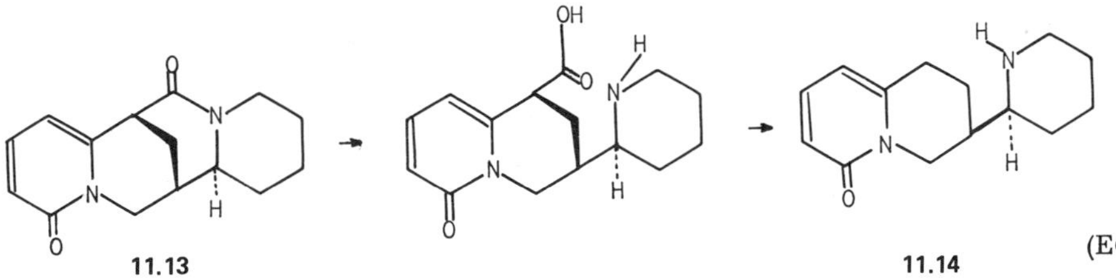

11.13 → → 11.14 (Eq. 11.4)

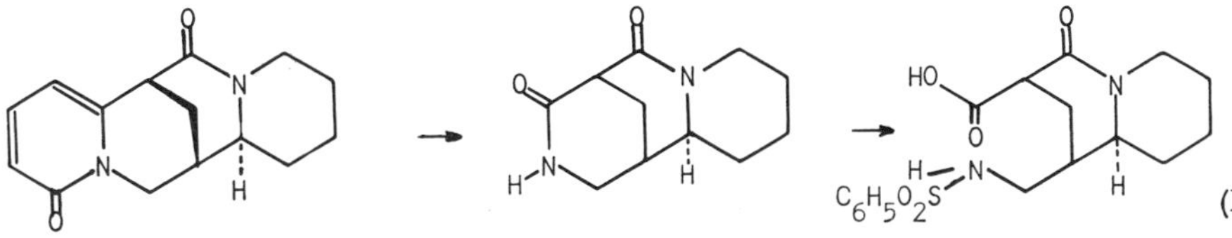

11.13 → 11.15 → 11.16 (Eq. 11.5)

reaction with ferric chloride, the bromination reaction, and the reduction to "hexahydrodeoxyanagyrine," pointed to the presence of an α-pyridone ring.

3. When the new lactam (11.15) is hydrolyzed, the resulting aminocarboxylic acid forms a sulfonamide (presumably 11.16, although other formulations might be written) which loses carbon dioxide at its melting point, indicating that a β-ketocarboxylic acid (11.16) had been formed (Equation 11.5).

4. Finally, when anagyrine (11.12) itself is subjected to sequential Hofmann elimination reactions and hydrogenations, a weakly basic oil, $C_{15}H_{23}NO$ (11.17), is obtained (Equation 11.6). Ozonolysis of this oil yields a lactam ($C_{11}H_{21}NO$; 11.18) which on

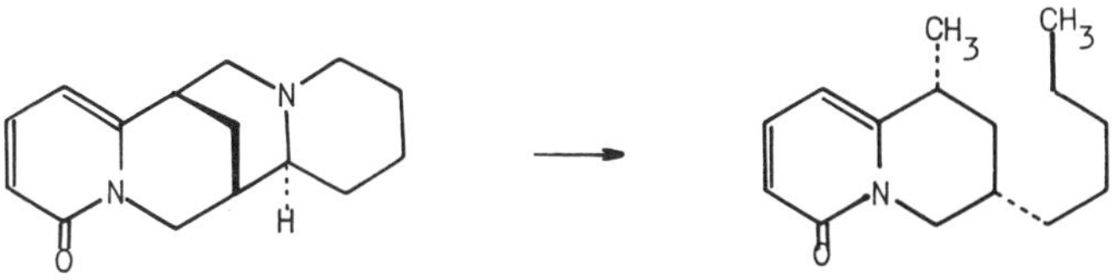

11.12 → 11.17 (Eq. 11.6)

hydrolysis and oxidation provides an amorphous dicarboxylic acid (11.19; Equation 11.7). The dicarboxylic acid (11.19) forms a crystalline imide, identical to a synthetic mixture of cis- and trans-α-methyl-α'-n-amylglutaric acid imides, thus confirming its structure as 11.19. The question which remained, i.e., whether the D ring of anagyrine (11.12) was a piperidine ring or an α-methylpyrrolidine ring was resolved by synthesis.

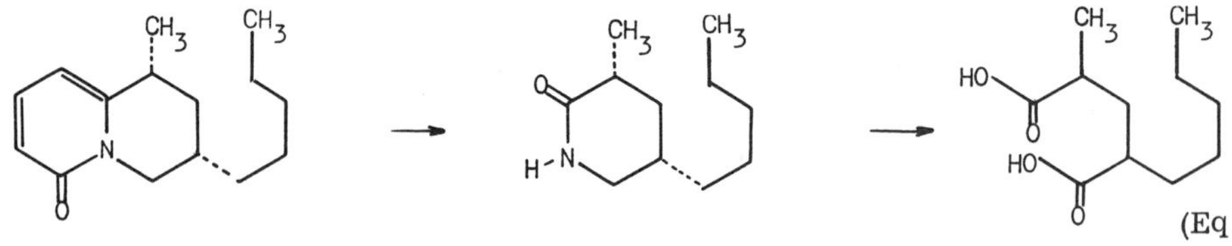

CH3 CH3 CH3 CH3 CH3 CH3 HO HO N H-N

11.17 11.18 11.19

(Eq. 11.7)

A number of syntheses of (±)-sparteine (11.11 and its mirror image) are available, including a biogenetically patterned one in which acetone, formaldehyde, and piperidine are allowed to condense and the resulting Mannich base oxidized with mercuric acetate to 8-ketosparteine (11.20) from which Wolff-Kishner reduction generates (±)-sparteine (11.11; Scheme 11.1; [35]). Although the configurations of sparteine (11.11) and its congeners had been laboriously worked out, x-ray crystallography was eventually used to establish their relative configurations and confirm the chemical work done [36].

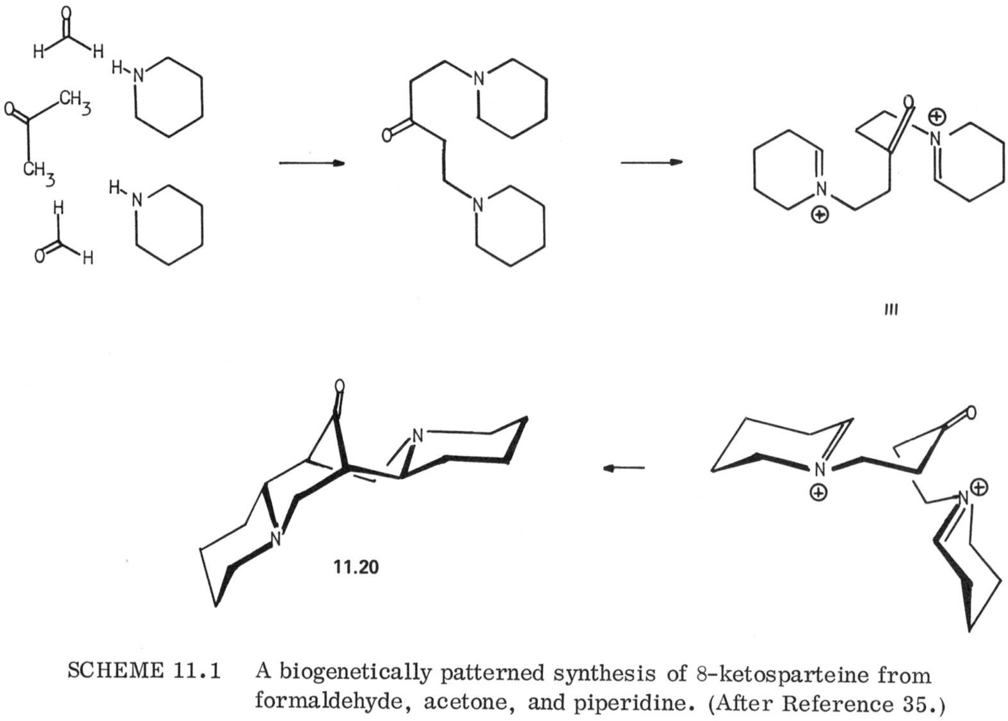

H H CH3 CH3 H-N H-N H H N N N N N III N N N

11.20

SCHEME 11.1 A biogenetically patterned synthesis of 8-ketosparteine from formaldehyde, acetone, and piperidine. (After Reference 35.)

III. MATRINE

Matrine (11.21) corresponds to $C_{15}H_{24}N_2O$ and is known to exist as a series of at least four more or less readily interconvertible forms which yield isomeric methiodides. Until its synthesis and resolution of its diasteromers, the structures were not firmly established.

The correct molecular formula and the nature of the nitrogen atoms, but not their disposition, was readily established. Thus, the nitrogens were recognized as different from each other; one, a tertiary basic nitrogen through which salts could form, and the other present in a lactam which could be hydrolyzed with potassium hydroxide solution to the salt

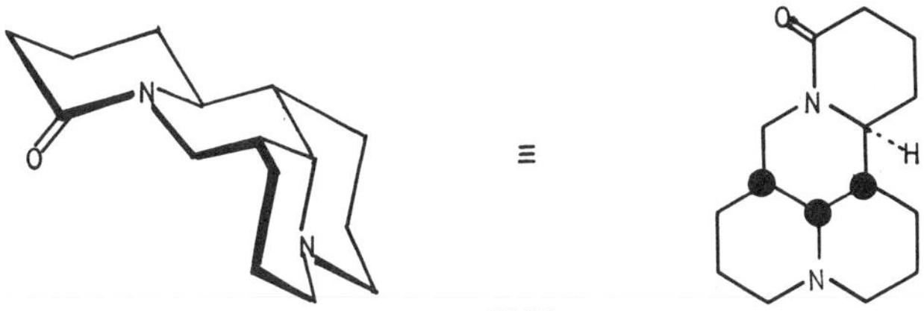

11.21

of a carboxylic acid called matrinic acid (11.22), an amino acid. Matrinic acid (11.22) clearly contained a "new" secondary amino group.

Zinc dust distillation of matrinic acid hydrochloride (11.22-HCl) gives a variety of products, among which is an optically inactive tertiary amine, $C_{10}H_{19}N$, identified as 1-methylquinolizidine (β-lupinane, 11.23; thus establishing a part of the skeleton; Equation 11.8). Similarly, when potassium matrinate (11.22, salt) is distilled from sodium hydroxide,

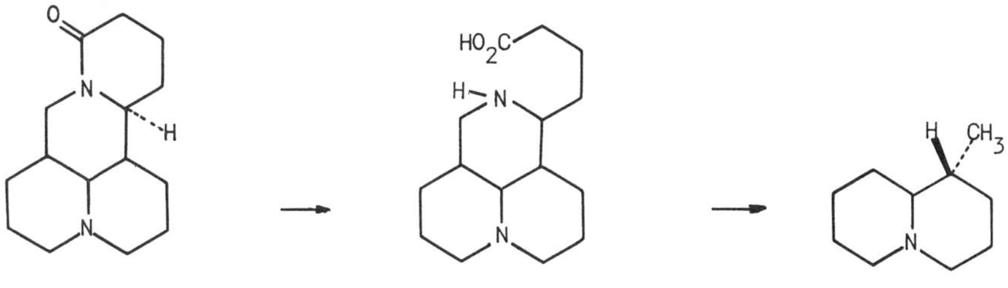

(Eq. 11.8)

11.21 11.22 11.23

two C_{12} products are isolated: α-matrinidine, $C_{12}H_{16}N_2$ (11.24), and dehydro-α-matrinidine, $C_{12}H_{16}N_2$ (11.25). Hydrogenation of α-matrinidine (11.24) to dihydro-α-matrinidine,

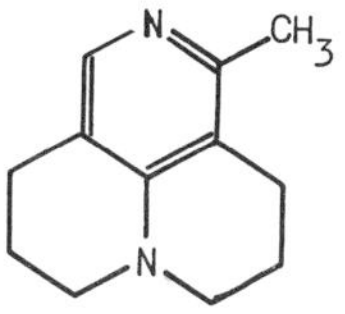

11.25

$C_{12}H_{22}N_2$ (11.26), is easily effected, while both α-matrinidine (11.24) and dihydro-α-matrinidine (11.26) undergo dehydrogenation on zinc dust distillation to dehydro-α-matrinidine (11.25) and β-lupane (11.23).

The difference of six hydrogens between dihydro-α-matrinidine (11.26) and dehydro-α-matrinidine (11.25) suggests aromaticity is being introduced, and although dihydro-α-matrinidine (11.26) readily undergoes N-acylation, α-matrinidine (11.24) itself, on attempted acetylation, adds the elements of water along with the acyl function (with the formation of a ketone, 11.27, as well as the amide; Equation 11.9). This fixes the structure of α-matrinidine as an imine. The fact that dehydro-α-matrinidine (11.25) possesses a methyl side chain was suggested by the formation of a benzal derivative.

Finally, when matrine (11.21) itself is heated with palladium on asbestos at 280°C, a liquid ditertiary amine, $C_{14}H_{20}N_2$ (11.28), is obtained, along with other products. This same compound is also obtained from dehydro-α-matrinidine (11.25) by lithiation and condensation with ethyl bromide. That the final carbon of matrine which forms the lactam should be placed

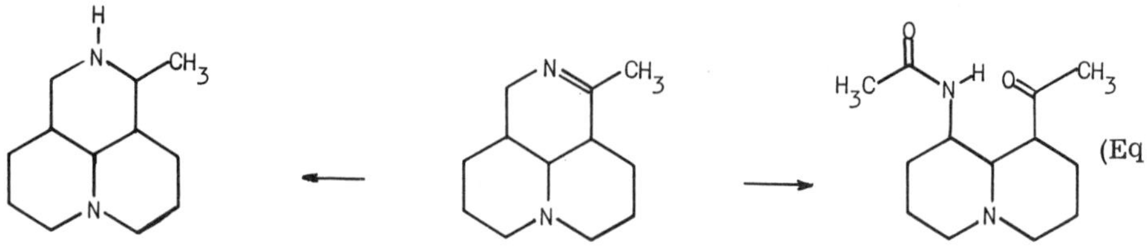

11.26 11.24 11.27 (Eq. 11.9)

at the end of the propyl side chain in 11.28 was deduced from a series of experiments on matrinic acid (11.22) in which the carboxy group is converted by the Curtius degradation into an amine (11.29; with loss of carbon dioxide) and, ultimately, elimination to an alkene which yields the same n-propylated derivative (11.28) found in the earlier dehydrogenation reaction (see above; Scheme 11.2).

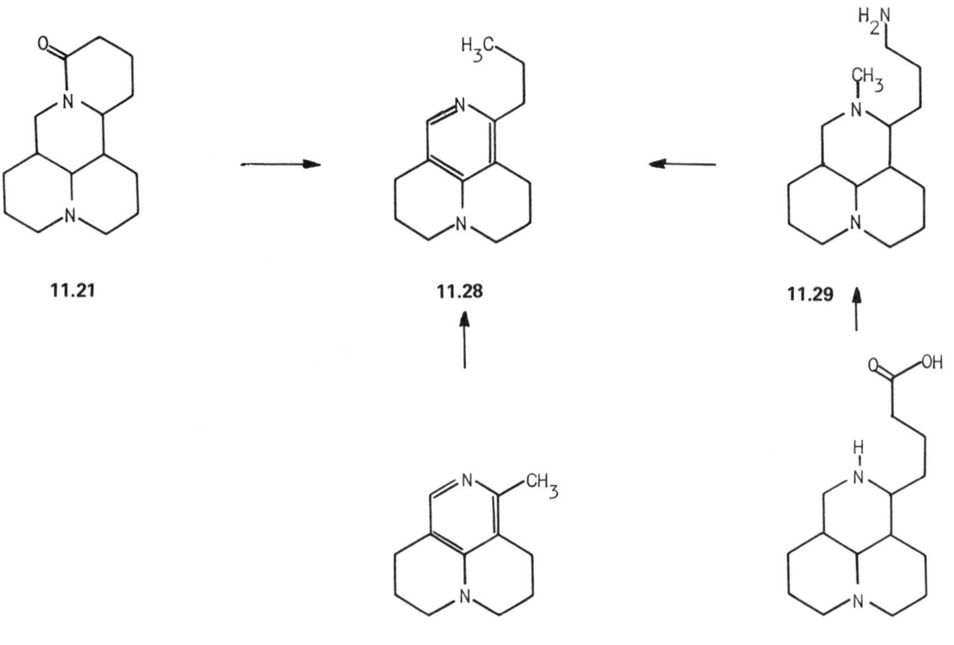

11.21 11.28 11.29 11.25 11.22

SCHEME 11.2 Establishment of the nature of the side chain of matrine.

For the synthesis of matrine (11.21; [37]; Scheme 11.3), diethyl 3-oxopimelate (11.30) is condensed with ethyl β-allinate (11.31) to the enamine (11.32), which in turn is reduced and cyclized to the diester 11.33. Dieckmann cyclization then affords a crystalline β-ketoester which was not completely characterized and is presumably 11.34a or 11.34b. Decarboxylation is effected in moderate yield by refluxing the β-ketoester in glacial acetic acid, and the product is bisalkylated with acrylonitrile via the enamine to give the dinitrile (11.35), which since it is formed under conditions of thermodynamic control, presumably has the most stable (all equatorial substituents on a "trans-decalin") configuration. Reductive

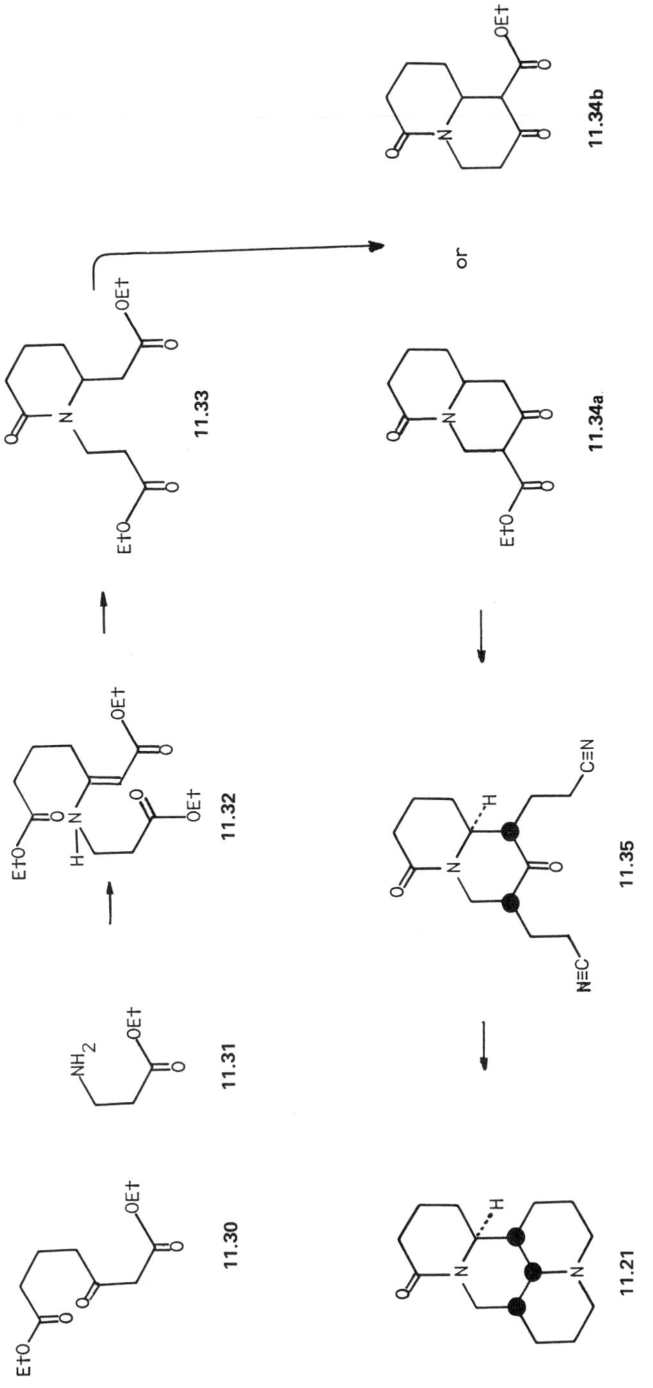

SCHEME 11.3 The synthesis of matrine. (After Reference 37.)

cyclization of 11.35 with 5% palladium on carbon in glacial acetic acid at 50 psi hydrogen pressure yields (±)-matrine (11.21 and its mirror image).*

SELECTED READING

F. Bohlman and D. Schumann, in The Alkaloids, Vol. 9, R. H. F. Manske (ed.). Academic, New York, 1967, pp. 175 ff.

N. J. Leonard, in The Alkaloids, Vol. 7, R. H. F. Manske (ed.). Academic, New York, 1960, pp. 253 ff.

N. J. Leonard, in The Alkaloids, Vol. 3, R. H. F. Manske (ed.). Academic, New York, 1953, pp. 119 ff.

*That these compounds may be easily isomerized (p. 107) is evidenced by the fact that should the reduction be carried out with 10% Pd/C, an isomer, epimeric at three of the four chiral centers, is obtained!

12

CHEMISTRY OF THE *LYCOPODIUM* ALKALOIDS

> The Night is Mother of the Day
> The Winter of the Spring
> And ever upon old Decay
> The greenest mosses cling.
>
> J. G. Whittier, A Dream of Summer

Annotinine (12.1) and lycopodine (12.2) are only several of the alkaloids of *Lycopodium* (club mosses), and while some of the alkaloids are toxic (to frogs), their relative inaccessibility has apparently preserved mankind from toxicity studies. The spores of *Lycopodium clavatum* L. (vegetable sulfur) have been used medicinally as an adsorbant dusting powder, but other

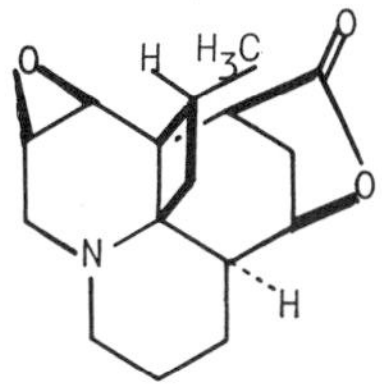

12.1

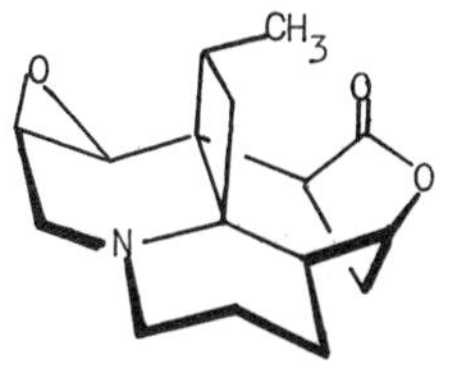

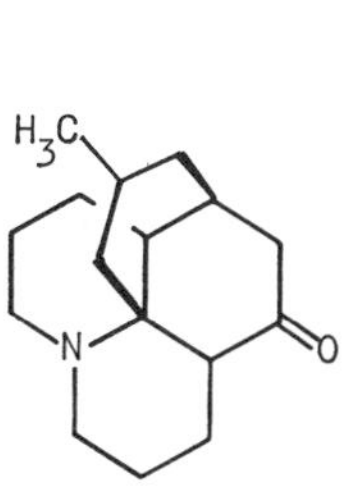

12.2

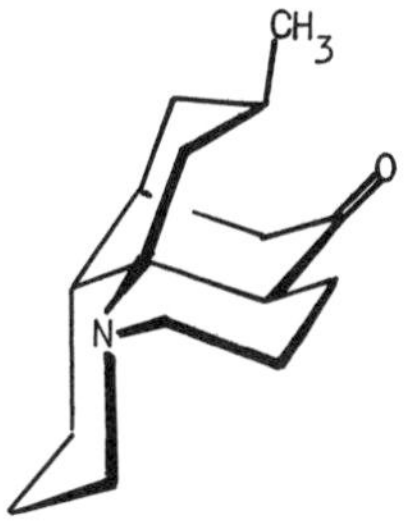

uses (as diverse as additives to gun powder and suppository coatings) have been recorded. Annotinine (12.1), one of the major alkaloids, the first upon which significant work was done, and among the most complex, has a unique ring system. Lycopodine (12.2), on the other hand, is more typical.

I. ANNOTININE

Several comprehensive review and other complete articles have appeared detailing the massive amount of work required to unravel the complete structure of annotinine (12.1). Some of the salient reactions will be discussed here, but for more information the interested reader is directed to the work of Wiesner [38].

Oxidation of the base annotinine (12.1), $C_{16}H_{21}NO_3$, with potassium permanganate gives a neutral compound, $C_{16}H_{19}NO_4$, which although originally considered to be a ketone, is no longer basic and thus must be a cyclic lactam with a six-membered or larger ring (ν_{max} 1645 cm^{-1}; 12.3). This amide readily yields a chlorohydrin (12.4; $C_{16}H_{20}NO_4Cl$) with hydrogen chloride, which reversibly generates the starting material on treatment with base, as does annotinine (12.1) itself when similarly treated, implying the presence of an epoxide ring. Treatment of the lactam chlorohydrin (12.4) with phosphorous oxychloride causes dehydration to a vinyl chloride (12.5), $C_{16}H_{18}ClNO_3$, in which the double bond is conjugated with the carbonyl of the lactam and which can be oxidized to an amino acid (12.6), $C_{14}H_{19}NO_4$, indicating that the epoxide is dialkylated. When the lactone-lactam (12.5) is treated with alcoholic alkali, a hydroxyacid (12.7) is obtained, which on heating in benzene with p-toluenesulfonic acid, loses water to form a new lactone (12.8) isomeric with 12.5. From the infrared spectrum (ν_{max} 1760 and 1660 cm^{-1}) it is clear that the lactam functionality is no longer in conjugation with the double bond. These reactions are indicated in Scheme 12.1.

The carboxylic acid (12.6) provided a great deal of information useful to the structure determination of annotinine (12.1). Thus, when 12.6 is dehydrogenated over palladium on carbon at 270° C, three compounds are obtained, namely, 12.9 ($C_{13}H_{13}NO$), 12.10 ($C_{13}H_{15}NO$), and 12.11 ($C_{14}H_{15}NO_3$; Equation 12.1). Compound 12.9 is identical to the product obtained

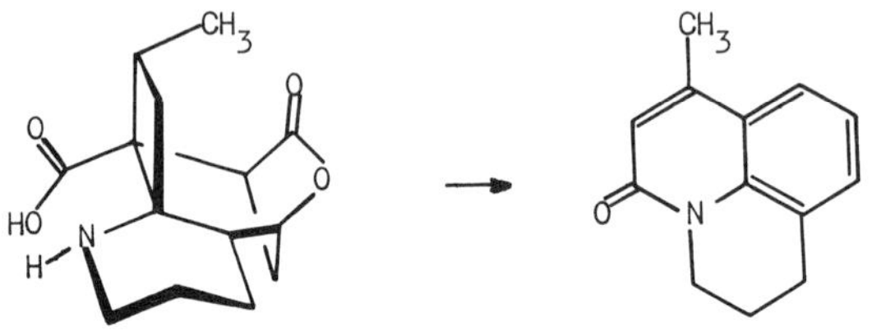

12.6 → 12.9 + 12.10 + 12.11 (Eq. 12.1)

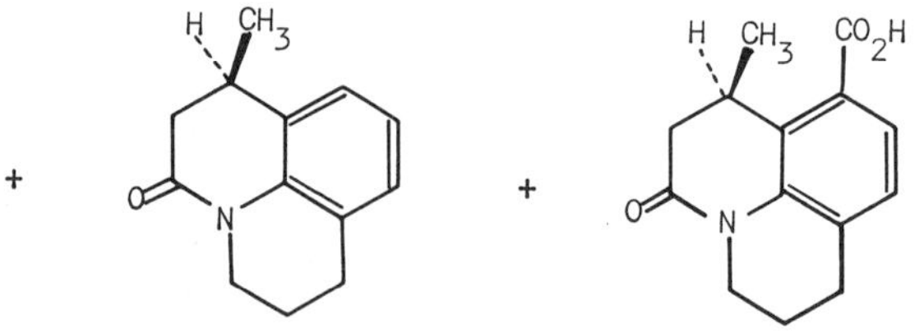

on treating tetrahydroquinoline with ethyl acetoacetate. Reduction of synthetic 12.9 with sodium amalgam yields a racemic material identical (by infrared) to the chiral 12.10 (except for rotation), while 12.11, which undergoes copper-catalyzed decarboxylation to 12.10 and whose ultraviolet spectrum is similar to m-acetylaminobenzoic acid, could also be synthesized. Although a number of pathways (one is shown) for the conversion of 12.6 to 12.11 can be written, this facile process was much misunderstood until the configuration of the methyl

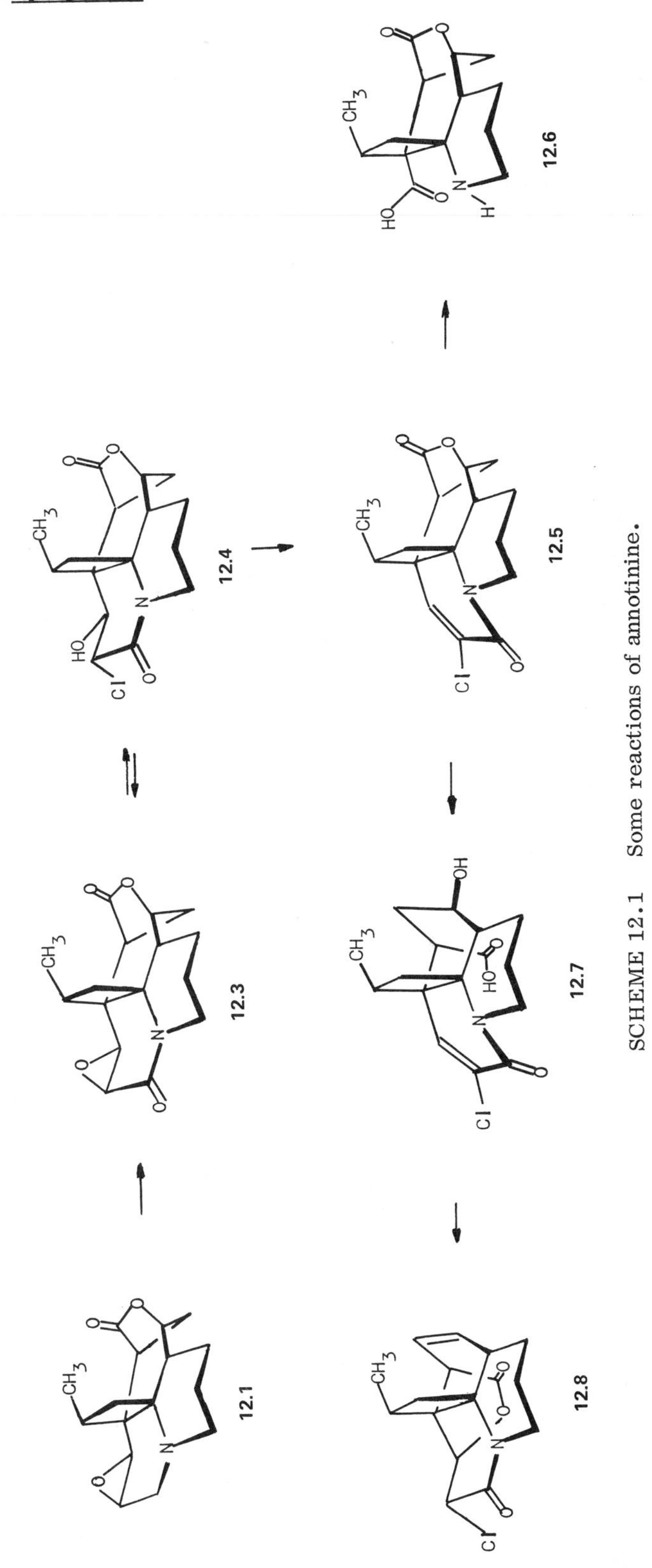

SCHEME 12.1 Some reactions of annotinine.

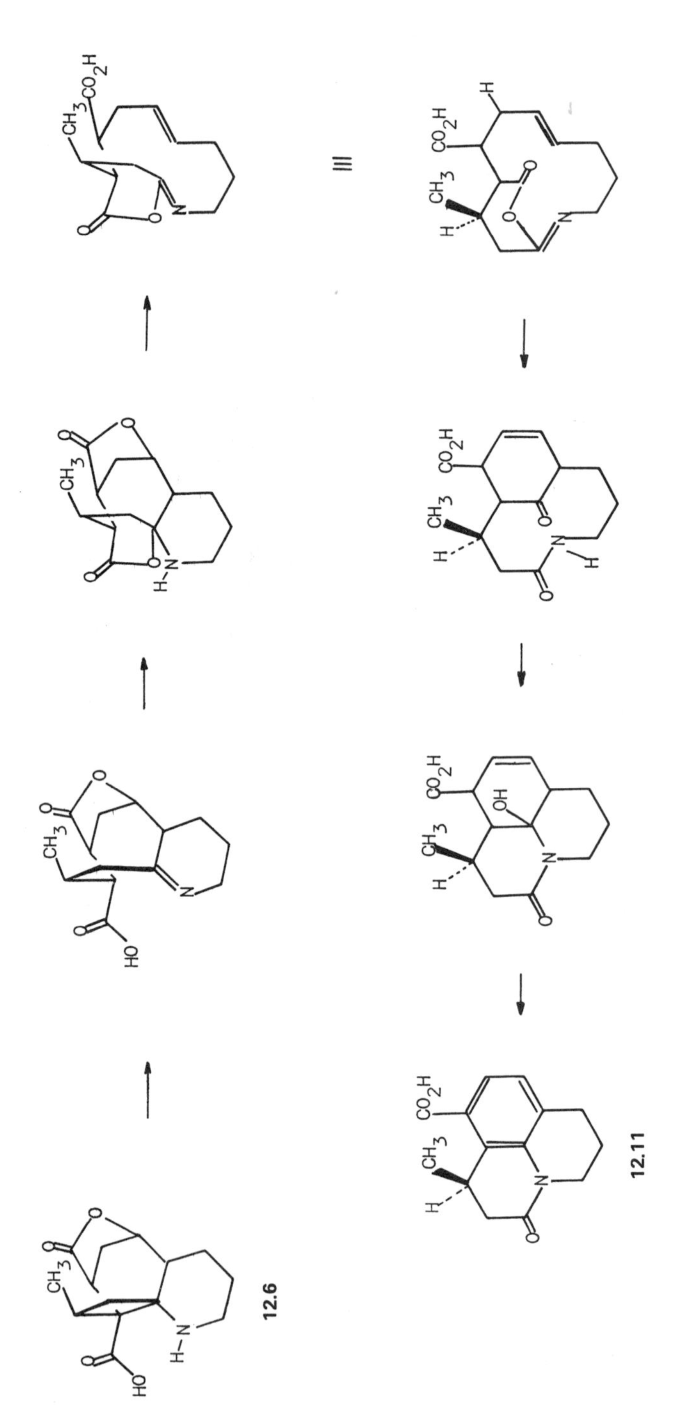

SCHEME 12.2 A potential pathway for the conversion of the acid 12.6 to the quinolone 12.11.

in annotinine (12.1) had been established by x-ray crystallography [39]. A possible pathway is shown in Scheme 12.2.

Treatment of the hydroxy acid 12.7 with phosphorus pentoxide in refluxing xylene generates an alkene (12.12). This is apparently a simple dehydration reaction since there is no change in the ultraviolet spectrum of 12.7. Dehydrogenation of 12.12 over palladium on carbon leads to the formation of a quinolone acid (12.13) which yields the quinolone (12.14; whose structure had been established) on decarboxylation (Equation 12.2). The acid 12.13 has all

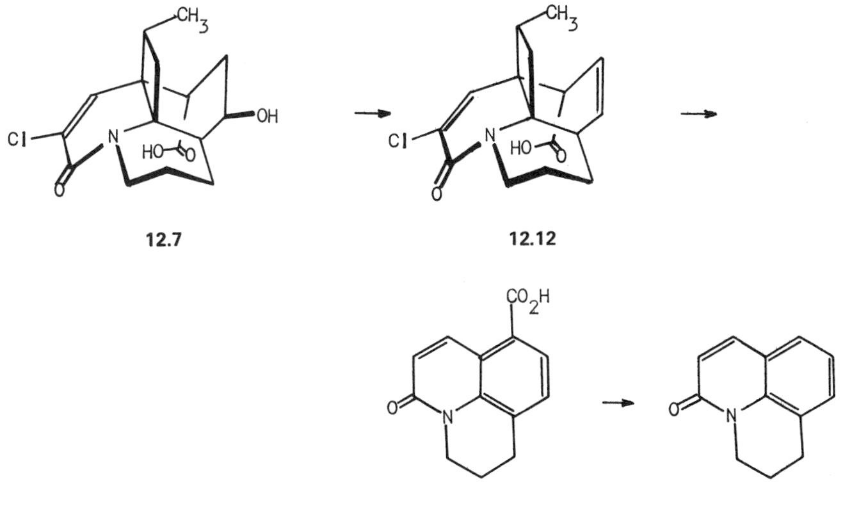

12.7 12.12

(Eq. 12.2)

12.13 12.14

the carbon atoms of annotinine save three. That these are fused in a bridge attached at one end to C_{12} was deduced from the fact that the methyl ester of 12.6 undergoes hydrolysis with difficulty, as expected for the ester of a quaternary carboxylic acid.

The site of attachment of the other end of the bridge was deduced as follows: reductive hydrogenolysis of the chlorolactam 12.5 generates the saturated lactam 12.15, which on hydrolysis and oxidation yields a ketoacid (12.16). When the ester corresponding to 12.16 is subjected to selenium dioxide oxidation, a variety of products is isolated; among them are the epimeric diols 12.17 and the α,β-unsaturated ketone 12.18. Oxidation of the diols (12.17) with periodic acid generates, among other products, succinic acid (12.19; Scheme 12.3). Since this presumably arises from carbon atoms 1 through 4, only C_{12}, which is already quaternary, and C_{13} are vailable for bridging. Finally, a gem-dimethyl structure is excluded because formation of 12.9, 12.10, and 12.11 favors the methylcyclobutane structure. The stereochemistry of the methyl on the bridge, which has not been defined in the above discussion but which nevertheless was in concert with the formulation of annotinine as shown (12.1), was confirmed by x-ray crystal analysis [39].

The synthesis of annotine (12.1; [40]; Scheme 12.4) requires the photochemical $_{\pi}2_s + _{\pi}2_s$ addition of allene to an appropriate alkene. Thus, the tricyclic lactone resulting from the alkylation of the vinylogous amide 12.20 ($\Delta^{8,9}$-octahydroquinol-7-one) with acrylic acid (12.21) yields the tetracyclic intermediate 12.22 and its mirror image on photolysis in the presence of allene at low temperature. Only one product is obtained in this reaction, the isomer in which the double bond lies over the less encumbered side. Formation of the ethylene ketal, hydrogenation, and hydrolysis yield the desired C_{15} epimer (12.23) with the carbonyl at C_7 being regenerated in the last step. Reduction of the carbonyl (sodium borohydride), conversion of the resulting 7-β-ol to its mesylate, and elimination with t-butoxide in dimethyl sulfoxide

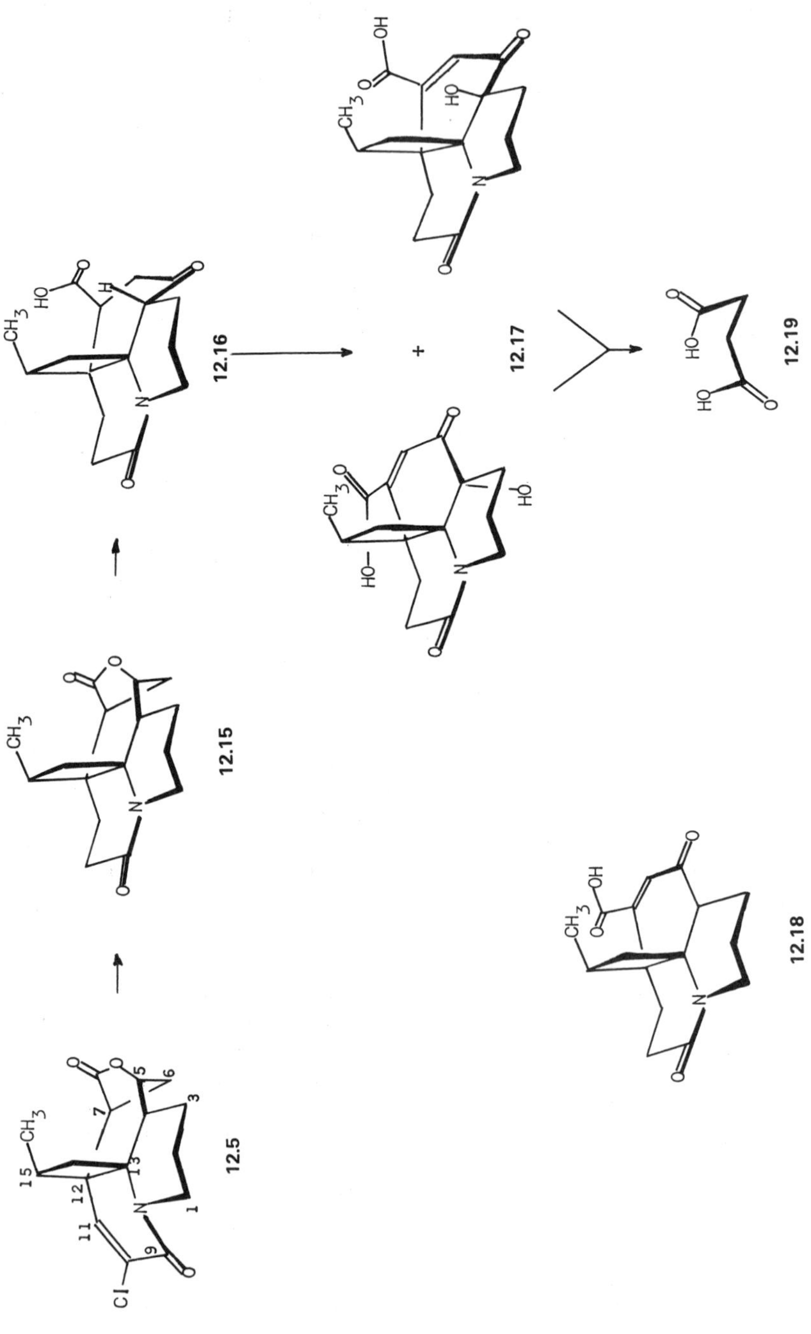

SCHEME 12.3 The degradation of annotinine to succinic acid.

gives the alkene (12.24) which is converted into a mixture of C_5 acetates with selenium dioxide in acetic acid and thence, after hydrolysis, to the C_5 ketone 12.25. Treatment of the α,β-unsaturated ketone (12.25) with hydrogen cyanide followed by methanolysis yields a racemic ester which is capable of resolution via the brucine salt of the corresponding carboxylic acid. An enantiomer so obtained was identical, on esterification, to the ester of the keto acid (12.16).

The acid 12.16 was converted to its enol acetate (12.26), esterified with methanol, and the latter reduced with sodium borohydride to a mixture of epimeric hydroxy esters.* Conversion of the mixture of the alcohols to the corresponding acids (12.27) and refluxing with toluenesulfonic acid in benzene permits equilibration at C_7 and lactonization of the appropriate epimer to 12.28. Finally, when 12.28 is treated with N-bromosuccinimide in carbon tetrachloride, bromination and dehydrohalogenation occur, and the vinyl bromide obtained is hydrated with aqueous hydrogen bromide to the bromohydrin 12.29. Then, aqueous bicarbonate converts the bromohydrin to the epoxide, oxoannotinine (12.30), which readily undergoes acid-catalyzed hydrogenolysis to annotinine (12.1) in the presence of platinum.

II. LYCOPODINE

Although the alkaloid lycopodine (12.2) was isolated quite early (1881), its formula, $C_{16}H_{25}NO$, was established some 50 years later. Among the bits of information available for structural elucidation were: (a) there are no N-methyl or O-methyl groups or active

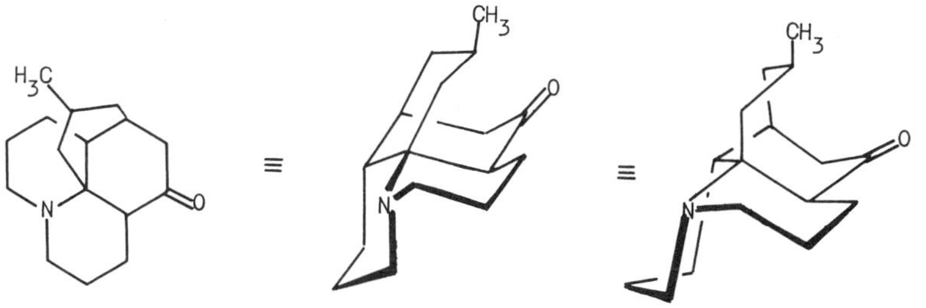

12.2

hydrogens; (b) the oxygen is present as a ketone carbonyl (ν_{max} 1693 cm^{-1}) as evidenced by reduction to an alcohol with lithium aluminum hydride and formation of a hydrazone; and (c) selenium and palladium dehydrogenations afford 7-methylquinoline (12.31) and 5,7-dimethylquinoline (12.32; Equation 12.3). When the Hofmann degradation was attempted on lycopodine (12.2) a rearrangement, not understood until much later, occurred (Equation 12.4). Although more than one pathway can be written for the process and the one shown involves the intermediacy of a species which marginally violates Bredt's rule, it appears to accommodate the

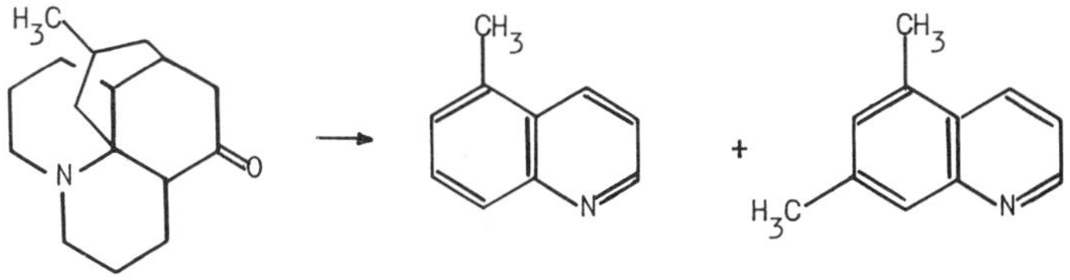

(Eq. 12.3)

12.2 12.31 12.32

*It was necessary to prepare the enol acetate since direct reduction of the keto function failed to epimerize the C_4 position, and the wrong isomer resulted.

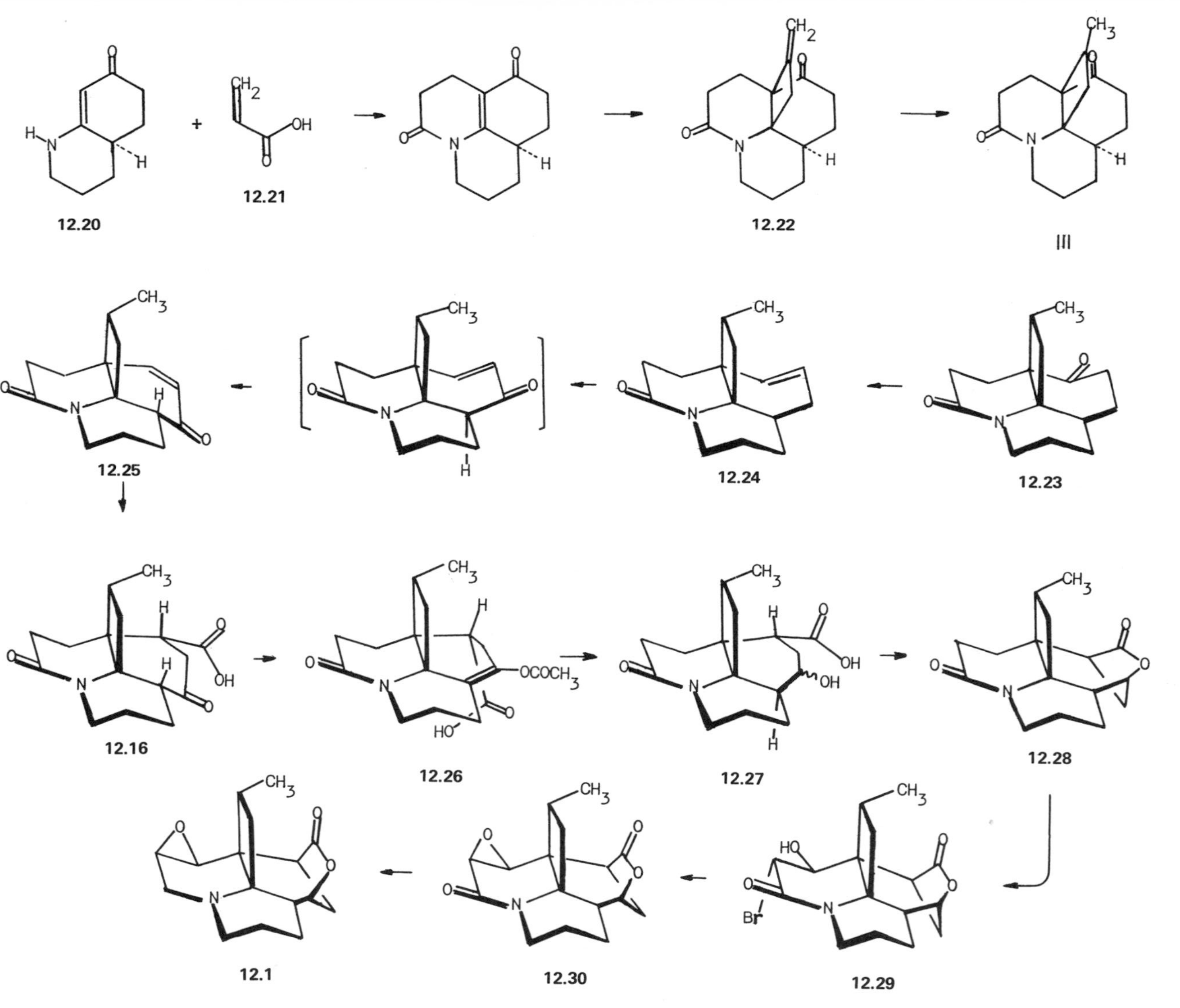

SCHEME 12.4 The synthesis of annotinine. (After Reference 40.)

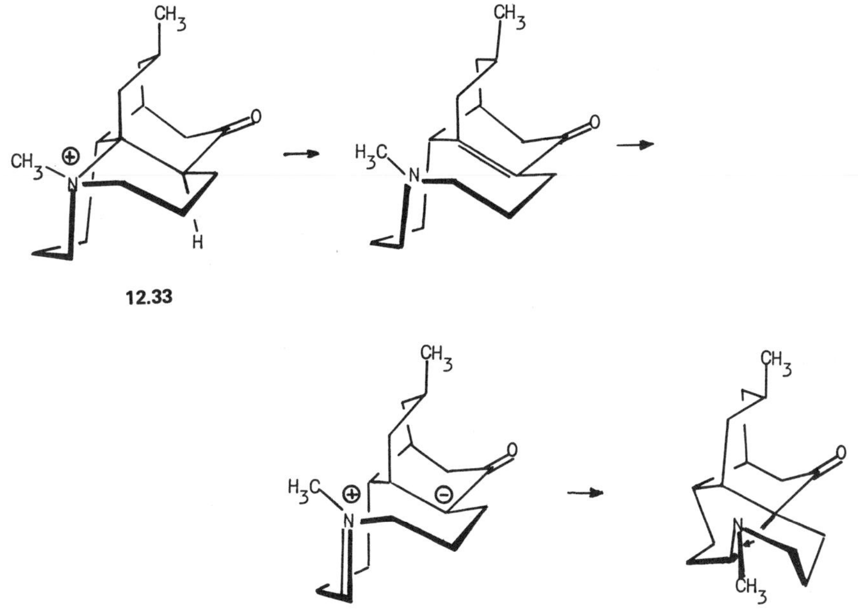

12.33

(Eq. 12.4)

12.34

stereochemical requirements and is the least-motion route. Thus, 12.2 on methylation yields 12.33 which rearranges with an intramolecular hydride transfer to 12.34.

Of greater import was the work describing the reaction of cyanogen bromide with lycopodine (12.2). Two products (12.35 and 12.36) are isolated from this reaction. Each of these is converted, via the corresponding acetate (displacement of bromine by acetate), to their respective primary alcohols and thence to the corresponding carboxylic acids 12.37 and 12.38, respectively. Hydrolysis of the cyano groups and treatment of the acids with diazomethane leads to the lactams 12.39 and 12.40, and each of these, on reduction with lithium aluminum hydride, generates dihydrolycopodine (12.41), which had previously been obtained from lycopodine (12.2) itself. These reactions are shown in Scheme 12.5

The products derived from treatment of lycopodine (12.2) with cyanogen bromide, α-bromocyanolycopodine (12.35), and β-bromocyanolycopodine (12.36; see above) were useful in other ways, too: (a) Both 12.35 and 12.36, on hydrogenolysis (each under slightly different conditions) yield, respectively, α-cyanolycopodine (12.42) and β-cyanolycopodine (12.43). Each of these derivatives, separately, undergoes borohydride reduction to the corresponding alcohol, which on modified Kuhn-Roth oxidation provides acetic, propanoic, and butanoic acids. This indicates that both α-cyanolycopodine (12.42) and β-cyanolycopodine (12.43)

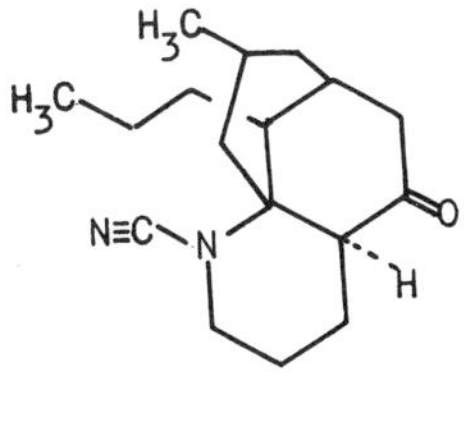

12.42

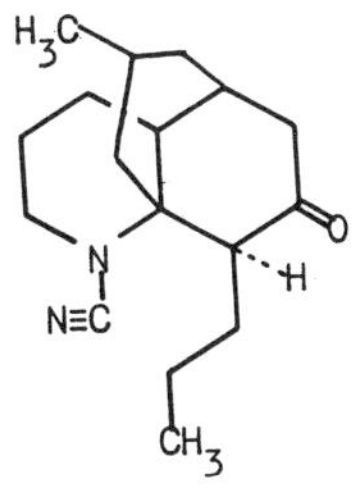

12.43

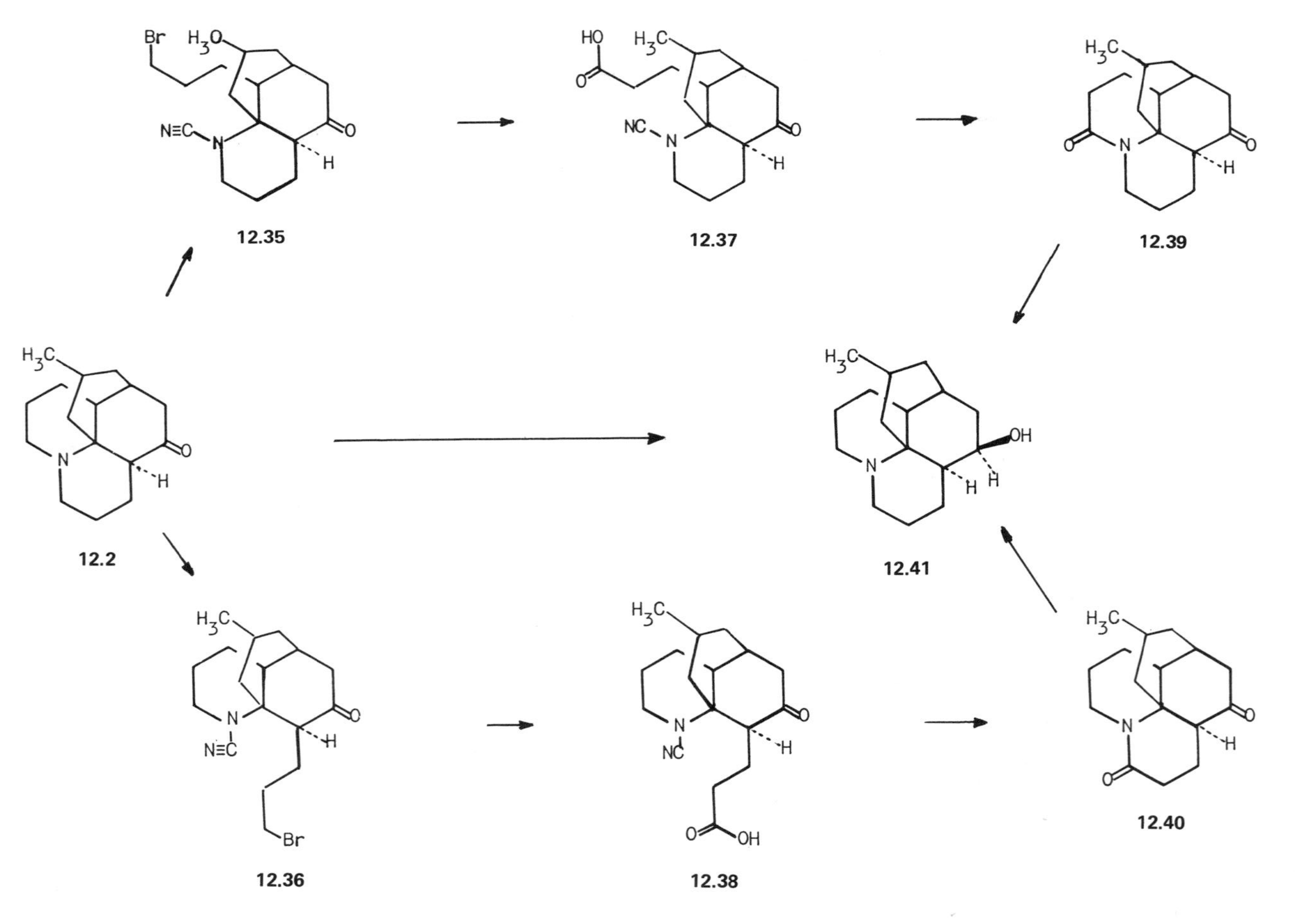

SCHEME 12.5 Some reactions of lycopodine.

possess n-propyl side chains. (b) In the conversion of β-bromocyanolycopodine (12.36) to the primary alcohol necessary for oxidation to the carboxylic acid 12.38, a neutral compound, formulated as the enol ether 12.44, is occasionally isolated from the acetolysis of 12.36. This is not true for the isomer 12.35 (α-bromocyanolycopodine). (c) Reduction of the carboxylic acid 12.38 derived from β-bromocyanolycopodine (12.36; but not the one, 12.37, derived from α-bromocyanolycopodine, 12.35) with sodium borohydride results in the formation of a lactone (12.45). (d) Both α- and β-cyanolycopodines (12.42 and 12.43, respectively) and

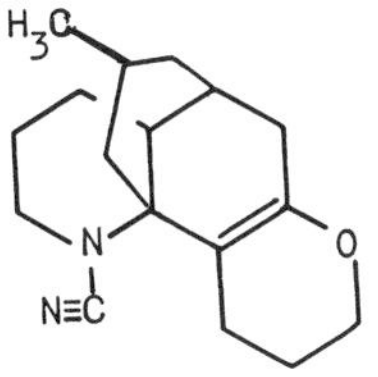

12.44 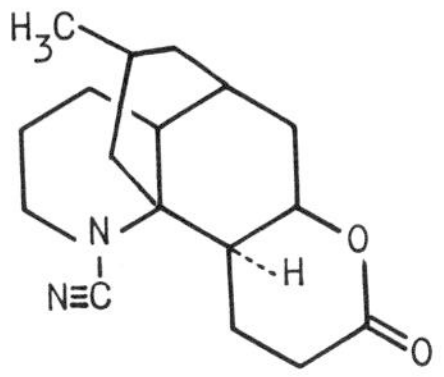 12.45

lycopodine (12.2) itself undergo deuterium exchange in basic medium with incorporation of three deuterium atoms per molecule, indicating that there are three enolizable hydrogens. (e) The benzylidine derivative (12.46) of α-cyanolycopodine (12.42) on selenium dioxide oxidation provides a mixture of compounds from which 12.47 can be isolated. It was recognized that the hydroxyl group in 12.47 was tertiary and α to the carbonyl. Ozonolysis of 12.47 provides a mixture from which a yellow compound, to which the structure 12.48 is assigned, is isolated. The assignment is made on the basis of its spectral properties (λ_{max} 420 nm, ϵ = 25, unaffected by alkali, ν_{max} 1724 cm^{-1}, and hydrogen bonded —OH). The formation of 12.48 is shown in Equation 12.5.

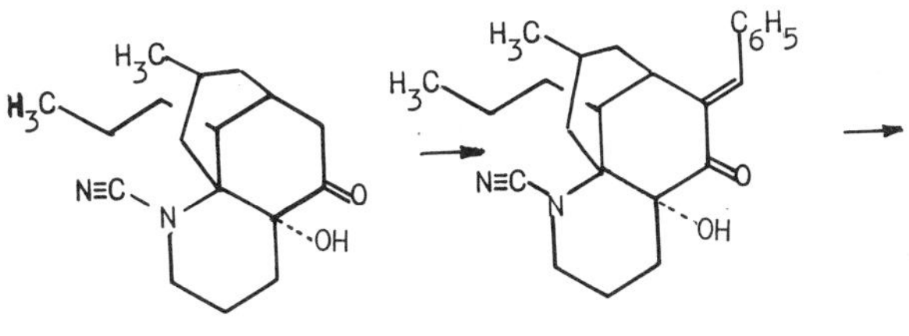

12.42 12.46

(Eq. 12.5)

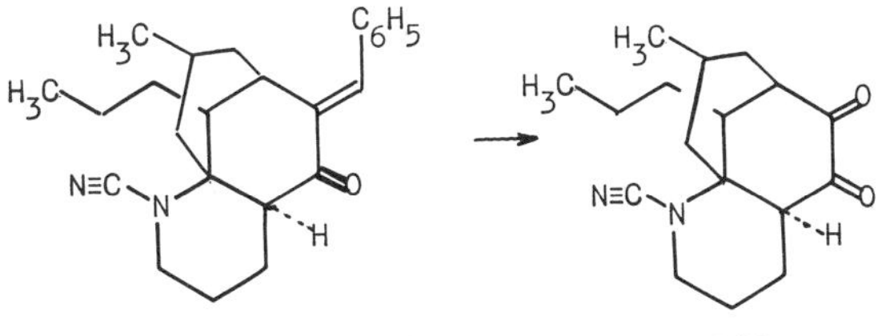

12.47 12.48

Now about this time the structure determination of annotinine (12.1) was completed, and since both compounds contain sixteen carbon atoms and one nitrogen, a hydrojulolidine (12.49) skeleton seemed likely. Additionally, the evidence provided in (e) above implies that C_7 is the position for attachment of one end of the bridge while C_{13} is, reasonably, the other, i.e., evidence in (a) above excludes either of the other two ring junctions as points of attachment

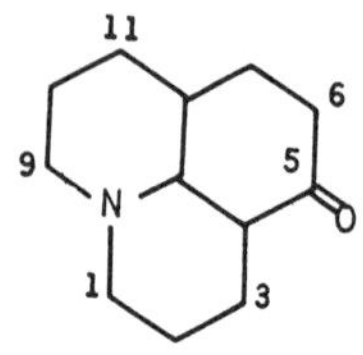

12.49

and neither carbon α to the carbonyl is available for bridging from the evidence in (d). That the ring should be six-membered with a single, symmetrically arranged methyl group was deduced from the proton magnetic resonance (pmr) spectrum and the dehydrogenation data (i.e., to 5-methyl- and 5,7-dimethylquinoline, 12.31 and 12.32, respectively).

Finally, the stereochemistry (except for the methyl at C_{15} which was deduced later through the interrelationship of lycopodine, 12.2, with other *Lycopodium* alkaloids) was established as shown since (a) α-bromocyanolycopodine (12.35) yields a cyclized product (12.50)

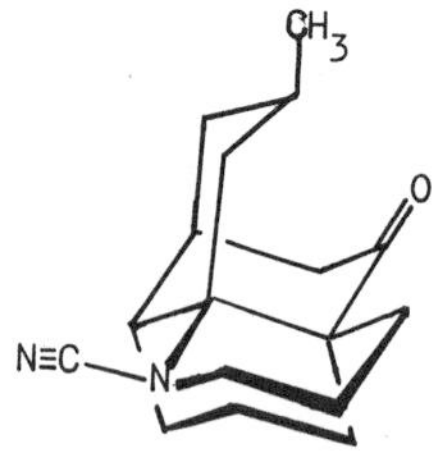

12.50

on treatment with hydroxide in ethanol and (b) epimerization (considered likely if the hydrogen α to the carbonyl in lycopodine, 12.2, at the ring juncture, were equatorial, i.e., a cis decalin results) fails to occur in acid or in base.

A synthesis of lycopodine (12.2; [41] begins with the copper-catalyzed 1,4-addition of m-methoxybenzylmagnesium bromide (12.51) to 5-methyl-2-cyclohexenone (12.52). This addition results in the formation of an enolate salt (12.53) in which it is pointed out that the m-methoxybenzyl and the methyl groups are arranged trans to one another on the cyclohexene ring. Alkylation of the enolate with allyl bromide, followed by oxidative hydroboration to a primary alcohol, conversion of the alcohol to the corresponding carboxylic acid, and esterification of the latter, results in a keto ester which undergoes aminolysis to the lactam 12.54. When the lactam (12.54) is treated with phosphoric acid in formic acid, cyclization occurs and a tetracyclic intermediate (12.55) results (Scheme 12.6).

Lithium aluminum hydride reduction of the amide function in 12.55, followed by Birch reduction of the aromatic ring and isomerization of the resulting enol ether, generates the amine 12.56. Then, acylation of the amine 12.56 with trichloroethyl chlorocarbonate and ozonolysis of the resulting carbamate yield an acid, esterification of which results in 12.57. Baeyer-Villager-type oxidation of 12.57 generates an enol ester which forms the ketone 12.58 on hydrolysis. Removal of the N-protecting group, lactam formation, and hydride reduction give (±)-dihydrolycopodine (12.59 and its mirror image), which is then oxidized to (±)-lycopodine (12.2 and its mirror image).

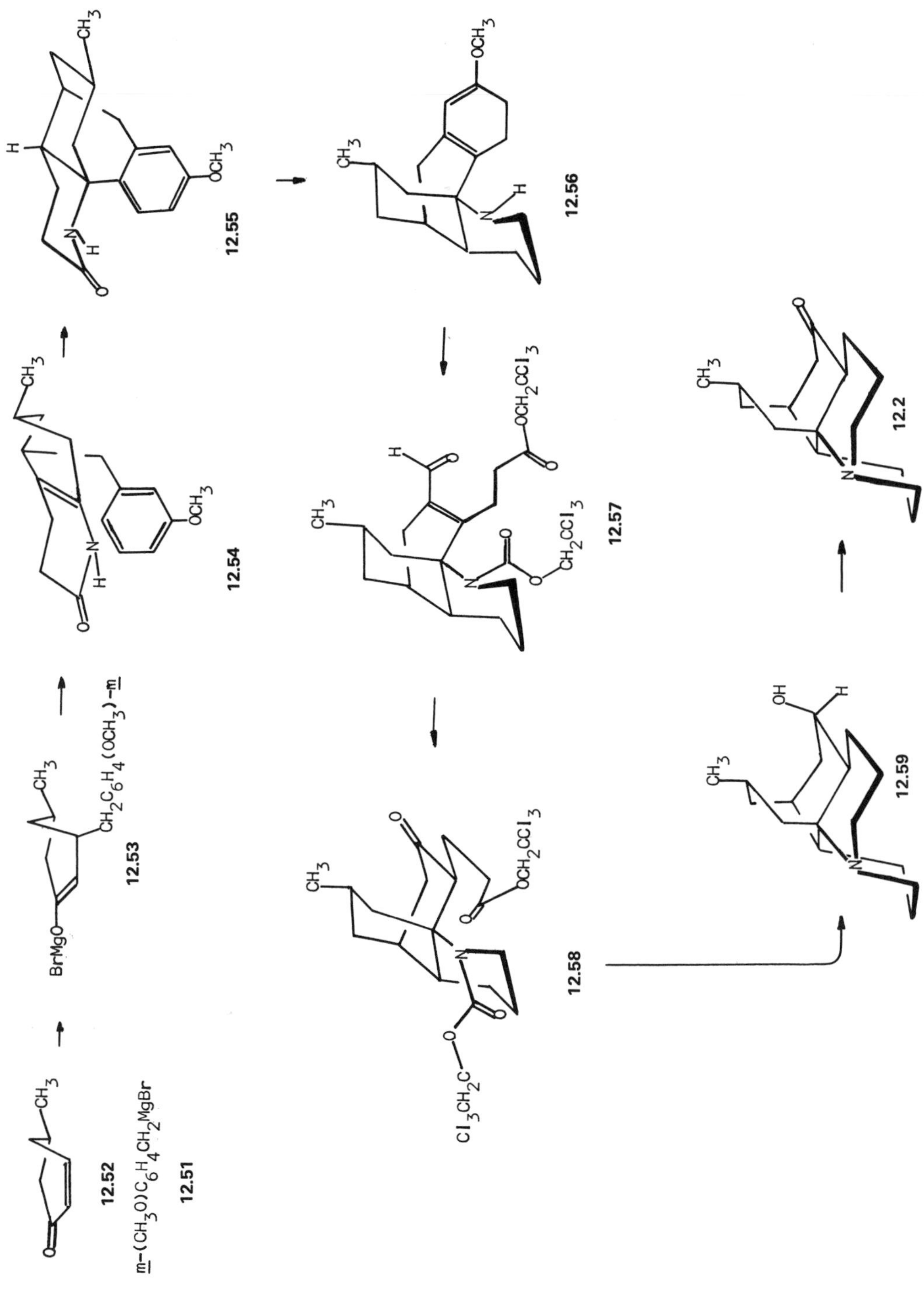
12.52
m-(CH3O)C6H4CH2MgBr
12.51
BrMgO
CH2C6H4(OCH3)-m
12.53
12.54
12.55
12.56
12.57
12.58
12.59
12.2

SCHEME 12.6 The total synthesis of lycopodine. (After Reference 41.)

SELECTED READING

D. B. MacLean, in The Alkaloids, Vol. 10, R. H. F. Manske (ed.). Academic, New York, 1968, pp. 306 ff.

R. H. F. Manske (ed.), in The Alkaloids, Vol. 7. Academic, New York, 1960, pp. 505 ff.

R. H. F. Manske (ed.), in The Alkaloids, Vol. 5. Academic, New York, 1955, pp. 295 ff.

Part 4

ALKALOIDS DERIVED FROM NICOTINIC ACID

The relatively small number of alkaloids derived from nicotinic acid (13.1) are obtained from plants of significant commercial value and have been extensively studied. Ricinine (13.2), which is a substituted pyridone, is easily isolated in high yield as the only alkaloid from the castor bean (Ricinus communis L.). The oil (castor oil, a gentle cathartic, rich in fatty acids), obtained by pressing the bean, serves as a starting material for many industrial endeavors. The highly toxic alkaloid (-)-nicotine (13.3), and related tobacco bases which

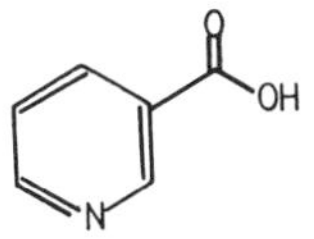

OH N | OCH_3 C≡N N O CH_3 | H N CH_3 N

13.1 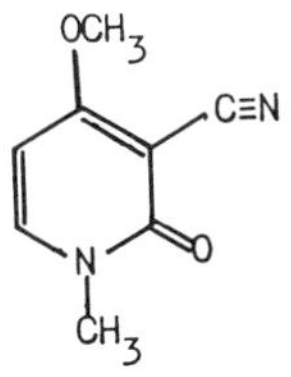 | 13.2 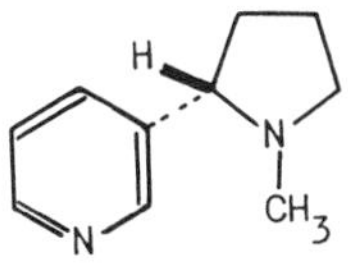 | 13.3

include (-)-anabasine (13.4) and (-)-anatabine (13.5), are obtained from, among others, the commercially grown tobacco plant Nicotiana tabacum L. Various tobaccos have differing amounts of these and other bases (as well as different flavoring constituents), some of which

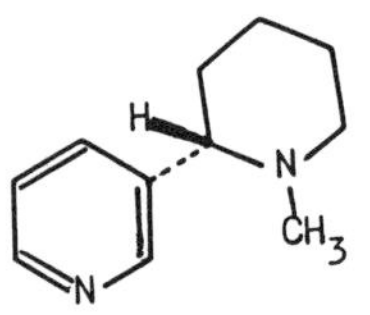

13.4

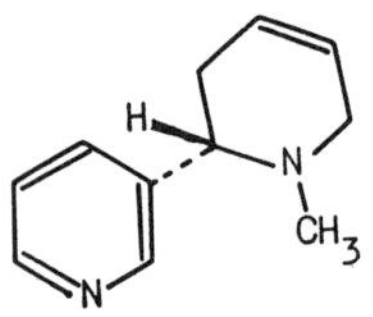

13.5

are apparently habituating to some individuals. Currently, assay of the (-)-nicotine (13.3) content in tobacco (annual world production in excess of three million tons) is desirable (in some countries mandatory), although the toxicity of the unassayed plant bases may be as high or higher than that of (-)-nicotine (13.3). As is apparent from their structures, these compounds share the feature of a modified or substituted pyridine ring, which although it could presumably have come from oxidation of a piperidine nucleus,* arises instead from nicotinic acid (13.1).

*It is interesting to note that the formation of alkaloids containing a pyridine ring derived by formal dehydrogenation of a piperidine appears to be rare.

13

BIOSYNTHESIS OF NICOTINIC ACID

Two roads diverged in a wood, and I—
I took the one less traveled by
And that has made all the difference.

Robert Frost, "The Road Not Taken" (1910)

I. THE TRYPTOPHAN PATHWAY TO NICOTINIC ACID

Nicotinic acid (13.1) is derived in mammalian systems and some microorganisms catabolically from tryptophan (13.6). In the higher order plants, this route does not appear to operate. Nevertheless, because catabolism of tryptophan (13.6) to nicotinic acid (13.1) occurs via anthranilic acid (13.7), which, it will be recalled (Schemes 2.20 and 2.21) was used in the generation of tryptophan (13.6), and because the subject has not been examined exhaustively in the higher plants, the known catabolic pathway is presented here.

As shown in Scheme 13.1, tryptophan (13.6) undergoes oxidation to N-formylkynurenine (13.8). This oxidation appears to utilize molecular oxygen and a metal-bearing enzyme which may mediate in what would otherwise be a "forbidden" electrocyclic process. Deformylation to kynurenine (13.9), followed by cleavage to anthranilic acid (13.7) and alanine (13.10), serves to provide the pathway (since anthranilic acid is excreted by mammals) for removal of kynurenine (13.9) which was not oxidized further to 3-hydroxykynurenine (13.11). Cleavage of 3-hydroxykynurenine (13.11) to alanine (13.10) and 3-hydroxyanthranilic acid (13.12), followed by oxidation (again by molecular oxygen) to 2-acroleyl-3-aminofumarate (13.13) and isomerization of the fumarate 13.13 to the corresponding maleate, generates a precursor suitable for cyclization to quinolinic acid (13.14). Quinolinic acid (13.14) is known to serve as a precursor to nicotinic acid (13.1) and readily undergoes loss of the α-carboxylate. While loss of the α-carboxylate can be readily accomplished thermally in vitro, in vivo loss takes a different pathway, and it appears that the ribonucleotide 13.15 is first created from quinolinic acid (13.14) and that this results in the ribonucleotide of nicotinic acid which then goes on to nicotinamide adenine dinucleotide (NAD^+). Degradation of NAD^+ provides nicotinamide (13.17) while nicotinic acid (13.1) itself can presumably be obtained by hydrolysis of 13.16.

II. THE NONTRYPTOPHAN PATHWAY TO NICOTINIC ACID

The experimental observation that nicotinic acid (13.1) serves as a precursor of the pyridine ring in the alkaloids (-)-nicotine (13.3), (-)-anabasine (13.4), and ricinine (13.2), whereas tryptophan (13.6) does not, suggests the absence of a pathway from tryptophan (13.6) to

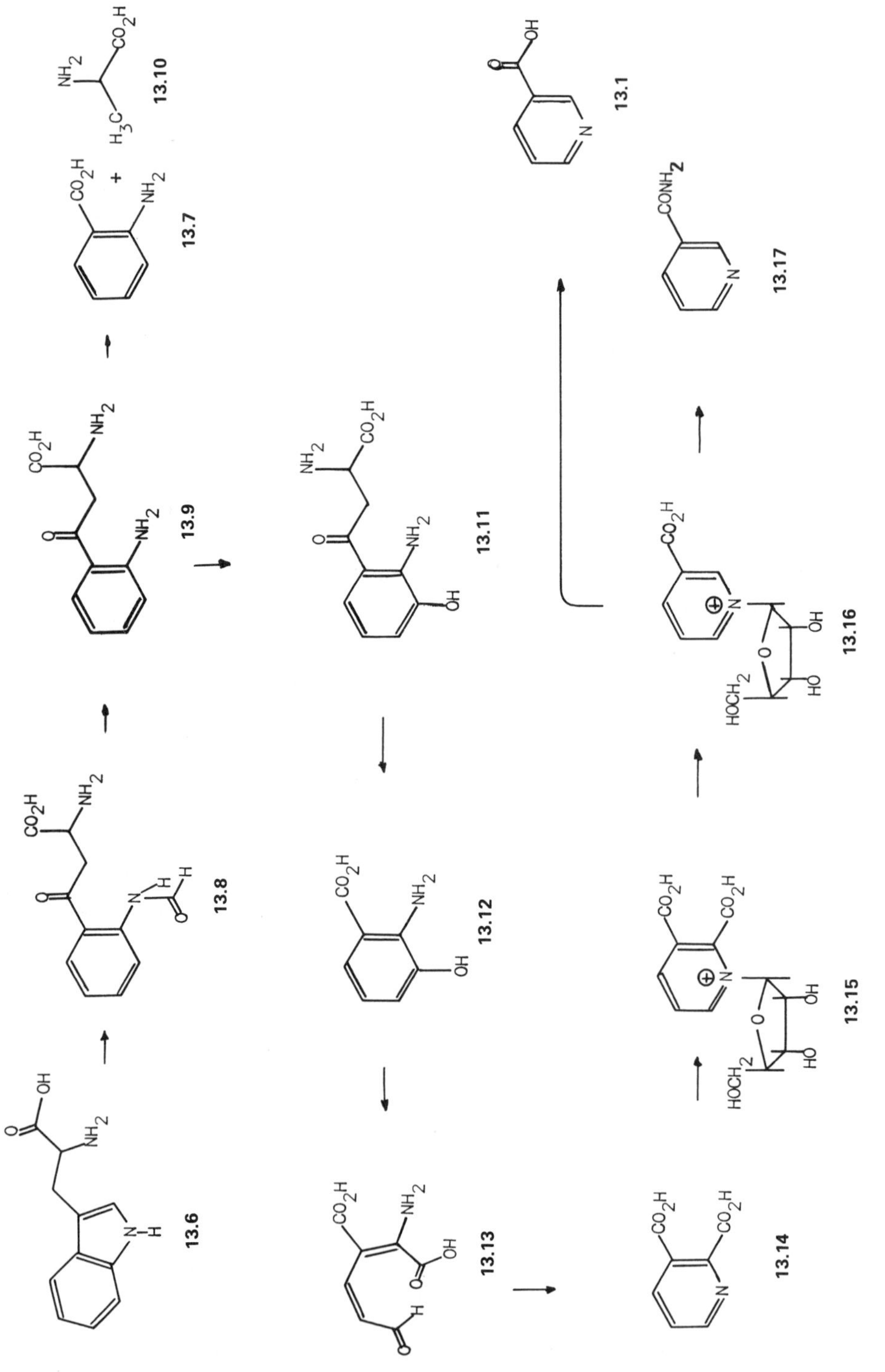

SCHEME 13.1 The catabolism of tryptophan to nicotinic acid.

nicotinic acid (13.1; incorporation into the plant and/or transport to the site of alkaloid synthesis, etc., may have also failed) and the presence of some other route.

Large numbers of feeding experiments on various plant species elaborating the abovementioned alkaloids were carried out by various research groups with somewhat inconsistent results, but a general picture has nevertheless emerged. As outlined in Scheme 13.2, a three-carbon fragment such as 3-phosphoglyceraldehyde (13.18) or its equivalent is presumed to serve as half the molecule while a four-carbon fragment such as aspartic acid (13.19) or its equivalent, serves to provide the nitrogen and the other half of the molecule. The quinolinic acid (13.14) resulting from the condensation can then, as in Scheme 13.1, go on to nicotinic acid (13.1). The simple elegance of this scheme suffers from the problem that while many feeding experiments are consonant with it, others are discordant. Thus, the pathway shown does account for nonrandom incorporation of variously labeled samples of glyceraldehyde (13.18) and aspartic acid (13.19), but it does not account readily for nonrandom incorporation of, for example, bicarbonate and glutamate (13.20), nor does it account for the results of studies involving the use of labeled nitrogen (see below). The obligatory nature of the pathway shown (Scheme 13.2) is therefore still an open question.

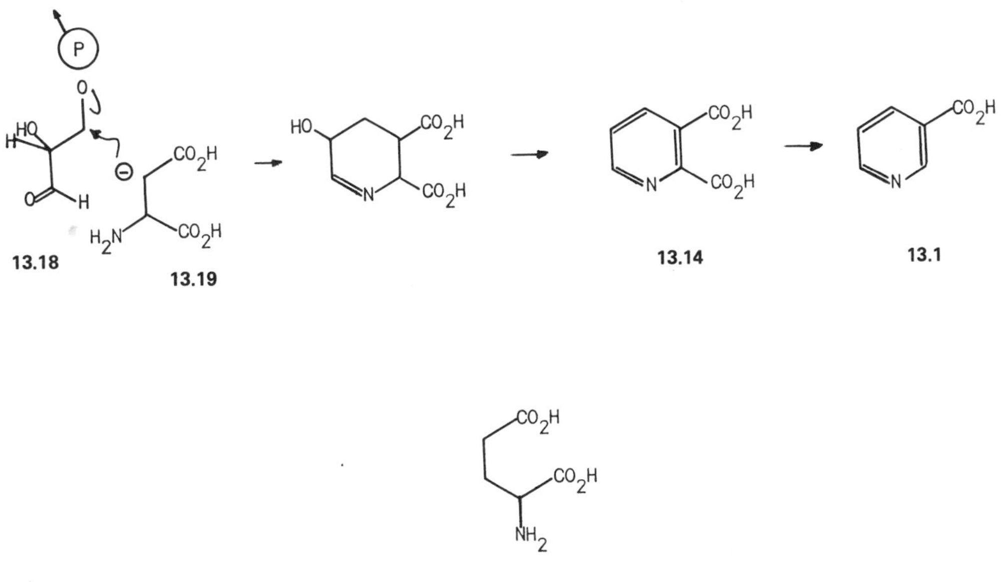

P O HO H O H ⊖ CO_2H H_2N CO_2H HO CO_2H N CO_2H N CO_2H CO_2H N CO_2H

13.18 13.19 13.14 13.1

CO_2H CO_2H NH_2

13.20

SCHEME 13.2 The formation of nicotinic acid via quinolinic acid from aspartic acid and 3-phosphoglyceraldehyde.

14

BIOSYNTHESIS OF RICININE AND THE TOBACCO ALKALOIDS

God offers to every mind its choice between truth and repose.

Ralph Waldo Emerson

I. RICININE

In intact plants and in excised roots of Ricinus communis L., nicotinic acid (14.1) and nicotinamide (14.2) serve as direct precursors of ricinine (14.3). The nitrile carbon and the nitrile nitrogen atoms of ricinine (14.3) are specifically derived from the carboxyl carbon of nicotinic acid (14.1) and from the amide nitrogen of nicotinamide (14.2), respectively. Double-labeled (^{3}H and ^{14}C or ^{15}N) precursors have also been utilized to confirm the studies with single labels which gave rise to the information provided above.

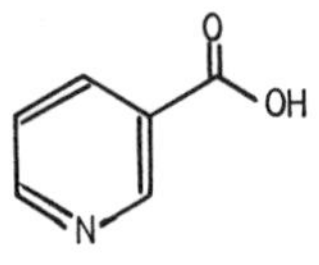

14.1 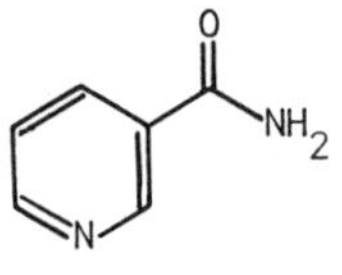 14.2 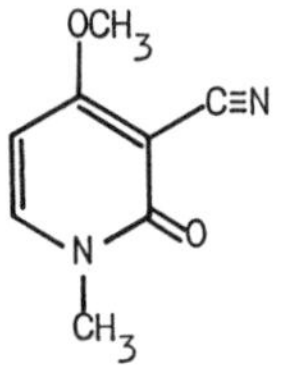 14.3

In concert with feeding experiments outlined earlier for nicotinic acid (p. 149), glycerol (14.4) is incorporated into ricinine (14.3) as C_4 to C_6 in a nonsymmetrical fashion. Aspartic acid (14.5) is also incorporated. However, when aspartic acid (14.5) labeled in the nitrogen

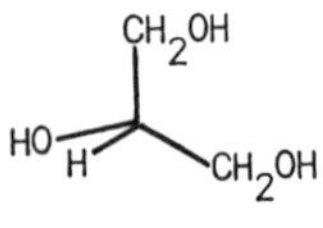

14.4 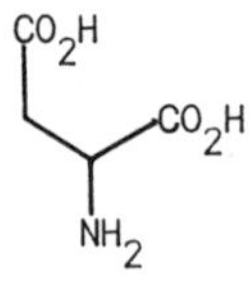 14.5

(^{15}N) atom was fed, both nitrogen atoms of ricinine (14.3) contained ^{15}N, and thus nitrogen exchange must have occurred prior to introduction of aspartic acid (14.5) into ricinine (14.3), or a symmetrical intermediate (e.g., 14.6) was involved. The first possibility is unlikely given the normal size of the nitrogen pool, while the latter is excluded since ricinine (14.3) derived from [4-^{14}C]aspartic acid (14.5) is labeled exclusively at the nitrile carbon. Nevertheless, quinolinic acid (14.7), the presumed intermediate, serves as a specific precursor of ricinine (14.3). Additional anomalies include (a) the incorporation of glycerol (14.4) into C_2 and C_3 of ricinine (14.3) and (b) the nonrandom incorporation of bicarbonate and [1-^{14}C]glutamic acid (14.8). Thus, despite much effort, the problem of the biosynthesis of ricinine (14.3) remains clouded.

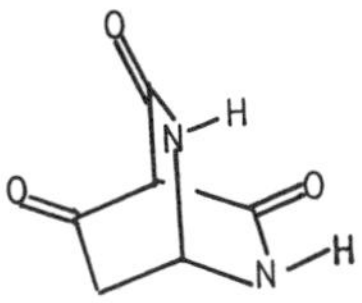

14.6 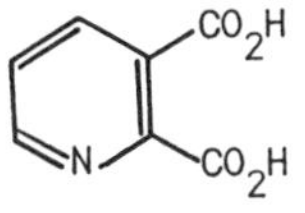 14.7 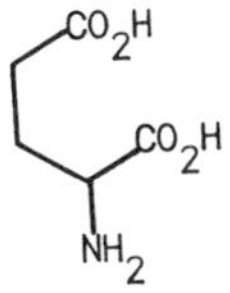 14.8

II. THE TOBACCO ALKALOIDS

A. (-)-Nicotine

The pyridine nucleus of five carbons and one nitrogen found in (-)-nicotine (14.9) is derived, intact, from nicotinic acid (14.1). The hydrogen atoms at carbons 2, 4, and 5 are also retained in the conversion. Only the carboxylic acid function at C_3 and the hydrogen at C_6 are lost in the process of going from nicotinic acid (14.1) to (-)-nicotine (14.9).

The pyrrolidine portion of four carbons and one nitrogen in (-)-nicotine (14.9) is provided by ornithine (14.10) or its equivalent. Contrary to what was found earlier in alkaloids derived from ornithine (14.10; Part 2, Chapter 3), this amino acid appears to be incorporated

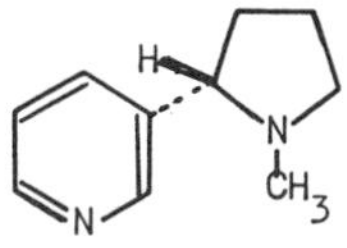

14.9 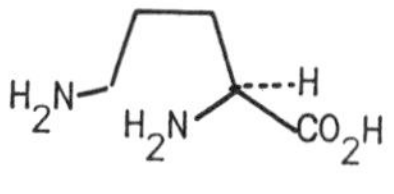 14.10

into the alkaloid in a symmetrical fashion. Thus, radioactivity from [2-^{14}C]ornithine (14.10) is equally distributed between C_2 and C_5 of the pyrrolidine ring of (-)-nicotine (14.9). However, the nitrogen in [δ-^{15}N, 2-^{14}C]ornithine (14.10) is the one retained in the isolated (-)-nicotine (14.9), while that in [α-^{15}N, 2-^{14}C]ornithine (14.10) is not [42].

The methyl group of (-)-nicotine (14.9) is derived from S-adenosylmethionine (14.11) by methyl transfer. However, there may still be some question as to exactly when methylation occurs. It seems clear that (-)-nicotine (14.9) is converted into, and is thus the precursor for, (-)-nornicotine (14.12), so that methylation must occur early. Second, it has been demonstrated [43] that the N-methyl-Δ^1-pyrrolinium ion (14.13) is incorporated into nicotine (14.9) without equilibration (i.e., $C_2 \neq C_5$). Thus, methylation must occur after ornithine (14.10) but before the N-methyl-Δ^1-pyrrolinium ion (14.13) is generated. Now, if ornithine were methylated to yield either α-N-methylornithine (14.14) or δ-N-methylornithine (14.15),

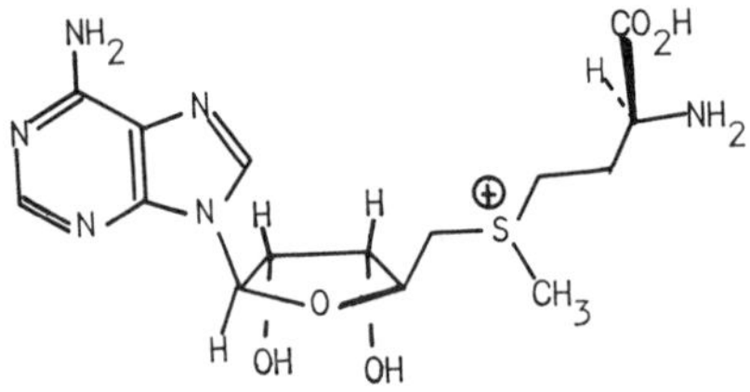

14.11

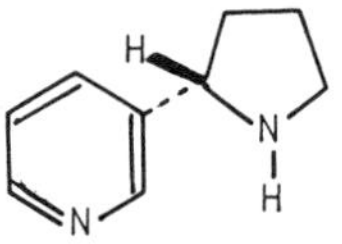

14.12 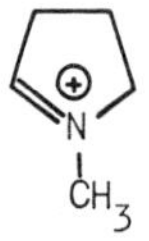 14.13 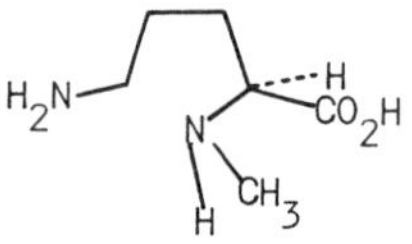 14.14 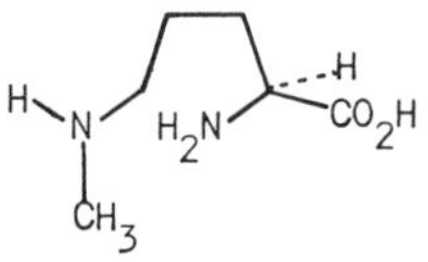 14.15

the symmetry of the decarboxylation product of ornithine (14.14; i.e., putrescine, 14.16) would have been destroyed, the nitrogens of N-methylputrescine (14.17) being distinguishable by one having been methylated before decarboxylation to a symmetrical intermediate.

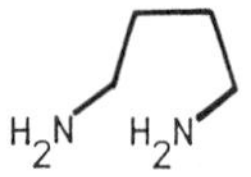

14.16 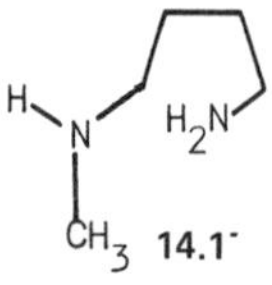 14.1˙

Presumably, therefore, since ornithine (14.10) is incorporated in a symmetrical fashion, neither α- nor δ-N-methylornithine (14.14 and 14.15, respectively) should be incorporated.

When α- and δ-N-[methyl-^{14}C]ornithine[2-^{14}C] (14.14 and 14.15, respectively) were fed to Nicotiana tabacum [44], both (!) were incorporated, but to different extents and in different ways. Thus, δ-N-[methyl-^{14}C]ornithine[2-^{14}C] (14.15) was incorporated better than the corresponding α-isomer (14.14). Second, 14.15 was incorporated to provide nicotine (14.9) in which only C_2 of the pyrrolidine ring (not C_5) bore label. Third, the ratio of activity in the pyrrolidine ring to that in the N-methyl group in the starting ornithine (14.15) was identical to that in the product nicotine (14.9). It was concluded that δ-N-methylornithine (14.15) is directly incorporated in a nonsymmetrical fashion. On the other hand, feeding α-N-[methyl-^{14}C]ornithine[2-^{14}C] (14.14) under the same conditions resulted in isolation of nicotine (14.9) in which both C_2 and C_5 of the pyrrolidine ring were equally labeled and the relative level of activity at the N-methyl of nicotine (14.9) had declined to about one-half of that expected for direct incorporation. It was therefore concluded that both N-methyl ornithines (14.14 and 14.15) were "false precursors" and were incorporated via different "aberrant" pathways utilizing enzymes which were not specific for ornithine (14.10) as substrate. The overall biosynthesis of the N-methyl-Δ^1-pyrrolinium ion (14.13) is thus suggested [44], as shown in Scheme 14.1, to proceed from ornithine (14.10) to putrescine (14.16) and then, after methylation to N-methylputrescine (14.17), to the product ion (14.13). The demand that the nitrogen atoms be differentiated (i.e., the δ-nitrogen but not the α-nitrogen being retained) can be accommodated by assuming that normal transamination of ornithine (14.10) occurs much more rapidly than decarboyxlation to putrescine (14.16).

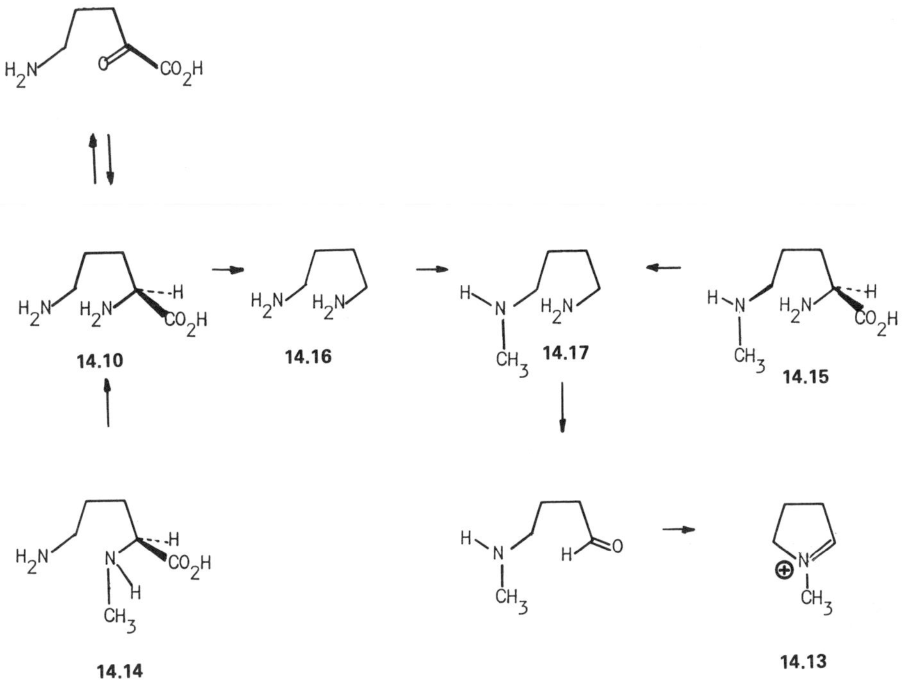

SCHEME 14.1 A pathway for the formation of an N-methyl-Δ^1-pyrrolinium ion from ornithine.

The aberrant incorporation of the N-methylornithines 14.14 and 14.15, indicating nonspecific enzymatic processes, and fundamental studies on the incorporation of carbon dioxide into various Nicotiana (see below) have, in part, resulted in questioning of the status of ornithine (14.10) as a "normal" precursor for the pyrrolidine portion of the nicotine (14.9) nucleus [45]. That is, although there is no question that ornithine (14.10) is incorporated, the question is: Is ornithine (14.10) an "obligatory" precursor?

(-)-Nicotine (14.9) isolated from short-term biosynthesis (Nicotiana glutinosa) utilizing only $^{14}CO_2$ as the carbon precursor, provides alkaloid in which complete and specific degradation of the pyrrolidine ring shows an unsymmetrical labeling pattern which is inconsistent with the use of ornithine (14.10) as a preformed precursor (however, there is some question about the validity of this work, as will be discussed later). Thus, the pattern in which C_4 and C_5 of the pyrrolidine ring are found to be extensively and equally labeled suggests that a two-carbon fragment is serving as the precursor of the pyrrolidine ring. Such a two-carbon fragment is available from hydrolysis of "active glycoaldehyde" (14.18); Chapter 2, p. 13), and the glycolic aldehyde (14.19) or glycolic acid (14.20) which results from its use would be expected to be equally labeled after a single pass (but not from the first cycle unless some active glycoaldehyde were available from isomerized tetrose) through the Calvin-Bassham cycle (Scheme 2.4). A potential pathway from "active glycoaldehyde" (14.18) to the N-methyl-Δ^1-pyrrolinium salt (14.13) is shown as Scheme 14.2 [43].

In concert with the hypothesis outlined above is the observation that glycolic acid (14.20) is an early product of photosynthesis in tobacco leaves [46]. Thus, according to Scheme 14.2, glycolic aldehyde (14.19) serves to generate 3,4-dihydroxybutyric acid (14.21) on condensation with acetate or malonate followed by decarboxylation. Dehydration to succinic acid semialdehyde (14.22), transamination, and methylation to N-methyl-4-aminobutyric acid (14.23),

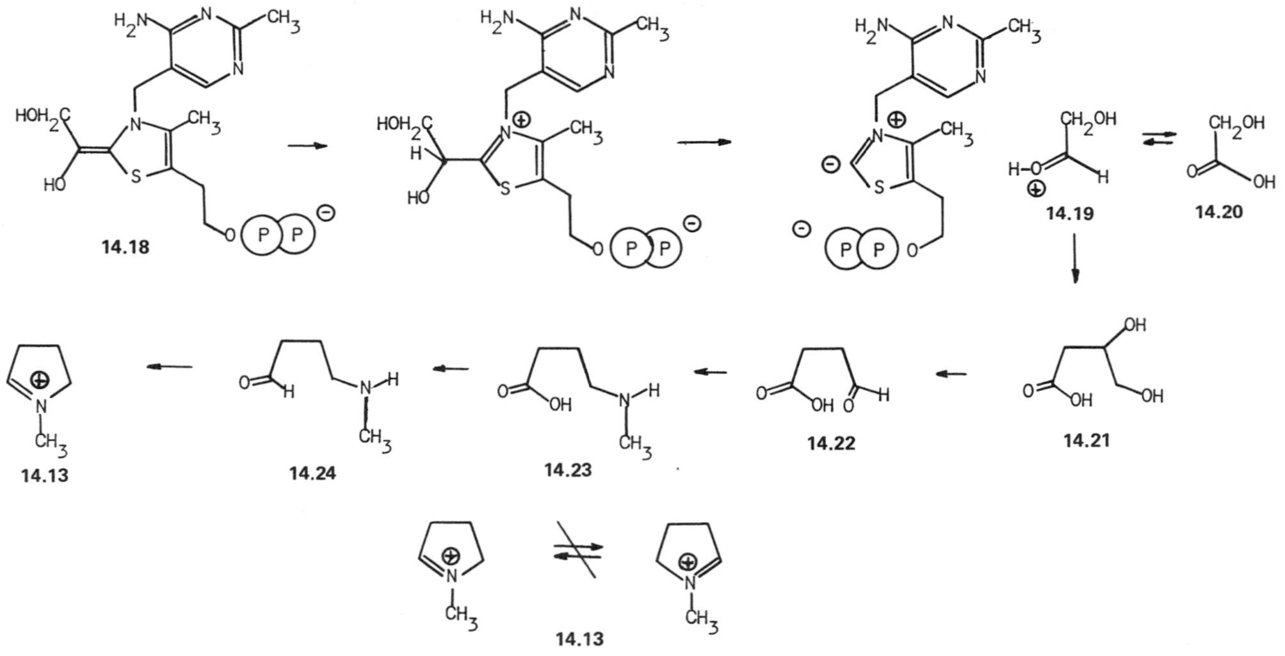

SCHEME 14.2 A pathway to N-methyl-Δ^1-pyrrolinium ion from glycolic acid. (After Reference 43.)

followed by reduction to N-methyl-4-aminobutanal (14.24) and cyclization to the N-methyl-Δ^1-pyrrolinium salt (14.13) which does not equilibrate, then completes the picture.*

Now, it has been pointed out [46] that at low CO_2 concentrations, an alternate pathway to glycolic acid (14.20; other than that shown in Scheme 14.2) appears to exist. Not much is yet known about this pathway except that the carbon balance appears to exclude generation of glycolic acid (14.20; which has the same specific activity as CO_2 available to the plant) from any three-carbon precursor. It was also concluded that at CO_2 concentrations close to those found in air, glycolic acid (14.20) is synthesized in the light by a carboxylation reaction different than that involving "active glycoaldehyde" (14.18). More recent experiments [48] do appear, however, to indicate that using $^{13}CO_2$ at normal atmospheric concentrations results in symmetrical rather than unsymmetrical labeling, and even the degradative work [45, 49] is now suspect. However, despite the struggle over the details, it appears safe to suggest that (-)-nicotine (14.9) forms from nicotinic acid (14.1) and an N-methyl-Δ^1-pyrrolinium ion (14.13) as shown in Scheme 14.3.

B. (-)-Anabasine

The pyridine ring of (-)-anabasine (14.27) is derived from nicotinic acid (14.1), but the piperidine ring is not. Thus, for example [50], when [6-^{14}C]nicotinic acid (14.1) is fed to 3-month-old Nicotiana glutinosa plants, the (-)-anabasine (14.27) isolated has all its activity in the pyridine ring. This was demonstrated by oxidation of the (-)-anabasine (14.27) to nicotinic acid (14.1) and decarboxylation of the latter to pyridine (14.28). In each case, the

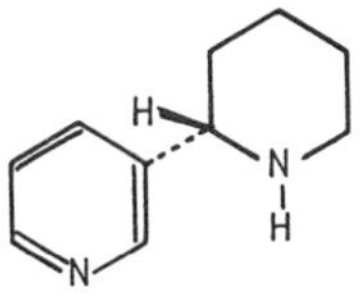

14.27 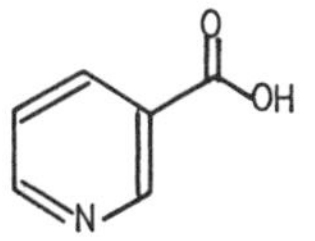 14.1 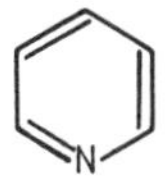 14.28

specific activity, (-)-anabasine (14.27) ⟶ nicotinic acid (14.1) ⟶ pyridine (14.28) is, within experimental error, the same.

On the other hand, when [1,5-^{14}C]cadaverine (14.29) or (±)-[2-^{14}C]lysine (14.30 and its antipode) is fed to Nicotiana glauca, all the activity in (-)-anabasine (14.27) is confined to the piperidine nucleus. Although the result from the cadaverine (14.29) feeding suggests that a symmetrical intermediate lies on the pathway to the piperidine portion of (-)-anabasine (14.27), the result from the (±)-lysine (14.30) feeding suggests that cadaverine (14.29) may not be an obligatory precursor, or, as recently suggested [18], it may serve as an intermediate bound to the enzyme surface but in equilibrium (K[bound]/[unbound] $\gg$ 1) with free cadaverine (14.29), since the intermediate Δ^1-piperidine (14.32), in contrast to the case of the N-methyl-Δ^1-pyrrolinium (14.13) ion in nicotine (14.9), remains nonsymmetrical.

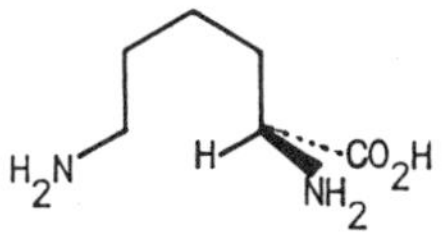

14.30 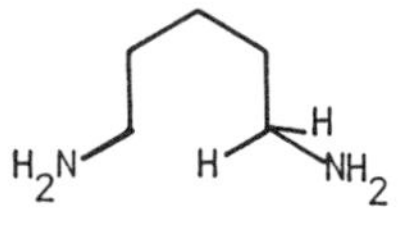 14.29 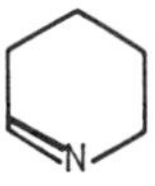 14.32

*With regard to this proposed pathway, it should be noted [47] that N-methyl-4-aminobutyric acid (14.23) is not incorporated into nicotine (14.9).

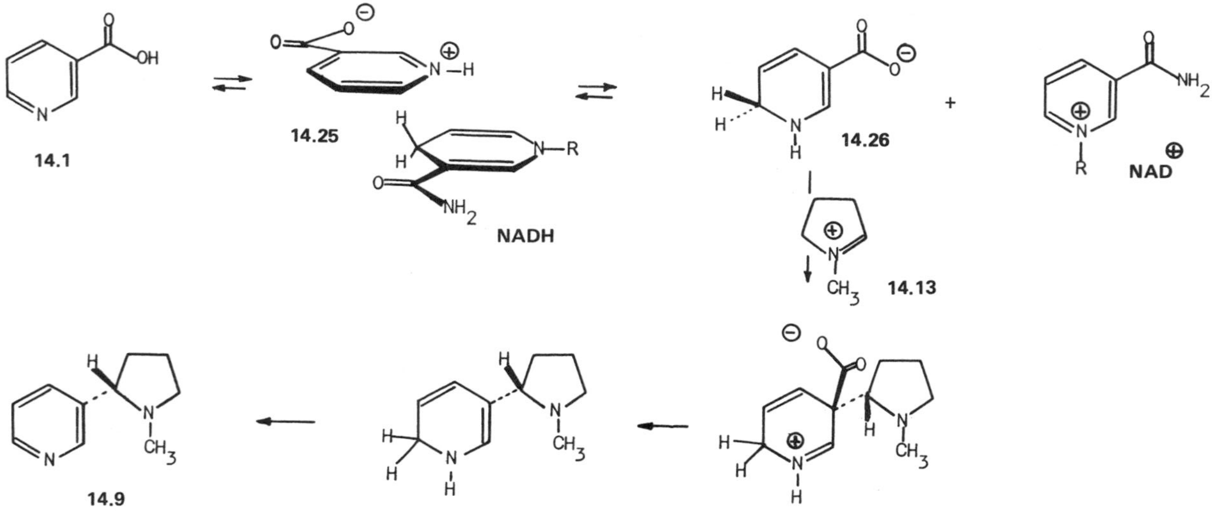

SCHEME 14.3 A pathway for the formation of (-)-nicotine (14.9) from nicotinic acid and an N-methyl-Δ^1-pyrrolinium ion.

The idea of bound cadaverine (14.29) as expressed in Scheme 14.4 is particularly attractive since it is in concert with the preferential incorporation of L-lysine (14.30) and the retention of ^{15}N from (±)-[ε-^{15}N,2-^{14}C]lysine (14.30) but not (±)-[α-^{15}N,2-^{14}C]lysine (14.30), and the retention of tritium from (±)-[2-^{3}H]lysine (14.30) observed in the case of sedamine (14.31). Reaction of the nonequilibrating Δ^1-piperideine (14.32) with nicotinic acid (14.1; Scheme 14.5), by analogy with that already outlined for nitotine (14.9), yields (2S)-(-)-anabasine (14.27).

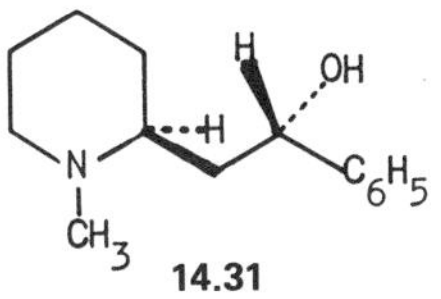

14.31

C. (-)-Anatabine

(-)-Anatabine (14.33) is (-)-$\Delta^{4'}$-dehydroanabasine, and it was expected, as recently pointed out [50], to be formed in a fashion analogous to (-)-anabasine (14.27). However, feeding of (±)-[2-^{14}C]lysine (14.30) to Nicotiana glutinosa generates (-)-anabasine (14.27) containing all of its activity in the piperidine ring and (-)-anatabine (14.33) with no activity [51]. On the other hand, when [6-^{14}C]nicotinic acid (14.1) is fed under the same conditions, it is found that the (-)-anatabine (14.33) isolated possesses activity in both rings and that the activity is equally divided between C_6 and $C_{6'}$. The degradative scheme which permitted such a determination is shown in Scheme 14.6 and may be described as follows: (-)-anatabine (14.33) is converted to its N-benzoyl derivative (14.34) with benzoic anhydride. Permanganate oxidation of 14.34 yields nicotinic acid (14.1) and N-benzoylglycine (14.35). The nicotinic acid (14.1) is then converted to β-methylpicoline (14.36) by treatment, in sequence, with thionyl chloride and hydrogen (in the presence of palladium on calcium carbonate). Further reduction and methylation of the β-methylpicoline (14.36) to the N,N-dimethylpiperidinium iodide (14.37) is accomplished with hydrogen on palladium in the presence of hydrogen chloride and then exhaustive methylation with methyl iodide. Hofmann elimination on 14.37 generates the two N,N-dimethylamino alkenes, 14.38 and 14.39, which are separated by gas-liquid phase chromatography and the desired isomer (14.39) oxidized first with osmium tetroxide and then with periodate to excise the terminal methylene group as formaldehyde. The label is there as expected.

At the same time, the hippuric acid (14.35), which possesses the other carbon of interest, is oxidized with lead tetraacetate and the resulting N-benzoyl-O-acetate (14.40) hydrolyzed with 2 N sulfuric acid to formaldehyde. Again, the activity is there, thus confirming the use of 2 equivalents of nicotinic acid (14.1) in the genesis of (-)-anatabine (14.33). A potential biosynthetic scheme which permits such a result is shown in Scheme 14.7. Here it is assumed that two reduced nicotinic acid systems (14.41 and 14.42), one of which (14.42) has already undergone decarboxylation, react together, and that subsequently a second equivalent of carbon dioxide is lost and further reduction to (-)-anatabine (14.33) occurs (NADH ⟶ NAD^+). Confirmation of this general scheme has recently been elegantly set forth [52].

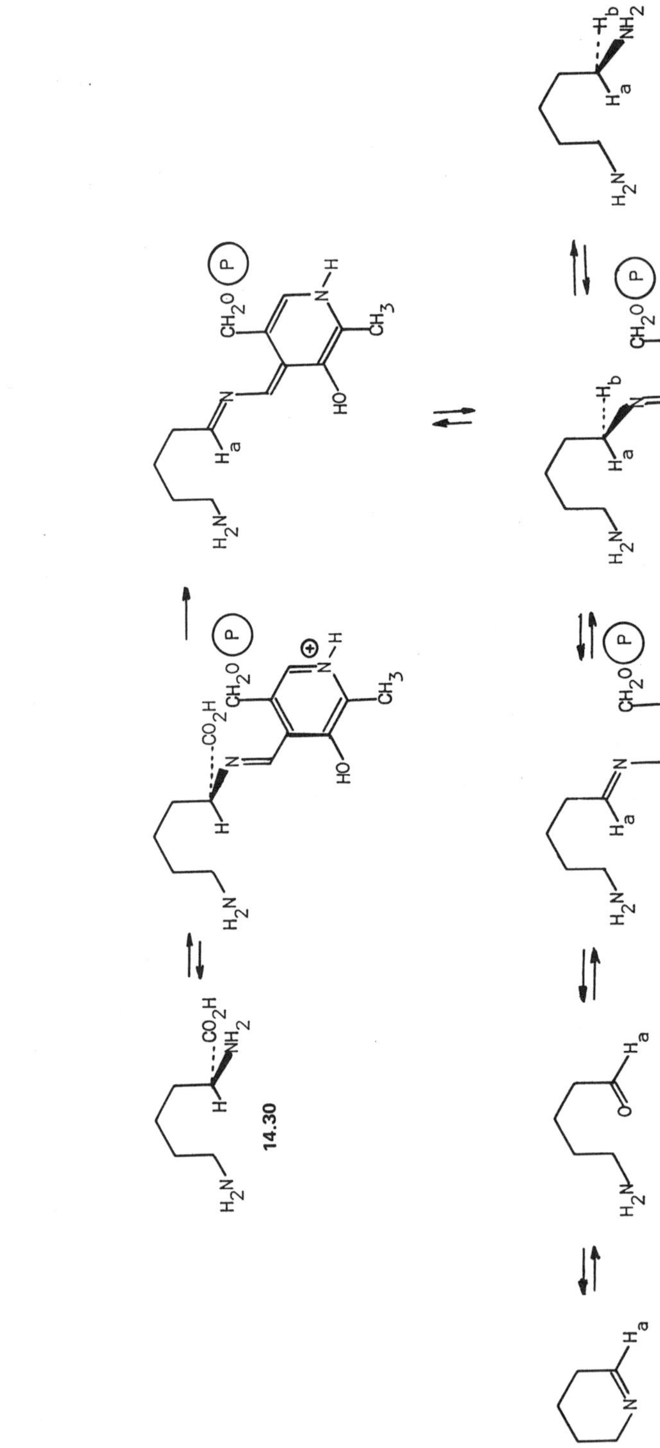

SCHEME 14.4 A pathway for the production of Δ^1-piperideine consonant with available evidence. (After Reference 18.)

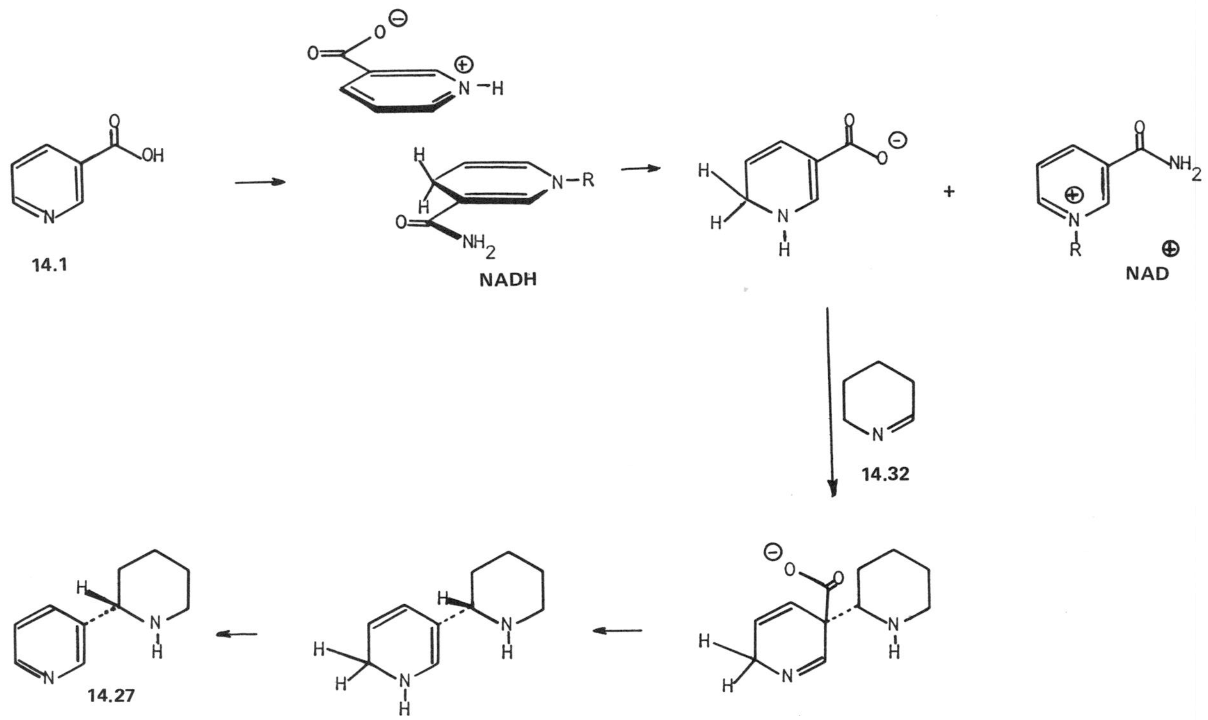

SCHEME 14.5 A pathway for the formation of (-)-anabasine from a Δ^1-piperideine and nicotinic acid (which has been reduced to a dihydronicotinic acid by NADH). The final oxidation could be used to reconvert NAD^+ to NADH.

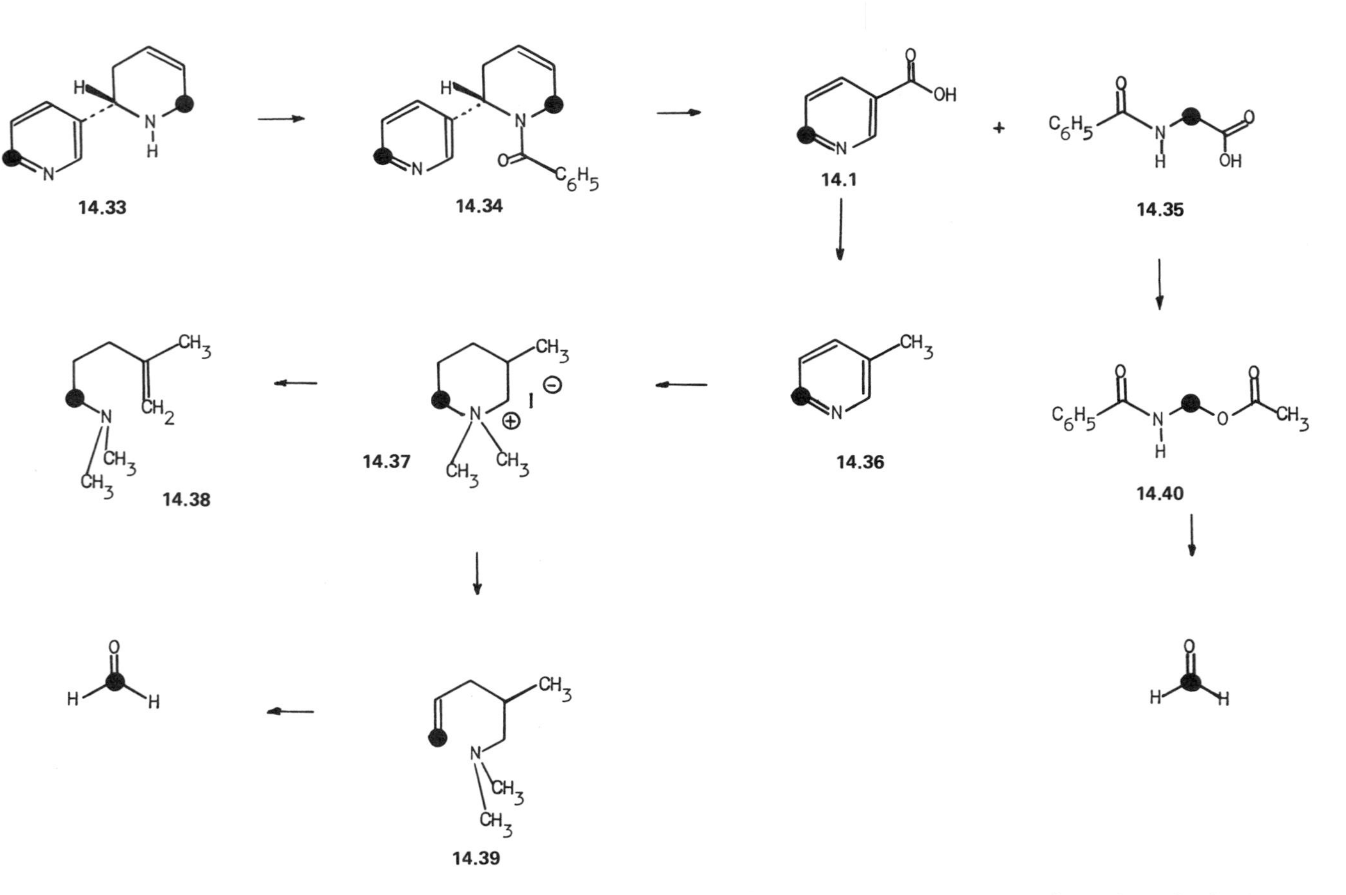

SCHEME 14.6 The degradative scheme of (-)-anatabine which determined its genesis from 2 equivalents of nicotinic acid. (After References 50 and 51.).

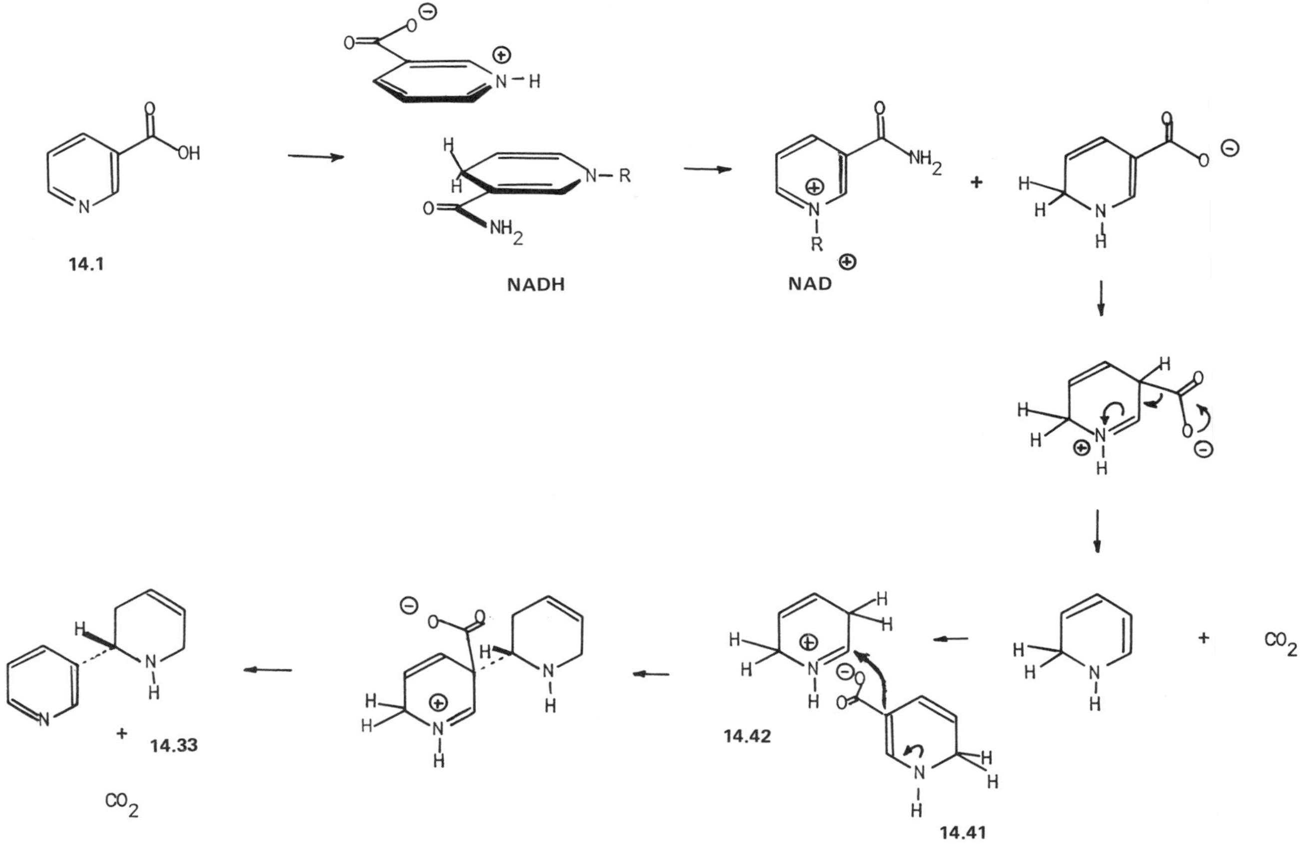

SCHEME 14.7 A potential pathway for the formation of (-)-anatabine from 2 equivalents of nicotinic acid.

15

CHEMISTRY OF RICININE

Even from the body's purity the mind receives a secret sympathetic aid.

James Thomson (ca. 1748)

Ricinine, $C_8H_8O_2N_2$ (15.1), is obtained as the only alkaloid from Ricinus communis L., occurring both in the seeds (0.3%) and the total plant (1.1%). The oil derived from the pressed seeds of this plant, castor oil, a purgative, is very different from other fatty plant oils in that it is soluble in alcohols and other polar solvents but insoluble in hydrocarbons. The use of castor oil obtained directly from the plant (as an extender, etc.), as well as when hydrogenated (castrowax), has made the plant of considerable commercial value.

The optically inactive base, ricinine (15.1), mp 201.5°C, undergoes hydrolysis in sodium hydroxide solution to yield methanol and ricininic acid (15.2), $C_6H_6O_2N_2$, mp 320°C. Treatment of ricininic acid (15.2) with hydrochloric acid at 150°C causes its decomposition to carbon dioxide, ammonium chloride, and a hydrochloride, $C_6H_7O_2N \cdot HCl \cdot H_2O$, mp 155-160°C, the free base of which, $C_6H_7O_2N$ (15.3), melts at 170-171°C when free of solvent (Equation 15.1). When the base 15.3 is treated with phosphorus pentachloride, a dichloride,

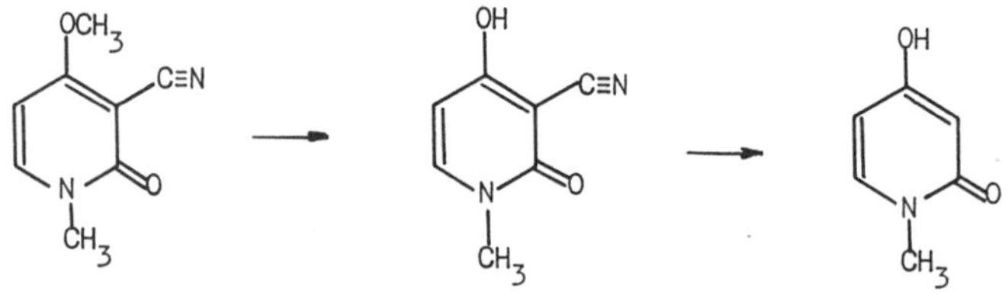

OCH3, C≡N, N, O, CH3 → OH, C≡N, N, O, CH3 → OH, N, O, CH3 (Eq. 15.1)

15.1 15.2 15.3

$C_5H_3NCl_2$ (15.4), bp 90°/18 torr, is produced, which on reduction with hydrogen iodide and phosphorus yields pyridine (15.5; Equation 15.2). Pyridine (15.5) has also been obtained from ricinine (15.1) on zinc dust distillation.

When ricininic acid (15.2) is treated with phosphorus oxychloride, a chloride (15.6) is produced which regenerates ricinine (15.1) on reaction with sodium methoxide in methanol (Equation 15.3). If, on the other hand, the chloride 15.6 is reduced catalytically (hydrogenolysis), desmethoxyricinine (15.7) is generated, and the latter, on warming with potassium hydroxide solution, yields 1-methyl-2-pyridone-3-carboxylic acid (15.8) and ammonia (Equation 15.4). The formation of 15.8 (Equation 15.4) on the one hand and the generation of hydrogen cyanide from ricinine (15.1) on treatment with chromic acid, on the other, leads to the conclusion that ricinine (15.1) contains a cyano group in the 3-position of the pyridine ring.

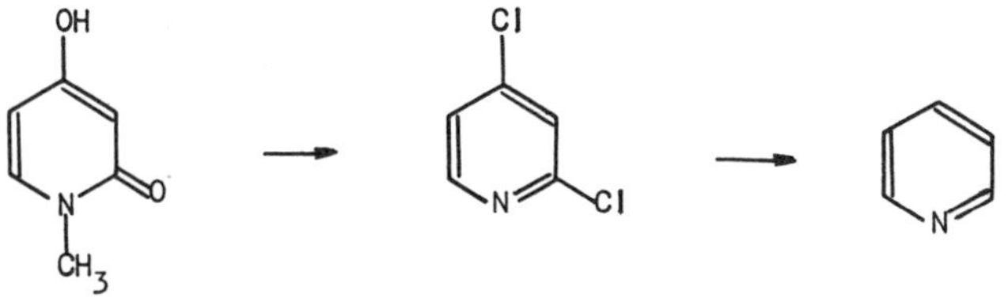

15.3 → 15.4 → 15.5 (Eq. 15.2)

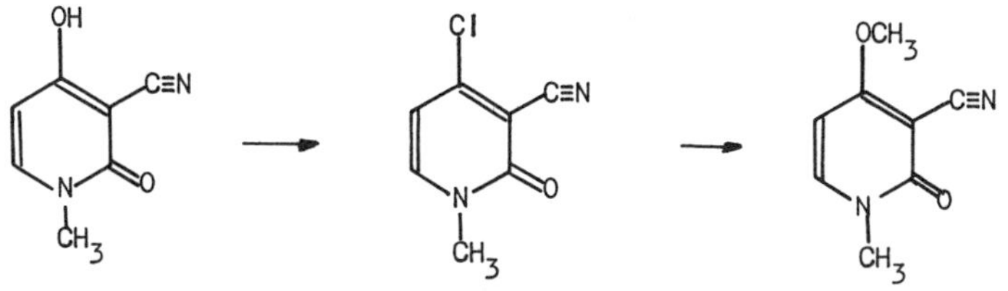

15.1 → 15.6 → 15.1 (Eq. 15.3)

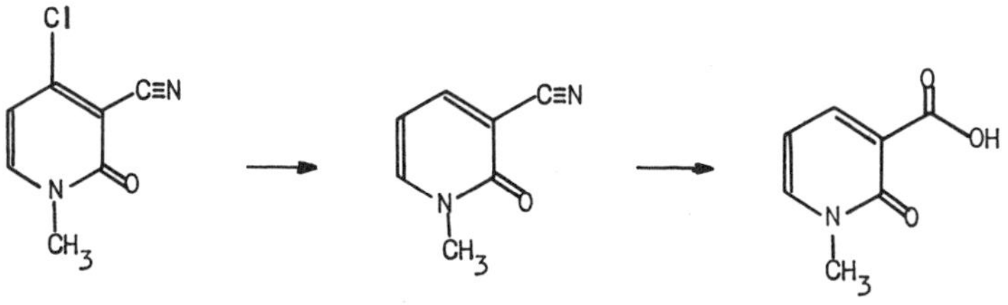

15.6 → 15.7 → 15.8 (Eq. 15.4)

That ricinine (15.1) is 1-methyl-3-cyano-4-methoxy-2-pyridone was established by synthesis [53]; Scheme 15.1). Thus, 4-chloroquinoline (15.9) on oxidation yields 4-chloroquinolinic acid (15.10), the anhydride of which on treatment with ammonia generates the amide

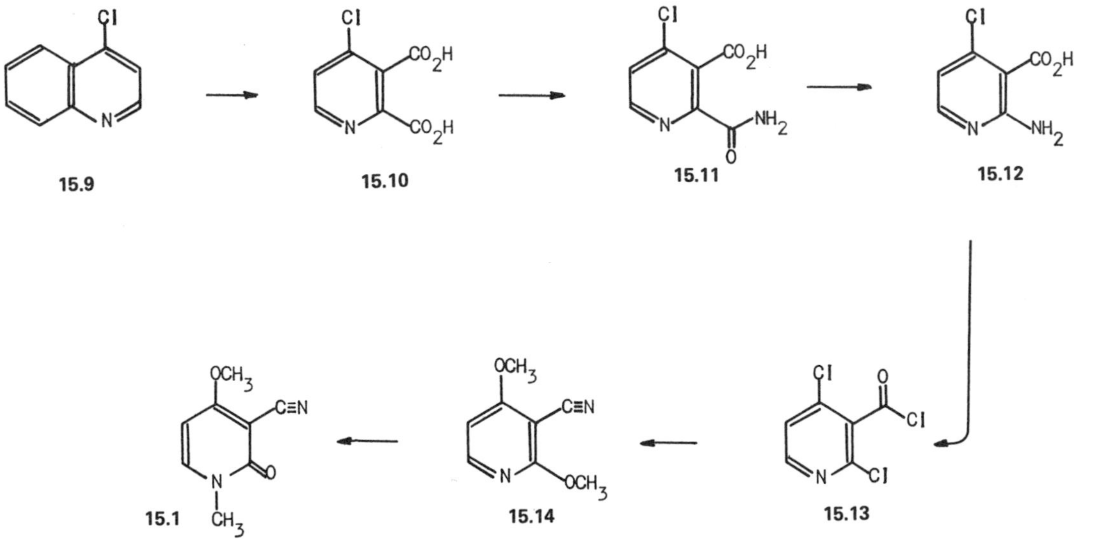

15.9 → 15.10 → 15.11 → 15.12 → 15.13 → 15.14 → 15.1

SCHEME 15.1 A synthesis of ricinine. (After Reference 53.)

15.11. Conversion of the amide 15.11 to 2-amino-4-chloronicotinic acid (15.12) is accomplished via the Hofmann rearrangement on the amide. Diazotization of 15.12 and hydrolysis and treatment of the resulting pyridone with phosphorus pentachloride in phosphorus oxychloride yield 2,4-dichloronicotinic acid chloride (15.13). The acid chloride 15.13 is then converted, via the corresponding amide, to 2,4-dimethoxy-3-cyanopyridine (15.14) with sodium methoxide. Methylation of 15.14 with methyl iodide yields ricinine (15.1).

SELECTED READING

L. Marion, in The Alkaloids, Vol. 1, R. H. F. Manske (ed.). Academic, New York, 1950, pp. 126 ff.

L. Marion, in The Alkaloids, Vol. 6, R. H. F. Manske (ed.). Academic, New York, 1960, p. 206.

16

CHEMISTRY OF THE TOBACCO ALKALOIDS

Warning: The Surgeon General Has Determined That Cigarette Smoking Is Dangerous to Your Health

---

The tobacco alkaloids, although originally considered taxonomic indicators of Nicotiana species, have since been found to occur much more widely. Therefore, as might be expected, the total base content in Nicotiana (including nicotine, 16.1) varies considerably with the variety of tobacco, and numerous analytic methods have been developed which permit nicotine (16.1) assay. Generally, cultivation of tobacco appears to increase the alkaloid content (from which it can be argued that increased alkaloid content has insured survival of a particular variety), and innumerable hybrids have been created and breeding experiments carried out. Much information is available which suggests that some of the bases present in tobacco are generated specifically in the root system and translocated to the leaves where they accumulate. Other bases are synthesized in the leaves, and while many bases (besides ammonia) have been found in various Nicotiana, only three typical bases will be considered here. These are (-)-nicotine (16.1), (-)-anabasine (16.2), and (-)-anatabine (16.3).

H N N CH_3

16.1

H N N H

16.2

H N N H

16.3

I. NICOTINE

(-)-Nicotine (16.1), $C_{10}H_{14}N_2$, $[\alpha]_D^{20}$ -169.3°, is a toxic, colorless liquid, bp 124-125°C/18 torr, which rapidly turns yellow in air and is quite hygroscopic. Oxidation of (-)-nicotine (16.1) with chromic acid in sulfuric acid yields an amino acid (16.4) which forms pyridine (16.5) on decarboxylation (Equation 16.1). The amino acid (16.4) was named "nicotinic acid" and was identified as pyridine-β-carboxylic acid. The nature of the $C_5H_{10}N$ residue attached to the pyridine (16.5) nucleus at the β-position was established by a study of the bromination of (-)-nicotine (16.1) and confirmed in other degradative and synthetic studies.

The original work [54] which examined the bromination (Equation 16.2) of (-)-nicotine (16.1) in acetic acid indicated that, among other materials, a tractable compound, "dibromocotinine," $C_{10}H_{10}N_2OBr_2$, to which the structure 16.6 was assigned, could be isolated. Although this structure has since been revised to 16.7 [55], the reaction of dibromocotinine

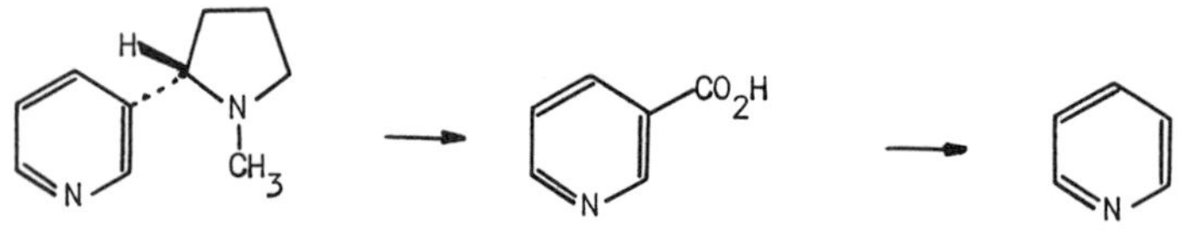

16.1 → 16.4 → 16.5 (Eq. 16.1)

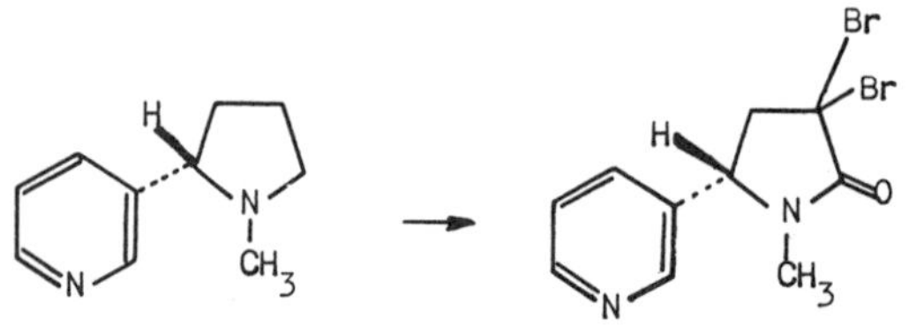

16.1 → 16.7 (Eq. 16.2)

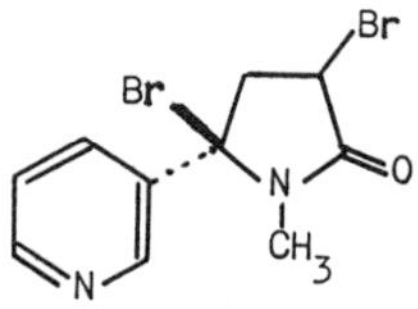

16.6

(16.7) with a mixture of sulfuric and sulfurous acids at 130-140°C to generate methylamine (16.8), oxalic acid (16.9), and methyl-β-pyridyl ketone (16.10) usefully permitted the structure 16.1 to be written for (-)-nicotine (Equation 16.3).

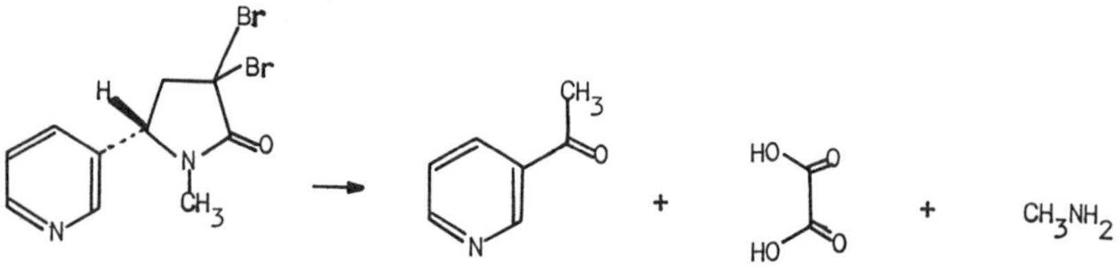

16.7 → 16.10 + 16.9 + CH_3NH_2 16.8 (Eq. 16.3)

In addition, (-)-nicotine (16.1) affords two isomeric monomethiodides. One of them, nicotine isomethiodide (16.11), which is formed by treating (-)-nicotine (16.1) with acid and then methyl iodide, is a quaternary salt whose hydroxide yields trigonelline (16.12) on oxidation with permanganate (Equation 16.4) and L-hygrinic acid (16.3; the absolute configuration

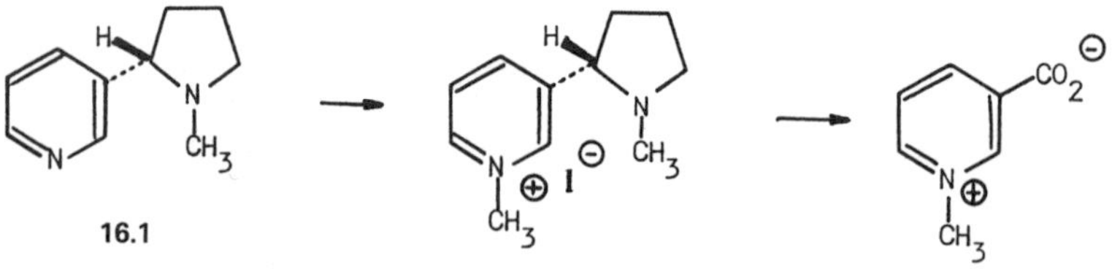

16.1 → 16.11 → 16.12 (Eq. 16.4)

of which is known) on oxidation with alkaline ferricyanide and then chromic acid (Equation 16.5). The second methiodide, nicotine methiodide, is therefore 16.14.

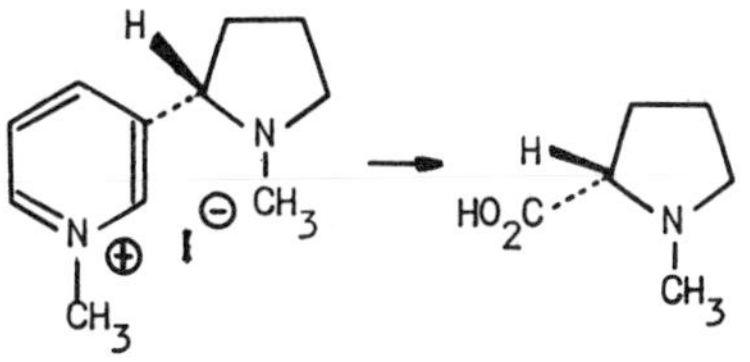

16.11 16.13 (Eq. 16.5)

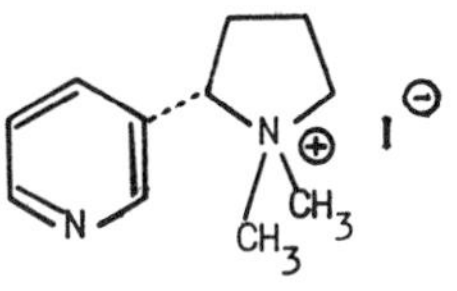

16.14

The structure of nicotine (16.1) has been confirmed by many syntheses, among which one of the first begins (Scheme 16.1; [56]) with the condensation reaction between ethyl nicotinate (16.15) and 1-methylpyrrolid-2-one (16.16) in the presence of sodium ethoxide. The product obtained is β-pyridyl-β'-(N'-methyl-α'-pyrrolidenyl)ketone (16.17). Treatment of the ketoamide (16.17) with hydrochloric acid at 130°C causes hydrolysis and decarboxylation, and this is followed by reduction with zinc in alcoholic sodium hydroxide to generate the amino alcohol 16.18. This secondary alcohol (16.18) is converted into the corresponding iodide with hydrogen iodide at 100°C, and cyclization to (±)-nicotine (16.1 and its antipode) occurs on alkali treatment. Finally, as pointed out earlier [45], a degradation for (-)-nicotine (16.1) has been developed which permits isolation of each carbon of the pyrrolidine ring.

The degradation, which is shown in Scheme 16.2, proceeds as follows: N-benzoylmetanicotine (16.19) is prepared by treatment of (-)-nicotine (16.1) with benzoyl chloride for 20 hr in refluxing xylene. Oxidation of N-benzoylmetanicotine (16.19) with sodium periodate and potassium permanganate generates N-benzoyl-N-methyl-β-alanine (16.20) and nicotinic acid (16.4), which yields pyridine (16.5) and $C_{2'}$ of the pyrrolidine ring as carbon dioxide on decarboxylation. Acid hydrolysis of N-benzoyl-N-methyl-β-alanine (16.20) followed by reductive formylation yields N,N-dimethylamino-β-alanine (16.21), which on conversion to the corresponding acid chloride with thionyl chloride and treatment with aluminum trichloride in benzene is converted into β-dimethylaminopropiophenone (16.22). Chromic acid oxidation of the propiophenone (16.22) generates benzoic acid ($C_{3'}$ of the pyrrolidine ring) and N,N-dimethylglycine (16.23), which on oxidation with lead tetraacetate in benzene at 55°C generates carbon dioxide ($C_{4'}$ of the pyrrolidine ring), formaldehyde ($C_{5'}$ of the pyrrolidine ring; isolated as its dimidone, 16.24, derivative), and N,N-dimethylamine (N-methyl of nicotine, 16.1; isolated as its p-bromobenzenesulfonamide, 16.25, derivative).*

*A possible analytic problem exists here. The oxidation of N,N-dimethylglycine (16.23; Scheme 16.2) with lead tetraacetate to generate carbon dioxide, formaldehyde, and N,N-dimethylamine requires, for meaningful utility in labeling experiments, that no formaldehyde come from the N,N-dimethylamine (either before or after fragmentation of N,N-dimethylglycine (16.23). It is not certain that this is the case [57].

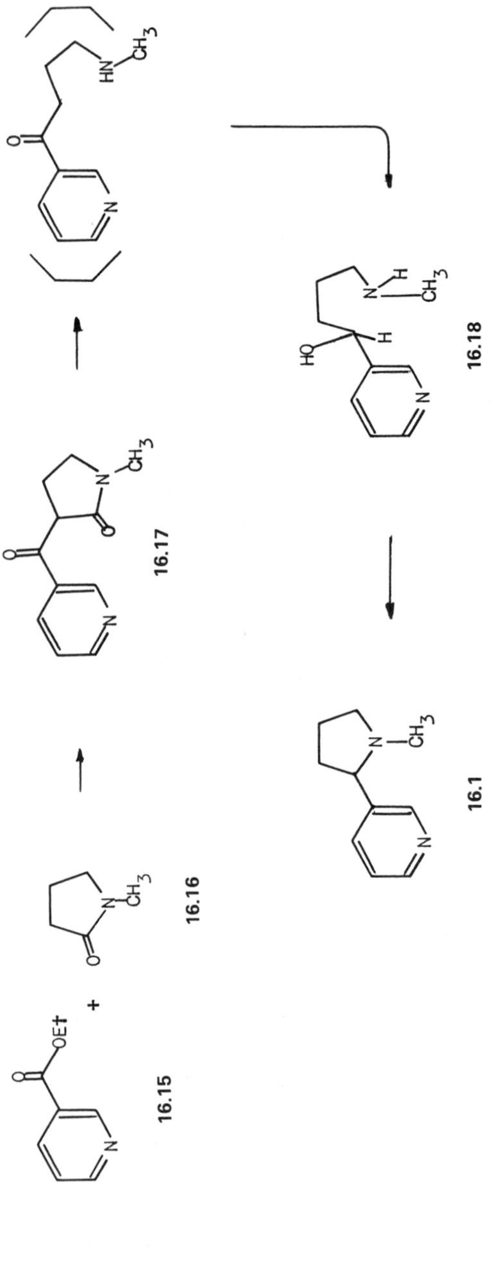

SCHEME 16.1 A synthesis of (±)-nicotine. (After Reference 56.)

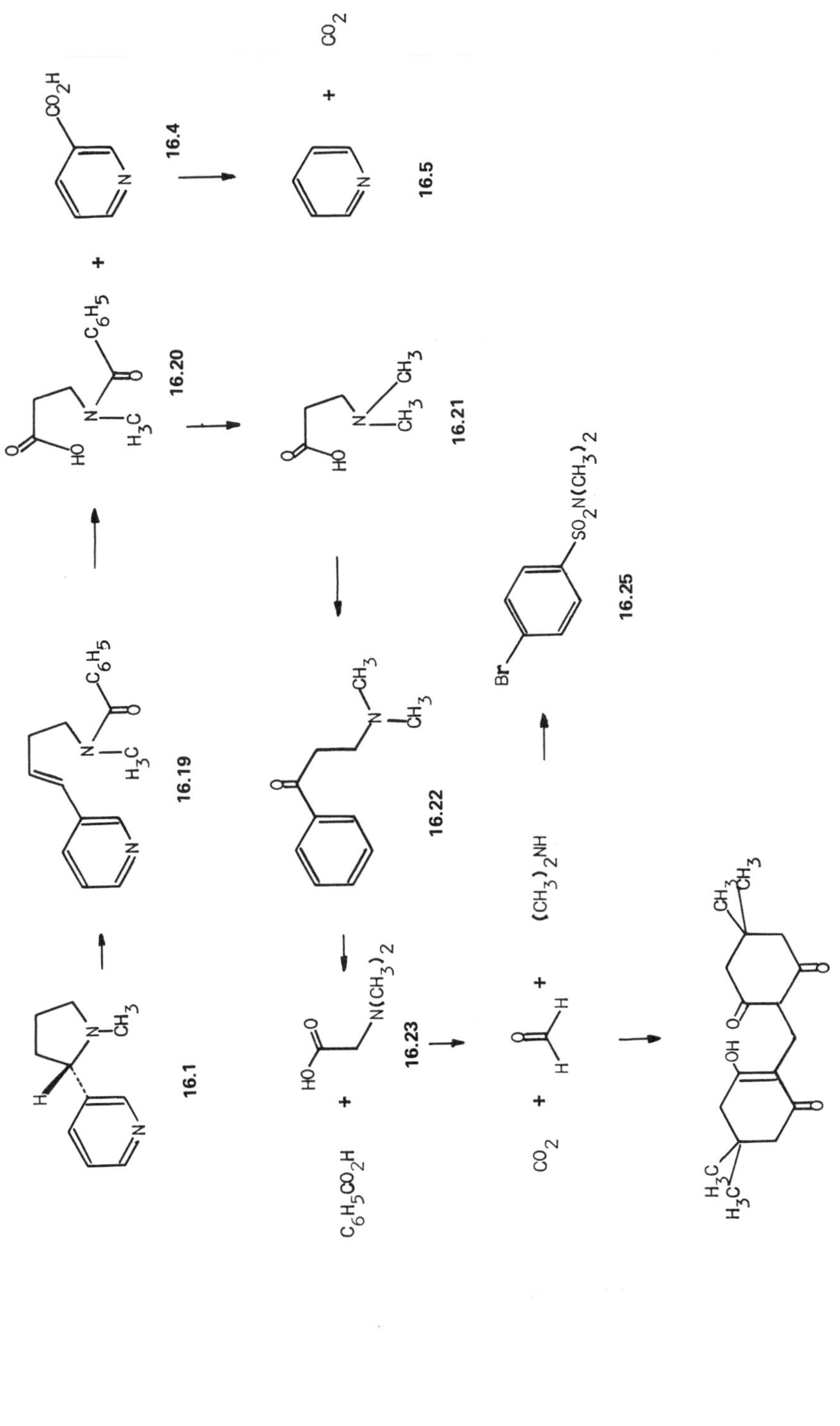

SCHEME 16.2 A degradation of (–)-nicotine. (After Reference 45.)

II. ANABASINE

(-)-Anabasine (16.2), $C_{10}H_{14}N_2$, supplants nicotine (16.1) as the major alkaloid in some species of *Nicotiana*,* and its large-scale isolation from *Nicotiana* and other genera has been extensively studied since (-)-anabasine (16.2) was at one time widely used as an insecticide. Many derivatives have been prepared from (-)-anabasine (16.2) with an eye to improving its pharmacologic action as an insecticide, but it has been replaced because of problems with toxicity toward users.

Since it was found quite early that on oxidation (-)-anabasine (16.1) generates nicotinic acid (16.4; Equation 16.6) and on catalytic dehydrogenation 3',2-dipyridyl (16.26; Equation 16.7) is formed, the conclusion was reached that (-)-anabasine (16.2) was 2-(3-pyridyl)piperidine. Further, when N-methylanabasine methiodide (16.27) is oxidized with potassium

16.2 → 16.4 (Eq. 16.6)

16.2 → 16.26 (Eq. 16.7)

ferricyanide, N,N-dimethylanabasone (16.28) is produced, and oxidation of the latter with chromic anhydride generates (-)-N-methylpipecolic acid (16.29; Equation 16.8); the absolute configuration of which is known. Thus, (-)-anabasine is (2S)-2-(3-pyridyl)piperidine (16.2).

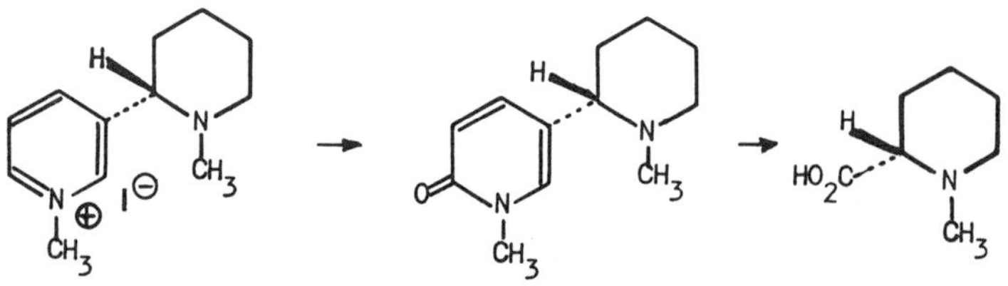

16.27 → 16.28 → 16.29 (Eq. 16.8)

The synthesis of (±)-anabasine (16.2; Scheme 16.3; [58]) is directly analogous to the synthesis of (±)-nicotine (16.1) already outlined above (p. 168), save that ethyl nicotinate (16.15) is condensed with N-benzoyl-2-piperidone (16.30) [instead of N-methylpyrrolid-2-one, 16.16, as was the case with nicotine, 16.1 (see Section I)] to yield the pyridyl ketone 16.31, which on hydrolysis (hydrogen chloride in a sealed tube) yields the dehydroanabasine (16.33). This compound (16.33) was originally called "anabaseine" and formulated as 16.34 [55]. Catalytic hydrogenation of 16.33 yields (±)-anabasine (16.2 and its mirror image).

---

*Racemic anabasine (16.2 and its mirror image) appears to be the major alkaloid in *Nicotiana glauca* [477].

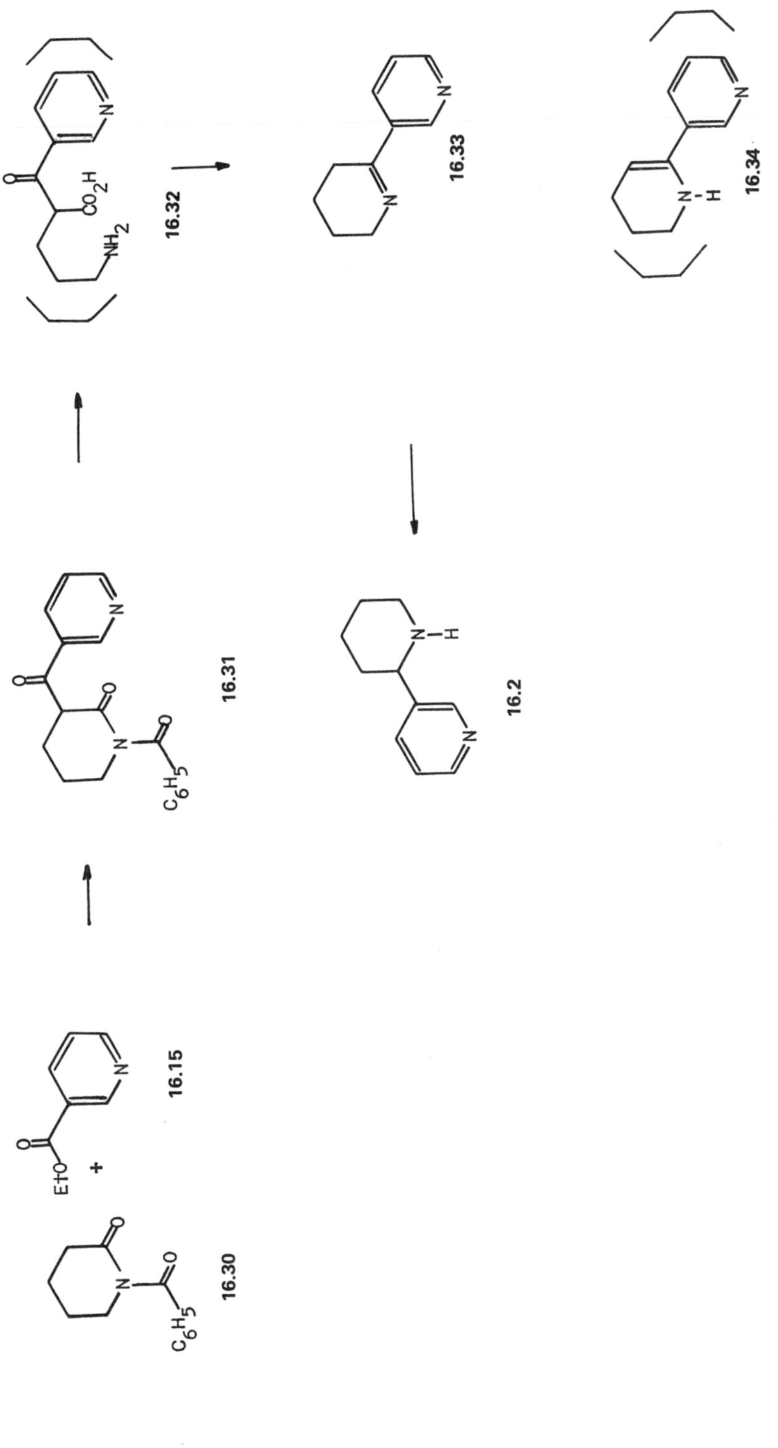

SCHEME 16.3 The synthesis of (±)-anabasine. (After Reference 58.)

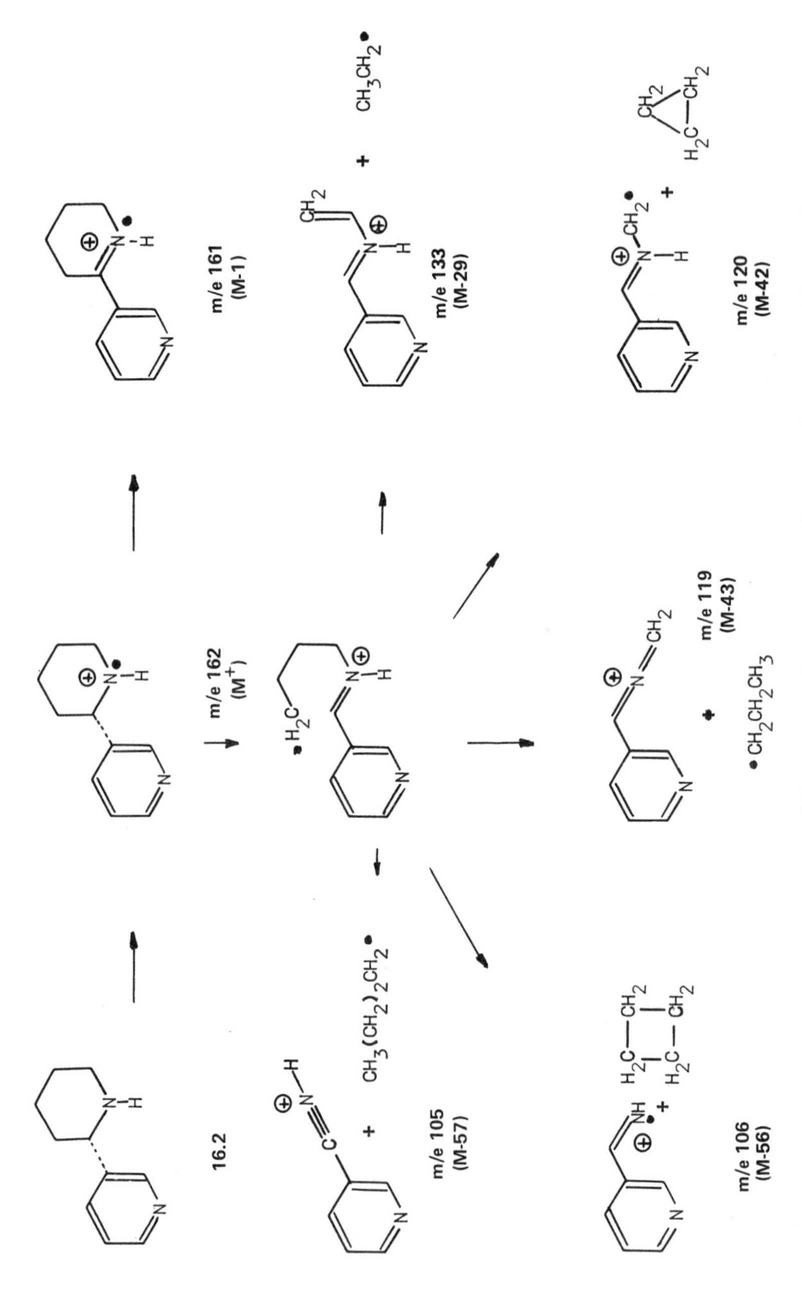

SCHEME 16.4 The mass spectral fragmentation pattern of (±)-anabasine. (After Reference 55.)

The mass spectra of (±)-anabasine (16.2) and several deuterated analogs have been examined [55] in some detail. The base peak in the spectrum of (±)-anabasine (16.2) occurs at m/e 84 (M-78) and corresponds to loss of the pyridine ring through α-cleavage in the molecular ion (16.35) to yield the fragment 16.36 (Equation 16.9). The deuterated analogs, (±)-anabasine-N-d (16.2) and (±)-anabasine-2-d_1 (16.2; prepared by deuteration of 16.33) are in

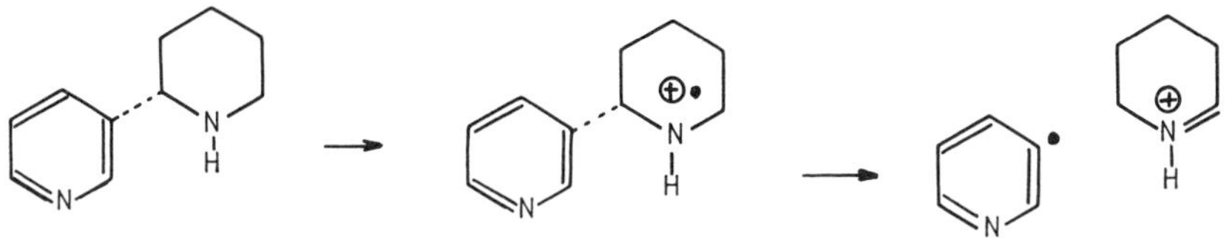

16.2 → 16.35 m/e 162 → 16.36 m/e 84 (Eq. 16.9)

concert with this fragmentation since on electron impact the m/e originally found for (±)-anabasine (16.2) is quantitatively displaced to m/e 85 for each of the monodeuterated derivatives. Further fragmentation patterns which emerged and can be recognized as being viable on the basis of deuterium labeling (at least to the extent of 40-50% or more of the ion arising from the fragment shown) are presented in Scheme 16.4.

III. ANATABINE

(-)-Anatabine (16.3) was originally isolated from the lower boiling fractions of crude nicotine (16.1) preparations as a colorless oil, $C_{10}H_{12}N_2$, bp 145-146°C/10 torr, $[\alpha]_D^{17}$ -177.8° (neat). It was quickly recognized as a secondary base (benzoylation to 16.37) and, in fact, as a "dehydroanabasine," since (-)-anabasine (16.2) is formed on hydrogenation (1 equivalent of hydrogen being consumed) of (-)-anatabine (16.3). Since (-)-anatabine (16.3) has none of the properties of an imine or an enamine, only 16.38 and 16.3 are available structural representations, and the former is excluded since mild oxidation of benzoyl anatabine (16.37) generates nicotinic acid (16.4) and hippuric acid (16.39; Equation 16.10).

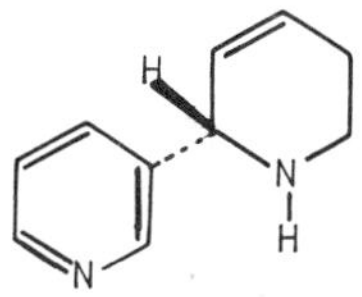

16.38

An interesting synthesis of (±)-anatabine (16.3) has been recorded [59] and is shown in Scheme 16.5. Here, the biscarbamate of 3-formylpyridine (16.40), when heated with butadiene (16.41) in acetic acid - boron trifluoride solution, yields the N-carboethoxy derivative of (±)-anatabine (16.42), which on hydrolysis affords (±)-anatabine (16.3 and its mirror image).

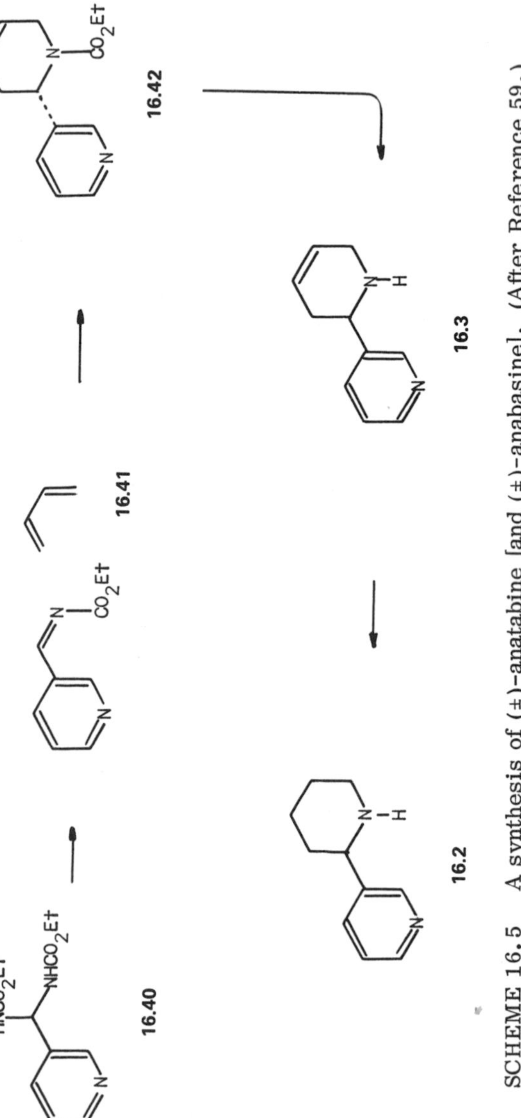

SCHEME 16.5 A synthesis of (±)-anatabine [and (±)-anabasine]. (After Reference 59.)

16.3 → 16.37 → 16.39 + 16.4

(Eq. 16.10)

SELECTED READING

A. A. Ayer and T. E. Habgood, in The Alkaloids, Vol. 11, R. H. F. Manske (ed.). Academic, New York, 1968, pp. 477 ff.

L. Marion, in The Alkaloids, Vol. 1, R. H. F. Manske (ed.). Academic, New York, 1950, pp. 128 ff.

L. Marion, in The Alkaloids, Vol. 6, R. H. F. Manske (ed.). Academic, New York, 1960, pp. 228 ff.

Part 5

ALKALOIDS DERIVED FROM TYROSINE

As noted earlier, chorismic acid (17.1) serves as the precursor to prephenic acid (17.2; Chapter 2, Section I.B) and anthranilic acid (17.3; Chapter 2, Section II.E). Anthranilic acid (17.3) gives rise to tryptophan (17.4) and alkaloids derived from tryptophan (see Part 6). Contrary to what is true for mammals, the conversion of phenylalanine (17.5) into tyrosine (17.6), in plants, seems to be of little importance (but see [60, 61]). Tyrosine (17.6) may be converted to its more highly hydroxylated homologs (i.e., dihydroxy- and trihydroxyphenylalanine, 17.7 and 17.8, respectively; see Part 1; [1]) wherein the second hydroxyl group (e.g., in 17.7) enters ortho to the one already present, and in 17.8, the third hydroxyl also enters ortho to one of the two already present.* Phenylalanine (17.5), on the other hand, does not commonly [60, 61] undergo hydroxylation. Instead, phenylalanine (17.5) is converted to cinnamic acid (17.9), the parent of the nonnitrogenous phenylpropanoids, and then cinnamic acid (17.9) undergoes hydroxylation to derivatives which are incorporated into the alkaloids. The analogous deamination of tyrosine (17.6) is not common, although in some systems the phenylalanine lyase enzyme which converts phenylalanine (17.5) into cinnamic acid (17.9) has been shown to catalyze the corresponding deamination of tyrosine (17.6; [60]).

Therefore, as we shall see, tyrosine (17.6) and its higher hydroxylated derivatives serve as the nitrogen-carrying C_6-C_2 or C_6-C_3 portion of a wide variety of alkaloids. The nonnitrogenous portion of these systems comes from tyrosine (17.6) or phenylalanine (17.5; via, for the latter, cinnamic acid, 17.9) for C_6-C_1, C_6-C_2, and C_6-C_3 bases or other readily available precursors, e.g., formate (17.10) or glyoxylate (17.11) for C_1; pyruvate (17.12) for C_2; mevalonate (17.13) for C_5; and loganin (17.14) for C_9 and C_{10} alkaloids.

Finally, many alkaloids to be encountered in this part have O-methyl, N-methyl and/or methylenedioxy groups. These are almost invariably derived, as has been noted (p. 38) from S-adenosylmethionine (17.15) by methyl transfer.

---

*It appears that if hydroxylation occurs prior to elaboration of the amino acid into an alkaloid, the third hydroxyl also enters ortho to the first (i.e., a 3,4,5-trihydroxy unit is created), while if hydroxylation is a late-stage process, the particular ortho site is less certain.

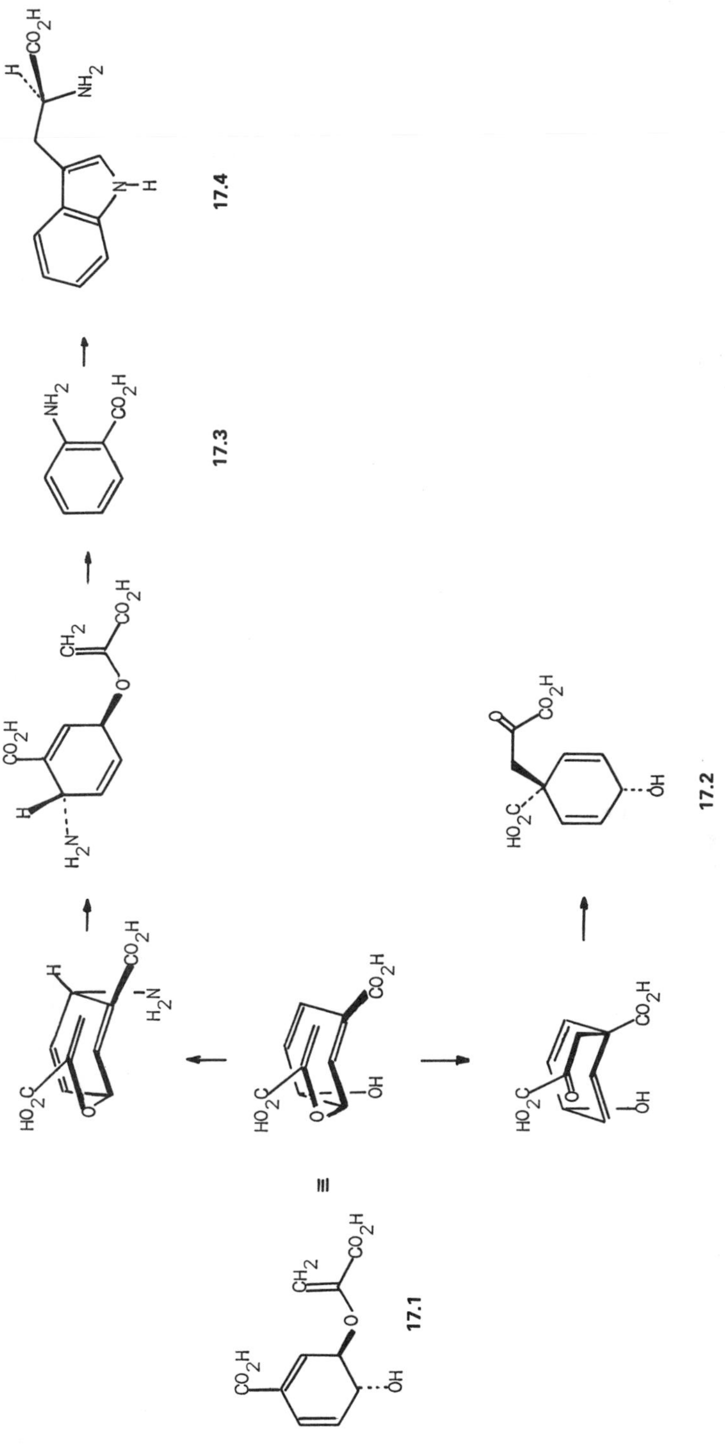

17.1

17.2

17.3

17.4

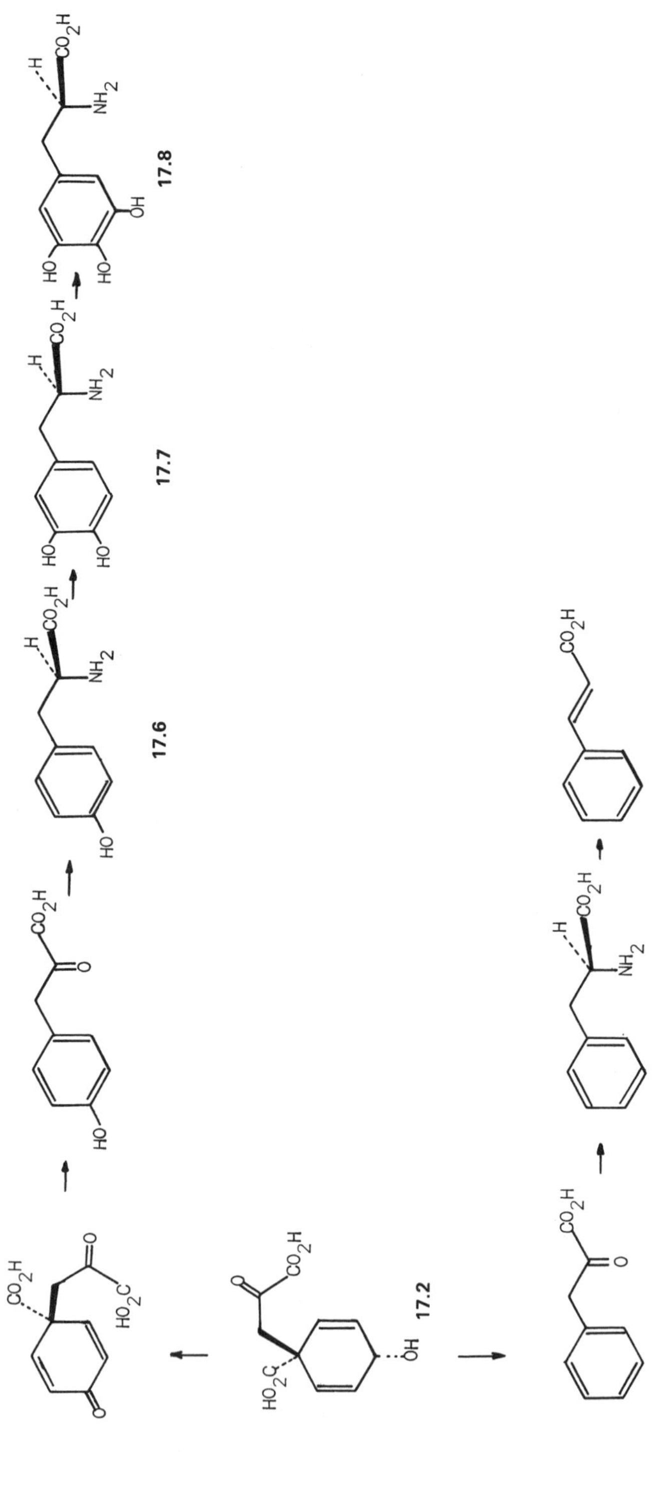
17.2
17.5
17.9
17.6
17.7
17.8
HO
OH
NH2
CO2H
H
HO2C
O

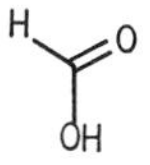

17.10

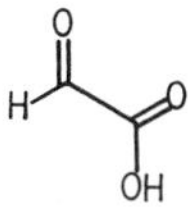

17.11

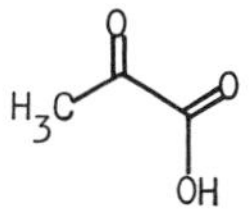

17.12

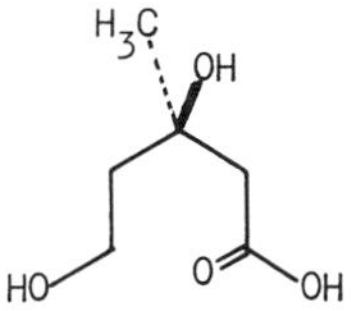

17.13

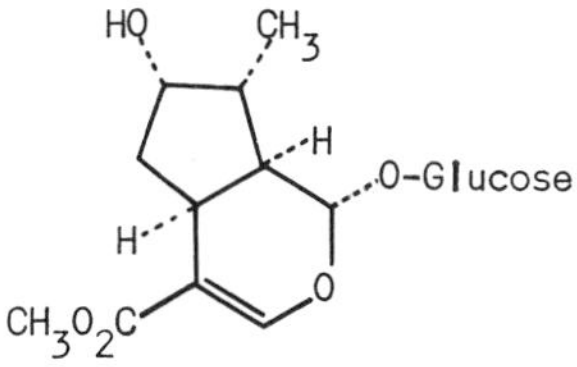

17.14

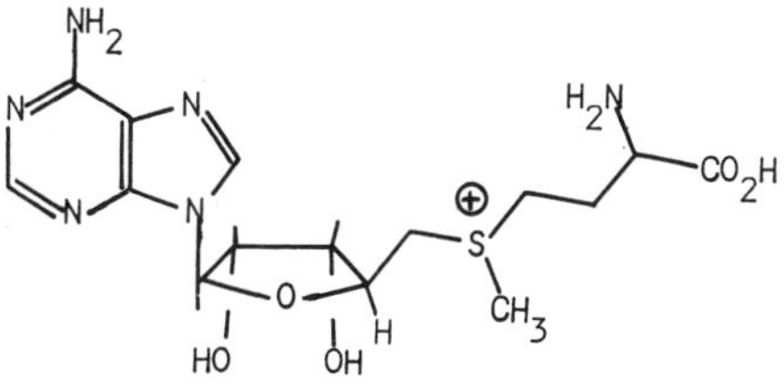

17.15

17

BIOSYNTHESIS OF β-PHENETHYLAMINE, SIMPLE TETRAHYDROISOQUINOLINE, AND 1-ALKYLTETRA-HYDROISOQUINOLINE ALKALOIDS

All that we see or seem
Is but a dream within
a dream.

Edgar Allan Poe (1827)

There is a relatively large number of alkaloids (about seventy) which fall into the various groups to be considered in this chapter. Typical of these are: (a) the simple phenethylamine, mescaline (17.16), from the small wooly "peyotyl" cactus Lophophora williamsii (Lemaire) Coult.; (b) anhalamine (17.17), anhalonidine (17.18), and lophocerine (17.19) from other Cactaceae; and (c) the important antamebic alkaloids (-)-protoemetine (17.20), (-)-ipecoside (17.21), and (-)-emetine (17.22), from the South American straggling bush Cephaelis ipecacuanha (Brotero) Rich. In each case, as we shall see, tyrosine (17.6) provides a C_6-C_2 portion while the remainder of the alkaloid is derived from "normal" plant constituents which might be expected to react with available amine.

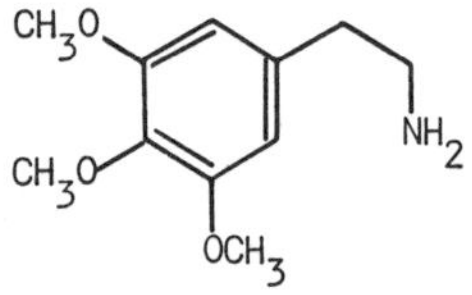

17.16 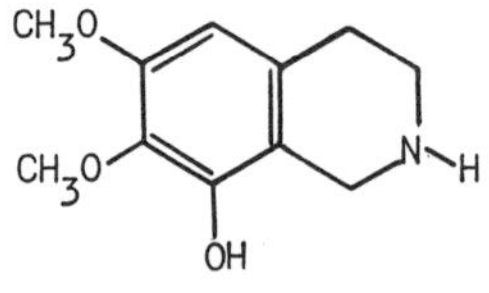 **17.17** 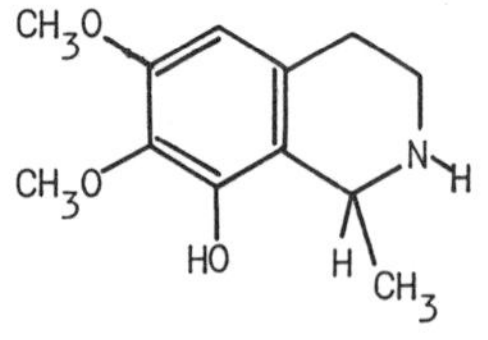 **17.18**

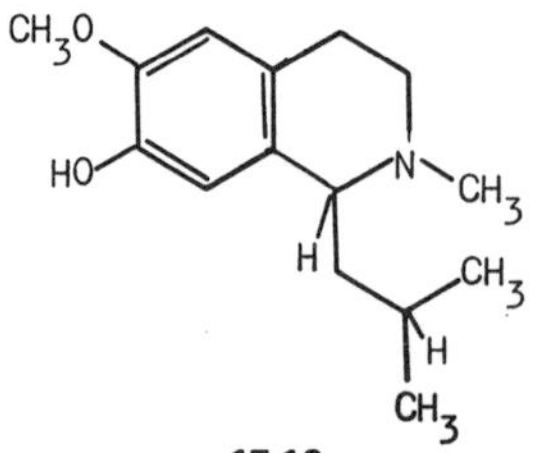

17.19

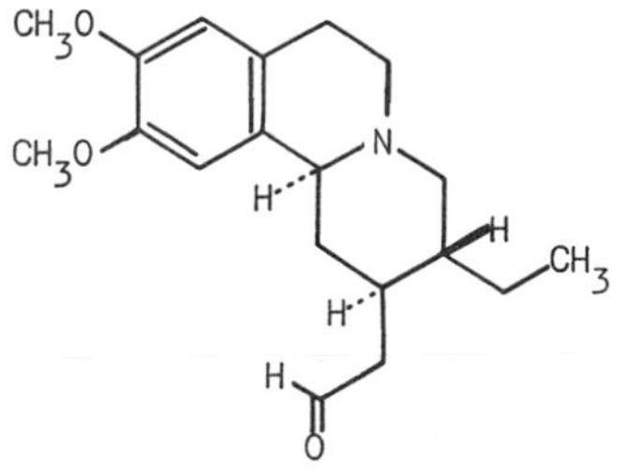

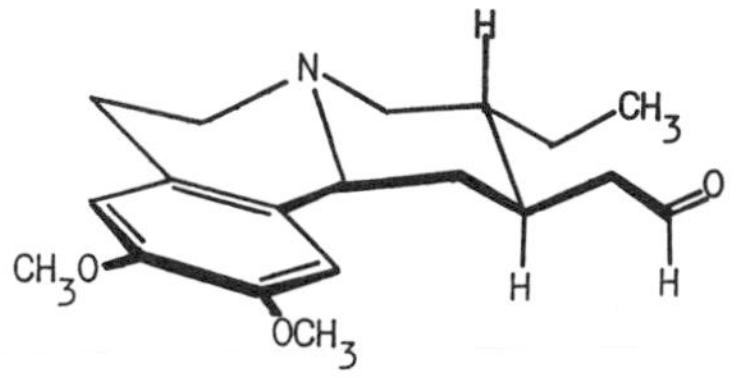

17.20

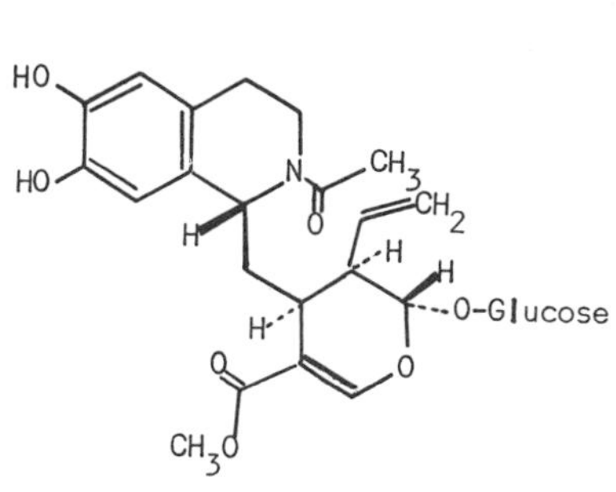

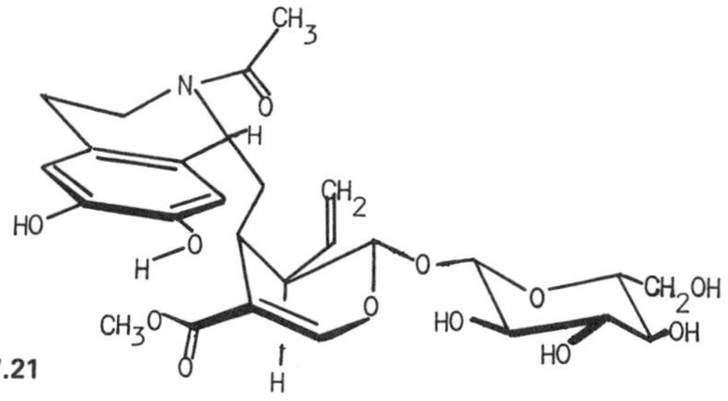

17.21

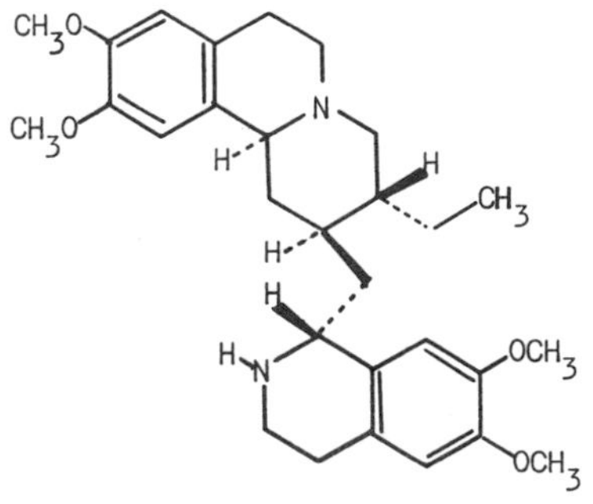

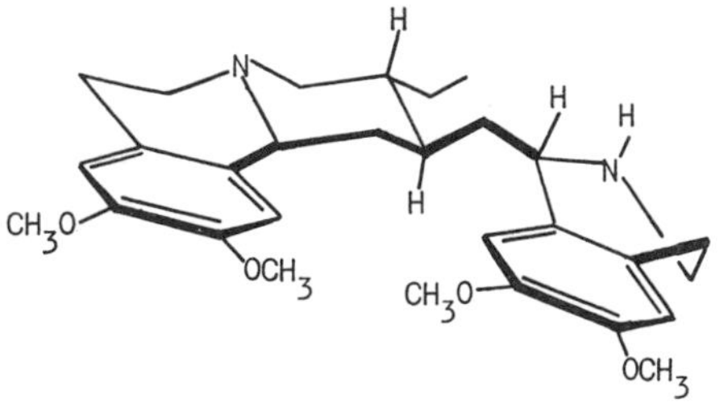

17.22

I. MESCALINE, ANHALAMINE, ANHALONIDINE, AND LOPHOCERINE

The biosynthesis of mescaline (17.16) and the simple tetrahydroisoquinolines anhalamine (17.17), anhalonidine (17.18), and lophocerine (17.19) from the Cactaceae has been examined in some detail. For all these compounds, it has been shown that tyrosine (17.6) serves as a specific precursor. Additionally, since 3,4-dihydroxyphenethylamine (17.23) is also incorporated well, it is assumed that the pathway from tyrosine (17.6) to the alkaloids involves hydroxylation of tyrosine (17.6) to 3,4-dihydroxyphenylalanine (17.7) and then decarboxylation to 17.23. [Alternatively, the decarboxylation might precede hydroxylation (Equation 17.1).] The decarboxylation of tyrosine (17.6) or 3,4-dihydroxyphenylalanine (17.7) can be formulated as occurring via a pyridoxyl - amino acid intermediate (e.g., 17.24, Scheme 17.1) as is usual, although specific evidence for this process here is lacking.

If O-methylation (via S-adenosylmethionine, 17.15) of 3,4-dihydroxyphenethylamine (17.23) occurs to yield 4-hydroxy-3-methoxyphenethylamine (17.25), the hydroxyl groups have been differentiated, and further hydroxylation ortho to the single hydroxyl originally

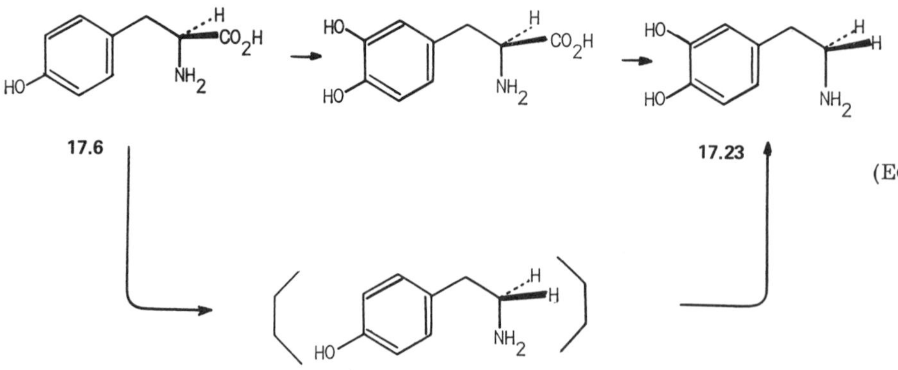

(Eq. 17.1)

present can be effected (generating 3,4-dihydroxy-5-methoxyphenethylamine, 17.26). Further O-methylation (again using S-adenosylmethionine, 17.15) can then generate either 3,5-dimethoxy-4-hydroxyphenethylamine (17.27) or 3,4-dimethoxy-5-hydroxyphenethylamine (17.28; Scheme 17.2).

Lophophora williamsii (Lemaire) Coult. has yielded an O-methyltransferase [62] which utilizes S-adenosylmethionine (17.15) to convert hydroxyphenethylamines to their O-methyl ethers. Using the O-methyltransferase, it has been found that 3,4-dihydroxyphenethylamine (17.23) is converted to 4-hydroxy-3-methoxyphenethylamine (17.25) exclusively. (No 3-hydroxy-4-methoxyphenethylamine, 17.29, is formed.) Second, the O-methyltransferase catalyzes methylation of 3,4-dihydroxy-5-methoxyphenethylamine (17.26) to a 50:50 mixture of 3,5-dimethoxy-4-hydroxyphenethylamine (17.27) and 3,4-dimethoxy-5-hydroxyphenethylamine (17.28). Finally, the same O-methyltransferase catalyzes methylation of 3,5-dimethoxy-4-hydroxyphenethylamine (17.27) exclusively (not 3,4-dimethoxy-5-hydroxyphenethylamine, 17.28) to yield mescaline (17.16; Scheme 17.2).

Since the tetrahydroisoquinolines anhalamine (17.17) and anhalonidine (17.18) possess the 3,4-dimethoxy-5-hydroxy pattern, it is presumed that they arise from 3,4-dimethoxy-5-hydroxyphenethylamine (17.27) which could not go on to mescaline (17.16).

Anhalamine (17.17) was thought, until recently, to arise by interaction of a one-carbon unit such as formate (17.10) with the 3,4-dimethoxy-5-hydroxyphenethylamine (17.27) to generate the formamide 17.30 which underwent cyclization, elimination to the imine 17.31, and reduction to anhalamine (17.17; Scheme 17.3). It is now believed [63] that a two-carbon fragment, e.g., glyoxylate (17.11), is more likely. Thus, reaction of 3,4-dimethoxy-5-hydroxyphenethylamine (17.27) with glyoxylate (17.11) would be expected to generate the imine (17.32) before cyclization to 17.33. Decarboxylation of 17.33 reductively or oxidatively followed by an overall reduction then yields anhalamine (17.17; Scheme 17.4), and, indeed, feeding of labeled peyoxylic acid (17.33), a constituent of peyote, lent credence to this suggestion. The pathway for decarboxylation of peyoxylic acid (17.33) remains uncertain, although it is likely that it is oxidative, rather than reductive. Thus, two possibilities are shown in Scheme 17.4 for this process.

The first of these involves generation of a suitable leaving group ("L") attached to nitrogen. It is presumed that the group L is electron deficient prior to attachment to nitrogen and electron rich as it leaves during the decarboxylation process to generate the ultimate anhalamine (17.17) precursor, the imine 17.31. Although it is apparently without precedent, there appears no reason that this group could not involve a pyridoxyl enzyme (Scheme 17.5) or conversion to the N-oxide, phosphorylation on the oxygen just introduced, and then elimination. Alternatively, the conversion of NAD^+ to NADH could be used to accomplish the same purpose, and this conversion would occur as the CO_2 left. The second possibility involves the generation of 17.31 from peyoxylic acid (17.33) by oxidation to the diradical 17.34 and

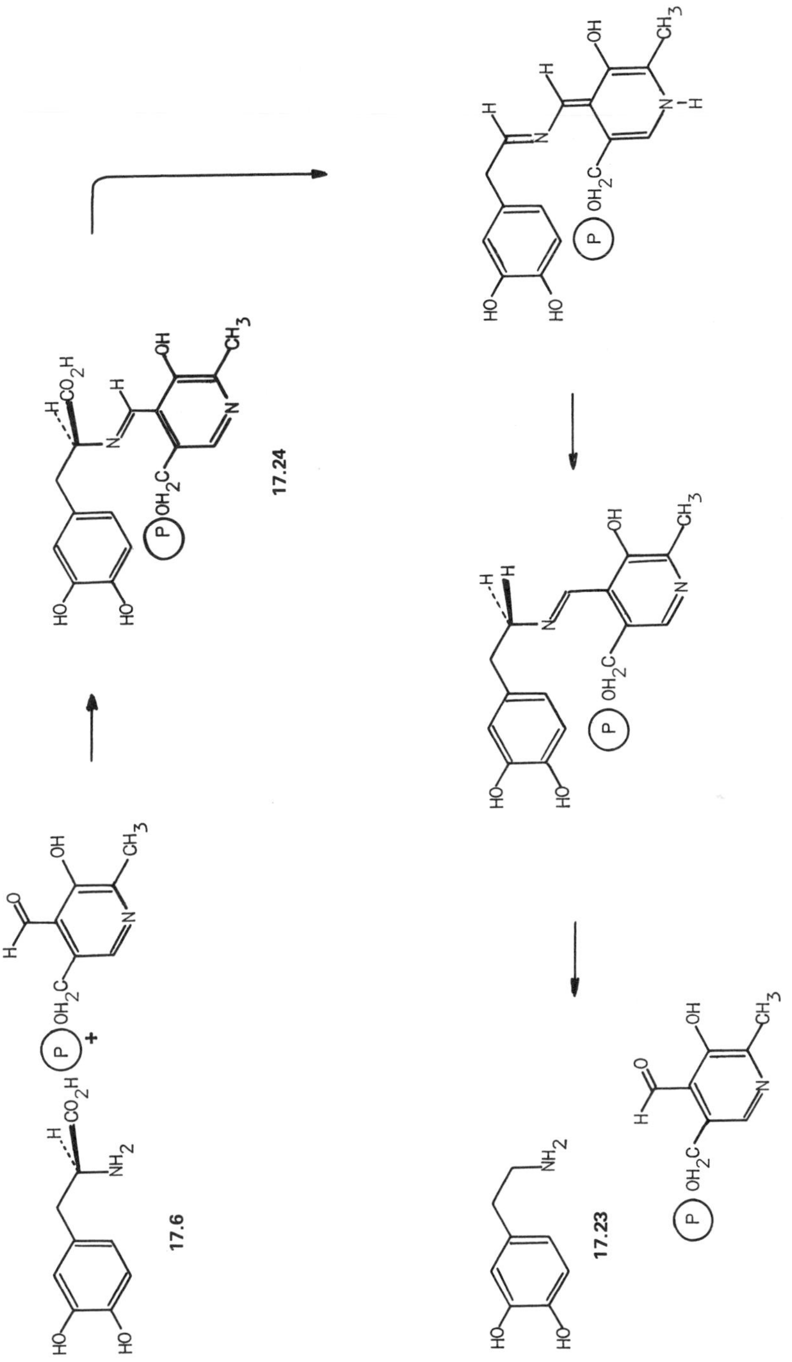

17.6

17.24

17.23

SCHEME 17.1 A possible pathway for the formation of 3, 4-dihydroxyphenethylamine from tyrosine.

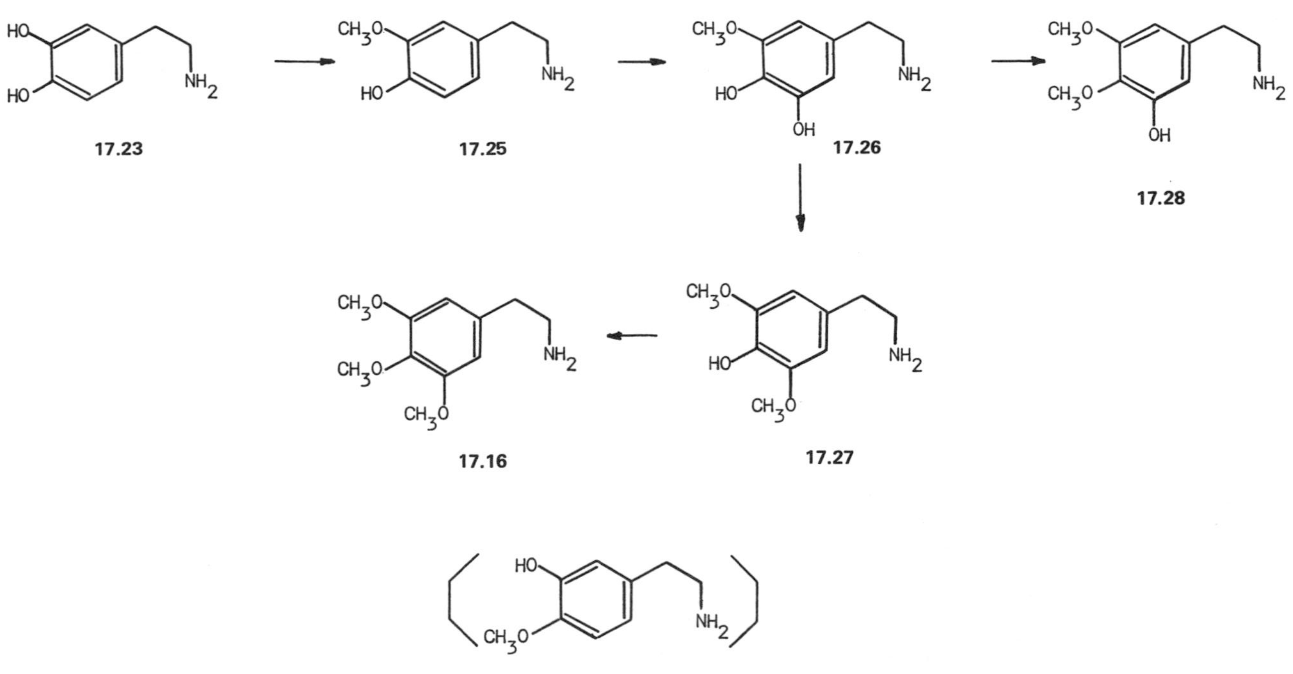

SCHEME 17.2 A pathway for the formation of mescaline in concert with O-methyltransferase results.

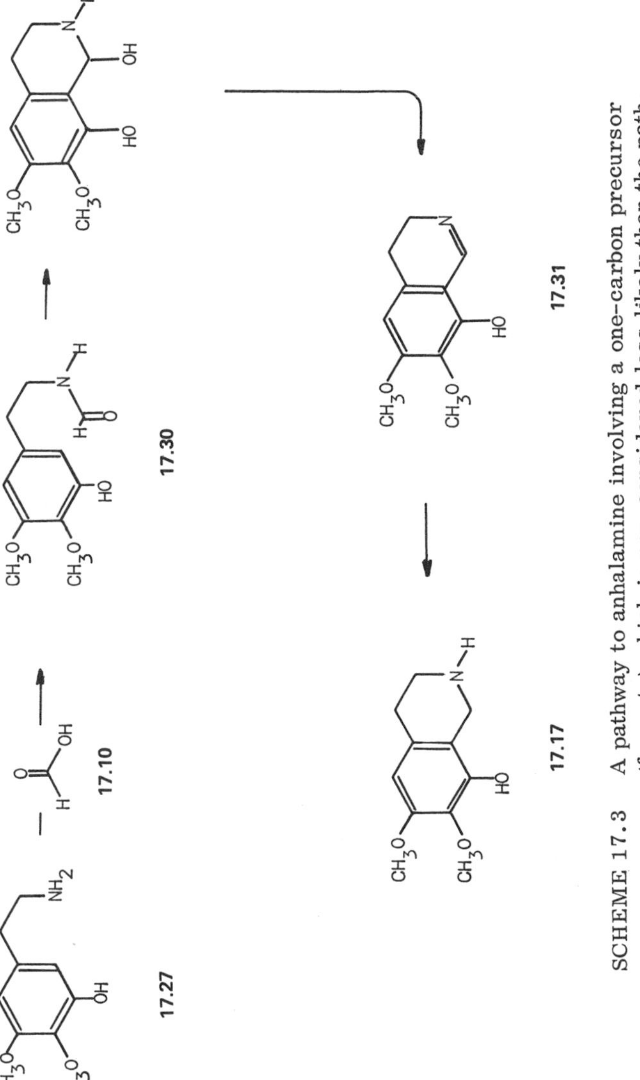

SCHEME 17.3 A pathway to anhalamine involving a one-carbon precursor (formate) which is now considered less likely than the pathway (Scheme 17.4) utilizing a two-carbon (glyoxylate) precursor.

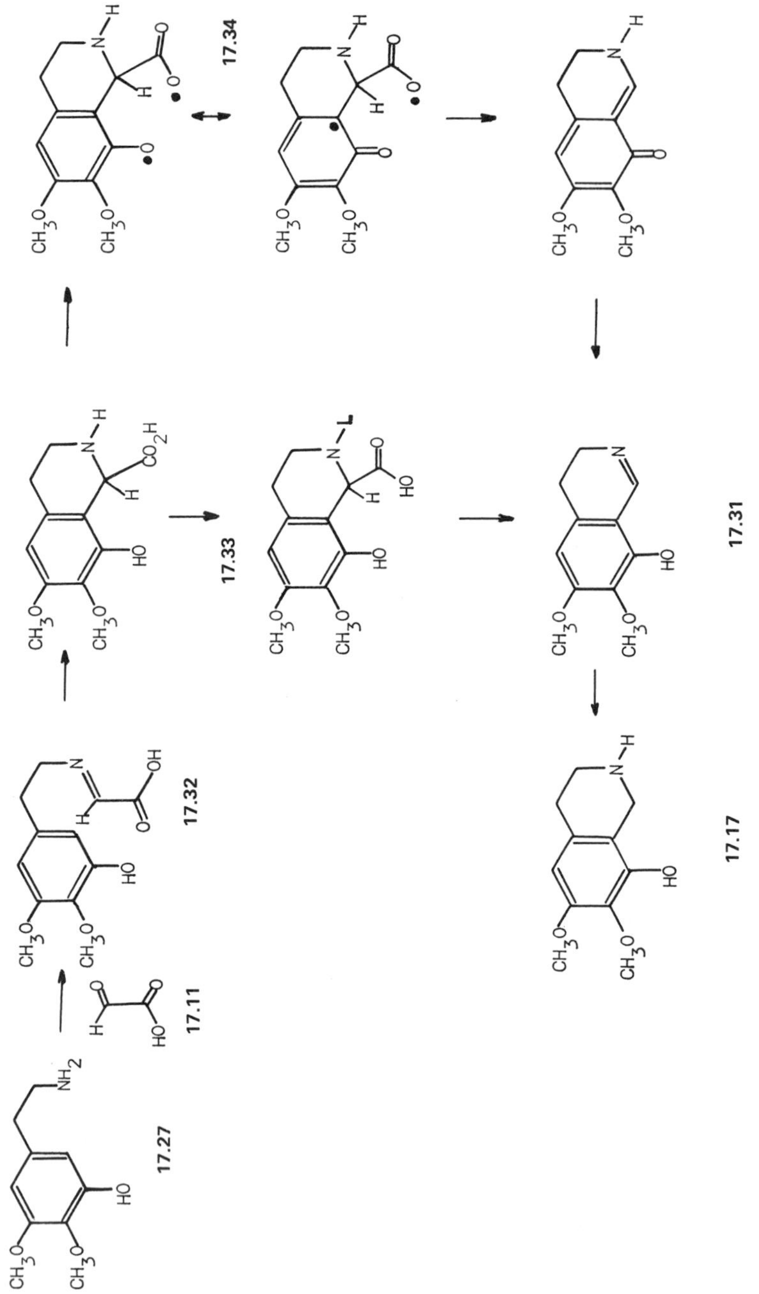

17.27 17.11 17.32 17.33 17.34 17.17 17.31

SCHEME 17.4 A pathway to anhalamine involving glyoxylic acid. Two routes are shown for the decarboxylation of peyoxylic acid.

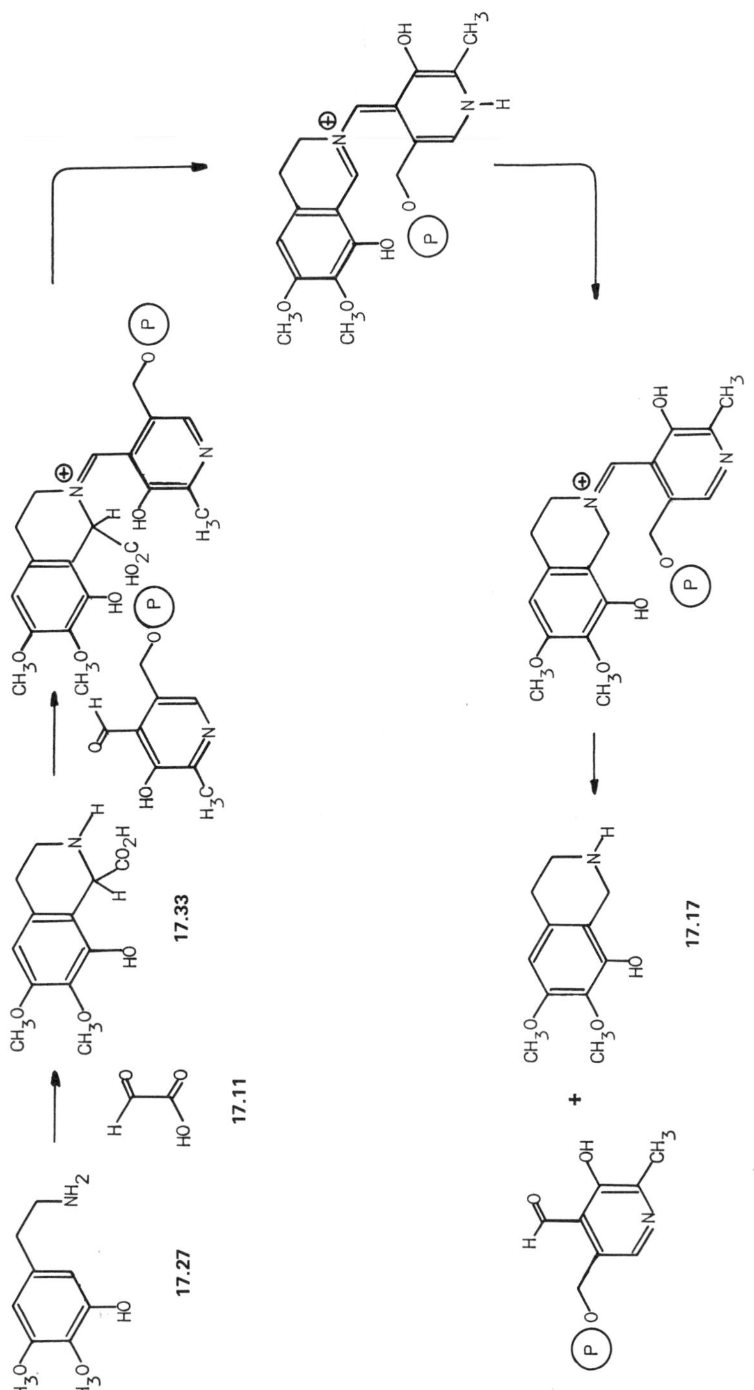

17.27

17.11

17.33

17.17

SCHEME 17.5 A possible (without precedent) decarboxylation route from peyoxylic acid to anhalamine.

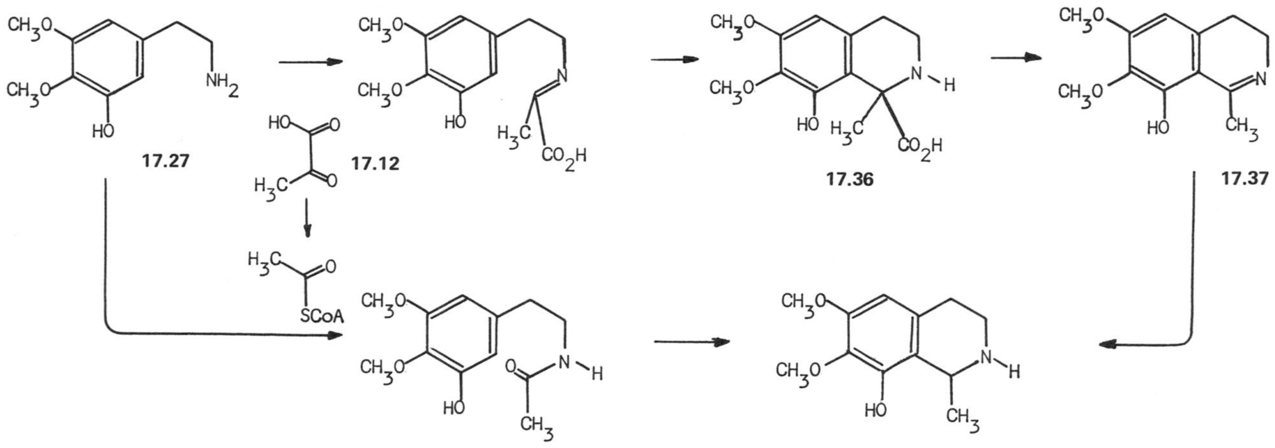

17.27

17.12

17.35

17.36

17.37

17.18

SCHEME 17.6 Several possible pathways to anhalonidine from 3, 4-dimethoxy-5-hydroxyphenethylamine. The pathway via peyoruvic acid appears preferred and is in concert with labeling experiments.

subsequent decarboxylation of the diradical. The generation of the radical ortho to the only unsubstituted oxygen is presumptive (but anthropomorphic) evidence for the reasonable nature of the pathway.

The route to anhalonidine (17.18;* Scheme 17.6) has been extensively tested by feeding experiments with presumed metabolic intermediates lying between tyrosine (17.6) and the final product (17.18). Although the usual cautionary words on the obligatory nature of the intermediates must be voiced, feeding of the amide 17.35, which had been doubly labeled, indicated that while radioactive anhalonidine (17.18) is generated, the distribution of label is not in accord with its (i.e., 17.35) intact incorporation. However, when doubly labeled peyoruvic acid (17.36), a normal constituent of peyote, is fed, efficient nonrandom incorporation into anhalonidine (17.18) occurs. The imine 17.37, analogous to the postulated imine 17.31, has also been isolated from in vivo experiments, and the decarboxylation of peyoruvic acid (17.36) to generate this imine (17.37) could, presumably, proceed via either of the pathways already proposed for anhalamine (17.17; see above).

Lophocerine (17.19)† is derived from tyrosine (17.6) which has undergone decarboxylation, hydroxylation, and presumably methylation to 4-hydroxy-3-methoxyphenethylamine (17.25). The methylation is accomplished using S-adenosylmethionine (17.15) and may be late stage rather than early (as noted above), accounting for the presence of only two hydroxyl groups rather than three, as found in mescaline (17.16), anhalamine (17.17), and anhalonidine (17.18), alkaloids obtained from the same plant. Intermediate feeding experiments which might clarify this point are lacking. Thus, either 4-hydroxy-3-methoxyphenethylamine (17.25) or 3,4-dihydroxyphenethylamine (17.23) serves as a portion of lophocerine (17.19). The remaining five carbons of lophocerine (17.19) can be derived both from mevalonic acid (17.13) and leucine (17.38).

Feeding of doubly labeled [3',4-^{14}C]mevalonic acid (17.13; [64]) to Pachycereus marginatus (DC) Britt. et Rose generates doubly labeled lophocerine (17.19). The labeling can be accounted for by assuming oxidation of mevalonic acid to the corresponding aldehyde (17.39), condensation of the aldehyde (17.39) with 3,4-dihydroxyphenethylamine (17.23), cyclization to 17.40, decarboxylation with loss of water to 17.41, and finally, reduction to lophocerine (17.19; Scheme 17.7).

Interestingly, (±)-[2-^{14}C]leucine (17.38) also serves as a specific precursor to lophocerine (17.19; Scheme 17.8). By analogy to anhalamine (17.17) and anhalonidine (17.18), it may be assumed that leucine (17.38) undergoes transamination to 4-methyl-α-ketopentanoic acid (17.42), which then condenses with the 3,4-dihydroxyphenethylamine (17.23) and undergoes decarboxylation. If this is correct, and if the decarboxylation is oxidative, it implies (by analogy to anhalamine, 17.17, and anhalodine, 17.18) that either methylation occurs at a late stage, permitting use of a para radical in the decarboxylation step or that the picture utilizing a leaving group (L) on nitrogen (or the reduction of NAD^+ to NADH) is correct, and methylation is early or late.

---

*In principle, of course, anhalonidine (17.18) must be a mixture of epimers or one chiral form. The naturally occurring form appears to be racemic or racemization may have occurred during isolation.

†As pointed out for anhalonidine (17.18), lophocerine (17.19) must also be a mixture of epimers or one chiral isomer. The naturally occurring form here, too, appears to be racemic or racemization may have occurred during isolation.

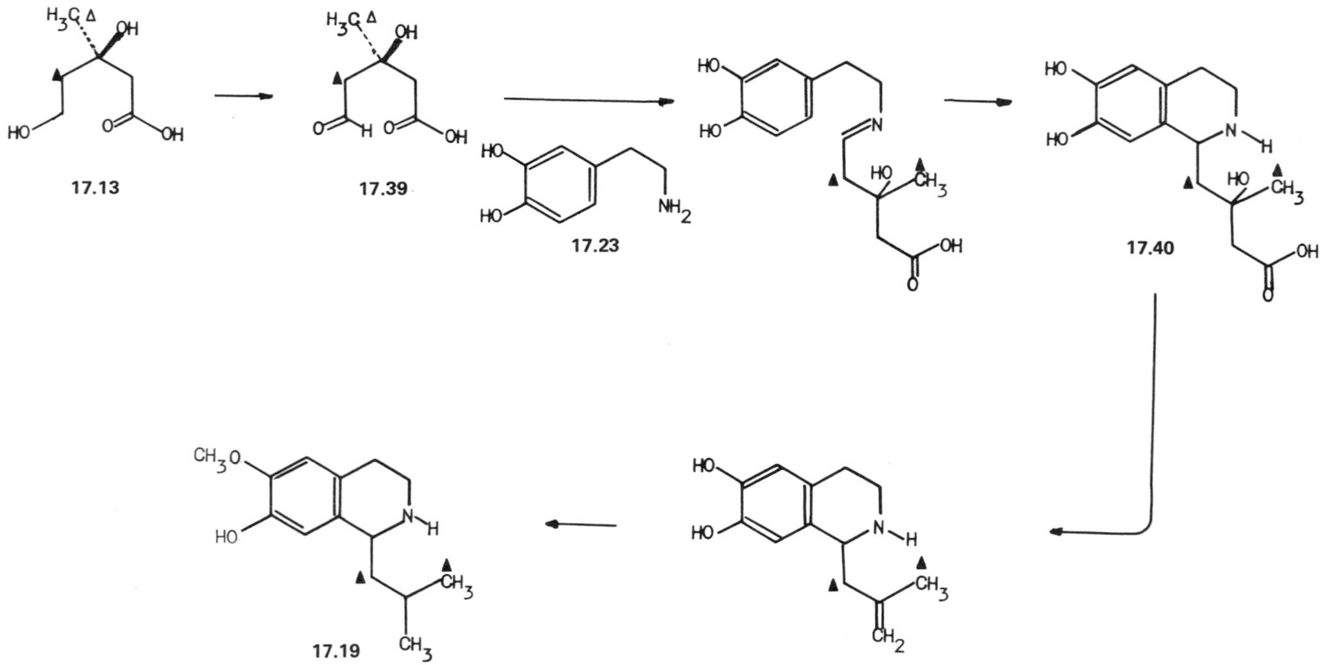

SCHEME 17.7 A potential pathway for the biosynthesis of lophocerine from 3, 4-dihydroxyphenethylamine and mevalonic acid (▲ = label).

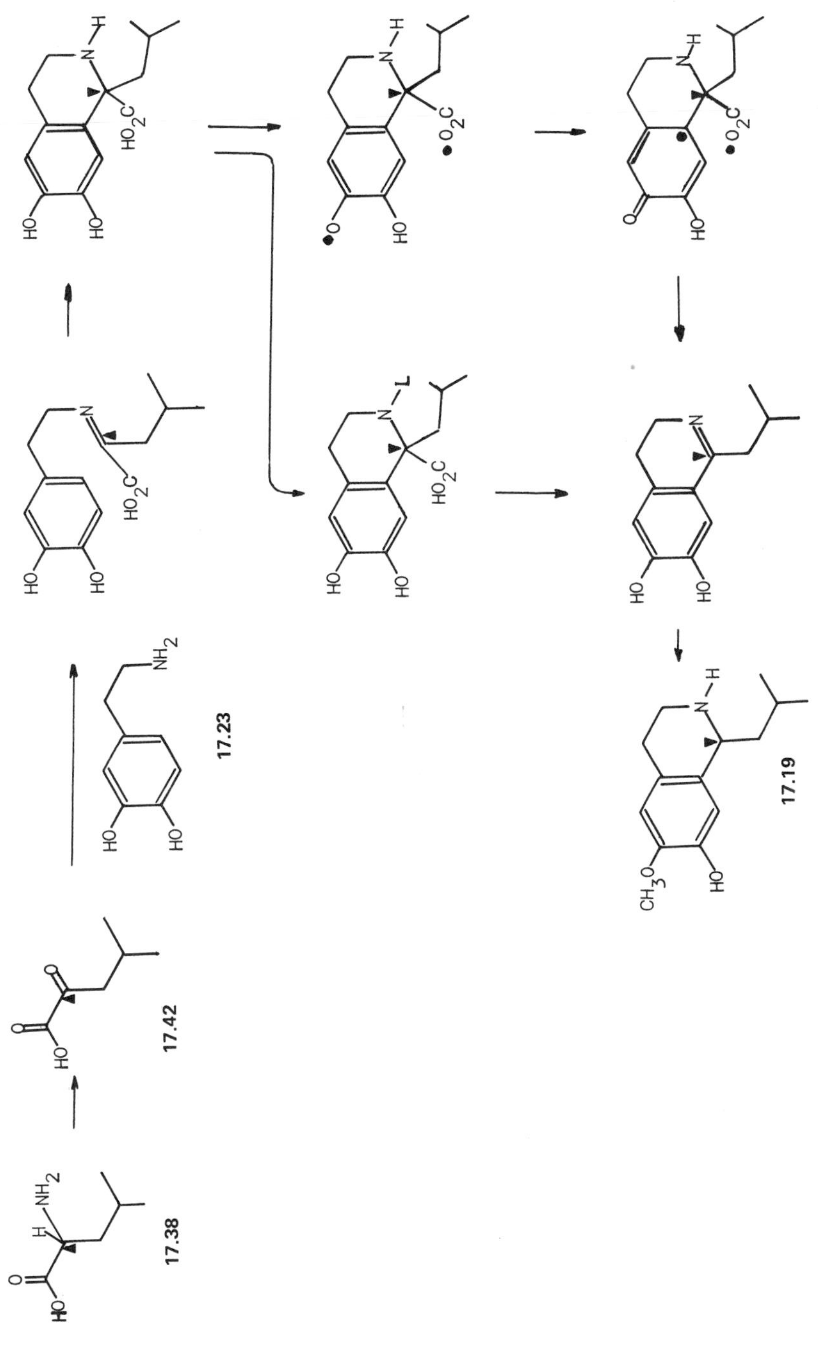

SCHEME 17.8 Pathways for the biosynthesis of lophocerine utilizing leucine as the precursor for the five-carbon fragment (▲ = label).

II. PROTOEMETINE, IPECOSIDE, AND EMETINE

Feeding experiments [65] demonstrate that (-)-ipecoside (17.21) precedes (-)-protoemetine (17.20) and (-)-emetine (17.22). The formation of (-)-ipecoside (17.21) has been established as occurring via a condensation reaction between tyrosine-derived (17.6) 3,4-dihydroxyphenethylamine (17.23) and secologanin (17.43; Chapter 2, p. 29 and Chapter 30, p. 435). Indeed, feeding experiments during which [O-methyl-^{3}H, 6-$^3H_2$]secologanin (17.43) was administered to Cephaelis ipecacuanha [65] yielded, among other products, (-)-ipecoside (17.21) in which, by degradation, the secologanin (17.43) was shown to have been incorporated intact (Scheme 17.9).

17.23

17.43

17.21

SCHEME 17.9 The formation of (-)-ipecoside from 3,4-dihydroxyphenethylamine and secologanin.

A most interesting feature of the biosynthetic pathway from secologanin (17.43) and 3,4-dihydroxyphenethylamine (17.23) to (-)-emetine (17.22; presumably via (-)-protoemetine, 17.20) has been investigated [65]. When [3'-^{14}C]desacetylipecoside (17.44) is fed to Cephaelis ipecacuanha, it is incorporated well and specifically into (-)-emetine (17.22). The epimer, [3'-^{14}C]desacetylisoipecoside (17.45), is not. Since desacetylisoipecoside (17.45) has the configuration extant in (-)-emetine [17.22; C_5 in the ipecoside becomes C_{11} in (-)-emetine, 17.22], this result implies that an inversion must occur.

The preferential cyclization of desacetylipecoside (17.44) to (-)-emetine (17.22) might be rationalized by the initial formation of an intermediate such as 17.46 (Scheme 17.10) from desacetylipecoside (17.44), rather than from its C_5 epimer (17.45) because of an expressed enzymatic preference for one epimer. Should the intermediate 17.46 then undergo ring opening and reclosure, i.e., a cis to trans-azadecalin conversion, hydrolysis, and decarboxylation yield desmethyldehydroprotoemetine (17.47) with the observed stereochemistry at C_{11}.

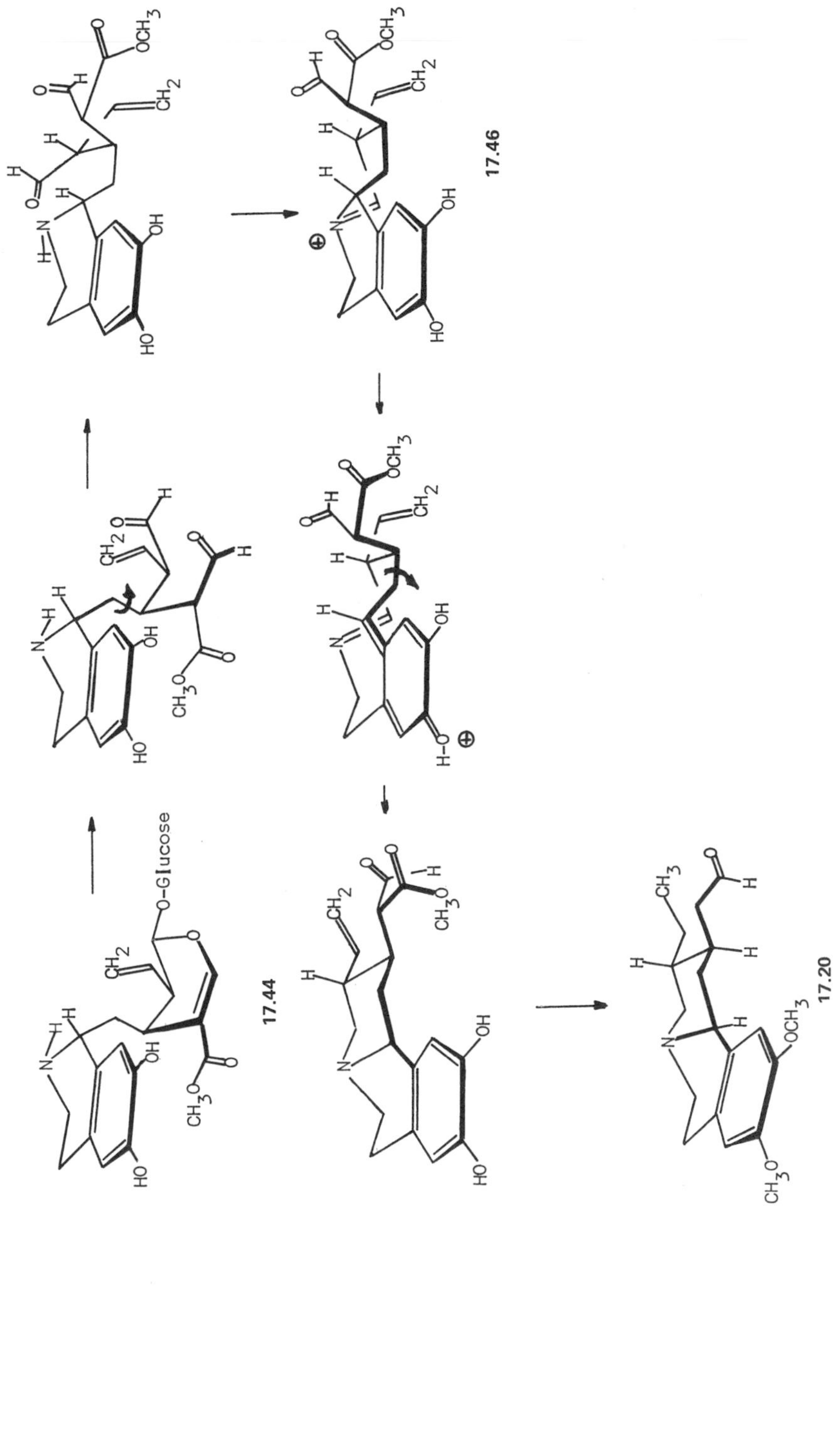

SCHEME 17.10 The preferential incorporation of desacetylipecoside into (-)-protoemetine.

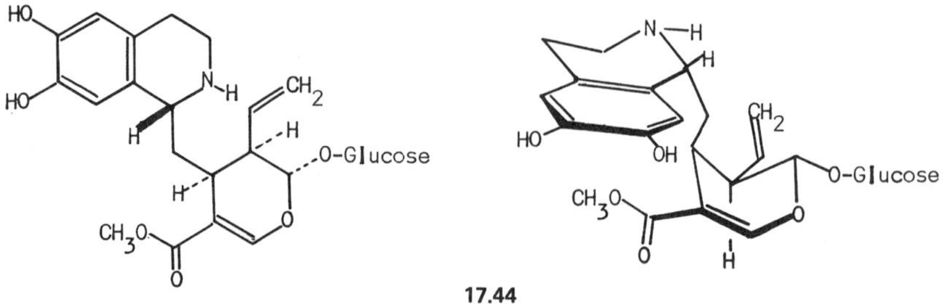

17.44

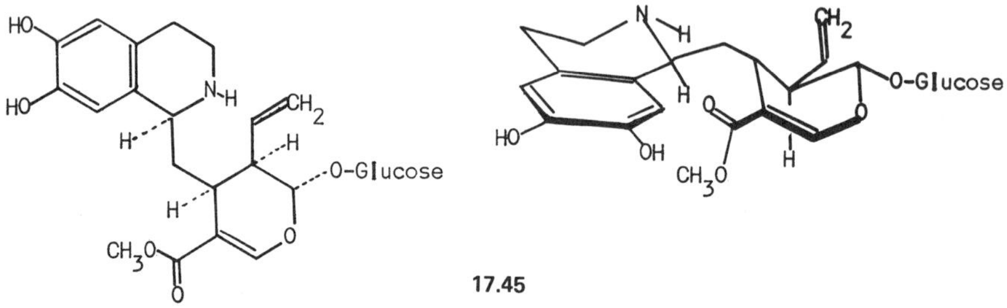

17.45

Methylation (S-adenosylmethionine, 17.15) and reduction of the vinyl side chain then generates (-)-protoemetine (17.20), while reaction with another equivalent of 3,4-dihydroxyphenethylamine (17.23) and reduction of the vinyl side chain leads to (-)-emetine (17.22).

Finally, it was shown that direct reduction of the vinyl side chain in the conversion of desacetylipecoside (17.44) to (-)-emetine (17.22) does not occur and that an isomerization to an ethylidene side chain is involved. This was deduced from the observation that the proton originally present at C_2 in loganin (17.14) was carried through into (-)-ipecoside (17.21) but not into (-)-emetine (17.22).

18

BIOSYNTHESIS OF 1-PHENYLTETRAHYDROISOQUINOLINE AND OTHER C_6-C_2 + C_6-C_1 ALKALOIDS DERIVED FROM TYROSINE

Daffodils, that come before the swallow dares
and take
The winds of March
with beauty.

William Shakespeare, The Winters Tale, Act IV, Scene 3

Only two groups of alkaloids appear to be derived from tyrosine (18.1) and a C_6-C_1 fragment. The first group, a rather small one, consists of alkaloids found only in Orchidaceae. (+)-Cryptostyline I (18.2) is an example. The second, much larger group, is composed of the

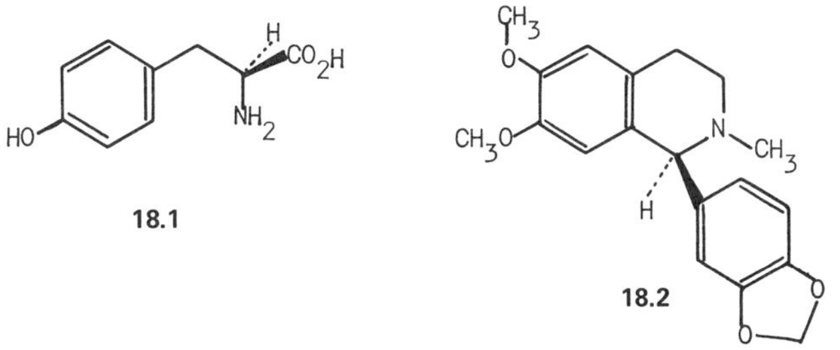

18.1

18.2

alkaloids of the Amaryllidaceae. This cosmopolitan family of related compounds includes over 100 isolated and characterized members of known structure, all of which may be derived from tyrosine (18.1) and a C_6-C_1 unit! This large number of alkaloids is in part accounted for by individual variations which result from late-stage oxidation reactions and rearrangements among the bases.

It is now believed that a single pregenitor, derived from tyrosine (18.1) and a C_6-C_1 unit and allowed to explore free-radical coupling reactions [66], provides for the structural variety ultimately obtained. Based upon the concept of a single pregenitor and observable differences and similarities in structure, the majority of known Amaryllidaceae alkaloids are divided into eight structural classes. These eight, and alkaloids typical of them, are as follows: (a) the simple N-benzyl-N-(β-phenethyl)amine derivative belladine (18.3), which in principal and perhaps in practice in its demethylated forms serves as the ultimate pregenitor of the remainder of the Amaryllidaceae alkaloids; (b) the pyrrolo[de]phenanthridines, of which lycorine (18.4) is perhaps the most ubiquitous of the thirty or so members; (c) the [2]benzopyrano[3,4g]indole alkaloids (about twenty-five), of which homolycorine (18.5) is a typical

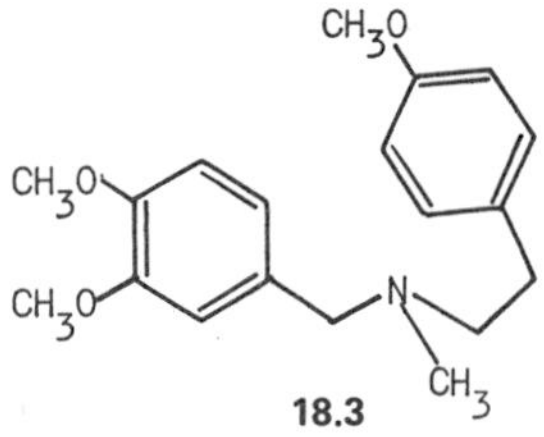

18.3

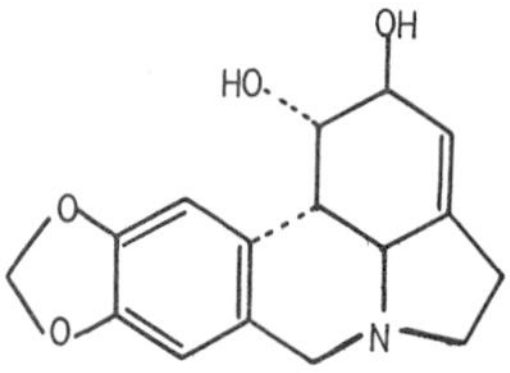

18.4

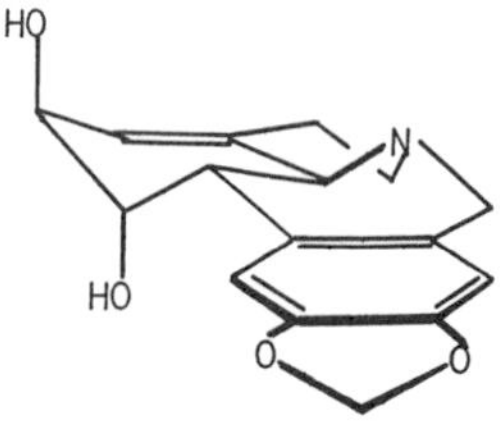

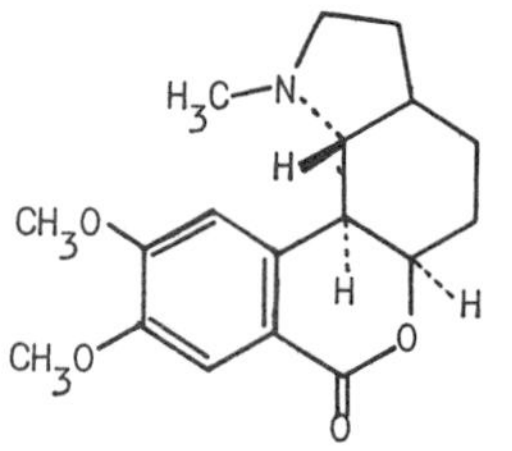

18.5

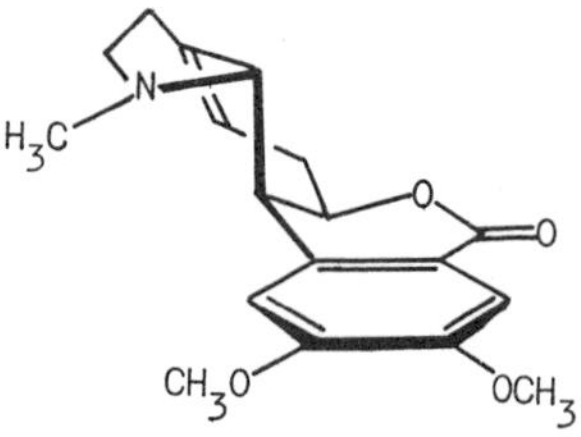

member, and which by relatively simple transformations may be derived from suitably substituted pyrrolo[de]phenanthridine alkaloids [type (b) above]; (d) the dibenzofuran bases (about ten) such as galanthamine (18.6); (e) the 5,10b-ethanophenanthridine bases (about thirty) such

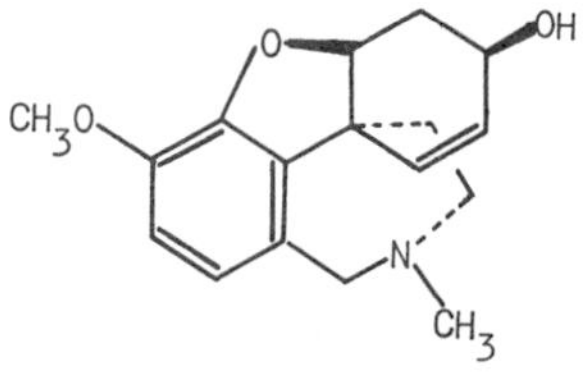

18.6

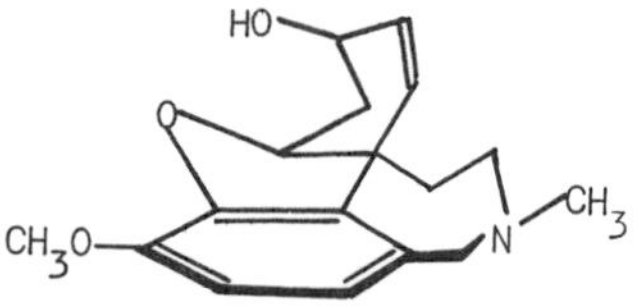

as crinine (18.7); (f) the [2]benzopyrano[3,4c]indole alkaloids (about five) such as tazattine (18.8), which can rationally be derived from the 5,10b-ethanophenanthridine bases; (g) the 5,11-methanomorphanthridine bases such as manthine (18.9) of which there are about ten known members and which can also, in principle (and in practice!) be obtained from suitably

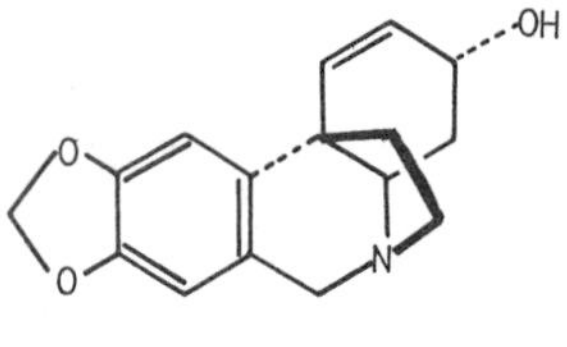

18.7

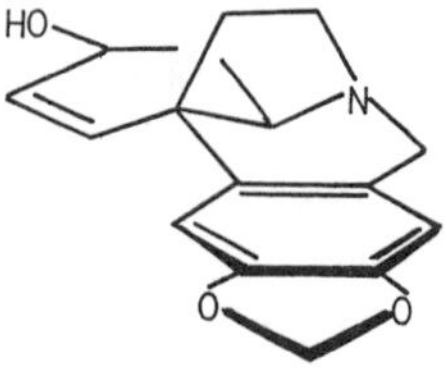

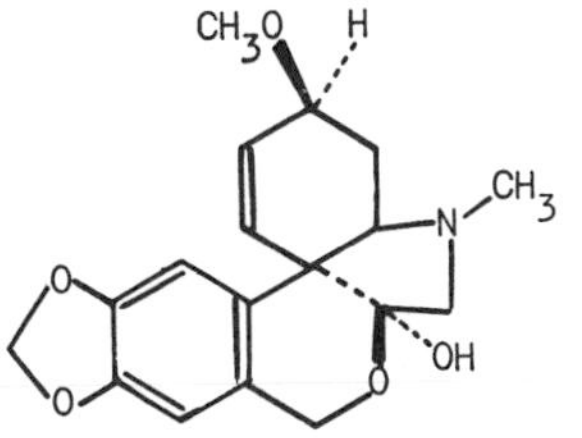

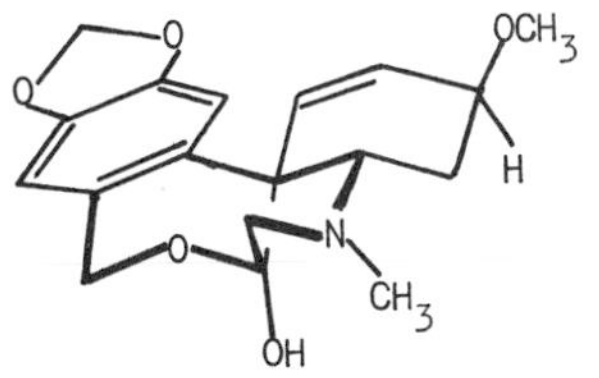

18.8

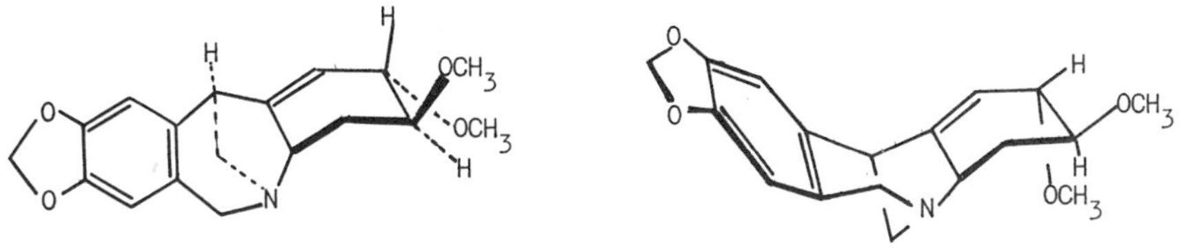

18.9

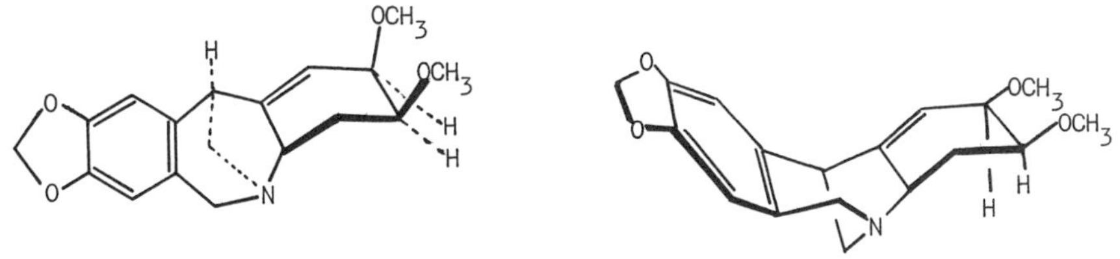

substituted 5,10b-ethanophenanthridines; and finally, (h) the simple base ismine (18.10) which is presumably generated by extensive degradation of a suitable member of the pyrrolo[de]phenanthridine or 5,10b-ethanophenanthridine group. Finally, we shall see that, as was the case with the peyote (Chap. 17) alkaloids, the methylation pattern of the precursors to belladine (18.3) presumably helps to determine which bases are elaborated.

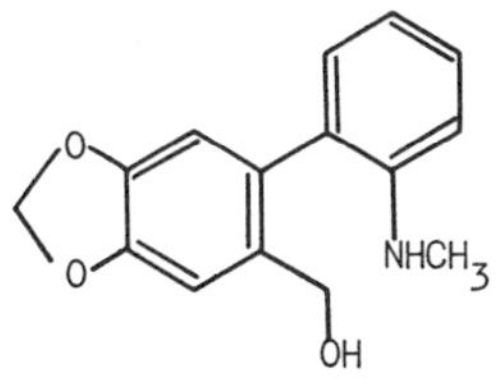

18.10

I. ORIGIN OF THE C_6-C_1 UNIT

The pathway for the generation of the C_6-C_1 unit in the Orchidaceae has not yet been examined in detail although some preliminary work has been done [67] which indicates that an appropriately hydroxylated benzaldehyde might serve this function. In the Amaryllidaceae, phenylalanine (18.11), not tyrosine (18.1), is known to provide the C_6-C_1 unit. Presumably,

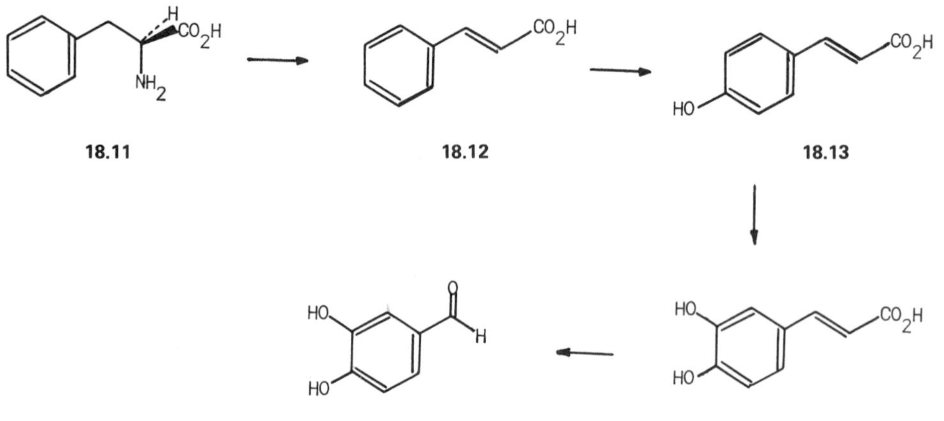

18.11 → 18.12 → 18.13 → 18.14 → 18.15

SCHEME 18.1 A pathway for the formation of protocatechuic aldehyde (a C_6-C_1 unit) from phenylalanine via cinnamic acid.

therefore, the phenylalanine lyase which operates in this family is incapable of utilizing tyrosine (18.1).

In the fully elaborated alkaloids which have been examined in some detail, the route appears to be that outlined in Scheme 18.1. Thus, phenylalanine (18.11) undergoes deamination to yield trans-cinnamic acid (18.12), which on oxidation (presumably by molecular oxygen) generates p-coumaric acid (18.13). Subsequent hydroxylation of p-coumarin acid (18.13) results in the formation of caffeic acid (18.14) which is cleaved to protocatechuic aldehyde (18.15), the actual C_6-C_1 fragment.

II. CRYPTOSTYLINE I

A reasonable scheme for the formation of (+)-cryptostyline I (18.2) can be written as follows (Scheme 18.2). The decarboxylation of tyrosine (18.1), involving, as usual, pyridoxal (18.16), yields 4-hydroxyphenethylamine (tyramine; 18.17; Scheme 18.3). Hydroxylation of tyramine to 3,4-dihydroxyphenethylamine (18.18) and reaction of this amine with protocatechuic aldehyde (18.15; Scheme 18.1) results in the formation of the imine 18.19. Then, cyclization to 18.20, methylation of 18.20 by S-adenosylmethionine (18.21) to 18.22 and cyclization to the methylenedioxy function results in the generation of (+)- or (-)-cryptostyline I (18.2 or the enantiomer of 18.2, respectively).

This pathway, or reasonable variations on it, is supported by the little experimental evidence available and by analogy to what is known in the Amaryllidaceae alkaloids (see below). Among the variations which might be reasonably considered are: (a) hydroxylation of tyrosine (18.1) to dihydroxyphenylalanine (18.23) prior to decarboxylation. By analogy to the Amaryllidaceae alkaloids, where such precursors have been tested, it is suggested here that oxidation occurs after decarboxylation, but in order to direct cyclization, before reaction with protocatechuic aldehyde (18.15). (b) Conversion of protocatechuic aldehyde (18.15) to vanillin (18.24) and/or isovanillin (18.25) prior to condensation with 3,4-dihydroxyphenethylamine (18.18). Vanillin (18.24), but not isovanillin (18.25), is incorporated [67] into (-)-cryptostyline I (enantiomer of 18.2). This difference may be due to difficulties encountered in introducing the aldehyde into the plant or to a real preference for one isomer over the other.

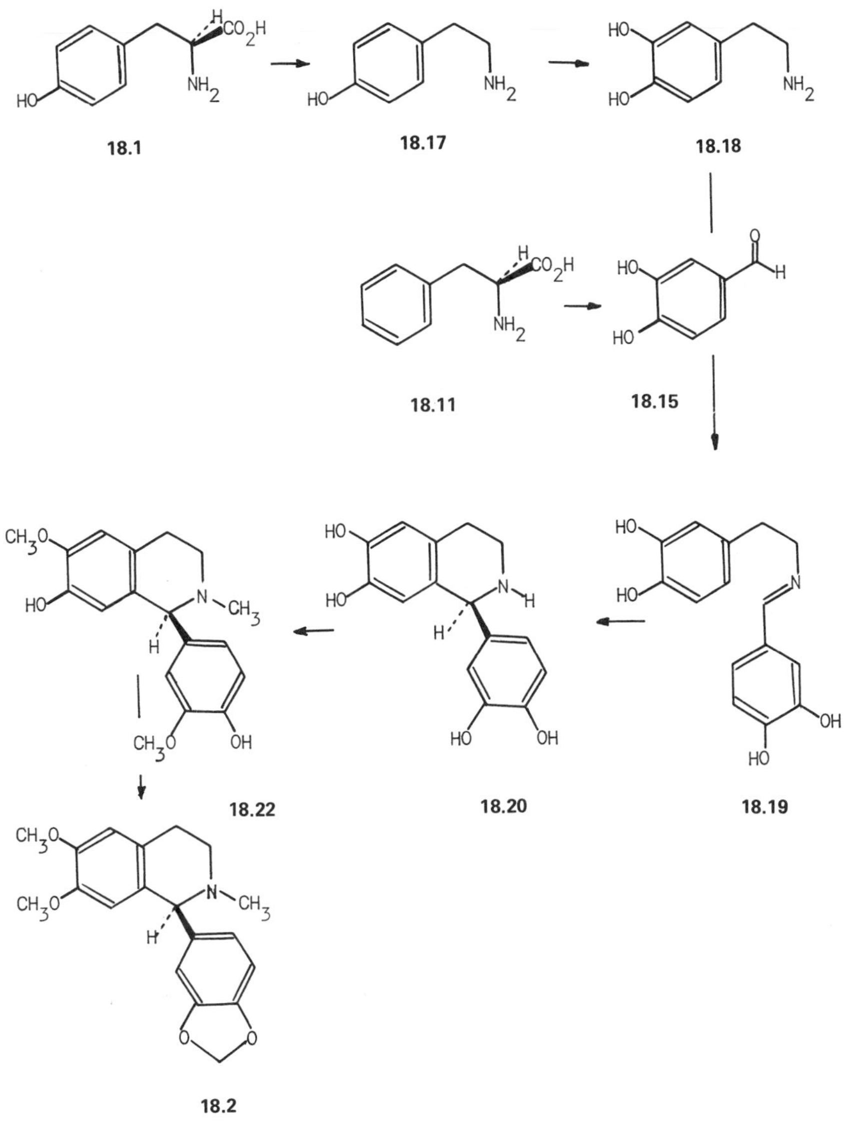

18.1
18.17
18.18
18.11
18.15
18.22
18.20
18.19
18.2

SCHEME 18.2 A pathway for the formation of cryptostyline I from tyrosine and protocatechuic aldehyde.

18.1

18.16

18.17

SCHEME 18.3 A pathway for the decarboxylation of tyrosine involving pyridoxyl.

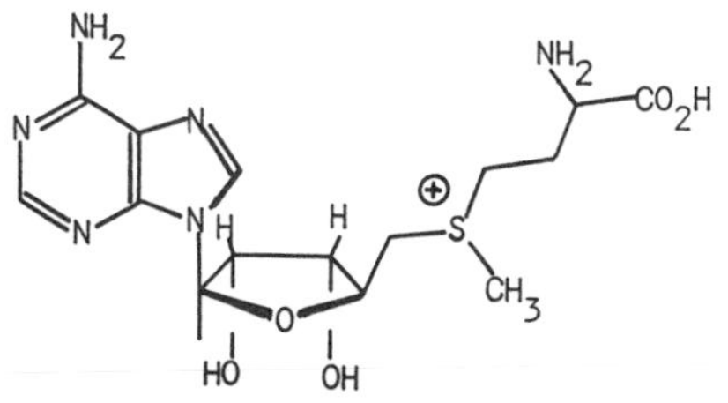

18.21

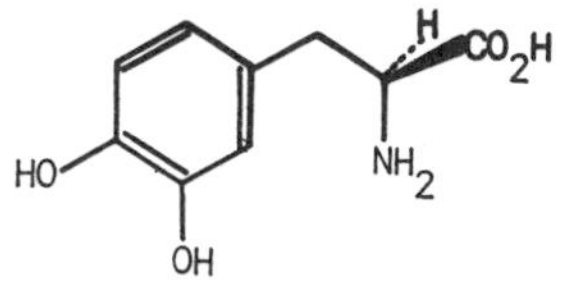

18.23

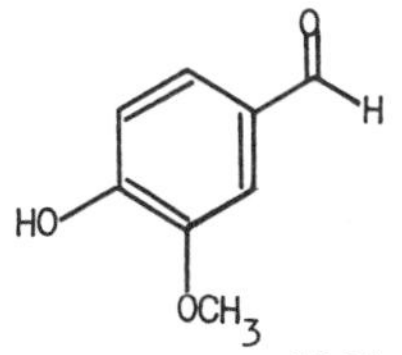

18.24

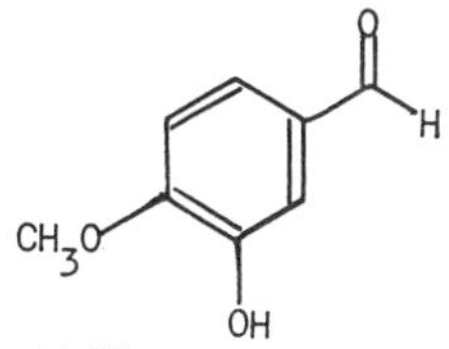

18.25

III. BELLADINE

In Nerine bowdenii, activity from [3-^{14}C]tyrosine (18.1) entered the C_6-C_2 portion of belladine (18.3) but not the C_6-C_1 unit. As it must be the case for the other alkaloids of the Amaryllidaceae for which belladine (18.3) in its desmethyl forms (e.g., norbelladine, 18.26) has been shown to be a precursor, phenylalanine (18.11) and trans-cinnamic acid (18.12) are incorporated well and serve specifically as the precursors to the C_6-C_1 unit of belladine (18.3). Presumably, therefore, protocatechuic aldehyde (18.15) is again generated from phenylalanine (18.11; Scheme 18.1) and the aldehyde reacts with tyramine (18.7; Scheme 18.3), which arises from the decarboxylation of tyrosine (18.1). The product of this reaction (18.27), on reduction, yields N-3,4-dihydroxybenzyl-N-(β-4-hydroxyphenethyl)amine (norbelladine, 18.26). Methylation, by S-adenosylmethionine (18.21), then results in the formation of belladine (18.3; Scheme 18.4).

It is interesting to note that long before belladine (18.3) was found, norbelladine (18.26) had been postulated as the precursor to the remainder of the Amaryllidaceae alkaloids [68]. Indeed, oxidative ortho and para coupling of phenoxy radicals derived from norbelladine (18.26) as a route to the other Amaryllidaceae alkaloids (Schemes 18.5, 18.7, and 18.9) had proved such an attractive idea that norbelladine (18.26) and various O-methyl derivatives of norbelladine (18.26) were synthesized and shown to be incorporated, intact, into belladine (18.3). Analogs of norbelladine (18.26) such as N-4-hydroxybenzyl-N(β-phenethyl)amine (18.28) and N-3,4-dihydroxybenzyl-N(β-3,4-dihydroxyphenethyl)amine (18.29) are not incorporated at all into belladine (18.3) nor the other Amaryllidaceae alkaloids.

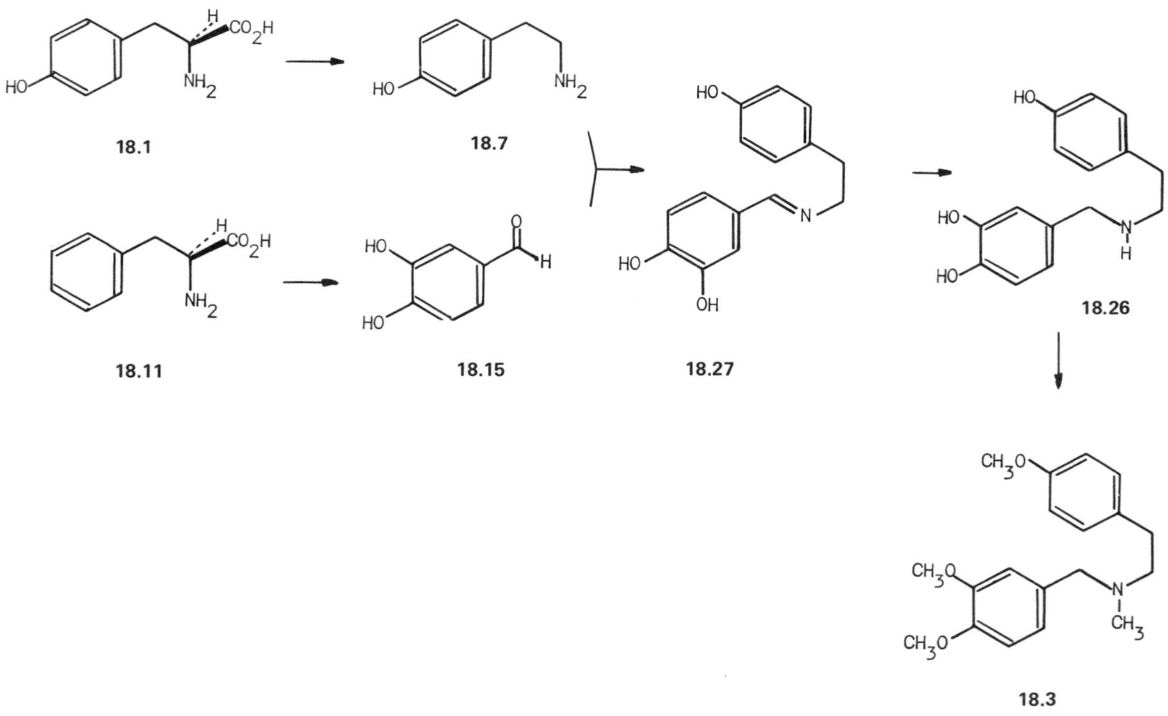

SCHEME 18.4 A pathway for the formation of belladine from tyrosine and phenylalanine.

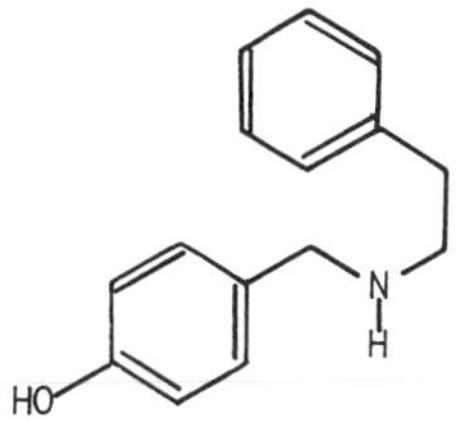

18.28

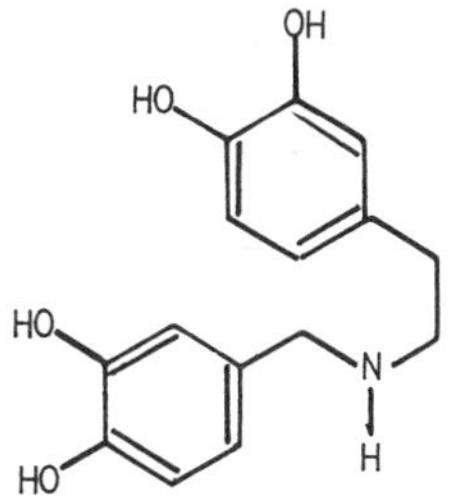

18.29

IV. LYCORINE

Because it is, perhaps, the most ubiquitous of the Amaryllidaceae alkaloids, the biosynthesis of lycorine (18.4) has been extensively examined. Norbelladine (18.26) is an intermediate, and accordingly it has been found that the phenethylamine portion of lycorine (18.4) is derived from tyrosine (18.1) via tyramine (18.17; Scheme 18.3) while the C_6-C_1 portion comes from protocatechuic aldehyde (18.15) derived from phenylalanine (18.11; Scheme 18.1).

The generation, oxidatively, of radicals ortho to the hydroxyl in the C_6-C_2 portion of norbelladine (18.26) and para to the hydroxyl in the C_6-C_1 portion of the same molecule and the coupling of the same radicals to each other (Scheme 18.5) lead directly to 18.30, which on conversion to 18.31, reduction, and methylation (by S-adenosylmethionine, 18.21) generates the naturally occurring norpluviine (18.32). Alternatively, since one of the hydroxy groups in norbelladine (18.26) is not utilized during the process outlined above at all, early methylation [i.e., norbelladine, 18.26, is converted to N-3-hydroxy-4-methoxybenzyl-N(β-4-hydroxyphenethyl)amine, 18.33, prior to cyclization] could occur, and thus norpluviine (18.32) could be formed directly.

The conversion of norpluviine (18.32) to carinine (18.34) and ultimately to lycorine (18.4; Scheme 18.6) has been established by demonstrating that (a) the O-methyl group of norpluviine (18.32) becomes the methylenedioxy group of lycorine (18.4; [69]) and (b) N-3,4-dihydroxybenzyl-N(β-3,4-dihydroxyphenethyl)amine (18.29), in which a second hydroxyl

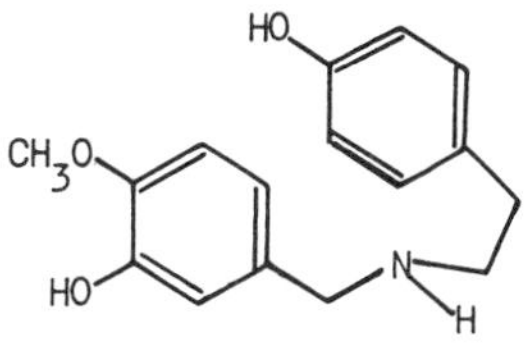

18.33

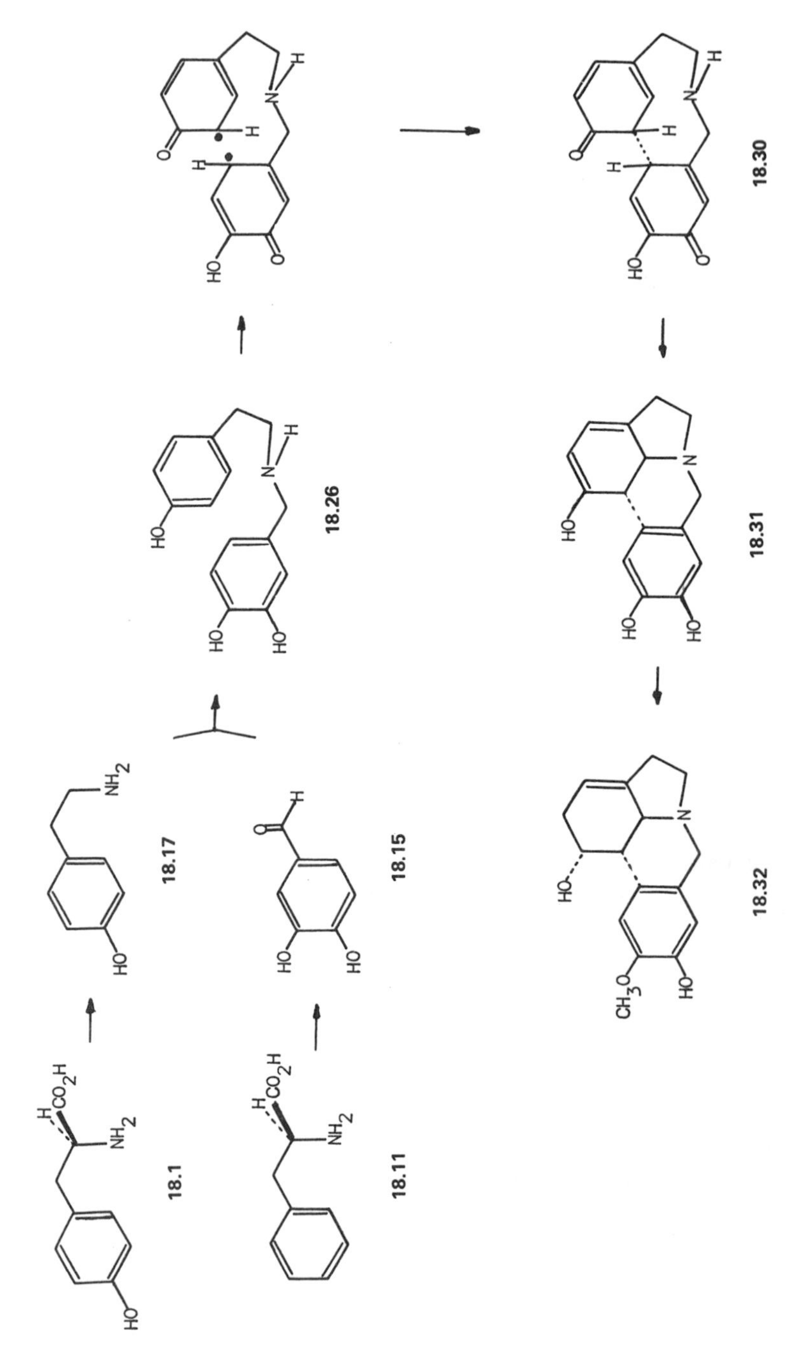

SCHEME 18.5 A pathway for the formation of norpluviine by coupling of radicals derived from norbelladine.

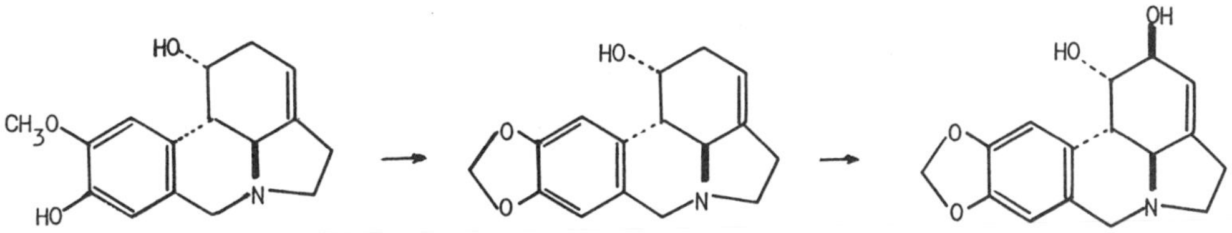

18.32 18.34 18.4

SCHEME 18.6 A pathway for the conversion of norpluviine to lycorine (via carinine).

present in the C ring of lycorine (18.4) has been incorporated at an early stage, is not itself incorporated into lycorine (18.4).

V. HOMOLYCORINE

The details of the biosynthesis of homolycorine (18.5) have not yet been worked out, but some of those of its likely precursor, lycorenine (18.35), have recently been illuminated. Thus, it is now known that phenylalanine-derived (18.11) protocatechuic aldehyde (18.15) is incorporated into the C_6-C_1 unit of lycorenine (18.35). In addition, the route N-3-hydroxy-4-methoxybenzyl-N-(β-4-hydroxyphenethyl)amine (18.33) from methylation of norbelladine (18.26)* to norpluviine (18.32), during which tritium originally present on the aldehydic carbon of protocatechuic aldehyde (18.15) is retained, can be extended to lycorenine (18.35), which continues to retain the original tritium [70]. Scheme 18.7 presents a pathway for the formation of lycorenine (18.35) consonant with the available information. Oxidation of lycorenine (18.35) then generates homolycorine (18.5; Equation 18.1).

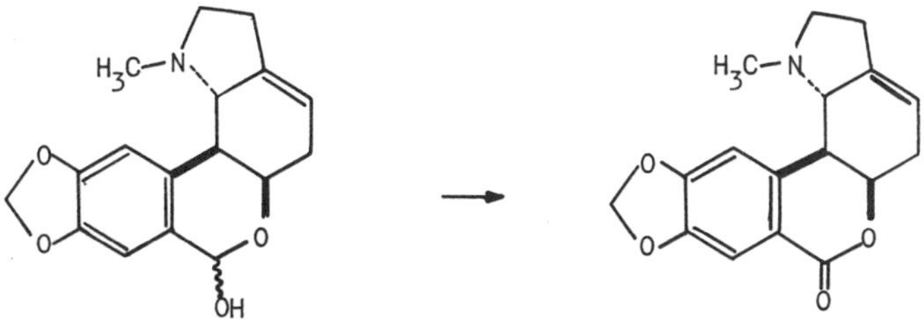

(Eq. 18.1)

18.35 18.5

*Presumably, as indicated in Scheme 18.7, norbelladine (18.26) is derived from protocatechuic aldehyde (18.15) and tyramine (18.17).

SCHEME 18.7 A pathway for the formation of lycorenine from tyrosine and protocatechuic aldehyde. $\underline{H}$ indicates ^3H label.

VI. GALANTHAMINE

In daffodils, norbelladine (18.26) generated, as usual, from phenylalanine-derived (18.11) protocatechuic aldehyde (18.15) and tyramine (18.17) obtained from tyrosine (18.1), is specifically incorporated into galanthamine (18.6). Furthermore, label from the N-methyl group of N-p'-O-dimethylnorbelladine (18.36) is exclusively retained as the N-methyl of galanthamine (18.6), and, indeed, it has been suggested that N-methylation of norbelladine (18.26) is a prerequisite to formation of the dibenzofuran bases of which galanthamine (18.6) is a member. Thus, N-methylation results in formation of the dibenzofuran alkaloids while lack of a methyl on nitrogen allows the other reactions to occur which generate the pyrrolo[de]-phenanthridine and 5,10b-ethanophenanthridine bases. Finally, the oxidative generation of radicals ortho to the hydroxyl in the C_6-C_1 portion of norbelladine (18.26) and para to the hydroxyl in the C_6-C_2 portion of the same molecule, followed by coupling of the radicals and a final ether-forming cyclization of the Michael type, leads to galanthamine (18.6; Scheme 18.8).

VII. CRININE

Crinine (18.7), which has been isolated in moderate amounts from a number of Amaryllidaceae (e.g., *Buphane distich* Herb, *Crinum laurentic* Durand and DeWild, etc.), and which has proved itself as a reference compound which interrelates the 5,10b-ethanophenanthridine bases subsequently isolated with each other, has not yet been studied from a biosynthetic point of view. Indeed, although crinine (18.7) is the archetypal transformation product of phenol oxidative coupling of N-3-hydroxy-4-methoxybenzyl-N(β-4-hydroxyphenethyl)amine (18.33)* in the position para to each of the oxygens (Scheme 18.9), it is usually the more highly oxidized enantiomeric product, haemanthamine (18.37), which is studied.

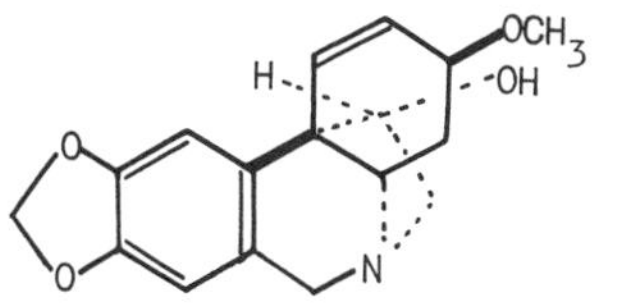

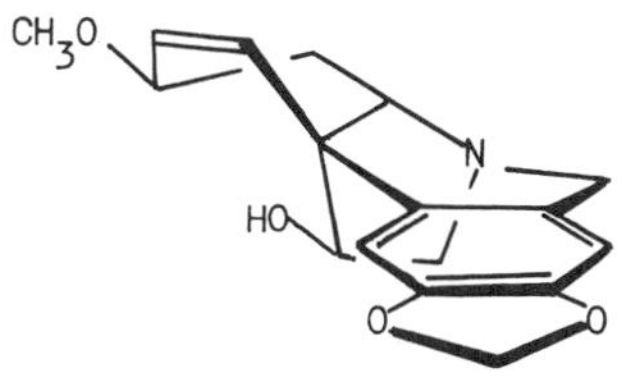

18.37

With regard to haemanthamine (18.37), therefore, N-3-hydroxy-4-methoxybenzyl-N(β-4-hydroxyphenethyl)amine (18.33), labeled with ^{14}C in the O-methyl, was incorporated into haemanthamine (18.37) without loss of label. When [formyl-^3H]protocatechuic aldehyde (18.15) is administered to "Twink" daffodils, stereospecifically labeled 6R- or 6S-[6-^3H]haemanthamine (18.37) is obtained. Additionally, synthetic (±)-[3-^3H]crinine (18.7 and its mirror image) has been converted in vivo into haemanthamine (18.37), and (±)-octopamine (18.38) is not incorporated as well as tyramine (18.17; [71]). These results argue for a late-stage hydroxylation in the formation of haemanthamine (18.37) with, perhaps, involvement of the nitrogen.

---

*This amine (18.33) is presumably formed in the usual way from tyrosine-derived (18.1) tyramine (18.17; Scheme 18.3) and phenylalanine-derived (18.11) protocatechuic aldehyde (18.15; Scheme 18.1) followed by reduction (Scheme 18.4) and methylation by S-adenosylmethionine (18.21).

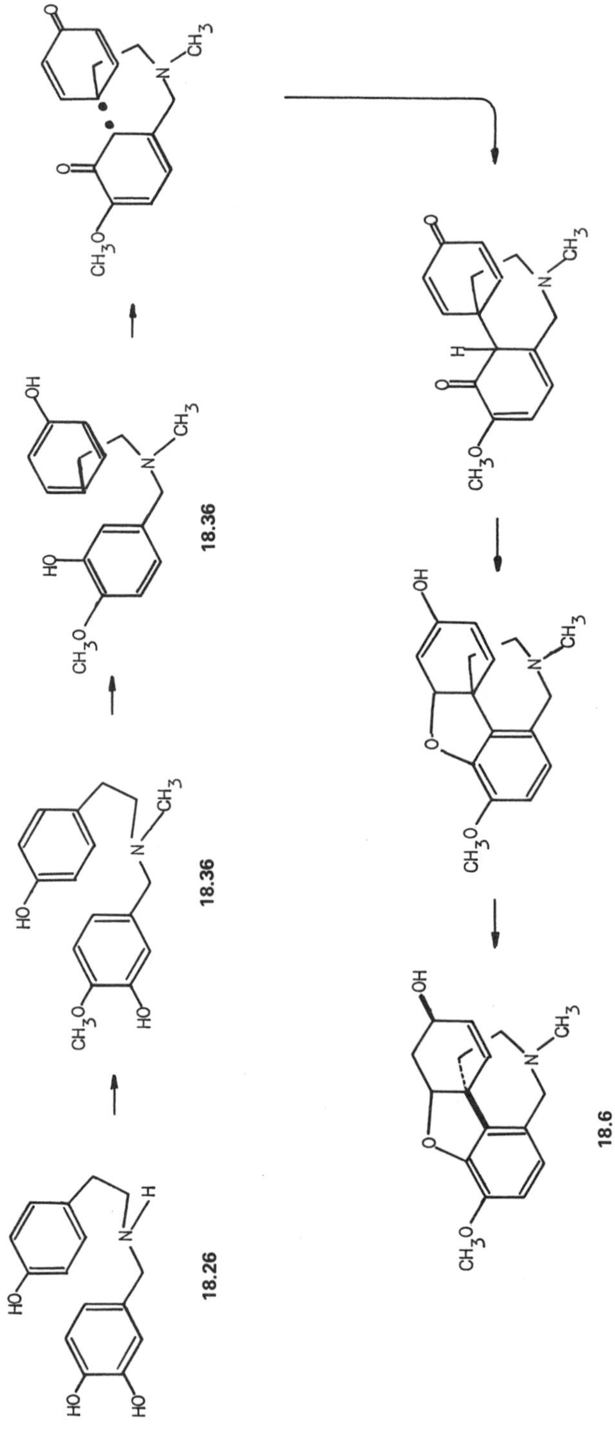

SCHEME 18.8 A pathway for the formation of galanthamine from norbelladine.

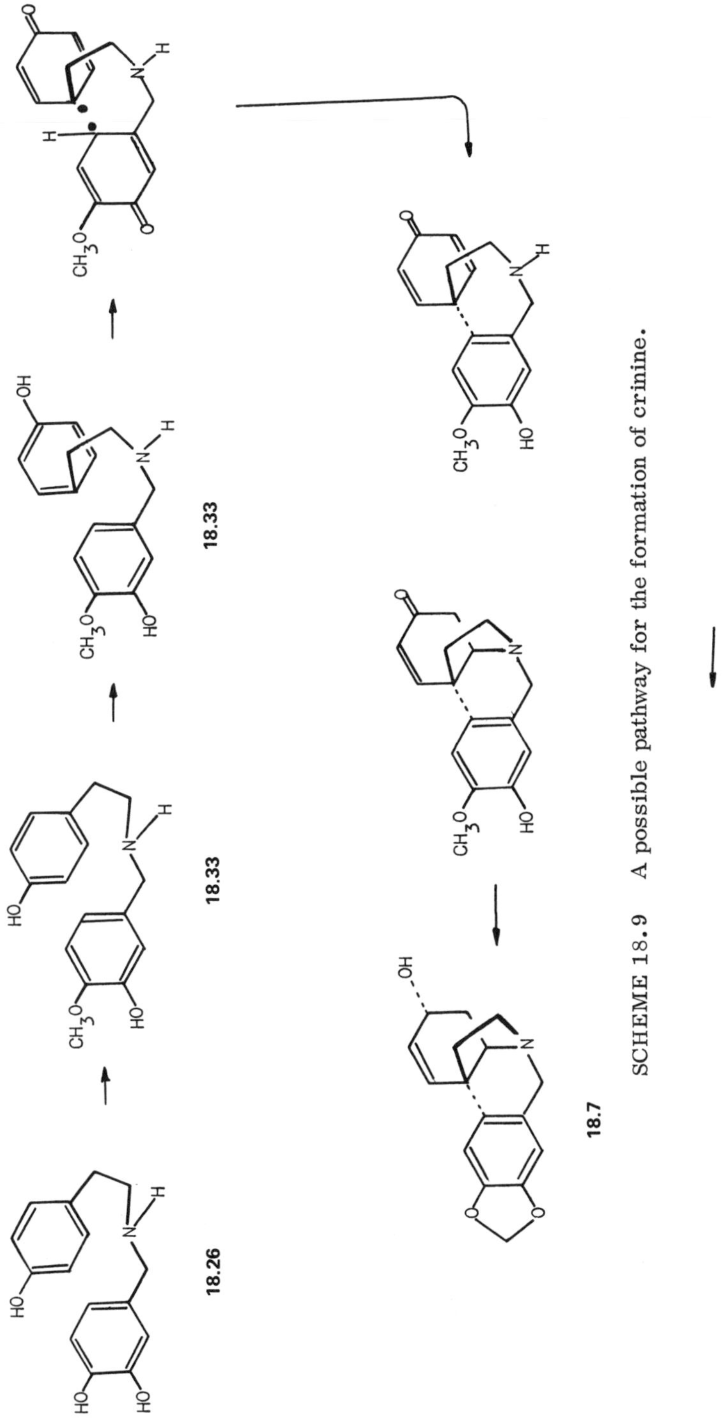

SCHEME 18.9 A possible pathway for the formation of crinine.

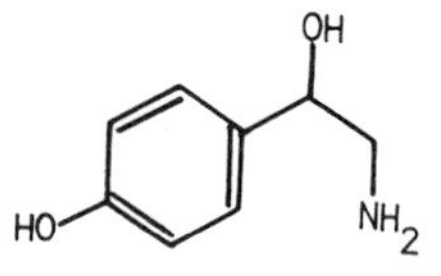

18.38

If the pathway (Scheme 18.9) shown is correct, the C_3 methylene of tyrosine (18.1) becomes the C_2 methylene of N-3-hydroxy-4-methoxybenzyl-N(β-4-hydroxyphenethyl)amine (18.33), which in turn becomes the C_{11} methylene of crinine (18.7) and haemanthamine (18.37). Thus, tritium introduced at C_3 of tyrosine (18.1) should eventually be replaced by oxygen in haemanthamine (18.37). Feeding of 2S-[3S-^{3}H]tyrosine (18.39) which had been mixed with (±)-[2-^{14}C]tyrosine (18.39) to "Texas" daffodils, yields ^{14}C-labeled haemanthamine (18.37) in which about 90% of the tritium label has been retained. Feeding of the

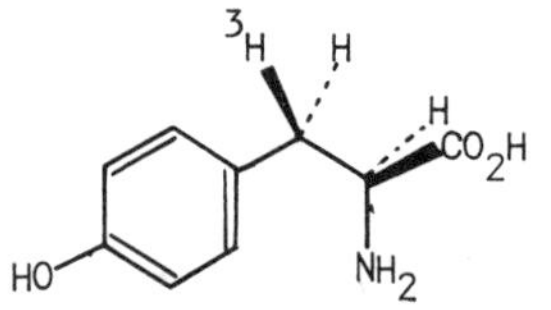

18.39

enantiomer of 18.39 yields the opposite result [72]. Thus, late-stage hydroxylation proceeds with retention at C_{11}, with the pro-R hydrogen being removed and might thus argue for involvement of, and perhaps initial oxidation at, the nitrogen and subsequent suprafacial transfer of oxygen from nitrogen to carbon.

VIII. TAZETTINE

The alkaloids tazettine (18.8) was incorrectly shown to be derived, in vivo, in Sprekelia formosissima, from haemanthamine (18.37) via haemanthadine (18.40) (Equation 18.2; [73]). The biosynthetic route pictured (Equation 18.2) was in accord with feeding experiments and did provide a reasonable pathway for generation of the alkaloid. However, very careful laboratory work, in which the strongly basic conditions usually employed in alkaloid isolation were avoided, showed that tazettine (18.8) is an artifact of isolation and that it is readily generated from its precursor, pretazettine (18.41).* Later [74], haemanthidine methosalts (18.42) were found, in aqueous solution, to be in equilibrium with salts of protonated pretazettine (18.41). Basification of such aqueous solutions of the salts yields either tazettine (18.8) or pretazettine (18.41), depending upon the amount of time elapsed between basification and isolation (Scheme 18.10). Appropriate labeling experiments indicate that the required hydride transfer in the conversion of pretazettine (18.41) into tazettine (18.8) is intramolecular and that the same hydride transfer reaction is also the slowest step in the entire sequence of transformations shown in Scheme 18.10.

Thus, the biosynthesis of the haemanthidine methosalts (18.42), precursors of the artifact tazettine (18.8), is that already shown for the enantiomer of crinine (18.7; i.e., vittatine, 18.43; Scheme 18.9) followed by (a) late-stage oxidation of vittatine (18.43) to haemanthamine (18.37), (b) benzylic oxidation of haemanthamine (18.37) to haemanthidine (18.40), and (c) methylation (presumably by S-adenosylmethionine, 18.21) to the appropriate methosalt (18.42; Scheme 18.11).

*It is interesting to speculate about the number of similar, as yet undetected, cases of the same phenomenon.

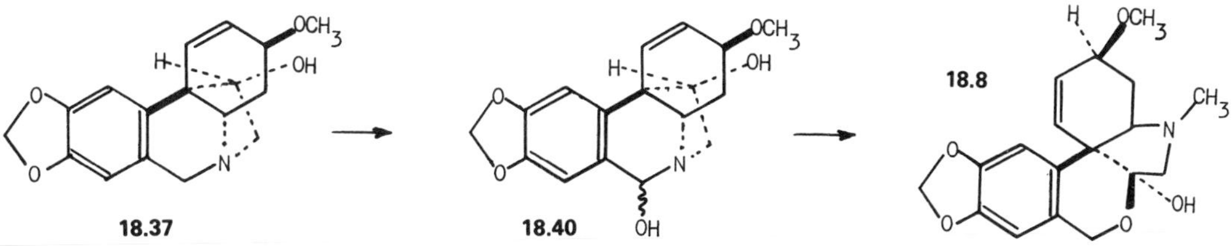

18.37 18.40 18.8

(Eq. 18.2)

IX. MANTHINE

Manthine (18.9) and similar minor bases are frequently found in South African Haemanthus species, and although the alkaloid manthine (18.9) has been known for some time, controversy about its structure is still rife: either of the structures 18.9 can still be written [75]. It is currently believed, on the basis of in vitro studies (although an in vivo study reporting the contrary exists [76] that alkaloids such as manthine (18.9) are derived from haemanthamine (18.37). The basis for the belief rests upon the outcome of the treatment of the methanesulfonate of haemanthamine (18.44) with methanolic sodium methoxide, from which manthine (18.9; [76]) is isolated.

Manthine (18.9) has also been related to other alkaloids of the same basic ring structure, and as recently pointed out [75], no single structure is compatible with all the data currently available. Regardless, should the belief that manthine (18.9) is derived from haemanthamine (18.37) by a process similar to that outlined in Equation 18.3 be correct, it would follow that the biosynthesis applicable is that already shown for crinine (18.7) and its enantiomer vittatine (18.43; Scheme 18.9) from which haemanthamine (18.37) is derived by late-stage oxidation.

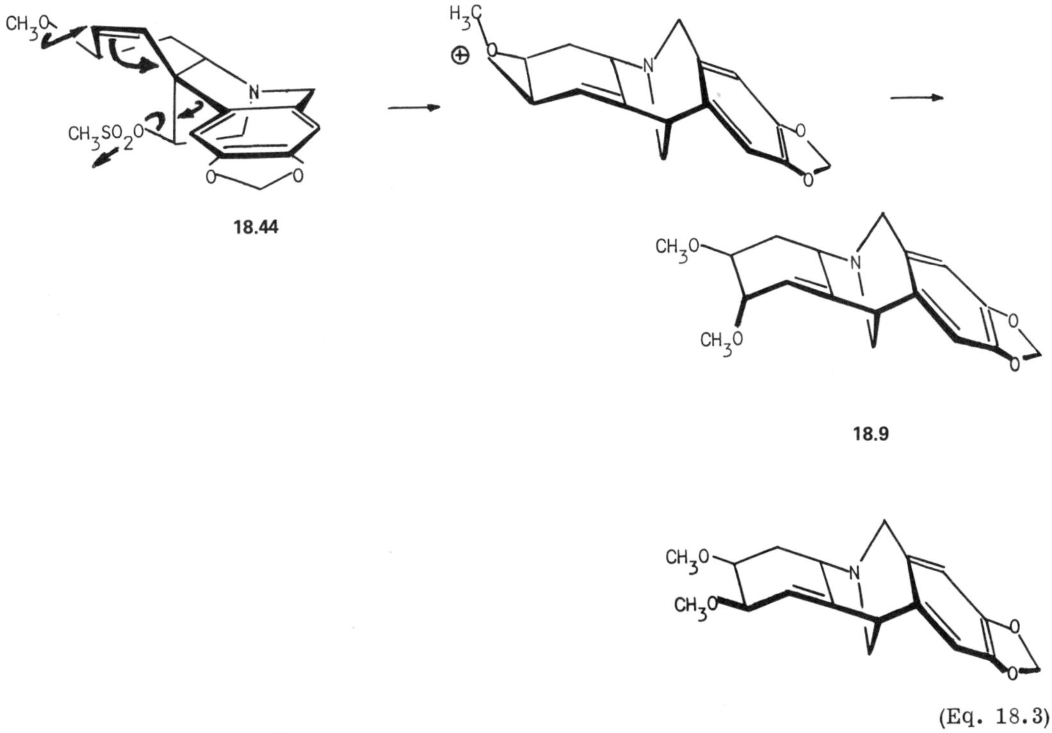

18.44

18.9

(Eq. 18.3)

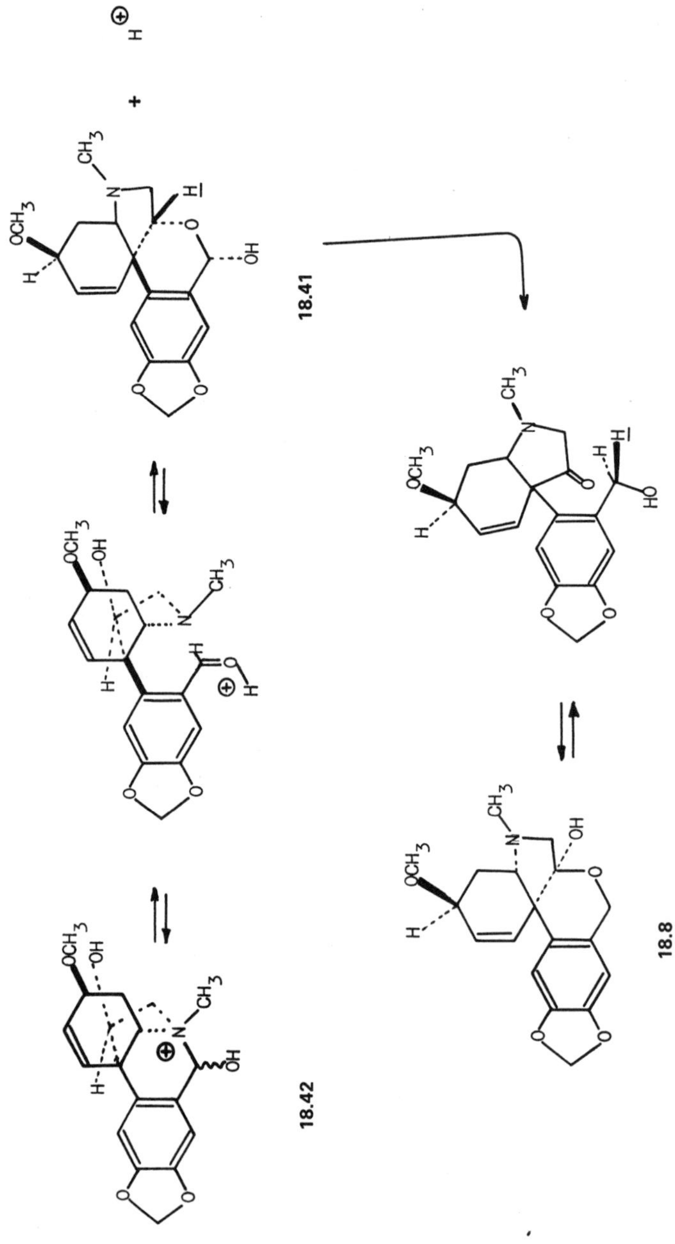

SCHEME 18.10 A pathway for the formation of tazettine and pretazettine from an haemanthadine methosalt. H indicates ^{3}H label.

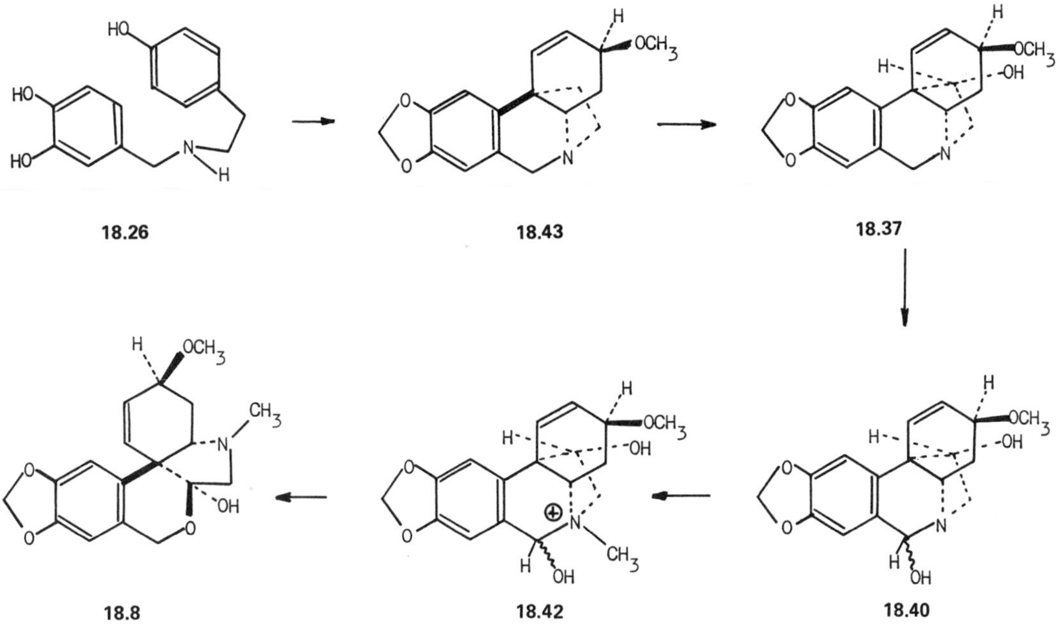
HO
HO
HO
N
H
OCH3
OH
CH3
18.26
18.43
18.37
18.8
18.42
18.40

SCHEME 18.11 A pathway for the biosynthesis of a haemanthidine methosalt precursor to the artifact tazettine.

X. ISMINE

Ismine (18.10), isolated from, for example, Sprekelia formosissima, occurs along with haemanthamine (18.37) and haemanthadine (18.40) and has long been considered a degradation product of one or more of the classes of the more typical alkaloids. Thus, in concert with this idea, although the incorporation was low, [2,3,3-$^3H_3$]oxocrinine (18.45) yielded specifically labeled ismine (18.10; Equation 18.4; [77]). Any number of pathways can be written to

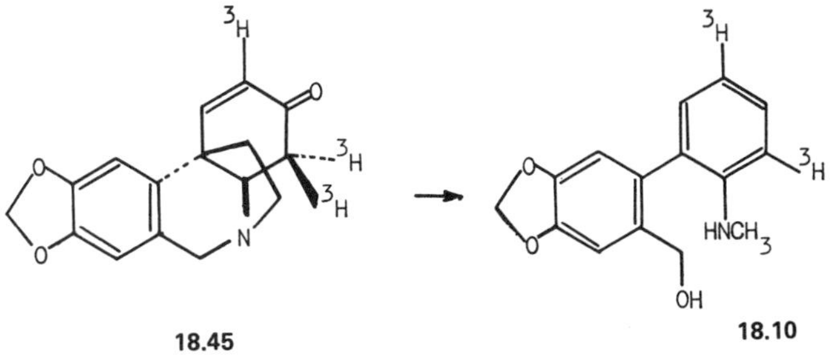
3H
O
N
HNCH3
OH
18.45
18.10

(Eq. 18.4)

ismine (18.10). All of them involve oxidative conversion of crinine (18.7) or its enantiomer vittatine (18.43) to haemanthamine-like (18.37) species and thence to haemanthadine (18.40) analogs and products of further oxidation, culminating in loss of the ethano bridge originally present and aromatization of the remaining ring.

19

BIOSYNTHESIS OF 1-BENZYLTETRAHYDROISOQUINOLINE ALKALOIDS AND OTHER C_6-C_2 + C_6-C_2 ALKALOIDS DERIVED FROM TYROSINE

> There is some soul of goodness in things evil
> Would men observingly distil it out.
>
> William Shakespeare, King Henry V, Act IV, Scene 1

The alkaloids falling into the broad category with which this chapter deals comprise one of the most abundant groups to be considered in this volume. In the time since the discovery of morphine (19.1), when the nineteenth century dawned, more than 400 related (defined here as C_6-C_2 + C_6-C_2 bases derived from tyrosine), naturally occurring bases have been isolated, derivatized, and characterized. Many of these compounds have been synthesized by unequivocal routes and others fall daily. All this large number of alkaloids are derived from tyrosine (19.2), and all are rationalized as coming from norlaudanosoline (19.3) as the penultimate

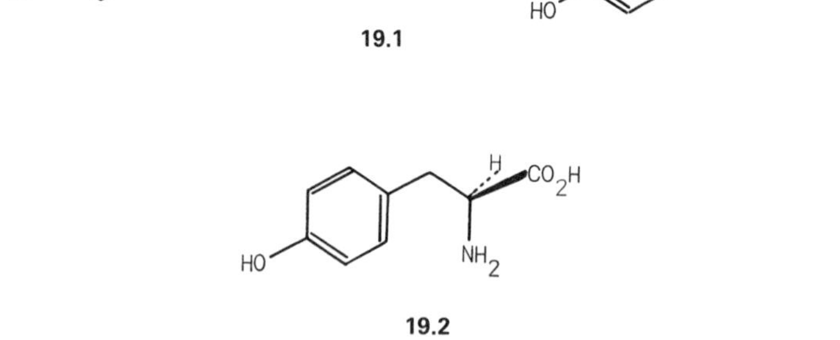

19.1

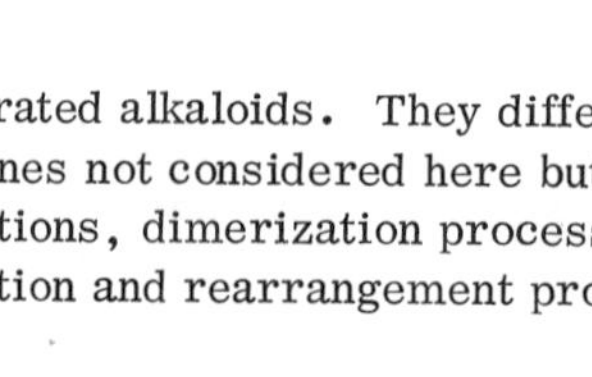

19.2

precursor to the elaborated alkaloids. They differ, in forming the 11 <u>major</u> groups listed below (and some minor ones not considered here but derivable in the same fashion) in details of initial cyclization reactions, dimerization processes, methylation patterns (O and N), and finally, late-stage oxidation and rearrangement processes which make and break bonds.

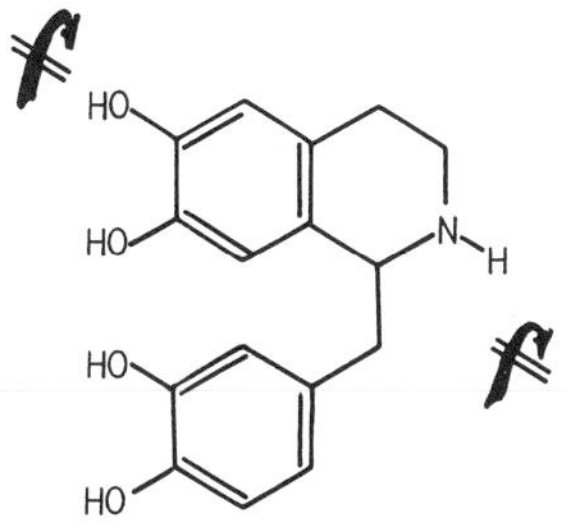

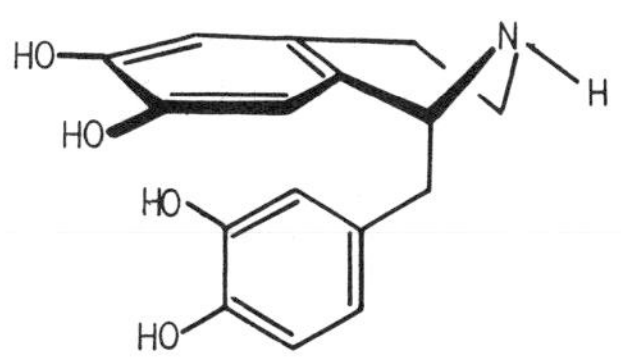

19.3

The specific compounds representative of the major groups corresponding to C_6-C_2 + C_6-C_2 alkaloids which will be considered here (followed by the approximate number of members in that group) are:

1. The simple 1-benzyltetrahydroisoquinoline alkaloids laudanosine (19.4) and its aromatic 1-benzylisoquinoline analog, papaverine (19.5), both of which are clearly related to norlaudanosoline (19.3; about twenty-five members).

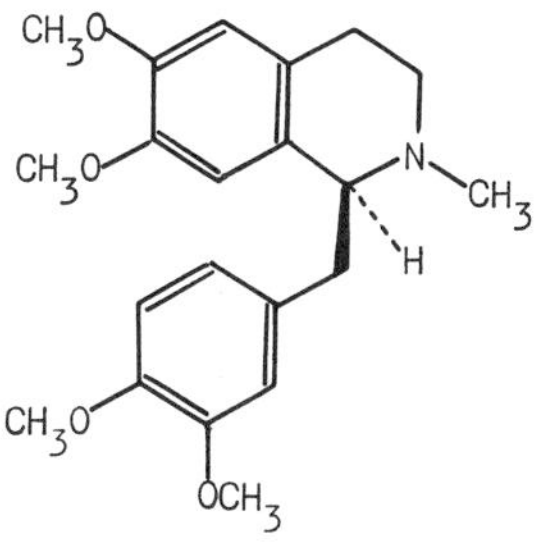

19.4

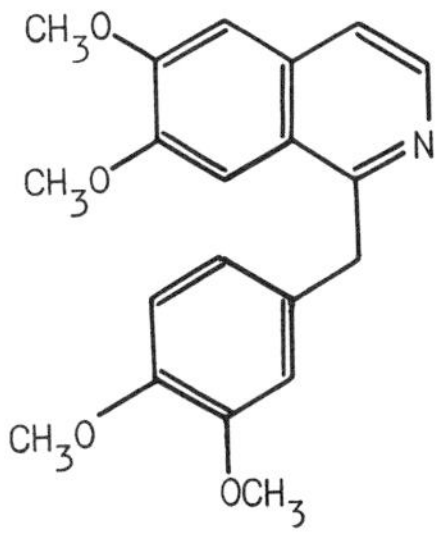

19.5

2. The bisbenzylisoquinoline alkaloids (+)-oxyacanthine (19.6) and (+)-tubocurarine (19.7), which presumably arise by oxidative coupling reactions of suitable norlaudanosoline-like (19.3) derivatives (over 100 such compounds are known).

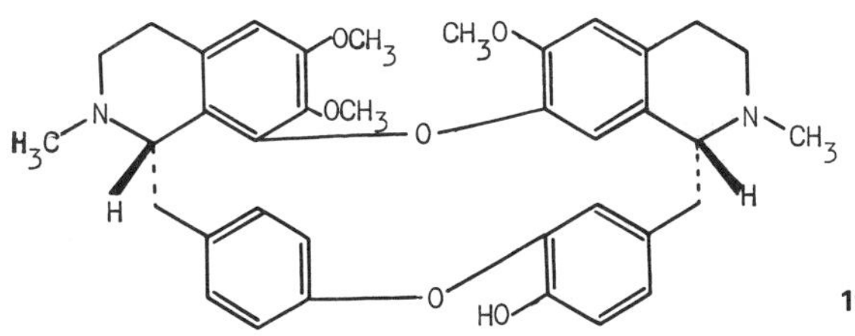

19.6

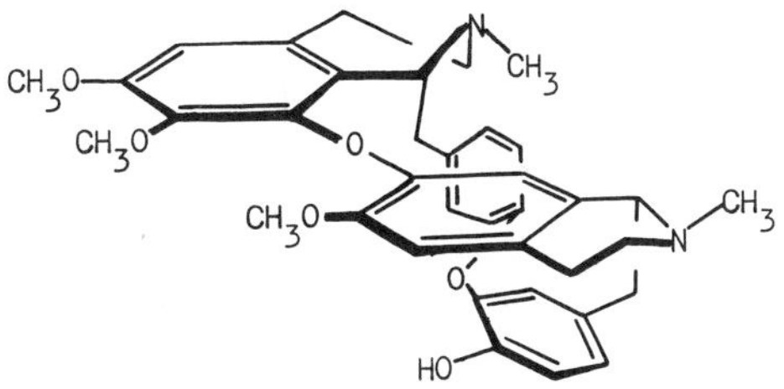

3. The proaporphine alkaloids related to (-)-orientalinone (19.8), which may arise from intramolecular oxidative coupling in norlaudanosoline (19.3; about fifteen have been reported), and the aporphine alkaloid (+)-isothebaine (19.9), which appears to be derived from (-)-orientalinone (19.8; more than seventy aporphine alkaloids have been isolated).

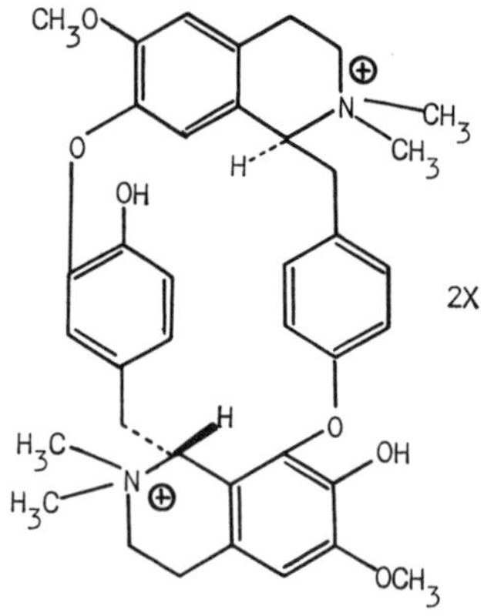

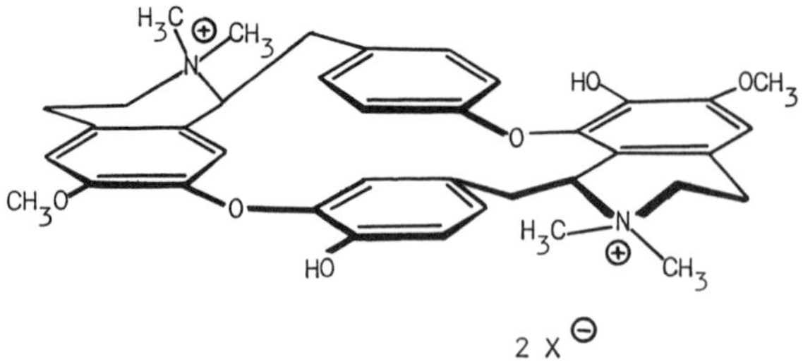

19.7

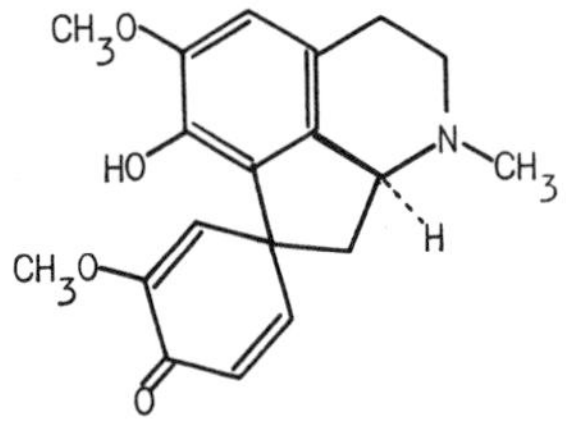

19.8

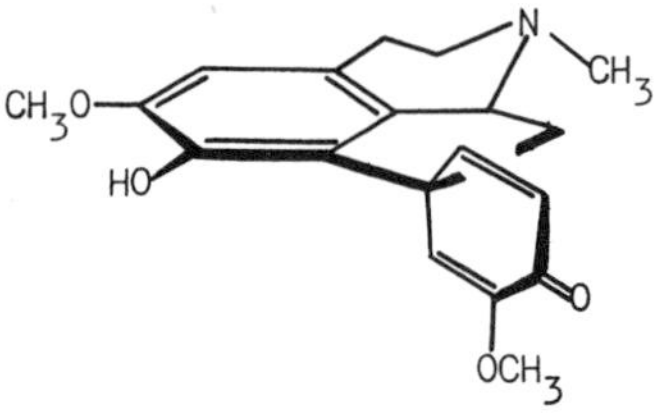

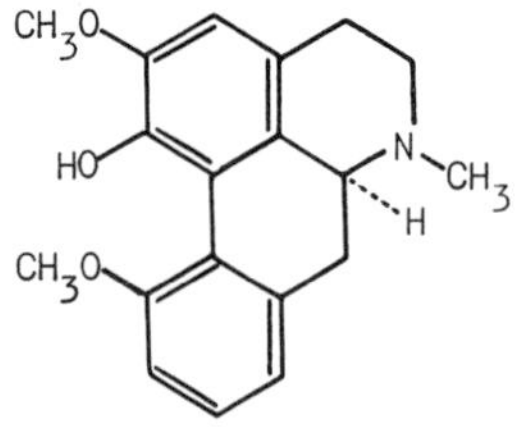

19.9

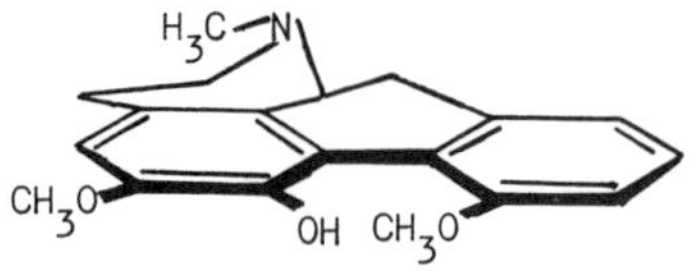

4. The morphine bases, morphine (19.1), codeine (19.10), and thebaine (19.11), also derived by intramolecular coupling reactions of norlaudanosoline (19.3) or norlaudanosoline-like (19.3) intermediates, but in a different sense than for the proaporphines (about fifteen bases are grouped along with morphine).

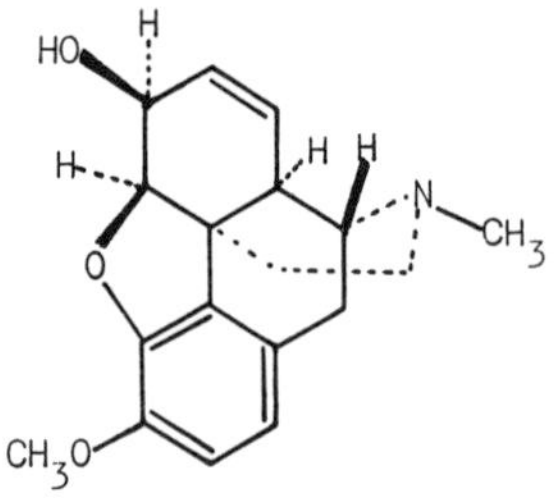

19.10

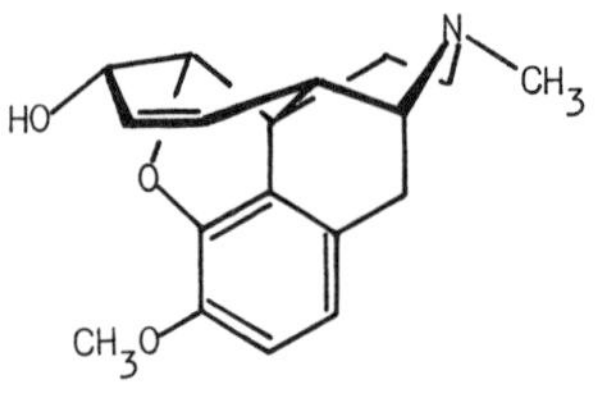

5. The erythrina alkaloid erythraline (19.12) from still another intramolecular coupling pattern (fifteen similar bases are known).

6. The pavine, argemonine (N-methylpavine, 19.13), which can be considered a rearrangement product of an oxidized laudanosoline-like (19.3) derivative (ten such materials have been found).

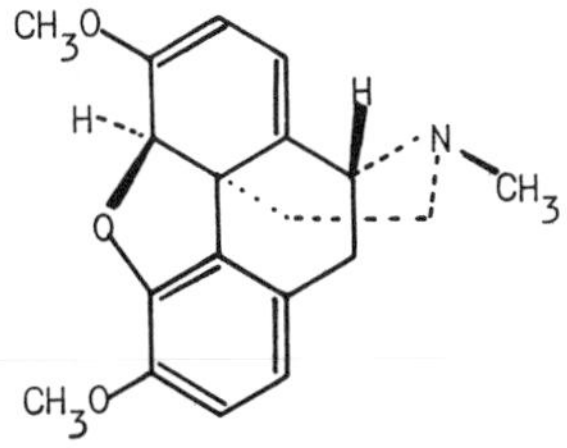

19.11

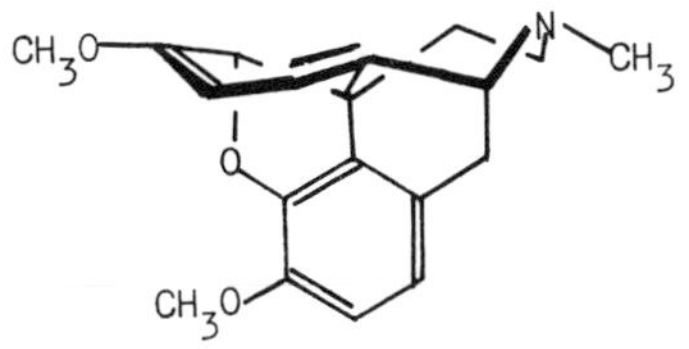

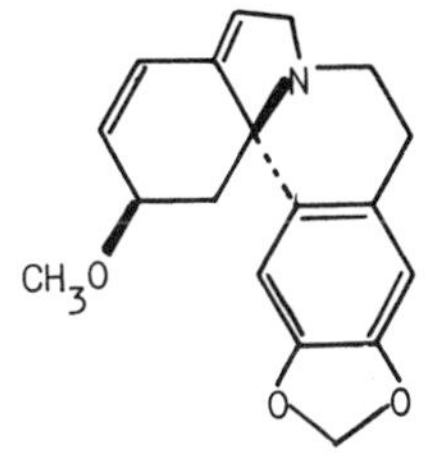

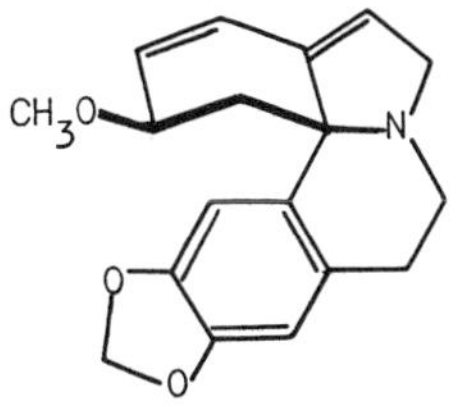

19.12

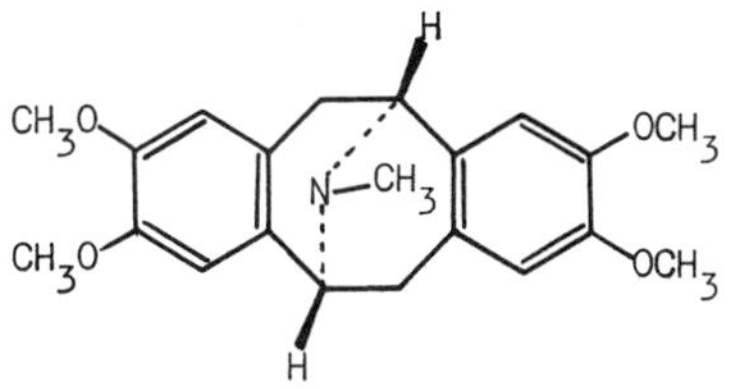

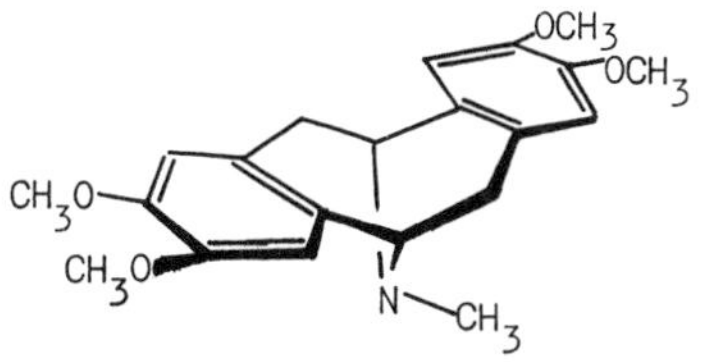

19.13

7. The protoberberine alkaloid, berberine (umbellatine, 19.14) in which a different kind of late-stage oxidation has been effected (about seventy analogs have been isolated).

8. The alkaloid protopine (19.15), a member of a class which arises by further transformations of the protoberberine group (almost fifteen of these are now known).

9. The benzophenanthridine bases sanguinarine (19.16) and (+)-chelidonine (19.17), where rearrangement of the protoberberine skeleton has taken a different course (about twenty-five members of this group have been found).

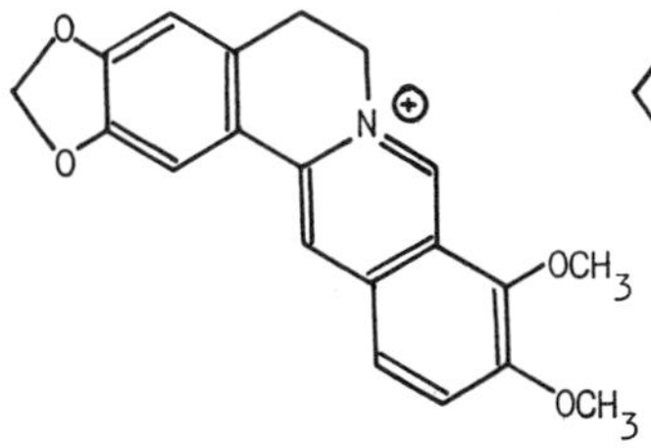

19.14 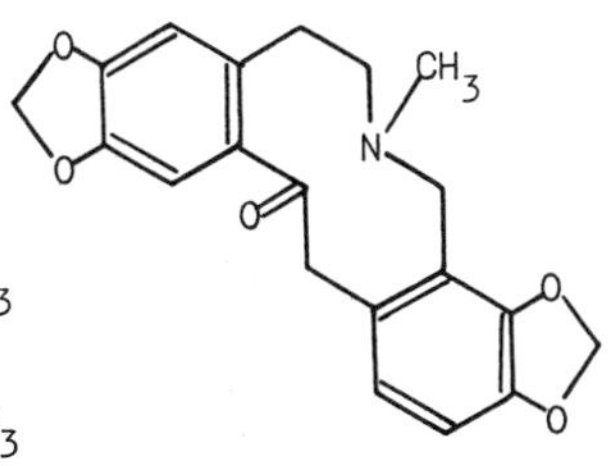 19.15 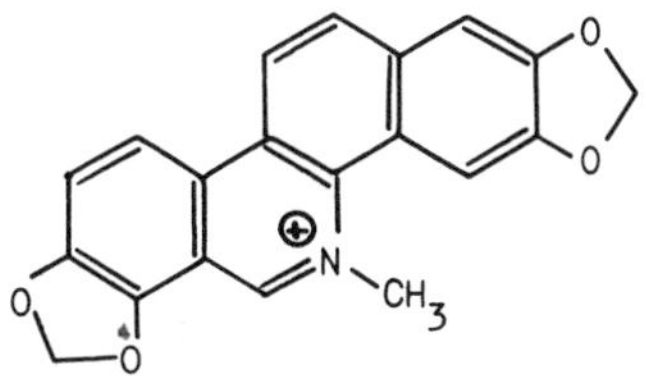 19.16

10. The highly oxidized phthalid isoquinoline narcotine (19.18; for which fifteen analogous compounds are known).

11. Rhoeadine (19.19) and its analogs, which also comprise late-stage rearrangement products presumably derived from the phthalid isoquinoline system (about twelve analogs are known).

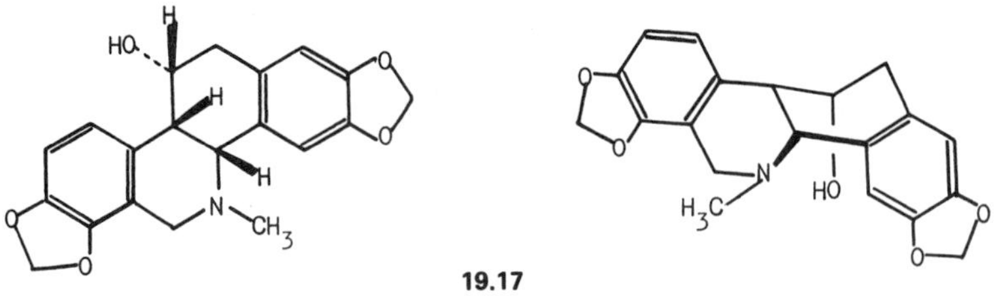

19.17

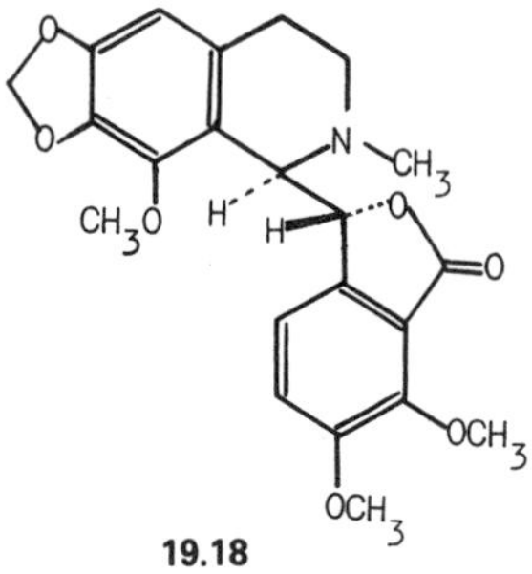

19.18

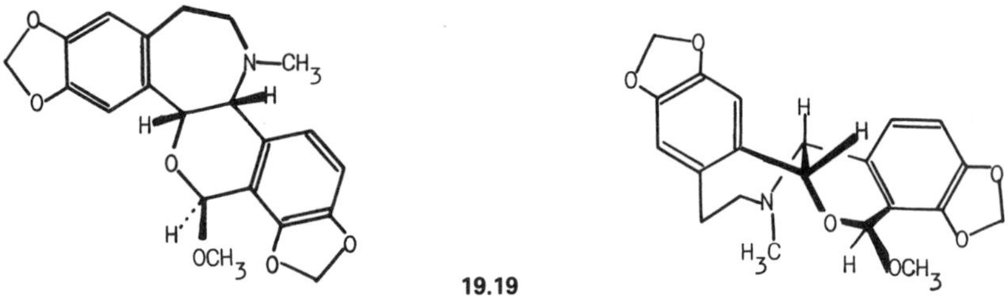

19.19

Finally, as was the case with the peyote (Chapter 17) and the Amaryllidaceae (Chapter 18) alkaloids, which are also derived from tyrosine (19.2), we shall see that the methylation pattern present in the 1-benzyltetrahydroisoquinoline system helps determine which alkaloids are elaborated; that the methylation to form the intermediate methylated 1-benzyltetrahydroisoquinolines may be early or late stage; and that the stereochemistry (axial or equatorial) of the benzyl (or hydrogen) at C_1 in the 1-benzyltetrahydroisoquinoline parent, are important considerations in controlling which specific alkaloid product develops in the plant.

I. LAUDANOSINE AND PAPAVERINE

(-)-Tyrosine (19.2) serves as a specific precursor to (+)-laudanosine (19.4) and papaverine (19.5) in Papaver somniferum L. Labeling experiments indicate that 2 equivalents of [2-^{14}C]tyrosine are incorporated into, for example, papaverine (19.5). The incorporation is specific but is not identical for each tyrosine (19.2), implying that the tyrosine (19.2) is partitioned into two different daughter molecules which then recombine to yield the alkaloids. The construction of the bases (+)-laudanosine (19.4) and papaverine (19.5), in concert with the above observations, can be accomplished in a number of ways. Three such acceptable pathways are as follows.

First, the oxidation (or hydroxylation) of tyrosine (19.2; Chapter 1, p. 3 and the introduction to this part, p. 178; Equation 19.1) is followed by partitioning of the latter between 3,4-dihydroxyphenethylamine (19.21) and 3,4-dihydroxyphenylpyruvic acid (19.22): The

19.2 → 19.20 (Eq. 19.1)

phenethylamine 19.21 would presumably be formed via the usual pyridoxyl-mediated (19.23) decarboxylation (Scheme 19.1), while the pyruvic acid (19.22) would be generated by means of a transamination reaction with an enzyme-bearing pyridoxyl (19.23) involved again (Scheme 19.2). Where they have been sought, the requisite enzymes are known to be present [78]. The combination of these fragments (Scheme 19.3) to the corresponding imine, cyclization to 19.24, and decarboxylation either oxidatively, perhaps via the N-oxide, to the dihydrobenzylisoquinoline 19.25, or without oxidation to norlaudanosoline (19.3), then follows. Methylation of norlaudanosoline (19.3) by S-adenosylmethionine (19.26) to (+)-N-norlaudanosine

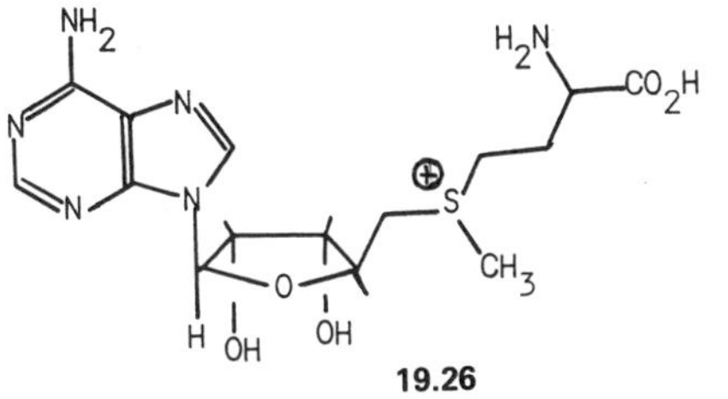

19.26

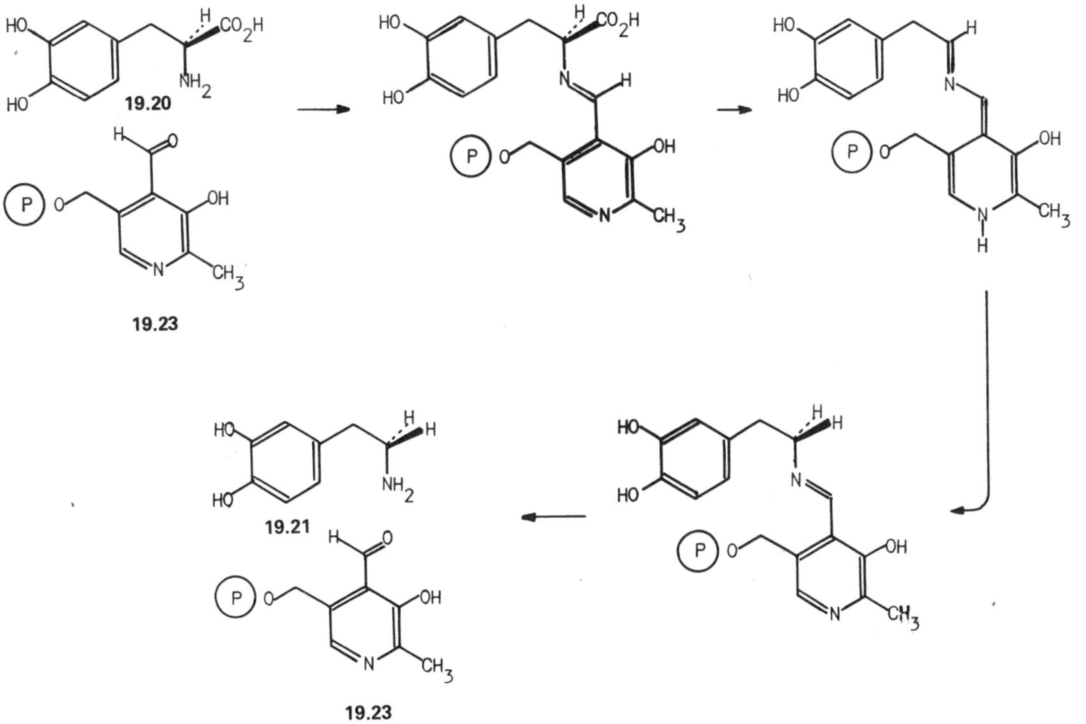

19.20 19.23 → → → 19.21 19.23

SCHEME 19.1 A pathway for the decarboxylation of 3,4-dihydroxyphenylalanine to 3,4-dihydroxyphenethylamine utilizing pyridoxyl catalysis.

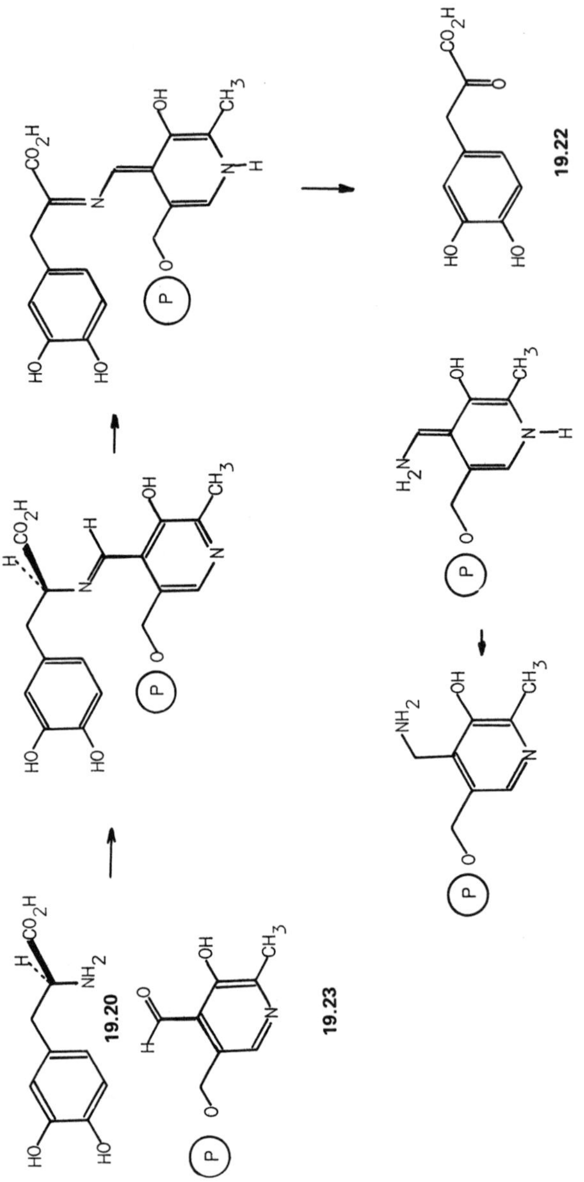

SCHEME 19.2 A pathway for the pyridoxyl-mediated transamination of dihydroxyphenylalanine to the corresponding dihydroxypyruvic acid.

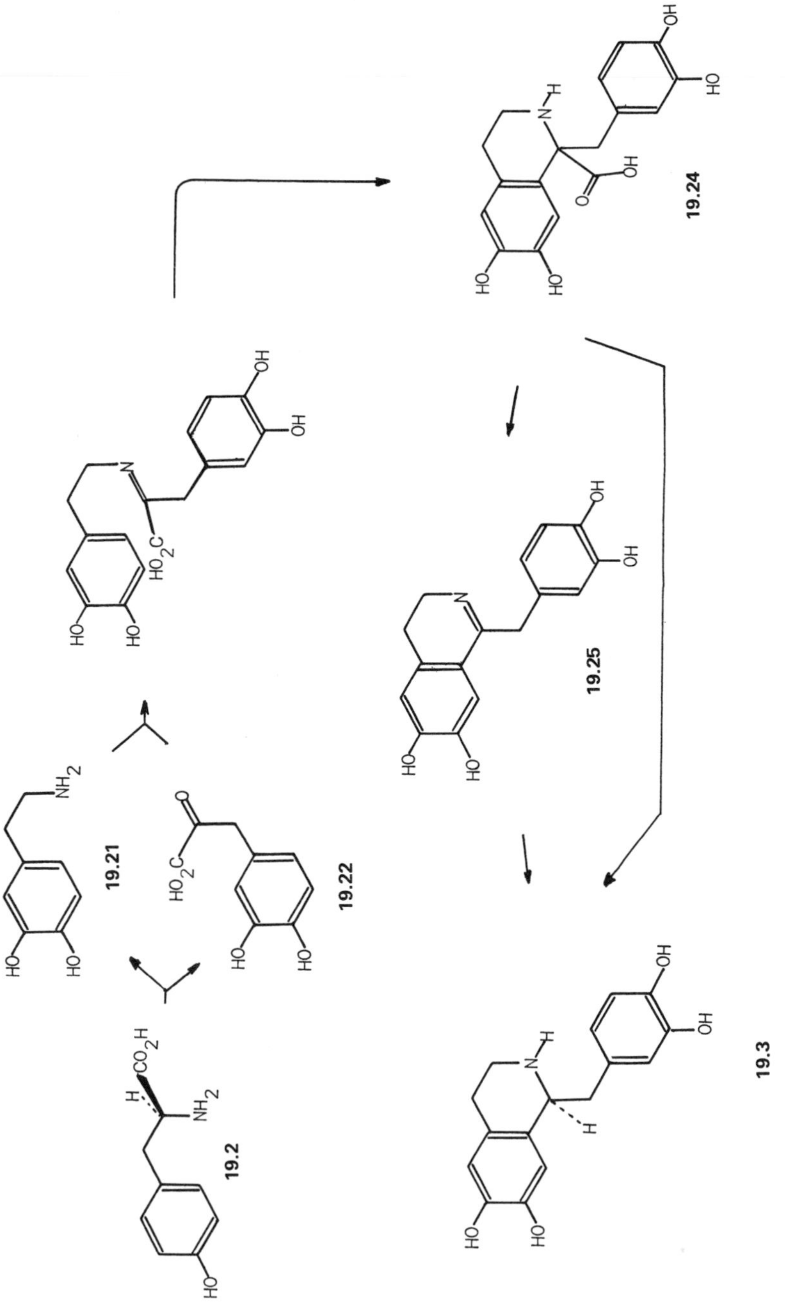

SCHEME 19.3 A pathway to norlaudanosoline from tyrosine.

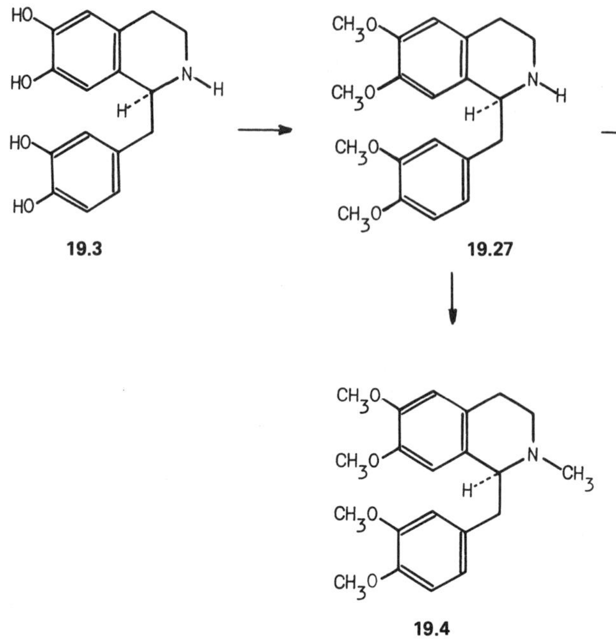

SCHEME 19.4 A pathway to papaverine and laudanosine from norlaudanosoline.

(tetrahydropapaverine, 19.27), oxidation to papaverine (19.5), or further methylation to (+)-laudanosine (19.4) would then finish the process (Scheme 19.4). The formation of a dehydrobenzylisoquinoline (19.25) along the path (Scheme 19.3) has been proposed on the grounds that 1,2-dehydronorreticuline (19.28) is incorporated well into morphine (19.2; see below),

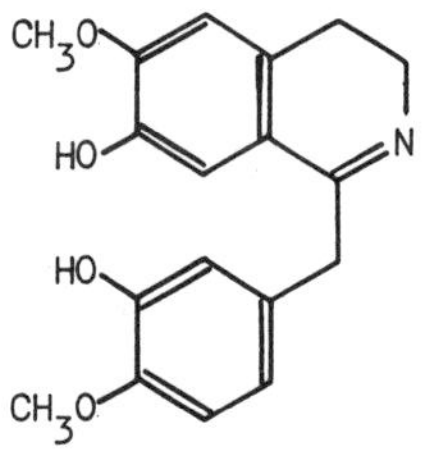

but it does not appear to operate here since 19.28 is not incorporated into papaverine (19.5). Instead, a pathway potentially available, but as yet without precedent, again utilizing a pyridoxyl-bearing (19.23) enzyme, may apply to generate norlaudanosoline (19.3) directly from 19.24 (Scheme 19.5).

Second, while norlaudanosoline (19.3) is a recognized precursor of (+)-reticuline (19.29) and tetrahydropapaverine (N-norlaudanosine, 19.27), as well as papaverine (19.5), recent feeding experiments [79] indicate that N-norreticuline (19.30) is very well incorporated into papaverine (19.5), and this suggests that early- rather than late-stage methylation may prevail. Thus, as indicated in Scheme 19.6, methylation of dihydroxyphenylalanine (19.20) by S-adenosylmethionine (19.26), followed by the reactions outlined in Schemes 19.1, 19.2, and e.g., 19.5, but this time on the methylated analogs (19.31 and 19.32), would generate norreticuline (19.30) directly without ever passing through norlaudanosoline (19.3) itself.

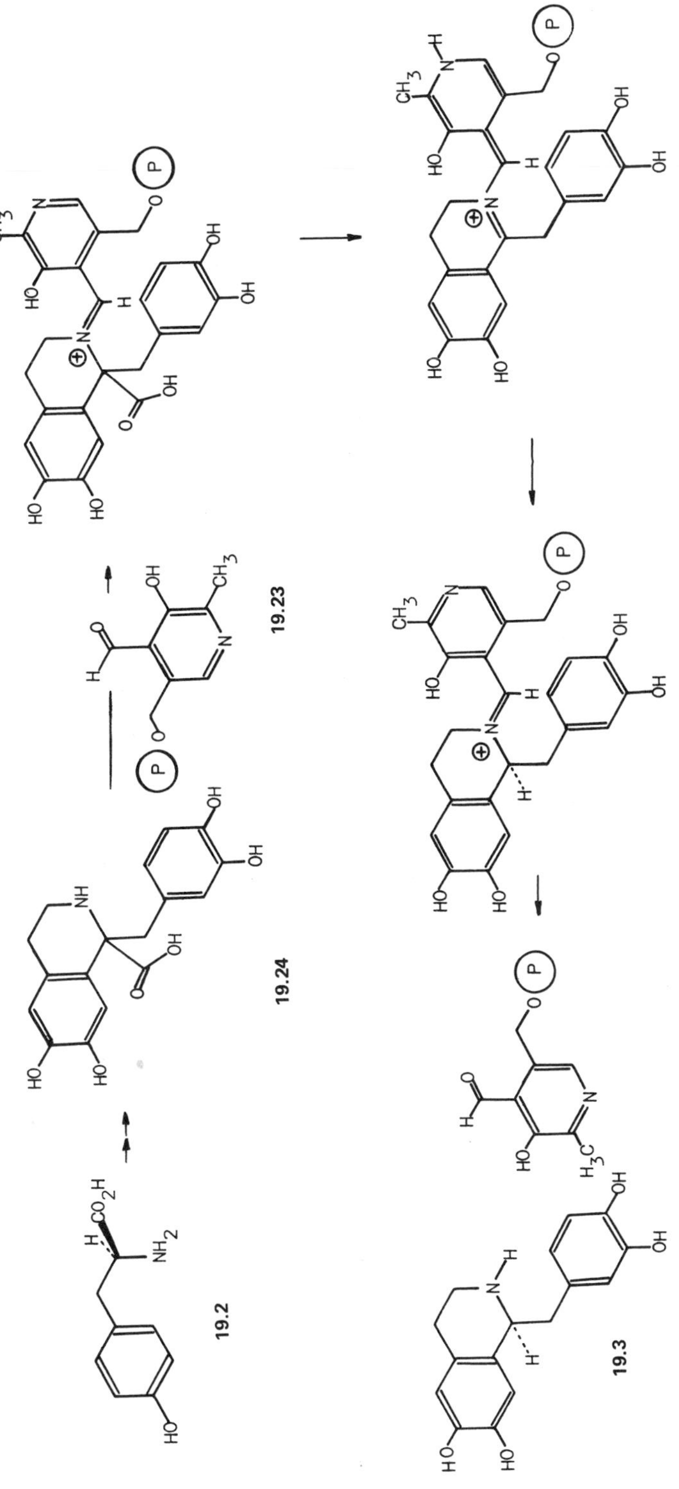

19.2

19.24

19.23

19.3

SCHEME 19.5 A pathway (without precedent) for the generation of norlaudanosoline from tyrosine with the mediation of a pyridoxyl-bearing enzyme.

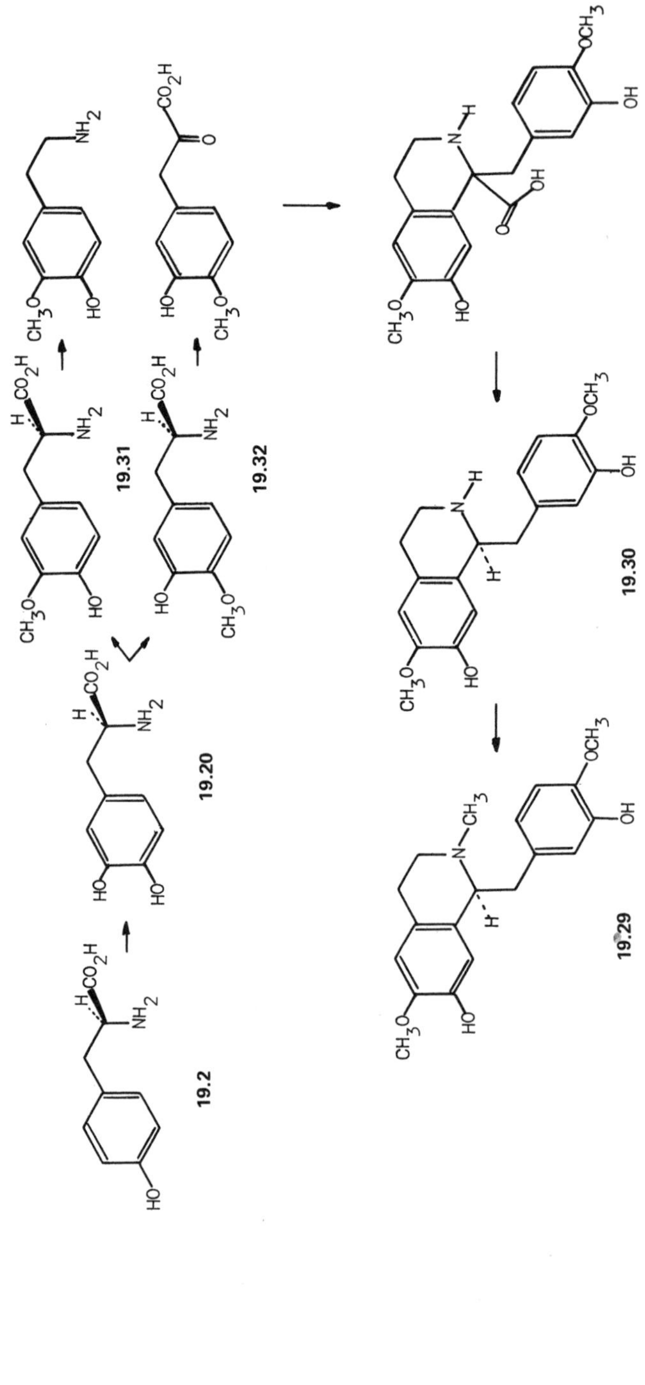
19.2
19.20
19.31
19.32
19.30
19.29
HO
CO2H
NH2
CH3O
OCH3
OH
N–H
N–CH3

SCHEME 19.6 A pathway to reticuline (involving norreticuline) which does not go through norlaudanosoline.

Methylation of norreticuline (19.30) would lead to tetrahydropapaverine (N-norlaudanosine, 19.27) which could be further methylated to (+)-laudanosine (19.4) or oxidized to papaverine (19.5). Such a scheme implies that norlaudanosoline (19.3) is not an obligatory intermediate on the route to reticuline (19.29), laudanosine (19.4), or papaverine (19.5), and it offers the advantage of early differentiation, by a methylation process, for the two different routes of tyrosine (19.2) utilization.

A third commonly written pathway may exist. It too is consonant with the available information. In this scheme of things, tyrosine (19.2) is converted to 3,4-dihydroxyphenethylamine (19.21; Scheme 19.1), which is then oxidized and hydrolyzed to 3,4-dihydroxyphenylacetaldehyde (19.33). [Alternatively, 19.33 might come (Scheme 19.7) from the thiamine pyrophosphate-catalyzed (19.34) decarboxylation of the pyruvic acid 19.22 (Scheme 19.2)]. Combination of 3,4-dihydroxyphenethylamine (19.21) and 3,4-dihydroxyphenylacetaldehyde (19.33) generates the imine 19.35 which then undergoes cyclization to norlaudanosoline (19.3; Scheme 19.8), the methylation of which generates N-norlaudanosine (19.27). As noted before (Scheme 19.4), N-norlaudanosine (19.27) may be converted to (+)-laudanosine (19.4) and to papaverine (19.5). This pathway can also accommodate early methylation to permit differentiation of the two diverse routes tyrosine (19.2) must take.

Most recent feeding experiments [80, 81] appear to favor the first route.

II. OXYACANTHINE AND TUBOCURARINE

(+)-Oxyacanthine (19.6) and (+)-tubocurarine (19.7) are representative members of different subgroups of bisbenzylisoquinoline alkaloids. About twenty such subgroups exist, and together they contain over 100 known bisbenzylisoquinoline alkaloids. This large number of kinds of dimeric benzylisoquinoline bases comes about because of the possible variations in the joining (conceptually, if not actually) of two benzyltetrahydroisoquinoline-derived fragments and their subsequent methylation. In all cases known, the ether bridge joining the two benzylisoquinoline-derived fragments is fixed to one benzylisoquinoline in a position dictated by the derivation of that fragment from tyrosine (19.2) or its further transformation products, while the other end is fixed to a carbon ortho to that demanded by derivation from tyrosine (19.2) or its further transformation products.

Although feeding experiments have not been reported, it is generally accepted that S-adenosylmethionine (19.26) serves as the source of the N- and O-methyl (or methylenedioxy) groups. Thus, (+)-oxyacanthine (19.6) might arise, for example, by: (a) oxidative dimerization of 2 equivalents of 1-(p-hydroxybenzyl)-1,2,3,4-tetrahydro-6,7-dihydroxyisoquinoline (19.36), derived, as per Scheme 19.9, from 3,4-dihydroxyphenethylamine (19.21) and p-hydroxyphenylpyruvic acid (19.37), followed by methylation by S-adenosylmethionine (19.26; Scheme 19.10); or (b) by S-adenosylmethionine methylation of 3,4-dihydroxyphenethylamine (19.21) derived from tyrosine (19.2) and reaction of that fragment with the pyruvic acid (19.37) to yield N-methylcoclaurine (19.38): dimerization of N-methylcoclaurine and further methylation yields (+)-oxyacanthine (19.6; Scheme 19.11); or (c) simply by methylation of 1-(p-hydroxybenzyl)-1,2,3,4-tetrahydro-6,7-dihydroxyisoquinoline (19.36) to N-methylcoclaurine (19.38), oxidative dimerization and a final methylation. Interestingly, the dimerization proceeds, whatever the pathway, in such a way that one new carbon-oxygen bond is formed ortho to an oxygen already present in one half, and the second new carbon-oxygen bond is formed in the reverse sense (i.e., ortho to an oxygen already present in the second half).

The curarine alkaloids are different in the way dimerization occurs. One scheme is shown (Scheme 19.12), although others may be written. In this group of alkaloids, both new carbon-oxygen bonds are formed ortho to oxygen substituents in the same half. Thus, as shown in Scheme 19.12, coupling occurs ortho to two oxygens in 19.39b, a resonance form

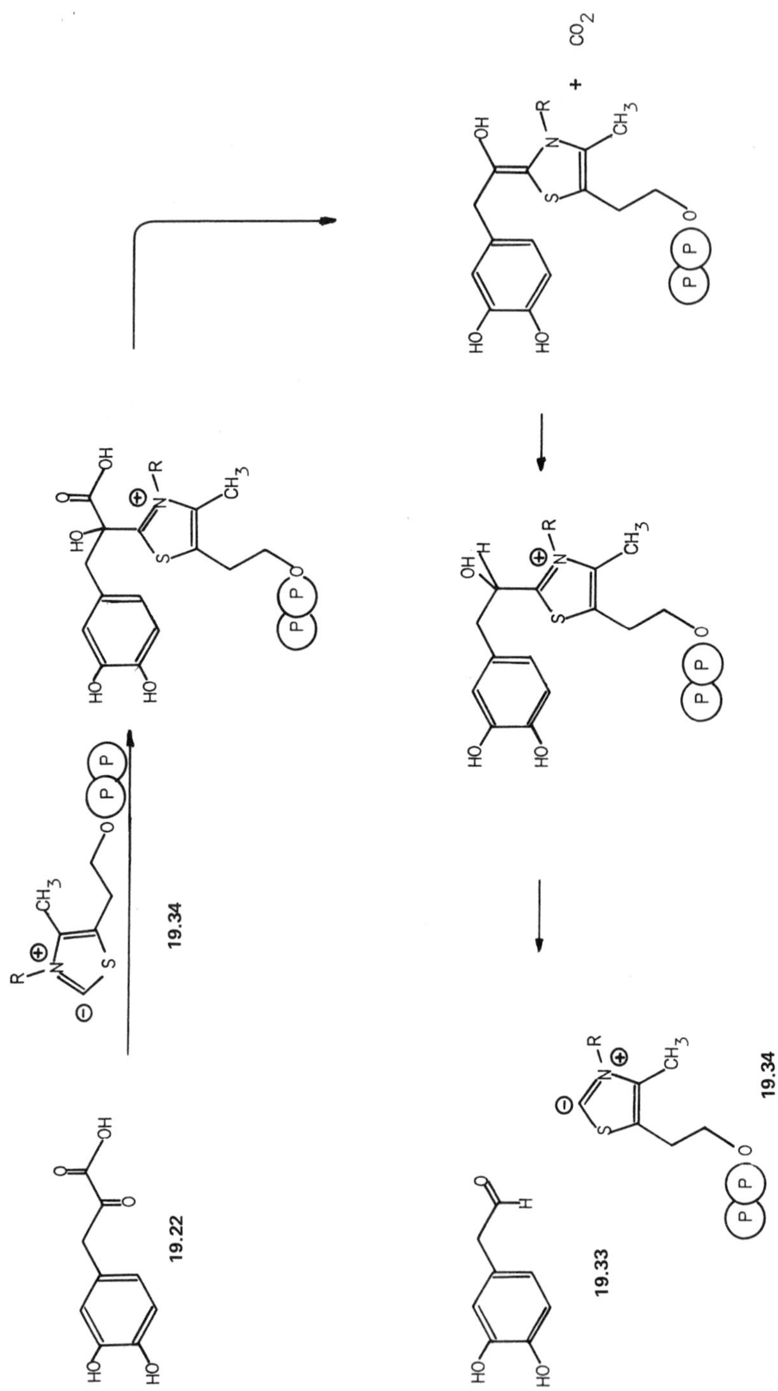

19.22

19.34

19.33

19.34

SCHEME 19.7 A possible pathway to 3,4-dihydroxyphenylacetaldehyde by the decarboxylation (thiamine pyrophosphate-catalyzed) of phenylpyruvic acid.

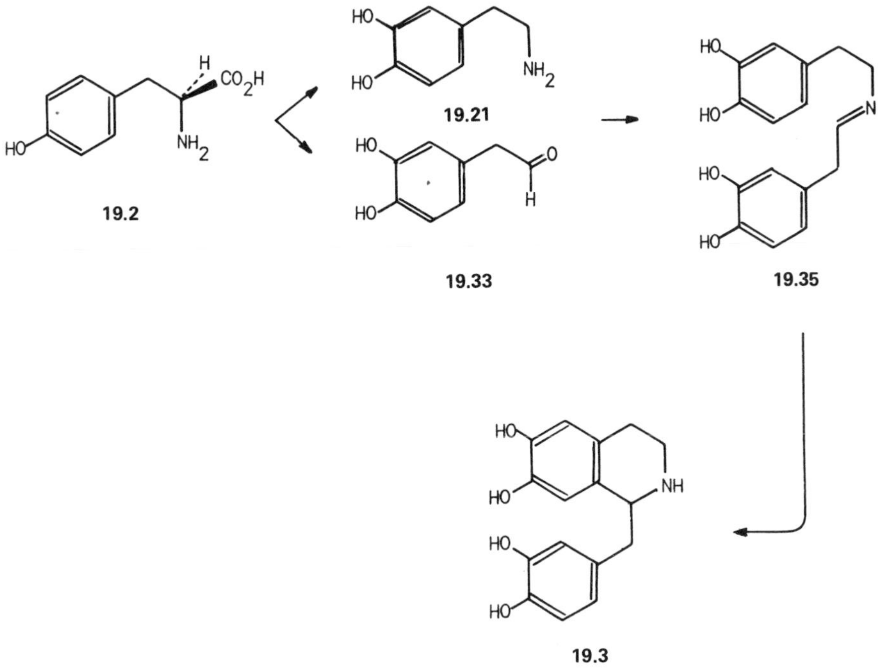

SCHEME 19.8 A pathway to norlaudanosoline involving 3,4-dihydroxyphenethylamine and 3,4-dihydroxyphenylacetaldehyde.

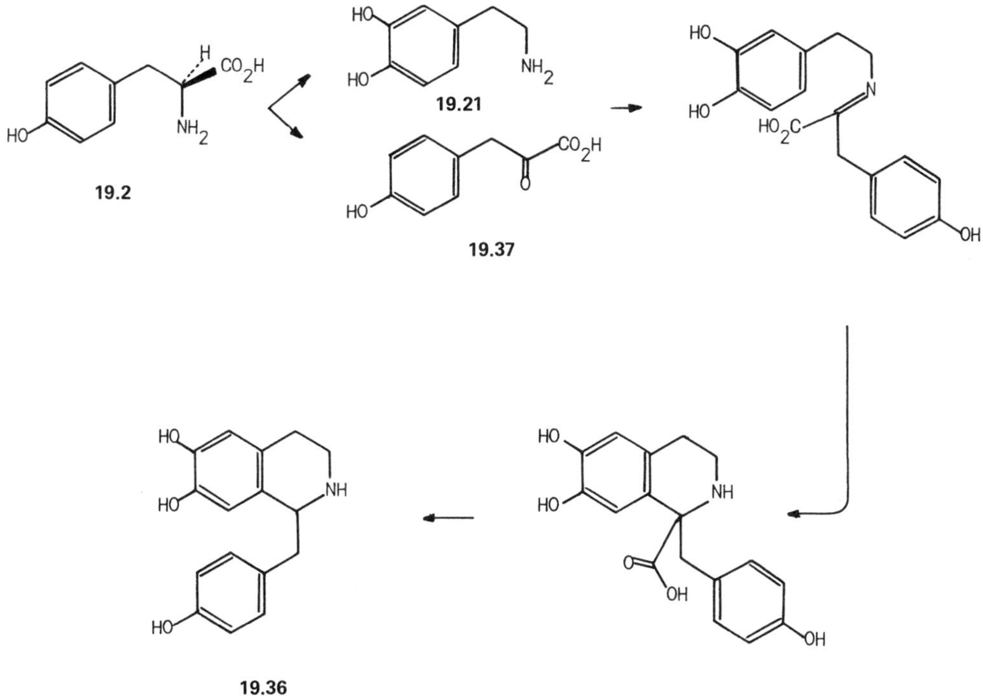

SCHEME 19.9 A pathway for the formation of 1-(p-hydroxybenzyl)tetrahydroisoquinoline from tyrosine.

SCHEME 19.10 A pathway for the formation of oxyacanthine from dimerization of 2 equivalents of 1-(p-hydroxybenzyl)-1, 2, 3, 4-tetrahydro-6, 7-dihydroxyisoquinoline.

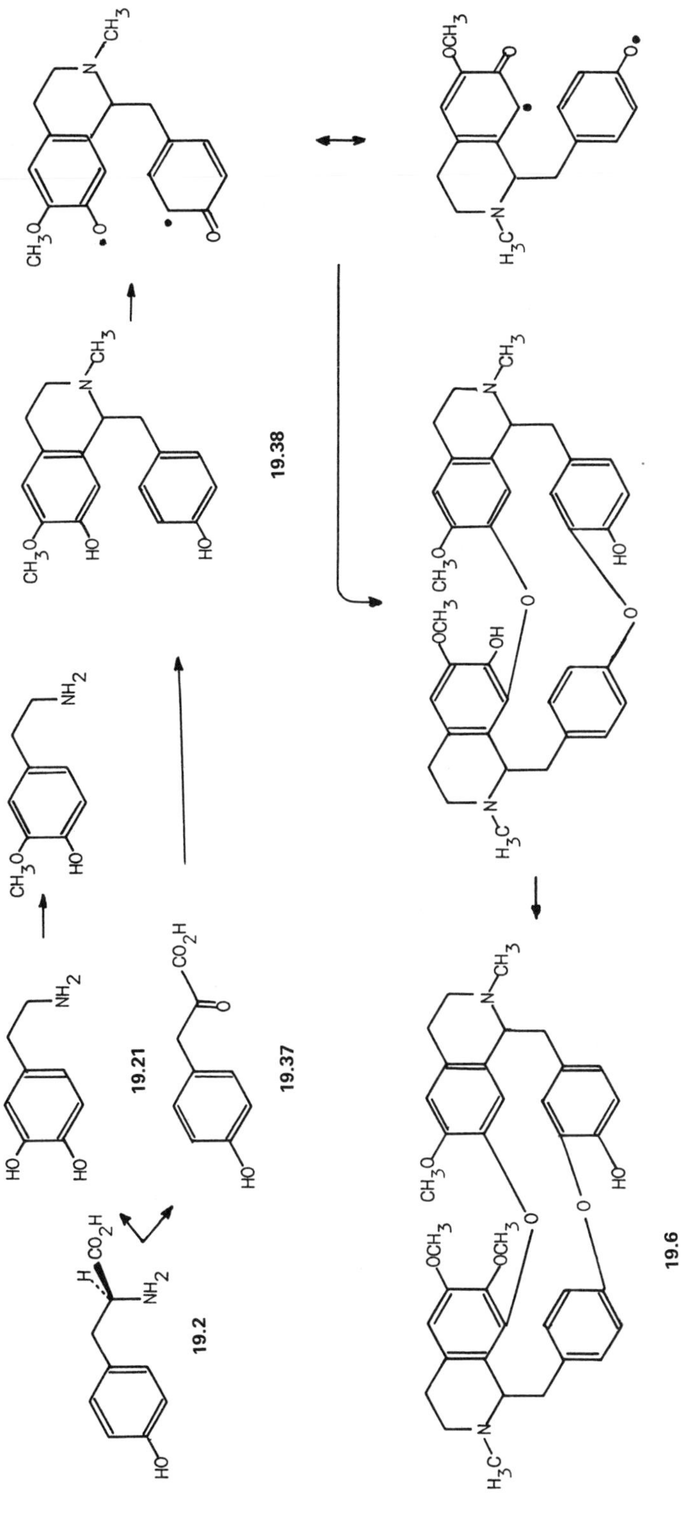

SCHEME 19.11 A pathway to oxyacanthine via N-methylcoclaurine.

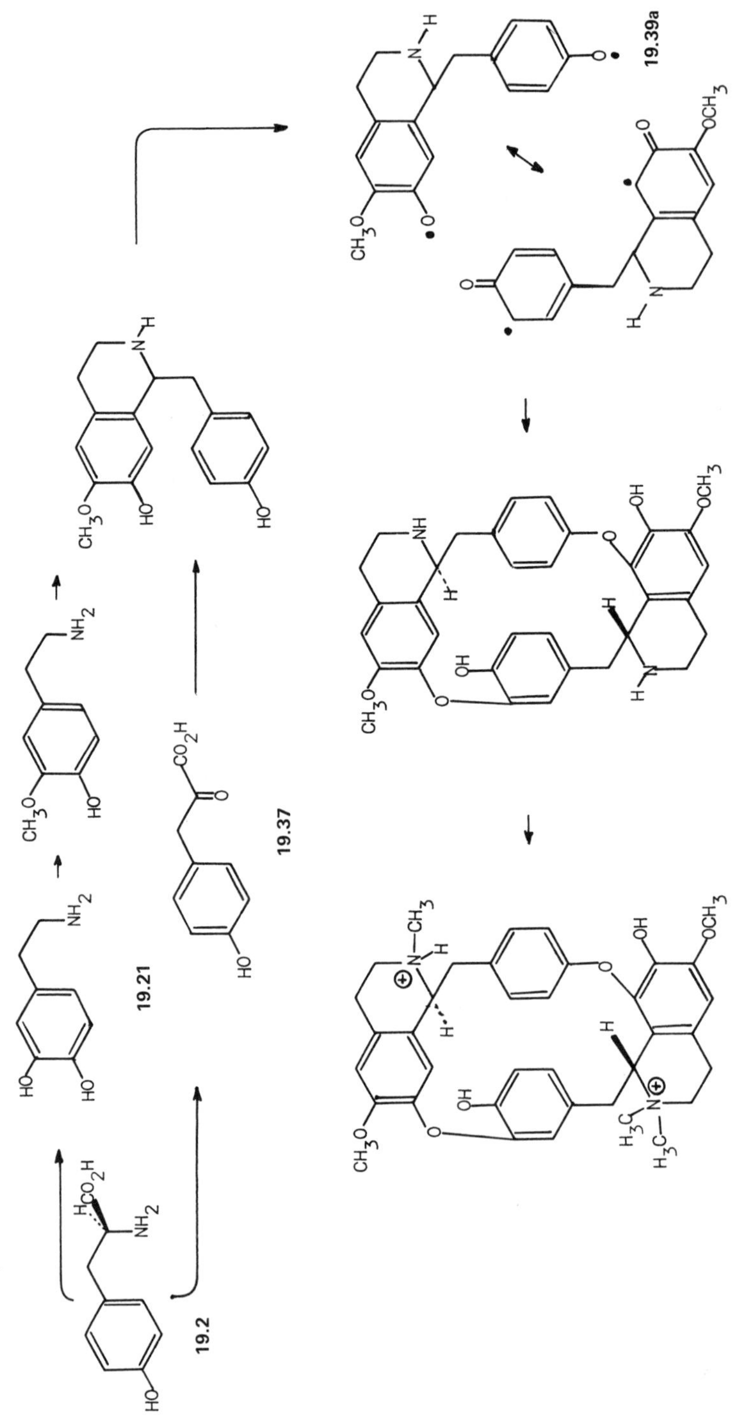
19.2
19.21
19.37
19.39a
19.39b
19.7
CH3O
HO
NH2
CO2H
OCH3
OH
N—H
NH
N—CH3
H3C
H

SCHEME 19.12 A pathway to tubocurarine by dimerization of two tetrahydroisoquinoline precursors.

of 19.39a, while 19.39a, a resonance form of 19.39b, could be considered as providing the oxygens to form the bridges. Clearly, however, since these cannot be individual reacting species, (+)-tubocurarine (19.7) must form in a more subtle fashion.

III. ORIENTALINONE AND ISOTHEBAINE

(-)-Orientalinone (19.8) is derived from (+)-orientaline (19.40) in Papaver orientale L. (Scheme 19.13). As is true for the other tetraoxygenated tetrahydrobenzylisoquinoline alkaloids already discussed, i.e., laudanosine (19.4), reticuline (19.29), and N-norreticuline (19.30), the derivation of (+)-orientaline (19.40) from tyrosine (19.2) presumably occurs via decarboxylation of 3,4-dihydroxyphenylalanine (19.20), before or after condensation with 3,4-dihydroxyphenylpyruvic acid (19.22) or 3,4-dihydroxyphenylacetaldehyde (19.33). Cyclization, reduction to norlaudanosoline (19.3), and, finally, methylation by S-adenosylmethionine (19.26) yield (+)-orientaline (19.40). Alternatively, methylation at the 3,4-dihydroxyphenylalanine (19.20), dihydroxyphenethylamine (19.21), and/or pyruvic acid (19.22) stage might occur before condensation and cyclization to (+)-orientaline (19.40).

Regardless, however, the establishment of (-)-orientalinone (19.8) as the pregenitor of (+)-isothebaine (19.9) served to demonstrate that the spirodienone system of proaporphines lay between the tetrahydrobenzylisoquinolines and the aporphine alkaloids. Indeed, although it is still possible that direct coupling of some benzyltetrahydroisoquinolines to aporphines may occur, the evidence appears to militate against such a pathway. As is shown in Scheme 19.13, the currently held view of the biogenesis of aporphine alkaloids is that they are formed via proaporphines [65], which in turn arise oxidatively from 1-benzyltetrahydroisoquinolines [68].

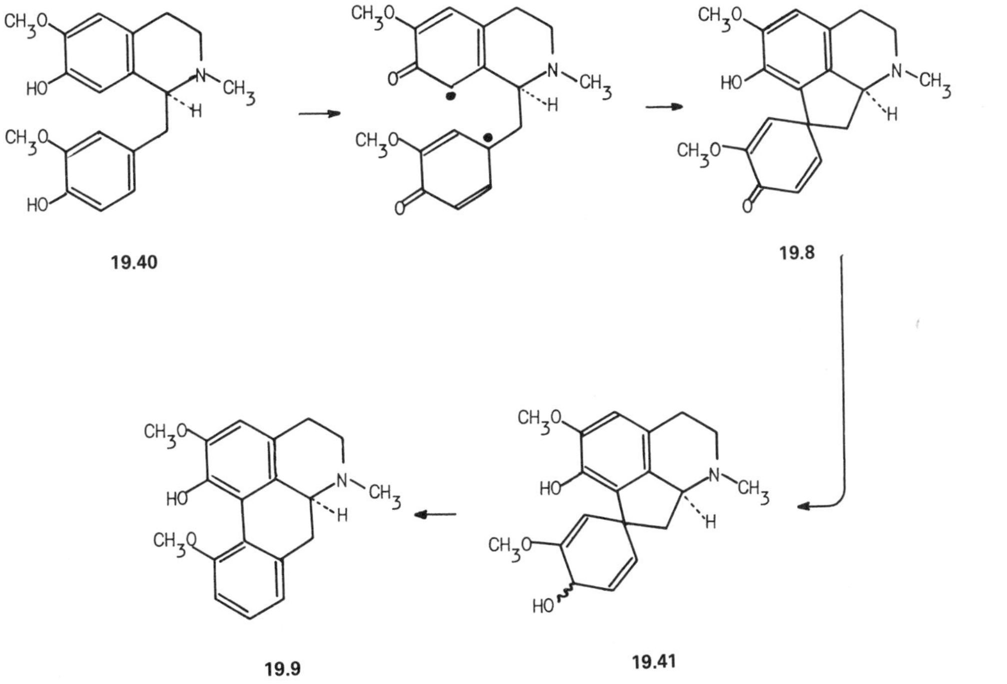

19.40

19.8

19.9

19.41

SCHEME 19.13 A pathway for the production of (+)-isothebaine from (+)-orientaline via (-)-orientalinone.

According to this scheme of things (Scheme 19.13), (+)-orientaline (19.40) is converted to (-)-orientalinone (19.8) via intramolecular radical coupling. Reduction of (-)-orientalinone (19.8) to orientalinol (19.41) and the dienol - benzene rearrangement, with loss of water, to generate the aporphine (+)-isothebaine (19.9) follow and serve to account for the relationship surmised from feeding experiments. Thus, tritium-labeled (at the N-methyl and at C_{10}) epimeric orientalinoles (19.41) were fed separately to *Papaver orientale* L. One of the two orientalinoles (19.41) was incorporated well into (+)-isothebaine (19.9) with retention of tritium, while the other was only poorly utilized. Since the configuration of the epimer utilized best is not yet known, further details of the process cannot be resolved [82].

Finally, it is interesting to note, since it demonstrates the controlling effect of the methylation pattern and stereochemistry at C_1 in the tetrahydrobenzylisoquinoline system, that of the alkaloids considered here *and found in the same plant* (*Papaver orientale* L.), *neither* (-)-reticuline (19.42) *nor* (+)-reticuline (19.29) will serve to generate (+)-isothebaine

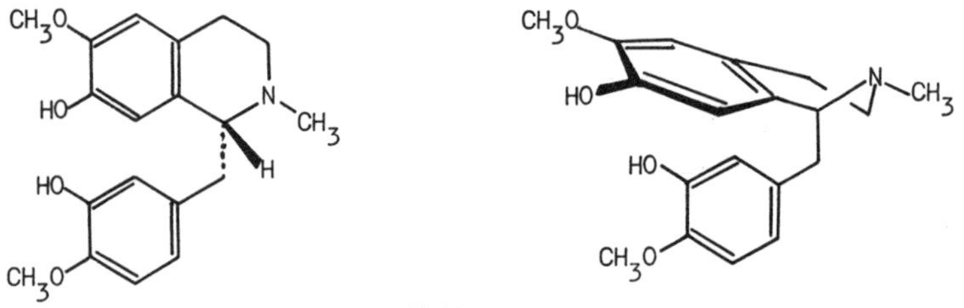

19.42

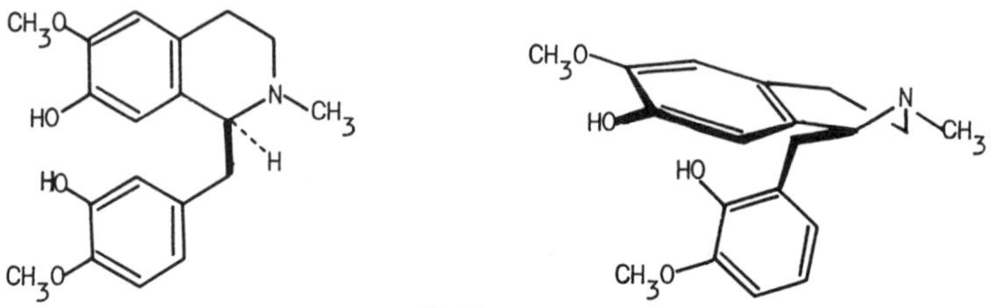

19.29

(19.9) (which comes *only* from (+)-orientaline, 19.40) and that (-)-reticuline (19.42) yields *only* the morphine alkaloids (see below) while (+)-reticuline (19.29) yields *only* narcotine (19.18) and the other alkaloids (e.g., berberine, 19.14) from which narcotine (19.18) is derived (see below).

IV. MORPHINE, CODEINE, AND THEBAINE

Of the four possible O,O,N-trimethyllaudanosolines, i.e., (±)-reticuline (19.43), (±)-orientaline (19.44), (±)-protosinomenine (19.45), and the unnamed and as yet unreported isomer (±)-(19.46), *only* (-)-reticuline (19.42) serves as the precursor to thebaine (19.11), codeine (19.10), and morphine (19.1) in *Papaver orientale* L.

Pathways to (±)-reticuline (19.43) from tyrosine (19.2) have already been presented (Section I). Although 1,2-dehydroreticuline (19.28) has been demonstrated to be a precursor to thebaine (19.11), it appears unlikely, given the specificity for (-)-reticuline (19.42) incorporation, that 19.28 is obligatory. Indeed, the presently accepted biosynthesis of the morphine bases, in concert with $^{14}CO_2$ feeding experiments [83], as well as the earlier work, up

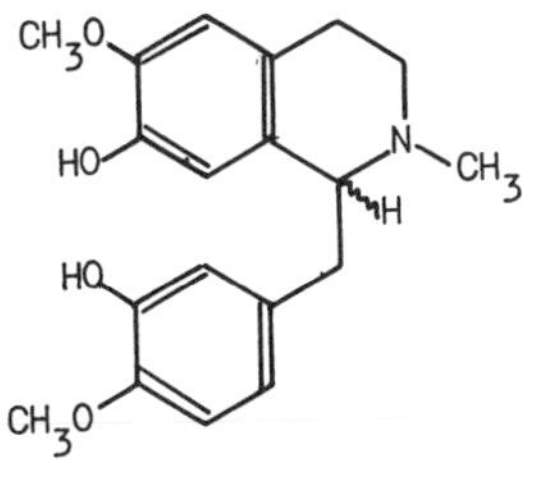

19.43 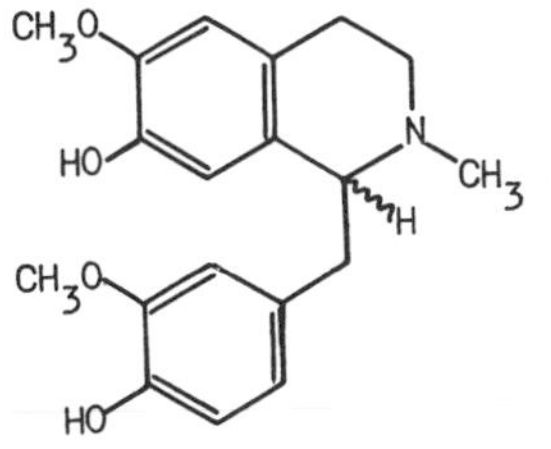 **19.44**

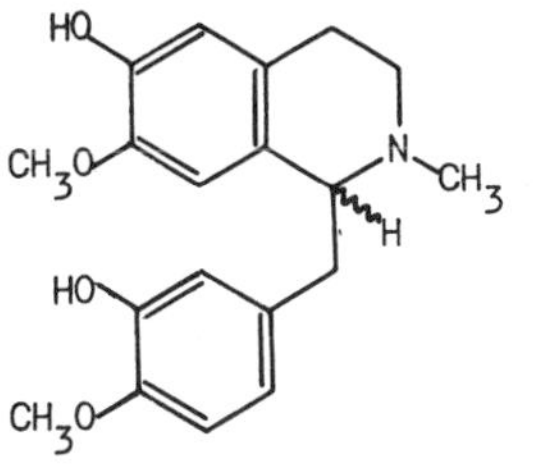

19.45 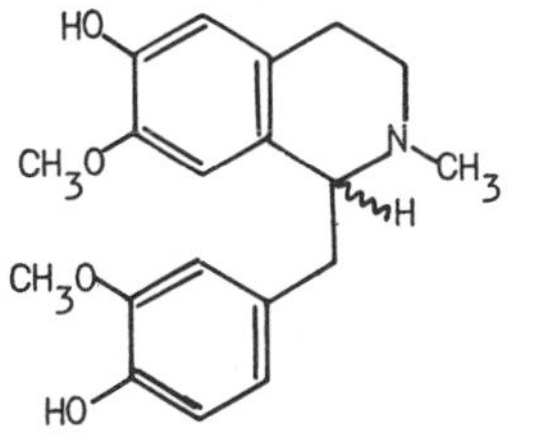 **19.46**

to and including that on (-)-reticuline (19.42), which is outlined in Scheme 19.14, would appear to indicate that 1,2-dehydroreticuline (19.28) might be incorporated, but that it is not necessary. In Scheme 19.14, it is seen that (-)-reticuline (19.42) undergoes oxidative coupling, reduction, and the loss of the elements of water, to yield thebaine (19.11) directly. Then (Scheme 19.15), hydrolysis of the enol ether of thebaine (19.11) yields neopinone (19.47), a natural constituent of Papaver somniferum L., from which it has been isolated, which subsequently rearranges to codeinone (19.48). Reduction of codeinone (19.48) yields codeine (19.10) and demethylation of the latter, morphine (19.1).

There are two further points of interest in the biosynthesis of these bases. First, codeinone (19.48), which goes on to generate codeine (19.10), does not yield thebaine (19.11). Thus, the process is apparently irreversible. Second, morphine (19.1) does not accumulate in the plant. Thus, the degradation of thebaine (19.11) which yielded codeine (19.10) and morphine (19.1) continues. Little work has been reported on the potentially interesting products of the in vivo catabolism of morphine (19.1; [84]).

V. ERYTHRALINE

The role of (+)-N-norprotosinomenine (19.49), a dimethyl derivative of norlaudanosoline (19.3) in the genesis of erythraline (19.12), has recently been investigated [82, 85, 86]. (+)-N-Norprotosinomenine (19.49), which is incorporated into erythraline (19.12) in Erythrina crista galli much more significantly than its enantiomer, is presumably generated from tyrosine (19.2) via 3,4-dihydroxyphenylalanine (19.20) and 3,4-dihydroxyphenylpyruvic acid (19.22). Late-stage methylation of the norlaudanosoline (19.3), resulting from the above condensation by S-adenosylmethionine (19.26), then provides (+)-N-norprotosinomenine (19.49). The discussions already provided above for potential pathways to (±)-laudanosine (19.4 and its mirror image), (±)-reticuline (19.43), and (±)-orientaline (19.44), are applicable here too for (+)-N-norprotosinomenine (19.49), and, in particular, the methylation by S-adenosylmethionine (19.26) may be early or late stage.

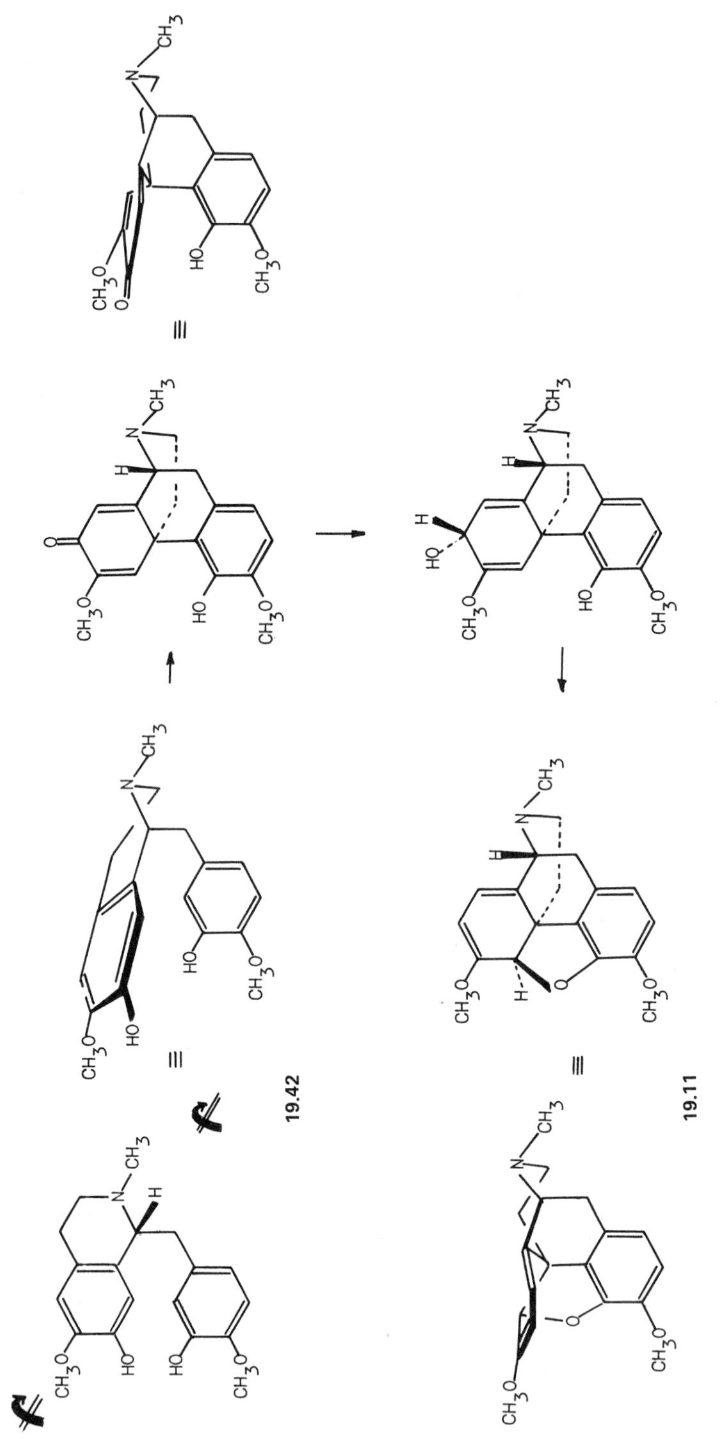

SCHEME 19.14 A pathway for the formation of thebaine from (-)-reticuline.

SCHEME 19.15 A pathway for the formation of morphine from thebaine (via codeine).

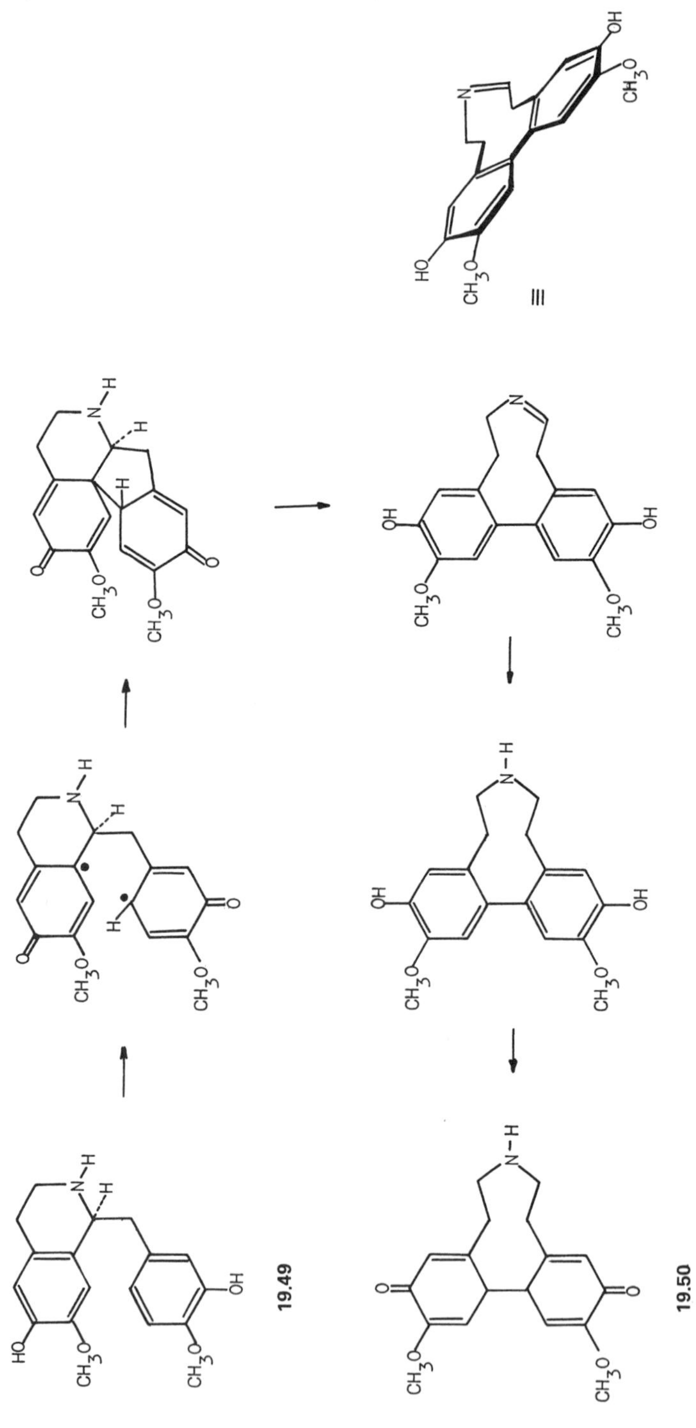

19.49

19.50

SCHEME 19.16 A pathway to erythraline intermediates from (+)-N-norprotosinomenine.

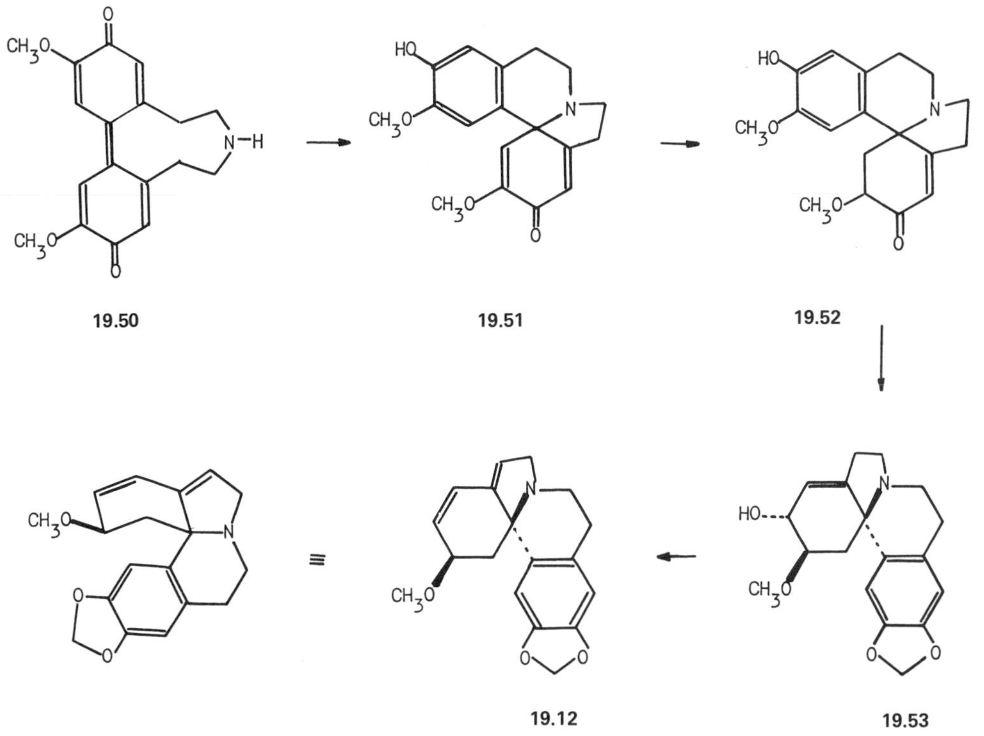

19.50 19.51 19.52

19.12 19.53

SCHEME 19.17 A pathway to erythraline from an N-norprotosinomenine-derived symmetrical intermediate.

As shown in Scheme 19.16, the incorporation of (+)-N-norprotosinomenine (19.49) into erythraline (19.12) occurs via a symmetrical intermediate (19.50). Thus, (+)-[5-^{3}H,3-^{14}C]N-norprotosinomenine (19.49) is incorporated as above into erythraline (19.12) with the distribution of label predicted on the basis of the symmetrical intermediate 19.50. Further (Scheme 19.17), when [O,O-dimethyl-^{14}C,1,17-$^3H_2$]erysodienone (19.51), derived from the symmetrical intermediate (19.50) is fed to the same plant, erythraline (19.12), in which there has been no scrambling of label, is isolated.

Additional evidence for the intermediacy of 19.50 is provided by good incorporation of this material, without randomization of label, and from the fact that when (+)-[4'-methoxy-^{14}C]N-norprotosinomenine (19.49) is administered to Erythrina crista galli, the erythraline (19.12) produced has activity equally distributed between the methoxy and the methylenedioxy groups. This last point demonstrates that a symmetrical intermediate is present, and that, as is the case with the Amaryllidaceae, the methylenedioxy group is generated from an ortho hydroxy-methoxy combination [87]. Other intermediates are also presumed present, and some of these materials, e.g., the amine 19.52 and erythratine (19.53), have been fed in labeled form and shown to lead to erythraline (19.12). They may not be, however, obligatory precursors.

VI. ARGEMONINE

Very little firm information exists on the genesis of the pavine (-)-argemonine (19.13) or, indeed, any of the pavines. In the absence of evidence, speculation has taken two courses. The earliest suggestion (Scheme 19.18; [88]) presumed that (+)-reticuline (19.29), a normal

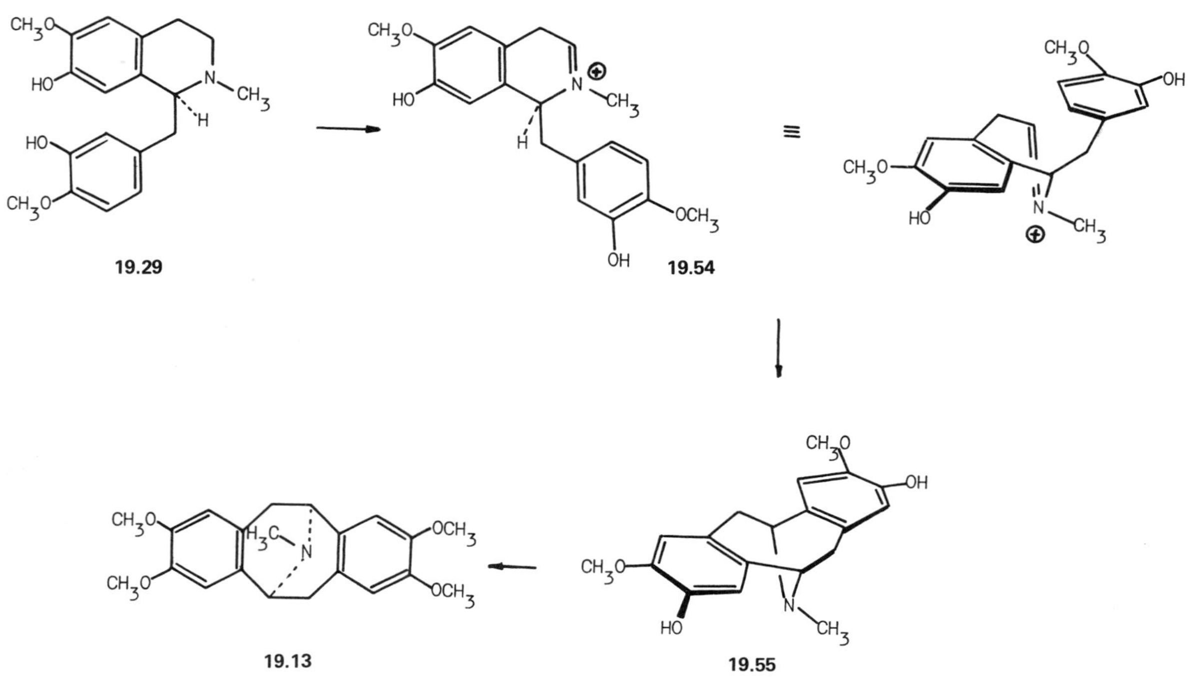

19.29

19.54

19.55

19.13

SCHEME 19.18 A pathway to erythraline from an N-norprotosinomenine-derived symmetrical intermediate.

constituent of Argemone mexicana L., serves as the precursor of an imine (19.54) which undergoes cyclization to (-)-bis-norargemonine (19.55) and sequential methylations to (-)-argemonine (19.13). Labeled alkaloids are not generated from feeding experiments with labeled (+)-reticuline (19.29), but it is not known if this is simply due to poor uptake by the plant. The second suggestion [89] is based upon the observation that attempted preparation of 4-hydroxytetrahydroisoquinolines, e.g., 19.56, under acidic conditions, fails and generates isopavines, e.g., (±)-amurensinine (19.57) instead (Equation 19.2). It is suspected that dehydration and rearrangement to the appropriate imine (e.g., 19.54) would lead to pavines. The genesis of the unknown, but required, 4-hydroxytetrahydroisoquinoline (19.56) which might, a priori, come from early oxidation (Equation 19.3; via noradrenaline, 19.58, for example) or late oxidation of the already elaborated 1-benzyltetrahydroisoquinoline (i.e., 19.59), has not been, for this system, expostulated upon (Equation 19.4).

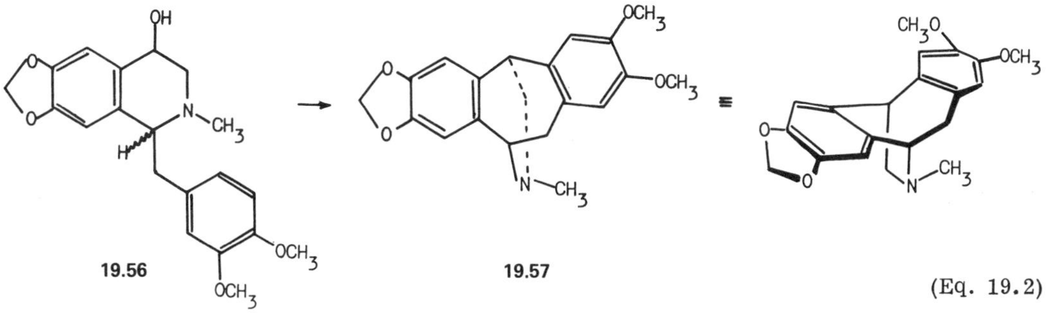

19.56 → 19.57 (Eq. 19.2)

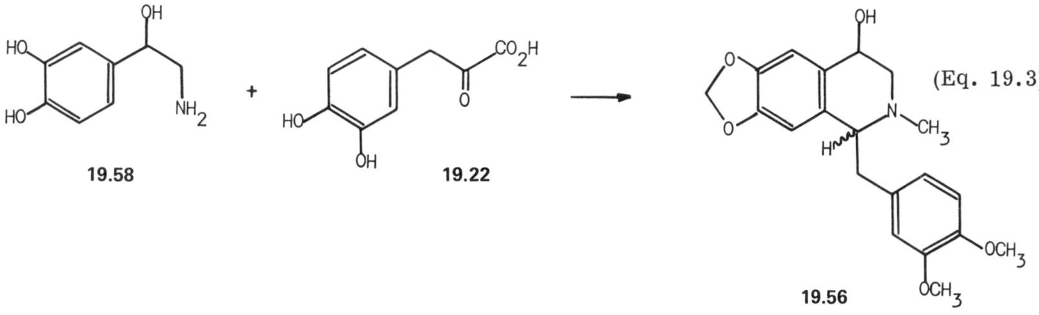

19.58 + 19.22 → 19.56 (Eq. 19.3)

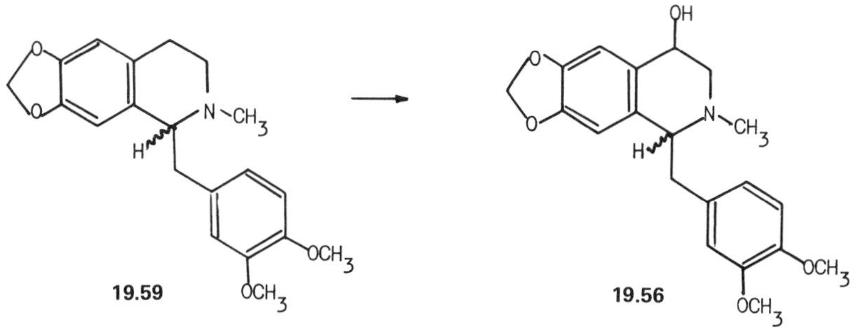

19.59 → 19.56 (Eq. 19.4)

VII. BERBERINE (UMBELLATINE)

It has been shown [90] that two tyrosine-derived (19.2) fragments are incorporated into berberine (19.14). However, since 3,4-dihydroxyphenethylamine (19.21) is only incorporated once, it is assumed that two nonidentical fragments, one of which is either 3,4-dihydroxyphenethylamine (19.21) or is derivable from 3,4-dihydroxyphenethylamine (19.21), while the other one is not, are required. In the same vein, it is known [91, 92] that (+)-reticuline (19.29), labeled at the nitrogen and oxygen methyls, yields in Hydrastis canadensis L., specifically labeled berberine (19.14). In addition to demonstrating that (+)-reticuline (19.29) is incorporated, these results also show that it is the N-methyl of (+)-reticuline (19.29) which becomes C_8 of berberine (19.14) and the O-methyl, as we have come to expect, which becomes the methylenedioxy group. In the same plant, (±)-protosinomenine (19.45) is not incorporated at all.

Scheme 19.19 provides a pathway, consonant with the available data, for the generation of berberine (19.14). The requirement for (+)-reticuline (19.29) appears to dictate cyclization to a suitable intermediate, i.e., (-)-scoulerine (19.60), before oxidation. (+)-Reticuline (19.29) can be derived, as indicated (p. 226) from condensation between 3,4-dihydroxyphenylpyruvic acid (19.22) and 3,4-dihydroxyphenalanine (19.20), both derived from tyrosine (19.2), to norlaudanosoline (19.3), which is also incorporated into berberine (19.14), and late-stage methylation by S-adenosylmethionine (19.26) or by early methylation and then condensation directly to (+)-reticuline (19.29).

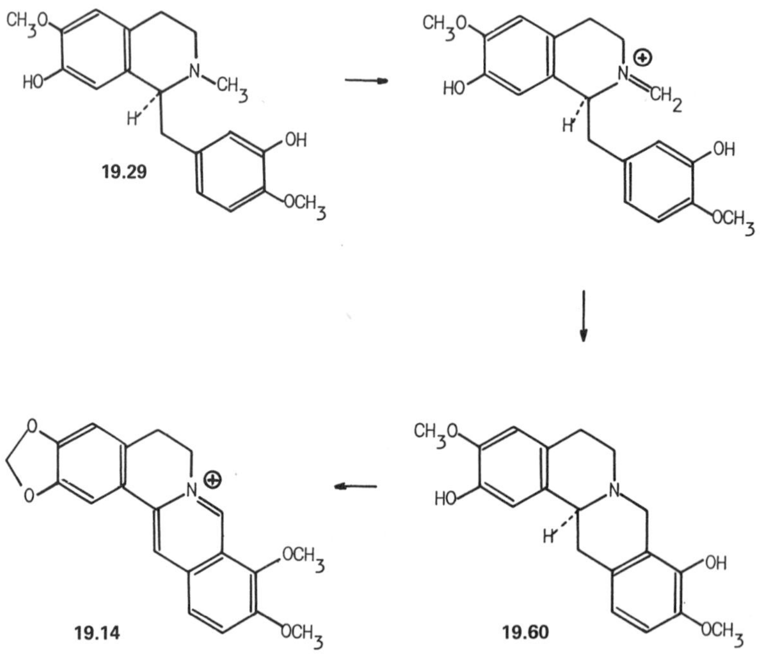

CH3O HO N CH3 H OH OCH3 19.29 CH3O HO N ⊕ CH2 H OH OCH3 O O N ⊕ OCH3 OCH3 19.14 CH3O HO N H OH OCH3 19.60

SCHEME 19.19 A pathway to berberine (umbellatine) from (+)-reticuline.

VIII. PROTOPINE

It has been suggested that protopines in general, and protopine (19.15) in particular, since this is the system to which the available experimental data refers, arise by oxidation of appropriate protoberberines. When labeled (+)-reticuline (19.29; [93]; labeled at nitrogen and oxygen methyls, and at C_3 with ^{14}C and at C_1 with tritium, as shown in Scheme 19.20) is administered to Cheledonium majus L., significant incorporation is obtained in the protoberberines (-)-scoulerine (19.60) and (-)-stylopine (19.61; with some loss of tritium); the protopine (19.15; with complete loss of tritium); and (+)-chelidonine (19.17; a benzophenanthridine alkaloid; Chapter 19, Section IX). Scheme 19.20 accounts for the production of protopine (19.15) by assuming that it is formed from (-)-stylopine (19.61) by oxidation to the imine (19.62), hydration to 19.63, a final methylation, and ring opening of the alcohol amine. The alternative possible imine (19.64) leads to (+)-chelidonine (19.17; see Section IX). As already noted for the protoberberine alkaloids (p. 242), the formation of (+)-reticuline (19.29) is presumed to occur from tyrosine (19.2).

IX. SANGUINARINE AND CHELIDONINE

As pointed out for protopine (19.15) and as incorporated into Scheme 19.20, (+)-reticuline (19.29) serves as a precursor for (-)-scoulerine (19.60) and (-)-stylopine (19.61) in Cheledonium majus L. In the same plant, along with protopine (19.15), labeled (+)-reticuline (19.29) generates labeled (+)-chelidonine (19.17). Thus, as shown in Scheme 19.21, of the two imines, 19.62 and 19.64; the former leads to protopine (19.15) by a pathway such as that shown in Scheme 19.20 while the latter leads to (+)-chelidonine (19.17). The labeling pattern found in the (+)-chelidonine (19.17) is in concert with that suggested by Scheme 19.21. It is not yet known if sanguinarine (19.16) is derived from (+)-chelidonine (19.17) or if it comes from an isomer (e.g., the epimer at C_6) of chelidonine (19.17).

X. NARCOTINE

Protopine (19.15) is generated by oxidation of a (-)-scoulerine (19.60) transformation product (19.61) to an imine (19.62) in which a double bond between C_{14} and the nitrogen undergoes hydrolysis, a subsequent methylation, and ring opening (Equation 19.5 and Scheme 19.20).

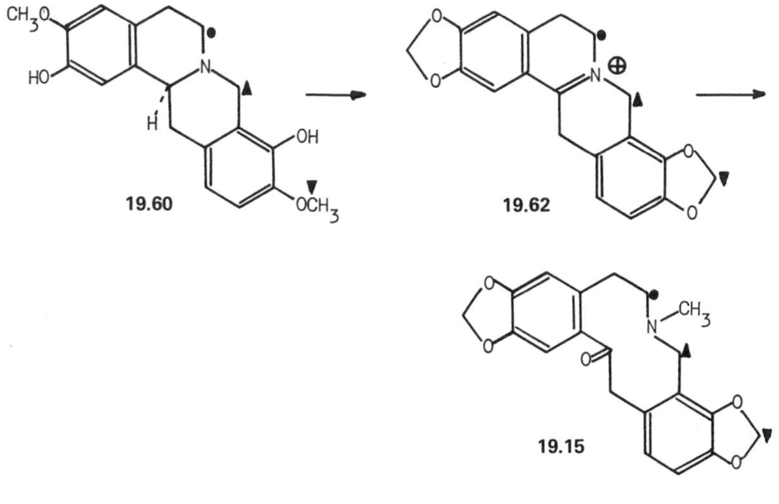

CH3O HO H N OH OCH3 19.60 → O O N⊕ O O 19.62 → (Eq. 19.5)

O O N CH3 O O O 19.15

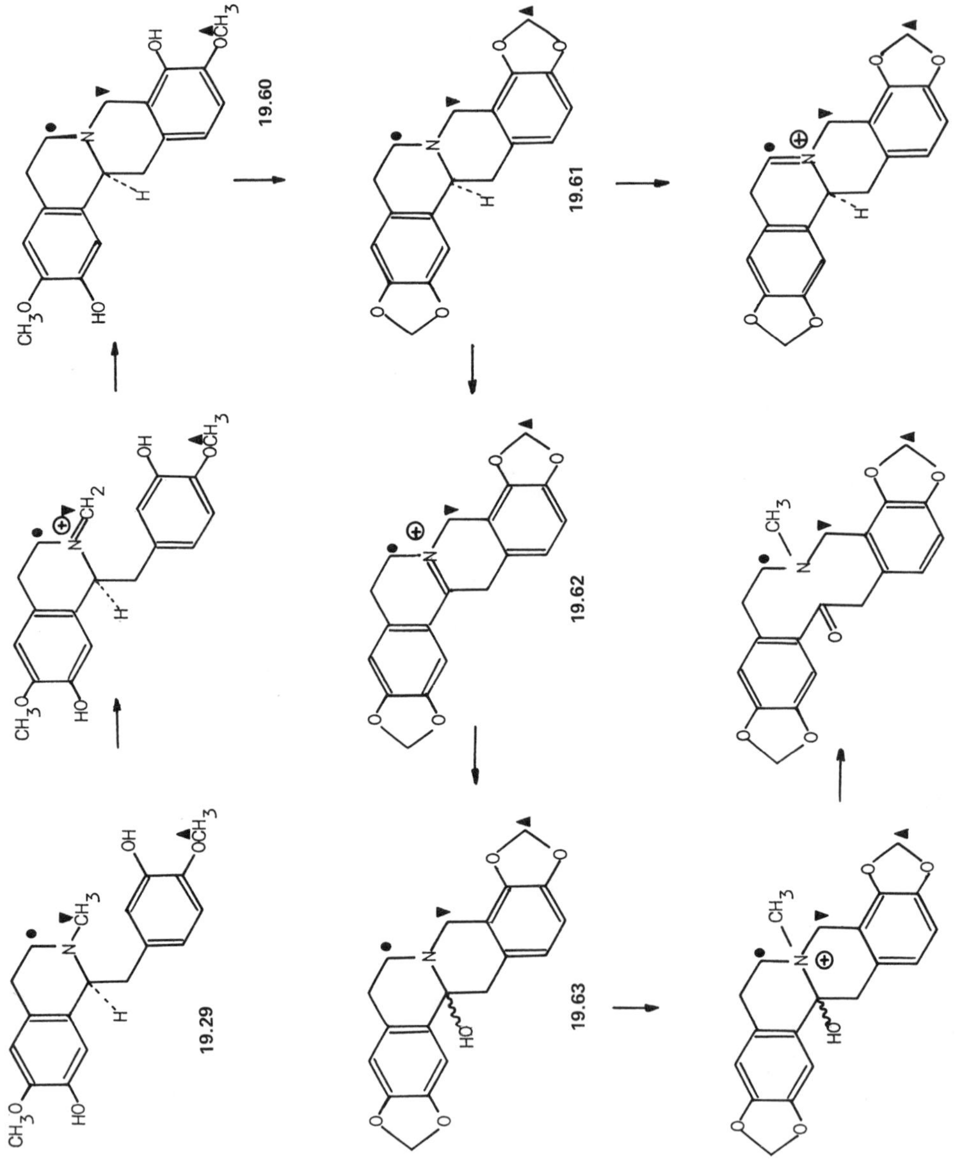

SCHEME 19.20 A pathway for the formation of protopine from (+)-reticuline.

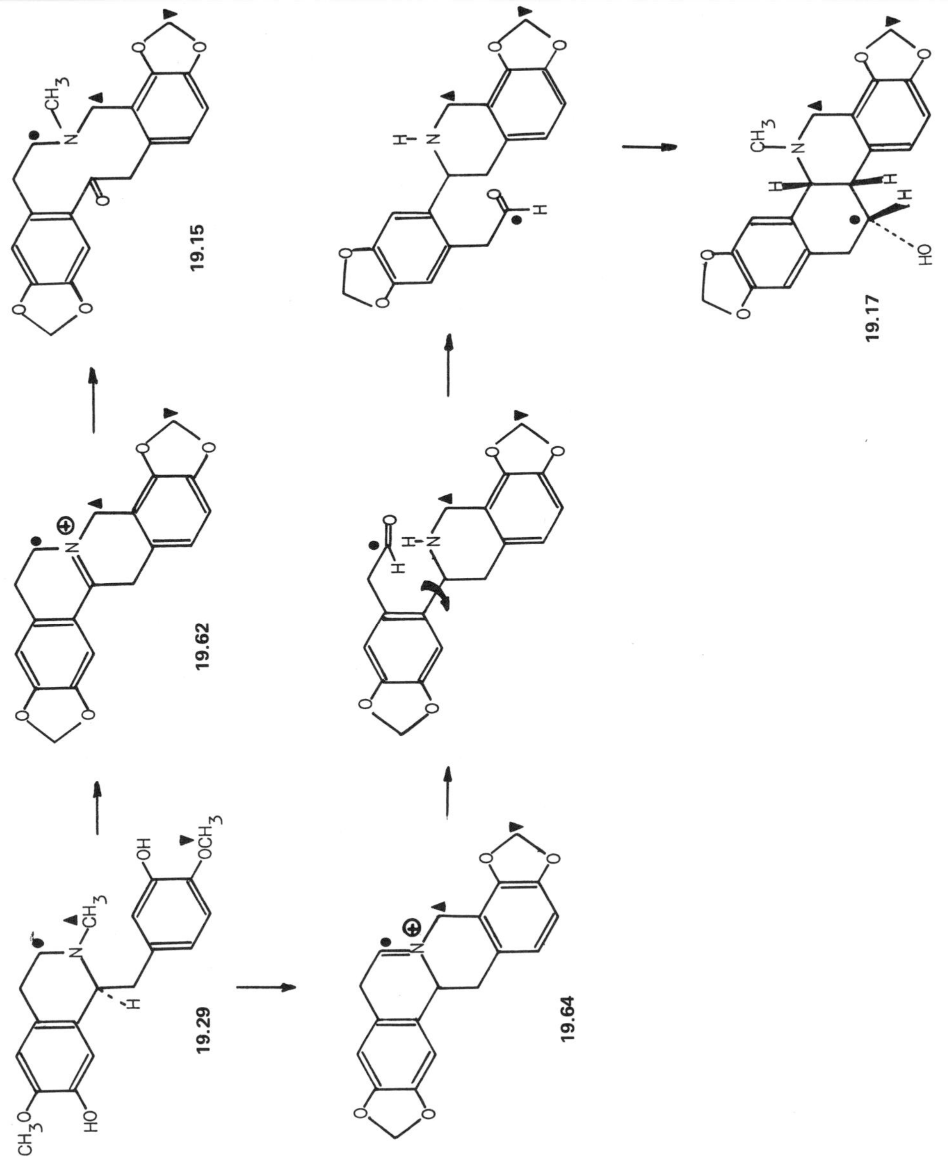

19.29

19.62

19.15

19.64

19.17

SCHEME 19.21 A pathway to (+)-chelidonine (and to protopine) from (+)-reticuline.

Oxidation of the same intermediate (19.61) between C_6 and the nitrogen generates a different imine (19.64), which on hydrolysis, methylation, and subsequent rearrangement yields (+)-chelidonine (19.17; Equation 19.6 and Scheme 19.21). The last possibility for imine formation (Scheme 19.22) from (-)-scoulerine (19.60) results from oxidation between C_8 and the nitrogen, i.e., 19.65 (Equation 19.7). In accord with this suggestion, (+)-reticuline (19.29) is incorporated better than its enantiomer into narcotine (19.18) in Papaver somniferum L. [94], and (-)-scoulerine (19.60) lies on the path between tyrosine (19.2) and narcotine (19.18).

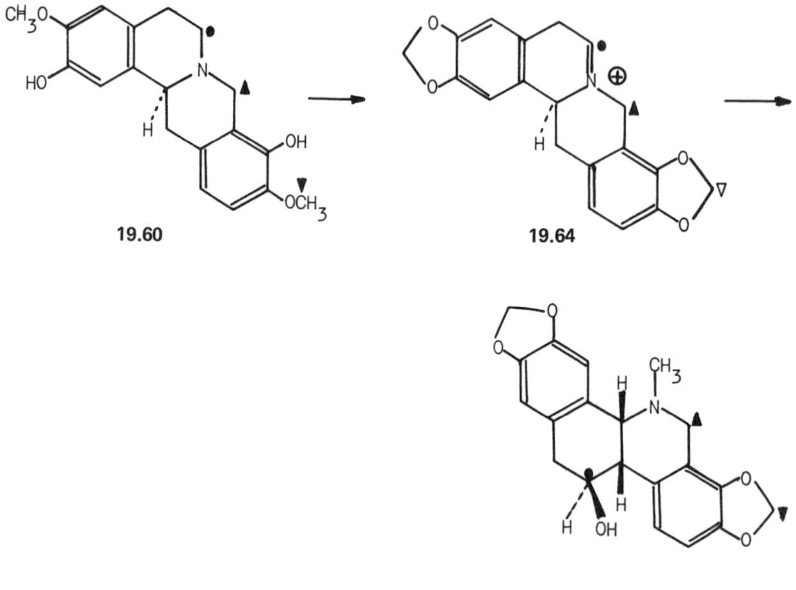

CH_3O HO N H OH OCH_3

19.60

O O N ⊕ H O O

19.64

(Eq. 19.6)

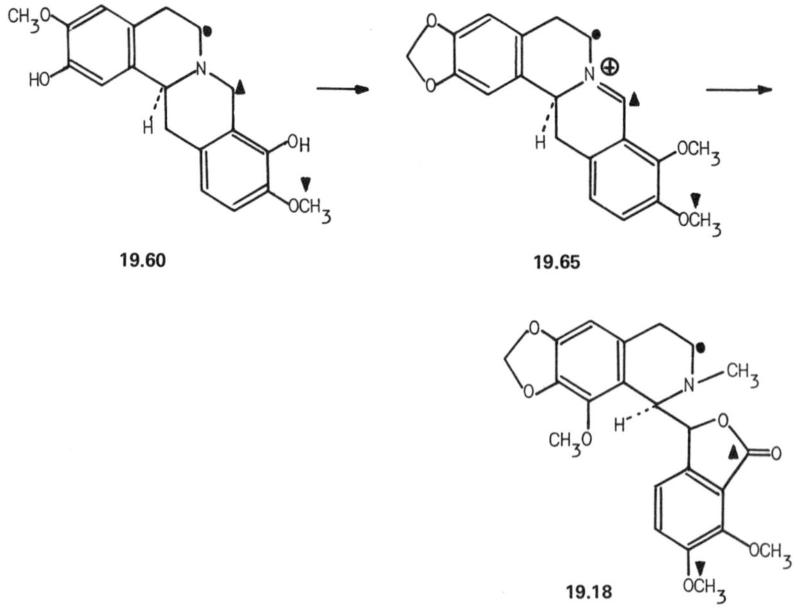

O O CH_3 H N H H OH O O

19.17

CH_3O HO N H OH OCH_3

19.60

O O N ⊕ H OCH_3 OCH_3

19.65

(Eq. 19.7)

O O N CH_3 H CH_3O O O OCH_3 OCH_3

19.18

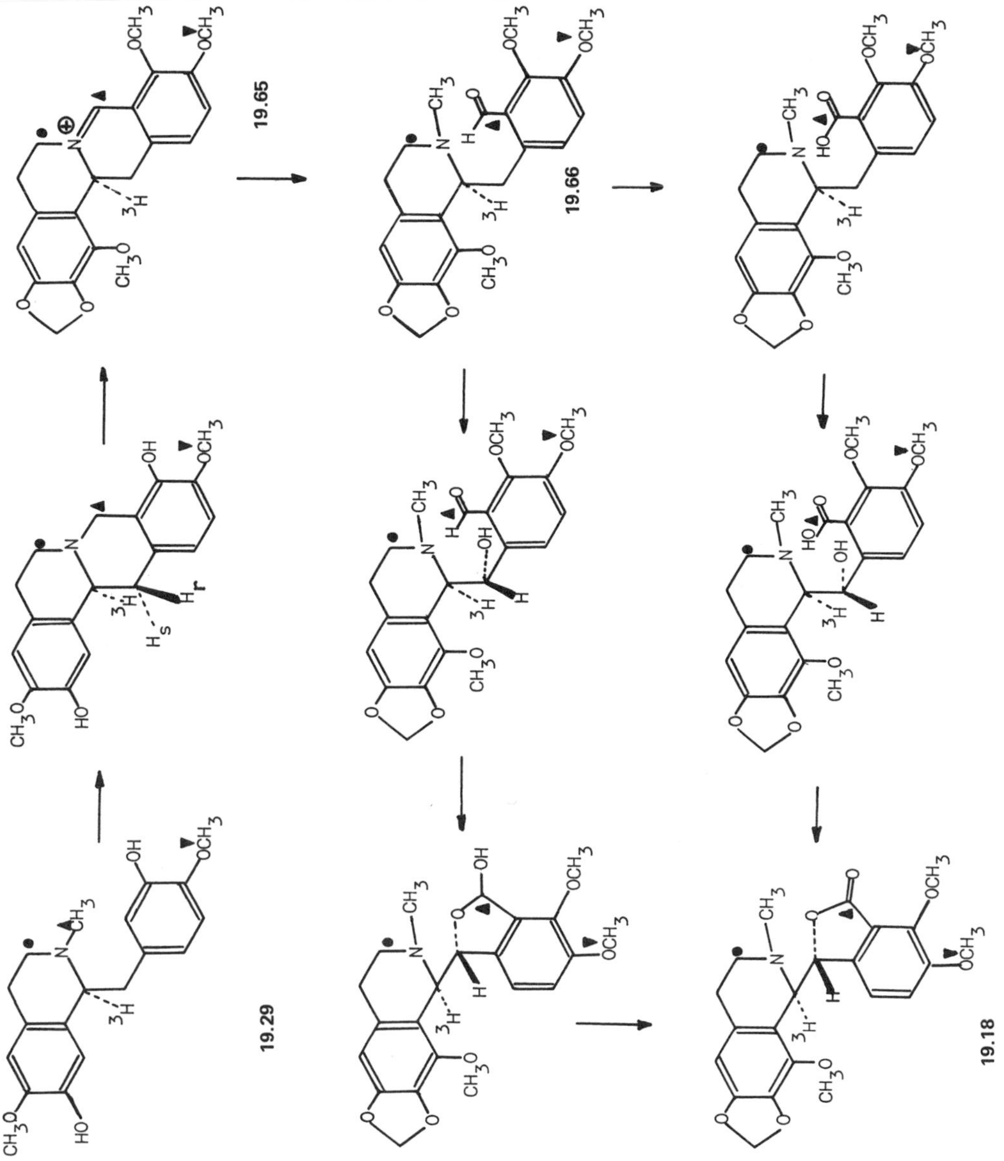

SCHEME 19.22 Two possible pathways to narcotine from reticuline.

The labeling patterns shown in Scheme 19.22 are in accord, as is shown, with the carbonyl carbon of narcotine (19.18) coming from the N-methyl of (+)-reticuline (19.29). The imine (19.65) and the aldehyde 19.66 therefore are presumed to arise from late-stage oxidation and methylation processes. Two pathways are shown in the Scheme 19.22. The first involves conversion of the aldehyde 19.66 into a hemiacetal which then undergoes oxidation to narcotine (19.18), while the second allows oxidation of 19.66 early and formation of the lactone in narcotine (19.18) later. Interestingly [95], it has been observed that the pro-S hydrogen at C_{13} of (-)-scoularine (19.60) is stereospecifically removed in this process and narcotine (19.18) formed. Finally, the additional methoxyl must also come about rather late in the process (i.e., had the oxidation and methylation occurred early, it is likely that the incorporation of (+)-reticuline (19.29) would have been less facile).

XI. RHOEADINE

Biosynthetic studies on the rhoeadine alkaloids (corresponding to 2-phenyl-1,2,4,5-tetrahydro-2H-3-benzazapins) have begun, and it has already been shown that 2 equivalents of [3-^{14}C]tyrosine (19.2) are incorporated into an analog of rhoeadine (19.19), i.e., (+)-alpinigenine (19.67; [96]). Aside from that experimental information, speculation has run [97] to

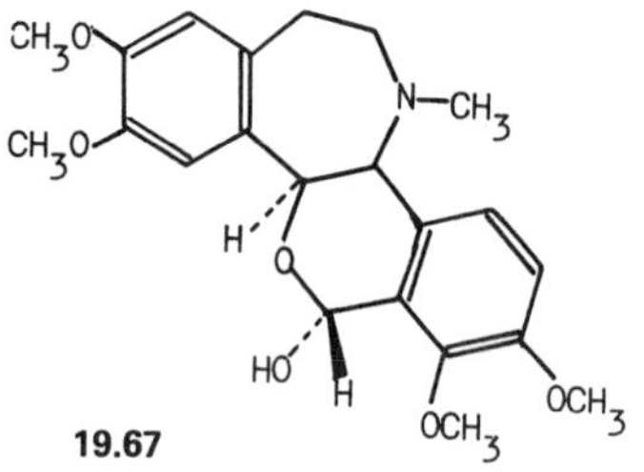

19.67

the derivation of, e.g., rhoeadine (19.19), from a phthalid isoquinoline. While it is possible that oxidation to the phthalid isoquinoline state had in fact been effected prior to rearrangement to (+)-rhoeadine (19.19), it is not necessary that this be the case, and indeed, oxidation at the benzylic carbon and hemiacetal formation on the aldehyde which arises directly from hydrolysis of the imine 19.65, rearrangement, and generation of the appropriate acetal, accomplishes the same goal. Scheme 19.23 outlines these possibilities.

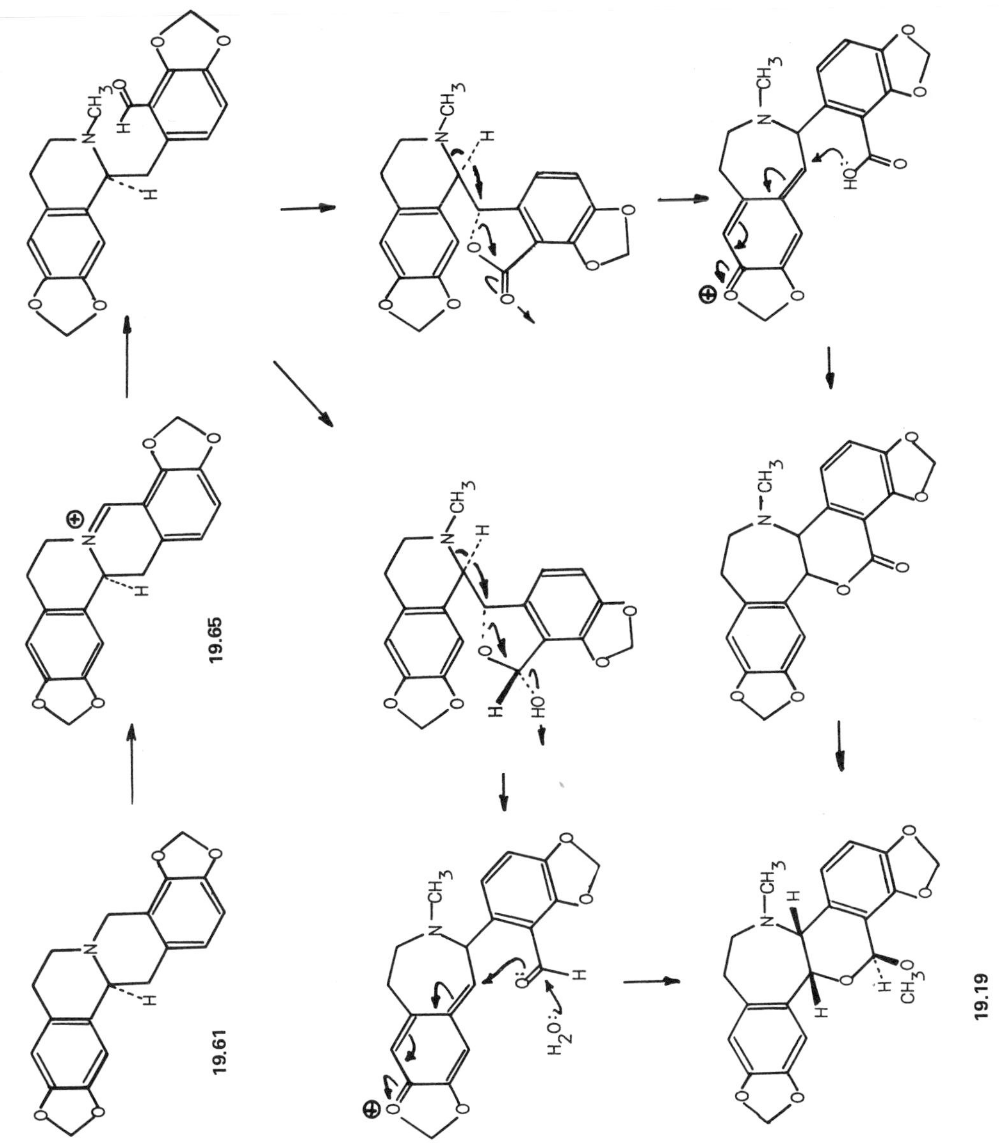

SCHEME 19.23 Two possible pathways to rhoeadine.

20

BIOSYNTHESIS OF 1-PHENETHYLTETRAHYDROISOQUINOLINE ALKALOIDS AND OTHER C_6-C_2 + C_6-C_3 ALKALOIDS DERIVED FROM TYROSINE

And in his dim, uncertain sight
Whatever wasn't must be right
From which it follows he had strong
Convictions that what was, was wrong.

G. W. Carryl, "The Iconoclastic Rustic and the Apropos Acorn"

There are a number of compounds, or small groups of compounds, which are reasonably, or indeed have already been shown to be, derived from tyrosine (20.1) as a C_6-C_2 fragment, and phenylalanine (20.2) as a C_6-C_3 fragment. All the compounds which appear to be derived by the combination of the C_6-C_2 tyrosine-derived (20.1) fragment and the C_6-C_3 phenylalanine-derived (20.2) fragment also appear to be related to each other.

H CO2H HO NH2

20.1

H CO2H NH2

20.2

The compounds to be considered here are the 1-phenethyltetrahydroisoquinoline, autumnaline (20.3), found in Colchicum cornigerum (Schweinf.) (the only true monomeric 1-phenethyltetrahydroisoquinoline known), the homoproaporphine, kreysiginone (20.4; about 10 of

CH3O HO N CH3 H CH3O OH CH3O

20.3

CH3O HO N CH3 CH3O O

20.4

this type have been detected) found in the same plant; the homoaporphine alkaloid (-)-kreysigine (20.5), which accompanies the above bases (about six of these analogs of the populous aporphine alkaloids have been found); and the toxic principle of the autumn crocus (Colchicum autumnale), colchicine (20.6). The studies on the biosynthesis of colchicine (20.6) have

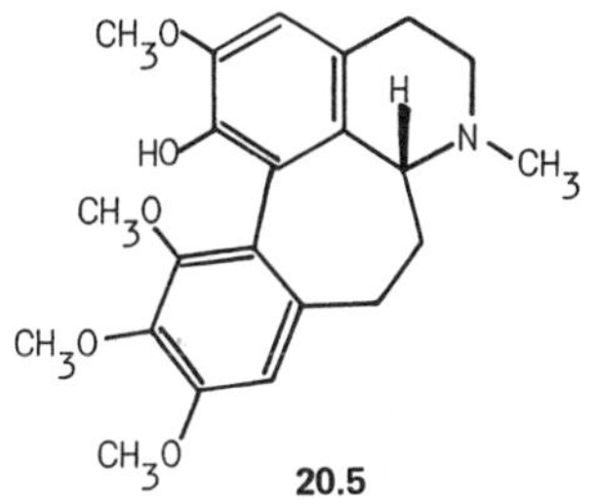

20.5

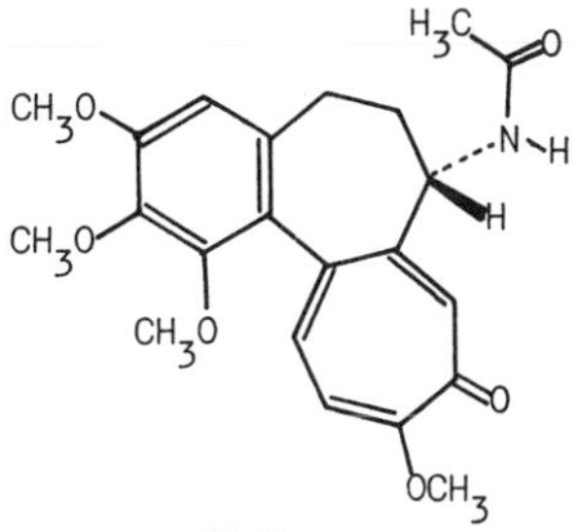

20.6

served as a guide to the more recent work on the other C_6-C_2 + C_6-C_3 bases. Finally, it will be noted that there is a curious pattern of a single methoxyl and a single hydroxyl (in the appropriate positions) in the tyrosine-derived portion of each of these molecules. It is amusing to speculate that this pattern may be required for elaboration into the alkaloids while other possibilities provide different, as yet undetected, bases.

I. ORIGIN OF THE C_6-C_3 UNIT

The anti (antarafacial) elimination of ammonia to generate trans (E)-cinnamic acid (20.7; [60]; Equation 20.1) from phenylalanine (20.2) permits the latter to serve as the precursor to a large number of nonnitrogenous "phenylpropanoids" as well as the alkaloids considered

(Eq. 20.1)

20.2 20.7

here. It is presumed that hydroxylation of the C_6-C_3 fragment occurs at a late stage, i.e., after ammonia loss, since it is rarely found that the ammonia lyase utilizing phenylalanine (20.2) will also operate upon tyrosine (20.1). Thus, even in cases where phenylalanine (20.2) incorporation has not been demonstrated, the implication is that it will prove to be involved.

The elimination of the pro-S hydrogen (Equation 20.1) along with the nitrogen, rather than the pro-R hydrogen, has been demonstrated for the conversion of phenylalanine (20.2) to (E)-cinnamic acid (20.7), but the concertedness of the process has not. Much effort has been expended in an attempt to learn further details of the elimination process [60], and much has been learned, but little is currently known about the activation required at nitrogen. Thus, although the original concept of the conversion of the amino function to a better leaving group through the intermediacy of a pyridoxyl phosphate (20.8) prosthetic group within the cavity of the enzyme is not supported by spectral studies, some carbonyl-like properties appear to be present which might be put to a similar use.

II. AUTUMNALINE

Although no biosynthetic work has yet been reported on autumnaline (20.3), it has been postulated [98] that it arises by condensation between a tyrosine-derived (20.1) 3,4-dihydroxyphenethylamine (20.9) fragment and a C_6-C_3 unit from phenylalanine (20.2) which has undergone deamination to E-cinnamic acid (20.7), subsequent hydroxylation, methylation, and reduction, although not necessarily in that order (Scheme 20.1). It is also assumed that the methyl groups arise by methyl group transfer from S-adenosylmethionine (20.10). Although

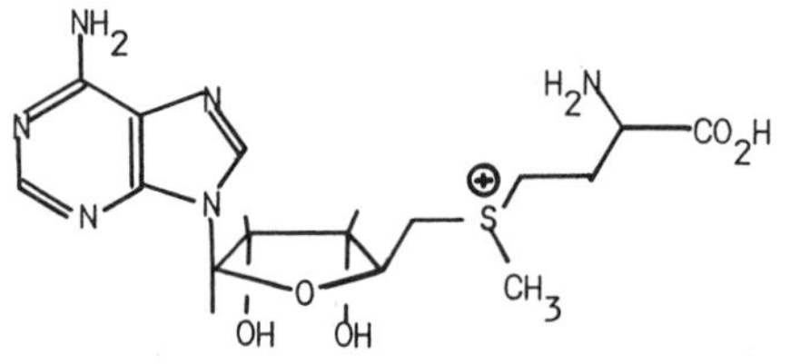

20.10

it is not shown in Scheme 20.1, it is presumed that tyrosine (20.1) is converted to 3,4-dihydroxyphenylalanine (20.11), oxidatively (Chapter 1, p. 3; Equation 20.2) and that, in accord

20.1 → 20.11 (Eq. 20.2)

with Scheme 20.2, which has been seen before, 3,4-dihydroxyphenylalanine (20.11) undergoes pyridoxyl-mediated (20.8) decarboxylation to 3,4-dihydroxyphenethylamine (20.9).

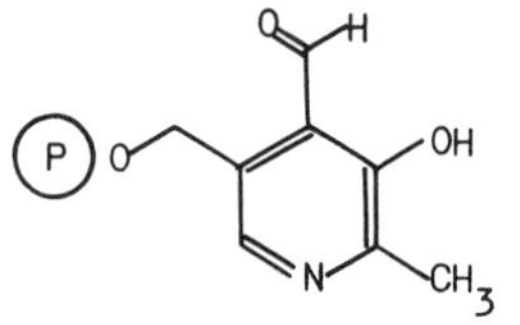

20.8

III. KREYSIGINE AND KREYSIGINONE

Since (-)-kreysigine (20.5) and its analog kreysiginone (20.4) are both found in the same plant (Colchicum cornigerum, Schweinf.) it is reasonable to assume, as in the case of the aporphine-proaporphine bases (Chapter 19) that a biogenetic relationship exists. One possible variation on the proposed linking between the homoaporphine-homopropaporphine bases has been tested, and although it failed to show the expected relationship, others, postulating for example a later methylation sequence, might not. When [3-^{14}C]autumnaline (20.3) is administered to Kreysigia multaflora Reichb. [99] and the total alkaloid extract methylated with

20.2

20.7

20.1

20.9

20.3

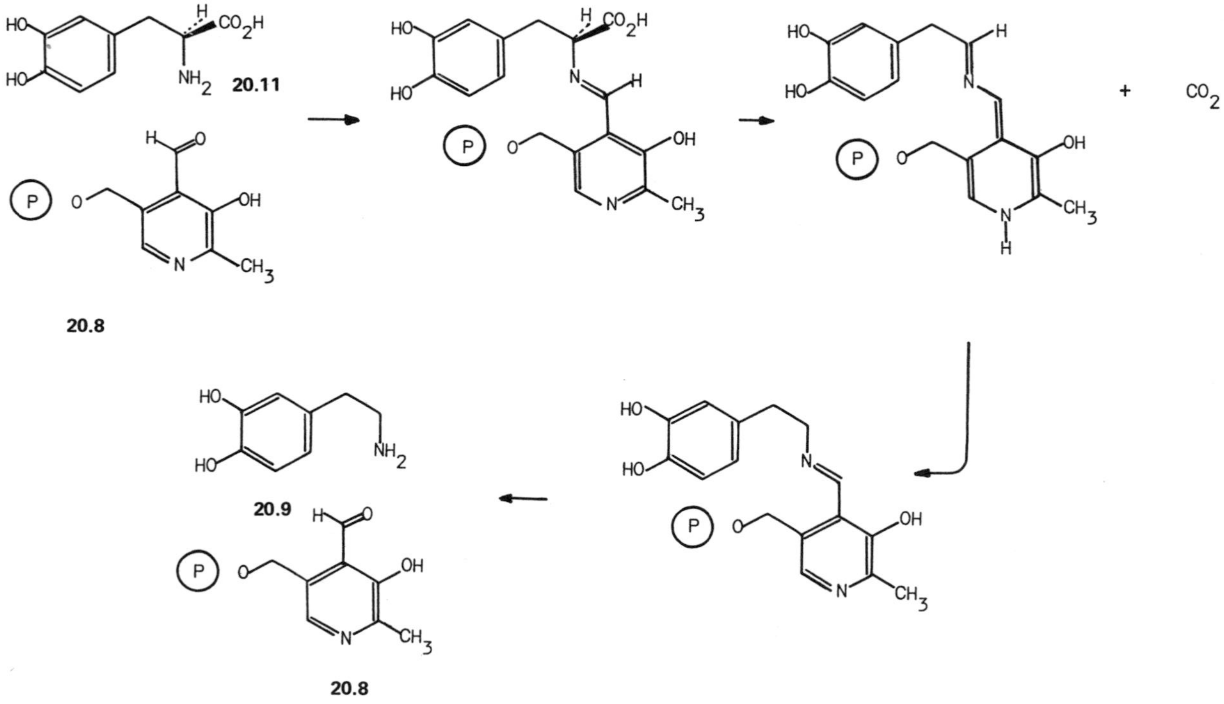

SCHEME 20.2 A pathway for the decarboxylation of 3, 4-dihydroxyphenylalanine to 3, 4-dihydroxy-phenethylamine utilizing pyridoxyl catalysis.

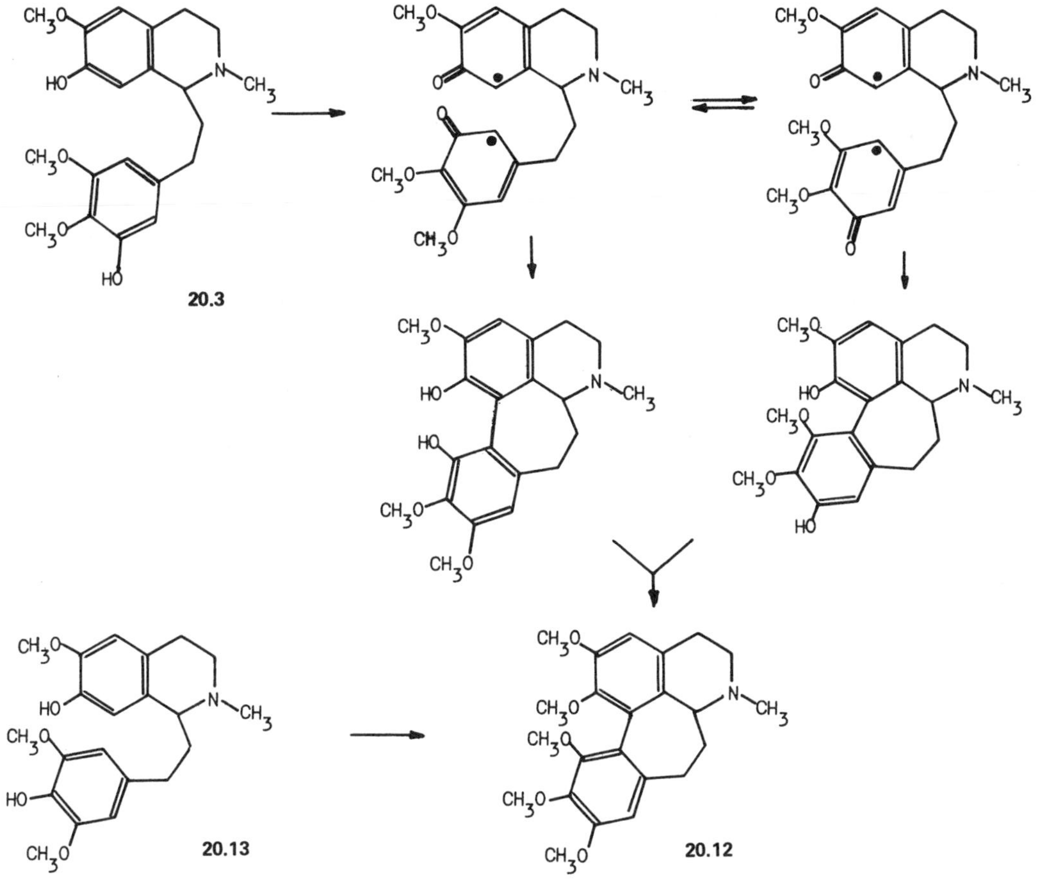

SCHEME 20.3 A pathway to kreysigine involving phenyl-phenyl coupling consonant with labeling experiments.

diazomethane, significant incorporation (1.6%) into O-methylkreysigine (20.12) is obtained. On the other hand, when the 3-^{14}C-labeled isomer (20.13) of autumnaline (20.3) is administered under the same conditions, a smaller (>0.015%) net incorporation into O-methylkreysigine (20.12) is obtained. This suggests that if, indeed, autumnaline (20.3) were an obligatory intermediate on the path to (-)-kreysigine (20.5), direct oxidative phenyl-phenyl coupling ortho to the available hydroxyl in the A ring and either ortho or para to that in the D ring would be necessary. This is outlined in Scheme 20.3.

IV. COLCHICINE

Colchicine (20.6) has been investigated more thoroughly than any of the other C_6-C_3 + C_6-C_3 bases hitherto discussed. Numerous postulates were advanced regarding its biosynthesis but not until it was found [100] that [3-^{14}C]phenylalanine (20.2) is incorporated almost exclusively into C_5 of colchicine (20.6) and nowhere else, and that phenylalanine (20.2) and tyrosine (20.1) are not interconvertible, that such speculations bore fruit. Thus, since the original work it has been found that: (a) C_3 and C_2 of cinnamic acid (20.7) become C_5 and C_6, respectively, of colchicine (20.6; [101]); (b) [3-^{14}C]tyrosine (20.1) enters C_{12} of colchicine (20.6; [102]);

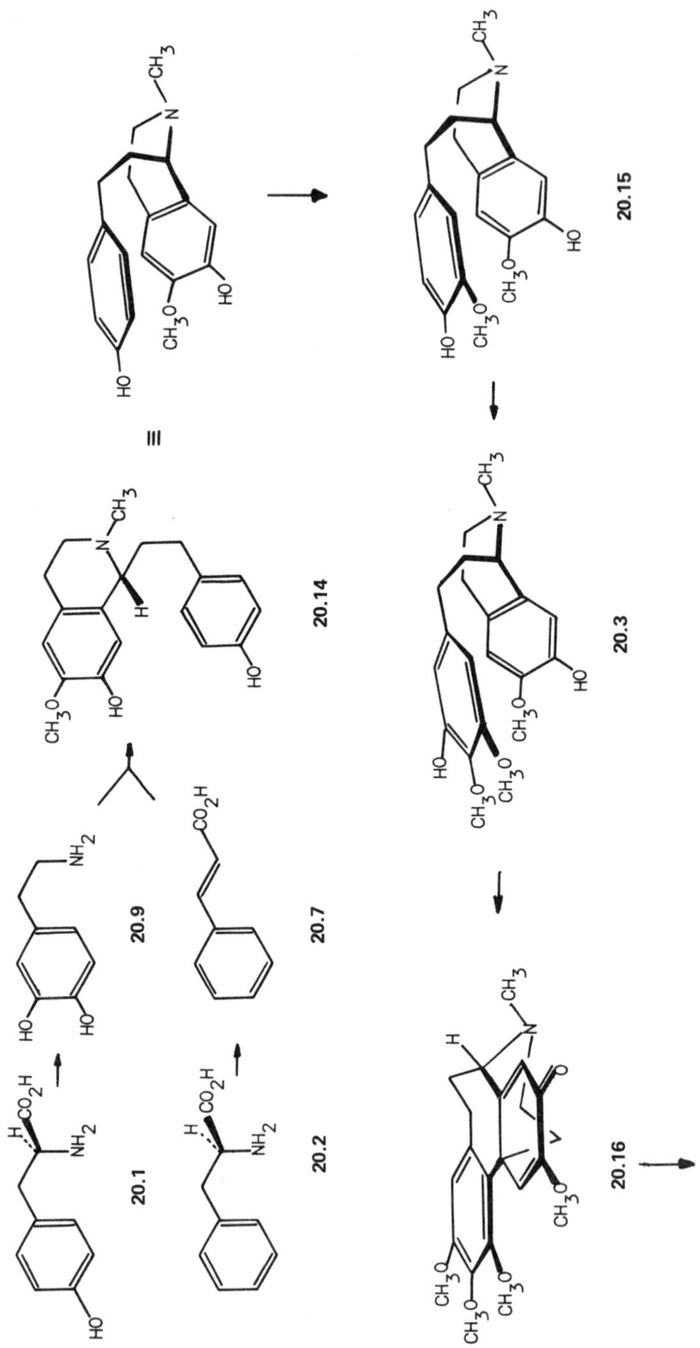
20.1
20.2
20.9
20.7
20.14
20.15
20.3
20.16
HO
CO2H
NH2
H
N
CH3
CH3O

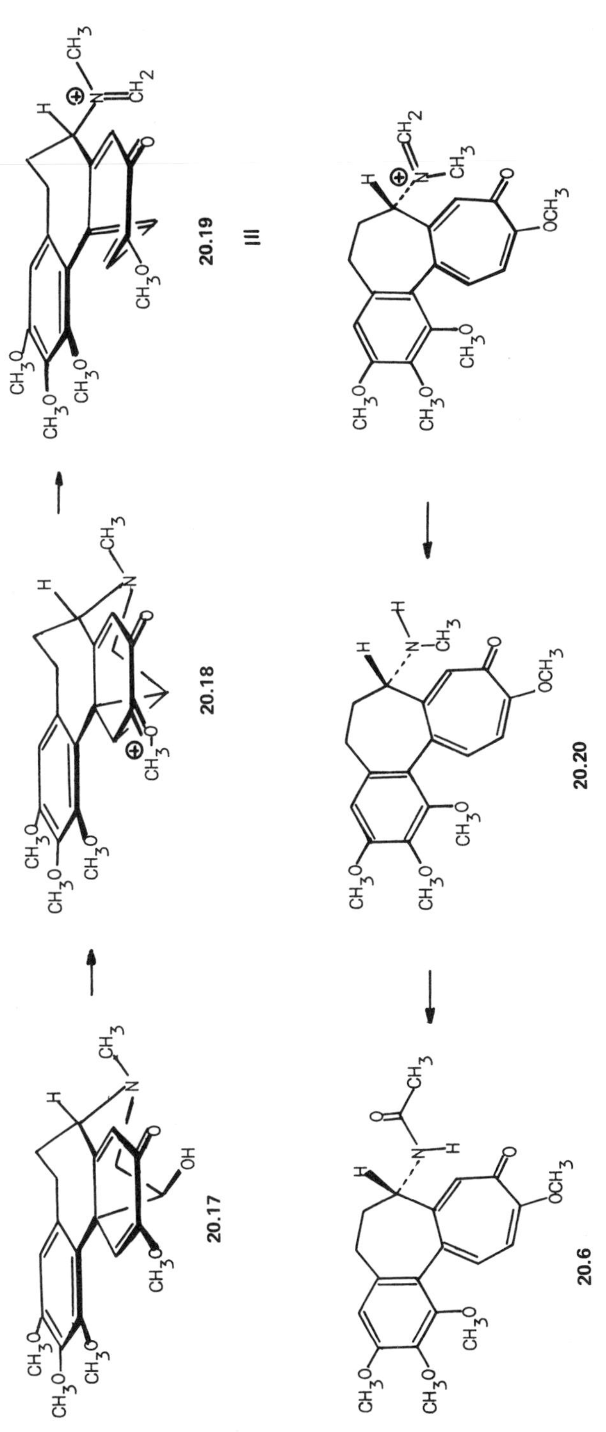

SCHEME 20.4 The biosynthesis of colchicine.

(c) C_4 of the aromatic nucleus of tyrosine (20.1) enters C_9 of colchicine (20.6; [103]); (d) ^{14}C- and ^{3}H-labeled 1-phenethyltetrahydroisoquinolines 20.13, 20.14, and 20.15, the latter labeled with both ^{14}C and ^{15}N, are incorporated intact into colchicine (20.6; [104]); and, finally (e) ^{13}C-labeled autumnaline (20.3) is incorporated into colchicine (20.6; [105]). Scheme 20.4 incorporates this data, and two further observations, namely, the methyl groups arise from S-adenosylmethionine (20.10) and the acetyl side chain from acetate, into the currently accepted pathway for the genesis of colchicine (20.6).

Thus, as shown in Scheme 20.4, tyrosine (20.1) serves as a C_6-C_2-N fragment, while phenylalanine (20.2) serves as a C_6-C_3 pregenitor, which when combined yield the 1-phenyltetrahydroisoquinoline (20.14). Further hydroxylation (oxidation) to 20.15 and 20.3 (autumnaline), followed by cyclization and methylation, yields O-methylandrocymbine (20.16), the O-desmethyl derivative of which is also naturally occurring and which also serves as a specific precursor to colchicine (20.6; [104]). Then, hydroxylation of O-methylandrocymbine (20.16) yields the alcohol 20.17 which undergoes elimination to 20.18 and ring expansion to 20.19. Finally, the immonium salt 20.19 undergoes hydrolysis to demecolchicine (20.20), demethylation, and acetylation to colchicine (20.6). Although this pathway successfully accounts for production of colchicine (20.6), the hydroxylation to the alcohol 20.17 is only one way of producing the appropriate oxidation level—oxidation at nitrogen would serve the same end.

21

BIOSYNTHESIS OF BETACYANINS: C_6-C_3 + C_6-C_3 ALKALOIDS DERIVED FROM TYROSINE

The dissenting opinions of one generation become the prevailing interpretations of the next.

B. J. Hendrick (1937)

The red-violet pigments of the fruits or flowers of a number (about a dozen) families of the Centrospermae are betacyanins (or betaxanthins, which are related and which may be generated from the betacyanins; see below). The betacyanins are invariably acetals or ketals of sugars and one of two aglycone fragments called betanidin (21.1) and isobetanidin (21.2; the

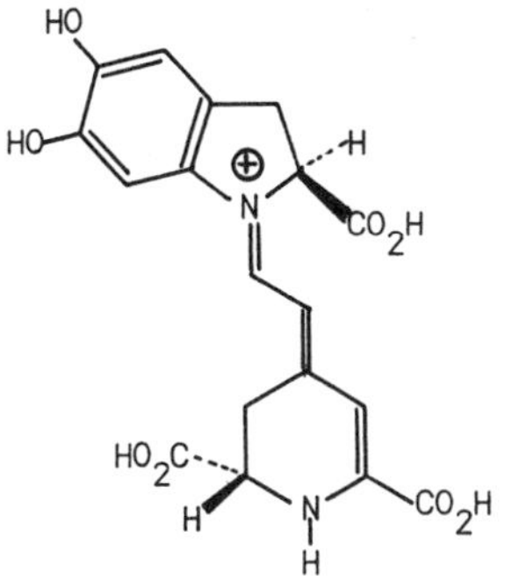

HO HO N ⊕ H CO_2H HO_2C H N H CO_2H

21.1

HO HO N ⊕ H CO_2H H HO_2C N H CO_2H

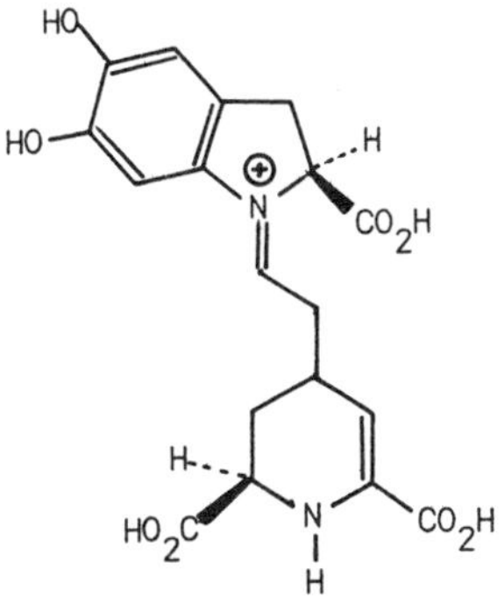

21.2

sugar being attached at the 5-hydroxy of the dihydroindolecarboxylic acid portion of the betacyanin). It is clear from the structures of these compounds that they should, since they are epimers, probably be derived from the same precursor, and although it was presumed that they were generated from 3,4-dihydroxyphenylalanine (21.3), firm evidence was lacking until relatively recently [106].

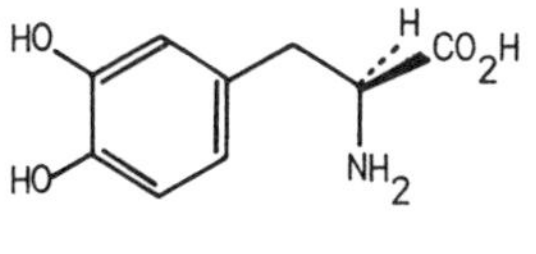

HO HO H CO_2H NH_2

21.3

It is now believed, although incorporation is limited, that 5,6-dihydroxy-2,3-dihydroindole-2-carboxylic acid (21.4) arises from tyrosine (21.5) via 3,4-dihydroxyphenylalanine (21.3) and "dopaquinone" (21.6) and goes on to 5,6-dihydroxyindole (21.7; Scheme 21.1).

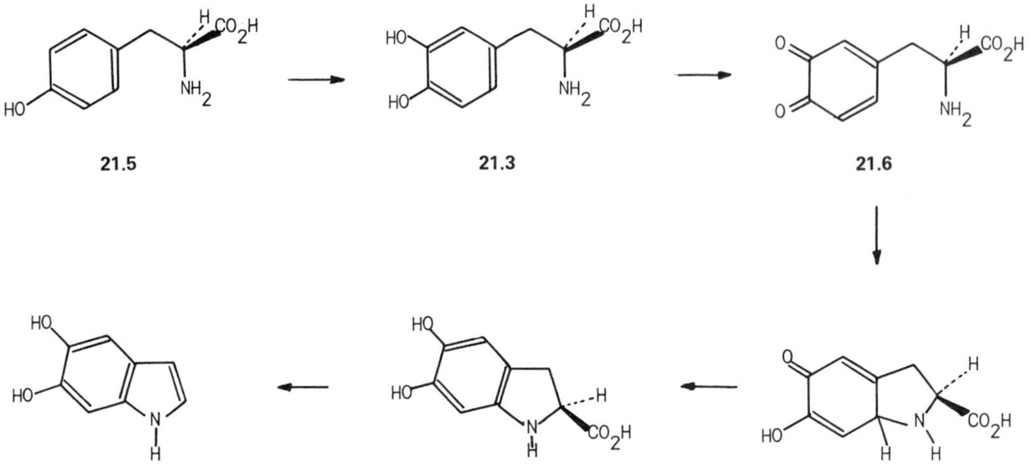

21.5 21.3 21.6

21.7 21.4

SCHEME 21.1 A pathway to 5,6-dihydroxy-2,3-dihydroindole-2-carboxylic acid from tyrosine.

Combination of 5,6-dihydroxy-2,3-dihydroindole-2-carboxylic acid (21.4) with the hydropyridine, (+)- or (-)-betalamic acid (21.8), yields the betacyanins betanidin (21.1) and isobetanidin (21.2) directly (Equation 21.1). Thus, all betacyanins or betaxanthins are simply derivatives of imines of (+)- or (-)-betalamic acid (21.8), the former utilizing only the dihydroindole carboxylic acid, 21.4, while the latter use any other amino acid.

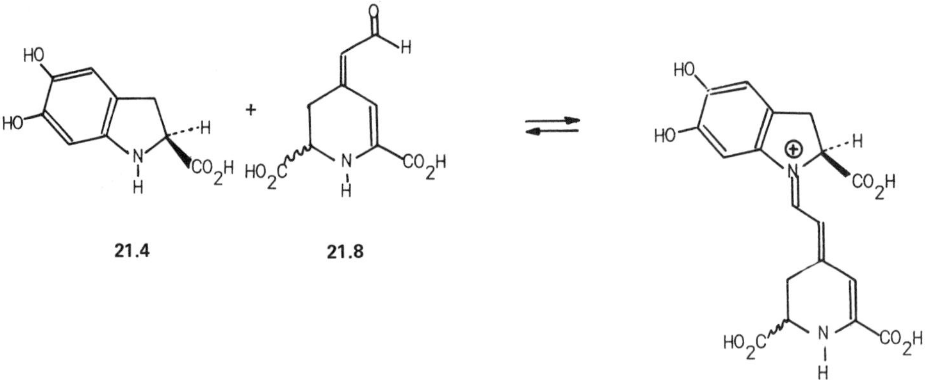

21.4 21.8

(Eq. 21.1)

Good incorporation into the betalamic acid (21.8) portion of the betacyanins has been obtained in cactus [88] on feeding both [1-^{14}C]- and [2-^{14}C]dihydroxyphenylalanine (21.3). This result is accounted for in Scheme 21.2, which, it will be noted, assumes further oxidation of 3,4-dihydroxyphenylalanine (21.3) to the trihydroxyphenylalanine (21.9) and a ring cleavage,

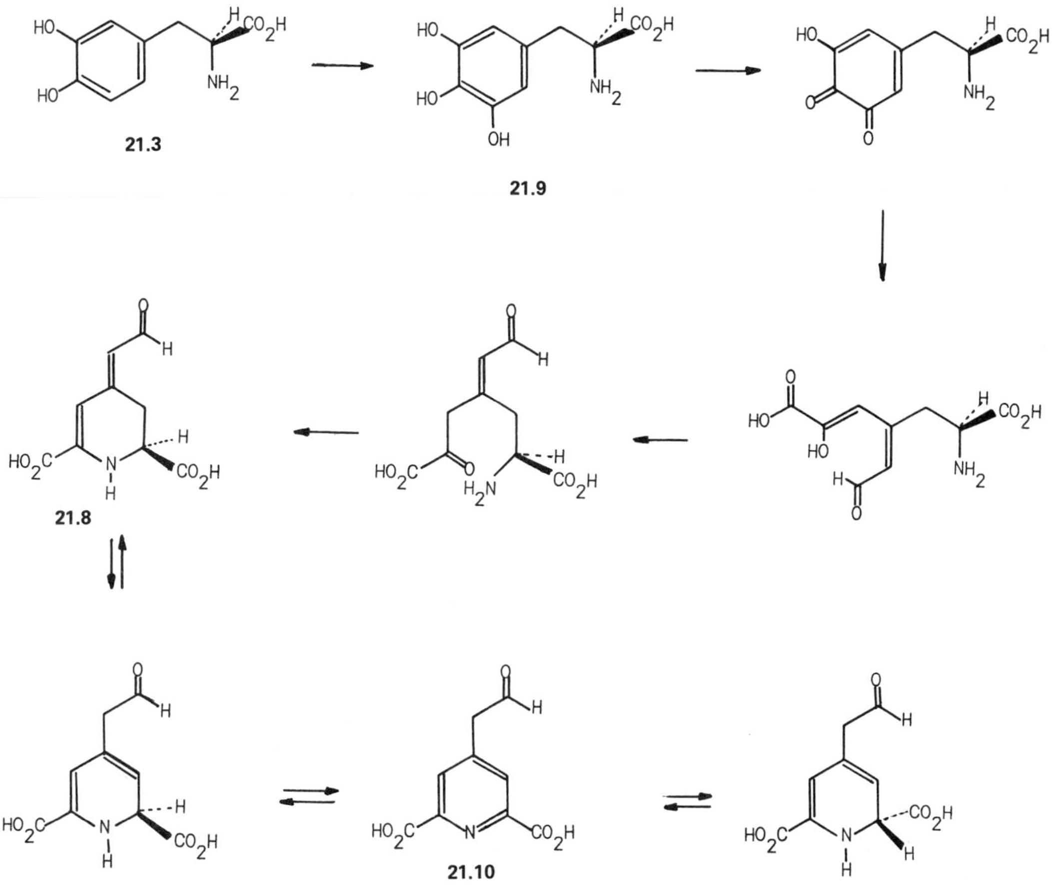

SCHEME 21.2 A pathway for the formation of chiral betalamic acid and its subsequent racemization.

followed by recyclization to optically active betalamic acid (21.8). Racemization of betalamic acid (21.8) might subsequently occur through a low-lying barrier involving an oxidation to the corresponding symmetrical pyridine (21.10) followed by a reduction to either the racemate or the enantiomer of the original betalamic acid (21.8).

22

CHEMISTRY OF β-PHENETHYLAMINE, SIMPLE TETRAHYDROISOQUINOLINE, AND 1-ALKYL-TETRAHYDROISOQUINOLINE ALKALOIDS

In nature, there are neither rewards nor punishments—
there are consequences.

R. G. Ingersoll, Some Reasons Why

There are, broadly speaking, two truly different kinds of bases to be considered in this chapter. On the one hand, mescaline (22.1), anhalamine (22.2), anhalonidine (22.3), and lophocerine (22.4) are relatively simple derivatives of β-phenethylamine. They all occur in cacti, but it is apparently true that only mescaline (22.1) is an "intoxicant." The chemistry of these materials is, correspondingly, simpler than that of the Ipecac bases which are the other kinds to be discussed.

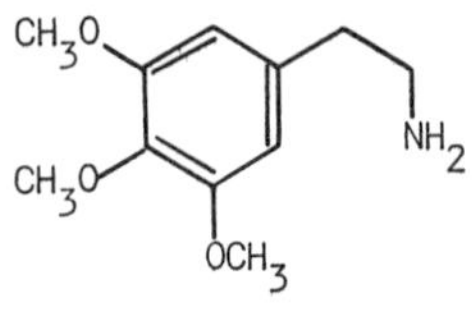

22.1 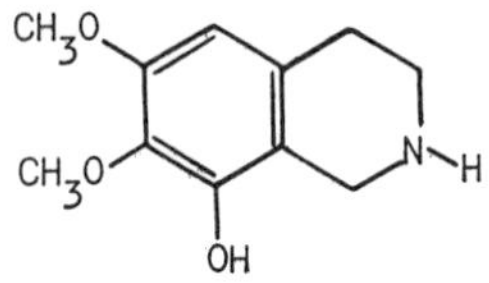 22.2 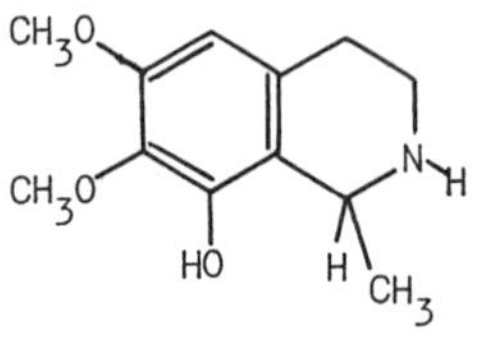 22.3

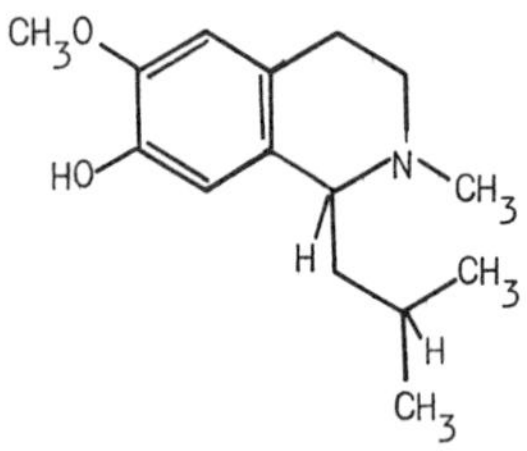

22.4

The rather more elaborate ipacec alkaloids, protoemetine (22.5), ipecoside (22.6), and emetine (22.7) are all presumably present in the crude ipacec extracts which have been used as emetics, and emetine (22.7), specifically, has been used as a drug against pathogenic ameba in amebic dysentery.

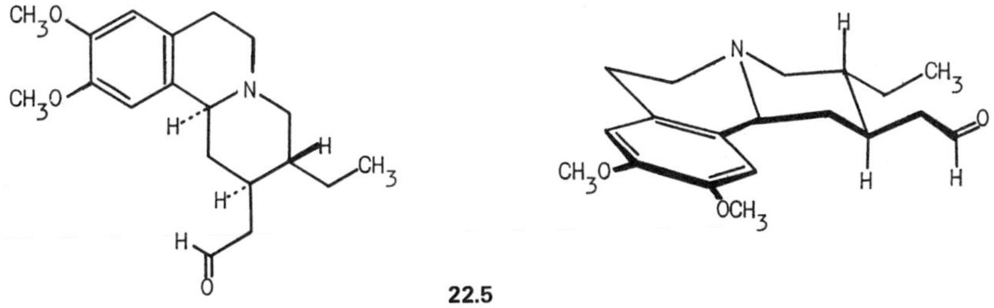

22.5

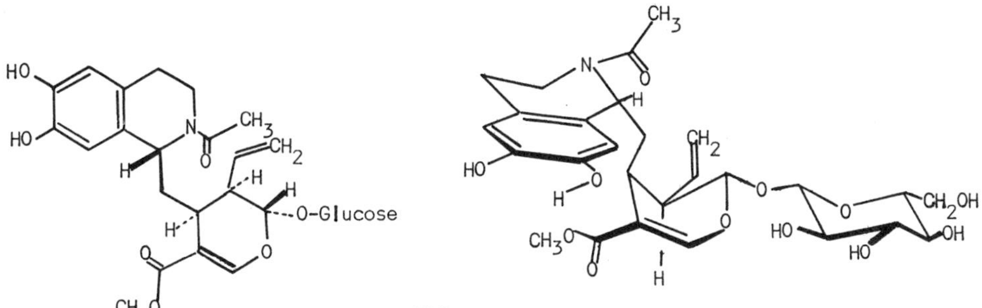

22.6

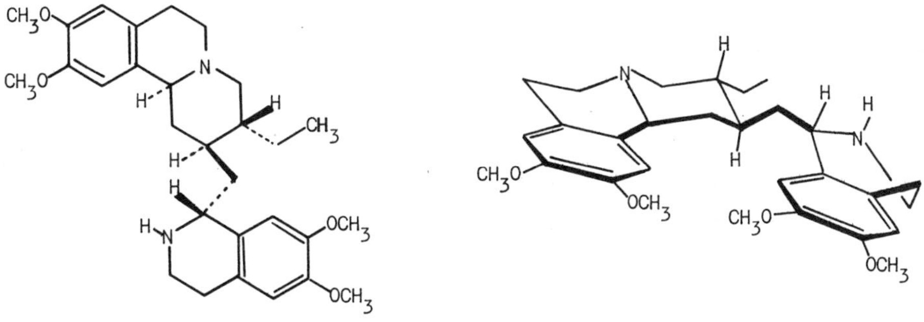

22.7

I. MESCALINE

Crude preparations of mescaline (22.1) from peyote were first reported by the Spanish as they learned of its use from the natives of Mexico during the Spanish invasion of that country in the sixteenth century. The colorful history of mescaline (22.1) has drawn attention to its use as an hallucinogen, and even today it is in use among natives of North and South America. Although in connection with drug abuse complaints, mescaline (22.1) is considered dangerous, it has been reported [107] that it is not a narcotic, nor is it habituating. It was also suggested that its sacramental use in the Native American Church of the United States be permitted since it appears to provoke only visual hallucination while the subject retains clear consciousness and awareness. The base itself was isolated in 1896 from "mescal buttons" (Anhalonium lewinii, Hennings, now called Lophophora williamsii), mp 35-36° C, bp 180° C, (12 mm). The sulfate $(C_{11}H_{17}O_3N)_2 \cdot H_2SO_4 \cdot 2H_2O$ and other derivatives, as well as the free base, yield trimethylgallic acid (22.8) on oxidation, so that the presently accepted structure (i.e., 3,4,5-trimethoxy-β-phenethylamine, 22.1) could be written.

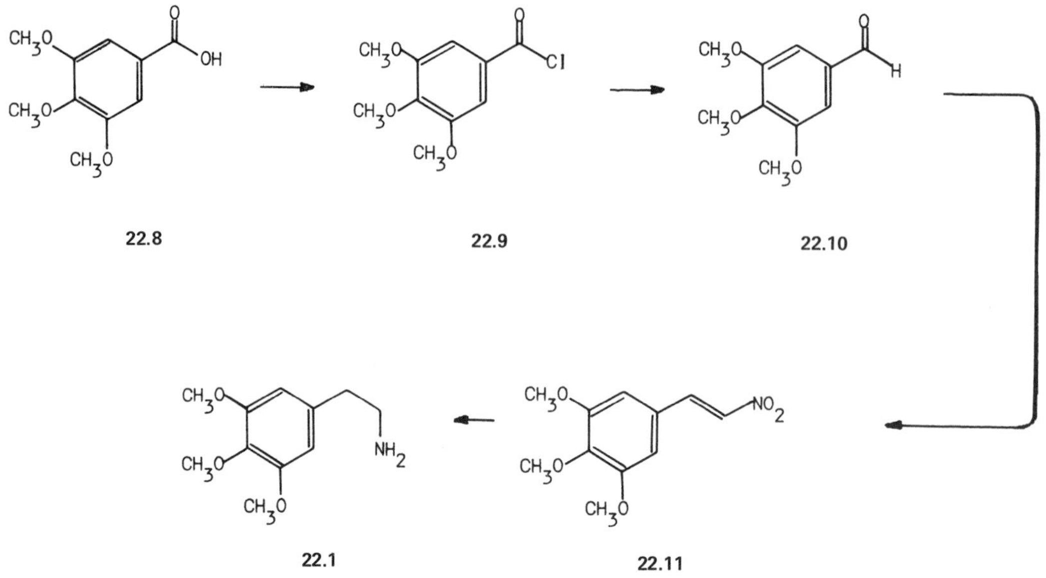

SCHEME 22.1 A synthesis of mescaline. (After Reference 108.)

Although numerous syntheses are currently available, the first, which was actually used as the proof of structure [108], is typical of this class of bases. Thus, the acid chloride of trimethylgallic acid (22.9) was converted to the corresponding benzaldehyde (22.10) by the Rosenmund method and the aldehyde (22.10) condensed with nitromethane to the ω-nitrostyrene (22.11). Reduction, first with zinc dust and acetic acid to the oxime and then with sodium amalgam to mescaline (22.1) completed the synthesis. The reactions involved are shown in Scheme 22.1.

II. ANHALAMINE

The isolation of anhalamine (22.2), $C_{11}H_{15}NO_3$, mp 189-191°C, from Lophophora williamsii was recorded in 1899 and its structure inferred on the basis of its similarity to mescaline (22.1) and experimental work related to mescaline (22.1) and its derivatives [109].* A recent synthesis [110] is interesting. As shown in Scheme 22.2, when mescaline (22.1) is converted into its carbamate derivative (22.12) with ethylchloroformate and the latter treated with polyphosphoric acid, conversion to the 1-oxo-6,7,8-trimethoxy-1,2,3,4-tetrahydroisoquinoline (22.13; ν_{max} 1670 cm^{-1}) occurs in about 40% yield. In this derivative, the methoxy group at C_8 is particularly labile, and simply heating 22.13 with concentrated hydrochloric acid is sufficient to cause demethylation to the corresponding 8-hydroxylactam (22.14; ν_{max} 1650 cm^{-1}). When the sodium salt of the phenol 22.14 is benzylated to 22.15 and the O-benzyl ether reduced in tetrahydrofuran with lithium aluminum hydride, a basic material (22.16) is isolated which in turn undergoes palladium-catalyzed debenzylation on hydrogenolysis in acetic acid to yield anhalamine (22.2).

---

*When the structure of mescaline (22.1) had been resolved, its pharmacological properties stimulated (and continues to stimulate) intensive searches for analogous compounds. Some of the synthetic materials, whose structures were reasonably well established, appeared later as natural substances.

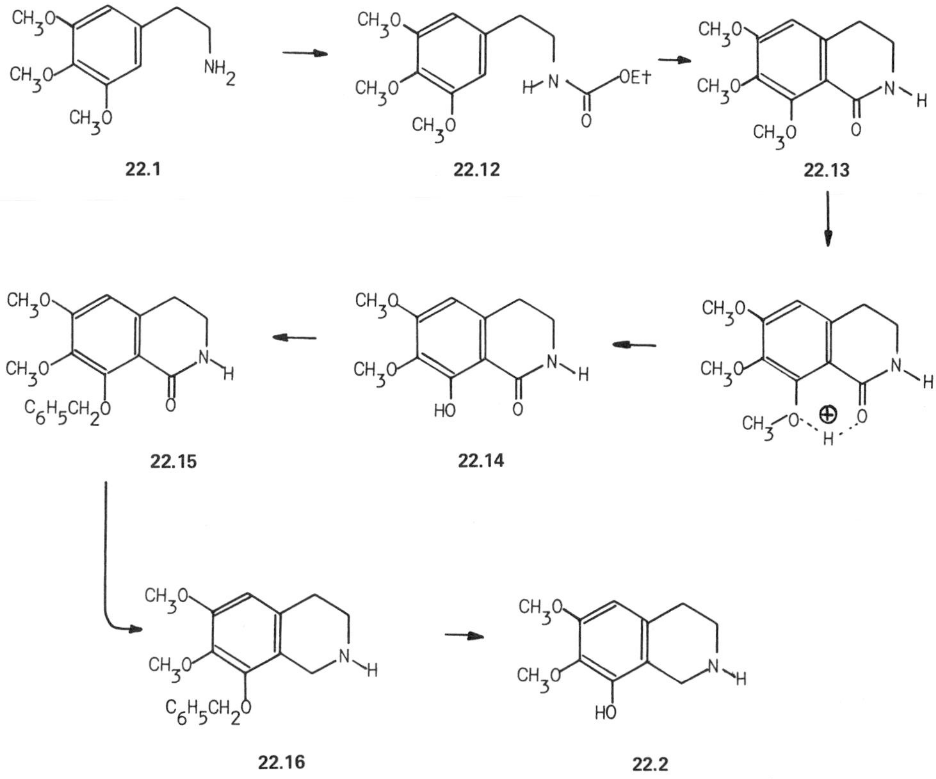

SCHEME 22.2 A synthesis of anhalamine. (After Reference 110.)

III. ANHALONIDINE

Anhalonidine (22.3), optically inactive, mp 160-161°C, is also isolated from Lophophora williamsii. Again, it is among those compounds which had already been prepared in a search for mescaline-like (22.1) activity elsewhere. Interestingly, the N-methyl derivative of anhalonidine (22.17), also naturally occurring and called pellotine (22.17), was prepared synthetically and at least partially resolved with (+)-tartaric acid [111] before anhalonidine (22.3) had been isolated. (-)-Pellotine (22.17) undergoes rapid racemization (!) in solution, from which it is concluded that even if the naturally occurring alkaloid (22.3) were optically active, racemization would occur on work-up. A pathway for the racemization process has not yet been suggested.

A recent synthesis of racemic anhalonidine (22.3) utilizes the same intermediate, i.e., 8-benzyloxy-6,7-dimethoxy-1,2,3,4-tetrahydroisoquinoline (22.16; Scheme 22.2) already generated from mescaline (22.1; [111]). As outlined in Scheme 22.3, 8-benzyloxy-6,7-dimethoxy-1,2,3,4-tetrahydroisoquinoline (22.16), on treatment with sodium hypochlorite, yields the corresponding 3,4-dihydroisoquinoline (22.18), which in turn undergoes benzylation to the O,N-dibenzyl derivative (22.19) and alkylation at C_1 on treatment with methyl magnesium iodide in ether to yield 22.20. Again, catalytic hydrogenolysis serves to effect debenzylation, and (±)-anhalonidine (22.3) results.

An interesting degradation of anhalonidine (22.3), isolated from feeding experiments with (-)-[2-^{14}C]tyrosine (22.21), has been reported [112] and is shown in Scheme 22.4. (±)-Anhalonidine (22.3) is methylated exhaustively to yield O,N-dimethylanhalonidine

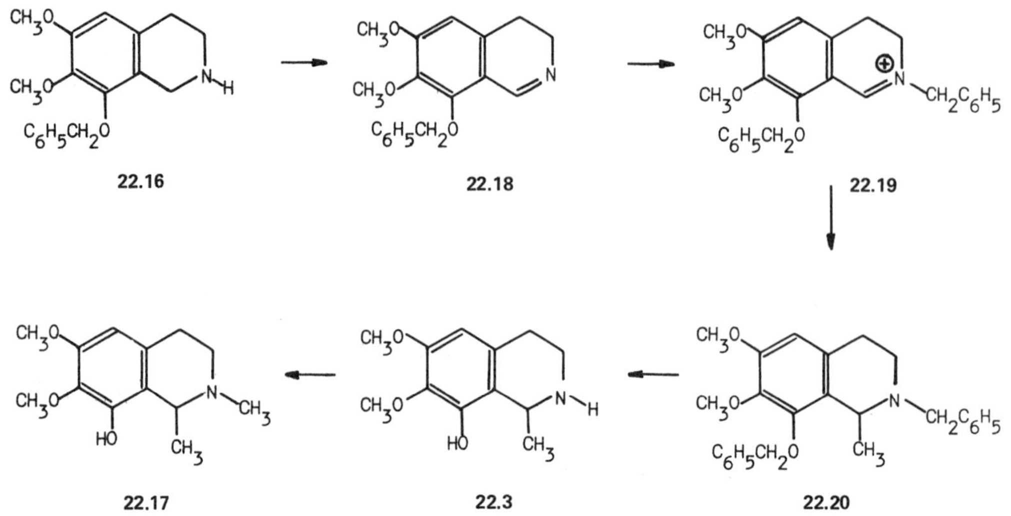

22.16 22.18 22.19

22.17 22.3 22.20

SCHEME 22.3 A synthesis of (±)-anhalonidine. (After Reference 111.)

methiodide (22.22), the corresponding chloride of which undergoes an Emde-type degradation (sodium amalgam) followed by hydrogenation to a mixture of amines which yields the ethyltrimethoxy-α-phenethylamine derivative 22.23 and the ethyltrimethoxy-β-phenethylamine (22.24) on methylation. The α- and β-phenethylamine derivatives, 22.23 and 22.24, respectively, are separated, and Hofmann degradation on 22.23 generates 2-ethyl-3,4,5-trimethoxystyrene (22.25) which produces inactive 2-ethyl-3,4,5-trimethoxybenzaldehyde (22.26) and active formaldehyde, isolated as its dimedone (22.27) derivative, on oxidation with osmium tetroxide and sodium metaperiodate.

IV. LOPHOCERINE

Racemic lophocerine (22.4) was isolated as its methyl ether after diazomethane treatment of the alkali-soluble fraction of the total alkaloid extract of Lophocereus schotti. It is presumed that lophocerine (22.4) occurs naturally as a racemate, although, as reported for

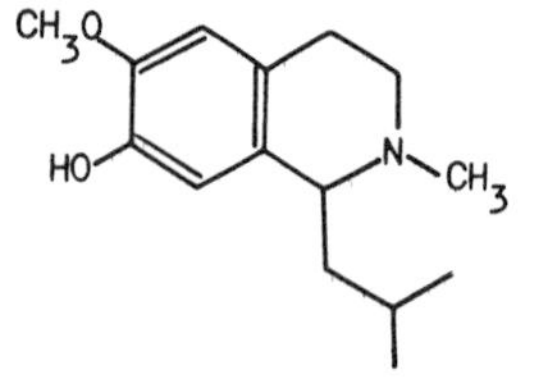

22.4

pellotine (22.17), it is possible that racemization might have occurred during work-up from the isolation. The structure of lophocerine (22.4) is based upon its synthesis (see below) and the identity of its methyl ether to a degradation product obtained during the structural elucidation of the trimeric alkaloid pilocerine (22.28) which occurs along with lophocerine (22.4) in Lophocereus schotti [113-15].

SCHEME 22.4 A degradation of anhalonidine. (After Reference 112.)

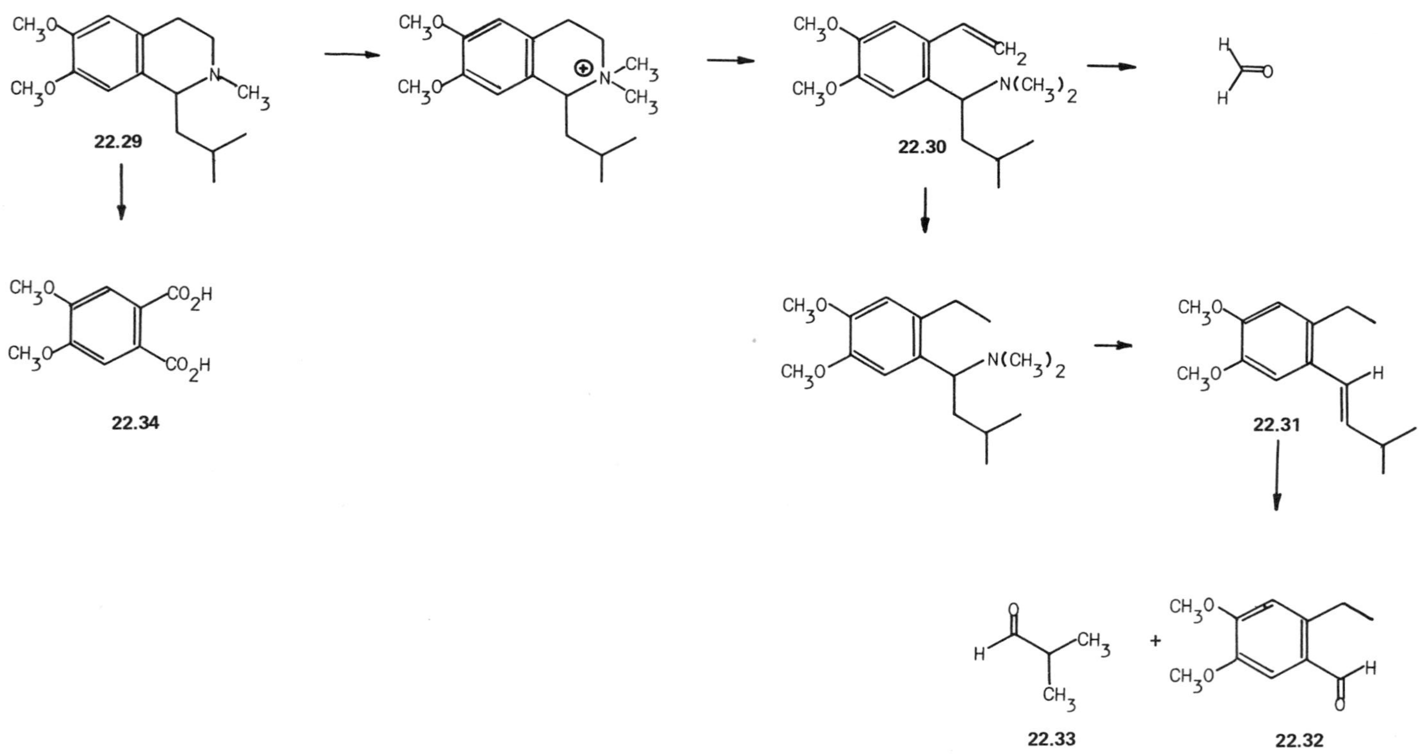

SCHEME 22.5 A degradation of O-methyllophocerine. (After Reference 114.)

When pilocerine (22.28) is cleaved with potassium and liquid ammonia, a phenol is obtained which generates an oil on methylation with diazomethane. This oil is identical (as shown later, [116]) to O-methyllophocerine (22.29) and was degraded as follows (Scheme 22.5).

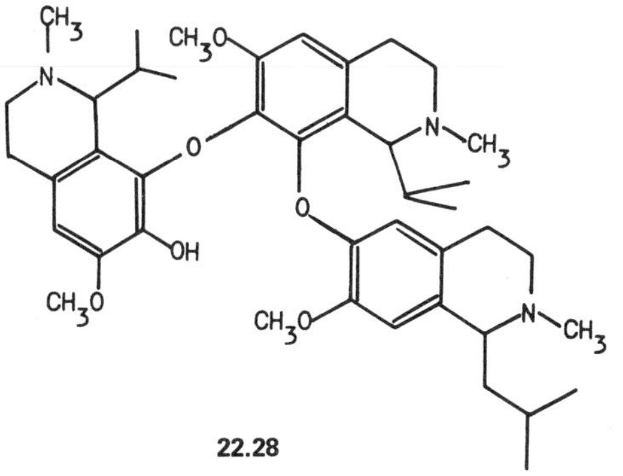

22.28

Exhaustive methylation of O-methyllophocerine (22.29) and Hofmann degradation yields a methine (22.30) which furnishes formaldehyde on ozonolysis. Hydrogenation of the methine (22.30), further methylation, and another Hofmann degradation result in formation of trimethylamine and a nitrogen-free neutral material assigned the structure 22.31. On ozonolysis, the styrene 22.31 yields 2-ethyl-4,5-dimethoxybenzaldehyde (22.32) and isobutyraldehyde (22.33). Finally, permanganate oxidation of the original O-methyllophocerine (22.29) generates a mixture of metahemipinic (22.34) and isobutyric and isovaleric acids. Structure 22.29 for O-methyllophocerine is demanded by these data. The position of the hydroxyl group in lophocerine (22.4) is assigned on the basis of biogenetic considerations (!) [116]; see p. 184) and the fact that use of diazoethane in place of diazomethane in the original alkylation of pilocerine (22.28) followed by degradation as outlined above and in Scheme 22.5 yields 2-ethyl-5-ethoxy-4-methoxybenzaldehyde (22.35).

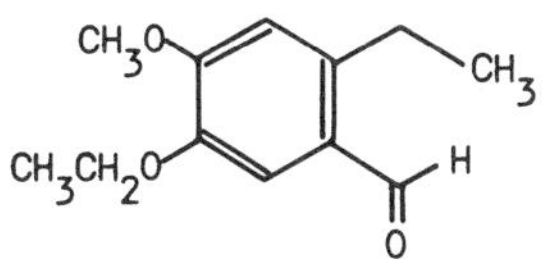

22.35

Lophocerine (22.4) has been synthesized [116]. Vanillin (22.36) is benzylated to 3-methoxy-4-benzyloxybenzaldehyde (22.37) which yields 3-methoxy-4-benzyloxy-ω-nitrostyrene (22.38) on reaction with nitromethane. Reduction of 22.38 with lithium aluminum hydride and treatment of the resulting amine with isovaleryl chloride in the presence of a basic anion exchange resin generates the corresponding isovaleryl amide, 22.39, which undergoes cyclization to 1-isobutyl-6-methoxy-7-benzyloxy-3,4-dihydroisoquinoline (22.40) on treatment with phosphorus pentachloride. Methylation with methyl iodide, followed by hydrogenation over Adams catalyst, yields 1-isobutyl-2-methyl-6-methoxy-7-benzyloxy-1,2,3,4-tetrahydroisoquinoline (22.41). Further reduction with H_2 in the presence of palladium on charcoal causes hydrogenolysis to lophocerine (22.4). Scheme 22.6 presents this synthesis.

As pointed out earlier, feeding experiments [117] have demonstrated that the phenethylamine portion of lophocerine (22.4) is derived from tyrosine (22.21), while both leucine

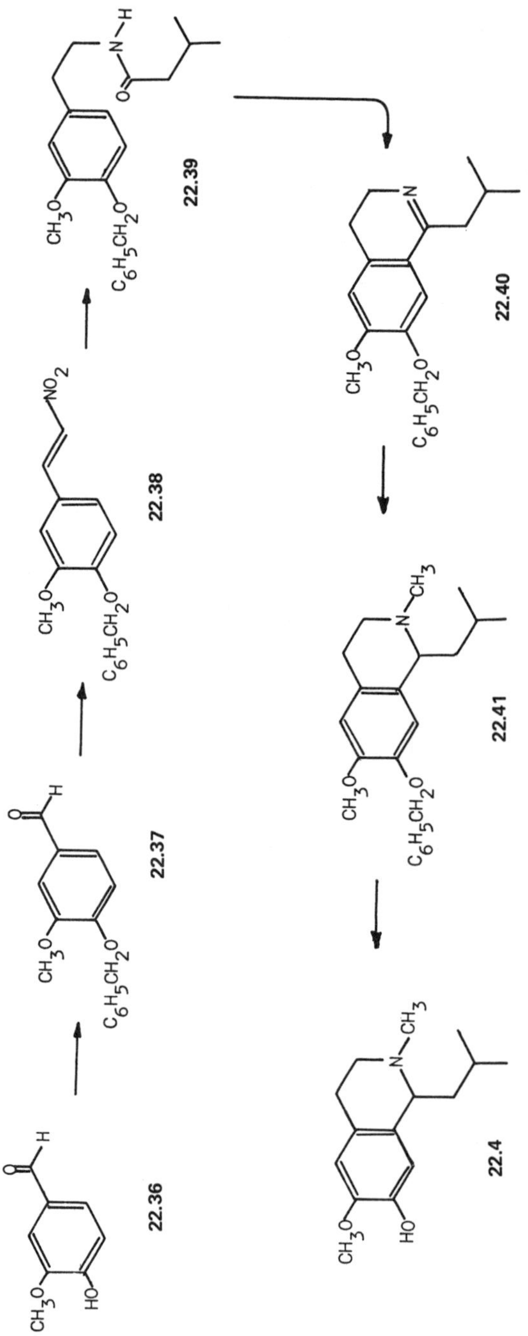

22.36 → 22.37 → 22.38 → 22.39 → 22.40 → 22.41 → 22.4

SCHEME 22.6 A synthesis of lophocerine. (After Reference 116.)

(22.42) and mevalonate (22.43) serve as precursors of the remaining C_5 unit. The degradative scheme which permits these conclusions is much the same as that used for the structure proof and is shown as Scheme 22.5.

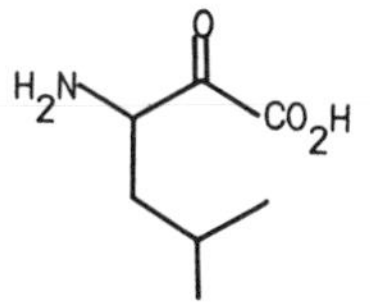

22.42

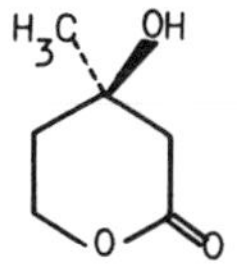

22.43

V. EMETINE

Ipecac root from a variety of sources has been used therapeutically for over 200 years as an emetic. Emetine (22.7), a major constituent of ipecac (e.g., in Cephaelis ipecachuanha) was isolated, shown to have the properties of the crude extract, and work on the structure was begun in 1817. The gross structure of emetine (22.7) was established by degradation and the stereochemistry, largely by synthesis. Some of the degradative work, now classic, which led to the structure of emetine (22.7) is described below and is shown in the schemes and reactions which follow.

When emetine (22.7) as its hydrochloride salt is treated with potassium permanganate in aqueous solution, 6,7-dimethoxyisoquinoline-1-carboxylic acid (22.44) is isolated from the oxidation products (Equation 22.1). In alkaline solution, however, the same oxidizing

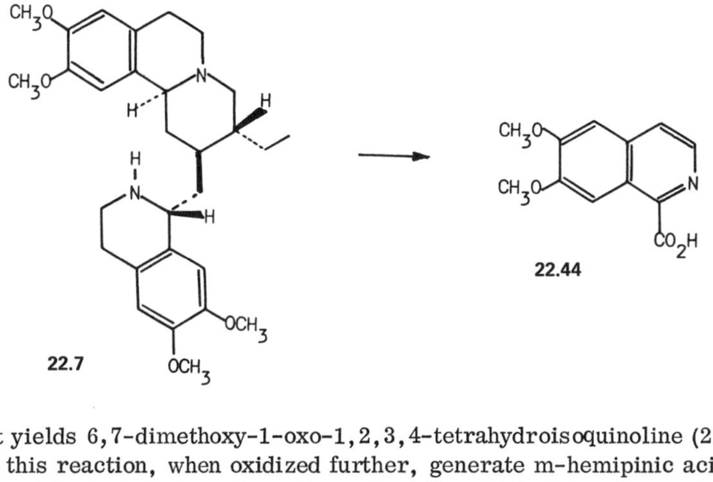

22.7 → 22.44 (Eq. 22.1)

agent yields 6,7-dimethoxy-1-oxo-1,2,3,4-tetrahydroisoquinoline (22.45), and the residues from this reaction, when oxidized further, generate m-hemipinic acid (22.34; Equation 22.2). The combined yield of products from the latter (96% of theory, assuming one dimethoxytetrahydroisoquinoline, or equivalent unit, per molecule of alkaloid) is approximately two times what precedent dictates is normally found in similar cases, which suggests that two dimethoxytetrahydroisoquinoline (or equivalent units) are present in emetine (22.7). Along the same lines, when cephaeline (22.46), an alkaloid which accompanies emetine (22.7) and which on O-methylation yields emetine (22.7), is subjected to O-ethylation, oxidation yields both m-hemipinic acid (22.34) and 4-ethoxy-5-methoxyphthalic acid (22.47; Scheme 22.7).

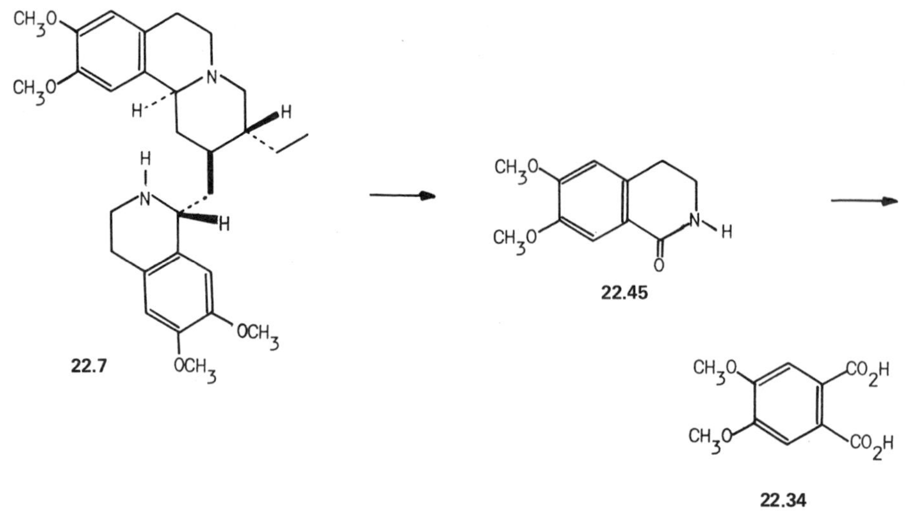

22.7 → 22.45 → 22.34 (Eq. 22.2)

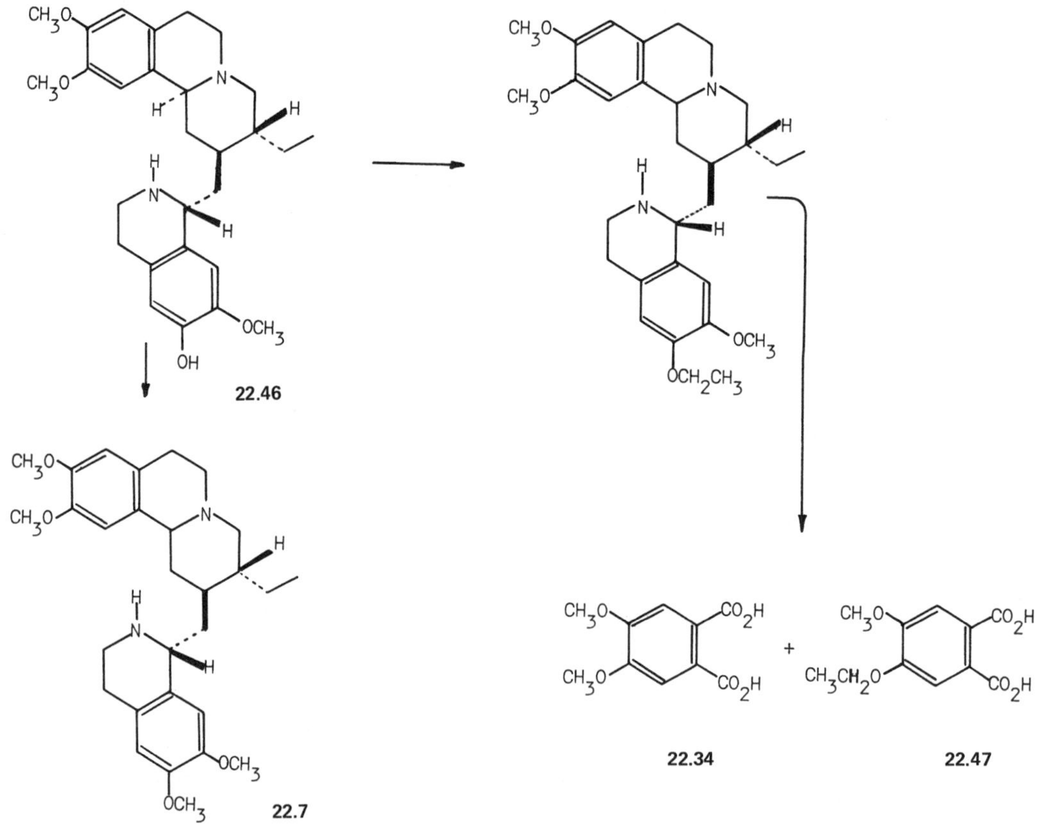

22.46 → 22.7

22.46 → ethylated product → 22.34 + 22.47

SCHEME 22.7 Conversion of cephaline to emetine and degradation of cephaline to prove the presence of two tetrahydroisoquinoline systems.

These results all demonstrate and confirm the presence and nature of the two dioxygenated aromatic rings present in emetine (22.7).

Exhaustive methylation and Hofmann degradation reactions and the preparation of the products of these degradation reactions, provided most of the information about the structure of emetine (22.7). Thus [118], the dimethiodide of N-methylemetine (22.48) can be converted to its bismethine (22.49) which is hydrogenated to 22.50 (Scheme 22.8). Since this molecule contains an N, N-dimethyl-p-methoxybenzylamine functionality, it is not surprising that quaternary salt formation at the amino groups results in ready elimination of trimethylamine from one site, and reduction of the monomethine which results over palladium catalyst yields 22.51 (Scheme 22.8). Two more Hofmann degradation sequences, without intermediate hydrogenation, yield the doubly unsaturated derivative 22.52.

The diene 22.52 undergoes ozonolysis (Scheme 22.9) to generate 4,5-dimethoxy-2-ethylbenzaldehyde (22.53) and its corresponding acid and a second aldehyde which is formulated as 22.54, since its oxidation results in formation of β-(4,5-dimethoxy-2-ethylphenyl)propanoic acid (22.55) and on ozonolysis it yields methyl ethyl ketone (22.56; Scheme 22.9). Other alternatives to the aldehyde (22.54) were eliminated by two different, equally ingenious, routes. The first of these (Scheme 22.10; [119]) actually proved the structure of emetine (22.7), while the second confirmed it.

When N-acetylemetine (22.57) is exhaustively methylated to 22.58, and the Hofmann degradation carried out, the alkene 22.59 results. Subsequent and repetitive hydrogenations, methylations, and Hofmann degradations culminating in a deacetylation reaction, result in isolation of the amine 22.60. When this amine (22.60) is subjected to the same reactions outlined above and the nitrogen excised, the alkene 22.61 is obtained. Ozonolysis of this alkene (22.61) yields an aldehyde (22.62) identical to the product obtained by hydrogenation of 22.54. However, the aromatic ring in 22.62 and that in the product obtained on hydrogenation of 22.54 are from the opposite ends of emetine (22.7)! The second, and confirming work on the structure of emetine (22.7) was also reported in 1948 [120] and is shown as Scheme 22.11.

γ-Methyl-n-caproic acid (22.63) is converted to its α,β-unsaturated analog via bromination and dehydrohalogenation and the latter esterified to 22.64. Michael addition of diethyl malonate (22.65), hydrolysis, and decarboxylation then yield the symmetric diacid 22.66, which undergoes a Friedel-Crafts acylation onto 2,3-dimethoxyethylbenzene (22.67) to yield the ketoacid 22.68. Clemmensen reduction of the keto group, followed by conversion of the acid to an amide of 3,4-dimethoxy-β-phenethylamine (22.69), cyclization, and reduction, yield the secondary amine 22.70. When the amine 22.70 is subjected to exhaustive methylation, Hofmann degradation, hydrogenation, further methylation, and elimination, 22.61 is formed (Equation 22.3). The alkene 22.61, and its ozonolysis product (22.62) are

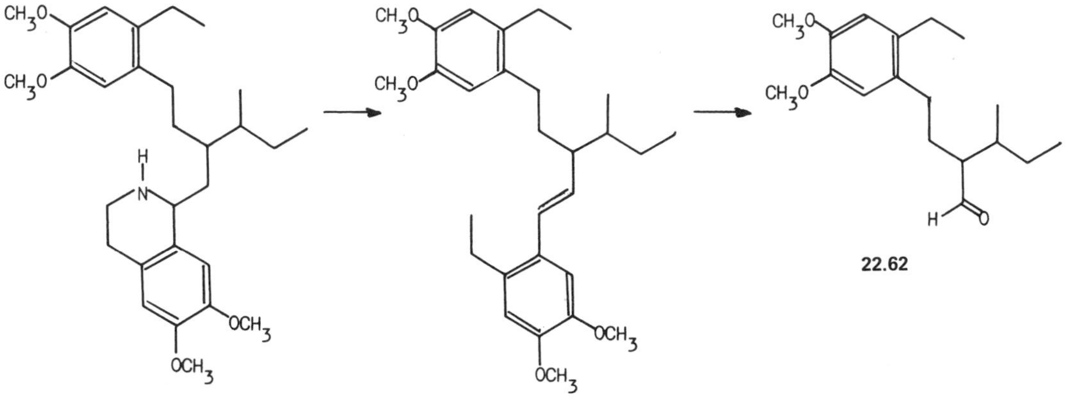

22.70 → 22.61 → 22.62 (Eq. 22.3)

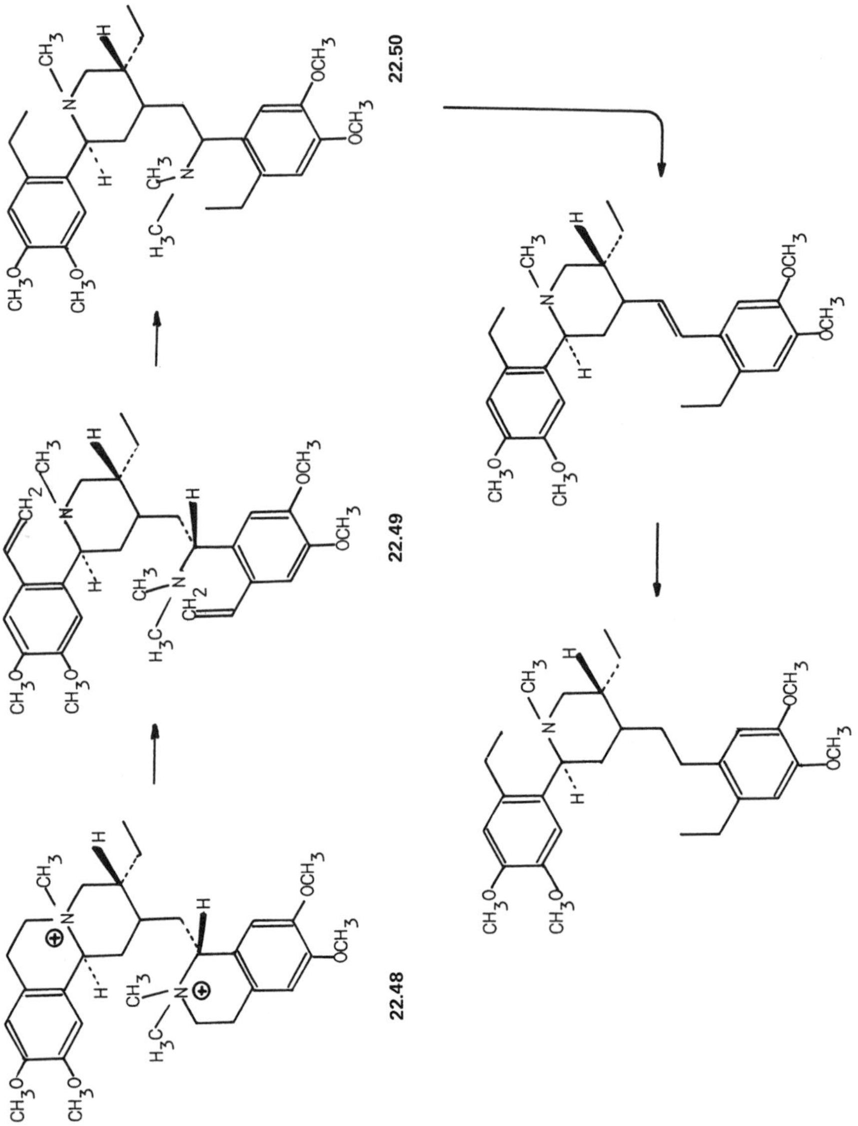

SCHEME 22.8 The degradation of emetine. (After Reference 118.)

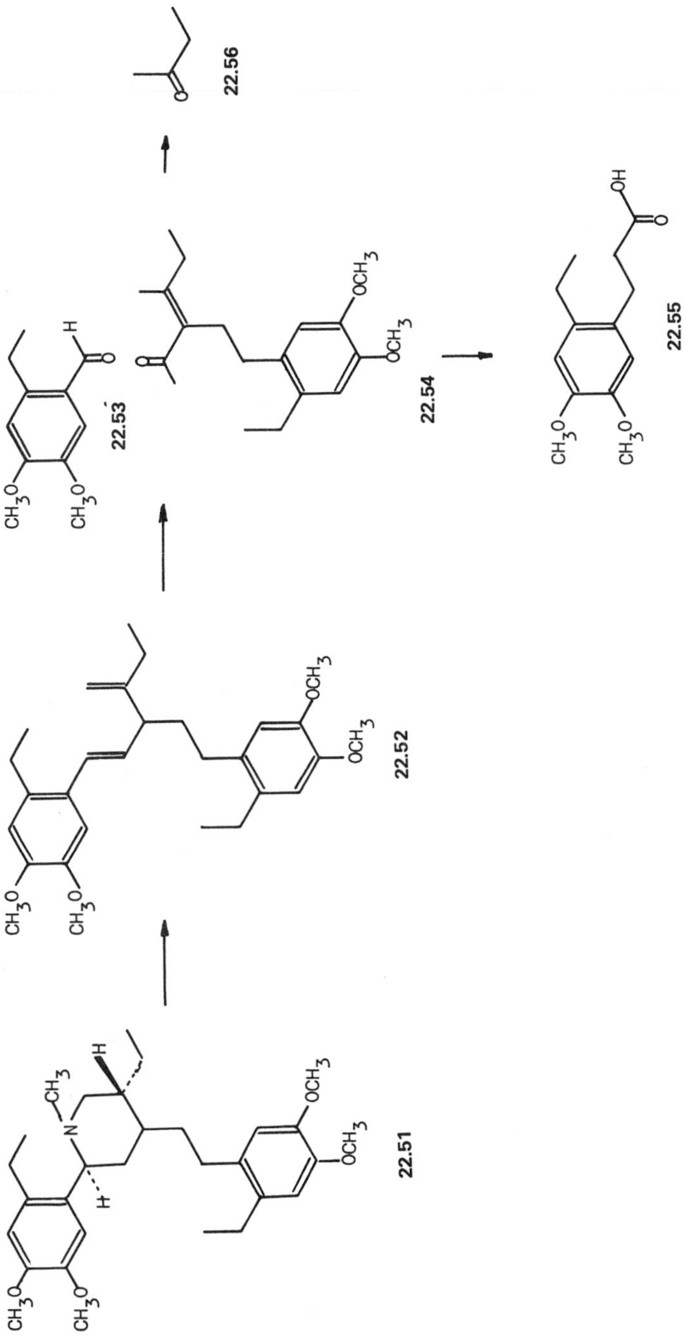

SCHEME 22.9 The degradation of emetine. (After Reference 118.)

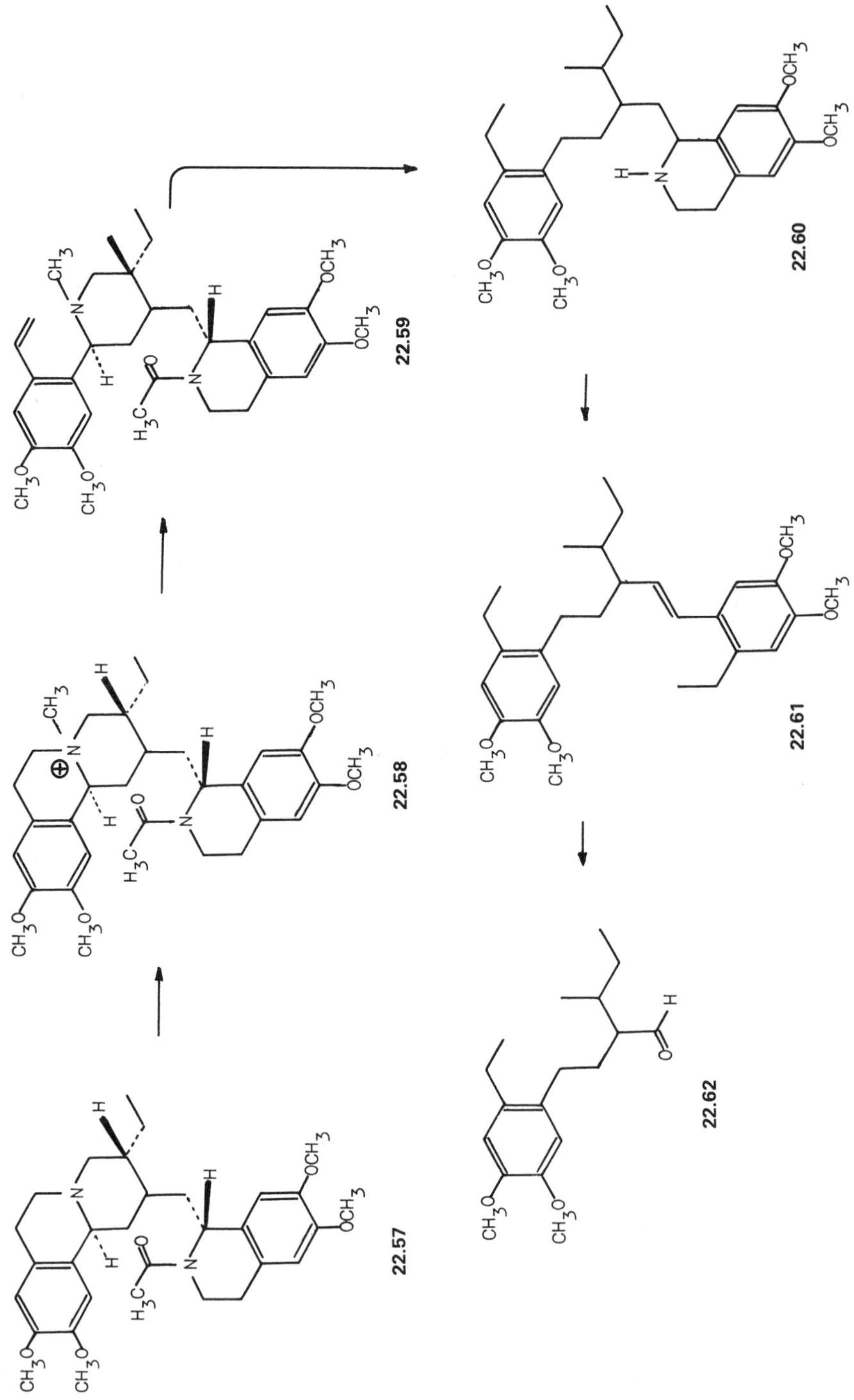

22.57

22.58

22.59

22.60

22.61

22.62

SCHEME 22.10 A portion of the structure proof of emetine. (After Reference 119.)

SCHEME 22.11 A partial synthesis of emetine. (After Reference 120.)

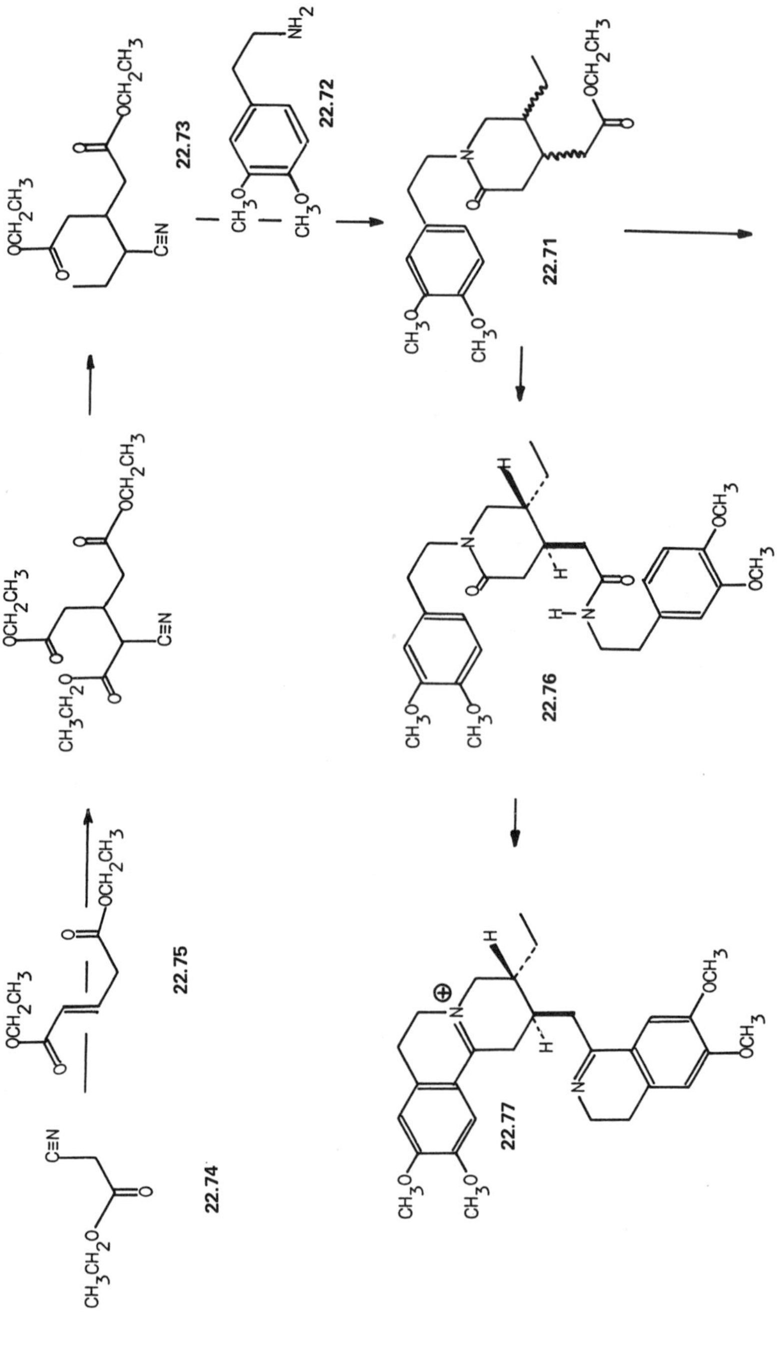

22.74
22.75
22.73
22.72
22.71
22.76
22.77

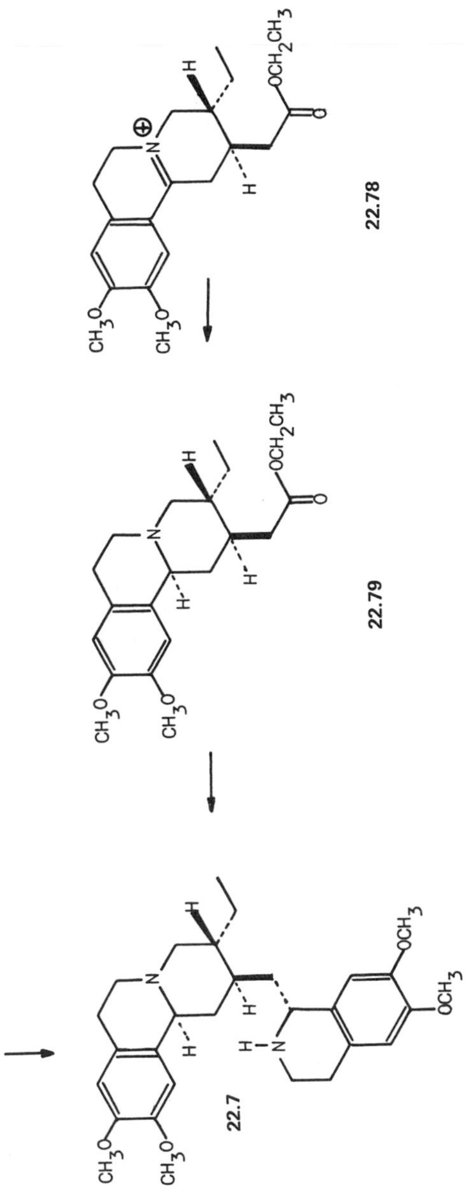

SCHEME 22.12 Syntheses of emetine. (After Reference 121.)

identical to the materials obtained (see above) by degradation of naturally occurring emetine (22.7).

Although emetine (22.7) is synthesized commercially, the first total synthesis of emetine (22.7) was announced before the detailed stereochemistry of the alkaloid had been established, and, indeed, the synthesis was useful in deciding what interrelationship existed between the four asymmetric centers present in the molecule. The synthesis (Scheme 22.12; [121]) involves the formation of 22.71 from the reductive cyclization of 3,4-dimethoxy-β-phenethylamine (22.72) with the cyano ester 22.73 obtained from the reaction of cyanoacetic ester (22.74) with diethylglutaconate (22.75), followed by ethylation, selective hydrolysis, and decarboxylation to 22.73. One of the isomers of 22.71 is then converted to the amide 22.76 with more 3,4-dimethoxy-β-phenethylamine (22.72), the imine 22.77 on cyclization, and to (±)-emetine (22.7 and its mirror image) on reduction. Alternatively, cyclization of one of the isomers of 22.71 to the imine 22.78, reduction to 22.79, and then reaction with 3,4-dimethoxy-β-phenethylamine (22.72), cyclization, and a last reduction, also yield (±)-emetine (22.7 and its mirror image). The stereochemistry of the required isomer of 22.71 was not known, nor indeed was the stereochemistry of either of the two imine reductions carried out either concomitantly on 22.77, or sequentially, from 22.71. The key was, however, available [122] and is shown in Scheme 22.13.

Carefully controlled reduction of the isomer of 22.71 which went on to yield emetine (22.7) with lithium borohydride generates the alcohol 22.80. Desulfurization of the corresponding isothiouronium salt (22.81; made from the tosylate of 22.80) yields the lactam 22.82 with, of course, retention of configuration at the two sites bearing alkyl side chains. This same lactam (22.82) is prepared by N-alkylation of 3,4-dimethoxy-β-phenethylamine (22.72) with racemic ethyl threo-3,4-diethyl-5-bromovalerate (22.83) and subsequent cyclization. Thus, the ethyl groups in 22.82 are trans (and almost certainly, diequatorial).

The third chiral center to be considered is that generated in, for example, the conversion of 22.78 to 22.79. This site was deduced as being S (the incoming hydrogen being axial, as in 22.7) since (a) reduction of the imine 22.78 under many different sets of conditions resulted in the same product and (b) Bohlmann bands [123], typically associated with quinolizidines in which the hydrogen in question is axially disposed, are present.

The final chiral center was established as <u>opposite</u> to the center in the benzoquinolizine ring by noting that while the ORD curve of emetine hydrobromide (22.7) remained constant in the range 350-700 nm, suggesting that the individual contributions of the chiral centers adjacent to the UV chromophore were about equal and opposite, the same region of the spectrum of isoemetine (22.84) hydrobromide rose to a maximum at about 300 nm, suggesting the sense of the asymmetric centers was alike.

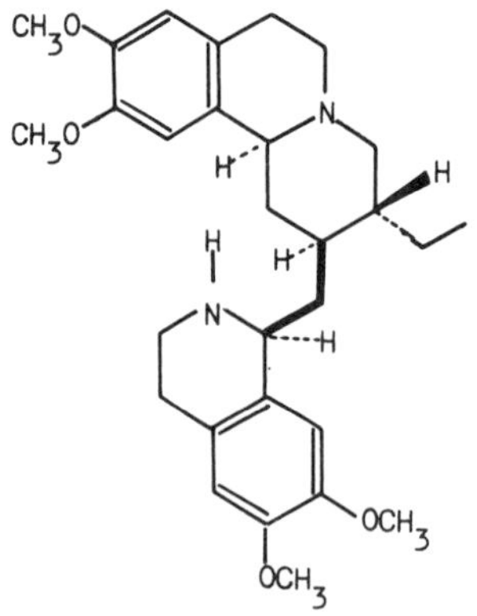

CH_3O CH_3O N H H H H N H OCH_3 OCH_3

22.84

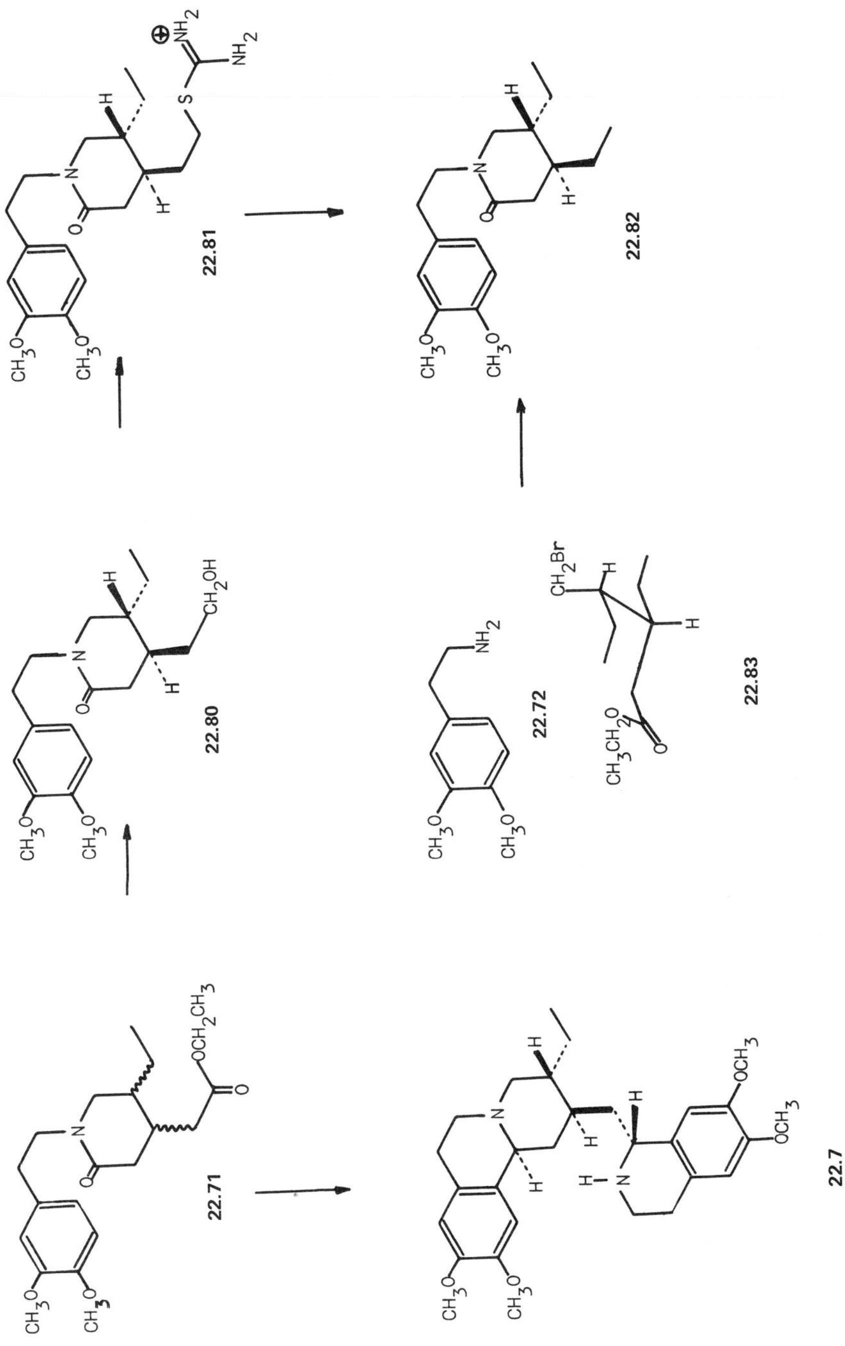

SCHEME 22.13 Establishment of the relative configuration of the alkyl substituents present in emetine.

VI. PROTOEMETINE

Aside from emetine (22.7) and a few closely related bases (e.g., cephaline, 22.46, and psychotrine, 22.85) which are readily isolated from ipecac, careful examination of the nonphenolic residues was required to find additional compounds which are present. Among the

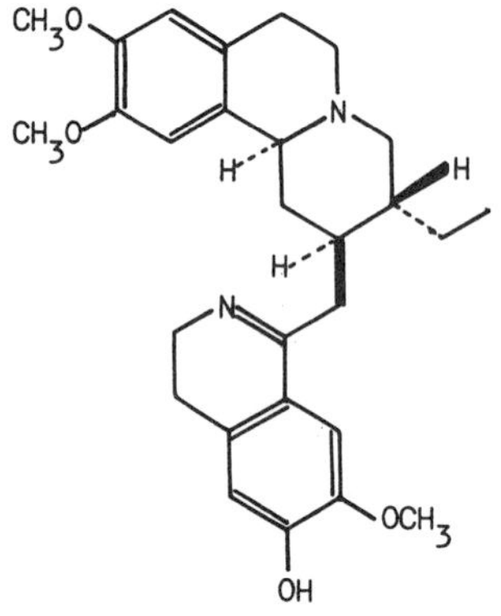

22.85

latter is an unstable base called protoemetine (22.5; [124]). This base (22.5), which can be converted to emetine (22.7), possesses spectral characteristics indicating the presence of a 6,7-dimethoxytetrahydroisoquinoline system and an aldehydic unit. By way of confirmation of the latter, the base (22.5) gives a Tollens test and forms a crystalline oxime which generates, without loss of carbon, a nitrile on dehydration. The nitrile (22.86) is capable of hydrolysis to the corresponding carboxylic acid (22.87) which can be converted to the amide 22.88 on treatment with 3,4-dimethoxy-β-phenethylamine (22.72). This amide (22.88) undergoes cyclization to 22.89, identical to (±)-O-methylpsychotrine (Scheme 22.14). (±)-O-Methylpsychotrine (22.89 and its mirror image) is capable of reduction to (±)-emetine (22.7 and its mirror image), thus confirming the suggested relationship.

VII. IPECOSIDE

The neutral compound, ipecoside (22.6), was characterized first in 1952, and in 1967, mostly on the basis of spectroscopic studies, a structure was proposed [125]. The maximum in the ultraviolet spectrum of ipecoside (22.6) is unchanged in acid but shifts to λ_{max} 306 nm in alkali. When the spectrum of 1-ethyl-2-acetoxy-6,7-dihydroxy-1,2,3,4-tetrahydroisoquinoline (22.90) is subtracted from the original spectrum of ipecoside (22.6), the remainder

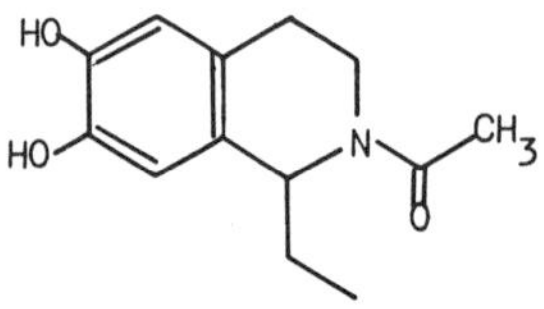

22.90

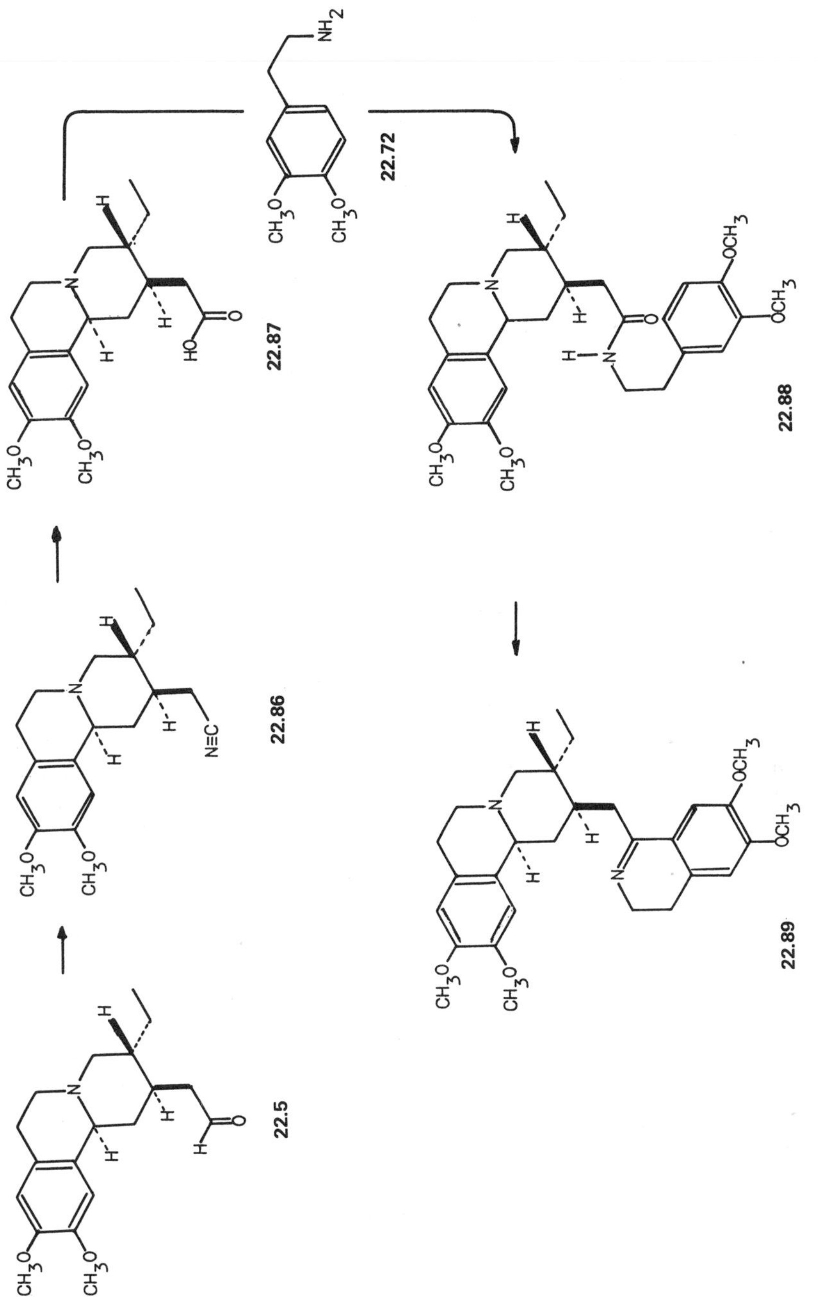

SCHEME 22.14 A synthesis of (±)-O-methylpsychotrine from protoemetine.

(λ_{max} 238 nm, ϵ 9.1 × 10^3) suggests an α,β-unsaturated ester, substituted β by oxygen (i.e., a fragment such as 22.91). The infrared spectrum confirms the presence of the β-oxygenated α,β-unsaturated ester (i.e., the fragment 22.91; 1690 cm^{-1}, 5.91 μ) and also shows the presence of an amide (1630 cm^{-1}, 6.12 μ). Mass spectrometry (EI) confirms the original molecular formula, $C_{27}H_{35}NO_{12}$, although no parent ion is detectable. A fragmentation pattern characteristic of N-acetyl-1-substituted-1,2,3,4-tetrahydroisoquinolines is detected.

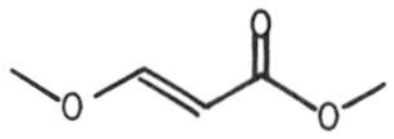

22.91

Additionally, (a) Mild acidic hydrolysis of ipecoside (22.6) generates (+)-glucose, while both acetic acid and (+)-glucose form under more vigorous conditions. (b) Hydrogenation of ipecoside (22.6) yields dihydroipecoside, $C_{27}H_{37}NO_{12}$ (22.92), for which the UV spectrum

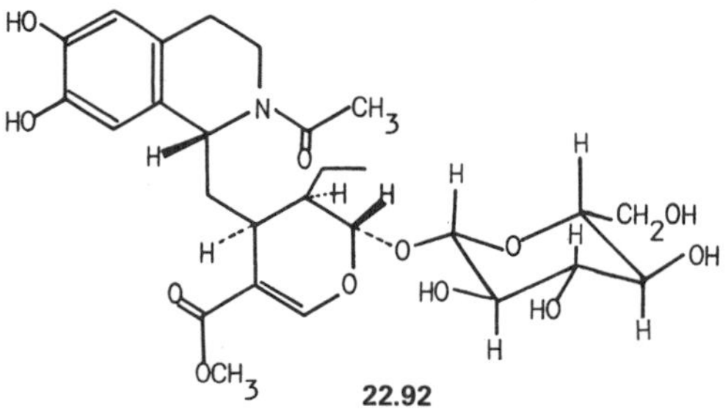

22.92

is unchanged (from that of ipecoside, 22.6), implying that there is a second double bond which has not undergone reduction, and therefore, ipecoside (22.6) is tetracyclic. (c) Kuhn-Roth oxidation of ipecoside (22.6) yields only acetic acid, but both acetic and propanoic acids are produced from dihydroipecoside (22.92). (d) Both ipecoside (22.6) and O,O-dimethyldihydroipecoside (22.93) are hydrolyzed by β-glucosidase, permitting the assignment of a β-glucosidic linkage of glucose to the aglucone fragment.

Proton magnetic resonance (pmr) confirms the loss of the vinylic side chain on conversion of ipecoside (22.6) to dihydroipecoside (22.92) and the gross structure of the aglucone 22.94 obtained on β-glucosidase hydrolysis of O,O-dimethyldihydroipecoside (22.93), although the details of the stereochemistry at the chiral centers were deduced only by inference and by the conversion via acid epimerization and then oxidation of 22.94 to 22.95 (Equation 22.4).

When O,O-dimethyldihydroipecoside (22.93) is hydrolyzed vigorously with acid and the intermediate (presumably 22.96) reduced, dihydroprotoemetine (22.97), identical with a sample produced by the reduction of protoemetine (22.5), is generated (Equation 22.5). Thus, the stereochemistry at each center is presumed identical to that in protoemetine (22.5; see below).

Attempted correlation of ipecoside (22.6) with other monoterpene glucosides related to indole alkaloids (Part 6) failed, and x-ray crystal analysis [126] was necessary to establish the structure as shown (22.6). It is interesting to note that there must, therefore, be an inversion at C_1 of the tetrahydrobenzylisoquinoline nucleus during conversion of ipecoside (22.6) to dihydroprotoemetine (22.97). This isomerization is written as occurring early, although there is no definitive evidence that it does so.

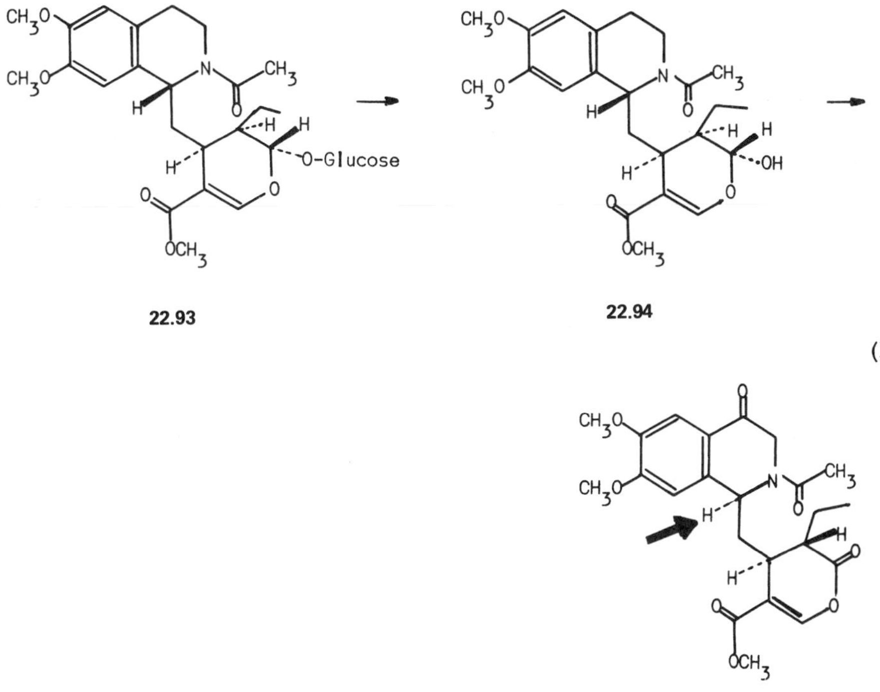

22.93

22.94

(Eq. 22.4)

22.95

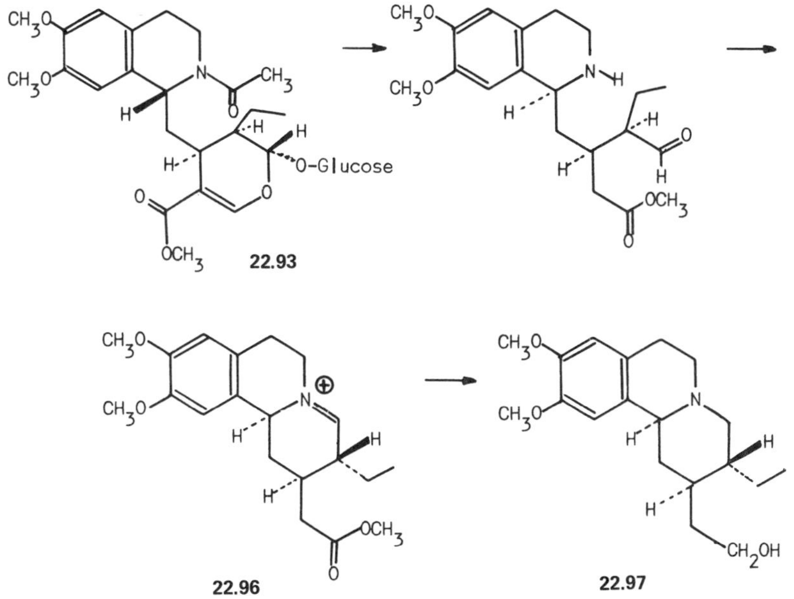

22.93

(Eq. 22.5)

22.96

22.97

SELECTED READING

A. Brossi, S. Teitel, and F. V. Parry, in The Alkaloids, Vol. 13, R. H. F. Manske (ed.). Academic, New York, 1971, pp. 189 ff.

M. M. Hanot, in The Alkaloids, Vol. 3, R. H. F. Manske (ed.). Academic, New York, 1953, pp. 363 ff.

R. H. F. Manske (ed.), in The Alkaloids, Vol. 7. Academic, New York, 1960, pp. 419 ff., 423 ff.

L. Reti, in The Alkaloids, Vol. 3, R. H. F. Manske (ed.). Academic, New York, 1953, pp. 313 ff.

23

CHEMISTRY OF 1-PHENYLTETRAHYDROISOQUINOLINE AND OTHER C_6-C_2 + C_6-C_1 ALKALOIDS DERIVED FROM TYROSINE

> Though analogy is often misleading, it is the least misleading thing we have.
>
> S. Butler, Notebooks (1912)

The C_6-C_2 + C_6-C_1 alkaloids derived from tyrosine (23.1) are unique in that the 1-phenyltetrahydroisoquinoline cryptostyline I (23.2) and more highly oxidized relatives are only found in the Orchidaceae, while the other major bases to be considered here, namely, belladine (23.3), lycorine (23.4), homolycorine (23.5), galanthamine (23.6), crinine (23.7), tazettine (23.8), manthine (23.9), and ismine (23.10), as representative of more than 100 related compounds, occur only in the Amaryllidaceae.

23.1

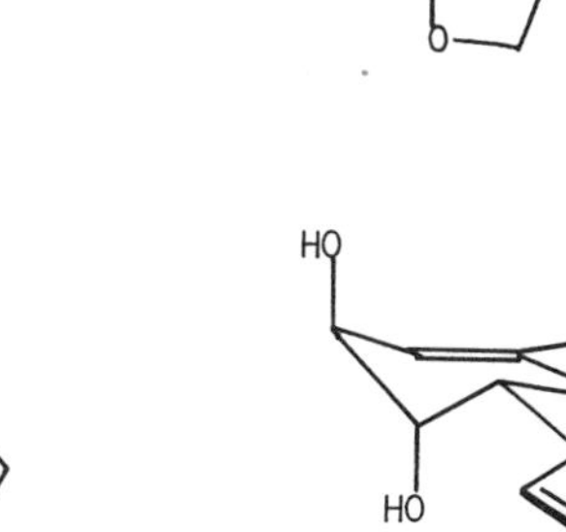

23.2

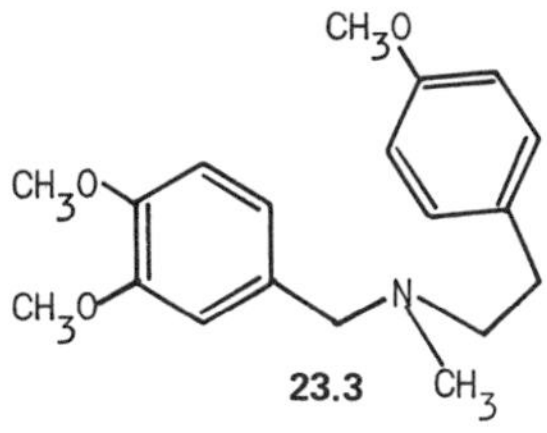

23.3

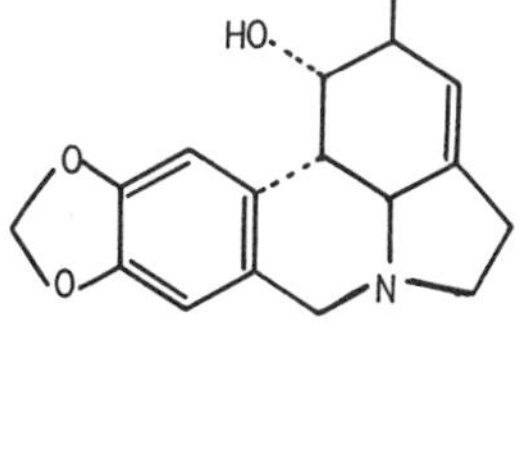

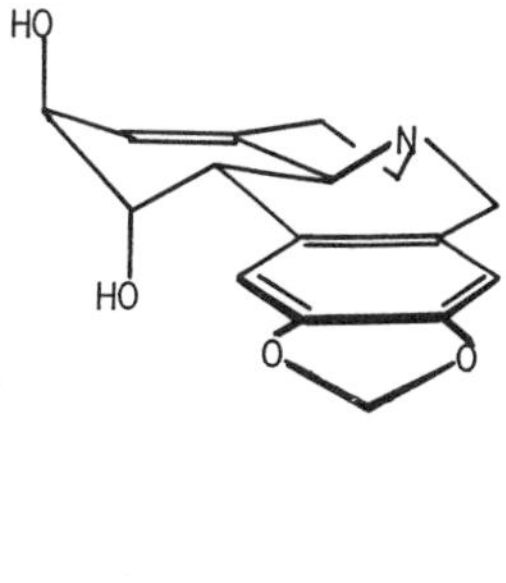

23.4

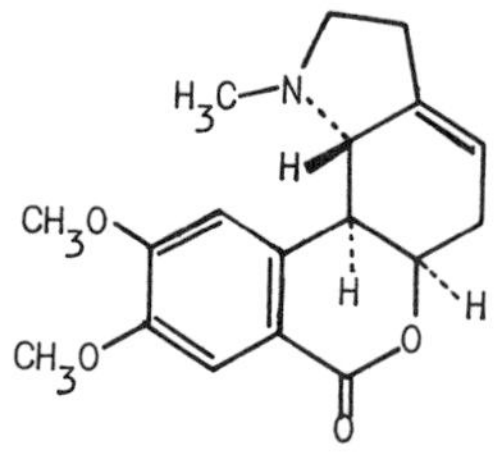

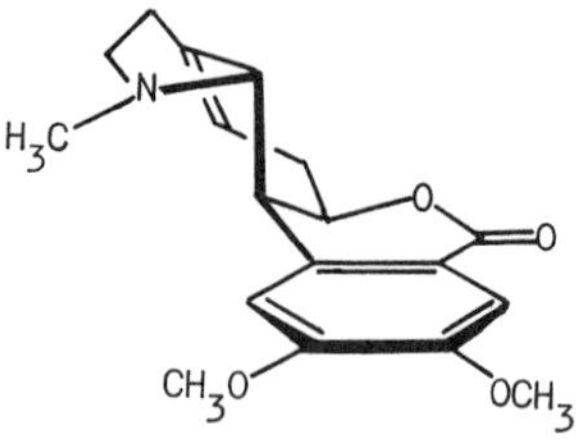

23.5

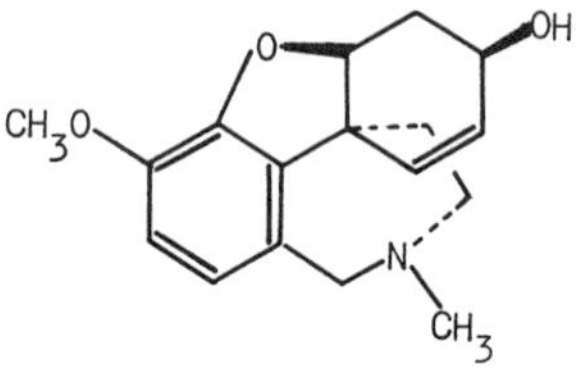

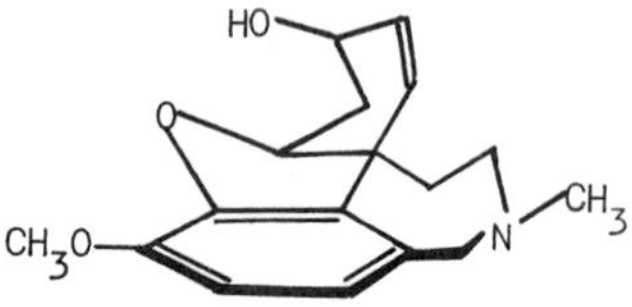

23.6

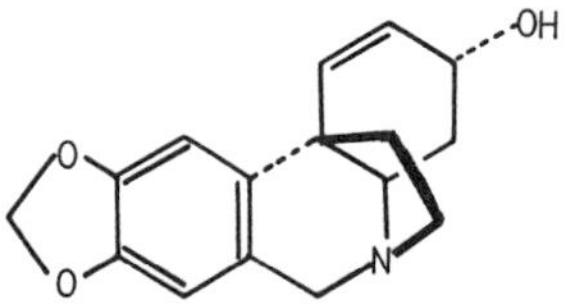

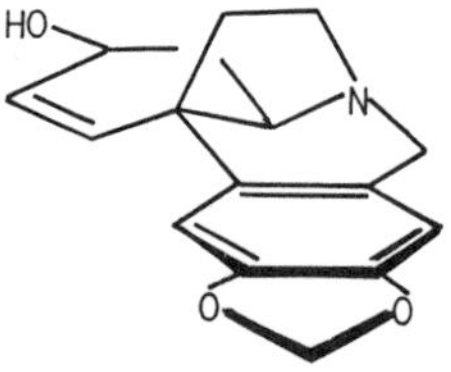

23.7

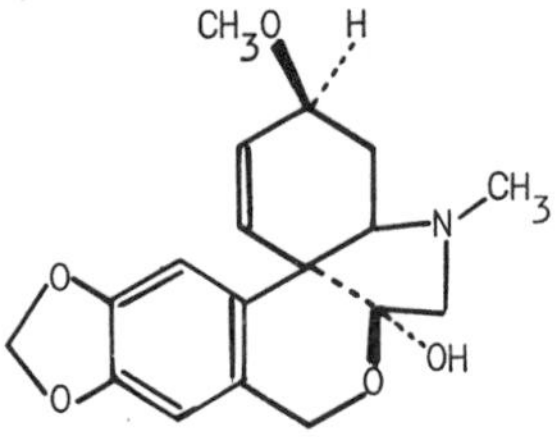

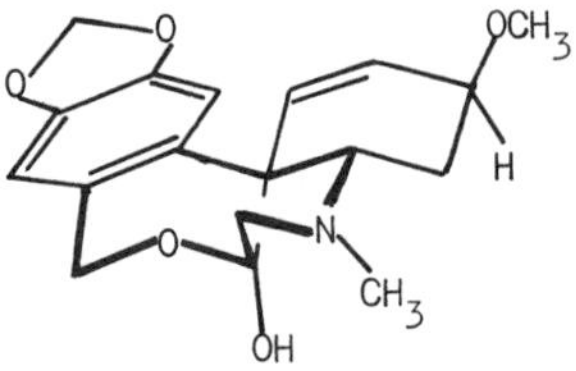

23.8

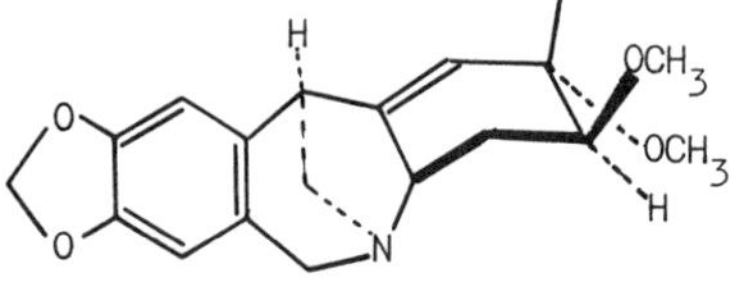

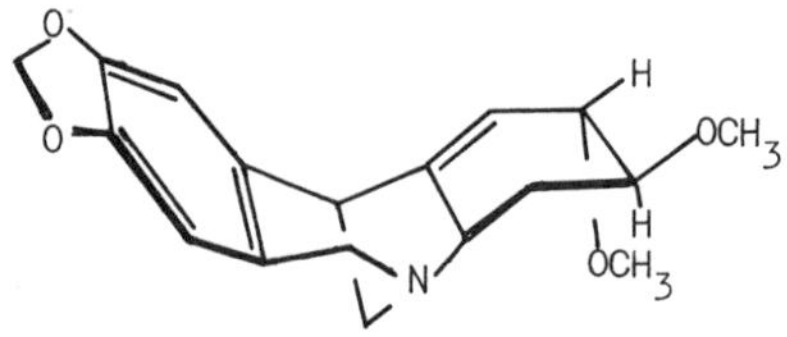

23.9

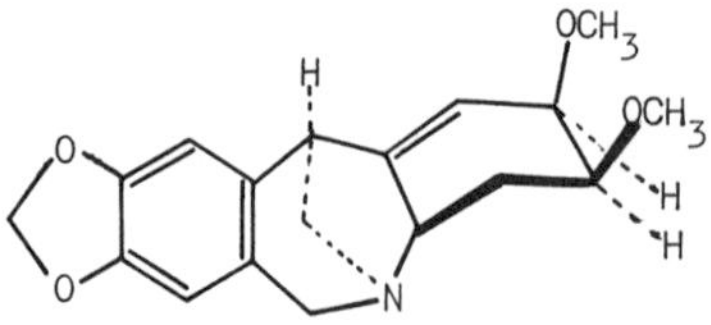

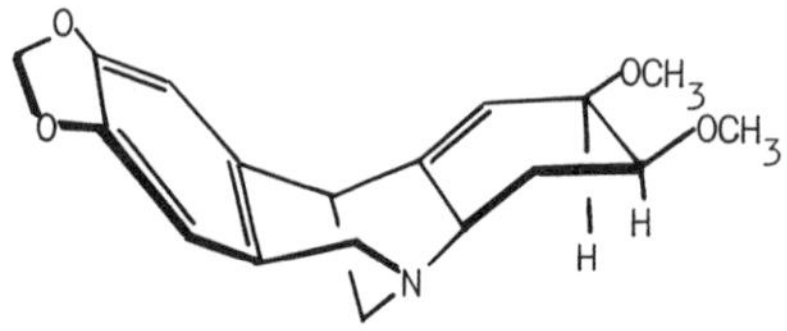

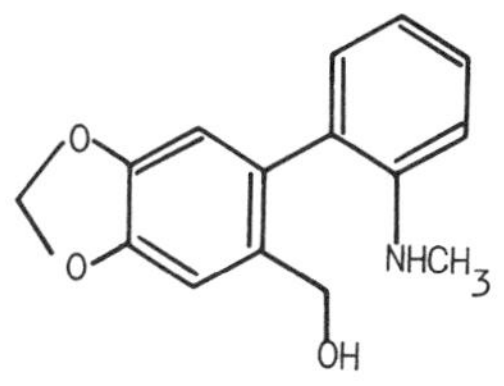

23.10

I. CRYPTOSTYLINE I

Optically active cryptostyline I (23.2) was isolated from Cryptostylis fulva Schltr. and was quickly identified [127] as a tetrahydroisoquinoline by ultraviolet (λ_{max} 235 nm, log ϵ 4.37 and 286 nm, log ϵ 3.89), mass (M^+, m/e 327, base peak, m/e 206), and proton magnetic resonance spectroscopy (C_1 proton at δ = 4.2 ppm, TMS = 0.00). It was then synthesized as outlined in Scheme 23.1. Thus, reaction of 3,4-dimethoxy-β-phenethylamine (23.11) with the acid chloride of piperonylic acid (23.12) yields the amide 23.13, which in turn forms the imine 23.14 on phosphorus oxychloride treatment. Methylation and reduction of the imine generate (±)-cryptostyline I (23.2 and its mirror image). The configuration of the (+)-enantiomer has been established as S on the basis of examination of the x-ray crystal structure of a related base [128].

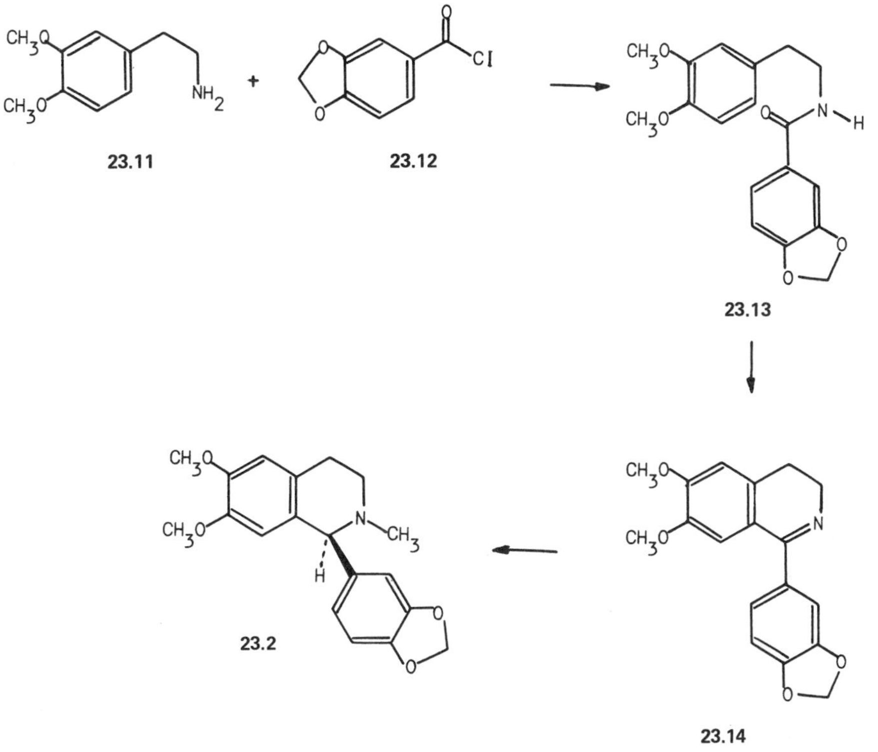

23.11 23.12 23.13 23.14 23.2

SCHEME 23.1 A synthesis of (±)-cryptostyline I. (After Reference 127.)

II. BELLADINE

As pointed out earlier (Chapter 18), a belladine (23.3) precursor (i.e., belladine, 23.3, prior to complete N- and/or O-alkylation) has been suggested [68] as the penultimate precursor of the remainder of the Amaryllidaceae alkaloids. The suggestion preceded isolation of belladine (23.3) from an Amaryllis hybrid [129]. The structure of belladine (23.3) was readily established by (a) its lack of chirality; (b) analysis, showing three O-methyl, one N-methyl, and no C-methyl groups; and (c) Hofmann degradation of belladine methiodide (23.15) to N,N-dimethylveratrylamine (23.16) and 4-methoxystyrene (23.17) (Equation 23.1). Belladine (23.3) and a number of similar phenethylamine derivatives had been synthesized (e.g., by reduction of amides such as 23.13, Scheme 23.1) for testing in pharmacological programs prior to their isolation.

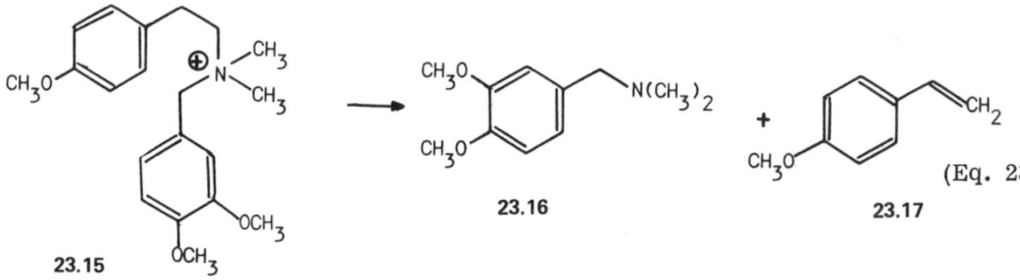

23.15 → 23.16 + 23.17 (Eq. 23.1)

III. LYCORINE

Lycorine (23.4), a levorotatory base, $C_{16}H_{17}NO_4$, mp 276-280°C (dec.) from ethanol, was recognized as a potent emetic and a moderately toxic base from the time of its initial isolation from Lycoris radiata Herb. (about 1877). Since that time, its isolation from many other Amaryllidaceae has served to establish it as the most cosmopolitan alkaloid of the family. Typically, as much as 1% of the dry weight of daffodil bulbs may consist of lycorine (23.4), which has been reported to crystallize, as colorless prisms, directly from the extract after basification.

Although the structure and absolute configuration of lycorine (23.4) have been established by x-ray crystallography [130], the degradative work carried out over a span of about 50 years is illustrative of the attack on more complicated structures which shall be encountered. It was recognized quite early that lycorine (23.4) is a tertiary base containing one ethylenic double bond, two nonphenolic hydroxyls, and (by a color test with potassium dichromate in sulfuric acid) a methylenedioxy, but no methoxy groups. Oxidation yields hydrastic acid (23.18; Equation 23.2), and both lycorine (23.4) and dihydrolycorine (23.19),

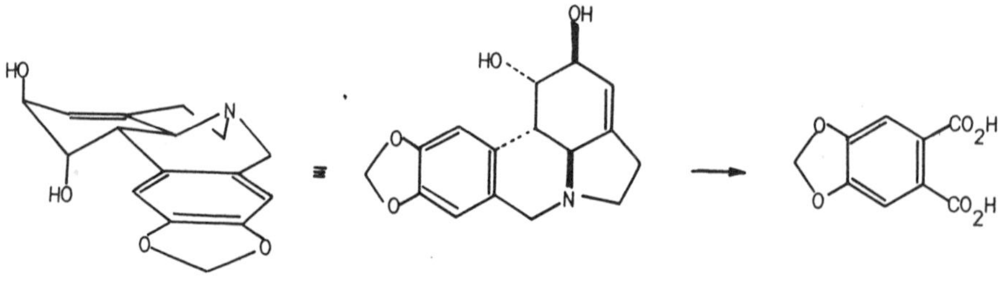

23.4 → 23.18 (Eq. 23.2)

which form on catalytic reduction of lycorine (23.4; Equation 23.3), yield diacetyl derivatives with acetic anhydride.

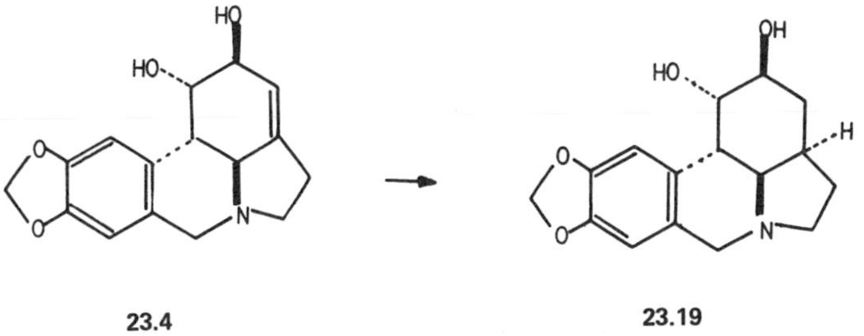

23.4 23.19

(Eq. 23.3)

Neither Hofmann nor Emde degradations were particularly useful on N-methyllycorine salts (23.20) since dehydration was easily effected during the course of these reactions. Nevertheless, the anhydromethine (23.21) formed from lycorine methiodide (23.20) under the conditions of the Hofmann reaction does yield, on oxidation, 6,7-methylenedioxy-N-methylphenanthridone (23.22), the structure of which had been established by synthesis (Equation 23.4); and the Emde reaction yields what was later recognized as 7-(3,4-methylenedioxy-6-methyl)phenyl-1-methyldihydroindole (23.23; Equation 23.5). When lycorine (23.4) is

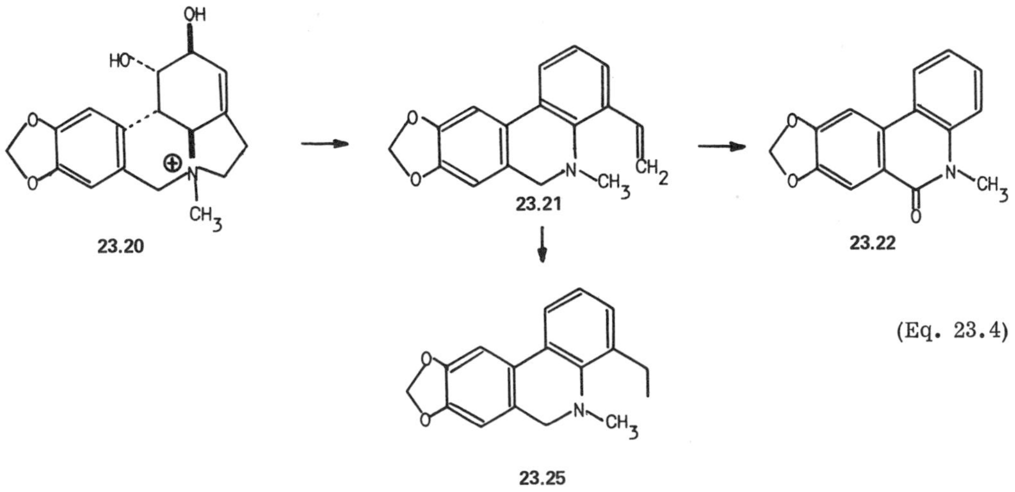

23.20 23.21 23.22 23.25

(Eq. 23.4)

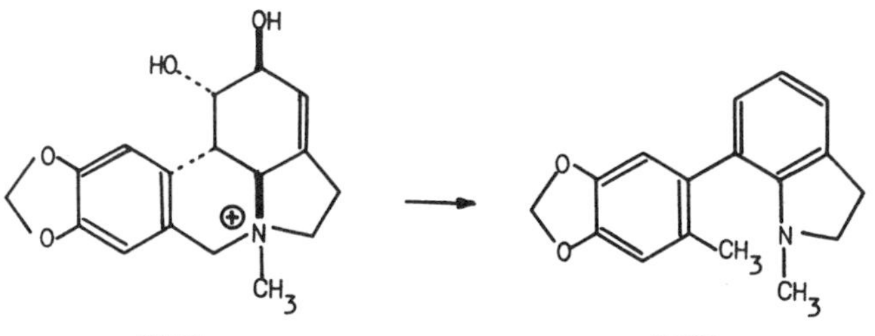

23.20 23.23

(Eq. 23.5)

distilled from zinc dust (Equation 23.6), phenanthridine (23.24) is formed. Distillation of the anhydromethine dihydride 23.25 yields phenanthridine (23.24), 1-methylphenanthridine (23.26), and 1-ethyl-6,7-methylenedioxyphenanthridine (23.27; Equation 23.7), the latter accounting for all the carbon atoms of lycorine (23.4).

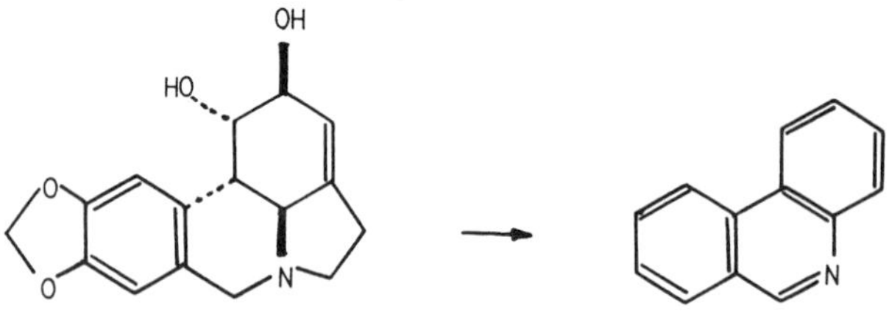

23.4 → 23.24 (Eq. 23.6)

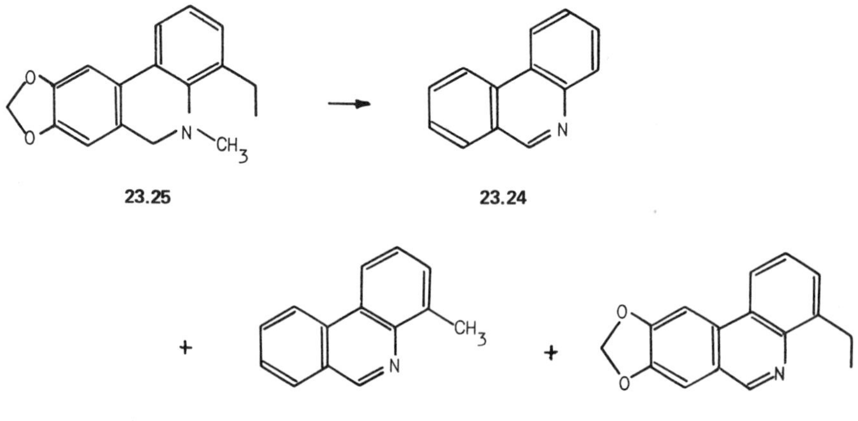

23.25 → 23.24 + 23.26 + 23.27 (Eq. 23.7)

Although much confusion reigned for some time regarding the disposition of the two hydroxyl functions, the weight of evidence indicated that they were vicinal and that the problem with lead tetraacetate and sodium periodate cleavage reactions (both of which gave anomalous results) lay in the ready oxidation of lycorine (23.4) to betaines such as 23.28 (Equation 23.8) on the one hand, and oxidation of the products of initial cleavage of the glycol, on the other. Of the possible structures which could have been written for lycorine (23.4) on the basis of these data, those in which the double bond was either conjugated to the aromatic ring or the nitrogen (i.e., an enamine) were rejected on the basis of ultraviolet and pKa measurements,

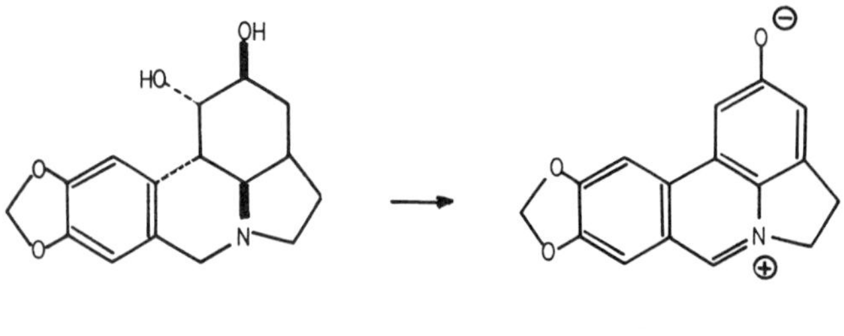

23.4 → 23.28 (Eq. 23.8)

respectively. Finally, since neither of the hydroxyl groups is enolic, and since 1-acetyllycorine (23.29), prepared by acid hydrolysis of diacetyllycorine (23.30; Equation 23.9), is oxidized by manganese dioxide to 1-acetyl-2-lycorinone (23.31) as expected for an allylic alcohol, the gross structure of lycorine (23.4) is established.

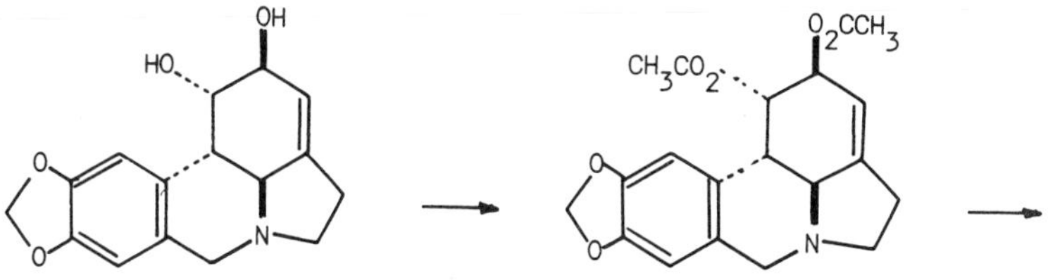

23.4 23.30

(Eq. 23.9)

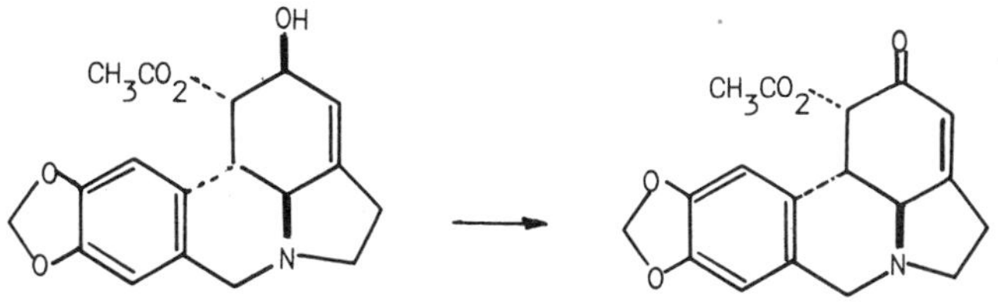

23.29 23.31

Now, when diacetyldihydrolycorine (23.32) is treated with cyanogen bromide, diacetyl-ω-bromo-N-cyanodihydrosecolycorine (23.33) forms. Treatment of the latter with ethanolic potassium hydroxide provides, among others, anhydro-N-cyanodihydrosecolycorine (23.34; Equation 23.10). Then, since tosylation of dihydrolycorine (23.19) yields a monotosylate (23.35) which generates an epoxide (23.36) with methanolic potassium hydroxide, and since

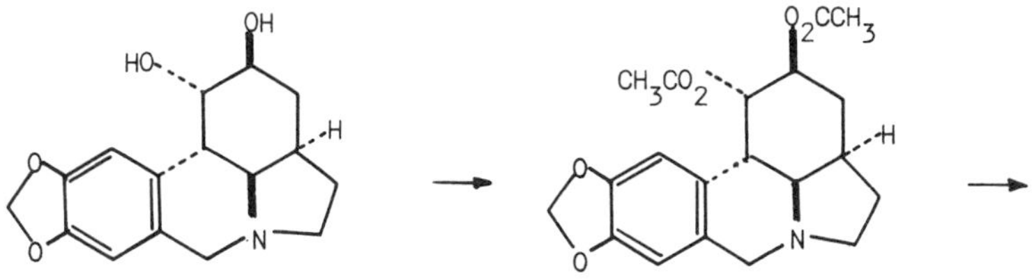

23.19 23.32

(Eq. 23.10)

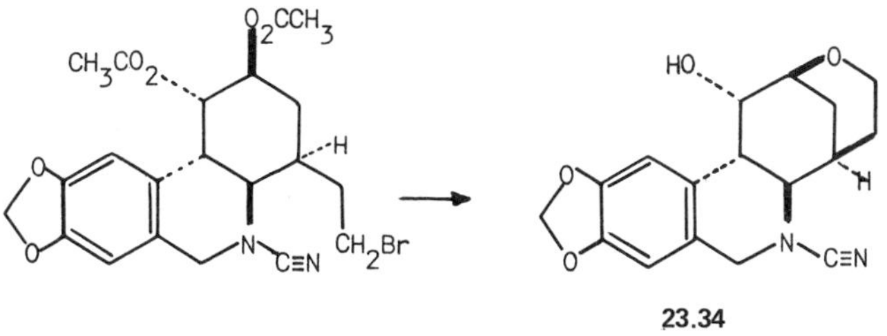

23.33 23.34

the epoxide (23.36) can be hydrolyzed with acid back to dihydrolycorine (23.19) or reduced with lithium aluminum hydride to α-dihydrocaranine (23.37; Equation 23.11; which yields an alkene, 23.38, where in the double bond is conjugated to the aromatic nucleus on treatment

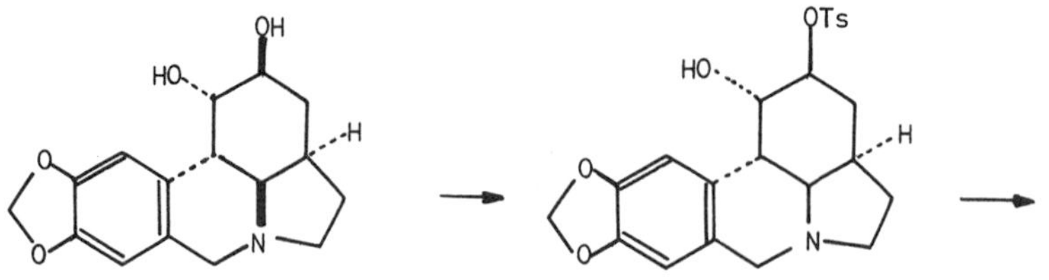

23.19 23.35 (Eq. 23.11)

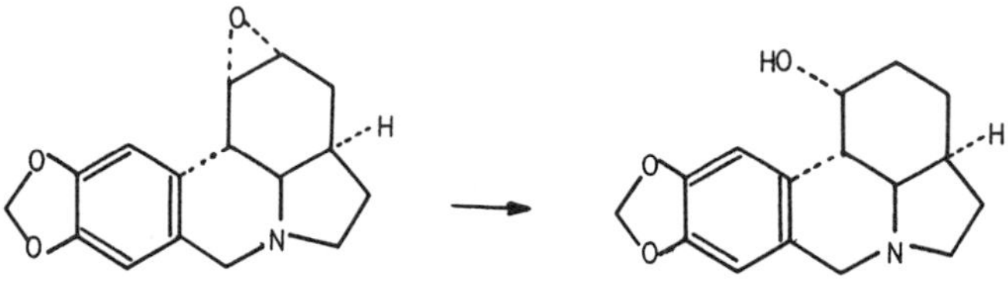

23.36 23.37

with phosphorus pentachloride, Equation 23.12), the hydroxyl groups are defined as vicinal, trans, and diaxial. The ready formation of the anhydro-N-cyanodihydroseco base (23.34; see above) suggests that the ethyl side chain must be cis and axial to the 2-hydroxyl. Finally,

23.37 23.38 (Eq. 23.12)

since a trans diaxial ring fusion of rings C and D is impossible, the nitrogen must be equatorial on ring C and rings B and C trans fused. This defines the structure of dihydrolycorine (23.19), and therefore the structure of lycorine (23.4) as shown <u>or</u> as the mirror image of that shown.

Lycorine (23.4) has been approached by many different synthetic pathways [131]. Dihydrolycorine (23.19) fell to synthesis [132] some 5 years before lycorine itself [133]. As shown in Scheme 23.2, the synthesis of dihydrolycorine (23.19; [132]) begins with 1,2,5,6-tetrahydro-3-(3,4-methylenedioxyphenyl)phthalic anhydride (23.39), which had been prepared from 1-(3,4-methylenedioxyphenyl)-1,3-butadiene (23.40) and maleic anhydride (23.41). Treatment of the anhydride (23.39) with methanol generates a mixture of the corresponding half-esters, and on treatment of the latter with stannic chloride in dichloromethane, a separable mixture of tetralone (23.42) and indanone (23.43). The indanone (23.43) is converted to the corresponding keto alcohol 23.44 through a series of oxidations and reductions, and treatment of the latter (23.44) with sodium azide in trichloroacetic acid leads to the trichloroacetamide derivative of the lactam 23.45. Hydrolysis and conversion to the corresponding carboxylic acid 23.46 (via tosylation, treatment with cyanide, and hydrolysis), followed by acetic anhydride treatment, generate the imide (23.47) which is reduced to the known (±)-γ-lycorane (23.48) with lithium aluminum hydride and then hydrogen and platinum oxide. Alternatively, the acid 23.46 can be used to generate the iodolactone 23.49 (in aqueous sodium bicarbonate with iodine-potassium iodide). The iodolactone 23.49 is unstable and yields the imide 23.50 on treatment with acetic anhydride. Dehydrohalogenation with lithium chloride in dimethyl formamide yields the alkene 23.51, epoxidation of which produced 23.52, and, finally, reduction of the latter with lithium aluminum hydride/zinc chloride yields (±)-dihydrolycorine (23.19 and its mirror image), spectroscopically identical to the chiral material obtained from natural sources. The yield in the last step is 5% theory.

The synthesis of lycorine (23.4; [133]) is shown in Scheme 23.3. In a fashion similar to that for dihydrolycorine (23.19; see above), a cycloaddition reaction is used to begin. Thus, heating fumaric acid (23.53) with 3,4-methylenedioxyphenylallylcarbinol (23.54) in acetic anhydride gives the two stereoisomeric trans anhydrides 23.55 and 23.56 as the major products and a small amount of the thermodynamically favored cis isomer 23.57. When 23.55 and 23.56 are converted into the corresponding acids by hydrolysis and the latter heated for 12 hr, only 23.57 is obtained. Then, a 3:7 mixture of 23.58 (R = H, R' = CH_3):23.59 (R = CH_3, R' = H) is generated from 23.57 on methanolysis, and the major isomer is converted into the urethane 23.60 via the corresponding acid chloride, the azide, and the isocyanate. The free acid of the ester 23.60 is then homologized to 23.61 through methanolysis of the corresponding diazo ketone.

Cyclization of 23.61 to the lactam 23.62 is accomplished with phosphorus oxychloride (in the presence of stannic chloride), and the bislactam 23.63 is generated from 23.62 with hydrogen chloride in acetic acid. Careful reduction of 23.63 with lithium aluminum hydride at 0°C for 1 hr yields the tetracyclic intermediate 23.64. Epoxidation of 23.64 with m-chloroperbenzoic acid yields the epoxylactam 23.65, which in chiral form can also be obtained from lycorine (23.4). Thus, a resolution is avoided and the chiral epoxylactam 23.65 can be used for the remainder of the synthesis. Treatment of optically active 23.65 in ethanol with sodium borohydride and diphenyl diselenide followed by sodium periodate yields the alcohol 23.66 which is converted, after acetylation, into the epoxide 23.67 with m-chloroperbenzoic acid. Then, treatment with diphenyl diselenide and sodium periodate again, generates lycorine lactam (23.68; the acetate having been hydrolyzed). Reduction of the lactam (23.68) diacetate with lithium aluminum hydride yields lycorine (23.4).*

IV. HOMOLYCORINE

The chemistry of homolycorine (23.5), $C_{18}H_{21}NO_4$, mp 175°C $[\alpha]_D^{19}$ +65.1° (ethanol) is closely tied to its reduced homolog, lycorenine (23.69), $C_{18}H_{23}NO_4$, mp 200-202°C, $[\alpha]_D^{22}$ +149.3° (methanol). Indeed, although both homolycorine (23.5) and lycorenine (23.69) are isolated from Lycoris radiata Herb., and the former was reported earlier, the latter is apparently more tractable. Both homolycorine (23.5) and lycorenine (23.69) were recognized as tertiary bases bearing one N-methyl and two methoxyls. Although little concrete evidence was available, lycorenine (23.69) was assumed to have a structure similar to that of lycorine (23.4). Then, after the oxidation of lycorenine (23.69) to homolycorine (23.5) was recognized as the conversion of a hemiacetal to a δ-lactone ([134]; ir, 5.68 μ, 1760 cm^{-1}; Equation 23.13) much of the chemistry of both of these materials became clearer. Thus, reduction of both homolycorine (23.5) and lycorenine (23.69) with lithium aluminum hydride yields tetrahydrohomolycorine (23.70), which in turn generates the ether 23.71 on dilute acid treatment (Equation 23.14).

---

*Note added in proof: A much more facile, higher yield synthesis of (±)-lycorine (23.4 and its mirror image) has been developed [473].

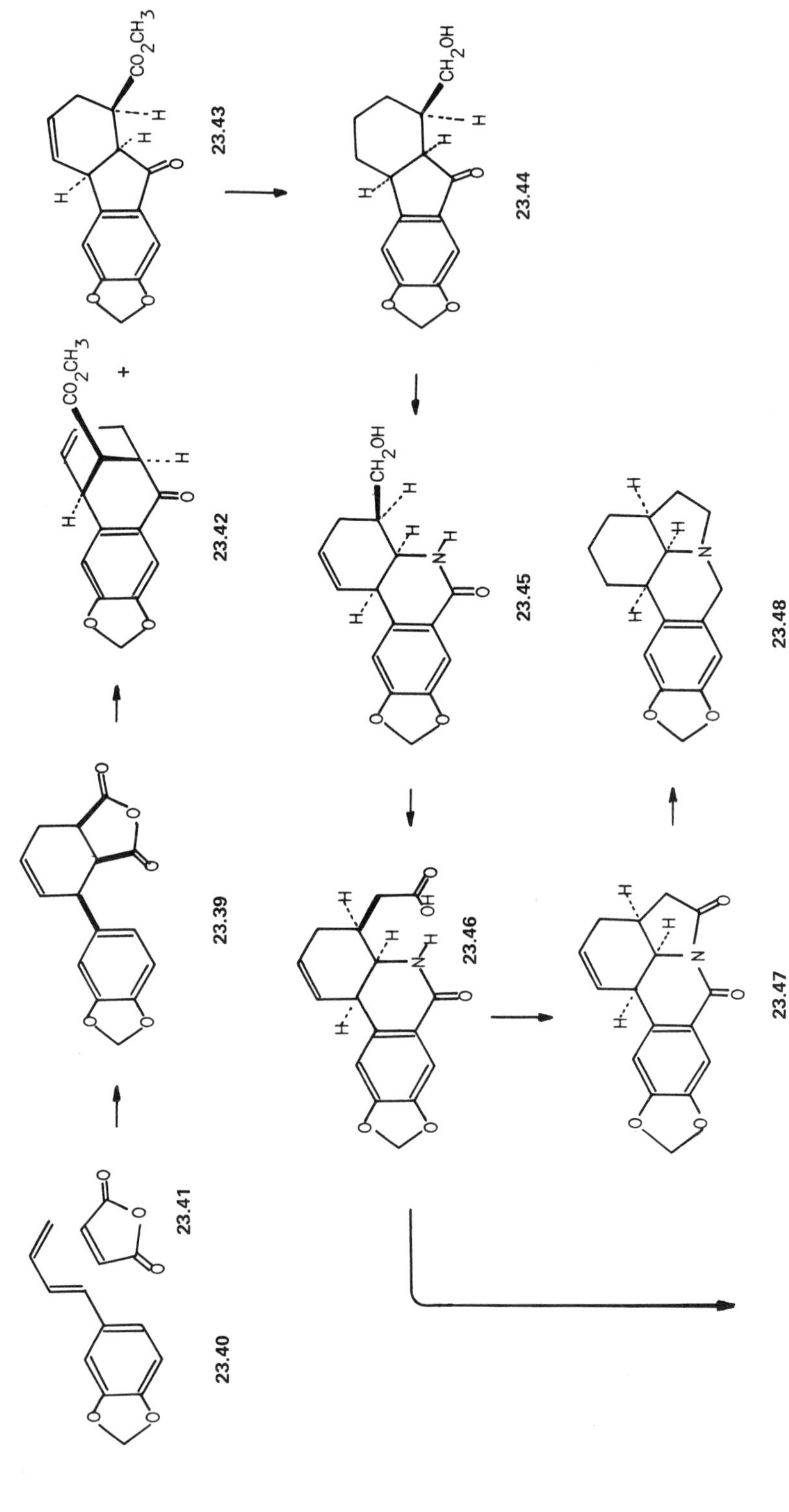

23.40
23.41
23.39
23.42
23.43
23.44
23.45
23.46
23.47
23.48

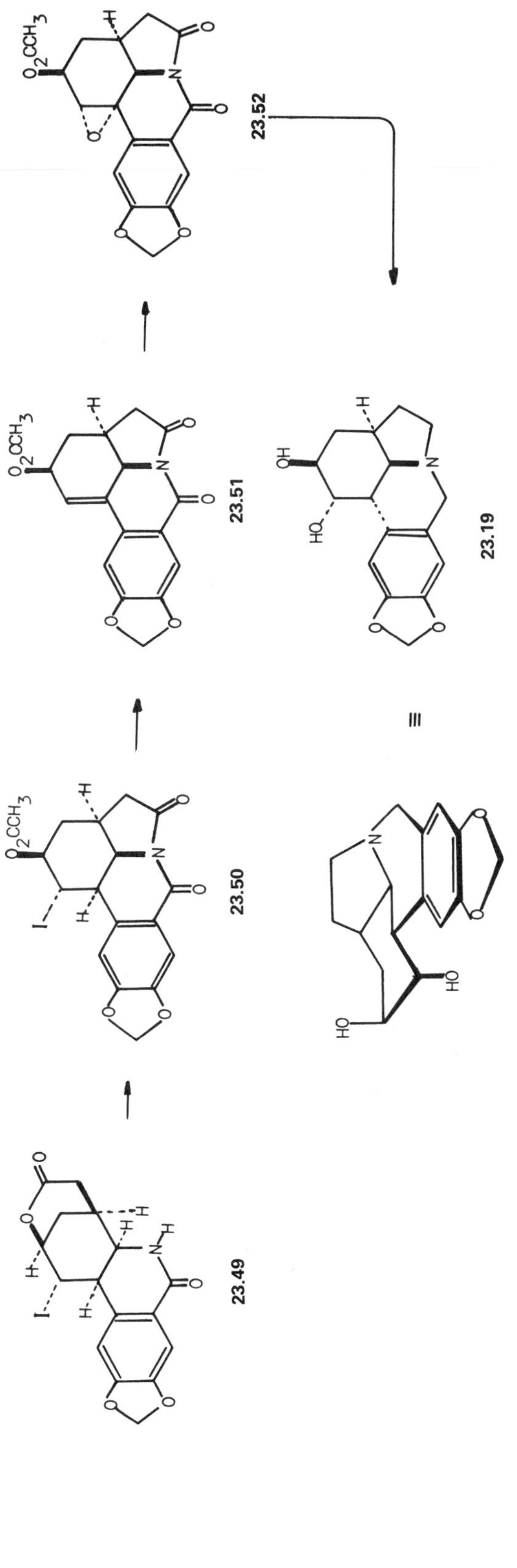

SCHEME 23.2 The synthesis of (±)-dihydrolycorine. (After Reference 132.)

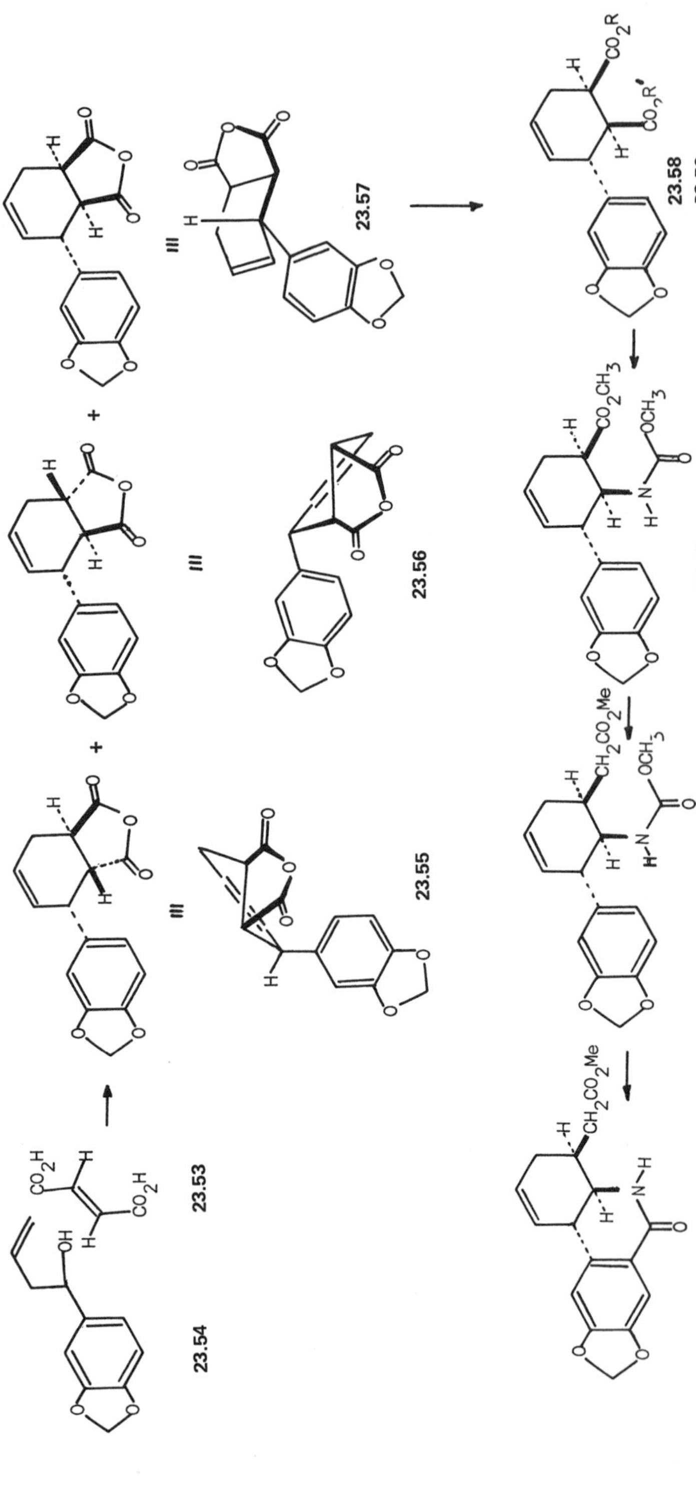
23.54
23.53
23.55
23.56
23.57
23.58
23.59
23.60
23.61
23.62
CO2H
CO2R
CO2R'
CO2CH3
OCH3
CH2CO2Me

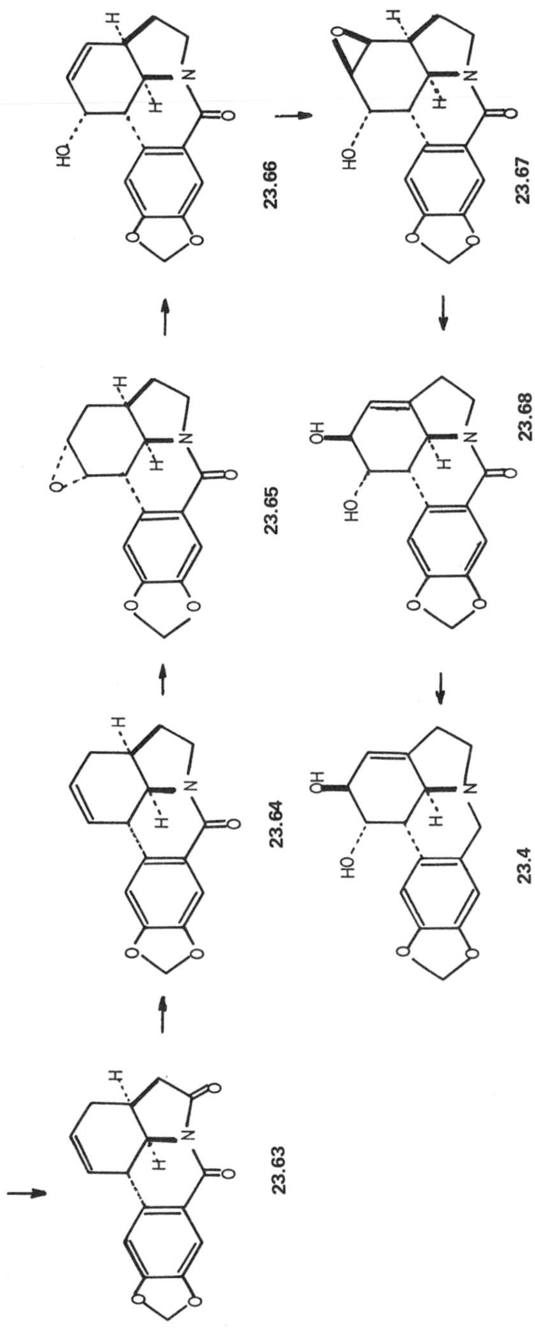

23.63 23.64 23.65 23.66 23.67 23.68 23.4

SCHEME 23.3 The synthesis of lycorine. (After Reference 133.)

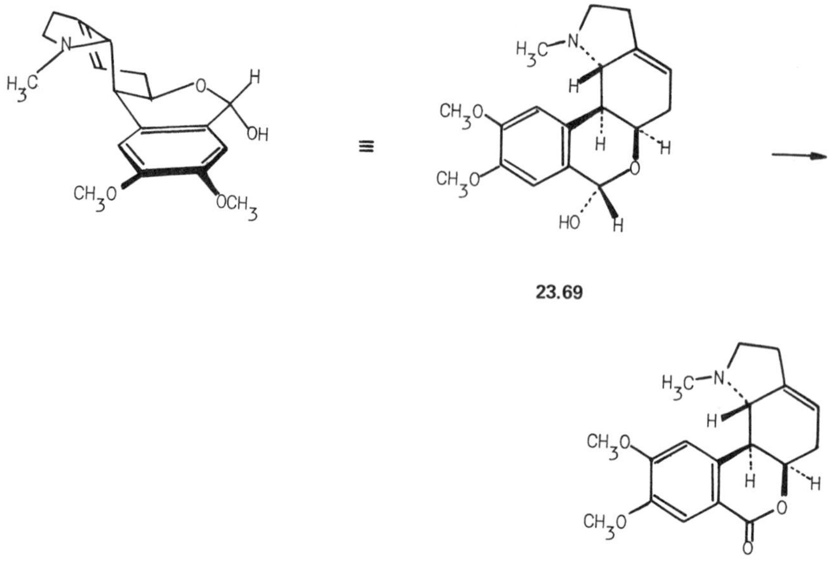

23.69

(Eq. 23.13)

23.5

23.70

23.71

(Eq. 23.14)

Lycorenine (23.69), but not homolycorine (23.5), can be monoacetylated, and lycorenine (23.69) yields an oxime with hydroxylamine hydrochloride. Other reasonable reactions could also be effected, but ultimately it was the interrelationship of lycorenine (23.69) with lycorine (23.4) which firmly established the gross structure and the stereochemistry of lycorenine (23.69) and thus of homolycorine (23.5).

For the first, lycorenine (23.69) yields a secondary alcohol (23.72) on Wolff-Kishner reduction. Aromatization of the alcohol yields, in turn, two indole derivatives, namely, 23.73 and 23.74 (Equation 23.15). The first of these, 23.73, establishes the position of hydroxyl attachment in lycorenine (23.69), while the second (23.74) is identical to the product obtained from the Emde base (23.23) of lycorine (23.4; Equation 23.5) after the latter has been treated with aluminum trichloride in refluxing chlorobenzene (to open the methylenedioxy ring), methylation with diazomethane, and dehydrogenation with palladium on carbon (Equation 23.16). Additionally, α-dihydrocaranine (23.37) obtained from lycorine (23.4; Equation 23.11), when treated under the von Braun reaction conditions with cyanogen bromide, yields two seco bases, namely, 23.75 and 23.76 (Equation 23.17). The latter (23.76) yields the corresponding secondary amine on lithium aluminum hydride reduction and, subsequently, the N-methyl derivative (23.77) with formaldehyde and formic acid. Treatment of the tertiary amine 23.77 with ethanolic potassium hydroxide in a sealed tube under an atmosphere of nitrogen ([135]; Equation 23.18) serves to cleave the methylenedioxy function so that methylation

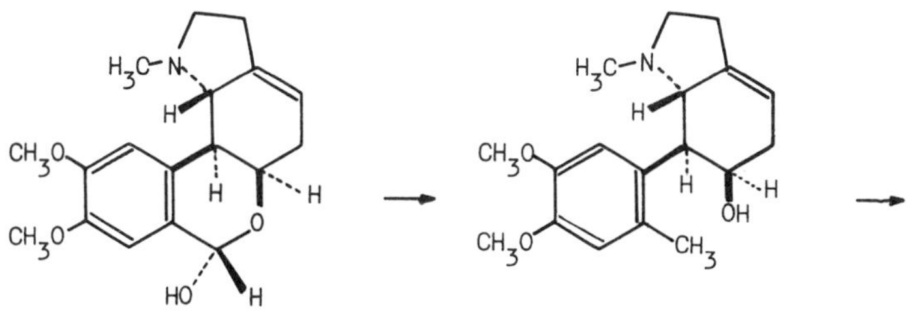

23.69

23.72

(Eq. 23.15)

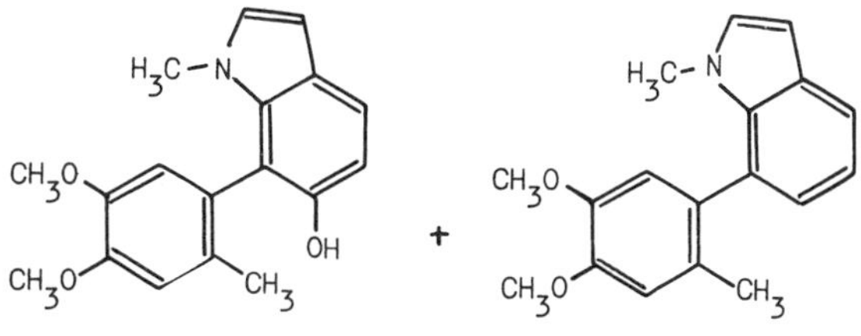

23.73

23.74

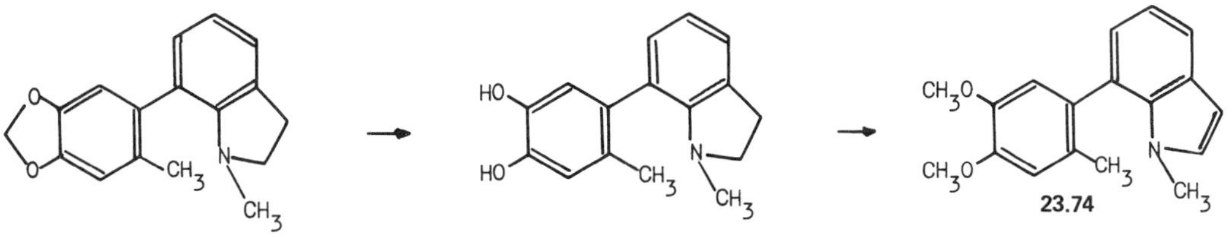

23.23

23.74

(Eq. 23.16)

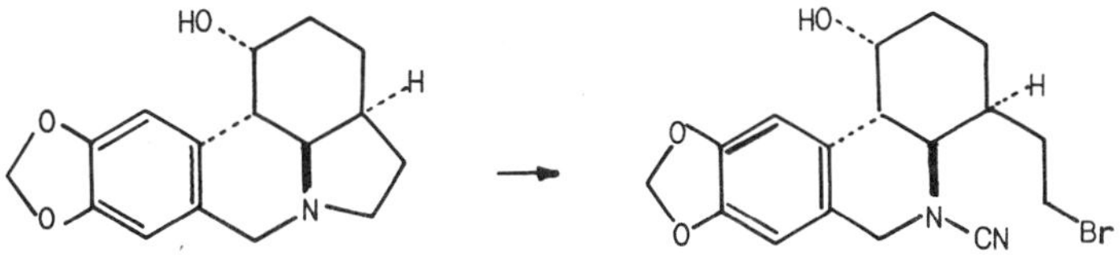

23.37

23.75

(Eq. 23.17)

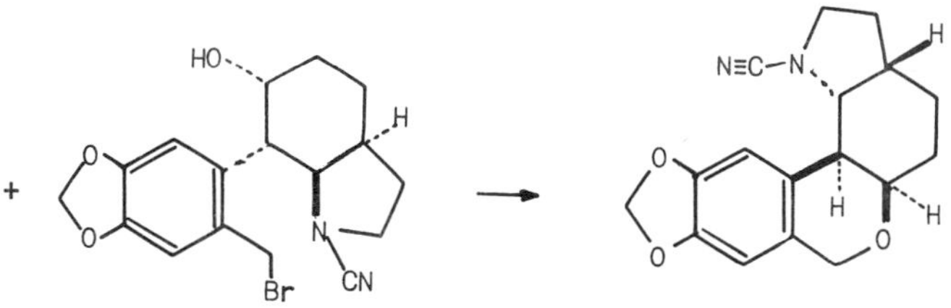

23.76

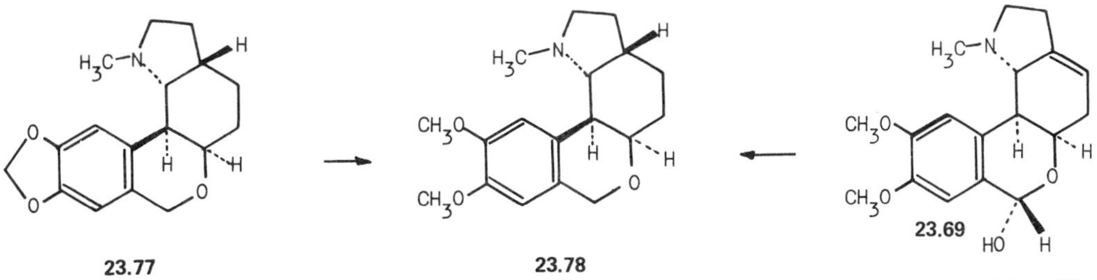

23.77

23.78

23.69

(Eq. 23.18)

with diazomethane generates α-deoxydihydrolycorenine (23.78), one of the two products of catalytic hydrogenation (and accompanying hydrogenolysis) of lycorenine (23.69).

The position of the double bond in lycorenine (23.69) and homolycorine (23.5) was assigned originally on the basis of reasoning similar to that outlined already for lycorine (23.4) and, ultimately, was established by further interrelationships in the series and completion of the x-ray crystallographic analysis of the corresponding methiodide [136].

The synthesis of neither homolycorine (23.5) nor lycorenine (23.69) has been reported, but although it has not been commented upon as such, the preparation of (±)-dihydrolycorine (23.19 and its mirror image) also completes the synthesis of α-dihydrohomolycorine (23.79), one of the hydrogenation products of homolycorine (23.5; Equation 23.19). This conclusion

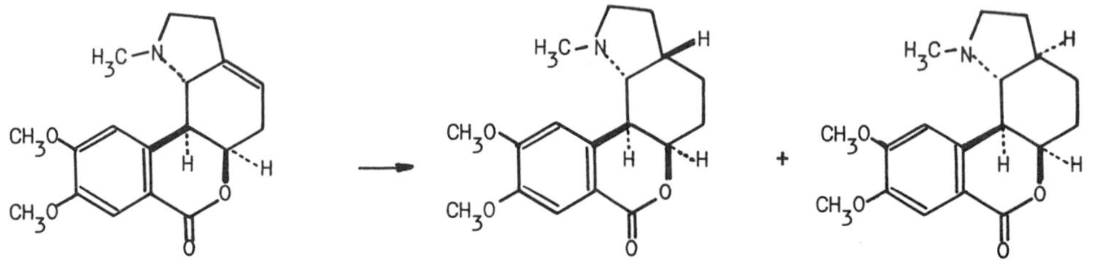

23.5 23.79 (Eq. 23.19)

follows from the known [137] conversion of α-deoxydihydrolycorenine (23.78), which can be obtained from (±)-dihydrolycorine (23.19 and its mirror image; see above) to α-dihydrohomolycorine (23.79) on oxidation with potassium dichromate in sulfuric acid (Equation 23.20).

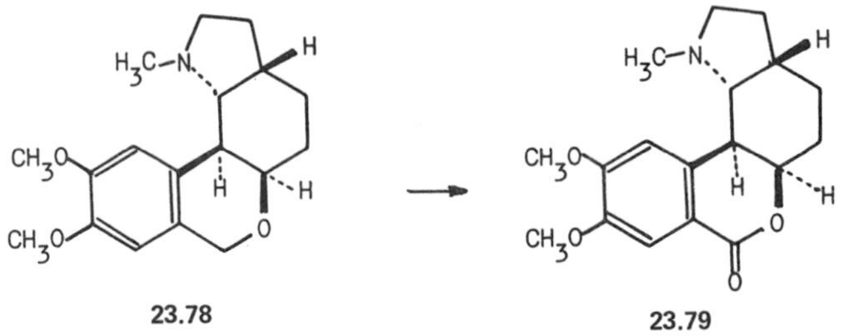

(Eq. 23.20)

23.78 23.79

V. GALANTHAMINE

Galanthamine (23.6), $C_{17}H_{21}NO_3$, was originally isolated from the Caucasian snowdrop, Galanthus woronowii, and is a tertiary, optically active base, possessing one methoxyl, one N-methyl, and one nonphenolic hydroxyl group. Hydrogenation of galanthamine (23.6) yields a dihydrobase (23.80; Equation 23.21) which is also found to be naturally occurring (lycoramine). With two of the oxygens accounted for (one methoxyl and one hydroxyl group), the third oxygen is placed as an ether linkage. The ether linkage in galanthamine (23.6) is labile.

Hydrobromic acid treatment of galanthamine (23.6) results in formation of apogalanthamine (23.81), a dibasic phenol (Equation 23.22). On the other hand, dihydrogalanthamine (lycoramine; 23.80) proved more resistant to hydrogen bromide, and the O-methyl is cleaved while the secondary hydroxyl is replaced by a bromine on treatment with this reagent; the bromophenol 23.82 (Equation 23.23) is the product. Although this difference in reactivity

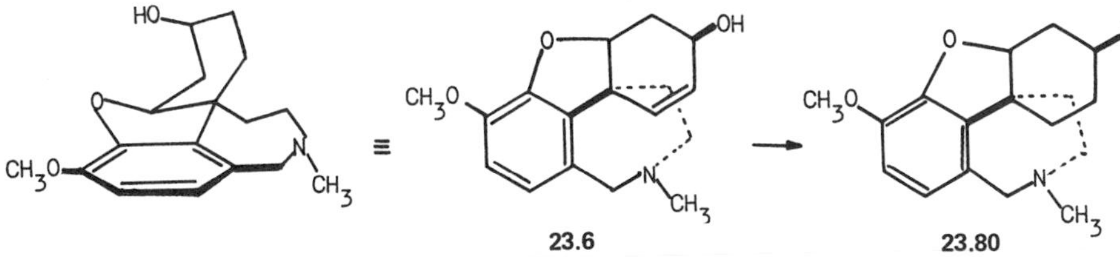

(Eq. 23.21)

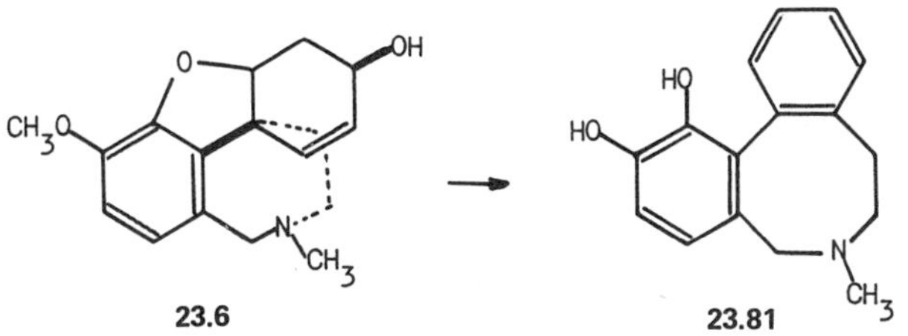

23.6 23.80

(Eq. 23.22)

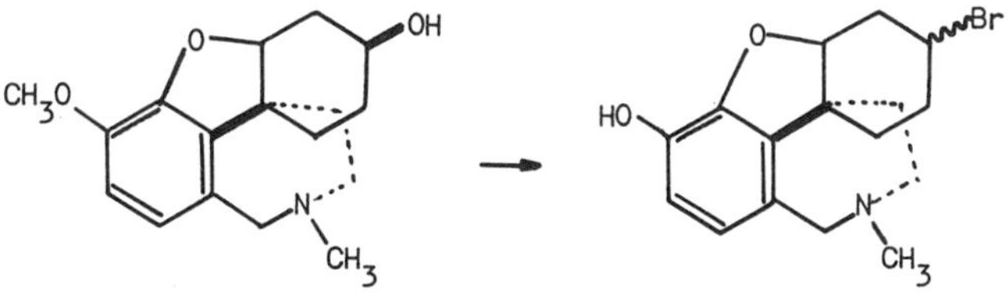

23.6 23.81

(Eq. 23.23)

23.80 23.82

was not immediately understood, when it was realized that apogalanthamine (23.81) possessed two aromatic rings, it was concluded that the driving force for rearrangement was aromatization accompanying ether cleavage.

When galanthamine (23.6) is cleaved with hydrochloric acid instead of hydrobromic acid, O-methylapogalanthamine (23.83) is formed, and treatment of the latter with methyl iodide results in generation of O,O-dimethylapogalanthamine methiodide (23.84; Equation 23.24).

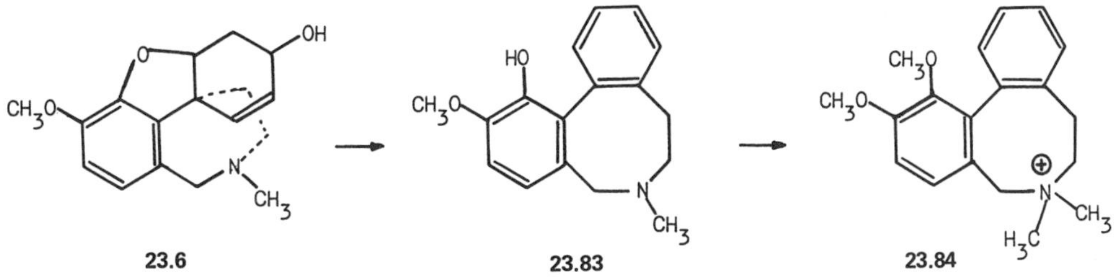

23.6 23.83 23.84

(Eq. 23.24)

Emde degradation (Equation 23.25) of O,O-dimethylapogalanthamine methochloride (23.84) yields a new base (23.85) which is oxidized by potassium permanganate to a mixture of two acids, shown by synthesis to be 2,3-dimethoxybiphenyl-6,2'-dicarboxylic acid (23.86) and 2,3-dimethoxy-6-methylbiphenyl-2'-carboxylic acid (23.87). O,O-Dimethylapogalanthamine (23.88) was synthesized [138] as shown in Scheme 23.4 via a copper-catalyzed Ulmann condensation between 2-bromo-3,4-dimethoxymethyl benzoate (23.89) and the methyl ester of α-iodophenylacetic acid (23.90). The resulting biphenyl (23.91) is converted, via the diol

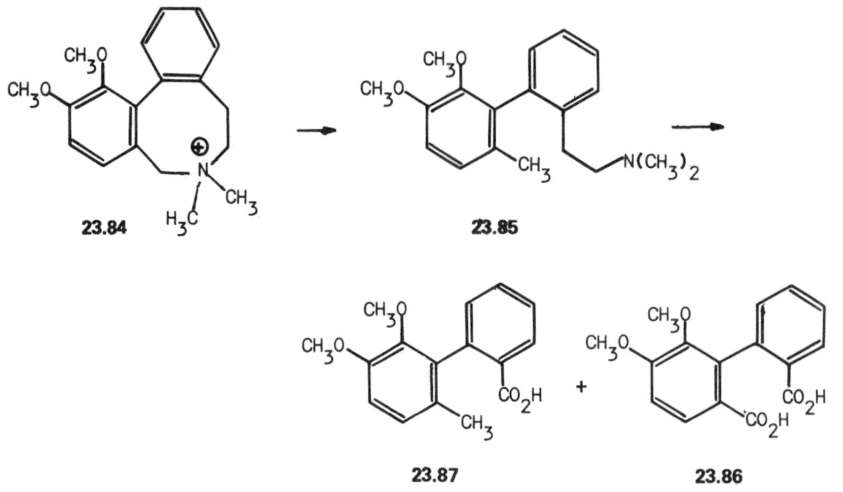

(Eq. 23.25)

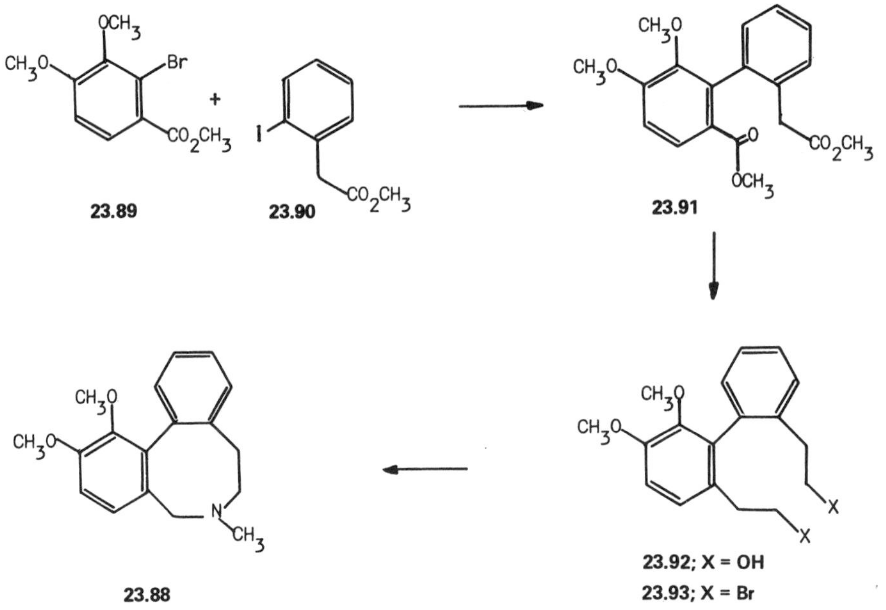

SCHEME 23.4 The synthesis of O,O-dimethylapogalanthamine. (After Reference 138.)

23.92, into the dibromide 23.93, which on treatment with methyl amine yields O,O-dimethylapogalanthamine (23.88).

Now, the hydroxyl group in galanthamine (23.6) must be allylic since manganese dioxide readily affects its oxidation to galanthaminone (23.94; which is also naturally occurring and is called narwedine). Narwedine (23.94; Equation 23.26) is converted to hydroxyapogalanthamine (23.95) on treatment with mineral acid. The structure of the latter was established by reactions similar to those outlined above for O,O-dimethylapogalanthamine (23.88).

Finally, gross structures which might have been written for galanthamine (23.6) were limited to that shown when it was recognized that galanthaminone (narwedine; 23.94) formed

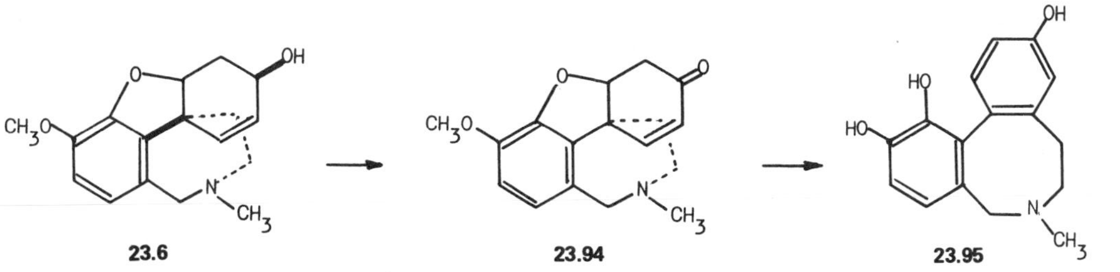

23.6 → 23.94 → 23.95

(Eq. 23.26)

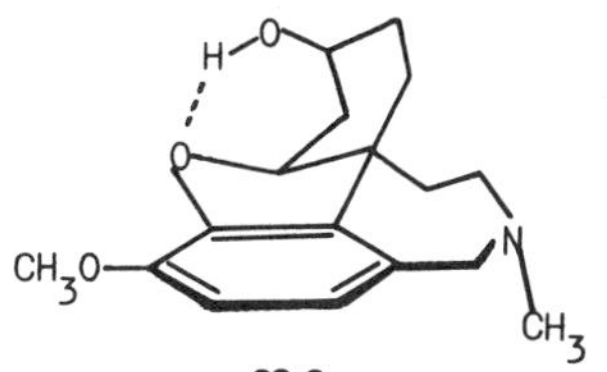

23.6

derivatives that were characteristic of a ketone flanked by a methylene unit and that a dienone-phenol rearrangement would account for the skeletal changes on ether cleavage.

The relative stereochemistry of the hydroxyl and the ether bridge was correctly defined as cis by (a) the observation of hydrogen bonding between the two groups, and (b) epimerization studies. An x-ray crystallographic analysis [139] was used to confirm the conclusion. (-)-Galanthamine (23.6) has been synthesized [140] by a "biomimetic" route (Scheme 23.5).

The cyanohydrin of p-hydroxybenzaldehyde (23.96) is converted by reduction and hydrolysis to p-hydroxyphenylacetic acid (23.97) and, after O-benzylation, to the corresponding acid chloride, 23.98. This acid chloride (23.98), when allowed to react with the N-methyl-p-methoxy-m-O-benzylamine (23.99) prepared from O-benzylisovanilline (23.100), yields the amide (23.101). Reduction of 23.101 with lithium aluminum hydride and palladium on charcoal-catalyzed hydrogenolysis to the norbelladine derivative (23.102) is then accomplished. The latter undergoes oxidative cyclization to (±)-narwedine (galanthaminone, 23.94 and its mirror image) with a variety of oxidizing agents (e.g., potassium ferricyanide) in a maximum yield of 1.4%. When (±)-narwedine (galanthaminone, 23.94)* is reduced with lithium aluminum hydride, (±)-galanthamine (23.6)* and (±)-epigalanthamine (23.103)* are formed. However, it had been noted that in the presence of trace quantities of (-)-galanthamine (23.6), (±)-narwedine (23.94)* was spontaneously resolved and (+)-narwedine (antipode of 23.94) preferentially crystallized. This was attributed to adsorption of traces of (-)-galanthamine (23.6) on the surface of developing narwedine crystals [140], and thus, since reduction of (-)-narwedine (23.94) with lithium aluminum hydride gives a separable mixture of (-)-galanthamine (23.6) and (-)-epigalanthamine (23.103; Equation 23.27), the total synthesis is complete.

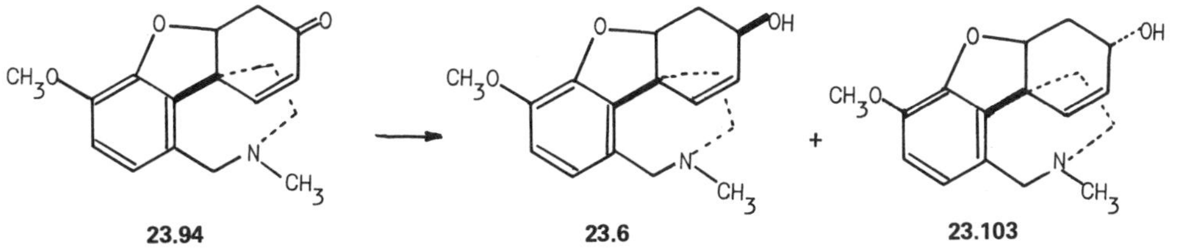

23.94 → 23.6 + 23.103

(Eq. 23.27)

*Shown as one optical isomer.

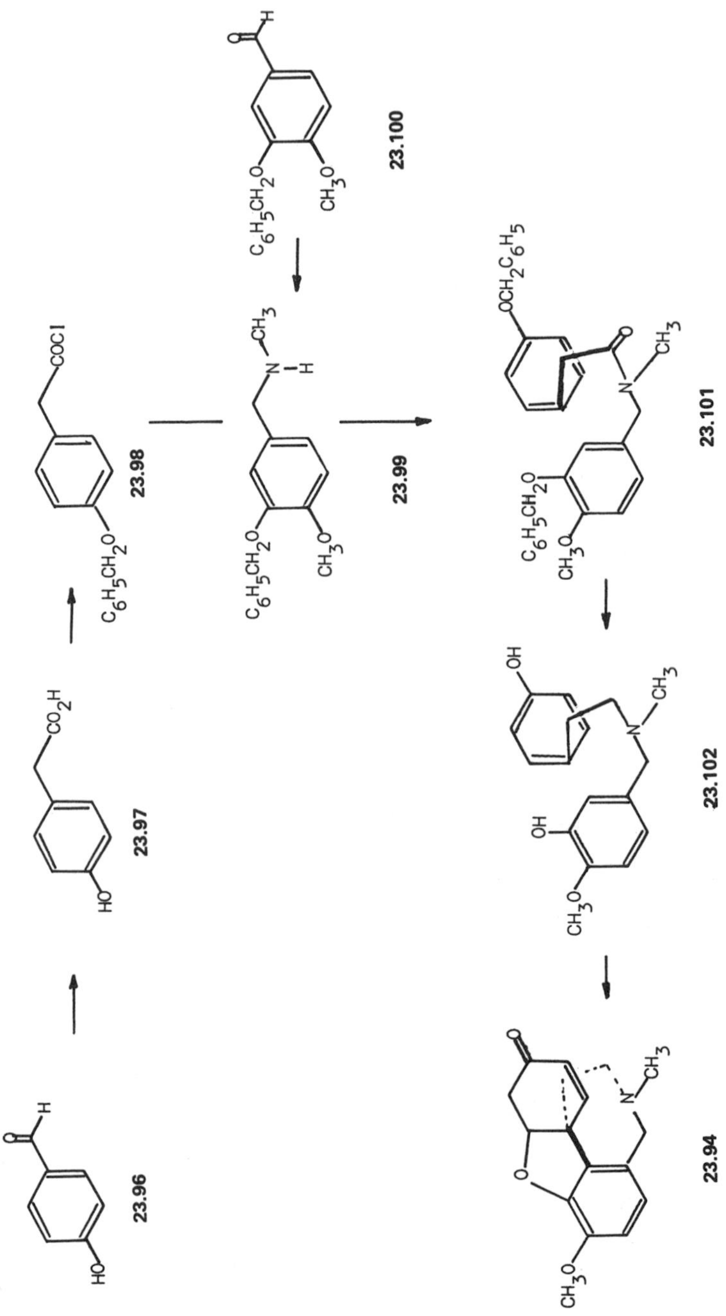

SCHEME 23.5 A synthesis of (±)-narwedine (and therefore galanthamine). (After Reference 140.)

VI. CRININE

Crinine (23.7), $C_{16}H_{17}NO_3$, the isolation of which was first reported in 1955, is a levorotatory base, isomeric with caranine (23.104; α-dihydrocaranine, 23.37, has already been discussed as a derivative of lycorine, 23.4). Like caranine (23.104), crinine (23.7) contains

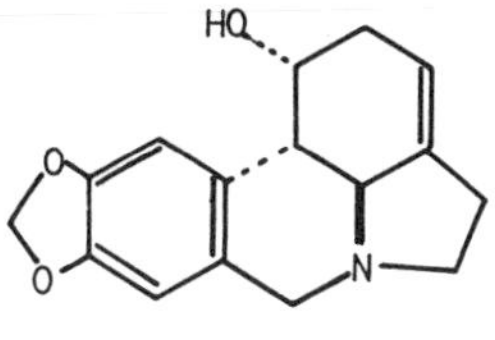

23.104

one hydroxyl, one double bond, a methylenedioxy function, but no N-methyl. Unlike caranine (23.104), however, crinine (23.7) is not dehydrogenated easily by palladium on charcoal (nor, indeed, is it dehydrogenated even at 200°C), and the betaine forms (e.g., 23.28) characteristic of pyrrolo[de]phenanthridine systems do not form on oxidation. Thus, it was concluded that caranine (23.104) and crinine (23.7) have different ring systems.

Oxidation of crinine (23.7) with manganese dioxide readily (indicating an allylic alcohol) yields oxocrinine (23.105) which can be reduced with sodium borohydride to a mixture of crinine (23.7) and epicrinine (23.106), and either epimer is capable of reoxidation to oxocrinine (23.105; Equation 23.28). Catalytic reduction of oxocrinine (23.105) affords dihydrooxocrinine (23.107), which on Wolff-Kishner reduction is converted to (-)-crinane (23.108; Equation 23.19). (±)-Crinane (23.108 and its mirror image), the infrared spectrum of which is identical to (-)-crinane (23.108), was synthesized as outlined in Scheme 23.6 [141].

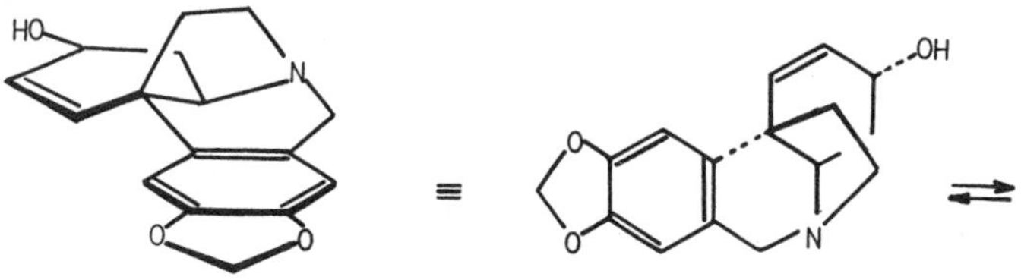

23.7

(Eq. 23.28)

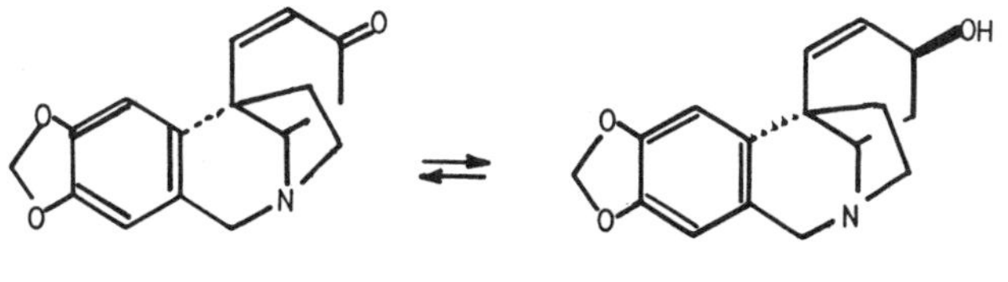

23.105 23.106

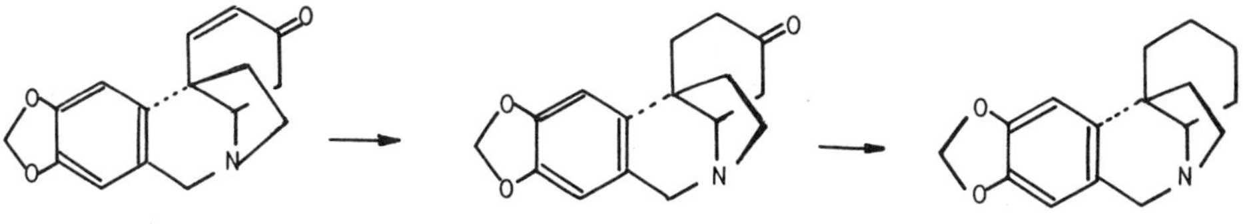

23.105 23.107 23.108

(Eq. 23.29)

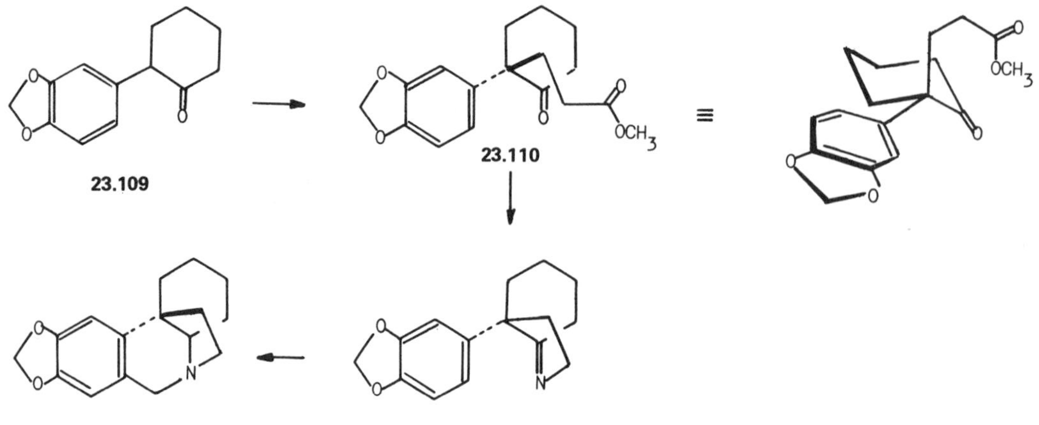

SCHEME 23.6 A synthesis of (±)-crinane. (After Reference 141.)

As shown, 2-(3,4-methylenedioxyphenyl)-cyclohexanone (23.109) undergoes cyanoethylation and methanolysis to the ester 23.110. Treatment of the ester (23.110) with hydrazine generates the corresponding hydrazone-hydrazide which, on nitrous acid-catalyzed decomposition, provides the hexahydroindole (23.111). Catalytic hydrogenation and treatment with formaldehyde in the presence of acid yields (±)-crinane (23.108 and its antipode).

Of the possible structures which might be written for crinine (23.7), the fact that (a) oxocrinine (23.105) yields a methiodide (23.112) which readily generates a dienone (23.113) on elimination (Equation 23.30); (b) dihydrooxocrinine (23.107) yields a methiodide (23.114)

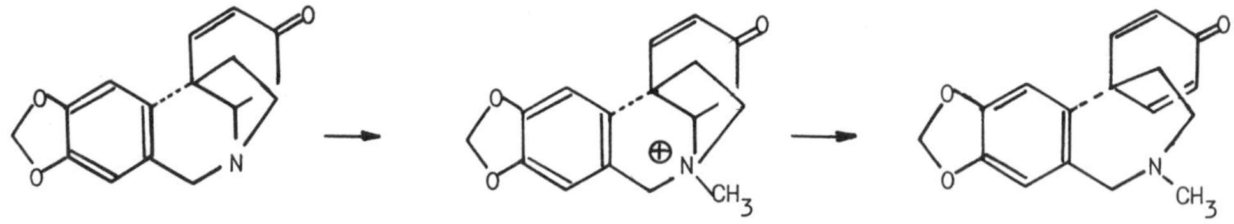

(Eq. 23.30)

which undergoes elimination to an α,β-unsaturated ketone (23.115; Equation 23.31) which is optically active and which, on hydrogenation, yields an optically inactive dihydro derivative (23.116); and (c) the dihydro derivative (23.116) is identical to the tetrahydro derivative obtained from the dienone (23.113), all point to the correct structure being assigned (without stereochemistry) as 23.7. The stereochemistry shown for (-)-crinine (23.7) is consistent with numerous interrelationships between crinine-type alkaloids, and the absolute configuration has been determined by relating (-)-crinine (23.7) to (+)-tazettine (23.8), which inter alia was degraded to (+)-R-3-methoxyadipate (23.117; [142]).

(±)-Crinine and (±)-epicrinine (23.7 and 23.106 and their mirror images, respectively) have been synthesized [143], and the synthesis of these epimeric alkaloids is shown in Scheme 23.7 where, although only one enantiomer is shown, racemic materials were used.

Ethylpiperonylacetate (23.118) is nitrosated with sodium nitrite in acetic acid and the product reduced and acetylated directly with zinc in acetic acid-acetic anhydride to yield ethyl N-acetylpiperonylglycinate (23.119). Condensation of the amide (23.119) with methyl vinyl ketone and cyclization of the Michael adduct yields the α,β-unsaturated cyclohexenone

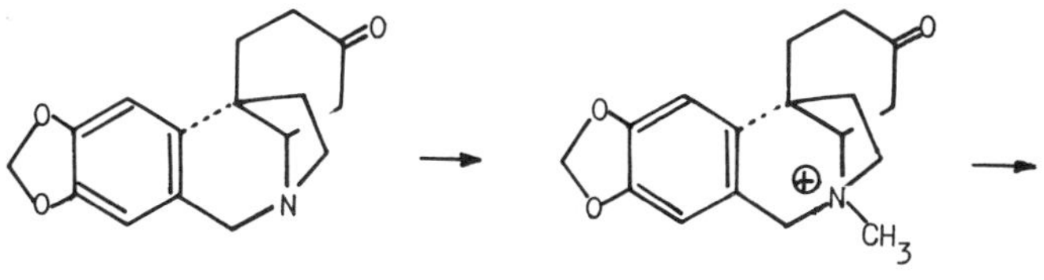

23.107 23.114

(Eq. 23.31)

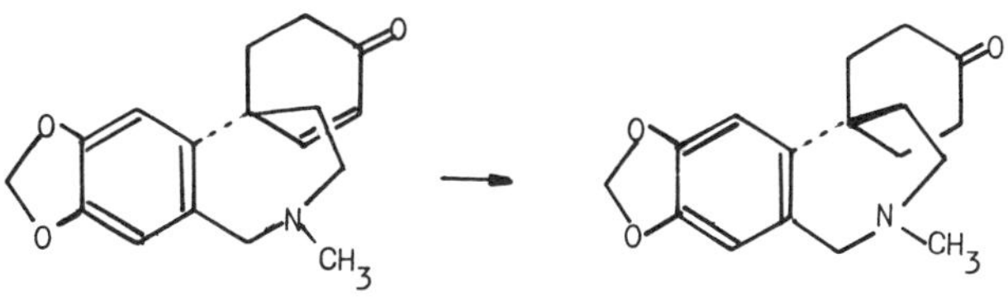

23.115 23.116

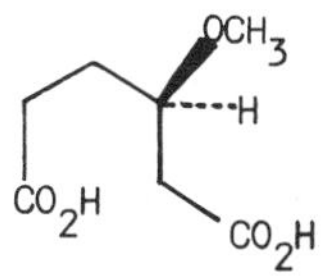

23.117

23.120 which undergoes decarboxylation (after hydrolysis of the ester) to yield the N-acetylcyclohexenone 23.121. When 23.121 is reduced with sodium borohydride the mixture of epimeric alcohols 23.122 and 23.123 is obtained, and this mixture is converted to the amides 23.124 and 23.125 by a modified Claisen rearrangement on treatment with 1,1-dimethoxy-1-dimethylaminoethane (23.126). Treatment of this mixture of diamides (23.124 and 23.125) with dilute sodium hydroxide solution yields a single lactam (23.127) and some unreacted diamide 23.125. That the cyclization had proceeded in the desired fashion was demonstrated by catalytic reduction of the lactam 23.127 to the corresponding dihydro derivative and then to the base 23.128 with lithium aluminum hydride. Treatment of 23.128 with formalin and hydrochloric acid then generates (±)-crinane (23.108 and its mirror image). The lactam 23.127 is reduced directly with lithium aluminum hydride to the base 23.129, which with formalin and hydrochloric acid yields (±)-α-desoxycrinine (23.130). Selenium dioxide oxidation in acetic acid-acetic anhydride at reflux generates (±)-crinine which is identical in all respects to a mixture of (-)-crinine (23.7) and its optical antipode (+)-crinine (mirror image of 23.7; also naturally occurring and called vittatine).

Dihydrocrinine (23.131) is also naturally occurring (elwesine). A synthesis of dihydrocrinine (23.131) has been reported [144], and because the method of synthesis is rather novel and incorporates several interesting reactions, it is recorded as Scheme 23.8.

Piperonyl cyanide (23.132) is allowed to react with ethylenedibromide in the presence of lithium amide in glyme solution to generate 1-cyano-1-(3,4-methylenedioxyphenyl)cyclopropane (23.133) which is then converted to the corresponding aldehyde (23.134) by reduction with diisobutylaluminum hydride. The aldehyde (23.134) yields the corresponding imine 23.135 on treatment with benzylamine in the presence of calcium chloride in benzene. This imine (23.135) then undergoes thermal isomerization to the corresponding pyrroline (23.136) in the presence of ammonium chloride. The hydrochloride of 23.136 is converted to the cis-octahydroindole (23.137) by treatment of its acetonitrile solution with methyl vinyl ketone. Since debenzylation occurs with elimination, the ketone 23.137 is hydrogenated to a mixture

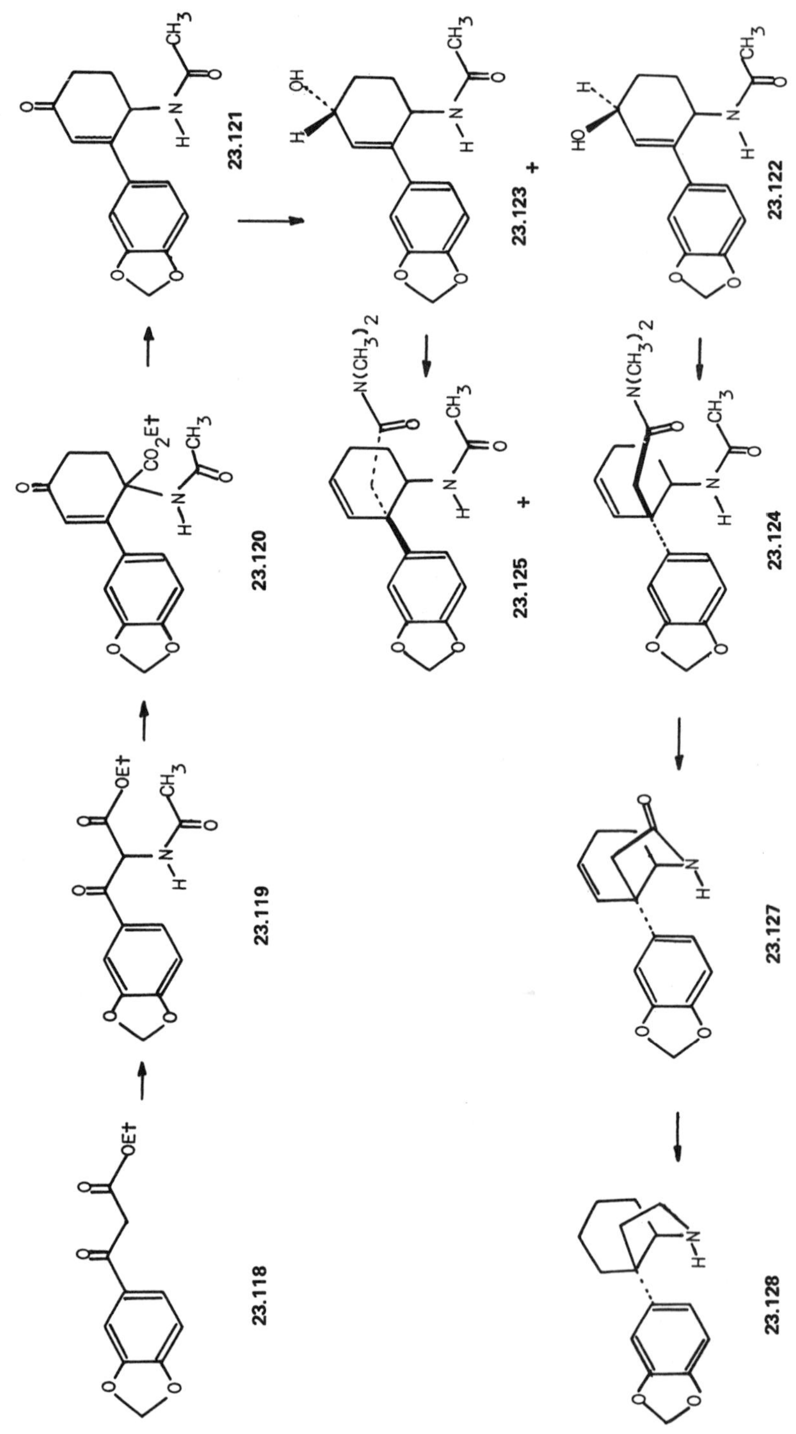

23.118

23.119

23.120

23.121

23.123 + 23.122

23.125 + 23.124

23.127

23.128

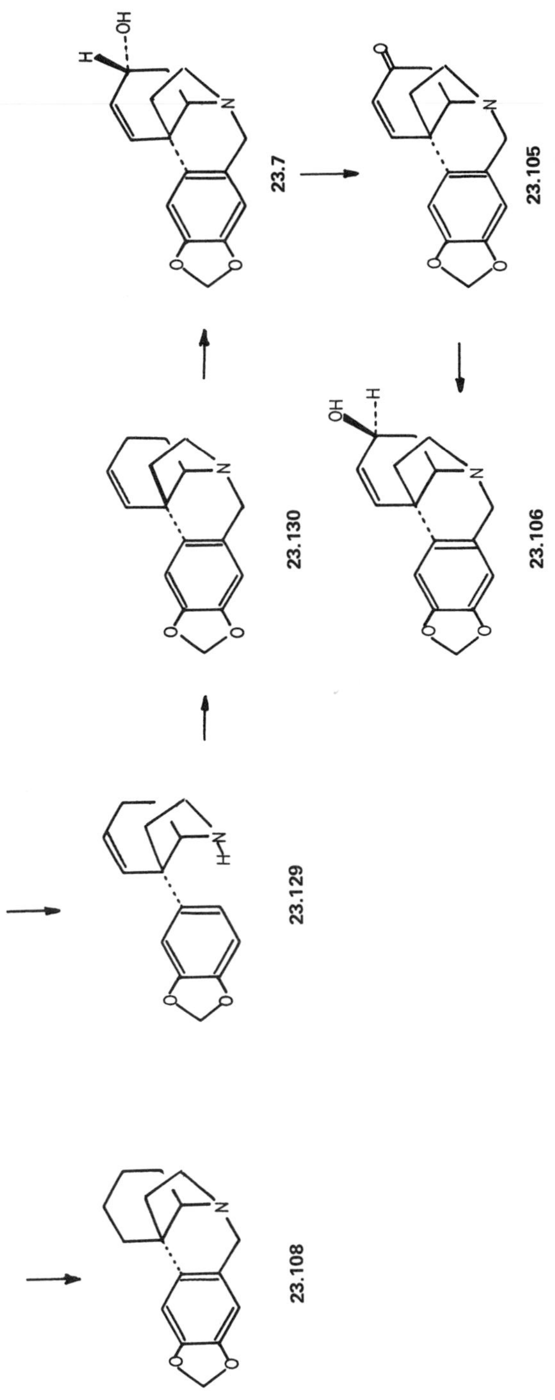

SCHEME 23.7 The synthesis of (±)-crinine and (±)-epicrinine. (After Reference 143.)

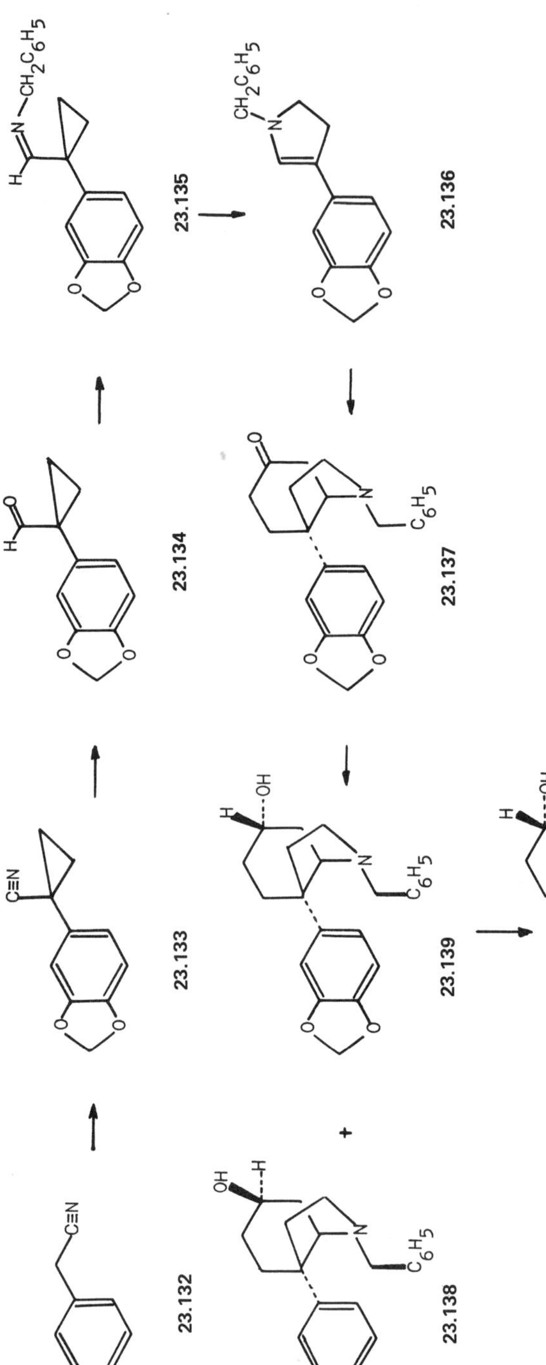

SCHEME 23.8 A synthesis of (±)-elwesine (dihydrocrinine). After Reference 144.)

of the epimeric alcohols 23.138 and 23.139 in the presence of palladium on carbon in isopropyl alcohol. The desired epimer (23.139) on debenzylation and cyclization with formalin in hydrochloric acid generates (±)-elwesine (23.131 and its mirror image). The two key steps in the synthesis of (±)-elwesine that will almost certainly be employed in other syntheses of note are (a) the acid-catalyzed thermal rearrangement of the cyclopropyl imine (23.135) to the pyrroline 23.136, and (b) the annelation of the Δ^2-pyrroline hydrochloride (23.136) with methyl vinyl ketone to the cis-octahydroindole (23.137).

VII. TAZETTINE

The work on unraveling the structure of tazettine (23.8) began in the 1930s with its isolation from Narcisus tazetta and the determination of its molecular formula ($C_{18}H_{21}NO_5$), some physical properties (mp 212°C, $[\alpha]_D^{18}$ +165.8, $CHCl_3$), and a little chemistry, i.e., it contains one methoxyl, a methylenedioxy, an N-methyl (a tertiary amine), and a double bond not conjugated to the aromatic ring. Oxidation of tazettine (23.8) with potassium permanganate yields hydrastic acid (23.18), and thus the methylenedioxy and the methoxyl groups are not on on the same ring. Zinc dust distillation of tazettine (23.8) yields phenanthridine (23.24). Tazettine (23.8) forms a methiodide (23.140) which undergoes Hofmann degradation with elimination of the methoxyl group and formation of a new aromatic ring (benzoic acid on oxidation) to a methine (23.141). Subsequent methylation and Hofmann degradation of the methine (23.141) yield 6-phenylpiperonyl alcohol (23.142; Scheme 23.9). About 25 years later, the structure of the methine (23.141) was elucidated by isolation and synthesis of the unstable methine methiodide (23.143) which, since all of the carbons of tazettine (23.8) save the methoxyl were present, meant that both the methoxyl and the double bond were present in the ring which became aromatic. Although many studies on the Hofmann degradation were undertaken in an effort to prevent methoxyl loss during the degradation, most of these were fruitless. However, other information was gained during these studies. Thus, it was found that when tazettine (23.8) is treated with dimethyl sulfate and then potassium iodide, methyltazettine methiodide (23.144) forms and that two methoxyl groups are now present. Several sequential Hofmann degradations on methyl tazettine methiodide (23.144) yield a plethora of nonnitrogenous products, among which is found 6-(4-methoxyphenyl)piperonyl alcohol (23.145; Scheme 23.10). This means that the methoxy group is present in the position para to the junction with the methylenedioxyphenyl, and since tazettine (23.8) is not an enol nor an enol ether, the double bond must be vinylogous to it.

When tazettine (23.8) is reduced with lithium aluminum hydride tazettadiol (23.146; $C_{18}H_{23}NO_5$) is formed, and dehydration of this diol leads to an ether, deoxytazettine (23.147; Equation 23.32). Hofmann degradation of the ether 23.147 results in the formation of deoxytazettine methine (23.148), which subsequently rearranges to deoxytazettine neomethine

23.8 → 23.146 → 23.147

(Eq. 23.32)

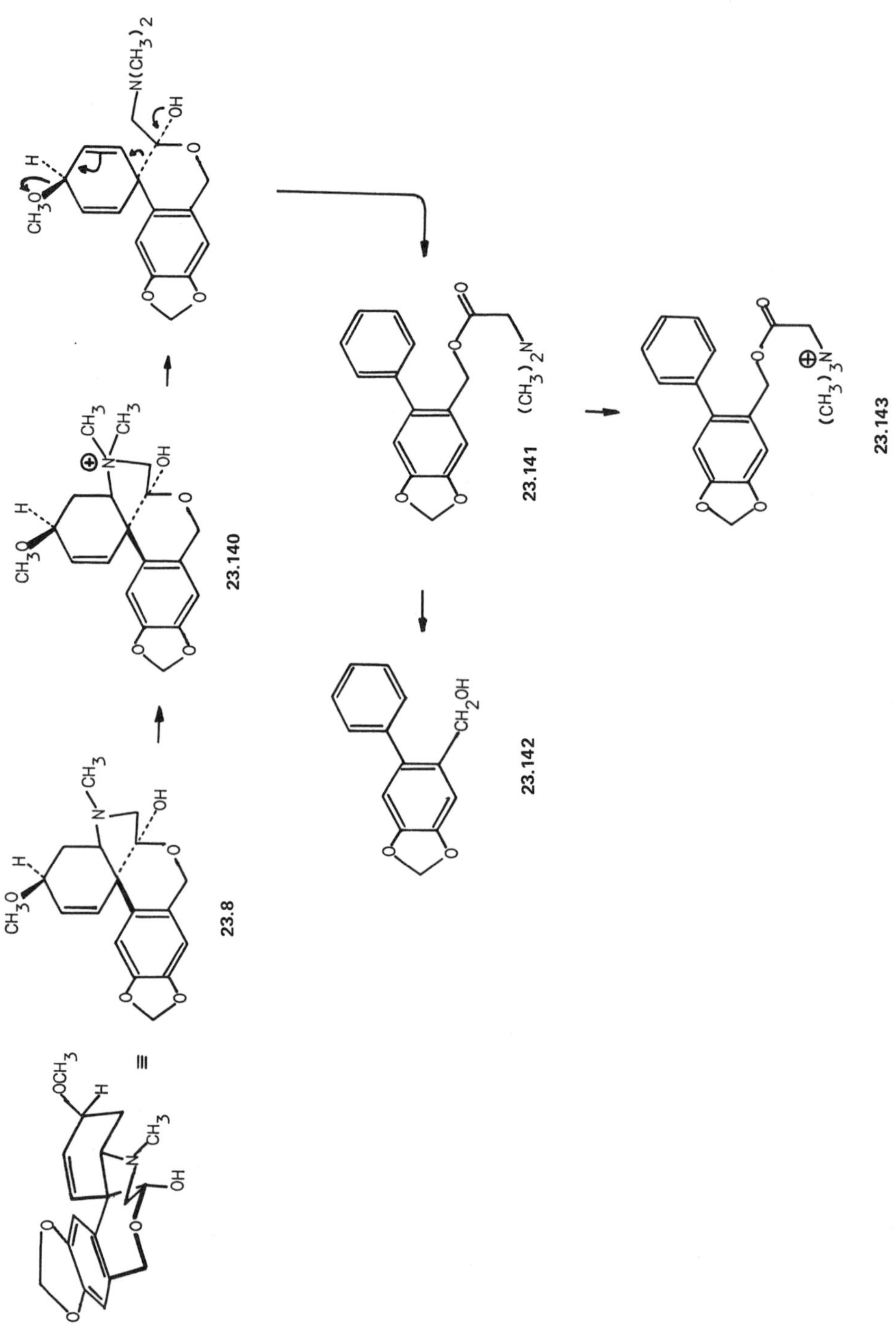

SCHEME 23.9 The degradation of tazettine.

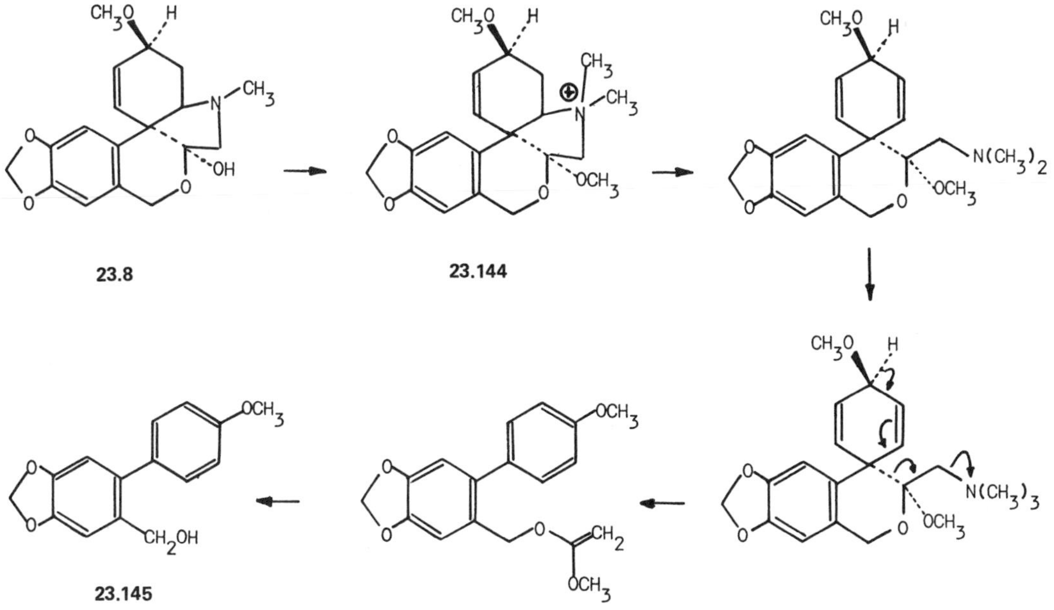

23.8

23.144

23.145

SCHEME 23.10 The Hofmann degradation of methyltazettine methiodide.

(23.149) in the presence of acid and the enol ether (23.150) on further Hofmann degradation. Permanganate oxidation of 23.150 finally leads to the lactone 23.151 (Equation 23.33). When the Hofmann degradation of deoxytazettine (23.147) is carried out and the acidic work-up previously used avoided, the methine 23.148 is isolated, and tazettine (23.8) is therefore best accommodated by the structure shown (23.8).

Numerous other reactions are consonant with the structure shown for tazettine (23.8), and significant advances in the stereochemical relationships of the functional groups one to another were made, piecemeal, by noting reactions undergone. For example, it was

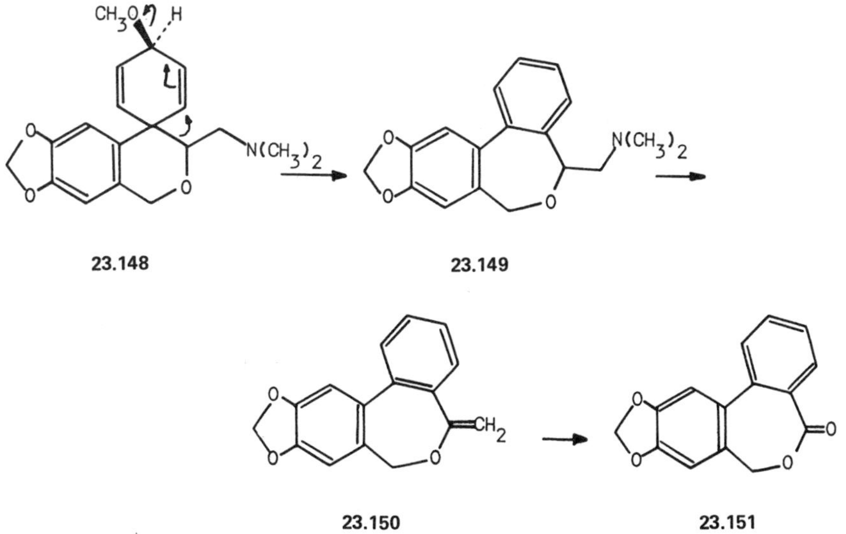

23.148 23.149 (Eq. 23.33)

23.150 23.151

appreciated that the allylic methyl ether of tazettine (23.8) is cleaved with hydrochloric acid to a mixture of epimeric allylic alcohols, tazettinol, and isotazettinol (23.152 and 23.153, respectively; Equation 23.34) and that only one of these will subsequently yield tazettine

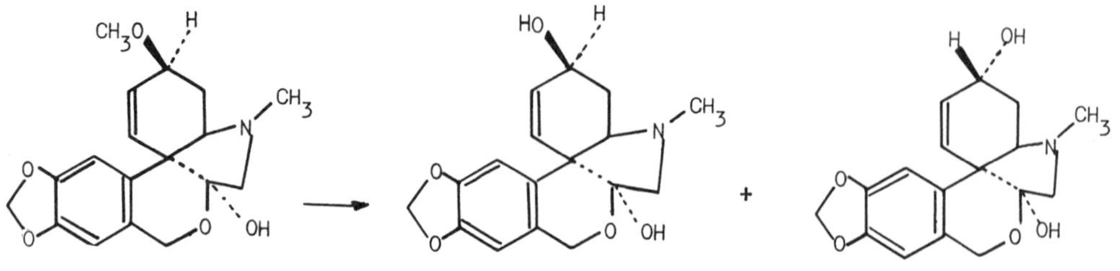

23.8 23.152 23.153

(Eq. 23.34)

derivatives. However, rearrangements from related Amaryllidaceae alkaloids whose structures were on a more firm basis served to establish the stereochemistry (prior to the x-ray crystal analysis of tazettine methiodide, 23.140; [145]).

The structure of 6-hydroxycrinamine (23.154) has been established by x-ray crystal analysis [146]. The C-3 epimer of 6-hydroxycrinamine (23.154) is haemanthidine (23.155).

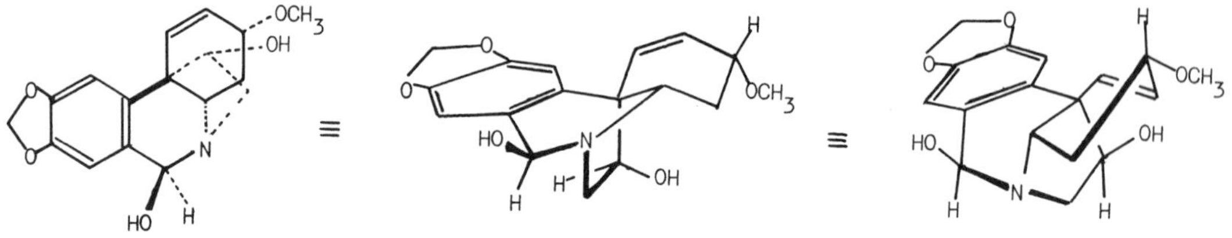

23.154

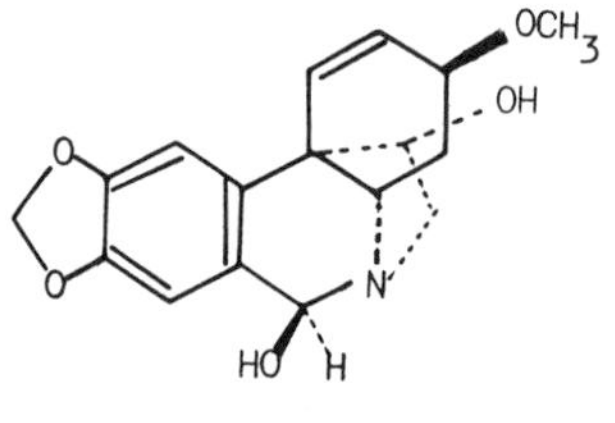

23.155

Although the C-6 hydroxyl in the former is known [146] in the crystal to be trans (or anti) to the pyrrolidine ring, in solution both epimers at C-6 are in equilibrium (Equation 23.35; [147]). Treatment of haemanthidine (23.155) with methyl iodide, followed by base in acetone solution, generates the rearranged methyl ketal (23.156; Scheme 23.11), which on mild hydrolysis yields pretazettine (23.157; Equation 23.36) whose absolute configuration is therefore also established. Pretazettine (23.157) is slowly converted to tazettine (23.8) on standing (Equation 23.37).

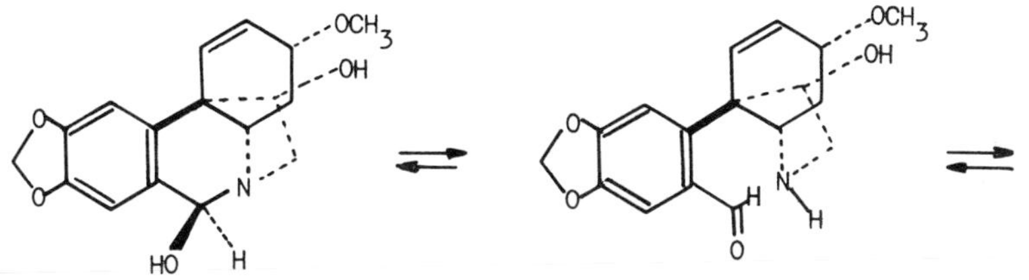

(Eq. 23.35)

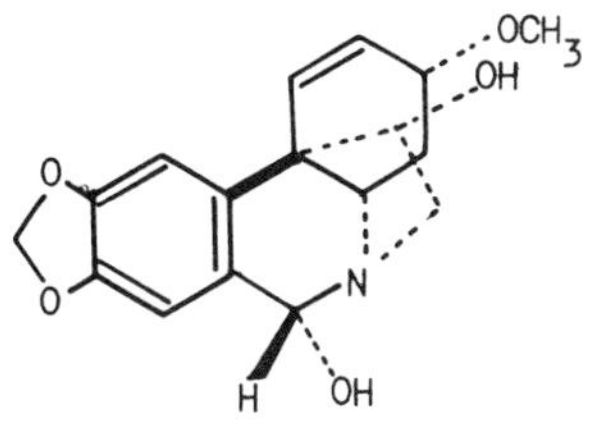

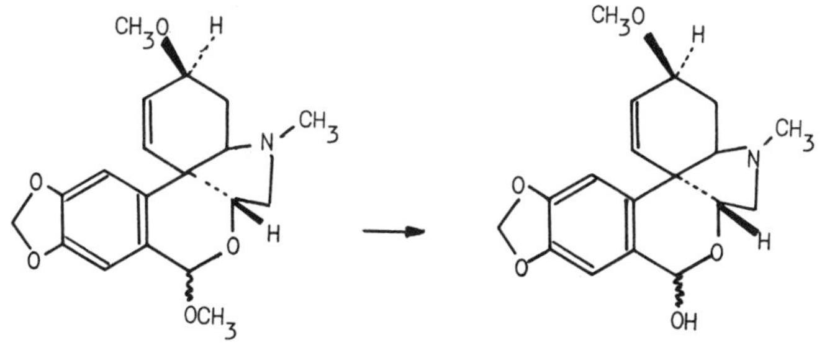

23.154

(Eq. 23.36)

23.156

23.157

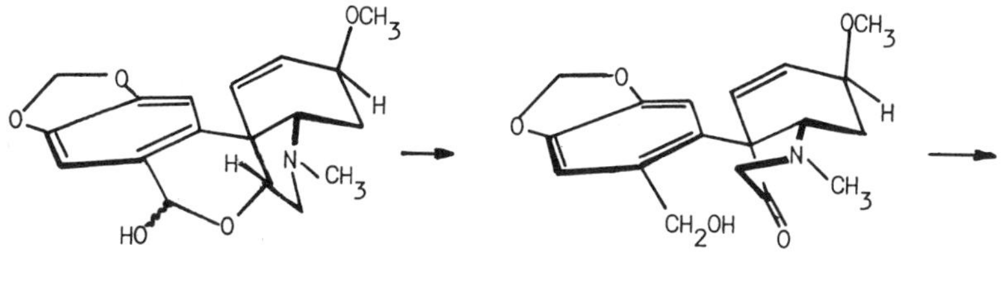

(Eq. 23.37)

23.157

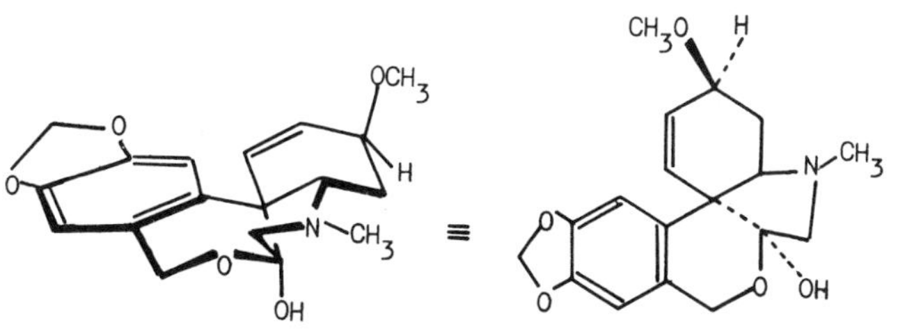

23.8

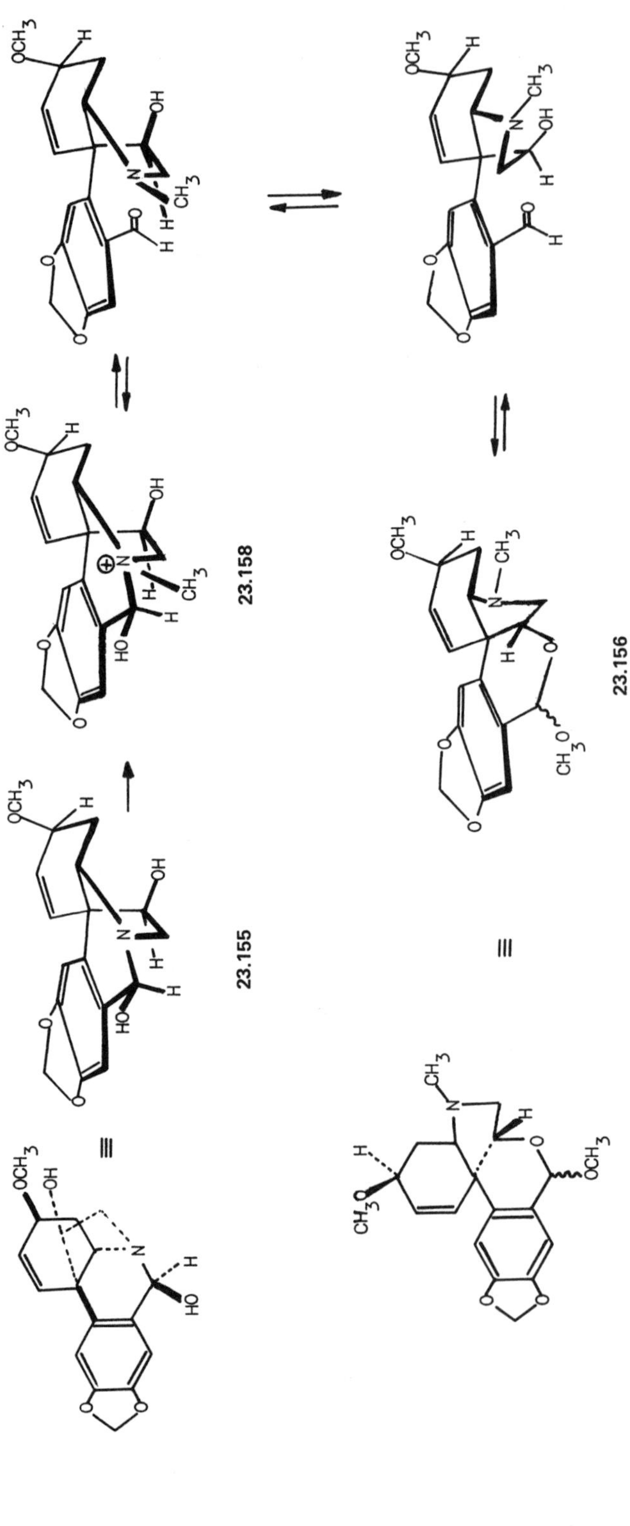

23.155

23.158

23.156

SCHEME 23.11 The rearrangement of haemanthidine on treatment with methyl iodide and base.

It has been demonstrated [148] by use of deuterium labeling that the rearrangement of N-methylhaemanthidine (23.158) to tazettine (23.8) involves a facile intramolecular hydride transfer (Equation 23.37; [148]. Additionally [149], attempted isolation of tazettine (23.8) under conditions which avoid strong base, alumina chromatography, or prolonged standing of the crude alkaloids fraction fails. Indeed, it appears that tazettine (23.8) may be an artifact and that either pretazettine (23.157) or another precursor (e.g., N-methylhaemanthidine, 23.158) may be the actual alkaloid present in the plant.

Haemanthidine (23.155; in its racemic form), and therefore tazettine (23.8), have fallen to synthesis [150]. The synthesis is outlined in Scheme 23.12.

Piperonal (23.159) is converted, via piperonylic acid, to the corresponding acid chloride (23.160), which on treatment with sodium azide and refluxing t-amyl alcohol yields the t-amyl carbamate 23.161. The carbamate (23.161) undergoes hydrolysis and decarboxylation to the methylenedioxy aniline 23.162. Nitrous acid treatment of the aniline 23.162 converts it to the corresponding diazonium salt which is used to arylate maleic anhydride. The Diels-Alder reaction of methylenedioxyphenylmaleic anhydride (23.163) with butadiene proceeds to yield the cyclohexene 23.164. Direct treatment of the anhydride (23.164) with sodium methoxide in methanol yields the half ester 23.165 (the thermodynamic product) which subsequently is converted to the corresponding acid chloride (with oxalyl chloride) to the azide (with sodium azide) and thence to the isocyanate (23.166). Treatment of 23.166 with polyphosphoric acid for one-half hour at room temperature causes cyclization, and saponification of the ester serves to generate the lactam 23.167.

When the potassium salt of the acid (from saponification of the above) (23.167) is allowed to react with a mixture of iodine and potassium iodide in bicarbonate solution, the iodolactone 23.168 is formed. The lactone 23.168 is smoothly transformed to the epoxyacid 23.169 with aqueous alkali and from there to the methoxylactone 23.170 wherein the correct stereochemistry at C-3 has been created. Conversion of the methoxylactone 23.170 into the p-bromophenacyl-mesylate ester (23.171) prevents relactonization of the lactone and prepares the ground for elimination to the requisite double bond. Then, mild hydrolytic conditions allow conversion of the carboxylate ester to the corresponding acid, while the mesylate remains and treatment with thionyl chloride, diazomethane, and cyclization to the "quasi-amide" (23.172) is possible. All the carbons and the appropriate functionality for the final conversion to haemanthidine (23.155) are now present. Reduction of 23.172 with disiamylborane and acetylation results in a mixture of the acetates 23.173 and 23.174, and direct treatment of the mixture with 1,5-diazabicyclo[4.3.0]nona-5-ene (DBN; 23.175) causes elimination. A final reduction of the mixture with lithium aluminum hydride yields (±)-haemanthidine (23.155 and its mirror image) and its C-11 epimer (6%). The overall yield for thirty sequential reactions is 0.4%.

VIII. MANTHINE

Manthine (23.9) occurs, along with several homologs, in South African _Haemanthus_ species. Manthine (23.9) contains one methylenedioxyphenyl ring, a double bond not conjugated with the aromatic ring, two aliphatic methoxyls, and a tertiary amino function, but no N-methyl. Although information about manthine (23.9) was mostly deduced from cooccurring analogs which lacked one of the methoxyls (having a hydroxyl in place of this), most of the information was ultimately obtained from the observation that treatment of haemanthamine (23.176), a crinine-type (23.7) alkaloid, with methanesulfonyl chloride followed by aqueous alkali, yields a base isomeric with haemanthamine (23.176; the base was called isohaemanthamine, 23.177), which on methylation yields manthine (23.9; Equation 23.38). Alternatively, manthine (23.9) can also be prepared by mesylation of haemanthamine (23.176) and treatment of the resulting product with methanolic sodium methoxide (Equation 23.39).

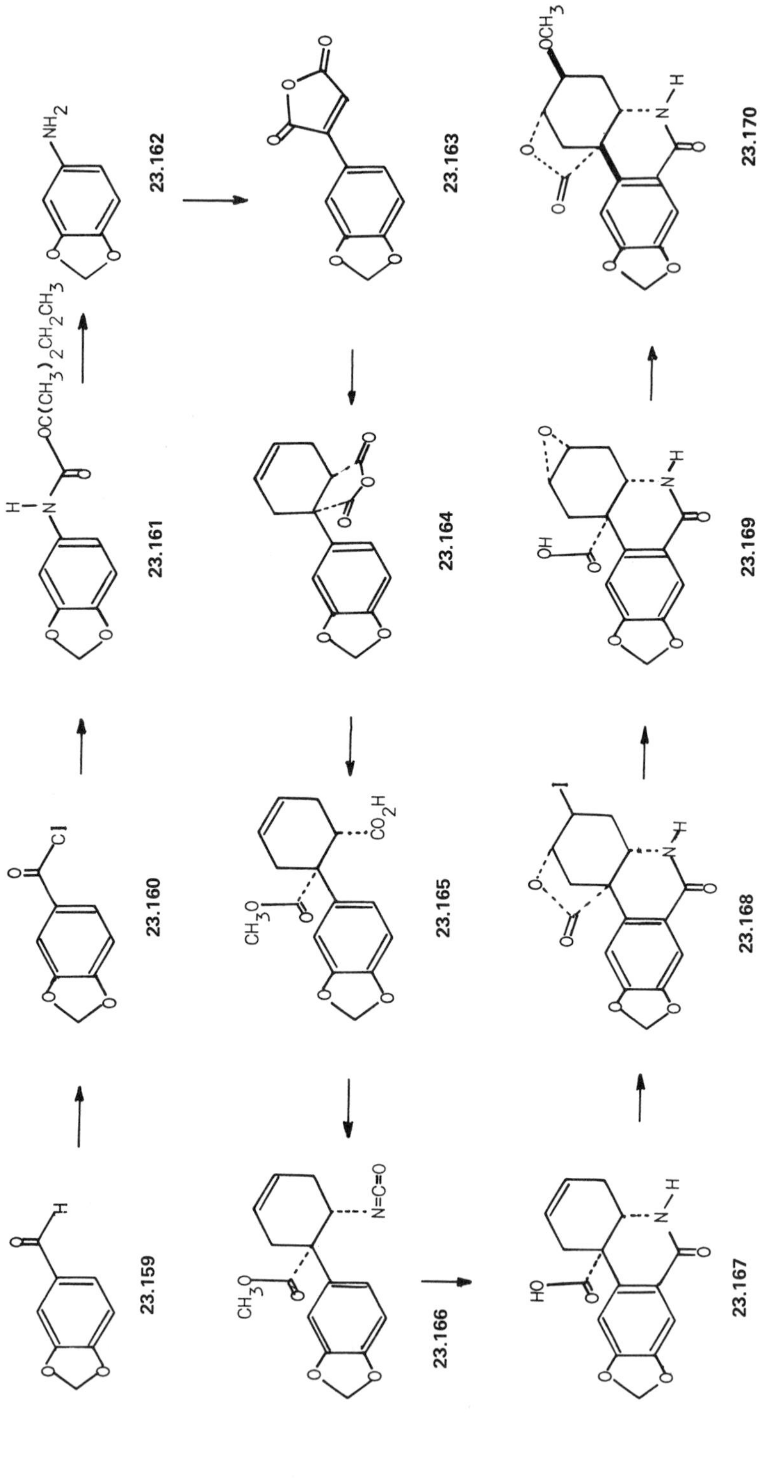

23.159

23.160

23.161

23.162

23.163

23.164

23.165

23.166

23.167

23.168

23.169

23.170

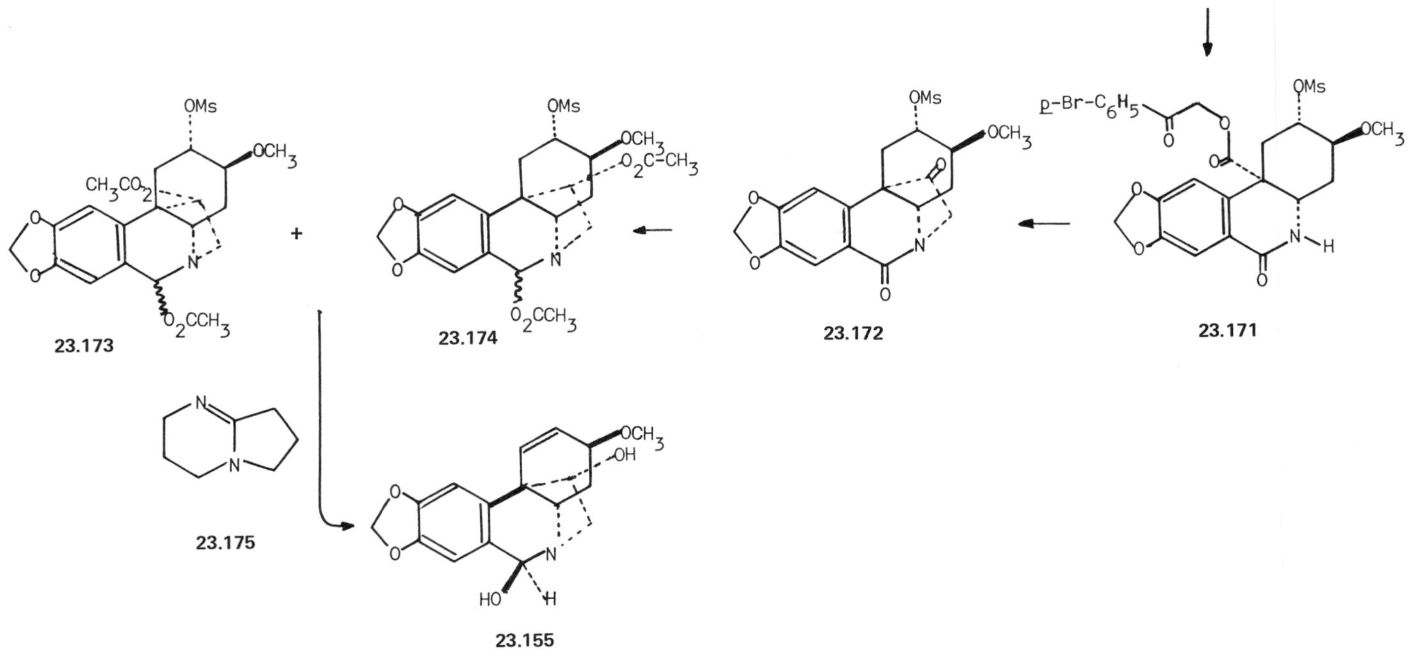

SCHEME 23.12 A synthesis of (±)-haemanthidine. (After Reference 150.)

23.176

(Eq. 23.38)

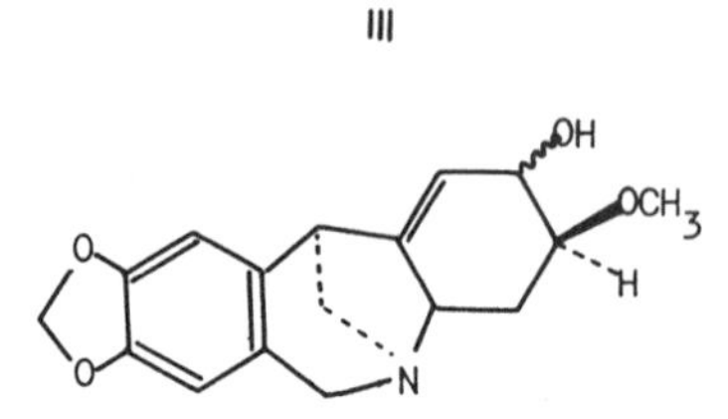

23.177

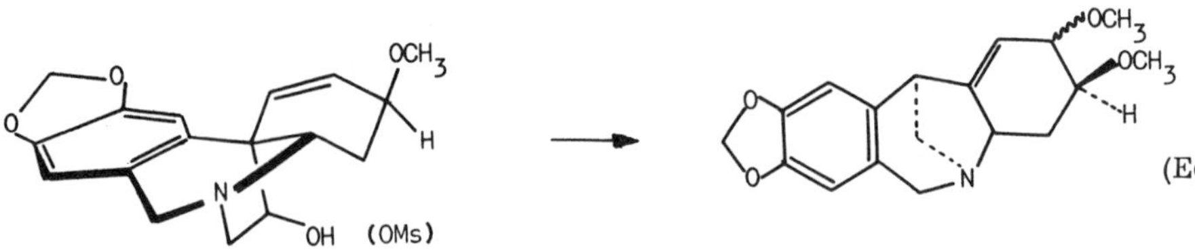

23.176

23.9

(Eq. 23.39)

The absolute configuration of haemanthamine (23.176) has been determined by x-ray crystal analysis [151] of its p-bromobenzoate.

In principle, the attack of hydroxide or methoxide during the rearrangement of haemanthamine (23.176) could occur from either above or below the plane of the cyclohexene ring. Since the methoxy group is quasiaxial in haemanthamine (23.176), and since isohaemanthamine (23.177), which is methylated to manthine (23.9), shows nonbonded hydroxyl absorption in the infrared, the assignment of the incoming hydroxyl entering in the trans (to the methoxyl) sense was made. However, two different molecules, both of structure 23.178, have been reported, and until this has been resolved the structure of manthine (23.9) remains in question.

23.178

The synthesis of manthine (23.9) has been accomplished because haemanthamine (23.176) has been synthesized. Although the synthesis of haemanthidine (23.155) is potentially useful in the preparation of haemanthamine (23.176), since removal of the benzylic hydroxyl is all that is required, the conversion has not yet been reported. An analogous conversion is known. Dihydroapohaemanthadine 23.179 [152] has been treated with thionyl chloride to generate the chloride 23.180, which on lithium aluminum hydride reduction yields dihydroapohaemanthamine (23.181; Equation 23.40). A more direct route was used to synthesize the parent compound, haemanthamine (23.176; [153]; Scheme 23.13).

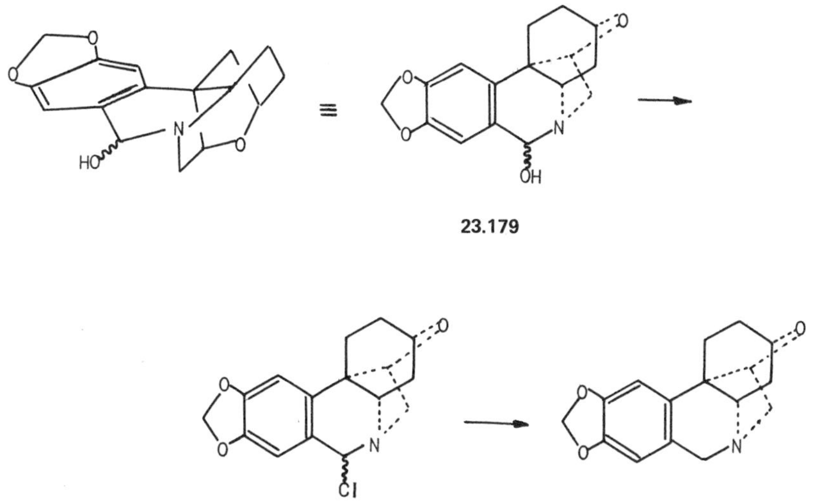

23.179

(Eq. 23.40)

23.180 23.181

The condensation of 3,4-methylenedioxybenzyl cyanide (23.182) with diethyloxylate yields the pyruvate derivative 23.183. Reduction of the latter (23.183) with hydrogen in the presence of Raney nickel yields the unstable imine 23.184, which on heating in ethanol is converted to the pyrrolidone 23.185. The Diels-Alder reaction between 23.185 and butadiene in dimethyl formamide at 160°C yields the cis-hydroindole intermediate 23.186 (presumably via reaction of the arylpyrroline-3,4-dione, 23.187, formed on elimination of ethanol from 23.185). When the Diels-Alder adduct 23.186 is treated with N-bromoacetamide the bromohemiketal (23.188) and the bromohydrin (23.189), both form, and without separation, both of these afford the same epoxide (23.190) on reaction with methanolic methoxide. When the epoxide (23.190) is opened with BF_3-methanol, the hemiketal methyl ether (23.191) is formed, and this undergoes reduction to a single product (23.192) with lithium aluminum hydride. Cyclization with methanolic formaldehyde in acetic acid yields the 11-hydroxycrinine derivative 23.193, which in turn forms a monotosylate that undergoes elimination to (±)-haemanthamine (23.176 and its mirror image).

Finally, it is noteworthy that if coccinine (23.194), which is clearly related to manthine (23.9), is oxidized under Oppenauer conditions, dehydrococcine (23.195) results (Equation 23.41). Dehydrococcine (23.195) bears a striking resemblance to (-)-cherylline (23.196), a rare phenolic Amaryllidaceae alkaloid of known [154] absolute configuration.

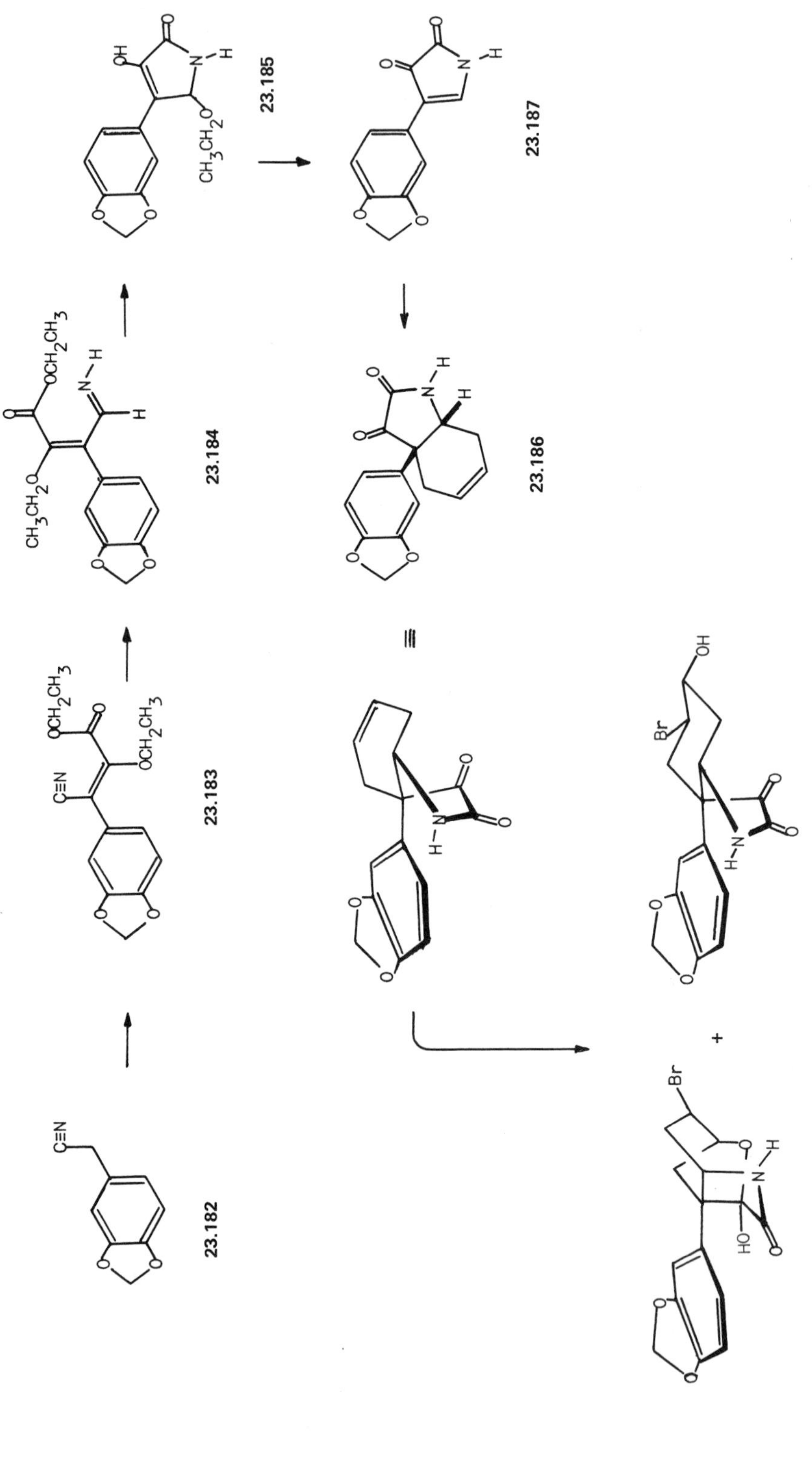

23.182

23.183

23.184

23.185

23.187

23.186

23.189

23.188

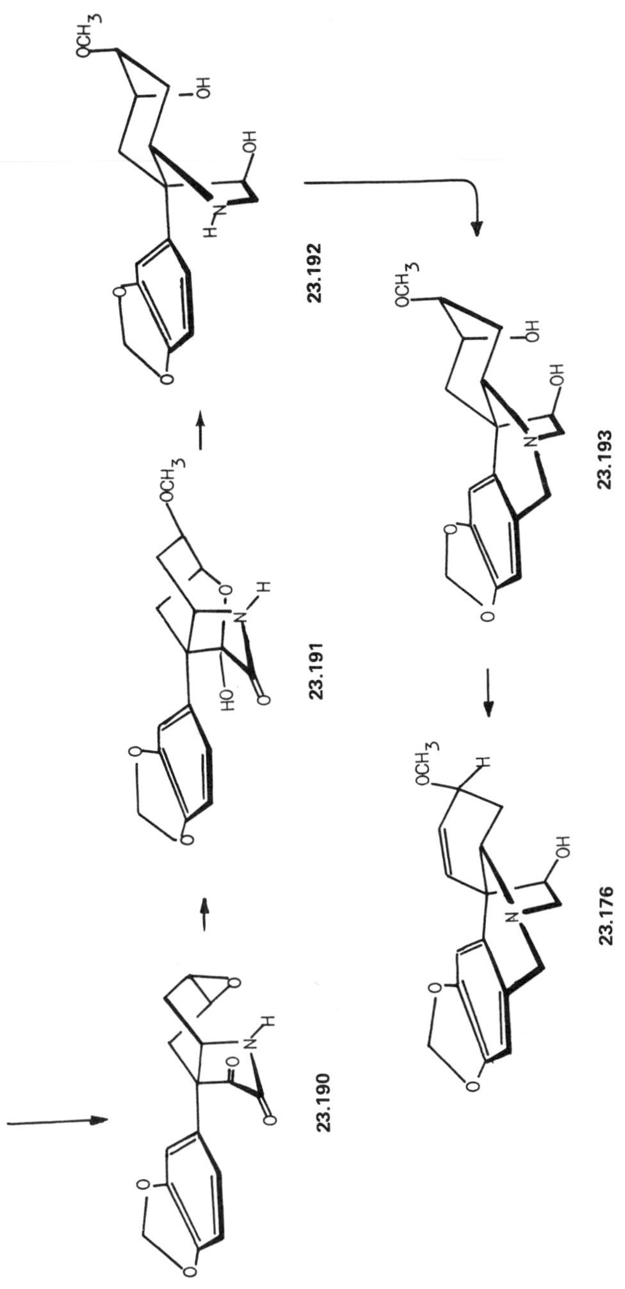

23.190 23.191 23.192 23.193 23.176

SCHEME 23.13 A synthesis of (±)-haemanthamine. (After Reference 153.)

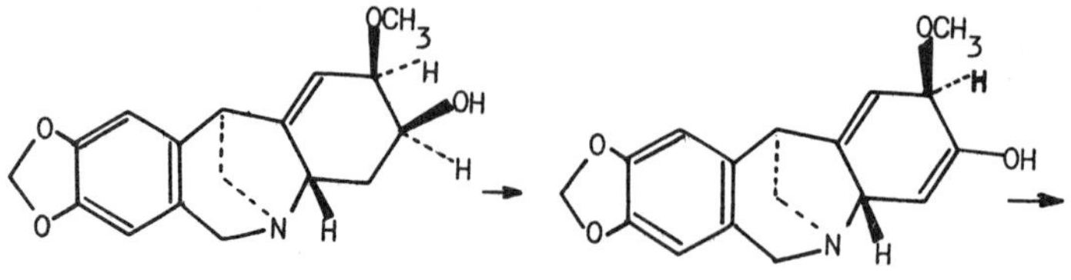

23.194

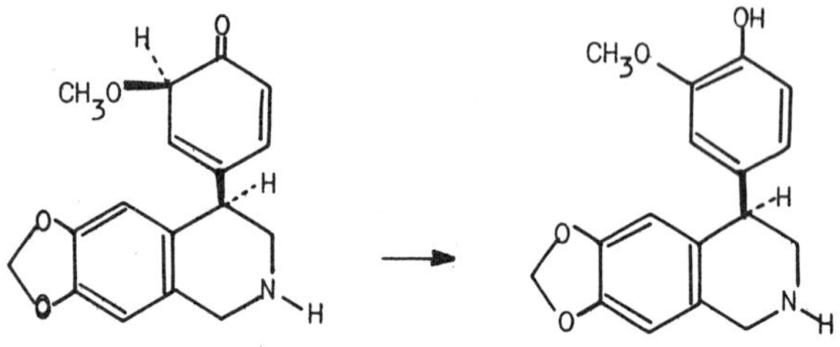

(Eq. 23.41)

23.195

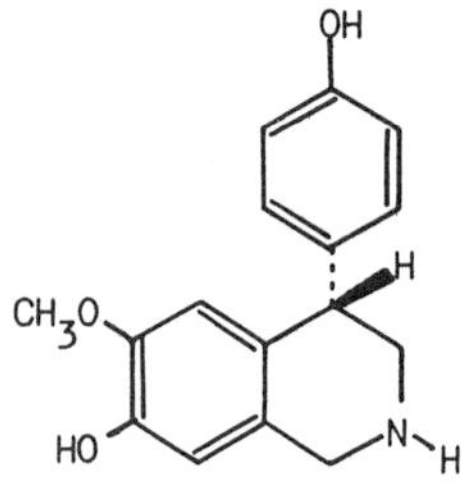

23.196

IX. ISMINE

Ismine (23.10) has been isolated in very low yield from a few Amaryllidaceae. It contains a methylenedioxy group and one N-methyl. Acetylation yields a diacetate (23.197), and treatment of ismine (23.10) with acid and then potassium ferricyanide generates 8,9-methylenedioxy-5-methyl-6-phenanthridone (23.198; Equation 23.42). Ismine has been synthesized [155]; Scheme 23.14) by chloromethylation of 3,4-methylenedioxy-2'-nitrobiphenyl (23.199)

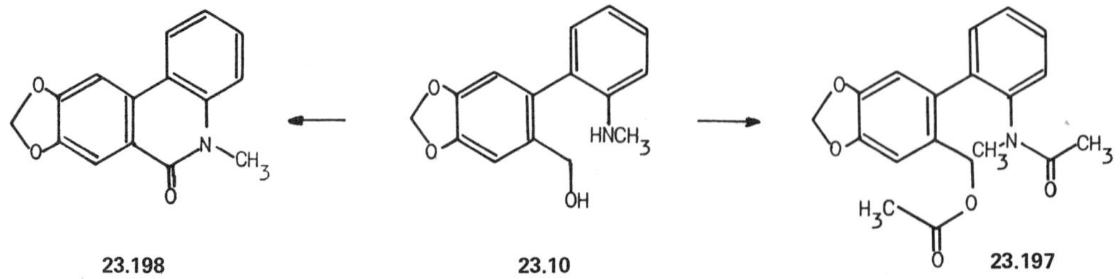

23.198 23.10 23.197

(Eq. 23.42)

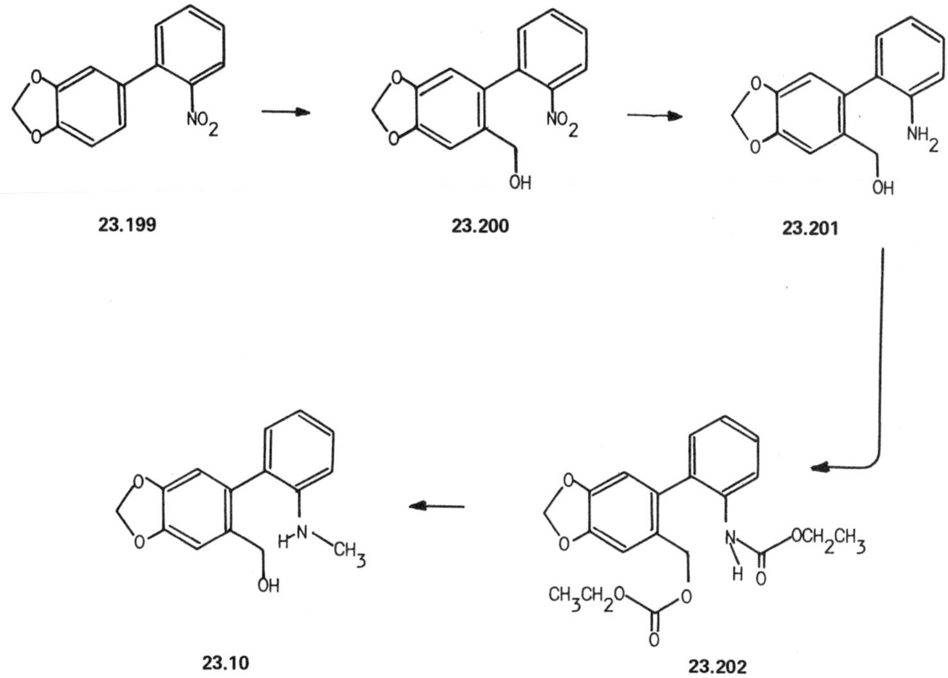

23.199 23.200 23.201

23.10 23.202

SCHEME 23.14 A synthesis of ismine. (After Reference 155.)

and hydrolysis to the alcohol (23.200). Catalytic reduction of 23.200 then affords the amino alcohol 23.201, which on reaction with ethyl chloroformate to the amido ester 23.202 and reduction with lithium aluminum hydride generates ismine (23.10).

SELECTED READING

J. W. Cook and J. D. Loudon, in The Alkaloids, Vol. 2, R. H. F. Manske (ed.). Academic, New York, 1952, pp. 1 ff.

W. C. Wildman, in The Alkaloids, Vol. 6, R. H. F. Manske (ed.). Academic, New York, 1960, pp. 289 ff.

W. C. Wildman, in The Alkaloids, Vol. 11, R. H. F. Manske (ed.). Academic, New York, 1968, pp. 307 ff.

24

CHEMISTRY OF 1-BENZYLTETRAHYDROISOQUINOLINE ALKALOIDS AND OTHER C_6-C_2 + C_6-C_2 ALKALOIDS DERIVED FROM TYROSINE

In Flanders field the poppies blow
Between the crosses, row on row. . . .

J. McCrae, "In Flanders Fields" (1915)

The first nitrogenous base isolated from plants (1805) was morphine (24.1). As a member of the general category of C_6-C_2 + C_6-C_2 alkaloids derived from tyrosine (24.2) via nor-laudanosoline (24.3), morphine (24.1) and its relatives have stimulated a massive amount of

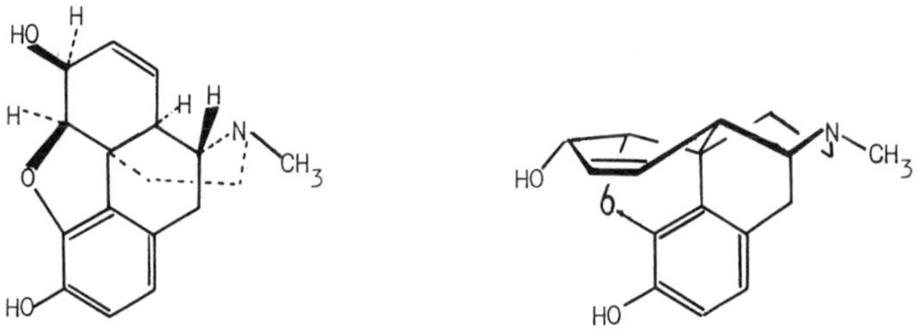

24.1

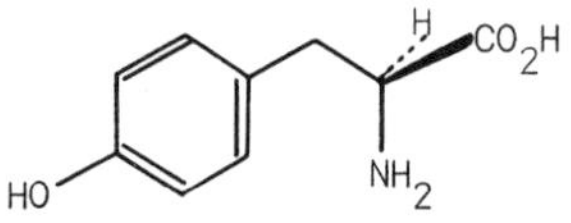

24.2

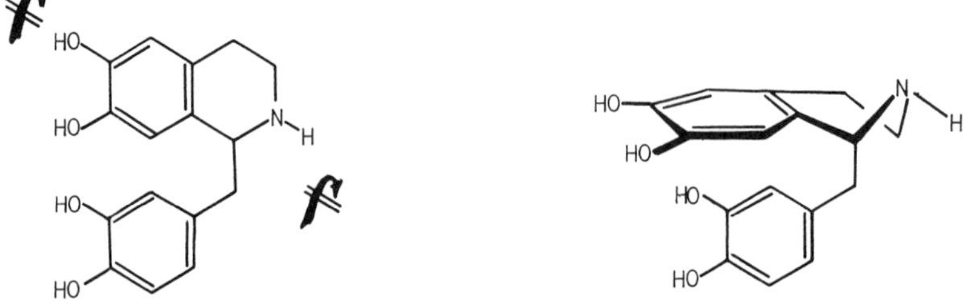

24.3

chemical work from the point of view of (a) degradation for structural elucidation; (b) synthesis, per se, as well as in the search for nonaddicting relatives; and (c) investigations into rearrangements of the initial skeletons of these systems into new and unusual forms. Although there are literally hundreds of naturally occurring compounds which could be included in the discussion of these bases, the 17 (see below) chosen are representative of the major varieties available, and the chemistry involved in their respective cases demonstrates much of the work that has been and continues to be done in the field.

For some of the compounds to be discussed, minor structural modification, capable of being carried out only after the structures were known and the compounds themselves had been synthesized, has proven of major therapeutic value. Indeed, even today, structural variants are under examination with an eye toward improved physiologic properties.

Finally, many of the bases within this group have been interrelated, either directly or through a series of derivatives, and this further serves to establish the rationale for their inclusion within the same biogenetic rubric. The alkaloids to be discussed are: laudanosine (24.4), papaverine (24.5), oxyacanthine (24.6), tubocurarine chloride (24.7), orientalinone

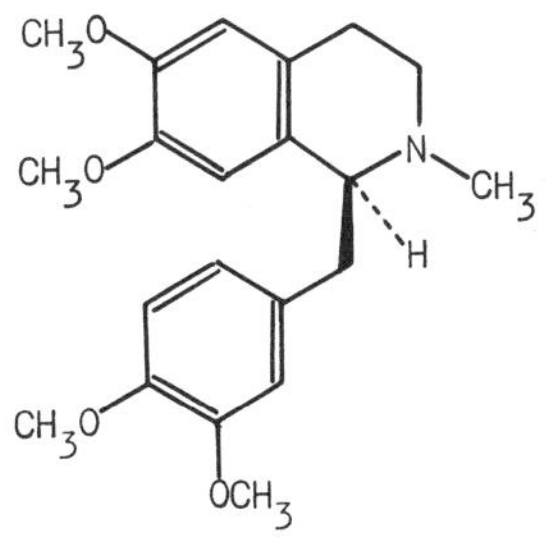

24.4

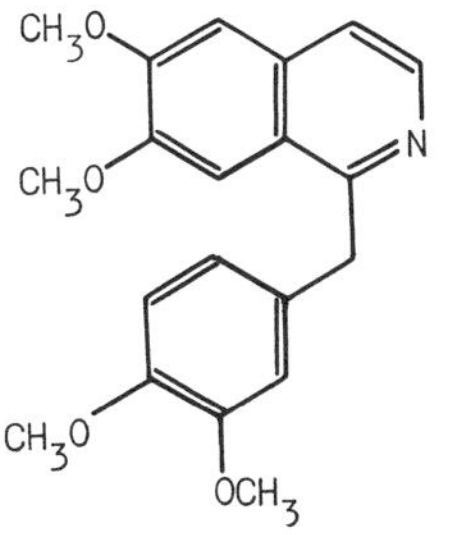

24.5

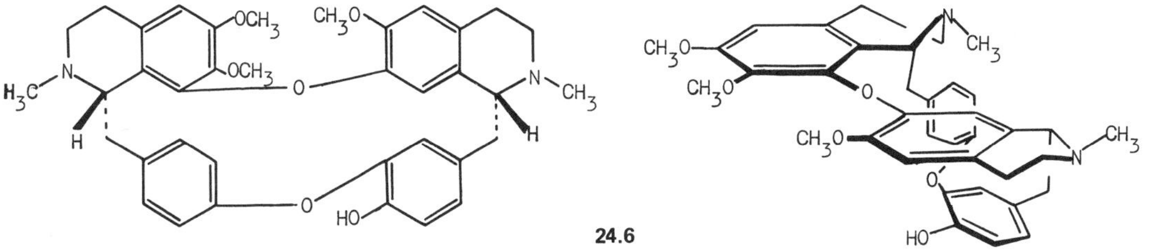

24.6

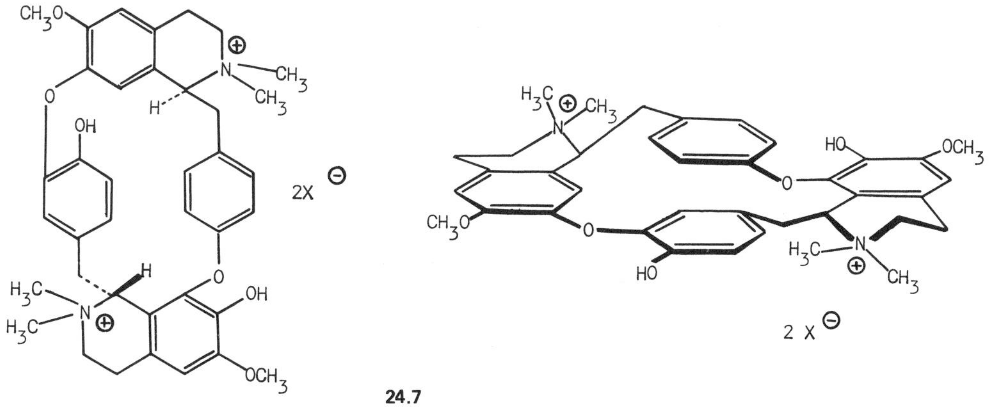

24.7

(24.8), isothebaine (24.9), morphine (24.1), codeine (24.10), thebaine (24.11), erythraline (24.12), argemonine (24.13), berberine (24.14), protopine (24.15), sanguinarine (24.16), chelidonine (24.17), α-narcotine (24.18), and rhoeadine (24.19).

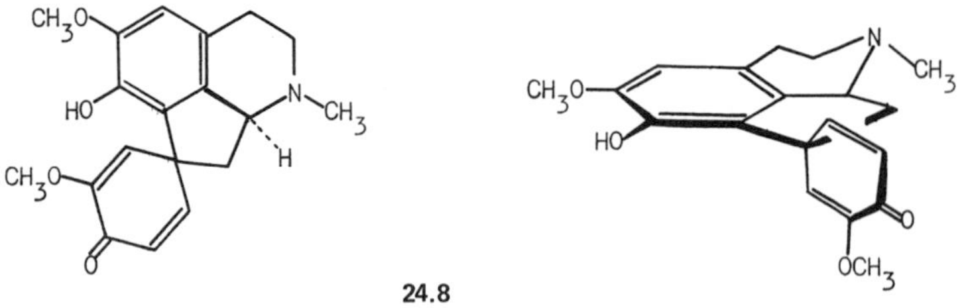

24.8

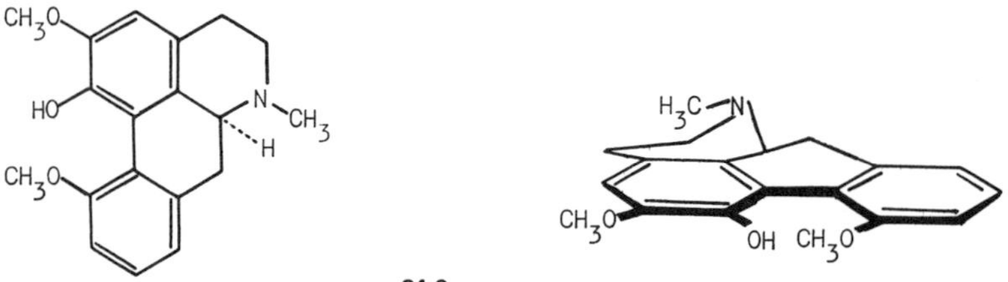

24.9

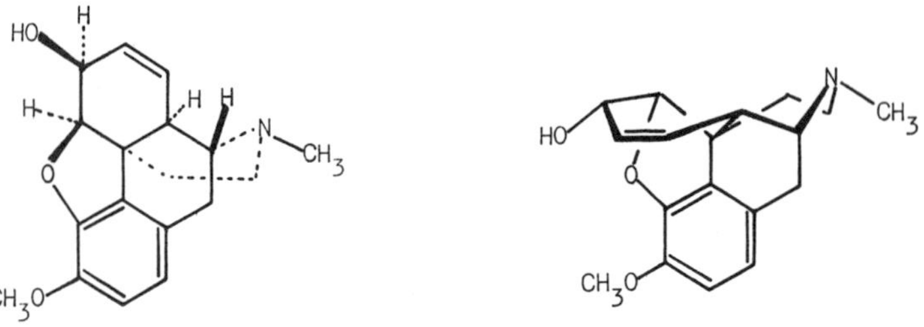

24.10

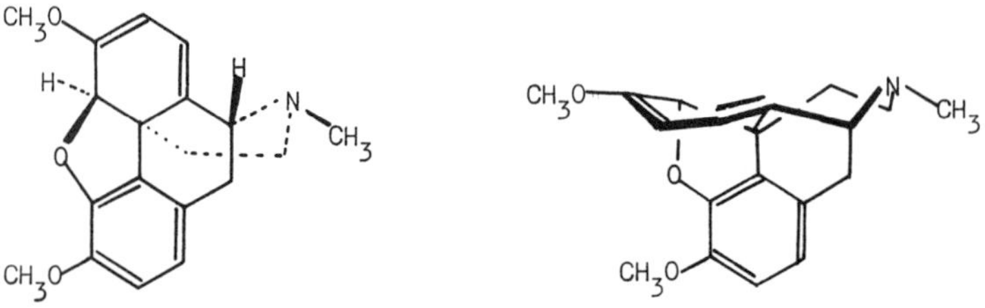

24.11

24.12

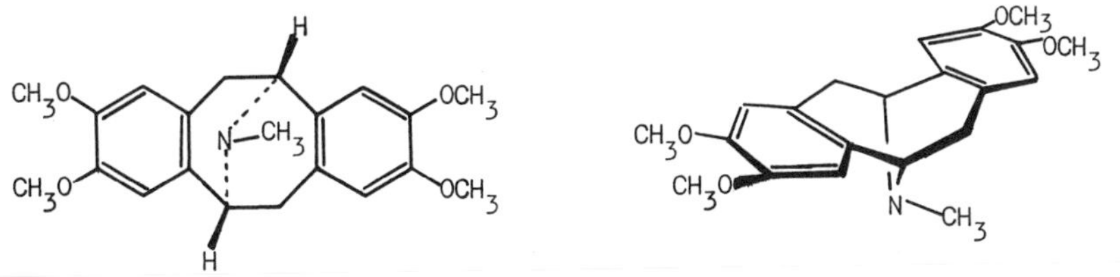

24.13

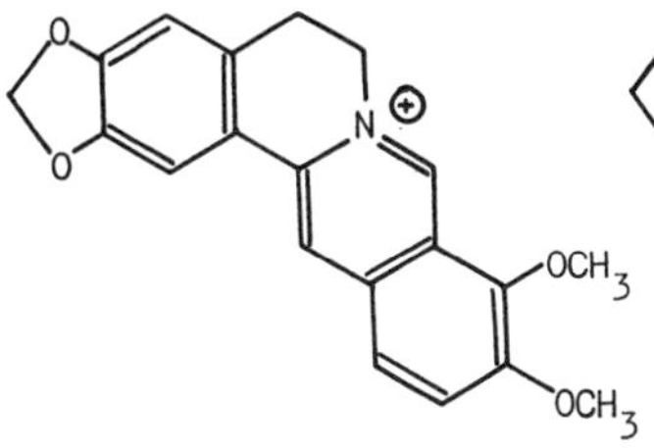

24.14 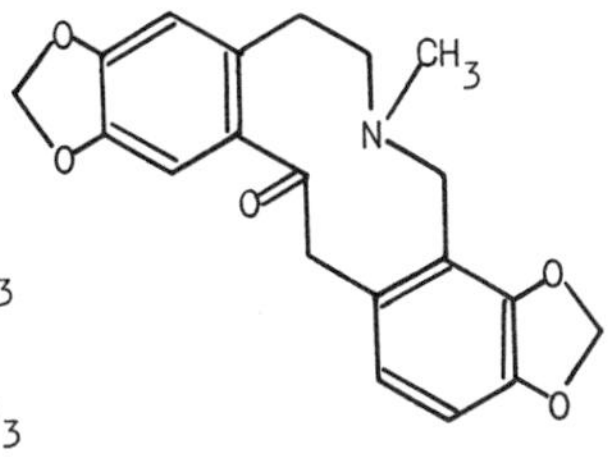 **24.15** 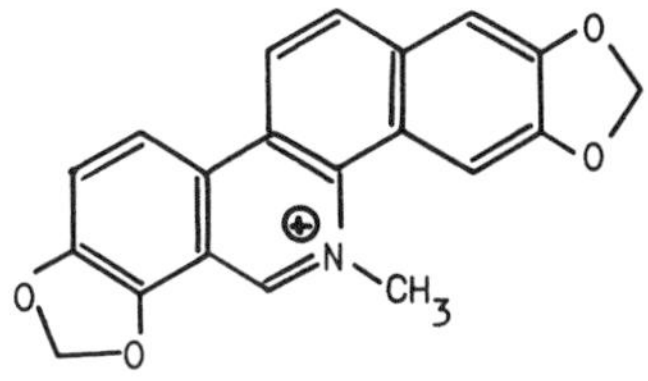 **24.16**

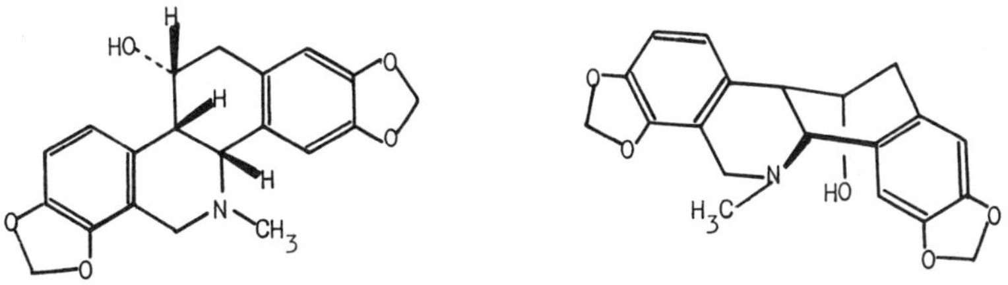

24.17

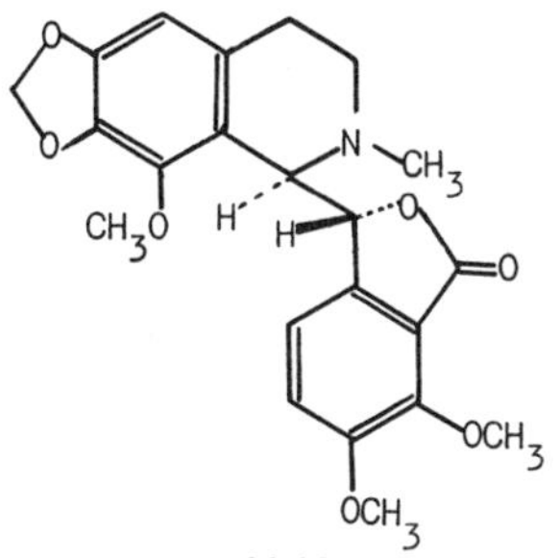

24.18

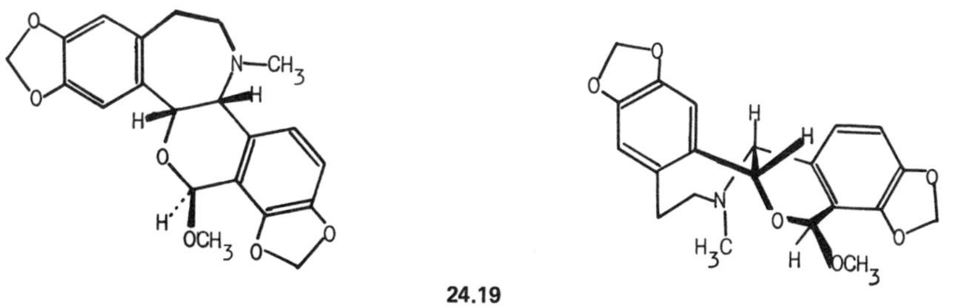

24.19

I. LAUDANOSINE AND PAPAVERINE

Laudanosine (24.4) and papaverine (24.5) are considered together here since it was recognized quite early that N-methylpapaverinium salts (e.g., 24.20) were reduced by a variety of reagents to (±)-laudanosine (24.4; Equation 24.1). Papaverine (24.5) crystallized readily from the mother liquors remaining after morphine (24.1) extraction from opium. Laudanosine (24.4) was found about 25 years after papaverine (24.5) and occurs in much lower yield in the plant.

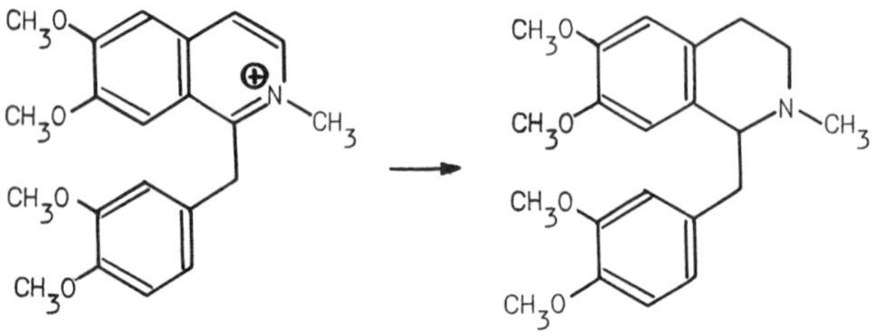

24.20 → 24.4 (Eq. 24.1)

Papaverine (24.5) usually precipitates from the mixed hydrochloride opium bases with narcotine (24.18) on sodium acetate treatment. It can be conveniently purified by chromatography, is optically inactive, melts at 147-148°C (from ethanol), and corresponds to $C_{20}H_{21}NO_4$. When papaverine (24.5) is boiled with hydroiodic acid, four equivalents of methyl iodide are liberated, indicating the presence of at least four methoxyls. Oxidation of papaverine (24.5) with a variety of reagents first generates papaveraldine (24.21), and further oxidation yields 6,7-dimethoxyisoquinoline-1-carboxylic acid (24.22; Equation 24.2).

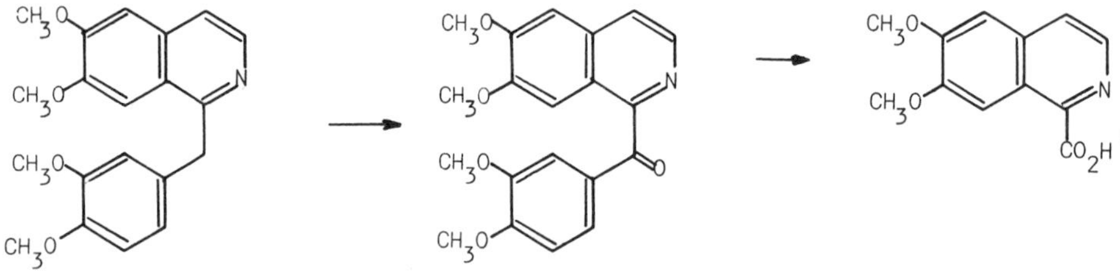

24.5 → 24.21 → 24.22 (Eq. 24.2)

Under other conditions, papaverine (24.5) yields m-hemipinic acid (24.23), veratric acid (24.24), papaverinic acid (24.25), and pyridine-2,3,4-tricarboxylic acid (24.26; Equation 24.3). Finally, very careful oxidation of papaverine (24.5) leads to papaverinol (24.27; Equation 24.4; papaverinol, 24,27, can also be obtained by reduction of papaveraldine, 24.21). Papaverinol (24.27) can be reduced back to papaverine (24.5). Alkali fusion causes cleavage of papaverine (24.5) into 6,7-dimethoxyisoquinoline (24.28) and 3,4-dimethoxytoluene (24.29; Equation 24.5). Chemical reduction (zinc and hydrochloric acid) of papaverine (24.5) leads to two compounds, namely, (±)-1,2,3,4-tetrahydropapaverine (N-norlaudanosine; 24.30) and pavine (N-norargemonine, 24.31; Equation 24.6).

Papaverine (24.5) has been synthesized by a number of different routes, among which the first [156] is quite general and is shown as Scheme 24.1. Acetoveratrone (24.32) is nitrosated and reduced to aminoacetoveratrone (24.33) with tin and hydrochloric acid.

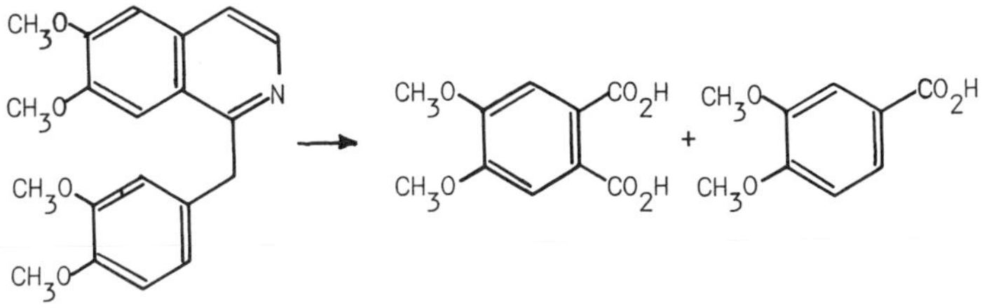

24.5 → 24.23 + 24.24

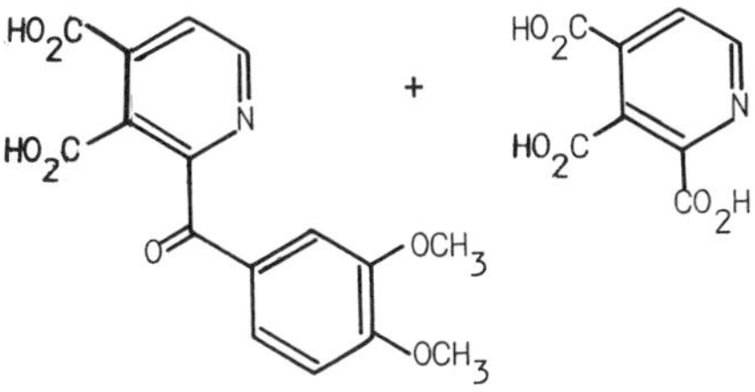

+ 24.25 + 24.26

(Eq. 24.3)

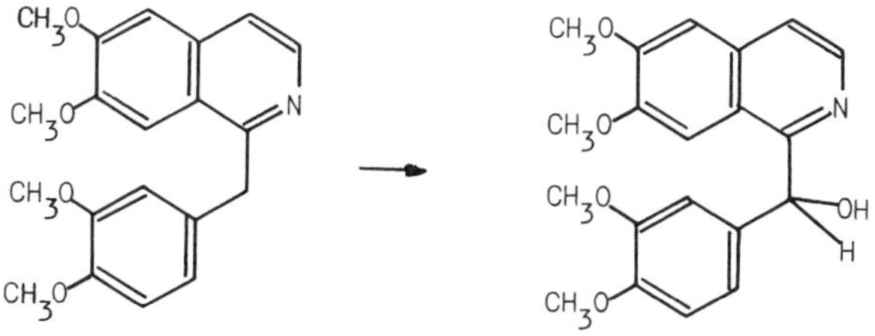

24.5 → 24.27

(Eq. 24.4)

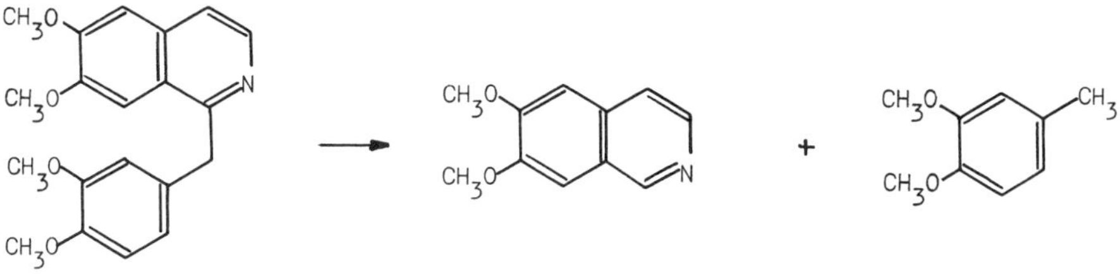

24.5 → 24.28 + 24.29

(Eq. 24.5)

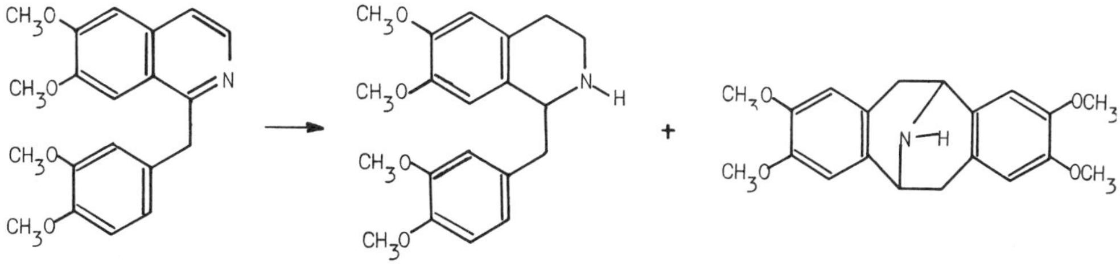

24.5 → 24.30 + 24.31

(Eq. 24.6)

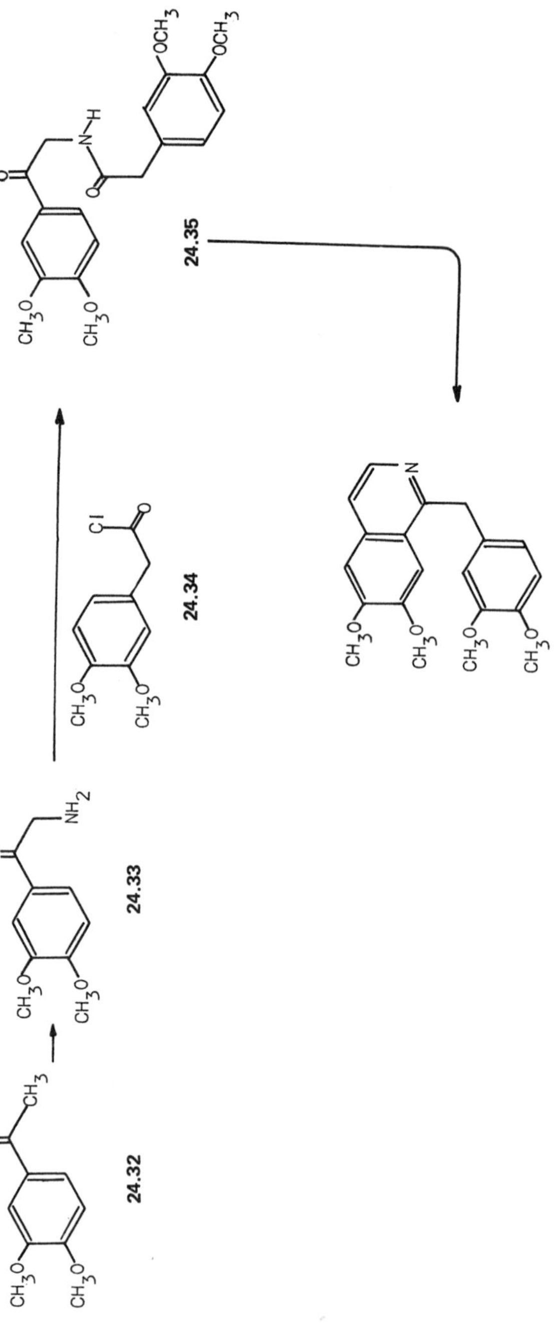

24.32 → 24.33 → (24.34) → 24.35 → 24.5

SCHEME 24.1 A synthesis of papaverine. (After Reference 156.)

Condensation of 24.33 with homoveratroyl chloride (24.34) yields the amide (24.35). Reduction to the corresponding benzylic alcohol and cyclization (with dehydration) on treatment with phosphorus pentoxide generates papaverine (24.5). Alternatively, palladium-catalyzed dehydrogenation of 1,2,3,4-tetrahydropapaverine (N-norlaudanosine, 24.30) in high-boiling solvents provides a potentially useful pathway to papaverine (24.5) since the tetrahydroisoquinoline (24.30) is readily generated biomimetically by the condensation of homoveratrylamine (24.36) with homoveratraldehyde (24.37) in ether at room temperature and then treatment with dilute hydrochloric acid (Equation 24.7).

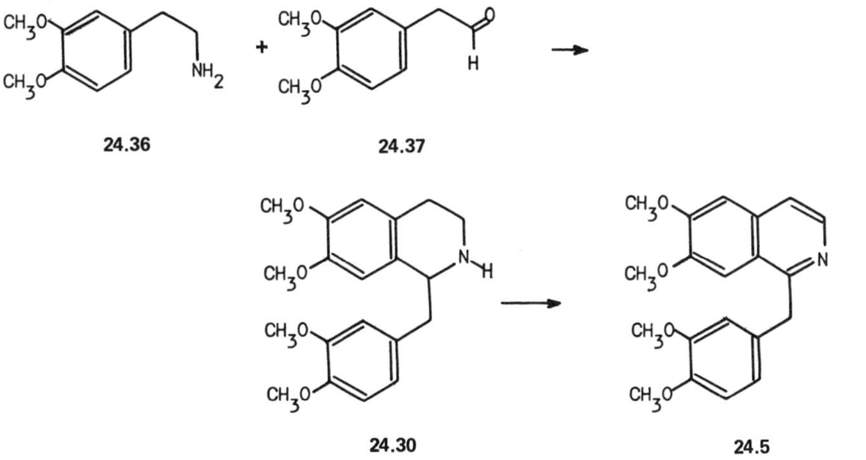

24.36 24.37

24.30 24.5

(Eq. 24.7)

Most of the chemistry of laudanosine (24.4) has been carried out on racemic material obtained from reduction of N-methylpapaverinium salts (Equation 24.1). The structural elucidation of laudanosine (24.4) was straightforward since it was recognized that laudanosine (24.4), $C_{21}H_{27}NO_4$, contains four methoxyl groups and a tertiary nitrogen (N-methyl) and that exhaustive methylation and Hofmann degradations led sequentially through the aminostilbene (24.38) to the nonbasic alkene 24.39 (Equation 24.8).

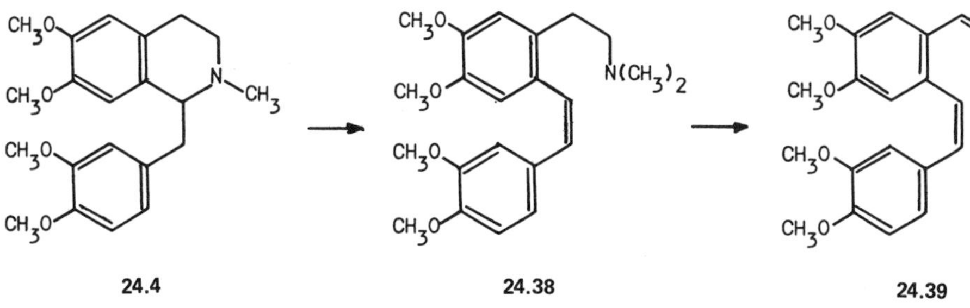

24.4 24.38 24.39

(Eq. 24.8

Among the most interesting reactions of a laudanosine (24.4) derivative is the attempted Hofmann degradation of α- or β-hydroxylaudanosine [157] which are presumably the diasteromeric alcohols of general form 24.40. The hydroxylaudanosines (24.40) cannot be prepared by oxidation of laudanosine (24.4). Instead, they are generated by reduction of N-methylpapaverinolium salts (24.41) made by methyl halide treatment of papaverinol (24.27; Equation 24.9). When hydroxylaudanosine methochloride (24.42) is warmed with aqueous sodium

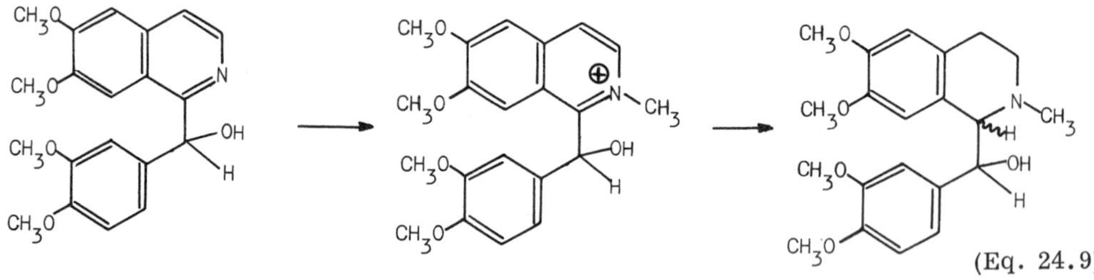

(Eq. 24.9)

24.27 24.41 24.40

hydroxide (or silver oxide), veratraldehyde (24.43) and N,N-dimethyl-2-vinyl-4,5-dimethoxybenzylamine (24.44) are formed (Equation 24.10). A suitable explanation for this unusual cleavage reaction has not yet appeared.

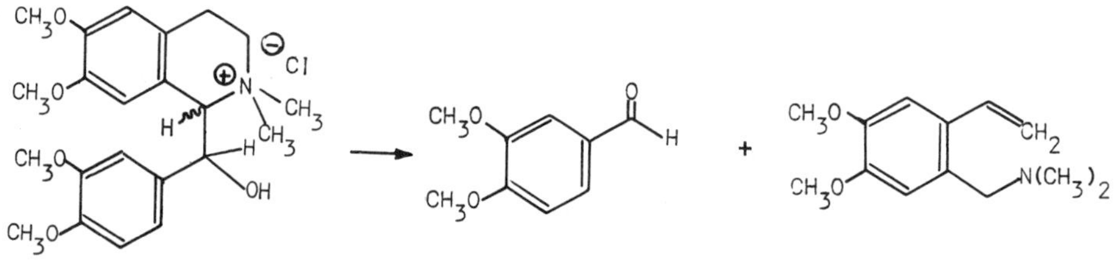

24.42 24.43 24.44 (Eq. 24.10)

(-)-N-Norlaudanosine (24.30) has served to establish the absolute configuration of the other benzyltetrahydroisoquinolines which can be related to it, since ozonolysis [158] and oxidation yield N-β-carboxyethyl-L-aspartic acid (24.45; Equation 24.11) of known absolute configuration.

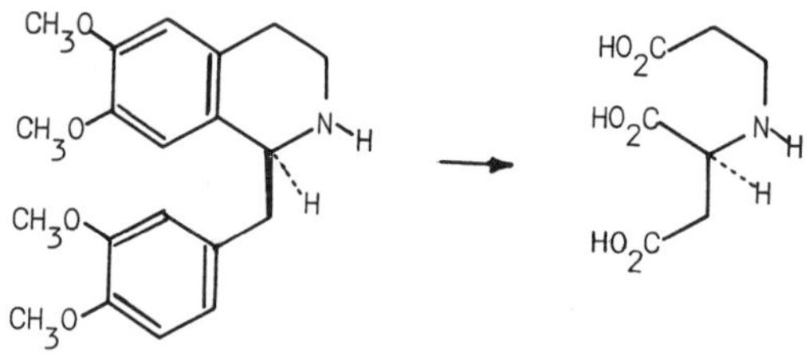

(Eq. 24.11)

24.30 24.45

II. OXYACANTHINE

Oxyacanthine (24.6) and related bisbenzylisoquinoline alkaloids from the roots of, e.g., Berberis vulgaris L., have been known for many years, but it was not until about 1925 that the dextrorotatory base corresponding to $C_{37}H_{40}N_2O_6$ was obtained in a relatively pure form.

Of the six oxygen atoms in oxyacanthine (24.6), four were brought to account through noting the presence of three methoxyls and a hydroxyl (phenolic). Methylation of oxyacanthine (24.6; O and N) followed by a single Hofmann degradation yields a methine (24.46) which undergoes permanganate oxidation to 4',5-dicarboxy-2-methoxydiphenyl ether (24.47; Equation

24.12). Ethylation of oxyacanthine (24.6) and similar degradation yields 4',5-dicarboxy-2-ethoxydiphenyl ether (24.48) which establishes the position of the phenol as ortho to the oxygen bridging the phenyl rings. When the methine (24.46) is subjected to ozonolysis, 4',5-diformyl-2-methoxydiphenyl ether (24.49), which is capable of oxidation to the corresponding

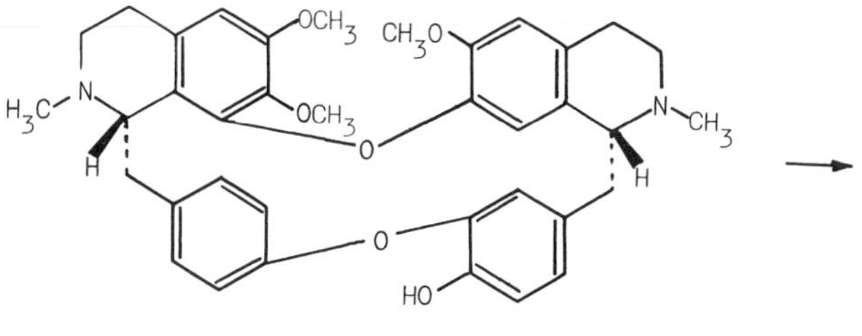

24.6

(Eq. 24.12)

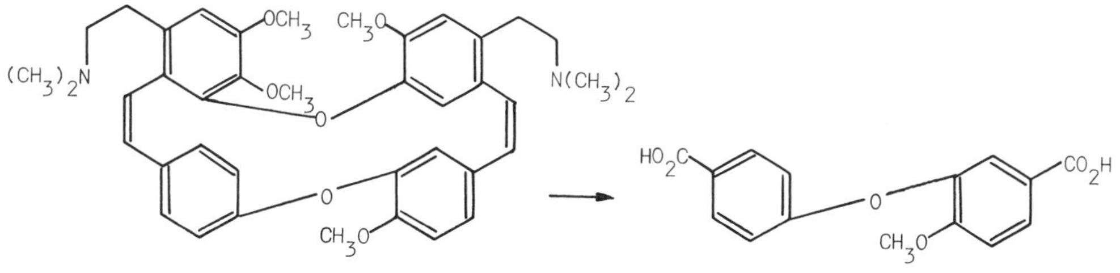

24.46

24.47

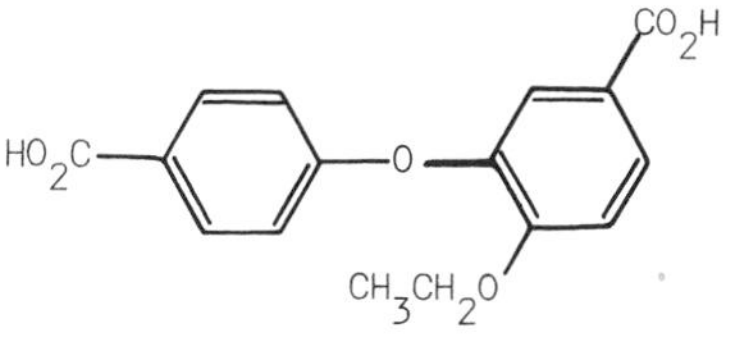

24.48

dicarboxylic acid, 24.47, and a nitrogenous fragment, presumably 24.50, capable of undergoing Hofmann degradation to a dialdehyde which can be reduced catalytically to 3,4'-diethyl-2,5'-diformyl-2',5,6-trimethoxydiphenyl ether (24.51), are formed (Equation 24.13). The carbonyl groups of 24.51 are removed on Clemmensen reduction, and 3,4,4'-trimethoxy-6,6'-diethyl-1,1'-dimethyl-2,3'-diphenyl ether (24.52), identical to the product of the Ulmann condensation between 1,2-dimethoxy-4-ethyl-5-methyl-6-bromobenzene (24.53) and the sodium salt of 2-methoxy-4-ethyl-5-methylphenol (24.54), is formed (Equation 24.14).

Later [159], oxyacanthine (24.6) was subjected to treatment with sodium in liquid ammonia to yield two phenolic products, 24.55 and 24.56 (Equation 24.15). The first (24.55) was isolated and identified as (+)-1-(p-hydroxybenzyl)-6,7-dimethoxy-2-methyl-1,2,3,4-tetrahydroisoquinoline [(+)-armepavine], while the second on methylation yields (-)-1-(p-methoxybenzyl)-6,7-dimethoxy-2-methyl-1,2,3,4-tetrahydroisoquinoline (24.57), thus establishing the stereochemistry at each of the chiral centers.

It should be noted that the sodium in liquid ammonia cleavage occurs specifically to generate those products in which the position para to the oxygen is occupied and not the other possible products. This is a general observation for the reductive cleavage reaction.

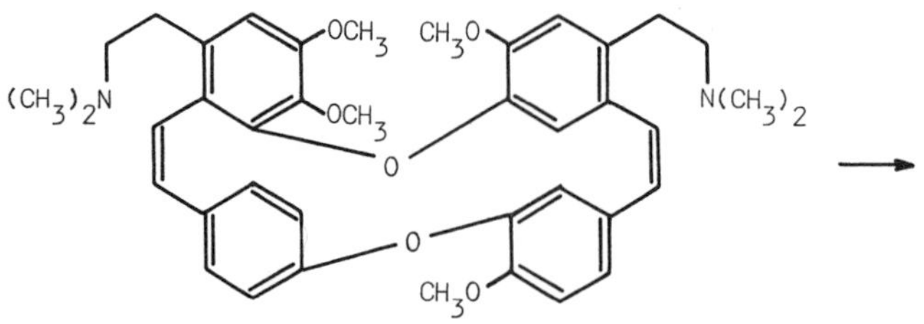

24.46

(Eq. 24.13)

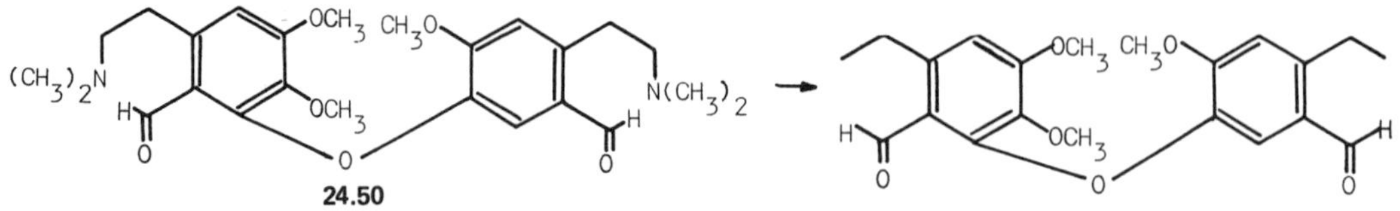

24.50

24.51

24.49

24.47

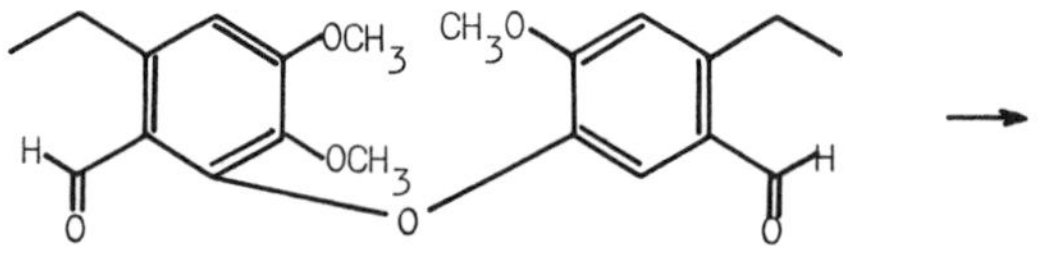

24.51

(Eq. 24.14)

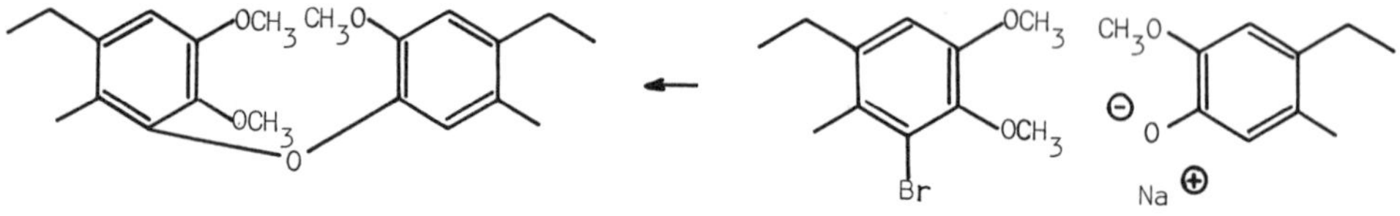

24.52

24.53

24.54

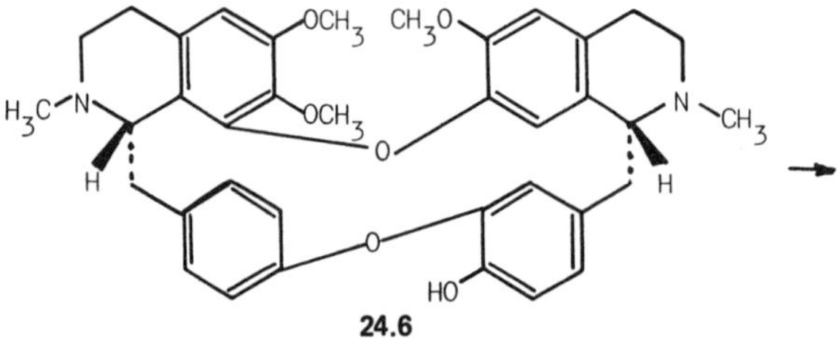

24.6

(Eq. 24.15)

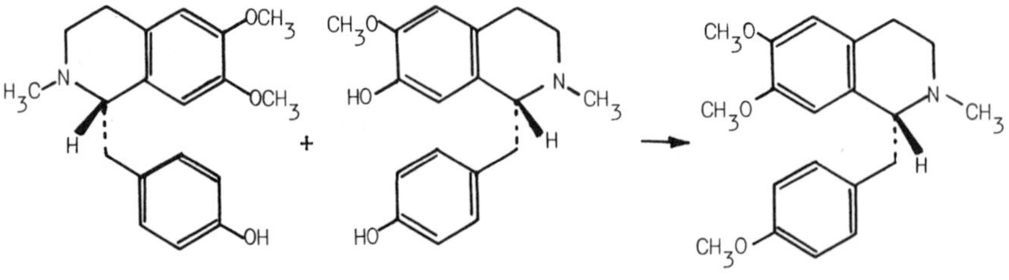

24.55

24.56

24.57

Oxyacanthine (24.6) may have been synthesized [160] although minor spectroscopic differences were noted in the racemic and chiral materials. The general synthetic pathway is, nevertheless, instructive and is shown as Scheme 24.2.

A mixture of 5-bromo-3,4-dimethoxybenzaldehyde diethyl acetal (24.58) and acetylvanillin diethyl acetal (24.59) in pyridine is heated with cupric oxide, copper powder, and potassium carbonate to effect the Ullmann condensation and generate 4-(5-formyl-2,3-dimethoxyphenoxy)-3-methoxybenzaldehyde diethyl acetal (24.60). Hydrolysis of the diethylacetal (24.60) to the corresponding dialdehyde, aldol condensation with nitromethane to the ω-nitrostyrene, and reduction with lithium aluminum hydride yield 4-[5-(2-aminoethyl)-2,3-dimethoxyphenoxy]-3-methoxyphenethylamine (24.61). Reaction of the amine (24.61) with the bis acid chloride of 4-benzyloxy-3,4'-oxydiphenyldiacetic acid (24.62) yields the two bis amides 24.63 and 24.64 which are separable and which were individually carried on through the Bischler-Napieralski cyclization with phosphorus oxychloride in chloroform, reduction to the respective secondary amines, and methylation and debenzylation to the respective amino phenols. Since 24.63 can yield 24.65 or 24.66, while 24.64 can result in 24.67 or 24.68 and the infrared spectra of the synthetic materials are different from that of oxyacanthine (24.6), the synthesis of the latter cannot yet be said to have been accomplished. Apparently the insolubility of the compounds obtained precluded pmr examination.

III. TUBOCURARINE

(+)-Tubocurarine (24.7) is a water-soluble quaternary ammonium base which was obtained about the beginning of the twentieth century from tubocurare (the red resinous mass packed in tubes and used as an arrow poison by Indian tribes living in the upper Amazon basin). It is obtained from, among others, Chondodendron tomentosam, and occurs, along with (-)-curine (24.69) whose gross structure was first elucidated and then compared (as the O,O-dimethylmethochloride) with (+)-tubocurarine (24.7). Curine (24.69) corresponds to $C_{36}H_{38}N_2O_6$ and yields a dimethyl ether (24.70) which on Hofmann degradation and oxidation generates 4',5,6-tricarboxy-2,3-dimethoxydiphenyl ether (24.71) and the isomeric 3,3',4-tricarboxy-2,2'-dimethoxydiphenyl ether (24.72). These two fragments together define the gross structure of the dimethyl ether as 24.70 (Equation 24.16).

The phenolic positions were established by ethylation (instead of methylation) and Hofmann degradation and oxidation as described above. Since ethylation of (+)-tubocurarine chloride yields the same nitrogen-free compound as (-)-curine (after Hofmann degradation) the two alkaloids must have the same orientation of methoxyl and phenolic groups and, on

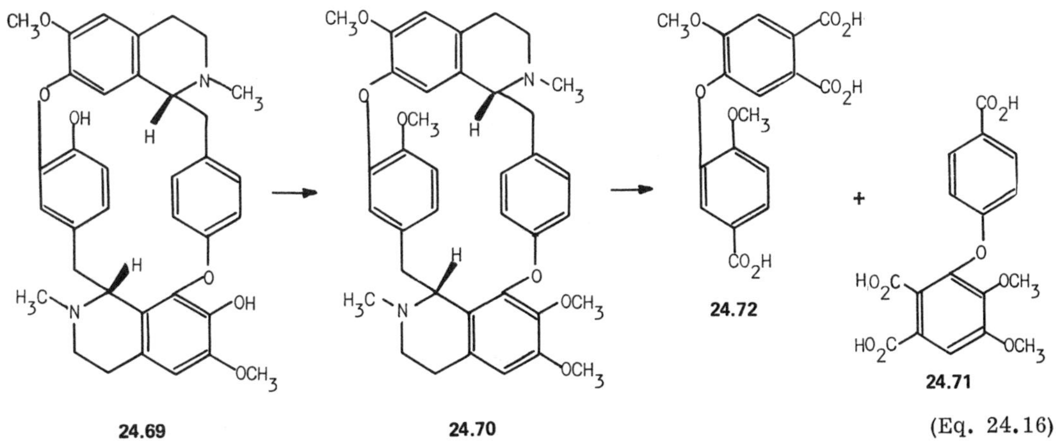

24.69 → 24.70 → 24.72 + 24.71

(Eq. 24.16)

24.58

24.59

24.60

24.61

24.62

24.63

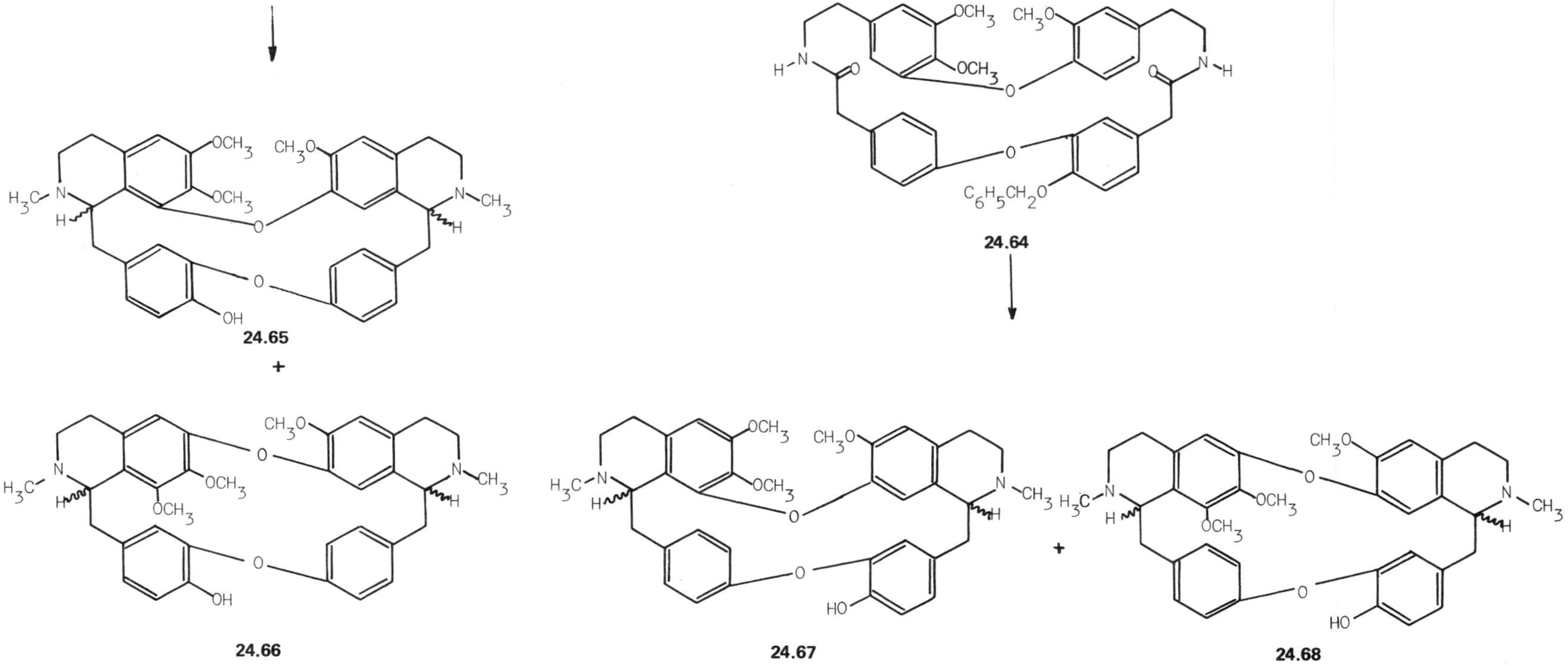

24.64

24.65

+

24.66

24.67

+

24.68

SCHEME 24.2 A possible synthesis of oxyacanthine. (After Reference 160.)

the basis of optical rotation measurements, must differ only in the sense of the chiral centers. Although (+)-tubocurarine (24.7) had been thought for many years to be a bis-quaternary salt, nmr measurements demonstrate that only three N-methyl groups are present [161]. Sodium in liquid ammonia cleavage of O,O-dimethyl-(+)-tubocurarine chloride (24.73) yields the inactive Emde product (24.74) and (+)-1-(4-hydroxybenzyl)-2-methyl-6-methoxy-7-hydroxy-1,2,3,4-tetrahydroisoquinoline (24.75), which demonstrates which of the two nitrogens is quaternary (Equation 24.17).

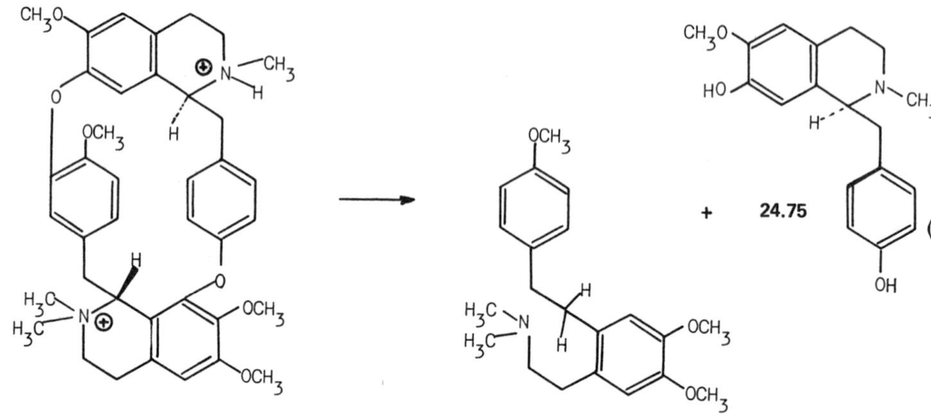

24.73 → 24.74 + 24.75 (Eq. 24.17)

A synthesis of (+)-tubocurarine (24.7) has not yet been effected although a general scheme such as that shown for oxyacanthine (24.6; Scheme 24.2) might be applicable. It should be noted, however, in contradistinction to oxyacanthine (24.6), one and only one of the methyl groups in tubocurarine (24.7) must be quaternary while the other remains tertiary. This introduces complications not present in the attempted synthesis of oxyacanthine (24.6).

IV. ORIENTALINONE

The isolation of (-)-orientalinone (24.8) from natural sources was a direct tribute to the insight of Barton and Cohen [68] who had postulated the intermediacy of such a dienone, lying between the tetrahydroisoquinoline bases and the aporphine alkaloids, on the basis of a presumed oxidative coupling pathway.

The alkaloid, $C_{19}H_{21}NO_4$, mp 230-232°C (dec.), $[\alpha]_D^{20}$ -76° ($CHCl_3$), was synthesized in its racemic form by oxidation of the (±)-benzyltetrahydroisoquinoline base (±)-orientaline (24.76) with potassium ferricyanide (Equation 24.18; [162]). The crude ketonic fraction (4%

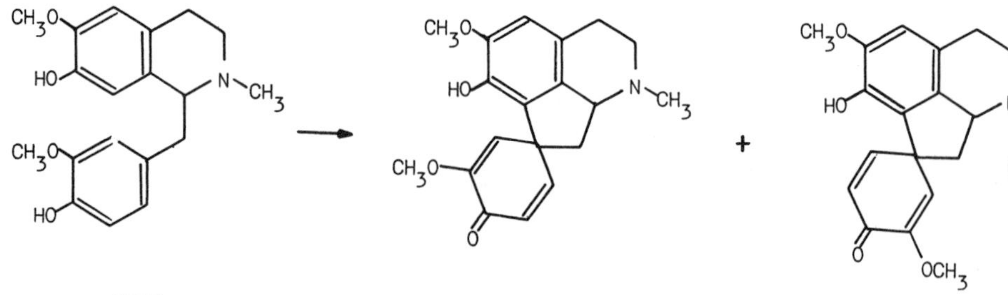

24.76 → 24.8 + 24.77 (Eq. 24.18)

yield) which presumably also contains other products, e.g., the diasteromer of (±)-orientalinone (24.77), yields a mixture of dienols (e.g., 24.78) which undergo the dienol-benzene rearrangement with acid to yield (±)-isothebaine (24.9; Equation 24.19). Resolution of

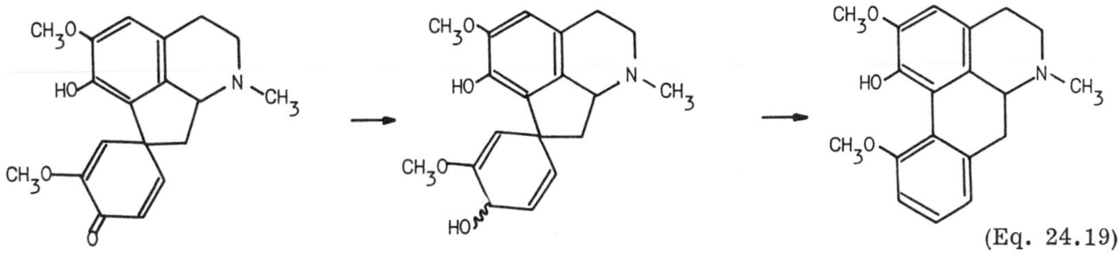

(Eq. 24.19)

24.8 24.78 24.9

(±)-orientaline (24.76) yields (-)-orientalinone (24.8) and when treated with acid (+)-isothebaine (24.9 shown as its racemate). Interestingly, when (±)-orientalinone (24.8 and its mirror image) is treated with hydrogen chloride in acetic acid, (±)-isocorytuberine (24.79) is obtained while HCl in methanol yields (±)-corydine (24.80; Equation 24.20).

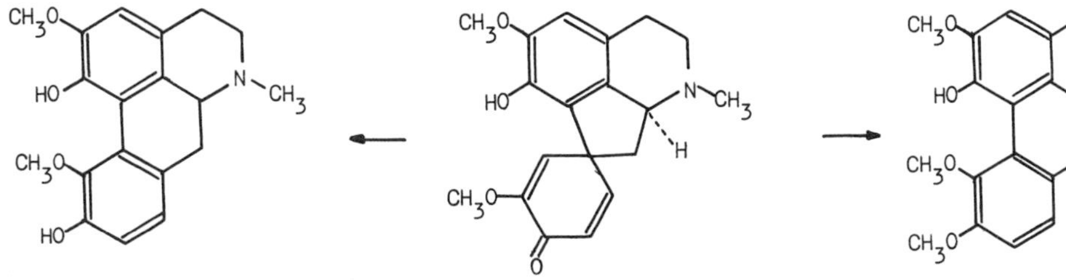

(Eq. 24.20)

24.79 24.8 24.80

The route of the dienol-benzene rearrangement followed by the dienol mixture 24.78 to yield (±)-isothebaine (24.9) is interesting since there are, in fact, four possible isomers (i.e., 24.9, 24.81, 24.82, and 24.83) which can be pictured as arising from the suitable movement of bonds in 24.78 (Scheme 24.3). However, since the most electron-rich bond is presumably most favored to migrate, 24.82 and 24.83 should be missing. Second, since the methoxyl is either protonated or hydrogen bonded to the protonated hydroxyl which is lost, the double bond bearing the methoxyl is the more electrophilic, and thus (±)-isothebaine (24.9) is the favored product. The a priori rigor of these arguments is tempered by the a posteriori isolation of (±)-isothebaine (24.9).

V. ISOTHEBAINE

Isothebaine (24.9), which may be derived from orientalinone (24.8), is isolated from the roots of Papaver orientale after the period of active growth of the aerial parts and the production of thebaine (24.11) has ceased. Hofmann degradation of the O-methyl ether of isothebaine (24.84) yields an optically inactive methine and an optically active isomethine (presumably 24.85 and 24.86, respectively) which subsequently proceed, via a trimethoxyvinylphenanthrene, to a carboxylic acid which yields 3,4,5-trimethoxyphenanthrene (24.87) on decarboxylation (Equation 24.21).

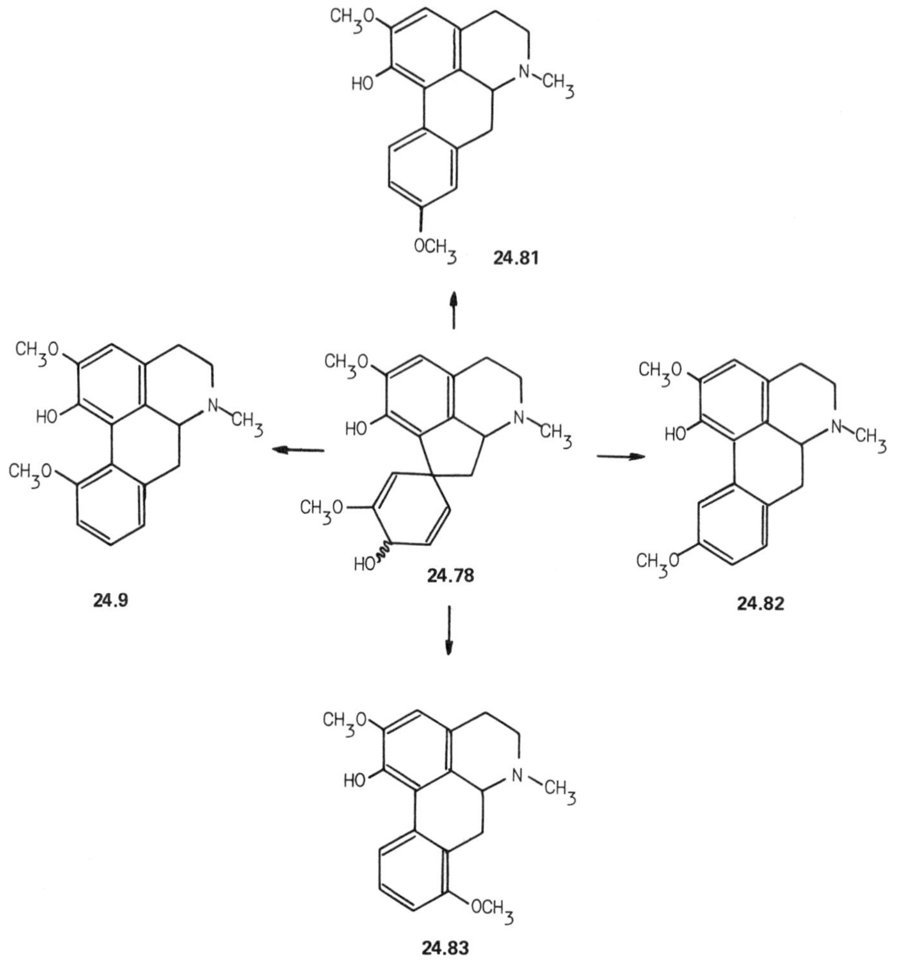

CH_3O
HO
$N-CH_3$
OCH_3
24.81

CH_3O
HO
$N-CH_3$
CH_3O

24.9

CH_3O
HO
$N-CH_3$
CH_3O
HO
24.78

CH_3O
HO
$N-CH_3$
CH_3O

24.82

CH_3O
HO
$N-CH_3$
OCH_3
24.83

SCHEME 24.3 Possible products from the dienol-benzene rearrangements of the dienols resulting from reduction of (±)-orientalinone.

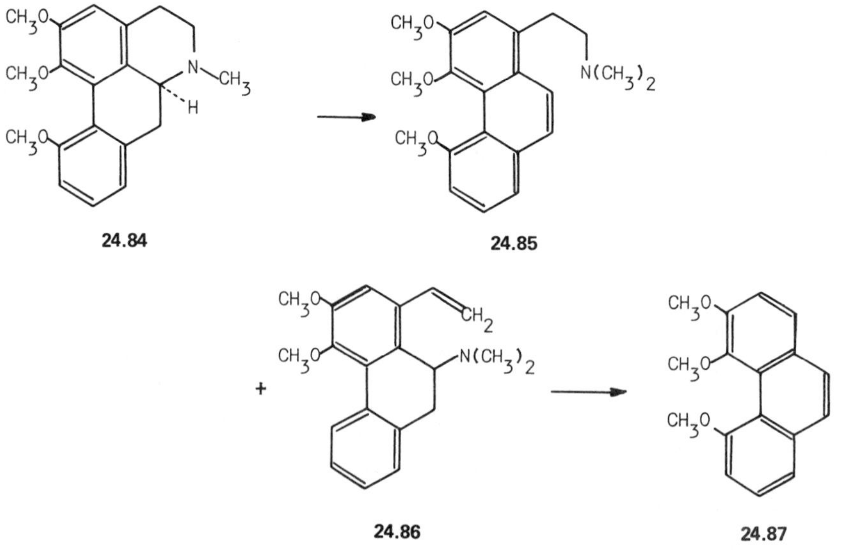

CH_3O
CH_3O
$N-CH_3$
H
CH_3O

24.84

CH_3O
CH_3O
$N(CH_3)_2$
CH_3O

24.85

(Eq. 24.21)

CH_3O
CH_3O
CH_2
$N(CH_3)_2$
+

24.86

CH_3O
CH_3O
CH_3O

24.87

After an involved argument (Reference 163) concerning the position of the phenolic hydroxyl, (±)-6, 8-dimethoxy-7-ethoxyaporphine (24.88) was synthesized and compared with the product of ethylation of isothebaine (24.9). The solution infrared and ultraviolet spectra of the synthetic material were identical to those obtained from the natural product, thus establishing the position of the hydroxyl and the two methoxyl groups in isothebaine (24.9).

The synthesis of the O-ethyl aporphine (24.88) is shown in Scheme 24.4, where it is seen that the reaction of 2-nitro-3-methoxy-ω-diazoacetophenone (24.89), prepared by diazomethane treatment of the corresponding acid chloride, yields on treatment with β-(3-methoxy-4-ethoxyphenyl)ethylamine (24.90) the corresponding 3'-methoxy-2'-nitrophenylaceto-3-methoxy-4-ethoxyphenethyl amide (24.91).

Cyclization of the amide (24.91) to the dihydroisoquinoline (24.92) is effected with phosphorus pentachloride in cold chloroform, and then, after N-methylation and without isolation of the intermediates, the nitro and imino functions are reduced with zinc and hydrochloric acid, the resulting amine diazotized with sodium nitrite in dilute sulfuric acid, and the Pschorr cyclization consummated by the addition of copper powder. A poor yield of dimethoxyethoxyaporphine (24.88) results.

Although the racemic material melts lower than the ethyl ether of isothebaine (24.9) itself, and resolution was not apparently effected, the identity of the infrared spectra and the degradative work reported above sufficed to establish the gross structure of isothebaine (24.9). The absolute configuration of (+)-isothebaine (24.9) can be deduced as shown by the following considerations (Scheme 24.5): (+)-isothebaine (24.9) is derived from (-)-orientalinone (24.8), which in turn comes from (+)-orientaline (24.76) by reactions which should

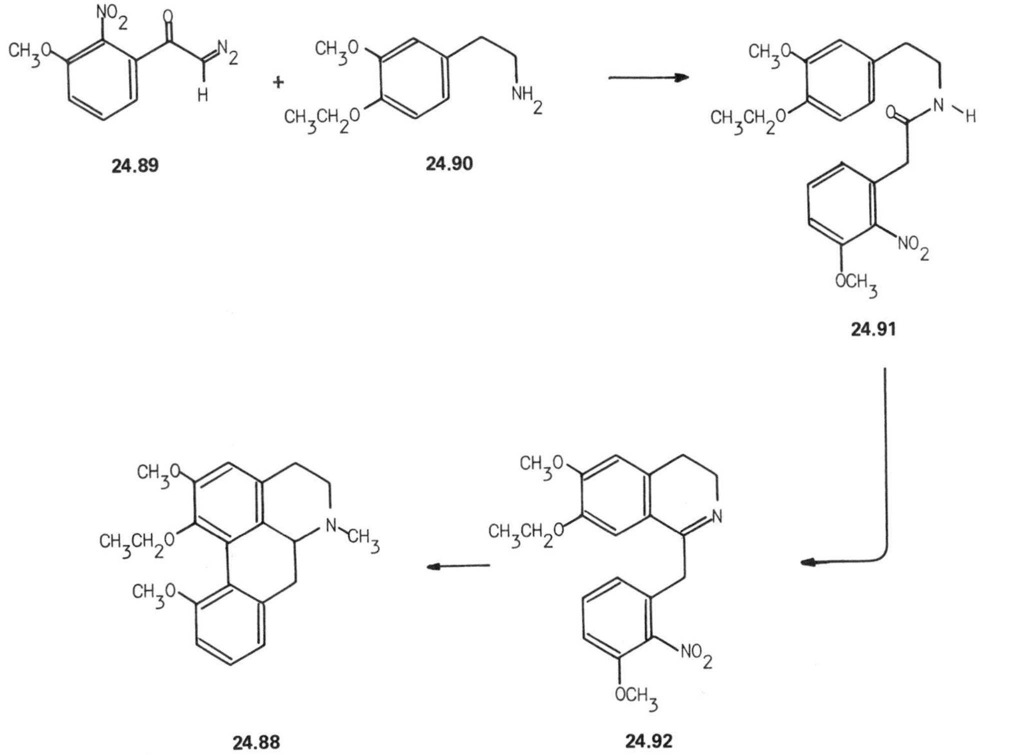

SCHEME 24.4 A synthesis of (±)-6, 8-dimethoxy-7-ethoxyaporphine. (After Reference 163.)

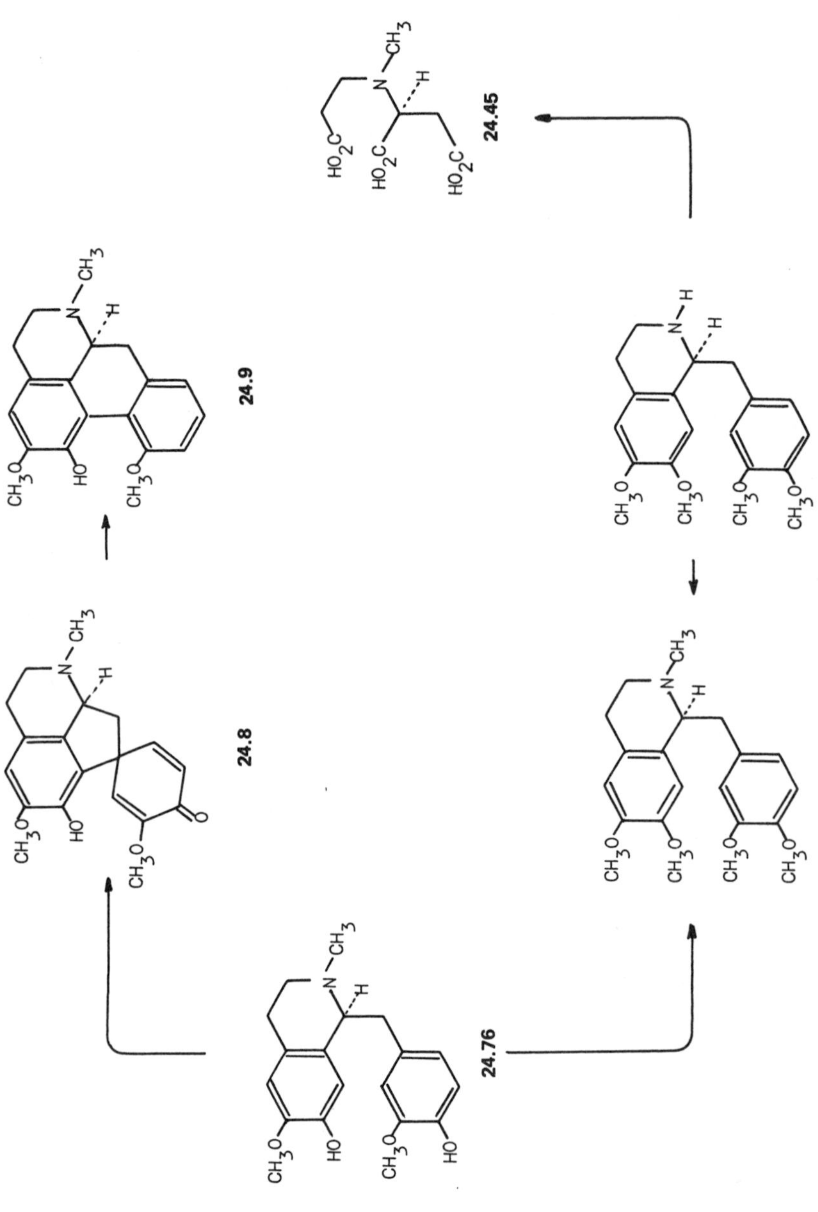

SCHEME 24.5 Interrelationships which allow deduction of the absolute configuration of (+)-isothebaine.

not affect the chiral center. Since the latter (i.e., 24.76) can be compared by methylation to (-)-laudanosine (24.4) which, it will be remembered, was of established absolute configuration, i.e., (-)-N-norlaudanosine (24.30) yields N-β-carboxyethyl-L-aspartic acid (24.45; Equation 24.11) of known absolute configuration.

An interesting oxidation of isothebaine (24.9) has been reported [164]. When isothebaine (24.9) is treated with iodine, a green salt, identical to an isolable alkaloid called PO-3 (24.92) is obtained. The salt (24.92) can be hydrogenated to (±)-7-hydroxyisothebaine (24.93; Equation 24.22). In a similar vein, the yellow color of the heartwood of the tulip tree, Liriodendron tulipifera L., is due, in part, to the alkaloid liriodenine (24.94) which is presumed to be generated by aerial oxidation of anonaine (24.95; Equation 24.23).

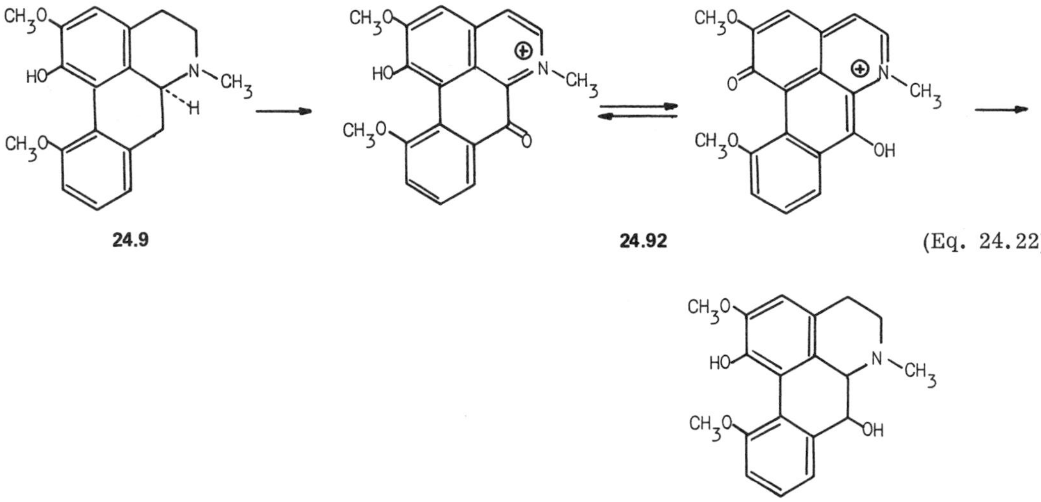

24.9 → 24.92 → 24.93 (Eq. 24.22)

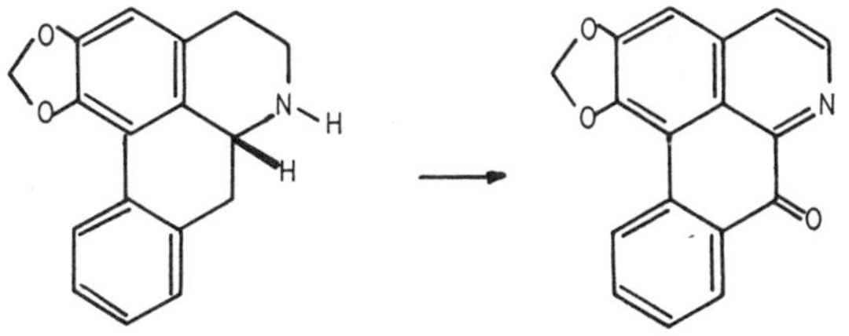

24.95 → 24.94 (Eq. 24.23)

VI. MORPHINE, CODEINE, AND THEBAINE

The viscous exudate of the unripe seeds of the opium poppy, Papaver somniferum, is opium. A tincture made from opium has been used for thousands of years medicinally, the use varying with the real and supposed needs of the culture, and the pharmacologically active constituents, often in purer form, or their laboratory-modified derivatives, find use today as analgesics. The degradative reactions involved in the structure proof of morphine (24.1), codeine (24.10), and thebaine (24.11), which were recognized as being closely related, were often misleading since numerous rearrangements were encountered; nevertheless, the

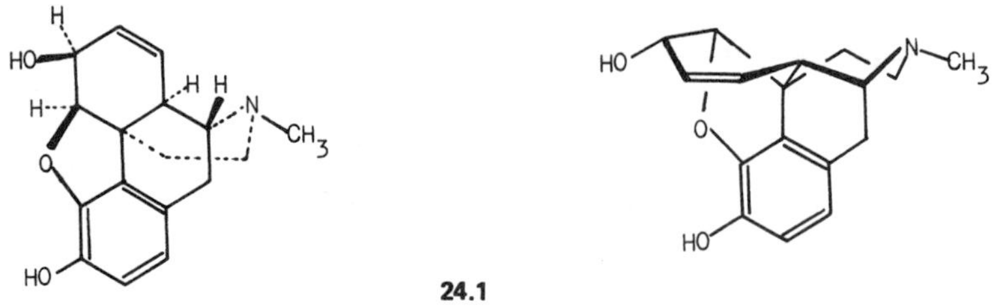

24.1

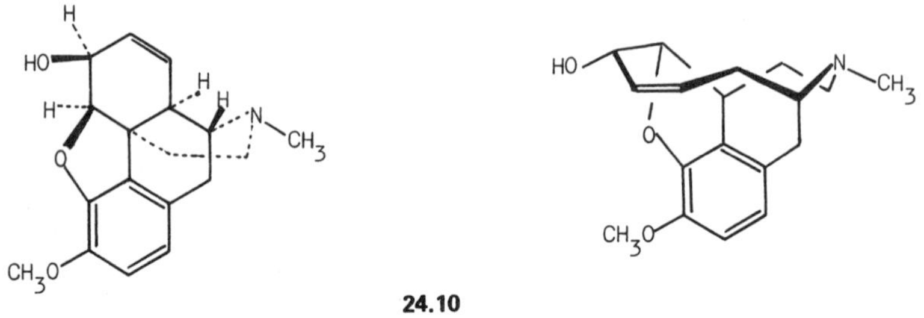

24.10

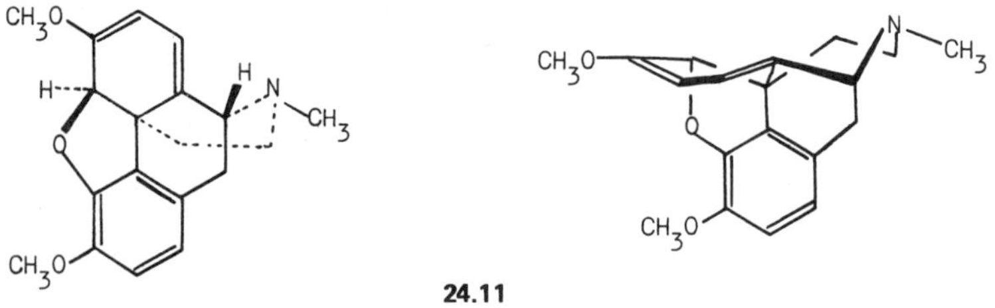

24.11

problem was solved and a great deal of chemistry done. However, for this reason, only a small fraction of the chemistry of this group of alkaloids can possibly be considered here and only some of the key reactions and rearrangements discussed.

The major constituent of opium is morphine (24.1) which was first isolated in 1805 and identified as $C_{17}H_{19}NO_3$. Codeine (24.10) was obtained by O-methylation of the phenolic hydroxyl in morphine (24.1), while thebaine (24.11) was recognized as an enol ether of, and capable of hydrolysis to, codeinone (24.96), the oxidation product of codeine (24.10; Equation 24.24).

The almost correct structures for these compounds were proposed in 1923 on the basis of the following information [165]:

1. Thebaine (24.11), as well as its methiodide (24.97), is converted to O-acetylthebaol (24.98) by refluxing acetic anhydride, while thebaol (24.99) itself is converted to phenanthrene (24.100) on zinc dust distillation (Scheme 24.6) and is capable of oxidation to a methoxyphthalic acid and a product with an odor resembling that of vanillin. This information suggested there was a methoxy group present in each of the terminal rings of the phenanthrene (24.100) system, which was the fundamental nucleus of the alkaloids. To corroborate this deduction, 4-methoxyphenylacetic acid sodium salt (24.101) was allowed to react with 3-acetoxy-4-methoxy-2-nitrobenzaldehyde (24.102) and the resulting cinnamic acid (24.103) reduced to the corresponding amine, diazotized, and cyclized to obtain the carboxylic acid 24.104. Pyrolysis of 24.104 caused loss of carbon dioxide, and the resulting product was identical to acetylthebaol (24.98; Scheme 24.6).

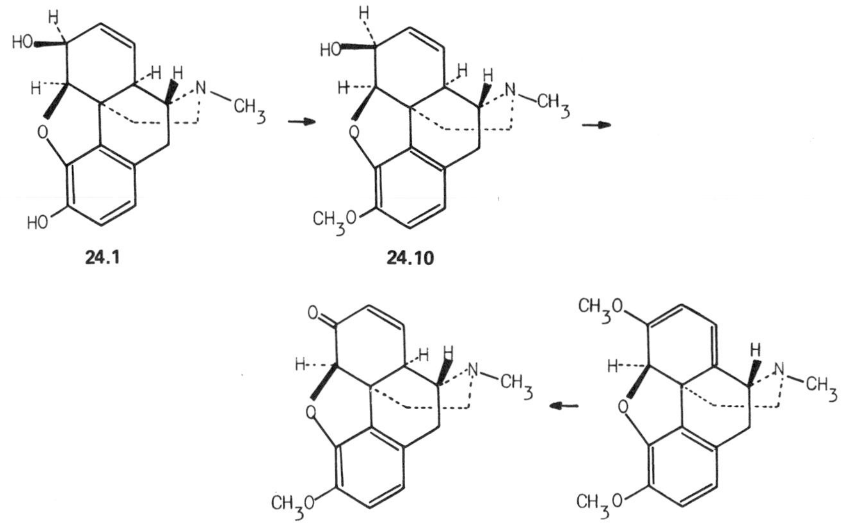

24.1 → 24.10 → 24.11 → 24.96

(Eq. 24.24)

2. Acid hydrolysis of thebaine (24.11) yields codeinone (24.96), which had previously been prepared by the oxidation of codeine (24.10) with chromic acid or potassium permanganate in acetic acid. Methylation of morphine (24.1) with, e.g., trimethylphenylammonium hydroxide or diazomethane, yields codeine (24.10).

3. Codeine (24.10) can be converted into its methiodide and thence to a pair of methines, i.e., α-codeimethine (24.105) and β-codeimethine (24.106; see Scheme 24.7). Either or both of these methines can be further methylated (i.e., 24.107) and degraded by heating with alkoxide (e.g., sodium amylate) to methylmorphenol (24.108; Scheme 24.7). Now, methylmorphenol (24.108) can be converted with hydrogen iodide to morphenol (24.109), and fusion of the latter with potassium hydroxide gives 3,4,5-trihydroxyphenanthrene (24.110; Equation 24.25), the trimethyl ether of which

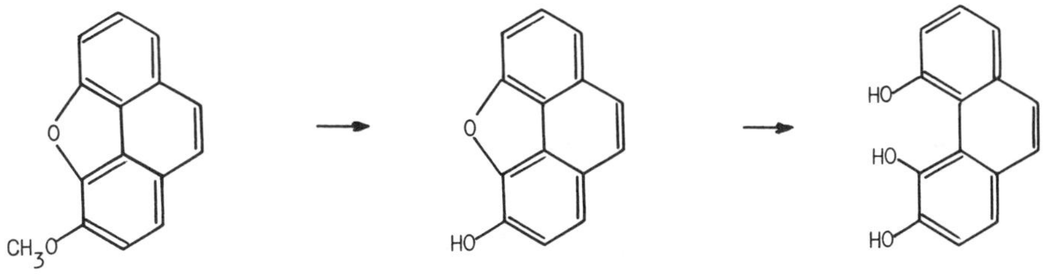

24.108 → 24.109 → 24.110

(Eq. 24.25)

was synthesized. The presumption that the second oxygen of morphenol (24.109; the first being that of a phenol) is present as ether bridge appears to have crept into the thinking of the various workers in the field at this time. The presumption has, of course, ultimately been vindicated by x-ray crystal analysis of codeine (24.10; [166]) and degradation of thebaine (24.11) to the dicarboxylic acid 24.111 of known absolute stereochemistry ([167]; Scheme 24.8). Thus, as is shown in Scheme 24.8, when codeine (24.10) is reduced to the corresponding dihydro derivative and the latter oxidized, dihydrocodeinone (24.112) is obtained. Two Hofmann degradations with a

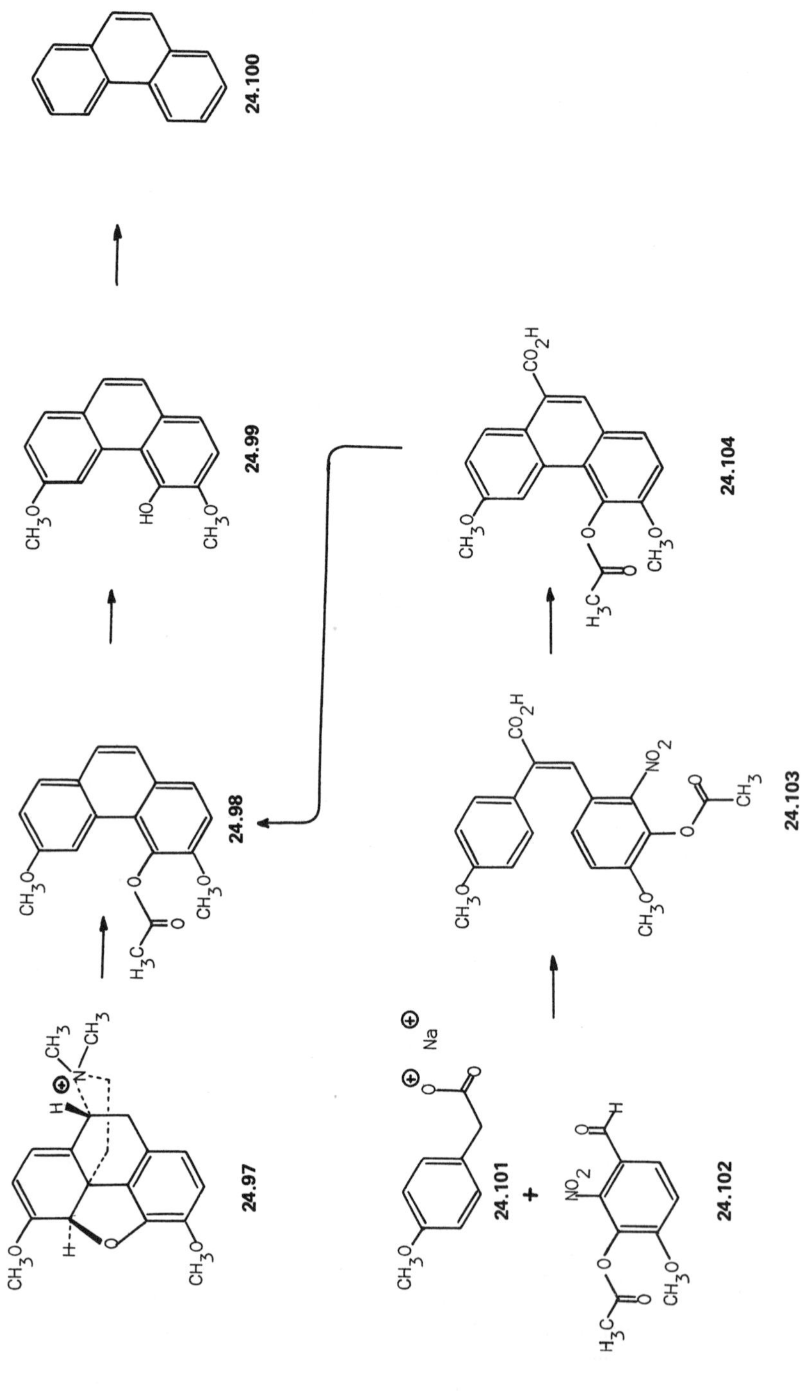

SCHEME 24.6 The degradation of thebaine to thebaol and the synthesis of thebaol [165].

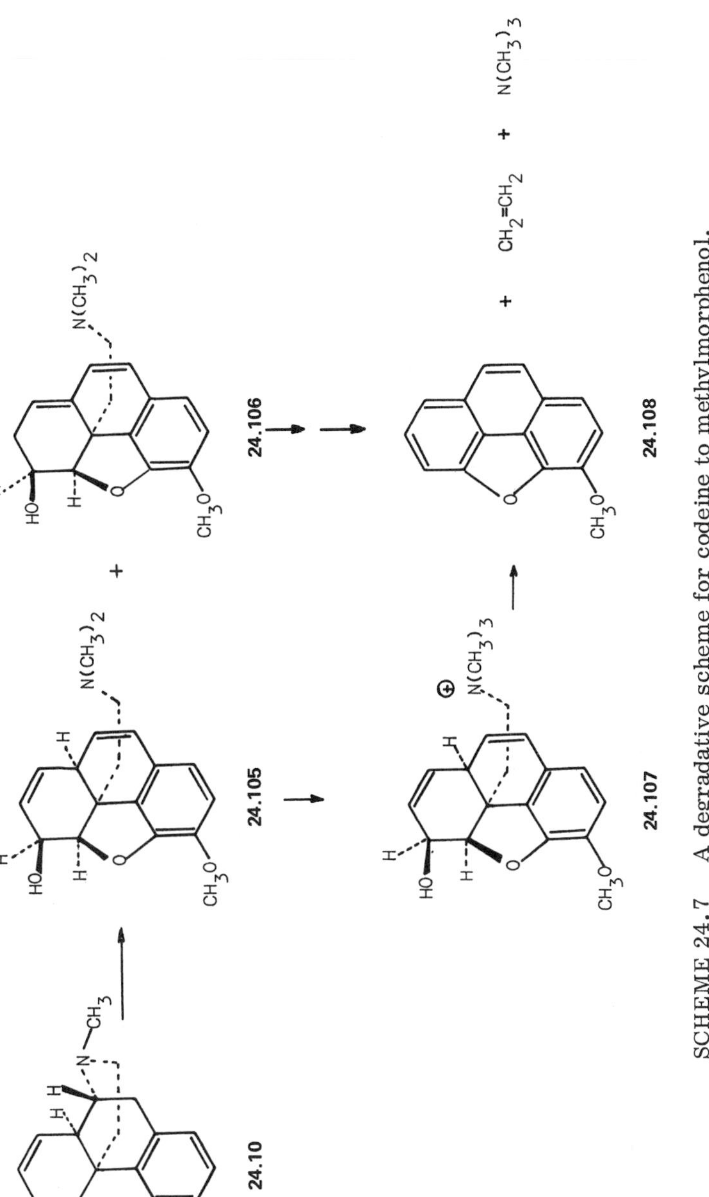

SCHEME 24.7 A degradative scheme for codeine to methylmorphenol.

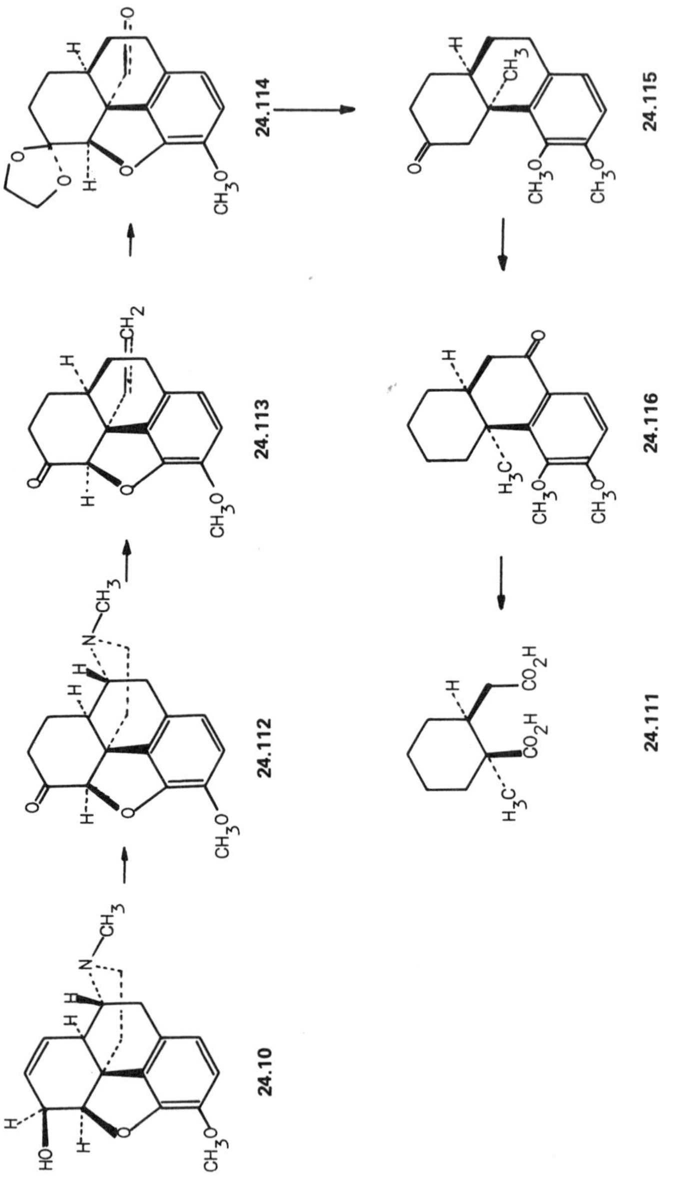

24.10

24.112

24.113

24.114

24.115

24.116

24.111

SCHEME 24.8 A degradation of codeine which established the absolute stereochemistry. (After Reference 167.)

reduction after the first gives the vinyl derivative 24.113 which is converted to the corresponding ethylene ketal and oxidized with osmium tetroxide and lead tetraacetate to the aldehyde 24.114. Reduction with hydrazine (Huang-Minlon modification of the Wolff-Kishner reduction), removal of the ketal blocking group, and reductive hydrogenolysis of the oxide bridge, followed by methylation of the phenol so generated, yield the ketodimethyl ether 24.115. The keto group is removed via reduction of the corresponding thioketal and the benzylic carbon oxidized with chromic anhydride in acetic acid to yield the ketone 24.116. Oxidative ozonolysis then generates 24.111.

4. When morphine (24.1) is heated with concentrated hydrochloric acid at 130-140°C in a sealed tube for 3 hr and the hydrochloride salt which results neutralized with sodium bicarbonate, apomorphine (24.117) is generated. The structure of the dimethyl ether (24.118) of apomorphine (24.117) was first established by oxidative degradation to the dimethoxyphenanthrene carboxylic acid 24.119 (Equation 24.26)

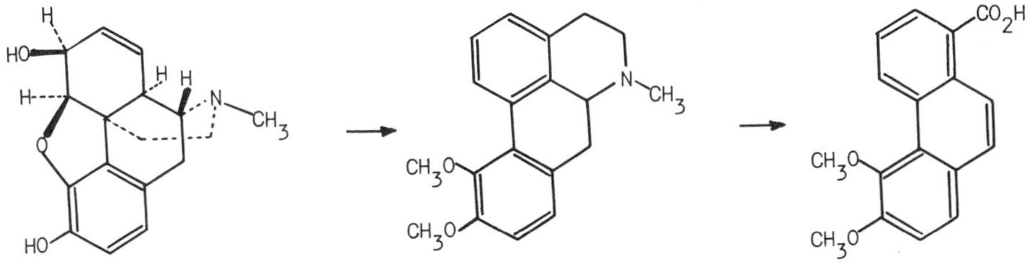

24.1 → 24.118 → 24.119

(Eq. 24.26)

and later by a synthesis using the reactions already outlined for isothebaine (24.9; see above, Scheme 24.4). Two different routes can be written for the conversion of morphine (24.1) into apomorphine (24.117). They essentially differ only in the distinction between migrating one or another group while the ether ring opens. These pathways, for which little real evidence is available, are shown in Scheme 24.9 and and Scheme 24.10. The similar transformation of thebaine (24.11) into morphothebaine (24.120), which is induced by warming the former in 40% hydrochloric acid on the steam bath for about 3 hr, can also be formulated in two different ways, analogous to those already outlined. These are shown as Schemes 24.11 and 24.12.

5. When thebaine (24.11) or codeinone (24.96) is warmed for only 2 min with dilute hydrochloric acid, the hydrochloride of an amorphous base, thebenine (24.121), is isolated. Contrary to what has been shown above, this transformation is accommodated by only one migratory pathway (Scheme 24.13). However, thebenine (24.121), as its hydrochloride, is obtained in only about 50% yield (!) from the rearrangement reaction. It is amusing to speculate that the alternative dienone-phenol rearrangement and resulting cleavage reaction (Scheme 24.14) is also occurring, but since cyclization to the phenanthrene cannot occur, the syrupy mass remaining after thebenine (24.121) hydrochloride is removed results from further transformation products of the intermediate aldehyde 24.122.

6. The position of the attachment of the nitrogen in morphine (24.1), codeine (24.10), and thebaine (24.11) was assigned on the basis of oxidative studies, the rearrangements shown above to aporphine bases, and on biogenetic grounds [165]. The position of the attachment of the other end of the bridge (one end being to nitrogen) must have been one which assured extrusion of the bridge during the aromatization reactions.

SCHEME 24.9 A possible pathway from morphine to apomorphine.

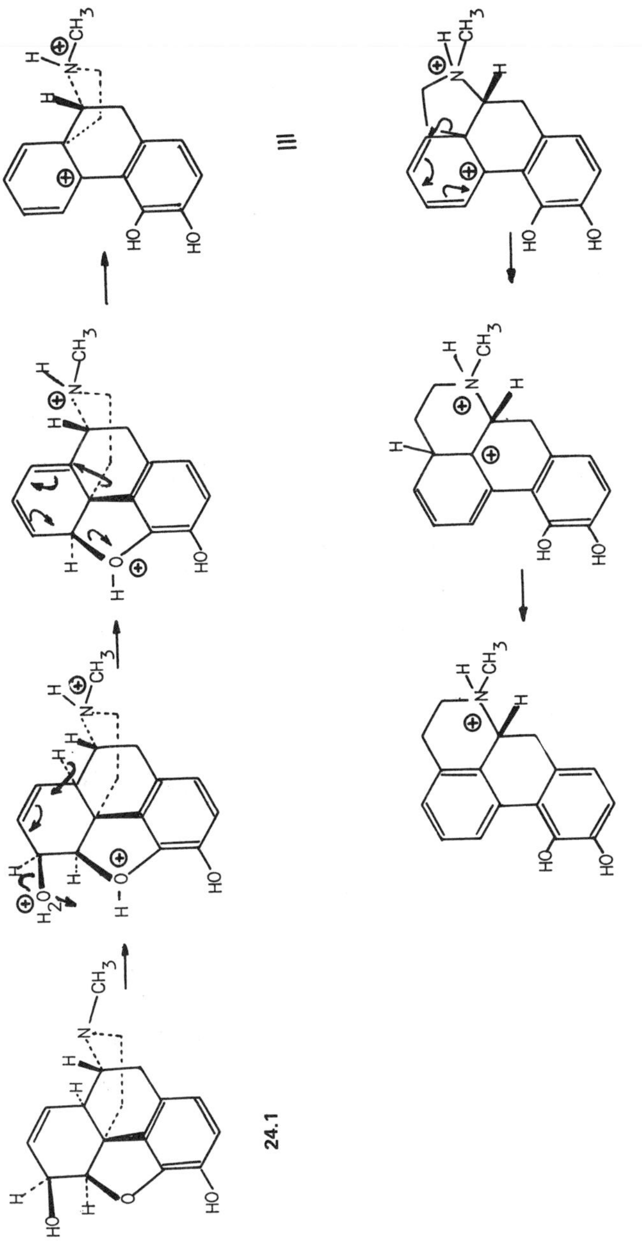

24.1

SCHEME 24.10 A possible pathway from morphine to apomorphine.

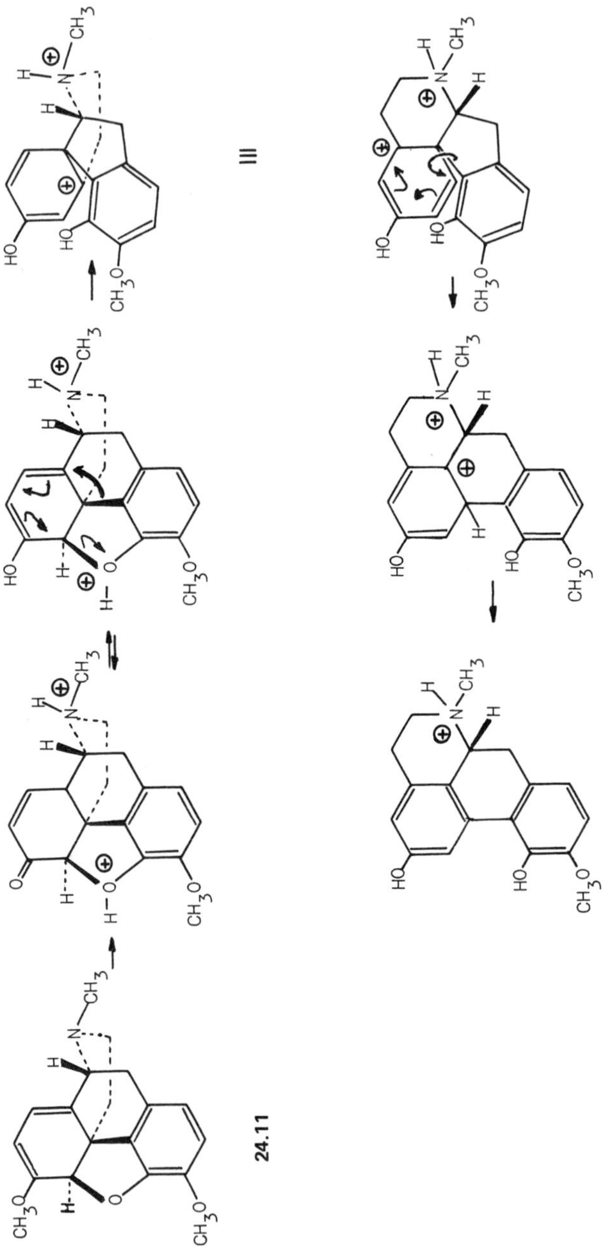

24.11

24.120

SCHEME 24.11 A possible pathway from thebaine to morphothebaine.

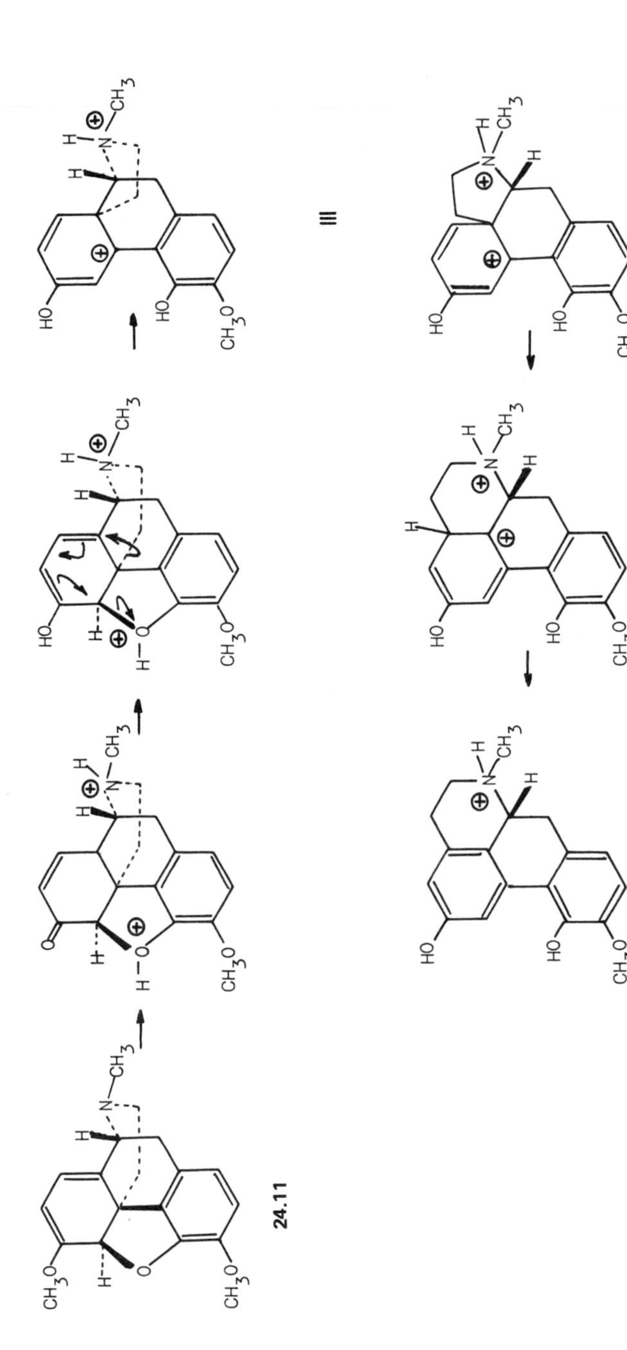

SCHEME 24.12 A possible pathway from thebaine to morphothebaine.

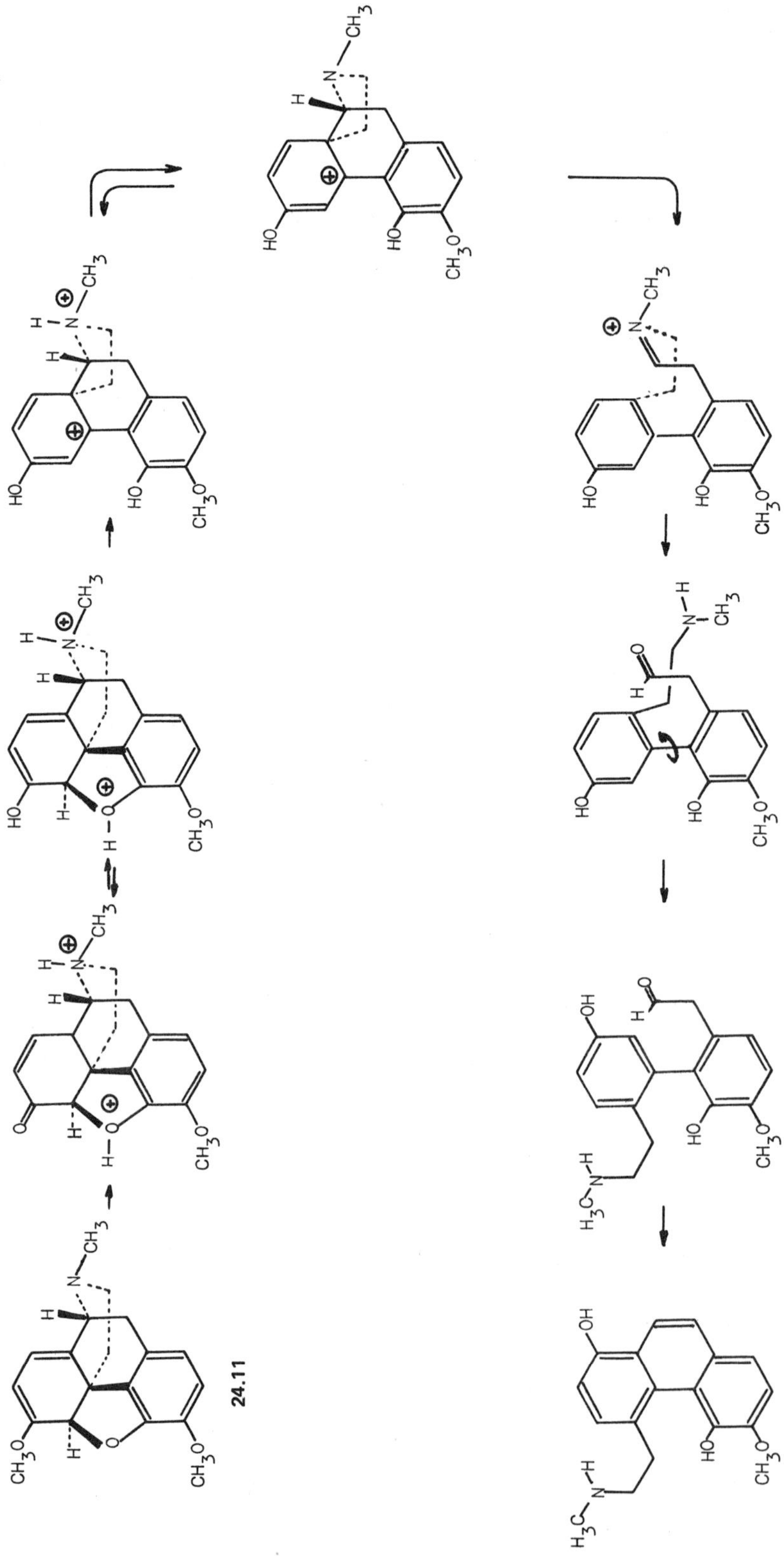

SCHEME 24.13 A possible pathway from thebaine to thebenine.

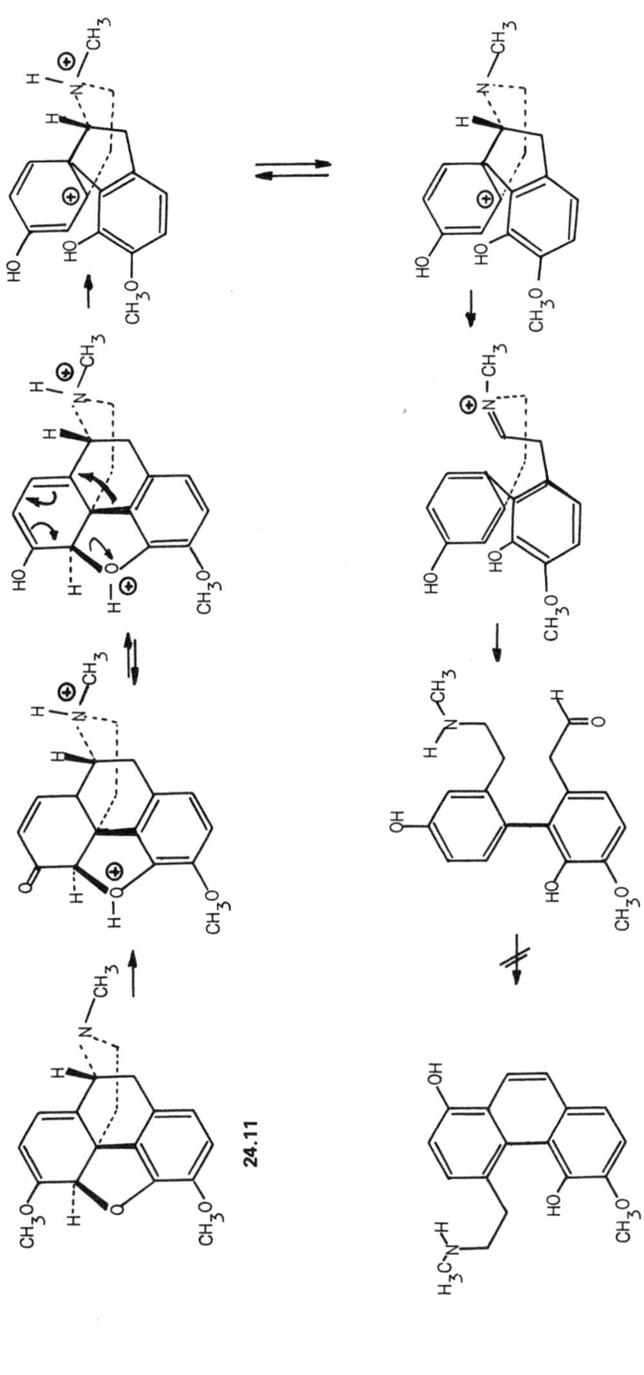

SCHEME 24.14 A possible cleavage reaction of thebaine which does not lead to thebenine.

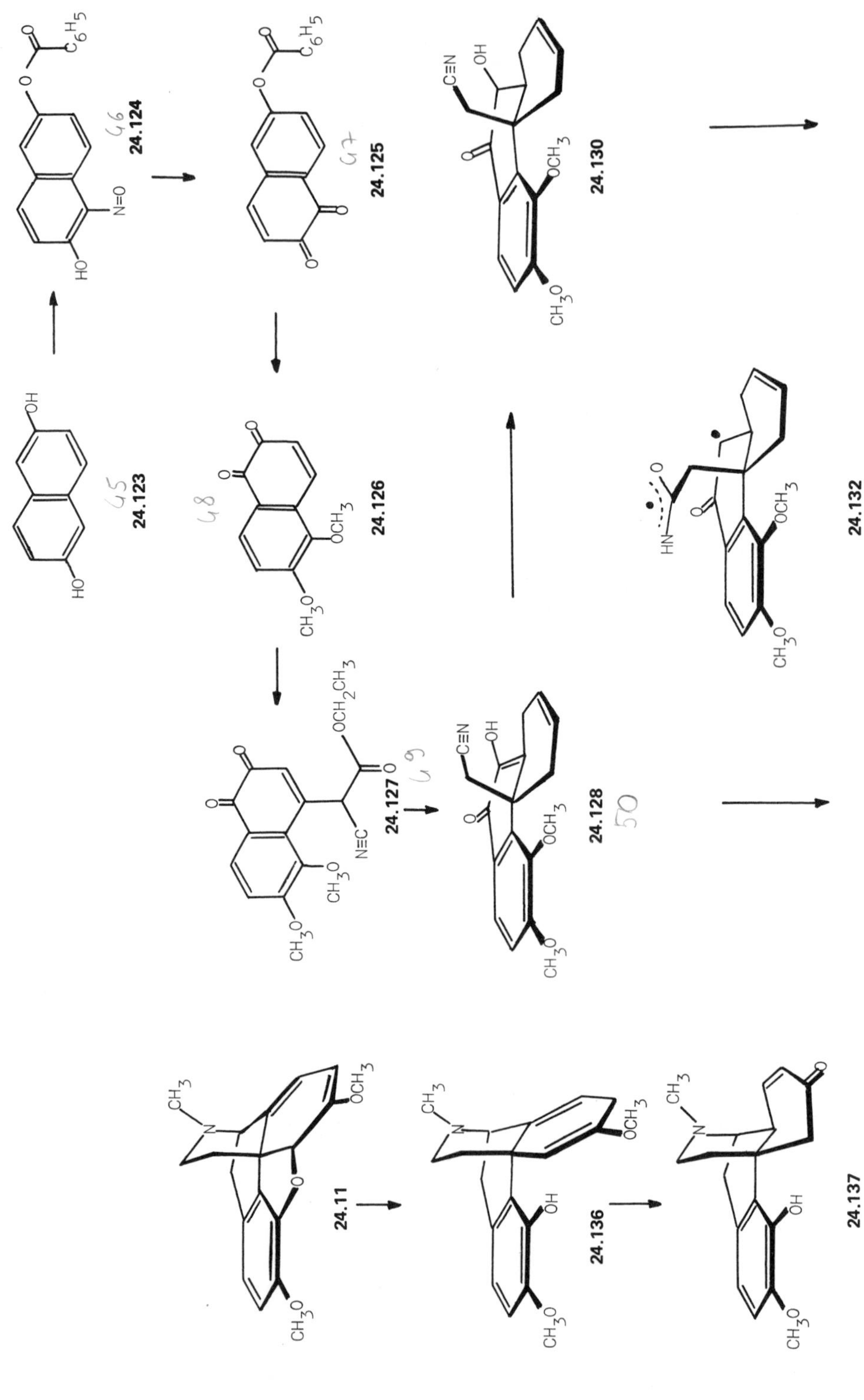

24.123
24.124
24.125
24.126
24.127
24.128
24.130
24.132
24.11
24.136
24.137
HO
OH
N=O
O
C6H5
OCH3
CH3O
OCH2CH3
N≡C
C≡N
HN
N
CH3

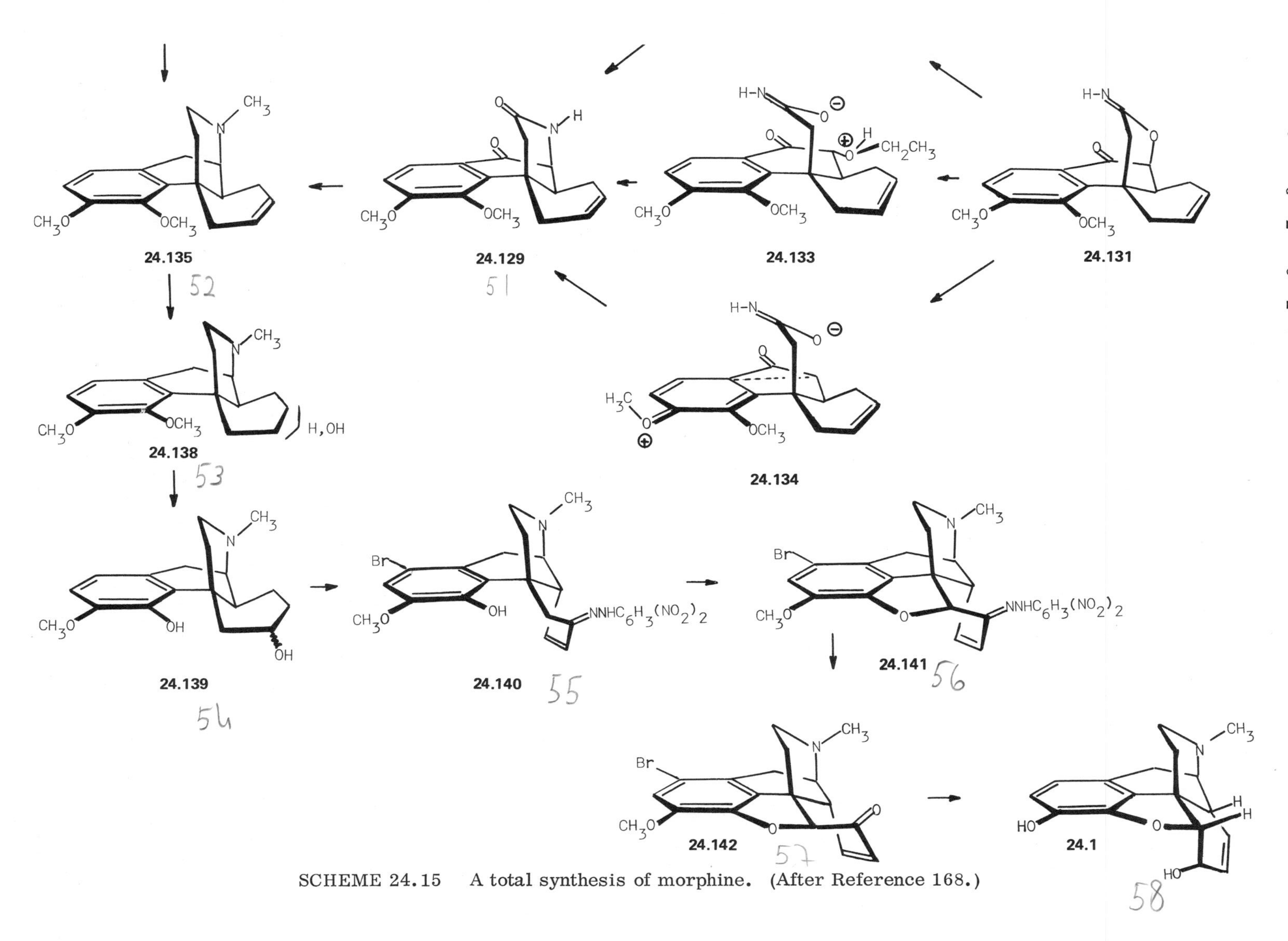

SCHEME 24.15 A total synthesis of morphine. (After Reference 168.)

Thus, until the total synthesis of morphine (24.1; see below) and codeine (24.10) and the later x-ray and comprehensive degradative studies to determine the absolute configuration (see above), the structures which are shown were only "best guesses." The total synthesis of morphine (24.1) has been accomplished [168,169], and the first synthesis completed is set forth below (Scheme 24.15).

2,6-Dihydroxynaphthalene (24.123) is converted into its monobenzoate which is nitrosated to 24.124. Reduction to the corresponding α-aminophenol and oxidation with ferric chloride yields the naphthaquinone 24.125. After reduction of the quinone (24.125) to the corresponding catechol with sulfur dioxide and methylation, the benzoate blocking group is removed and the same sequence of reactions indicated above repeated on the other end of the molecule to yield (25% overall) the dimethoxynaphthaquinone 24.126. A Michael condensation of this quinone with ethyl cyanoacetate and ferricyanide oxidation to regenerate the quinone yield 24.127, which then undergoes hydrolysis and decarboxylation to the corresponding nitrile which possesses the "ene" portion of a system capable of reaction with butadiene (Diels-Alder) to generate the hydrophenanthrene 24.128. Reduction of 24.128 in ethanol with hydrogen over copper chromite catalyst leads to a mixture of products which includes the ketolactam 24.129.

Several possible pathways, in the absence of evidence, can be written for the process observed. All, however, presume reduction to the dihydro intermediate 24.130 which then adds to the nitrile to generate the imino ether 24.131. Then, either via the radical pair 24.132, solvent participation and redisplacement (i.e., 24.133), or homoallylic participation (e.g., 24.134) and readdition, 24.129 is generated. Reduction of the keto group of 24.129 by the Wolff-Kishner procedure, N-methylation with methyl iodide (sodium hydride), and finally, lithium aluminum hydride reduction, lead to the amine 24.135 which is capable of resolution with dibenzoyltartaric acid to provide the (+)-enantiomer. This amine (24.135) is identical to a degradation product obtained from thebaine (24.11), i.e., by reduction of thebaine (24.11) with sodium in ethanol to 24.136 and hydrolysis with aqueous acid to (among others) 24.137 which can then be reduced to an alcohol that undergoes phenolic O-methylation, tosylation, and elimination to 24.135. Chiral 24.135 is then hydrated (aqueous sulfuric acid) to a mixture of alcohols (24.138), and after separation, partial demethylation to 24.139 is accomplished by heating 24.138 with potassium hydroxide in diethylene glycol at 225°C. Oxidation using a modified Oppenauer method and bromination yield a dibromide which forms a 2,4-dinitrophenylhydrazone identical to that obtained from 1-bromothebaine (24.140). This means that after elimination of hydrogen bromide, the α,β-unsaturated ketone is capable of epimerization to the required enantiomer. Exchange of the dinitrophenylhydrazine with acetone (hydrochloric acid catalysis), reduction of the double bond, and rebromination yield a dibromide which in turn generates a 2,4-dinitrophenylhydrazone derivative of 1-bromocodeinone (24.141). Again, acid-catalyzed exchange of the dinitrophenylhydrazine with acetone generates the free ketone (24.142) which undergoes reduction with lithium aluminum hydride and a final demethylation with pyridine-hydrogen chloride complex at 220°C to morphine (24.1).

Innumerable derivatives of morphine (24.1), codeine (24.10), and thebaine (24.11) have been prepared and rearrangements within them and the parent compounds studied. Some of the fundamental rearrangements of the alkaloids themselves have been discussed above. Among the more interesting that remain are the plethora of reactions which occur on treatment of codeine (24.10) and its epimers with thionyl chloride. This system demonstrates a remarkable mixture of S_N2, S_N2', and S_Ni reactions typical of α,β-unsaturated alcohols can occur, and subtle differences in solvent, stereochemical requirements, etc., influence exactly which reaction occurs and which chlorocodide is produced. Scheme 24.16 sets forth the major products obtained from codeine (24.10), allopseudocodeine (24.143), isocodeine (24.144), and ψ-codeine (24.145) with thionyl chloride, i.e., α-chlorocodide (24.146) and β-chlorocodide (24.147).

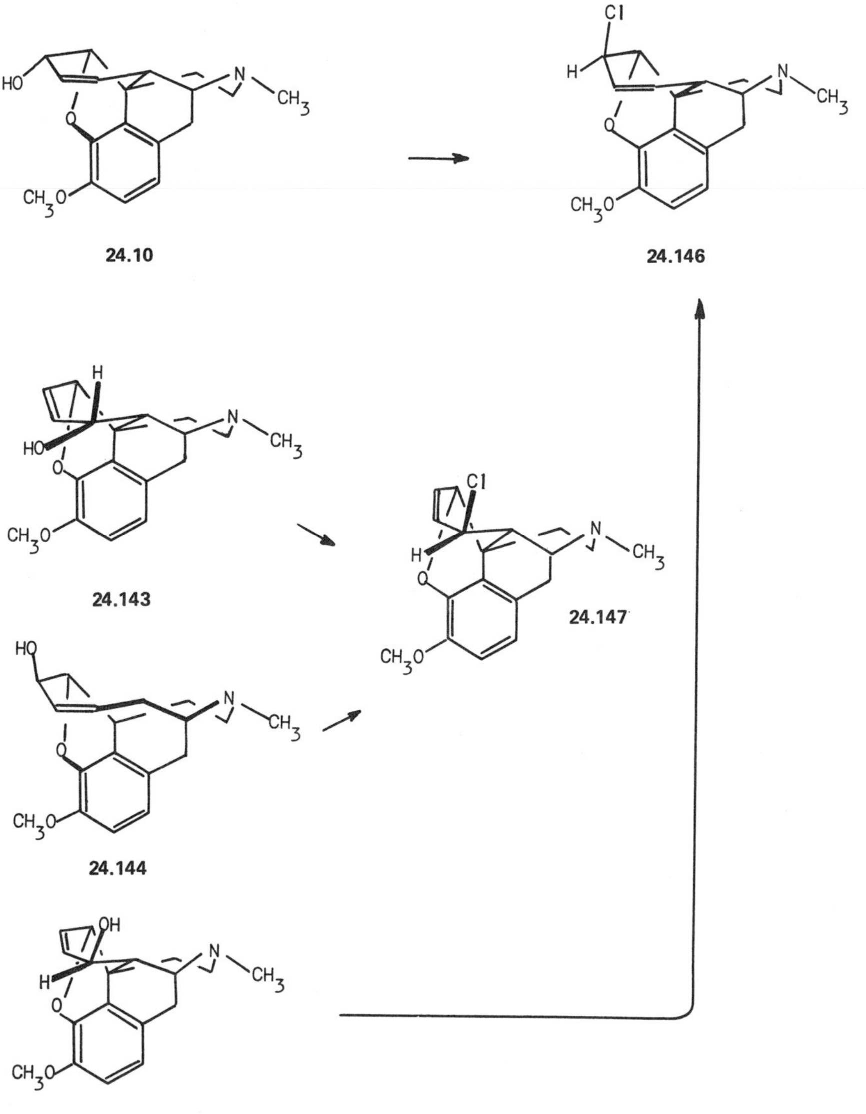

24.10

24.146

24.143

24.147

24.144

24.145

SCHEME 24.16 The chlorocodides derived from codeine isomers on treatment with thionyl chloride.

Finally, when thebaine (24.11) is treated with phenylmagnesium bromide, phenyldihydrothebaine (24.148) is obtained as the major product. To account for this transformation, it is assumed that the bridging-ether oxygen is complexed by the Grignard reagent and that this serves as the driving force for ring opening and migration of the appropriate bonds (Scheme 24.17). Since the structure of the product (i.e., 24.148) has been proved by oxidation (to the dicarboxylic acid 24.149) of the alkene which results from successive exhaustive methylation and Hofmann degradation reactions (after formation of the methyl ether), i.e., 3,4-dimethoxy-2(5-methoxy-2-vinylphenyl)stilbene (24.150), there is little question as to which

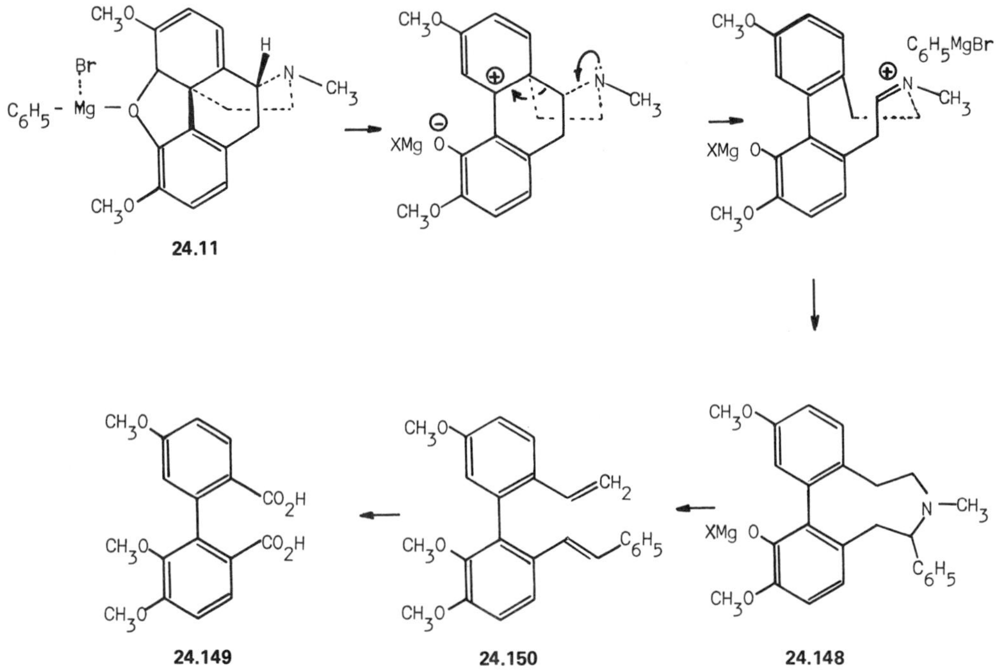

CH_3O Br C_6H_5 - Mg — O H N CH_3 CH_3O **24.11** → CH_3O ⊕ N CH_3 ⊖ XMg O CH_3O → CH_3O C_6H_5MgBr ⊕ N CH_3 XMg O CH_3O

↓

CH_3O CO_2H CH_3O CO_2H CH_3O **24.149** ← CH_3O CH_2 CH_3O C_6H_5 CH_3O **24.150** ← CH_3O N —CH_3 XMg O C_6H_5 CH_3O **24.148**

SCHEME 24.17 A possible pathway for the formation of phenyldihydrothebaine from thebaine and phenylmagnesium bromide.

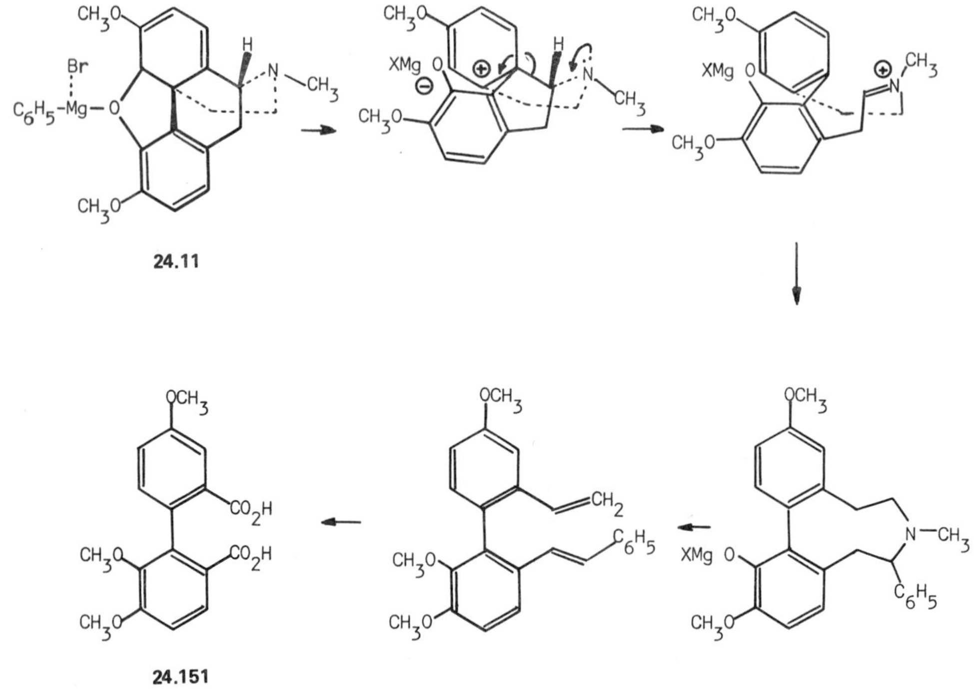

CH_3O Br C_6H_5-Mg— O H N CH_3 CH_3O **24.11** → CH_3O XMg O ⊖ ⊕ H N CH_3 CH_3O → CH_3O XMg O ⊕ CH_3 N CH_3O

↓

OCH_3 CO_2H CH_3O CO_2H CH_3O **24.151** ← OCH_3 CH_2 C_6H_5 CH_3O CH_3O ← OCH_3 N—CH_3 XMg O C_6H_5 CH_3O

SCHEME 24.18 An alternate pathway for bond migration during the formation of phenyldihydrothebaine from thebaine. The pathway is apparently not followed to any appreciable extent.

bond migrates during the dienone-phenol rearrangement. The alternate pathway (Scheme 24.18) would lead to the acid 24.151, which is not obtained [170].

VII. ERYTHRALINE

The alkaloids found in all plant parts of Erythrina species have been intensively studied because they produce smooth muscle paralysis much like tubocurare (see Section III). Although there are two broad categories, i.e., the aromatic alkaloids, of which erythraline (24.12) is an example, and second, the lactone alkaloids such as α-erythroidine (24.152), our discussion will be limited to the former.

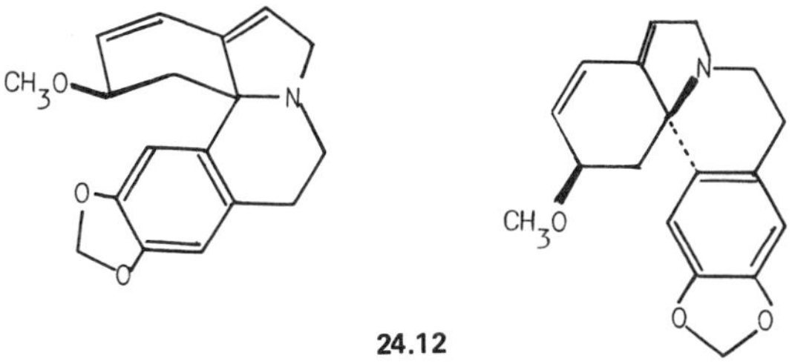

24.12

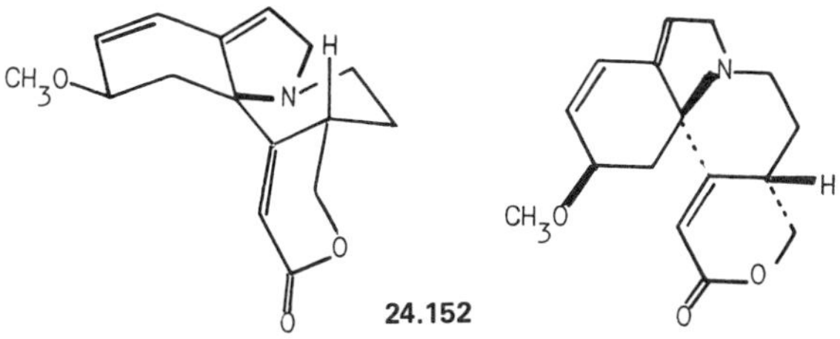

24.152

Erythraline (24.12), $C_{18}H_{19}NO_3$, $[\alpha]_D$ +212° (ethanol), mp 106-107°C, has been isolated from more than 10 Erythrina species and was recognized as a tertiary amine lacking an N-methyl group but possessing an aromatic methylenedioxy function and a methoxyl group.

Partial hydrogenation of erythraline (24.12) yields dihydroerythraline (erythramine; 24.153) and then tetrahydroerythraline (24.154; Equation 24.27). On oxidation, erythraline (24.12) yields hydrastic acid (24.155; Equation 24.28), and on fusion with potassium hydroxide, indole (24.156) forms (Equation 24.29). When erythraline (24.12) is heated with mineral acid, apoerysopine (24.157) results (Scheme 24.19). On mild acid treatment, however,

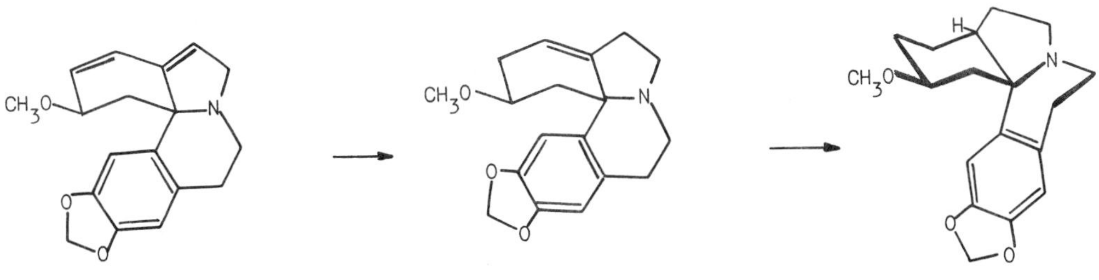

24.12 → 24.153 → 24.154

(Eq. 24.27)

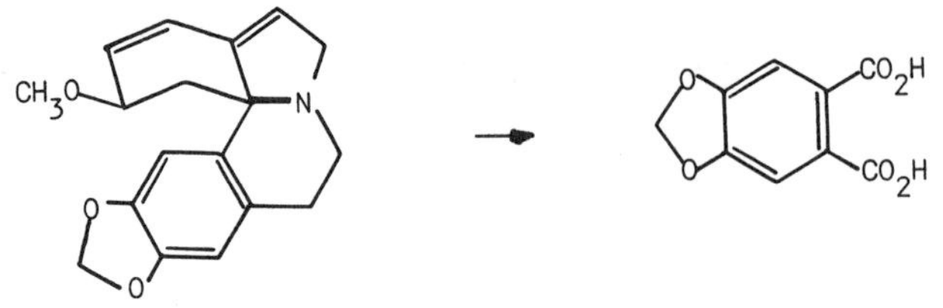

24.12 → 24.155 (Eq. 24.28)

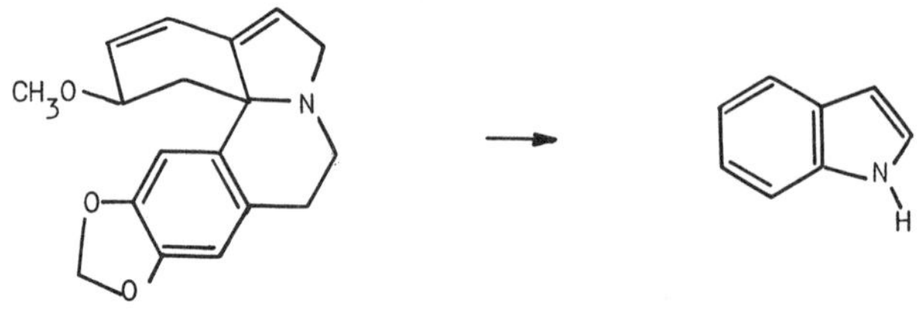

24.12 → 24.156 (Eq. 24.29)

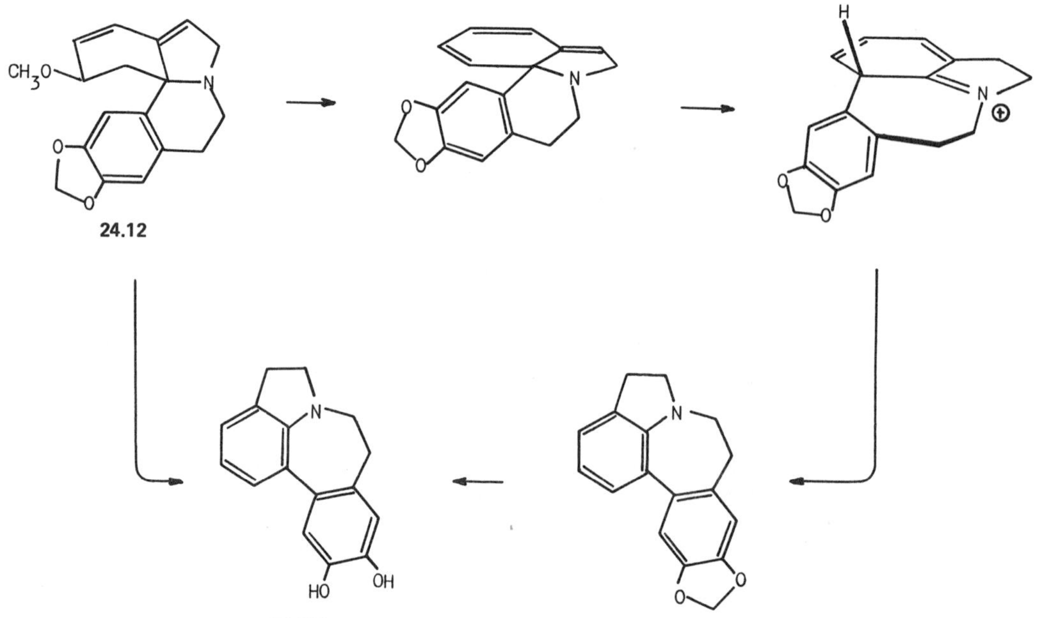

SCHEME 24.19 The rearrangement of erythraline to apoerysopine and apoerythraline. (After Reference 171.)

apoerythraline (24.158) is formed, and further acid treatment of apoerythraline (24.158) then generates apoerysopine (24.157). This rearrangement was recognized [171] as the key to the structure of the erythrina alkaloids, and the spiroamine (erythrinane) skeleton was postulated on the basis of the rearrangement after the degradation and synthesis of apoerysopine (24.157) had established its structure. Thus, as shown in Scheme 24.20, Hofmann

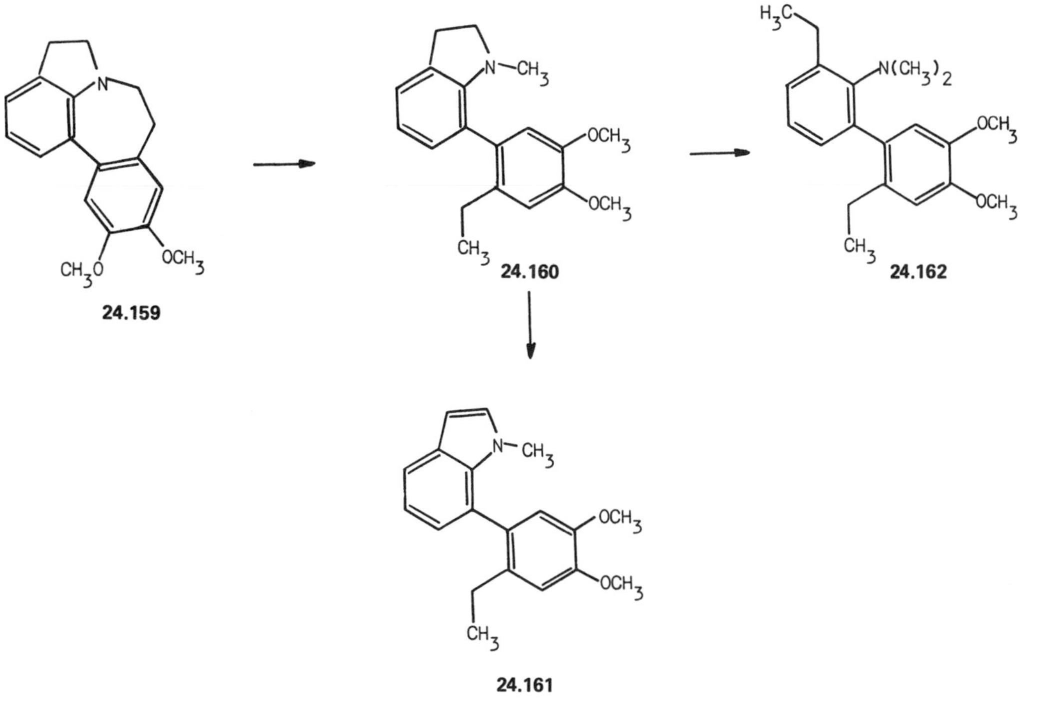

24.159 24.160 24.162 24.161

SCHEME 24.20 The degradation of dimethylapoerysopine. (After Reference 171.)

degradation of dimethylapoerysopine (24.159) with hydrogenation of the methine base leads to 24.160 which can be dehydrogenated to the indole 24.161 or subjected to further Hofmann degradation and hydrogenation to the dimethylaniline 24.162.

Dimethylapoerysopine (24.159) was then synthesized as shown in Scheme 24.21 [172]. Here, condensation of dimethoxyphenethylamine (24.36) with the hydroxycyclohexanone acetic acid (24.163) generates the tetrahydroindolanone 24.164. Bromination of the latter with N-bromosuccinimide followed by treatment with phosphorus pentoxide and formic acid results in the unprecedented cyclization (with double-bond migration and dehydration) to the bromolactam 24.165 which undergoes dehydrohalogenation to 24.166. Aromatization with palladium on carbon at 250°C followed by reduction of the lactam with lithium amalgam generates dimethylapoerysopine (24.159).

Although the stereochemistry of erythraline (24.12) was correctly inferred from extensive degradative and synthetic studies of Boekelheide [173] and Hill [174] on related compounds, the x-ray crystal structure of erythraline (24.12) as its hydrobromide [175] allows the present structure to be drawn with certainty.

Erythraline (24.12) has not yet been synthesized, but its close relative erysotrine (24.167), which is not naturally occurring but which is obtained by O-methylation of several of other erythrina alkaloids, has been prepared. The synthesis of erysotrine (24.167) is shown in Scheme 24.22 [176] and, with the possible exception of the acid-catalyzed ether cleavage shown, should also be applicable to the synthesis of erythraline (24.12).

As is shown in Scheme 24.22, the oxalyl derivative of 4-methoxycyclohexanone (24.168) is treated with 3,4-dimethoxyphenethylamine (24.36) to generate the lactam 24.169. Hydrogenation followed by treatment with concentrated sulfuric acid produces the bridged ether 24.170, in which the ring juncture demands the cis fusion shown. An elimination reaction with toluenesulfonic acid in the presence of acetic anhydride produces, among others, the

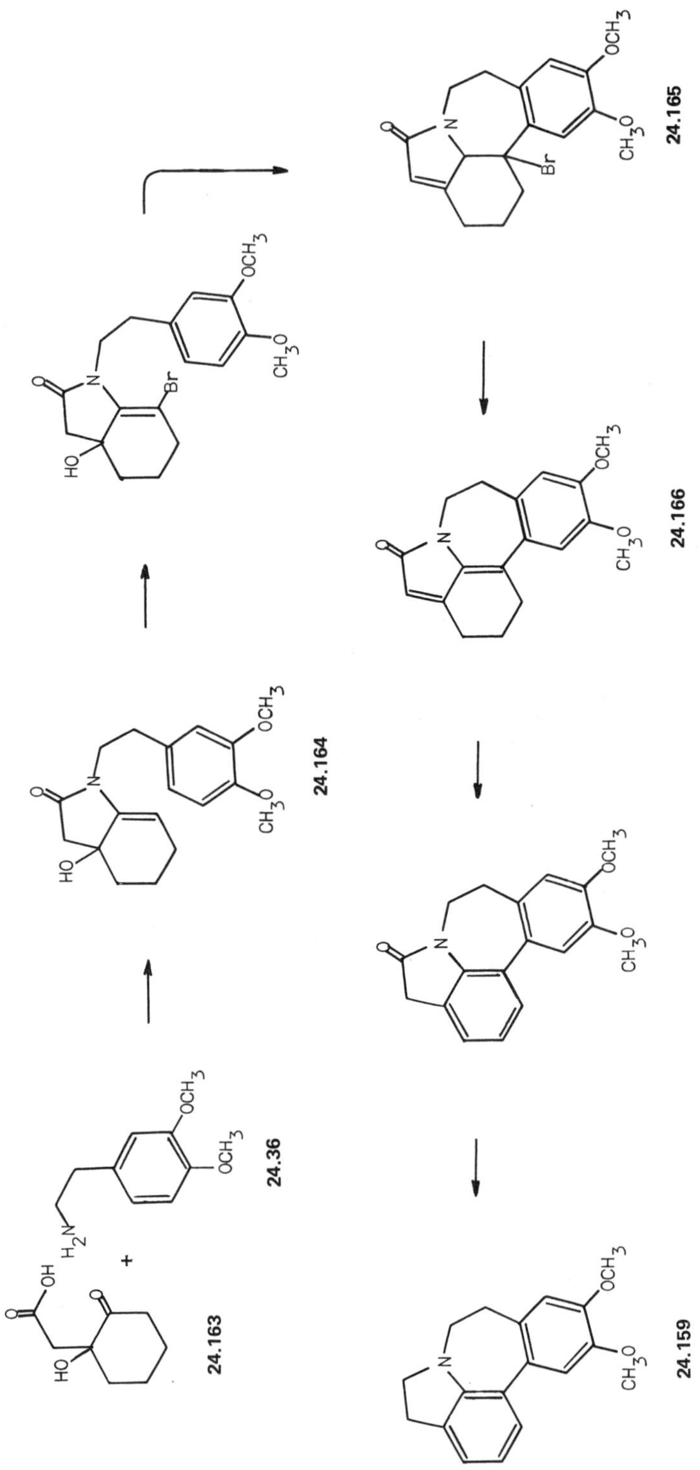

SCHEME 24.21 A synthesis of dimethylapoerysopine. (After Reference 172.)

SCHEME 24.22 A synthesis of erysotrine. (After Reference 176.)

alkene 24.171. Epoxidation of 24.171 with perbenzoic acid and ring opening with dimethylamine to the aminoalcohol 24.172, followed by a Cope elimination, yield the allylic alcohol 24.173. When 24.173 is treated with methanolic hydrogen chloride, a separable mixture of epimeric ethers 24.174 is obtained. The correct isomer is reduced with lithium aluminum hydride to 24.175, which forms a mesylate and undergoes elimination to (±)-erysotrine, resolution of which to (+)-erysotrine (24.167) is accomplished with dibenzoyl tartaric acid.

VIII. ARGEMONINE

(-)-Argemonine (24.13), from several Argemone species, $C_{21}H_{25}NO_4$, $[\alpha]_D^{24}$ -189° (chloroform) was originally considered to be an aporphine alkaloid (with an unusual methoxyl substitution pattern) since exhaustive methylation and Hofmann degradation led, ultimately, to a

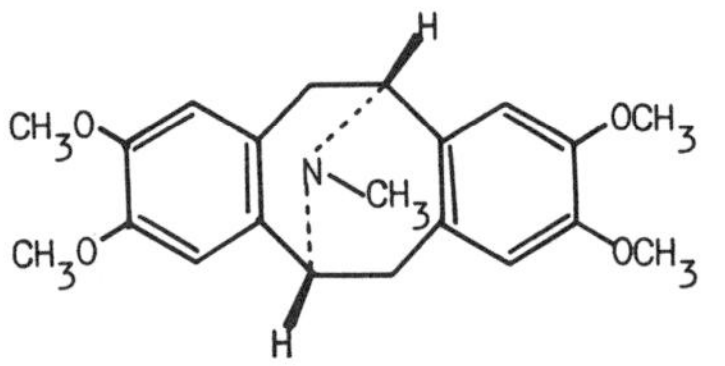

24.13

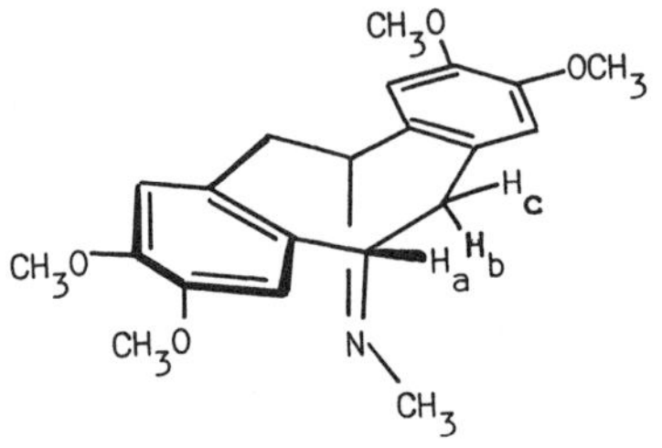

nitrogen-free product which is subsequently converted to what was presumed to be tetramethoxybiphenyltricarboxylic acid. However [177], the ultraviolet spectrum [λ_{max} 287 nm (log ϵ 4.01)] is not at all like that of aporphines [178], which have three typical absorbances in the 300-310, 268-282, and 220 nm regions. Second, in addition to a molecular ion at m/e 355, the mass spectrum of argemonine (24.13) possesses a large fragment (base peak) at m/e 204, which is typical of an N-methyldimethoxyisoquinolinium fragment (e.g., 24.176),

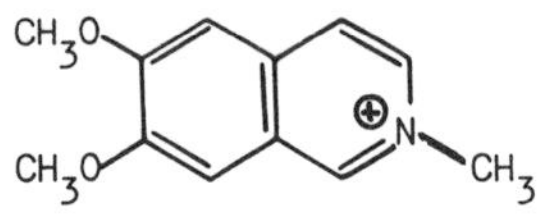

24.176

not normally obtained from an aporphine. Finally, the pmr spectrum of argemoine (24.13) possesses four methoxyl (two singlets at δ 3.84 and 3.92 ppm) and four aromatic proton signals (two singlets at δ 6.39 and 6.76; as well as the N-methyl) arranged in a symmetrical pattern not found in aporphine systems and an ABX system wherein the bridgehead protons H_a are coupled to one adjacent proton (H_b) as a doublet (centered at δ 2.62 ppm, J_{ab} = 6 Hz) and H_b is vicinally coupled to H_a and geminally coupled to H_c (two doublets at δ 3.42 and 3.53) while H_c is geminally coupled to H_b (center at δ 2.62, J_{bc} = 17 Hz).

The suggested structure, N-methylpavine (24.13), was confirmed by the methylation of pavine (24.31) whose structure had been established [179] by degradation. Thus (Scheme 24.23), pavine (24.31), which had been obtained from reduction of papaverine (24.5), was degraded to 2,2'-dicarboxy-4,5,4',5'-tetramethoxydibenzyl (24.177) by sequential Hofmann degradations and hydrogenations to a nitrogen-free compound and subsequent oxidation with potassium permanganate to give a product (24.177) identical to that obtained from oxidation of the corresponding diketone 24.178.

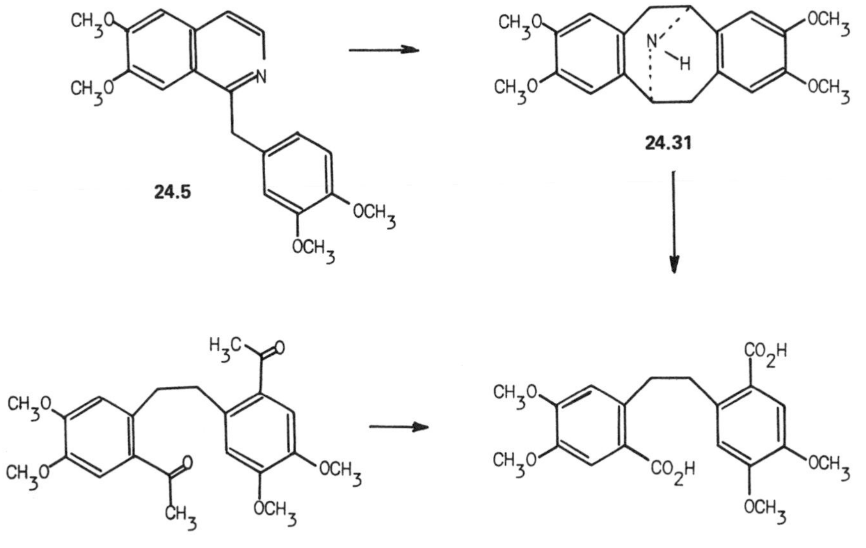

CH3O, CH3O, N, 24.5, OCH3, OCH3 → 24.31 → 24.177; 24.178 → 24.177

SCHEME 24.23 A pathway for the degradation of pavine. (After Reference 174.)

The absolute configuration [180] of (-)-argemonine (24.13) has been determined (Scheme 24.24) by conversion of N-benzyl-(-)-argemonine chloride (24.179) to the methine (24.180), catalytic debenzylation and formylation to 24.181 with formamide, and finally, oxidative ozonolysis to a mixture of aspartic and N-methylaspartic acids. Since N-methylation and esterification (with n-propanol) of the aspartic acid mixture yields (-)-N,N-dimethyldi-n-propyl aspartate (24.182) of known absolute configuration, the absolute configuration of (-)-argemonine (24.13) is defined.

Among the more interesting reactions of (-)-argemonine (24.13) is its conversion to a resonance-stabilized dibenzocycloheptatrienyl cation (24.183; Scheme 24.25). Thus, N-methylation and Hofmann degradation of (-)-argemonine (24.13) to the methine (24.184) and further N-methylation and silver oxide treatment to the alcohol 24.185 are followed by a Wagner-Meerwein reorganization on reaction with hydrogen chloride to the chloride 24.186. Then, elimination with sodium hydroxide to the exomethylene derivative 24.187 and protonation with perchloric acid in acetic acid lead to 24.183.

IX. BERBERINE

Although berberine (24.14), which despite its toxicity has been used as an antimalarial, is reported to have been isolated as early as 1826, it was not until 1862 [181] that the correct molecular formula of its hydroxy salt was determined as $C_{20}H_{19}NO_5$.

The classic structural examination was carried out by W. H. Perkin, Jr. [182], who obtained m-hemipinic acid (24.188) on vigorous permanganate oxidation, and under more mild conditions, a mixture of products which were: oxyberberine, $C_{20}H_{17}NO_5$ (24.189); dioxyberberine, $C_{20}H_{17}NO_6$ (24.190); berberal, $C_{20}H_{17}NO_7$ (24.191); anhydroberberilic acid, $C_{20}H_{19}NO_9$ (24.192); and berberilic acid, $C_{20}H_{19}NO_9$ (24.193), for which "two forms" were known. All these products are shown as Scheme 24.26.

24.13 → 24.179 → 24.180 → 24.181 → 24.182

SCHEME 24.24 The establishment of the absolute configuration of (-)-argemonine. (After Reference 180.)

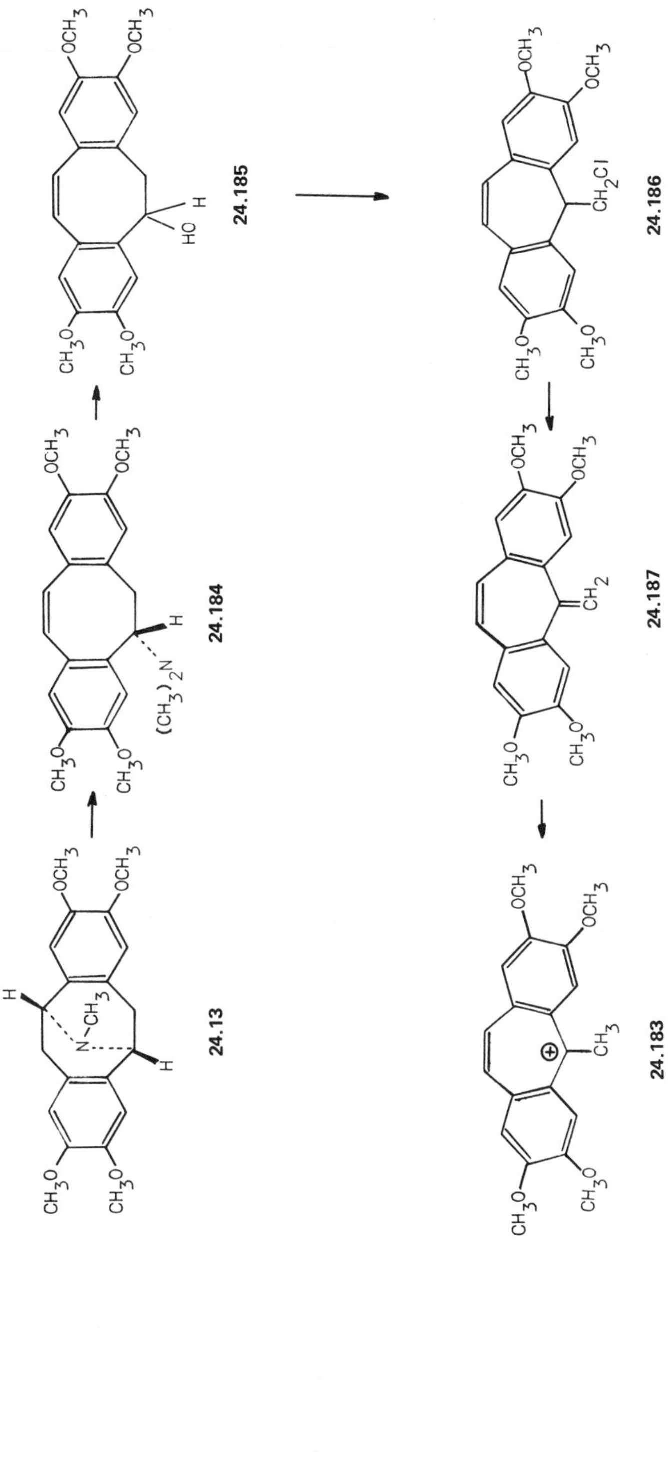

SCHEME 24.25 The generation of a resonance-stabilized cation from (-)-argemonine.

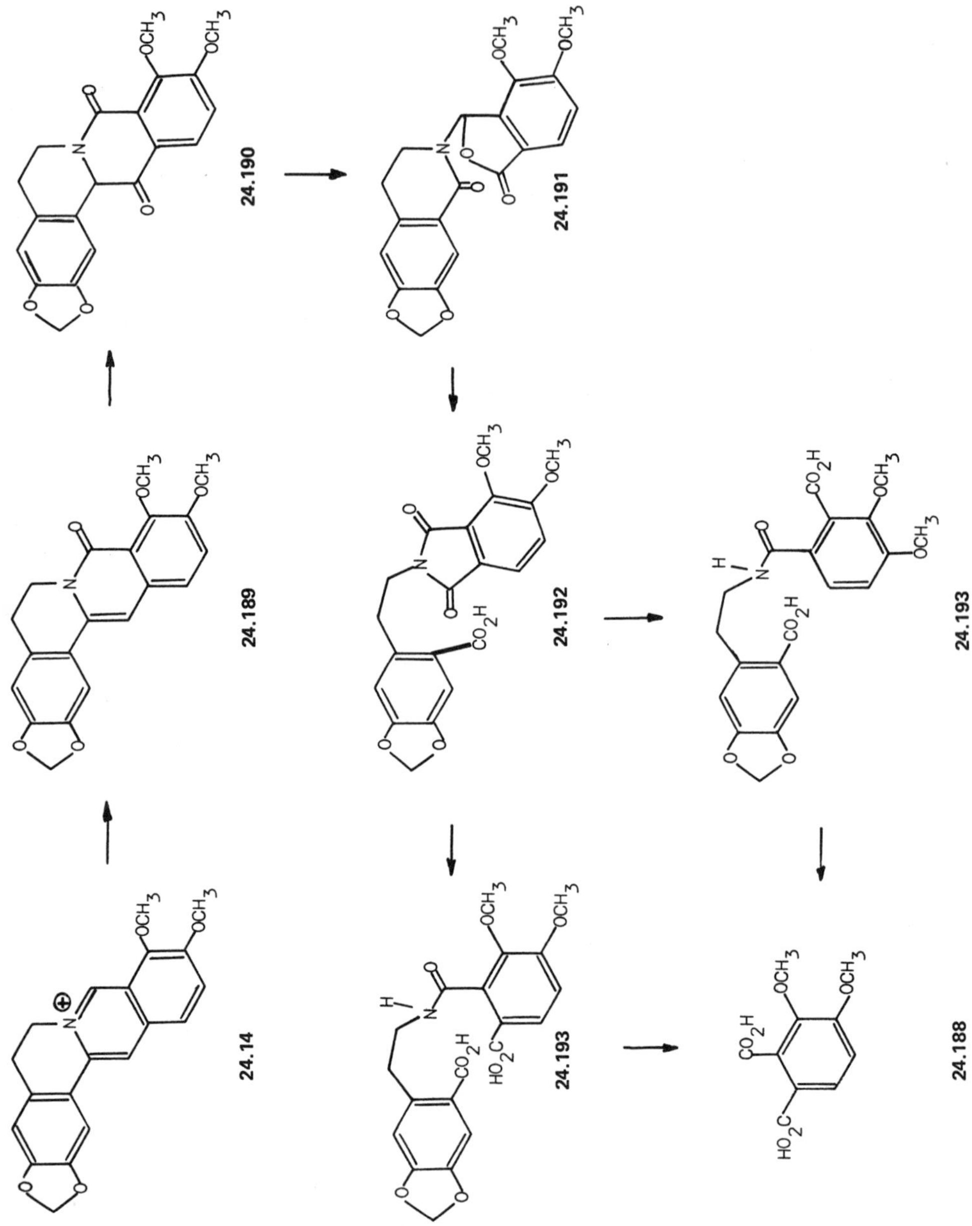

SCHEME 24.26 The oxidation products of berberine. (After Reference 182.)

When anhydroberberilic acid (24.192) is subjected to hydrolysis with sulfuric acid, m-hemipinic acid (24.188) and an amino acid (24.194) are formed (Equation 24.30). Heating the amino acid (24.194) results in lactam formation, and the latter yields an N-nitroso derivative (24.195) which decomposes in alkali to a hydroxy acid (24.196; Equation 24.31).

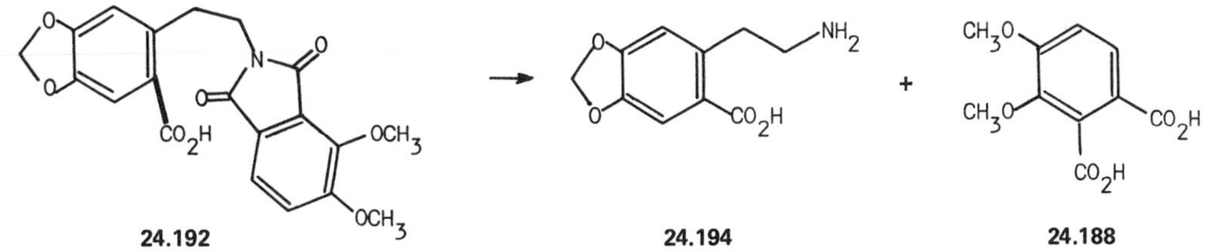

24.192 → 24.194 + 24.188

(Eq. 24.30)

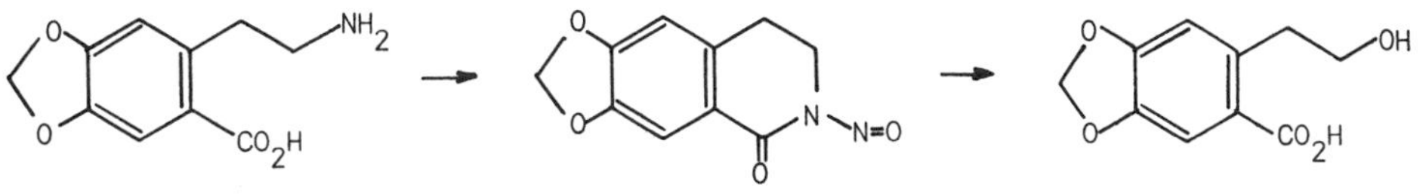

24.194 → 24.195 → 24.196

(Eq. 24.31)

The lactone (24.197) of the acid 24.196 is converted to the corresponding ω-chloroethyl acid chloride (24.198) with phosphorus pentachloride in chloroform and then, via the corresponding methyl ester, to oxyhydrastine (24.199), on heating with methylamine, whose structure is known (Equation 24.32). Thus, it is not surprising that when the amino acid 24.194 is heated with m-hemipinic acid (24.188), anhydroberberilic acid (24.192) is formed (Equation 24.33). Finally, when berberal (24.191) is hydrolyzed with dilute sulfuric acid, desmethyl-oxyhydrastine (24.200) and ψ-opianic acid (24.201) are formed. Heating a mixture of 24.200

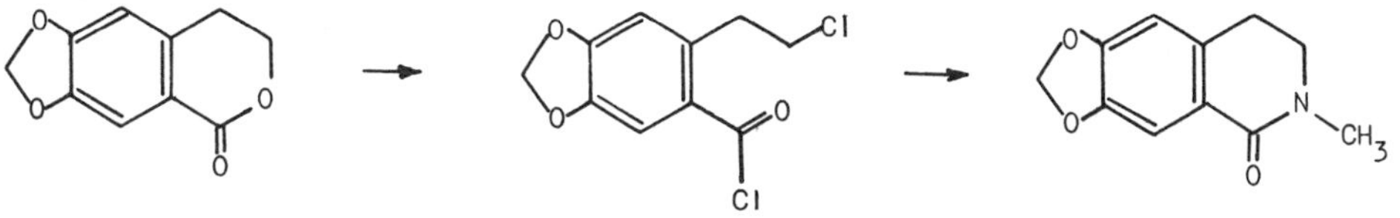

24.197 → 24.198 → 24.199

(Eq. 24.32)

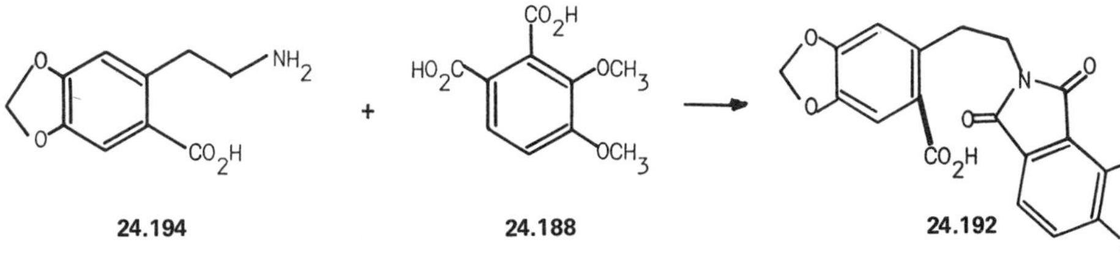

24.194 + 24.188 → 24.192

(Eq. 24.33)

and 24.201 causes their recombination to berberal (24.191; Equation 24.34). This information provided the structure of berberine (24.14), and a synthesis followed. The first synthesis is shown in Scheme 24.27 [183, 184].

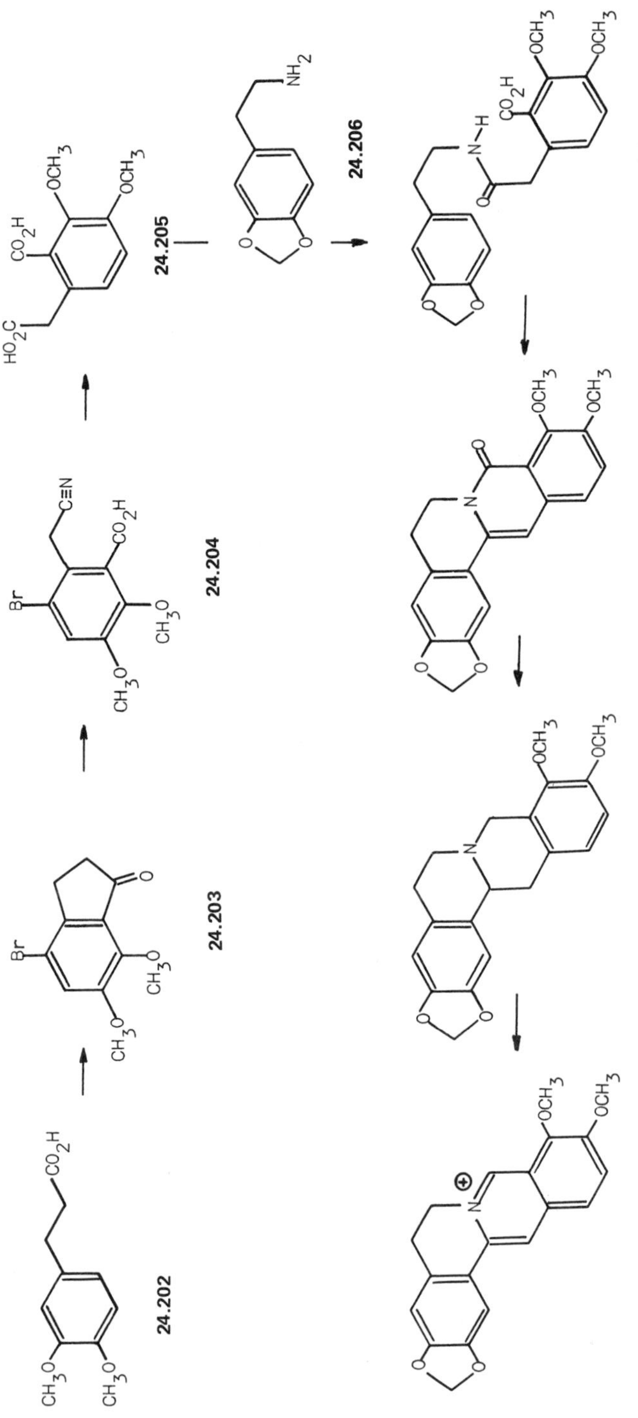

SCHEME 24.27 A synthesis of berberine. (After Reference 183.)

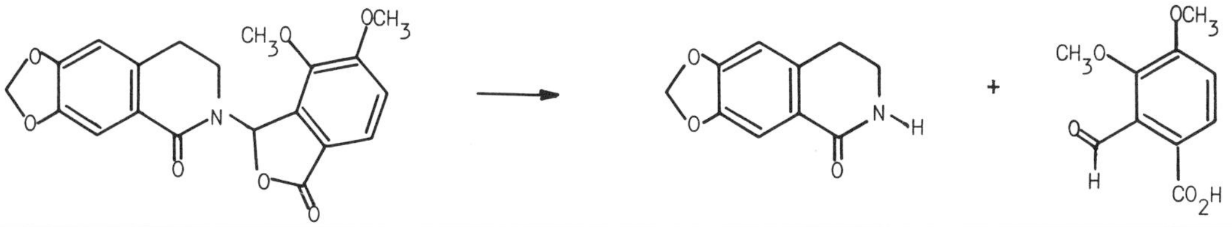

24.191 24.200 24.201

(Eq. 24.34)

Bromination of β-3,4-dimethoxyphenylpropanoic acid (24.202) followed by phosphorus pentoxide dehydration yields the bromoindanone (24.203), which is converted with methyl nitrite to the corresponding isonitroso derivative, and thence, with phosphorus pentachloride, to the o-carboxyphenylacetonitrile 24.204. Hydrolysis of the latter and reductive (sodium amalgam) debromination yield the carboxylic acid 24.205. Condensation of 24.205 with 3,4-methylenedioxyphenethylamine (24.206) followed by partial hydrolysis yields the amide 24.207, which as the methyl ester undergoes cyclization and dehydration on phosphorus pentachloride treatment to oxyberberine (24.189). Reduction to tetrahydroberberine [24.208; (±)-canadine] at a lead cathode in alcoholic sulfuric acid, followed by oxidation by atmospheric oxygen, generates berberine (24.14).

A second synthesis of berberine (24.14) which appears much simpler was attempted by Pictet and Gams [185], and although it was reported to yield the desired berberine (24.14), it is now clear that ψ-berberine (24.209) was the actual product. In this synthesis (Scheme 24.28), reaction of the phenethylamine (24.206) with the dimethoxyphenylacetic acid 24.210 followed by reduction and cyclization leads to the 1-benzyltetrahydroisoquinoline 24.211. Then, condensation of 24.211 with the formaldehyde equivalent methylal in the presence of hydrogen chloride and oxidation generate 24.209, not, as originally proposed, berberine (24.14).

Among the more interesting reactions of berberine (24.14) are (Equation 24.35) its reaction with phenylmagnesium bromide to the 8-phenyl derivative (24.212; [186]) and (Equation 24.36) the Diels-Alder reaction with ethyl vinyl ether to yield 24.213 [187].

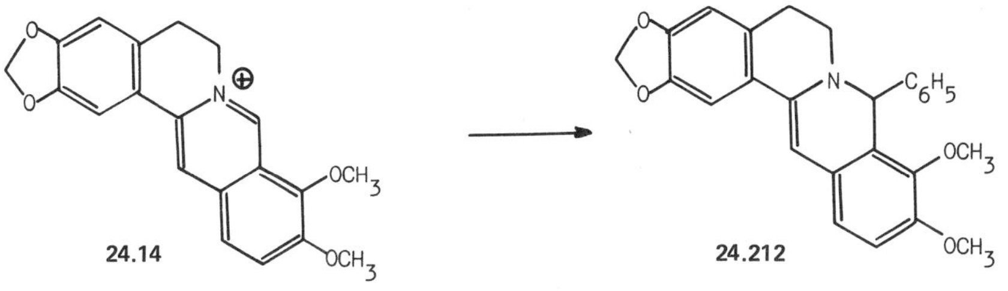

24.14 24.212 (Eq. 24.35)

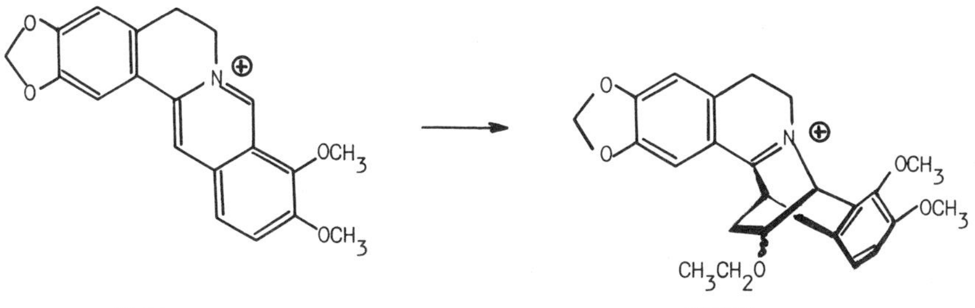

24.14 24.213 (Eq. 24.36)

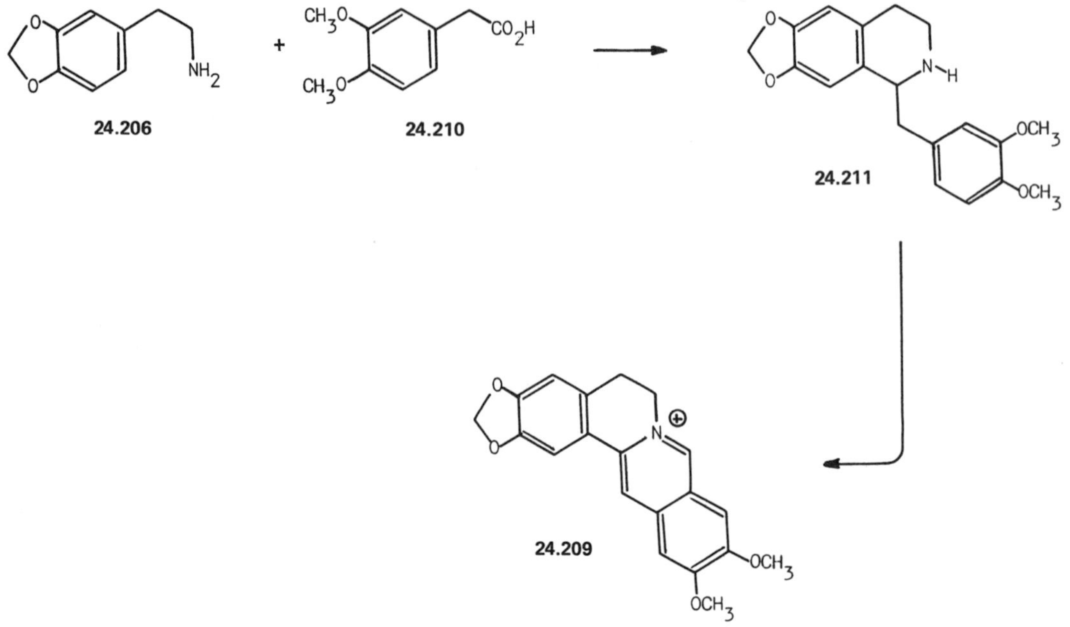

24.206 + 24.210 → 24.211 → 24.209

SCHEME 24.28 A synthesis of ψ-berberine. (After Reference 185.)

X. PROTOPINE

Protopine (24.15) is one of the most widespread of all alkaloids, having been isolated from many different plant families. However, although a number of reactions had been carried out it was the similarity of protopine (24.15) to cryptopine (24.214) which originally suggested its structure. Thus, cryptopine (24.214) on treatment with phosphorus oxychloride yields isocryptopine chloride (24.215), and the latter generates anhydrocryptopine (24.216) when allowed to react with methanolic potassium hydroxide (Equation 24.37). Exactly

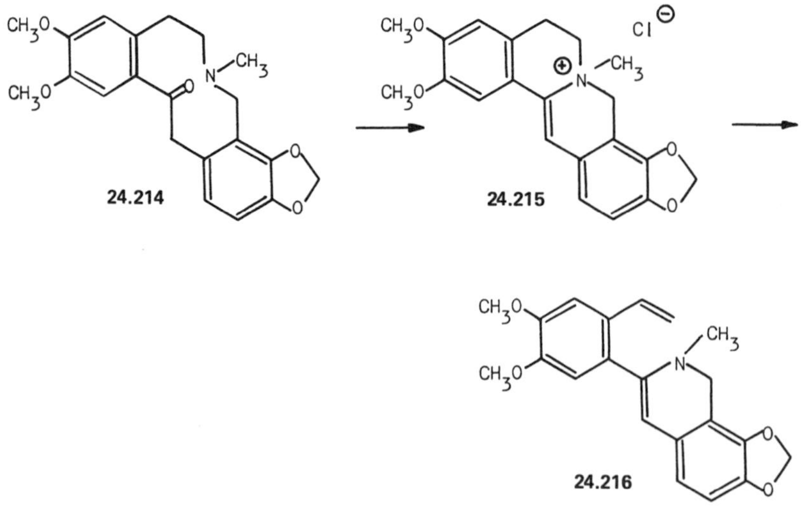

24.214 → 24.215 → 24.216

(Eq. 24.37)

analogous reactions were observed with protopine (24.15), which provided the corresponding isoprotopine chloride (24.217) and anhydroprotopine (24.216; Equation 24.38), respectively.

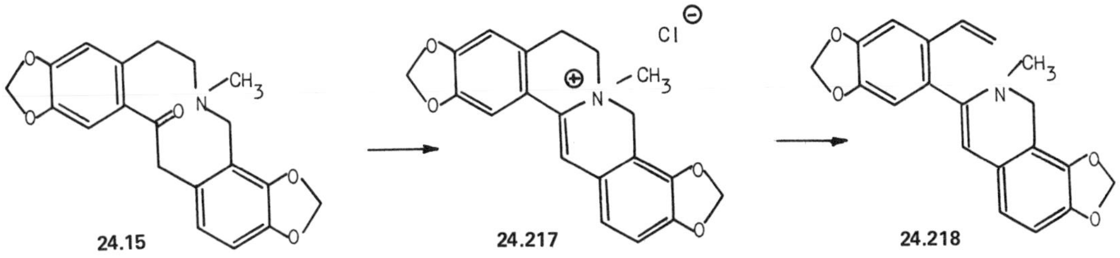

24.15 24.217 24.218

(Eq. 24.38)

The structural elucidation of cryptopine (24.214) was due to the now-classic work of W. H. Perkin, Jr. [188] who was able to show (Scheme 24.29) that on treatment with dimethyl sulfate, cryptopine (24.214) forms a methosulfate (24.219) which undergoes an Emde degradation, with sodium amalgam in acid, to yield tetrahydromethylcryptopine (24.220). The latter, with acetyl chloride, then dehydrated to anhydrotetrahydromethylcryptopine (24.221). Oxidation of 24.221 with potassium permanganate (Scheme 24.30) yields a variety of products which include 4,5-dimethoxy-2-(β-dimethylaminoethyl)benzaldehyde (24.222), 2-methyl-3,4-methylenedioxybenzaldehyde (24.223), and their further respective transformation products, N-formyl-4,5-dimethoxy-2-(β-dimethylaminoethyl)benzoic acid (24.224) and 2-methyl-3,4-methylenedioxybenzoic acid (24.225). These could reasonably be ascribed to structure 24.221, but it was not until careful reduction of cryptopine (24.214) with sodium amalgam

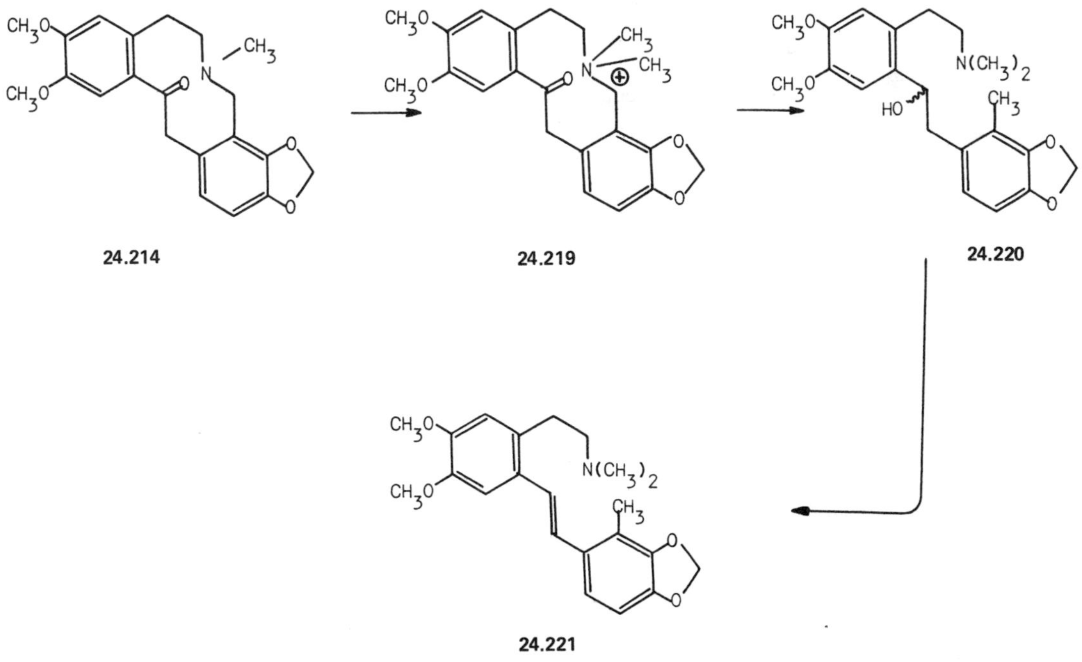

24.214 24.219 24.220

24.221

SCHEME 24.29 The degradation of cryptopine. (After Reference 188.)

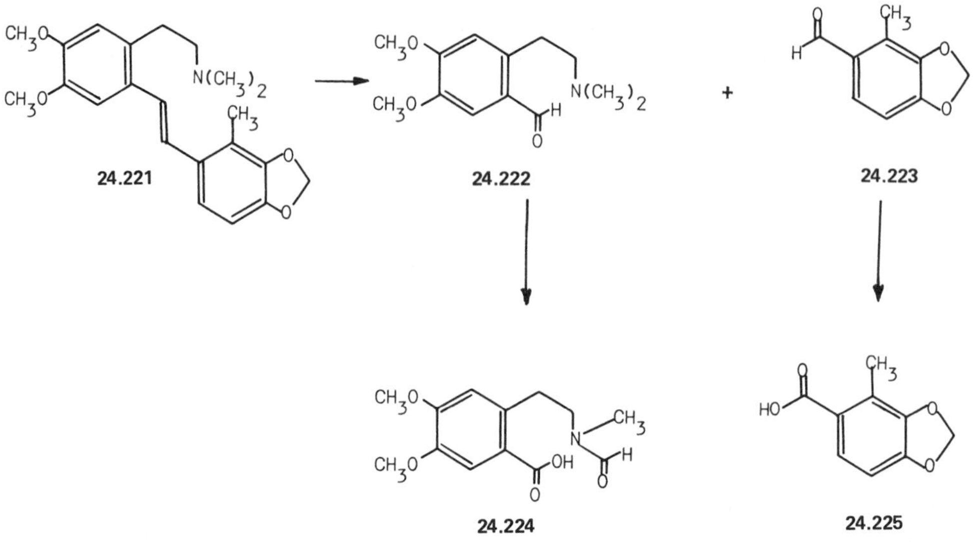

24.221 24.222 24.223

24.224 24.225

SCHEME 24.30 Oxidation of anhydrotetrahydromethylcryptopine with permanganate.

could be shown to generate a dihydro compound (24.226) which underwent cyclization with acetyl chloride (or phosphorus oxychloride) to a separable mixture of diasteromeric quaternary methochlorides, 24.227 and 24.228 (which could now be related to the tetrahydroprotoberberine alkaloids), that the position of the carbonyl in cryptophine (24.214) was assigned (Scheme 24.31).

Cryptopine (24.214), as well as protopine (24.15), has been synthesized. For the latter (Scheme 24.32), 3,4-methylenedioxy-β-phenethylamine (24.206) is allowed to condense with 3,4-methylenedioxyhomophthalic acid (24.229) (which had been prepared from homopiperonylic acid, 24.230, by the reactions already outlined in Scheme 24.27), the adduct partially hydrolyzed to 24.231, cyclized, and reduced to the tetrahydroprotoberberine 24.232. Conversion of the latter to the corresponding methochloride mixture leads, on treatment with silver oxide and heat, to anhydrodihydroprotopine (24.233). On careful oxidation of 24.333 with perbenzoic acid the N-oxide 24.234 is formed. The latter, with hydrogen chloride in acetic acid, yields protopine (24.15; Equation 24.39).

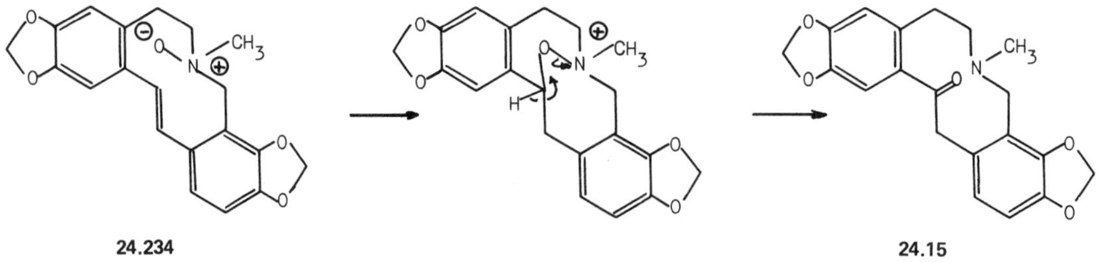

24.234 24.15

(Eq. 24.39)

Interestingly, although protopine (24.15) appears, a priori, capable of cyclization to the alcohol ammonium salt 24.235, the infrared spectrum indicates a carbonyl group at 1675 cm^{-1} (5.96 μ) rather than 1692 cm^{-1} (5.9 μ), which might be expected for a conjugated ketone, showing some transannular interaction. However, acid salts of cryptopine (24.214) have no carbonyl bands at all and therefore, presumably, exist as tetracyclic tautomers 24.236.

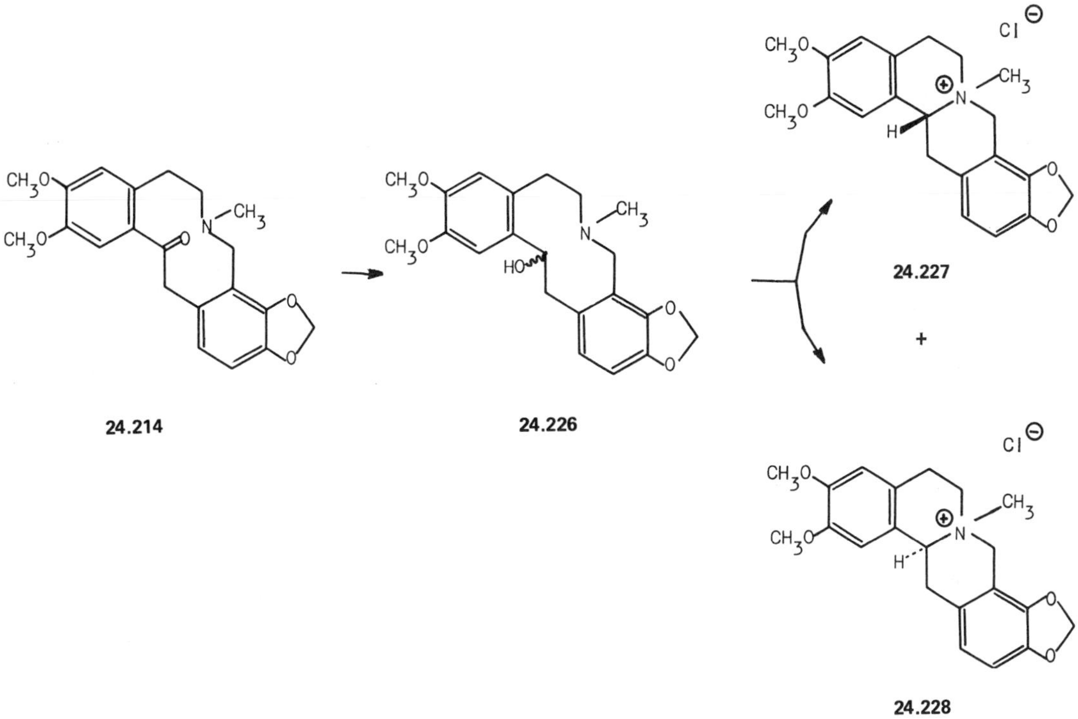

24.214

24.226

24.227

+

24.228

SCHEME 24.31 The reduction and cyclization products of cryptopine.

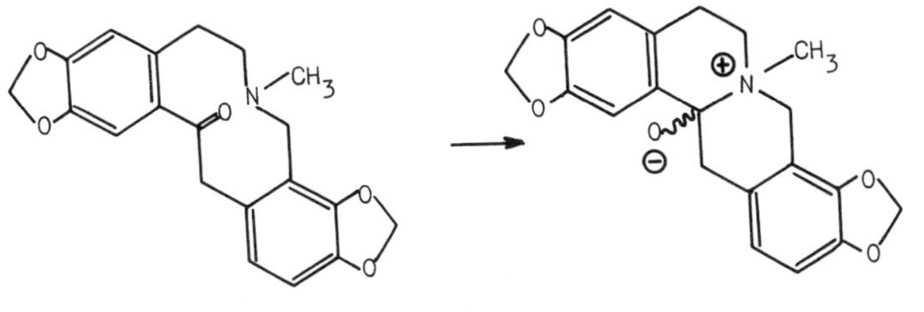

24.15

24.235

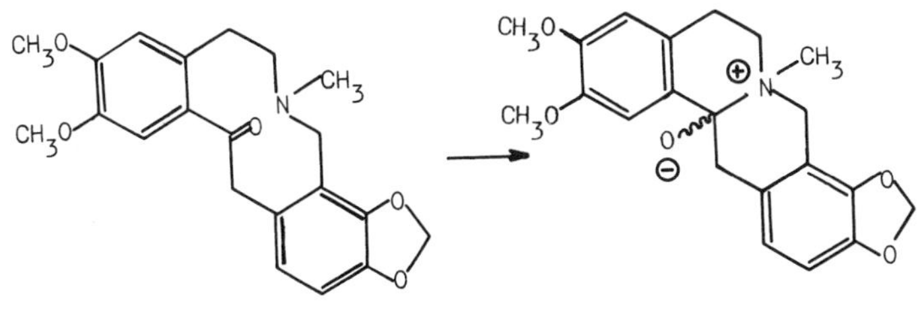

24.214

24.236

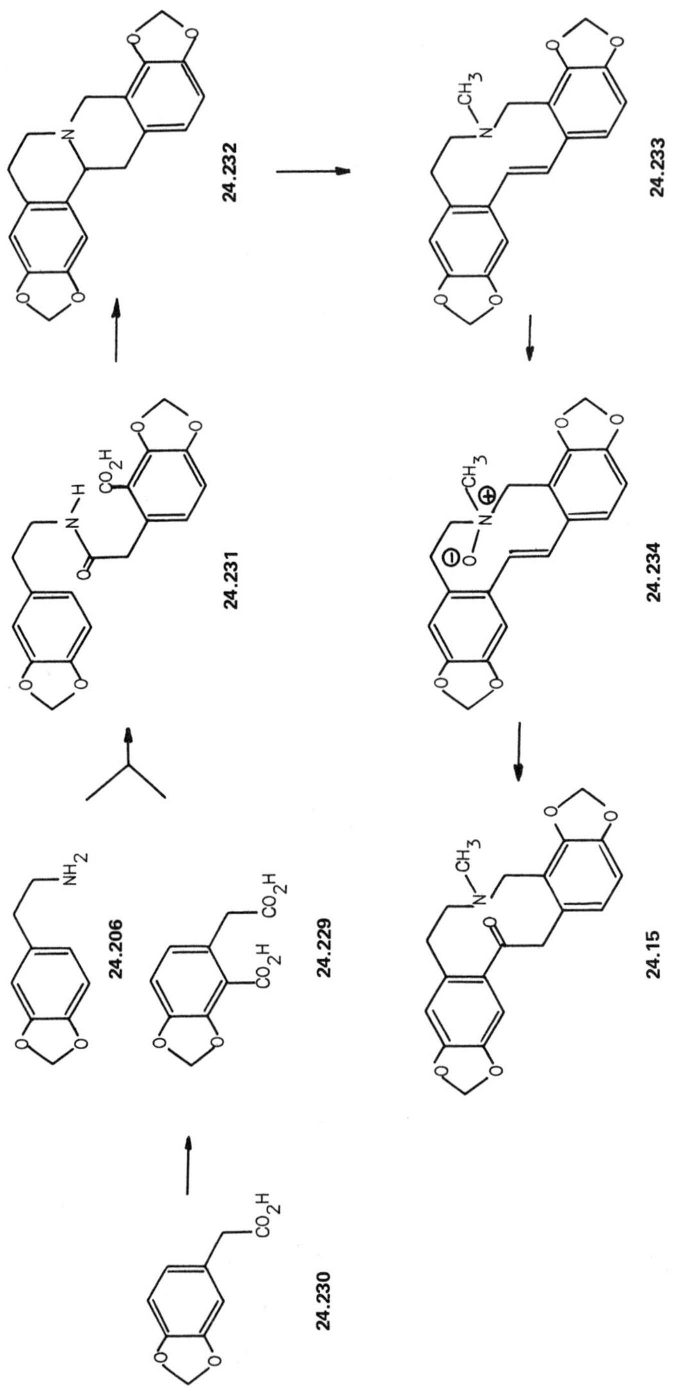

24.230

24.206

24.229

24.231

24.232

24.233

24.234

24.15

SCHEME 24.32 A synthesis of protopine.

XI. SANGUINARINE AND CHELIDONINE

The structure of chelidonine (24.17) was established by about 1930, and exploration of its chemistry elucidated its relationship to sanguinarine (24.16). Thus, both sanguinarine, $C_{20}H_{13}NO_4$ (24.16) and chelidonine, $C_{20}H_{19}NO_5$ (24.17) are isolated from *Chelidonium majus* L., and both yield benzo[c]phenanthridine (24.237) on zinc dust distillation. Chelidonine (24.17) can be converted into sanguinarine (24.16; Equation 24.40) by acetylation and oxidation with mercuric acetate.

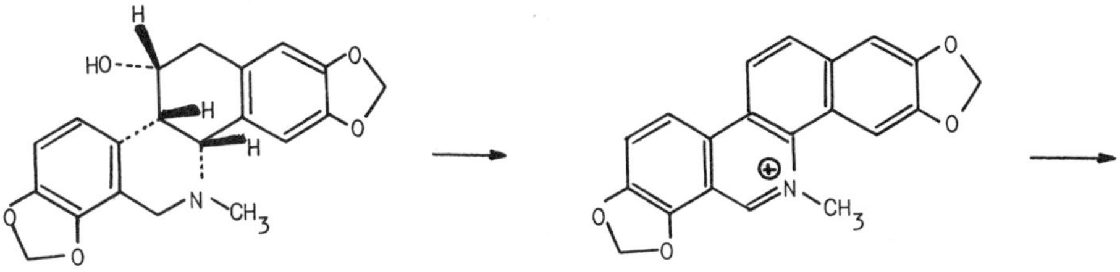

24.17 → 24.16 → (Eq. 24.40)

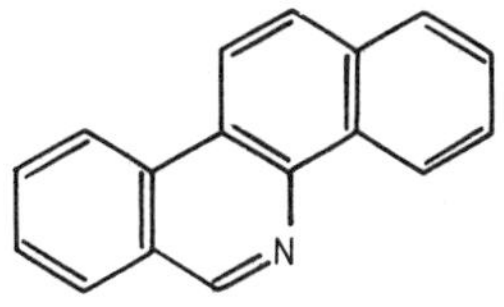

24.237

Oxidation of chelidonine (24.17) with potassium permanganate yields 3,4-methylenedioxyphthalic acid (24.238) and hydrastic acid (24.239) which confirms the presence of two differently situated methylenedioxy substituents (Equation 24.41).

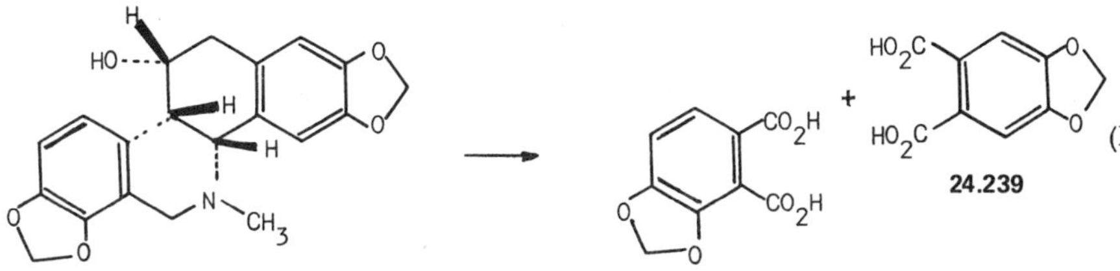

24.17 → 24.238 + 24.239 (Eq. 24.41)

When chelidonine (24.17) is converted to a quaternary base with methyl iodide and the latter subjected to Hofmann degradation, the base (24.240) which is isolated generates hydrastic acid (24.239) and 2-dimethylaminomethylpiperonylic acid (24.241; Equation 24.42) on oxidation. Additionally, further methylation of 24.240 and treatment with base yields an ether (presumably 24.242 or a double-bond isomer of 24.242) which establishes the position of the hydroxyl group in chelidonine (24.17).

Attempted oxidation of the hydroxyl group of chelidonine (24.17) to the corresponding keto compound fails. For example, treatment of chelidonine (24.17) with aluminum isopropoxide in cyclohexanone yields hydroxydihydrosanguinarine (24.243) and the hydroxysanguinarine betaine 24.244 (Equation 24.43), while mercuric acetate oxidation results in

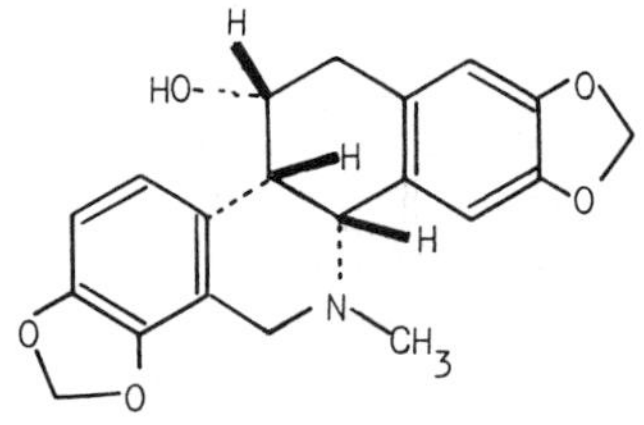

24.17

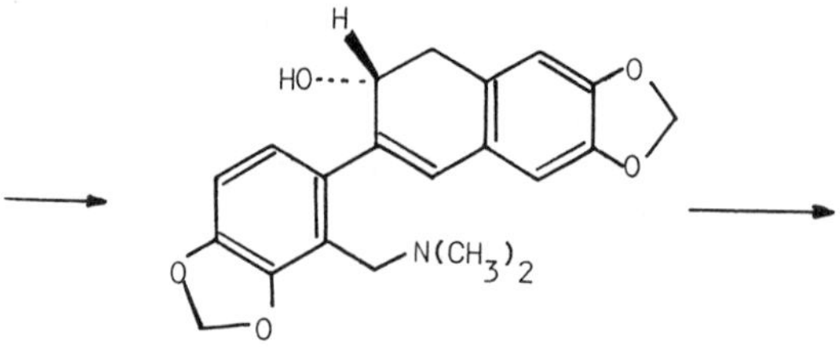

24.240

(Eq. 24.42)

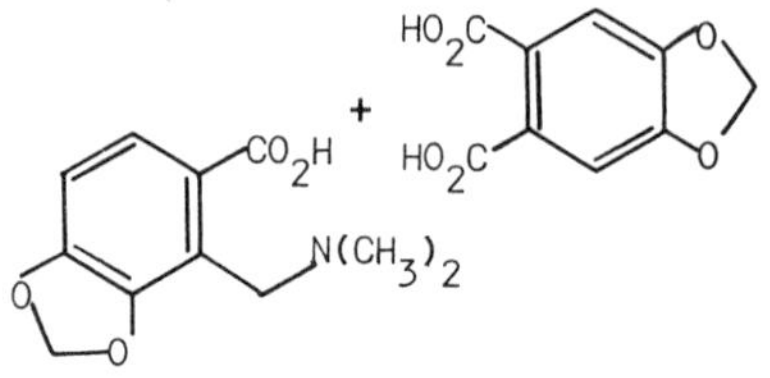

24.241

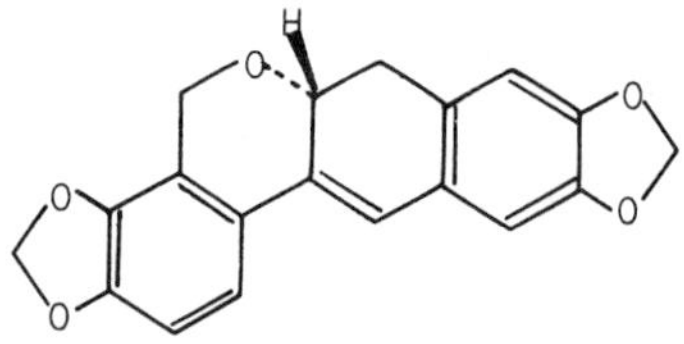

24.242

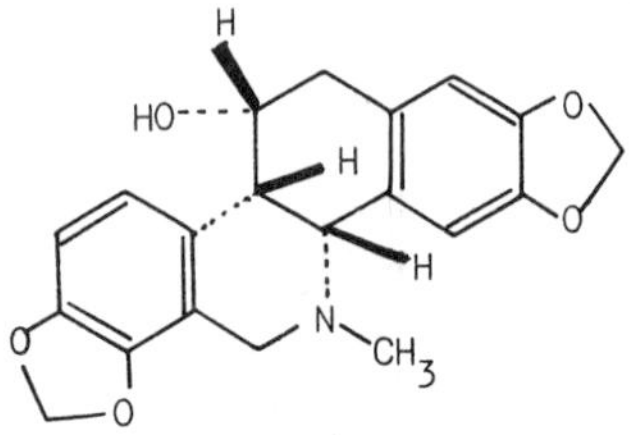

24.17

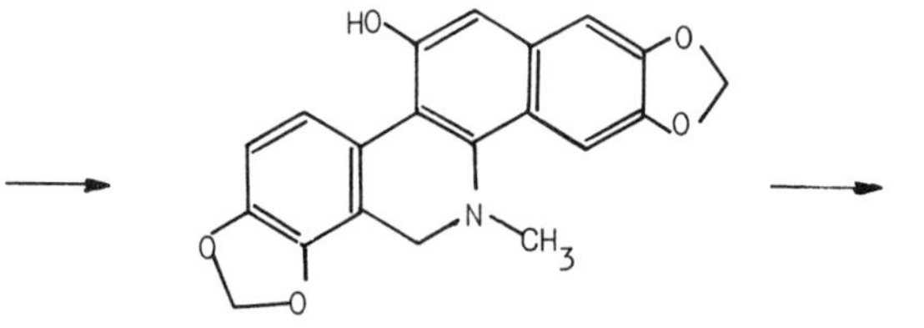

24.243

(Eq. 24.43)

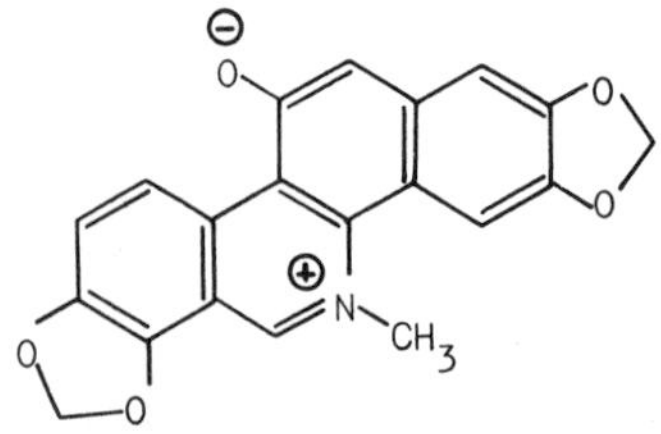

24.244

formation of sanguinarine (24.16), dihydrosanguinarine (24.245), oxysanguinarine (24.246), and didehydrochelidonine (2.247; Equation 24.44; [189]). Interestingly, the latter compound (24.247), which suggests the stereochemistry of the hydroxyl function in chelidonine (24.17), is converted into dihydrosanguinarine (24.245) on acid treatment.

(Eq. 24.44)

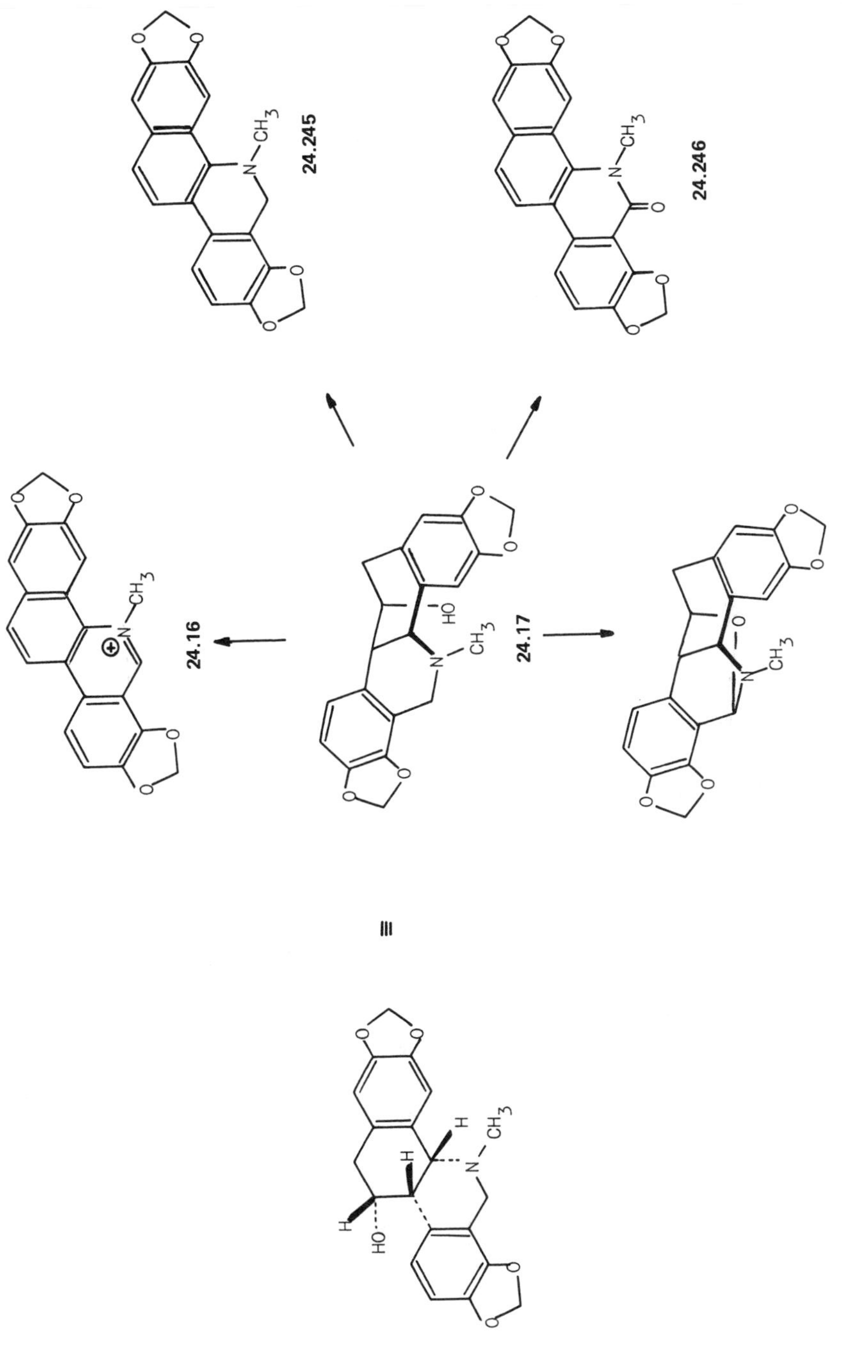

24.245

24.246

24.16

24.17

24.247

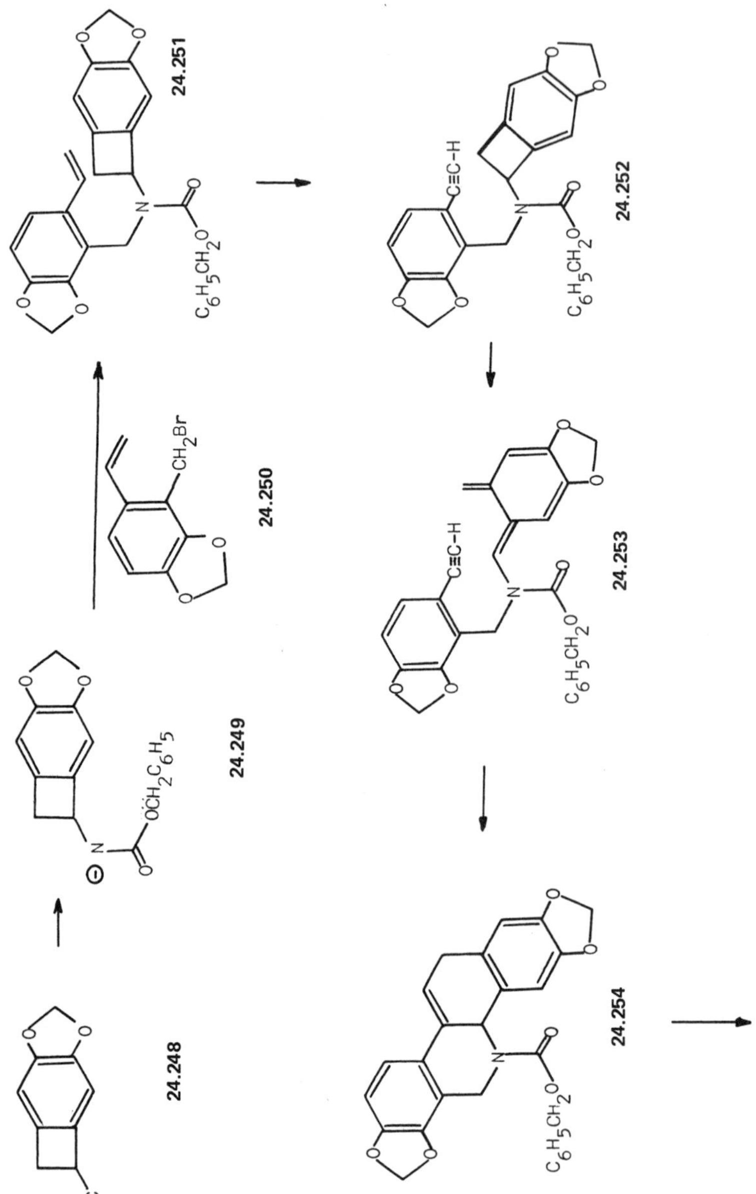

24.248

24.249

24.250

24.251

24.252

24.253

24.254

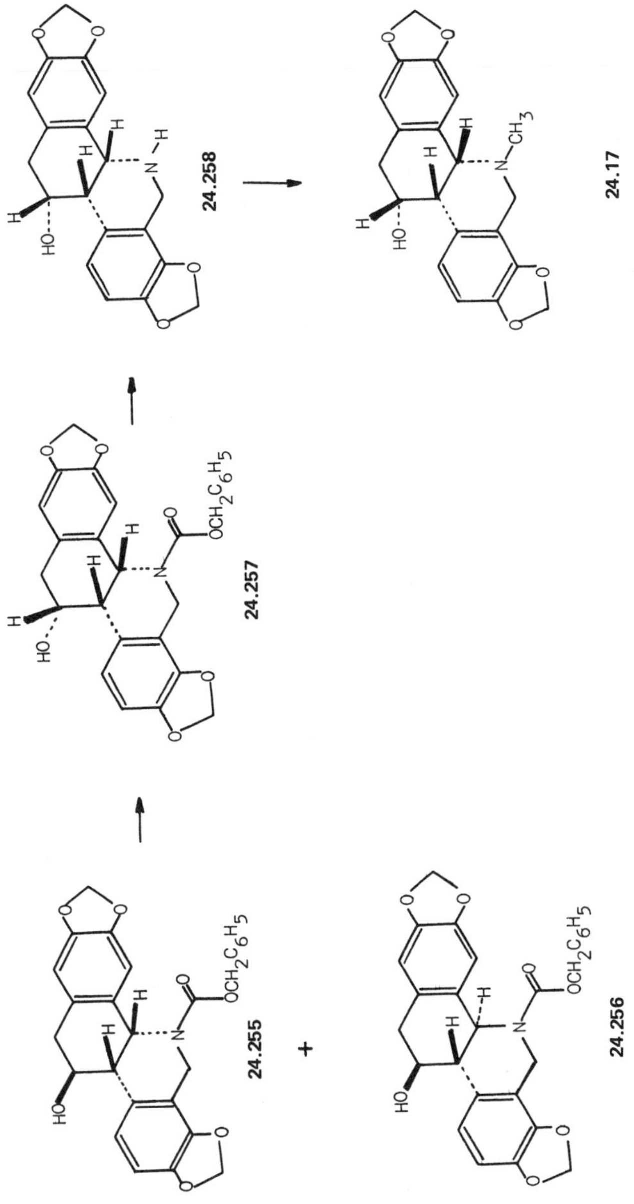

SCHEME 24.33 A synthesis of chelidonine. (After Reference 192.)

The relative stereochemistry of chelidonine (24.17) was postulated as a cis B/C ring fusion with the hydroxyl at the C_6 axial to account for the extensive hydrogen bonding observed in the infrared spectrum [190] and, based upon an empirical correlation involving esterification with (±)-α-phenylbutyric anhydride and the chirality of the remaining unreacted α-phenylbutyric acid, the absolute configuration of chelidonine (24.17) was assigned as shown [191].

An exciting synthesis of chelidonine (24.17) and thus of sanguinarine (24.16) has been developed [192] and is outlined below as Scheme 24.33. The cyanobenzocyclobutane 24.248 was hydrolyzed and converted (via the Curtius rearrangement) to the corresponding isocyanate, thence to the benzylurethane with benzyl alcohol, and finally to the anion 24.249. On reaction of the anion 24.249 with the benzyl bromide 24.250 in dimethyl formamide-containing sodium iodide, the tertiary carbamate 24.251 is formed. Low-temperature bromination and didehydrohalogenation of 24.251 yields the alkyne 24.252, which undergoes thermal rearrangement (presumably via 24.253) to the tetracyclic intermediate 24.254. Oxidative hydroboration yields the alcohol mixture 24.255 and 24.256. The former is converted to the alcohol 24.257 via oxidation with Jones reagent and reduction with sodium borohydride. Hydrogenolysis of the benzyl carbamate is accompanied by decarboxylation to (±)-norchelidonine (24.258), and N-methylation yields (±)-chelidonine (24.17 and its mirror image).

XII. NARCOTINE

(-)-α-Narcotine, $C_{22}H_{23}O_7N$, mp 132°C (24.18) cooccurs with the other opium alkaloids in Papaver somniferum L. With dilute sulfuric acid, (-)-α-narcotine (24.18) is cleaved into two fragments, namely, cotarnine (24.259) and opianic acid (24.260; Equation 24.45).

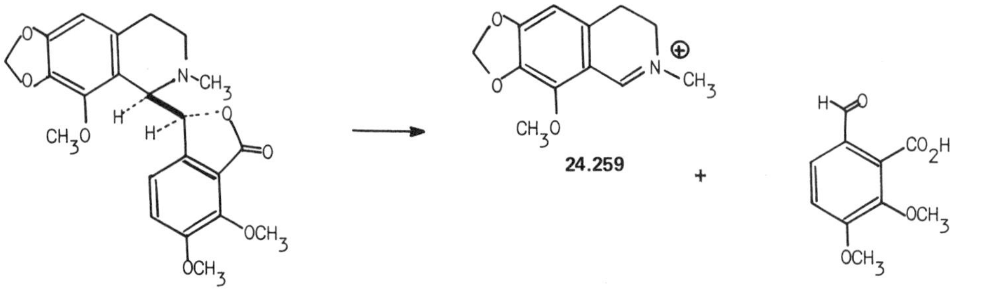

24.18 → 24.259 + 24.260

(Eq. 24.45)

Reductively, however, (-)-α-narcotine (24.18) yields hydrocotarnine (24.261) and meconine (24.262; Equation 24.46). Cotarnine (24.259) is readily reduced to hydrocotarnine (24.261) and reversibly adds a hydroxyl group to yield a ψ-base (24.263; Equation 24.47). Oxidation of cotarnine (24.259) results in the o-dicarboxylic acid cotarnic acid (24.264). On the other hand, opianic acid (24.260), which can also be represented as a lactol (Equation 24.48), is formed from meconine (24.262) oxidatively while meconine (24.262) itself is capable of generation by formylation of 2,3-dimethoxybenzoic acid (24.265). Recombination of the above fragments dictates the structure of narcotine (24.18) be written as shown.

When (Scheme 24.34) (-)-α-narcotine (24.18) is heated in methanolic potassium hydroxide, (-)-β-narcotine (24.266) is formed. Both (-)-α-narcotine (24.18) and (-)-β-narcotine (24.266) can be reduced with lithium aluminum hydride to form, respectively, α-narcotinediol (24.267) and β-narcotinediol (24.268). Then, since the respective monomesylates (formed at the benzylic carbon) undergo cyclization to the corresponding protoberberine bases which

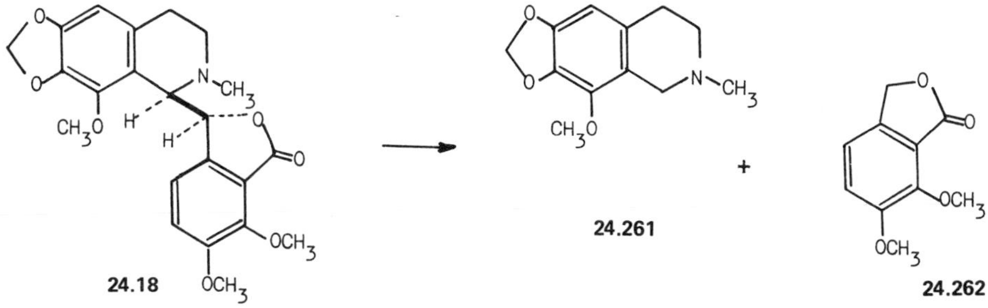

(Eq. 24.46)

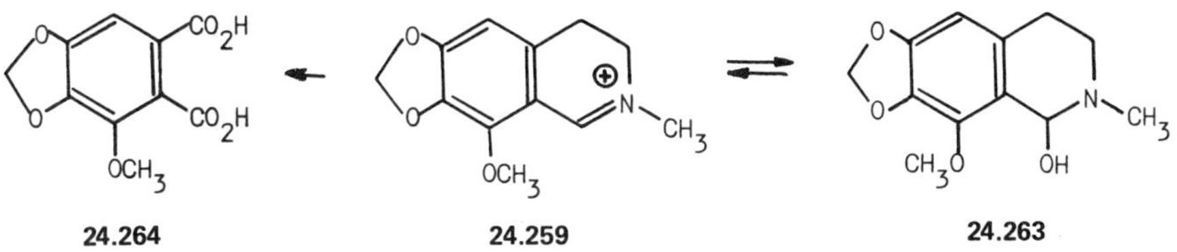

(Eq. 24.47)

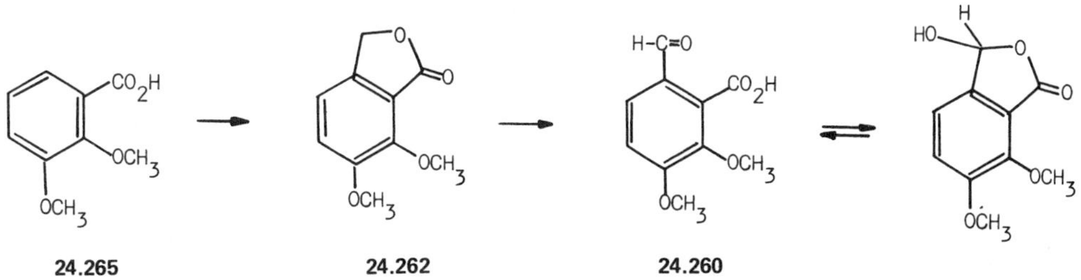

(Eq. 24.48)

can be thermally demethylated to 24.269 and 24.270, respectively, and reductively converted to the identical protoberberine (24.271) of known configuration, only the configuration at the center bearing the hydroxyl remained to be determined.

In one of these bases (24.269) the angle subtended by the protons at C-13 and C-14 is about 160° while in the other (24.270) the angle is about 60°. Thus, that isomer (24.269) in which the protons are trans or anti about the single bond is assigned to the form in which there is a coupling constant ($J_{13,14}$) of 9 Hz while the opposite arrangement (24.270) shows $J_{13,14}$ of 1.5 Hz [193].

(±)-Narcotine (24.18 and its mirror image) is readily synthesized [194]. As shown (Equation 24.49), heating cotarnine (24.259) and meconine (24.262) in ethanol results in the formation of (±)-narcotine (24.18 and its mirror image) which is capable of resolution.

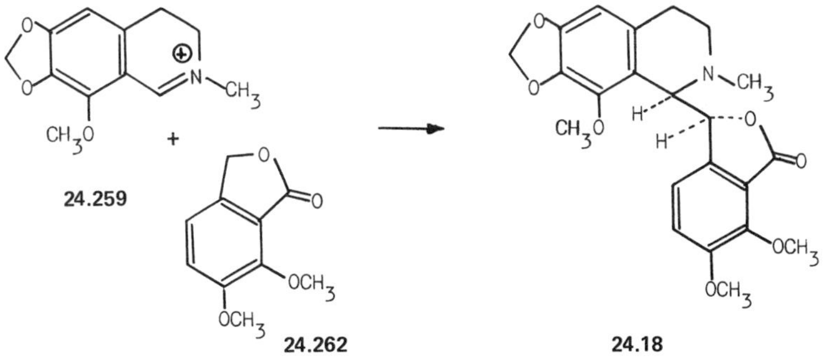

(Eq. 24.49)

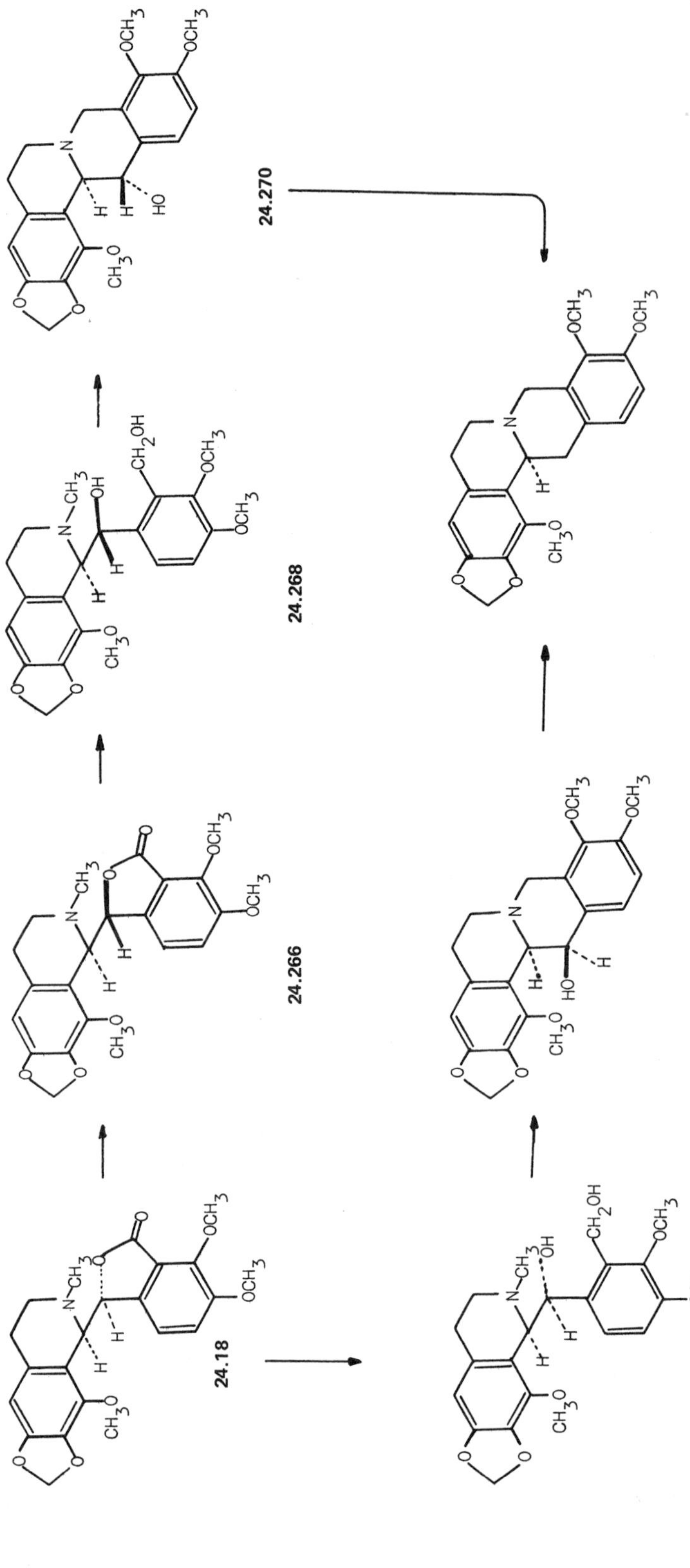

24.18 → 24.266 → 24.268 → 24.270 → 24.271

24.18 → 24.267 → 24.269 → 24.271

SCHEME 24.34 The degradation of both narcotinediols to the same protoberine base.

XIII. RHOEADINE

Rhoeadine, $C_{21}H_{21}NO_6$ (24.19), or its relatives (methoxy groups in place of the methylenedioxy groups) are widely found in plants of the genus Papaver. All these closely related bases are optically active, and they are all dextrorotatory.

Hofmann degradation of the methiodide of rhoeadine (24.19) yields desrhoeadine (24.271), and a further Hofmann degradation yields the bismethine (24.272; Equation 24.50).

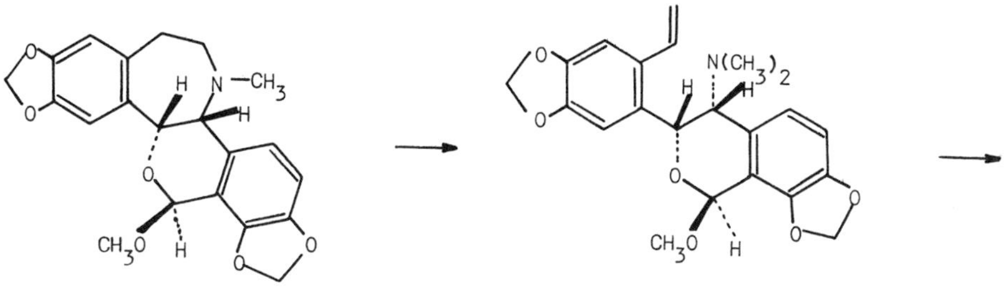

24.19 24.271 24.272 (Eq. 24.50)

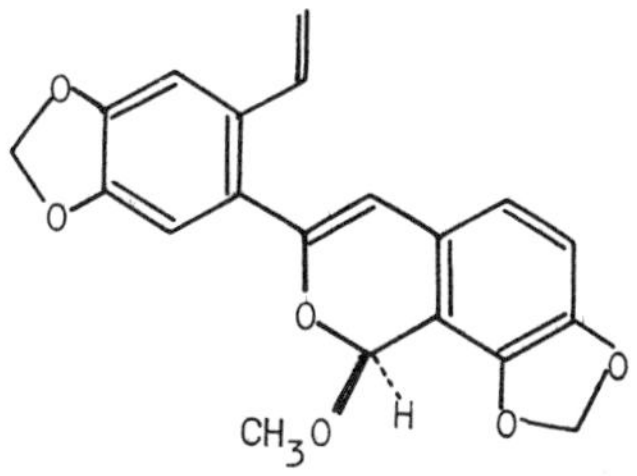

When rhoeadine (24.19) is treated with dilute mineral acid, rhoeagenine (24.273), also found naturally, is obtained (Equation 24.51). Zinc dust distillation of the latter yields

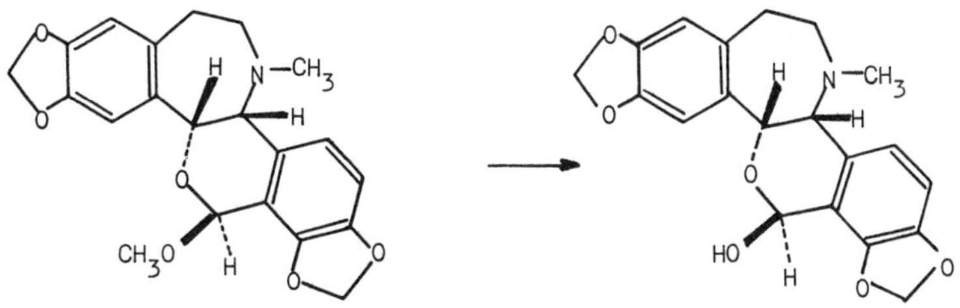

24.19 24.273 (Eq. 24.51)

isoquinoline (24.274; Equation 24.52) while basic permanganate affords, on acidification, hydrastic acid (24.239) and (just like chelidonine, 24.17; Equation 24.53) 3,4-methylenedioxyphthalic acid (24.238). However, lithium aluminum hydride reduction of rhoeagenine (24.273) yields rhoeageninediol (24.275), convertible on oxidation to a δ-lactone (ir, 1725 cm^{-1}, 5.80 μ) oxyrhoeagenine (24.276), the careful reduction of which regenerates rhoeagenine (24.273; Scheme 24.35). Finally, oxidation of rhoeageninediol (24.275) with dilute nitric acid generates hydrastinine (24.277) and 4,5-methylenedioxyphthalide (24.278; Equation

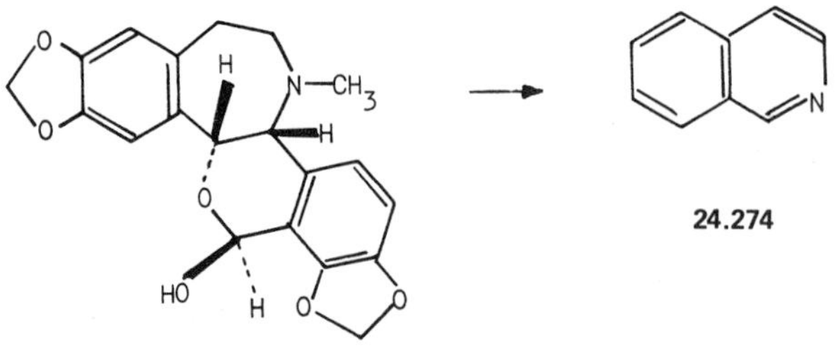

24.273 → 24.274

(Eq. 24.52)

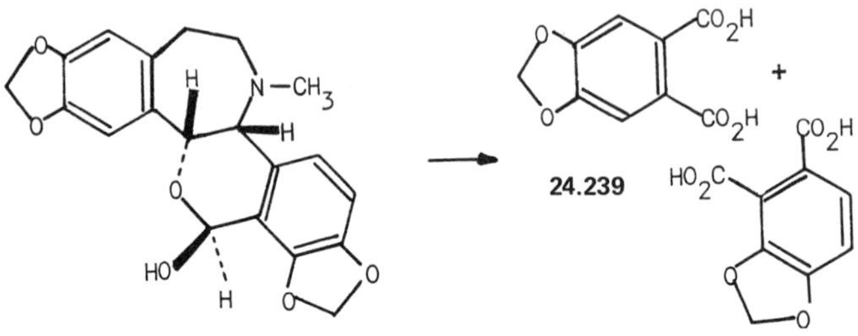

24.273 → 24.239 + 24.238

(Eq. 24.53)

24.54), and treatment [195] with an excess of boiling thionyl chloride affords an enamine (24.217) salt identical to the compound obtained from protopine (24.15) with phosphorus oxychloride (Equation 24.55). A potential pathway for this transformation is shown as Scheme 24.36.

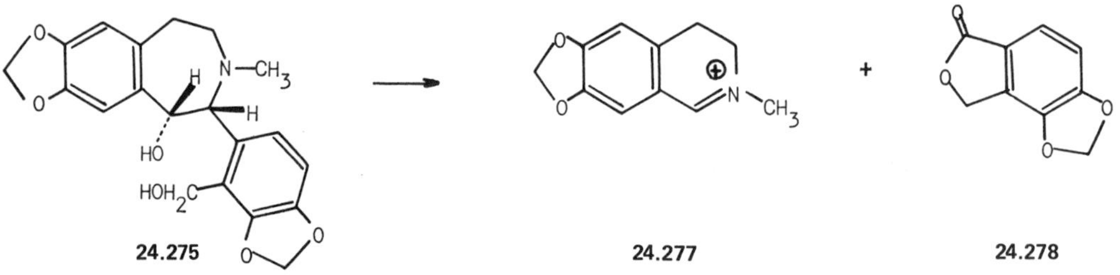

24.275 → 24.277 + 24.278

(Eq. 24.54)

Mass spectrometric examination [196] of rhoeadine (24.19) and rhoeagenine (24.273) confirmed the structures shown. Thus, the former possesses a spectrum characterized by the molecular ion (M^+) at m/e 383, an intense M-15 ion, m/e 368, and the base peak at m/e 177. A reasonable fragmentation pattern of rhoeadine (24.19) accounting for these fragments is shown as Equation 24.56. On the other hand, rhoeagenine (24.273) has intense peaks at m/e 206, 192, and 163, for which the fragments shown (Equation 24.57) can be written. The relative stereochemistry shown for rhoeadine (24.19) was derived by a series of arguments [197] in which it was pointed out that the relatively small coupling constant ($J_{1,2}$ = 2 Hz) for the hydrogens at the B/D ring juncture indicated they were cis to each other, while the chemical shift of the benzylic proton (C-1) was downfield from that observed in its epimer, a result of deshielding by the methoxyl. The structure of rhoeagenine (24.273) has been confirmed by x-ray crystal analysis [198].

(+)-Rhoeadine (24.19) has been synthesized [198] from the phthalideisoquinoline alkaloid (-)-bicuculline (24.279; which could itself be independently synthesized by methods shown

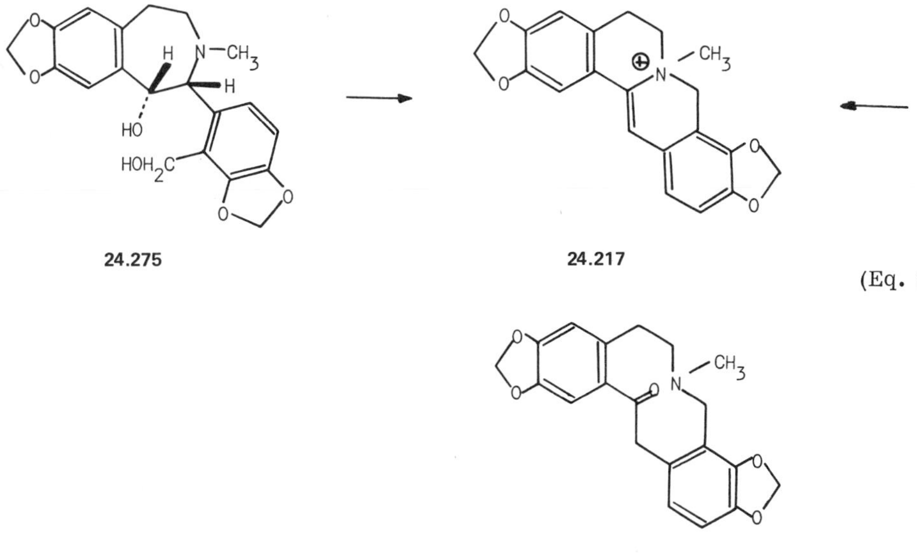

24.275 → 24.217 ← 24.15

(Eq. 24.55)

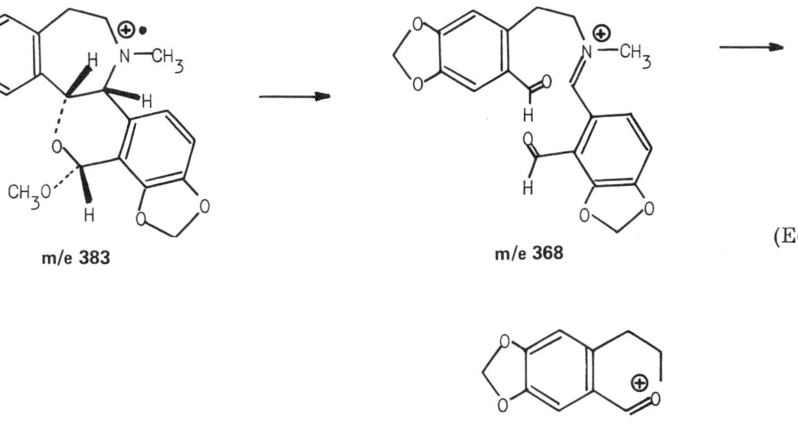

m/e 383 → m/e 368 → m/e 177

(Eq. 24.56)

above). The synthesis (Scheme 24.37) begins with a series of reactions which convert 24.279 into the amidolactone (24.280) through the use of phenylchloroformate and diisopropylethylamine, followed by dehydrohalogenation with dimethyl sulfoxide and the same amine. Then, on treatment with 2 N sodium hydroxide, 24.280 is converted to the acid salt 24.281, which on oxidation in acetic acid presumably leads, via 24.282, to the ketolactone 24.283. Reduction of 24.283 with lithium borohydride in tetrahydrofuran followed by acidification with acetic acid generates the acid 24.284, which spontaneously cyclizes to the lactone (±)-oxyrhoeagenine (24.285 and its mirror image). This lactone (24.285) was successfully resolved with (-)-10-camphorsulfonic acid in methanol and the dextrorotatory enantiomer reduced with sodium bis-(2-methoxyethoxy)-aluminum hydride to a mixture of anomeric lactols, which with trimethylorthoformate in methanol generates (+)-rhoeadine (24.19).

24.273

m/e 163

m/e 206

m/e 192

(Eq. 24.57)

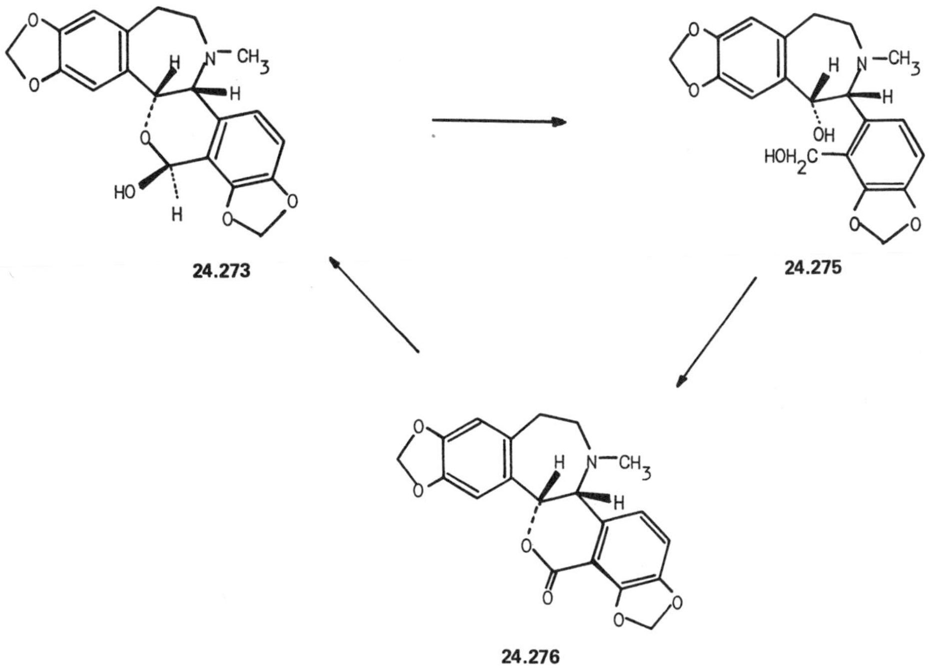

24.273

24.275

24.276

SCHEME 24.35 The interconversion of rhoeagenine, rhoeageninediol, and oxyrhoeagenine.

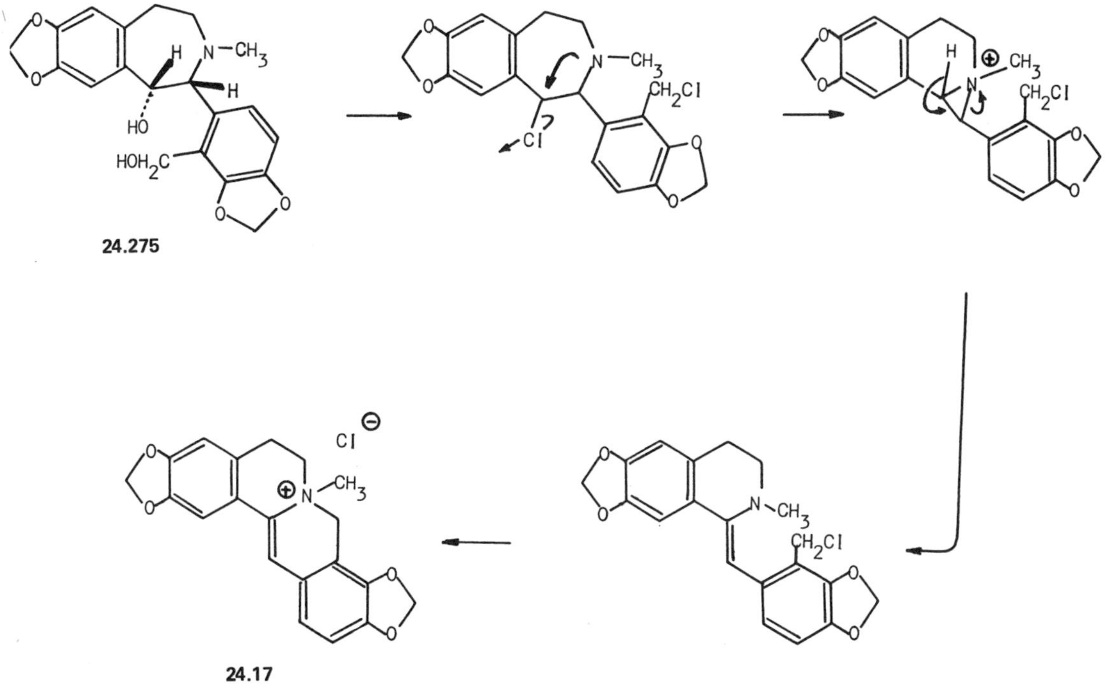

24.275

24.17

SCHEME 24.36 A possible route for the transformation of rhoeageninediol into a tetrahydroprotoberberine.

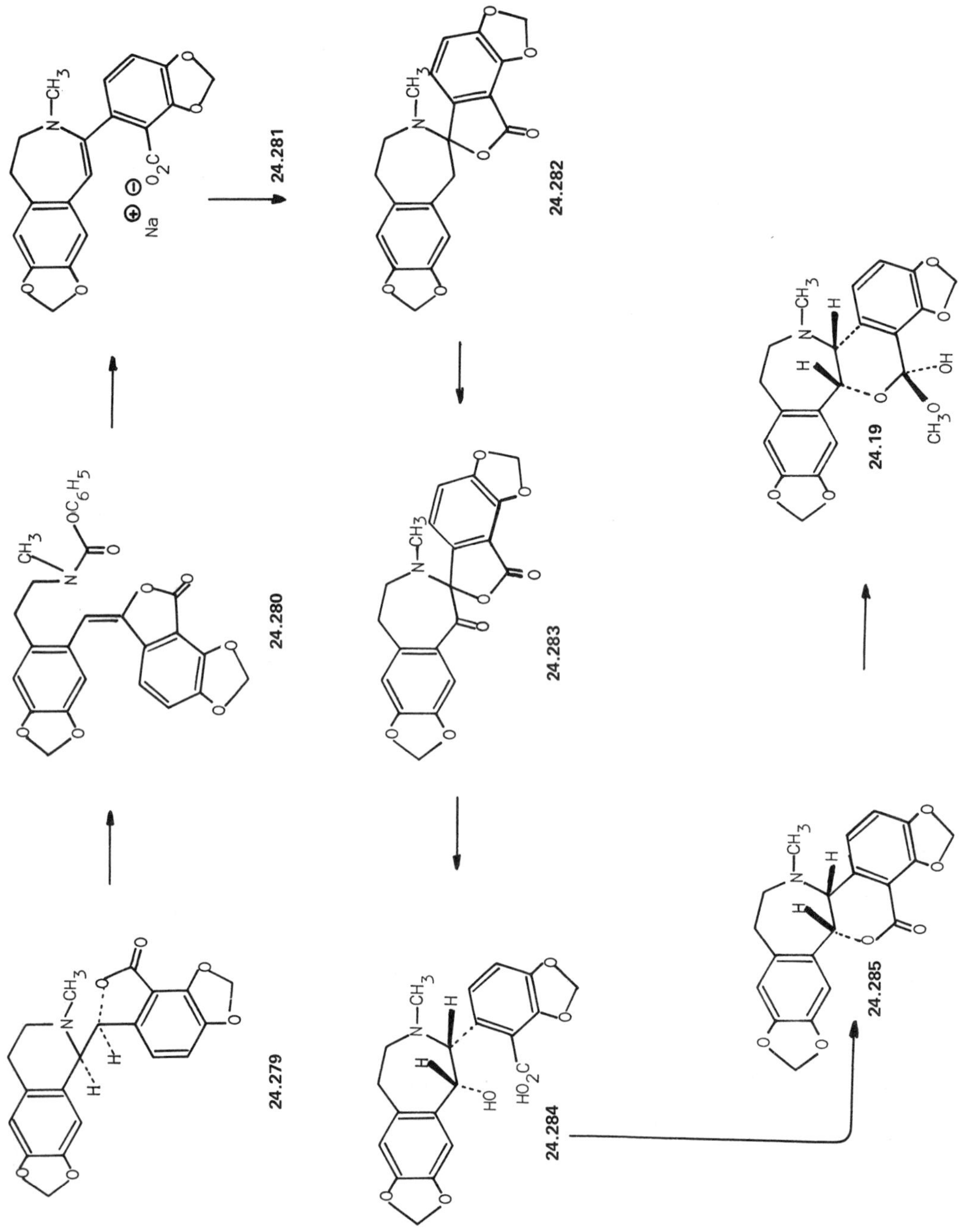

24.279

24.280

24.281

24.282

24.283

24.284

24.285

24.19

SCHEME 24.37 A synthesis of (+)-rhoeadine. (After Reference 198.)

SELECTED READING

K. W. Bentley, The Isoquinolone Alkaloids. Pergamon, New York, 1965.
K. W. Bentley, in The Alkaloids, Vol. 13, R. H. F. Manske (ed.). Academic, New York, 1971, pp. 3 ff.
A. Berger, in The Alkaloids, Vol. 4, R. H. F. Manske (ed.). Academic, New York, 1954, pp. 29 ff.
V. Boekelheide, in The Alkaloids, Vol. 7, R. H. F. Manske (ed.). Academic, New York, 1960, pp. 201 ff.
M. Curcumelli-Rodostamo, in The Alkaloids, Vol. 13, R. H. F. Manske (ed.). Academic, New York, 1971, pp. 304 ff.
M. Curcumelli-Rodostamo and M. Kulka, in The Alkaloids, Vol. 9, R. H. F. Manske (ed.). Academic, New York, 1967, pp. 133 ff.
V. Deulofeu, J. Comin, and M. J. Vernengo, in The Alkaloids, Vol. 10, R. H. F. Manske (ed.). Academic, New York, 1968, pp. 402 ff.
D. Ginsburg, The Opium Alkaloids. Wiley-Interscience, New York, 1962.
H. L. Holmes, in The Alkaloids, Vol. 2, R. H. F. Manske (ed.). Academic, New York, 1952, pp. 1 ff, 219 ff.
H. L. Holmes and G. Stork, in The Alkaloids, Vol. 2, R. H. F. Manske (ed.). Academic, New York, 1952, pp. 161 ff.
P. W. Jeffs, in The Alkaloids, Vol. 9, R. H. F. Manske (ed.). Academic, New York, 1967, pp. 41 ff.
T. Kametani, The Chemistry of the Isoquinolone Alkaloids. Hirokawa Publ., Tokyo, 1969.
M. Kulka, in The Alkaloids, Vol. 7, R. H. F. Manske (ed.). Academic, New York, 1960, pp. 439 ff.
M. Kulka, in The Alkaloids, Vol. 4, R. H. F. Manske (ed.). Academic, New York, 1954, pp. 199 ff.
R. H. F. Manske (ed.), in The Alkaloids, Vol. 4. Academic, New York, 1954, pp. 119 ff, 147 ff.
R. H. F. Manske (ed.), in The Alkaloids, Vol. 10, Academic, New York, 1968, pp. 467 ff.
R. H. F. Manske and W. R. Ashford, in The Alkaloids, Vol. 4, R. H. F. Manske (ed.). Academic, New York, 1954, pp. 77 ff.
F. Šantávy, in The Alkaloids, Vol. 12, R. H. F. Manske (ed.). Academic, New York, 1970, pp. 333 ff.
M. Shamma, in The Alkaloids, Vol. 9, R. H. F. Manske (ed.). Academic, New York, 1967, pp. 1 ff.
M. Shamma, in The Alkaloids, Vol. 13, R. H. F. Manske (ed.). Academic, New York, 1971, pp. 165 ff.
M. Shamma, The Isoquinoline Alkaloids. Academic, New York, 1972.
J. Stanek, in The Alkaloids, Vol. 7, R. H. F. Manske (ed.). Academic, New York, 1960, pp. 433 ff.
J. Stanek, in The Alkaloids, Vol. 9, R. H. F. Manske (ed.). Academic, New York, 1967, pp. 117 ff.
J. Stanek and R. H. F. Manske, in The Alkaloids, Vol. 4, R. H. F. Manske (ed.). Academic, New York, 1954, pp. 167 ff.
G. Stork, in The Alkaloids, Vol. 6, R. H. F. Manske (ed.). Academic, New York, 1960, pp. 219 ff.

25

CHEMISTRY OF 1-PHENETHYLTETRAHYDROISOQUINOLINE ALKALOIDS AND OTHER C_6-C_2 + C_6-C_3 ALKALOIDS DERIVED FROM TYROSINE

The melancholy days are come, the saddest of the year,
Of wailing winds, and naked woods, and meadows brown and sere.

W. C. Bryant, "The Death of Flowers"

The autumn crocus, Colchicum autumnaline L. (Lilaceae; meadow saffron), has been known for possessing an active principle which even in crude form could be used to advantage in the treatment of gout. As a consequence, other Liliaceae were examined for similar active materials, and recently [199] the proposed precursor of many of the alkaloids found in Colchicum autumnaline L. and relatives, namely, autumnaline (25.1), was isolated from Colchicum cornigerum. Other Liliaceae have yielded related compounds to be discussed in this chapter. These include the homoproaporphine alkaloid kreysiginone (25.2) and the homoaporphine alkaloid kreysigine (25.3). Finally, the active ingredient of Colchicum autumnaline L., colchicine (25.4), which was actually the first alkaloid (although nonbasic!) of the family

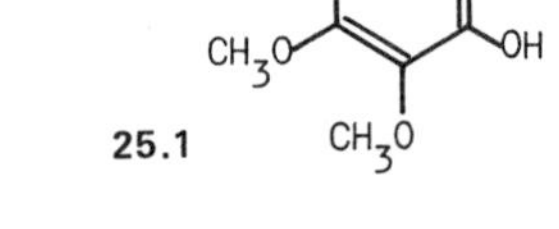

25.1

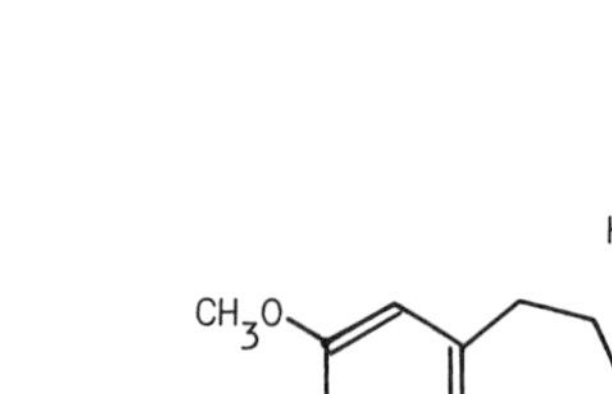

25.2

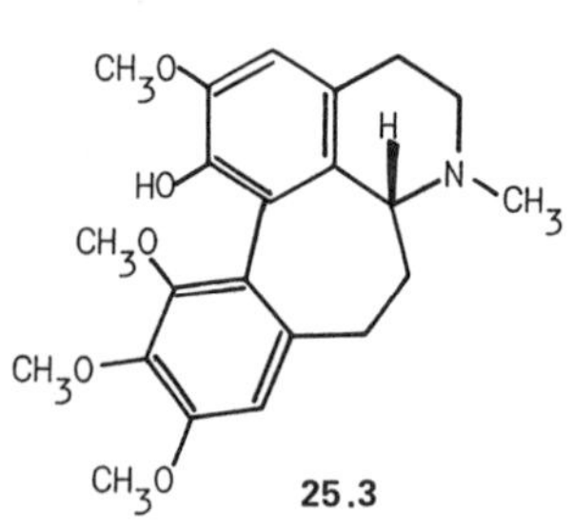

25.3

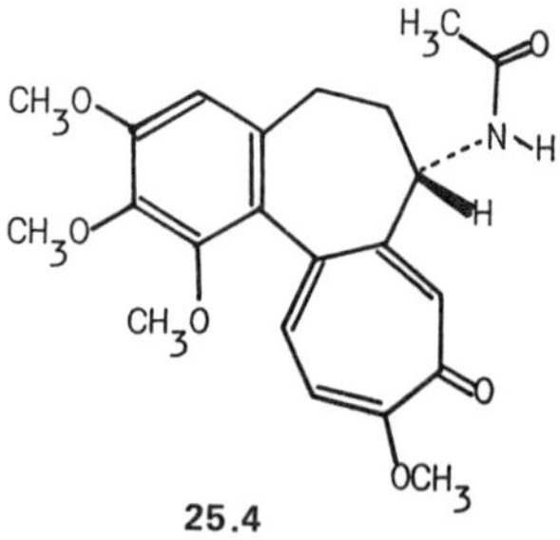

25.4

to have been isolated, and which apparently has the most pronounced physiological effects, will be discussed.

I. AUTUMNALINE

Laborious separation of the alkaloid mixture found in Colchicum cornigerum L. yielded the long sought (from a biogenetic point of view) autumnaline (25.1). The molecular formula of the natural base, $C_{21}H_{27}NO_5$, was measured mass spectrometrically and the major cleavages at m/e 192 (base peak) and m/e 167 expected on the basis of precedent were found (Equation 25.1). Proton magnetic resonance (pmr) confirmed the presence of one N-methyl, three

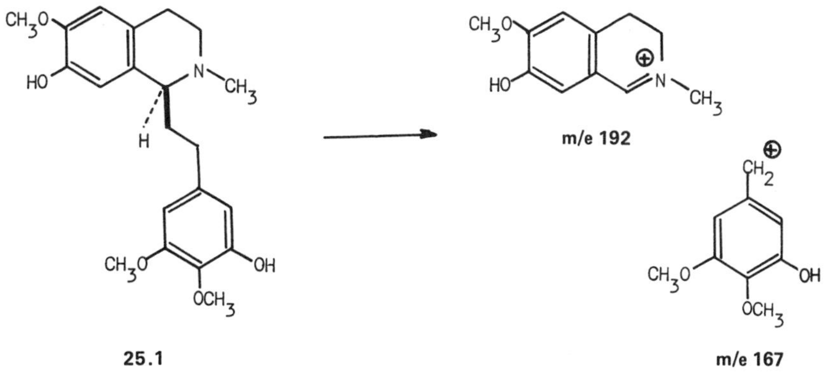

CH_3O HO N CH_3 H CH_3O OH OCH_3 25.1 → CH_3O HO ⊕ N CH_3 m/e 192 ⊕ CH_2 CH_3O OH OCH_3 m/e 167 (Eq. 25.1)

O-methyls, and four isolated aromatic protons, while the infrared and ultraviolet spectra produced evidence for the phenolic hydroxyls and the two independent aryl chromophors. The comparison of the natural levorotatory base with racemic synthetic material completed its identification.

The synthesis of autumnaline (25.1) has been reported and was carried out on labeled precursors [200], as shown in Scheme 25.1. Thus, reaction of 4-benzyloxy-3-methoxyphenethylamine (25.5) with the acid chloride derived from 3-benzyloxy-4,5-dimethoxyphenylpropanoic acid (25.6) yields the amide 25.7, which on Bischler-Napieralski cyclization generates the imine 25.8. N-Methylation, reduction of the imine, and catalytic debenzylation lead to autumnaline (25.1 and its mirror image).

II. KREYSIGINONE

Based upon the isolation of homoaporphine alkaloids (see below) from Kreysigia multiflora and biogenetic arguments analogous to those already discussed for the aporphine alkaloids, the presence of homoproaporphine alkaloids was predicted [199]. Thus, the diphenol 25.9 was synthesized (the method of synthesis was not specified, but a synthesis analogous to that shown as Scheme 25.1 with appropriate changes in substituents would suffice) and oxidized with ferricyanide (Equation 25.2) to a mixture of dienones (25.10) and (25.11). The two spirodienones were separated and characterized by mass, pmr, infrared, and ultraviolet spectrometry. One of them proved identical to kreysiginone (25.2), which was found only after the synthesis had been completed and the dienones could be used as comparison samples for the chromatography of the minor plant bases.

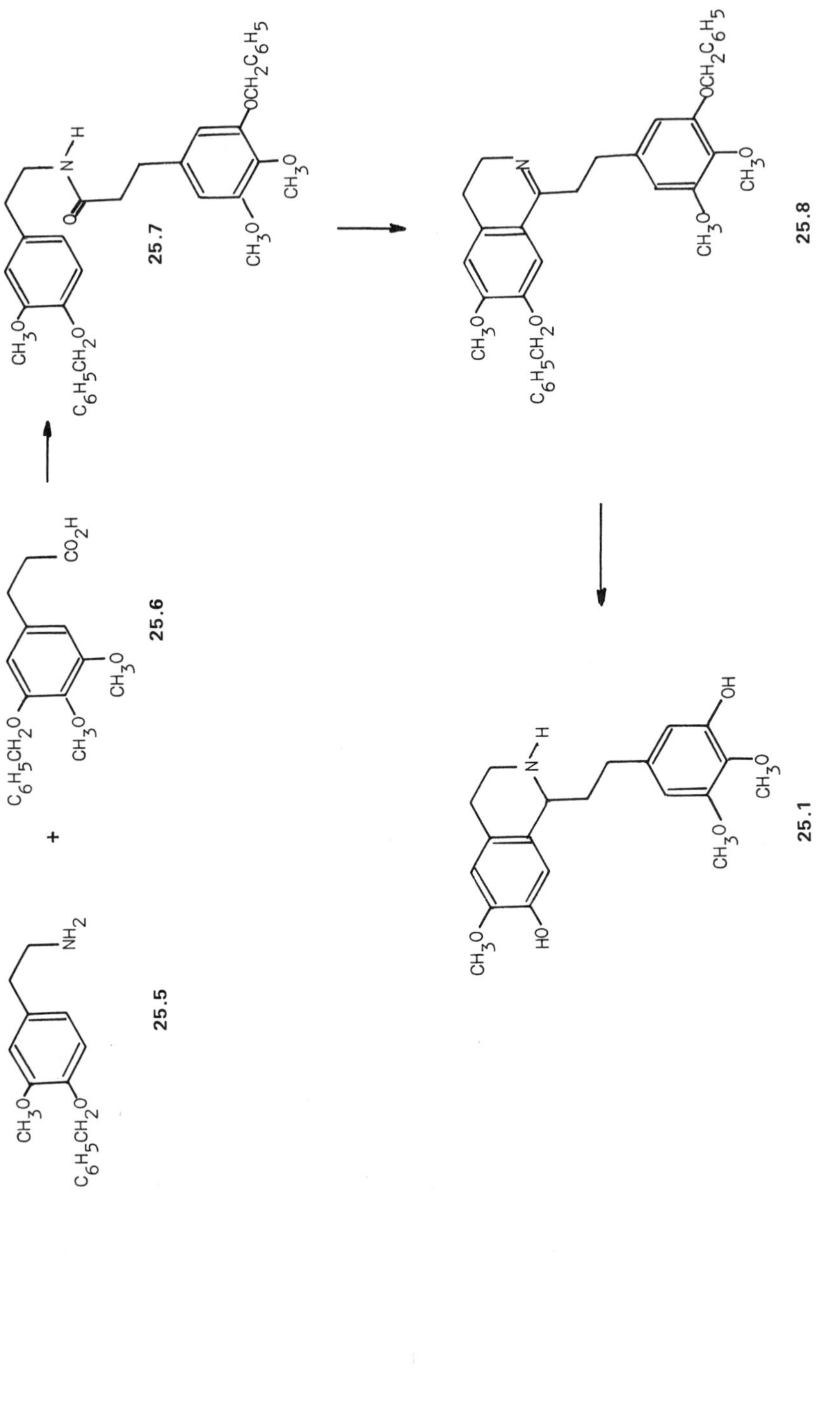

SCHEME 25.1 The synthesis of autumnaline. (After Reference 200.)

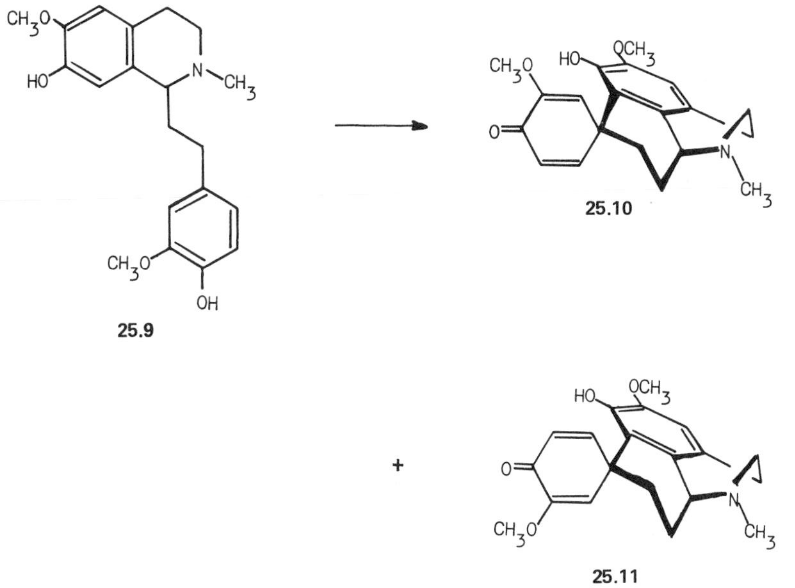

25.9

25.10

(Eq. 25.2)

+

25.11

III. KREYSIGINE

Kreysigine (25.3), $C_{22}H_{27}NO_5$, a phenol, was found as the (-)-enantiomer in Colchicum cornigerum but has also been isolated as a racemic base [201]. The pmr spectrum shows that kreysigine (25.3) possesses one N-methyl and four O-methyl groups as well as two aromatic protons. One of the methoxyl groups appears at high field (δ 3.59 ppm) relative to the normal methoxyl chemical shift (δ 3.83 and 3.86 ppm) for the others. The observation of a single high-field methoxyl is reminiscent of the methoxy-substituted aporphine alkaloids [202] where the methoxyl ortho to the bond between the A and D rings is routinely observed upfield of the other methoxyls. In addition, the ultraviolet spectrum of kreysigine (25.3) is similar to that of the aporphine alkaloids, while the mass spectrum shows the parent ion (M^+) at m/e 385 and the base peak (M-17) at m/e 368, the latter presumably corresponding to loss of the C-1 hydroxyl.

The structure of kreysigine (25.3) was confirmed by synthesis [201]; Scheme 25.2). The diphenolic phenethylisoquinoline 25.12 was oxidized with potassium ferricyanide to the homoproaporphine (25.13) which undergoes an acid-catalyzed dienone-phenol rearrangement to yield (±)-multifloramine (25.14), later found to be naturally occurring. Treatment of this homoaporphine (25.14) with diazomethane yields (±)-kreysigine (25.3) and O-methyl-(±)-kreysigine (25.15).

IV. COLCHICINE

The highly toxic nature of Colchicum has been reported to have been known to Dioscorides, a contemporary of Pliny, who served Nero as a physician and was the first to establish systematically the medicinal value for some 600 plants. The use of Colchicum extracts in the treatment of gout appears to have begun in the sixteenth century, although crystalline preparations of the active principal colchicine (25.4) were first described in 1884. Recent interest in colchicine (25.4) stems from its ability to bring cell division to an abrupt halt at a particular stage.

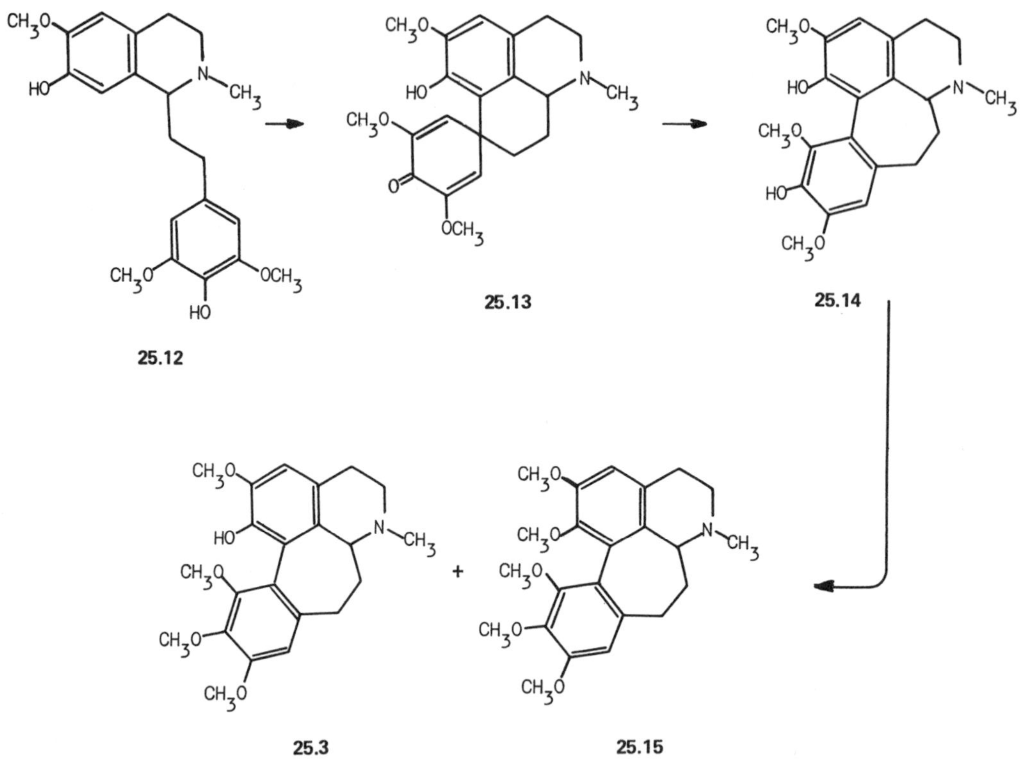

25.12 25.13 25.14

25.3 25.15

SCHEME 25.2 A synthesis of (±)-kreysigine and (±)-O-methylkreysigine. (After Reference 201.)

Solvent-free crystalline colchicine (25.4), mp 143-147°C, corresponds to $C_{22}H_{25}NO_6$, and since it is not basic it does not form the usual acid salts common among plant bases. Acid hydrolysis of colchicine (25.4) yields, depending upon the conditions, as many as five different products (Scheme 25.3). Thus, under very mild conditions, colchicine (25.4) yields colchiceine (25.16), $C_{21}H_{23}NO_6$ and methanol. On further acid treatment (HCl, 150°C, 6 hr), trimethylcolchicinic acid (25.17), $C_{19}H_{21}NO_5$, and acetic acid are formed. It was clear that trimethylcolchicinic acid (25.17) possessed a free amino group as well as the enolic hydroxyl present in colchiceine (25.16). Finally, colchiceine (25.16) can also be converted into dimethylcolchicinic acid (25.18), $C_{18}H_{19}NO_5$, and colchicinic acid (25.19), $C_{16}H_{15}NO_5$. In this way, colchicine (25.4) was recognized as having four methoxyl groups (one readily hydrolyzed) and an N-acetyl moiety. That the amino group of trimethylcolchicinic acid is primary was evidenced by exhaustive methylation which required three equivalents of methyl iodide.

Now, colchiceine (25.16), which gives a positive ferric chloride test, can on methylation with diazomethane be converted into a mixture of colchicine (25.4) and an isomer, isocolchicine (25.20; Equation 25.3). Additionally, although colchicine (25.4) does not give derivatives which might have been expected on the basis of the presence of a carbonyl group, a carbonyl group was deduced as being present on the basis of hydrogenation data. Thus, colchicine (25.4) on hydrogenation over platinum oxide generates a hexahydro derivative (hexahydrocolchicine, 25.21, $C_{22}H_{31}NO_6$; Equation 25.4) which contains a hydroxyl group and still possesses a double bond, since an epoxide ($C_{22}H_{31}NO_7$) can be formed on treatment of 25.21 with perbenzoic acid. Oxidation of colchicine (25.4) with warm alkaline permangante

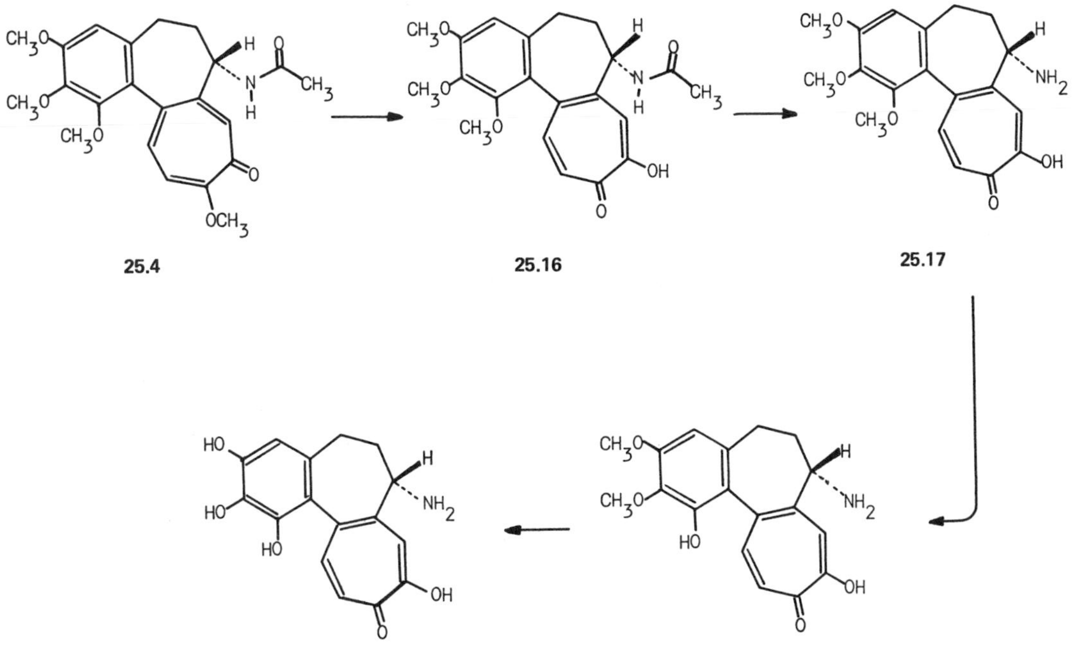

25.4 25.16 25.17

25.19 25.18

SCHEME 25.3 The products of acid hydrolysis of colchicine.

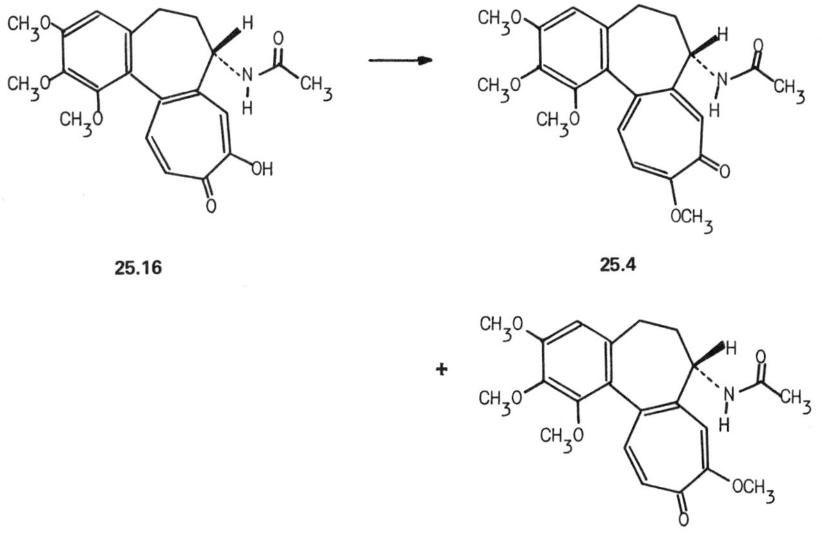

25.16 25.4 (Eq. 25.3)

25.20

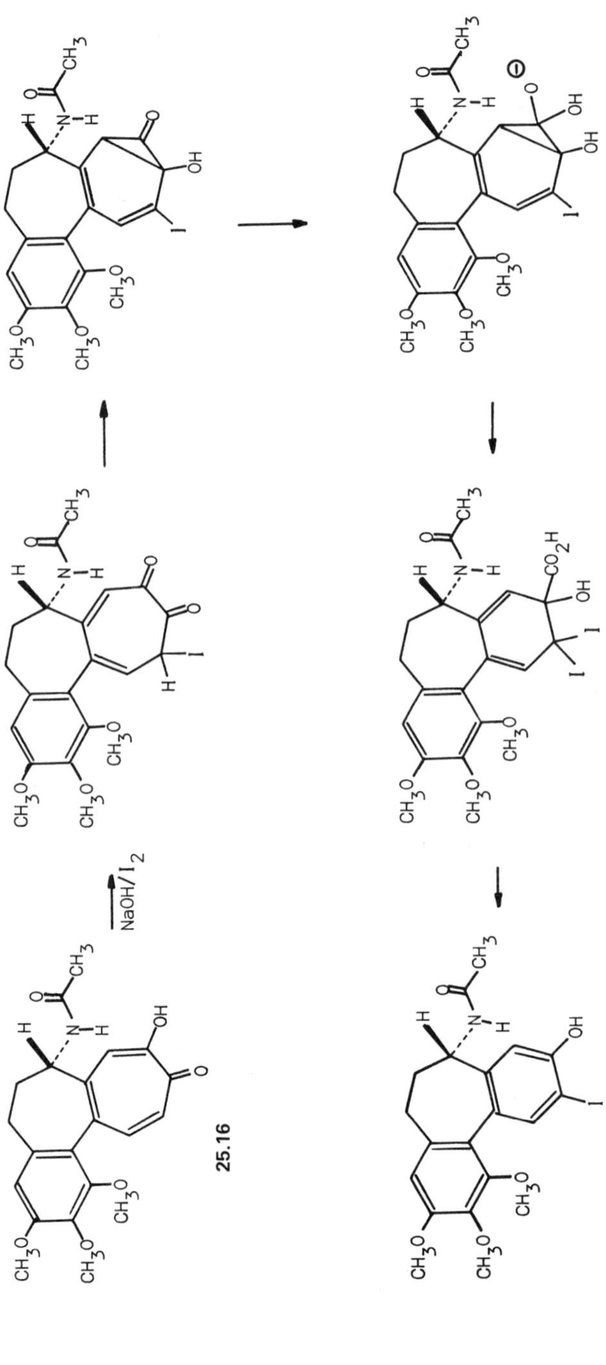

NaOH/I_2

25.16

25.23

SCHEME 25.4 A possible pathway for the formation of N-acetyliodocolchinol from colchicine.

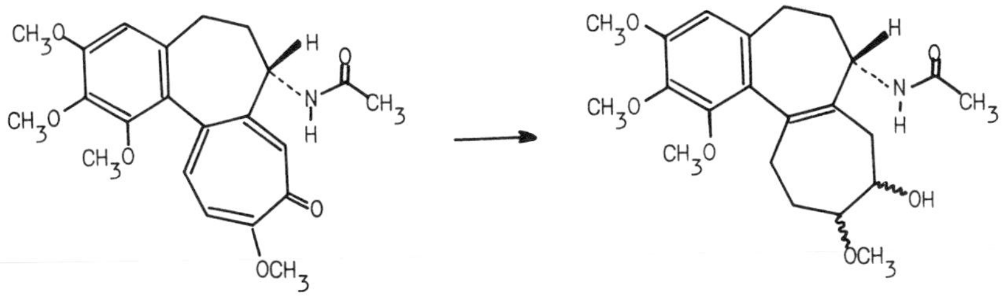

(Eq. 25.4)

25.4 25.21

(Equation 25.5) yields 3,4,5-trimethoxyphthalic acid (25.22). When colchiceine (25.16) is treated with iodine and base, an iodophenol, N-acetyliodocolchinol (25.23), is formed. A possible pathway for this process is shown as Scheme 25.4. N-Acetyliodocolchinol (25.23) proved a most valuable compound in the structure determination of colchicine (25.4).

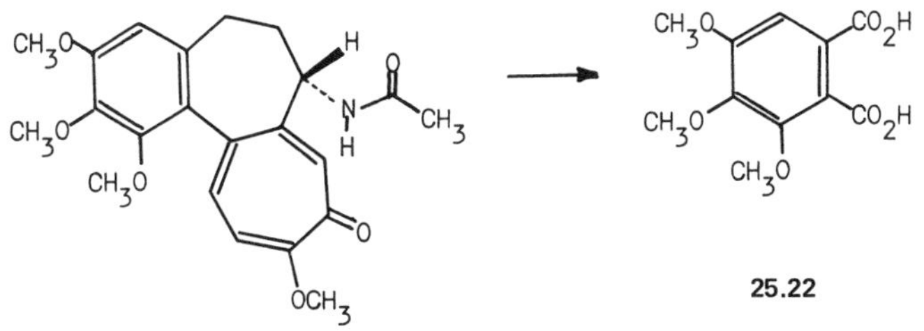

(Eq. 25.5)

25.4 25.22

As shown in Scheme 25.5, methylation of 25.23 yields N-acetyliodocolchinol methyl ether, which on oxidation, first with nitric acid and then with alkaline permanganate, yields 3-iodo-4-methoxyphthalic acid (25.24); while reduction of the iodophenol 25.23 with zinc and base gives N-acetylcolchinol (25.25) which forms a methyl ether (25.26), N-acetylcolchinol methyl ether. When 25.26 is heated with phosphorus pentoxide in xylene, acetamide is eliminated and the same product, namely, deaminocolchinol methyl ether (25.27), formed on hydrolysis of N-acetylcolchinol (25.25) followed by exhaustive methylation and Hofmann degradation, could be isolated. Then, deaminocolchinol methyl ether (25.27) yields a diol on oxidation with osmium tetroxide which can be converted, on further oxidation and treatment with alkali, to 2,3,4,7-tetramethoxy-10-phenanthraldehyde (25.28), the corresponding carboxylic acid (25.29) of which is identical to a synthetic sample. Since it was known that the hydrocarbon 25.30 undergoes an analogous series of transformations, the nature of ring B of colchicine (25.4) is established. The transformation of ring C into the iodophenol 25.25 was difficult to explain, and although it was postulated (on the basis of tropolone chemistry) that ring C was seven-membered, definitive proof was lacking until the x-ray crystal structure

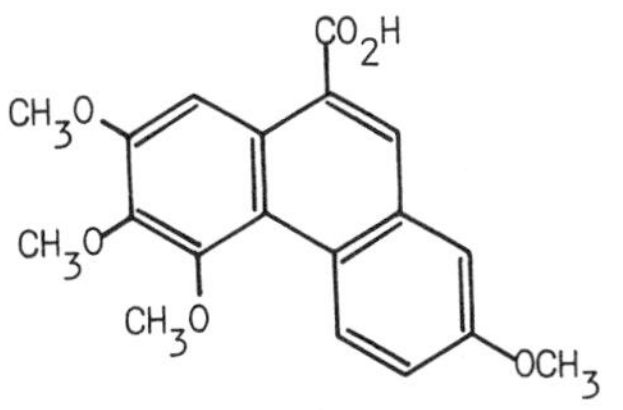

25.29 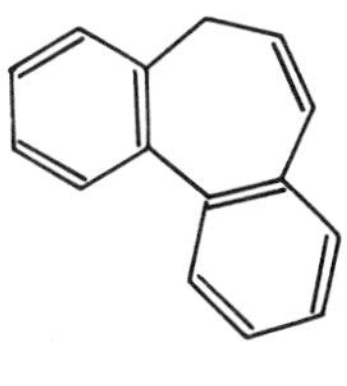 25.30

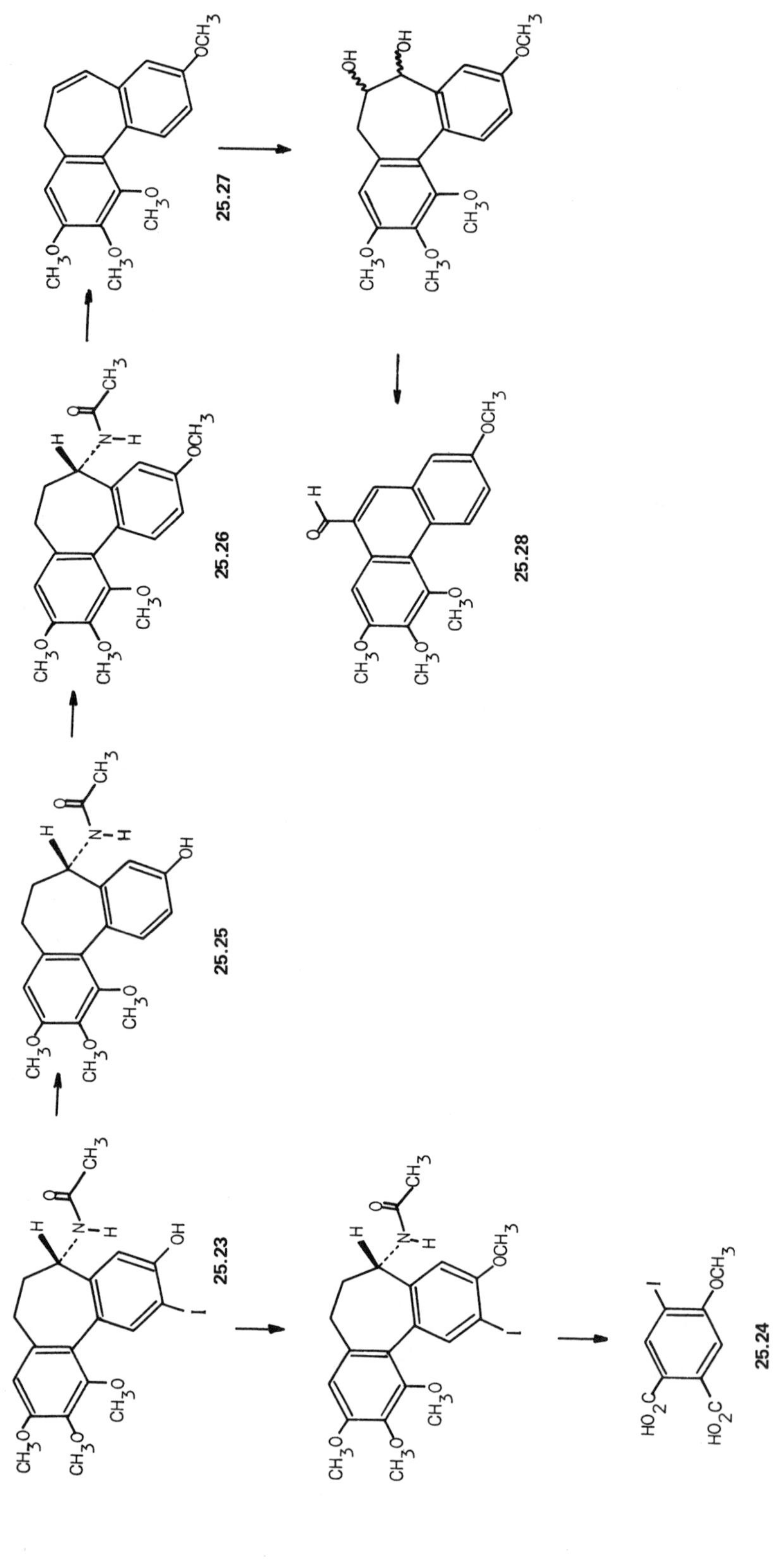

SCHEME 25.5 A degradation of N-acetylcolchinol.

of colchicine (25.4) was complete [203]. Once the structure was known, several very interesting rearrangements could be accounted for; thus, (a) alkaline hydrogen peroxide converts colchiceine (25.16) into N-acetylcolchinol (25.25; Scheme 25.6), while (b) treatment of colchicine (25.4) with methanolic sodium methoxide yields allocolchicine (25.31; Scheme 25.7). The absolute configuration of colchicine (25.4) at the single chiral center has been defined as "S" since strenuous oxidation of colchicine (25.4) results in the formation of N-acetyl-L-glutamic acid (25.32; Equation 25.6; [204]).

A number of syntheses of colchicine (25.4) have been reported since 1959. One of these [205] is shown as Scheme 25.8. The cycloheptanone derivative (25.33) is converted, with acrylonitrile, to the cyanoketone 25.34. A Reformatsky reaction on 25.34 yields tertiary alcohols 25.35, which after hydrolysis to the corresponding diacids can be converted to the lactones 25.36 with N,N-dicyclohexylcarbodiimide followed by diazomethane treatment. Only one of the two dicarboxylic acid diesters (25.36; the minor one) could be converted into the

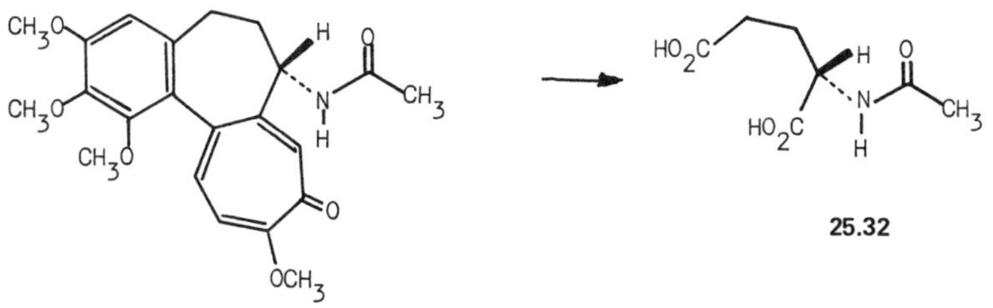

25.4 → 25.32 (Eq. 25.6)

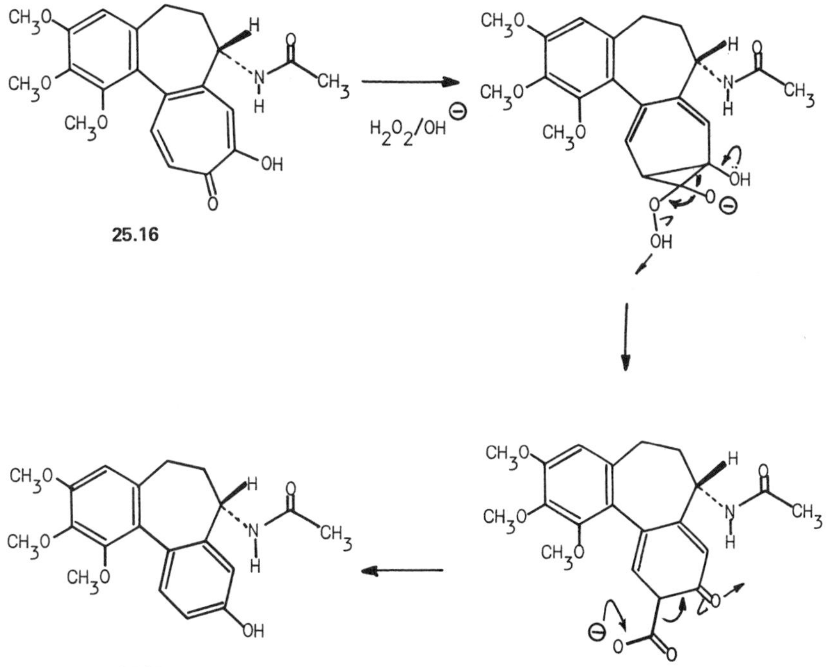

$H_2O_2/OH^{\ominus}$

25.16 → 25.25

SCHEME 25.6 A possible pathway for the formation of N-acetylcolchinol from colchiceine.

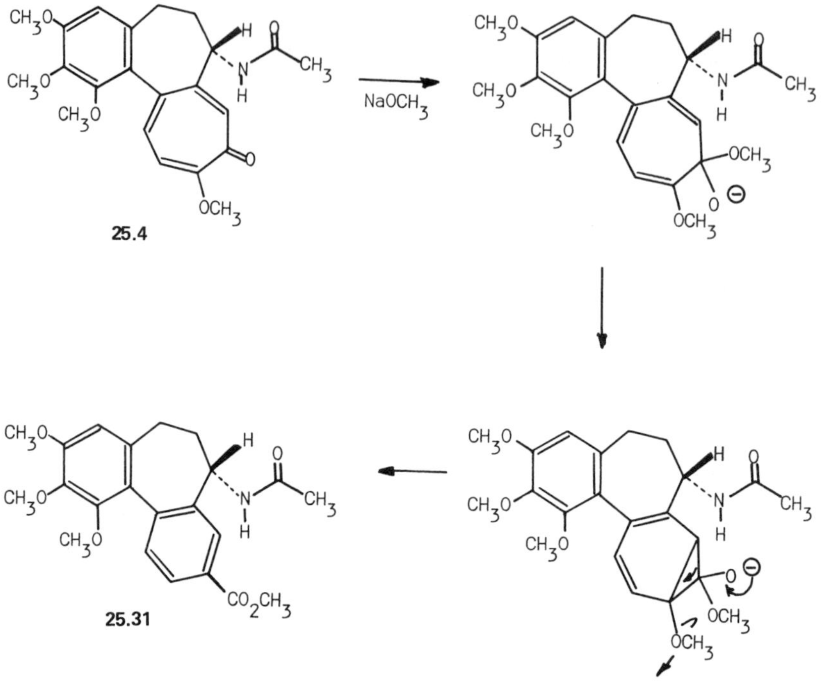

SCHEME 25.7 A pathway for the formation of allocolchicine from colchicine.

α-hydroxyhemiketal 25.37 via the acyloin condensation with sodium in xylene. Oxidation of the hydroxyl group in 25.37 to the corresponding carbonyl with copper acetate and acid-catalyzed (toluene sulfonic acid in benzene) elimination yield the α-hydroxyketone (25.38), which on bromination (NBS) and dehydrohalogenation affords desacetamidocolchiceine (25.39). Desacetamidocolchiceine methyl ether was brominated with N-bromosuccinimide (NBS), the bromine displaced by azide, and the latter catalytically reduced and gently hydrolyzed to the amine (±)-trimethylcolchicinic acid (25.17), which was, in turn resolved, acetylated, and methylated to yield colchicine (25.4).

Finally, several photo isomers of colchicine (15.4) are generated on exposure of the alkaloid to ultraviolet light by the two expected disrotatory modes of cyclization of the tropolone ring (Equation 25.7). Thus, β- and γ-lumicolchicine have been identified as 25.40 and 25.41, respectively, by careful examination of their pmr spectra and reduction of the respective carbonyl groups to the corresponding alcohols. (α-Lumicolchicine is a dimer of β-lumicolchicine, 25.40.)

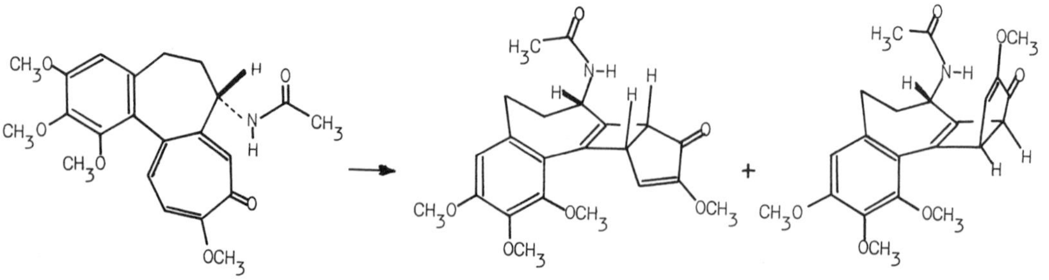

25.4 → 25.40 + 25.41 (Eq. 25.7)

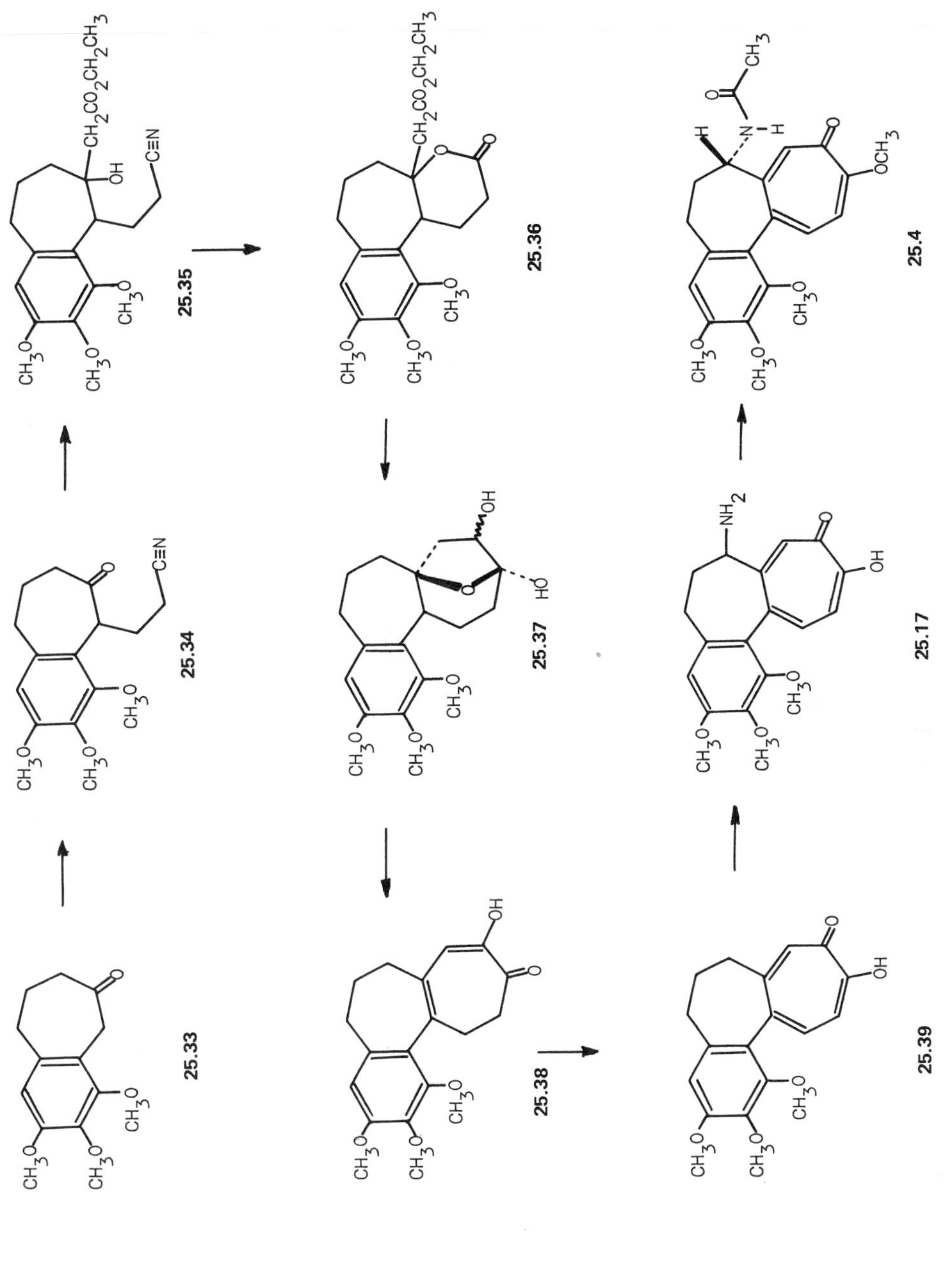

SCHEME 25.8 A total synthesis of colchicine. (After Reference 205.)

SELECTED READING

J. W. Cook and J. D. Loudon, in The Alkaloids, Vol. 2, R. H. F. Manske (ed.). Academic, New York, 1952, pp. 261 ff.

M. Shamma, The Isoquinolone Alkaloids. Academic, New York, 1972.

W. C. Wildman, in The Alkaloids, Vol. 2, R. H. F. Manske (ed.). Academic, New York, 1952, pp. 247 ff.

W. C. Wildman and B. A. Pursey, in The Alkaloids, Vol. 11, R. H. F. Manske (ed.). Academic, New York, 1968, pp. 407 ff.

26

CHEMISTRY OF THE BETACYANINS: C_6-C_3 + C_6-C_3 ALKALOIDS DERIVED FROM TYROSINE

> Always do right. This will gratify some people and astonish the rest.
>
> Mark Twain (Samuel Clemens), 1901

The structure of the brightly colored (red-violet and yellow) alkaloids found in the order Centraspermae (cacti, red beet, etc.) remained unknown until the 1960s. In part, this might have been due to the fact that these pigments are water soluble (they exist as Zwitterions), although, in the 1930s, a group under the direction of Sir Robert Robinson began an intensive investigation of these pigments (then called "nitrogenous anthoxyanins") which failed to reach fruition since problems in obtaining crystalline materials prevailed. Of the several dozen pigments, now called betacyanins, most of the structural work has been carried out on only two aglycones (which are obtained on hydrolysis of the betacyanins, now identified as β-glucosides). The chemistry described here will thus concentrate on these materials.

When the aglycones betanidin (26.1) or isobetanidin (26.2) as their respective hydrochloride salts, in methanol, are treated with diazomethane, the yellow crystalline di-O-methylneobetanidin trimethyl ester (26.3) is obtained (Equation 26.1). The five O-methyl groups in 26.3 which were absent in the precursors, are readily observed by proton magnetic resonance (pmr). In addition, betanidin (26.1) can be converted by acid-catalyzed esterification into an oily trimethyl ester (26.4; capable of hydrolysis to betanidin, 26.1, and

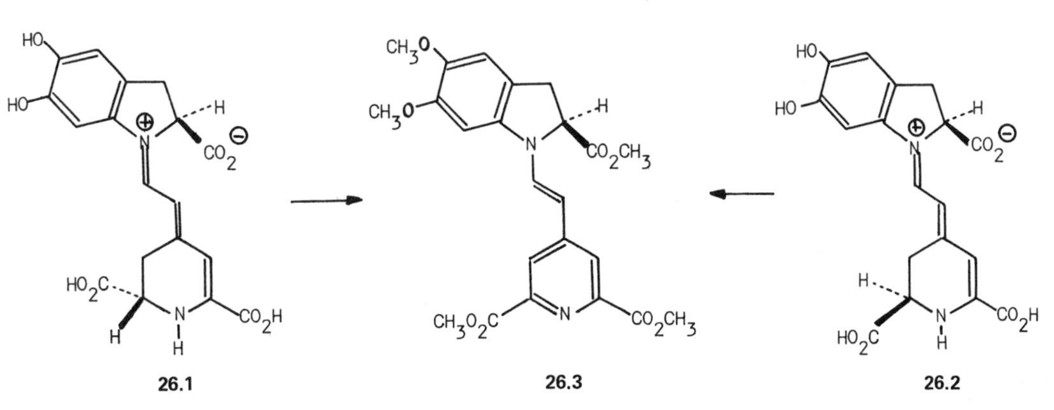

26.1 → 26.3 ← 26.2

(Eq. 26.1)

isobetanidin, 26.2, implying that racemization can occur in either step) which yields a diacetyl derivative (26.5), 5,6-di-O-acetylneobetanidin trimethyl ester (Equation 26.2).*

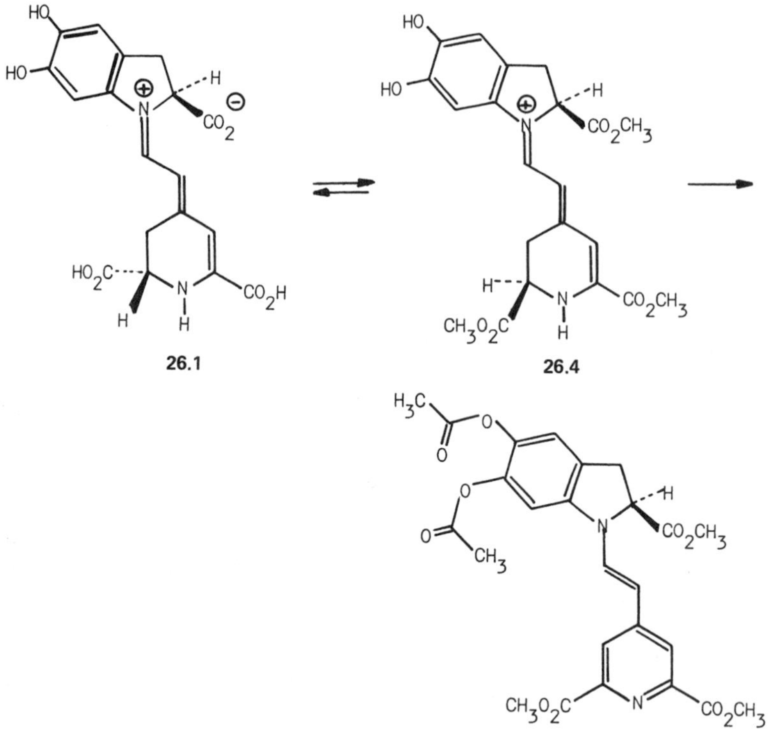

26.1 26.4

(Eq. 26.2)

26.5

Alkaline degradation of betanidin (26.1) yields 5,6-dihydroxy-2,3-dihydroindole-2-carboxylic acid (S-cyclodopa, 26.6), formic acid, and 4-methylpyridine-2,6-dicarboxylic acid (26.7; Equation 26.3). These fragments account for all the carbon atoms of betanidin (26.1).

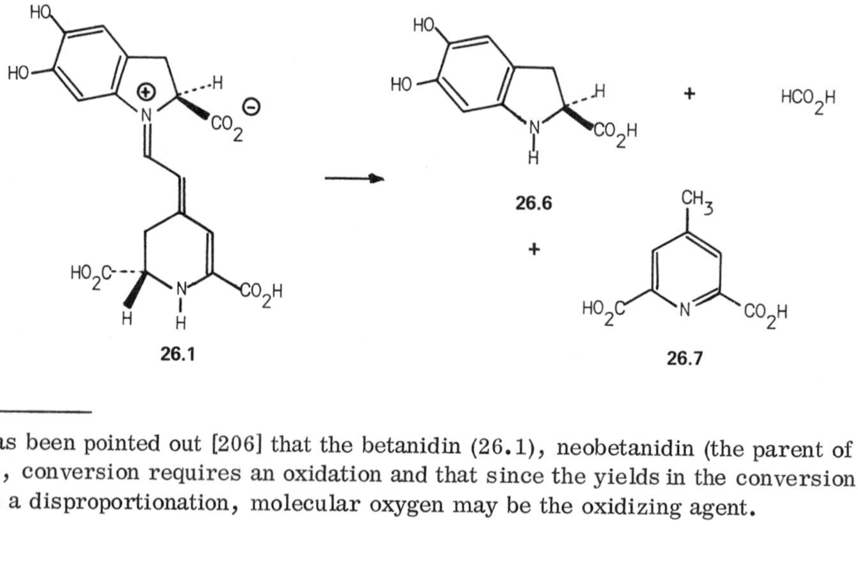

26.1 26.6 + HCO_2H + 26.7

(Eq. 26.3)

*It has been pointed out [206] that the betanidin (26.1), neobetanidin (the parent of 26.3 and 26.5), conversion requires an oxidation and that since the yields in the conversion often preclude a disproportionation, molecular oxygen may be the oxidizing agent.

When the pmr spectra of N-acetyl-5,6-diacetoxy-2,3-dihydroindole-2-carboxylic acid methyl ester (26.8) and 4-methylpyridine-2,6-dicarboxylic acid methyl ester (26.9) were compared to the spectrum of the diacetyl derivative 26.5, it was apparent that only 2

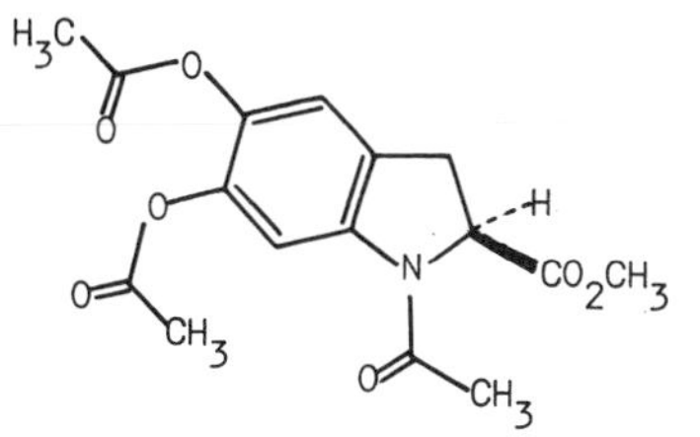

26.8

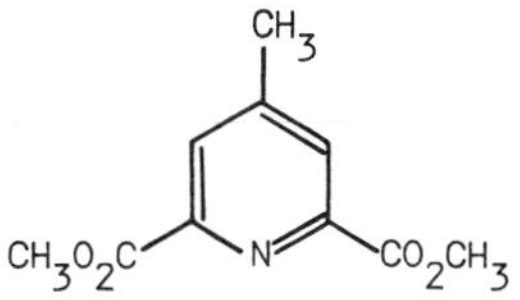

26.9

one-proton doublets assigned to the vinylene connecting piece were absent. Palladium-catalyzed disproportionation converts di-O-methylneobetanidin trimethyl ester (26.3) to the corresponding di-O-methyl-2,3-dehydro-11,12-dihydroneobetanidin trimethyl ester (26.10; Equation 26.4) which was readily identified by pmr. The same change had been observed in

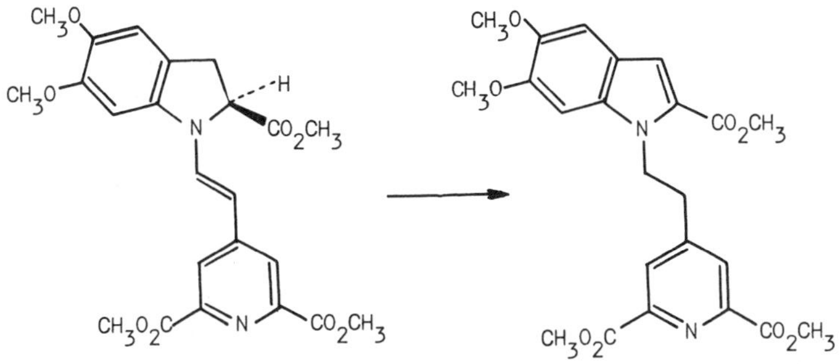

(Eq. 26.4)

26.3

26.10

the model compound N-stryrylindoline (26.11). Then, the formulation of betanidin (26.1) and isobetanidin (26.2) as containing reduced pyridine rings was deduced by careful examination of their respective pmr spectra (as hydrochlorides in trifluroracetic anhydride; [207]).

Now, since both betanidin (26.1) and isobetanidin (26.2) yield S-cyclodopa (26.6) on hydrolysis, it was concluded that they differ in the configuration at C-15, where another carboxylate residue was attached. Furthermore, since other betaxanthins, e.g., indicaxanthin (26.12), can be generated from betanidin (26.1) by hydrolysis (water-SO_2) followed by basification in the presence of the appropriate amino acid [i.e., for indicaxanthin, 26.12; L-proline (26.13) would be used; Equation 26.5), and since indicaxanthin (26.12) has been oxidized to L-aspartic acid (26.14) in hydrogen peroxide (Equation 26.6), betanidin must be 26.1 and isobetanidin (26.2) its C-15 epimer.

Finally, although the above interconversion can be carried out with a variety of amino acids, a total synthesis of betanidin (26.1) has not yet been recorded nor, indeed, is a synthesis of betalamic acid (26.15) known.*

*Note added in proof: A synthesis of betanidin [474] has been completed.

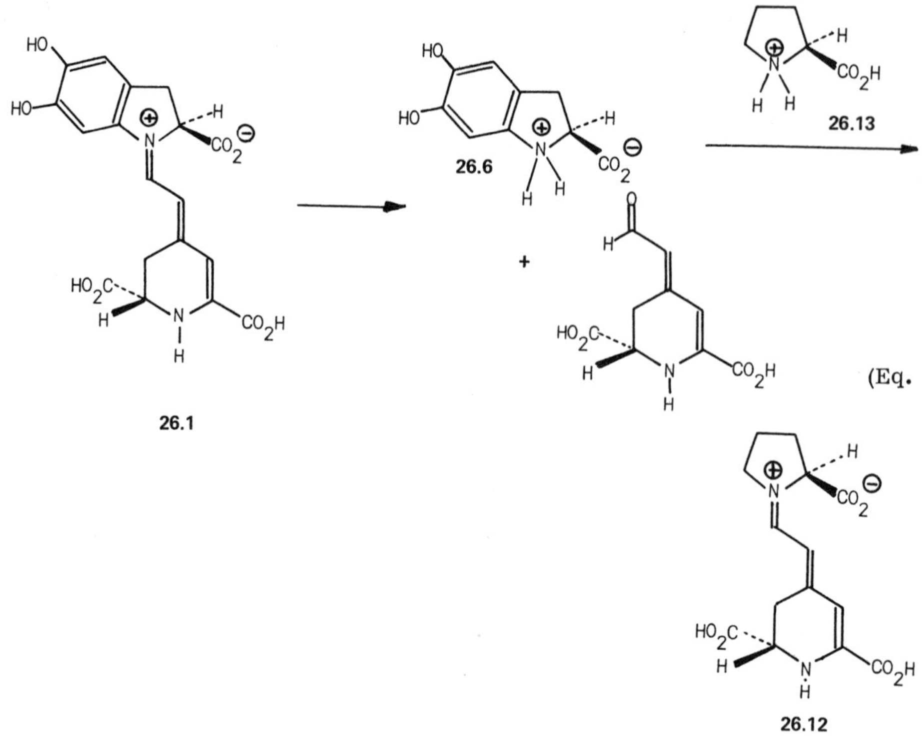

HO HO H $CO_2^{\ominus}$ N HO_2C H N H CO_2H

26.1

HO HO 26.6 N H H H $CO_2^{\ominus}$ + O H HO_2C H N H CO_2H

N H H H CO_2H 26.13

(Eq. 26.5)

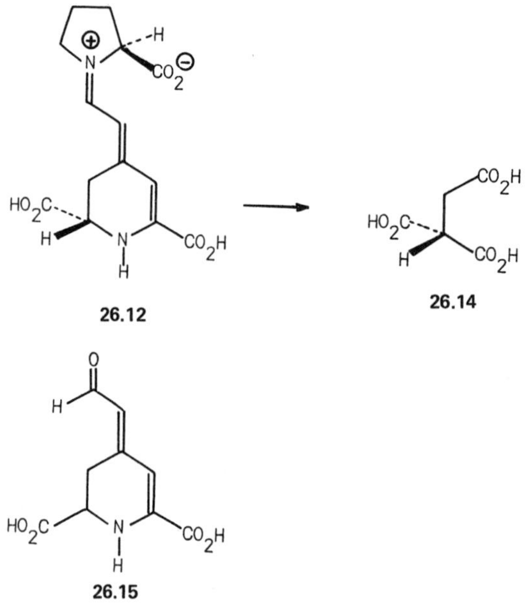

N H $CO_2^{\ominus}$ HO_2C H N H CO_2H

26.12

N H $CO_2^{\ominus}$ HO_2C H N H CO_2H

26.12

CO_2H HO_2C H CO_2H

26.14

(Eq. 26.6)

O H HO_2C N H CO_2H

26.15

SELECTED READING

T. J. Mabry, in Chemistry of the Alkaloids, S. W. Pelletier (ed.). Van-Nostrand-Reinhold, New York, 1970, pp. 367 ff.

Part 6

ALKALOIDS DERIVED FROM TRYPTOPHAN

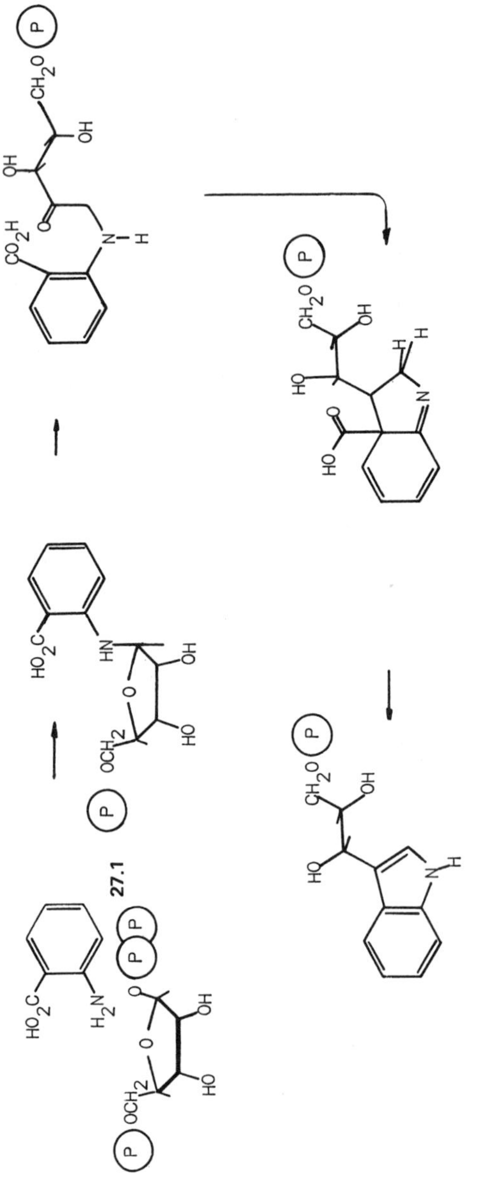

SCHEME 27.1 A pathway to indole-3-glycerol phosphate from anthranilic acid.

Anthranilic acid (27.1) serves, as has been pointed out (Chapter 2), as the precursor for tryptophan (27.2). Scheme 27.1 (from anthranilic acid to indole-3-glycerol phosphate) and Scheme 27.2 (from indole-3-glycerol phosphate to tryptophan) indicate the pathway utilized.

Tryptophan (27.2) may be directly hydroxylated (or partially degraded) to from a relatively simple group of plant bases which shall be considered first. Then, as we have already noted for other alkaloids we have examined, elaboration of the tryptophan (27.2) nucleus can occur by reaction of the amino acid with a C_2 (acetate) or C_3 (pyruvate) fragment. With tryptophan (27.2) this results in formation of the tricyclic β-carboline alkaloids. Second, a C_5-mevalonate-derived (27.3) or a C_{10}-loganin-derived (27.4) fragment can be combined with tryptophan (27.2) to form, respectively, the ergoline and the polycyclic β-carboline alkaloids. The last case to be considered in this part involves alkaloids which result from dimerization of tryptophan (27.2) and tryptophan derivatives.

Given the great complexity of the ring systems to be examined here, the propensity for rearrangement available in the elaborated skeletons, and the active investigation of the more elaborate bases by several capable groups, it is not surprising that more than 1000 tryptophan-derived alkaloids [208] have been characterized. We will actively consider only a small fraction of this number here but shall in passing point to others which lie between those to which detail will be given.

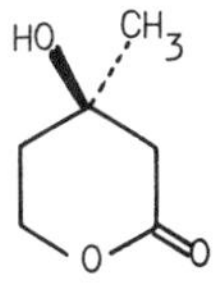

27.3

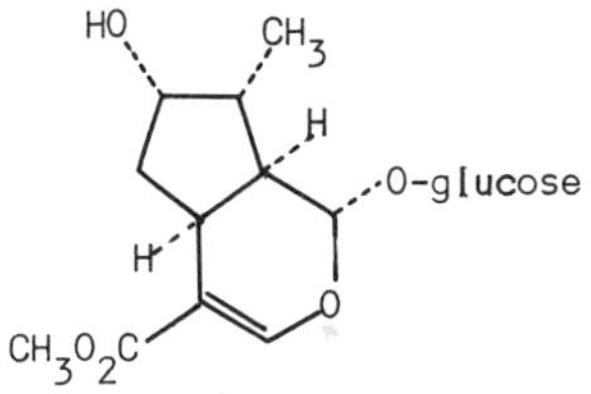

27.4

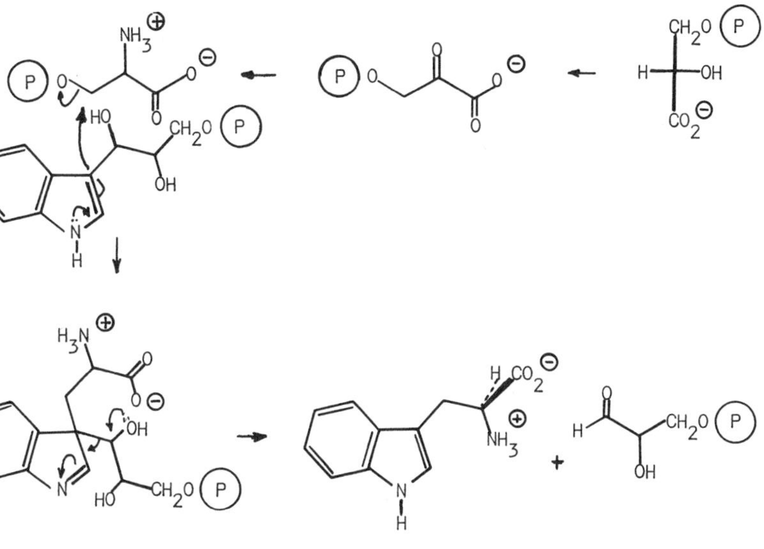

27.2

SCHEME 27.2 A pathway for the formation of tryptophan from indole-3-glycerol phosphate and serine.

Finally, while it is true, as usual, that the methylation pattern of the various bases serves to increase the number of such compounds, the majority nevertheless comes from alternate patterns of elaboration of the simple fragments which make them up. The methyl groups are presumably derived from methionine (27.5), probably via S-adenosylmethionine (27.6).

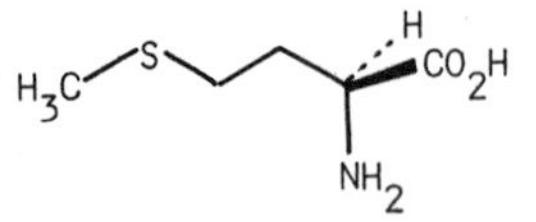

27.5

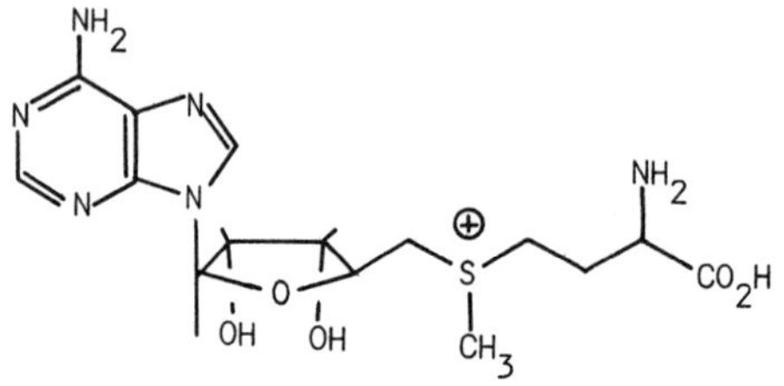

27.6

27

BIOSYNTHESIS OF THE SIMPLE INDOLE ALKALOIDS

I say the very things that make the greatest Stir
An' the most interestin' things,
are things that didn't occur.

S. W. Foss (ca. 1900)

The simpler derivatives of tryptophan (27.2) are widely distributed in the plant and animal kingdoms. Serotonin (5-hydroxytryptamine, 27.7) is found in the banana and stinging nettle, but only in mammals has it been shown to be derived from tryptophan (27.2). Thus, by hydroxylation of S-tryptophan (27.7), S-5-hydroxytryptophan (27.8) is formed, and decarboxylation yields serotonin (27.6; Scheme 27.3). Tryptamine (27.9) does not serve as a precursor of serotonin (27.7), and thus hydroxylation precedes decarboxylation. N,N-Dimethylserotonin (bufotenin; 27.10), a hallucinogen which is presumably generated by methylation of

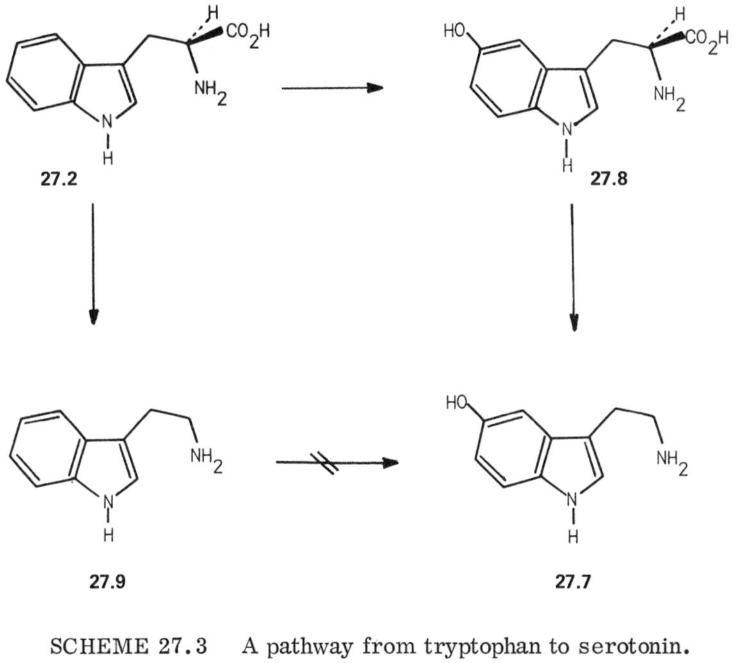

SCHEME 27.3 A pathway from tryptophan to serotonin.

27.2

27.12

SCHEME 27.4 A pathway for the production of gramine from tryptophan which accounts for retention of ^{3}H:^{14}C (▲) in the product. Nitrogen in the gramine (27.12) side chain is presumably derived from exogenous nitrogen (NH_3). (After Reference 210.)

serotonin (27.7) or a serotonin precursor (Equation 27.1) is isolated from some leguminous shrubs (Piptadenia, spp.), certain fungi (Amanita mappa), and secretions of the common toad (Bufo vulgaris). Interestingly, hydroxylation of tryptophan (27.2) in Psilocybe semperiva

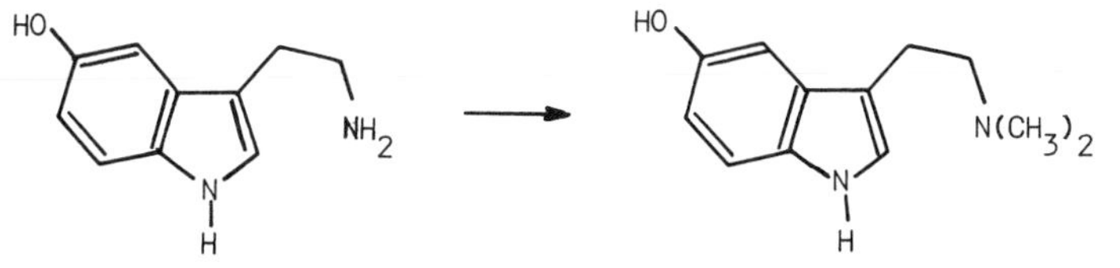

HO NH2 → HO N(CH3)2 (Eq. 27.1)

27.7 27.10

yields the phosphate derivative of N,N-dimethyl-4-hydroxytryptamine, psilocibin (27.11; Equation 27.2), another hallucinogenic compound. It does not yet appear to be known at what stage (i.e., before or after decarboxylation) 4-hydroxylation occurs.

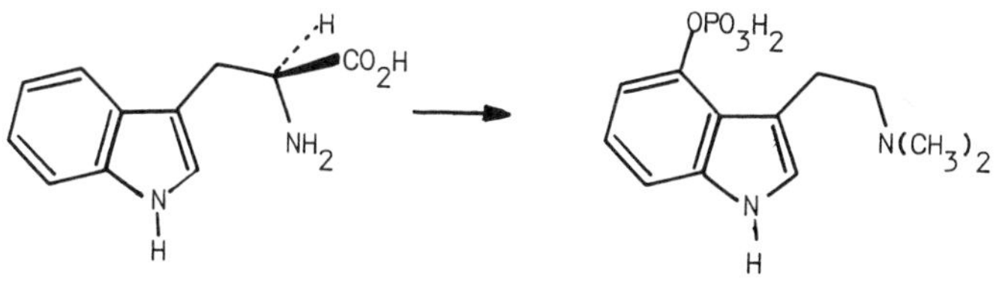

H CO2H NH2 → OPO3H2 N(CH3)2 (Eq. 27.2)

27.2 27.11

The first contribution of ^{14}C and ^{3}H feeding studies to alkaloid biosynthesis (except for methylation studies) dealt with the biosynthesis of gramine (27.12; [209]). A process was suggested (Scheme 27.4) for converting tryptophan (27.2) to gramine (27.12) which involved excision of a carbon atom from the side chain and incorporation of exogenous nitrogen. Retention of the ^{3}H:^{14}C ratio (both originally present at the β-carbon atom of the side chain in tryptophan, 27.2, and both found in the resulting gramine, 27.12, in the same ratio; [210]) confirmed the first portion of the scheme, but as was later shown [211], the second portion could not be correct since the terminal amino group in gramine (27.11) was also tryptophan-derived (27.2). Retention of both the β-carbon and the nitrogen of tryptophan (27.2) in the resulting gramine (27.11) can be rationalized by a different, oxidative, pathway, shown in Scheme 27.5. Alternatively, since the demonstration of nitrogen retention requires unusual quantities of nitrogen-labeled precursor be administered, the obligatory retention of nitrogen under more "normal" conditions remains open to question.

(Scheme 27.5 appears on p. 422.)

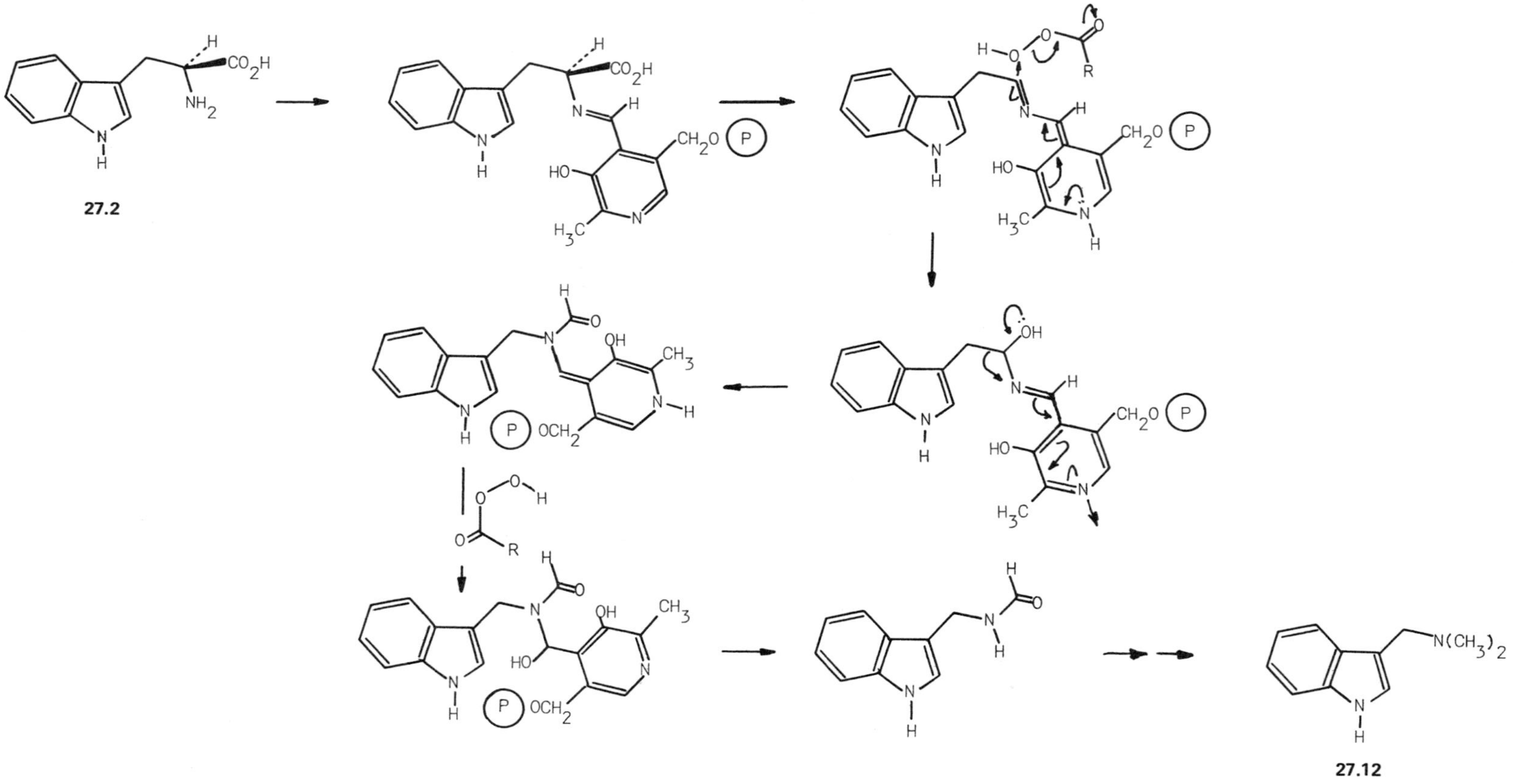

27.2

27.12

SCHEME 27.5 A proposed pathway to gramine from tryptophan which accounts for retention of the nitrogen.

28

BIOSYNTHESIS OF THE TRICYCLIC β-CARBOLINE ALKALOIDS: TRYPTOPHAN + C_2

Progress has been much more general than retrogression.

C. R. Darwin, The Descent of Man

About 20 alkaloids, differing in oxidation state, are known among the β-carboline alkaloids derived from tryptophan (28.1) and a C_2 fragment. We shall consider only two, namely, eleagnine (28.2) and harmine (28.3). These alkaloids, which occur in various plant families,

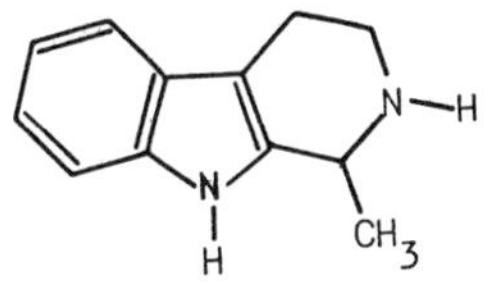

28.2

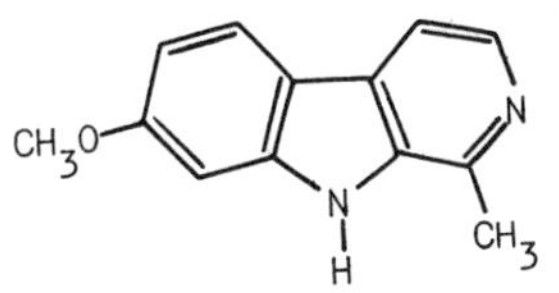

28.3

e.g., Leguminoseae and Rubiaceae, were at one time used therapeutically against tremors in Parkinson's disease, and the seeds of the African rue, Peganum harmala L., which contain harmine (28.3) and related alkaloids, have been used as a tapeworm remedy.

Eleagnine (28.2), isolated from Elagnus angustifolia to which [α-^{14}C]-(±)-tryptophan had been fed [212], indicated, as expected, incorporation of the tryptophan (28.1) moiety; feeding of [1-^{14}C]acetate provided, again as expected, label in the C_2 (Scheme 28.1). As is shown (Scheme 28.1), a pyridoxyl phosphate-catalyzed decarboxylation yields tryptamine (28.4) which condenses with, e.g., acetyl coenzyme A or some other suitable C_2-providing species, to ultimately yield eleagnine (28.2). Presumably, the decarboxylation could have occurred later since data is unavailable regarding the introduction of tryptamine into the alkaloid. On the other hand, in Peganum harmala, it has been shown [213] that harmine (28.3) is generated from [β-^{14}C-^{15}N]tryptamine (28.4) with retention of the ^{14}C:^{15}N ratio and that both [1-^{14}C]acetate and [2-^{14}C]- and [3-^{14}C]pyruvate served as the source of the

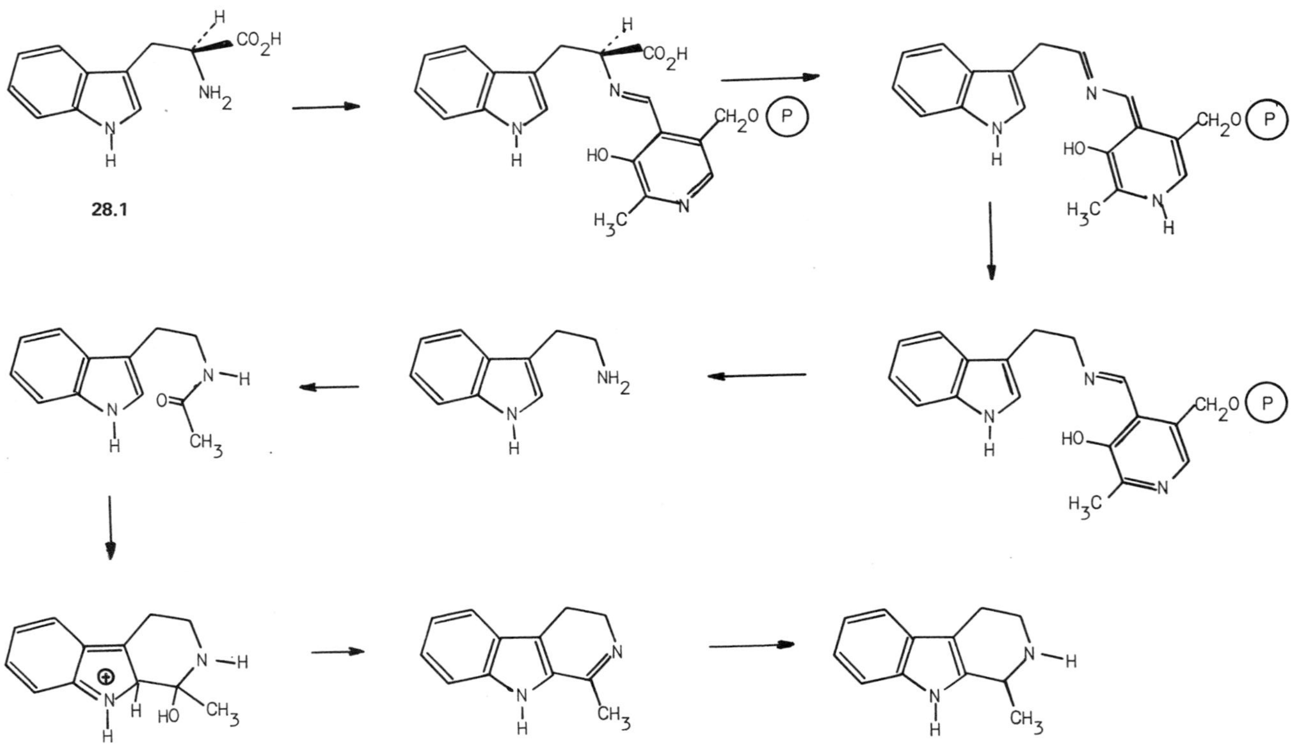

28.1

28.2

SCHEME 28.1 A potential pathway from tryptophan to eleagnine. (After Reference 212.)

two-carbon unit. Scheme 28.2 outlines possibilities which reflect the results obtained. Finally, although feeding of intermediates such as 28.5 demonstrate their incorporation, the obligatory nature of the pathway remains questionable.

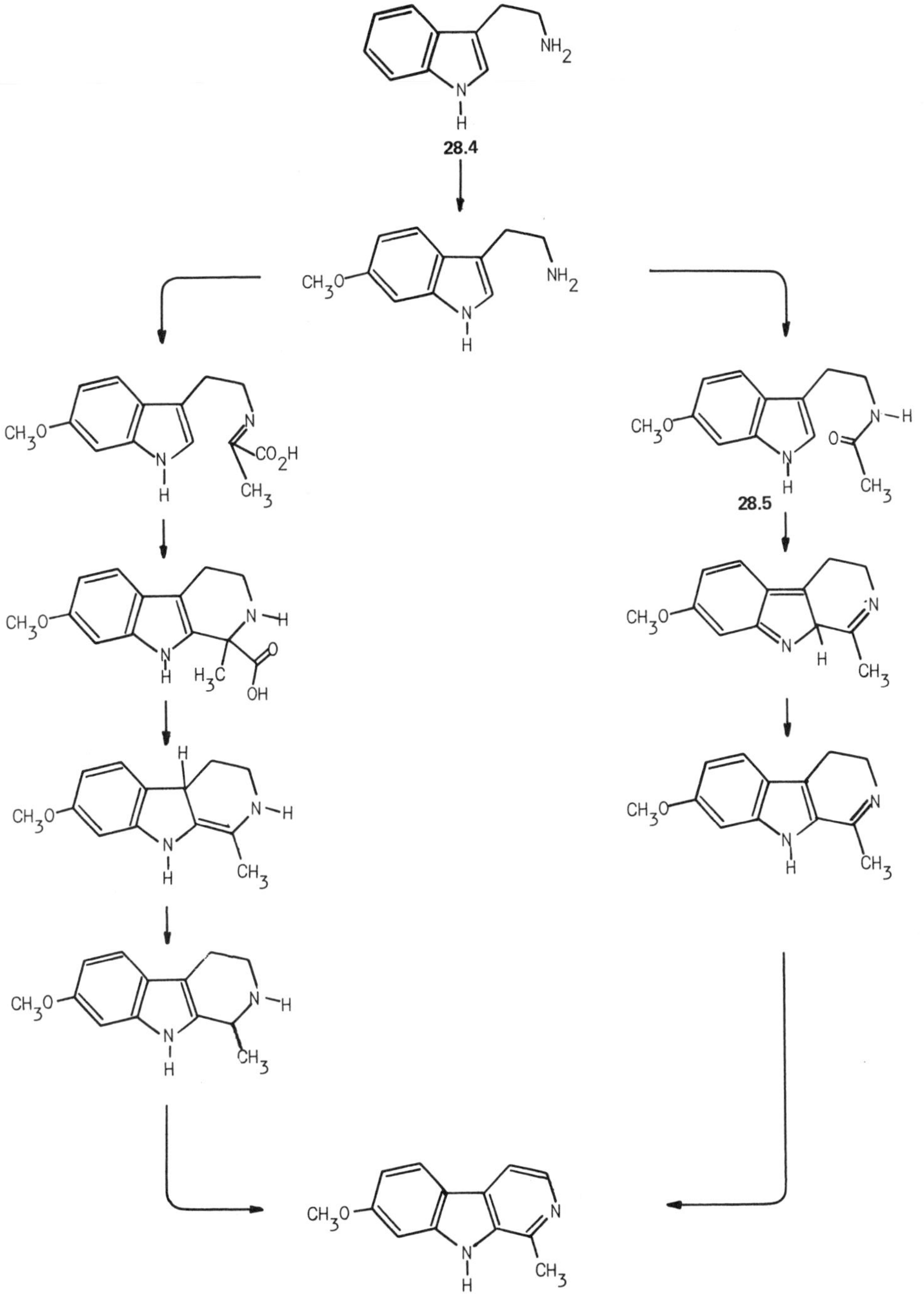

28.4

28.5

28.3

SCHEME 28.2 Potential pathways from tryptamine to harmine utilizing both acetate (or its equivalent) and pyruvate (or its equivalent).

29

BIOSYNTHESIS OF THE ERGOLINE ALKALOIDS: TRYPTOPHAN + C_5

> It oft falls out,
> To have what we would have, we speak not what we would mean.
>
> William Shakespeare, Measure for Measure, Act II, Scene IV

The ergoline alkaloids comprise a numerous family of bases and some related amino acids, the latter found bound to peptide fragments. The compounds we will consider here, namely, elymoclavine (29.1), agroclavine (29.2), and lysergic acid (29.3), are typical and, biosynthetically, the first appears to lie between the other two. These and related materials (see

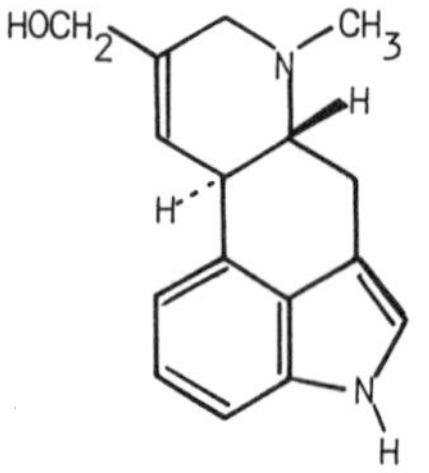

29.1 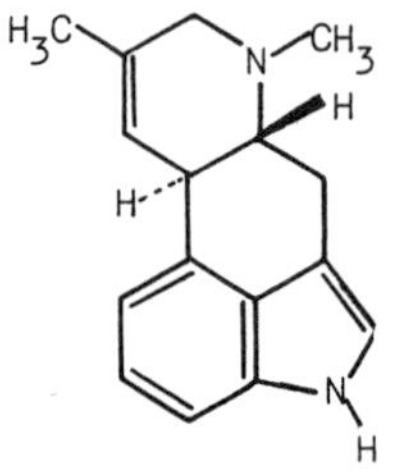 29.2 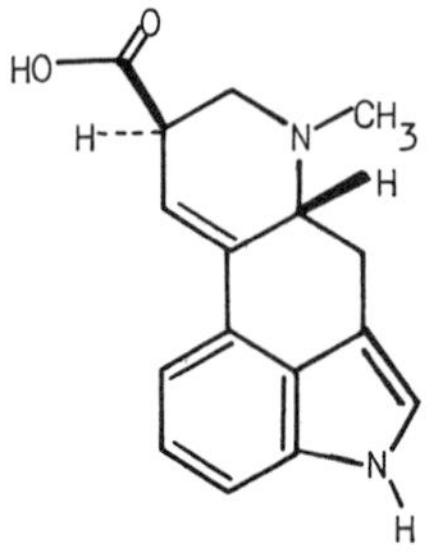 29.3

below) are routinely obtained from the fungus Claviceps purpurea which grows as a parasite on rye. This fungus, which also infests other grasses, can be grown in culture, and some of the ergoline alkaloids have also been reported to occur in higher plants.

As we have come to expect, the N-methyl group found in all these compounds is methionine-derived (via S-adenosylmethionine, 29.4), and, in the early 1960s [214] it was suggested that the ergot bases were possibly formed by a combination of tryptophan (29.5) and a

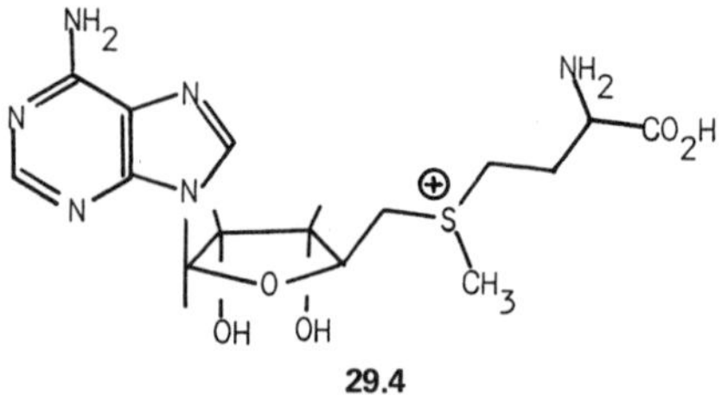

29.4 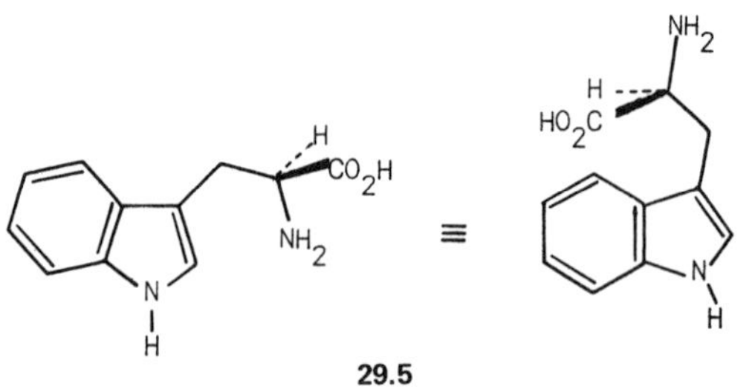 29.5

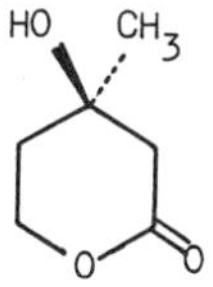

29.6

C_5 fragment derived from mevalonate (29.6). In support of this hypothesis, a number of doubly labeled (^{14}C and ^{3}H) tryptophan (29.5) feeding experiments have been carried out (on the one hand) and also some labeled mevalonate (^{14}C and ^{3}H) feeding experiments (on the other). As indicated below, although more complicated than originally supposed, the general concept has been supported. Thus, L-tryptophan (29.5) is incorporated into the ergot alkaloids with loss of the carboxylate carbon and retention of the α-hydrogen [215]. However, D-tryptophan (enantiomer of 29.5) is incorporated with decarboxylation and loss of the α-hydrogen. This may indicate a conversion of D- to L-tryptophan (29.5) but demands an inversion during alkaloid formation since the elaborated bases (and lysergic acid, 29.3) all have the D (or R) configuration at the chiral center derived from L-tryptophan (39.5).

As to the C_5 unit, the story is still incomplete, but some information is available. In the formation of elymoclavine (29.1), it has been demonstrated [216] that a C_5-mevalonate-derived (29.6) unit is incorporated. It is suggested that mevalonate (29.6) is incorporated via γ,γ-dimethylallylpyrophosphate (29.7), and it appears that the most reasonable course is reaction of 29.7 with L-tryptophan (29.5) to yield 4-(γ,γ-dimethylallyl)tryptophan (29.8; Equation 29.1). However, although it is known that 29.8 is present in the ergot fungus, it is not known if it is an obligatory precursor of the alkaloids.

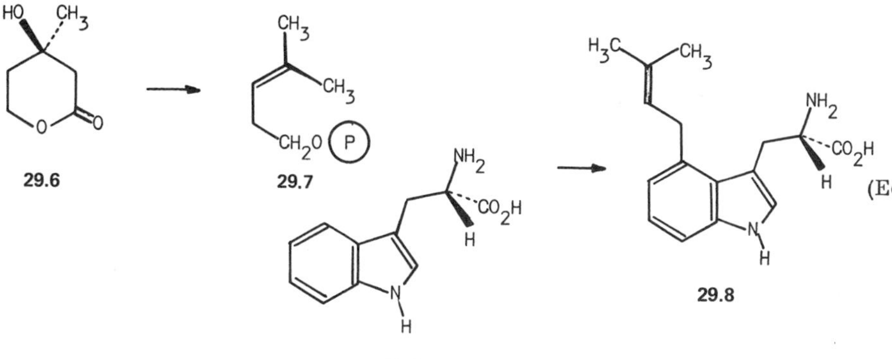

29.6 → 29.7 + 29.5 → 29.8 (Eq. 29.1)

The next course of events permits conversion of 4-(γ,γ-dimethylallyl)tryptophan (29.8) into a set of stereochemically diverse, isolable bases in which a new ring has been formed. These bases are isomers of chanoclavine I (29.9), which possesses a trans ring fusion (Equation 29.2). Of the possible ring fusion isomers, only chanoclavine I (29.9) is further elaborated into the alkaloids to be considered. Interestingly, chanoclavine I (29.9) carries the label from [2-^{14}C]mevalonate in the C-methyl group, not in the hydroxymethyl group, demonstrating that, as is expected on the basis of their cis-trans relationship to the methylene unit, the methyl groups of 4-(γ,γ-dimethylallyl)tryptophan are not identical. However, the tritium from 4R-[4-^{3}H]mevalonate (29.10) but not from 4S-[4-^{3}H]-mevalonate (29.11) is also incorporated into chanoclavine I (29.9). As shown in Scheme 29.1, decarboxylation of 4R-[2-^{14}C, 4-^{3}H]mevalonate (29.10) leads, formally, to the

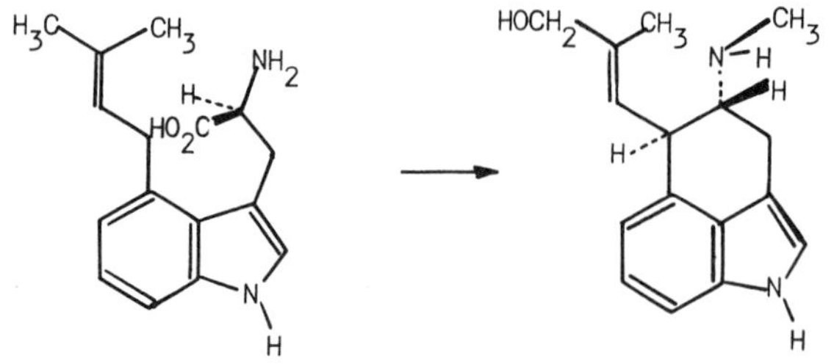

29.8 → 29.9 (Eq. 29.2)

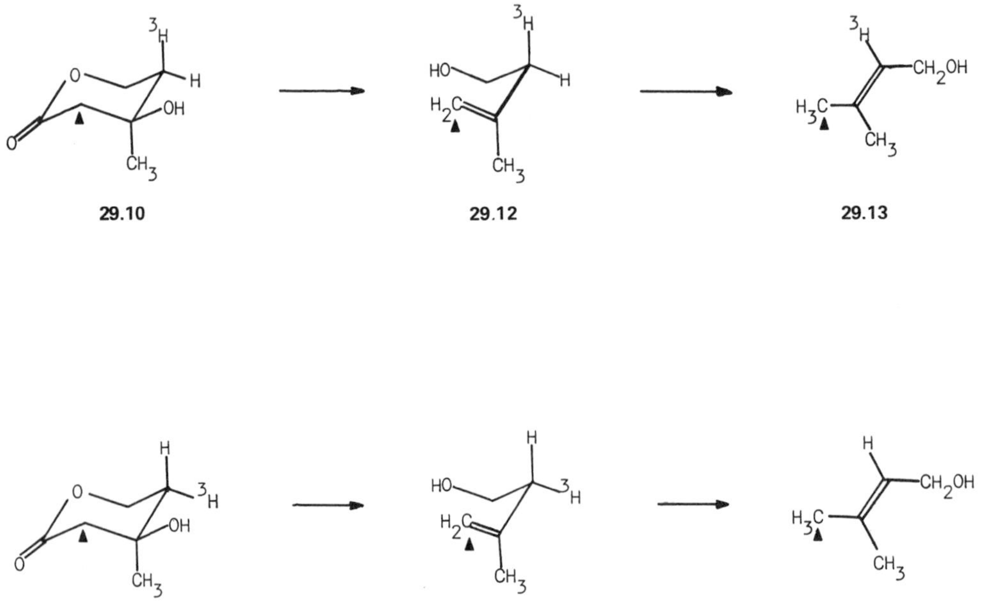

29.10 → 29.12 → 29.13

29.11 → 29.14 → 29.15

SCHEME 29.1 The conversion of mevalonate to γ,γ-dimethylallyl alcohol showing the expected positions of tritium and ^{14}C (▲) label.

specific 2-methyl-4-hydroxy-1-pentene 29.12, which formally isomerizes to the specific γ,γ-dimethylallyl alcohol 29.13; while 4S-[2-^{14}C, 4-^{3}H]mevalonate (29.11) leads, via 29.14, to the specific γ,γ-dimethylallyl alcohol 29.15. Thus, chanoclavine I (29.9), as shown in Equation 29.3, must form with two inversions: one at the tryptophan (29.5) center (L to D) and one in the C_5 side chain where the ^{14}C label starts out cis (Z) to the ^{3}H and ends up, in (29.9), trans (E). A hypothetical series of events to account for the conversion of 4-(γ,γ-dimethylallyl)tryptophan (29.8) into chanoclavine I (29.9) is shown in Scheme 29.2.

It is assumed that oxidation of 4-(γ,γ-dimethylallyl)tryptophan (29.8) to the corresponding allyl alcohol (29.16) leads the process. Dehydration of 29.16 results in generation of the diene 29.17, which on rotation and epoxidation of the terminal double bond, coupled with pyridoxal-catalyzed* decarboxylation, leads to the postulated intermediate 29.18. Cyclization, removal of the pyridoxal, and methylation then lead to chanoclavine I (29.9).

*There is no evidence that pyridoxal pyrophosphate is required for this process.

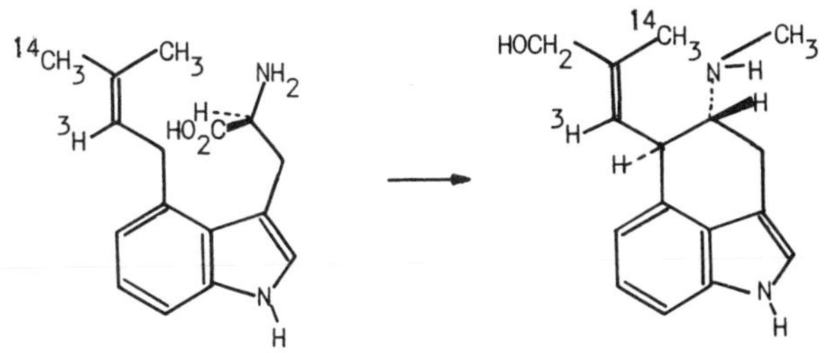

(Eq. 29.3)

29.8 29.9

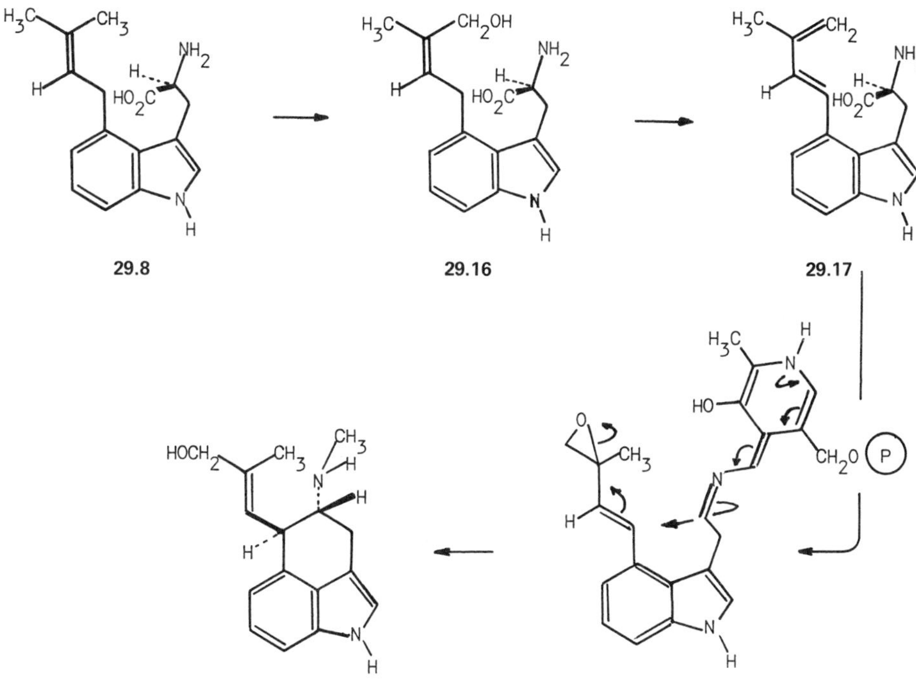

29.8 29.16 29.17

29.9 29.18

SCHEME 29.2 A postulated pathway from 4-(γ,γ-dimethylallyl)tryptophan to chanoclavine I.

Similarly, a potential pathway for the conversion of chanoclavine I (29.9) to elymoclavine (29.1) can be imagined. Such a pathway is shown in Scheme 29.3. Here it is postulated that isomerization of chanoclavine I (29.9) to the corresponding enol 29.19 occurs initially. Chanoclavine-I-aldehyde (29.20) has been shown to lead to elymoclavine (29.1). However, the isomerization of 29.9 to the enol 29.19 also accounts for the loss of one hydrogen from the hydroxymethylene unit and is in concert with feeding experiments with labeled chanoclavine-I-aldehyde (29.20; [217, 218]). Then, rotation, conversion to the aldehyde 29.21, and cyclization lead to the alcohol amine 29.22. Dehydration of 29.22 to the enamine 29.23 and

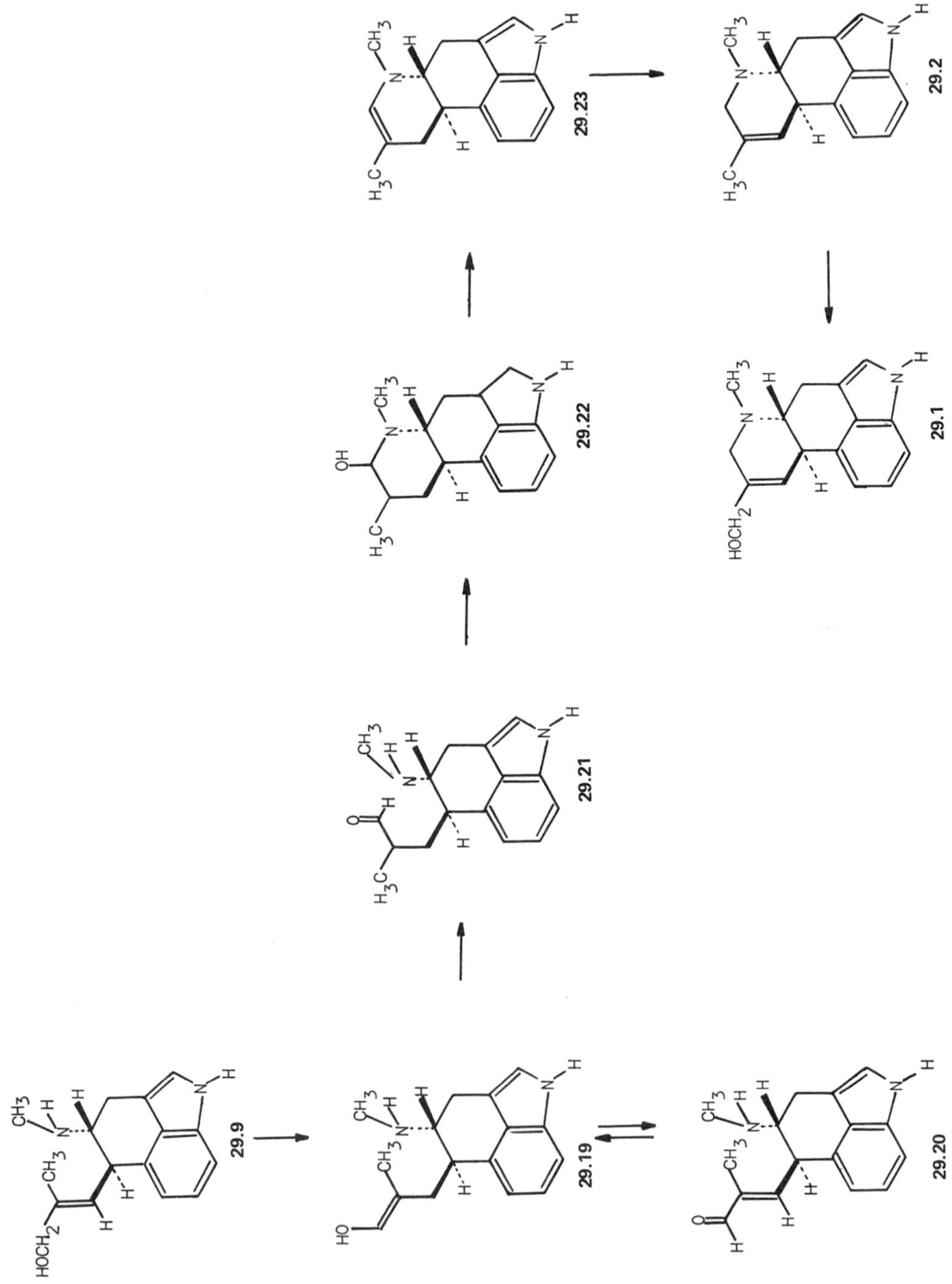

SCHEME 29.3 A postulated pathway for the conversion of chanoclavine I to elymoclavine (via agroclavine).

isomerization of the double bond result in the formation of agroclavine (29.2), the oxidation of which generates elymoclavine (29.1) and, ultimately, lysergic acid (29.3; Equation 29.4).

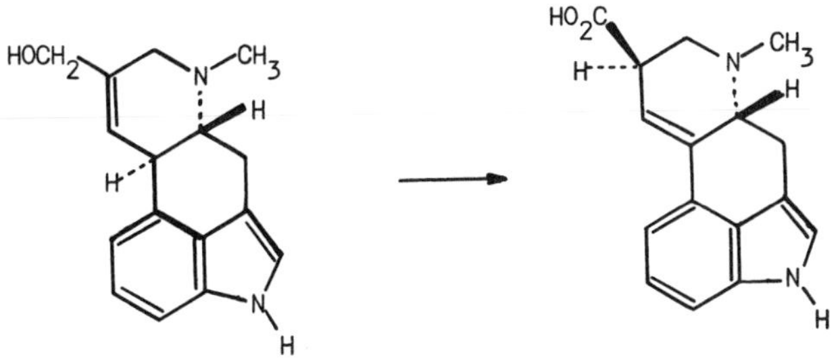

HOCH$_2$ N CH$_3$ H H N H

29.1

HO$_2$C H N CH$_3$ H N H

29.3

(Eq. 29.4)

30

BIOSYNTHESIS OF THE POLYCYCLIC β-CARBOLINE AND RELATED ALKALOIDS: TRYPTOPHAN + C_{9-10}

> One precedent creates another. They soon accumulate and constitute law.
>
> Janius (1772)

The alkaloids to be considered here are members of a group of bases numbering over 700 well-characterized compounds. The permuted structures of these alkaloids are often novel, and numerous single examples of types exist, but, broadly, five categories are widely recognized.* These are the Corynanthé-Strychnos bases such as ajmalicine (30.1), yohimbine (30.2), reserpine (30.3), ajmaline (30.4), sarpagine (30.5), echitamine (30.6), and

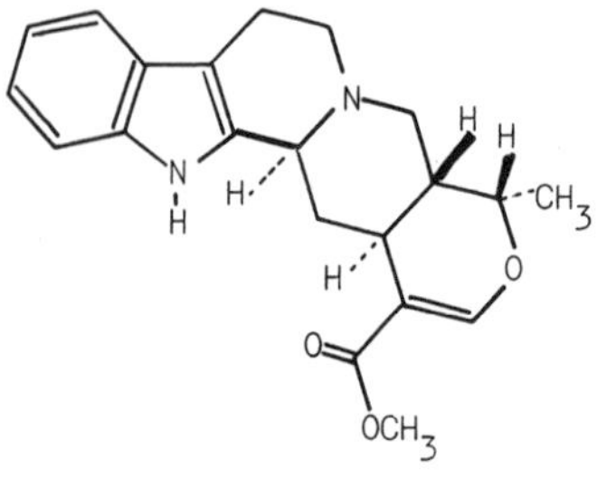

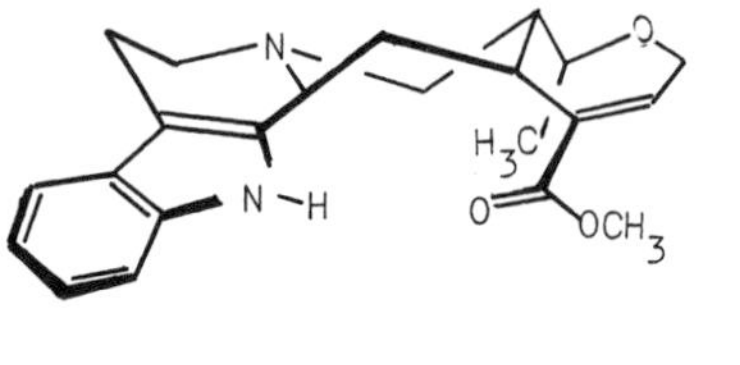

30.1

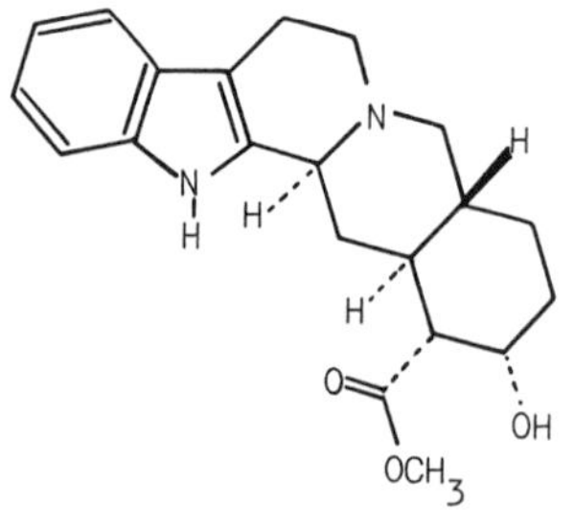

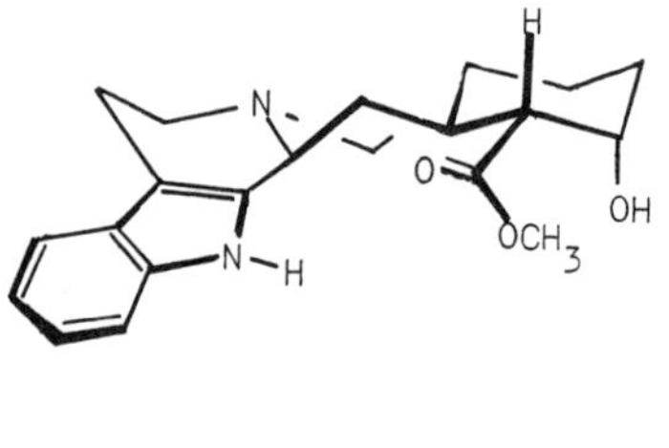

30.2

*A different classification system has been developed, but the details of some of the members of that system lie beyond the scope of this work [219].

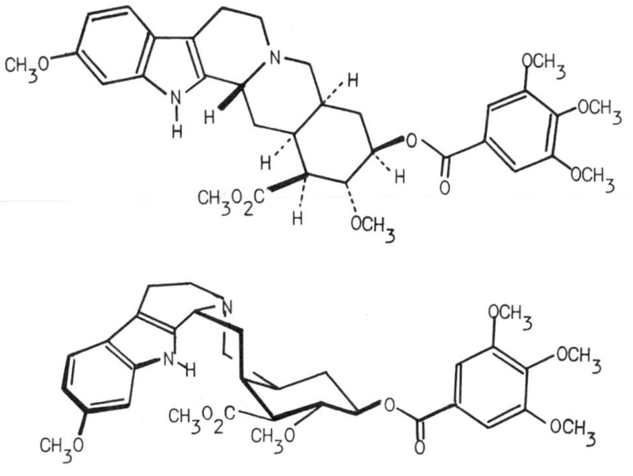

30.3

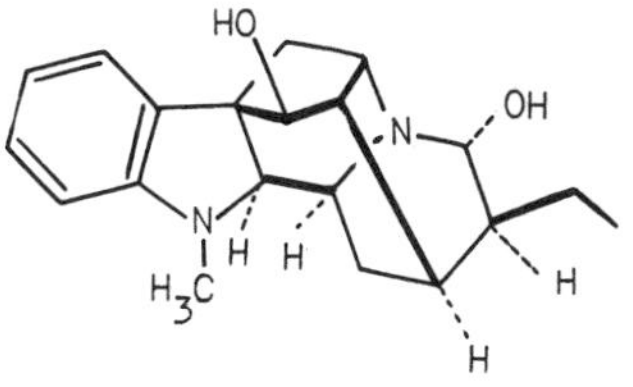

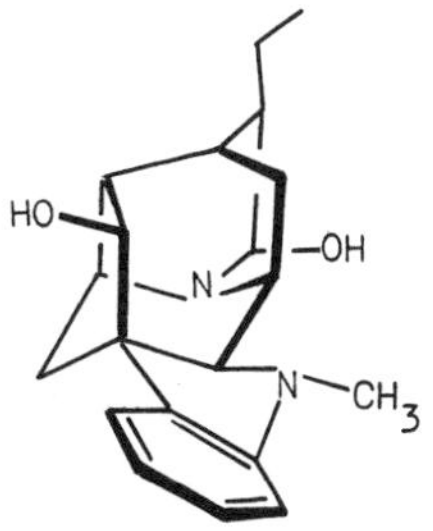

30.4

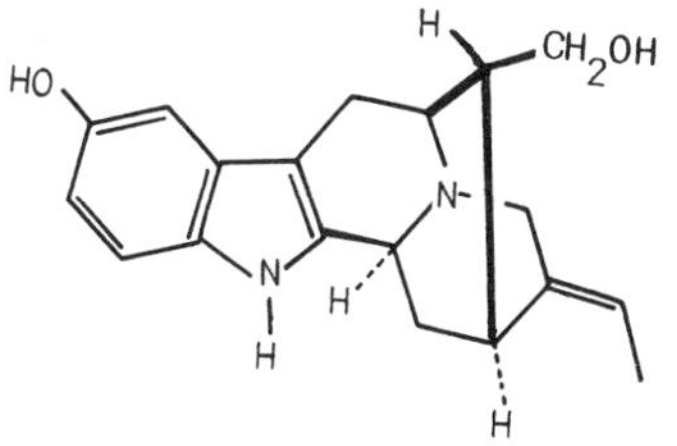

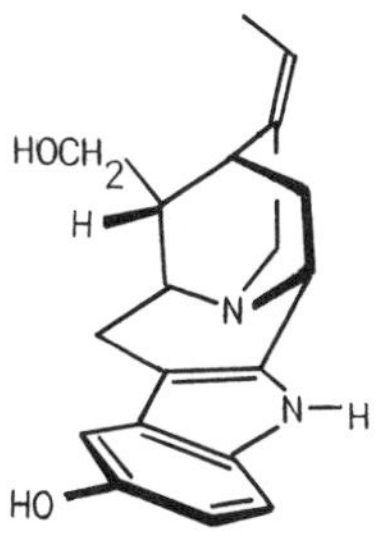

30.5

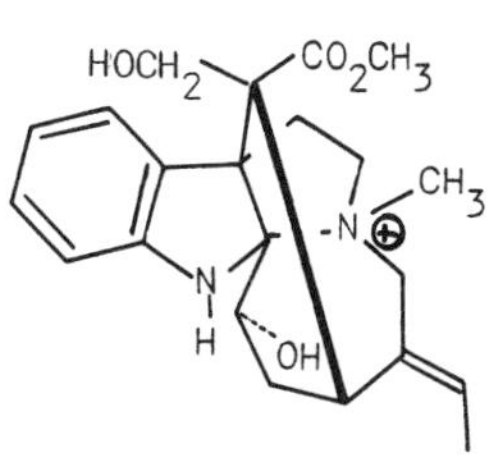

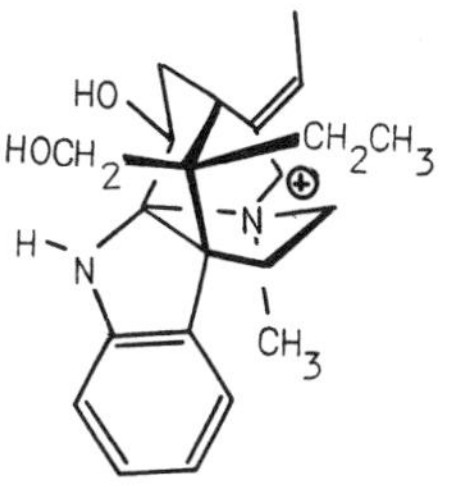

30.6

strychnine (30.7); the Cinchona alkaloids, of which quinine (30.8) is typical; the Iboga alkaloids, such as catharanthine (30.9); the Aspidosperma bases, typified by tabersonine (30.10); and the Eburna family, which includes vincamine (30.11).

Numerous feeding experiments (and, more recently, isolated crude enzyme preparations, [220] have demonstrated that tryptophan (30.12) serves, in all these different bases, as the source of the indole portion, and further, that the ten carbon and two nitrogen atoms of tryptamine (30.13) are carried through into the elaborated alkaloids. The conversion

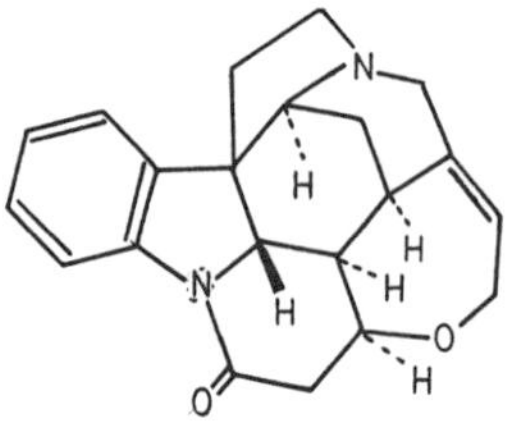

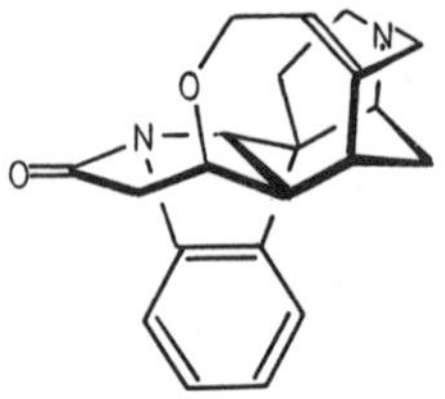

30.7

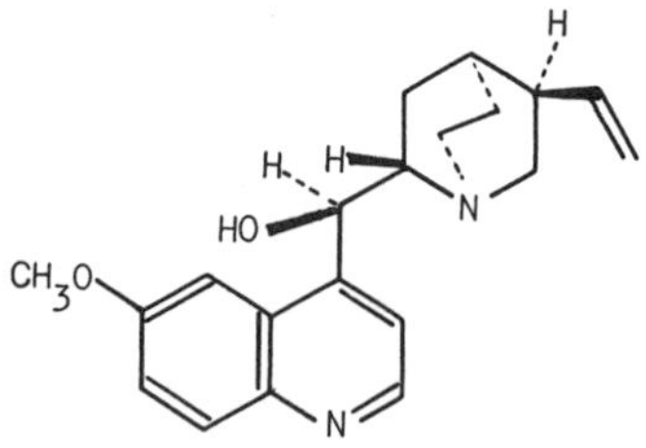

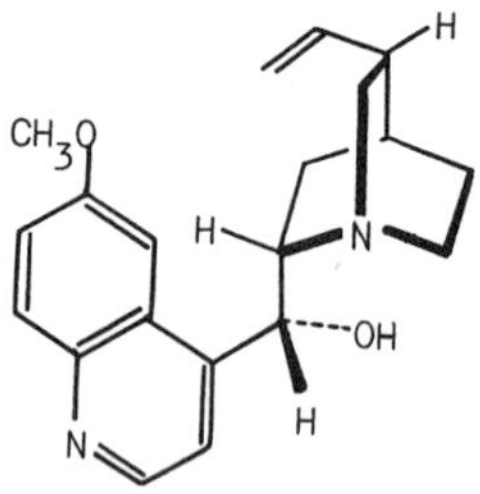

30.8

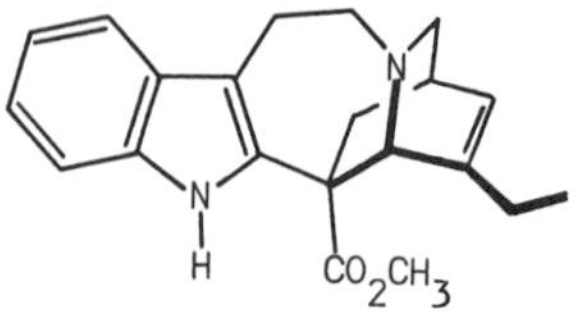

30.9

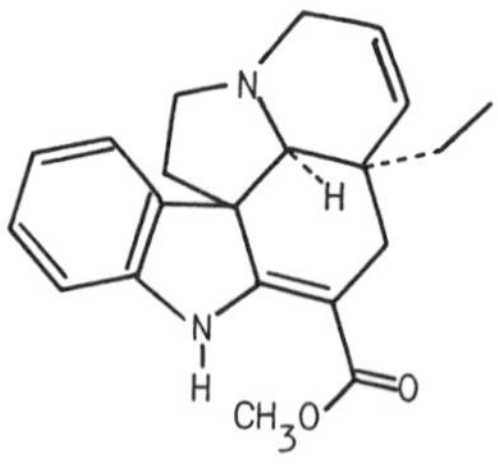

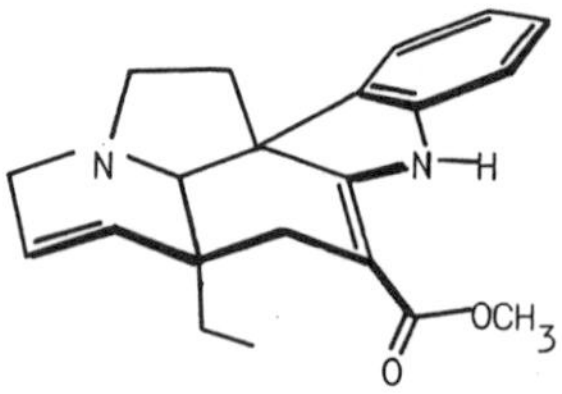

30.10

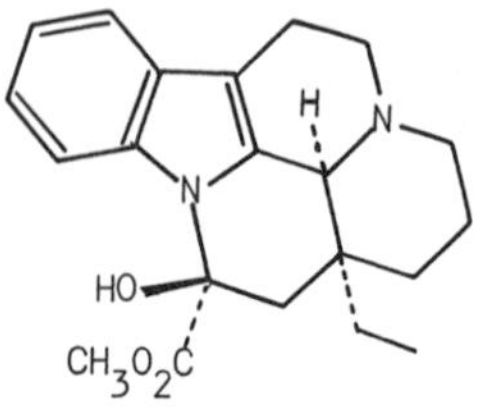

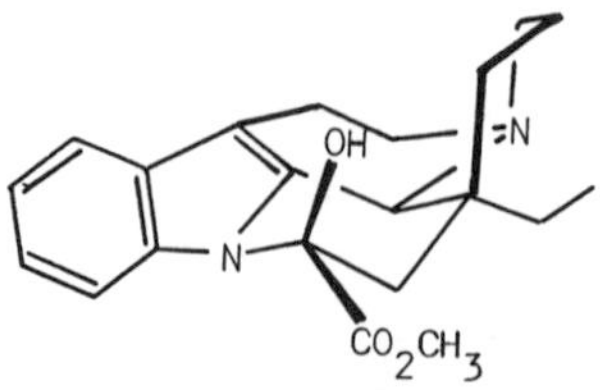

30.11

of tryptophan (30.12) into tryptamine (30.13) presumably involves decarboxylation via a pyridoxal-bound (30.14) intermediate and is shown as Scheme 30.1. Additionally, evidence has recently been accumulated [221] that, in at least one plant system, bases of three of these groups are formed sequentially. Thus, in Catharanthus roseus, it is possible to demonstrate that the combination of tryptophan (30.12) with secologanin (30.15), a monoterpene-derived fragment [222, 223] which we have previously encountered (pp. 29 and 194) leads first to vincoside (30.16) and isovincoside (30.17; Scheme 30.2), then, to the Corynanthé-Strychnos alkaloids, and finally to the Iboga and Asiposperma bases. Therefore, this chapter is arranged in what is presumed to be a more or less sequential elaboration of an initially formed tryptamine (30.13) plus secologanin (30.15) adduct (as vincoside, 30.16) to its more highly oxidized and rearranged offspring.

Before beginning with the bases themselves, however, it is necessary to first examine the origin of the C_{9-10} unit, which has been studied in some detail in connection with these bases, since many facets of the labeling pattern of this unit, derived from labeled mevalonate (30.18), are later utilized in arguments regarding the elaborated alkaloids. As has been pointed out previously, in an abbreviated fashion (Scheme 2.14 and Scheme 29.1), mevalonate (30.18) is converted via geraniol (30.19) and/or nerol (30.20) and loganin (30.21) into secologanin (30.15). These transformations have been investigated in various connections which

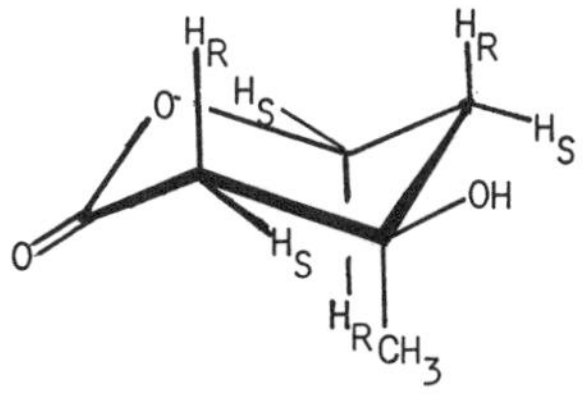

30.18

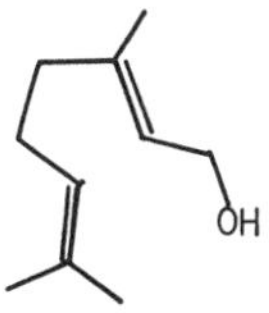

30.19

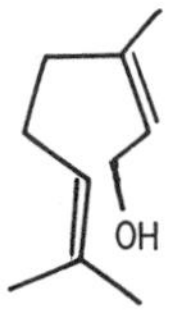

30.20

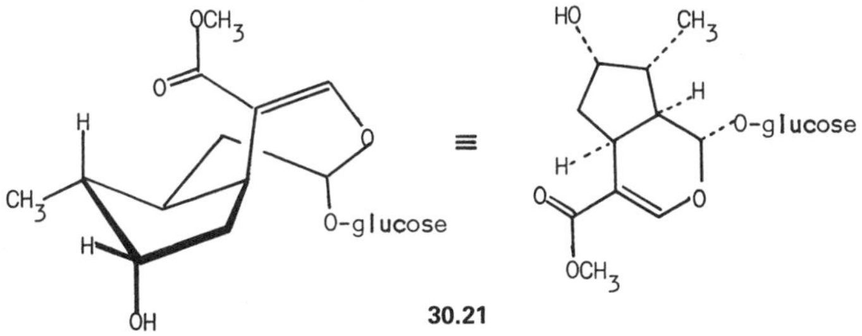

30.21

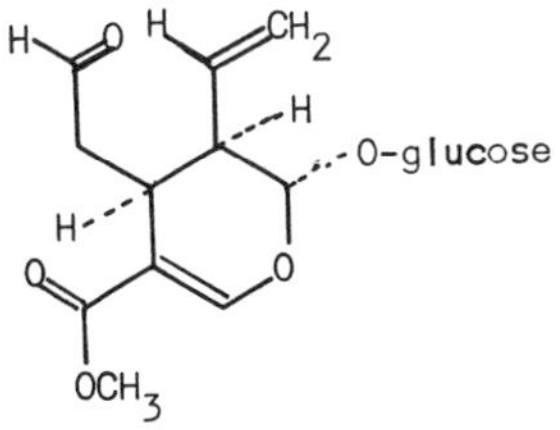

30.15

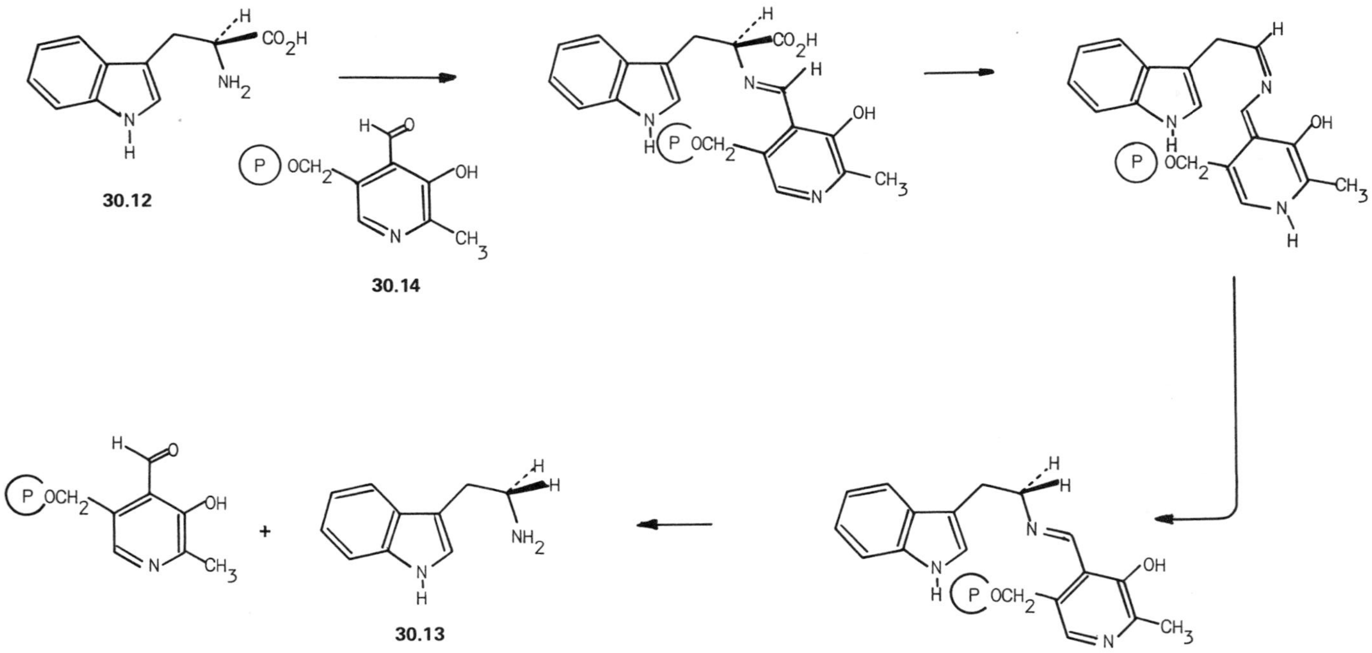

SCHEME 30.1 A pathway for the formation of tryptamine from tryptophan.

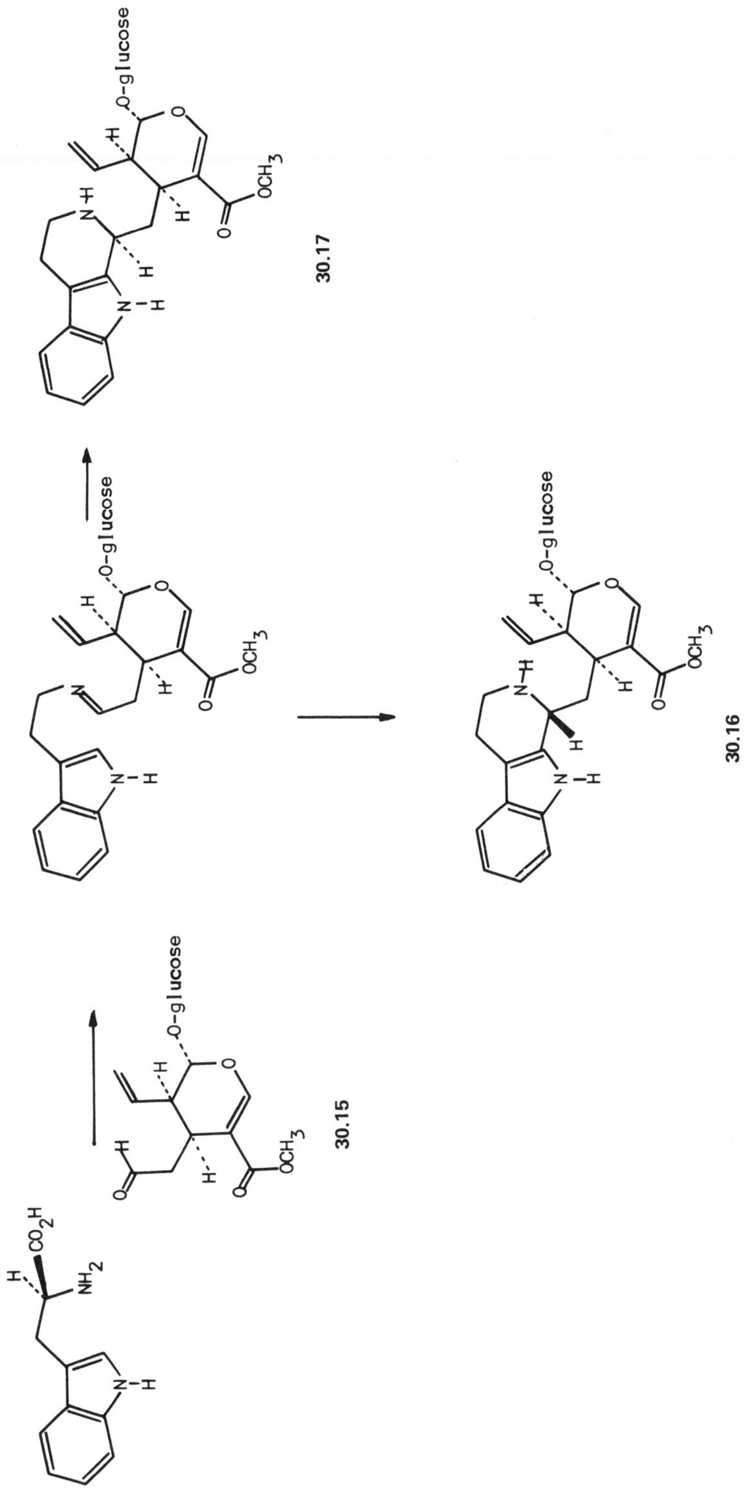
O-glucose
OCH3
30.15
30.16
30.17

SCHEME 30.2 A pathway to vincoside and isovincoside from tryptophan and secologanin.

can be considered in three distinct parts: (a) the conversion of mevalonate (30.18) to geraniol (30.19); (b) the transformation of geraniol (30.19) and/or nerol (30.20) into loganin (30.21); and (c) the cleavage of loganin (30.21) to secologanin (30.15).

I. THE CONVERSION OF MEVALONATE TO GERANIOL

The details of the biosynthesis of geraniol (30.19) have been worked out in connection with the genesis of squalene (30.22) from mevalonate (30.18; Equation 30.1; [224]) and may be applicable to the alkaloids to be considered here [221]. They are shown below (Scheme 30.3).

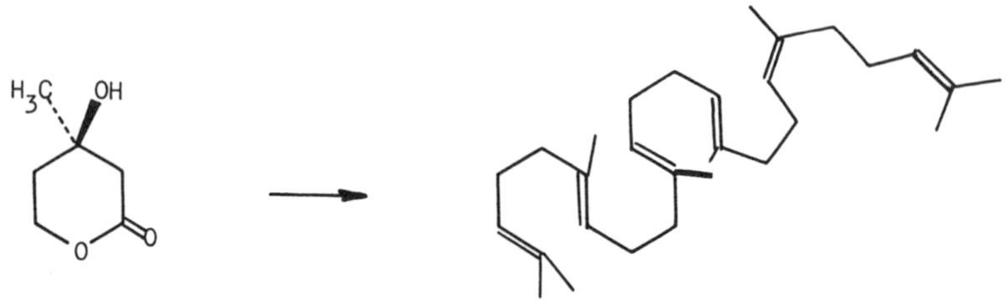

(Eq. 30.1)

30.18 30.22

First, mevalonate 5-pyrophosphate (30.23) undergoes loss of -OH and $-CO_2H$ (i.e., CO_2 and H_2O) in a trans (antarafacial) sense to yield isopentenyl pyrophosphate (30.24). When isopentenyl pyrophosphate (30.24) isomerizes to 3,3-dimethylallyl pyrophosphate (30.25), of the two possible protons at C_2 which might be lost, only the 2R proton (i.e., the 2S proton of mevalonate, 30.18) comes off, and the new proton is (presumably) picked up on the opposite face of the double bond.* The "new" methyl group thus created when the double bond is protonated is now E (i.e., trans to the methylenepyrophosphate group).

In the condensation of 3,3-dimethylallyl pyrophosphate (30.25) with isopentenyl pyrophosphate (30.24), the dimethylallyl group is added to that side of the double bond on which the groups $-CH_2CH_2O$(P), $-CH_3$, -H, and -H appear in a clockwise fashion, and in this process the 2R hydrogen of the isopentenyl pyrophosphate (30.24; the 2S hydrogen of mevalonate, 30.18) is lost, and the new carbon-carbon bond is formed with inversion of configuration to yield geraniol (30.19, shown as its pyrophosphate). It is reasonable to inquire if the loss of the 2R-hydrogen of the isopentenyl pyrophosphate (30.24), when it reacts with 3,3-dimethylallyl pyrophosphate (30.25), in the formation of the new carbon-carbon bond creating geraniol (30.19) could be a synchronous process. The current belief [224] is that the S_N2 displacement of phosphate on 3,3-dimethylallyl pyrophosphate (30.25) by isopentenyl pyrophosphate (30.24) involves addition of some cofactor to the double bond of the face opposite to that reacting (i.e., trans or antarafacial addition) and that a subsequent elimination of the cofactor and its corresponding anti hydrogen (i.e., the 2R hydrogen, trans or antarafacial elimination) obtains. The alternative, a concerted displacement and syn or suprafacial elimination has not been excluded.

Regardless of the details of the addition and elimination reactions themselves, the overall outcome of the set of processes is clear, and specific label introduced early (such as ^3H or ^{14}C) via mevalonate can be traced into geraniol (30.19) with certainty.

---

*This was not definitively shown [224] but has since been confirmed, is in concert with antiperiplanar (antarafacial) addition and, regardless, is not critical to the arguments to be used.

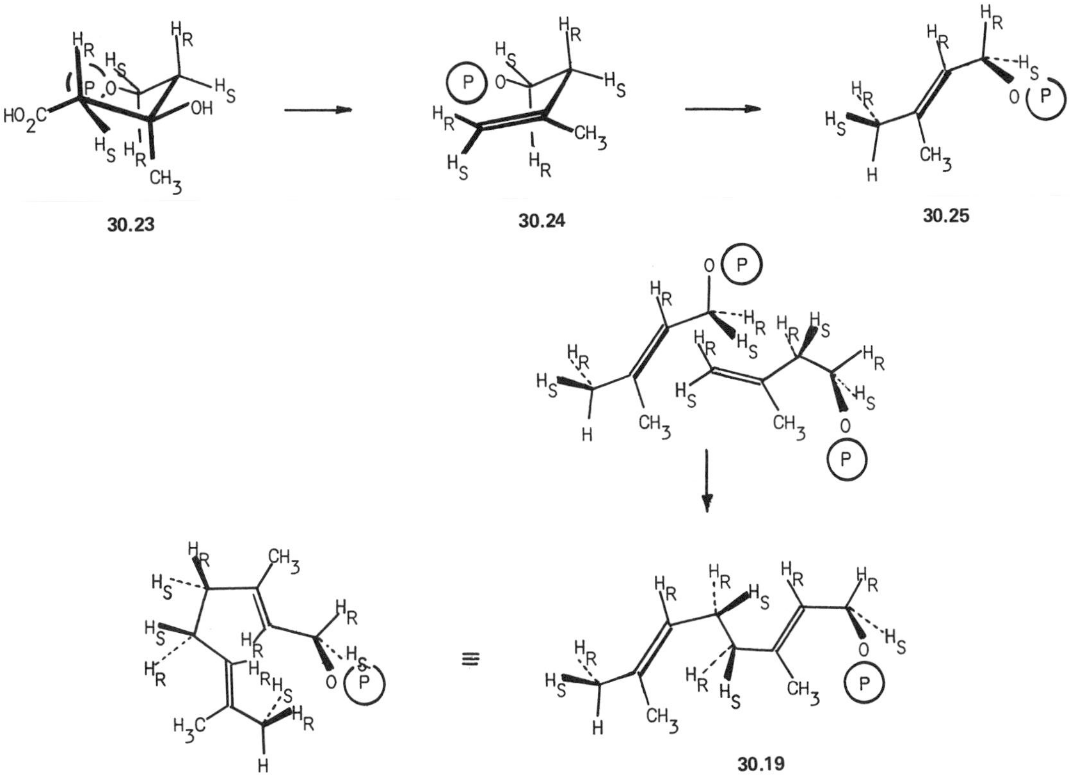

SCHEME 30.3 The fate of the hydrogens of mevalonate in its conversion to geraniol. (After Reference 224.)

II. THE CONVERSION OF GERANIOL AND NEROL TO LOGANIN

The randomization of C_9 and C_{10} of geraniol (30.19) and nerol (30.20) as demonstrated by ^{14}C-labeling [225] in the resulting loganin (30.21) has suggested that both of these compounds undergo oxidation to an intermediate (e.g., the dialdehyde 30.26) in which the same oxidation level obtains at both carbon atoms prior to cyclization. Second, since nerol (30.20) is incorporated slightly better than geraniol (30.19), isomerization about the double bond is presumed to precede eventual cyclization (or a dual pathway to loganin, 30.21, exists). Third, the C_{10} hydroxylated isomers 30.27 and 30.28 have been suggested [226] as lying on the pathway to the penultimate loganin (30.21) precursor, deoxyloganin (30.29), the aglycone of which is not, however, incorporated into loganin (30.21).*

*Since the conversion of deoxyloganin (30.29) to loganin (30.21) requires oxidation at an unactivated position, it is worthwhile considering oxidation to the appropriate intermediate, while the double bond, e.g., in 30.26 or its precursors, is properly (allylic) located. Such a supposition would entail the assumption that deoxyloganin (30.29) is not an obligatory precursor of loganin (30.21) even though the incorporation of deoxyloganin (30.29) without randomization into loganin (30.21) and a number of the indole alkaloids has been demonstrated [227].

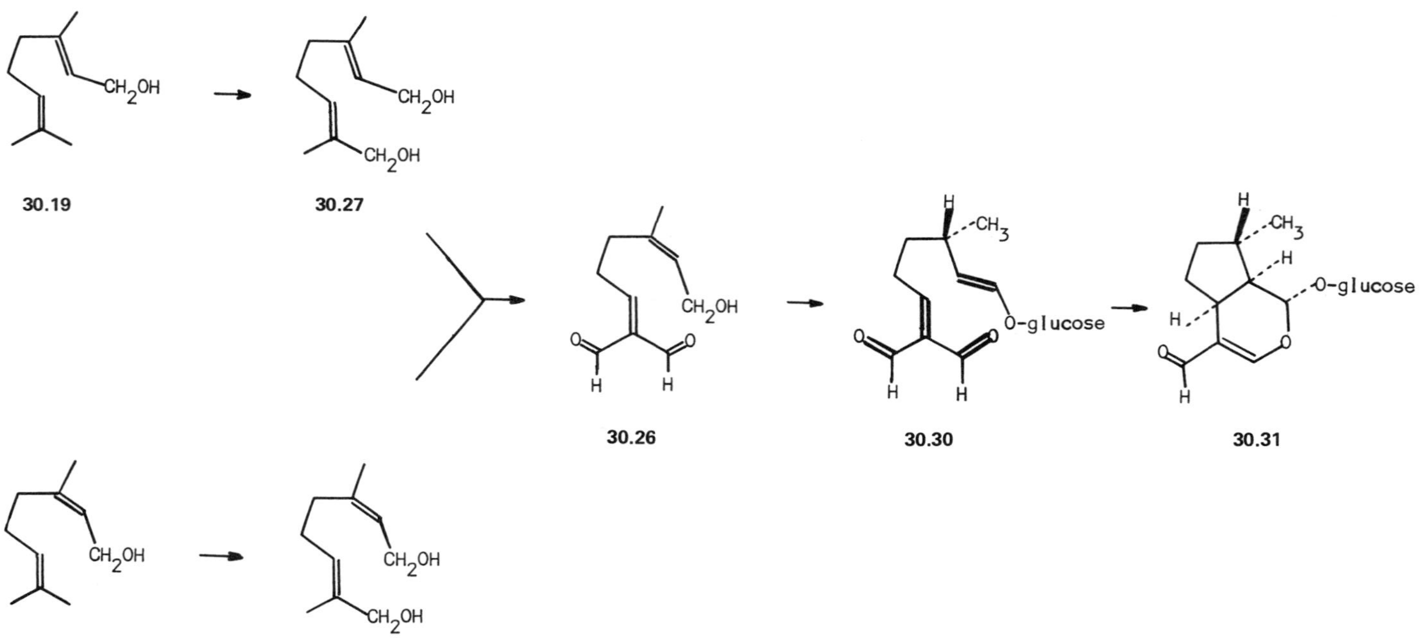

30.19 → 30.27

30.20 → 30.28

30.27 + 30.28 → 30.26 → 30.30 → 30.31

SCHEME 30.4 A potential pathway for the conversion of geraniol and nerol into a deoxyloganin precursor.

Although intermediates between the C_{10} hydroxylated isomers 30.27 and 30.28 have not been isolated, it is interesting to speculate that isomerization of the dialdehyde 30.26 and conversion to the enol ether 30.30 might first occur and that a $_{\pi}2_s + _{\pi}2_s + _{\pi}2_s$ reorganization would lead directly to the aldehyde 30.31 (Scheme 30.4). Oxidation of the aldehyde 30.31 would lead directly to deoxyloganic acid (30.32), which on methylation and further oxidation provides loganin (30.21; Equation 30.2). Deoxyloganin (30.29) and loganic acid (30.33) are known to occur in Catharanthus roseus, and the former occurs along with loganin (30.21) in Strychnos nux-vomica.

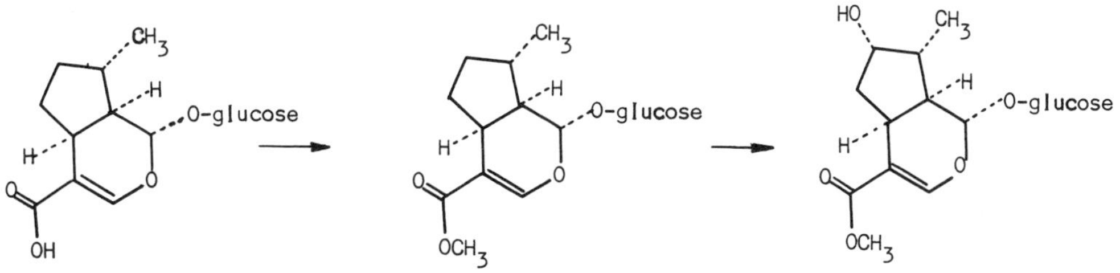

30.32 30.29 30.21

(Eq. 30.2)

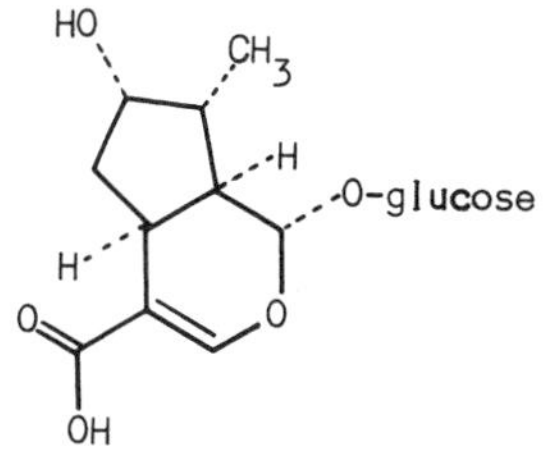

30.33

III. THE CLEAVAGE OF LOGANIN TO SECOLOGANIN

Although several glucosides bearing a cleaved loganin (30.21) fragment have been isolated from various families, details of the cleavage of loganin (30.21) to secologanin (30.15) remain obscure. Several possible pathways include (a) oxidation of loganin (30.21) to the unknown hydroxyloganin (30.34; [226]) and a cleavage reaction (Equation 30.3) and (b) introduction of

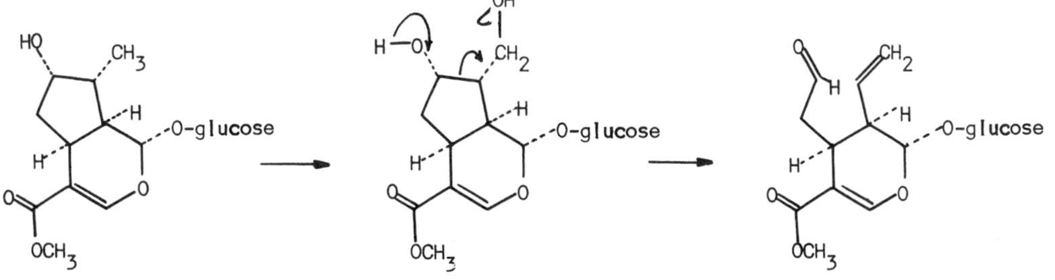

30.21 30.34 30.15

(Eq. 30.3)

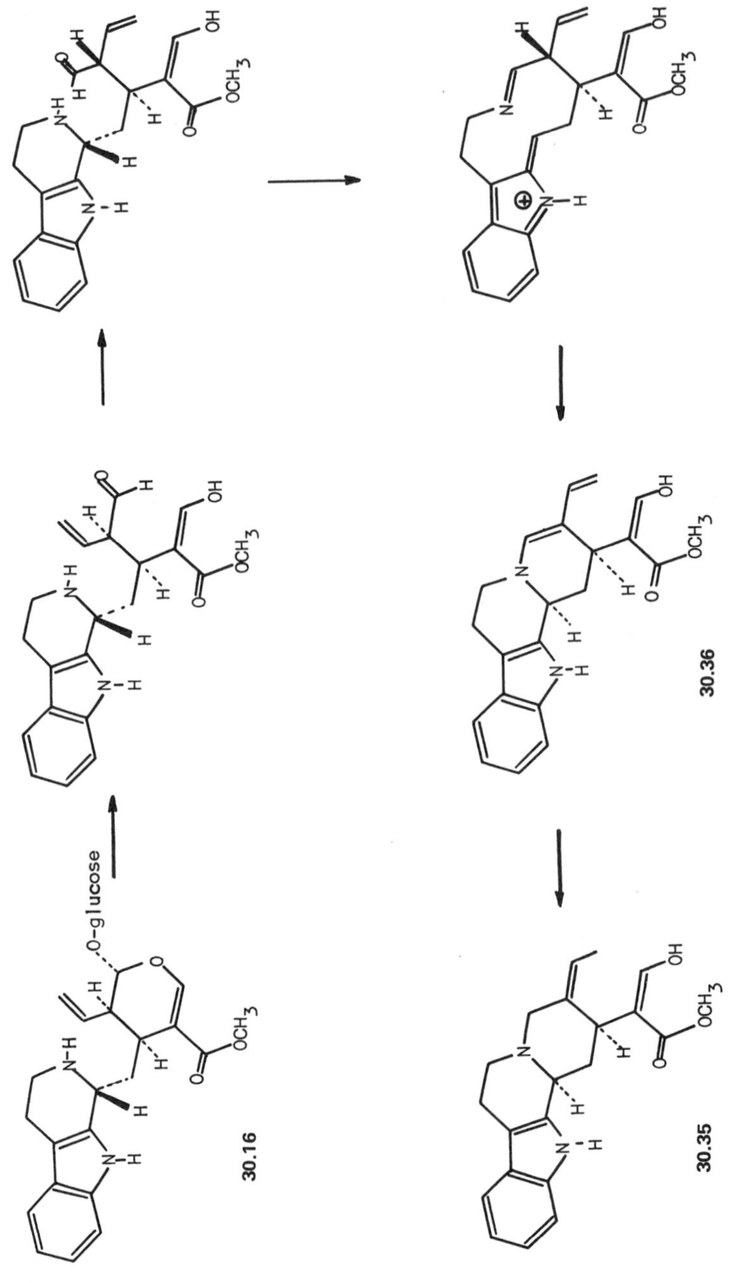

30.16

30.36

30.35

SCHEME 30.5 A possible pathway for the formation of geissoschizine (via a pregeissoschizine) from vincoside.

a suitable leaving group onto the hydroxyl (e.g., as a peroxide) and cleavage in the opposite sense to route (a) (Equation 30.4).

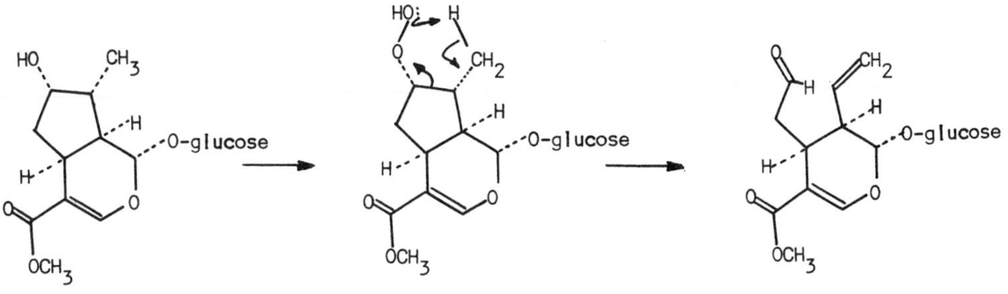

HO CH3 H O-glucose H O OCH3 O

30.21

HO: H O CH2 H O-glucose H O O OCH3

O CH2 H H O-glucose H O O OCH3

30.15

(Eq. 30.4)

IV. THE ALKALOIDS

As already pointed out (p. 435) much of the work on the elaboration of the molecules formed from an initial combination of tryptophan (30.12) with secologanin (30.15) has been done in a single plant (i.e., Catharanthus roseus; [221]), but complementary work is available in other plant systems as well [228, 229].

The initial steps in the elaboration of all these bases appears to be the formation of vincoside (30.16) and isovincoside (30.17; Scheme 30.2). Of the two intermediates, vincoside (30.16) but not isovincoside (30.17) is utilized in C. roseus for further alkaloid production (see footnote, p. 445). Additionally, in C. roseus, H_5 of loganin (30.12; i.e., $4H_R$ of geraniol, 30.19; $2H_R$ of mevalonate, 30.23) appears at C_3 in vincoside (30.16) and remains in the Corynanthé alkaloids subsequentially isolated, but the stereochemistry has been inverted at C_3 in going from vincoside (30.16) to the more elaborate bases. While it is not known at what stage the inversion occurs, the isolation of geissoschizine (30.35) and the demonstration that properly labeled geissoschizine (30.35) could be further elaborated [230] suggest that the inversion occurs before the alkaloids themselves are formed. Scheme 30.5 traces a possible pathway for the formation of geissoschizine (30.35) via a hypothetical "pregeissoschizine" (30.36), which is presumably in equilibrium with geissoschizine (30.35) itself.*

Now, when labeled (±)-[2-^{14}C]tryptophan was fed to young C. roseus [232], the amount of labeled geissoschizine (30.35) rapidly rose and then slowly declined while the amount of labeled ajmalicine (30.1) slowly rose over several days to a constant value, suggesting that ajmalicine (30.1), once formed, is not further transformed into other alkaloids under the short-term conditions employed. Scheme 30.6 outlines a path to ajmalicine (30.1) from geissoschizine (30.35). That ajmalicine (30.1) was derived in a manner approximately as shown was demonstrated by feeding of [2-^{14}C]-, [3-^{14}C]-, [4-^{14}C]-, and [5-^{14}C]mevalonolactone (30.18) and specific degradations [229] using Kuhn-Roth and Schmidt processes to excise the methyl groups as acetic and propanoic acids from ajmalicine (30.1) and its derivatives (e.g., ajmaliciol, 30.37, the product of hydrolysis, decarboxylation, and Wolff-Kishner reduction of ajmalicine, 30.1).

*It should be appreciated that the isomerization of "pregeissoschizine" (30.26) to geissoschizine (30.35) involves the loss of a proton which was originally at C_2 of geraniol (30.19; $2H_R$; C_5 [$5H_R$] of mevalonate, 30.32; [231]).

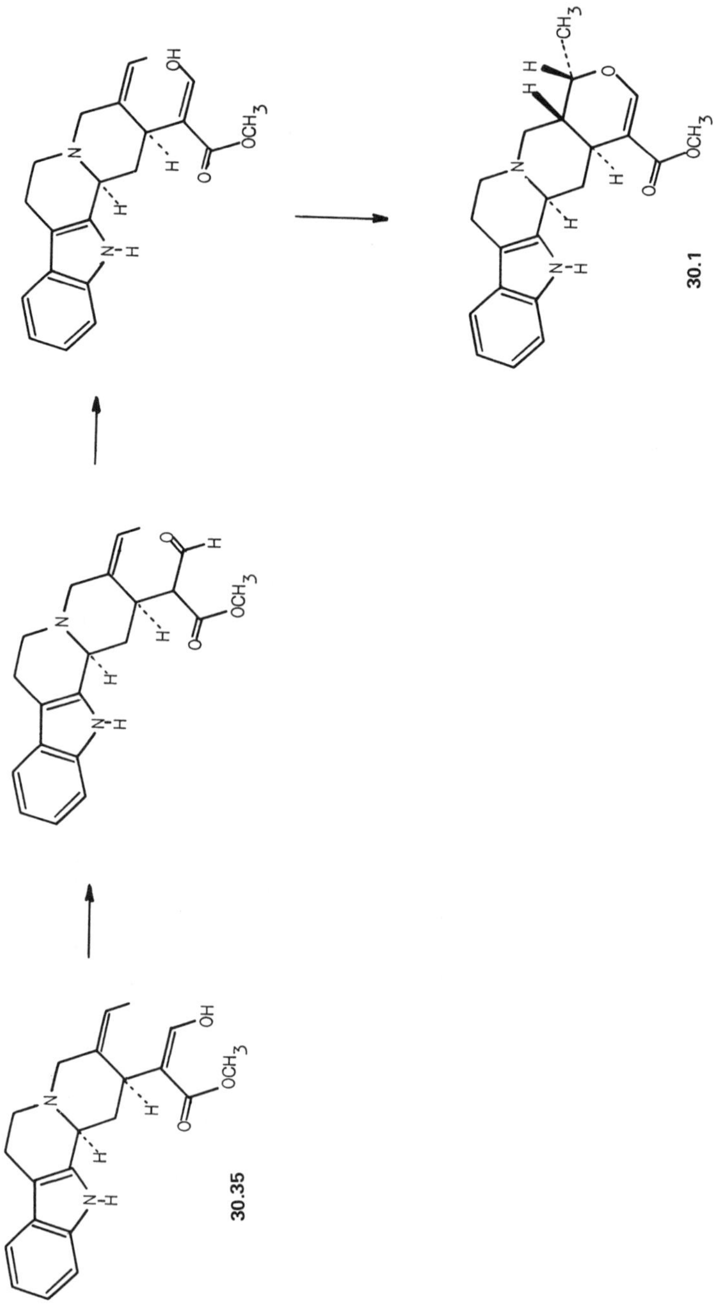

30.35

30.1

SCHEME 30.6 A possible pathway for the formation of ajmalicine from geissoschizine.

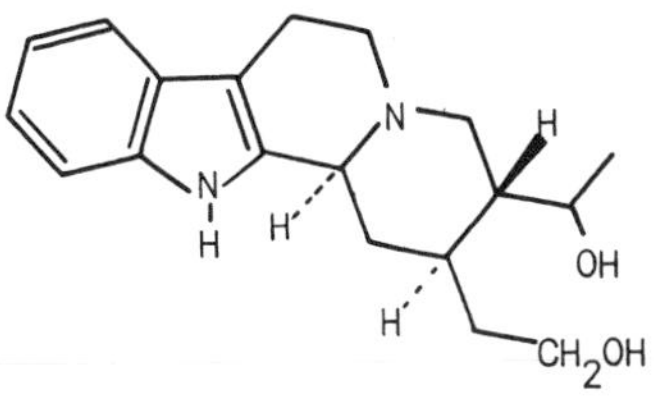

30.37

A similar pathway (Scheme 30.7) can be written for the formation of yohimbine (30.2). Although it appears that experimental evidence is not available to support or refute the hypothesis as written, the pattern of substitution of yohimbine (30.2) and the stereochemistry lend credence to the suggestion that vincoside (30.16) also provides yohimbine (30.2) via pregeissoschizine (30.36). Other pathways can also be written, and it would be interesting to know, in particular, if, as the scheme predicts, the hydrogen originally at C-2 (2 H_R of geraniol; 30.19) is lost in the final alkaloid.*

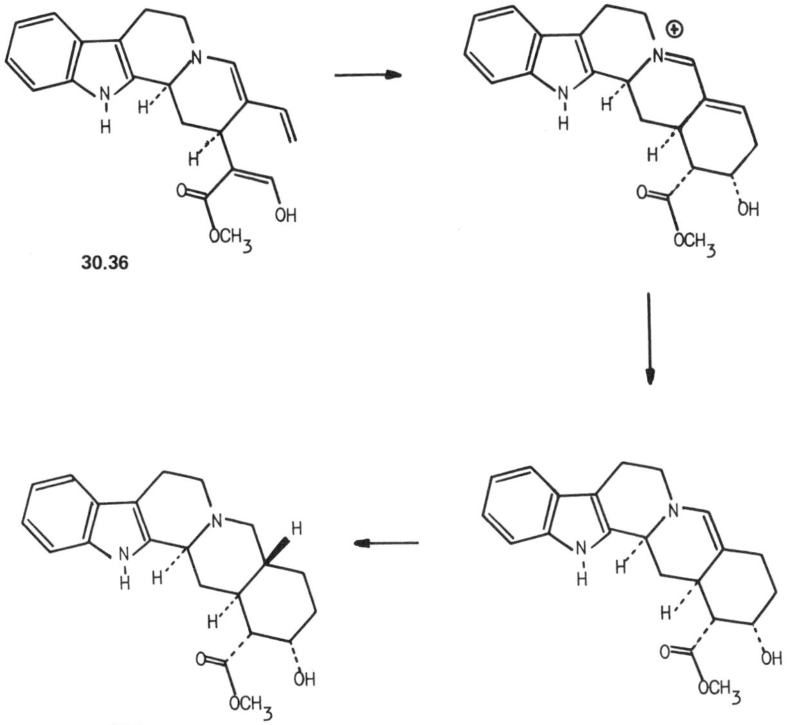

30.36

30.2

SCHEME 30.7 A pathway for the formation of yohimbine from a geissoschizine precursor.

*During the final preparation of this manuscript, it was pointed out that new work (A. I. Scott, private communication) indicates that, contrary to previous reports [221, 228, 229], it is isovincoside (30.17) and not vincoside (30.16) which is further elaborated. The early inversion (Scheme 30.5) at C-3 is therefore unnecessary, and isovincoside (30.17) can lead directly to pregeissoschizine (30.36) and geissoschizine (30.35) without isomerization. NOTE ADDED IN PROOF: See also [475, 476]. Isovincoside is now called strictosidine (30.17).

The administration of (±)-[2-^{14}C]tryptophan (30.12 and its mirror image) to Rauwolfia serpentina has been investigated [228], and both the reserpine (30.3), which was not degraded completely, and the ajmaline (30.4), which was more closely examined, were found to be labeled. However, it does not yet appear to be established that loganin (30.21) is, in fact, the source of the remainder of the carbon atoms in these alkaloids.

With regard to reserpine (30.3), since the stereochemistry at C_3 is opposite to that previously encountered (indeed, it is also opposite to that in ajmaline, 30.4, with which it cooccurs), it cannot be assumed, in the absence of suitable labeling experiments, if secologanin (30.15) is a pregenitor, that vincoside (30.16) rather than isovincoside (30.17) is the precursor. That is, either vincoside (30.16) is incorporated and epimerization does not occur, or isovincoside (30.17) is the precursor of this alkaloid and epimerization does occur (or, of course, both!). Second, it is not known at what stage the aromatic ring is oxidized. Thus, the conjectural Scheme 30.8 presumes early oxidation of the appropriate precursor to a methoxy derivative of the C_3 epimer of pregeissoschizine (30.38) and then a series of transformations, including an allylic oxidation, to the methyl ester of reserpic acid (30.39). Esterification of the methyl ester of reserpic acid with 3,4,5-trimethoxybenzoic acid (30.40)* then yields reserpine (30.3).

Ajmaline (30.4) and sarpagine (30.5) are apparently related to each other through the absolute sense of the cyclization of pregeissoschizine (30.36) intermediate. Thus (Scheme 30.9), through a series of proton migrations, pregeissoschizine (30.36) can be converted into an immonium ion (30.44), which on transformation into a pentacyclic derivative, e.g., 30.45, could have formed the new ring with the carbomethoxy group to the "right" or to the "left," in an absolute sense, of the imine nitrogen. If cyclization produced that molecule in which the carbomethoxy group is to the right, hydrolysis and decarboxylation followed by hydration of the resulting imine and cyclization of the aldehyde produce ajmaline (30.4).

---

*The genesis of this acid, in this case, also does not appear to be known. However, it is presumably derived by methylation of gallic acid (30.41) which can arise either from shikimic acid (30.42; see also p. 23) or phenylalanine (30.43) via deamination and oxidation (Equation 30.5).

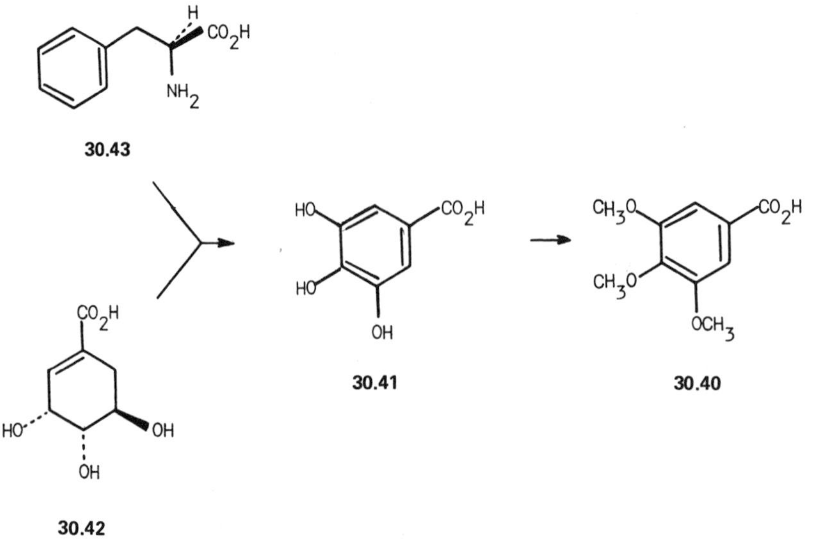

H
CO_2H
NH_2

30.43

HO CO_2H HO OH

30.41

CH_3O CO_2H CH_3O OCH_3

30.40

(Eq. 30.5)

CO_2H
HO OH
OH

30.42

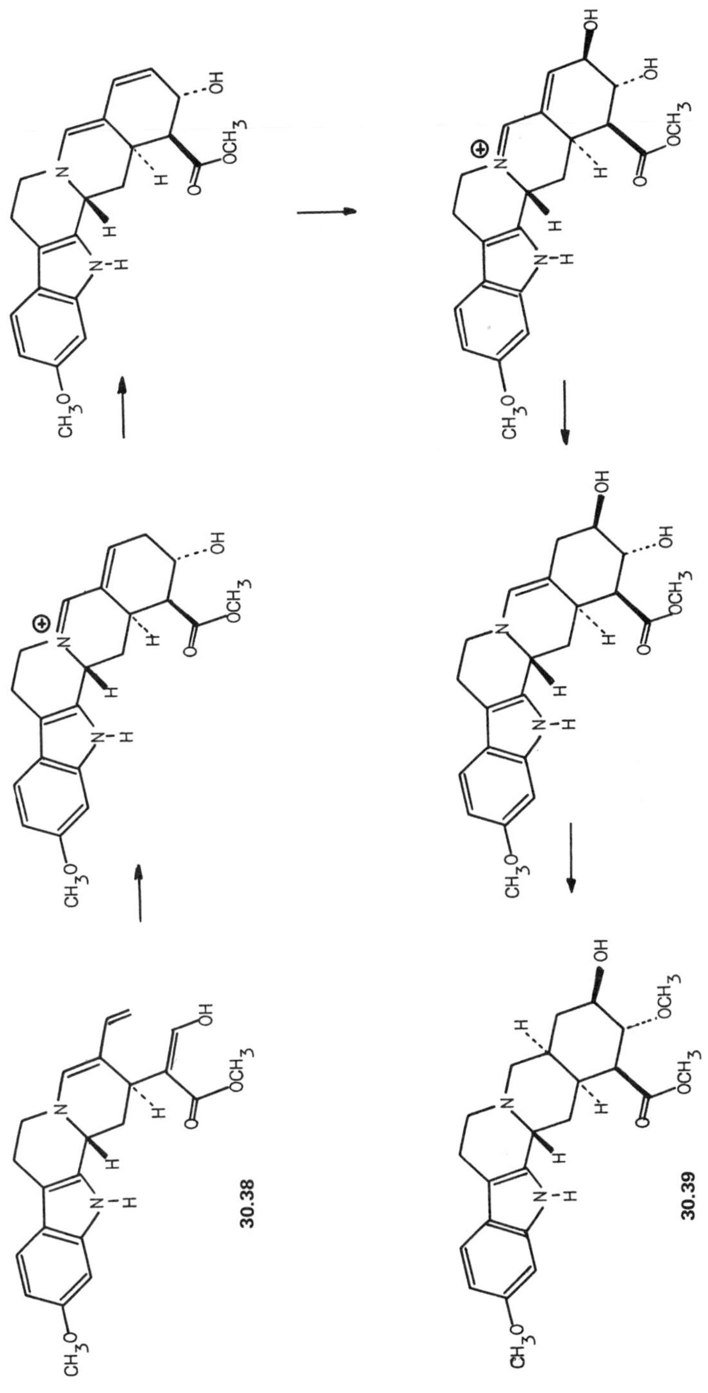

30.38

30.39

SCHEME 30.8 A possible pathway from a secologanin-tryptamine derivative (epipregeissoschizine) to reserpic acid methyl ester.

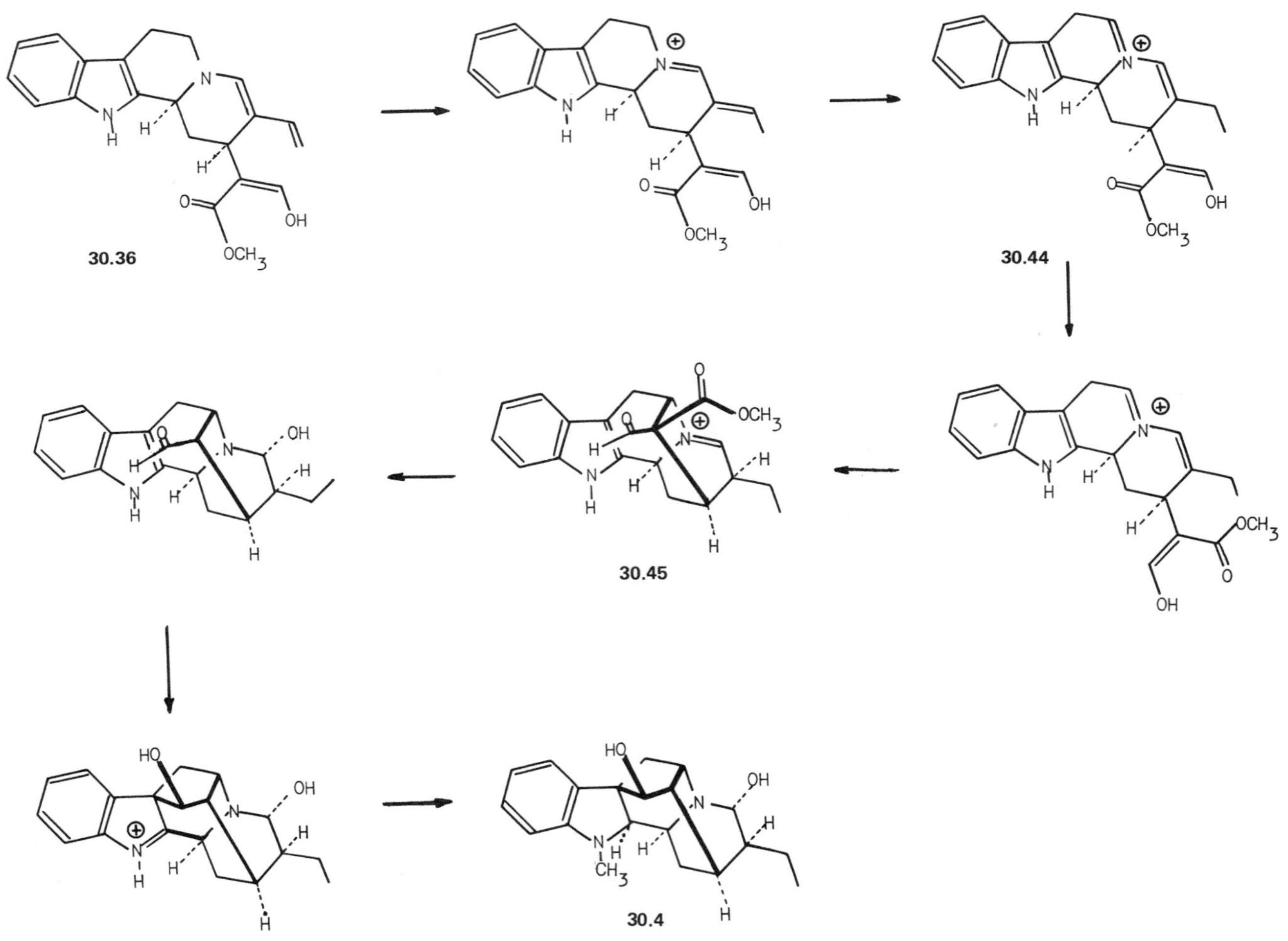

SCHEME 30.9 A possible pathway from "pregeissoschizine" to ajmaline.

Alternatively (Scheme 30.10), the same pregeissoschizine (30.36) in a form where early hydroxylation has occurred* (i.e., 30.46) can form a pentacyclic derivative, e.g., 30.47, in which the carbomethoxy group is found on the left, in an absolute sense, of the imine nitrogen. Here, cyclization of the aldehyde (30.48) which results after hydrolysis and decarboxylation is stereoelectronically impossible, and simple reduction leads directly to sarpagine (30.5). In still another variation (Scheme 30.11), pregeissoschizine (30.36) or its C_3 epimer can lead to a different immonium ion (30.49). Cyclization involving the indole portion of the nucleus, followed by oxidation at nitrogen to the N-oxide** and an elimination reaction generate an intermediate (30.50) which on hydration, recyclization, and reduction, ultimately leads to echitamine (30.6).

By a slightly more elaborate pathway, quinine (30.8) can also be shown to be derived from a pregeissoschizine derivative, namely, methoxypregeissoschizine (30.51). Thus, [2-^{14}C]tryptophan (30.12; [235]), [2-^{14}C]geraniol (30.19; [236]), and [3-^{14}C]geraniol (30.19; [237]) are incorporated, without randomization, into quinine (30.8) in Cinchona succiruba. As indicated in Scheme 30.12, it is presumed that early oxidation of a pregeissoschizine (30.36) derivative generates the corresponding methoxy compound (30.51)† which undergoes suitable prototropic shifts to the imine 30.52 and hydrolysis and decarboxylation to the dialdehyde 30.53. Then, cyclization to 30.54 is followed by reduction in the quinuclidine ring and oxidation in the indole portion to 30.55.‡ Ring cleavage of 30.55 to 30.56 is then followed by recyclization, dehydration, and reduction to quinine (30.8). The presence of the ^{14}C label from [2-^{14}C]tryptophan (30.12) at the α-carbon of the quinoline ring of quinine (30.8), the ^{14}C label from [2-^{14}C]geraniol (30.19) at C_{20} of quinine (30.8) and that from [3-^{14}C]geraniol at C_{19} of quinine (30.8) all help to confirm the broad outlines of the path shown as Scheme 30.12.

The elaboration of geissoschizine (30.35) into strychnine (30.7), catharanthine (30.9), tabersonine (30.10), and vincamine (30.11), as well as many related bases, has been thoroughly studied in C. roseus, and the work has been summarized [221]. The key appears to lie in the conversion of geissoschizine (30.35), via preakuammicine (30.57), into akuammicine (30.58), from which the later bases are elaborated. However, the path from geissoschizine (30.35) to akuammicine (30.58) has yet to be elucidated, and two suggestions (Schemes 30.13 and 30.14) are current [221, 234]. The first (Scheme 30.13) and oldest suggests the intermediacy of geissoschizine oxindole (30.59). Here, the oxidation of geissoschizine (30.35) to the β-hydroxyindoline (30.60) and thence to the "diol" 30.61 (written as the monophosphate) or the direct conversion of geissoschizine (30.35) into 30.61 is suggested. Then, ring opening leads to the oxindole 30.59 which undergoes cyclization to preakuammicine (30.57). Geissoschizine (30.35), β-hydroxyindoline (30.60), the diol (30.61), and the oxindole (30.59) have been reported to occur in C. roseus [221]. The major difficulty with this scheme is that oxindole derivatives such as 30.59 are apparently reluctant to undergo (in vitro) the further cyclization reaction demanded for conversion of 30.59 into preakuammicine (30.57).

---

*Early hydroxylation is suggested since the alkaloid missing the hydroxyl group is not apparently found, and evidence being unavailable, it is convenient to presume early oxidation.
**Use of oxidation at nitrogen, which we will encounter again, has been suggested in other systems as providing reasonable pathways for fragmentation [208, 233, 234].

†Although such oxidation may well have occurred at the tryptophan (30.12) state, evidence is lacking and introduction of the oxygen at the pregeissoschizine (30.36) level is convenient. There is no evidence that oxidation may not have occurred much later.

‡There is no evidence that reduction did not precede cyclization, in which case a displacement rather than a condensation reaction would be involved; nor is there evidence that transformation of 30.54 into 30.55 is not, somehow, concerted.

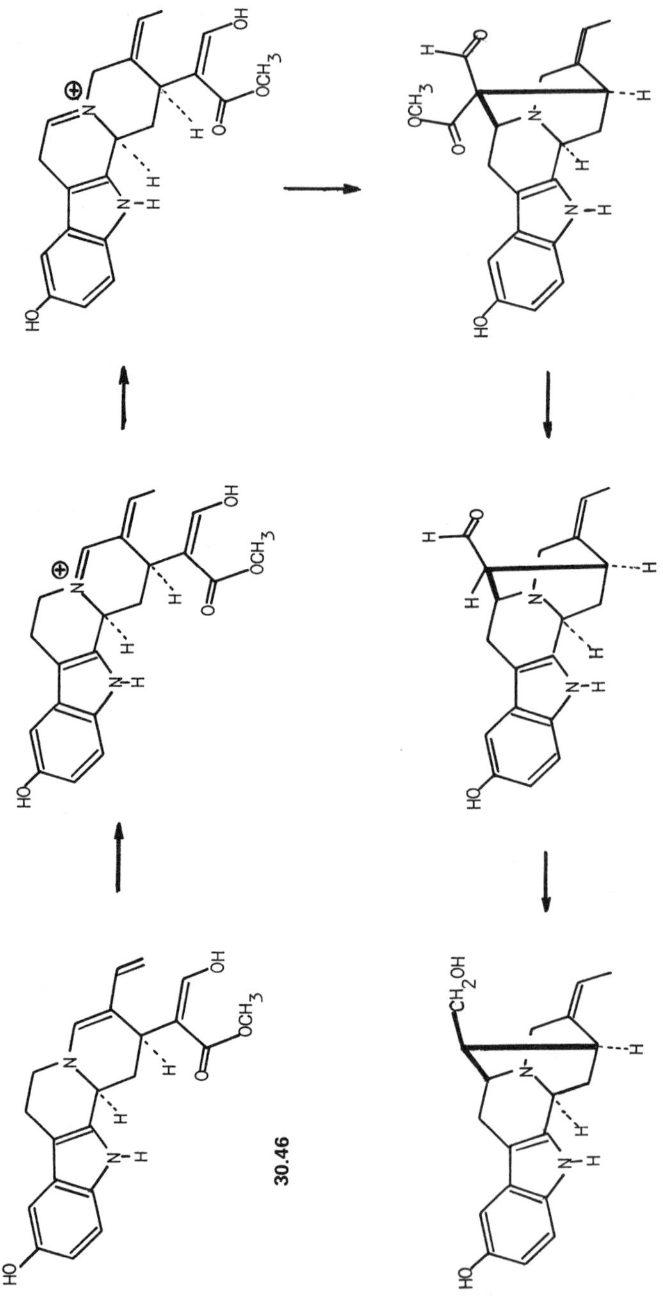

30.46

30.47

30.48

30.5

SCHEME 30.10 A possible pathway for the formation of sarpagine from a hydroxylated geissoschizine precursor.

SCHEME 30.11 A possible pathway to echitamine from pregeissoschizine or its epimer at C_3.

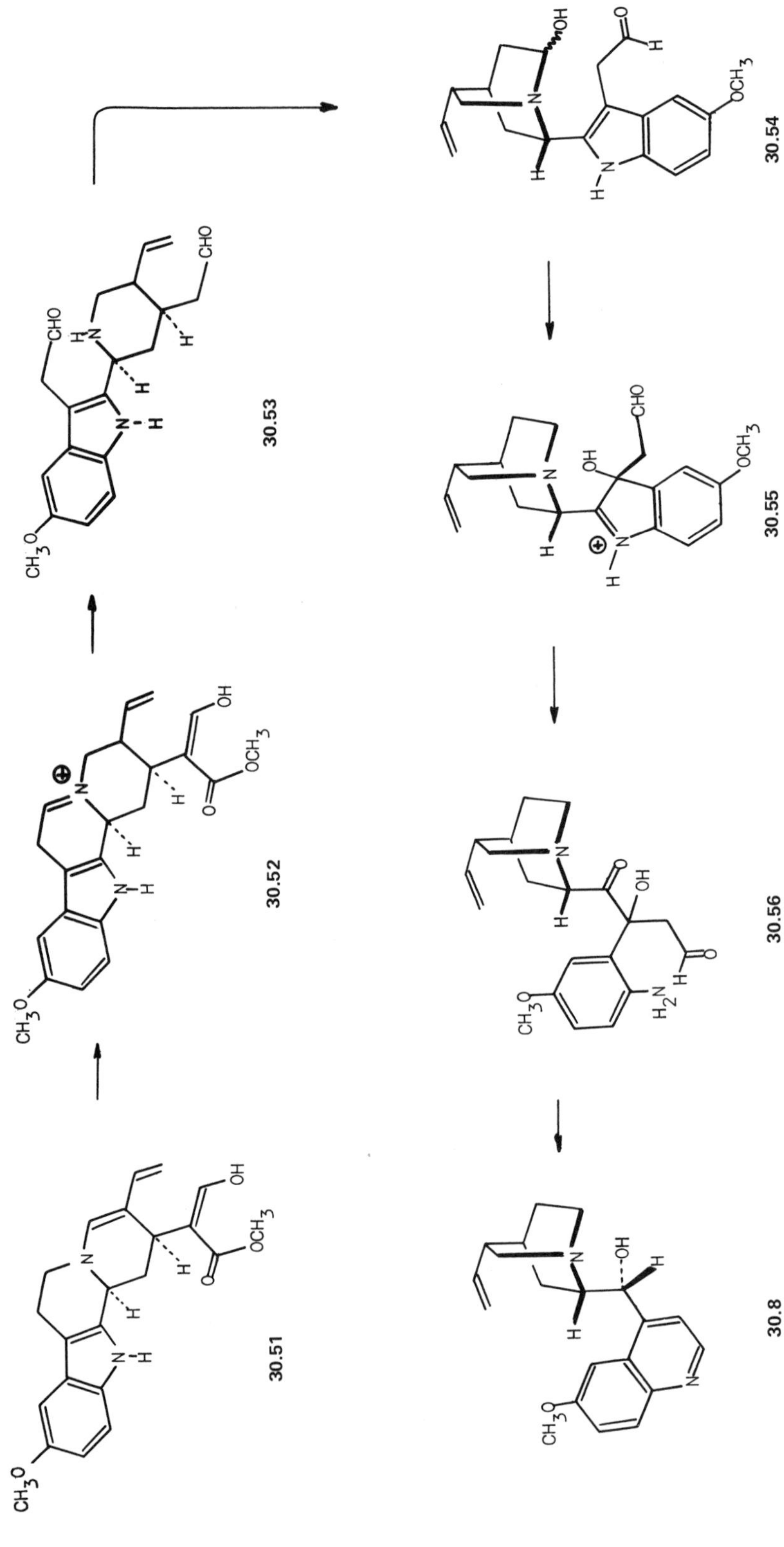

30.51

30.52

30.53

30.54

30.55

30.56

30.8

SCHEME 30.12 A possible pathway from a pregeissoschizine derivative to quinine which is in concert with established labeling patterns.

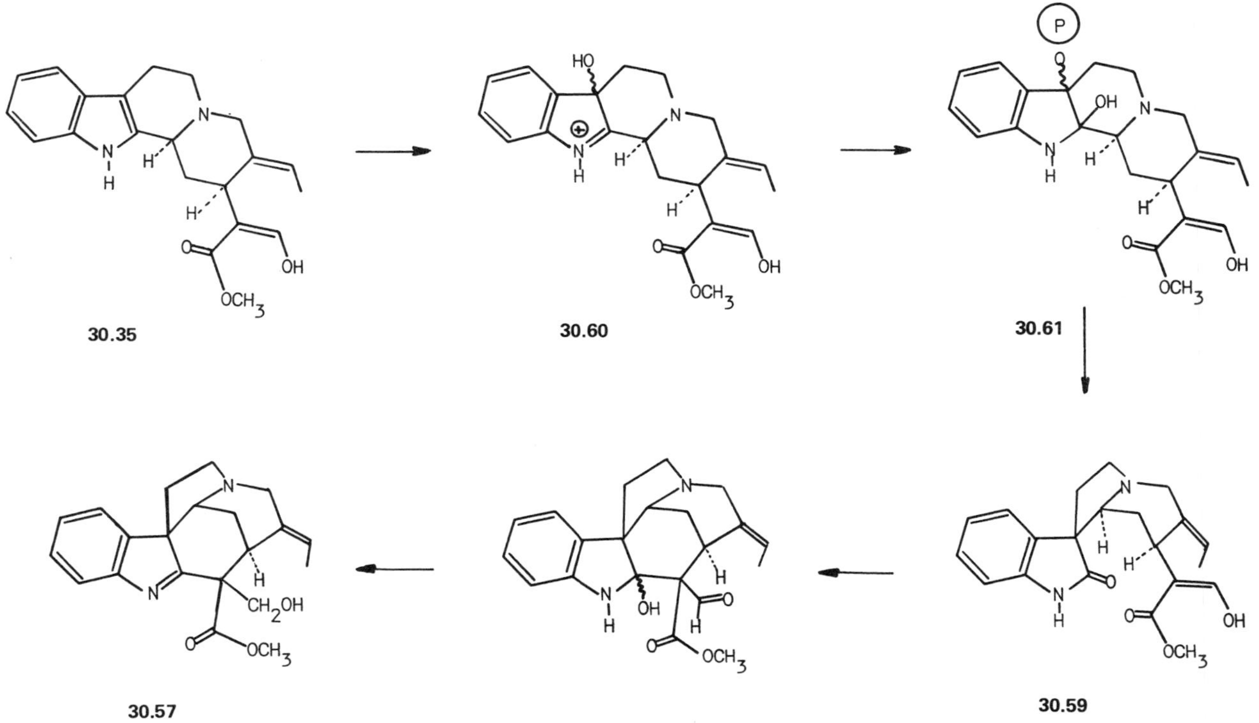

30.35

30.60

30.61

30.59

30.57

SCHEME 30.13 A possible pathway for the conversion of geissoschizine into preakuammicine.

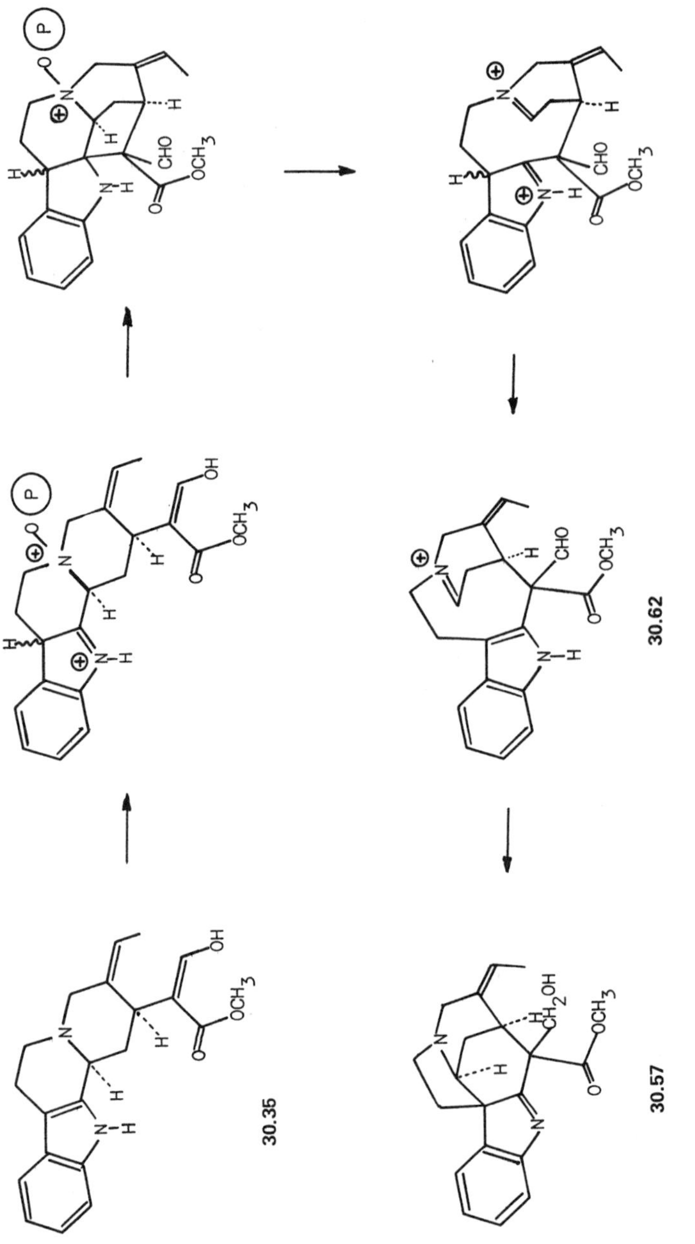

30.35

30.57

30.62

SCHEME 30.14 A possible pathway for the conversion of geissoschizine into preakuammicine.

The second scheme (Scheme 30.14; [234]) makes use of oxidation at nitrogen (see p. 449) to generate an appropriate leaving group. Such reactions are known to occur in vitro [233], but their in vivo counterpart is still lacking. The particular attraction of this proposal lies in the generation of 30.62 as an intermediate, since on reduction it provides the next observed intermediate, stemmadenine (30.63; Scheme 30.15).

Stemmadenine (30.63) has been implicated in the biosynthesis of the alkaloids which follow later and can, as shown (Scheme 30.15), be derived from preakuammicine (30.57) as well as from the intermediate 30.62. Preakuammicine (30.57) itself can be pictured as a precursor to strychnine (30.7), and although incorporation into *Strychnos nux-vomica* is slight,* it is convenient to consider preakuammicine (30.57) to which a two-carbon, acetate-derived fragment has been added as the actual pregenitor of strychnine (30.7). Thus, as shown in Scheme 30.16, preakuammicine (30.57) may be converted via a retroaldol process into akuammicine (30.58) which may, at least in *C. roseus*, which does not elaborate strychnine (30.7), lie on the pathway to the more complex alkaloids. Addition of a two-carbon, acetate-derived fragment to akuammicine (30.58), which has been oxidized at the allylic terminal carbon, and reductive cyclization then leads to strychnine (30.7). The in vivo conversion of [O-methyl-^{3}H, Ar-^{3}H]geissoschizine (30.35) into akuammicine (30.58; [239]) has been demonstrated in *C. roseus*.

The remaining three alkaloids to be considered, namely, catharanthine (30.9), tabersonine (30.10), and vincamine (30.11), may be derived from stemmadenine (30.63) by a ring cleavage reaction [240] with migration of the exo double bond. It is interesting to note, as demonstrated by feeding experiments [221], that the hydrogen lost when the double bond migrates in stemmadenine (30.63) is derived from C_1 of loganin (30.12) and corresponds specifically to the $5H_S$ of mevalonate (30.32). Thus, as shown in Scheme 30.17, stemmadenine (30.63) is presumably converted into the acrylic ester (30.65) which can then undergo one of two allowed (Diels-Alder) ${}_{\pi}2_{s} + {}_{\pi}2_{s} + {}_{\pi}2_{s}$ reactions to yield either catharanthine (30.9) or tabersonine (30.10).

Finally (Scheme 30.18), tabersonine (30.10), after reduction of the nonconjugated double bond to the dihydro derivative 30.66, can undergo oxidation, e.g., at nitrogen to the N-oxide, elimination to the immonium ion 30.67, recyclization, and a second ring cleavage to an α-ketoester, which on cyclization to the nitrogen of the indole nucleus yields vincamine (30.11).

The possible schemes (Schemes 30.5 through 30.18) from vincoside (30.16) to pregeissoschizine (30.36) and thence to the *Corynanthé-Strychnos* bases of which ajmalicine (30.1) is typical, or to the *Cinchona* alkaloid quinine (30.8) and from pregeissoschizine (30.36), via preakuammicine (30.57) and stemmadenine (30.63) to the *Iboga* alkaloid catharanthine (30.9) or to the *Aspidosperma* representative tabersonine (30.10) and finally vincamine (30.11), the *Eburna* alkaloid, are outlined in Scheme 30.19. In the same scheme, the

*The minimal incorporation of even the Wieland-Gumlich aldehyde (30.64) is presumptive evidence for permeability problems, and long-term feeding experiments may resolve this issue [234, 238].

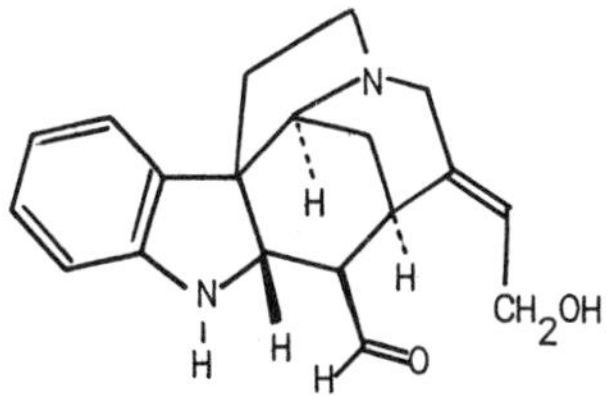

N H H H H H O CH$_2$OH

30.64

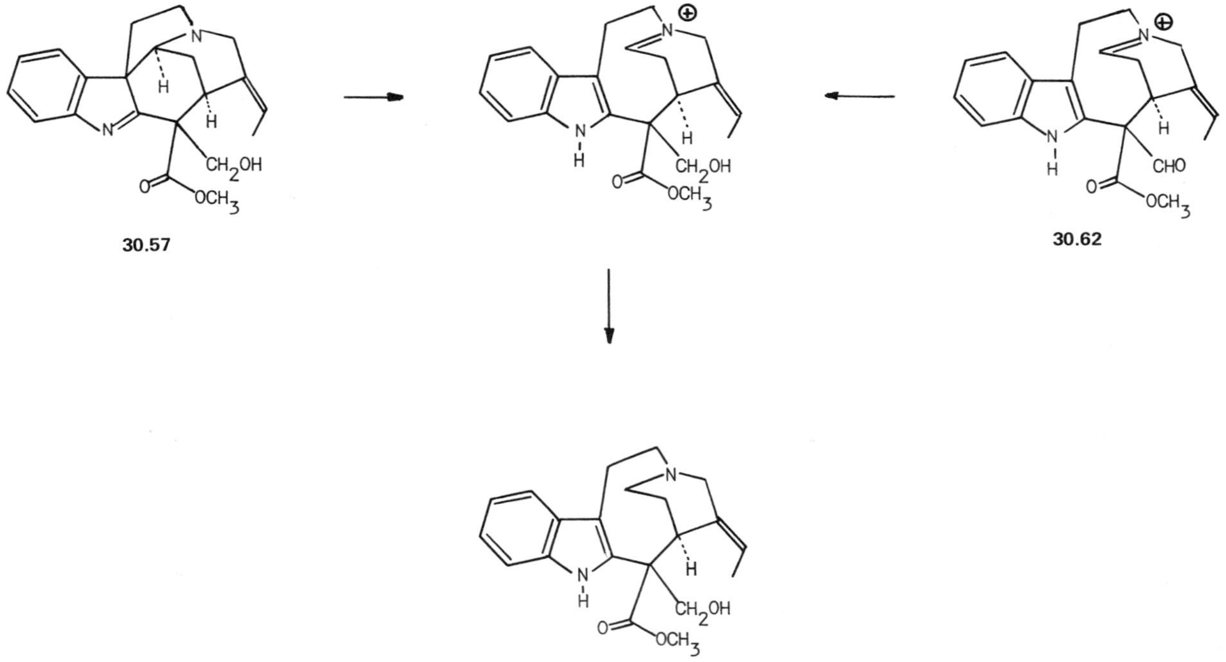

30.57

30.62

30.63

SCHEME 30.15 Possible pathways to stemmadenine from preakuammicine or an intermediate on the way to preakuammicine.

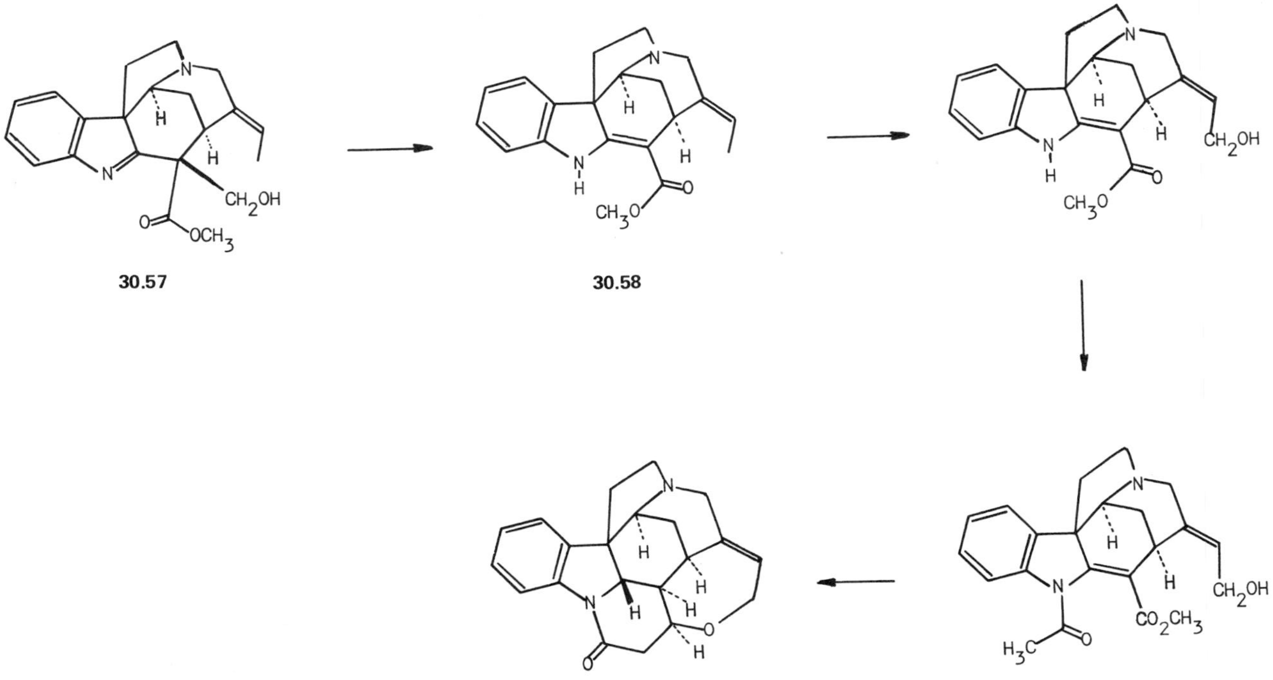

30.57

30.58

30.7

SCHEME 30.16 A possible pathway to strychnine from preakuammicine.

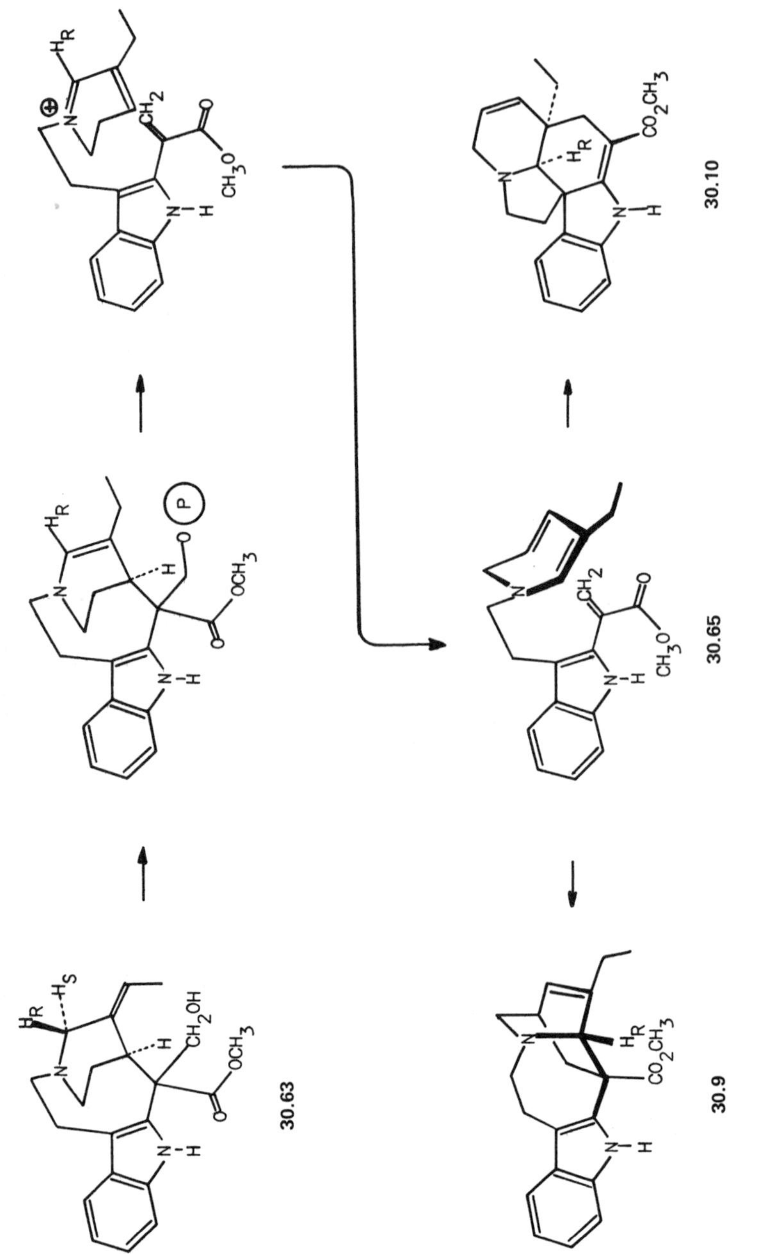

SCHEME 30.17 A possible route from stemmadenine to catharanthine and tabersonine.

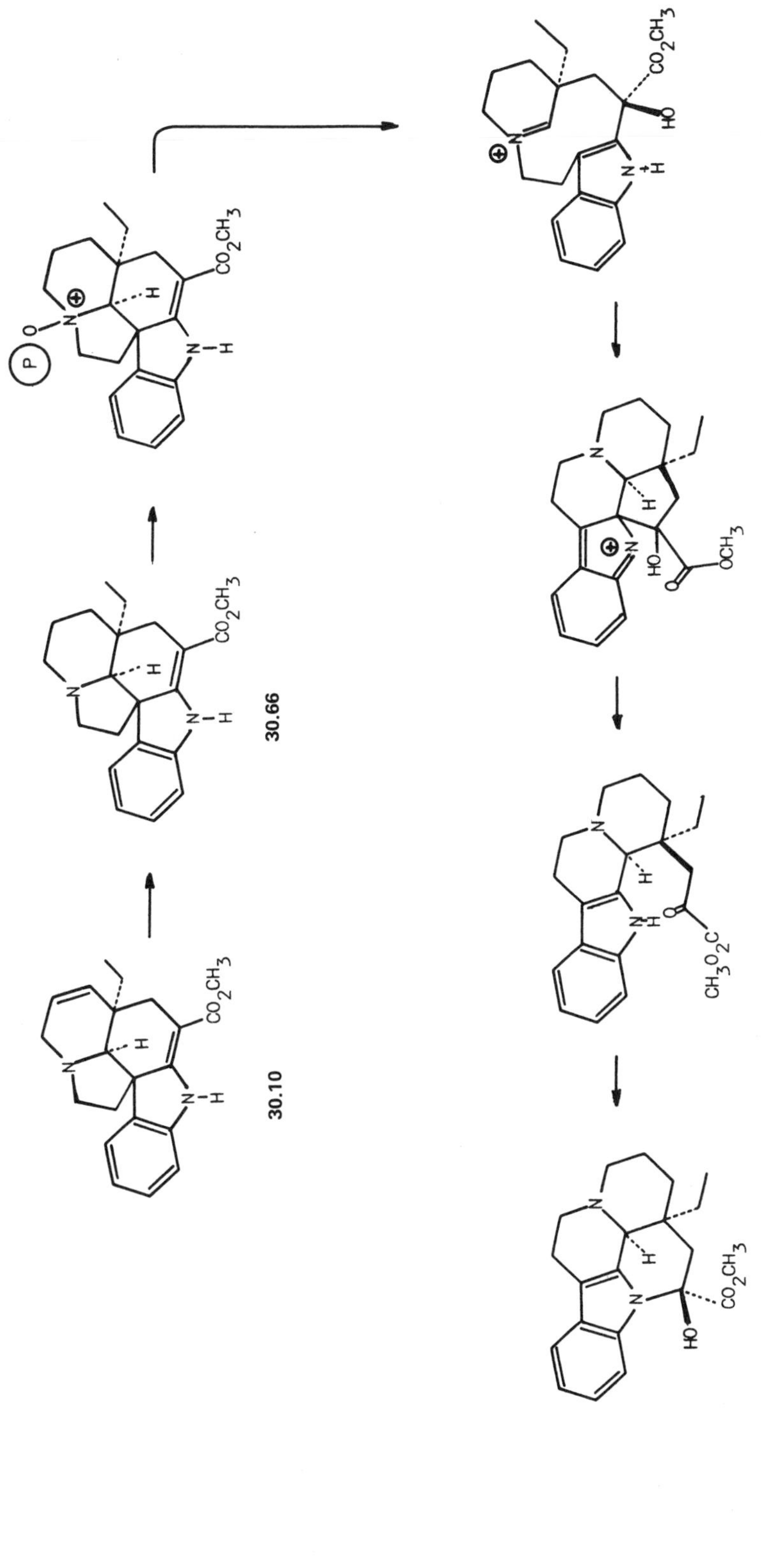

30.10

30.66

30.67

30.11

SCHEME 30.18 A possible pathway from tabersonine to vincamine.

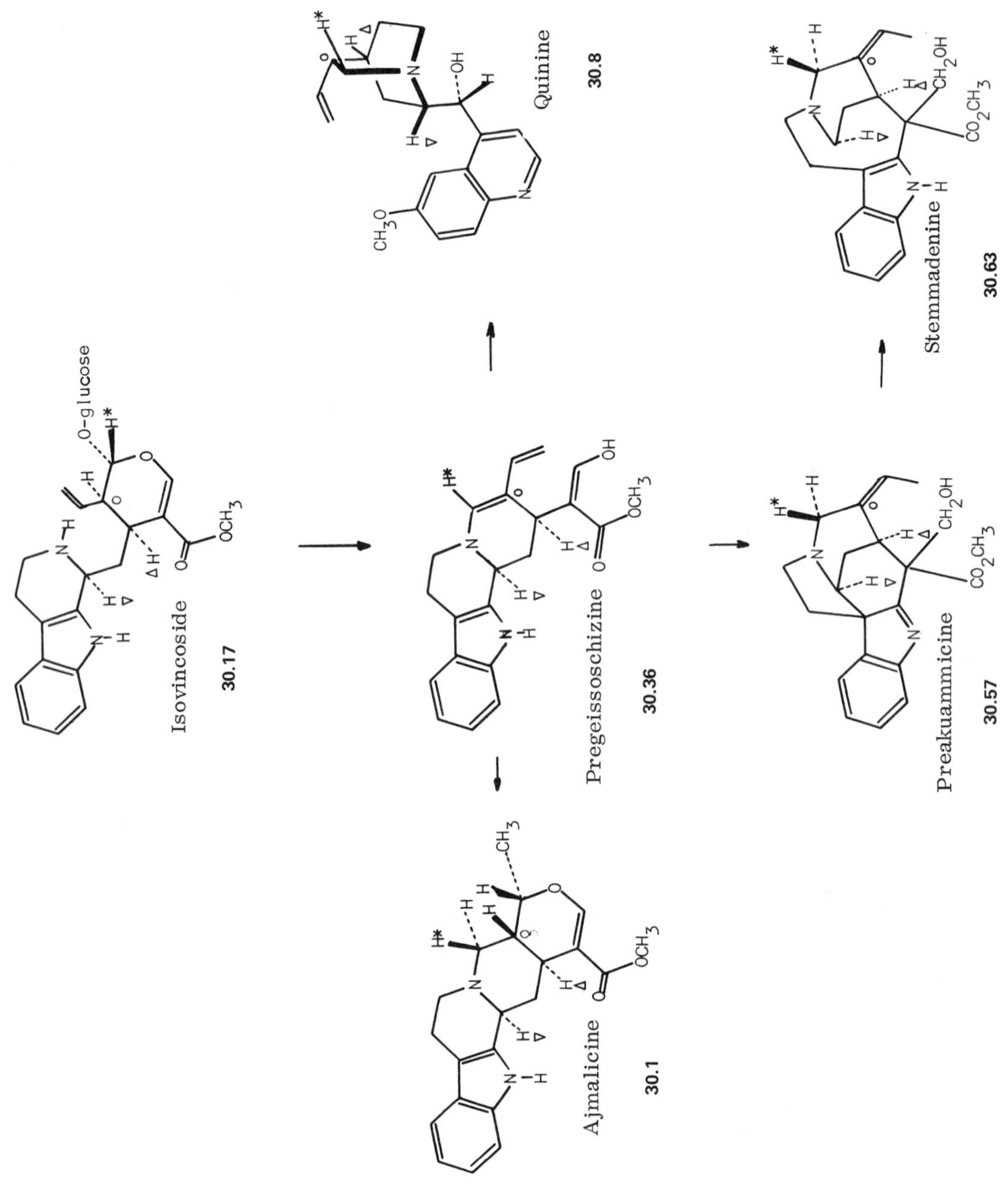

Isovincoside
30.17

Pregeissoschizine
30.36

Ajmalicine
30.1

Quinine
30.8

Preakuammicine
30.57

Stemmadenine
30.63

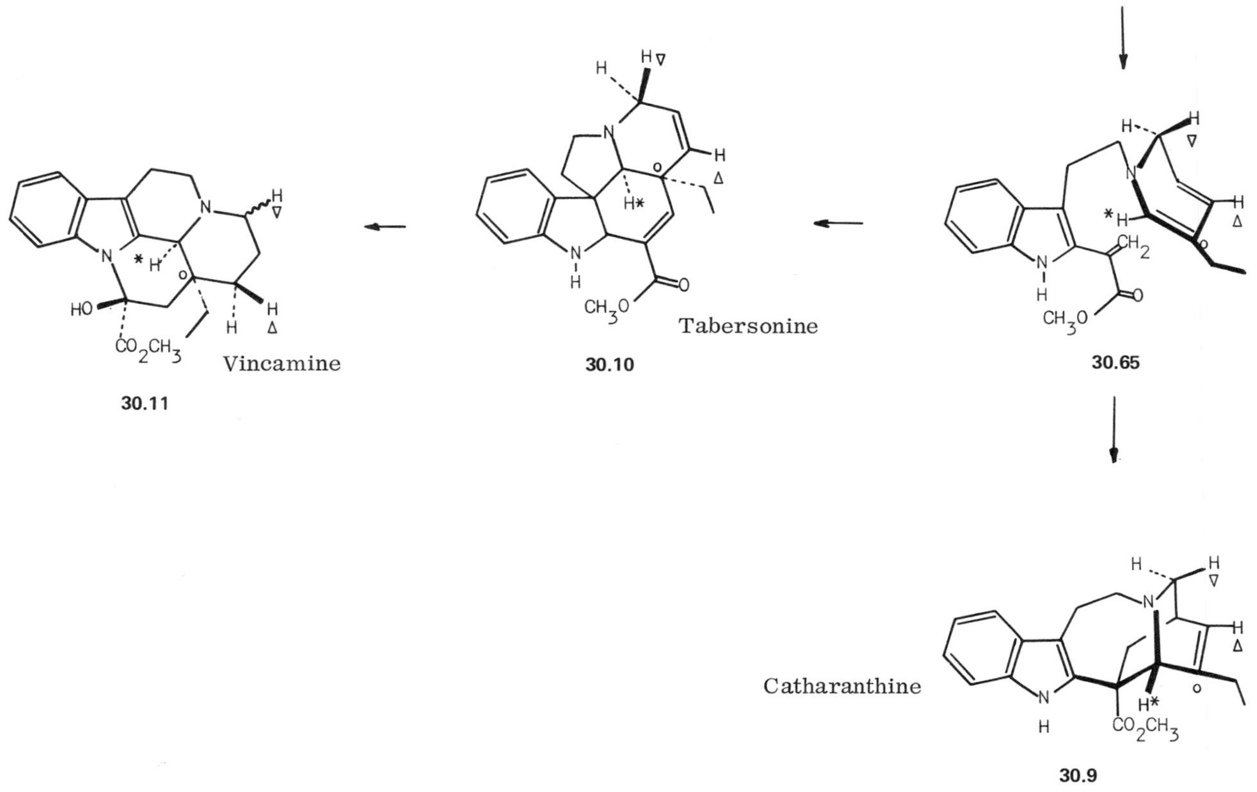

30.65

Tabersonine

30.10

Vincamine

30.11

Catharanthine

30.9

SCHEME 30.19 The fate of fed precursors in the elaboration of alkaloids derived from tryptophan. * = [1-$^{3}H_2$]geraniol (30.19) and [5-$^{3}H_R$]mevalonate (30.23); ○ = [2-^{14}C]geraniol (30.19); △ = [4-$^{3}H_R$]mevalonic acid (30.23); ▽ = [5-^{3}H]loganin (30.21).

predicted results of feeding [1-$^{3}H_2$]geraniol, 30.19 and [5-$^{3}H_R$]mevalonate, 30.23, *; [2-^{14}C]geraniol, 30.19, ○; [4-$^{3}H_2$]mevalonic acid, 30.23, △; and [5-^{3}H]loganin, 30.21, ▽, are traced through each of the above.

For the above bases (and many others we have not considered here) which actually occur in *C. roseus*, i.e., vincoside (30.16), geissoschizine (30.35), akuammicine (30.58), stemmadenine (30.63), catharanthine (30.9), and tabersonine (30.10), the labeling pattern predicted by Scheme 30.19 is actually observed although it must be added that the conversion of stemmadenine (30.63) into *either* tabersonine (30.10) or into catharanthine (30.9) may be a function of which hydrogen is abstracted in the formation of the acrylic ester 30.65 (the reaction appears stereoselective; [221]). Additionally, the formation of tabersonine (30.10) and catharanthine (30.9) may be reversible and [241] tabersonine (30.10) and its precursors have been shown to lead to vincamine (30.11) in *Vinca minor*.

31

BIOSYNTHESIS OF DIMERIC ALKALOIDS DERIVED FROM TRYPTOPHAN AND FROM POLYCYCLIC β-CARBOLINE ALKALOIDS

> Nine times out of ten . . . there is actually no truth to be discovered; there is only error to be exposed.
>
> H. L. Mencken, Prejudices (1922)

There are currently about 100 "bis-indole" or "dimeric" indole alkaloids known, and these can broadly be divided into two categories, namely, (a) those in which both "halves" are made up of closely related or identical groupings, and (b) those in which each "half," while formally derivable from tryptophan (31.1), is considerably different from the other [242]. In this chapter we shall consider only three such bases, i.e., calycanthine (31.2), found in both genera of Calycanthaceae; C-toxiferine (31.3), a major curare alkaloid found in Calabash curare _Strychnos_; and vinblastine (31.4), an antileukemic alkaloid of _Vinca_.

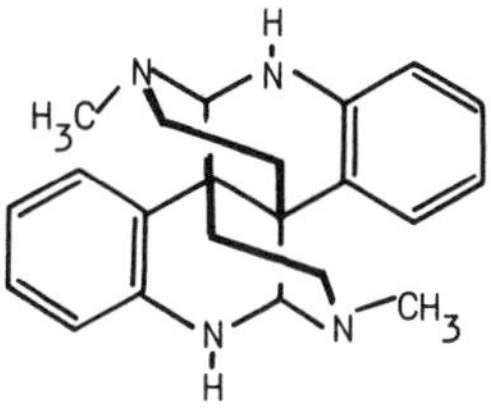

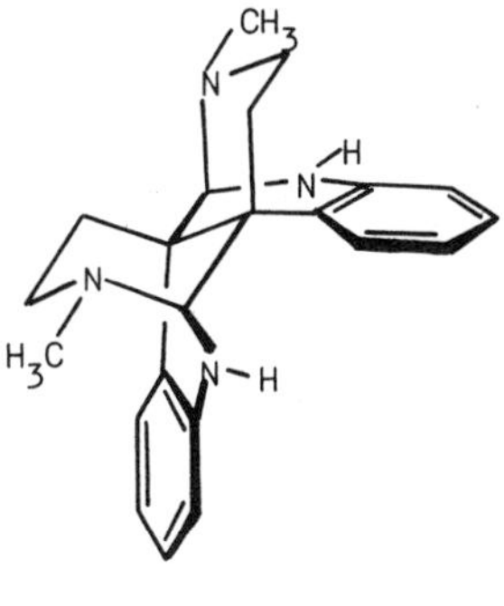

31.2

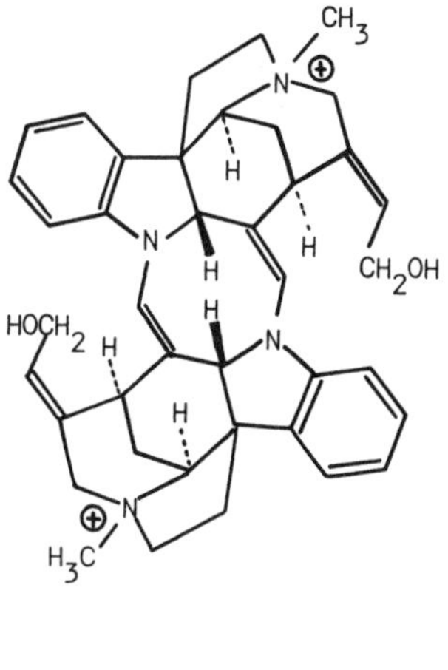

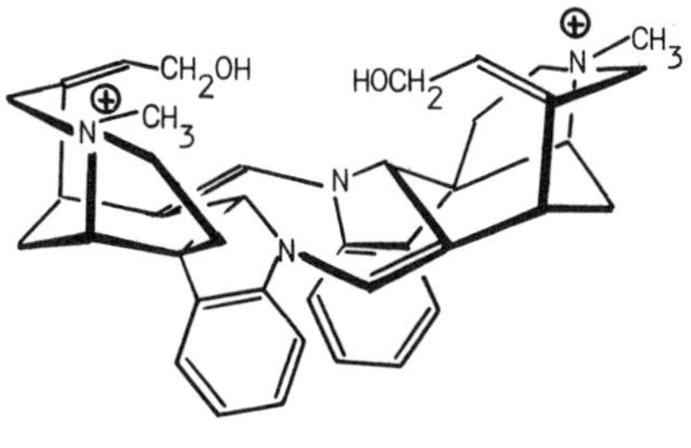

31.3

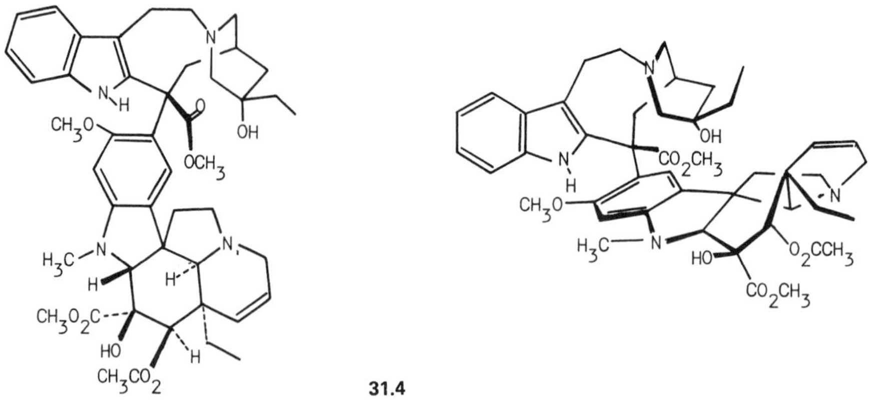

31.4

I. CALYCANTHINE

Calycanthine (31.2) is formed in Calycanthus floridus with incorporation of label from [^{14}C]-tryptophan (31.1; [243]). However, degradative work has yet to be reported. In the related base chimonanthine (31.5), also occurring in C. floridus, [β-^{14}C,2-^{3}H]tryptophan (31.1) is incorporated without change in the ^{3}H:^{14}C ratio [244]. Both calycanthine (31.2) and chimonanthine (31.5) are presumably [245] formed by oxidative dimerization of an appropriate tryptophan-derived (31.1) fragment. Thus, as shown in Scheme 31.2, it may be presumed that tryptophan (31.1) undergoes decarboxylation (Scheme 31.1) and methylation to N-methyltryptamine (31.6).* Then, oxidation of N-methyltryptamine (31.6) to a β-indolyl radical,

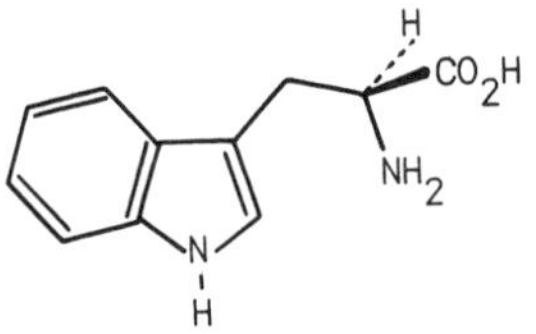

31.1

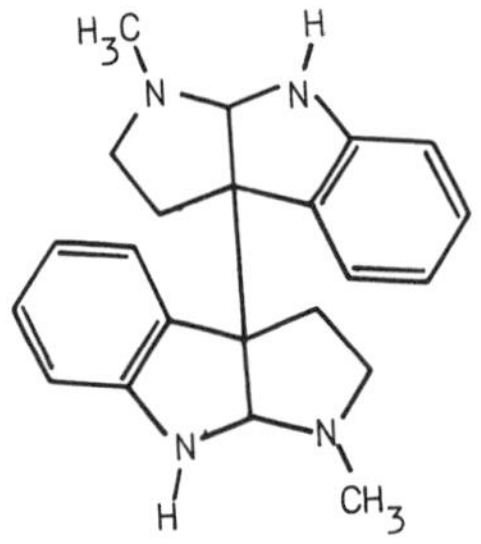

31.5

*Presumably, the methylating species is S-adenosylmethionine (31.7) and, at least in principle, methylation need not be this early in the biosynthesis. It is presumed that the decarboxylation of tryptophan (31.1) is, as usual, pyridoxal (31.8) catalyzed (Scheme 31.1).

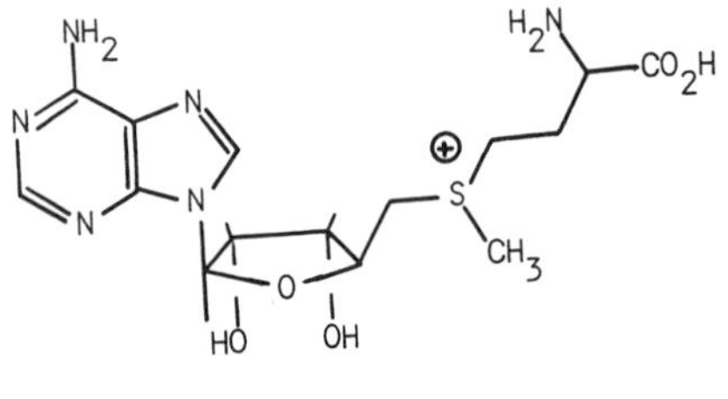

31.7

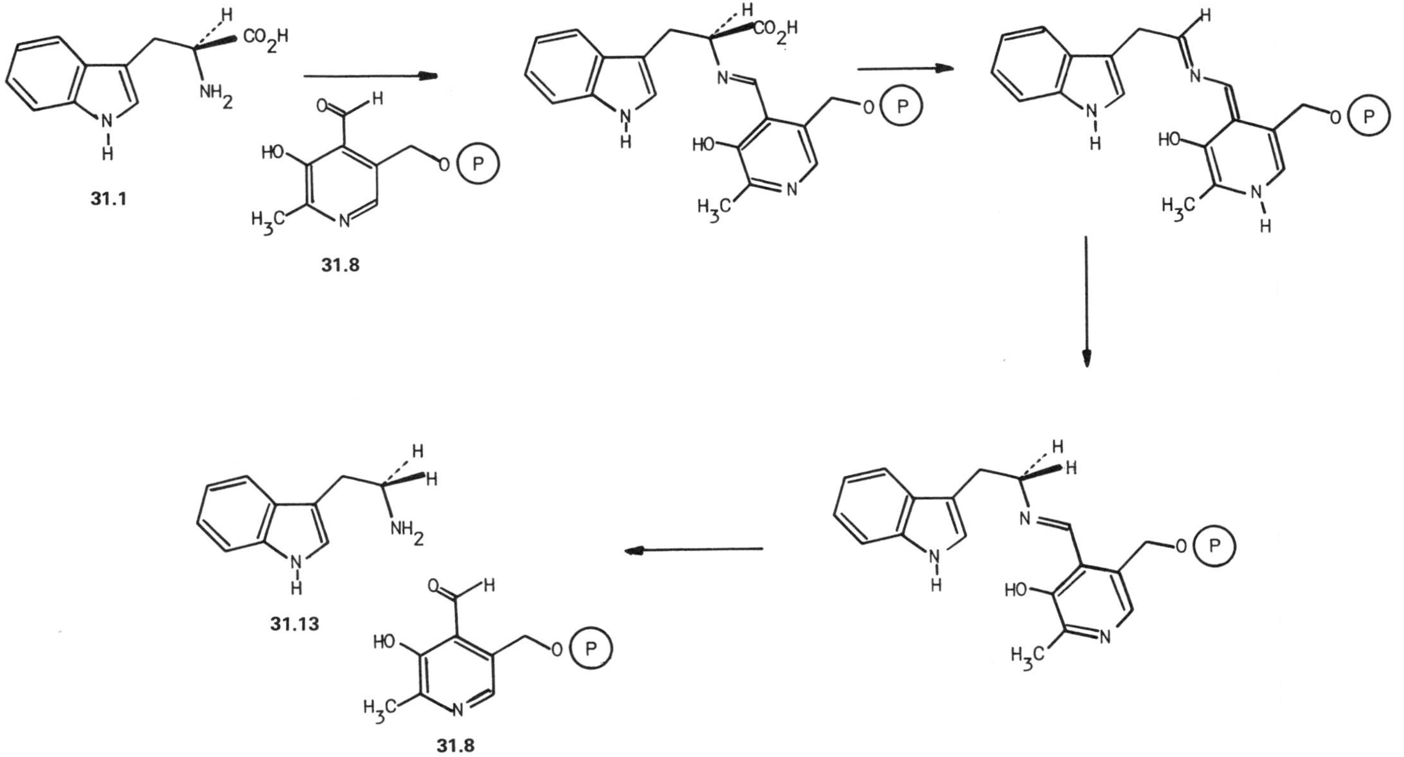

31.1

31.8

31.13

31.8

SCHEME 31.1 The presumed pathway for the formation of tryptamine from tryptophan utilizing pyridoxal catalysis.

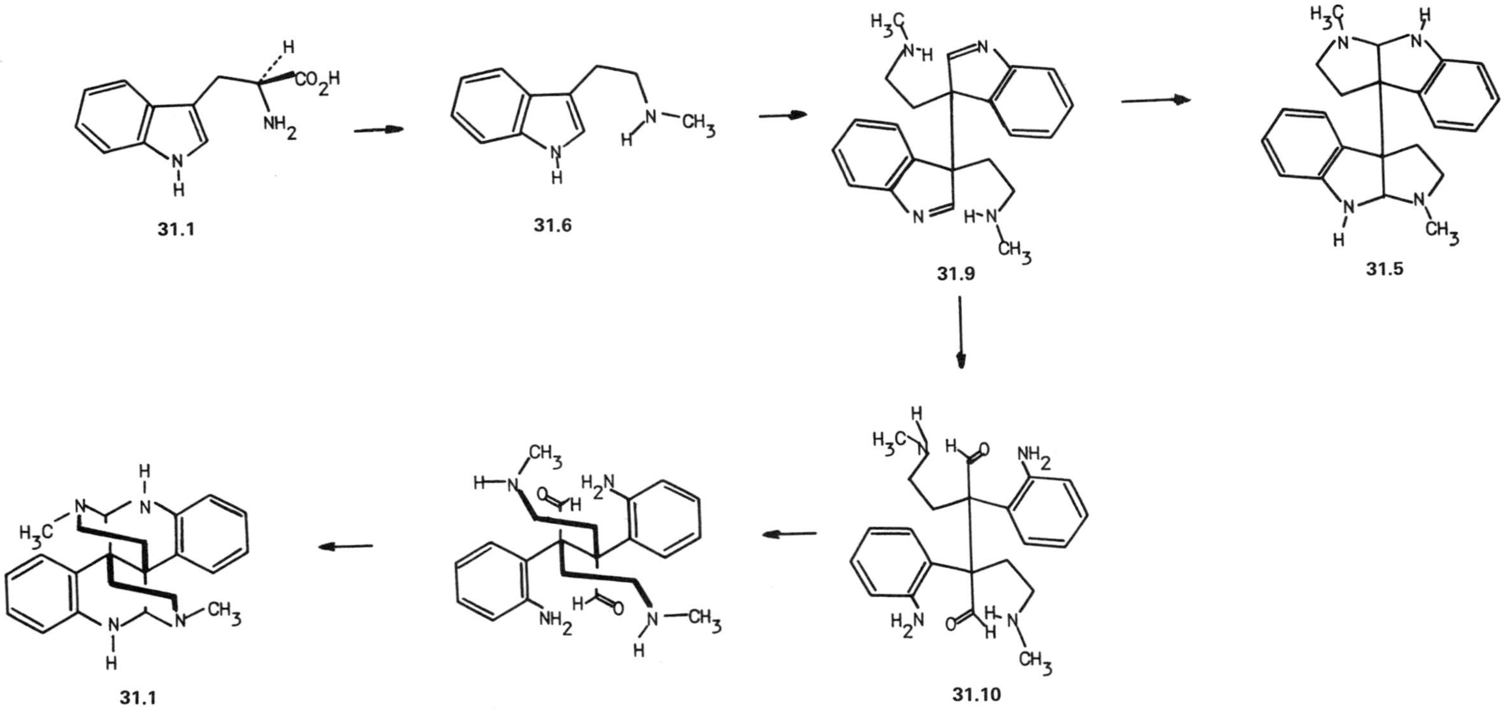

SCHEME 31.2 A possible pathway for the formation of calycanthine and chimonanthine based upon β-indolyl radical dimerization.

followed by self-coupling, generates a dimeric intermediate 31.9 which on simple cyclization leads to chimonanthine (31.5) and, potentially, other products. Alternatively, hydrolysis of 31.9 to the corresponding bis-aminoaldehyde (31.10) followed by recyclization can lead to calycanthine (31.1).

II. C-TOXIFERINE

In contrast to tubocurare, the arrow poison packed in tubes (p. 339) which is primarily made from plants of the genus Chondrodendron and which contains bisbenzylisoquinoline alkaloids, Calabash curare, the arrow poison packed in gourds, comes from the plants of several Strychnos species, and while other materials may also be present, the poisonous properties of this curare are presumably due to alkaloids.

Among the most common alkaloids of Calabash curare is C-toxiferine (31.3), and while it does not appear that biosynthetic work has been carried out, it is worthwhile to note that C-toxiferine (31.3) can be prepared in vitro from the Wieland-Gumlich aldehyde (31.11) by

31.11

acetic acid - sodium acetate catalyzed self-condensation and dehydration. Interestingly, the Wieland-Gumlich aldehyde (31.11) has been isolated (as caracurine VII, 31.11) from Strychnos toxifera Rob. Schomb., a plant that also serves as a source of C-toxiferine (31.3). Since the Wieland-Gumlich aldehyde (caracurine VII, 31.11) presumably has a genesis similar to that for strychnine (31.12), a hypothetical scheme (Scheme 31.3) from tryptophan-derived (31.1) tryptamine (31.13) and secologanin (31.14) via isovincoside (31.15), pregeissoschizine (31.16), the intermediate 31.17, preakuammicine (31.18), and caracurine VII (Wieland-Gumlich aldehyde, 31.11) to C-toxiferine (31.3) can be written.

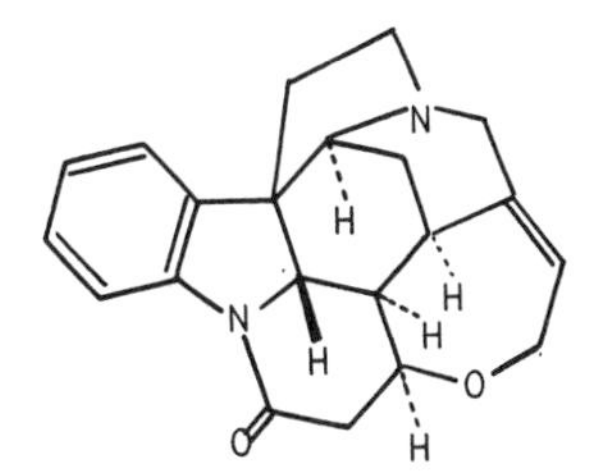

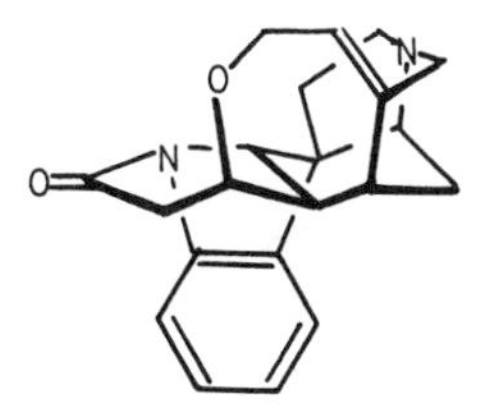

31.12

III. VINBLASTINE

Experiments in Catharanthus roseus have begun in an attempt to understand the origin of the antileukemic alkaloid vinblastine (31.4; [221, 234, 246].). Examination of the structure of vinblastine (31.4) indicates it is formally composed of an hydroxylated seco-catharanthine

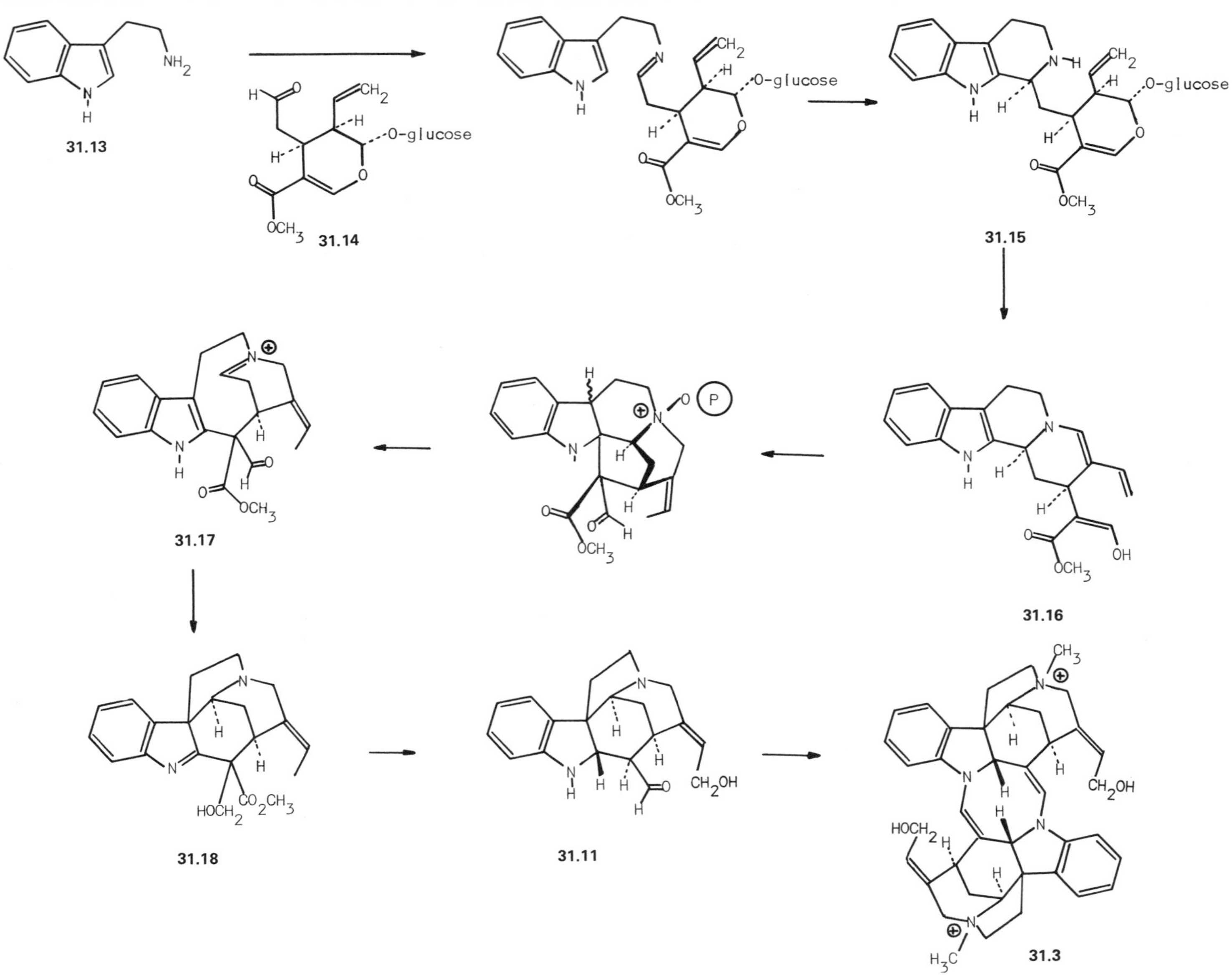

31.13

31.14

31.15

31.16

31.17

31.18

31.11

31.3

(31.18) and an oxidized and hydroxylated N-methyltabersonine (31.19) called vindoline (31.20), which is also isolated from C. roseus. Preliminary experiments [221, 234, 246] indicate that during a 5-week feeding period, [O-methyl-^{3}H]catharanthine (31.18) is not incorporated into vinblastine (31.4).

One possible pathway for the genesis of vinblastine (31.4) which is, despite the negative experimental report cited above, suggested by the similarity in structure of the two halves of vinblastine (31.4) to simpler alkaloids present in C. roseus is presented in Scheme 31.4 and Scheme 31.5. Here (Scheme 31.4), the intermediate 31.17 from Scheme 31.3 (generated, as shown there from tryptamine, 31.13, and secologanin, 31.14) is allowed to undergo reduction to 31.21 and isomerization to 31.22 (which, as shown in Scheme 30.17, is presumably stereoselective). Fragmentation to 31.23 and proton loss then generates the acrylic ester 31.24 (see p. 455), the presumed direct precursor to catharanthine (31.18) and tabersonine (31.19). Now (Scheme 31.5), if tabersonine (31.19) undergoes oxidation and methylation to 11-methoxy-N-methyltabersonine (31.25)* and then allylic oxidation, esterification, and hydration, vindoline (31.20) is formed. Then, since vindoline (31.20) is presumably capable of undergoing electrophilic aromatic substitution ortho to the methoxy group, attack by vindoline (31.20) on an oxidized and phosphorylated acrylic ester 31.26 could lead to 31.27. Reduction of the imine and hydration of the double bond in the Markownikoff sense leads to vinblastine (31.4).

(Scheme 31.4 appears on p. 470 and Scheme 31.5 appears on p. 471.)

---

*The oxidation and methylation could, of course, have been early rather than late stage processes.

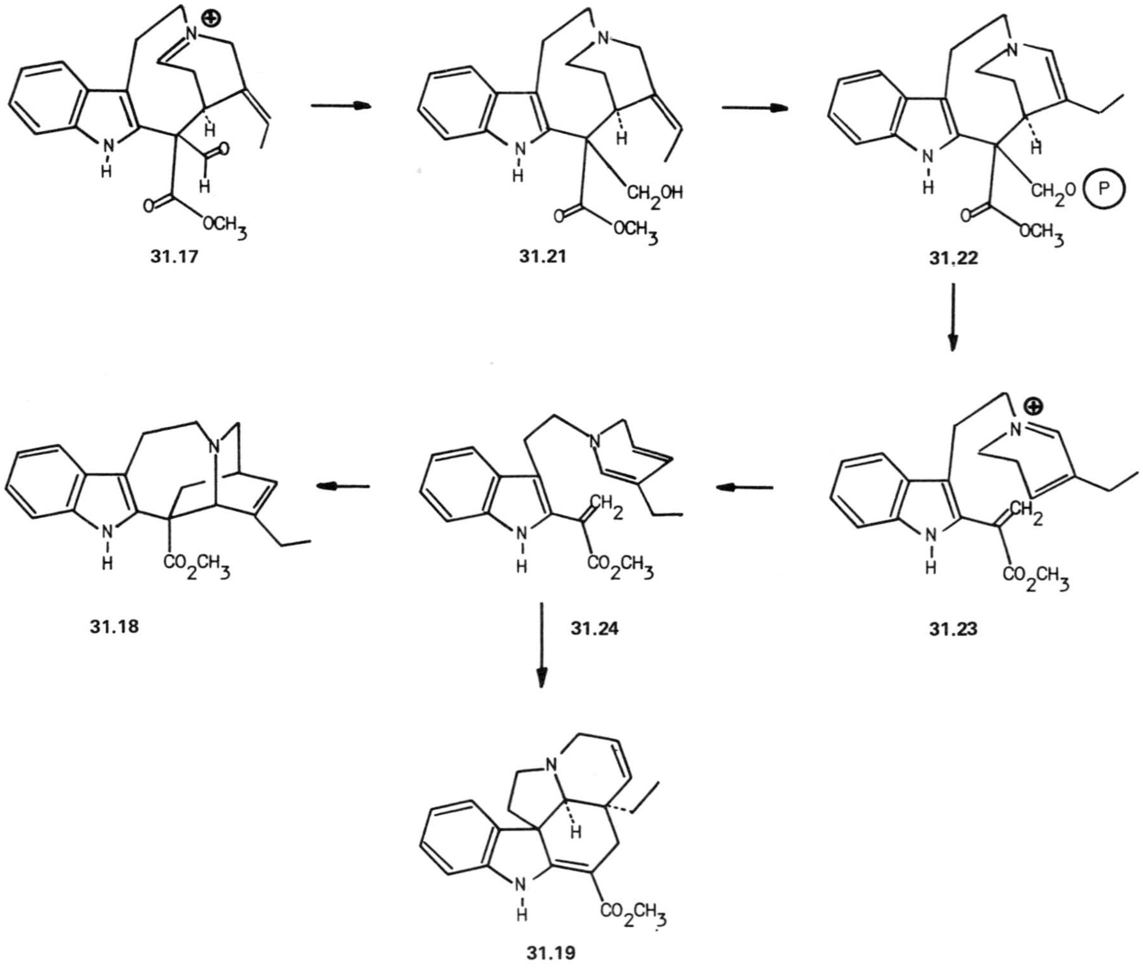
31.17
31.21
31.22
31.23
31.24
31.18
31.19
N-H
CH2OH
OCH3
CH2O
P
CH2
CO2CH3

SCHEME 31.4 A possible pathway to catharanthine and tabersonine and an intermediate which may lie on the route to vinblastine.

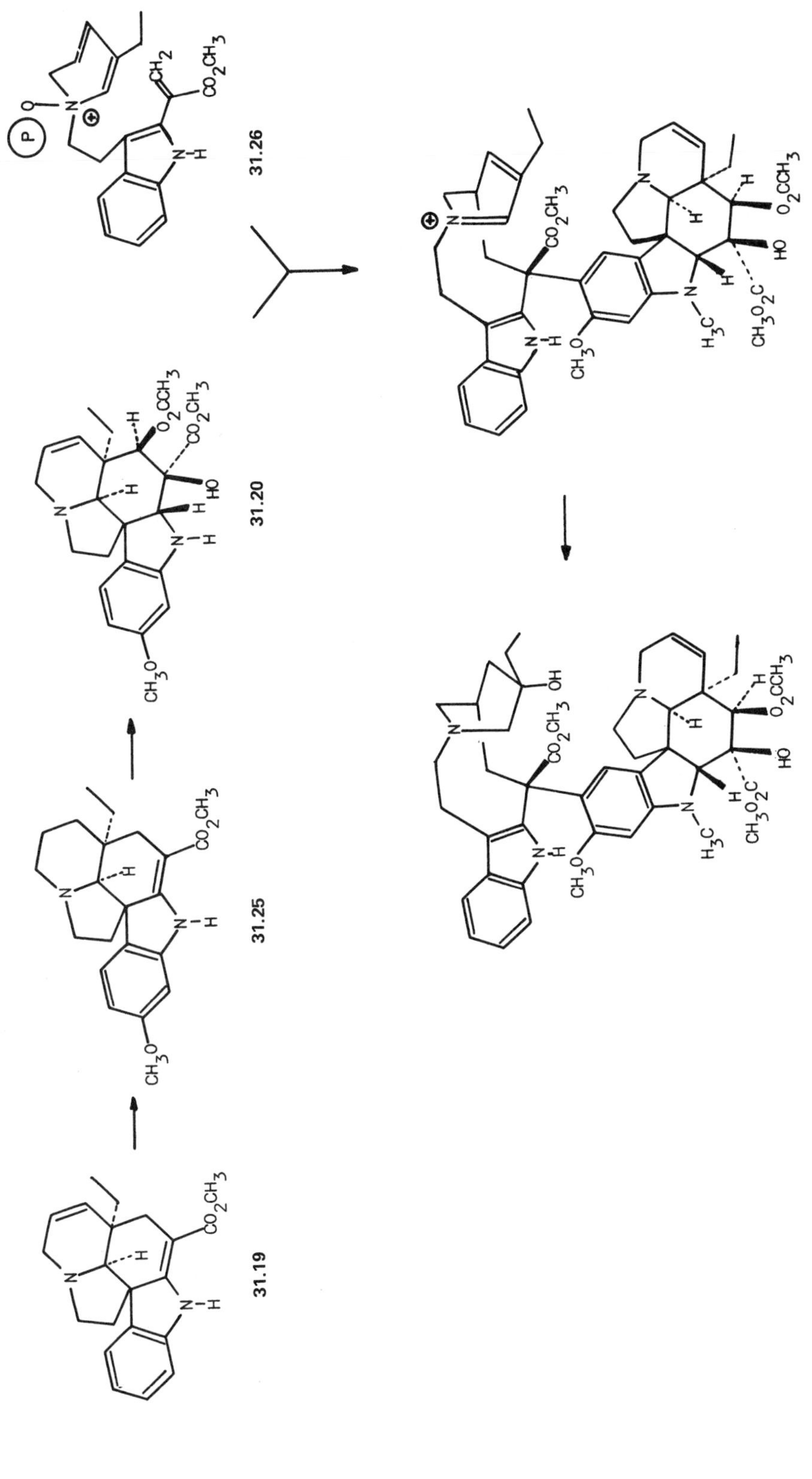

SCHEME 31.5 A possible pathway to vinblastine from intermediates similar to catharanthine and tabersonine.

32

CHEMISTRY OF THE SIMPLE INDOLE ALKALOIDS

It is as foolish to make experiments upon the constancy of a friend, as upon the chastity of a wife.

S. Johnson (1779)

Some of the simple indole alkaloids have been found in plants because they have been specifically sought, others have made their presence known through their physiological manifestations on ingestion, and still others were found "by accident." Thus, serotonin (32.1) was sought in plants after its isolation from beef serum and recognition of its neurohormonal activity. Bufotenine (32.2) and psilocybin (32.3), supposedly hallucination-causing alkaloids,

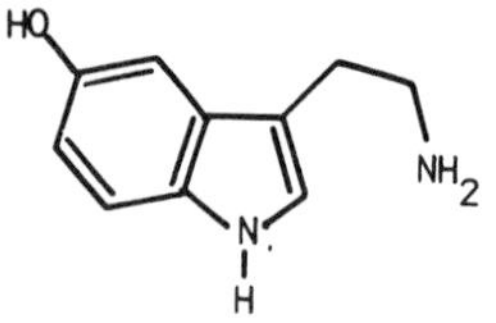

HO NH2 N H 32.1

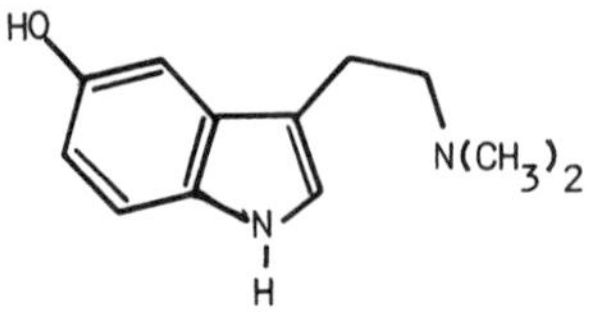

HO N(CH3)2 N H 32.2

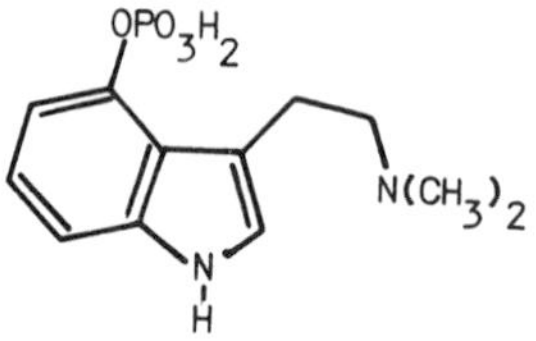

OPO3H2 N(CH3)2 N H 32.3

were found while attempting to isolate the active principles from seeds of Piptadenia peregrina and the fungus Psilocybe mexicana Heim., respectively. Finally, gramine (32.4) was orignally isolated from chlorophyll-deficient barley mutants in an attempt to understand the nature of chlorophyll deficiency.

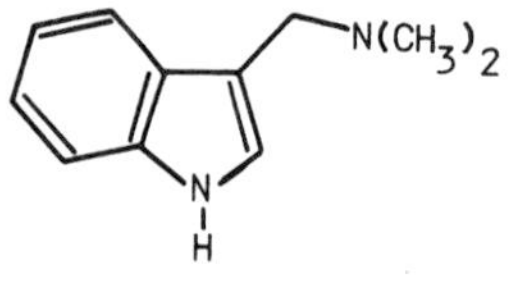

N(CH3)2 N H 32.4

I. SEROTONIN

The isolation of a vasoconstrictor substance from beef serum was first reported [247] in 1948. Later in the same year, an indolic base, $C_{10}H_{14}N_2O_2$, serotonin (32.1), was isolated (as a complex with creatinine, 32.5) as the sulfate salt, separated as a picrate, and tentatively

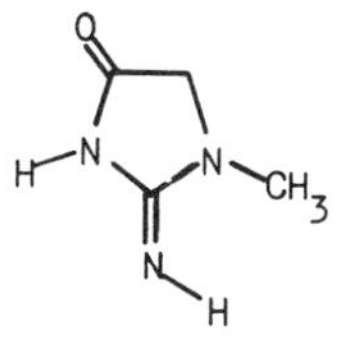

32.5

assigned, by ultraviolet spectroscopy and many spot tests, the presently accepted structure [248]. The presumed structure was established by comparison to a synthetic sample.

Serotonin (5-hydroxytryptamine, 32.1) has been prepared in many different ways. One of the first syntheses (Scheme 32.1; [249]) begins with 5-benzyloxyindole (32.6; which can be prepared via the Fischer indole synthesis) and involves its conversion to 5-benzyloxygramine (32.7) through a Mannich reaction. Formation of 5-benzyloxyindolacetonitrile (32.7) is then accomplished by cyanide treatment of the corresponding quaternary salt. Reduction yields the tryptamine 32.9, and catalytic debenzylation results in the formation of serotonin (32.1). Alternatively [250], the Fischer cyclization of the p-methoxyphenylhydrazone of 2,3-dioxopiperidine (32.10) is reported to yield 6-methoxy-1-keto-1,2,3,4-tetrahydro-β-carboline (32.11), which undergoes hydrolysis to the amino acid 32.12. Decarboxylation and demethylation with aluminum chloride yield serotonin (32.1; Scheme 32.2).

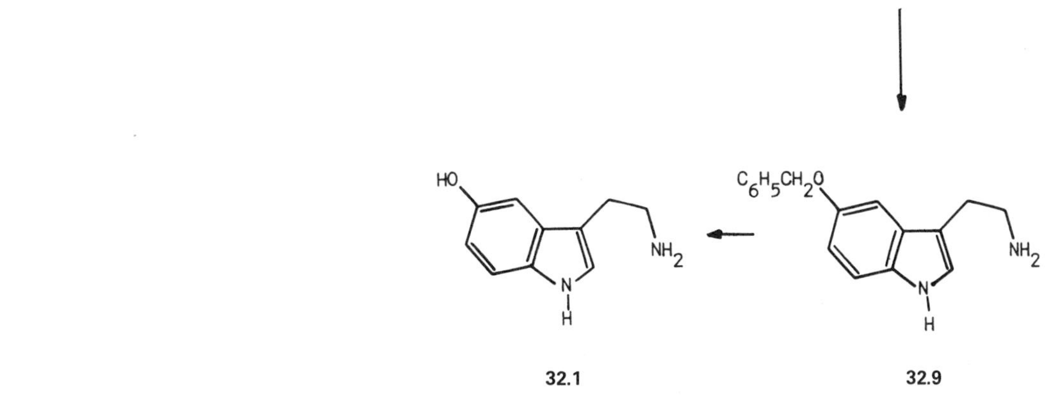

32.6 32.7 32.8

32.1 32.9

SCHEME 32.1 A synthesis of serotonin. (After Reference 249.)

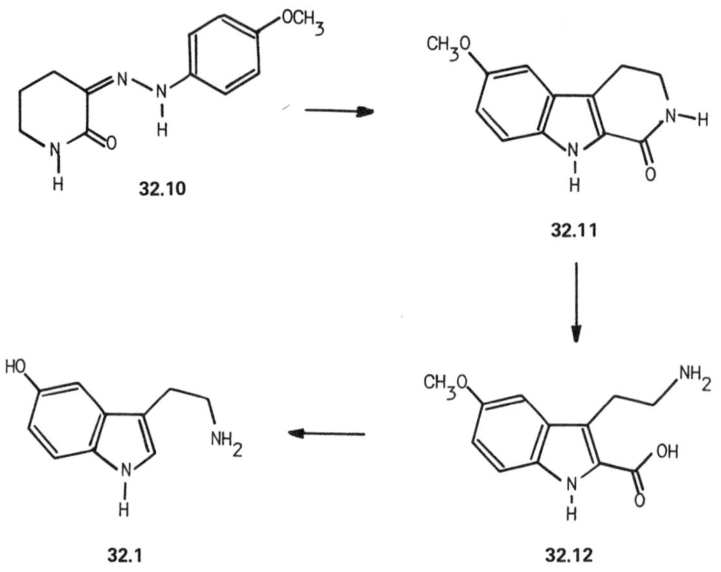

32.10

32.11

32.1

32.12

SCHEME 32.3 A synthesis of serotonin. (After Reference 250.)

II. BUFOTENIN

Bufotenin (32.2) has been isolated from such widely diverse sources as the shrub Piptadenia peregrina, the seeds of which are said to be the source of a ceremonial narcotic snuff, the parotid gland of the toad (Bufo vulgaris, Laur.), certain fungi (e.g., Amanita mappa, Batsch.), and human urine.

Although the first isolation of bufotenine (32.2) was reported before the beginning of the twentieth century, it was not until 1934 [251] that it was crystallized and the molecular formula established as $C_{12}H_{16}N_2O$. On the basis of color tests and other laboratory work in hand, bufotenine (32.2) was suspected to be a hydroxylated N,N-dimethyltryptamine. 5-Methoxy-N,N-dimethyltryptamine (32.13) was synthesized and compared, as its methiodide, with O-methylbufotenine methiodide (32.14); they were found to be identical.

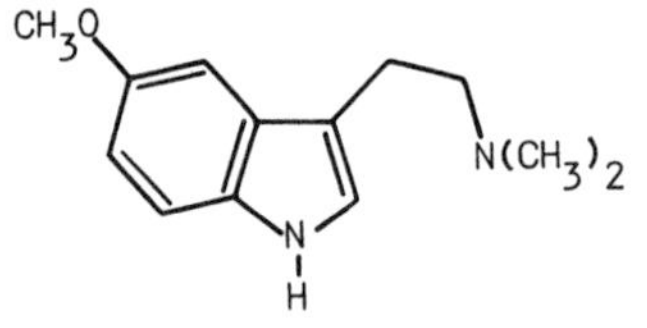

32.13

The synthesis of 5-methoxy-N,N-dimethyltryptamine methiodide (32.14) was accomplished [251]; Scheme 32.3) by condensation of the Grignard derivative of 5-methoxyindole (32.15) with chloroacetonitrile to 32.16, reduction of the latter with sodium and alcohol, and methylation with methyl iodide and thallium ethoxide. Shortly thereafter, the first synthesis of bufotenine (32.2) was reported [252]. As shown in Scheme 32.4, 5-ethoxyindole-3-acetic

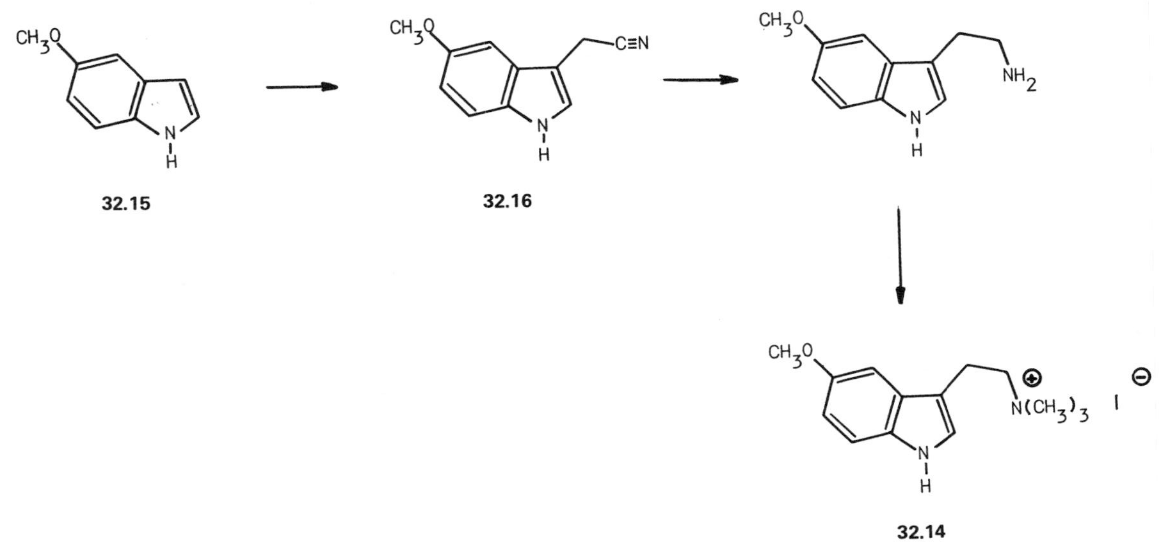

SCHEME 32.3 A synthesis of 5-methoxy-N,N-dimethyltryptamine methiodide (O-methylbufotenine). (After Reference 251.)

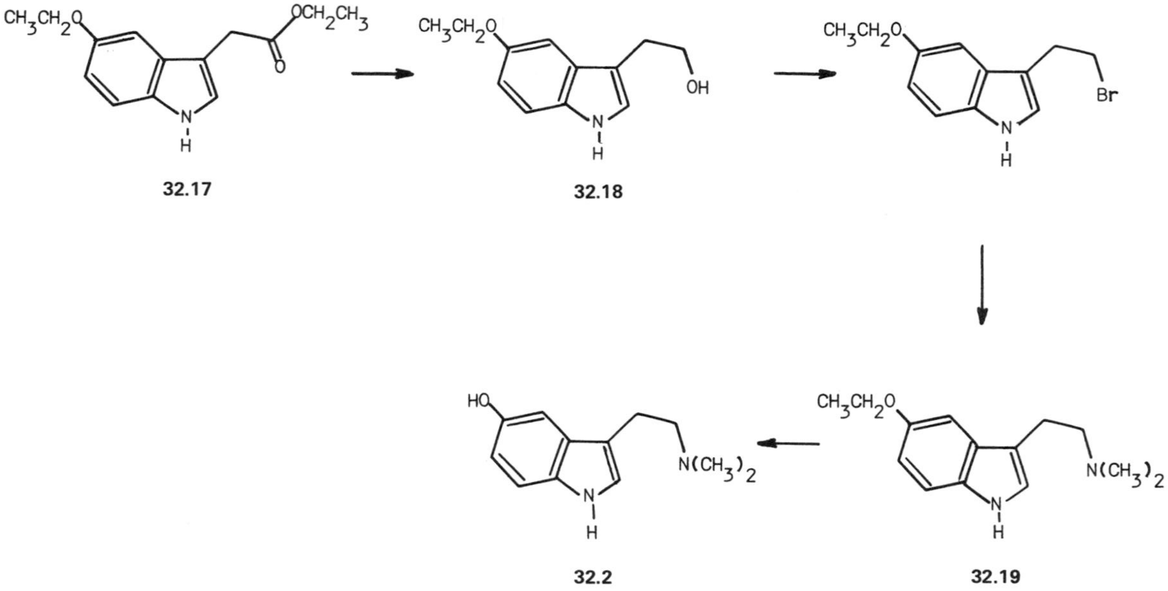

SCHEME 32.4 A synthesis of bufotenine. (After Reference 252.)

acid ethyl ester (32.17) is reduced with sodium in alcohol to the primary alcohol (32.18) which is converted to the corresponding primary bromide with phosphorus tribromide and thence to O-ethylbufotenine (32.19) with dimethylamine. Ether cleavage with aluminum chloride generates bufotenine (32.2).

III. PSILOCYBIN

Psilocybin (32.3) was first isolated in 1958 [253] from a Mexican fungus Psilocybe mexicana Heim. as an optically inactive, amphoteric solid, $C_{12}H_{17}N_2O_4P$, the ultraviolet spectrum of which was similar to that of 4-hydroxyindole and its derivatives. The halucinogen-causing effects of ingestion of psilocybin (32.3) itself were then demonstrated equivalent to that of its mushroom source.

Hydrolysis of psilocybin (32.3) leads to psilocin (32.20), which is also naturally occurring, and phosphoric acid (Equation 32.1). On treatment with diazomethane, psilocybin

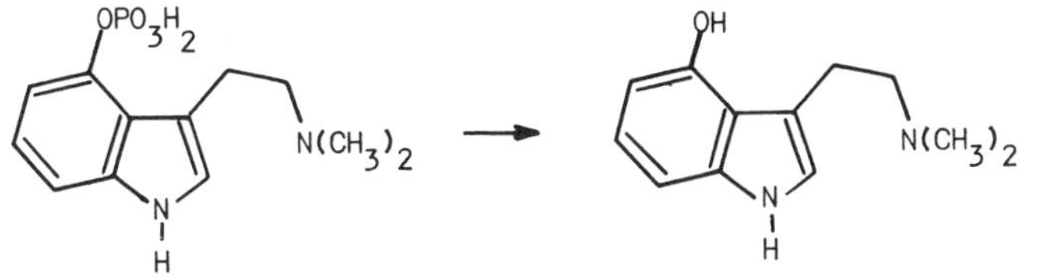

OPO_3H_2 $N(CH_3)_2$ N H **32.3** → OH $N(CH_3)_2$ N H **32.20** (Eq. 32.1)

(32.3) yields dimethylpsilocybin (32.21), and the latter affords trimethylamine on pyrolysis (Equation 32.2), a reaction not undergone by psilocybin (32.3).

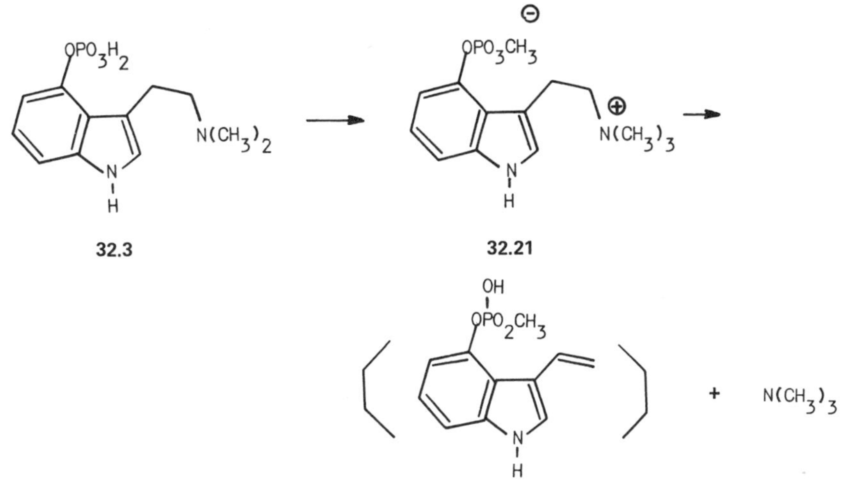

OPO_3H_2 $N(CH_3)_2$ N H **32.3** → ⊖ OPO_3CH_3 ⊕ $N(CH_3)_3$ N H **32.21** → OH OPO_2CH_3 N H + $N(CH_3)_3$ (Eq. 32.2)

On the basis of this information alone, psilocybin (32.3) was assigned its presently accepted structure and synthesized [254] as shown below (Scheme 32.5). Thus, 4-benzyloxyindole (32.22) is acylated with oxalyl chloride and the α-ketoacylchloride 32.23, converted to the corresponding N,N-dimethylamide (32.24) with dimethylamine. Reduction of 32.24 with lithium aluminum hydride leads to the O-benzyltryptamine 32.25 and then, on hydrogenolysis,

SCHEME 32.5 A synthesis of psilocybin. (After Reference 254.)

to psilocin (32.20). Conversion to psilocybin (32.3) is then accomplished by allowing psilocin (32.20) to react with dibenzylphosphochloridate (32.26) to the phosphate 32.27 and hydrogenolysis of the latter to 32.3.

IV. GRAMINE

Gramine (32.4), an optically inactive base, $C_{11}H_{14}N_2$, mp 134°C, was isolated [255] from chlorophyll-deficient barley mutants in a futile search for simple organic compounds related to the genetic deficiency of the mutants. It has since been isolated from normal sprouting barley and many other families of plants.

Gramine (32.4) was recognized as a tertiary base with two N-methyl groups and one acidic proton. Although its ultraviolet spectrum was typical of that of an indole derivative, suggestions for the structure were all incorrect until it was accidently synthesized by the reaction of indole magnesium iodide (32.27) with N,N-dimethylacetonitrile (32.28; Equation 32.3; [256]). Gramine (32.4) was later quantitatively prepared [257] by the Mannich condensation between indole, formaldehyde and dimethyl amine (Equation 32.4).

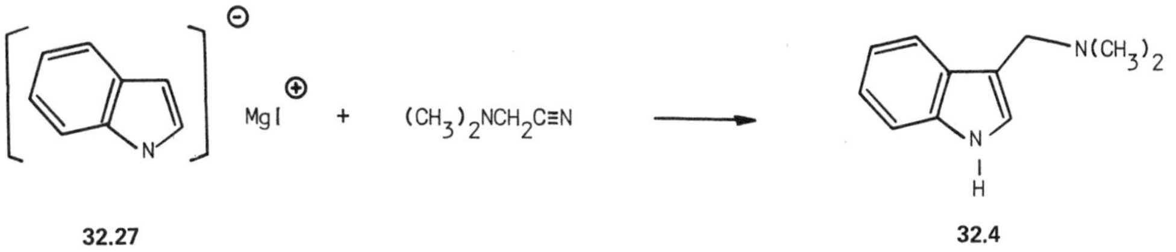

$$[C_8H_6N]^{\ominus}\ MgI^{\oplus} + (CH_3)_2NCH_2C{\equiv}N \longrightarrow \text{32.4}$$

32.27 32.4

(Eq. 32.3)

$$\text{indole} + CH_2O + HN(CH_3)_2 \longrightarrow \text{gramine}$$

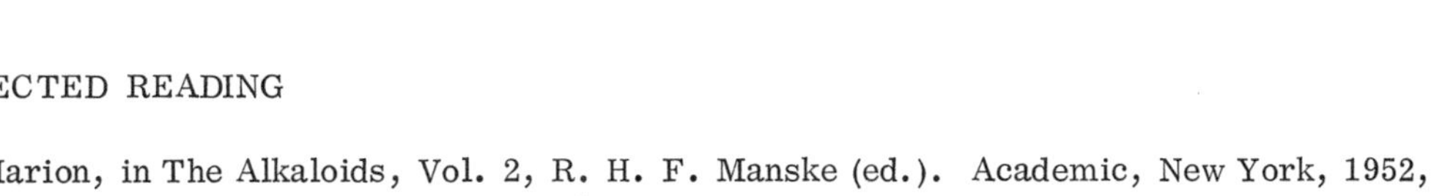

(Eq. 32.4)

SELECTED READING

L. Marion, in The Alkaloids, Vol. 2, R. H. F. Manske (ed.). Academic, New York, 1952, pp. 469 ff.

J. E. Saxton, in The Alkaloids, Vol. 7, R. H. F. Manske (ed.). Academic, New York, 1960, pp. 1 ff.

J. E. Saxton, in The Alkaloids, Vol. 8, R. H. F. Manske (ed.). Academic, New York, 1965, pp. 1 ff.

J. E. Saxton, in The Alkaloids, Vol. 10, R. H. F. Manske (ed.). Academic, New York, 1968, pp. 491 ff.

33

CHEMISTRY OF THE TRICYCLIC β-CARBOLINE ALKALOIDS: TRYPTOPHAN + C_2

Fire in each eye, and papers in each hand
They rave, recite and madden 'round the land.

A. Pope, The Satires, Prologue

The seeds of Peganum harmala L. and an extract from the South American liana (Banisteria caapi, Spruce) along with numerous others, have provided a large number of related tricyclic β-carboline alkaloids which, in their crude state, have been utilized as both anthelmintic and narcotic agents. Two alkaloids which are typical of the many bases present are harmine (33.1), from Peganum harmala L., and eleagine [(±)-tetrahydroharman (33.2)] from Arthrophytum leptocladum Popov. The former, isolated about 100 years [258] before the latter [259] suffered the structural work which made identification of eleagine (33.2) much simpler. Indeed, eleagine (33.2) was synthesized during structural work on harmine (33.1; see below) many years prior to its isolation.

Harmine (33.1), $C_{13}H_{12}N_2O$, is accompanied by a closely related base, harmaline (33.3), $C_{13}H_{14}N_2O$, both of which are readily crystallized from, e.g., an extract of the crushed seeds of Peganum harmala L., and both of which are optically inactive. Since both harmine (33.1) and harmaline (33.3) yield tetrahydroharmine (33.4) on reduction with sodium in ethanol and since harmaline (33.3) is oxidized to harmine (33.1) by a variety of reagents, it was concluded that harmaline (33.3) was dihydroharmine (Equation 33.1).

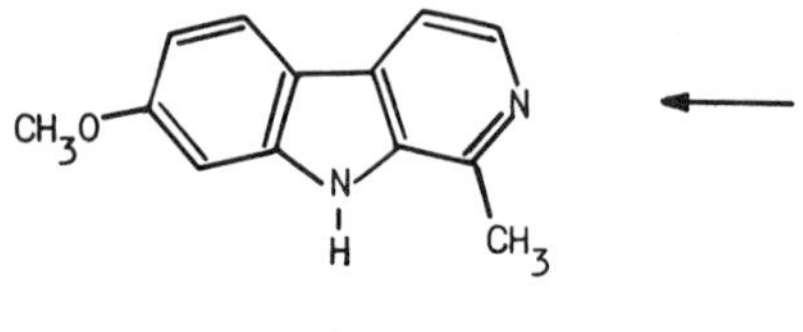

CH_3O N N H CH_3

33.1

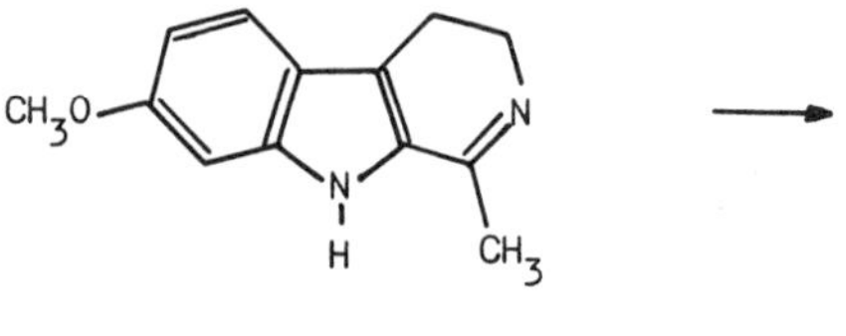

CH_3O N N H CH_3

33.3

(Eq. 33.1)

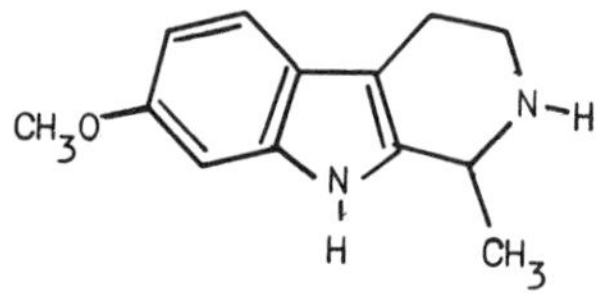

CH_3O N–H N H CH_3

33.4

When harmine (33.1) is heated with concentrated hydrochloric acid at 140°C, methyl chloride and harmol (33.5), $C_{12}H_{10}N_2O$, a phenol, are isolated (Equation 33.2). Harmol

33.1 33.5

(Eq. 33.2)

(33.5) can be converted, with zinc chloride, ammonium chloride, and ammonia, into aminoharman (33.6) and the latter into harman (33.7) on diazotization and reduction (Equation 33.3).

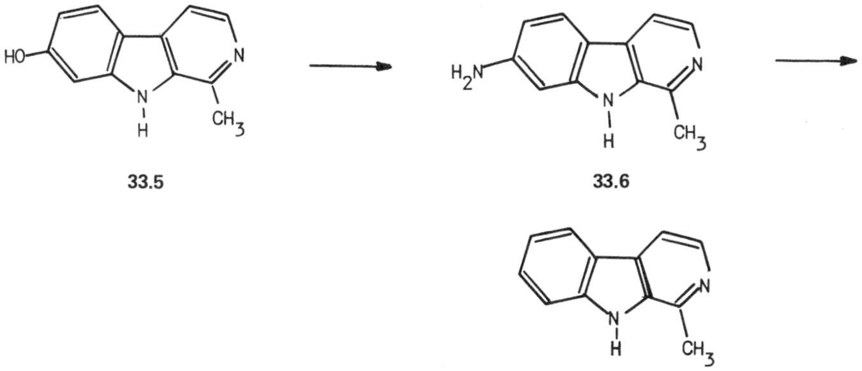

33.5 33.6

33.7

(Eq. 33.3)

Oxidation of harmaline (33.3) with fuming nitric acid yields m-nitroanisic acid (33.8; Equation 33.4), while oxidation of harmine (33.1) with chromic and sulfuric acids yields harminic acid (33.9), an ortho-dicarboxylic acid, which sequentially undergoes loss of two equivalents of carbon dioxide on pyrolysis to generate apoharmine (33.10), $C_8H_8N_2$ (Equation 33.5).

33.3 33.8

(Eq. 33.4)

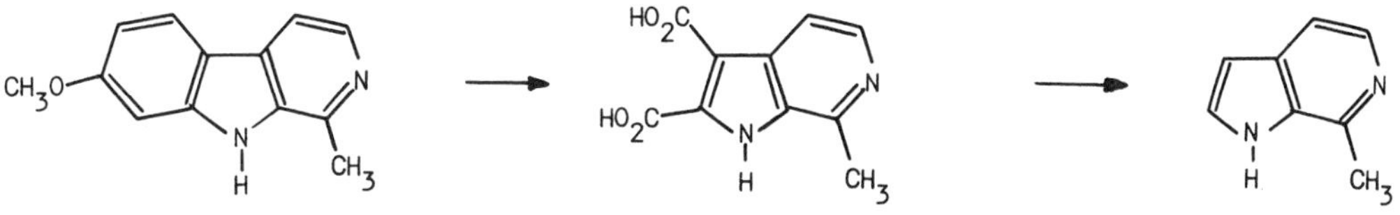

33.1 33.9 33.10

(Eq. 33.5)

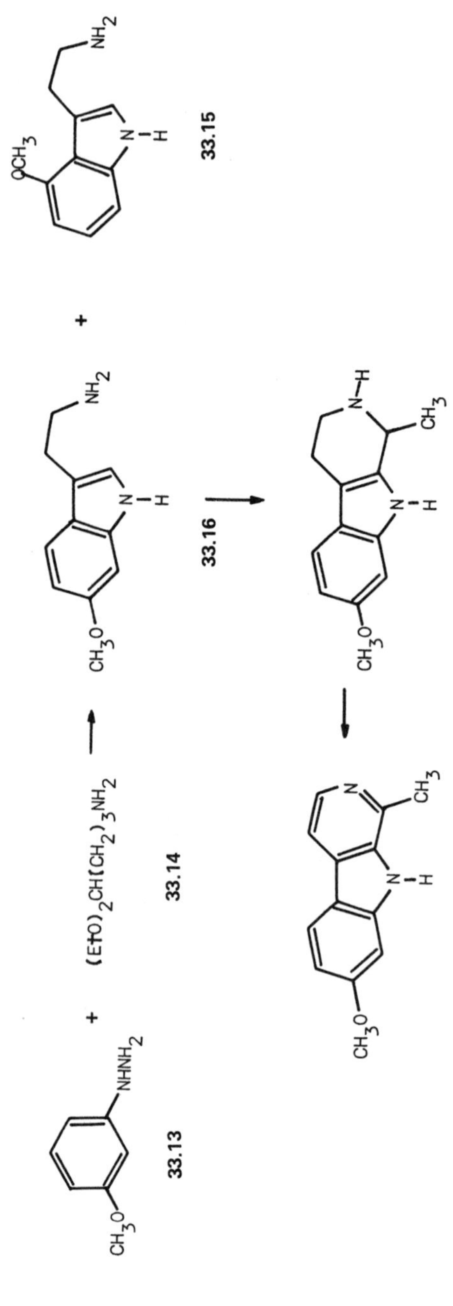

33.13 + $(EtO)_2CH(CH_2)_3NH_2$ (33.14) → 33.16 + 33.15

33.16 → 33.4 → 33.1

SCHEME 33.1 A synthesis of harmine. (After Reference 260.)

Of the two nitrogens present in harmine (33.1), harmaline (33.3), and apoharmine (33.10), only one is basic since salts are formed with only one equivalent of acid. Additionally, since both harmine (33.1) and harmaline (33.3) condense with benzaldehyde to yield benzylidine derivatives, a methyl group ortho to nitrogen on the pyridine ring was presumed present. Then, since the sum of the number of carbon atoms in the methoxyl, methyl, benzene, and pyridine rings is equal to the number of carbon atoms present, a pyrrole ring lying between the benzene and pyridine rings was suggested to accommodate the last nitrogen. This idea was confirmed by the formation of harman (33.7) by oxidation of the acid-catalyzed condensation product between tryptophan (33.11) and acetaldehyde with potassium dichromate (Equation 33.6) and by the fact that tetrahydroharman (eleagine, 33.2), which can be oxidized to harman (33.7) with palladium on carbon and air, is readily formed on condensation of tryptamine (33.12) with acetaldehyde at pHs between 5.2 and 6.2 (Equation 33.7).

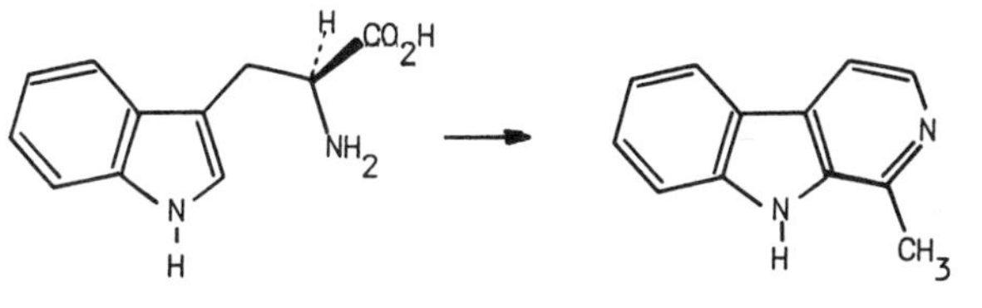

H CO_2H NH_2 N N H H CH_3

33.11 → 33.7 (Eq. 33.6)

NH_2 N H N—H N H CH_3 N N H CH_3

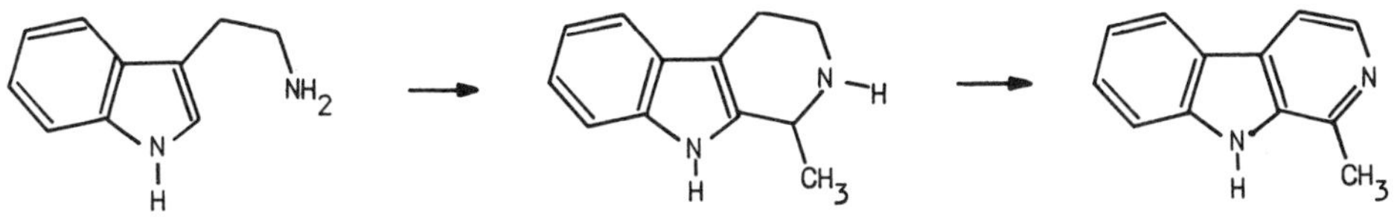

33.12 → 33.2 → 33.7

(Eq. 33.7)

Finally, harmine (33.1) itself [260] was synthesized (Scheme 33.1) by the condensation of m-methoxyphenylhydrazine (33.13) with γ-aminobutyraldehyde diethylacetal (33.14) in the presence of zinc chloride to a mixture of 4-methoxy- and 6-methoxy-3-β-aminoethylindoles (33.15 and 33.16, respectively). The latter (33.16), on treatment with acetaldehyde, yields tetrahydroharmine (33.4) which undergoes dehydrogenation to harmine (33.1) with maleic anhydride in the presence of palladium on carbon.

SELECTED READING

R. H. F. Manske (ed.), in The Alkaloids, Vol. 8. Academic, New York, 1965, pp. 47 ff.
N. Neuss, in Chemistry of the Alkaloids, S. W. Pelletier (ed.). Van Nostrand-Reinhold, New York, 1969, pp. 213-261.

34

CHEMISTRY OF THE ERGOLINE ALKALOIDS: TRYPTOPHAN + C_5

> And constancy lives in realms above;
> And life is thorny; and youth is vain;
> And to be wroth with one we love
> Doth work like madness in the brain.
>
> S. T. Coleridge, Cristabel, Part II

The pistil of rye and certain other grasses is capable of infection by the parasitic fungus Claviceps purpurea. Unless the infected grain is sieved, the fungus passes into the flour, and bread made from such contaminated flour apparently retains activity from some of the alkaloids elaborated by and present in the fungus. Thus, ingestion of the contaminated flour results in the disease called ergotism ("St. Anthony's Fire"). Outbreaks of ergotism have been recorded into the twentieth century. Convulsive ergotism (as distinct from gangrenous ergotism) has occasionally been looked upon as a form of spiritual or demonic possession during which violent muscle spasms, bending the sufferer into otherwise unattainable positions, occurred, and after passing left mental as well as physical scars.

Extracts of C. purpurea have long been used medicinally since they effect smooth muscle contraction, and even today, bases related to the alkaloid hydrolysis products, lysergic acid (34.1), elymoclavine (34.2), and agroclavine (34.3), are used for the same purpose. Additionally, it was accidently discovered in the early 1940s that the diethylamide of (+)-lysergic acid (34.4; LSD-25; as the tartrate salt) could, apparently, be absorbed through the skin with resulting inebriation. In a bold experiment, it was then demonstrated that oral ingestion resulted in symptoms characteristic of schizophrenia, which, although temporary, were quite dramatic [261].

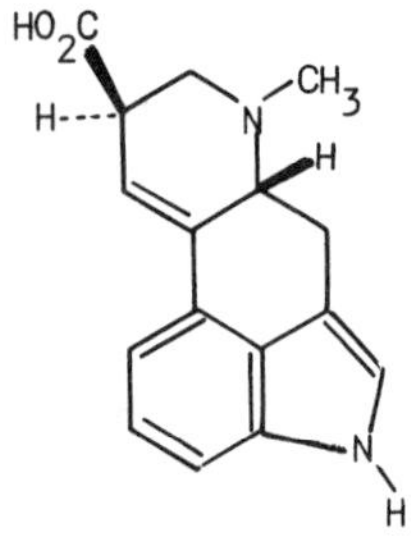

34.1

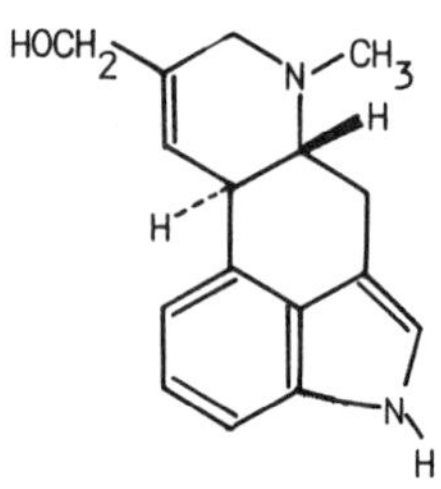

34.2

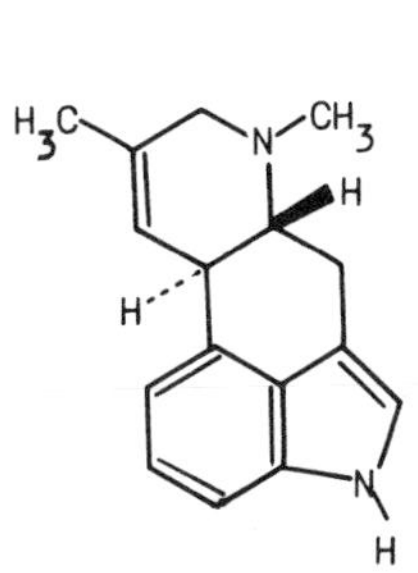

34.3

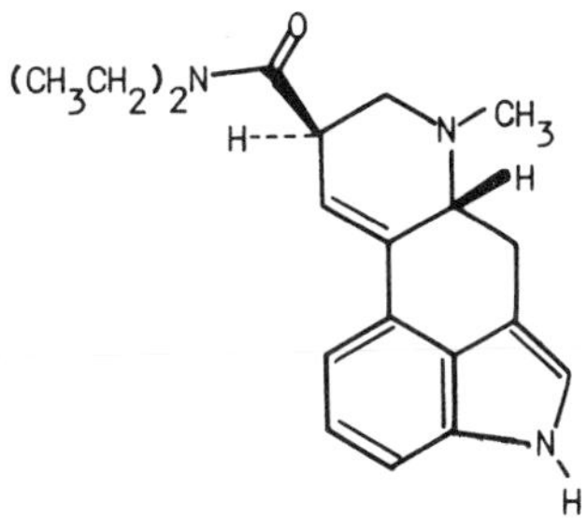

34.4

I. LYSERGIC ACID

The action of methanolic potassium hydroxide on the extracts of Claviceps purpurea results in cleavage of the peptide alkaloids and yields (+)-lysergic acid (34.1), $C_{16}H_{16}N_2O_2$, $[\alpha]_D$ + 40°, which was identified as possessing one N-methyl, a readily reducible (sodium in amyl alcohol) double bond (to yield dihdyrolysergic acid I, 34.5) and only one basic nitrogen. Since the methyl ester of lysergic acid (34.1) still possessed (Zerwittinov) one active hydrogen, a pyrrole or indole group was supposed present. This supposition was confirmed by the isolation of 3,4-dimethylindole (34.6) from potassium hydroxide fusion of dihydrolysergic acid I (34.5; Equation 34.1).

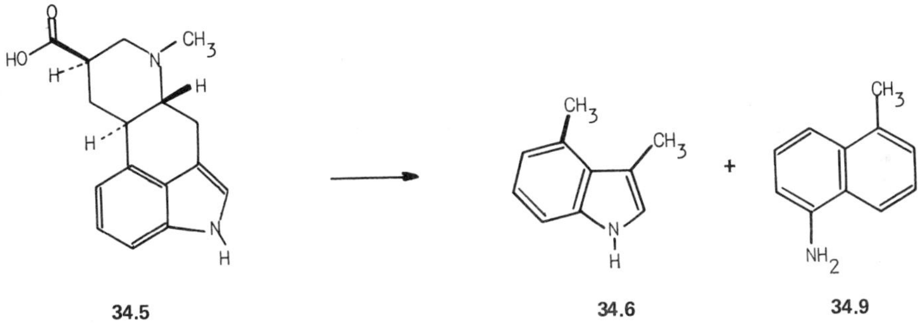

34.5 34.6 34.9

(Eq. 34.1)

In addition, heating lysergic acid (34.1) to 210-240°C causes decomposition with evolution of carbon dioxide and methylamine (Equation 34.2). This behavior is not mimicked in dihydrolysergic acid I (34.5), which at 350°C, yields instead a neutral unsaturated dehydration

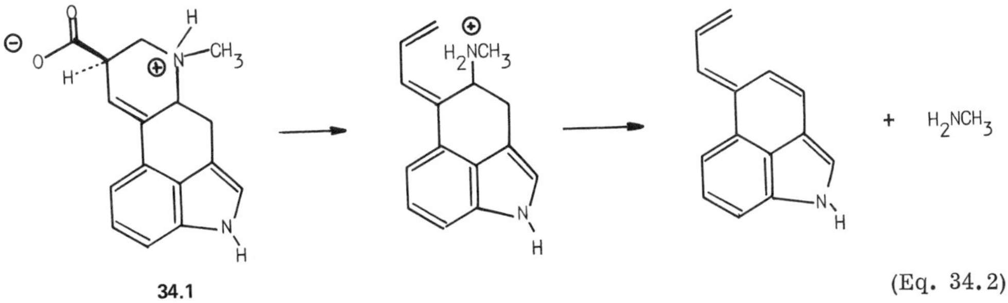

34.1

(Eq. 34.2)

product ($C_{16}H_{16}N_2O$, 34.7) which can, in turn, be hydrogenated to a neutral dihydro derivative, $C_{16}H_{18}N_2O$, 34.8 (Equation 34.3). These reactions permit formulation of lysergic acid (34.1) as a β-amino acid. On the basis of the presence of one readily reducible double

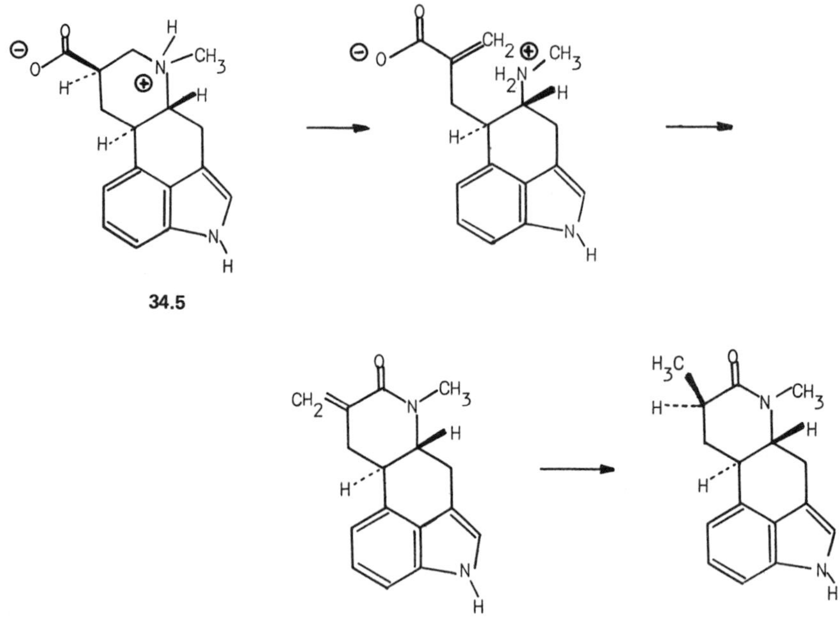

34.5

(Eq. 34.3)

34.7

34.8

bond, a carboxylic acid grouping, and the presence of two nitrogens, it was concluded that lysergic acid (34.1) must be tetracyclic. Then, when repetition of the potassium hydroxide fusion (Equation 34.1) yielded 3,4-dimethylindole (34.6) and methylamine as well as 1-methyl-5-aminonaphthalene (34.9), it was concluded that 34.1, or some similar structure (stereochemistry omitted) would be reasonable.

The position of the double bond, as lying between the carboxylate and the aromatic ring, could be tentatively decided on the basis of the following observations:

1. The double bond is, by ultraviolet spectroscopy, in conjugation with the aromatic system.
2. (+)-Lysergic acid (34.1) in methanol undergoes mutarotation to (+)-isolysergic acid (34.10);* and both (+)-lysergic acid (34.1) and (+)-isolysergic acid (34.10) can be converted to the same optically active lactam (34.14) by acetic anhydride (Equation 34.4; [262]). The "new" double bond in the lactam 34.14 is in conjugation (ultraviolet) with the aromatic system.

---

*Possible intermediates, 34.11, 34.12, and 34.13, in this conversion are shown in Scheme 34.1. Of these, the first (34.11) has been preferred [262] on the grounds of economy of explanation and the fact that the corresponding dihydroisolysergic acid I (34.16; see below) can also be epimerized (albeit less readily than lysergic acid, 34.1, itself). Interestingly, the third (i.e., 34.13) intermediate accounts for the reported racemization [see (3)] of (+)-lysergic acid (34.1) on treatment with barium hydroxide.

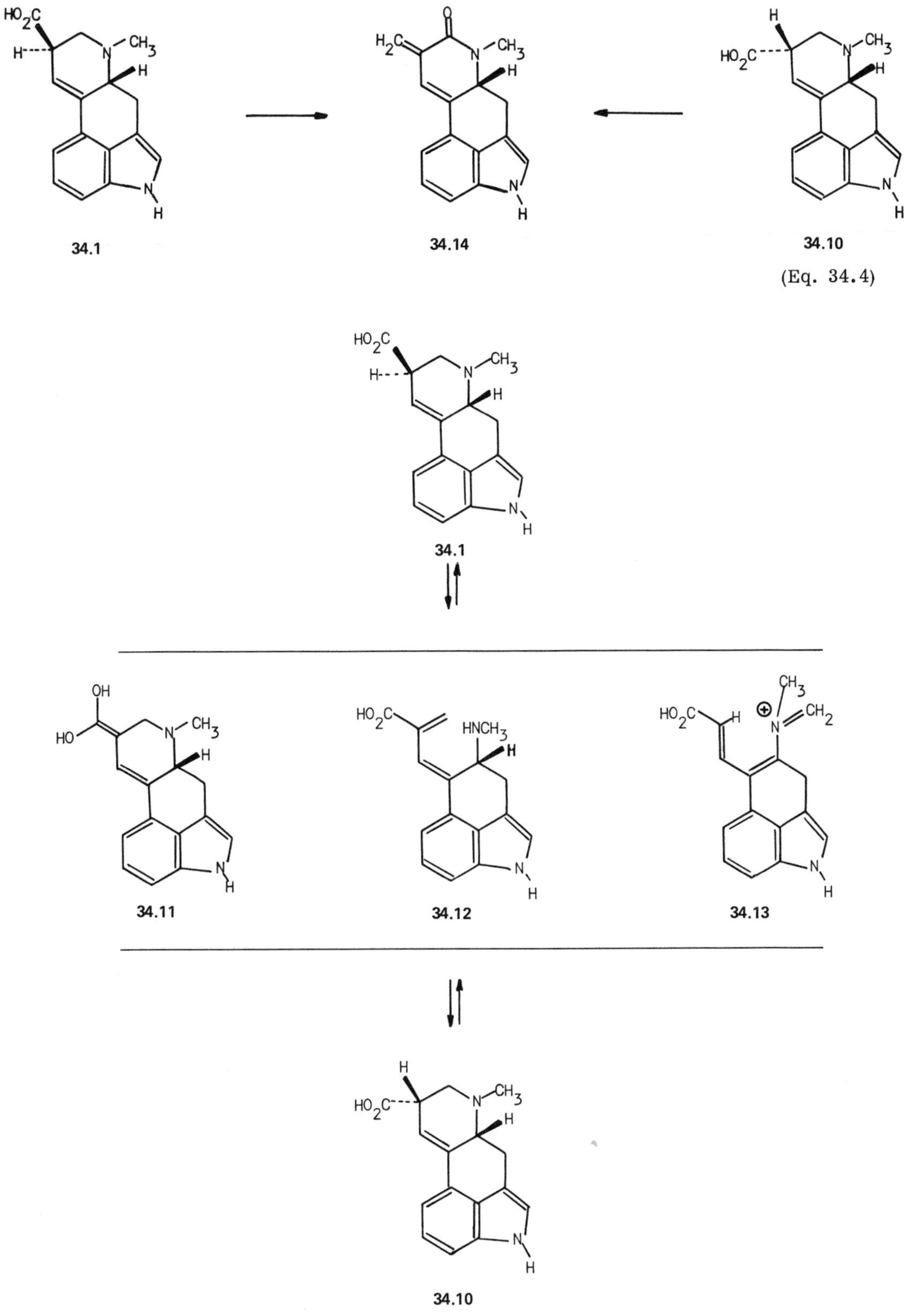

34.1

34.14

34.10

(Eq. 34.4)

34.1

34.11

34.12

34.13

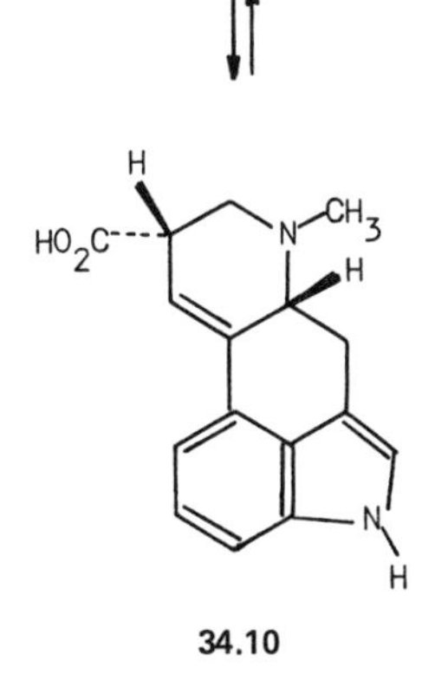

34.10

SCHEME 34.1 Possible intermediates which can be used to account for the mutarotation of lysergic acid.

3. When (+)-lysergic acid (34.1) or (+)-isolysergic acid (34.10) is heated with barium hydroxide solution, racemization occurs.

The position of the double bond, and indeed, the other structural information was subsequentially confirmed by synthesis, but the stereochemical details of the structure of (+)-lysergic acid (34.1) were obtained deductively.

Thus, based upon the position of the double bond as shown (+)-lysergic acid (34.1) and (+)-isolysergic acid (34.10) are epimeric at the carbon bearing the carboxyl group (C-8). Reduction of (+)-lysergic acid (34.1) yields a single dihydrolysergic acid, namely, dihydrolysergic acid I (34.5; Equation 34.5). However, reduction of (+)-isolysergic acid (34.10)

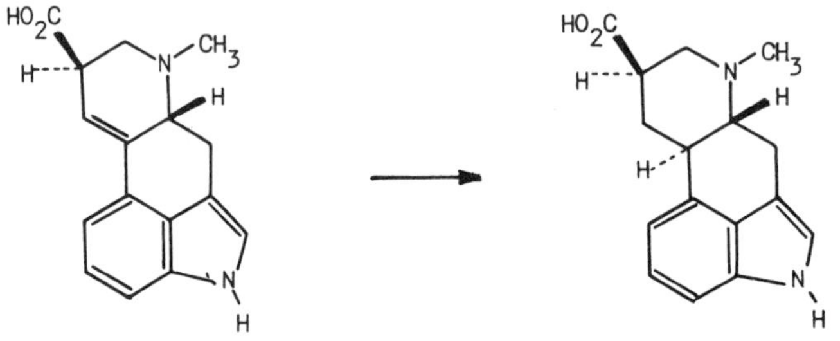

34.1 → 34.5 (Eq. 34.5)

yields a pair of dihydroisolysergic acids (Equation 34.6). One of these, dihydroisolysergic acid II (34.15), being formed on rapid (platinum catalysis in acetic acid solution) hydrogenation, and the other, dihydroisolysergic acid I (34.16) being formed along with dihydroisolysergic acid II (34.15) when the hydrogenation is carried out more slowly over palladium.

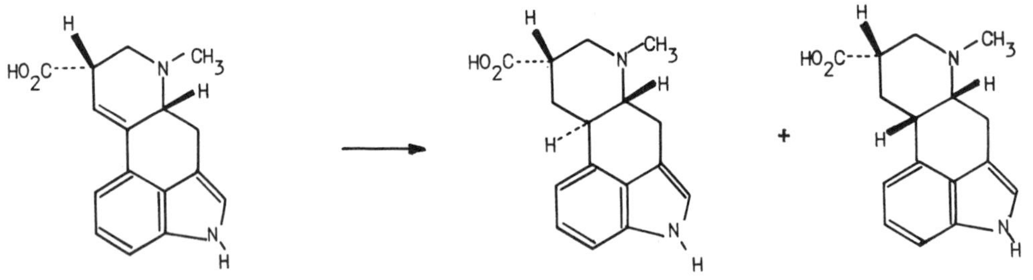

34.10 → 34.16 + 34.15 (Eq. 34.6)

Now, since catalytic hydrogenation of the double bond should result in a cis ring fusion which ought to be less stable than the trans isomer, it is presumed that dihydroisolysergic acid II (34.15) has a cis fusion where rings C and D are joined and dihydroisolysergic acid I (34.16) must have the trans C/D ring fusion. Additionally, dihydroisolysergic acid I (34.16) must have the same C/D ring fusion as dihydrolysergic acid I (34.5), and these differ only in the sense of the carboxyl stereochemistry since they can be equilibrated in alkali, and, on treatment with acetic anhydride, both yield the same lactam (34.7; Scheme 34.2).

Finally, since vigorous alkaline hydrolysis of the methyl ester of dihydroisolysergic acid I (34.17) yields dihydrolysergic acid I (34.5) and not dihydroisolysergic acid I (34.16), and since synthetically obtained dihydroisolysergic acid II methyl ester (34.18) yields

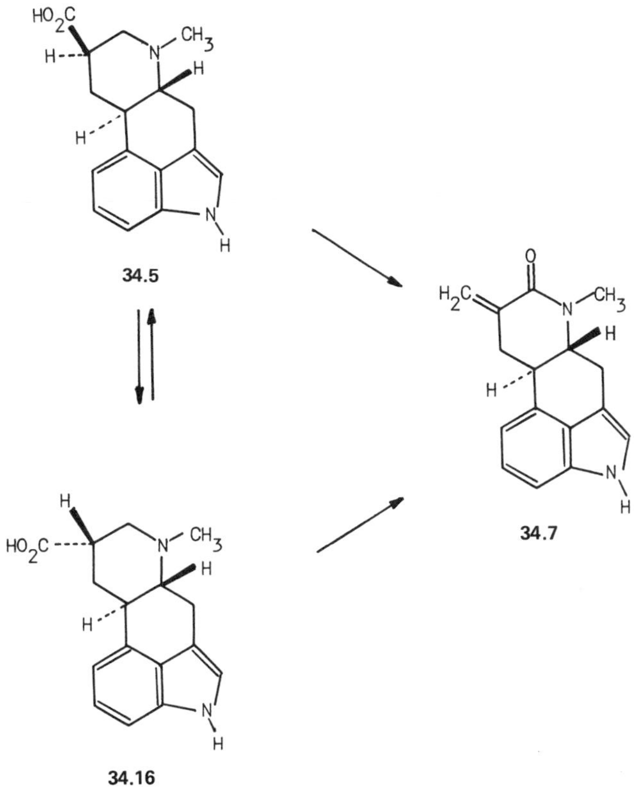

SCHEME 34.2 The conversion of dihydrolysergic acid I and dihydroisolysergic acid I to the same lactam, indicating the identity of the ring fusion in both isomers.

dihydroisolysergic acid II (34.15), it was assumed that both dihydrolysergic acid I (34.5) and dihydroisolysergic acid II (34.15) have the carboxylate groups equatorial. The hydrolysis reactions are shown in Equations 34.7 and 34.8, respectively.

With the relative configuration defined, i.e., the hydrogen at C-5 of lysergic acid (34.1) is β and the hydrogen at C-8 is α, attention was turned to the absolute stereochemistry [263]. The lactam 34.7 was subjected to oxidative ozonolysis and the resultant unstable tricarboxylic acid (34.19) hydrolyzed with dilute hydrochloric acid to the diacid 34.20. Esterification with n-propanol generates the di-n-propylester of D-(+)-N-methylaspartic acid (34.21), demonstrating that the configuration of (+)-lysergic acid (34.1) is 5R:8S. The reactions are shown in Scheme 34.3.

Finally, as shown in Scheme 34.4, (±)-lysergic acid has fallen to synthesis [264]. Thus, the N-benzoyl-2,3-dihydroindole-3-propanoic acid (34.22) is converted to the corresponding acid chloride and cyclized to the cyclohexenone (34.23. Bromination α to the carbonyl and displacement of the bromine with the ethylene ketal of N-methylaminoacetone (34.24) to yield 34.25 is followed by hydrolysis of the ketal and base-catalyzed aldol condensation to the α,β-unsaturated amino ketone 34.26. Reacetylation of 34.26, reduction with sodium borohydride, and conversion to the corresponding vinyl chloride with thionyl chloride are then followed by replacement of the chlorine with cyanide to yield 34.27. After the cyano group is hydrolyzed to the corresponding carboxylic acid and the protecting acetyl group removed, conversion to the indole, and thus to (±)-lysergic acid (34.1 and its mirror image), is consummated by dehydrogenation with deactivated Raney nickel in the presence of sodium arsenate.

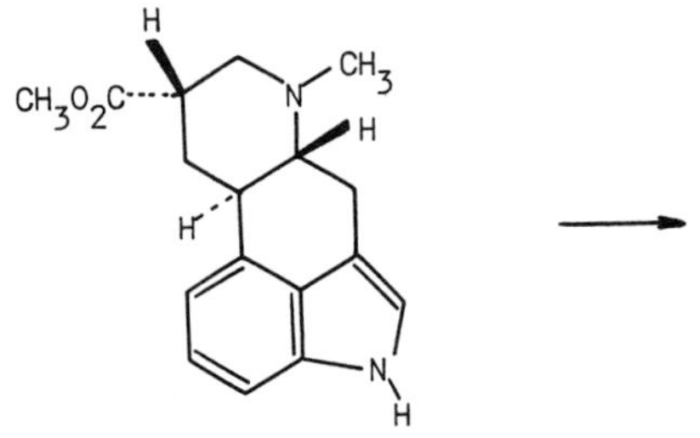

34.17

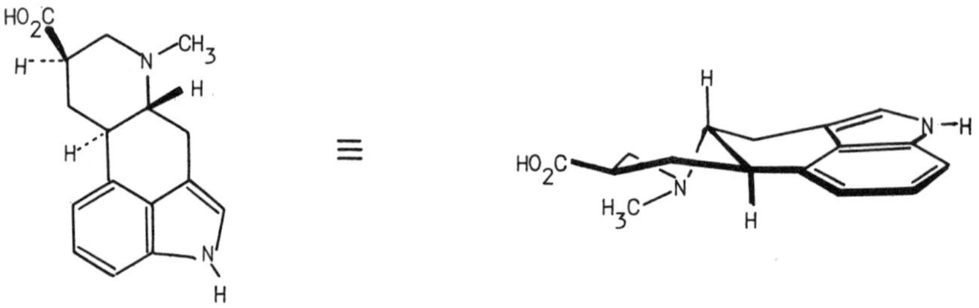

34.5

(Eq. 34.7)

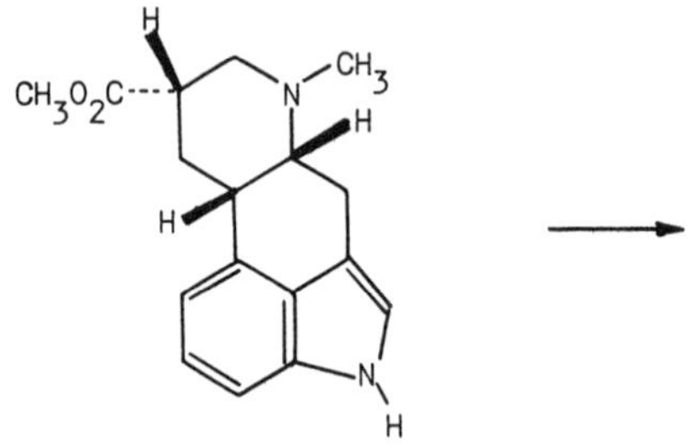

34.18

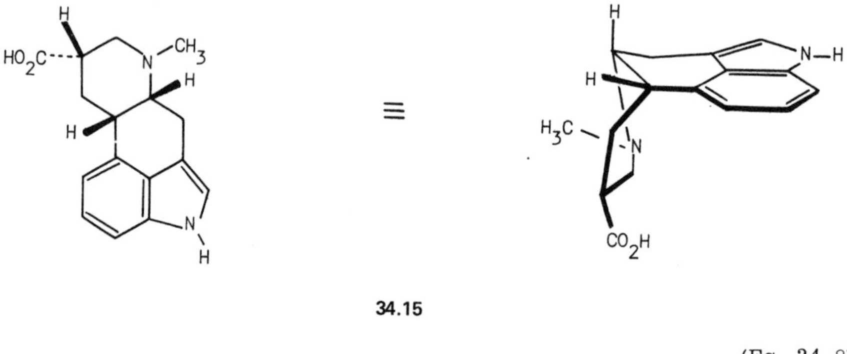

34.15

(Eq. 34.8)

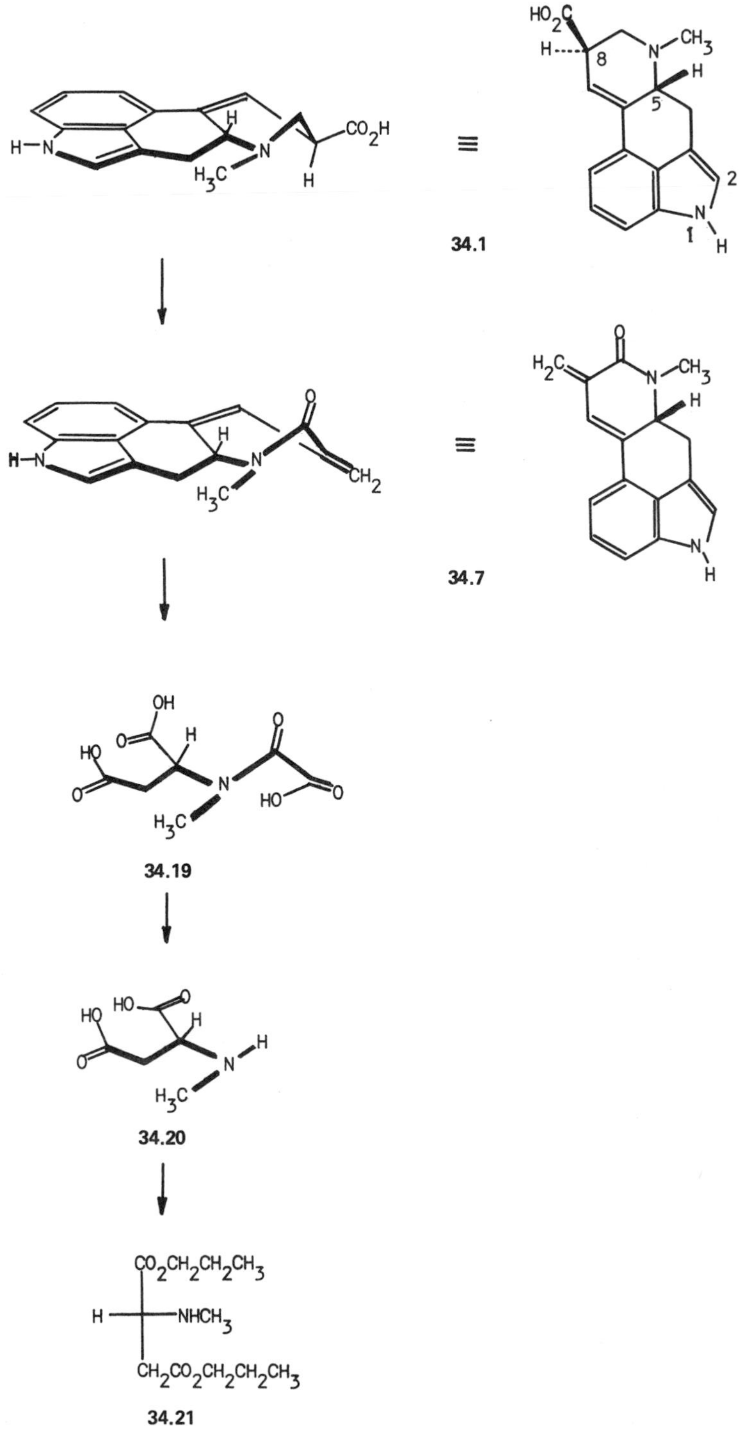

34.1

34.7

34.19

34.20

34.21

SCHEME 34.3 The degradation of lysergic acid to the di-n-propyl ester of (D)-(+)-N-methylaspartic acid. (After Reference 263.)

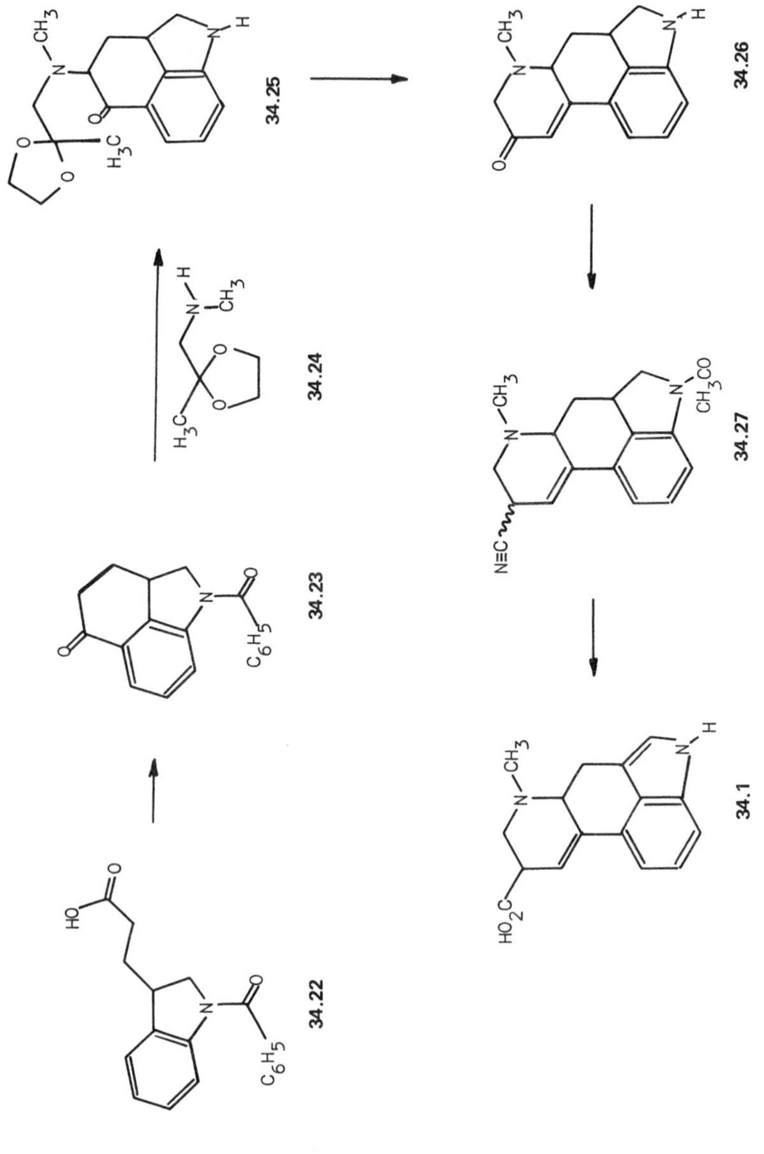

SCHEME 34.4 A total synthesis of (±)-lysergic acid. (After Reference 264.)

II. ELYMOCLAVINE AND AGROCLAVINE

The reduction of (+)-lysergic acid methyl ester (34.28) with sodium in butyl alcohol generates two alcohols, namely, dihydrolysergol I (34.29) and dihydrolysergol II (34.30; Equation 34.9). Similar reduction of the methyl ester of (+)-isolysergic acid (34.31) yields a single

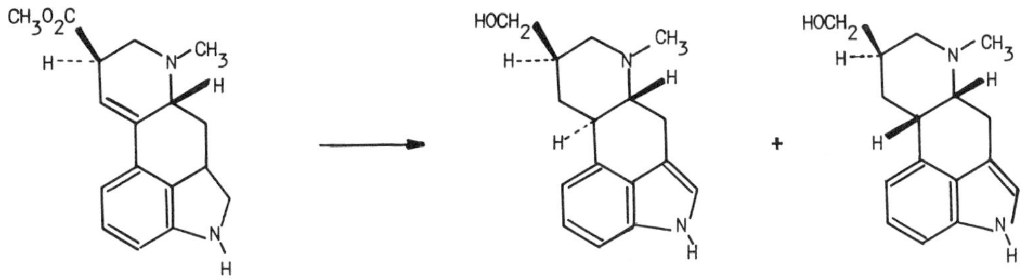

34.28 34.29 34.30

(Eq. 34.9)

dihydroisolysergol (34.32) different from the other two and called dihydroisolysergol I (34.32; Equation 34.10). Apparently the all-cis isomer dihydroisolysergol II (34.33) is not formed under these conditions.

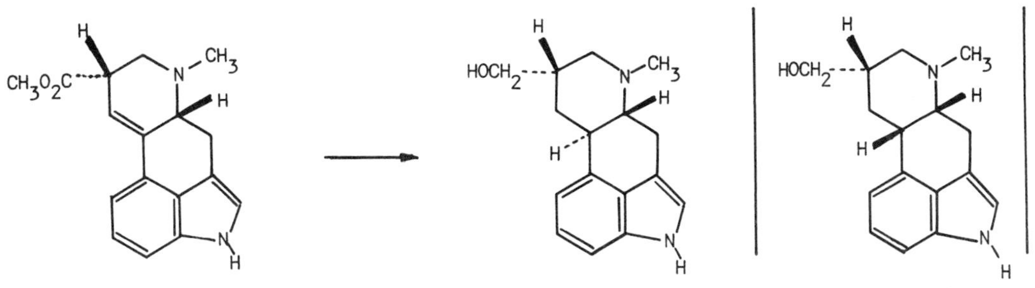

34.31 34.32 34.33

(Eq. 34.10)

When elymoclavine (34.2), $C_{16}H_{18}N_2O$, $[\alpha]_D^{20}$ -109° (ethanol), isolated from the action of the ergot fungus C. purpurea on Elumus mollis, Tri., is reduced catalytically, (+)-dihydrolysergol I (34.29) and (+)-dihydroisolysergol I (34.32) are formed (Equation 34.11). Since

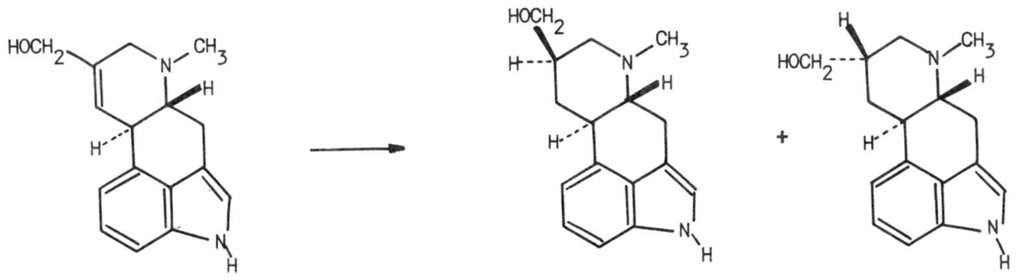

34.2 34.29 34.32

(Eq. 34.11)

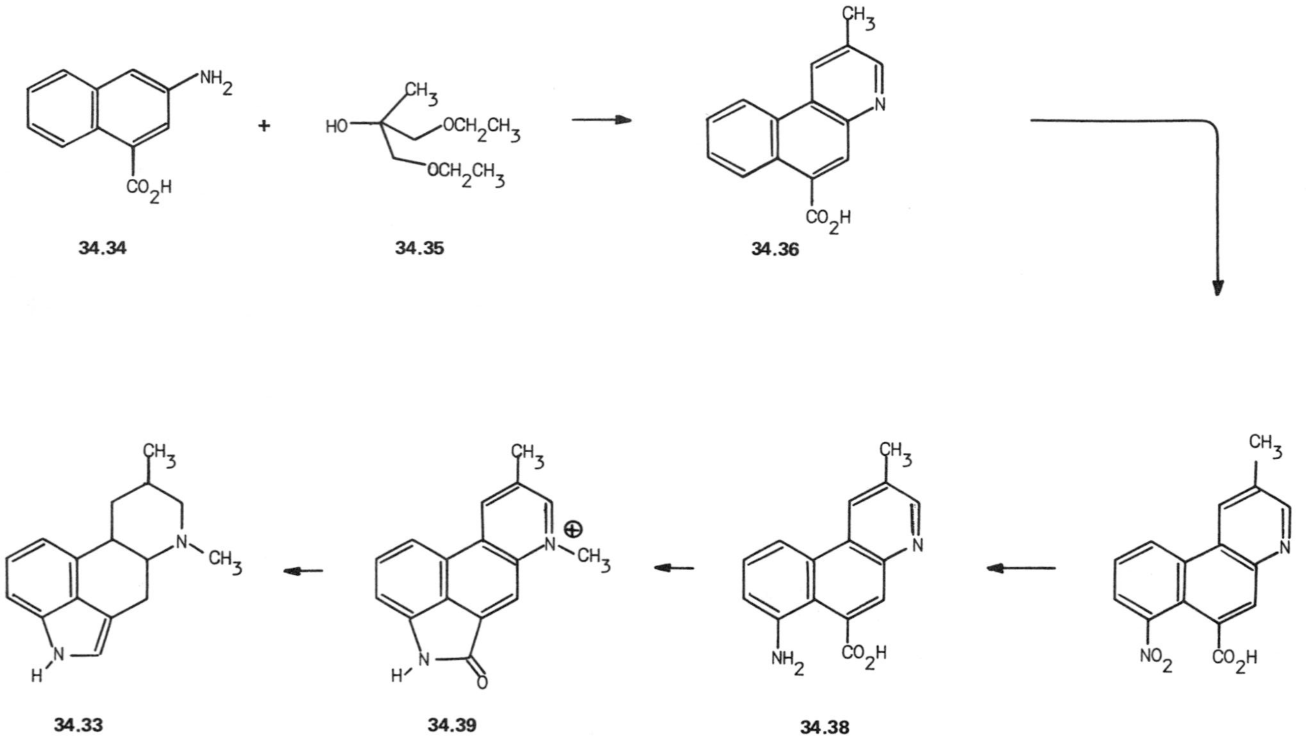

SCHEME 34.5 A synthesis of the mixture of diasteromeric 6, 8-dimethylergolines. (After Reference 265.)

the stereochemistry of the C/D ring juncture was known (conversion to the corresponding acids, see above), and since the double bond in elymoclavine (34.2) was clearly not conjugated with the aromatic ring (ultraviolet spectroscopy) nor was the double bond conjugated with the nitrogen (the pKas of the alkaloid 34.2 and the dihydro bases were almost the same), it was clear that proposed structure (34.2) was reasonable.

In the meantime, it was noted that reduction of elymoclavine (34.2) with sodium in butanol yields a wide variety of reduced and/or deoxygenated products which include agroclavine (34.3) and a mixture of the diasteromeric 6,8-dimethylergolines (34.33; Equation 34.12).

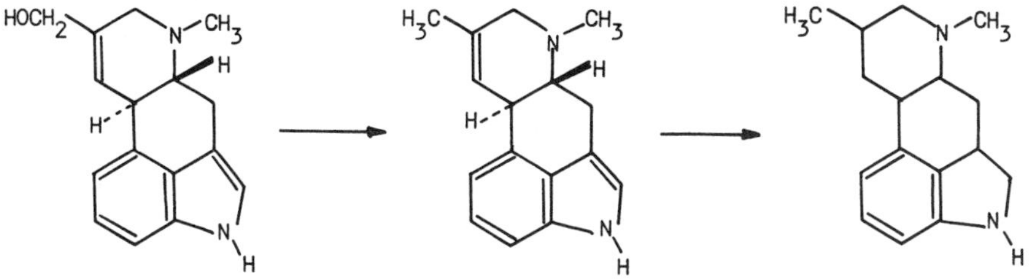

34.2 34.3 34.33

(Eq. 34.12)

Agroclavine (34.3), $C_{16}H_{18}N_2$, $[\alpha]_D^{20}$ -151° (chloroform), a base elaborated by the action of C. purpurea on Elymus mollis and other grasses, possesses an ultraviolet spectrum almost identical with elymoclavine (34.2; i.e., elymoclavine 227, 283, 293 nm; agroclavine 227, 284, and 293 nm). Since agroclavine (34.3) is stable to acid and cannot be isomerized under conditions which suffice for the lysergic (34.1) and isolysergic (34.10) acids, and since the double bond present is not conjugated with the aromatic ring (ultraiolet spectroscopy) or the nitrogen atom (again, pKa measurements on the initial base and the dihydro derivatives are essentially identical), the presently accepted structure was deduced.

Interestingly, a mixture of diasteromeric 6,8-dimethylergolines (34.33) derived from reduction of elymoclavine (34.2) and agroclavine (34.3) had been synthesized quite early [265] by classic methods and shown to be identical to the products obtained by lysergic acid (34.1) degradation (e.g., by reduction of the lactams 34.7, 34.8, and 34.14). Scheme 34.5 presents the synthesis of 6,8-dimethylergoline (34.33). Thus, as shown, 3-amino-1-naphthoic acid (34.34) is allowed, under the conditions of the Skraup reaction, to condense with the α,γ-diethyl ether of β-methylglycerol (34.35) to yield 3-methyl-5,6-benzoquinoline-7-carboxylic acid (34.36). On nitration, the 3'-nitro derivative, 34.37, is isolated, and this is reduced to 3'-amino-3-methyl-5,6-benzoquinoline-7-carboxylic acid (34.38). Lactam formation and generation of the methochloride (34.39) followed by catalytic reduction and then chemical reduction (sodium in butanol) generates the 6,8-dimethylergoline mixture (34.33).

SELECTED READING

A. Stoll and A. Hofmann, in The Alkaloids, Vol. 8, R. H. F. Manske (ed.). Academic, New York, 1965, pp. 726 ff.

A. Stoll and A. Hofmann, in Chemistry of the Alkaloids, S. W. Pelletier (ed.). Van Nostrand-Reinhold, New York, 1969, pp. 267-298.

35

CHEMISTRY OF THE POLYCYCLIC β-CARBOLINE AND RELATED ALKALOIDS: TRYPTOPHAN + C_{9-10}

Unity and diversity, and never one without the other.

A. Camus, The Wager of Our Generation (1957)

The alkaloids to be considered in this chapter, namely, yohimbine (35.1), quinine (35.2), ajmaliçine (35.3), reserpine (35.4), ajmaline (35.5), sarpagine (35.6), echitamine (35.7), strychnine (35.8), catharanthine (35.9), tabersonine (35.10), and vincamine (35.11) are

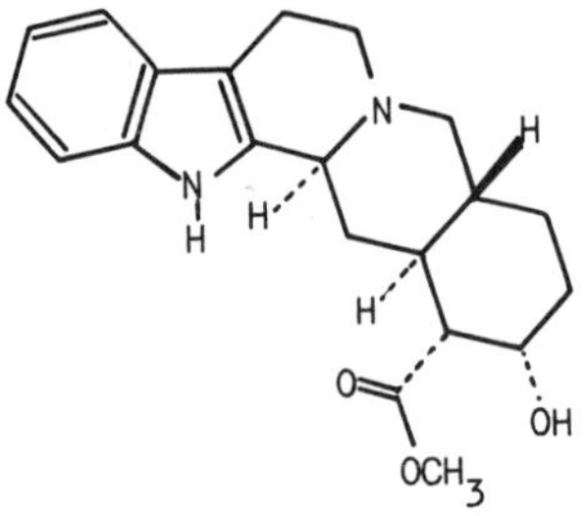

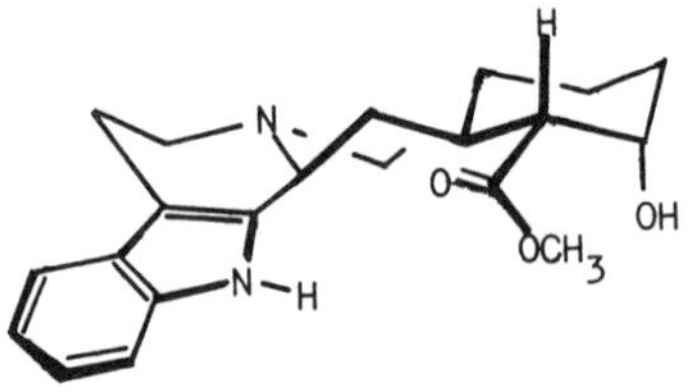

35.1

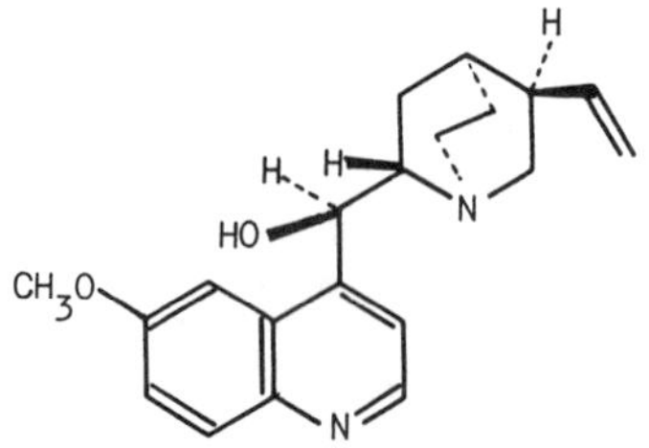

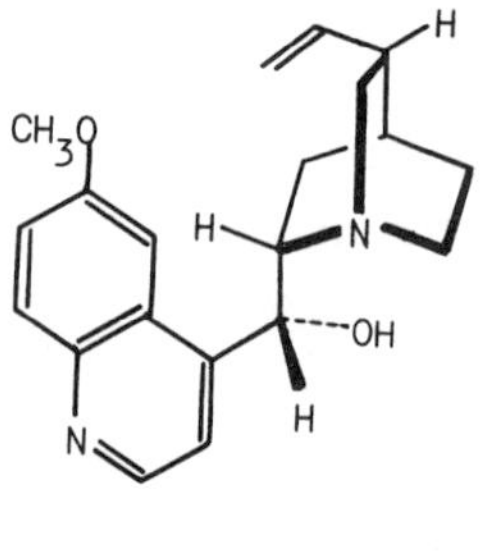

35.2

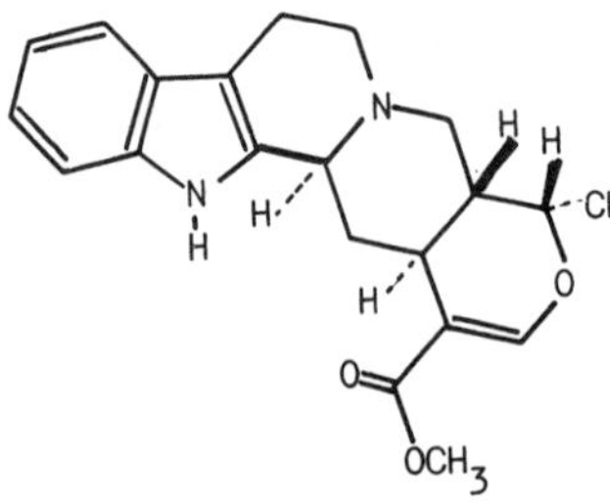

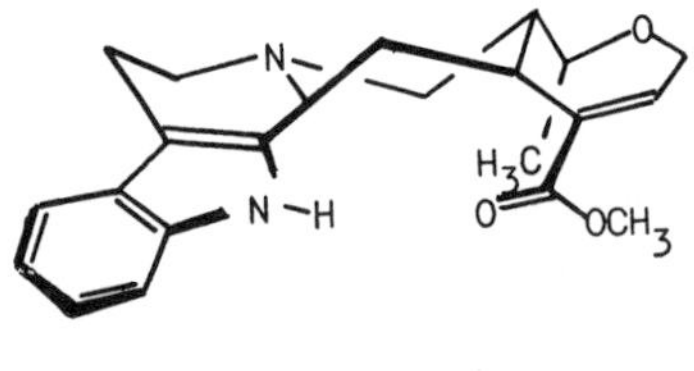

35.3

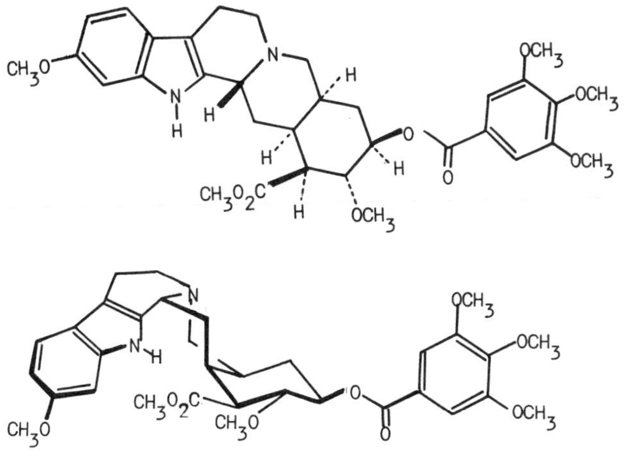

35.4

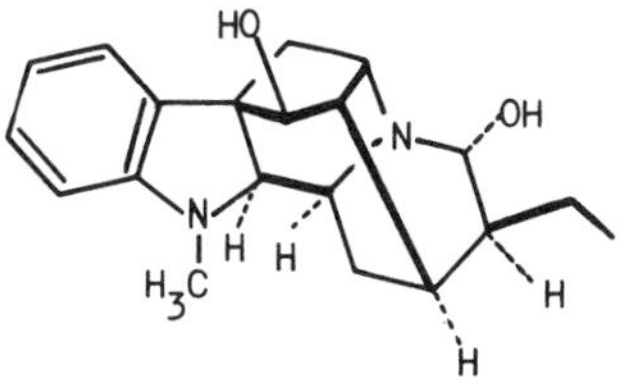

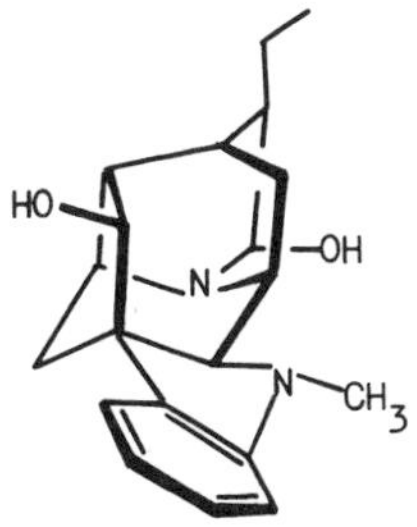

35.5

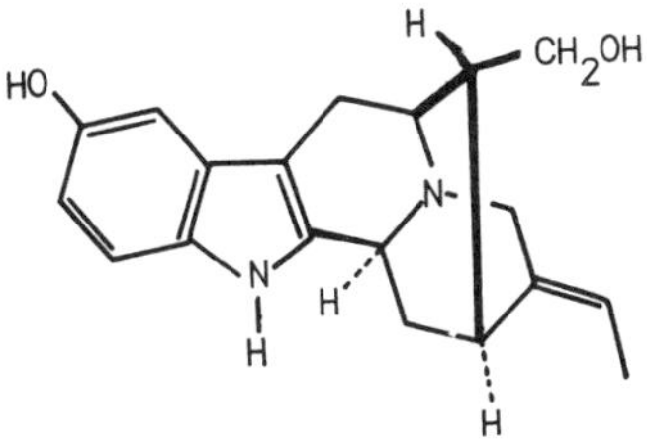

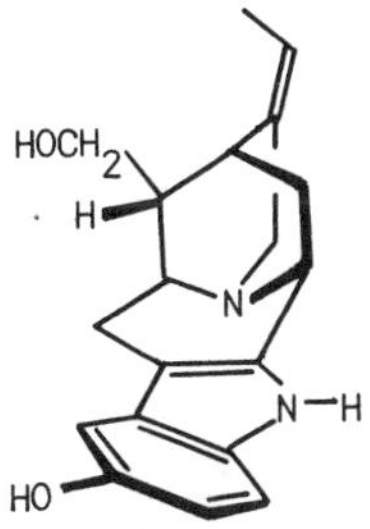

35.6

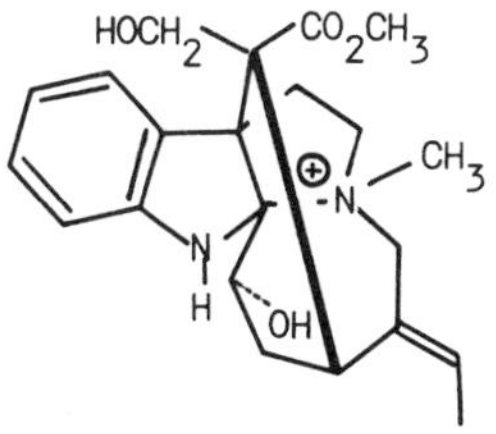

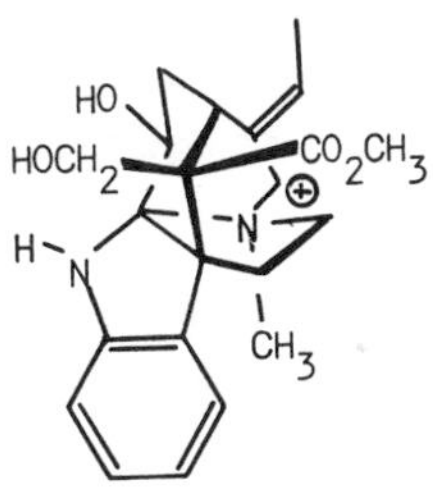

35.7

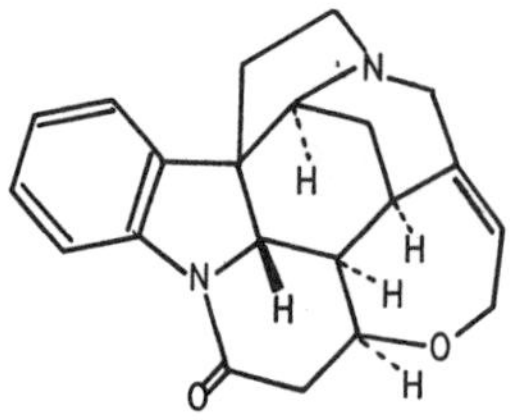

35.8

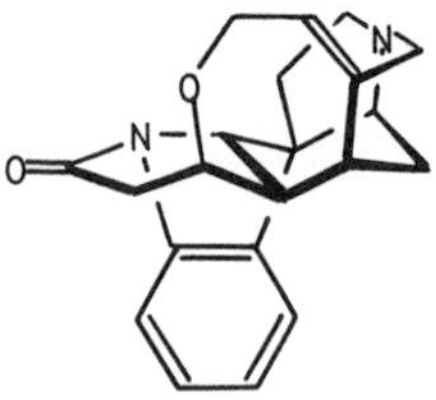

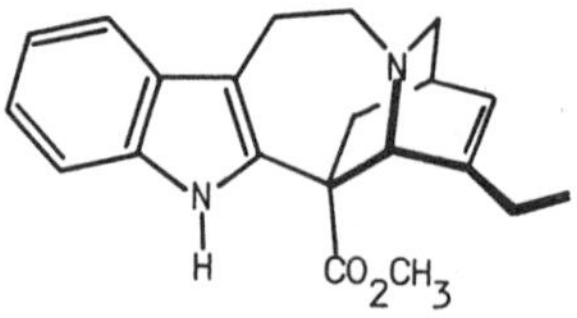

35.9

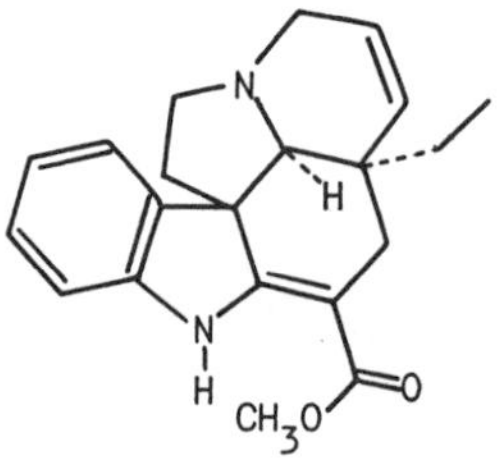

35.10

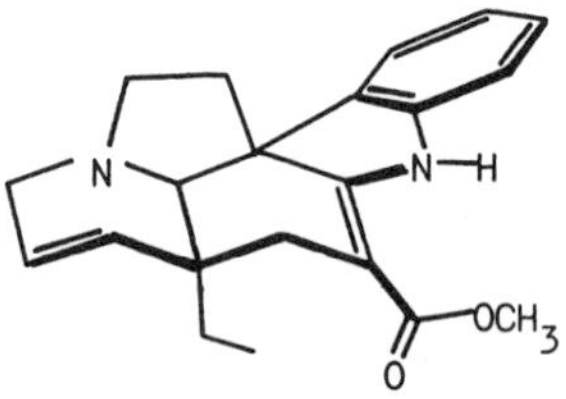

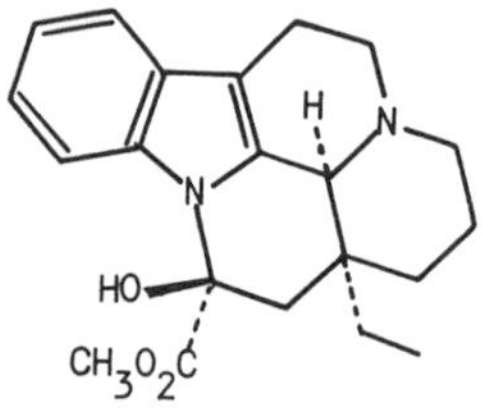

35.11

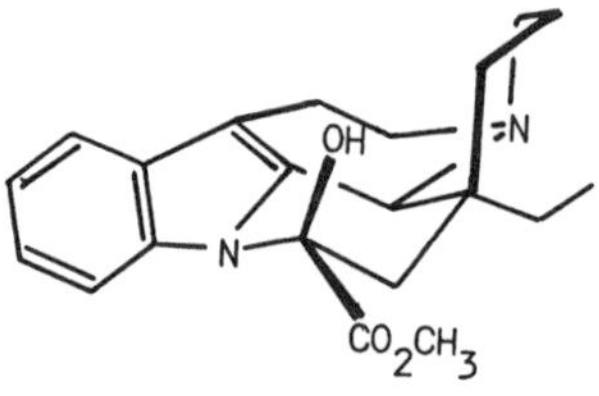

representative of more than 700 compounds of known structure derived from a wide variety of plants (but mostly from the family Apocynaceae). Because of their complexity, structural elucidation was often difficult, and syntheses were mounted as massive efforts. Indeed, there is more than passing evidence that progress in spectroscopy, beginning with ultraviolet and visible and continuing with infrared and nuclear magnetic resonance, have gone hand in glove with structural work and that without the development in the former, the latter would have languished. Clearly, along the same lines, x-ray crystallography, on suitable samples, coupled with computer-assisted data analysis and plotting, can be considered to have removed much of the challenge in structure determination so that more time is available for synthetic endeavors. As is usually the case, much of the work carried out on the bases to be described was stimulated by the observations, often recorded in folklore, of the activity demonstrated by crude plant extracts. As a result, benefits to medicine have accumulated, and many of these alkaloids and closely related compounds are currently used to alleviate the ills of mankind.

I. YOHIMBINE

Yohimbine (35.1) is primarily obtained from the bark of a tree (Corynanthe yohimbe K. Schum.) indigenous to the Congo and is generally isolated as its hydrochloride. The optically active ($[\alpha]_D^{20}$ +107.2°) base, $C_{21}H_{26}N_2O_3$, which readily forms a monoacetyl derivative, also undergoes saponification to methanol and yohimbic acid (35.12), $C_{20}H_{24}N_2O_3$ (Equation 35.1), demonstrating the presence of both carbomethoxy and hydroxyl groups.

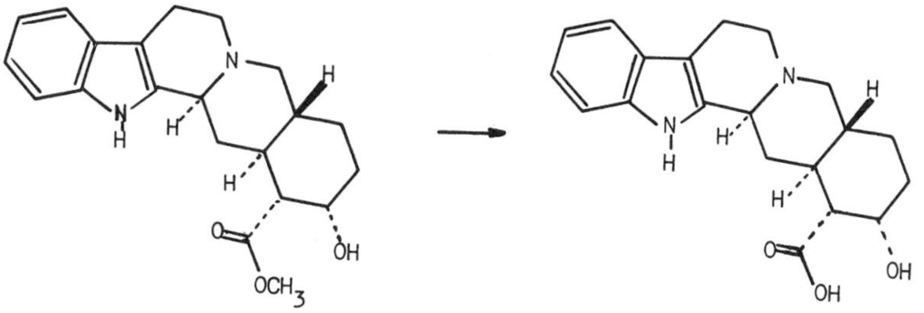

35.1 → 35.12 (Eq. 35.1)

Treatment of yohimbine (35.1) with sulfuric acid results in the formation of an unstable hydrogen sulfate ester which readily undergoes base-catalyzed elimination to apoyohimbine, $C_{21}H_{24}N_2O_2$ (35.13). The latter (35.13) can be hydrogenated to dihydroapoyohimbine (35.14; Equation 35.2).

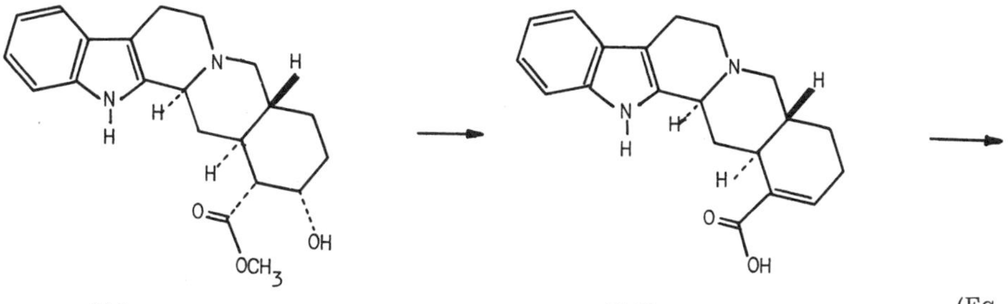

35.1 → 35.13 → (Eq. 35.2)

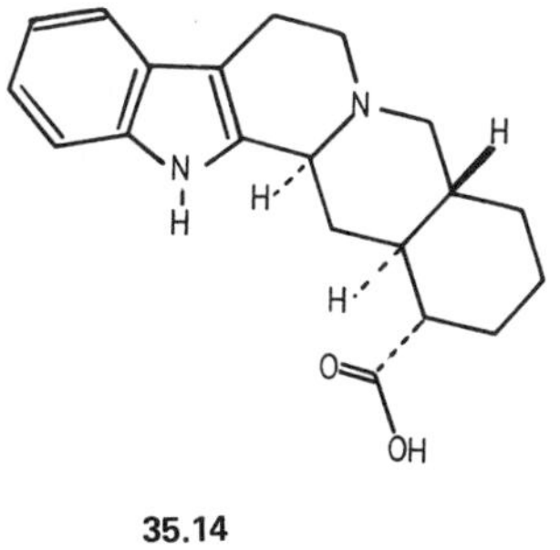

35.14

When yohimbic acid (35.12) is heated with soda lime, it undergoes decarboxylation with the formation of yohimbol, $C_{19}H_{24}N_2O$ (35.15), and yohimbone (35.16; Equation 35.3). Yohimbone (35.16) is also obtained on Oppenauer oxidation of yohimbine (35.1; Equation 35.4). Further, pyrolysis of yohimbine (35.1) with soda lime or zinc dust yields, among other

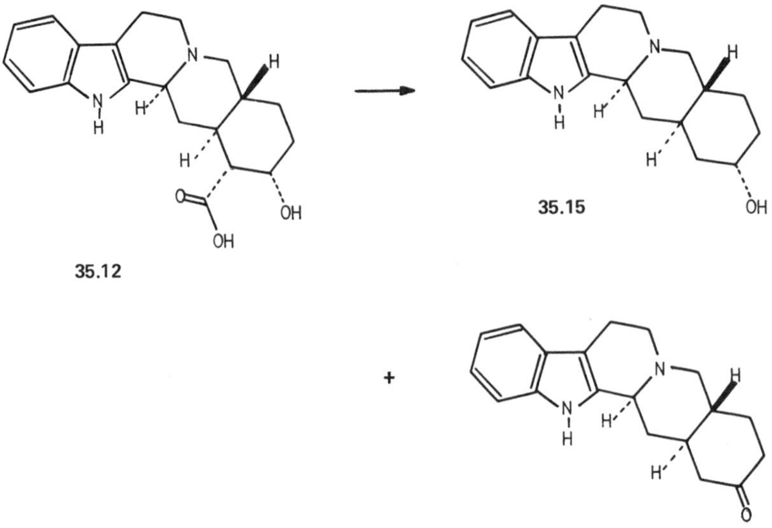

35.12 35.15

(Eq. 35.3)

+

35.16

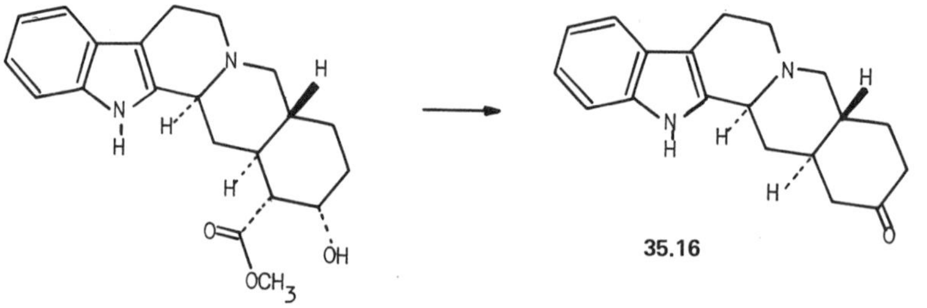

35.1 35.16

(Eq. 35.4)

products, 3-ethylindole (35.17), 3-methylindole (35.18), and isoquinoline (35.19), while destructive distillation generates indolyl-2-carboxylic acid (35.20) and harman (35.21; Scheme 35.1). Although these fragments were important in the structural elucidation, it was primarily thermal degradation in the presence of selenium which provided the greatest insight into the structure of yohimbine (35.1). Thus, selenium-catalyzed dehydrogenation of yohimbine (35.1) results in the formation of yobyrine, $C_{19}H_{16}N_2$ (35.22), tetrahydroisoyobyrine, $C_{19}H_{20}N_2$ (35.23), and ketoyobyrine, $C_{20}H_{16}N_2O$ (35.24) (Equation 35.5).

These materials were degraded further in order to determine their structures. Ozonolysis of tetrahydroisoyobyrine (35.23) yields a ketoamide (35.25) which undergoes hydrolysis to o-aminopropiophenone (35.26) and 5,6,7,8-tetrahydroisoquinoline-3-carboxylic acid (35.27; Equation 35.6), thus establishing its structure as 2-[3-(5,6,7,8-tetrahydroisoquinolyl)]-3-ethylindole. This was confirmed by a simple synthesis [266] in which the phenylhydrazone of 3-butyryl-5,6,7,8-tetrahydroisoquinoline (35.28) was allowed to undergo the Fisher indole reaction to yield tetrahydroisoyobyrine (35.23) directly (Equation 35.7).

On the other hand, yobyrine (35.22), which also contains an indole nucleus, is oxidized by chromic acid to phthalic acid (35.29) and o-toluic acid (35.30; Equation 35.8). The structure suggested by these fragments from yobyrine (35.22) was readily established by synthesis. Thus [267]; Scheme 35.2), condensation of tryptamine (35.31) with o-methylphenylacetic acid (35.32) leads to an amide (35.33) which, on treatment with phosphorus oxychloride, undergoes the Bischler-Napieralski cyclization to the imine 35.34. Palladium-catalyzed dehydrogenation of the latter generates yobyrine (35.22).

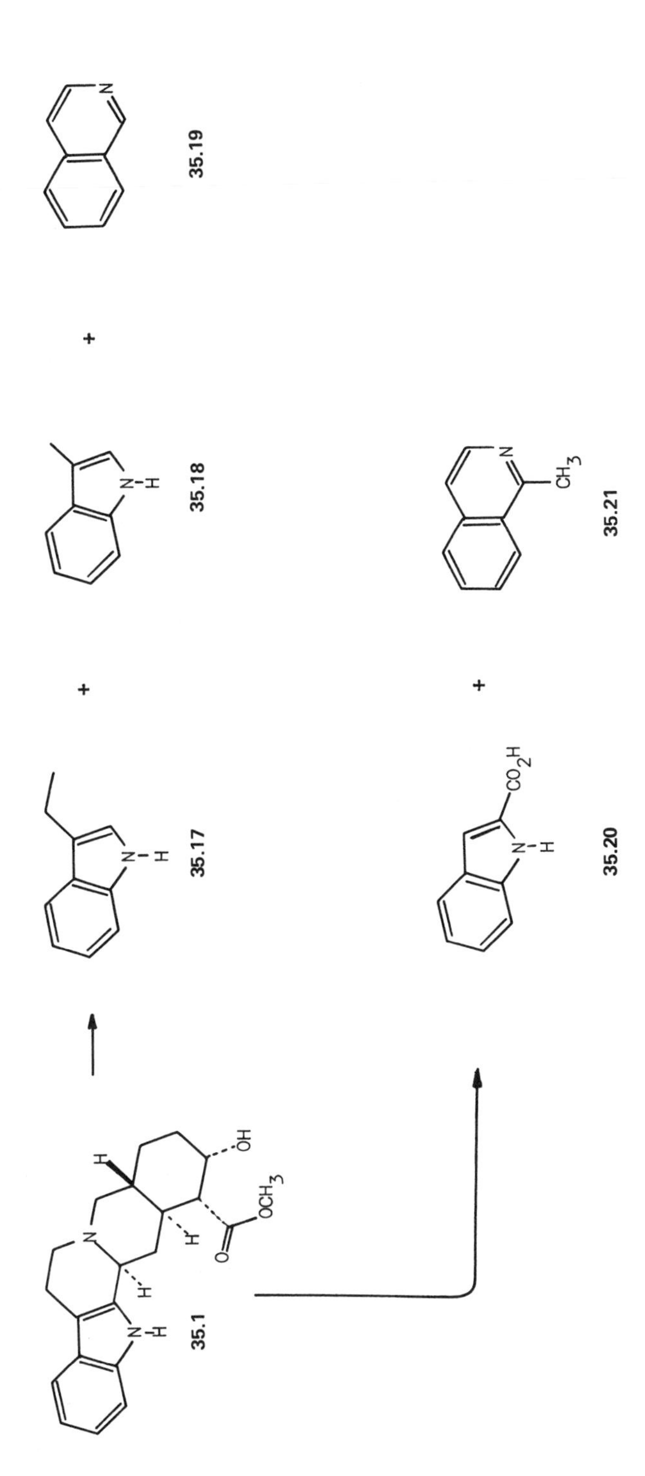

SCHEME 35.1 Products from the soda lime and zinc dust distillation of yohimbine.

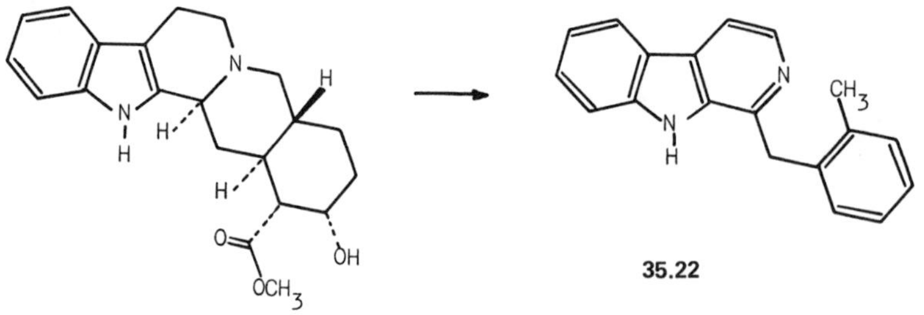

CH_3 OH OCH_3

35.1 35.22

(Eq. 35.5)

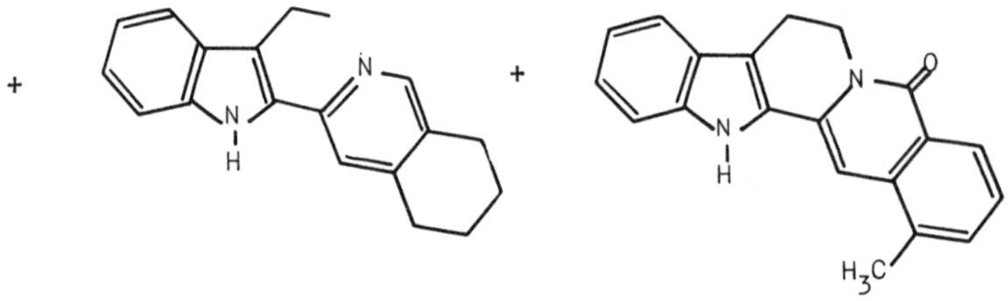

H_3C

35.23 35.24

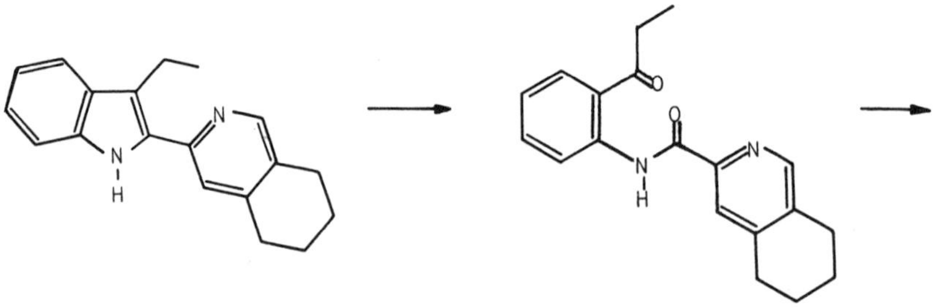

35.23 35.25

(Eq. 35.6)

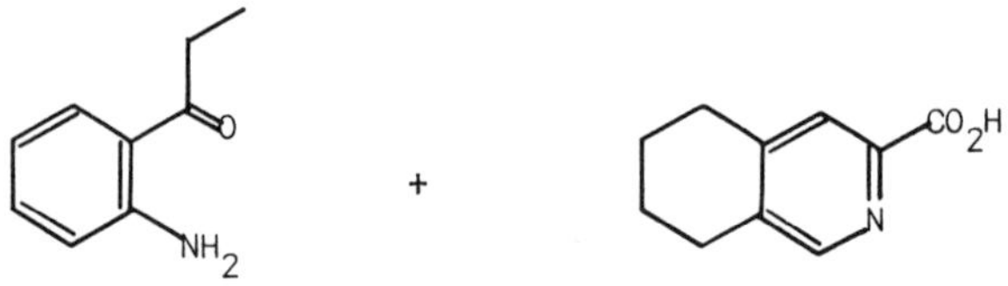

NH_2 CO_2H

35.26 35.27

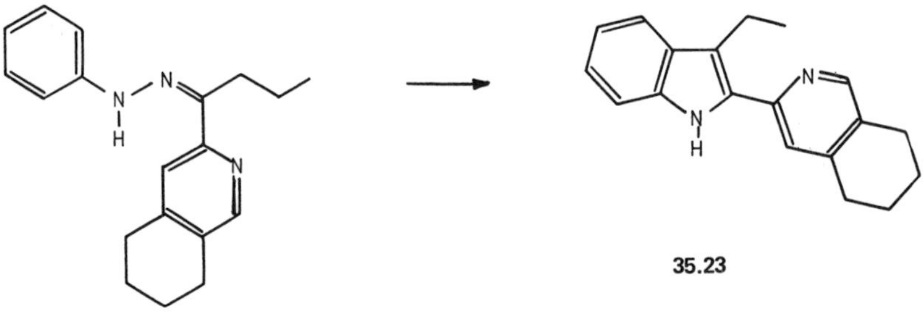

35.28 35.23

(Eq. 35.7)

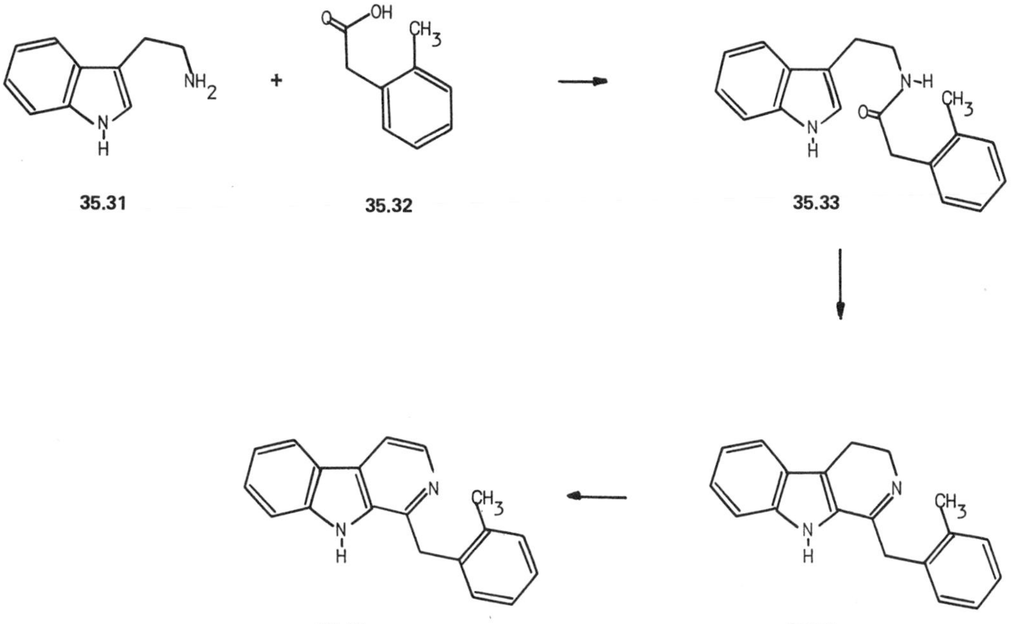

35.31 35.32 35.33

35.22 35.34

SCHEME 35.2 A synthesis of yobyrine, a degradation product of yohimbine. (After Reference 267.)

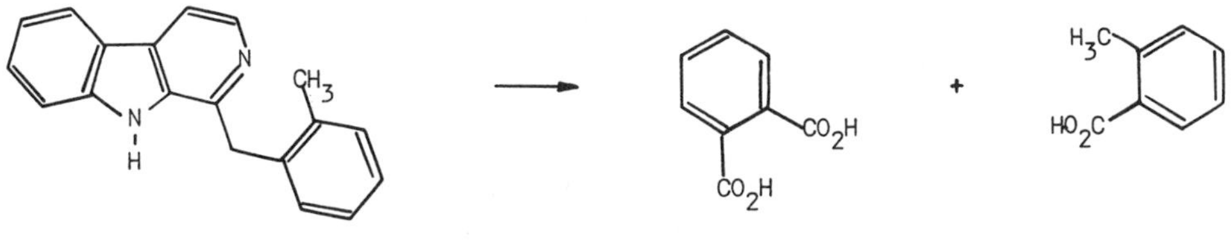

35.22 35.29 35.30

(Eq. 35.8)

Meanwhile, the formation of ketoyobyrine (35.24) establishes the position of the carbomethoxy group in yohimbine (35.1; Equation 35.9). Ketoyobyrine (35.24) was synthesized [268]; Scheme 35.3) by allowing tryptamine (35.31) to condense with 6-methylhomophthalic

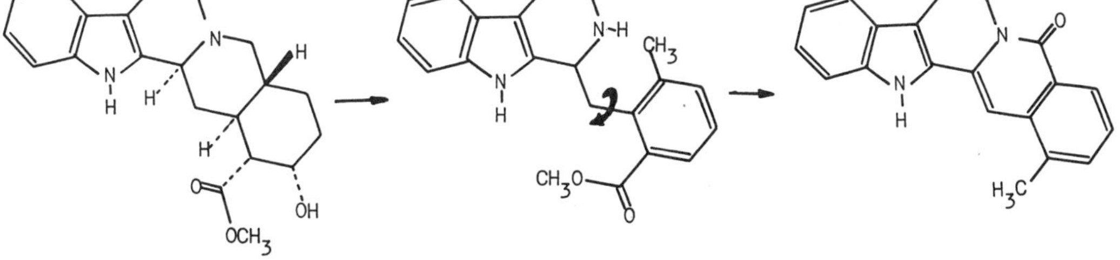

35.1 35.24

(Eq. 35.9)

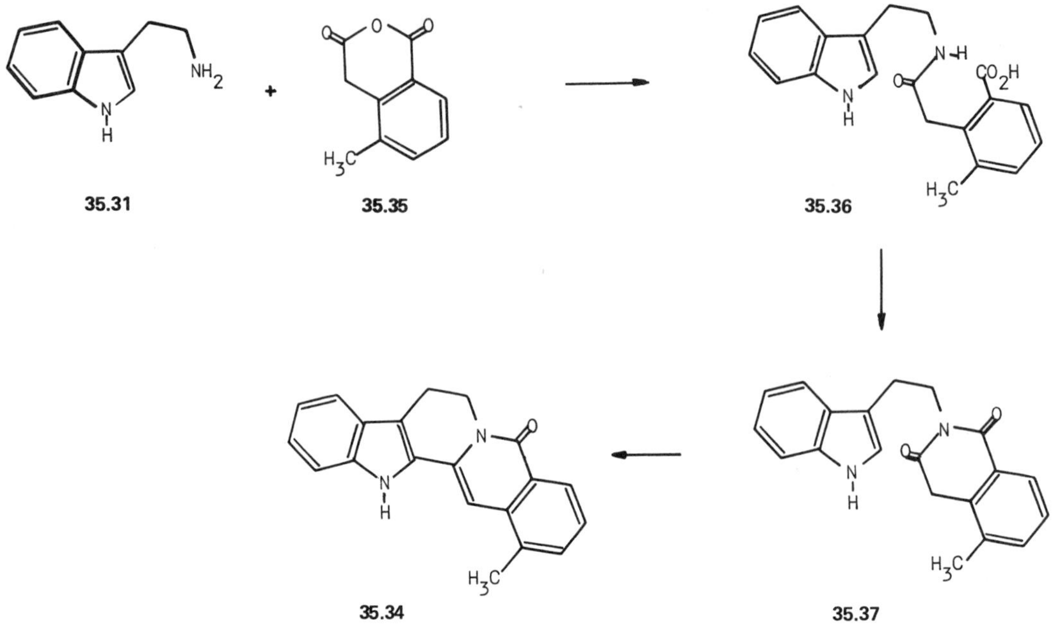

35.31 35.35 35.36

35.34 35.37

SCHEME 35.3 A synthesis of ketoyobyrine, a degradation product of yohimbine. (After Reference 268.)

acid anhydride (35.35) to the amide 35.36 which readily yields the imide 35.37. Phosphorus oxychloride treatment of 35.37 generates ketoyobyrine (35.24).

Finally, the position of the hydroxyl group in yohimbine (35.1) was deduced as being β to the carboxylic acid since (see above) yohimbine (35.1) undergoes ready decarboxylation to yohimbone (35.16) on oxidation and the structure assigned to the latter was confirmed by synthesis. As shown in Scheme 35.4 [269], condensation of tryptamine (35.31) with m-methoxyphenylpyruvic acid (35.38) followed by reaction with formaldehyde yields the pentacyclic ether 35.39. Birch reduction, followed by hydrolysis, produces 35.40 and a diasteromer. The isomer which predominates is hydrogenated to (±)-yohimbone (35.16 and its mirror image) which is then resolved with (-)-camphor-10-sulfonic acid (35.41) to generate (-)-yohimbone (35.16), identical with that derived from yohimbine (35.1) itself.

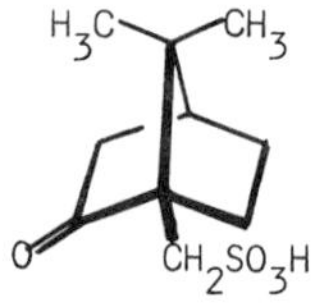

The relative stereochemistry of the five asymmetric centers in yohimbine (35.1) was determined deductively and established by synthesis [270]. First, the D/E ring fusion was established as being (almost certainly, see below) trans by degradation of yohimbic acid (35.12) (Scheme 35.5; [271]). Thus, when yohimbic acid (35.12) is distilled from thallous oxide (or carbonate), water and carbon dioxide are lost, and chanodesoxyyohimbol, $C_{19}H_{23}N_2$,

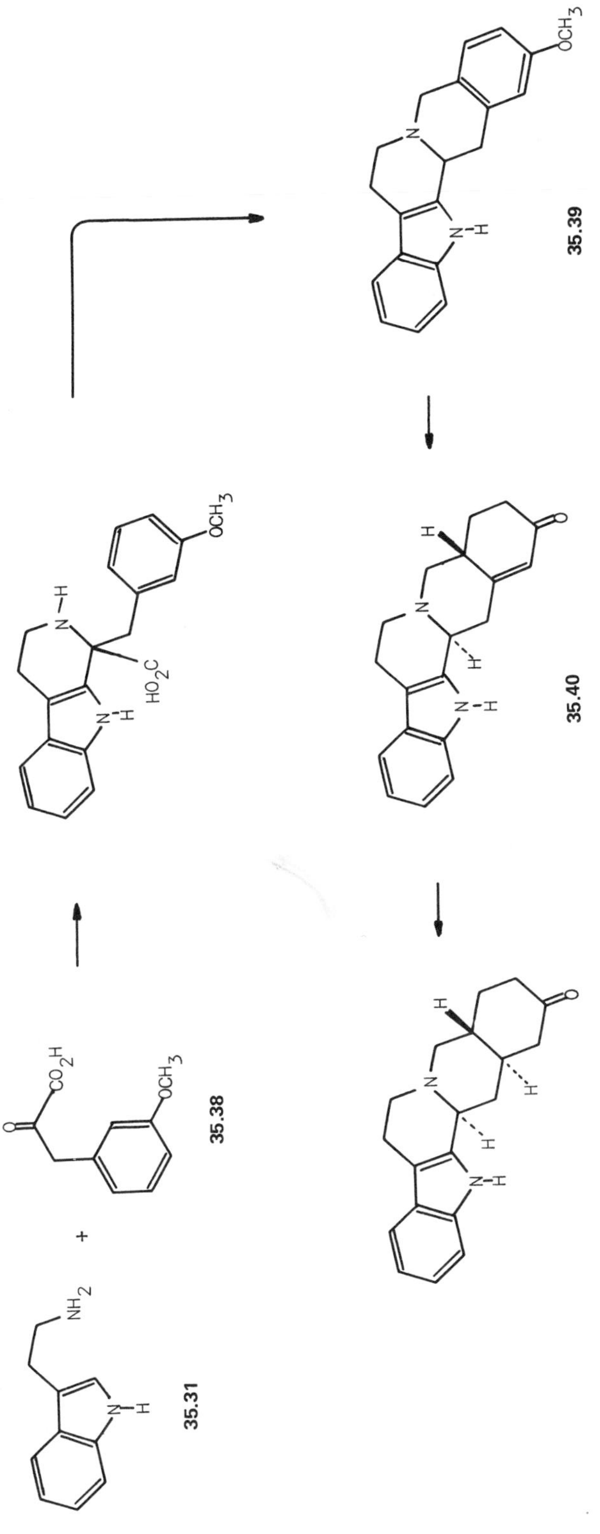

SCHEME 35.4 A synthesis of yohimbone. (After Reference 269.)

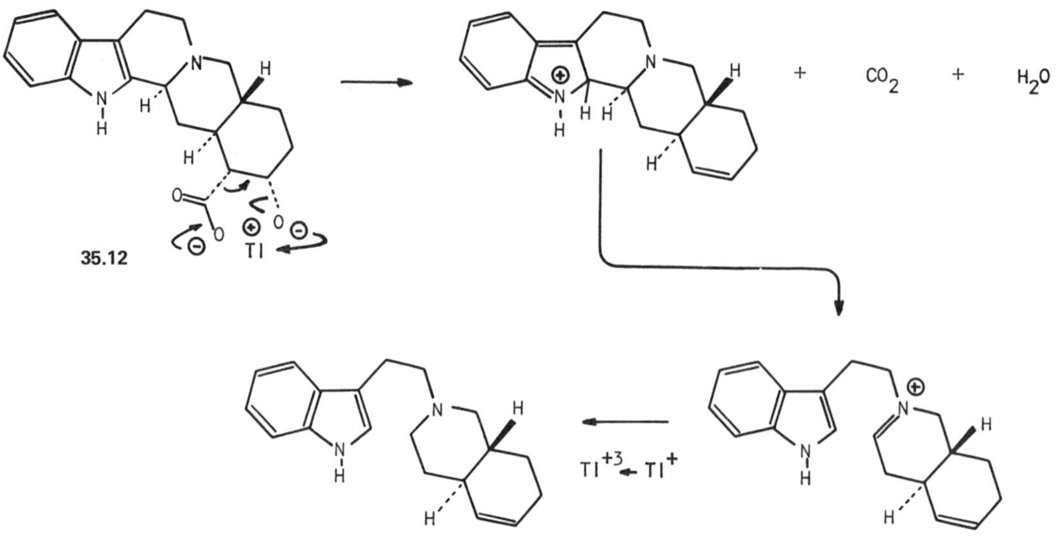

35.12

35.42

SCHEME 35.5 A possible pathway for the production of chanodesoxyyohimbol by the thallium-mediated thermal decomposition of yohimbic acid.

to which the structure 35.42 was assigned, was obtained.* Reduction of chanodesoxyyohimbol (presumably 35.42) with hydrogen and platinum in acetic acid, followed by Hofmann degradation, yields trans-N-methyldecahydroisoquinoline (35.43) which is optically active (Equation 35.10). Isolation of chiral trans-N-methyldecahydroisoquinoline (35.43) implies that the ring juncture had not been affected during the degradation and that the D/E ring fusion was the same (i.e., trans) as it had been in yohimbine (35.1).

Next, it was noted that gentle oxidation of yohimbine (35.1) with lead tetraacetate leads to a tetradehydroyohimbine (35.44) which can be reconverted, by reduction with hydrogen

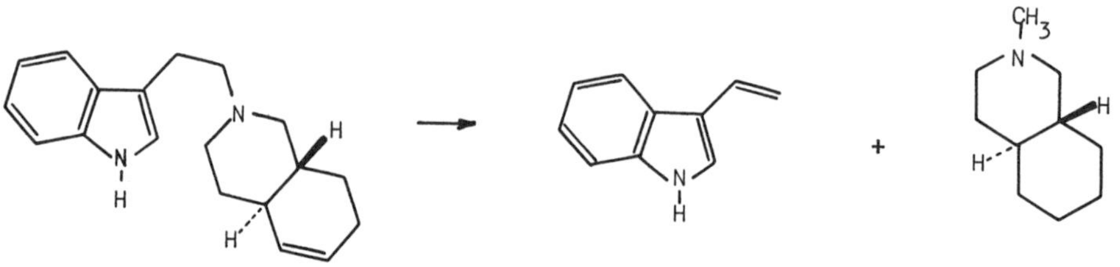

35.42 35.43

(Eq. 35.10)

*Only about 10% yield of chanodesoxyyohimbol (35.42) is found. A variety of other materials, some of which were not investigated further, are also isolated. The assignment of structure of chanodesoxyyohimbol (35.42) rests upon (a) color reactions indicating the indole is not substituted in the α-position; (b) its ready hydrogenation, indicating the presence of a double bond and the fact that if such a double bond were indeed present along with the indole nucleus, only four rings are allowed; and (c) subsequent degradation to trans-N-methyldecahydroisoquinoline (35.43).

over a platinum catalyst in the presence of base, to yohimbine (35.1). The naturally occurring ψ-yohimbine (35.45), on oxidation under the same conditions, yields the same tetradehydroyohimbine (35.44; Equation 35.11). Thus, it was assumed that yohimbine (35.1) and

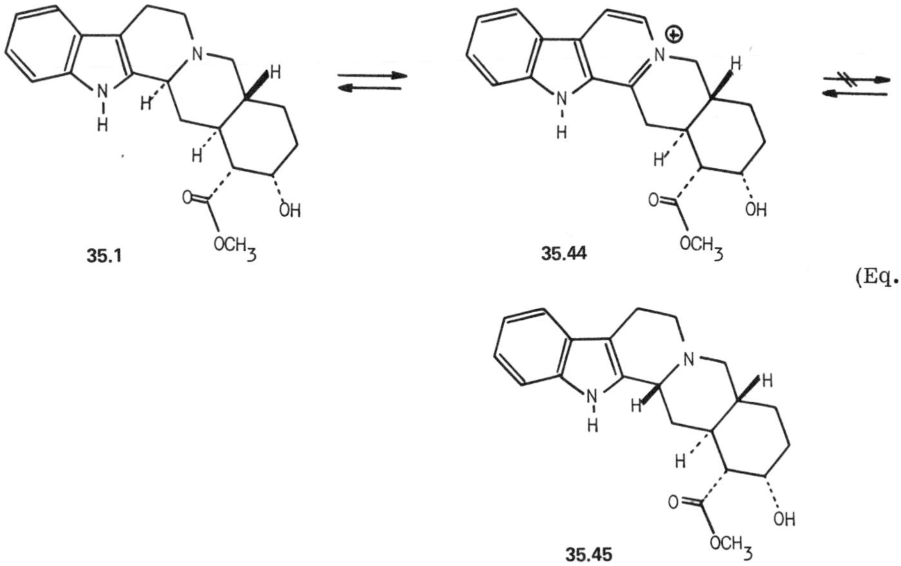

35.1 35.44

(Eq. 35.11)

35.45

ψ-yohimbine (35.45) are epimeric at C_3 and that reduction of the common tetradehydroyohimbine (35.44) yields the most stable isomer (i.e., 35.1).

Finally, since the elimination of water from yohimbine (35.1) to generate apoyohimbine (35.13; Equation 35.2) is so facile, it was concluded that the hydroxyl at C_{17} and the adjacent proton at C_{16} (i.e., the carbon bearing the carboxyl group) must be trans (or E) and diaxial, and thus the carboxyl group is equatorial.

The absolute configuration of yohimbine (35.1) was established by determining the absolute configuration of the hydroxyl at C_{17} [272]. The absolute configuration of the hydroxyl at C_{17} was established by conversion of the alcoholic function to its phenylglyoxylate ester and reaction of that derivative with methyl magnesium iodide. Hydrolysis of the resultant adduct yields S-(+)-atrolactic acid (35.46), demonstrating that the hydroxyl at C_{17} in yohimbine (35.1) is S (Scheme 35.6).*

---

*The same conclusion had been reached [273] from optical rotation studies. For a discussion of the ideas beyind the proof of the configuration at C_{17} in yohimbine (35.1) through the use of induction of optical activity in the atrolactic acid (35.46) isolated, the interested reader should consult References 274 and 275.

The lack of an experimental portion in the communication of these results [272] concerning the absolute configuration at C_{17} does not place the conclusion on as firm ground as would be desired. Second, the particular situation here, with the adjacent carbomethoxy group capable of complexation with, e.g., the magnesium in the Grignard reagent, may induce preferential formation of S-(+)-atrolactic acid (35.46). Regardless, the conclusion regarding the absolute stereochemistry of C_{17} and thus of the remainder of the chiral centers in yohimbine (35.1) has been accepted.

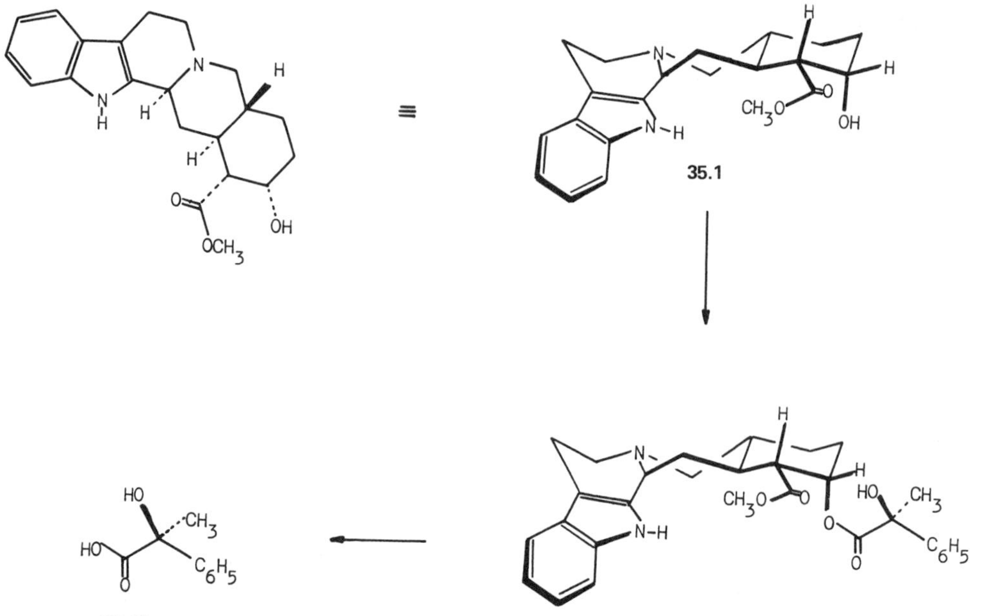

35.1

35.46

SCHEME 35.6 The determination of the absolute stereochemistry of the C_{17} hydroxyl in yohimbine. (After Reference 272.)

The total synthesis of yohimbine (35.1; Scheme 35.7; [270]) begins with the conversion of cis-Δ^6-octalin-1,4-dione (35.47), obtained from the zinc in acetic acid reduction of the Diels-Alder adduct of butadiene with 1,4-benzoquinone, into a mixture of diasteromeric trans-glycidic esters (35.48; only one of which is shown in the scheme). The glycidic esters (35.48) are formed from the dione 35.47 with potassium t-butoxide and ethyl chloroacetate. Hydrolysis of the mixture of esters 35.48 and decarboxylation yield the aldehyde 35.49 which is then oxidized with silver oxide to the acid 35.50. The carboxylic acid 35.50 is converted to its corresponding acid chloride with oxalyl chloride and thence to its amide 35.51 with tryptamine (35.31). Hydroxylation to the corresponding ketodiol with osmium tetroxide followed by platinum-catalyzed hydrogenation generates a triol (e.g., 35.52) which undergoes cleavage and cyclization to the lactam 35.53 on treatment with periodate followed by phosphoric acid. Conversion of 35.53 to the corresponding O-methylacetal with methanol and reduction with lithium aluminum hydride then yield the amino ether 35.54. When the O-acetate corresponding to 35.54 is pyrolyzed, elimination occurs, and oxidation of the alkene with osmium tetroxide yields a diol mixture 35.55. Cleavage of the diol with periodate results in formation of the O-formate ester of (±)-ψ-yohimbaldehyde (35.56), which on oxidation with chromic acid in methanol - acetone, yields (±)-ψ-yohimbine (35.45). The resolution of (±)-ψ-yohimbine (35.45) is accomplished with (-)-camphor-10-sulfonic acid (35.41), and since (+)-ψ-yohimbine (35.45) which results had already (see above) been epimerized to (+)-yohimbine (35.1; Equation 35.11), this resolution formally completes the synthesis.

II. QUININE

Quinine (35.2), $C_{20}H_{24}N_2O_2$, mp 177°C, $[\alpha]_D^{17}$ -117° ($CHCl_3$), which is triboluminescent, is one of a number of medicinally valuable bases isolated from the bark of Cinchona species indigenous to the Andes. A crude preparation from this source was introduced into use as a palliative for malaria in the seventeenth century, but several hundred years then elapsed before the first mixture of crystalline bases was obtained. Although more than 25 individual bases have since been isolated and characterized from Cinchona and related genera, the structural work on quinine (35.2) was enhanced by, and closely correlated with, work on three other co-occurring bases, i.e., quinidine (35.57), the C_9 epimer of quinine (35.2), cinchonidine (35.58), which is desmethoxyquinine, and cinchonine (35.59), which is desmethoxyquinidine.

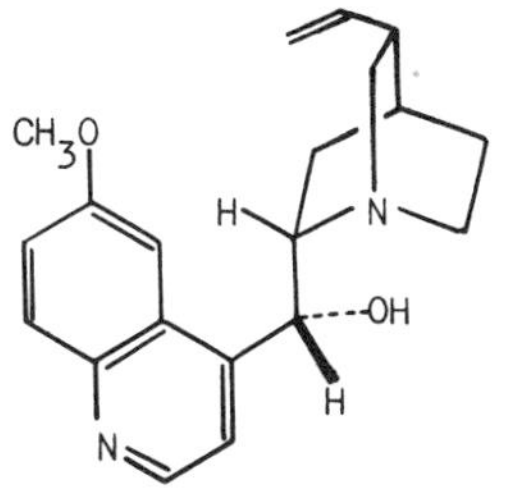

35.2

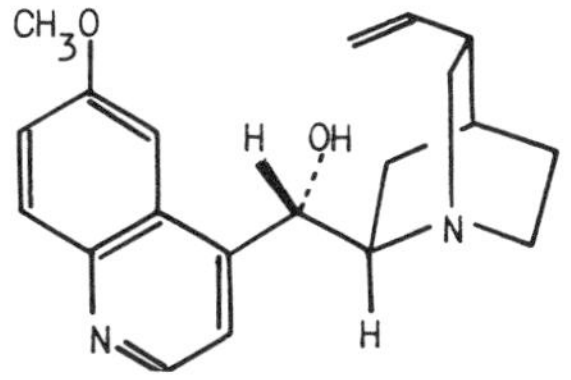

35.57

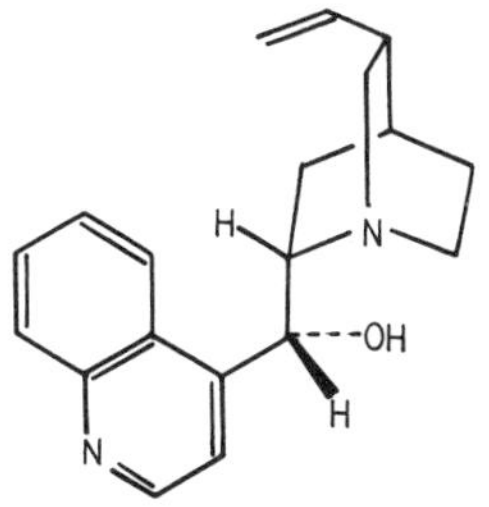

35.58

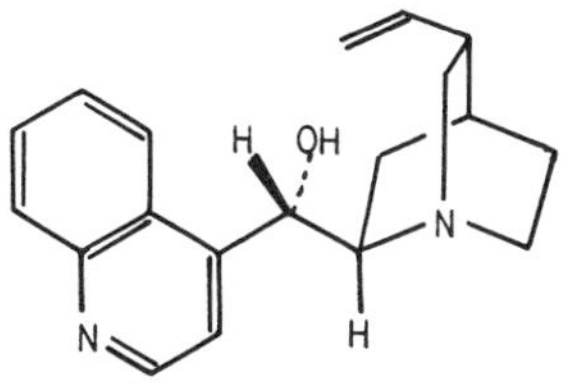

35.59

The functional groups present in this family of four were determined over a period of about 35 years. Thus, it was shown for all of them that two basic tertiary nitrogen atoms are present and the formation of an O-acetyl derivative on acetylation, which can be reconverted to starting material on hydrolysis, shows that a hydroxyl function needs to be accommodated in their structures. The presence of a vinylic double bond in each of the four, convertible by oxidation to formic acid and the corresponding alkaloid carboxylic acid (shown for the transformation quinine, 35.2, to quinitine, 35.60, Equation 35.12) was also confirmed by appropriate addition reactions.

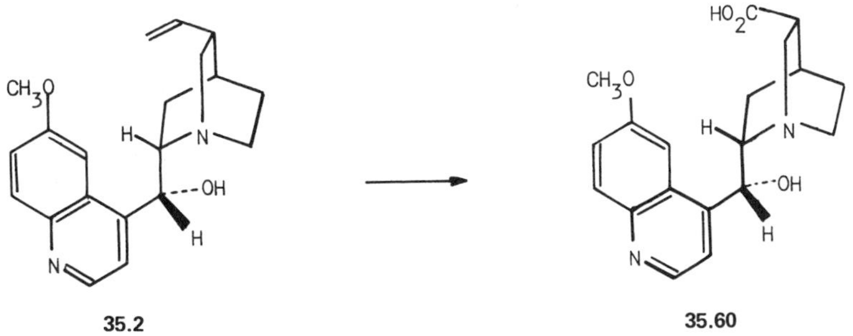

35.2 → 35.60 (Eq. 35.12)

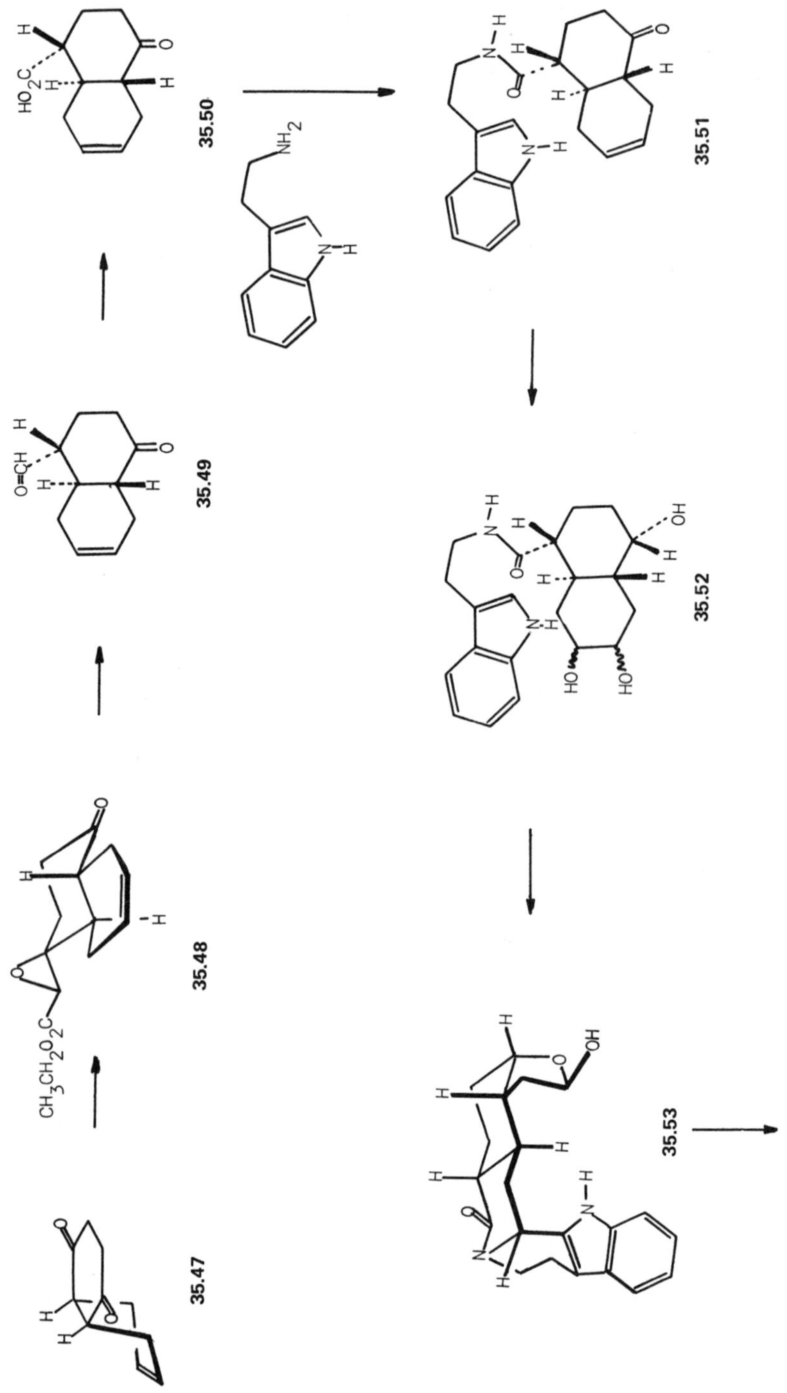

35.47

35.48

35.49

35.50

35.51

35.52

35.53

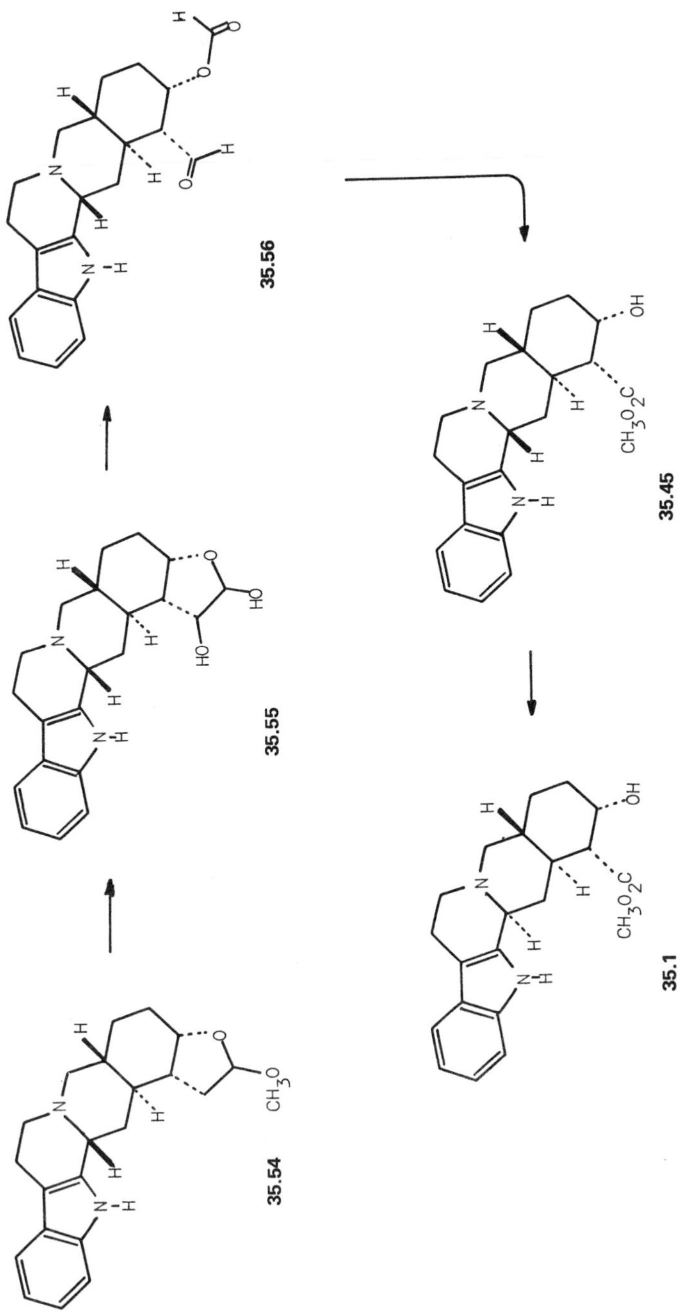
35.54
35.55
35.56
35.45
35.1
N-H
N
H
CH_3O
OH
HO
CH_3O_2C

SCHEME 35.7 The total synthesis of yohimbine. (After Reference 270.)

Fusion of cinchonidine (35.58) or cinchonine (35.59) with potassium hydroxide yields quinoline (35.61) (the first time quinoline (35.61) was obtained was in this fashion, as might be suspected from its name!), while quinine (35.2) and quinidine (35.57) yield 6-methoxy-quinoline (35.62) under the same conditions. Additionally, cinchonine (35.59) is simultaneously degraded to 4-methylquinoline (35.63) as well as 3-ethyl- and 3-ethyl-4-methylpyridine (35.64 and 35.65, respectively). These reactions are shown in Scheme 35.8.

Among the key pieces of evidence for the structures of quinine (35.2), quinidine (35.57), cinchonidine (35.58), and cinchonine (35.59) is their fragmentation into tractable pieces which contain all the atoms present before the degradation process is carried out. Thus, as shown in Scheme 35.9, treatment of cinchonidine (35.58) and cinchonine (35.59) with phosphorus pentachloride in chloroform yields different chlorides, each of which undergoes base-catalyzed dehydrohalogenation to the same alkene, cinchene (35.66). Analogous treatment of quinine (35.2) and quinidine (35.57) yields quinene (35.67).

Heating cinchene (35.66) with 20% aqueous phosphoric acid yields 4-methylquinoline (35.63) and meroquinene (35.68); similarly, quinene (35.67) generates 4-methyl-6-methoxy-quinoline (35.69) and, again, meroquinene (35.68). These reactions [276] are shown in Scheme 35.10. The relationship between quinine (35.2) and quinidine (35.57) as methoxyl-bearing derivatives of cinchonidine (35.58) and cinchonine (35.59) was inferred from this degradative evidence.

A reasonable structure for meroquinine (35.68) was established by a series of reactions (Scheme 35.11) which saw meroquinene (35.68) converted into 3-ethyl-4-methylpyridine (35.65) on heating with dilute hydrochloric acid at 240°C and, on oxidation with dilute acidic permanganate in the cold, into a new acid, cincholoiponic acid (35.70). Further permanganate oxidation of the latter yields lioponic acid (35.71), an epimer of which had been synthesized. Although other formulations might have been possible for meroquinene (35.68), the expression shown was generally accepted. The connection of the two fragments to reconstitute cinchene (35.66) or quinene (35.67) was deduced on the basis of the observation that both quinine (35.2) and cinchonine (35.59) are converted by aqueous acid into ketones (called toxines, e.g., 35.72 and 35.73, respectively) which contain a secondary amino function and that a suitably substituted 1-azabicyclo[2.2.2]octane system would account for this transformation ([277]; Scheme 35.12). The position of the hydroxyl group and thus the structure of cinchonine (35.59; as well as its epimer) were confirmed by oxidation to the corresponding ketone (35.74) and reduction back to the mixture of epimeric alcohols (Equation 35.13; [278]).

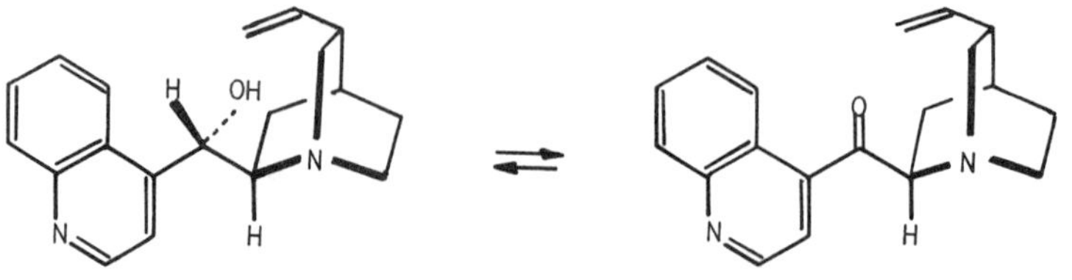

(Eq. 35.13)

35.59 35.74

Now, since all four bases, i.e., quinine (35.2), quinidine (35.57), cinchonidine (35.58), and cinchonine (35.59) give meroquinine (35.68) by transformations which do not affect C_3 and C_4 (in the bicyclic ring), they must all have the same configurations at C_3 and C_4. The acid obtained on degradation, cincholoiponic acid (35.70; Scheme 35.11) is optically active and was unstable with respect to a stereoisomer (35.75) to which it is converted on heating

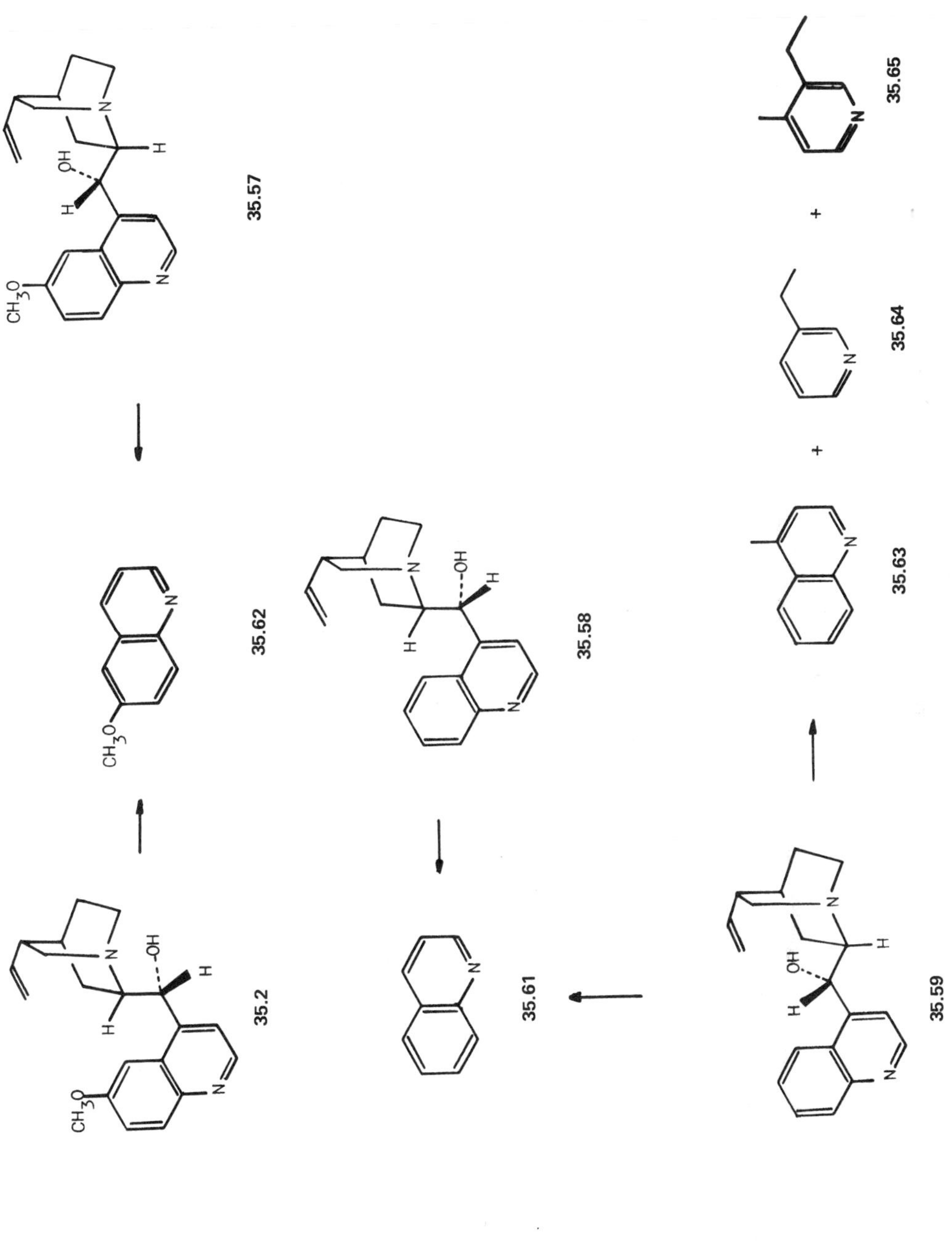

SCHEME 35.8 The potassium hydroxide fusion products of quinine, quinidine, cinchonine, and cinchonidine.

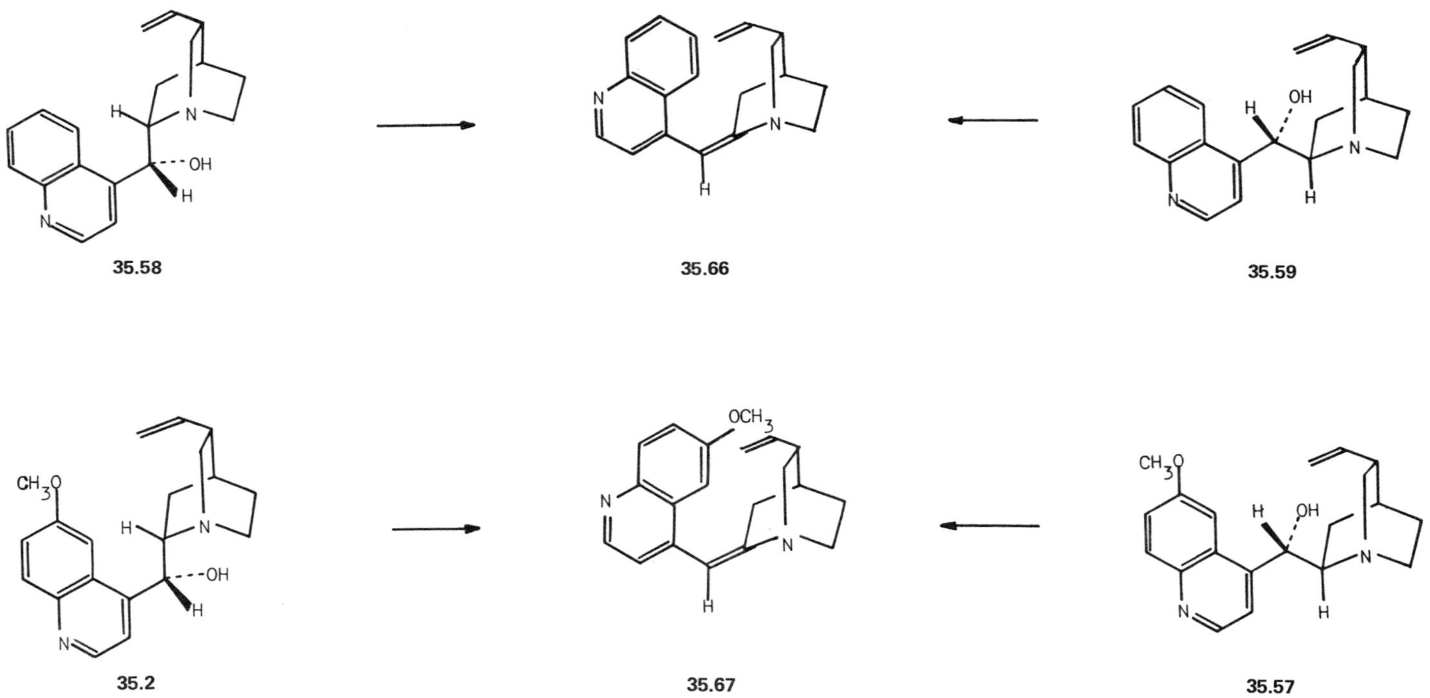

35.58

35.66

35.59

35.2

35.67

35.57

SCHEME 35.9 The formation of cinchene from cinchonine and cinchonidine and the formation of quinene from quinine and quinidine, showing the relationship between the pairs of bases as epimers at C_8.

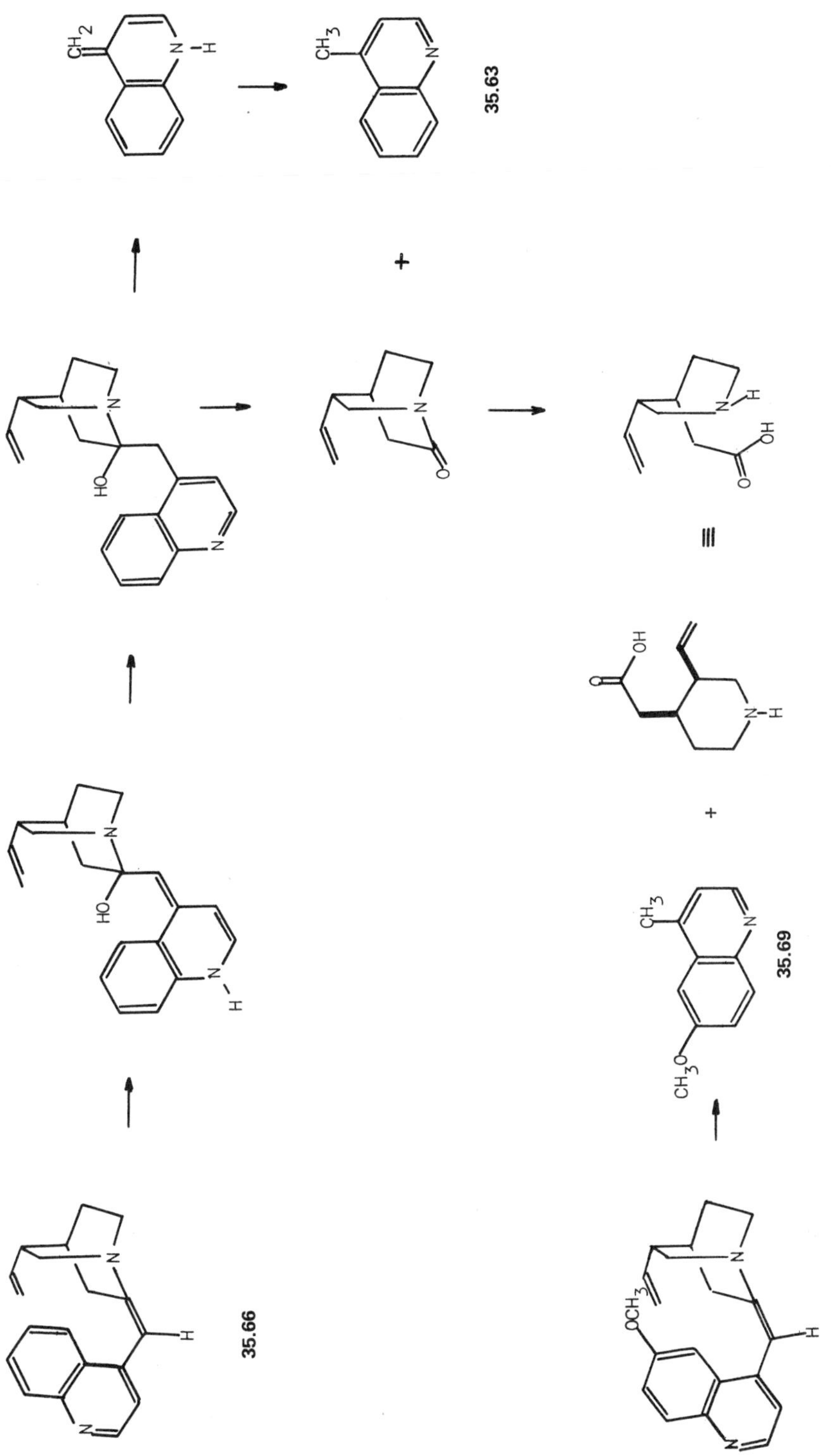

SCHEME 35.10 A possible pathway for the formation of isoquinolines and meroquinene from cinchene and quinene.

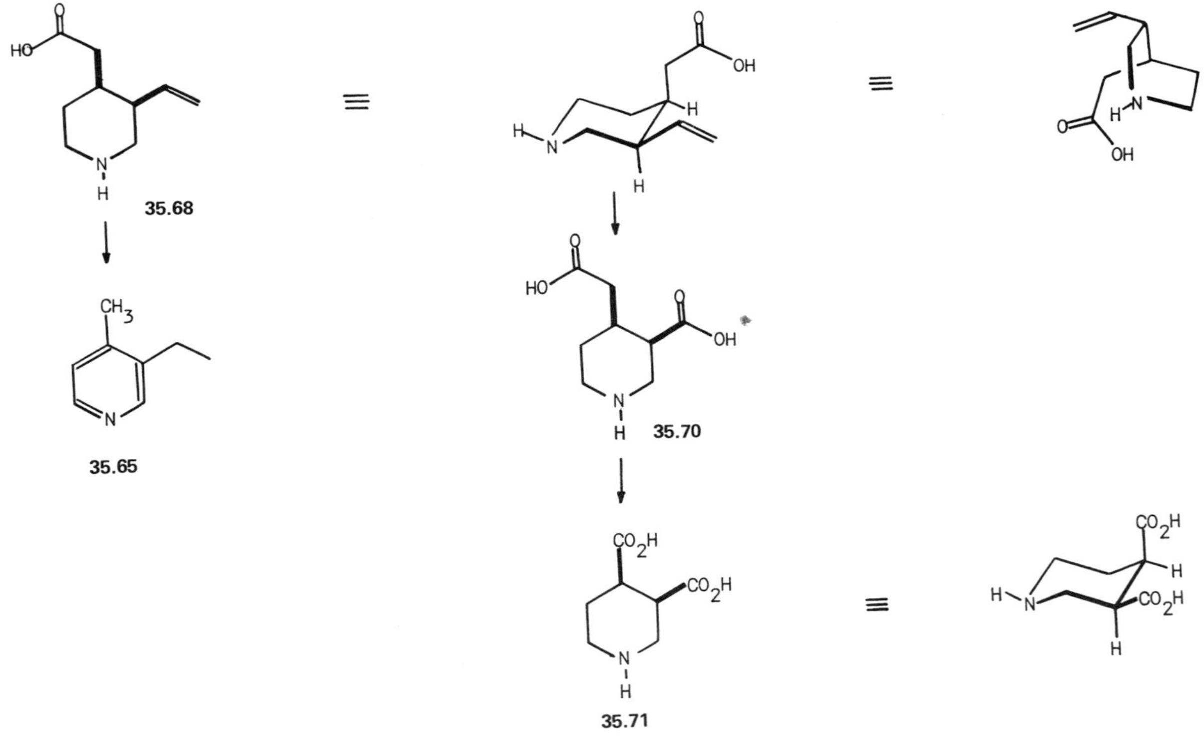

35.68 → 35.65

35.70 → 35.71

SCHEME 35.11 Some of the degradative reactions utilized to establish the structure of meroquinene.

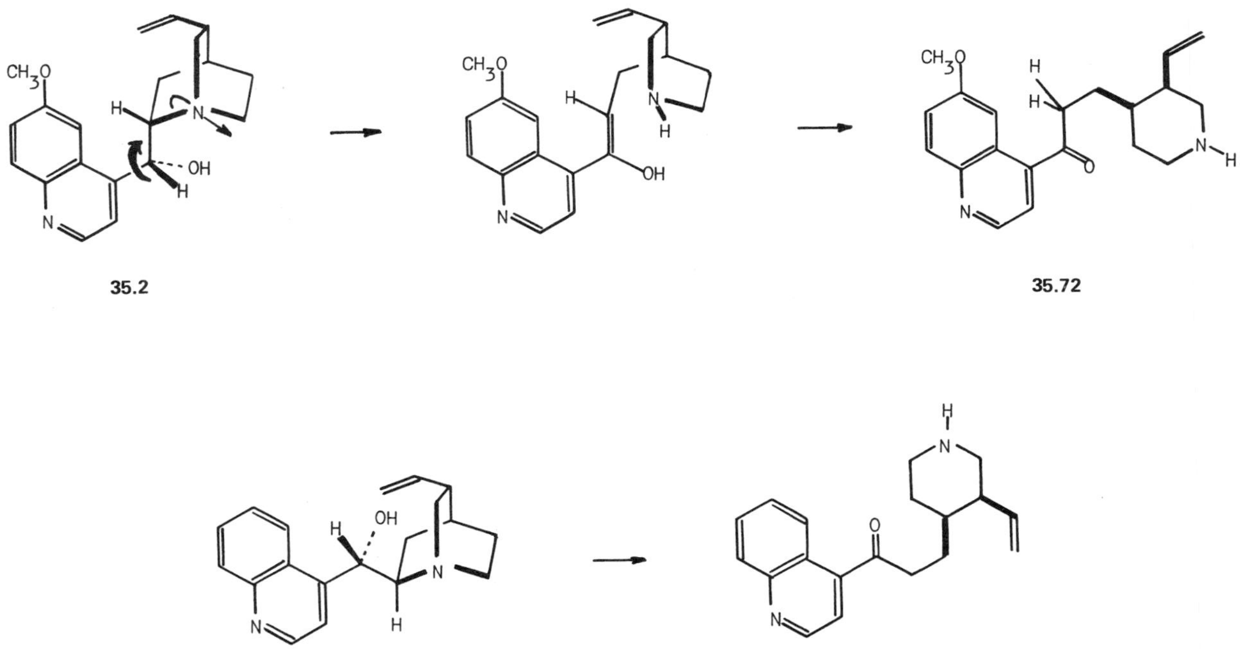

35.2 35.72

35.59 35.73

SCHEME 35.12 A possible scheme for the conversion of quinine and of cinchonine into their respective toxines.

with aqueous potassium hydroxide and reacidification (Equation 35.14). Second, when ethyl malonate (35.76) is added to the 3-cyano-Δ^3-piperidiene 35.77 and the reaction mixture hydrolyzed and decarboxylation effected, both racemic cincholoiponic acids 35.70 and 35.75 are obtained, with the latter predominating in the mixture (Equation 35.15).

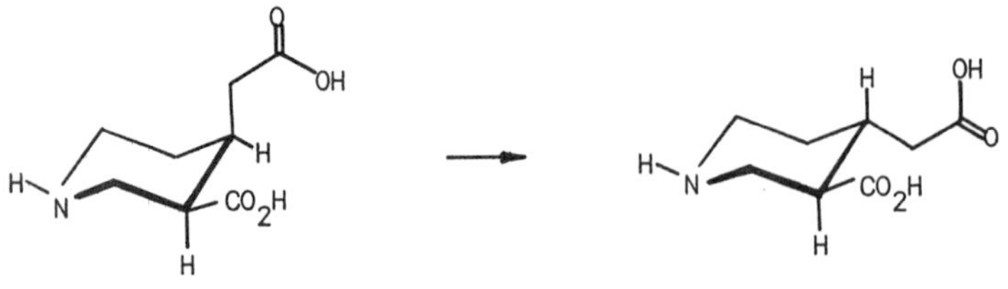

(Eq. 35.14)

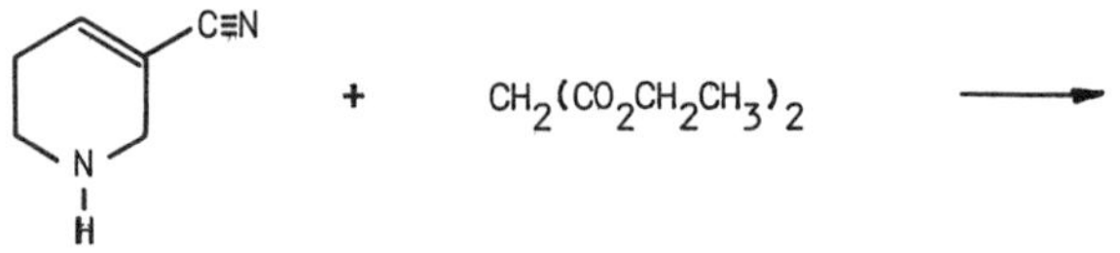

35.70 → 35.75

35.77 + $CH_2(CO_2CH_2CH_3)_2$ (35.76) → 35.75 + 35.70

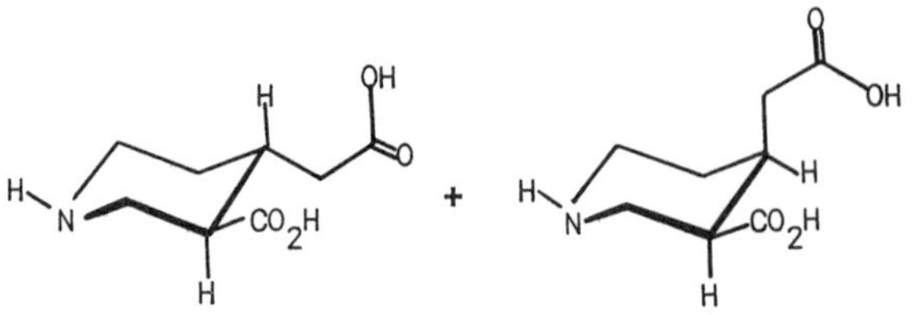

(Eq. 35.15)

It was thus concluded that the cincholoiponic acid isomer (35.70) derived from the alkaloids possesses a cis configuration [279]. This was confirmed [280] by conversion of the ethyl ester of dihydromeroquinene (35.78) to 3-ethyl-4(β-hydroxyethyl)-piperidine (35.79) with sodium in alcohol and thence, via the corresponding bromide, with zinc in acetic acid, to 3,4-diethylipiperidine (35.80). von Braun degradation to excise the nitrogen and generate 1,5-dibromo-2,3-diethylpentane (35.81) is then followed by condensation with ethyl malonate (35.76), and decarboxylation yields chiral 3,4-diethylcyclohexanecarboxylic acid (35.82). Hunsdiecker degradation of 35.82 followed by reductive dehalogenation generates optically inactive 1,2-diethylcyclohexane (35.83) which demands that the ethyl groups, and thus the substituents of cincholoiponic acid (35.70) derived from the alkaloids, must be cis. These reactions are shown in Scheme 35.13.

Additionally, the dibromide derived via the von Braun degradation (i.e., 35.81) is reduced to optically active (-)-3-methyl-4-ethylhexane (35.84) whose absolute configuration was correlated with R-(-)-2-methylbutyric acid (35.85; Equation 35.16). Next, the assignment of configuration at C_8 was considered, and the relative geometry was deduced from the observation that quinidine (35.57) but not quinine (35.2) and cinchonine (35.59) but not cinchonidine (35.58) are capable, on acid treatment, of conversion into cyclic ethers (35.86 and 35.87, respectively; Scheme 35.14). Since this is possible only if a cis relationship obtains between the vinyl side chain and the C_8-C_9 bonds, the opposite (i.e., trans) configuration is assigned to quinine (35.2) and cinchonidine (35.58). Finally, although the stereochemistry

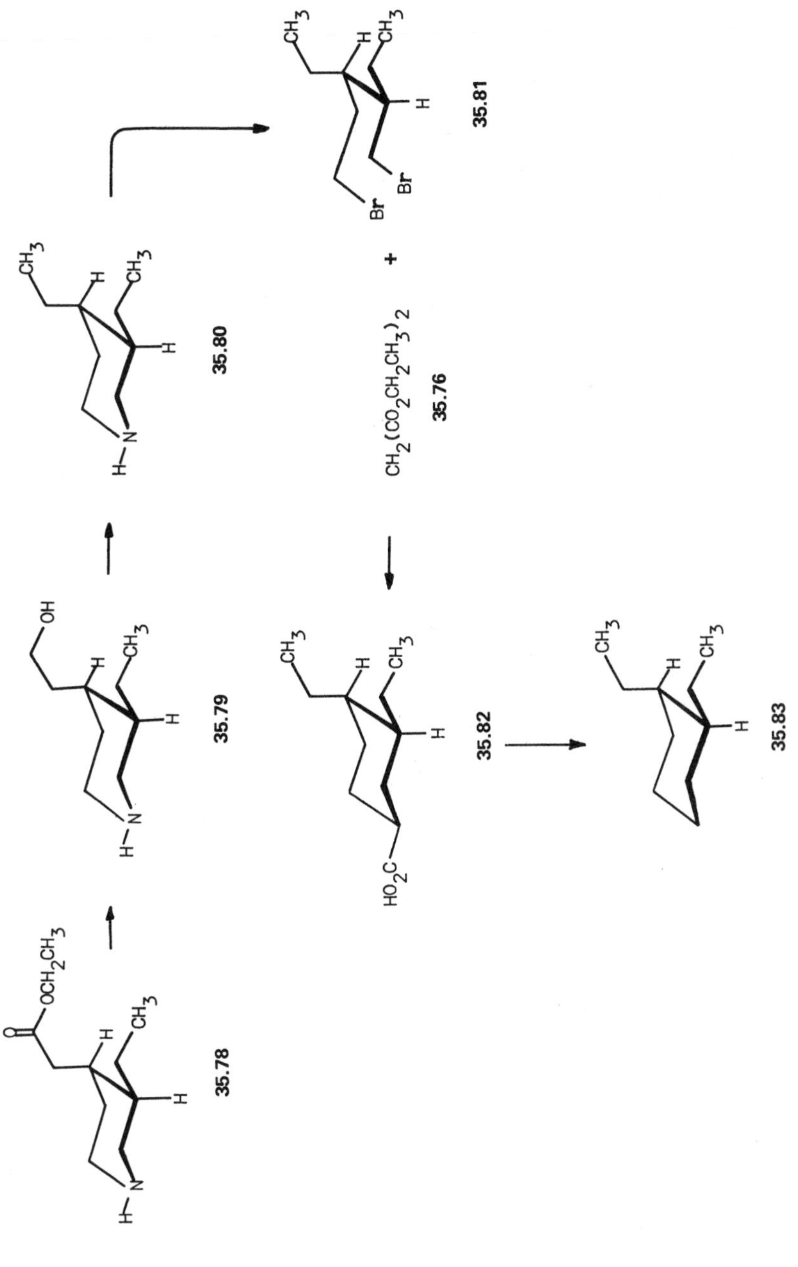

SCHEME 35.13 The degradation of ethyl dihydromeroquinene to optically inactive diethylcyclohexane. (After Reference 280.)

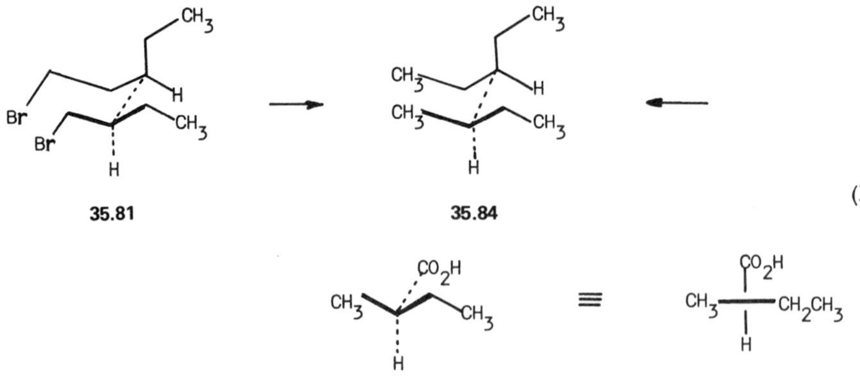

35.81 → 35.84 ← 35.85 (Eq. 35.16)

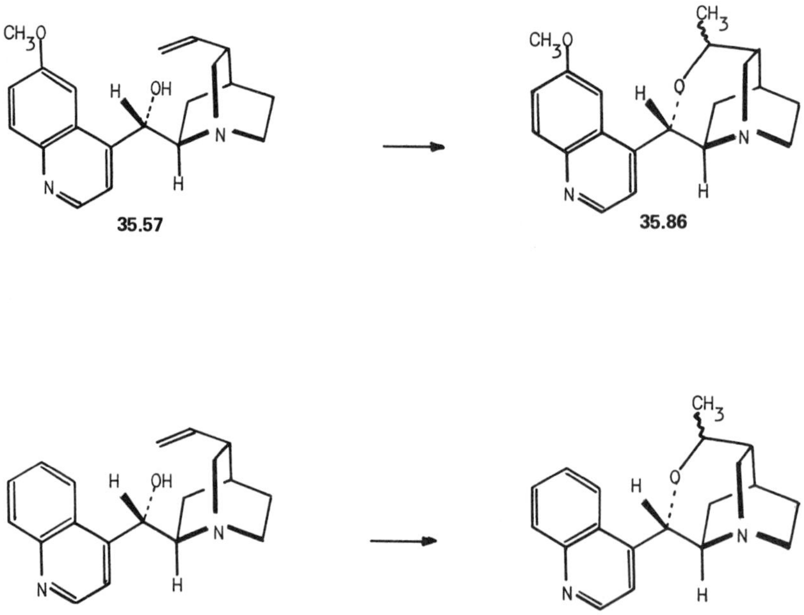

35.57 → 35.86

35.59 → 35.87

SCHEME 35.14 The conversion of quinidine and cinchonine into cyclic ethers.

at C_9 was accurately deduced by consideration of partial rotatory contributions of C_8 and C_9 and correlation of the bases with (-)-ephedine (35.88) and (+)-ψ-ephedrine (35.89) [281], determination of the absolute configuration of quinidine (35.57) by x-ray crystallography utilizing Bijvoet's anamolous dispersion method [282] substantiated the conslusions reached [283].

The first synthesis of quinine* (35.2; [284]), which still stands as a landmark on the path toward even more complicated bases to be considered later, was actually directed toward

*Additional syntheses of quinine (35.2) and related bases have recently been reported [285, 286].

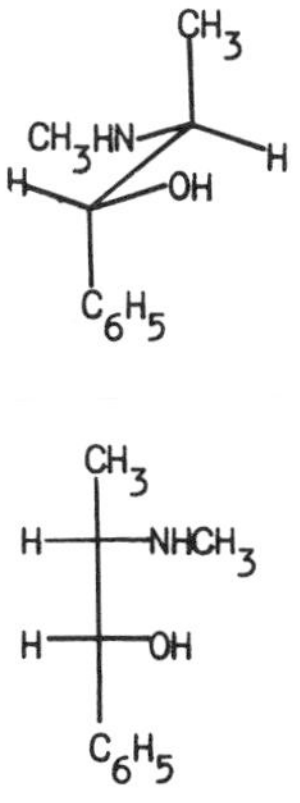

35.88

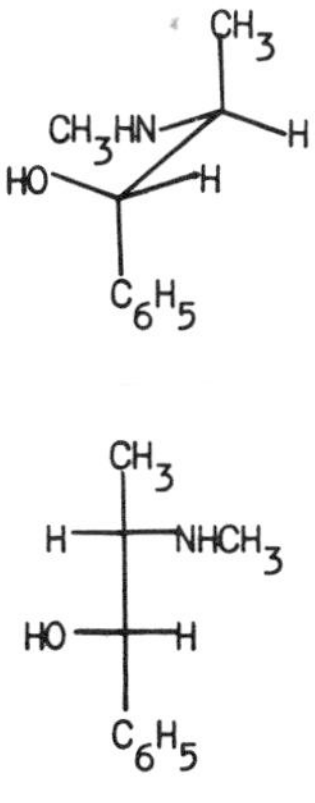

35.89

quinotoxine (35.72) can be converted by sodium hypobromite to an N-bromo derivative (35.90), which with base generates quininone (35.91; see Equation 35.13). Reduction of the latter (35.91) with aluminum powder and ethanolic sodium ethoxide yields a mixture from which quinine (35.2) and quinidine (35.57) can be isolated (Scheme 35.15) [287].

The synthesis of quinotoxine (35.72; Scheme 35.16) begins with the condensation of m-hydroxybenzaldehyde (35.92) and aminoacetal (35.93) to an imine and cyclization directly to a mixture of 7-hydroxyisoquinoline (35.94) and its 5-hydroxy isomer (35.95) with sulfuric acid. The sodium salt of 7-hydroxyisoquinoline (35.94) is less soluble than its isomer and thus separates. The purified 7-hydroxyisoquinoline (35.94) is permitted to condense in methanol with formaldehyde and piperidine, and the resulting 7-hydroxy-8-piperidinomethyl-isoquinoline (35.96) is converted by heating in methanolic sodium methoxide into 7-hydroxy-8-methylisoquinoline (35.97). When the phenol 35.97 is subjected to hydrogenation in acetic acid over platinum, 7-hydroxy-8-methyl-1,2,3,4-tetrahydroisoquinoline (35.98) results. Acetylation with acetic anhydride in methanol converts 35.98 into its N-acetyl derivative 35.99, which undergoes further reduction in ethanol with Raney nickel at 150°C and 3000 psi pressure to yield a mixture of stereoisomeric N-acetyl-7-hydroxy-8-methyldecahydroiso-quinolines. Then, oxidation with chromic anhydride in acetic acid leads to a mixture of isomeric N-acetyl-7-oxo-8-methyldecahydroisoquinoles from which the pure cis isomer (35.100) is isolated in 36% yield. The trans ketone (35.101) is also present.*

cis-N-Acetyl-7-keto-8-methyldecahydroisoquinoline (35.100), on treatment with ethyl nitrite and sodium ethoxide in ethanol, generates the oximinoester 35.102 which is reduced to the corresponding amine, N-acetyl-10-aminodihydrohomomeroquinene ethyl ester (35.103). Exhaustive N-methylation followed by elimination (with 60% sodium or potassium hydroxide at 140°C) and concomitant amide hydrolysis yields homomeroquinene (35.104) which is isolated as its N-uramido derivative (35.105) by warming the neutralized hydrolysate with potassium cyanate and subsequent acidification. When the N-uramido derivative (35.105) is treated with ethanolic hydrogen chloride and ammonium chloride, the hydrochloride of homomeroquinine ethyl ester forms. Direct treatment of this reaction mixture with benzoyl chloride and potassium carbonate yields N-benzoylmonomeroquinine ethyl ester (35.106). Condensation of the latter (35.106) with ethyl quininate (35.107) in the presence of sodium ethoxide yields (±)-quinotoxine (35.72) which is resolved with dibenzoyl-d-tartaric acid (35.108) to yield pure (+)-quinotoxine (35.72), identical with the naturally occurring material.

---

*It is worthwhile noting that the assignment of stereochemistry was based upon the successful conversion of 35.100 into quinotoxine (35.72)!

35.72 → 35.90 → 35.91 → 35.57 + 35.2

SCHEME 35.15 The formation of quinine and quinidine from quinotoxine. (After Reference 287.)

III. AJMALICINE

Ajmalicine (35.3), mp 257°C (dec.), $[\alpha]_D^{24}$ -58.1° ($CHCl_3$), has been isolated numerous times (occasionally under the name δ-yohimbine) from various species of Rauwolfia where it is often found to be the weakest base present. The alkaloid (35.3) corresponds to $C_{21}H_{24}N_2O_3$ and possesses an ultraviolet spectrum (λ_{max} 225 and 285 nm) characteristic of the chromophore present in a β-enol ester (e.g., 35.109). In the infrared, this particular grouping has been identified by maxima at 5.89 and 6.20 μ (1698 and 1614 cm^{-1}) of about equal intensity, and ajmalicine (35.3) is not an exception.

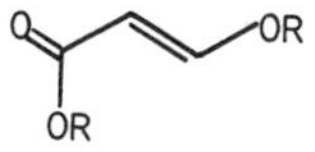

35.109

In addition to the observation that the isolated double bond in ajmalicine (35.3) is not readily reduced, it was noted that a methoxyl group is present in a carbomethoxy function since hydrolysis (with base; Equation 35.17) yields the corresponding carboxylic acid

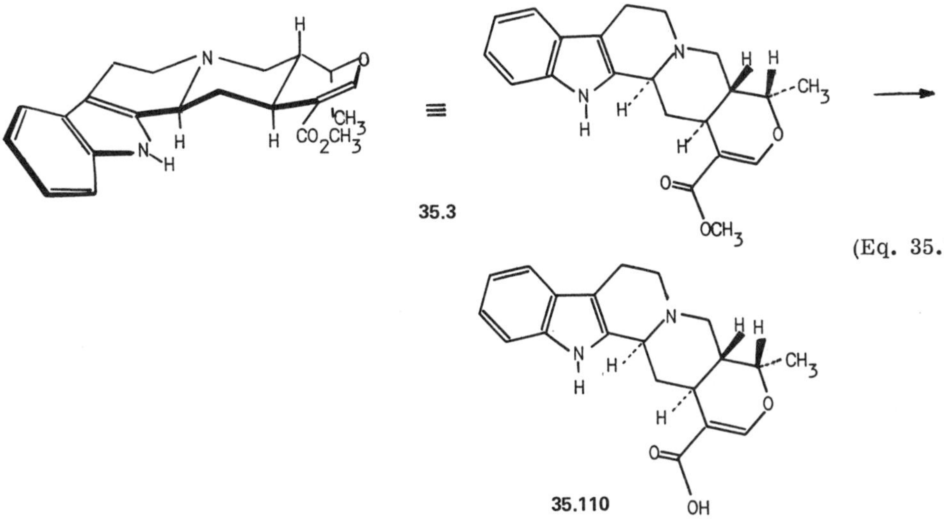

35.3

35.110

(Eq. 35.17)

(ajmalicic acid, $C_{20}H_{22}N_2O_3$; 35.110) and that ajmalicine (35.3) contains at least one C-methyl group (by Kuhn-Roth). Dehydrogenation of ajmalicine (35.3) with selenium yields alstyrine (35.111; Equation 35.18) which was identified as α-(α-diethyl-3,4-pyridine)-β-ethyl indole since oxidative ozonolysis, followed by hydrolysis of the resultant product (35.112), yields o-aminopropiophenone (35.26) and 3,4-diethylpyridine-6-carboxylic acid (35.113; Equation 35.19).

Based upon the molecular formula, the absence of a readily reducible double bond (but the presence of the chromophore 35.109), and the formation of alstyrine (35.111; which was similar to the formation of tetrahydroisoyobyrine, 35.23, from yohimbine, 35.1),* the currently accepted pentacyclic structure was suggested.

*It will be recalled that tetrahydroisoyobyrine (35.23) also yielded o-aminopropiophenone (35.26) and a substituted pyridine carboxylic acid (Equation 35.6).

35.92

$H_2NCH_2CH(OEt)_2$

35.93

35.95

35.94

35.96

35.97

35.98

35.99

35.100

35.101

35.102

35.103

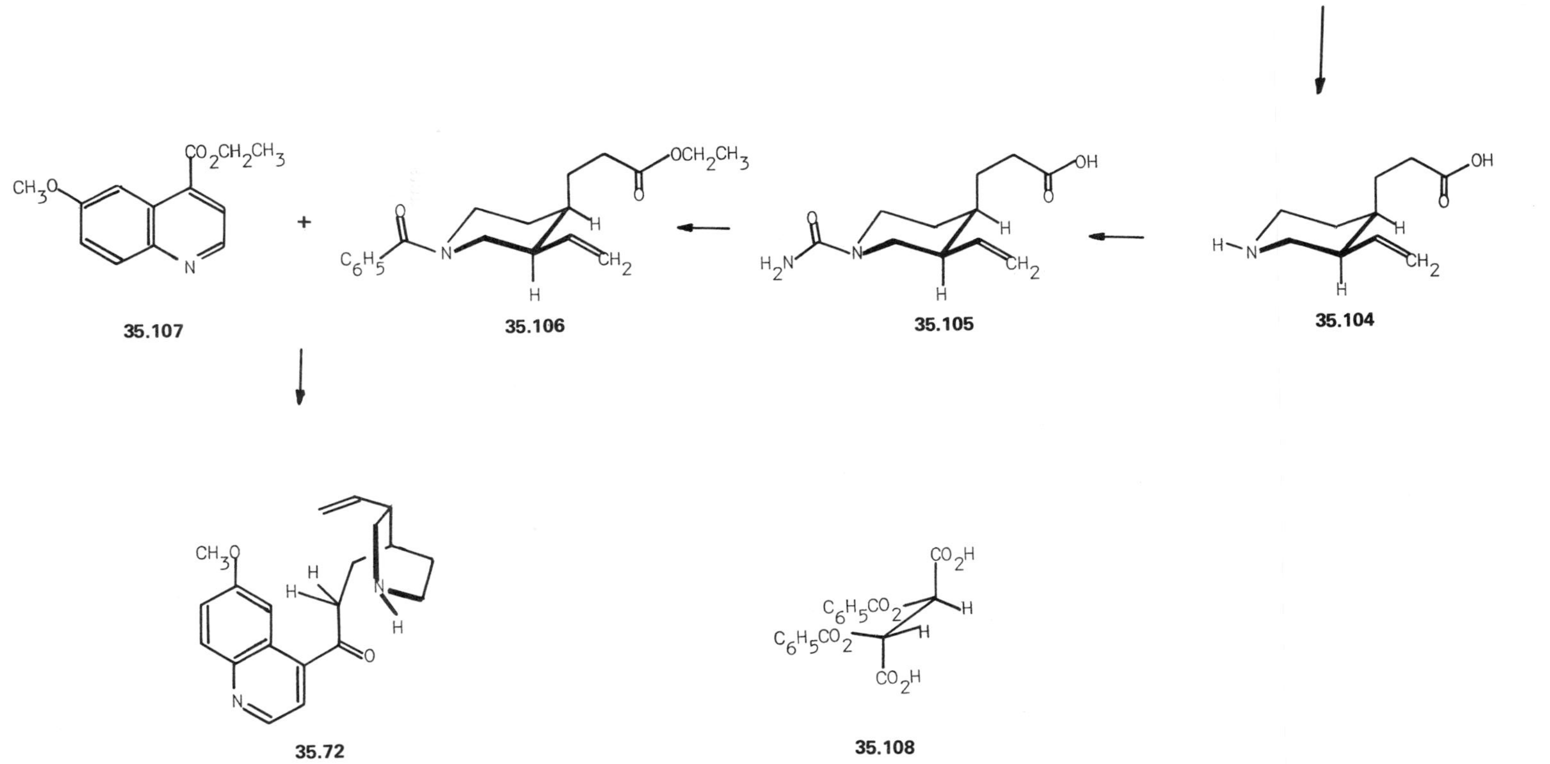

35.104

35.105

35.106

35.107

35.72

35.108

SCHEME 35.16 The total synthesis of quinotoxine. (After Reference 284.)

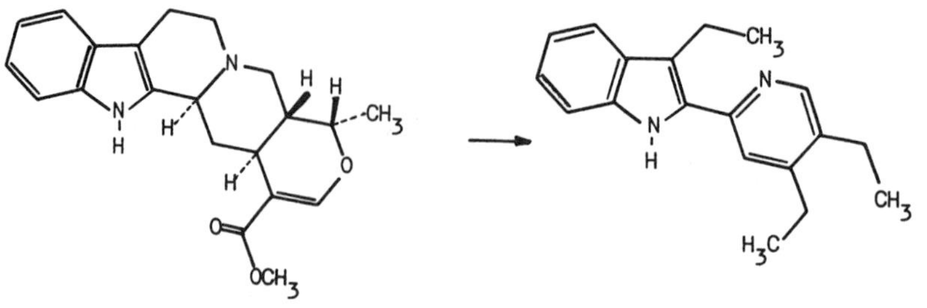

35.3 35.111 (Eq. 35.18)

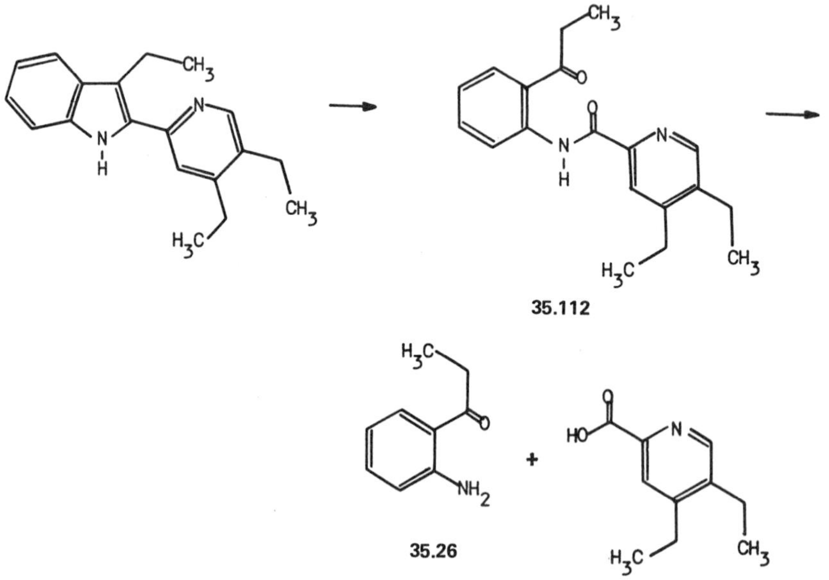

35.112 (Eq. 35.19)

35.26 35.113

Although the relative stereochemistry at three of the four asymmetric centers (i.e., C_3, C_{15}, and C_{20}) was originally assigned deductively on the basis of spectroscopic measurements and, inter alia, products of reduction of ring B aromatized products, it has proved possible to establish their absolute configurations through interrelation with cinchonine (35.59) whose absolute configuration now rests on x-ray crystallographic determination (see above) of quinidine (35.57). The stereochemistry of the fourth asymmetric center (C_{19}) has been decided (see below) on the basis of nmr measurements of compounds in which the absolute configuration at each of the other three centers is known.

A. Stereochemistry at C_{15}

The configuration at C_{15} [288] was demonstrated as follows. Acid-catalyzed hydration and decarboxylation of ajmalicic acid (35.110) yields what was thought to be an aldehyde, ajmalicial (35.114) but was identified, spectroscopically, as a hemiacetal. Wolff-Kishner reduction of ajmalicial (35.114) leads to ajmaliciol (35.115) which undergoes Oppenauer oxidation to 19-dihydrocorynantheone (35.116). These reactions are shown in Scheme 35.17. Since conditions of the Oppenauer oxidation are sufficient to have epimerized the ketone at C_{20} (i.e., α to the carbonyl), the thermodynamically stable isomer, with the acetyl function equatorial,

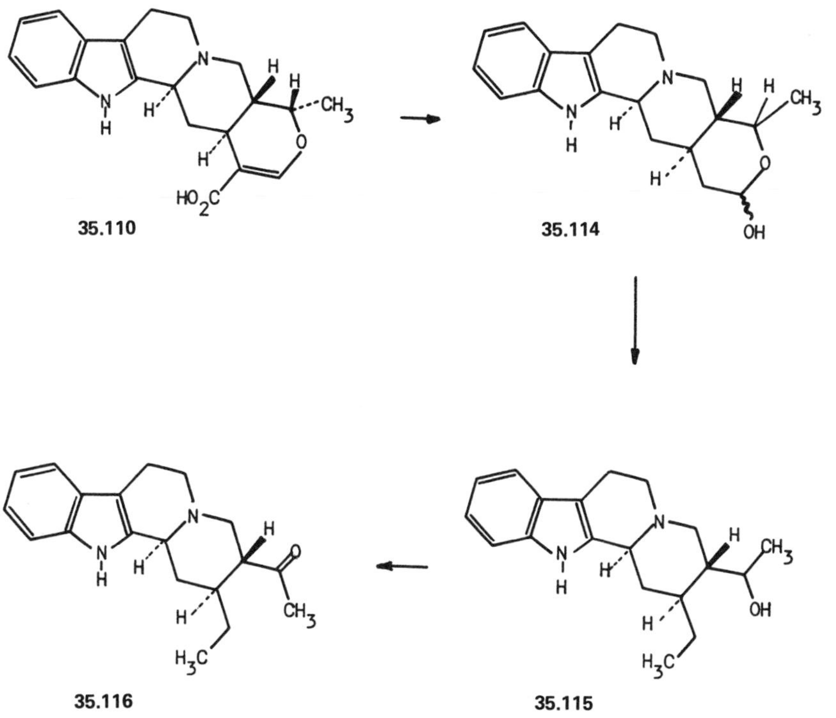

35.110 35.114 35.116 35.115

SCHEME 35.17 The degradation of ajmalicic acid to 19-dihydrocorynantheone. (After Reference 288.)

is obtained (Equation 35.20).* Wolff-Kishner reduction of 19-dihydrocorynantheone (35.116) yields dihydrocorynantheane (35.117), identical to that obtained by degradation of the alkaloid corynantheine† (35.118; Equation 35.21). This implies that the ethyl group (at C_{20}) in

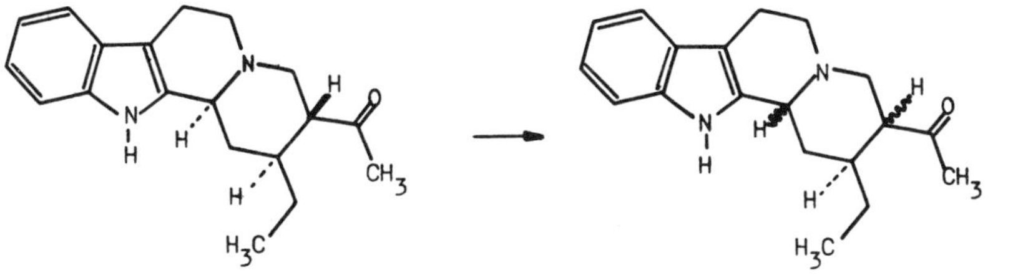

(Eq. 35.20)

*It is also pointed out [288] that the conditions of oxidation are capable of epimerizing C_3. Since this is possible and C_{20} could have been epimerized, only C_{15} is certain of having retained its original configuration.

†The chemistry of corynantheine (35.118), which we are not considering here beyond that necessarily discussed for the proof of structure of ajmalicine (35.3), is similar to that of ajmalicine (35.3). The interested reader is directed to the selected reading listed at the end of this chapter for the chemistry of many of the indole alkaloids not discussed here.

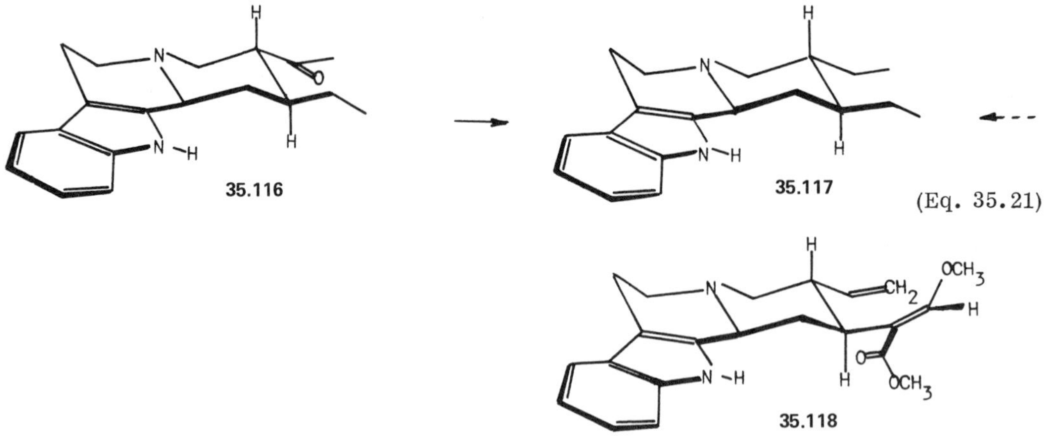

35.116

35.117

(Eq. 35.21)

35.118

dihydrocorynantheane (35.117) is equatorial (as shown) since this would be the thermodynamic isomer but, what is more important for our immediate purpose, that the substituent at C_{15}, which is the only one certainly not epimerized, is the same in both ajmalicine (35.3) and in dihydrocorynantheane (35.117).

Now, when corynantheine (35.118) is hydrolyzed with dilute base and then heated with acid, corynantheinal (35.119) results. Reduction of corynantheinal (35.119) with hydrogen on palladium in methanol generates dihydrocorynantheinal (35.120) which can be reduced further to dihydrocorynantheane (35.117; [289] or to the alcohol dihydrocorynantheol (35.121; [290]). These reactions are shown in Scheme 35.18. Since none of these transformations on corynantheine (35.118) could have affected C_{15} or C_{20} it is confidently assumed that both these centers retain their original geometries.

Cinchonine (35.59) yields a dihydro derivative, dihydrocinchonine (35.122), which, via the N-oxide (35.123), the unsaturated lactam (35.124), and its reduced derivative 35.125, can be converted to the benzoate 35.126 ([290]; Scheme 35.19). Then [291], cyanogen bromide (von Braun degradation) treatment of the benzoate 35.126 yields a bromocyanide (35.127). Reductive dehalogenation of 35.127 with Raney nickel in the presence of potassium hydroxide, and, without purification, subsequent treatment with 10% ethanolic potassium hydroxide causes hydrolysis of the benzoate, lactam, and nitrile, with decarboxylation of the carbamic acid intermediate and recyclization to the aromatic amino alcohol (35.128; 44% yield). Reduction of 35.128 with lithium aluminum hydride in tetrahydrofuran generates the diamine 35.129, and this oily base, on Oppenauer oxidation with lithium tert-butoxide and benzophenone, leads to an isomer (35.130) of dihydrocorynantheane (35.117). These reactions are shown in Scheme 35.20.

Now, as has been seen before (p. 507 and Equation 35.11) for the case of yohimbine (35.1), epimerization at C_3 via oxidation in ring C and subsequent reduction to the most stable isomer (which does not affect the stereochemistry at C_{15} or at C_{20}) can be effected. Thus, as in the case of yohimbine (35.1) (p. 507 and Equation 35.11), when 35.130 is oxidized with lead tetraacetate it is possible to obtain a tetradehydro derivative (35.131), and when the oxidation is effected with mercuric acetate at 120°C, a didehydro derivative (35.132) is generated. Reduction of the latter (35.132) with sodium borohydride generates dihydrocorynantheane (35.117; Scheme 35.21). Therefore, since the same dihydrocorynantheane (35.117) is generated from ajmalicine (35.3) and cinchonine (35.59) by processes which cannot affect C_{15}, the absolute configuration at C_{15} must be the same for both.

This conclusion was then confirmed as follows [280, 291]: The mono-N-oxide of dihydrocinchonine (35.123), whose absolute configuration is known (see Section II) was converted to the oxohexahydrocinchonine (35.133) by oxidation of the alcohol 35.125 [291]. Ethanolic

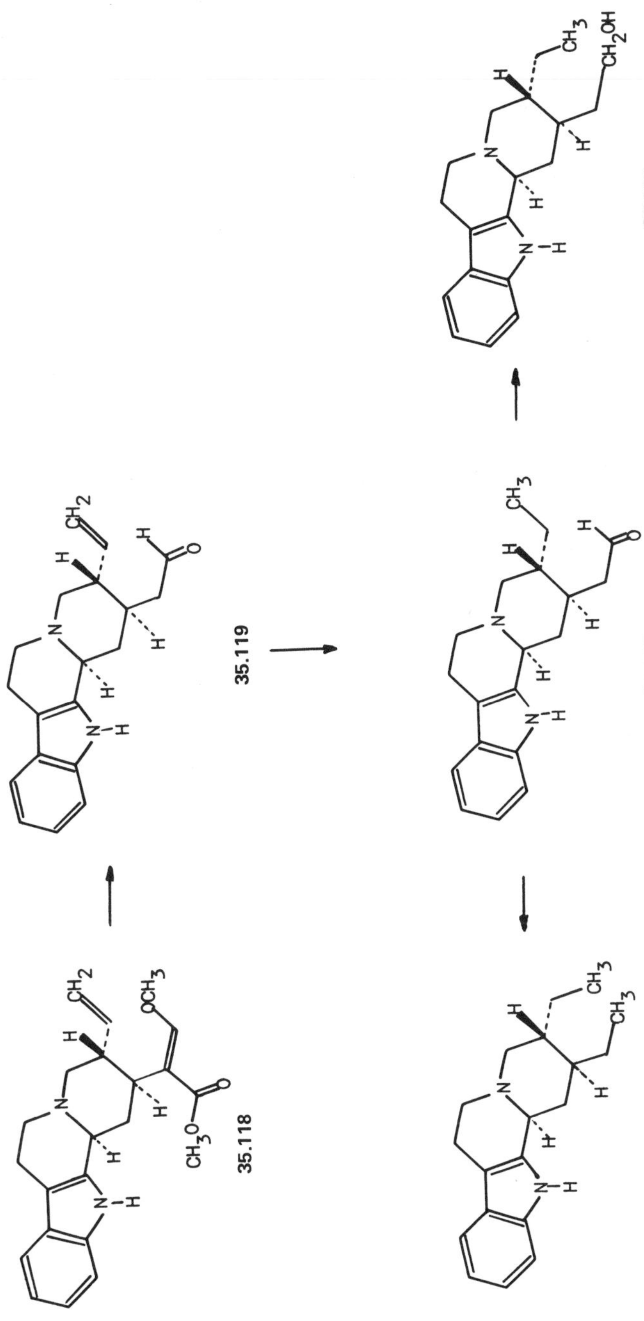

35.118 → 35.119 → 35.120 → 35.121

35.120 → 35.117

SCHEME 35.18 The conversion of corynantheine to dihydrocorynantheane and dihydrocorynantheol. (After Reference 289.)

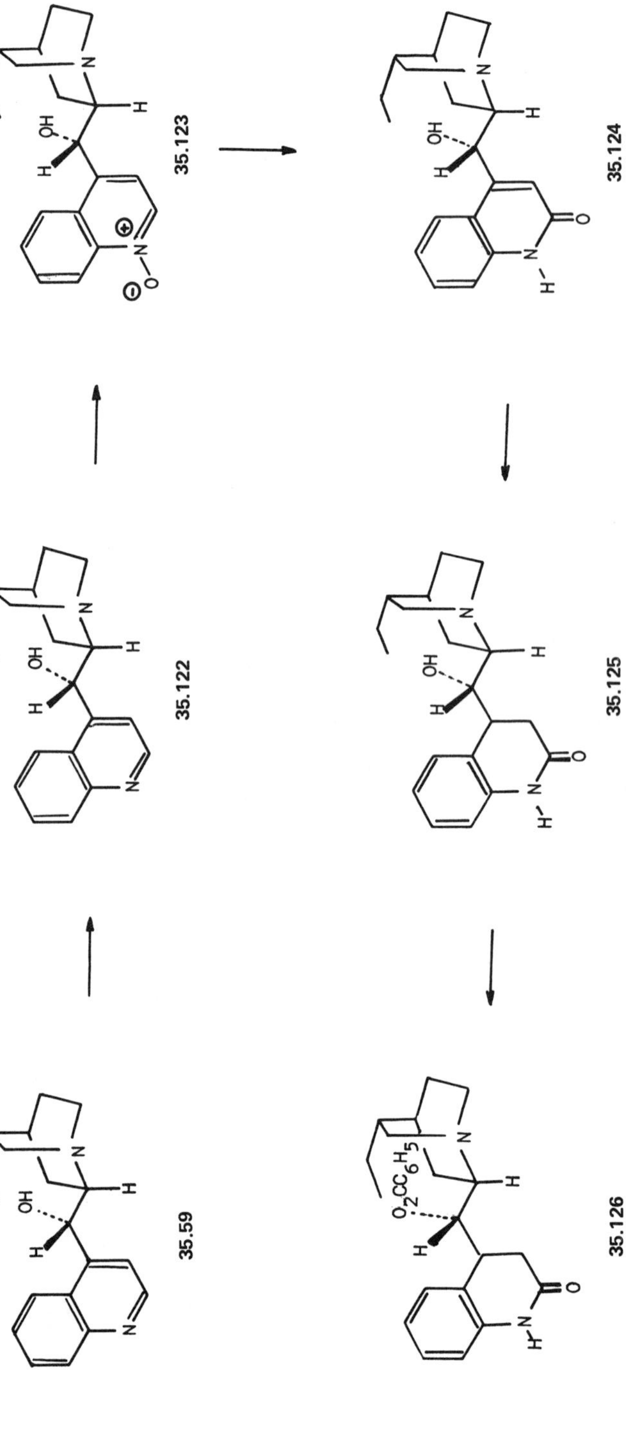

SCHEME 35.19 A partial degradation of cinchonine. (After Reference 290.)

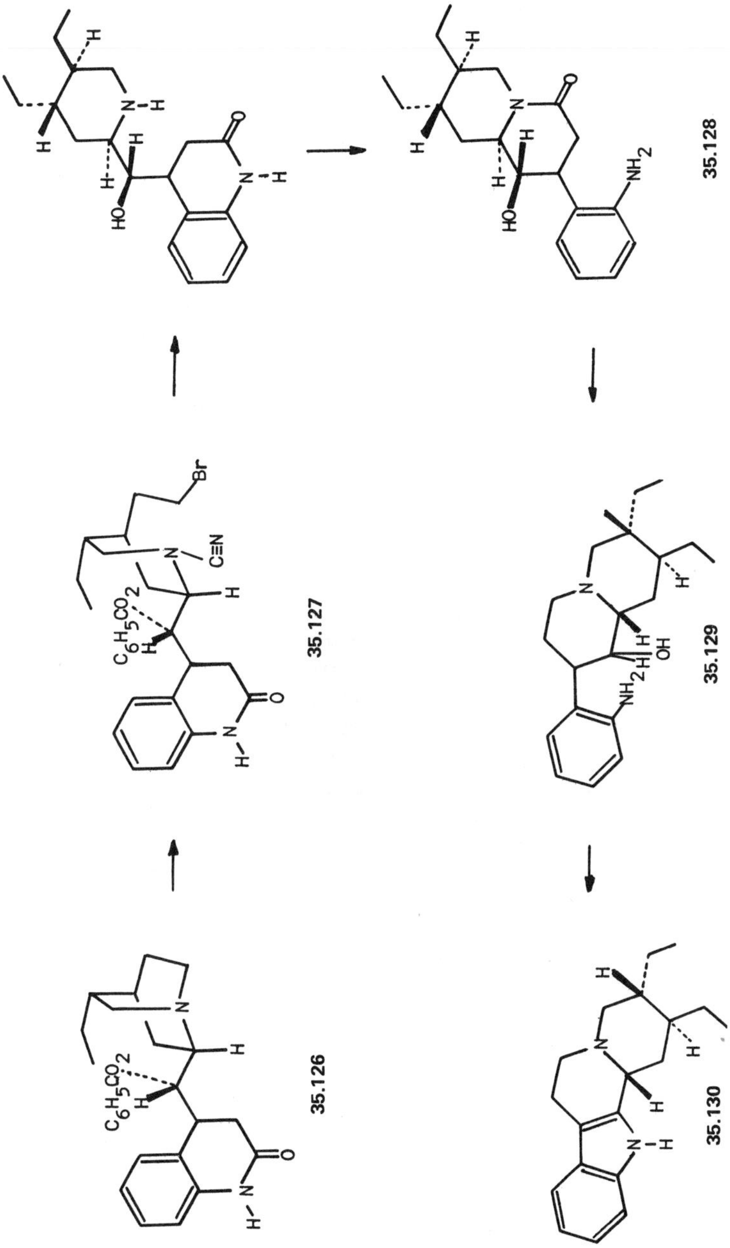
35.126
35.127
35.128
35.129
35.130
$C_6H_5CO_2$
Br
C≡N
NH_2
OH
HO
N–H

SCHEME 35.20 The formation of dihydrocorynantheane from a degradation product of cinchonine. (After Reference 291.)

hydrogen chloride treatment of 35.133 causes ring opening and recyclization to the ester 35.134 which yields, on reduction with lithium aluminum hydride, the alcohol, dihydrocinchonamine (35.135). The O-tosylate of dihydrocinchonamine (35.135) undergoes cyclization to the quaternary salt 35.136 (isolated as its tosylate), while the same tosylate is produced by tosylation and cyclization of the alcohol, dihydrocorynantheol (35.121), produced (see above) from corynantheine (35.118; [288]; Scheme 35.22).*

B. Stereochemistry at C_3

The stereochemistry at C_3 of ajmalicine (35.3) was demonstrated by a series of oxidation and reduction reactions similar to those already discussed for yohimbine (35.1; Equation 35.11; [292]) which illustrated that a C/D trans ring fusion with the C_3 hydrogen down (or α) was reasonable.†

---

*It must be pointed out that since the conversion of cinchonine (35.59) into dihydroinchonine (35.122) occurs with retention of configuration where the ethylidene side chain branches from the bicyclic system (i.e., at C_3 of cinchonine, 35.59, or C_{20} of corynantheine, 35.118) and since the further conversion of dihydrocinchonine (35.122) into dihydrocinchonamine (35.135; Scheme 35.22) also leaves this site unaffected, the absolute stereochemistry at C_{20} of corynantheine (35.118) is also demonstrated by the identity of the quaternary salt 35.136 from the two sources. However, because of the possibility of epimerization during Oppenauer oxidation of ajmaliciol (35.115) to 19-dihydrocorynantheone (35.116; Scheme 35.7), the identity of the salt 35.136 does not confirm the stereochemistry at C_{20} of ajmalicine (35.3). On the other hand, if it could be shown (see below) that epimerization during the Oppenauer oxidation of ajmaliciol (35.115) does not occur, the stereochemistry at C_{20} would be established.

†This work was facilitated by the availability of the various stereoisomers of ajmalicine (35.3) whose dehydrogenation products were identical, i.e., ajmalicine (35.3) and isoajmalicine (35.137) both yield serpentine (tetradehydroajmalicine; 35.138; Equation 35.22), while tetrahydroalstonine (35.139) and akuammagine (35.140) both yield alstonine (35.141; Equation 35.23). However, the reduction of these dehydrogenation products yields only one of the

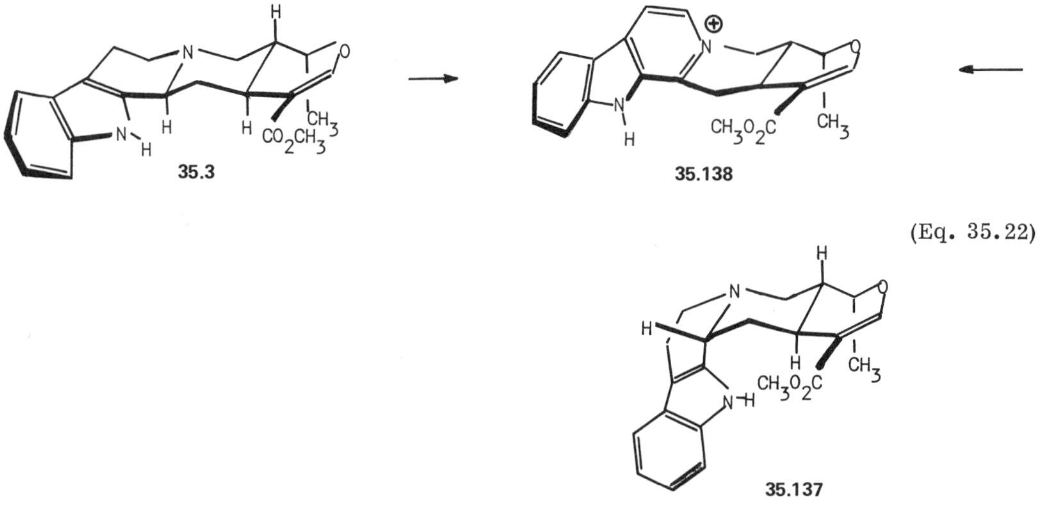

35.3 → 35.138 ← 35.137

(Eq. 35.22)

C. Stereochemistry at C_{19} and C_{20}

The stereochemical assignment at C_{20} is based upon epimerization from an axial substituent to an equatorial substituent [293]. Thus, tetrahydroalstonial (35.142) is generated by saponification of tetrahydroalstonine (35.139) followed by decarboxylation. As for ajmalicine (35.3), the hydrolysis product 35.142 was shown to be a hemiacetal which yields 19-corynantheidol (35.143) on Wolff-Kishner reduction. When 19-corynantheidol (35.143) is subjected to Oppenauer oxidation, 19-corynantheidone (35.144) is obtained, and this ketone, on treatment with sodium methoxide in methanol, yields 19-dihydrocorynantheone (35.116), which must therefore be the equatorial isomer, and 19-corynantheidone (35.144), the axial isomer. These reactions are shown in Scheme 35.23.

Now, when 19-dihydrocorynantheone (35.116) is reduced with sodium borohydride, ajmaliciol (35.115) and its C_{19} epimer result. Thus, ajmalicine (35.3), which had also yielded ajmaliciol (35.115), must have a trans-fused D/E ring system while tetrahydroalstonine (35.139) a D/E cis fusion. Additionally, since ajmalicine (35.3) and isoajmalicine (35.137) have coupling constants (J_{HH}) of 2.7 and 1.8 Hz, respectively, for the proton at C_{19} coupling to that at C_{20}, while tetrahydroalstonine (35.139) and akuammagine (35.140) have coupling constants (J_{HH}) of 10.3 and 5.8 Hz, respectively, for the same two protons [293], ajmalicine (35.3) and isoajmalicine (35.147) must bear the C_{19} and C_{20} protons in a cis relationship and tetrahydroalstonine (35.139) and akuammagine (35.140) must have the same two protons in a trans sense.

Racemic ajmalicine (35.155) has been synthesized ([294]; Scheme 35.24) by a pathway which, since formation of the most stable isomer is expected at each step, acts to substantiate the deductive reasoning behind the stereochemical assignments made above.

---

pair back again, i.e., serpentine (35.138) yields ajmalicine (35.3) while alstonine (35.141) generates tetrahydroalstonine (35.139). Further, since it can be shown, just as illustrated above for the relationship of ajmalicine (35.3) and cinchonine (35.59), that all four of the isomers (ajmalicine, 35.3, isoajmalicine, 35.137, tetrahydroalstonine, 35.139, and akuammagine, 35.140) bear the C_{15} substituent in the same absolute sense, it is clear that serpentine (35.138) and alstonine (35.141) can differ from each other at C_{19} and C_{20} only.

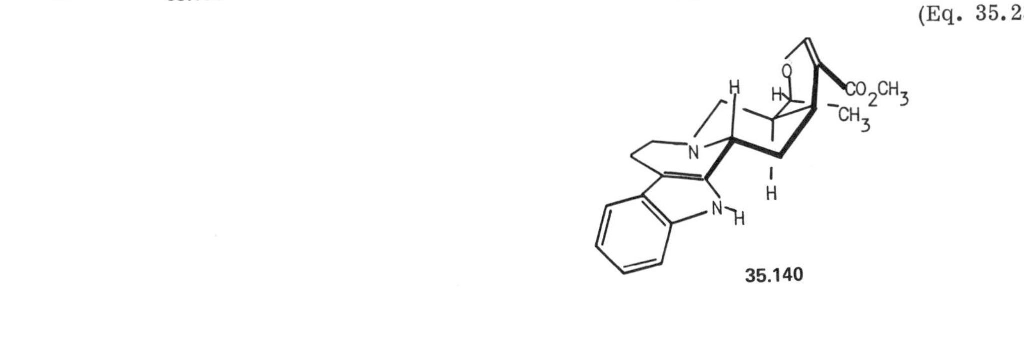

35.139 → 35.141 ← 35.140

(Eq. 35.23)

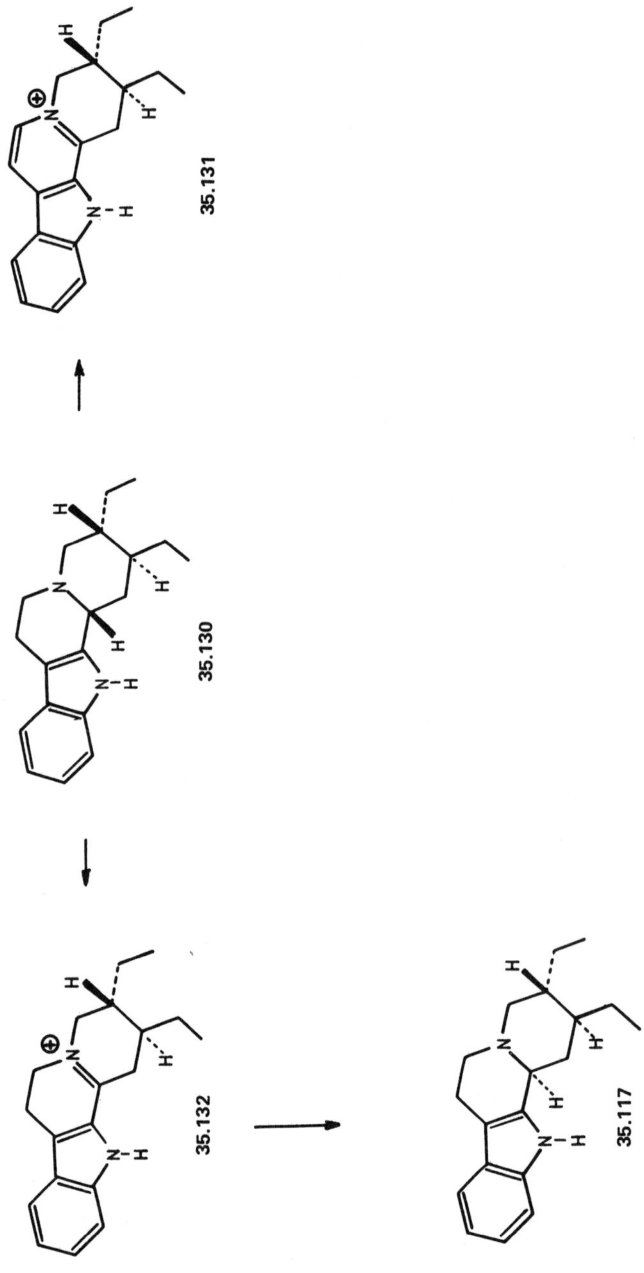

SCHEME 35.21 Epimerization of the cinchonine-derived isomer of dihydrocorynantheane to dihydrocorynantheane derived from corynantheine itself. (After Reference 290.)

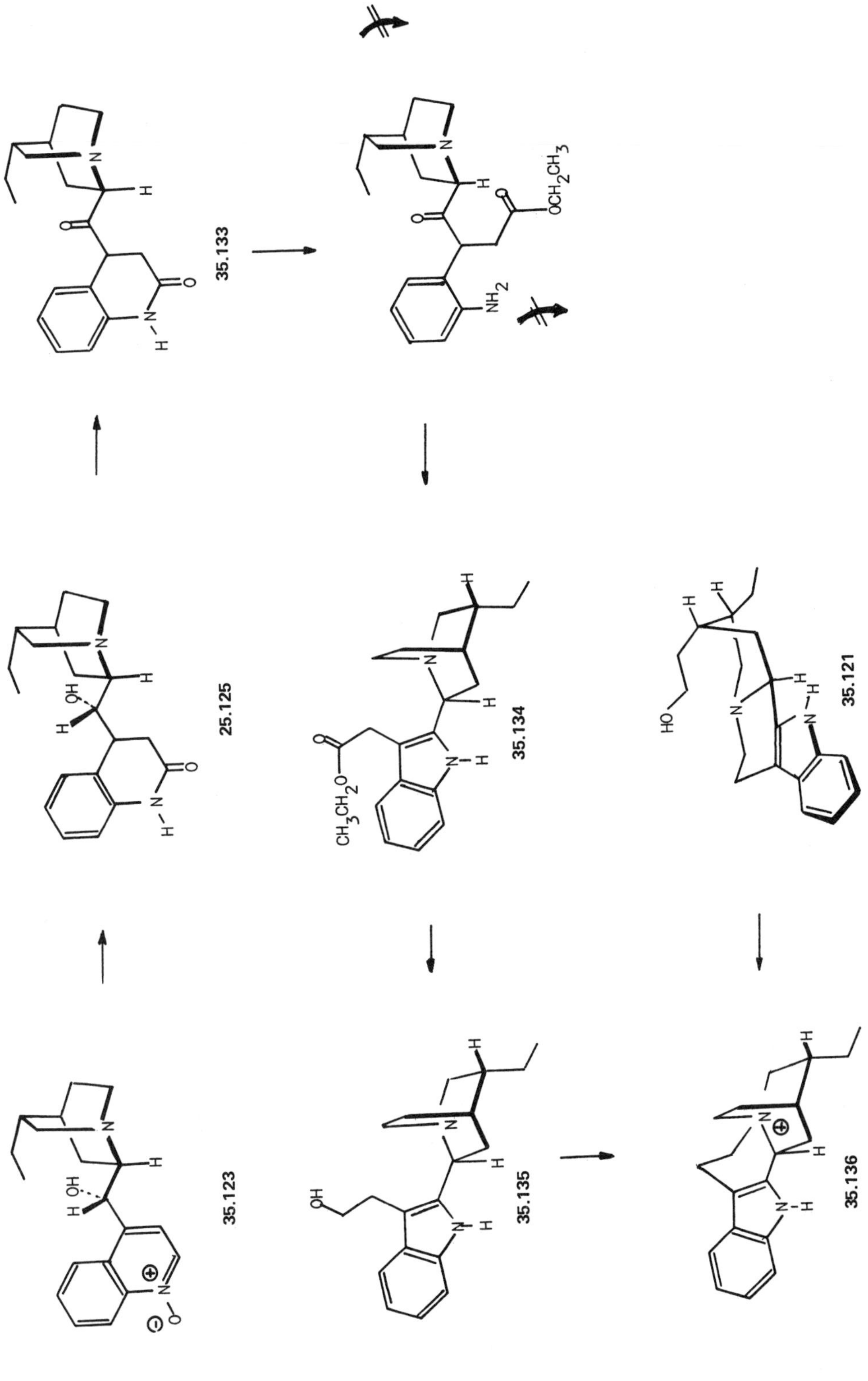

SCHEME 35.22 Conversion of cinchonine into a quaternary salt identical to one derived from corynantheine. (After Reference 288.)

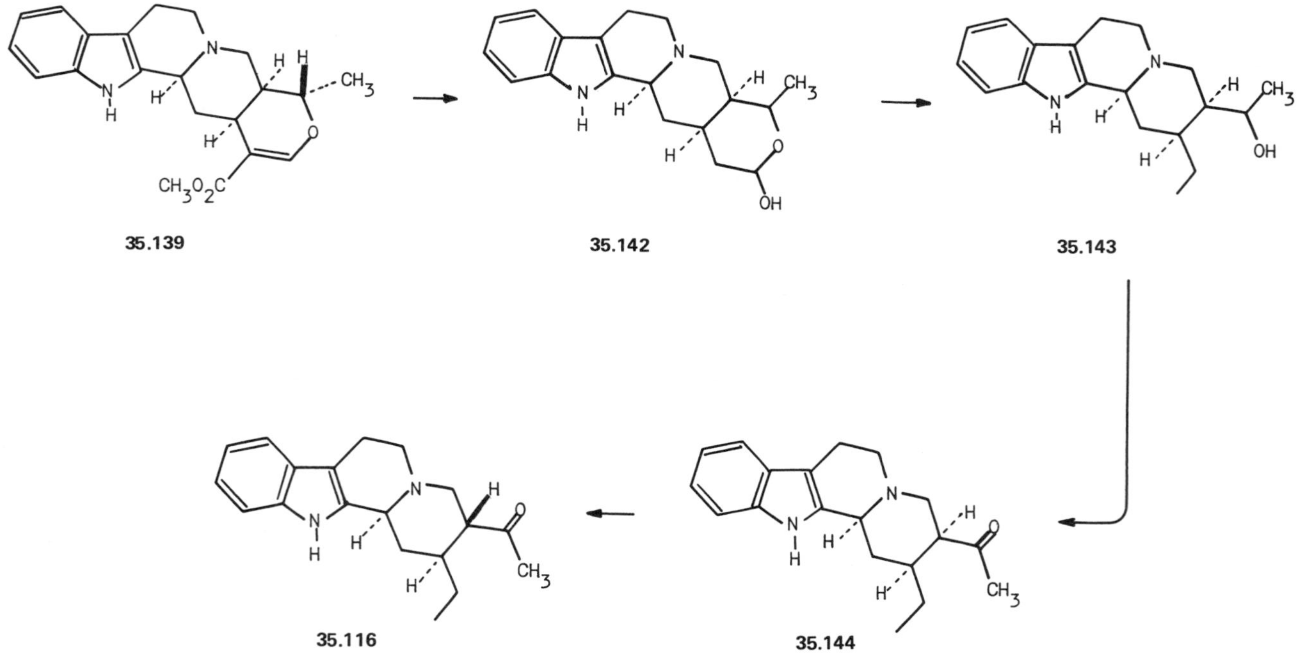

SCHEME 35.23 The conversion of tetrahydroalstonine into 19-dihydrocorynantheone. (After Reference 293.)

Methyl acetoacetate (35.145) is allowed to condense with dimethyl glutaconate (35.146) to yield the keto triester 35.147. Mannich condensation of the latter (35.147) with tryptamine (35.31) and formaldehyde in tert-butyl alcohol leads to the lactam 35.148. When 35.148 is treated with phosphorus oxychloride and the resulting imminium salt (35.149) reduced with hydrogen over a palladium on carbon catalyst in ethanol, the keto diester 35.150 results. Hydrolysis and decarboxylation of 35.150 to 35.151 is accomplished by heating with dilute hydrochloric acid, and the carbonyl group of the latter is reduced with sodium borohydride to generate the hydroxy acid 35.152. Cyclization of 35.152 to the corresponding lactone 35.153 is accomplished with N,N'-dicyclohexylcarbodiimide. The final, required carbon atom is attached by allowing the δ-lactone 35.153 to react with methyl formate in the presence of base (sodium triphenylmethane) to yield 35.154, and this is in turn induced to open and recyclize to (±)-ajmalicine (35.155) by heating in refluxing methanolic hydrogen chloride.

IV. RESERPINE

In 1952 [295], the principal hypotensive and sedative ingredient of Rauwolfia serpentina Benth. was isolated. Immediate utilization of reserpine (35.4), for so the agent was called, was made in the treatment of certain mental illnesses and high blood pressure, and its use is still extant. The crystalline, weakly basic alkaloid reserpine (35.4), $C_{33}H_{40}N_2O_9$, mp 263-264°C, $[\alpha]_D^{23}$ -117° (chloroform) proved to have several readily identifiable functional

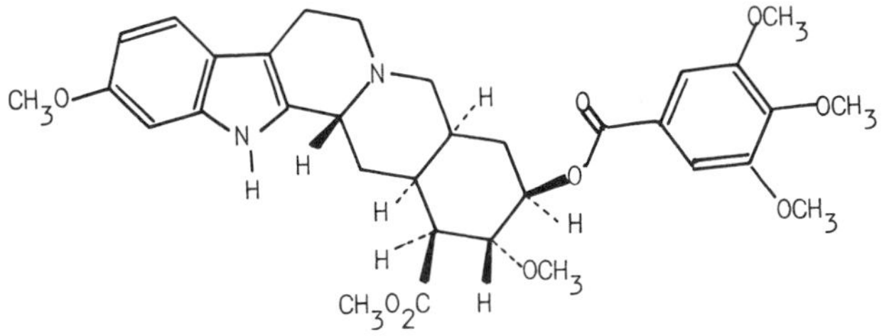

35.4

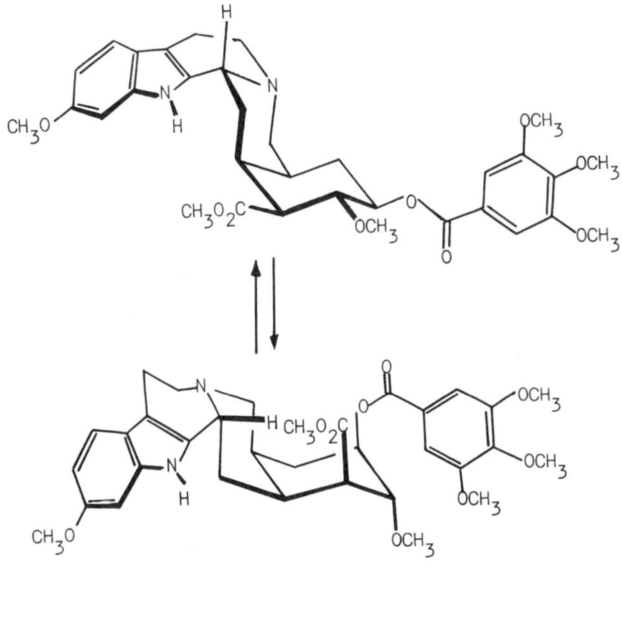

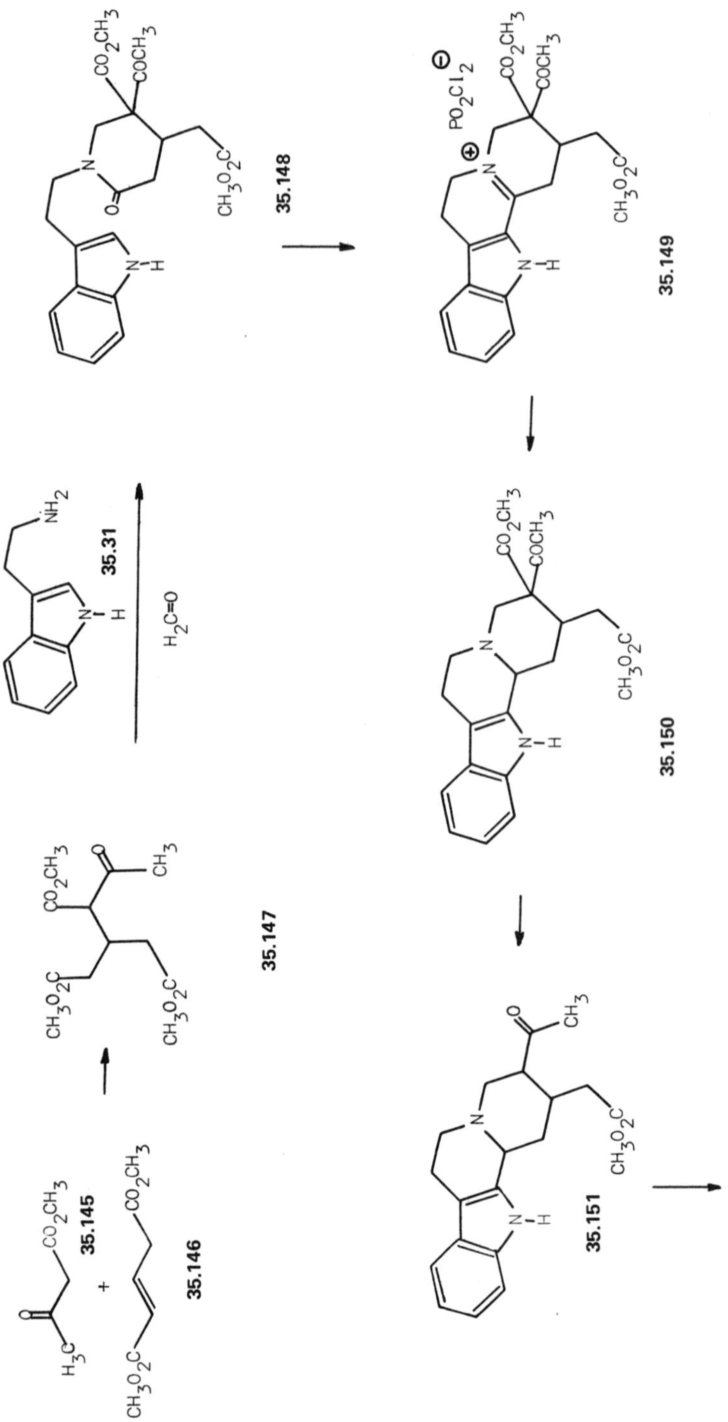

35.145 + 35.146 → 35.147

35.31

$H_2C{=}O$

35.148 → 35.149 → 35.150 → 35.151 →

$PO_2Cl_2^{\ominus}$

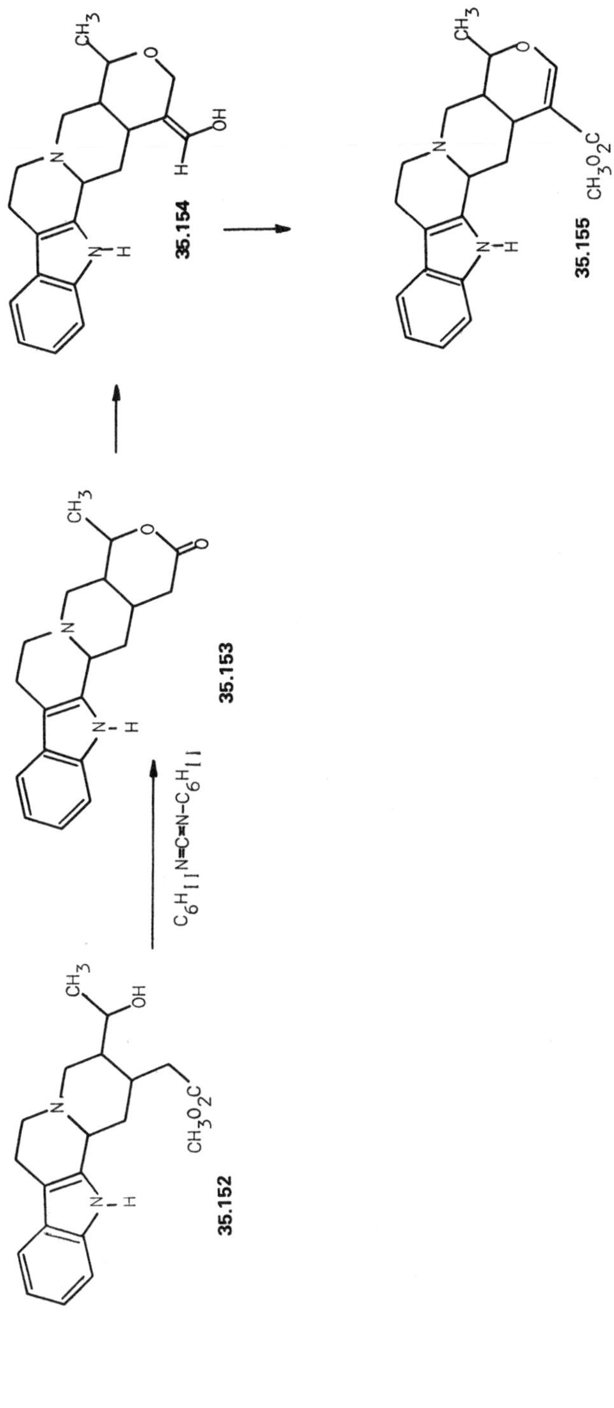

35.152 $\xrightarrow{C_6H_{11}N{=}C{=}N{-}C_6H_{11}}$ **35.153** $\rightarrow$ **35.154** $\rightarrow$ **35.155**

SCHEME 35.24 The total synthesis of (±)-ajmalicine. (After Reference 294.)

groups (Scheme 35.25) since basic hydrolysis yields reserpic acid (35.156), $C_{22}H_{28}N_2O_5$, methanol, and trimethylgallic acid (3,4,5-trimethoxybenzoic acid; 35.157). Reserpine (35.4) can be regenerated from reserpic acid (35.156) by treatment of the latter with diazomethane to yield methyl reserpate (35.158) and, subsequently, with 3,4,5-trimethoxybenzoyl chloride (the acid chloride of 35.157).

Since Rauwolfia species belong to the family Apocyanaceae, the alkaloid reserpine (35.4) was assumed to possess an indole nucleus, and it was found, by comparison of many ultraviolet spectra, that a close resemblance between the spectrum of reserpine [λ_{max} 216-218 nm (ϵ 58,500), 226-228 nm (sh, ϵ 43,200), 266-268 nm (ϵ 16,300), and 292-296 nm (ϵ 10,100)] and that of a mixture of 2,3-dimethyl-6-methoxyindole (35.159) and methyl-3,4,5-trimethoxybenzoate (methyl ester of 35.157; [296]) existed.

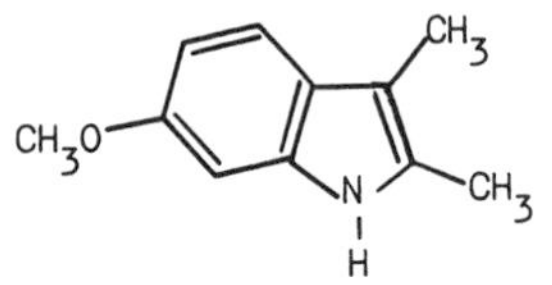

35.159

Now, oxidation of reserpic acid (35.156) with permanganate (Equation 35.24) generates, among other products, 4-methoxy-N-oxalylanthranilic acid (35.160). Selenium dehydrogenation of methyl reserpate (35.158) provides yobyrine (35.22) and a hydroxyyobyrine (35.161; Equation 35.25), the O-methyl ether of which (35.162) is identical to the product obtained by treatment of harmine (35.163) with lithium metal and then 2-methylcyclohexanone, dehydration

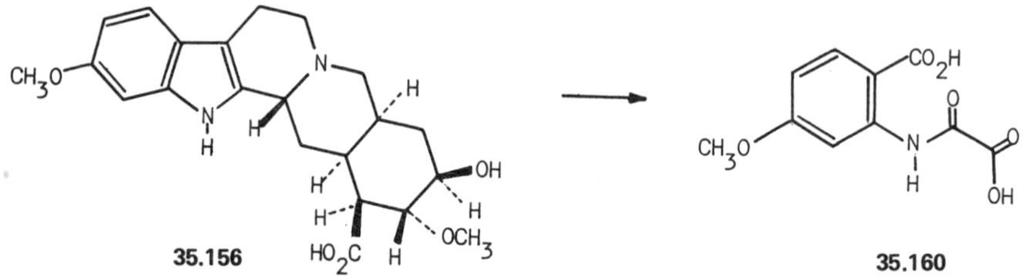

35.156 → 35.160 (Eq. 35.24)

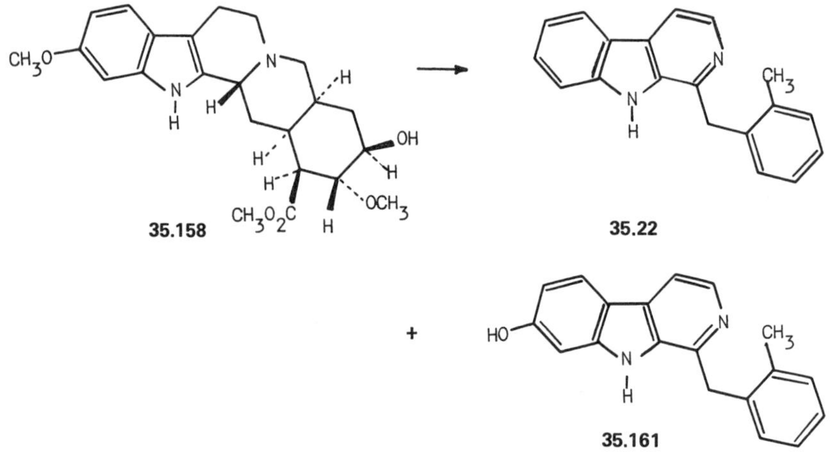

35.158 → 35.22 + 35.161 (Eq. 35.25)

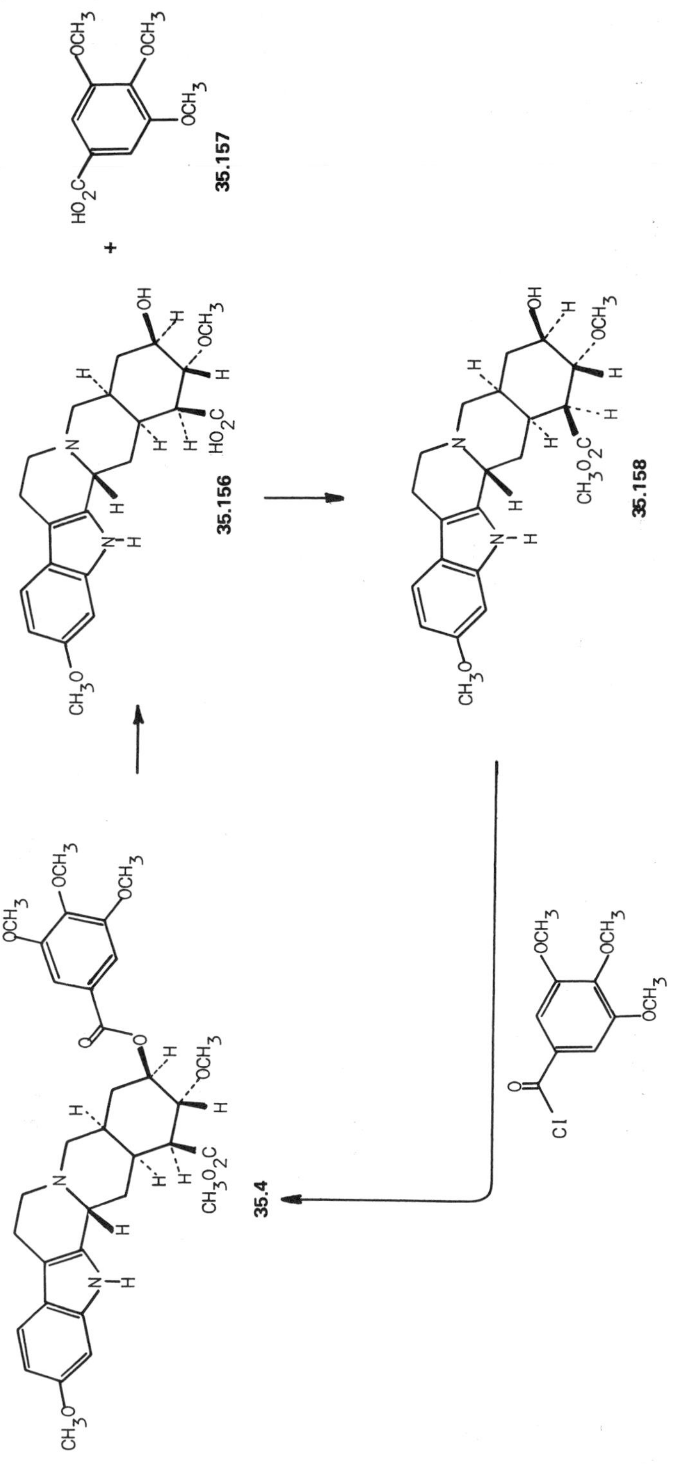

SCHEME 35.25 The hydrolysis of reserpine to reserpic acid and the regeneration of reserpine from reserpic acid.

of the resulting tertiary alcohol (35.165) with phosphorus pentoxide, and finally, dehydrogenation over palladium on carbon (Scheme 35.26; [297]). The reasonable suggestion that the functionality remaining to be placed in reserpic acid (35.156) should be in the E ring received confirmation from the isolation of 5-hydroxyisophthalic acid (35.166) from fusion of reserpic acid (35.156) with potassium hydroxide (Equation 35.26).

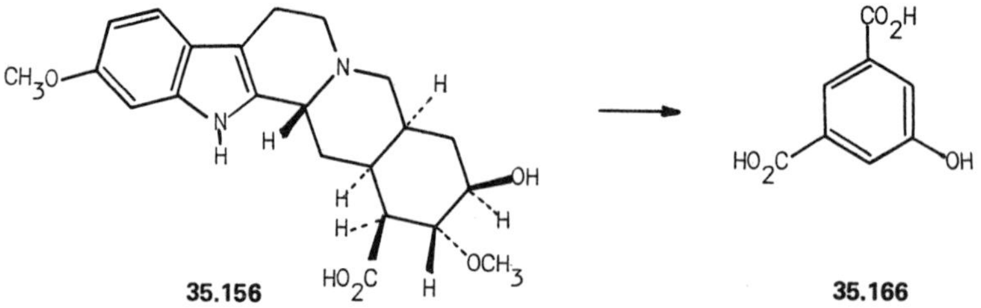

35.156 35.166

(Eq. 35.26)

When methyl reserpate (35.158) is treated with p-toluenesulfonic acid chloride (tosyl chloride) and the resultant tosylate (35.167) reduced with lithium aluminum hydride, a primary alcohol, reserpinol (35.168), is obtained. Selenium dehydrogenation of reserpinol (35.168) generates, inter alia, 11-hydroxy-6'-methylyobyrine (35.169; Scheme 35.27), the O-methyl ether of which was synthesized, as indicated above for the O-methyl ether of yobyrine (35.162), starting with harmine (35.164) and 2,6-dimethylcyclohexanone [297]. Additionally, since treatment of reserpic acid (35.156) with acetic anhydride leads to a γ-lactone (35.170; Equation 35.27), the carboxylate and hydroxyl functions in the E ring must bear a

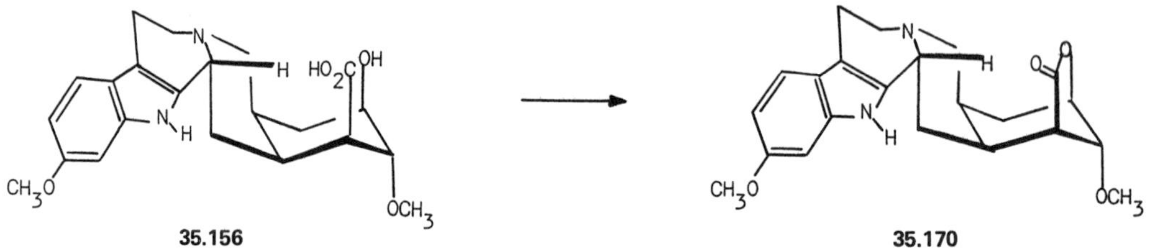

35.156 35.170

(Eq. 35.27)

cis-1,3-diaxial relationship. The methoxy group in the E ring must lie between the carboxyl function and the hydroxyl group of reserpic acid (35.156) since treatment of the tosylate 35.168 with collidine causes elimination to an enol ether which yields reserpone (35.171) and an isomer (35.172) on acid hydrolysis. The enol ether, methyl anhydroreserpate, is generally formulated as 35.173 rather than 35.174 which might have been expected to be the result

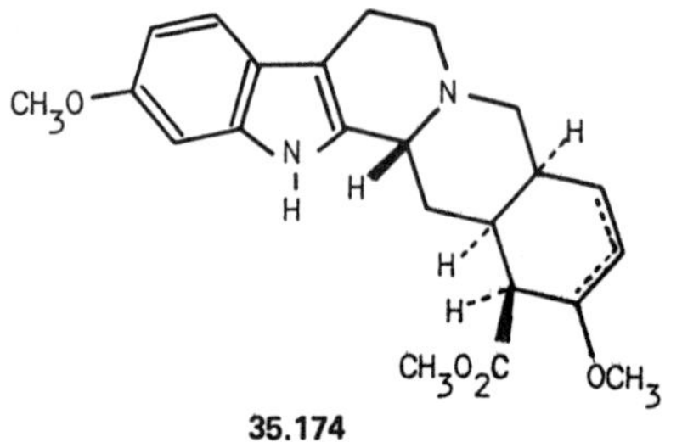

35.174

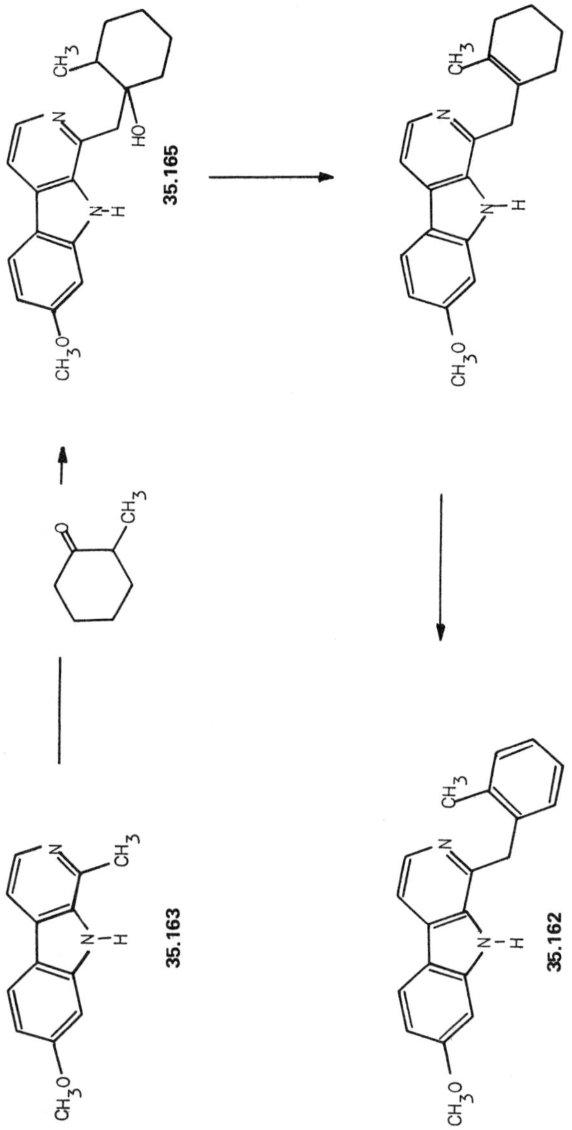

SCHEME 35.26 The synthesis of the O-methyl ether of the hydroxyyobyrine obtained on degradation of reserpic acid.

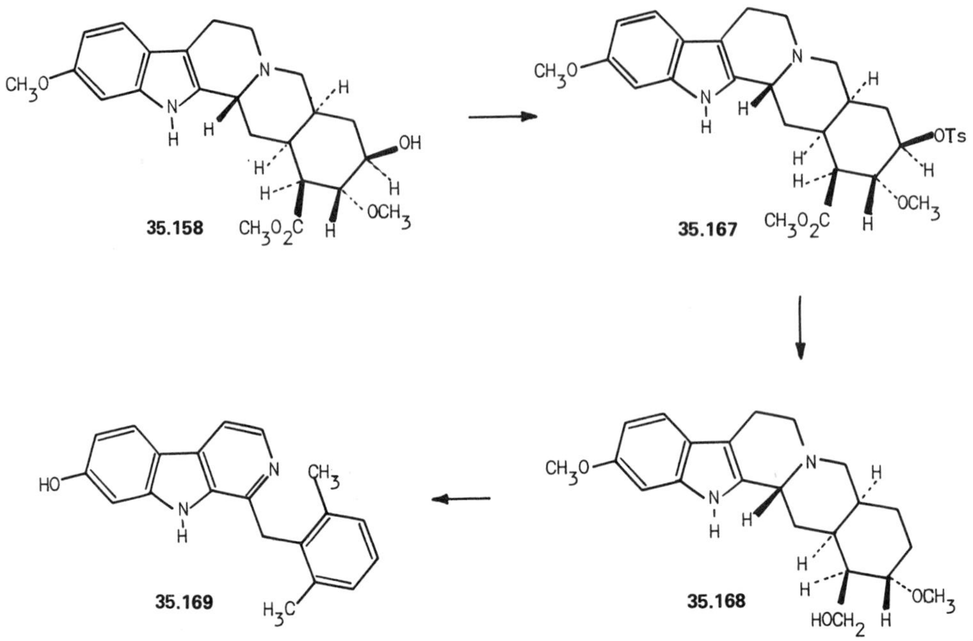

SCHEME 35.27 Degradation of reserpinol from the tosylate of methyl reserpate, 11-hydroxy-6'-methylyobyrine.

of elimination because of the presence of a conjugated carboxylate function (ultraviolet).* These reactions are shown in Scheme 35.28.

Although there was a good deal of evidence to suggest that the D/E ring junction in reserpine (35.4) was cis, the proof of this was based upon synthesis of further transformation products of reserpone (11-methoxyepialloyohimbone, 35.171) and its isomer, 11-methoxyalloyohimbone (35.172). Thus, conversion of reserpone (35.171) into the corresponding thioketal and Raney nickel desulfurization yield reserpane (11-methoxyepialloyohimbane, 35.175), while similar treatment of 11-methoxyalloyohimbone (35.172) generates 11-methoxyalloyohimbane (35.176; Scheme 35.29).

These could be synthesized ([298]; Scheme 35.30) by allowing harmine (35.164) to react with isopropoxymethylenecyclohexanone (35.177) to yield 11-methoxysempervirine (35.178) and reduction of the latter over a platinum catalyst with hydrogen to (±)-11-methoxyalloyohimbane (35.176). Long heating of (±)-11-methoxyalloyohimbane (35.176) in acetic acid causes it to undergo epimerization to (±)-reserpane (11-methoxyepialloyohimbane, 35.175). It is interesting to note that despite the transformation of 11-methoxyalloyohimbane (35.176) into 11-methoxyepialloyohimbane (35.175) and the formation of both reserpone (11-methoxyepialloyohimbone, 35.171) and its C_3 epimer 11-methoxyalloyohimbone (35.172) on acid hydrolysis of methyl anhydroreserpate (35.173), the presence of the additional functionality leads to a very large difference between reserpine (35.4) and its C_3 epimer 3-isoreserpine (35.177). The epimerization of reserpine (35.4) into 3-isoreserpine (35.177) occurs readily

*It is not quite obvious how 35.173 comes about unless a hydride shift from C_{17} to C_{18} is permitted, for which the geometry is wrong, while the neighboring methoxy assists in the tosylate leaving (see below).

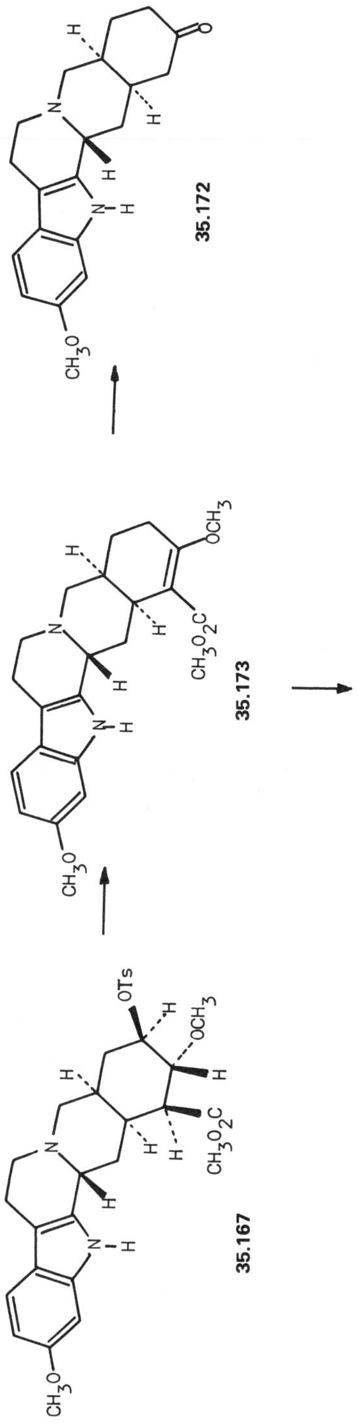

35.167

35.173

35.172

35.171

SCHEME 35.28 The degradation of the tosylate of methyl reserpate to reserpone and its C_3 isomer.

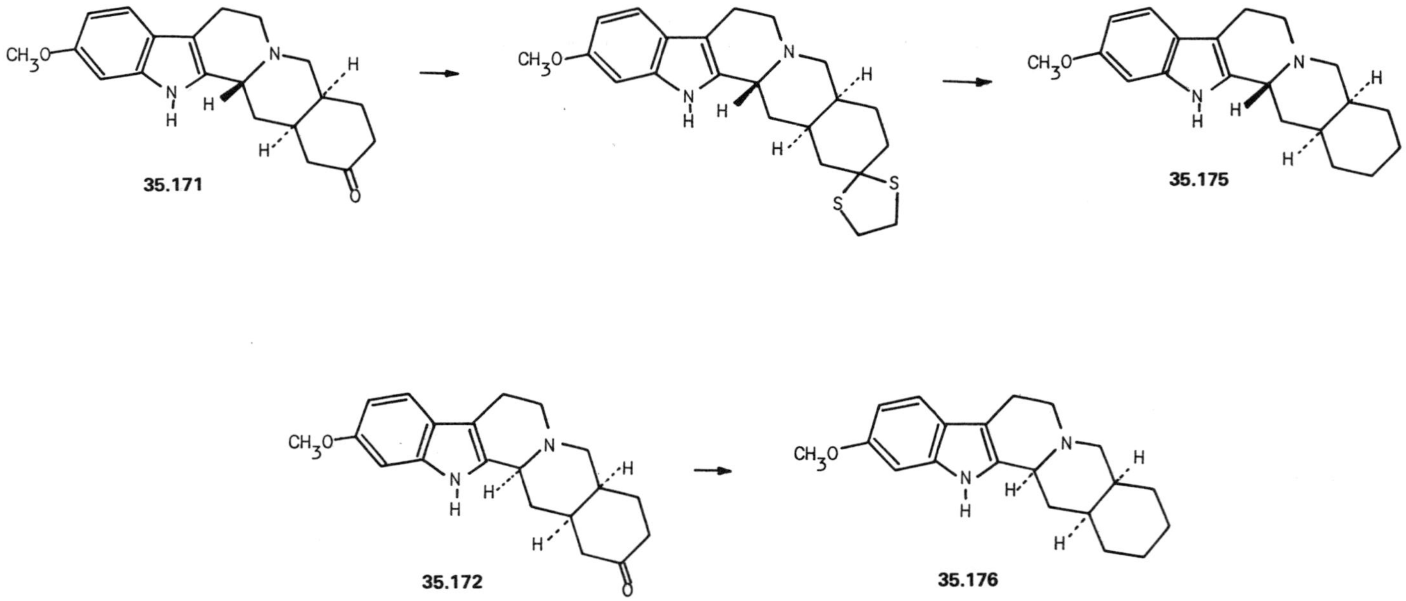

35.171 → → 35.175

35.172 → 35.176

SCHEME 35.29 The degradation of reserpone and 11-methoxyalloyohimbone to reserpane and 11-methoxyalloyohimbane, respectively.

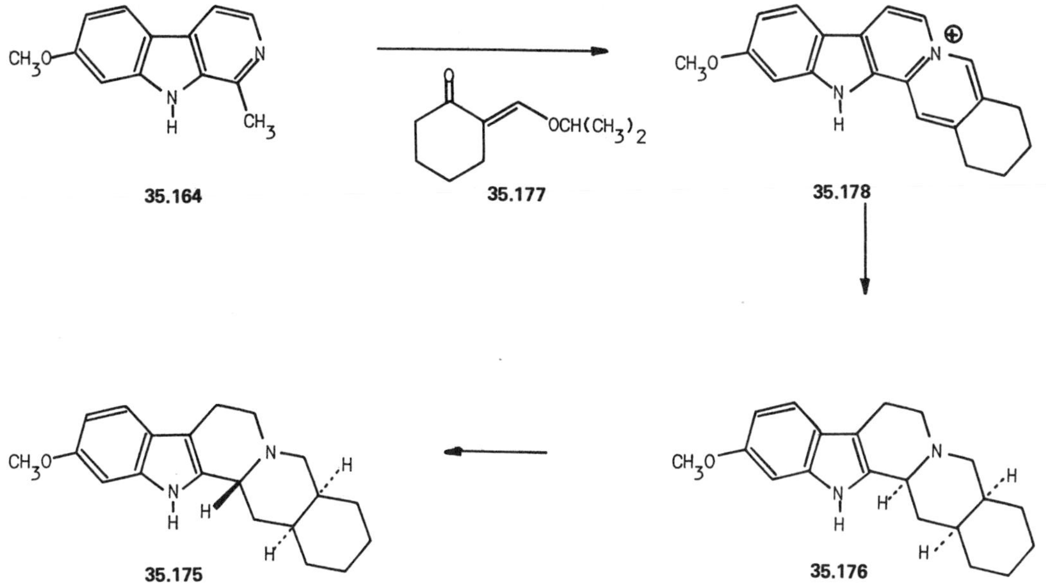

35.164

35.177

35.178

35.175

35.176

SCHEME 35.30 The synthesis of (±)-reserpane and (±)-11-methoxyalloyohimbane. (After Reference 298.)

and irreversibly on just heating in acetic anhydride (Equation 35.28). Further, while reserpine (35.4) readily forms the γ-lactone 35.170 (Equation 35.27), 3-isoreserpine (35.173) fails to do this. Since γ-lactone formation demands that the carboxyl group at C_{16} and the alcohol at C_{18} be diaxial on the E ring, the assignment of the proton at C_3 in reserpine (35.4)

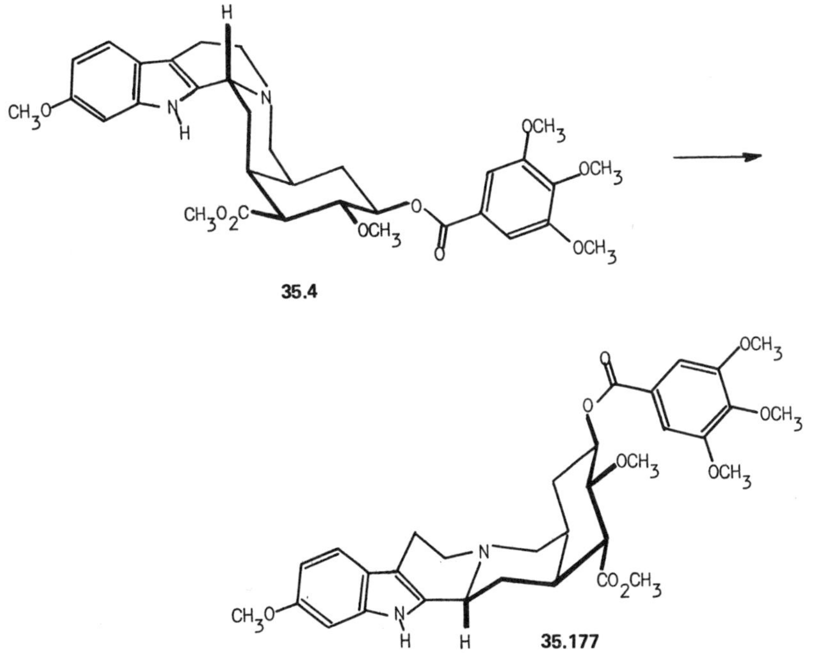

35.4

(Eq. 35.28)

35.177

as β (or up) and that in 3-isoreserpine (35.177) as α (or down), follows. Thus, for the hydroxyl and carboxylate functionalities to become cis diaxial in 3-isoreserpine (35.177) the D ring must also invert, putting the bulky methoxyindole group axial. Since this will create severe steric problems, the lactone formation is precluded.

Finally, the methoxy group at C_{17} is also placed axial since it was noted that the major product isolated during the formation of methyl anhydroreserpate (35.173; Scheme 35.28) from the tosylate 35.167 is a quaternary tosylate salt, formulated as 35.178 (Scheme 35.31). In order for such a product to form, given the α-proton at C_3 and the axial configuration already assigned to the hydroxyl at C_{18}, it is presumed (a) that the conformation of 35.167 with the substituents axial on ring E obtains; (b) that the methoxy participates as a neighboring group (for which the vicinal diaxial relationship is ideal) in the leaving of the tosylate (e.g., 35.179); (c) that epimerization at C_3 occurs to put the hydrogen there axial;* and (d) the re-displacement of the methoxonium ion occurs by the tertiary nitrogen in forming the salt 35.178. The relationships suggested by the chemical information thus provided are further substantiated by proton magnetic resonance spectroscopy [299], and the absolute configuration of reserpine (35.4) at C_{18} has been defined as R by the isolation of R-(-)-atrolactic acid (35.180) from the reaction of the phenyl glyoxylate ester of methyl reserpate (35.158) with methyl magnesium iodide (Scheme 35.32; [272]; see also p. 507).

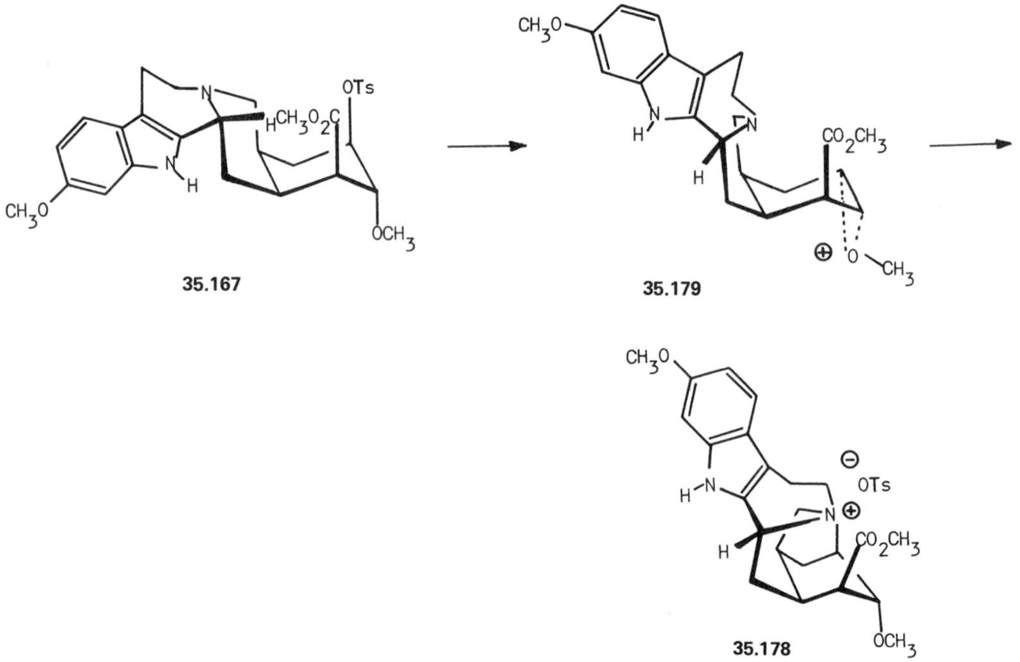

SCHEME 35.31 A pathway for the major product obtained on decomposition of the tosylate of methyl reserpate.

*This should suggest, as is in fact the case, that the same enol ether, formulated as methyl anhydroreserpate (35.173), is obtained from methyl reserpate tosylate (35.167) and from its C_3 epimer.

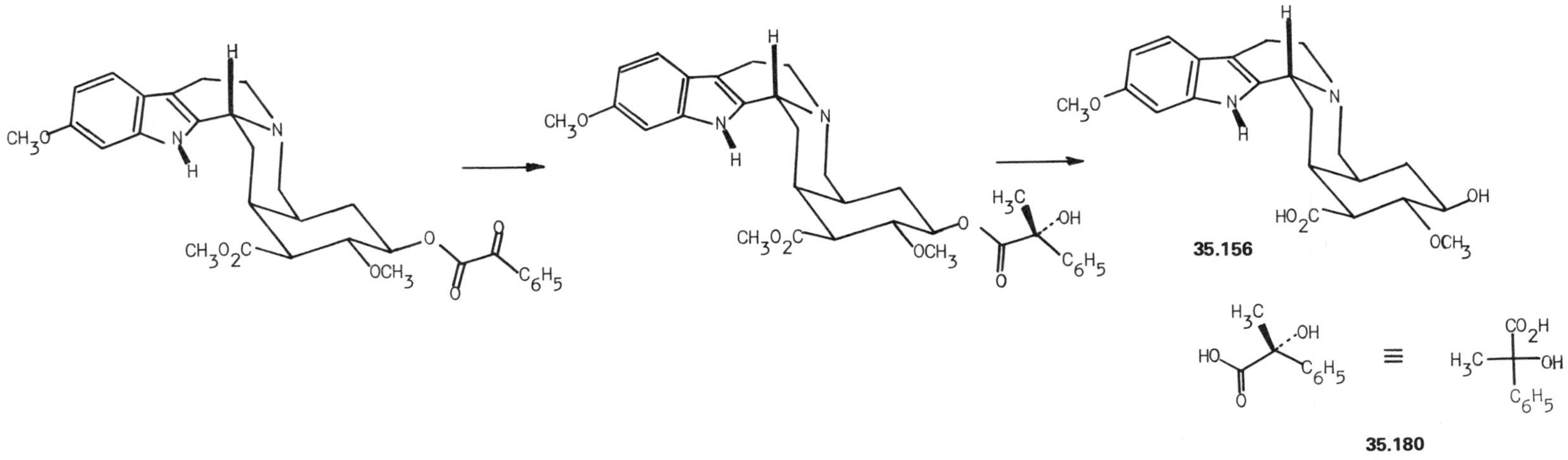

35.156

35.180

SCHEME 35.32 The isolation of R-(-)-atrolactic acid from the Grignard reaction of methyl magnesium iodide with the phenyl glyoxylate ester of methyl reserpate. (After Reference 272.)

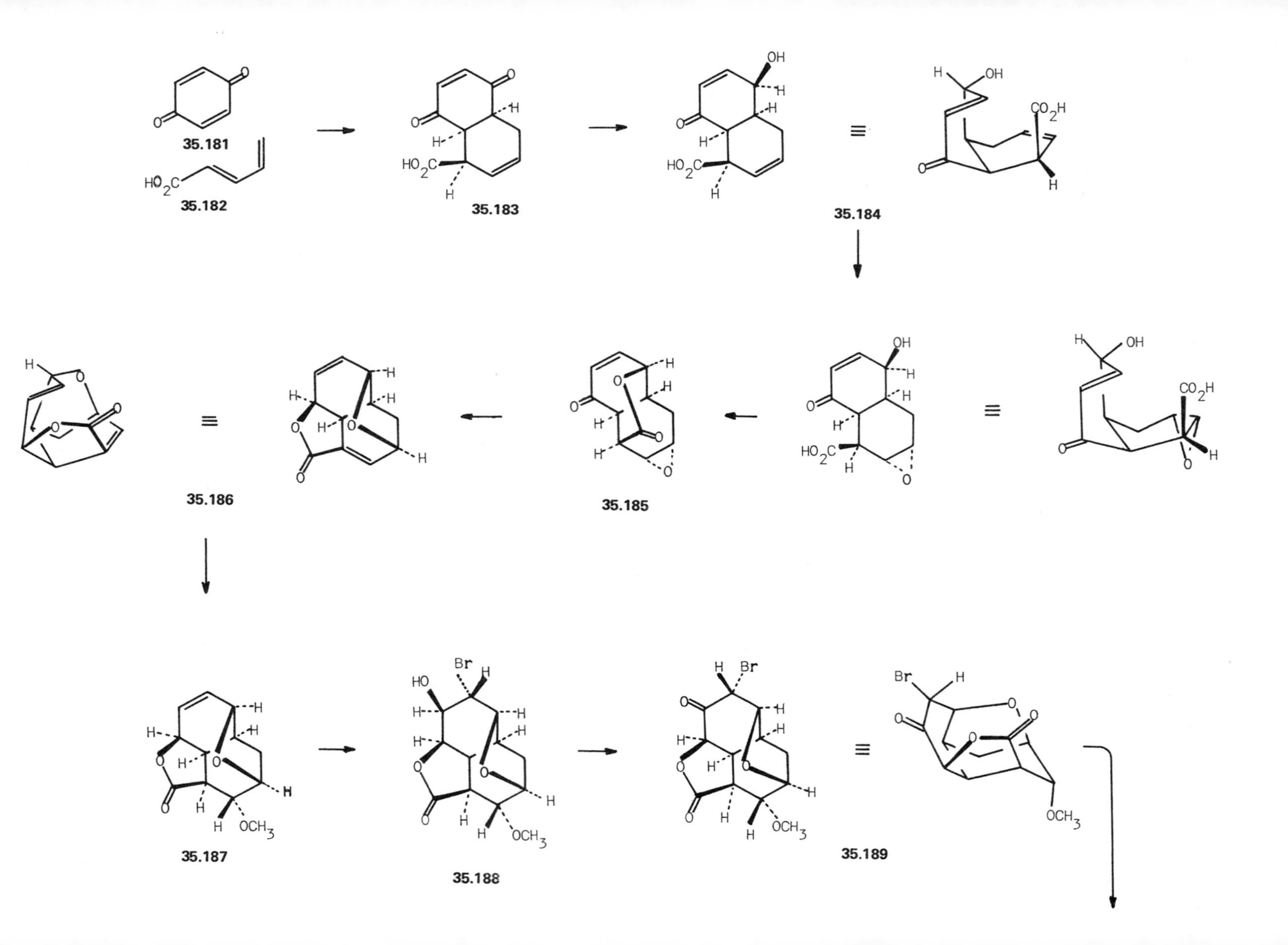
35.181
35.182
35.183
35.184
35.185
35.186
35.187
35.188
35.189

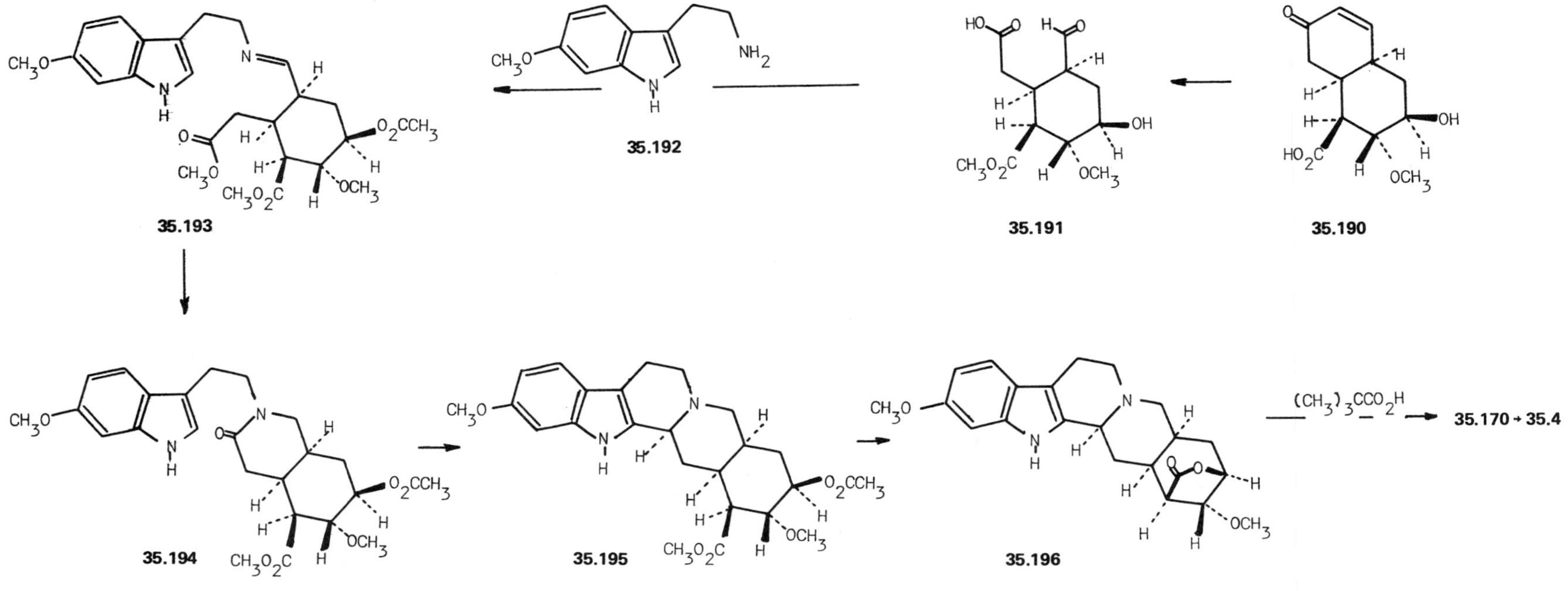

SCHEME 35.33 The total synthesis of reserpine. (After Reference 300.)

Finally, reserpine (35.4) has been synthesized ([300]; Scheme 35.33). p-Benzoquinone (35.181) undergoes a Diels-Alder reaction with vinyl acrylic acid (35.182) to yield the cis adduct 35.183. Reduction of 35.183 with sodium borohydride yields, specifically, the hydroxyketone 35.184. (This follows from the fact that epoxidation of 35.184 with perbenzoic acid followed by lactonization with acetic anhydride generates only 35.185. Had the epimer formed during reduction of 35.183, lactonization could not have occurred.) Next, reduction of the ketone carbonyl of 35.185 with aluminum isopropoxide in isopropyl alcohol is accompanied by opening of the lactone, which recloses to a new lactone at the hydroxyl group just formed, while the newly liberated hydroxyl oxygen opens the epoxide backside and the last-formed hydroxyl, being β to the carbonyl, is lost. The product of this series of transformations, therefore, is the alkene 35.186. Then, methoxide in methanol, attacking from the least hindered side, converts 35.186 into the lactone 35.187. Conversion of 35.187 into the corresponding bromohydrin (35.188) with N-bromosuccinimide in aqueous sulfuric acid followed by oxidation with chromic anhydride in acetic acid produces the bromoketone 35.189. When the ketone (35.189) is reduced with zinc in acetic acid, dehalogenation is accompanied by reduction of the lactone and opening of the ether: the unsaturated ketoacid 35.190 results. Esterification of the acid 35.190 with diazomethane and then acetic anhydride, followed by osmium tetroxide oxidation and periodate cleavage of the diol, yields the acid 35.191. Finally, esterification of the acid 35.191 with diazomethane and condensation of that ester with 6-methoxytryptamine (35.192) yields an imine (35.193) which is immediately reduced with sodium borohydride to the lactam 35.194. Bischler-Napieralski cyclization of 35.194 with phosphorus oxychloride followed by another reduction with sodium borohydride leads to (±)-methyl-O-acetylisoreserpate which is resolved with di-p-tolyl-(-)-tartaric acid to yield (-)-methyl-O-acetylisoreserpate (35.195), identical to that obtained from reserpine (35.14) itself. Hydrolysis of (-)-methyl-O-acetylisoreserpate (35.195) followed by heating with N,N-dicyclohexylcarbodiimide in pyridine yields the isoreserpic acid lactone (35.196), and this is epimerized by heating with pivalic acid (35.197) in xylene into reserpic acid lactone (35.170). Methanolysis of the lactone (35.170) then yields methyl reserpate (35.158) which (Scheme 35.25) can be converted, with 3,4,5-trimethoxybenzoyl chloride (acid chloride of 35.157), into reserpine (35.4) itself.

V. AJMALINE

Ajmaline (35.5), the major alkaloid of Rauwolfia serpentina Benth., was obtained originally in 1931 [301] but it was not until 1949 [302] that the first concrete information concerning its nature was forthcoming. Ajmaline (35.5), $C_{20}H_{26}N_2O_2$, usually crystallizes as a solvate

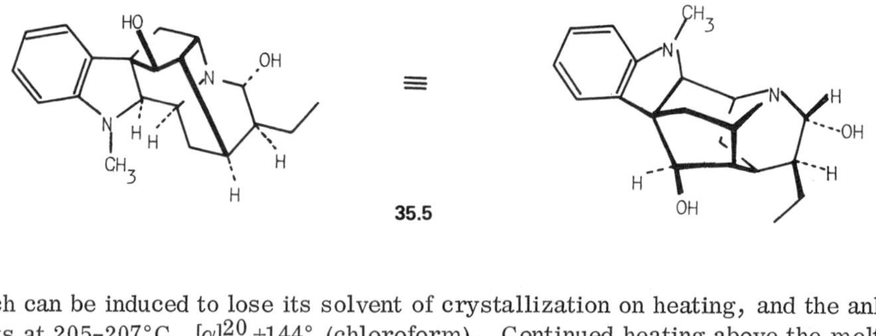

35.5

which can be induced to lose its solvent of crystallization on heating, and the anhydrous base melts at 205-207°C, $[\alpha]_D^{20}$ +144° (chloroform). Continued heating above the melting point or treatment of ajmaline (35.5) with refluxing ethanolic potassium hydroxide causes isomerization to isoajmaline (35.198), $C_{20}H_{26}N_2O_2$, mp 264-266°C, $[\alpha]_D^{20}$ +72.8° (ethanol; Equation

35.29), which may also be a naturally occurring base and which undergoes virtually all the same reactions as does ajmaline (35.5).

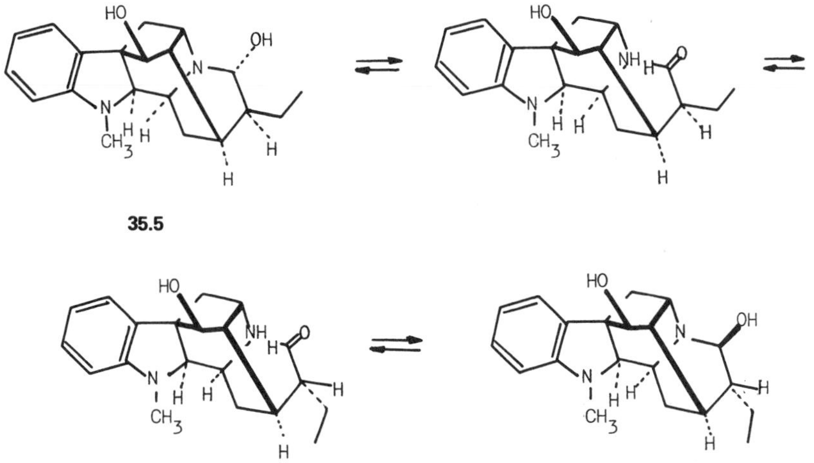

35.5

(Eq. 35.29)

35.198

Ajmaline (35.5; and, of course, its isomer, 35.198), which contains one N-methyl and at least one C-methyl, forms a dihydrochloride salt on treatment with hydrogen chloride and a monoquaternary methiodide (35.199; Equation 35.30) with methyl iodide. The quaternary

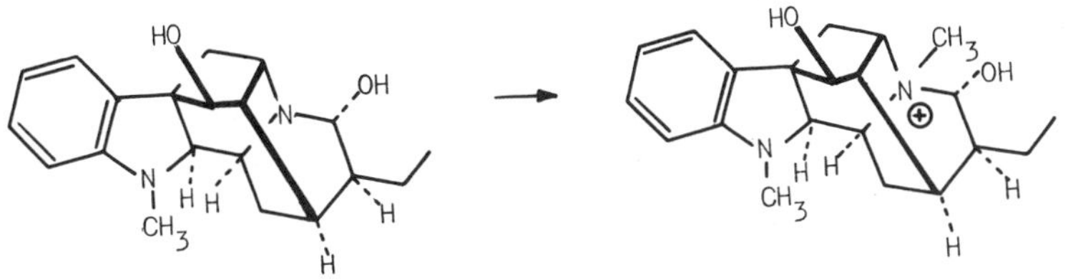

(Eq. 35.30)

35.5

35.199

iodide (35.199) is still basic, implying that both nitrogen atoms in ajmaline (35.5) itself are tertiary. The ultraviolet spectrum of ajmaline (35.5) shows absorbances [λ_{max} 248 nm (ϵ = 8,700) and 292 nm (ϵ = 3,100)] characteristic of ditertiary dihydroindole systems, and the ultraviolet spectra of isoajmaline (35.198) and the quaternary salt (35.199) are virtually the same as ajmaline (35.5) itself. Thus, it was deduced that isomerization of ajmaline (35.5) to isoajmaline (35.198) occurs away from the chromophore and that the indoline nitrogen is not involved in formation of the quaternary salt 35.199. That the original N-methyl group is present on the indoline nitrogen was established by isolation of (a) the N-methylharman 35.200 from soda lime distillation of ajmaline (35.5; Equation 35.31) and (b) the acetone adduct of N-methyl isatin (35.201) from permanganate oxidation of ajmaline (35.5) in acetone solution (Equation 35.32).

Apart from the presence of an aromatic ring which was rather easily reduced over Adam's catalyst, no other unsaturation could be detected, which means that ajmaline (35.5) must be hexacyclic. Additionally, since both nitrogen atoms are tertiary and since ajmaline (35.5) possesses two active hydrogens as well as hydroxyl absorption in the infrared, it was concluded that at least one, and perhaps two, hydroxyl groups must be present. That the

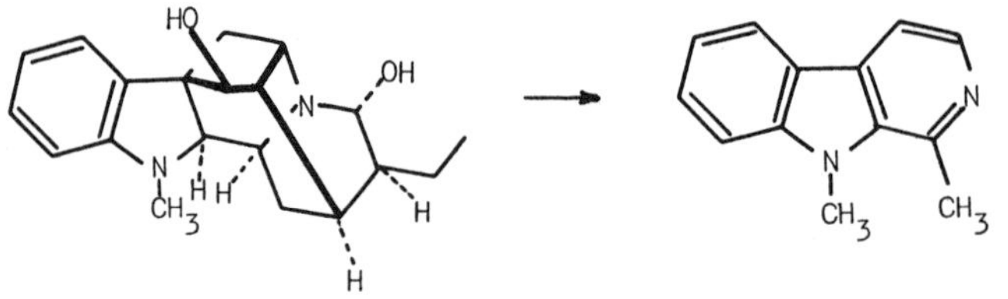

35.5 → 35.200 (Eq. 35.31)

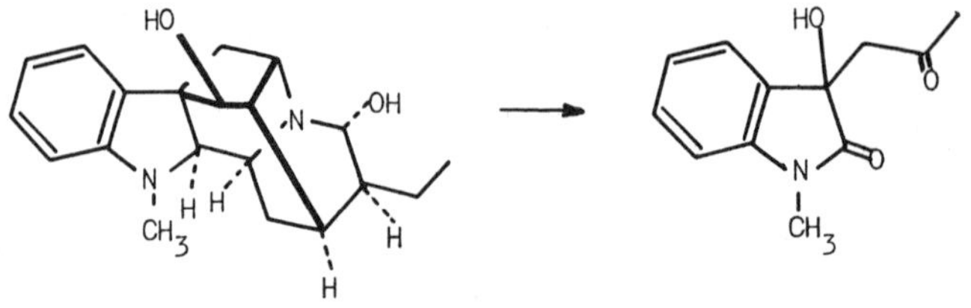

35.5 → 35.201 (Eq. 35.32)

latter is correct is confirmed by the formation of a diacetyl derivative which still yields a methiodide and possesses the intact dihydroindole nucleus. Hydrolysis of the diacetyl derivative regenerates ajmaline (35.5). It was, however, recognized that there was something unusual about the relationship between at least one of the hydroxyl groups and at least one of the nitrogen atoms since the pK_a of diacetylajmaline is only 4.9, while ajmáline itself has a pK_a of 8.15. Then, when it was found that ajmaline (35.5) forms an oxime (35.202; although it has no carbonyl absorption in the infrared, when the spectrum is taken in a nujol mull, and it cannot be reduced with lithium aluminum hydride) and reduces Tollen's reagent, the presence of a carbinolamine grouping was suggested. As to why this carbinolamine does not undergo loss of water to the corresponding imine, it was suggested that it is probable that the double bond which would have been introduced in that process is at a bridgehead. Subsequent transformations of ajmaline (35.5) confirmed the presence of the carbinolamine and led to the currently accepted structure.

The oxime of ajmaline (35.202, chanoajmaline oxime) undergoes dehydration to the nitrile 35.203 (Equation 35.33), and ajmaline (35.5) itself can be transformed into deoxydihydrochanoajmaline (35.204) by Wolff-Kishner reduction (Equation 35.34). Therefore, as

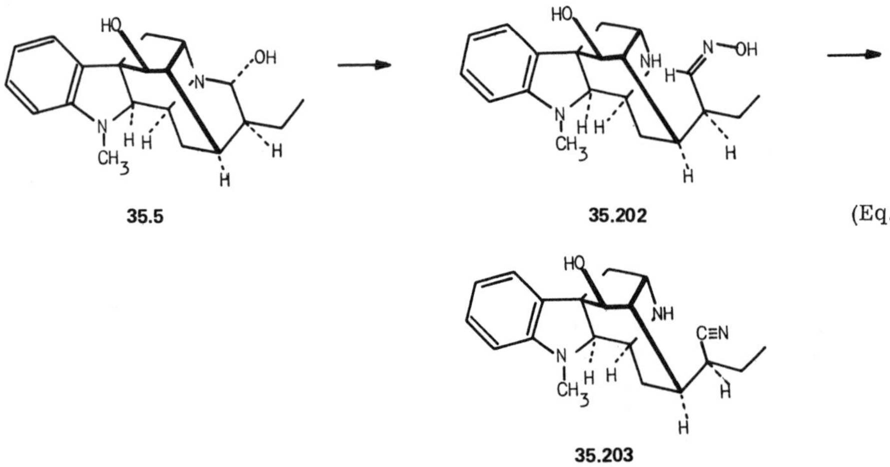

35.5 → 35.202 → 35.203 (Eq. 35.33)

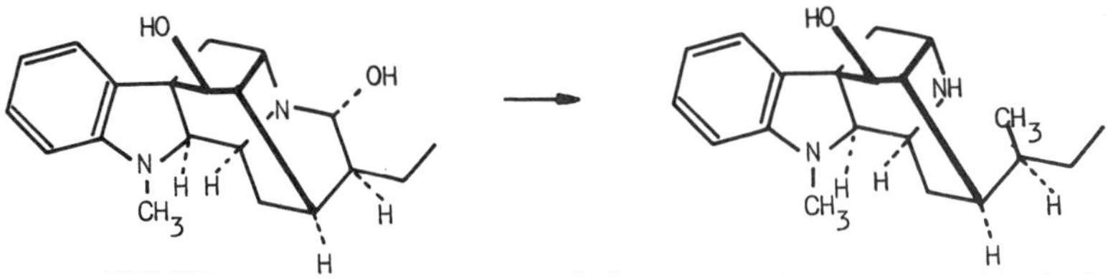

(Eq. 35.34)

35.5 35.204

is expected, ajmaline methiodide (35.199) can be converted to a free base (35.205) which possesses carbonyl absorption in the infrared (Equation 35.35). Finally, along these lines, ajmaline (35.5) can be reduced by borohydride to dihydrochanoajmaline (35.206), the

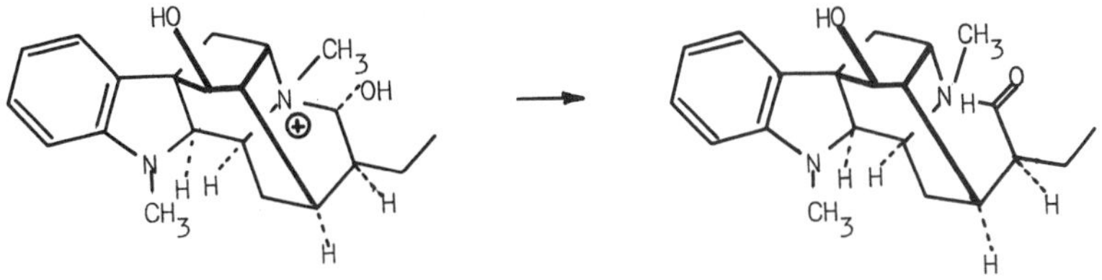

(Eq. 35.35)

35.199 35.205

hydrobromide of which forms 21-deoxyajmaline (35.207, as its hydrobromide) on pyrolysis, and ajmaline (35.5) can be decarbonolated on refluxing in xylene with Raney nickel to form decarbonochanoajmaline (35.208; Scheme 35.34).

Now, the other hydroxyl group (at C_{17}) of ajmaline (35.5) which had proved resistant to all but acetylation reactions, is capable, in 21-deoxyajmaline (35.207), of being oxidized. Under modified Oppenauer conditions (48 hr with potassium t-butoxide and benzophenone), a ketone (35.209) is formed, while with lead tetraacetate, an aldehyde (35.210) results. This aldehyde (35.210) possesses an indole chromophore (Scheme 35.35). Oxidation of dihydrochanoajmaline (35.206) with the same reagent generates a semiacetal (35.211; Equation 35.36).

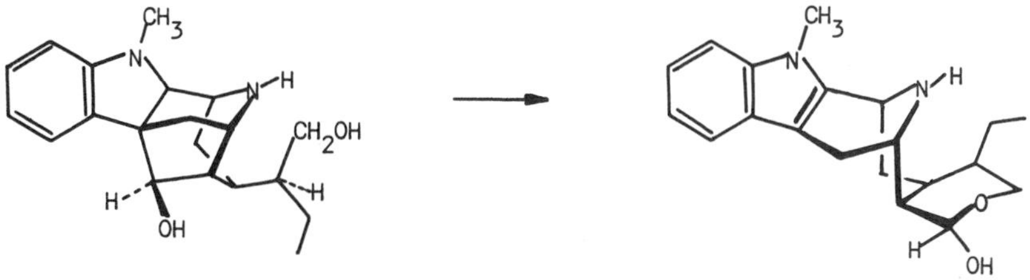

(Eq. 35.36)

35.206 35.211

Dehydrogenation of deoxydihydrochanoajmaline (35.204) with palladium on carbon yields a mixture of bases which was separated into N-methylharman (35.200) and two, more elaborate, bases: (a) ajarmine (35.212), $C_{20}H_{26}N_2$, and (b) ajmyrine (35.213), $C_{19}H_{22}N_2$ (Scheme 35.36; [303]). Ajarmine (35.212), which was recognized as a carboline derivative, (by ultraviolet spectroscopy) was synthesized (Scheme 35.37) by condensing the lithium derivative of N-methylharman (35.200) with ethyl isobutyl ketone (35.214) to the alcohol 35.215. Dehydration of 35.215 with phosphorus pentoxide yields an alkene that can be reduced with hydrogen over a platinum catalyst to yield ajarmine (35.212; [304]). Ajmyrine (35.213) was

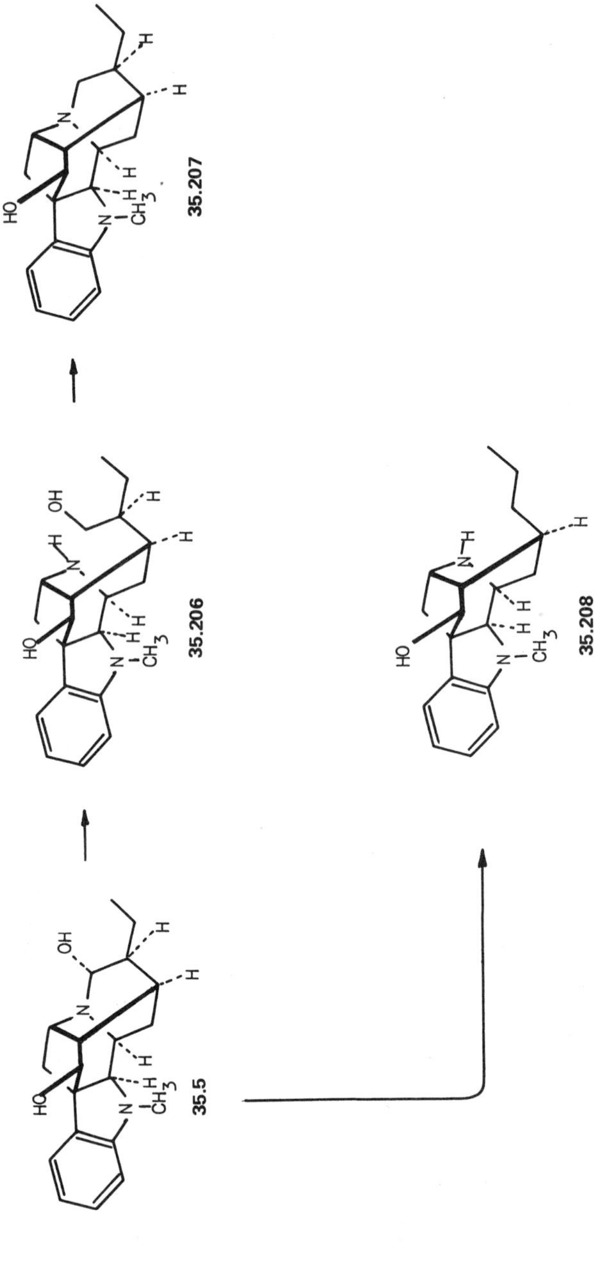

SCHEME 35.34 The transformation of ajmaline into 21-deoxyajmaline and into decarbonochanoajmaline.

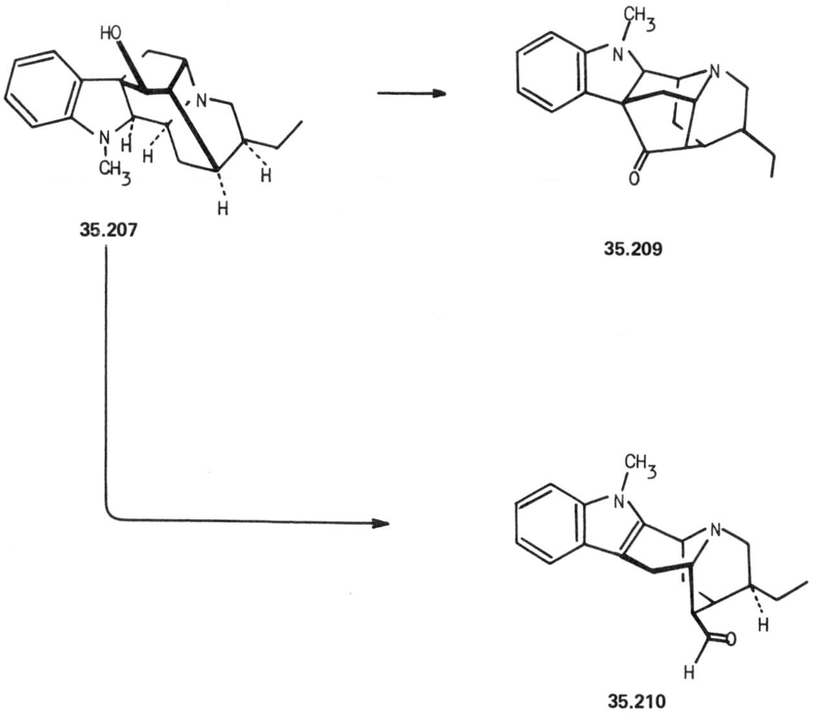

HO N H H CH3 H N H H 35.207

CH3 N N O 35.209

CH3 N N H H O 35.210

SCHEME 35.35 The oxidation of 21-deoxyajmaline to 21-deoxyajmalone and to deoxyajmalal-B.

also successfully synthesized (Scheme 35.38; [304]) by conversion of 4-nitropicoline N-oxide (35.216) to the corresponding nitrile (35.217), which after reduction to 4-cyano-2-methylpyridine is permitted to condense first with ethyl magnesium bromide and then with methyl magnesium iodide to generate the tertiary alcohol 35.218. Dehydration and reduction of 35.218 leads to 4-isobutylpyridine (35.219). Condensation of the latter (35.219) with N-methylisatin (35.220) in the presence of butyl lithium generates the oxyindole 35.221 in about 15% yield. Dehydration of 35.221 is accomplished with acetic anhydride, and the resultant alkene (35.222) is reduced first catalytically and then with lithium aluminum hydride to a mixture of indole and dihydroindole bases. The mixture, on palladium-catalyzed dehydrogenation, yields ajmyrine (35.213). The structure for ajmaline (35.5) suggested by this information is further substantiated by a detailed examination of its stereochemistry [305].

The absolute stereochemistry of ajmaline (35.5) was established by interrelation of ajmaline (35.5) with tetradehydrodihydrocorynantheane (35.131; see p. 528), the absolute stereochemistry of which was known. Thus, isoajmaline (35.198), as already noted, undergoes all the same reactions as does ajmaline (35.5) and is clearly epimeric with ajmaline (35.5) at C_{20}. Therefore, 21-deoxyajmaline (35.207) is epimeric at C_{20} with 21-deoxyisoajmaline (35.223). It is not surprising (Scheme 35.39), again by analogy with the chemistry of 21-deoxyajmaline (35.207; see, e.g., Scheme 35.35), that oxidation of 21-deoxyisoajmaline (35.223) with lead tetraacetate generates an aldehyde (35.224) epimeric at C_{20} with the aldehyde (35.210) mentioned earlier and shown in Scheme 35.35. The aldehyde (35.224) is called deoxyisoajmalal-A and can be readily isomerized to deoxyisoajmalal-B (35.225). Reduction of the isomerized aldehyde (deoxyisoajmalal-B, 35.225) with borohydride yields the corresponding alcohol, the tosylate of which (35.226) on heating in collidine undergoes fragmentation

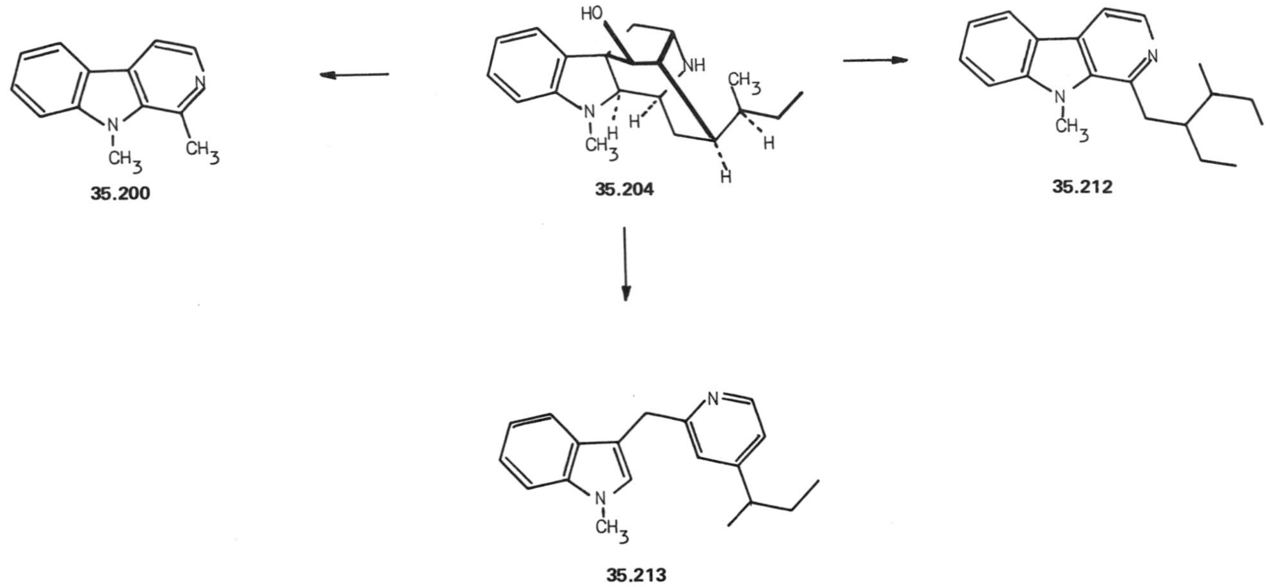
HO
NH
CH3
N
CH3
H
H
H
H
35.200
35.204
35.212
35.213

SCHEME 35.36 Products of the dehydrogenation of deoxydihydrochanoajmaline with palladium on carbon. (After Reference 303.)

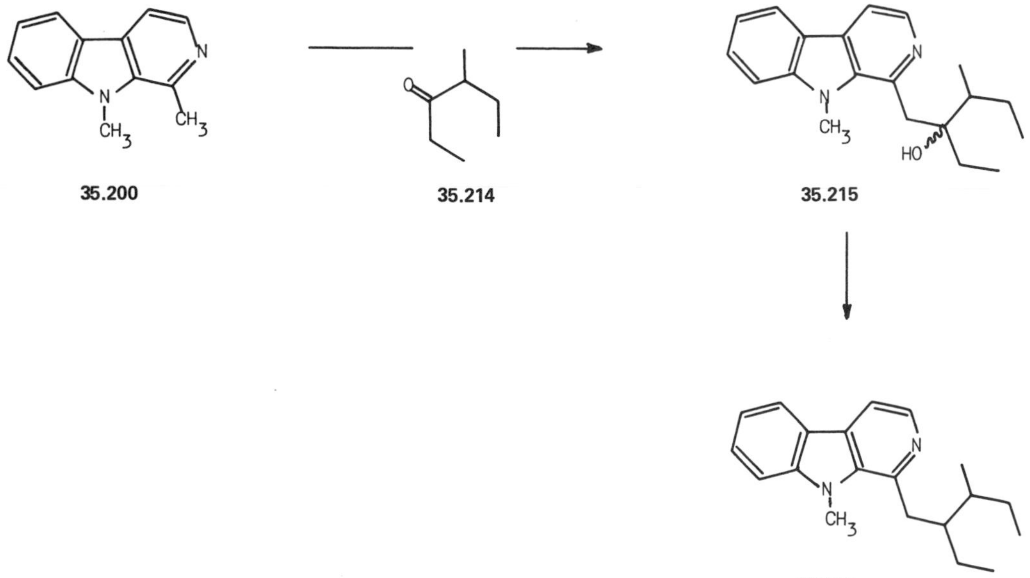

35.200 35.214 35.215 35.212

SCHEME 35.37 The total synthesis of ajarmine. (After Reference 304.)

and oxidation to the quinolizinium tosylate 35.227, which in turn can be catalytically reduced to N_a-methyltetradehydrodihydrocorynantheane (35.228), identical to the corresponding material prepared from tetradehydrodihydrocorynanthane (35.131). The methylated C_{20} epimer of 35.131 (N_a-methyltetradehydrocorynantheidane) was analogously prepared from ajmaline (35.5) itself.

The configuration of the hydroxyl at C_{17} of ajmaline (35.5) was established as shown by noting that deoxyajmalone (35.209) does not yield 21-deoxyajmaline (35.207) on reduction (catalytic or hydride). Instead, the epimeric alcohol, oxidizable back to deoxyajmalone (35.209), results. Thus, since hydride (or hydrogen) is doubtlessly delivered from the least hindered side of the carbonyl group, in 21-deoxyajmaline (35.207) the hydroxyl group must occupy that position.

Finally, the position of the hydrogen at C_2 of ajmaline (35.5) was demonstrated (Scheme 35.40) by noting that catalytic reduction of one of the chromic anhydride oxidation products of the O-acetate of 21-deoxyajmaline (35.207), i.e., 1-demethyl-Δ^1-deoxyajmaline-O-acetate, 35.229, yields 2-epi-21-deoxyajmaline (35.230) on catalytic reduction followed by N-methylation and hydrolysis. Since reduction to 35.230 doubtlessly proceeds to generate the isomer with the hydrogen at C_2 on the least hindered side (i.e., β) the α-C_2 epimer must represent ajmaline (35.5).

(±)-Ajmaline has been synthesized ([306]; Scheme 35.41). When N-methyl-3-indoleacetyl chloride 35.231 is allowed to condense with the magnesium chelate of ethyl hydrogen-Δ^3-cyclopentenylmalonate (35.232) the ketoester 35.233 results. The O-methyl oxime of 35.233 is prepared and reduced with lithium aluminum hydride to yield a mixture of epimeric amino alcohols, only one of which, 35.234, is shown. Treatment of the dibenzoyl derivative of 35.234 as a mixture of epimers with osmium tetroxide and sodium metaperiodate causes cleavage to a dialdehyde which spontaneously closes to the carbinolamide 35.235. Cyclization of 35.235 with acetic acid catalysis generates the aldehyde 35.236, and this, via the corresponding aldoxime, leads to the nitrile 35.237. Using triphenylmethyl sodium to generate the anion α to the cyano group, alkylation of 35.237 with ethyl iodide leads to 35.238, and this, after removal of the benzoate, produces the aldehyde 35.239 on

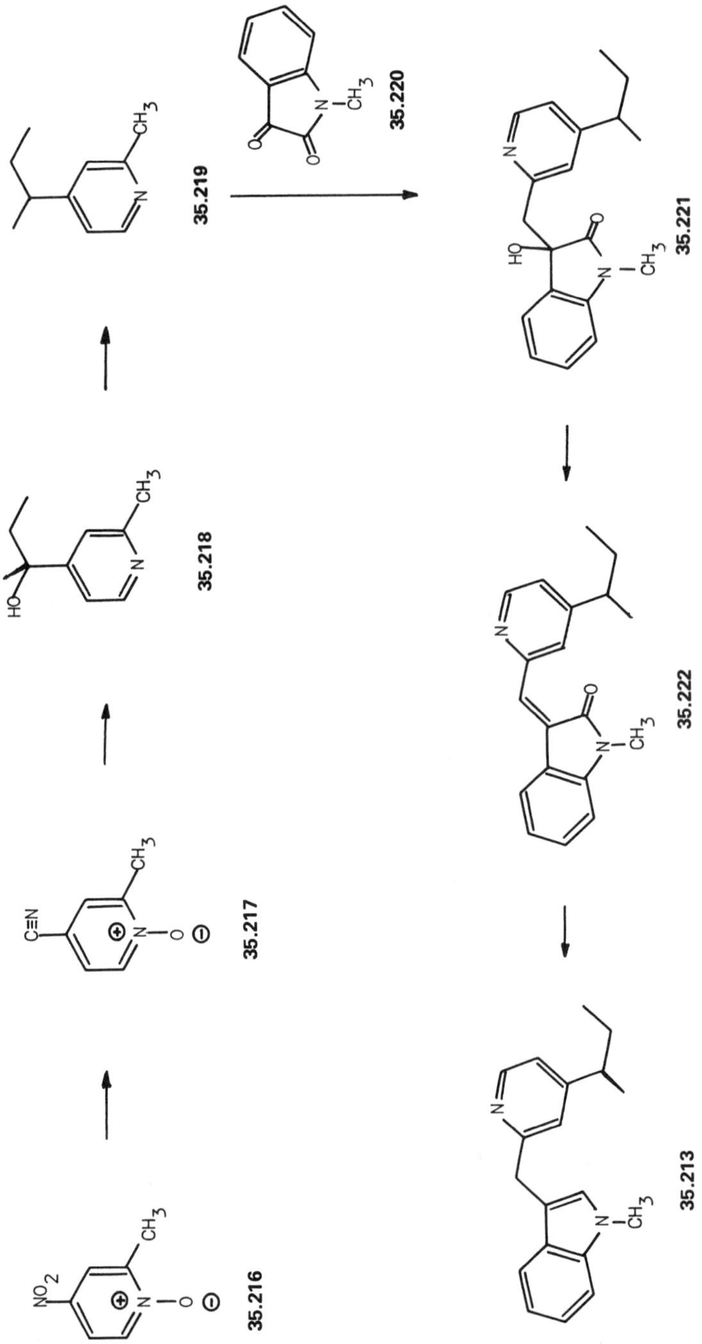

35.216 35.217 35.218 35.219 35.220 35.221 35.222 35.213

SCHEME 35.38 The total synthesis of ajmyrine. (After Reference 304.)

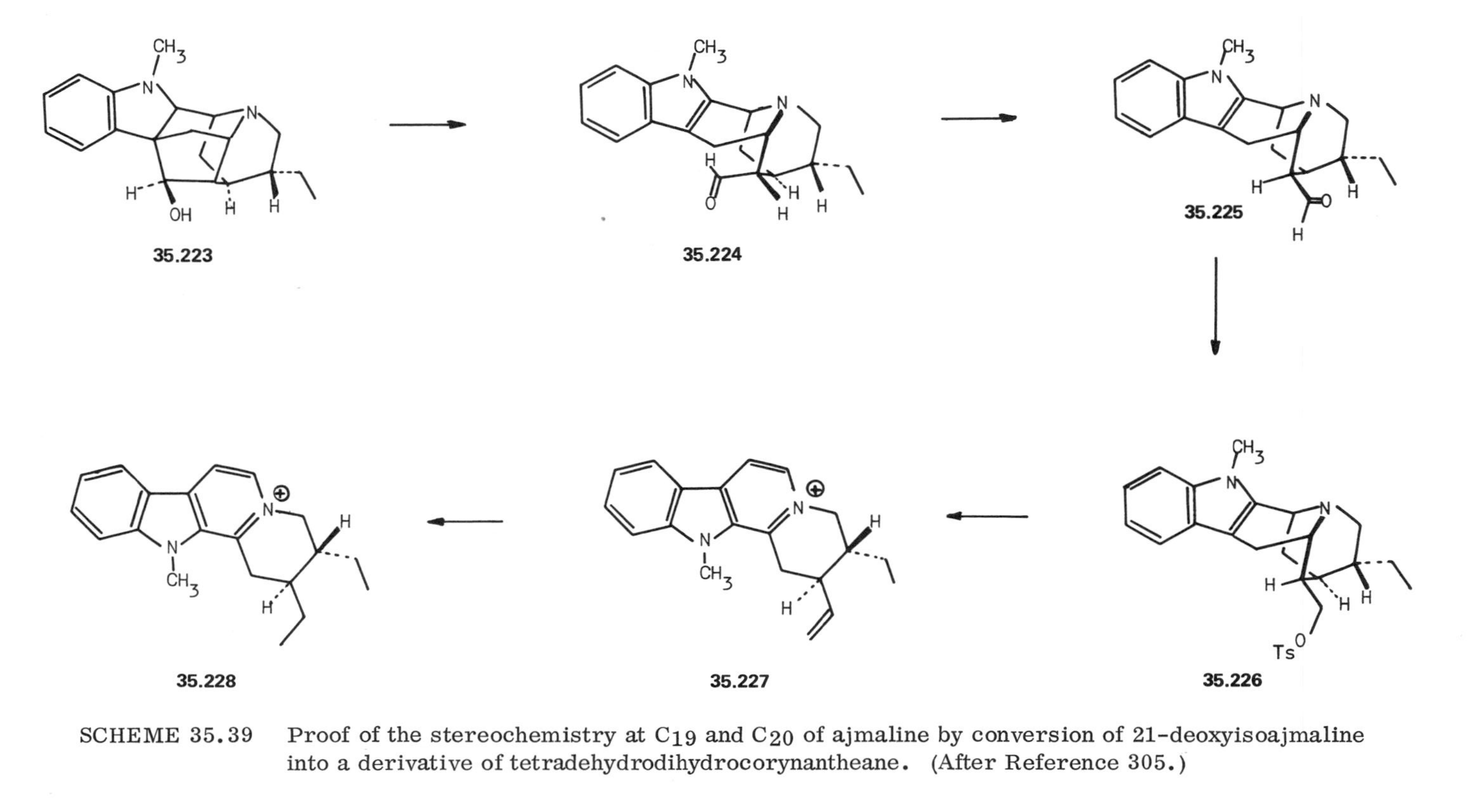

SCHEME 35.39 Proof of the stereochemistry at C_{19} and C_{20} of ajmaline by conversion of 21-deoxyisoajmaline into a derivative of tetradehydrodihydrocorynantheane. (After Reference 305.)

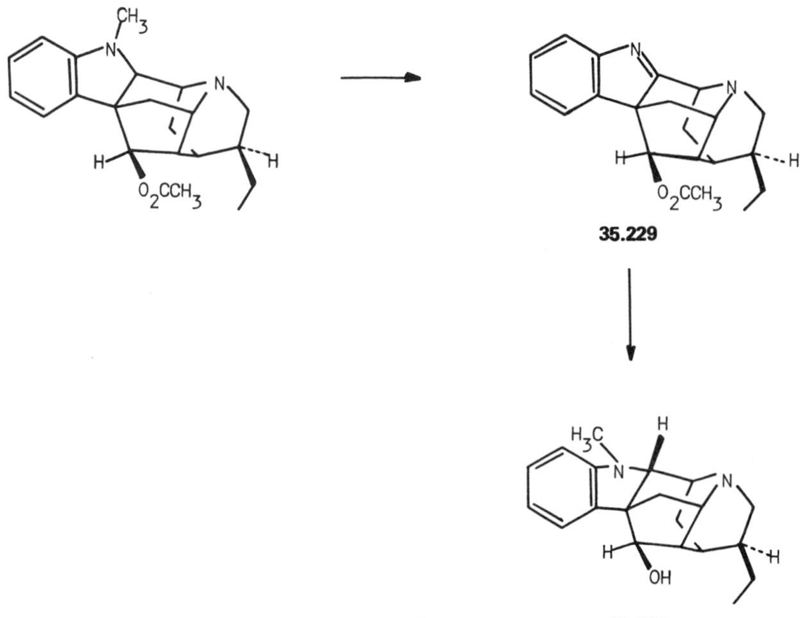

35.229

35.230

SCHEME 35.40 Demonstration that the hydrogen at C_2 of ajmaline is α by generation of the corresponding β isomer.

oxidation with dimethyl sulfoxide and acetic anhydride. When the aldehyde 35.239 is treated with acetic anhydride in a mixture of hydrochloric and acetic acids, cyclization to the acetate 35.240 occurs. Hydrogenolytic removal of the hydroxyl group at C_2 of 35.240, followed by reduction of the amide with lithium aluminum triethoxyhydride, generates the tertiary amine 35.241. After the N-benzyl group of 35.241 is removed by hydrogenolysis, the amino nitrile is reduced with lithium aluminum hydride, and (±)-ajmaline (35.5 and its mirror image) results on work-up.

VI. SARPAGINE

Although sarpagine (35.6), $C_{19}H_{22}N_2O_2$, mp 363-364°C, $[\alpha]_D^{24}$ +53.4° (pyridine) from R. serpentina Benth. possesses an ultraviolet spectrum [methanol; λ_{max} 226 nm (ϵ = 21,900); 277 nm (ϵ = 7,900)] and a shoulder at about 300 nm [307] very similar to that of a 5-methoxyindole, the functional groups present do not include an O-methyl. However, sarpagine (35.6) is phenolic and does yield an O-methyl ether (35.242) with diazomethane. The

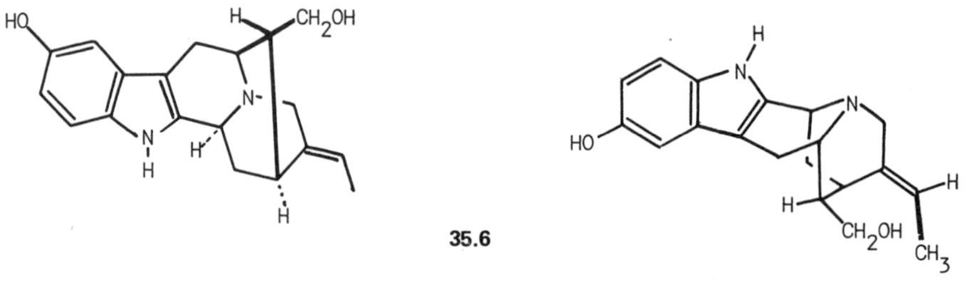

35.6

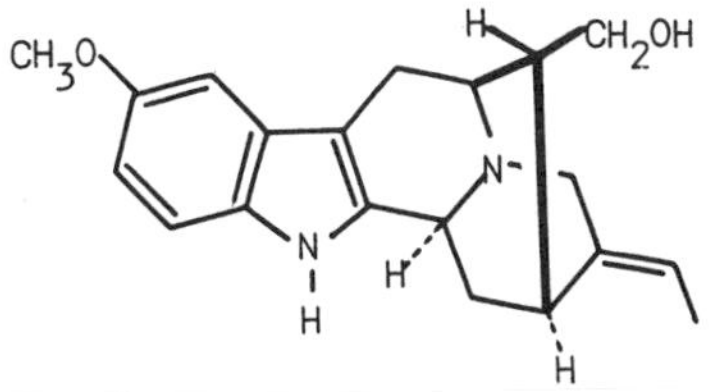

35.242

ultraviolet spectrum of the O-methyl ether (35.242) is similar to that of the base itself and thus it was correctly presumed that sarpagine (35.6) is a 5-hydroxyindole derivative [308].

Additionally, it was shown that the second oxygen present in sarpagine (35.6) is there as a primary alcohol function since (a) sarpagine (35.6) can be monoacetylated to 17-O-acetylsarpagine (35.243; Equation 35.37) simply by refluxing in acetic acid; (b) sarpagine

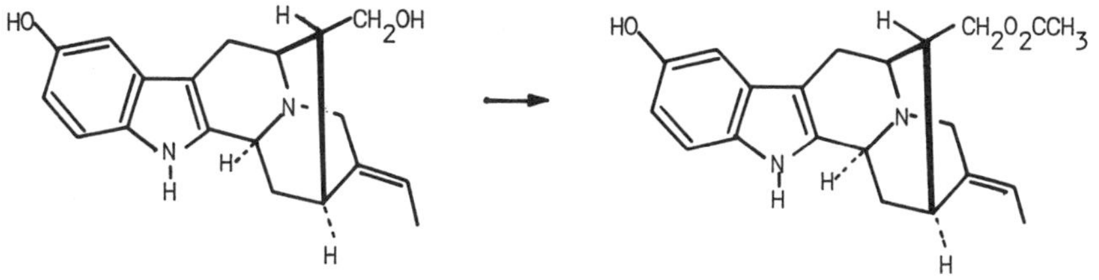

(Eq. 35.37)

35.6 35.243

(35.6) cannot be dehydrated with phosphoric acid since a primary carbocation would be generated; and (c) reduction of the tosylate of O-methylsarpagine (35.244), formed from O-methylsarpagine (35.242) and tosyl chloride with lithium aluminum hydride yields a deoxy derivative (35.245; Equation 35.38) which possesses two C-methyl groups (Kuhn-Roth) instead of only the one originally present in sarpagine (35.6) itself.

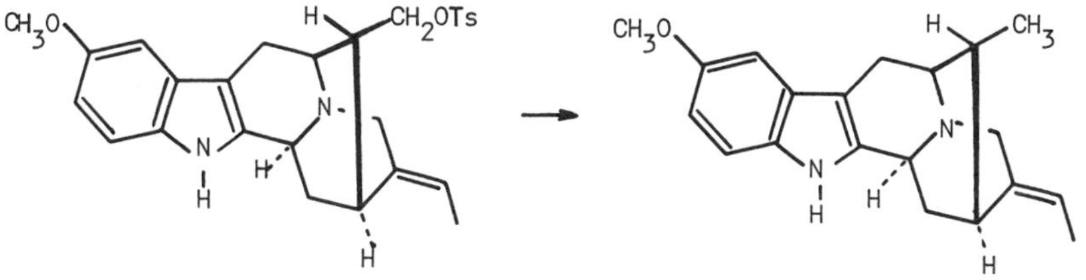

(Eq. 35.38)

35.244 35.245

When sarpagine (35.6) is treated with acetic anhydride in pyridine (Equation 35.39), O,O-diacetylsarpagine (35.246) is formed, and hydrolysis reconverts 35.246 into the original base. Chromatography of O,O-diacetylsarpagine (35.246) on alumina results in loss of an

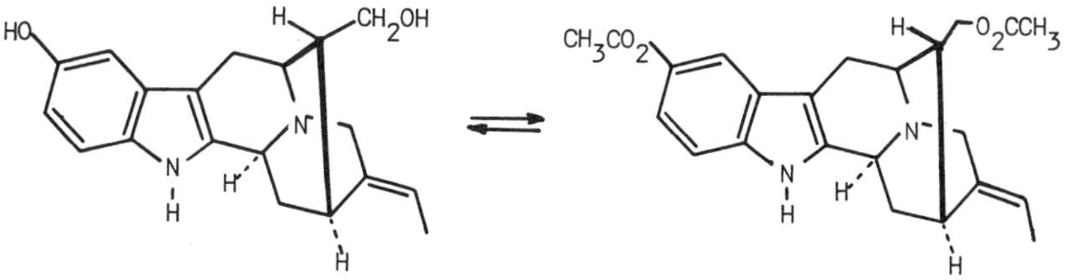

(Eq. 35.39)

35.6 35.246

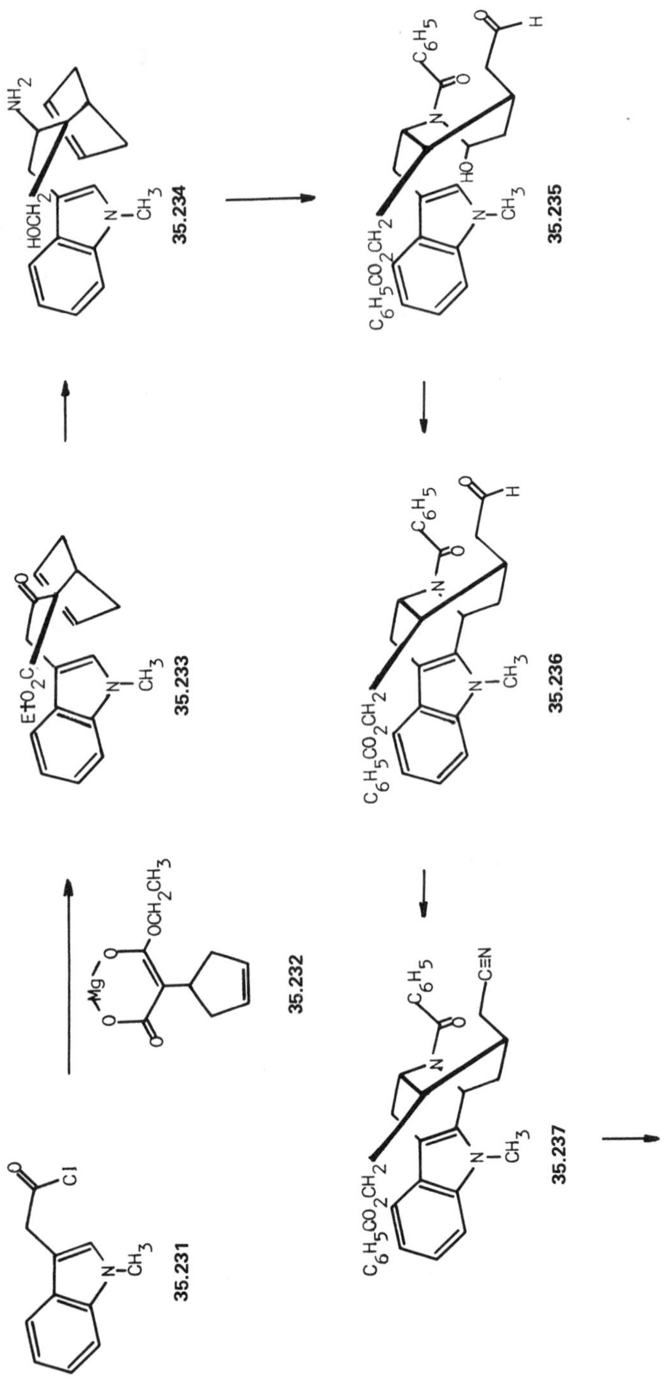

35.231

35.232

35.233

35.234

35.235

35.236

35.237

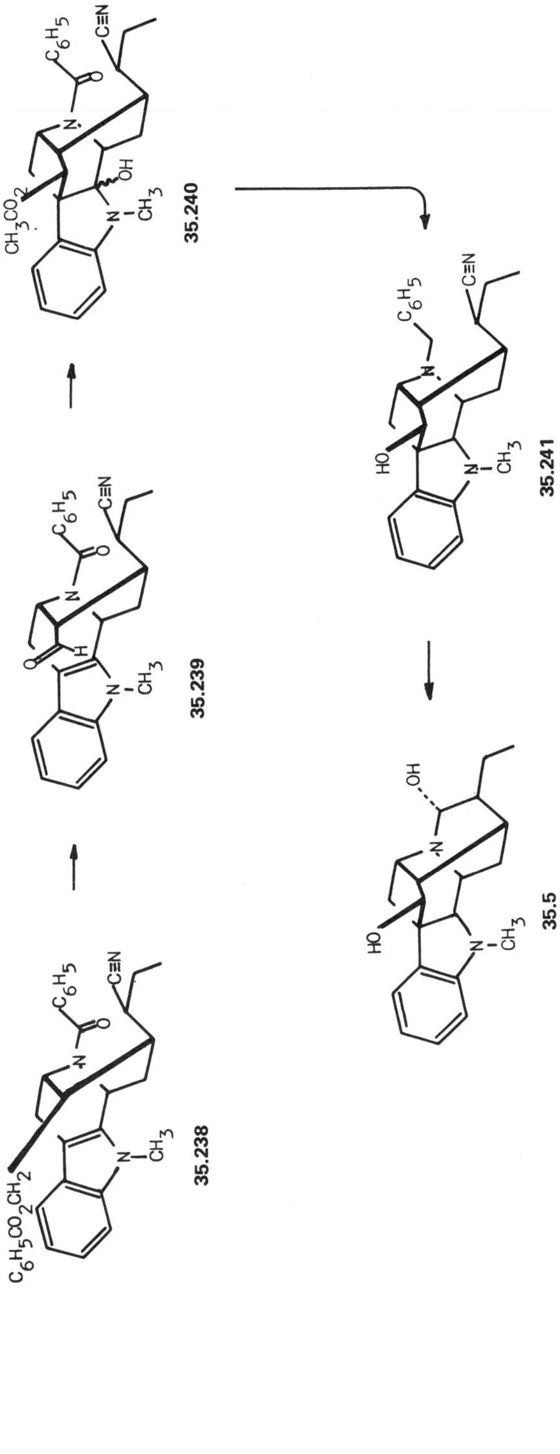

SCHEME 35.41 The total synthesis of (±)-ajmaline. (After Reference 306.)

acetyl function and isolation of the same monoacetyl derivative (35.243) formed directly from sarpagine (35.6) and acetic acid (see above.)

Acetylation of O-methylsarpagine (35.242) results in the formation of O-acetyl-O-methylsarpagine (35.247) and N,O-diacetyl-O-methylsarpagine (35.248; Equation 35.40). Since the

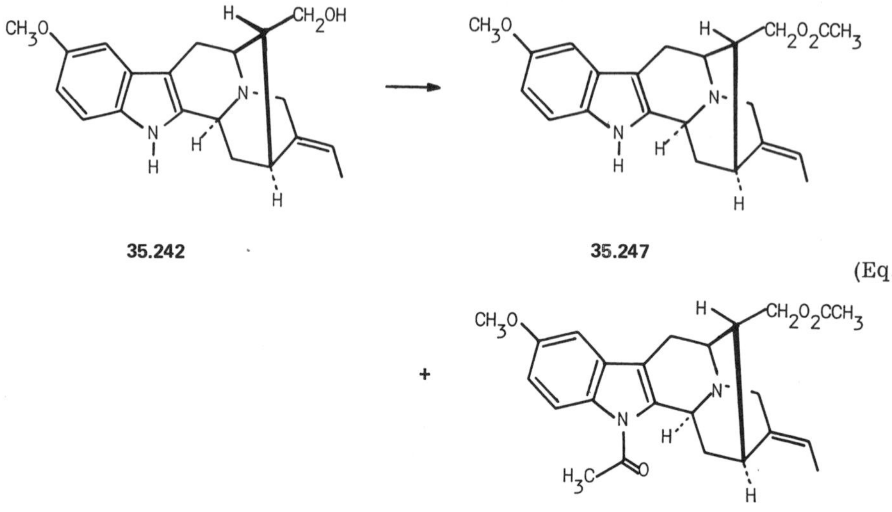

35.242 35.247

(Eq. 35.40)

35.248

latter (35.248) forms a quaternary methiodide with methyl iodide and has an ultraviolet spectrum characteristic of an N-acetylindole [309], the indole nucleus present in sarpagine (35.6) must be unsubstituted on nitrogen, and the second nitrogen atom present in sarpagine (35.6) must be tertiary.

Hydrogenation of O-methylsarpagine (35.242) yields, first, a dihydro derivative (35.249; Equation 35.41); continued reduction results in absorption of an additional four equivalents of hydrogen. That an ethylidene grouping is present to account for the one readily hydrogenated

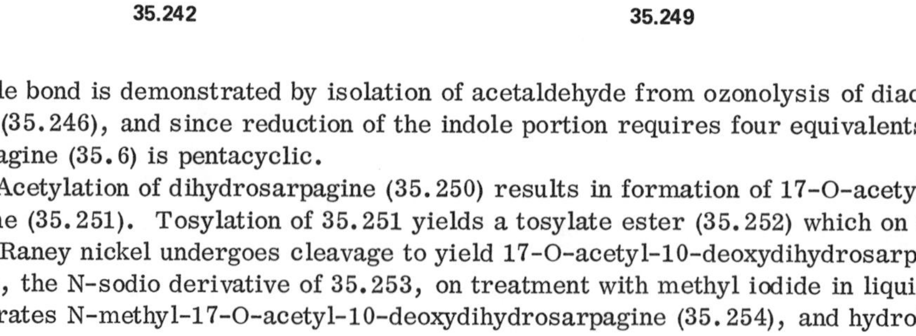

(Eq. 35.41)

35.242 35.249

double bond is demonstrated by isolation of acetaldehyde from ozonolysis of diacetylsarpagine (35.246), and since reduction of the indole portion requires four equivalents of hydrogen, sarpagine (35.6) is pentacyclic.

Acetylation of dihydrosarpagine (35.250) results in formation of 17-O-acetyldihydrosarpagine (35.251). Tosylation of 35.251 yields a tosylate ester (35.252) which on reduction with Raney nickel undergoes cleavage to yield 17-O-acetyl-10-deoxydihydrosarpagine (35.253). Then, the N-sodio derivative of 35.253, on treatment with methyl iodide in liquid ammonia, generates N-methyl-17-O-acetyl-10-deoxydihydrosarpagine (35.254), and hydrolysis of the acetate of 35.254 yields the alcohol (35.255), which is identical to the product obtained by borohydride reduction of 21-deoxyajmalal-B (35.210) and is thus 21-deoxyajmalol-B (35.255). The above reactions are shown in Scheme 35.42. The reactions in Scheme 35.42, which establish the absolute configuration at each of the asymmetric centers of sarpagine (35.6), do not, however, demonstrate the configuration about the double bond. That this configuration is E, as shown, was confirmed by x-ray analysis of akuammidine (35.256; [310]) which can be converted to 10-deoxysarpagine (35.257) by Oppenauer oxidation and accompanying decarboxylation followed by lithium aluminum hydride reduction back to the alcohol (35.257; Equation 35.42). 10-Deoxysarpagine (35.257) was generated from sarpagine (35.6) itself by tosylation of 17-O-acetylsarpagine (35.243) to 35.258 followed, first, by reductive cleavage of the tosylate with Raney nickel to 35.259, and then hydrolysis to the alcohol itself (Equation 35.43; [365]).*

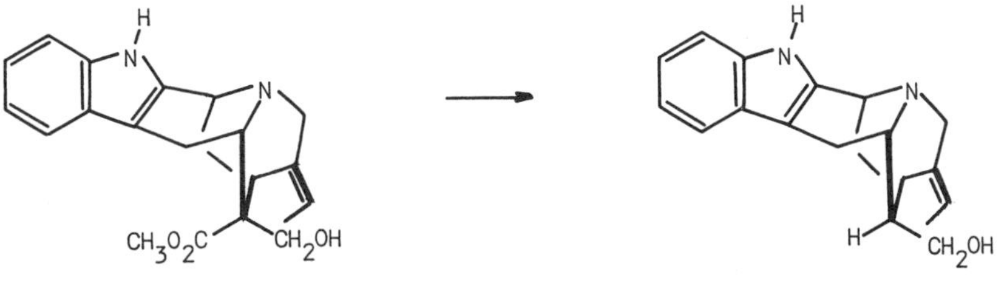

35.256 → 35.257 (Eq. 35.42)

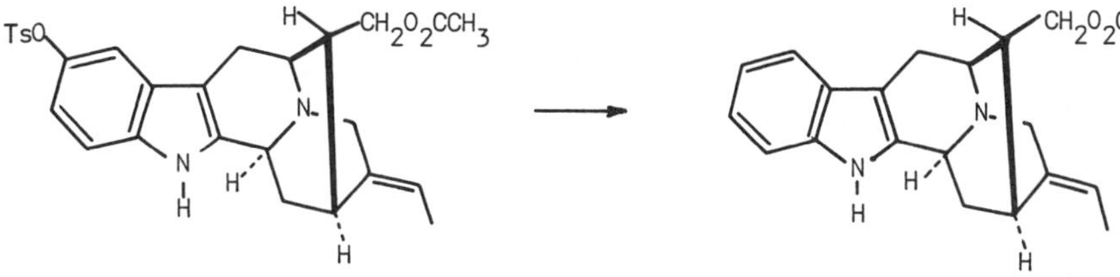

35.258 → (Eq. 35.43)

It does not yet appear that sarpagine (35.6) has fallen to synthesis. However, since 21-deoxyajmalal-B (35.210) has been synthesized (Scheme 35.43; [312]) it should in principle be possible, with minor modification, to prepare dihydrosarpagine (35.250) itself. The synthesis of 21-deoxyajmalal-B (35.210; Scheme 35.43; [312]) was begun by conversion of (±)-α-(Δ^3-cyclopentenyl)butyric acid (35.260) to the O-benzyl ether 35.261 by reduction of 35.260 to the corresponding alcohol with lithium aluminum hydride and O-benzylation with benzyl chloride. Oxidation of 35.261 with osmium tetroxide in pyridine generates the diol 35.262, the carbonate of which can be induced to undergo hydrogenolysis of the benzyl group to the corresponding primary alcohol, which with chromic anhydride affords the aldehyde 35.263. Reductive condensation of 35.263 with N-methyltryptophan (35.264) yields the amino acid 35.265. Hydrolysis of 35.265 and oxidation with periodate is presumed to lead to an intermediate dialdehyde which apparently undergoes spontaneous cyclization to the tetracyclic

*Although this clearly demonstrates the configuration about the carbon-carbon double bond, some doubt is apparently extant since a recent work [311] has presented the opposite conclusion.

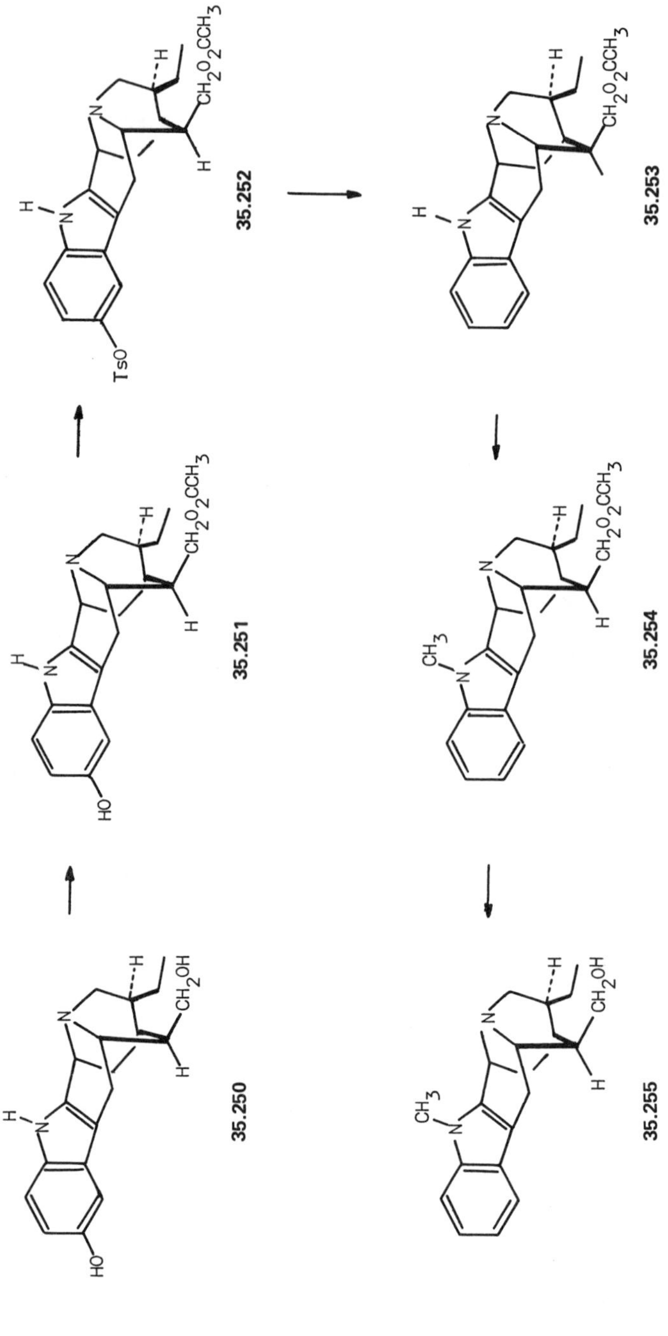

35.250 → 35.251 → 35.252 → 35.253 → 35.254 → 35.255

SCHEME 35.42 The conversion of dihydrosarpagine into 21-deoxyajmalol-B.

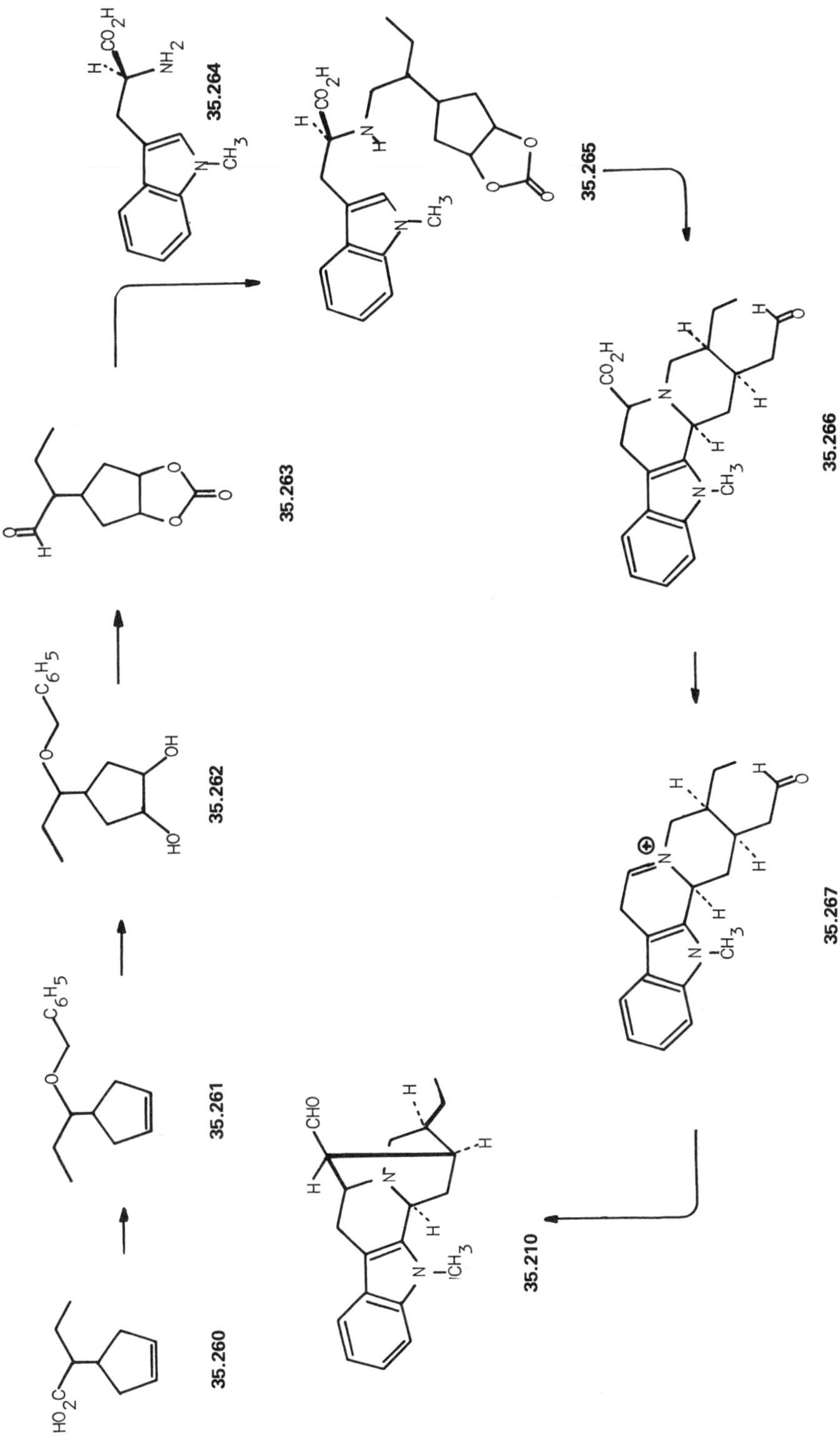

SCHEME 35.43 A total synthesis of 21-deoxyajmalal-B. (After Reference 312.)

monoaldehyde 35.266. Treatment of 35.266 with cyclohexylcarbodiimide in the presence of toluenesulfonic acid results in oxidative decarboxylation and, presumably, via the imine 35.267, spontaneous cyclization to (±)-deoxyajmalal-B (35.210 and its mirror image).

VII. ECHITAMINE

Echitamine (35.7), the major base obtained from the bark of the tree Alstonia scholaris R. Br., is contained in an extract which has been used for many years as an antimalarial drug. The base (35.7; [313]) was characterized originally as a chloride, $C_{22}H_{29}O_4N_2Cl$, $[\alpha]_D^{25}$ -57°,

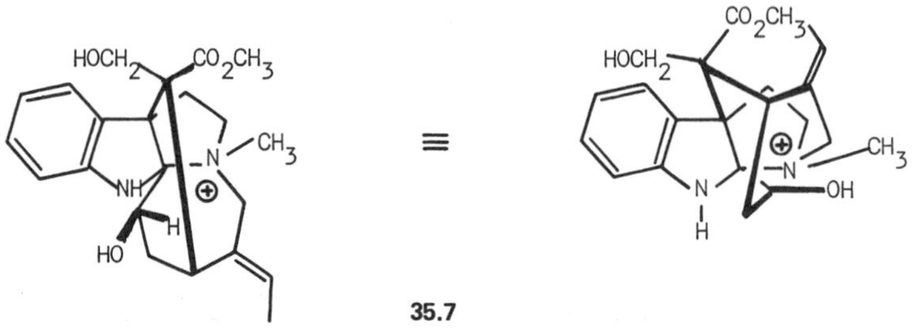

35.7

and it was also recognized that echitamine (35.7) was a quaternary amine whose hydroxide crystallized as a trihydrate. On heating echitamine hydroxide (35.7) above 105°C under vacuum, water is lost, and an anhydrous material, $C_{22}H_{28}O_4N_2$, of still uncertain structure (but very probably 35.268), is formed (Equation 35.44). The same material (35.268) can

(Eq. 35.44)

35.7 35.268

also be generated from echitamine chloride (35.7) on treatment with t-butoxide. Echitamine (35.7) can be regenerated from what is presumably 35.268 on treatment with aqueous acid (Equation 35.44). These early observations were later confirmed and extended [314] when it was also noted that echitamine (35.7) possesses one methoxyl, as a carbomethoxy ester, and one N-methyl group.

Over the next 30 years, much evidence for the complexity of echitamine (35.7) was compiled, but degradative and spectroscopic work failed to produce a completely satisfactory structure. During this period it was shown: (a) that echitamine (35.7) contains an indole nucleus, by ultraviolet spectroscopic examination of unidentified potassium hydroxide fusion products; (b) that hydroxyl, imino, ester, and aromatic functionality (based upon infrared spectra) are also present; (c) that alkyl nitrogen fission occurs on attempted hydrogenation, before the double bond (acetaldehyde on ozonolysis) is reduced; and (d) that at least one C-methyl group is present before two equivalents of hydrogen are absorbed and that at least two methyl groups are present afterwards (Kuhn-Roth, [315]).

When the x-ray crystal structure of echitamine (35.7) was completed [316, 317], reasonable explanations could be advanced for the reactions which had been observed.* Thus, it is now clear that catalytic hydrogenation of echitamine (35.7) in ethanol results in the formation of a dihydroechitamine (now called echitinolide, 35.269), $C_{21}H_{26}N_2O_3$, which is a δ-lactone and which has apparently come about from nitrogen - alkyl fission followed by lactonization (with loss of methanol; Scheme 35.44). The base echitinolide (35.269) can of course be hydrogenated, and on heating in hydrochloric acid, echitinolide (35.269) undergoes isomerization to isoechitinolide (3H, t, 0.62 δ, and 2H, q, 1.31 δ) formulated as 35.270 ([319]; Scheme 35.44).

VIII. STRYCHNINE

The seeds and leaves of Strychnos nux-vomica, L. contain the highly toxic alkaloid strychnine (35.8) and a number of close (A-ring methoxylated) relatives. The highly toxic alkaloid mixture, having found use as a rodenticide, was first isolated in the early nineteenth century, and strychnine (35.8) was purified and characterized, with a correct elemental analysis, as early as 1838. In the next 100 years, over 200 papers described attempts to elucidate the structure of strychnine (35.8) and its relatives, and an exciting story unfolded [320]. The account here, rather than being the historical description which has been well detailed elsewhere [321-323] attempts to delineate the salient features of the structural determination in a more concise fashion.

As shown in Scheme 35.45, the alkaline decomposition of strychnine (35.8; [324]) generates, among other products, seven isolable materials of known structure, namely, 3-ethylindole (35.17), 2-methylindole (35.18, skatol), tryptamine (35.31), β-collidine (35.65), carbazole (35.271), indole (35.272), and β-picoline (35.273). Not one of these compounds requires any carbon skeleton rearrangement during the degradation process, although at the time of their isolation, their relative placement in strychnine (35.8) could not be fathomed.

Strychnine (35.8), $C_{21}H_{22}N_2O_2$, mp 268-290°C (depending upon the rate of heating), $[\alpha]_D^{20}$ - 104° (c = 0.5 in absolute ethanol), which from the above information most probably contains at least an indole nucleus, was recognized as also containing one readily protonated nitrogen (called N_b) and a second nitrogen (called N_a), present in an amide linkage. Hydrolysis of strychnine (35.8) with alcoholic potassium hydroxide results in the formation of strychnic acid (35.274) which can be reconverted into strychnine (35.8; Equation 35.45). When

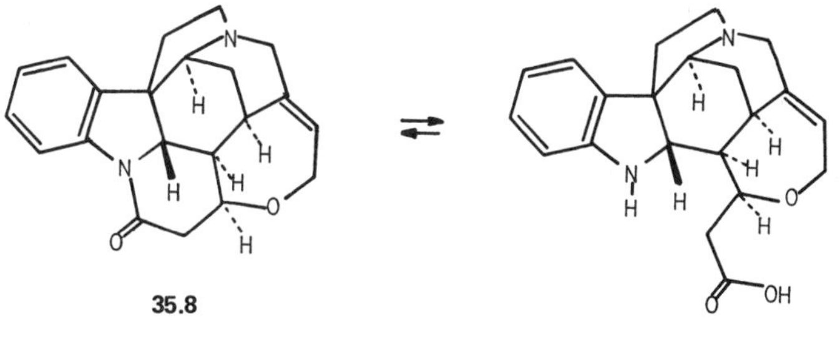

35.8 ⇌ 35.274 (Eq. 35.45)

*An interesting account which summarizes the work which preceded the final establishment of the structure of echitamine (35.7) is given in Reference 318.

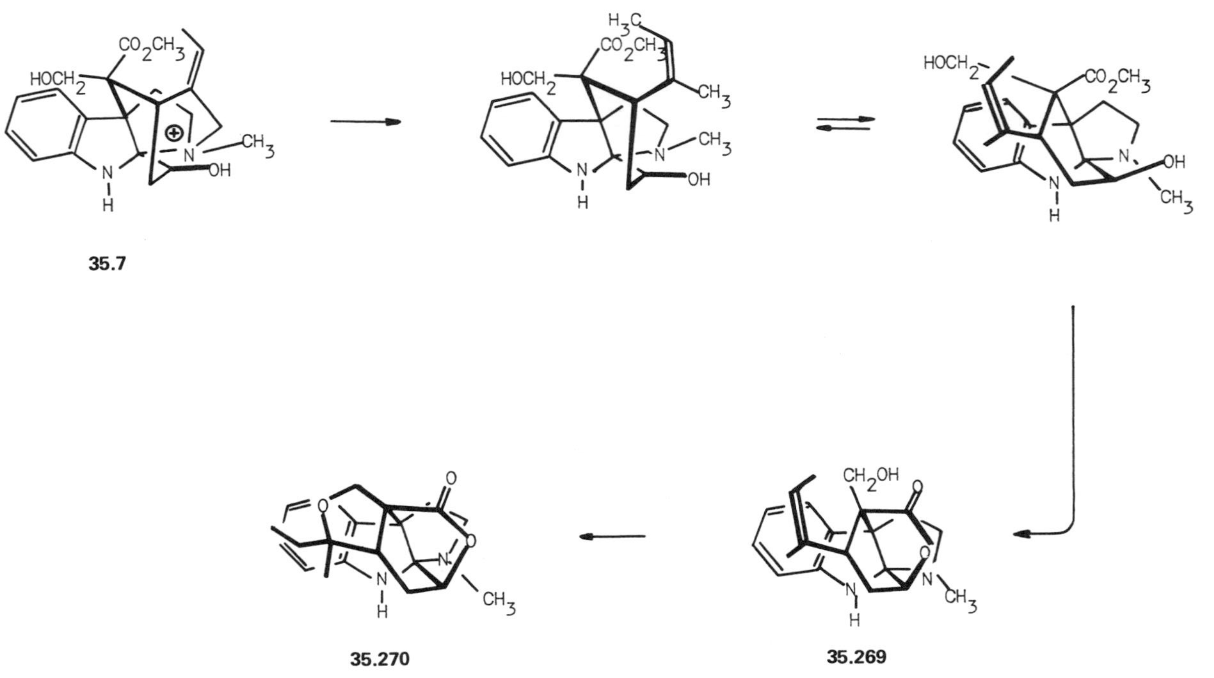

35.7

35.269

35.270

SCHEME 35.44 The products of the hydrogenation and rearrangement of echitamine. (After Reference 319.)

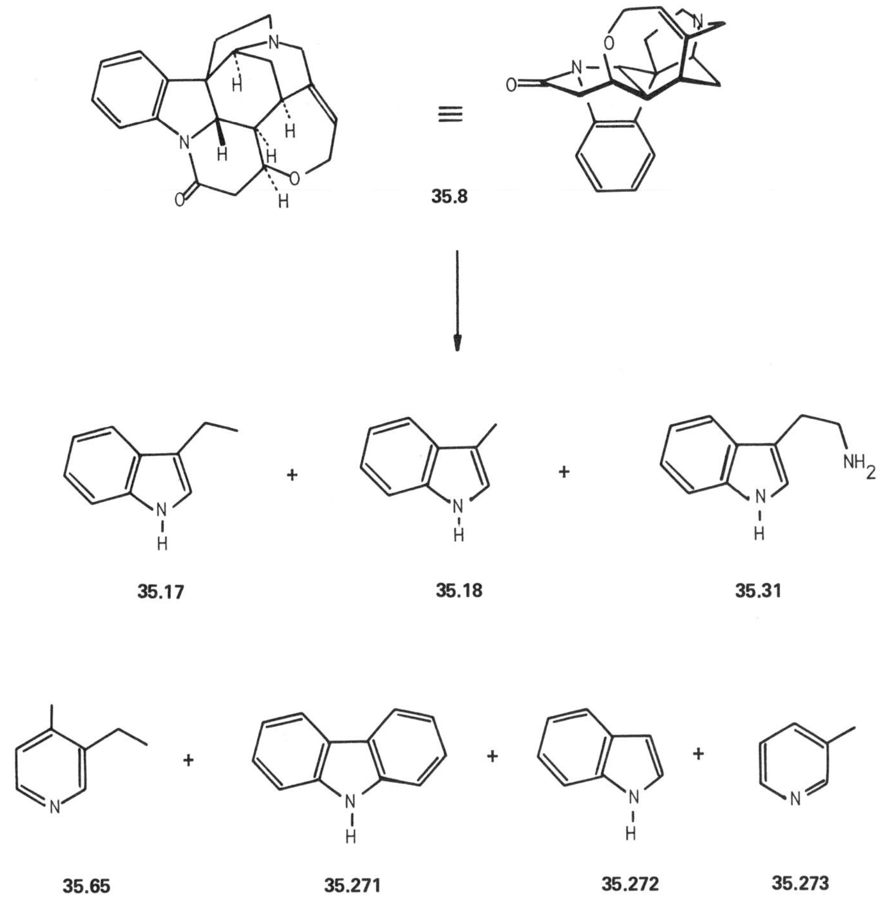

35.8

35.17 + 35.18 + 35.31

35.65 + 35.271 + 35.272 + 35.273

SCHEME 35.45 Products of the alkaline degradation of strychnine. (After Reference 324.)

strychnine (35.8) is treated with methyl iodide, a quaternary iodide, methylstrychninium iodide (35.275), is generated, and attempted base-catalyzed Hofmann degradation yields methylstrychnine (35.276), i.e., hydrolysis of the lactam to generate a betaine (35.276) occurs. Again (see above), the lactam can be reformed in acid or further methylation can be effected to yield dimethylstrychnine (35.277) which will not yield a lactam. Equation 35.46 depicts the processes occurring.

With one oxygen of strychnine (35.8) recognized as being present in the carbonyl group of the amide function, attention was turned to the second oxygen which was soon presumed to be in an ether linkage since it was unreactive.

Catalytic hydrogenation of strychnine (35.8) affords dihydrostrychnine (35.278), while heating strychnine (35.8) in refluxing xylene generates neostrychnine (35.279), which can also be reduced to dihydrostrychnine (35.278; Equation 35.47). On the other hand, electrolytic reduction of strychnine (35.8) leads to strychnidine (35.280), which can be further reduced (catalytically) to dihydrostrychnidine-A (35.281; Equation 35.48). The order of the reductive processes can be reversed with the same end result being obtained.

Oxidation of strychnine (35.8) proved among the most informative reactions. With aqueous nitric acid, strychnine (35.8) yields, among other products, dinitrostrycholcarboxylic

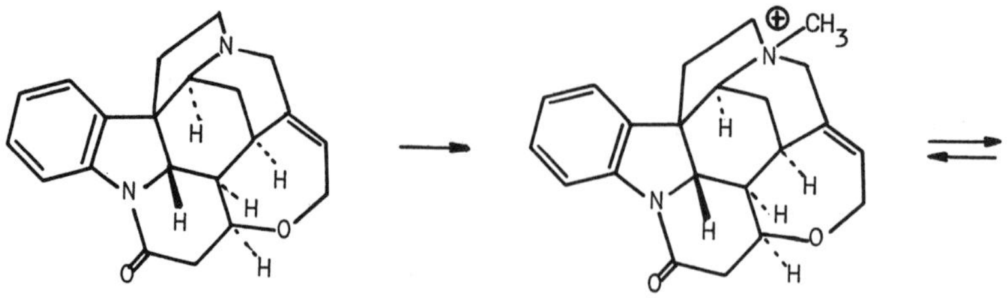

35.8 → 35.275 ⇌

(Eq. 35.46)

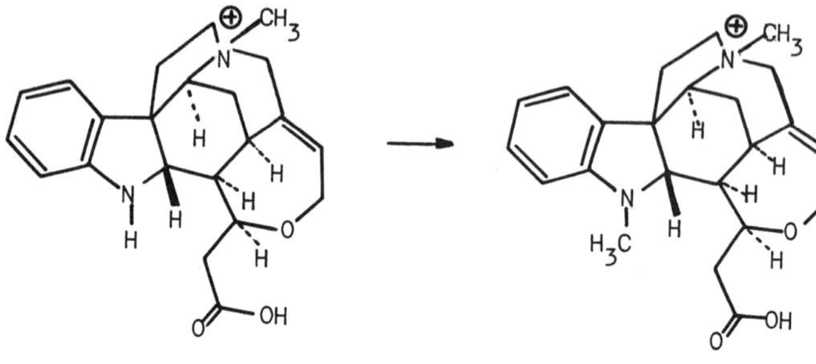

35.276 → 35.277

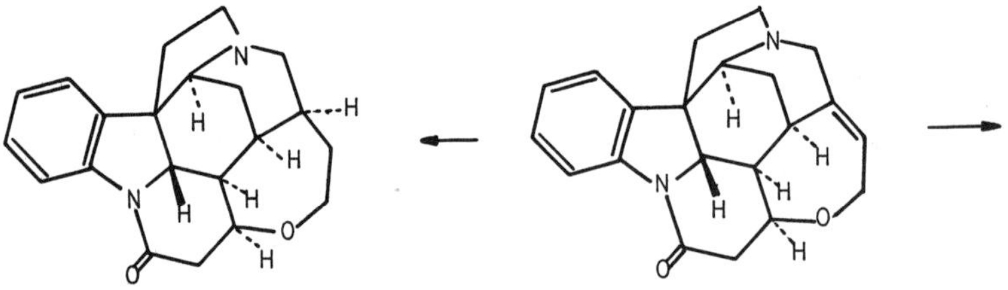

35.278 ← 35.8 →

(Eq. 35.47)

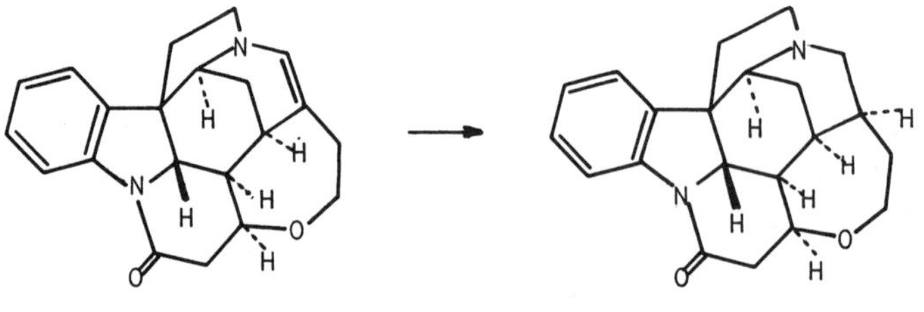

35.279 → 35.278

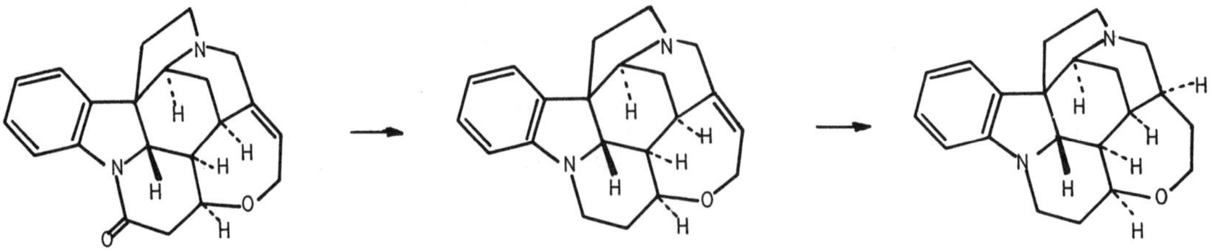

35.8 → 35.280 → 35.281

(Eq. 35.48)

acid (35.282; Equation 35.49), while with chromic acid, carboxyaponacine (35.283; Hanssen's C_{16}-acid; [325]; Equation 35.50) results. When the oxidation of strychnine (35.8) is carried out with potassium permanganate in acetone-chloroform solution, both strychninonic (35.284)

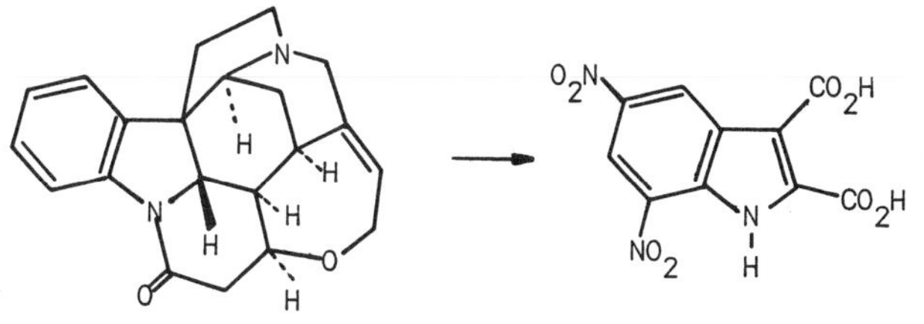

35.8 → 35.282 (Eq. 35.49)

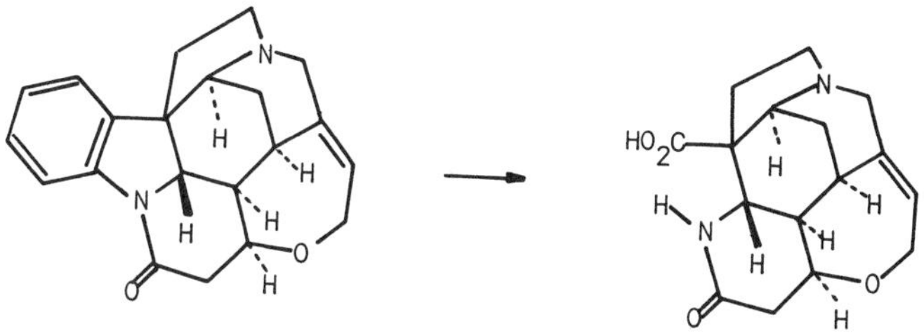

35.8 → 35.283 (Eq. 35.50)

and dihydrostrychninonic (35.285) acids are obtained (Equation 35.51; [326]). The latter can, of course, be further oxidized to the former, but reduction of strychninonic acid (35.284) catalytically generates strychninolic acid (35.286), the epimer of dihydrostrychninonic acid (35.285; Equation 35.52). Finally, along these lines, when strychninolic acid is warmed

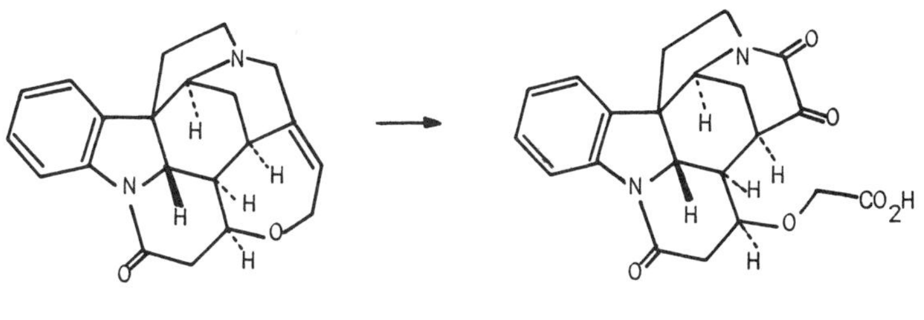

35.8 → 35.284 + 35.285 (Eq. 35.51)

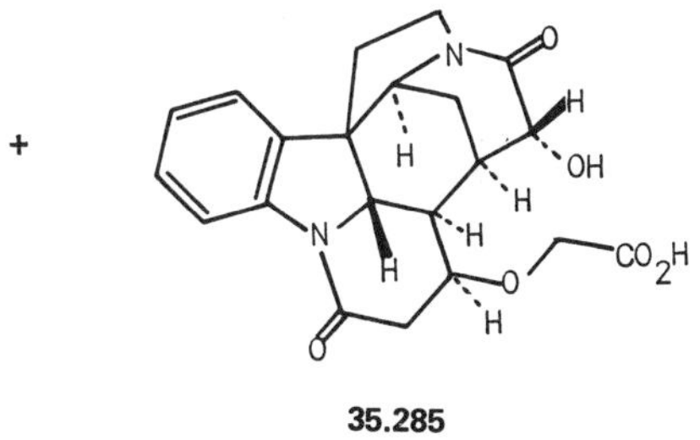

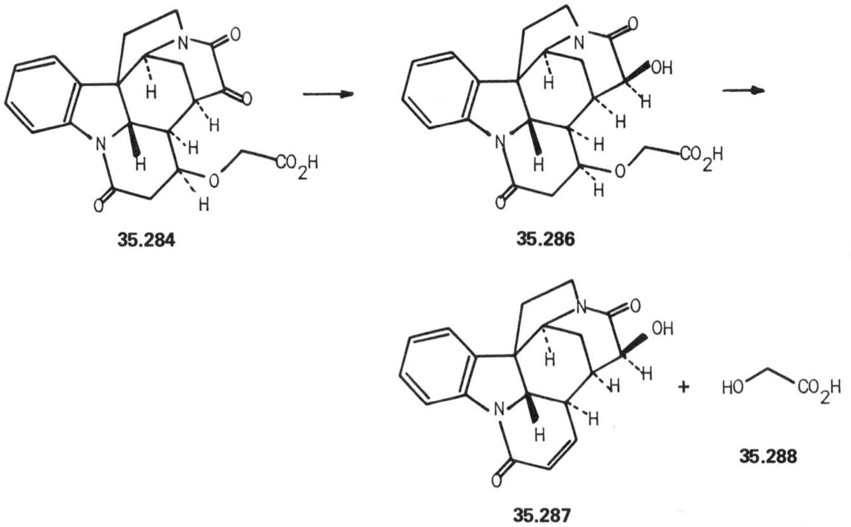

35.284 35.286

(Eq. 35.52)

35.287 35.288

with dilute aqueous sodium hydroxide, strychninolone-a (35.287) and glycolic acid (35.288) are produced.*

That a methylene unit lay next to the carbonyl group in strychnine (35.8) was confirmed by formation of a benzylidine derivative as well as oximinostrychnine (35.289 and 35.290) on oximination. Presumably, both the E and Z forms of oximinostrychnine (35.289 and 35.290, respectively) are formed and yield different products on attempted Beckmann rearrangement. Thus, treatment of the former (35.289) with thionyl chloride presumably yields the urea (35.291), which undergoes alkaline hydrolysis to norstrychninic acid (35.292; Equation 35.53).

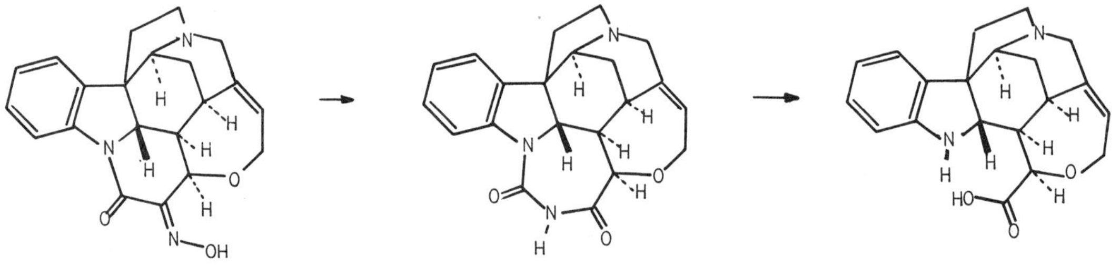

35.289 35.291 35.292

(Eq. 35.53)

On the other hand, it is presumably the Z isomer (35.290) which undergoes fragmentation via the carbamic acid 35.293 to the aldehyde 35.294 (the Wieland-Gumlich aldehyde), barium cyanide, and barium carbonate in the presence of barium oxide (Equation 35.54).†

---

*As expected, an epimer of strychninolone-a (35.287) and glycolic acid (35.288) result from treatment of dihydrostrychninonic acid (35.285) under the same conditions.

†Both sets of reactions (Equations 35.53 and 35.54) may be accounted for in terms of the E oxime (35.289) which should preferentially generate 35.292 (as well as yield 35.294). Alternatively, some preference for the Z oxime (35.290) should be expressed in the ground state because of hydrogen bonding. The oximes might be interconvertible under conditions of further reactions; it is unclear whether, in fact, two oximes are, or are not, present [327].

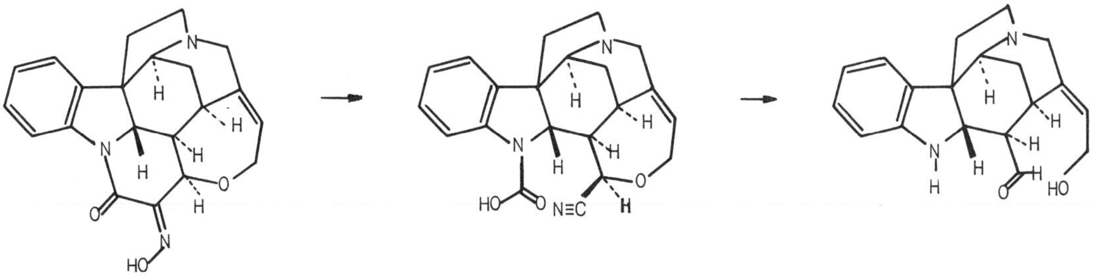

35.290 35.293 35.294

(Eq. 35.54)

When strychninolone-a (35.287) is treated further with base, strychninolone-b (35.295) and then strychninolone-c (35.296) are generated (Equation 35.55). Oxidation of the acetate of strychninolone-a (35.297) with permanganate yields an intermediate (presumably 35.298)

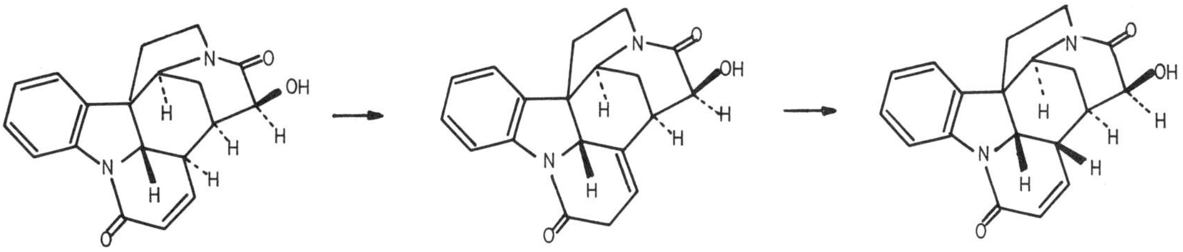

35.287 35.295 35.296

(Eq. 35.55)

which generates an amino acid (35.299) and oxalic acid on hydrolysis (Equation 35.56). On the other hand, the acetate of strychninolone-b (35.300 yields a keto carboxylic acid (35.301)

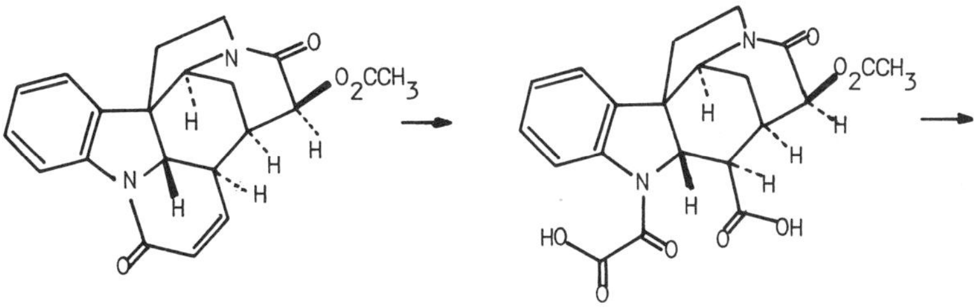

35.297 35.298

(Eq. 35.56)

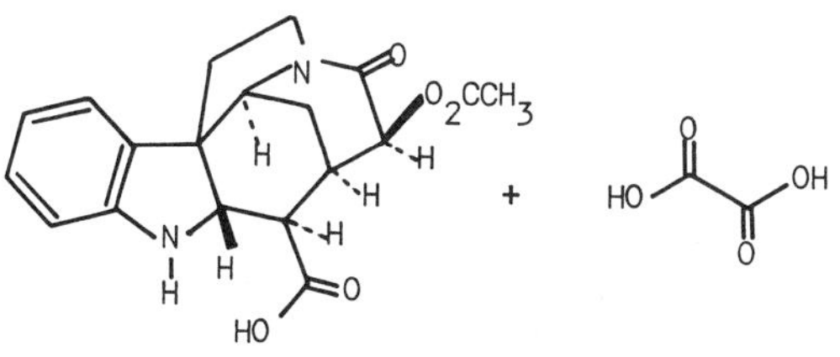

35.299

under the same conditions (Equation 35.57). The relationship suggested by the above information is confirmed by reduction of strychninolone-b (35.295) to a mixture of the dihydro derivatives of strychninolone-a (35.287) and strychninolone-c (35.296).

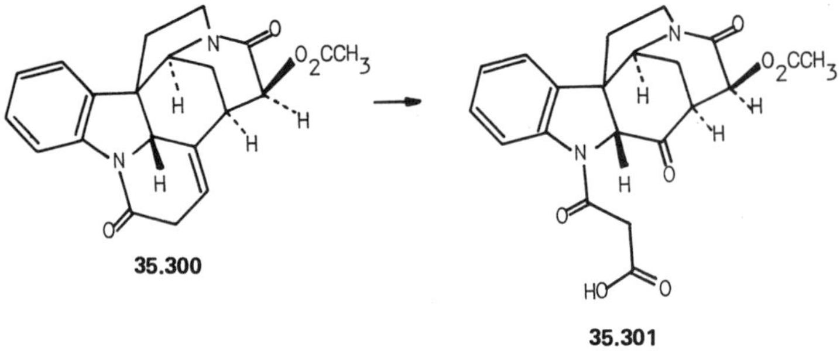

35.300 → 35.301 (Eq. 35.57)

The structure and stereochemistry of strychnine (35.8) deduced from the above and related reactions was amply confirmed by x-ray crystallographic examination of salts of strychnine (35.8; [328]), and the total synthesis of strychnine (35.8) was accomplished shortly thereafter [329]. The synthesis of strychnine (35.8; Scheme 35.46) begins with the condensation of phenylhydrazine (35.302) with acetoveratrone (35.303) in the presence of polyphosphoric acid to yield 2-veratrylindole (35.304). Conversion of 35.304 into 2-veratryltryptamine (35.306) is accomplished by allowing the indole (35.304) to undergo a Mannich reaction with formaldehyde and aqueous dimethylamine followed by treatment with methyl iodide to yield a quaternary ammonium iodide (35.305), displacing the trimethylamino leaving group with cyanide and finally reducing the nitrile with lithium aluminum hydride to generate 35.306. The amine (35.306) is then permitted to condense with ethyl glyoxylate to yield the imine 35.307, which with tosyl chloride in pyridine produces the indolenine 35.308. Borohydride reduction and acetylation converts 35.308 into the amide 35.309 which undergoes, in moderate yield, ozonolysis in aqueous acetic acid to the muconic ester 35.310. When the latter (35.310) is treated with boiling methanolic hydrogen chloride, hydrolysis of the amide, followed by lactamization and migration of the double bond, occurs, and the pyridone ester 35.311 results.

To avoid elimination reactions, the pyridone ester 35.311 is first hydrolyzed reductively to the amino diacid 35.312 with hydrogen iodide and red phosphorus and is then converted, by esterification and acetylation, into 35.313. Treatment of the latter (35.313) with sodium methoxide in methanol results in the occurrence of a Dieckmann cyclization and formation of the enol 35.314. Since the enolic oxygen (C_{15} - strychnine numbering) must be removed and carbonyl derivatives which might be used to that end fail to form, the enol (35.314) is converted into its tosylate ester with toluene sulfonyl chloride and the latter treated with the sodium salt of benzyl thiol to yield the benzylmercaptoester 35.315. Desulfurization to the unsaturated ester 35.316 is then accomplished with Raney nickel in hot ethanol.

When the double bond in 35.316 is reduced with hydrogen in the presence of palladium on charcoal, the cis unsaturated ester (35.317) is the major product and the trans isomer (35.318) accompanies it in minor amount. However, alkaline hydrolysis of 35.317 followed by reesterification with diazomethane yields 35.318. The acid intermediate in this process can be resolved with quinidine (35.57), and the resolved ester 35.318 is identical to a degradation product of strychnine (35.8).

Treatment of 35.318 with acetic anhydride in refluxing pyridine generates the enol acetate 35.319 which is hydrolyzed to the corresponding amino ketone (35.320) with hydrogen chloride in acetic acid at reflux. Oxidation of the ketone (35.320) with selenium dioxide in ethanol leads directly to dehydrostrychninone (35.321), identical, again, to that obtained from strychnine (35.8). Condensation of dehydrostrychninone (35.321) with sodium acetylide in tetrahydrofuran yields the carbinol 35.322, which on reduction, first with hydrogen over a palladium catalyst to generate the alcohol 35.323 and then with lithium aluminum hydride in refluxing ether, yields the amine 35.324. The isomerization of 35.234 to isostrychnine I (35.325) is accomplished by generation of a mixture of bromides from 35.324 with hydrogen bromide in acetic acid followed by acid-catalyzed hydrolysis. The final ring closure to strychnine (35.8) itself is then consummated by heating isostrychnine I (35.325) with ethanolic potassium hydroxide.

IX. CATHARANTHINE

Catharanthine (35.9), $C_{21}H_{24}N_2O_2$, mp 126-128°C, $[\alpha]_D^{20}$ +30° ($CHCl_3$), which is obtained from Catharanthus roseus, G. Don, was quickly identified as being related to coronaridine (35.326), ibogaine (35.327), and ibogamine (35.328) (other iboga alkaloids) by the similarity

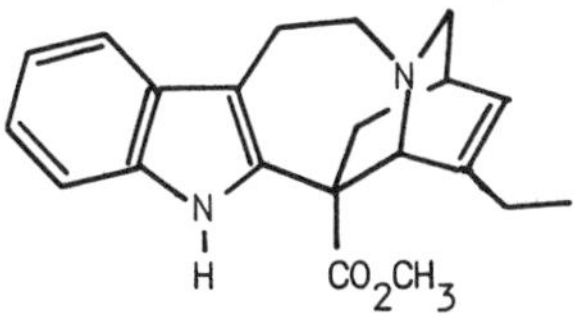

35.9

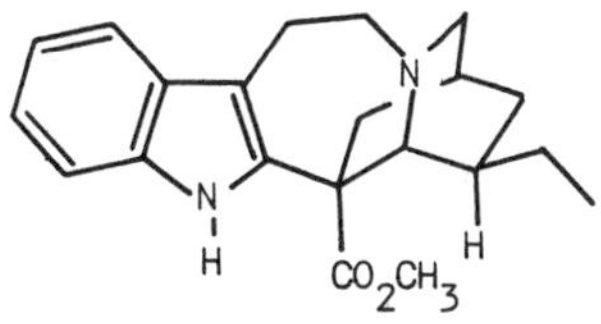

35.326

in their infrared spectra and mutual conversion of these bases into identical degradation products [330], the absolute configuration of one of which has been established, i.e., R-(+)-cleavamine (35.329) (see below; [331]).

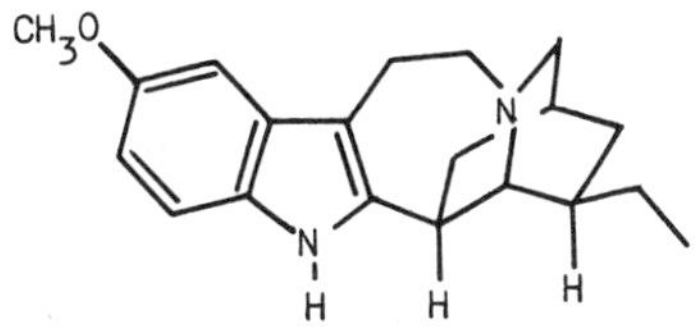

35.327

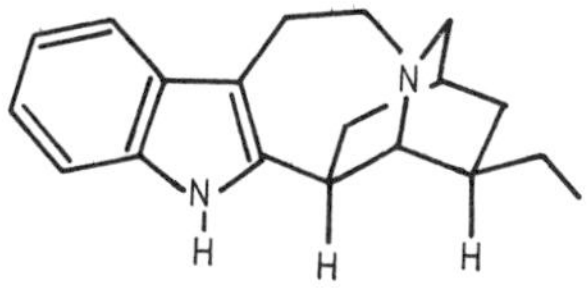

35.328

35.329

That catharanthine (35.9) contains an isolated double bond was demonstrated by its catalytic hydrogenation (PtO_2) in ethanol to dihydrocatharanthine (35.330; Equation 35.58). When the latter (35.330) is treated with hydrazine in absolute ethanol, decarboxylation occurs, and epiibogamine (35.331) results (Equation 35.59). That the ethyl group in dihydrocatharanthine (35.330) and in epiibogamine (35.331) is endo is suggested by the presumption that the hydrogenation occurs from the less hindered side [330] of the bicyclic system.

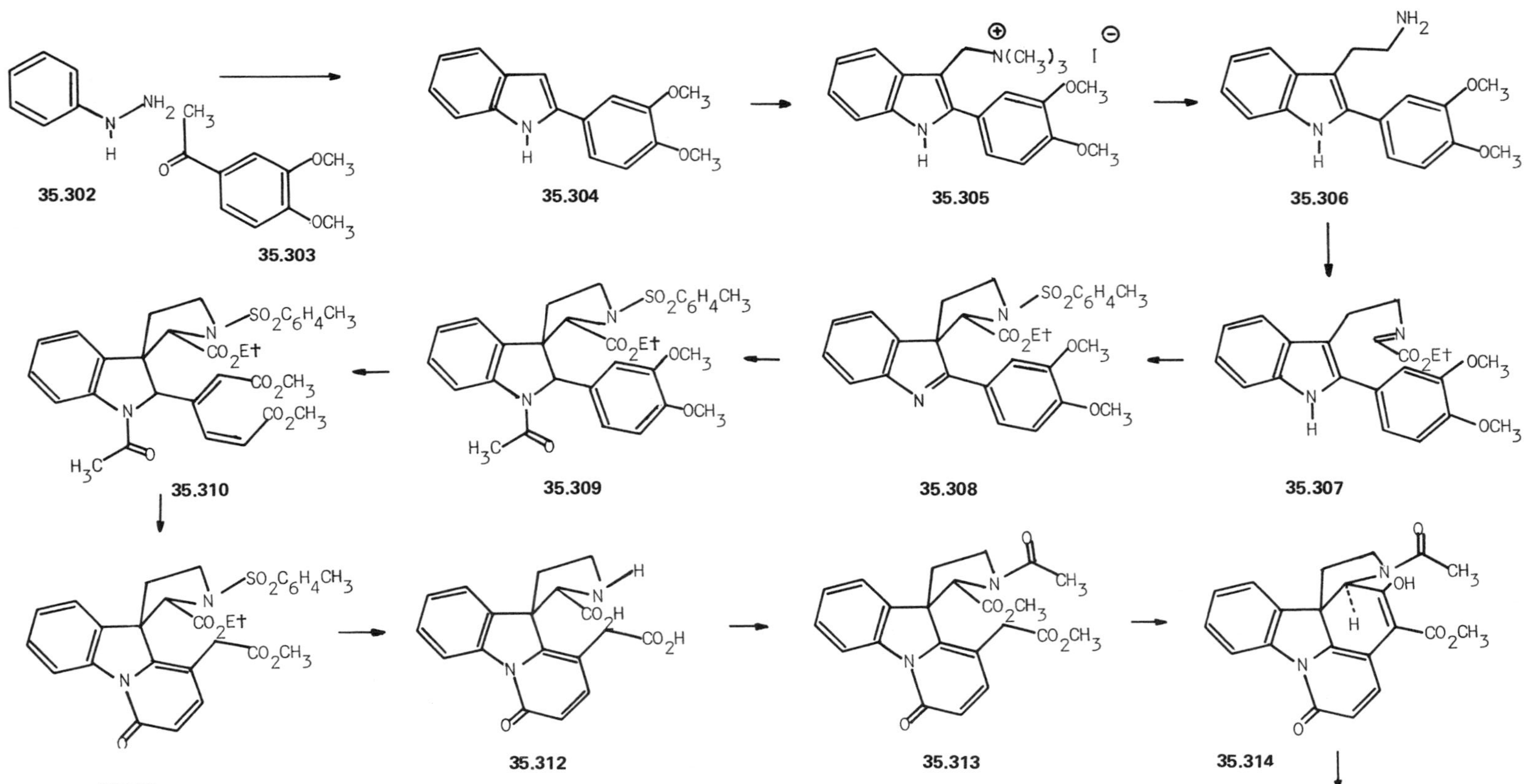

35.302

35.303

35.304

35.305

35.306

35.307

35.308

35.309

35.310

35.311

35.312

35.313

35.314

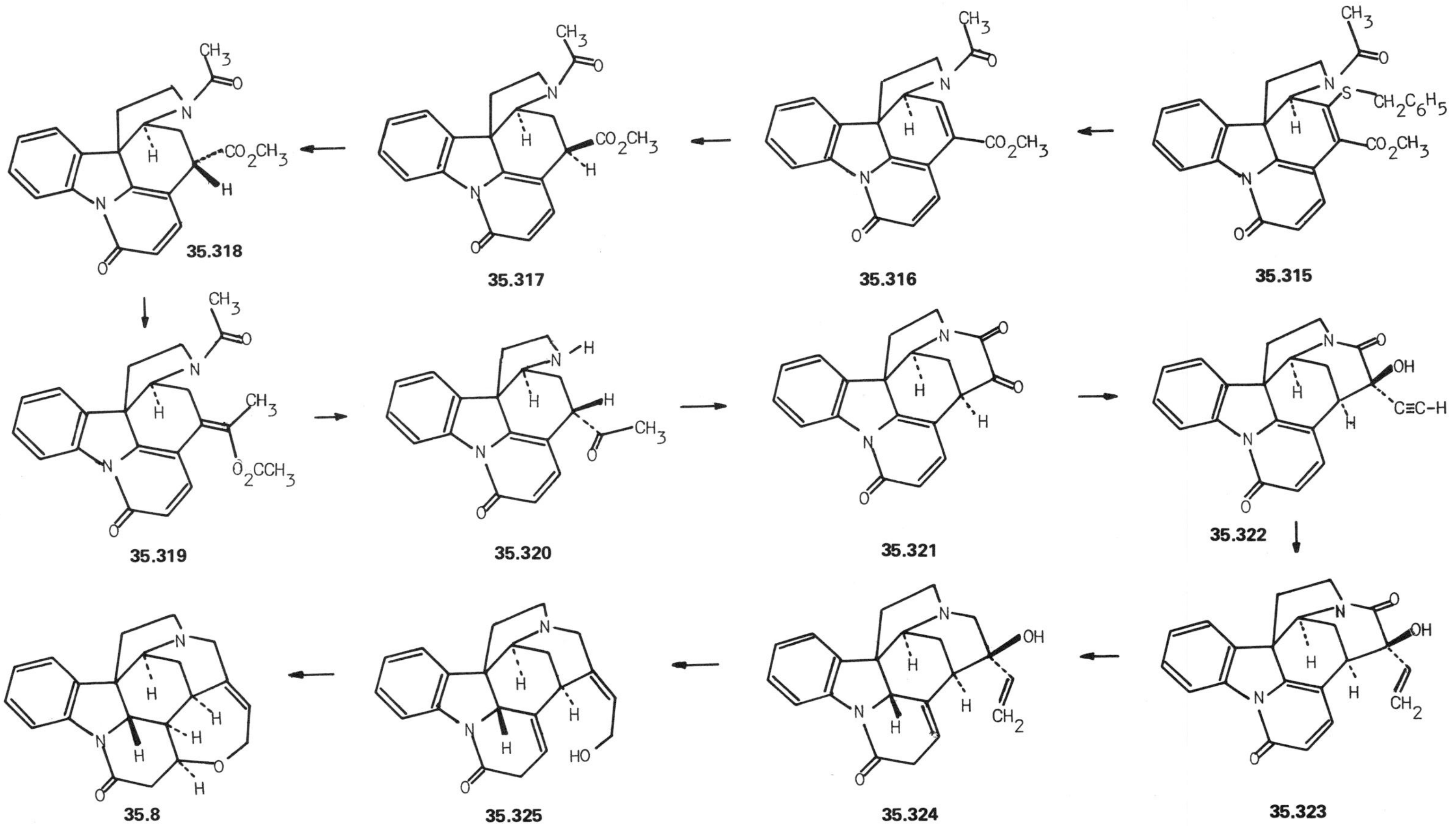

35.315 35.316 35.317 35.318

35.319 35.320 35.321 35.322

35.323 35.324 35.325 35.8

SCHEME 35.47 A possible pathway to Δ^3-ibogamine from catharanthine.

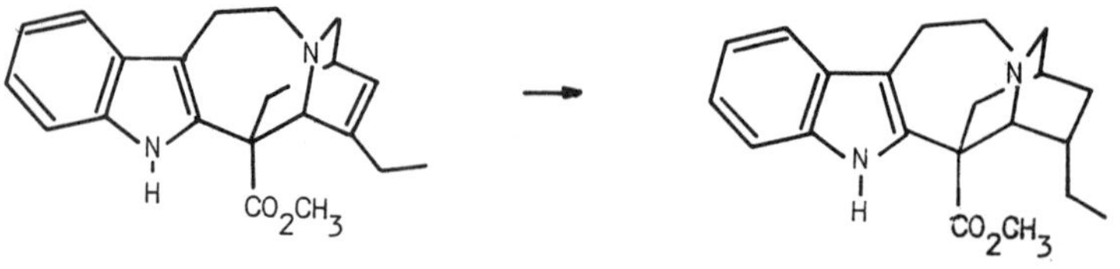

35.9 → 35.330 (Eq. 35.58)

35.330 → 35.331 (Eq. 35.59)

When catharanthine (35.9) is refluxed with concentrated hydrochloric acid, Δ^3-ibogamine (35.332) and cleavamine (35.329) are formed (Equation 35.60; [332]). A possible pathway for the formation of Δ^3-ibogamine (35.332) is shown in Scheme 35.47. Although cleavamine

35.9 → 35.332 + 35.329 (Eq. 35.60)

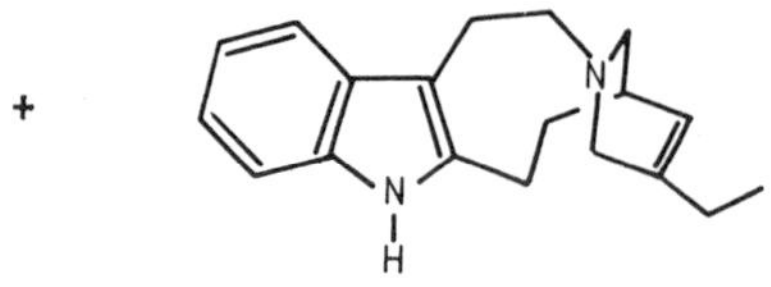

(35.329) is also formed in this reaction, the yield is low. The yield of cleavamine (35.329) can be increased, however, by the use of reducing conditions (e.g., stannous chloride and hydrochloric acid) for the reaction [333]. It may be speculated that two different pathways are involved in the formation of cleavamine (35.329). Thus (Scheme 35.48), when a reducing agent is present, generation of the imine 35.333, followed by reduction and decarboxylation, leads to 35.329. Alternatively, when a reducing agent is absent, the methyl group of the ester may serve to donate a hydride with acid-catalyzed hydrolysis of the oxonium salt (35.334), which then results in the cleaved acid 35.335. Decarboxylation of 35.335 then leads to cleavamine (35.329). This process is shown in Scheme 35.49.

The relative position of the carbomethoxy group in catharanthine (35.9) was established by reduction of the base with lithium aluminum hydride to the corresponding primary alcohol 35.336 and formation of a tetrahydro-1,3-oxazine (e.g., 35.337) on treatment of the alcohol 35.336 with acetone and hydrogen chloride (Equation 35.61), and the position of the double bond was established by pmr (H_3 was found at δ 5.9 ppm relative to TMS = 0.00; [330]).

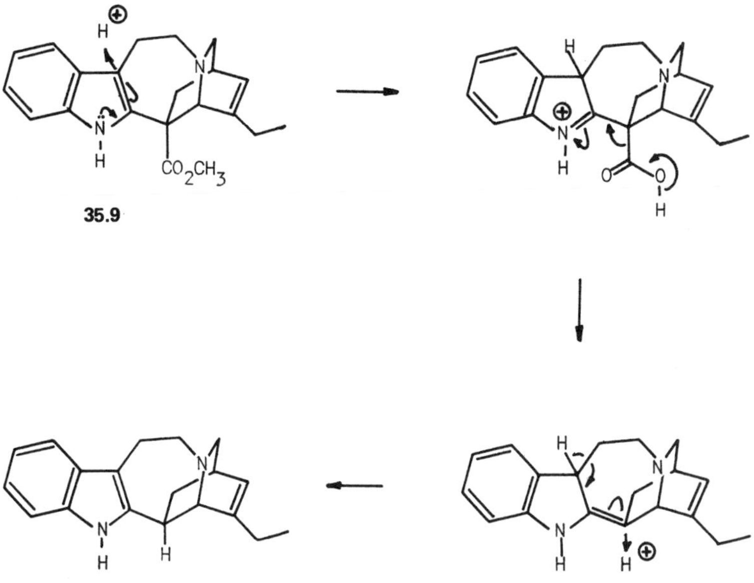

35.9

35.332

SCHEME 35.47 A possible pathway to Δ^3-ibogamine from catharanthine.

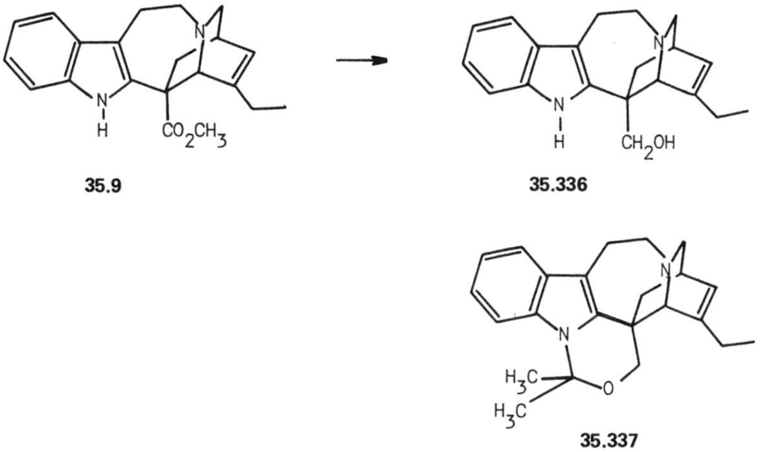

35.9

35.336

35.337

(Eq. 35.61)

When epiibogamine (35.331), generated (see above) from catharanthine (35.9) by reduction and decarboxylation, is subjected to selenium dehydrogenation [330, 334], the indoloquinoline (35.338) is obtained (Equation 35.62). This is the same material as that obtained

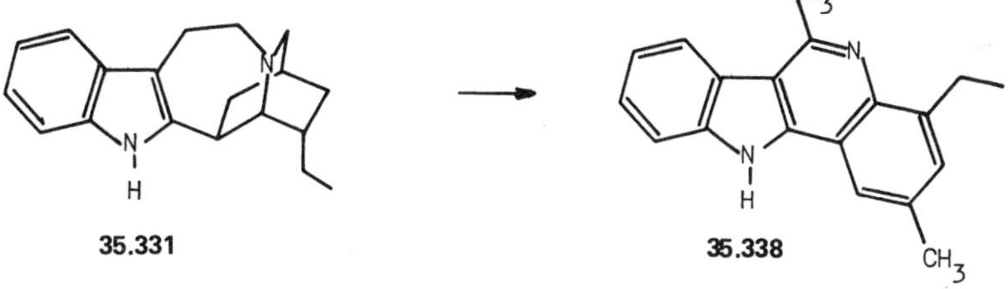

35.331

35.338

(Eq. 35.62)

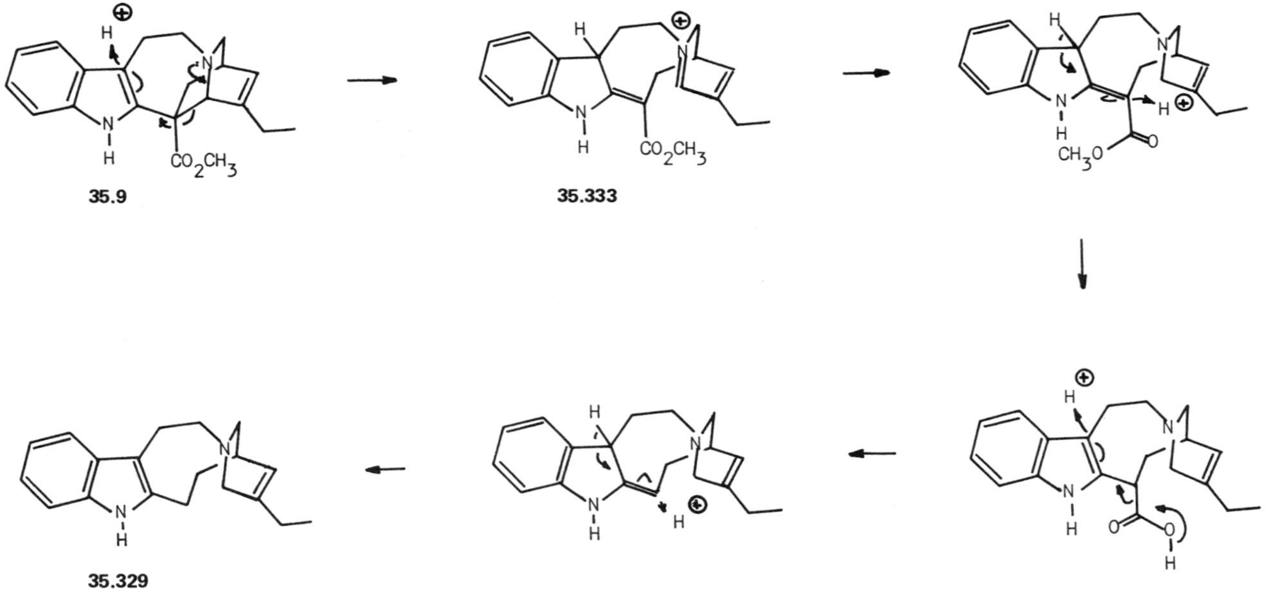

35.9

35.333

35.329

SCHEME 35.48 A possible pathway to cleavamine from catharanthine in the presence of a reducing agent.

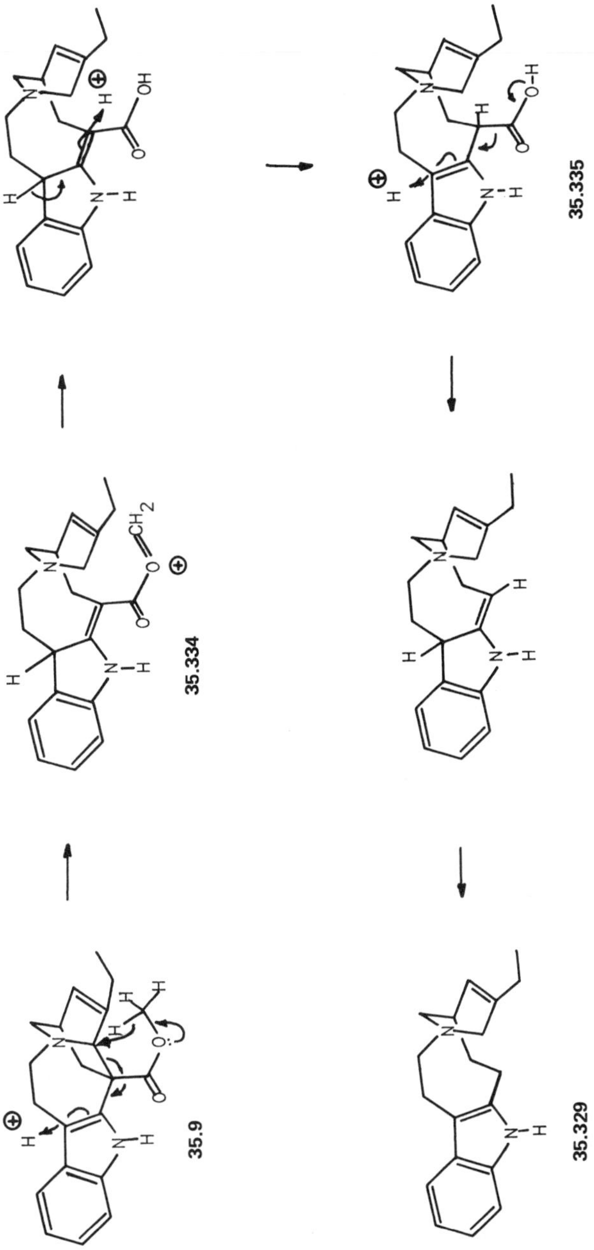

35.9

35.334

35.335

35.329

SCHEME 35.49 A possible pathway to cleavamine from catharanthine in the absence of a reducing agent.

from ibogamine (35.328) when the latter is treated, under the same conditions, with selenium, and thus, since the structure of ibogamine (35.328) is known, that of catharanthine (35.9) is also established.

Briefly, ibogamine (35.328) was shown to possess its currently accepted structure on the following degradative evidence [334].

1. On selenium dehydrogenation, ibogamine (35.328) yields the indoloquinoline (35.338) and the weakly basic secondary amine 35.339. Further treatment of 35.339 under the same dehydrogenation conditions generates 35.338 (Equation 35.63). Both these materials have been synthesized.

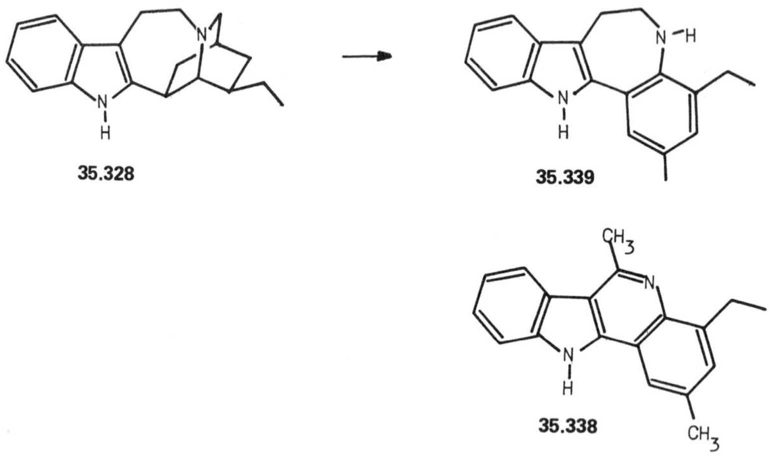

35.328 → 35.339 →

(Eq. 35.63)

35.338

2. Aerial oxidation of ibogamine (35.328) in ethanol presumably yields a hydroperoxyindolenine derivative, formulated as 35.340, which, on reduction to the corresponding hydroxyindoline followed by base treatment, generates demethoxyiboluteine (35.341; Equation 35.64).*

3. The oxime of demethoxyiboluteine (35.343), on refluxing with p-toluenesulfonyl chloride in pyridine, yields anthranilonitrile (35.344) and the tricyclic ketone (35.345; Equation 35.65).†

---

*Iboluteine (35.342) is obtained in an analogous fashion from ibogaine (35.327).

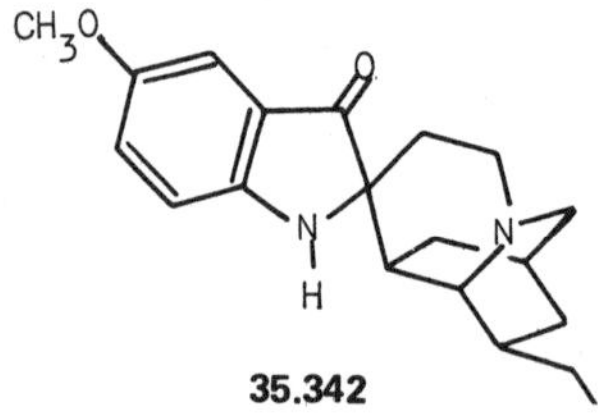

35.342

†The same ketone (35.345) is, of course, obtained from ibogaine (35.327), along with 5-methoxyanthranilonitrile (35.346).

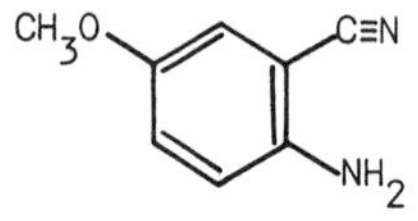

35.346

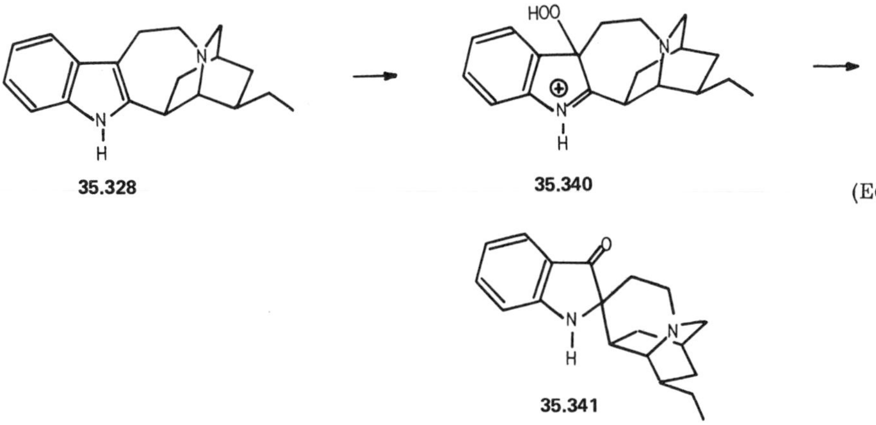

35.328 35.340 (Eq. 35.64)

35.341

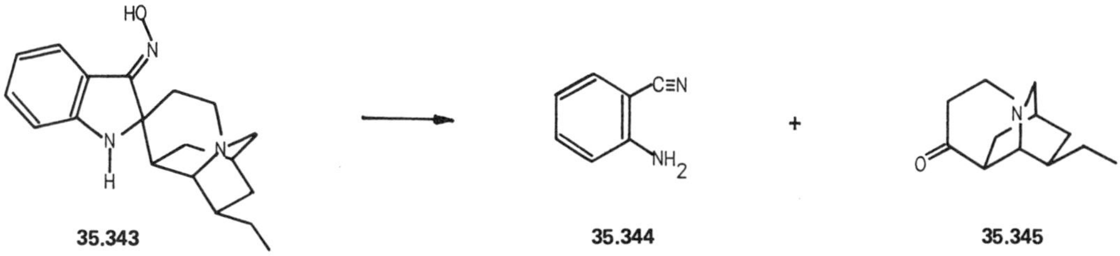

35.343 35.344 35.345

(Eq. 35.65)

4. The ketone, 35.345, reacts smoothly with cyanogen bromide to yield a crystalline N-cyanobromo derivative (35.347) which can be reduced with lithium aluminum hydride to the amino alcohol 35.348 (Equation 35.66).

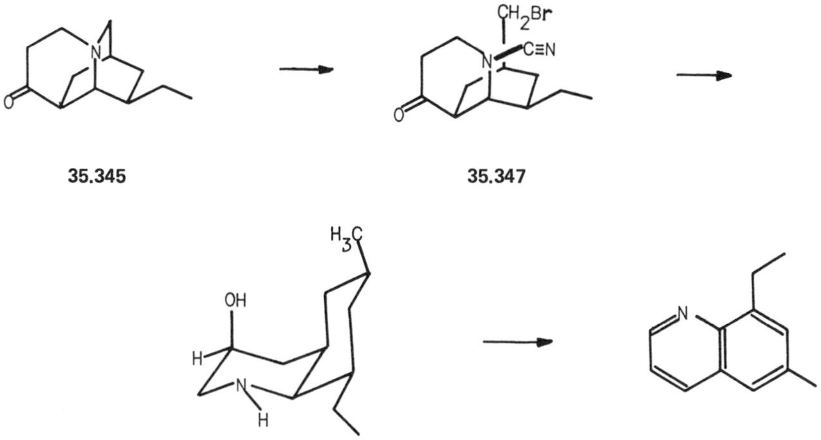

35.345 35.347

(Eq. 35.66)

35.348

5. The relative configuration of the ethyl group in ibogaine (35.327) and, by implication, in the demethoxy analog, ibogamine (35.328), was established by x-ray crystallography [335]. This, of course, also confirmed the deduced structures given above.

Finally, with the structure of catharanthine (35.9) established, a total synthesis was described [336, 337], and this synthesis is shown in Scheme 35.50.

Treatment of nicotinamide (35.350) with benzyl chloride generates N-benzyl-3-carboxamidopyridinium chloride (35.351) which can be reduced with sodium borohydride to what is presumably a mixture of the corresponding 1,6- and 1,2-dihydropyridines (35.352 and 35.353, respectively). Without separation, the mixture is treated, in chloroform, with methyl vinyl ketone (35.354) and the product, 35.355, is obtained on careful work-up in about 13% yield. Oxygenation of the ketone 35.355 in a solution of t-butyl alcohol and glyme saturated with potassium t-butoxide in the presence of triethyl phosphite at -20°C, yields the hydroxyketone 35.356 which is then reduced to a mixture of diasteromeric diols (35.357) with sodium borohydride in methanol. Cleavage of the diol mixture 35.357 with sodium metaperiodate yields the ketone 35.358 which can be converted into the corresponding dimethyl ketal (35.359) by treatment with trimethylorthoformate in hot methanol containing p-toluenesulfonic acid. When the dimethylketal (35.359) is oxidized with sodium hypochlorite and the resulting product mixture hydrolyzed with aqueous methanolic sodium carbonate, the ketone 35.360 results. Catalytic reductive debenzylation of the hydrochloride of 35.360 yields the amine 35.361, and this is followed by carbodiimide-catalyzed condensation of this secondary amine (35.361) with sodium indolacetate (35.362) to generate the amide 35.363. Then, when the amide 35.363 is treated with p-toluenesulfonic acid in hot benzene for 7 min, the isoquinuclidine 35.364 results. Reaction of 35.364 with vinyl magnesium bromide leads to a single vinylcarbinol (35.365) which may be catalytically reduced to the corresponding ethyl derivative and then converted into the amine (35.366) in 90% yield by effecting both lactam reduction and methoxyl hydrogenolysis with the hydride donor prepared from lithium aluminum hydride and aluminum chloride in tetrahydrofuran.

The conversion of the amine 35.366 to catharanthine (35.9) can then be completed by treatment of 35.366 with t-butyl hypochlorite to generate a mixture of 3-chloroindolenines (35.367), which with potassium cyanide in dimethyl acetamide affords the nitrile 35.368. Then, dehydration of 35.368 to the alkene with sulfuric acid, hydrolysis of the nitrile with potassium hydroxide in diethylene glycol to the corresponding carboxylic acid, and esterification with diazomethane, lead to racemic catharanthine (35.9 and its mirror image).

X. TABERSONINE

The amorphous base called tabersonine (35.10) was originally isolated from Amsonia tabernaemontana in 1954 [338], and some years later its correct molecular formula was established by mass spectrometry as $C_{21}H_{24}N_2O_2$. However, the fragmentation pattern of tabersonine (35.10) was somewhat different from that of many of the other Aspidosperma bases and was thus not directly definitive in establishing the structure of this particular base [339]. The optically active hydrochloride of tabersonine (35.10), $[\alpha]_D^{20}$ -310° (methanol), mp 196°C (dec.), has an ultraviolet spectrum (λ_{max} = 226 nm, ϵ = 10,700; λ_{max} = 297 nm, ϵ = 10,500;

35.10

λ_{max} = 329 nm, ϵ = 15,200) characteristic of a β-aminophenyl-α,β-unsaturated ester, and this information was confirmed by examination of the infrared spectrum of the same material in chloroform solution. However, although N-H and CO_2R groups are obviously present, the free base is resistant to both acetylation and hydrolysis. The pmr spectrum of the free base 35.10 possesses resonances assignable to two vinyl protons, four aromatic protons, one N-H proton and the typical triplet-quartet combination of a quaternary C-ethyl function.

Hydrogenation of tabersonine (35.10) results in formation of a dihydro base (35.369; Equation 35.67) in which the vinyl protons readily observed in the pmr spectrum of tabersonine (35.10) have vanished, and the ultraviolet spectrum of the product, 6,7-dihydrotabersonine (35.369), is essentially the same as that of tabersonine (35.10) itself, indicating that the

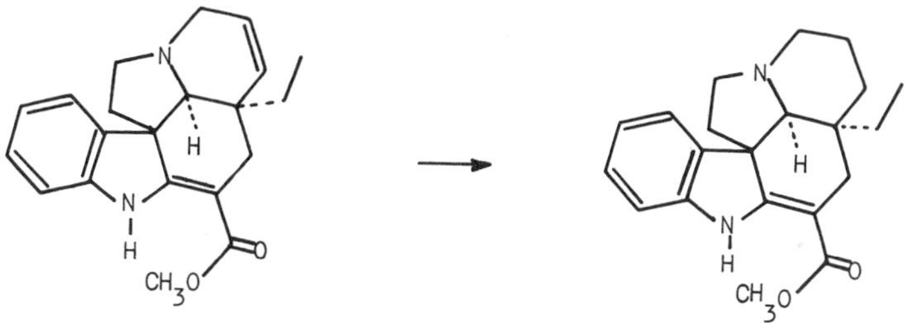

(Eq. 35.67)

35.10 35.369

double bond which has been reduced is not part of the chromophore. The mass spectrum of 6,7-dihydrotabersonine (35.369) has a very large m/e fragment 124, attributed to the grouping 35.370 (Equation 35.68; [339]).

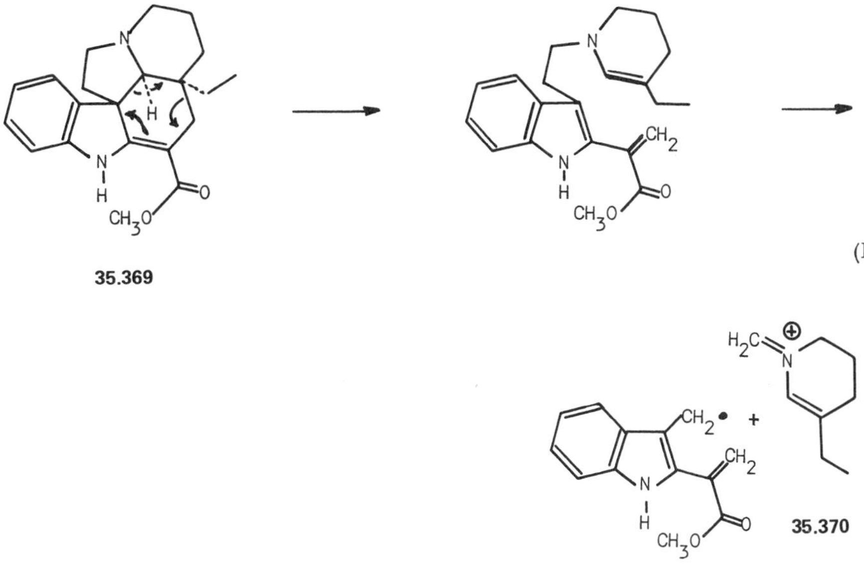

(Eq. 35.68)

35.369

35.370

Reduction of tabersonine (35.10) with zinc in hydrochloric acid yields 2,3-dihydrotabersonine (35.371; Equation 35.69), while with lithium aluminum hydride, tabersonine (35.10) leads to 2,3-dihydrotabersonol (35.372) and the corresponding deoxy base 35.373 (Equation 35.70). The ultraviolet spectra of 2,3-dihydrotabersonine (35.371), 2-3-dihydrotabersonol

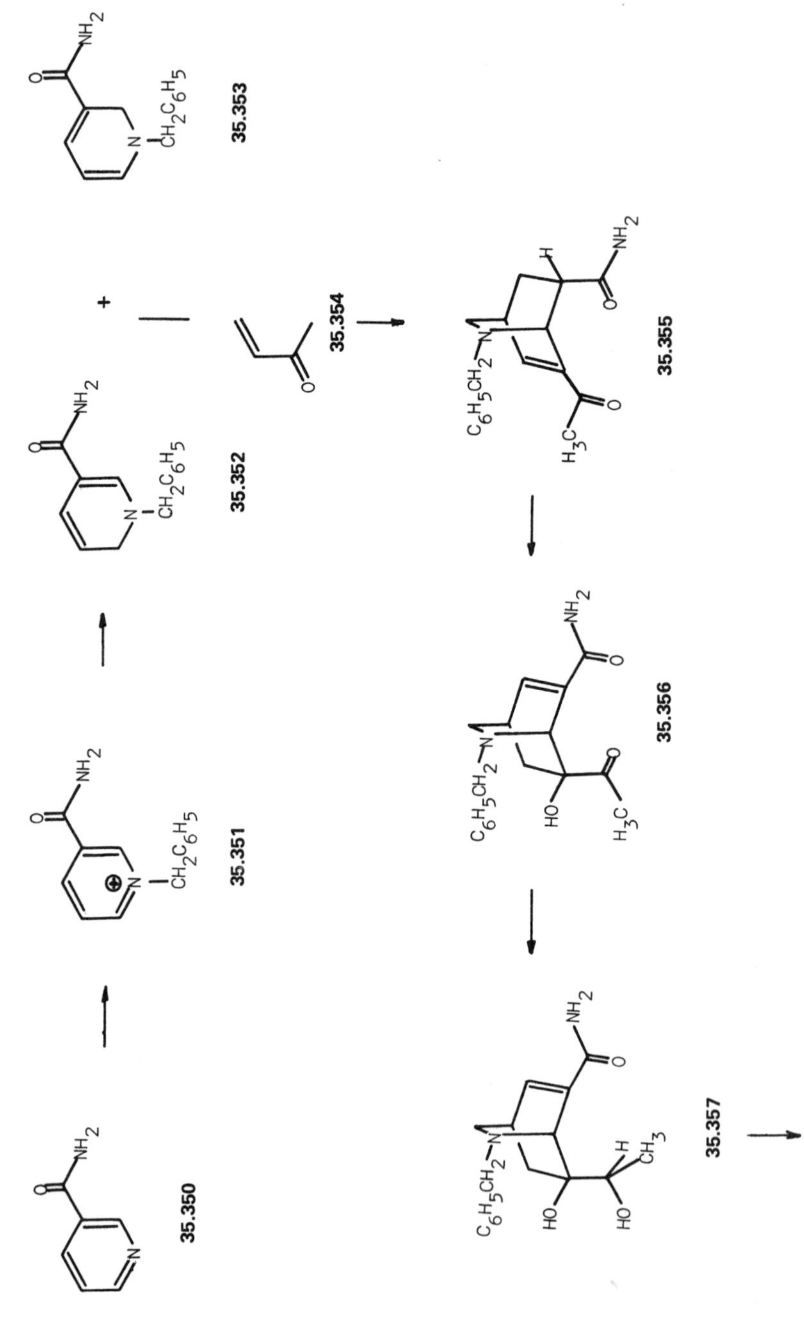

35.350 → 35.351 → 35.352 + 35.353

35.354 → 35.355 → 35.356 → 35.357 →

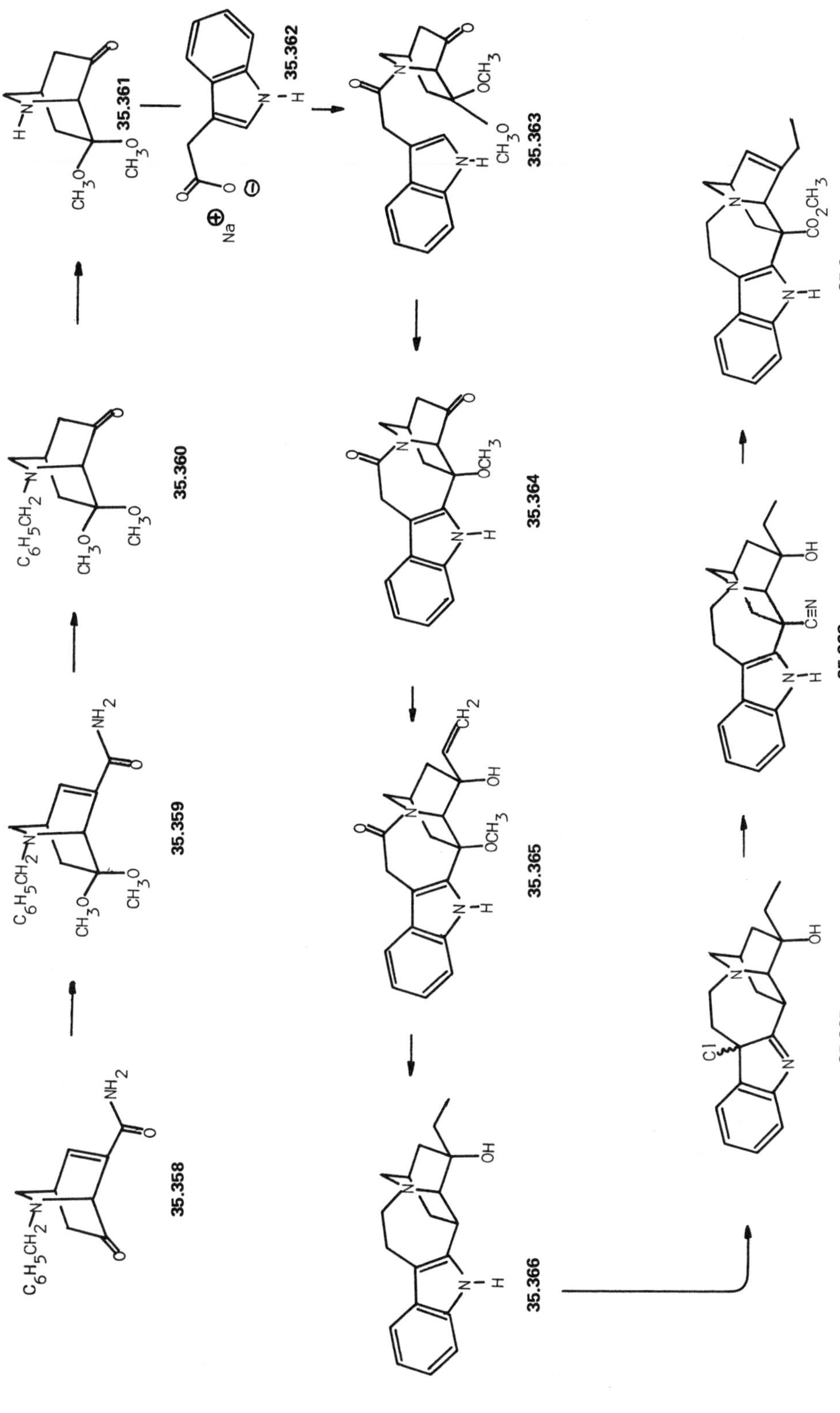

SCHEME 35.50 The total synthesis of catharanthine. (After References 336 and 337.)

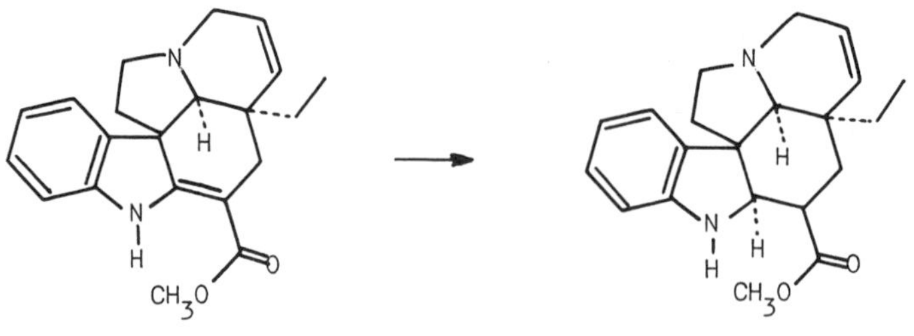

35.10 → 35.371

(Eq. 35.69)

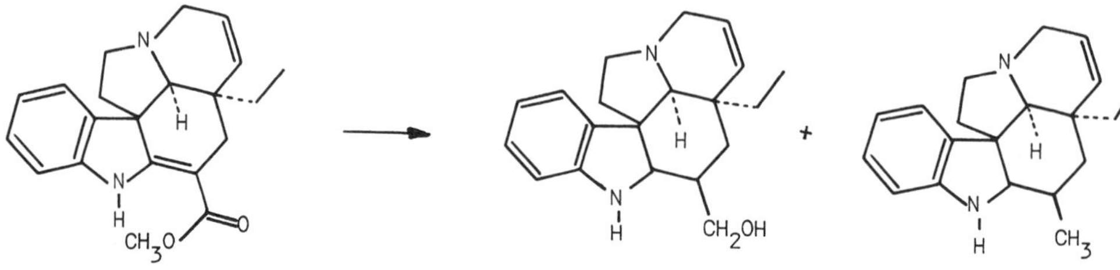

35.10 → 35.372 + 35.373

(Eq. 35.70)

(35.372), and the deoxy base (35.373) are all consonant with the reduction of the conjugated double bond, and the mass spectra also support the changes in the structures.

When 6,7-dihydrotabersonine (35.369) is hydrolyzed under forcing conditions, the indolenine (35.374) results (Equation 35.71). Borohydride reduction of 35.374 yields (+)-quebrachamine (35.375; Equation 35.72), whose structure had been previously assigned as follows:

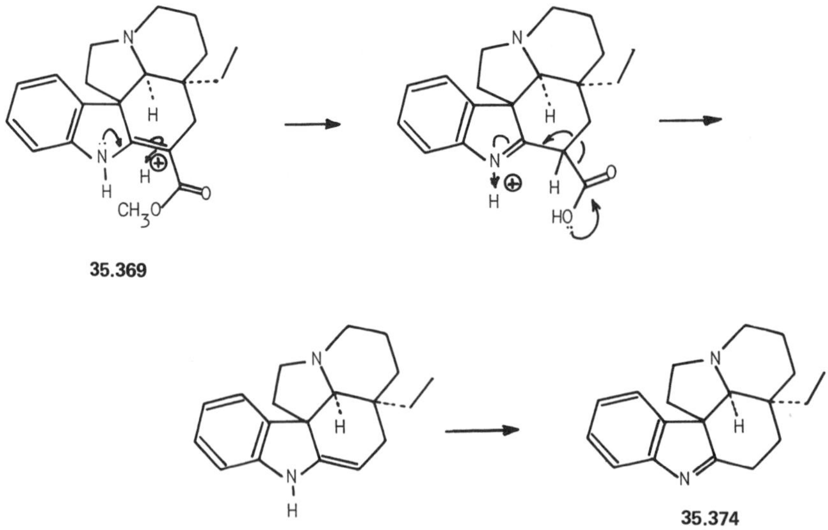

35.369 → → 35.374

(Eq. 35.71)

1. Zinc dust distillation of quebrachamine (35.375) yields 3-methylindole (35.18), 3-ethylpyridine (35.64), 3-ethyl-4-methyl pyridine (35.65), 3-methyl-5-ethylpyridine

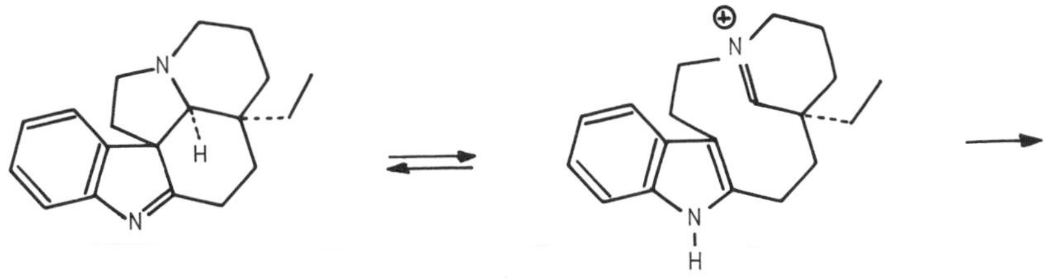

35.374

35.375

(Eq. 35.72)

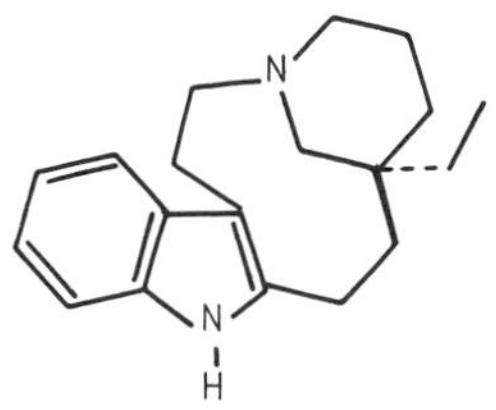

(35.376), 3,5-diethylpyridine (35.377), 2-ethylindole (35.378), 2,3-dimethylindole (35.379), and 2,3-diethylindole (35.380; see Scheme 35.51; [340]).

2. Mass spectral examination of quebrachamine (35.375) yields masses consonant with the fragments shown in Scheme 35.52.

3. Quebrachamine (35.375) was synthesized by modification of a route which had successfully produced aspidospermine (35.381; [341]), whose structure and absolute stereochemistry are known from x-ray crystallography [342].

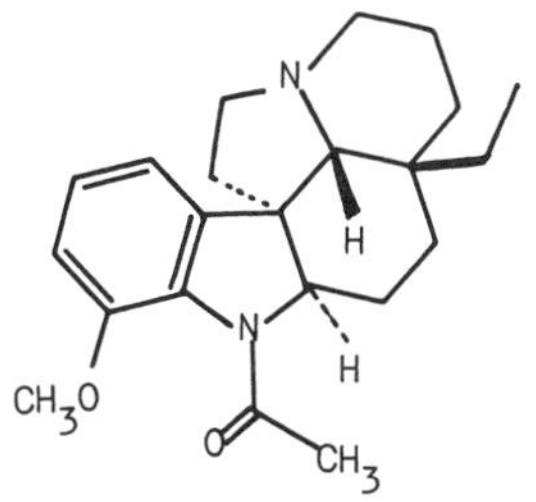

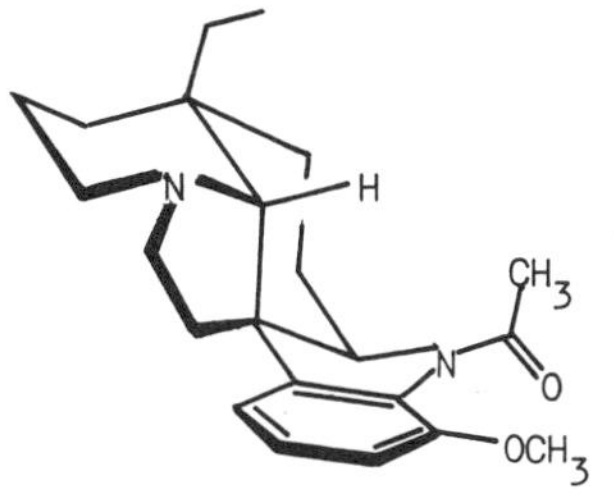

35.381

The synthesis of quebrachamine (35.375) is shown in Scheme 35.53 and was carried out as follows. n-Butyraldehyde (35.382) is condensed, via the enamine 35.383, with ethyl acrylate (35.384) to yield, after hydrolysis, the carbotheoxyaldehyde 35.385. Reaction of the corresponding enamine of 35.385 with methyl vinyl ketone (35.354) and hydrolysis leads, after internal condensation, to the isolabel cyclohexenone 35.386. Aminolysis of 35.386 results in formation of a mixture of amides 35.387 which, by protection of the keto group as its ketal, reduction with lithium aluminum hydride and hydrolysis, eventually generates the ketone 35.388.

When the latter (35.388) is allowed to react with the acid chloride of bromoacetic acid, the amide 35.389 results, and this is successfully cyclized to a mixture of tricyclic keto-amides 35.390 with t-butoxide in dimethyl sulfoxide. Then, protection of the ketone as a ketal, reduction of the amide, and removal of the protecting group generate the mixture of tricyclic ketones 35.391. Under conditions of the Fischer indole synthesis with phenylhydrazine,

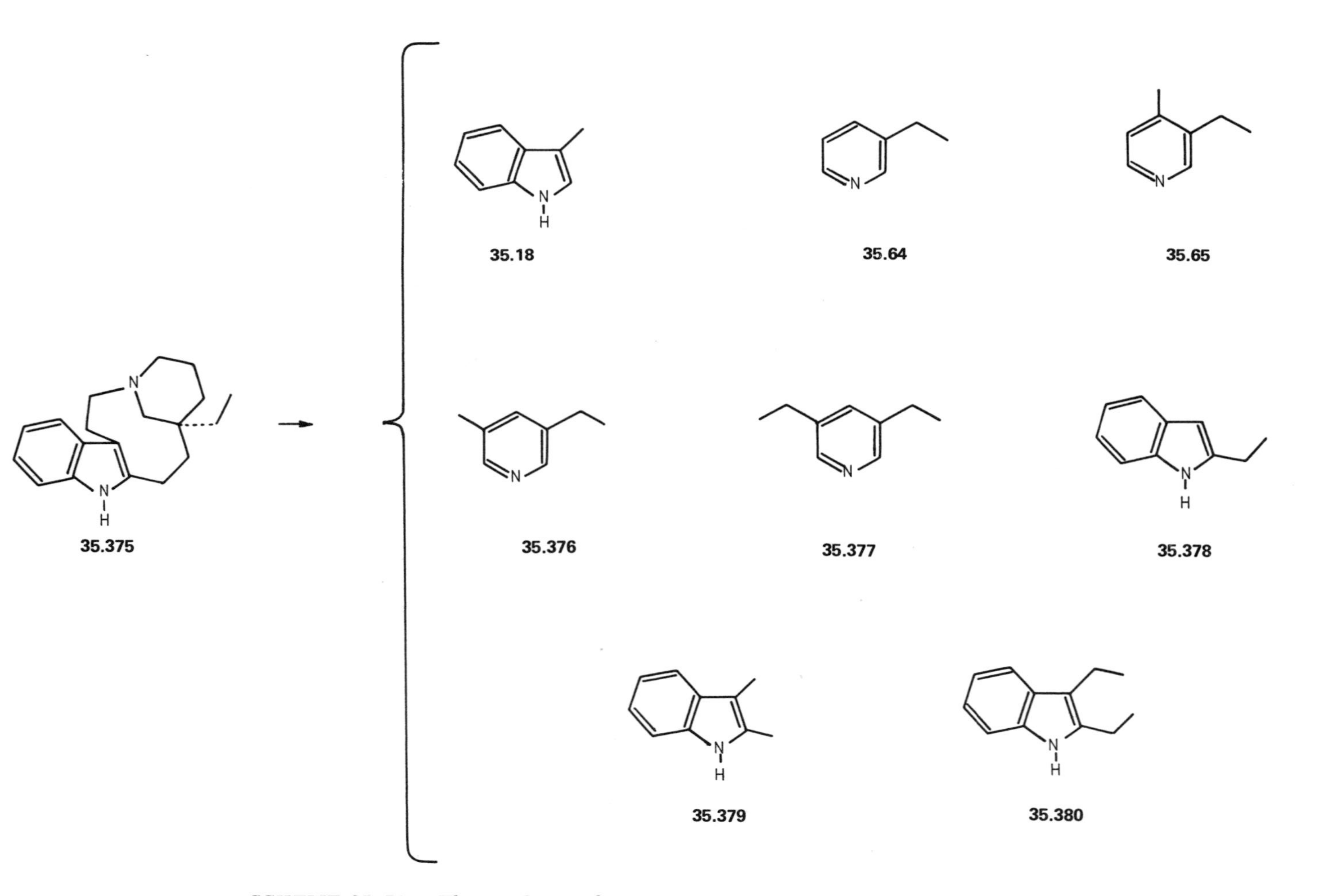

SCHEME 35.51 The products of zinc dust distillation of quebrachamine. (After Reference 340.)

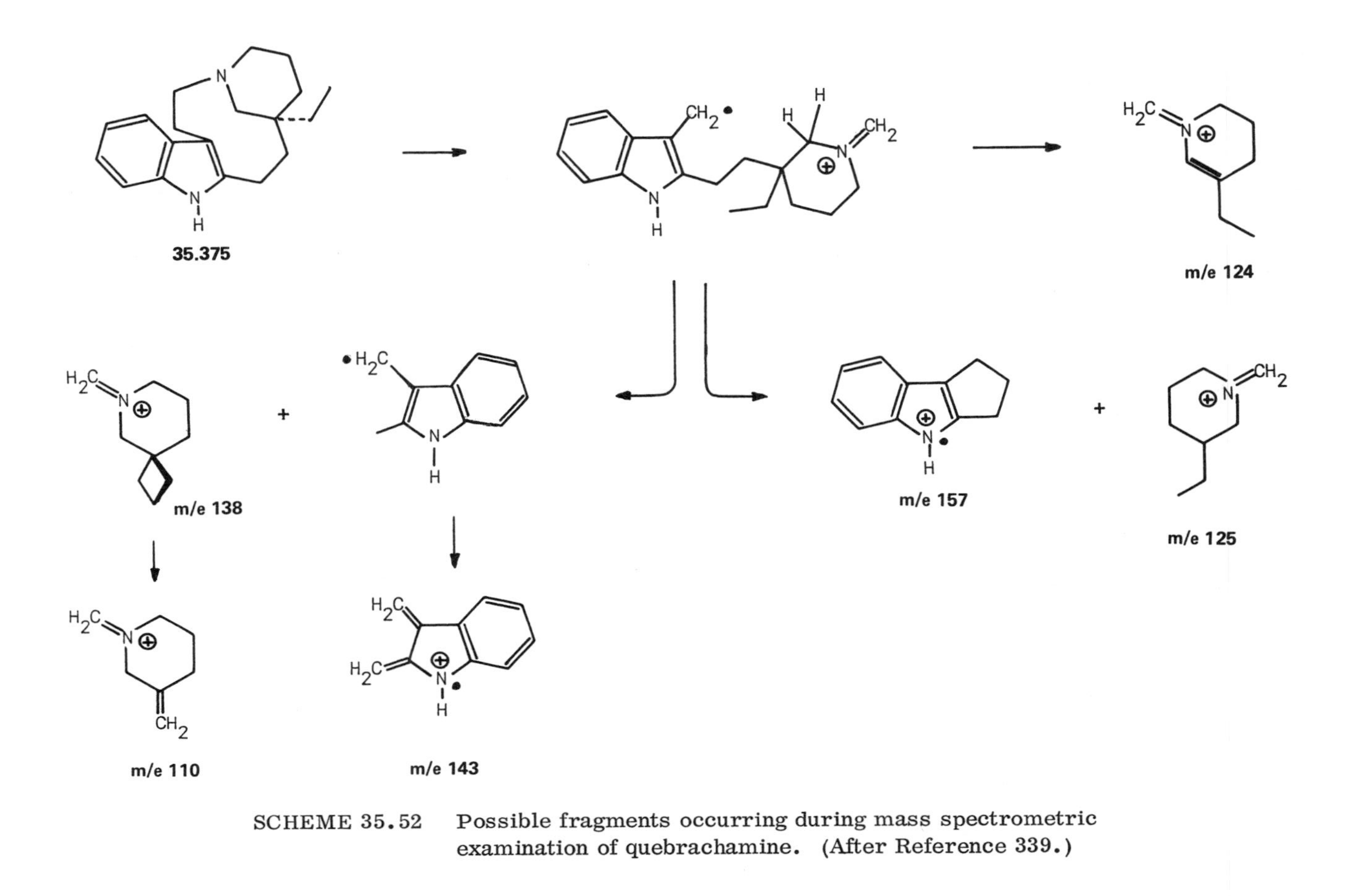

SCHEME 35.52 Possible fragments occurring during mass spectrometric examination of quebrachamine. (After Reference 339.)

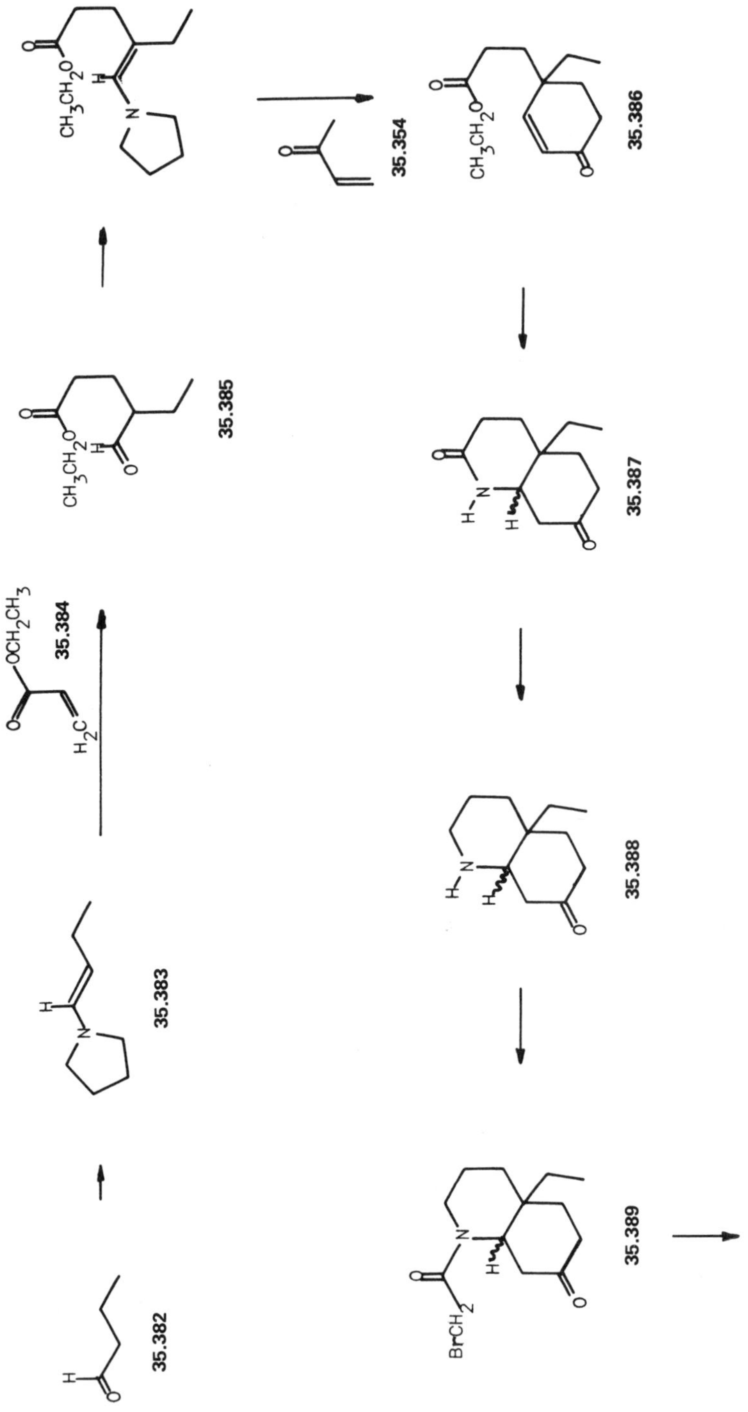

35.382

35.383

35.384

35.385

35.354

35.386

35.387

35.388

35.389

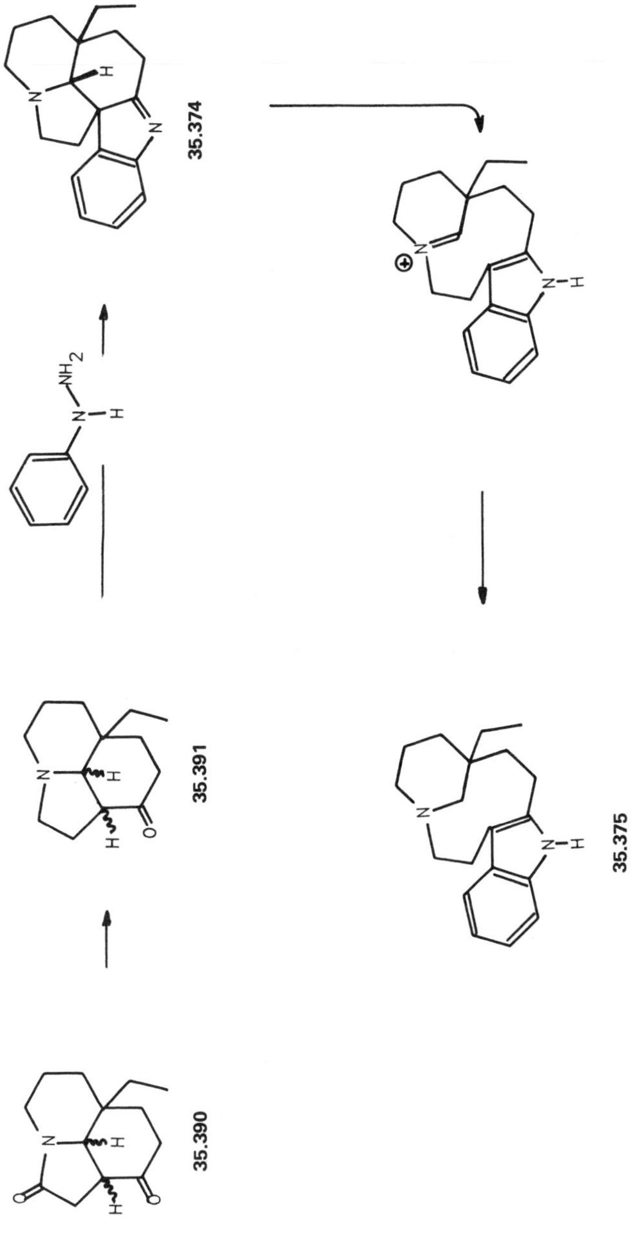

35.390

35.391

35.374

35.375

SCHEME 35.53 The total synthesis of quebrachamine. (After Reference 341.)

equilibration of the chiral centers occurs to generate the most stable isomer and the racemic indolenine 35.374 results; as noted previously (Equation 35.72), reduction of 35.374 generates racemic quebrachamine (35.375).

Tabersonine (35.10) itself has been synthesized [343], and the synthesis is shown as Scheme 35.54. When 5-bromonicotinamide (35.392) is treated with sodium hypobromite, 3-amino-5-bromopyridine (35.393) results. Diazotization of 35.393 followed by thermal acid-catalyzed decomposition yields the corresponding phenol, 3-bromo-5-hydroxypyridine (35.394), which is successfully methylated with diazomethane to the corresponding methyl ether. When the methyl ether is converted into the Grignard reagent (35.395) and treated with acetaldehyde, 3-(α-hydroxyethyl)-5-methoxypyridine (35.396) results. When the pyridine (35.396) is treated with benzyl chloride and then the resultant quaternary salt reduced with lithium aluminum hydride in tetrahydrofuran, a mixture of enol ether isomers 35.397 is found. Hydrolysis of the mixture with aqueous methanolic hydrochloric acid leads to the enone 35.398. When the enone (35.398) is reduced with lithium aluminum hydride in tetrahydrofuran, the corresponding allylic alcohol is produced, and heating of that material (35.399) with ethyl orthoacetate (35.400) effects a Claisen rearrangement and leads to the ester 35.401. The benzyl group of 35.401 is removed by treatment of the base with ethyl chloroformate in refluxing benzene, and the carbamate 35.402 is generated.

The tetracyclic lactam 35.403 is produced from the carbamate 35.402 by hydrolysis of the ester moiety with potassium hydroxide and decarboxylation of the carbamic acid which results, followed by conversion to a new amide with 3-indoleacetyl chloride (35.404), hydrolysis to free the carboxylate, and cyclization with polyphosphoric acid. When 35.403 is reduced with lithium aluminum hydride in tetrahydrofuran the corresponding amino alcohol mixture is produced, and this leads, on treatment with methanesulfonyl chloride-pyridine at 0°C, to the quaternary salt 35.405. The salt is reopened to a pair of nitriles (4:1 ratio) 35.406 on treatment with potassium cyanide in dimethylformamide. Then, the corresponding (35.407) methyl esters are prepared by saponification of the nitriles and subsequent treatment of each with diazomethane. Oxidation of the major isomer (i.e., α) with platinum-oxygen in ethyl acetate leads to racemic tabersonine (35.10 and its mirror image).

XI. VINCAMINE

Vincamine (35.11), $C_{21}H_{26}N_2O_3$, mp 231°C, $[\alpha]_D^{20}$ +39° (pyridine), is the major alkaloid of _Vinca minor_ L. The infrared spectrum of vincamine (35.11) suggests the presence of hydroxy and ester groups, while an N-substituted indole is indicated by the ultraviolet spectrum (λ_{max} 227 nm, ϵ = 26,200; λ_{max} 290 nm, ϵ = 4,700; methanol).

35.11

On basic hydrolysis, vincamine (35.11) yields a carboxylic acid (vincaminic acid, 35.408) and methanol; vincamine (35.11) can be regenerated from vincaminic acid (35.408) with diazomethane [344]. Oxidation of vincaminic acid (35.408) with lead tetraacetate yields (-)-eburnamonine (35.409; Equation 35.73; [345]). (-)-Eburnamonine (35.409) is the optical antipode of an alkaloid whose structure had already been determined (see below). When

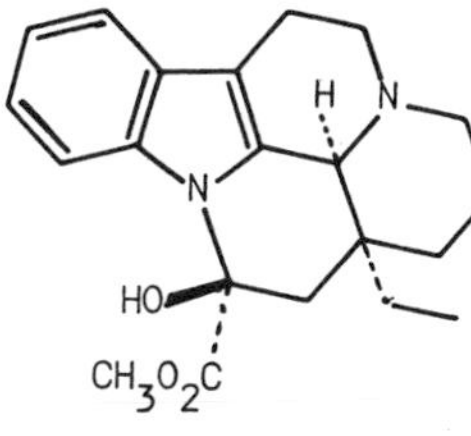

HO CH3O2C

35.11

HO HO2C

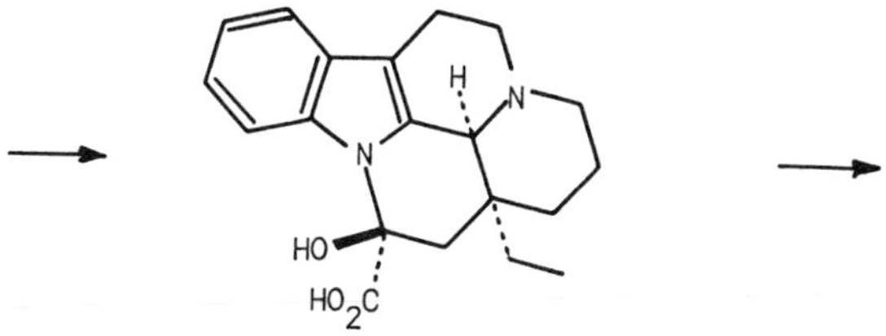

35.408

(Eq. 35.73)

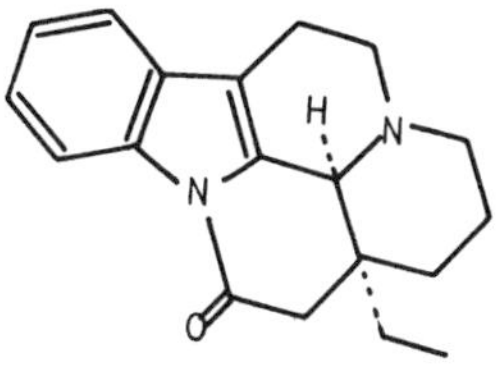

35.409

vincamine (35.11) is subjected to heating with acetic anhydride, apovincamine (35.410) is produced (Equation 35.74; [344]). Reduction of vincamine (35.11) with lithium aluminun hydride yields vincaminol (35.411), but if an acidic workup is employed, eburnamonine (35.409, chirality unknown) results (Scheme 35.55).

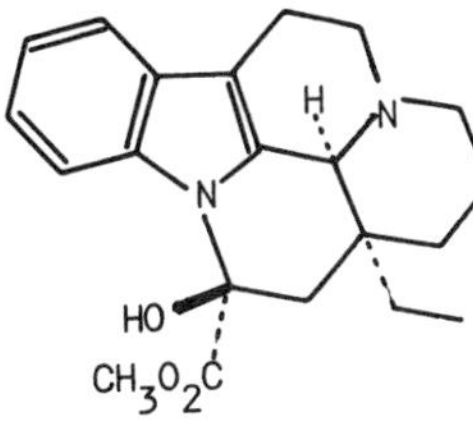

HO CH3O2C

35.11

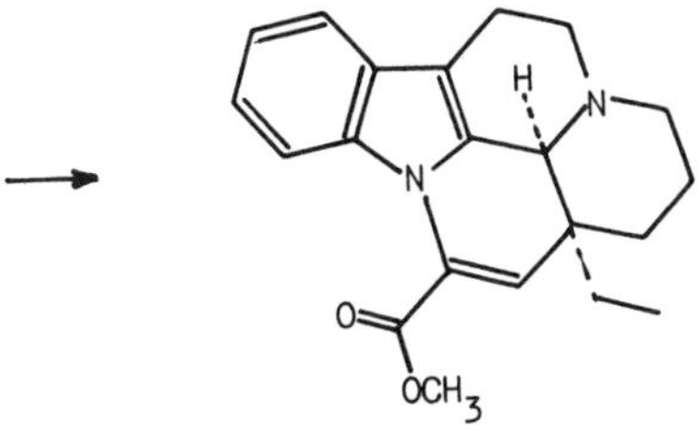

OCH3

35.410

(Eq. 35.74)

(+)-Eburnamonine (35.412; optical antipode of 35.409), $C_{19}H_{23}N_2O$, $[\alpha]^{20}$ +89°, mp 183°C, occurs naturally among the alkaloids of Hunteria eburnea Pichon and was recognized as a ketone since reduction with lithium aluminum hydride generates a mixture of two diastereomeric pentacyclic alkaloids, eburnamine (35.413) and isoeburnamine (35.414; Equation 35.75) which can be reoxidized to the parent alkaloid. On short treatment of (+)-eburnamonine (35.412) with selenium at 360°C, 4-ethyl-4-propyl-4,5-dihydrocanthin-6-one (35.415)

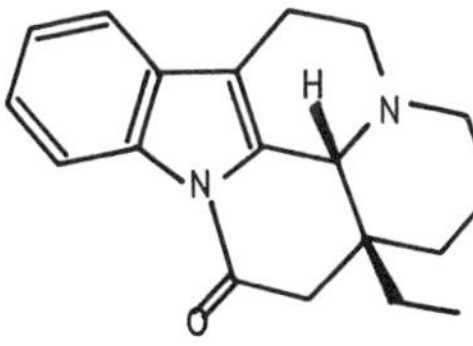

35.412

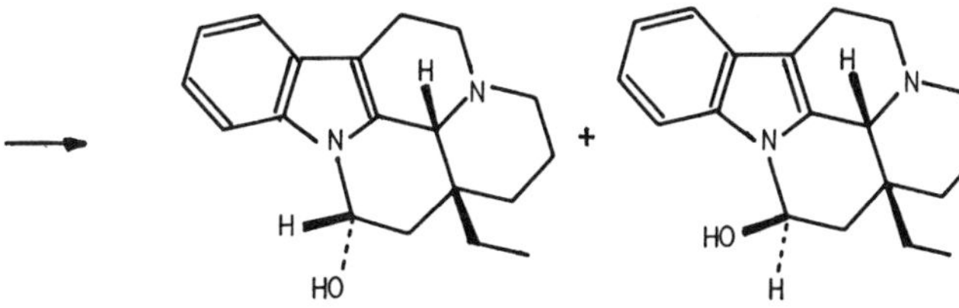

35.413

35.414

(Eq. 35.75)

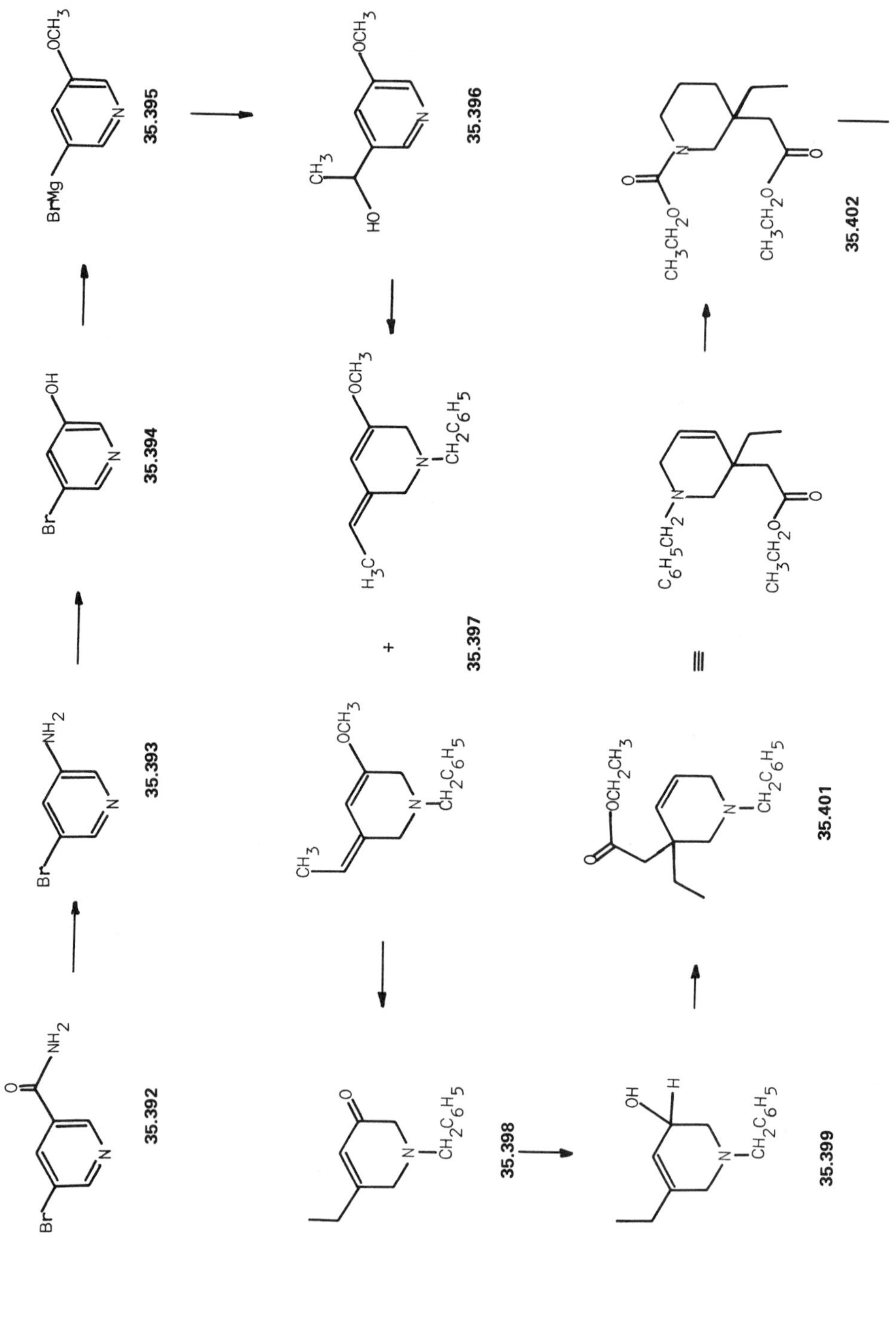

35.392

35.393

35.394

35.395

35.396

35.397

35.398

35.399

35.401

35.402

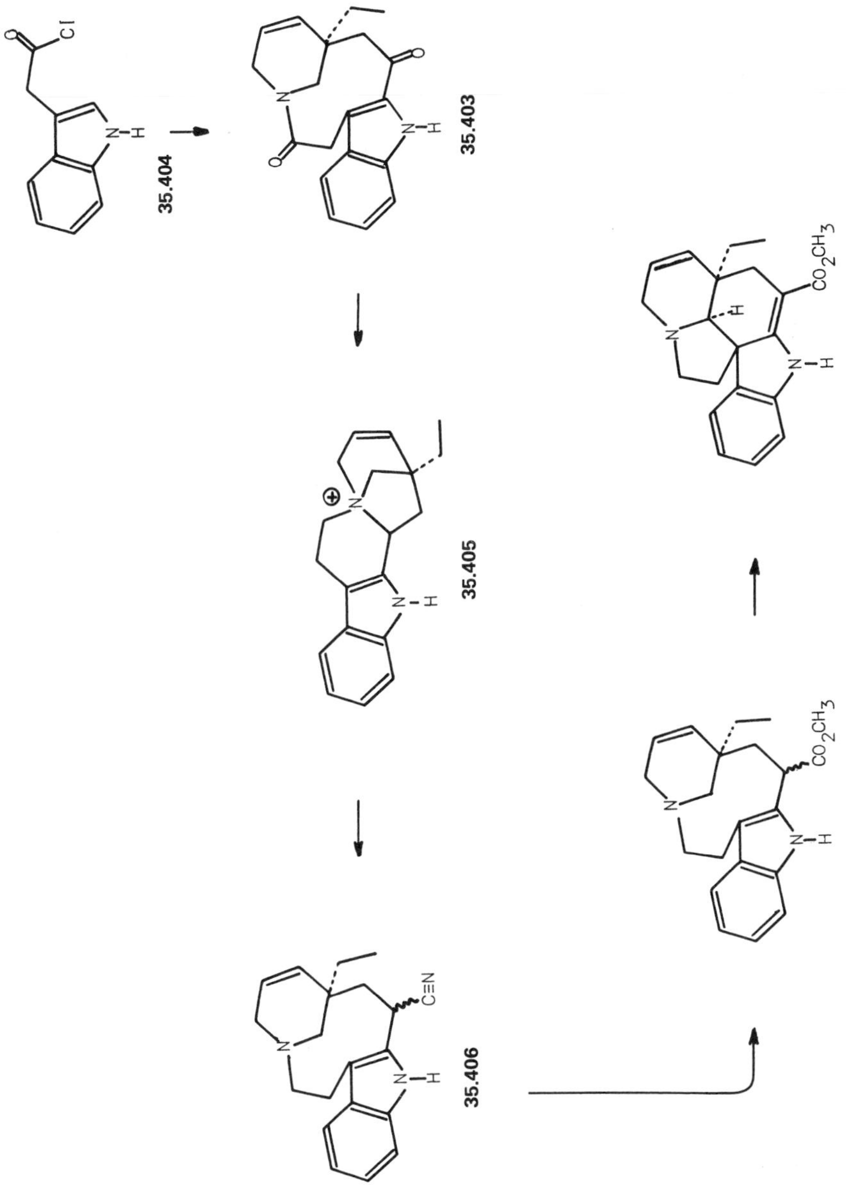

SCHEME 35.54 The total synthesis of tabersonine. (After Reference 343.)

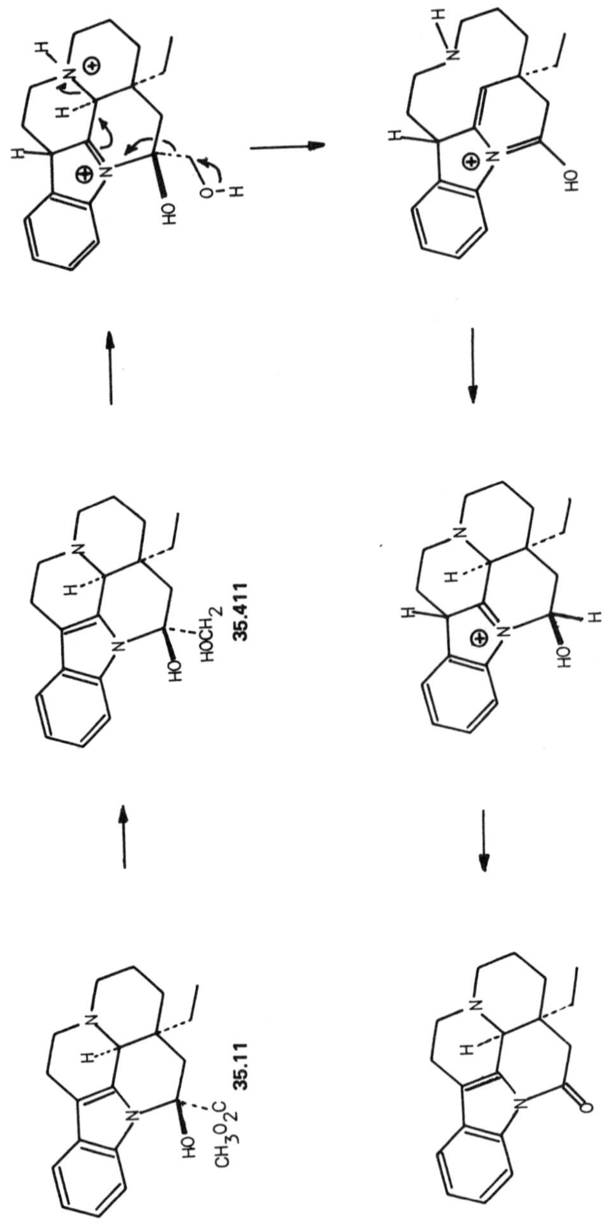

35.11

35.411

35.409

SCHEME 35.55 Conversion of vincamine to eburnamonine.

is formed; longer heating gives both ethyl- and propylcanthin-6-ones (35.416 and 35.417, respectively; Equation 35.76). Both 35.416 and 35.417 have been synthesized, as has been (±)-eburnamonine (35.409 and 35.412; [346]). The synthesis of (±)-eburnamonine (35.409

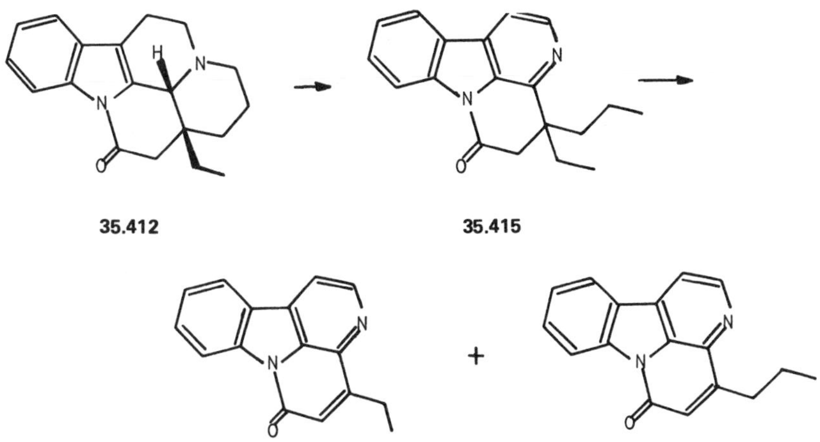

35.412 35.415 (Eq. 35.76)

35.416 35.417

and 35.412; Scheme 35.56) was designed not only to produce the desired alkaloid but also to establish the relative stereochemistry of hydrogen at C_3 and the ethyl group at C_{14}. Thus, diethylethylmalonate (35.418) was alkylated with 1,3-dibromopropane and the product treated with hydrogen bromide to effect decarboxylation to α-ethyl-δ-bromovaleric acid (35.419). Esterification with diazomethane followed by condensation with tryptamine (35.31) leads to the lactam 35.420 which undergoes a Bischler-Napieralsky cyclization on treatment with phosphorus oxychloride to the immonium salt 35.421, isolated as its perchlorate. The salt 35.421 is treated with base and then alkylated with ethyl iodoacetate to 35.422; cyclization to 35.423 is then induced by heating a buffered aqueous solution of the ester 35.422, and 35.423 is isolated as its perchlorate salt. Since reduction (sodium borohydride or catalytic hydrogenation) of 35.423 should lead to a trans system, i.e., 35.424 or its optical antipode, and when consummated affords an isomer of (±)-eburnamonine (35.409 and 35.412), it was concluded that (+)-eburnamonine (35.412) must be represented as shown or as the mirror image of that shown. However, when the absolute configuration (see below) of (+)-eburnamonine (35.412) became known, the problem was resolved. The conclusion that (+)-eburnamonine (35.412) must be as shown (i.e., the mirror image of the structure 35.409) was also reached on biogenetic grounds [347].

Vincamine (35.11) is thus represented as shown with the stereochemistry of the antipode of (+)-eburnamonine (35.412) and the hydroxyl group is placed axial because of the ease of dehydration (see above) to apovincamine (35.410). The absolute configuration at C_3 of vincamine (35.11) has been established [348] as shown in Scheme 35.57. When eburnamine (35.413; [349] is reduced under the conditions of the Wolff-Kishner reaction, (-)-1,1-diethyl-1,2,3,4,6,7,12,12b-octahydroindolo[2,3a]quinolizine (35.425) results. The optical rotatory dispersion curve (ORD) of the quinolizine 35.426 obtained from (-)-eburnamonine (35.409), which had been prepared by degradation of vincamine (35.11), is the mirror image of that obtained from (+)-eburnamonine (35.412), the naturally occurring base.

The ORD curve of (+)-1,2,3,4-tetrahydroharman (35.427) obtained by resolution of the corresponding racemate with (-)-dibenzoyltartaric acid is enantiomeric with that from the quinolizine 35.426 obtained from (-)-eburnamonine (35.409). Then, since 35.427 can be

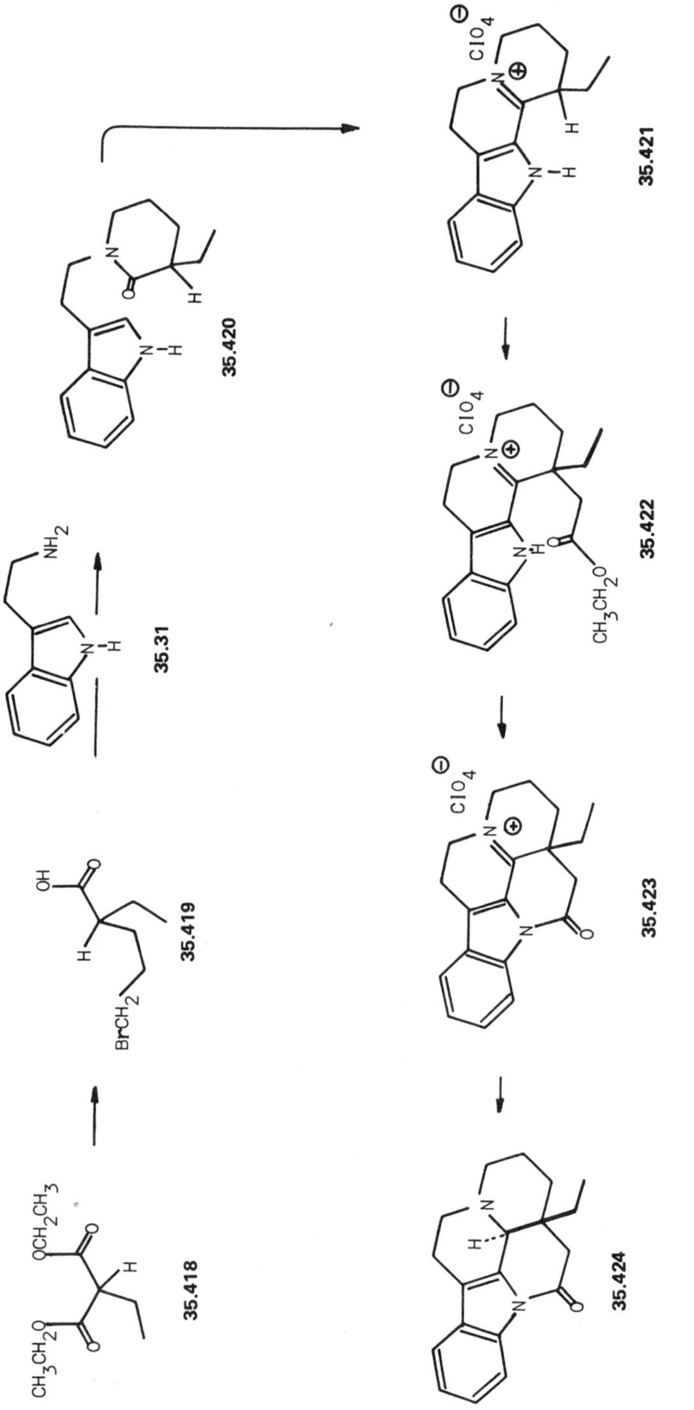

SCHEME 35.56 The total synthesis of (±)-eburnamonine. (After Reference 346.)

35.413 35.425

35.409 35.426

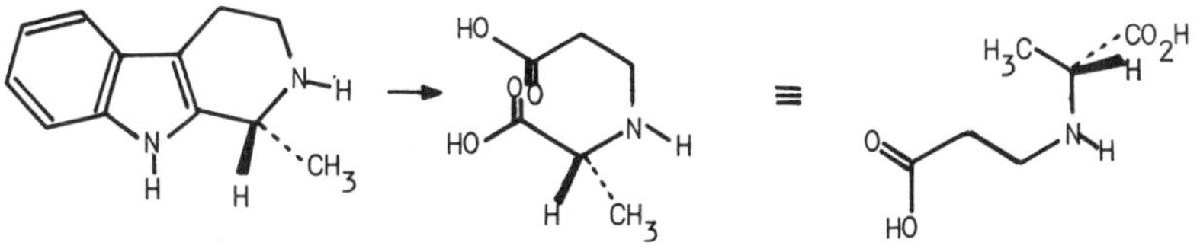

35.427 35.428

SCHEME 35.57 Determination of the absolute configuration of C_3 of vincamine. (After Reference 348.)

converted by formylation and ozonolysis into (-)-N-carboxyethyl-D-alanine (35.428), the R configuration is assigned to 35.427 and (-)-eburnamonine (35.409), and thus vincamine (35.11) must have the configuration shown.

Finally, vincamine (35.11) itself has been synthesized [350]. The synthesis, shown in Scheme 35.58, begins with the conversion of the pyrrolidine enamine of butyraldehyde (i.e., 35.383) to dimethyl 3-ethyl-3-formylpimelate (35.429) by condensation with two equivalents of methyl acrylate followed by hydrolysis. When the formylpimelate ester (35.429) is allowed to condense with tryptamine (35.31) and cyclization is induced, a mixture of the diasteromeric amides 35.430 and 35.431 is obtained. On treatment with phosphorus pentasulfide, these amides are converted into the corresponding thiolactams, which after chromatographic separation are individually converted to the corresponding amino acid esters 35.432 and 35.433, respectively, by Raney nickel desulfurization. Since the two amino acid esters (35.432 and 35.433) taken individually can be interconverted by epimerization at C_3 (through mercuric acetate oxidation followed by sodium borohydride reduction to a mixture of the two

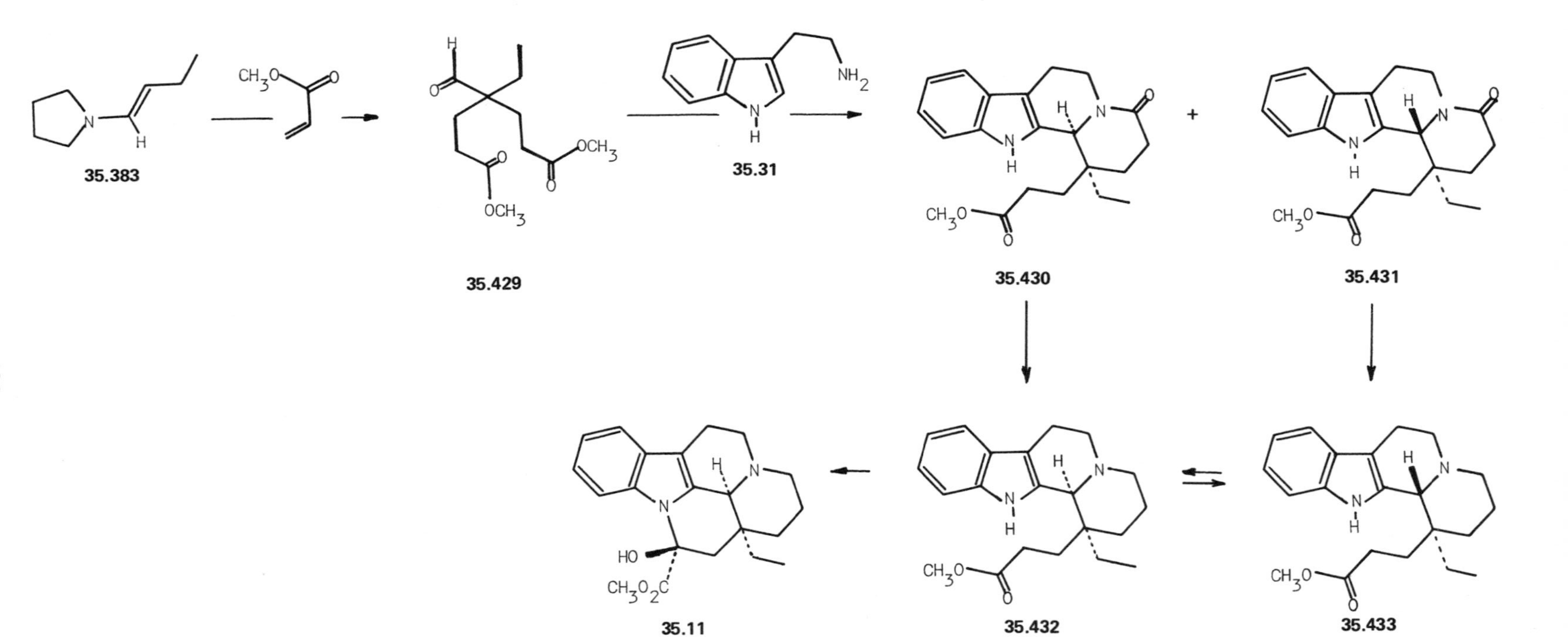

SCHEME 35.58 The total synthesis of vincamine. (After Reference 350.)

bases), either epimeric base could eventually be obtained. The oxidation of the appropriate ester (35.432) with p-nitrosodimethylaniline and triphenylmethyl sodium, followed by acidification, yields vincamine (35.11 and its mirror image).

SELECTED READING

A. R. Battersby and H. F. Hodson, in The Alkaloids, Vol. 8, R. H. F. Manske (ed.). Academic, New York, 1965, pp. 515 ff.

B. Gilbert, in The Alkaloids, Vol. 8, R. H. F. Manske (ed.). Academic, New York, 1965, pp. 336 ff.

B. Gilbert, in The Alkaloids, Vol. 11, R. H. F. Manske (ed.). Academic, New York, 1968, pp. 205 ff.

J. B. Hendrickson, in The Alkaloids, Vol. 6, R. H. F. Manske (ed.). Academic, New York, 1960, pp. 179 ff.

H. L. Holmes, in The Alkaloids, Vol. 2, R. H. F. Manske (ed.). Academic, New York, 1952, pp. 513 ff.

H. L. Holmes, in The Alkaloids, Vol. 1, R. H. F. Manske (ed.). Academic, New York, 1950, pp. 375 ff.

J. P. Kutney, in MTP International Review of Science, Series 1, Vol. 9: Alkaloids, K. Weisner (ed.). Butterworth's, New York, 1973, p. 27.

R. H. F. Manske (ed.), in The Alkaloids, Vol. 8. Academic, New York, 1965, pp. 697 ff.

R. H. F. Manske and W. A. Harrison, in The Alkaloids, Vol. 8, R. H. F. Manske (ed.). Academic, New York, 1965, pp. 679 ff.

H. J. Monteiro, in The Alkaloids, Vol. 11, R. H. F. Manske (ed.). Academic, New York, 1968, pp. 145 ff.

J. E. Saxton, in The Alkaloids, Vol. 8, R. H. F. Manske (ed.). Academic, New York, 1965, pp. 59 ff., 119 ff., 159 ff., 673 ff.

J. E. Saxton, in The Alkaloids, Vol. 10, R. H. F. Manske (ed.). Academic, New York, 1968, pp. 501 ff., 521 ff.

J. E. Saxton, in The Alkaloids, Vol. 12, R. H. F. Manske (ed.). Academic, New York, 1970, pp. 207 ff.

E. Schittler, in The Alkaloids, Vol. 8, R. H. F. Manske (ed.). Academic, New York, 1965, pp. 287 ff.

G. F. Smith, in The Alkaloids, Vol. 8, R. H. F. Manske (ed.). Academic, New York, 1965, pp. 592 ff.

W. I. Taylor, in The Alkaloids, Vol. 8, R. H. F. Manske (ed.). Academic, New York, 1965, pp. 203 ff.

W. I. Taylor, in The Alkaloids, Vol. 11, R. H. F. Manske (ed.). Academic, New York, 1968, pp. 41 ff., 79 ff., 99 ff., 125 ff.

W. I. Taylor and N. R. Farnsworth (eds.), The Catharanthus Alkaloids. Dekker, New York, 1975.

R. B. Turner and R. B. Woodward, in The Alkaloids, Vol. 3, R. H. F. Manske (ed.). Academic, New York, 1965, pp. 1 ff.

36

CHEMISTRY OF THE DIMERIC ALKALOIDS DERIVED FROM TRYPTOPHAN AND FROM POLYCYCLIC β-CARBOLINE ALKALOIDS

> Awake, O north wind; and come thou south; blow upon my
> garden, that the spice thereof may flow out. . . .
>
> The Song of Solomon IV, 7

There are only three alkaloids which shall be considered in this chapter. They are widely different, but share the common genesis of derivation from, ultimately, tryptophan (36.1).

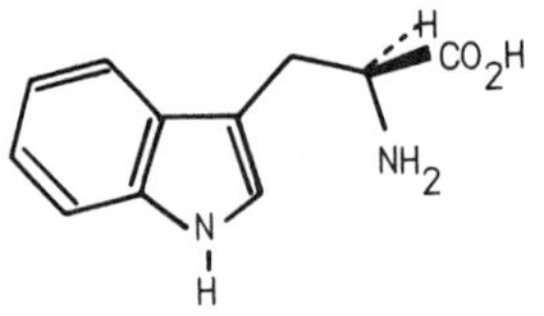

36.1

The first, calycanthine (36.2), is a representative of a rather small group of bases isolated from the seeds of the flowering aromatic shrubs, Carolina allspice (*Calycanthus floridus*) and the Japanese allspice (*Chimonanthus fragans*), and for many years calycanthine

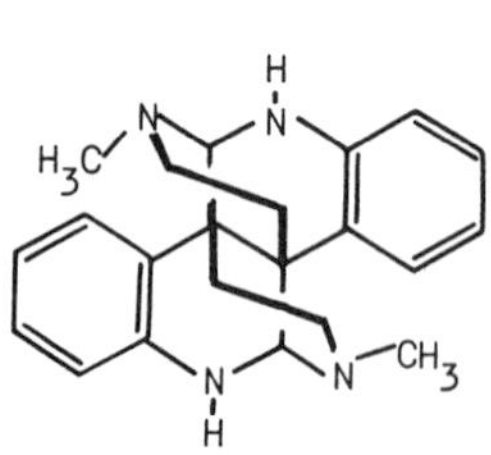

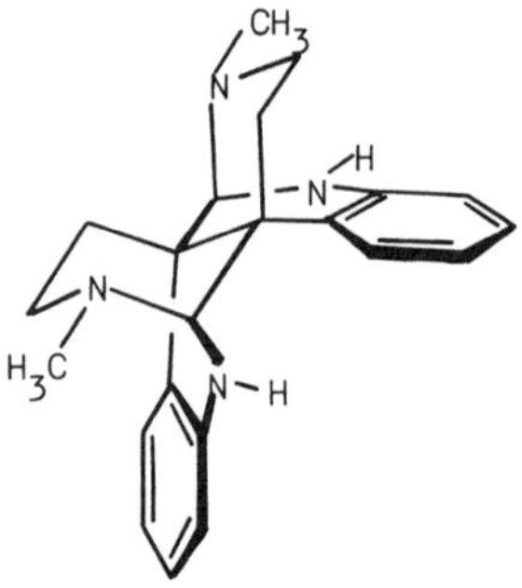

36.2

(36.2) was thought to be an indole alkaloid because of color reactions indicative of this nucleus. Indeed, as we have seen earlier (Chapter 31), calycanthine (36.2) is biogenetically related to bases containing the indole nucleus even though it is lacking in the elaborated alkaloid.

The second alkaloid to be discussed, C-toxiferine (36.3), is one of a very large number of bases isolated from a curare. Despite the fact that the word curare is a nonspecific term used to designate a group of arrow poisons (apparently introduced by the Caribs), two general

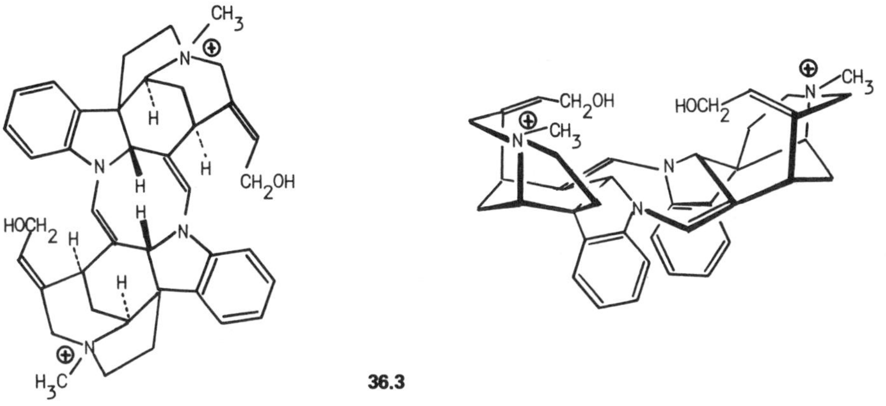

36.3

types of curare are recognized. The first, already discussed (Chapter 24) is called "tubocurare," is packed by its users in tubes, and consists largely of alkaloids possessing benzylisoquinoline and related structures. The second type, which concerns us here, is called "calabash curare," is packed in gourds, and consists largely of dimeric alkaloids related to strychnine (36.4).

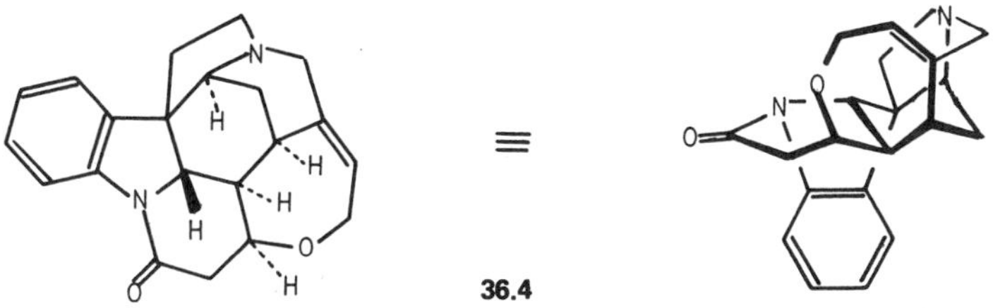

36.4

The final alkaloid to be considered in this chapter is vinblastine (36.5). This base is one of about 60 isolated from Catharanthus roseus, (L) G. Don, but is one of the few dimeric bases present. The search for vinblastine (36.5) was initiated largely on the basis of folklore, A brew made from Jamaican periwinkle had established itself in local medicine as a treatment for diabetes some years before the potent antileukemic alkaloid vinblastine (36.5) was found therein. Interestingly, no basic, alkaloidal material useful in treatment of diabetes has, apparently, yet been isolated from C. roseus.

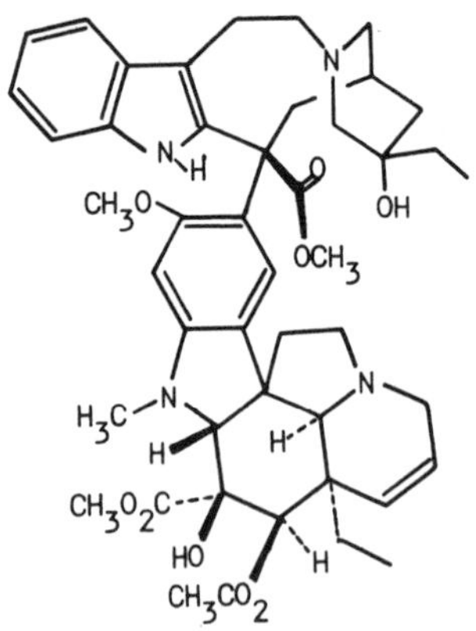

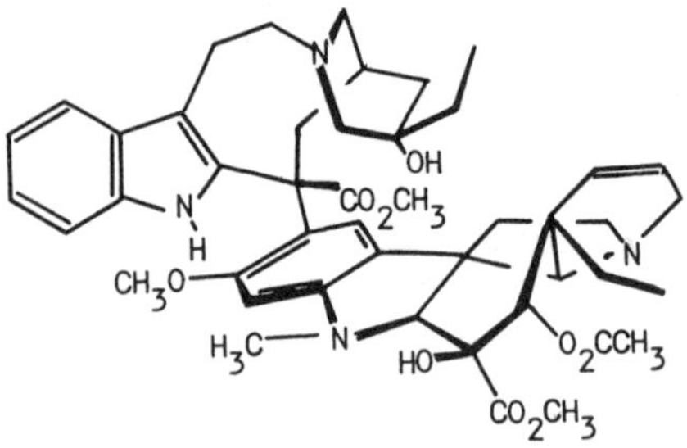

36.5

I. CALYCANTHINE

Although calycanthine (36.2), $C_{22}H_{26}N_4$, $[\alpha]_D^{18}$ +684° (ethanol), is reported to have been isolated as early as 1905 [351] from the seeds of Calycanthus glaucus Willd. its structure remained undetermined until it was revealed by x-ray crystallography of the corresponding dihydrobromide [352] in 1960.

A major problem encountered in the structure elucidation was the fact that when calycanthine (36.2) is boiled in ethanol with a trace of p-dimethylaminobenzaldehyde and hydrogen chloride (Erlich's reagent), the characteristic pink coloration attributed to monosubstituted indole nuclei develops. While this may indicate failure of the reagent to respond properly, it is possible that calycanthine (36.2) has rearranged under the conditions of the reaction to an indole-containing material. This conclusion derives from the observation [353] that when chimonanthine (36.6), as a racemate, is heated in acetic acid, a 1:4 mixture of chimonanthine (36.6) and calycanthine (36.2) results. This may occur as indicated in Scheme 36.1 via an intermediate such as 36.7. Among the other interesting reactions of calycanthine (36.2) are:

1. Benzoylation followed by gentle oxidation to yield N-benzoyl-N-methyltryptamine (36.8). This presumably also occurs via a species such as 36.7, which on benzoylation and oxidation (Scheme 36.2) can lead to 36.8.

2. Heating calycanthine (36.2) with phthalic anhydride is reported to yield the carboline derivative 36.9 [354]. Presumably (Scheme 36.3), this also occurs via an intermediate such as 36.7.

3. Pyrolysis (Scheme 36.4; [354]) of calycanthine (36.2) yields carboline (36.10), 3-methylindole (36.11), 3-ethylindole (36.12), 4-methylquinoline (36.13), quinoline (36.14), and calycanine (36.15), the structure of which was established unequivocally by a simple synthesis (Scheme 36.5) which involves condensation of o-aminoacetophenone (36.16) with o-nitrophenylpyruvic acid (36.17) in the presence of zinc chloride to yield, via 36.18, an isolable intermediate (36.19) which undergoes decarboxylation to 36.15 [355].

Finally, no satisfactory structure for calycanthine (36.2) being available, x-ray crystallography of the dihydrobromide provided the presently accepted structure [352]. The absolute configuration of calycanthine (36.2) has been determined by the coupled oscillator method [356] which involves analysis of the two noncoplanar aniline chromophores by means of a model in which the excitation dipoles of the two chromophores couple to give a resultant helical charge displacement.

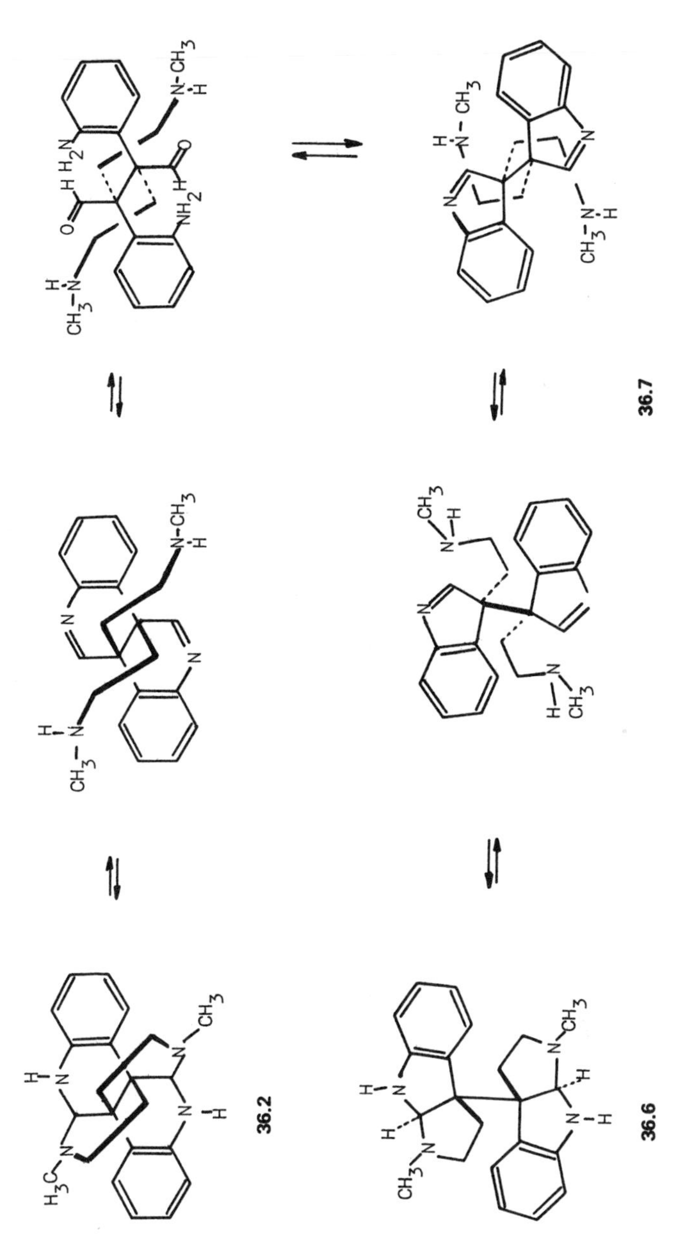

SCHEME 36.1 A possible pathway for the isomerization of calycanthine to chimonanthine.

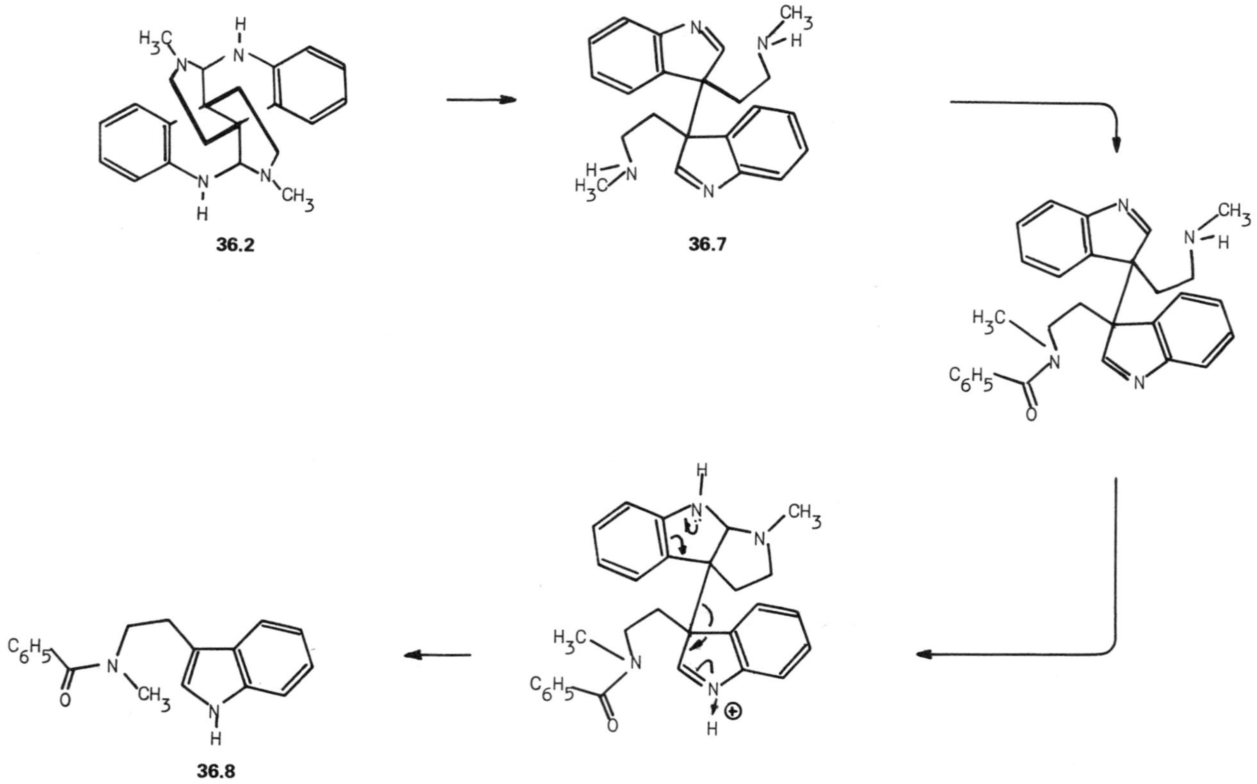

SCHEME 36.2 A possible pathway for the formation of N-benzoyl-N-methyltryptamine from the benzoylation of calycanthine.

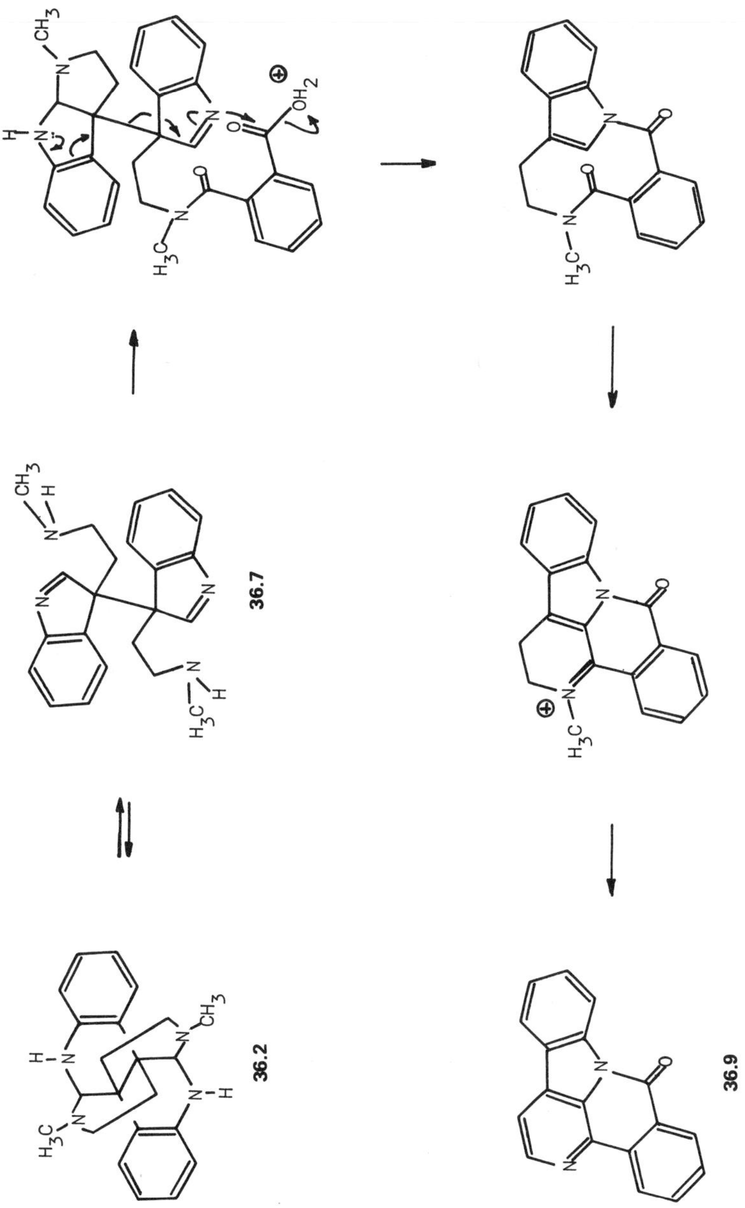

36.2

36.7

36.9

SCHEME 36.3 A potential pathway to the carboline formed on heating calycanthine with phthalic anhydride.

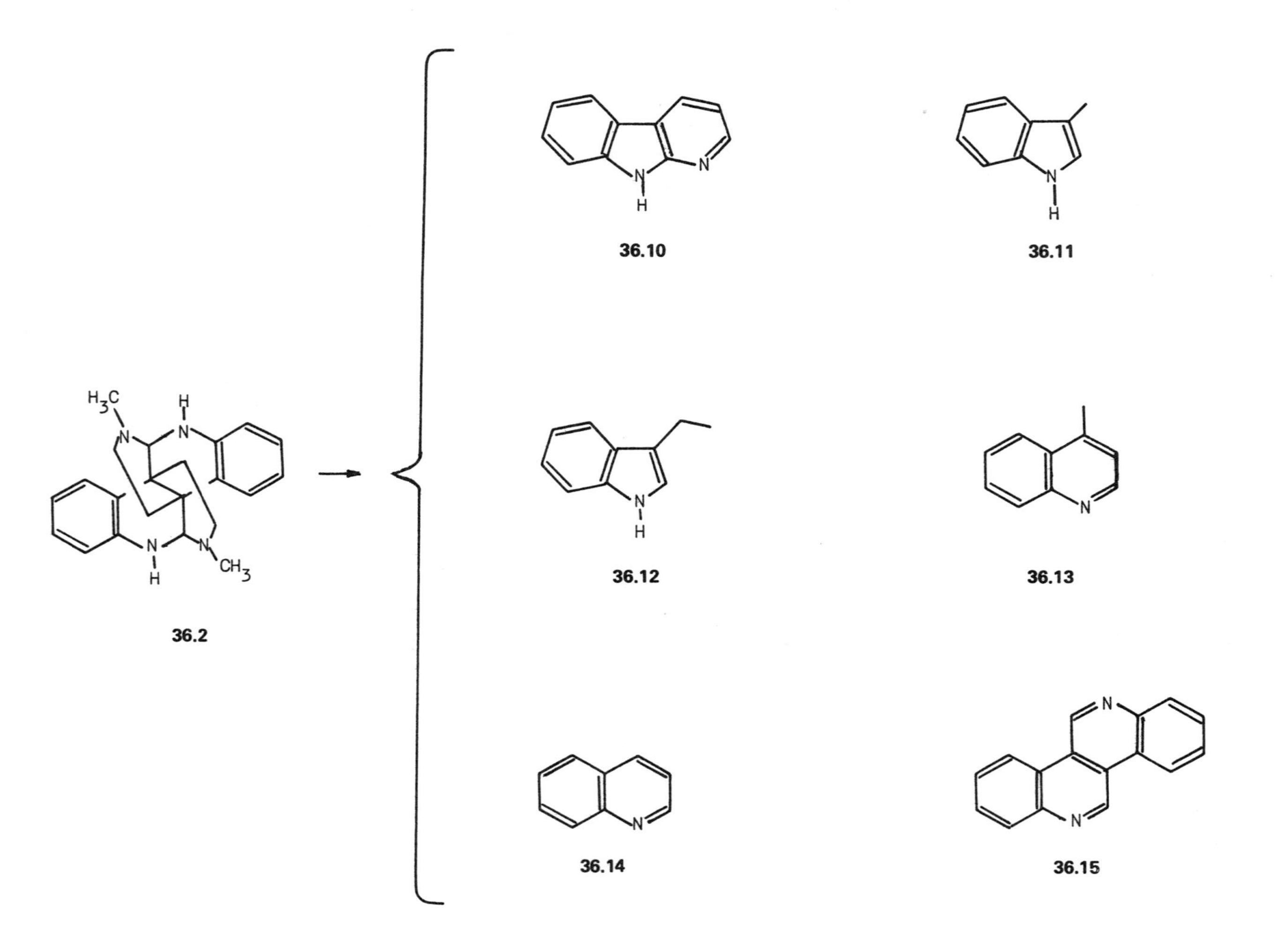

SCHEME 36.4 Products of the thermal decomposition of calycanthine. (After Reference 354.)

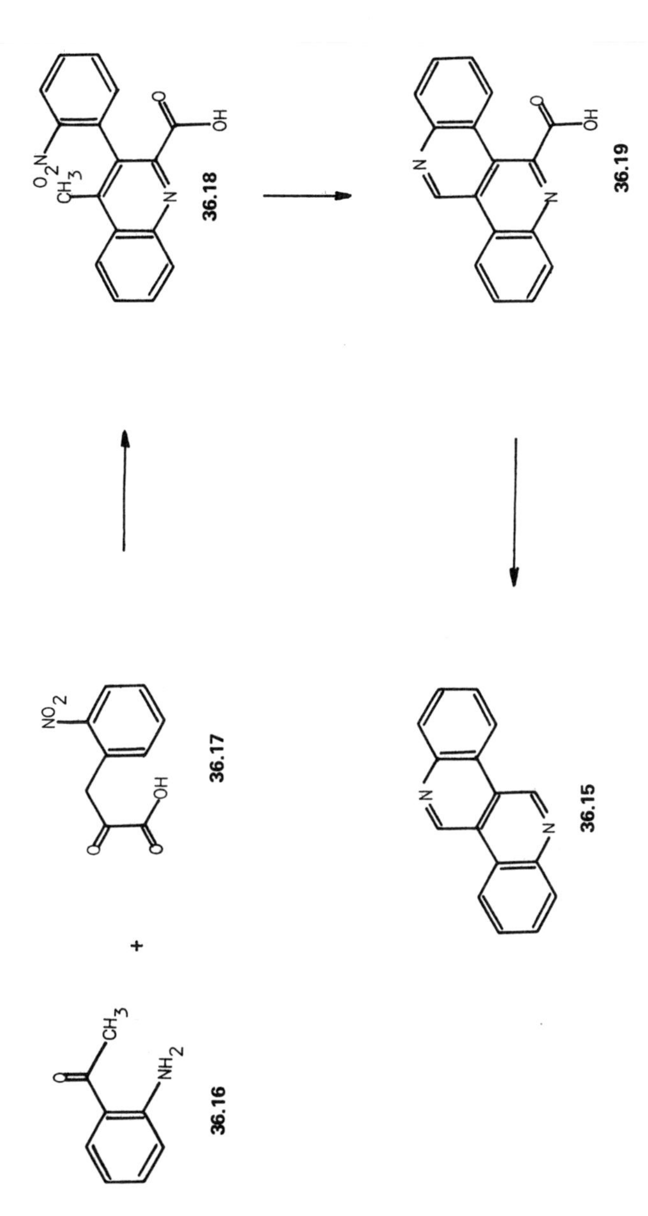

SCHEME 36.5 A synthesis of calycanine, a degradation product of calycanthine. (After Reference 355.)

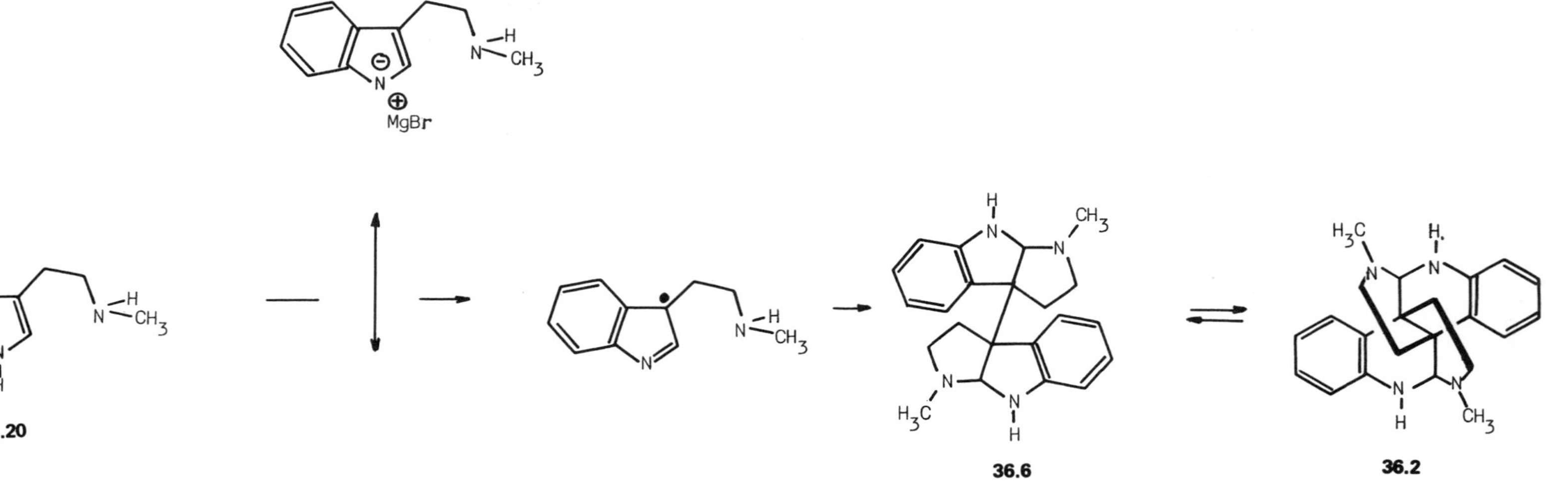

(Eq. 36.1)

Calycanthine (36.2) has been synthesized [357], and the synthesis, which also generates chimonanthine (36.6), is shown in Equation 36.1. Thus, when an ether solution of the Grignard derivative of $N_{(b)}$-methyltryptamine (36.20) is treated with ferric chloride and then hydrolyzed, unchanged $N_{(b)}$-methyltryptamine (36.20) and chimonanthine (36.6 and its optical antipode) along with other products are found. When (±)-chimonanthine (36.6 and its mirror image) is heated in acetic acid (Scheme 36.1) calycanthine (36.2 and its optical antipode) and unchanged racemic chimonanthine (36.6 and its mirror image) are obtained.

II. C-TOXIFERINE

The quaternary alkaloid C-tixoferine (36.3), occasionally called toxiferine I, was isolated in 1941 [358] from Strychnos toxifera and was demonstrated to have a high degree of curare-type activity. As its dichloride, the base, $C_{40}H_{46}N_4O_2Cl_2$, has $[\alpha]_D$ -540° (water), and the picrate

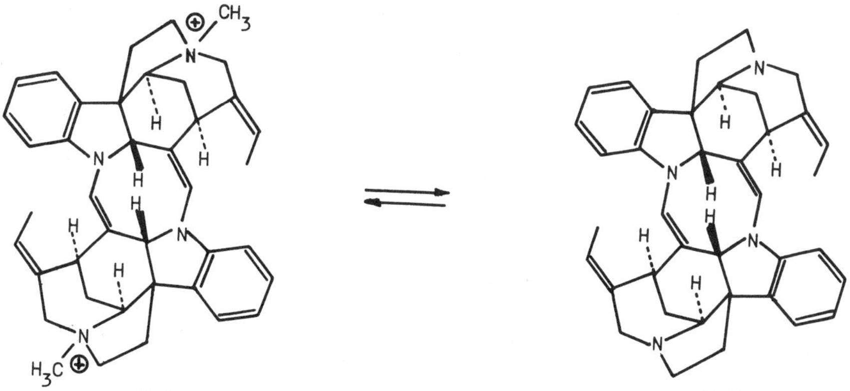

36.21 ⇌ 36.22 (Eq. 36.2)

has mp 278-280°C (dec.). Although a number of closely related bases accompany C-toxiferine (36.3), one other quaternary base and one tertiary base were particularly important in the elucidation of its structure.

The quaternary base of import is C-dihydrotoxiferine I (36.21), $C_{40}H_{46}N_4Cl_2$, as the dichloride. This base was capable of interconversion (Equation 36.2) with its corresponding tertiary base, nordihydrotoxiferine (36.22). Molecular distillation of the quaternary salt

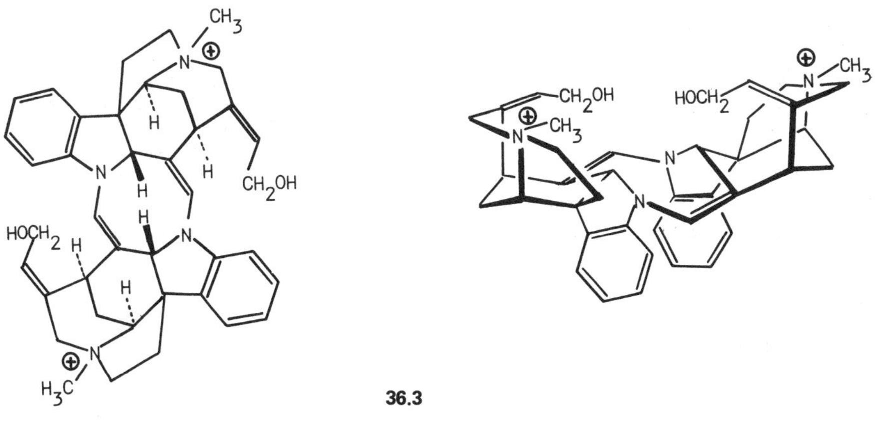

36.3

(36.21 as the dichloride) yielded methyl chloride and 36.22, while methylation of the tertiary base with methyl iodide yielded 36.21, as the diiodide [359].

Although much preliminary work had been done on the structure of these bases, it was not until about 1954 [360], when examination of the tertiary bases which accompany C-toxiferine (36.3) in S. toxifera was undertaken, that significant progress was made.

One of the tertiary bases, caracurine V, $C_{38}H_{40}N_2O_2$ (36.23), the important tertiary base to which reference was made above, on treatment with dilute aqueous acid (Scheme 36.6) rapidly yields an unstable base called caracurine Va (nortoxiferine I, 36.24). Then, more slowly, caracurine Va (36.24) is transformed into caracurine II, $C_{38}H_{38}N_4O_2$ (36.25), and caracurine VII, $C_{19}H_{22}N_2O_2$ (36.26), in the same acidic solution. Caracurine II (36.25), lacking two hydrogens, is still dimeric and is clearly a product of oxidation; caracurine VII, 36.26, on the other hand, can no longer be considered dimeric, and it proved identical to the Wieland-Gumlich aldehyde (36.26) isolated from a degradation of strychnine (36.4; see Chapter 35, Section VIII).

In a fashion similar to that shown above, C-dihydrotoxiferine (36.21), on treatment with dilute mineral acid, yields a product of oxidation and hemidihydrotoxiferine I (36.27), which in aqueous acetic acid is reconverted into C-dihydrotoxiferine (36.21; [360]; Equation 36.3). Under the same conditions, C-toxiferine (36.3) is converted into hemitoxiferine I (36.28; Equation 36.4).

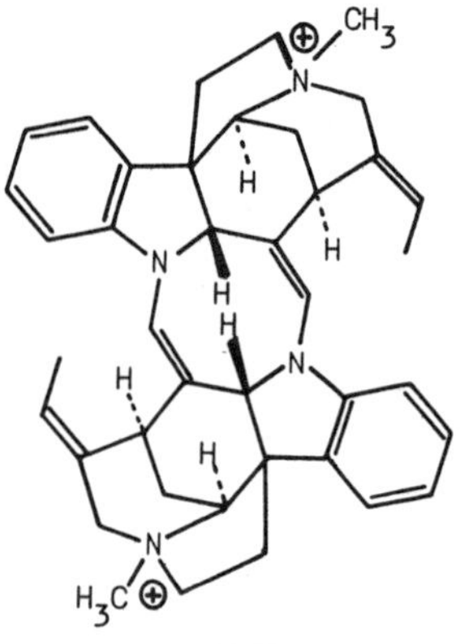

36.21 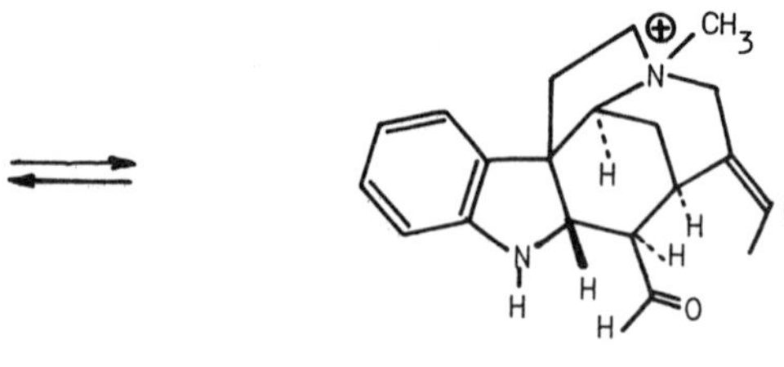 ⇌ 36.27 (Eq. 36.3)

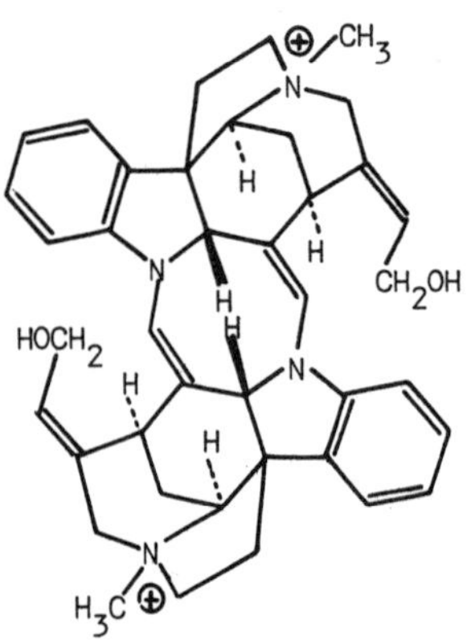

36.3 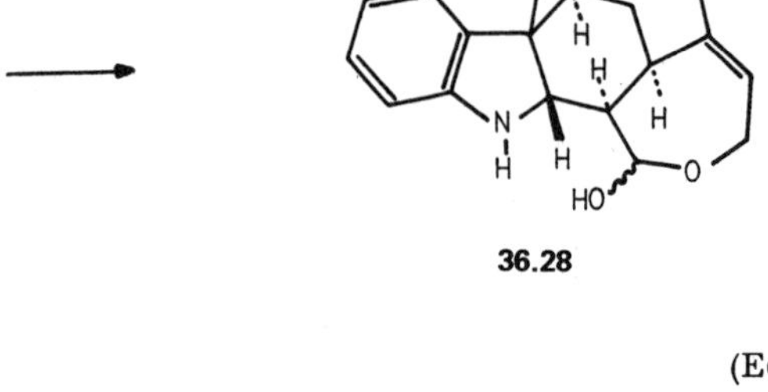 → 36.28 (Eq. 36.4)

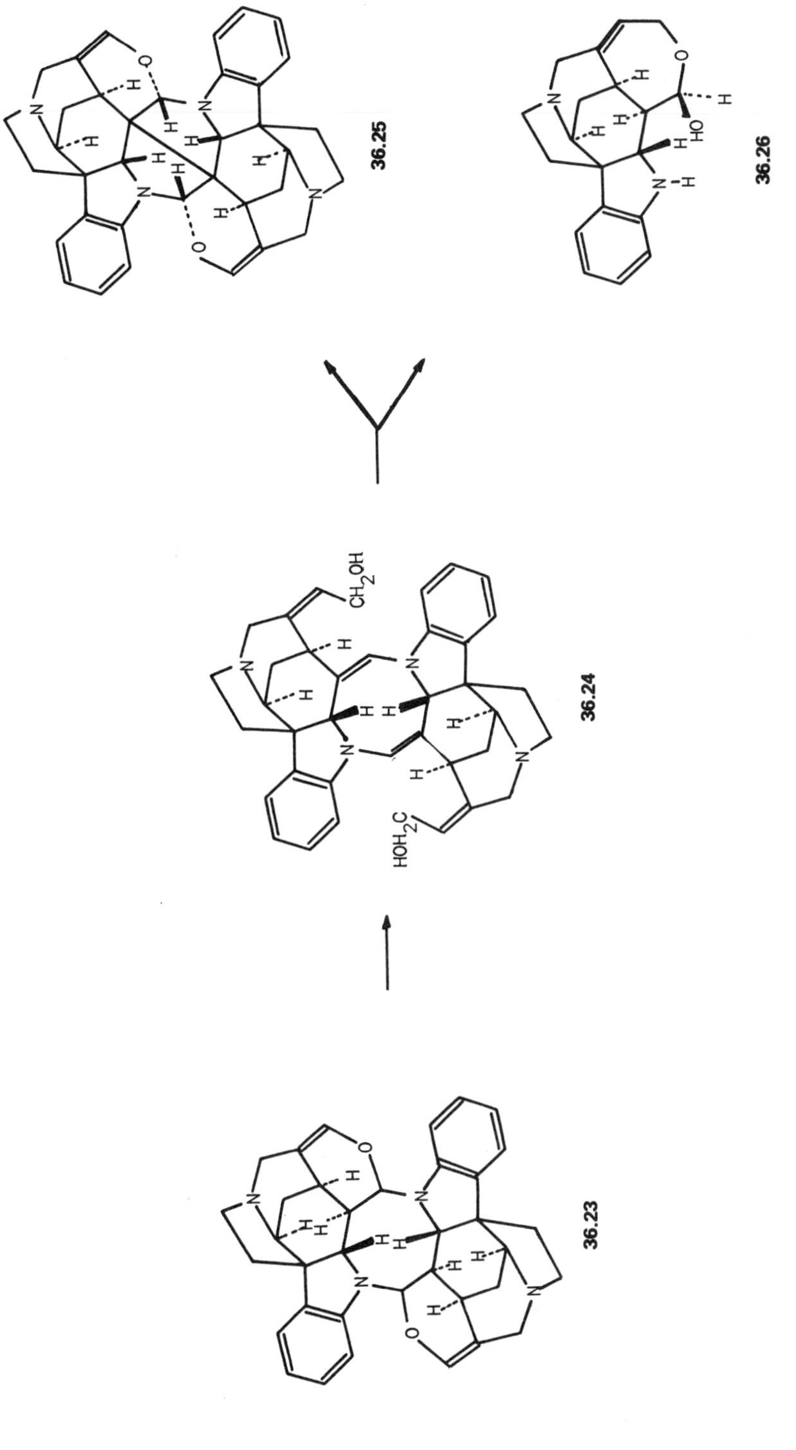

SCHEME 36.6 The products of the acid-catalyzed hydrolysis of caracurine V. (After Reference 360.)

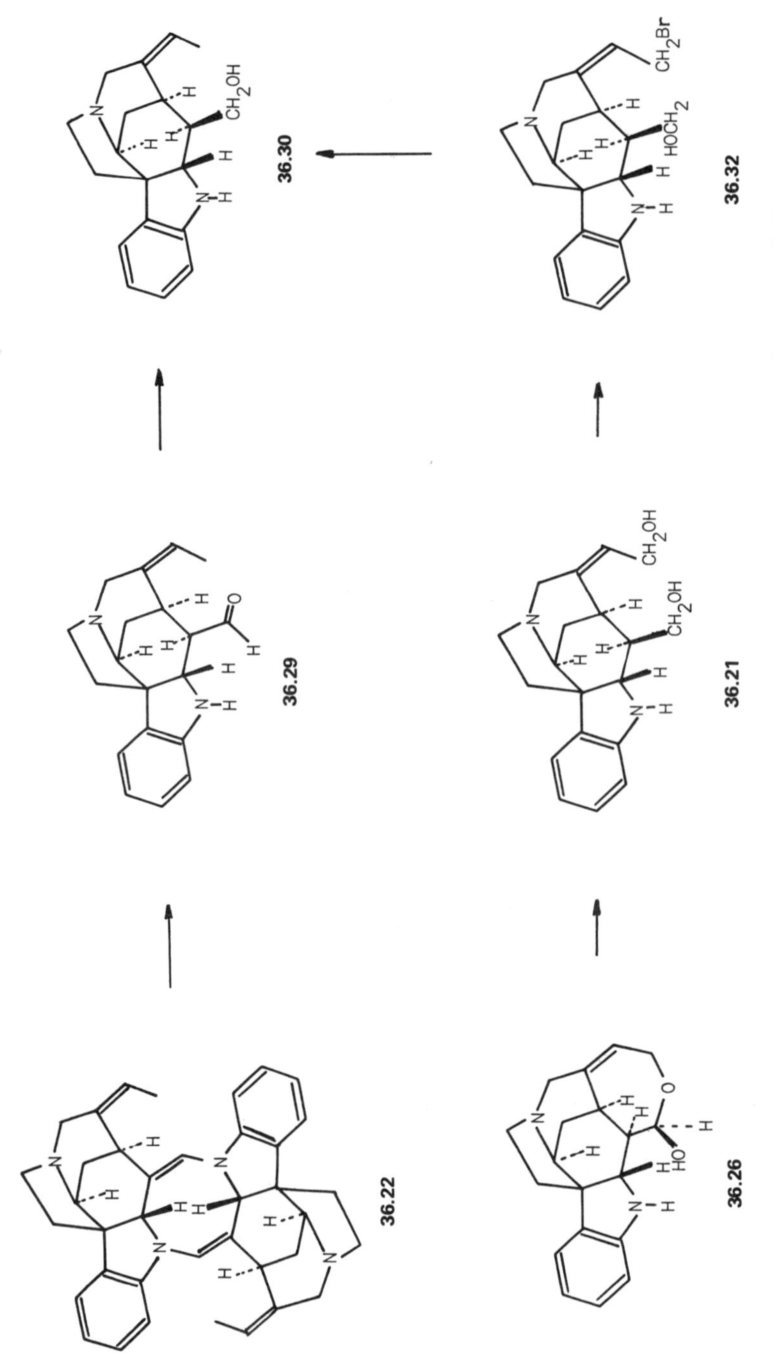

SCHEME 36.7 The interrelation of C-dehydrotoxiferine with the Wieland-Gumlich aldehyde. (After Reference 361.)

That these sequences are related was shown in the following fashion (Scheme 36.7): C-Dihydrotoxiferine (36.21) was converted into nordihydrotoxiferine I (36.22; see above), and the latter was treated with aqueous acid to yield norhemidihydrotoxiferine (36.29), which with sodium borohydride was reduced to the primary alcohol 36.30. This alcohol (36.30) is identical to that prepared from the Wieland-Gumlich aldehyde (36.26) by reduction of the latter with lithium aluminum hydride to the diol (36.31), selective allylic bromination to yield the bromide 36.32, and then reductive dehalogenation with zinc in acetic acid to 36.30 [361].

Thus, it was concluded that hemitoxiferine I (36.28) must be the $N_{(b)}$-methochloride of the Wieland-Gumlich aldehyde (36.26), and this was confirmed by a direct comparison which in turn allows for the condensation of two molecules of hemitoxiferine I (36.28) to yield C-toxiferine (36.3), two molecules of Wieland-Gumlich aldehyde (35.26) to yield caracurine V (36.23), and two molecules of norhemidihydrotoxiferine (36.29) to yield nordihydrotoxiferine (36.22).

The positions assigned to the double bonds in the dimers were made on the basis of the pmr spectra of these materials when a reasonable value for the chemical shifts of the methine protons at C_2 and C_2' was decided upon [362].

Finally, since the absolute configuration of strychnine (36.4) is known (p. 578) and the Wieland-Gumlich aldehyde (36.26) could be obtained from strychnine (36.4), the absolute stereochemistry of these dimeric bases is also known. The derivation of the Wieland-Gumlich aldehyde (36.26) from strychnine (36.4) can be carried out as shown in Scheme 36.8 [327]. Thus, strychnine (36.4) undergoes condensation with amyl nitrite in ethanolic sodium ethoxide to yield 11-oximinostrychnine (36.33), which is presumably a mixture of E and Z isomers (see p. 576). This oxime, with thionyl chloride, undergoes a Beckmann rearrangement to yield both the imide 36.34 and the carbamic acid 36.35. Hydrolysis of the latter yields the Wieland-Gumlich aldehyde (36.26). Then, Scheme 36.9, when heated in acetic acid - sodium acetate, two molecules of the Wieland-Gumlich aldehyde (36.26) condense and generate caracurine V (36.23), methylation of which leads to caracurine V dimethochloride (36.36) and the latter, on warming with p-toluenesulfonic acid in acetic acid, leads to C-toxiferine (36.3), thus formally completing a synthesis of this bisindole alkaloid [363].

III. VINBLASTINE

Before the structure and absolute stereochemistry of vinblastine (VLB, vincaleukoblastine; 36.5), a therapeutically useful antitumor alkaloid from Catharanthus roseus had been elucidated by x-ray crystallography [364], only a little degradative, but a great deal of spectroscopic (infrared, proton magnetic resonance, and mass spectrometry) work had been utilized

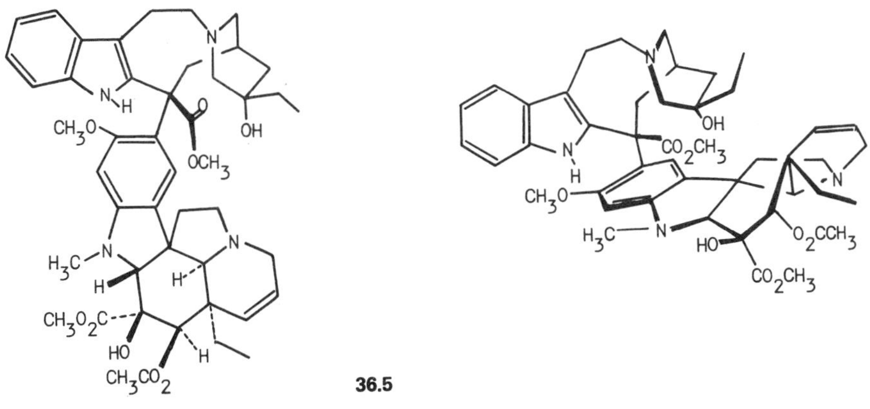

36.5

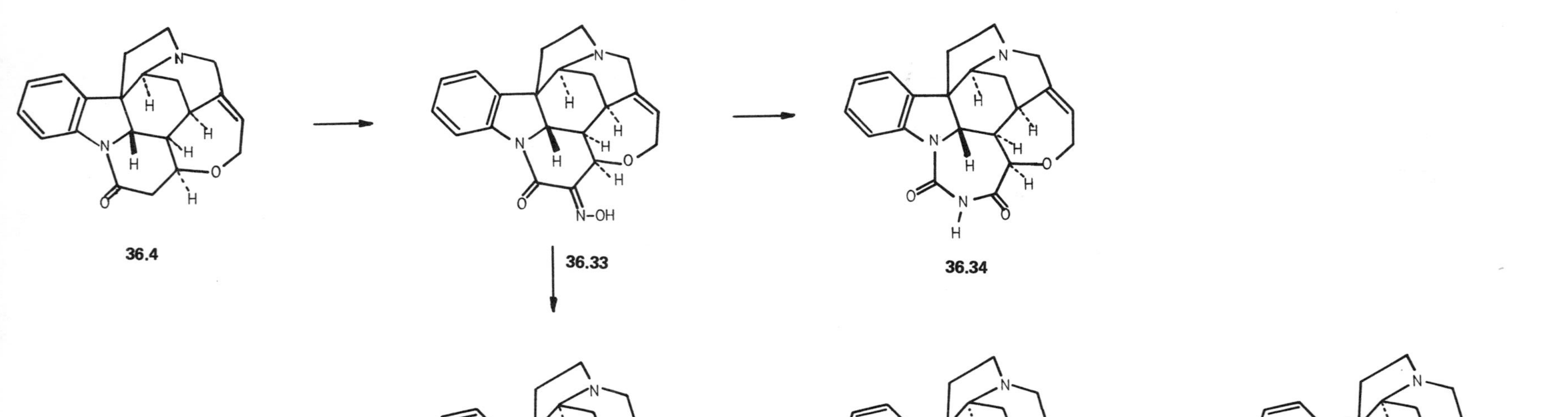

SCHEME 36.8 The conversion of strychnine into the Wieland-Gumlich aldehyde. (After Reference 327.)

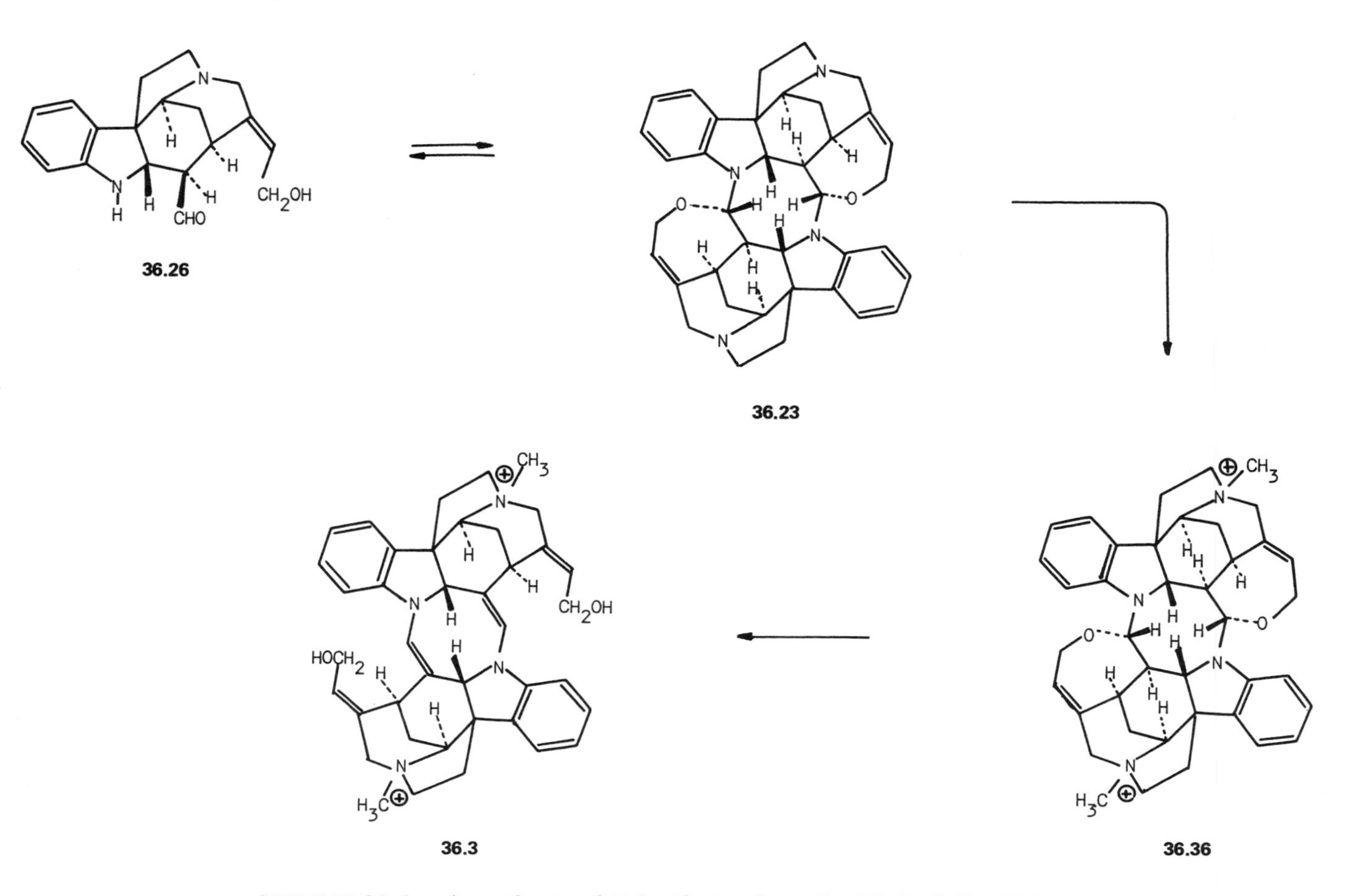

SCHEME 36.9 A synthesis of C-toxiferine from the Wieland-Gumlich aldehyde. (After Reference 363.)

to derive its structure [365]. The physical properties of vinblastine (36.5), $C_{46}H_{56}N_4O_9$ (etherate, mp 180-182°C, 201-211°C), $[\alpha]_D^{26}$ +42° (chloroform), were reported in 1958, when its biological activity was recognized [366].

Among the few reactions carried out was the treatment of vinblastine (36.5) with a mixture of concentrated hydrochloric acid, stannous chloride, and tin metal at reflux (Equation 36.5) which resulted in isolation of two basic "alkaloid" fragments (36.37 and 36.38), corresponding in their sum to the entire parent [367] less a carbomethoxy group.

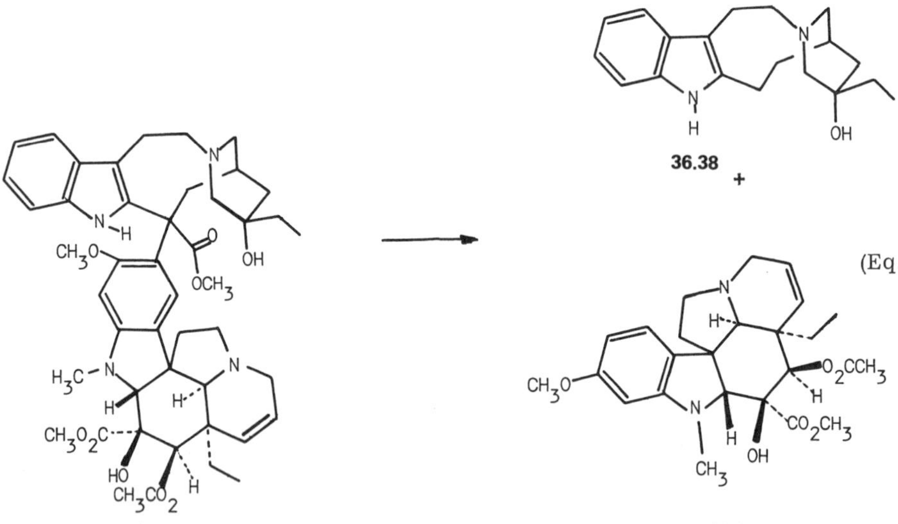

36.38 +

(Eq. 36.5)

36.5

36.37

The first of these fragments, identified as vindoline (36.37), $C_{25}H_{32}N_2O_6$, is a major alkaloid of C. roseus, while the second, named velbanamine, $C_{19}H_{26}N_2O$ (36.38), was recognized as a relative of cleavamine (36.39; p. 579). Velbanamine (36.38) could be converted into cleavamine (36.39; [365]; Equation 36.6), and the structure of cleavamine (36.39) had been established unequivocally by x-ray crystallographic analysis [331].

(Eq. 36.6)

36.38

36.39

Soda lime distillation of vindoline (36.37) affords N-methylnorharman (36.40; Equation 36.7), and hydrogenation of vindoline (36.37) yields dihydrovindoline (36.41; Equation 36.8). When dihydrovindoline (36.41) is pyrolyzed at 195-200°C in vacuo, a ketone (36.42), the result of dehydration, hydrolysis, and decarboxylation of the resulting β-ketoester (Equation 36.9), is obtained [368]. A complete analysis of the pmr spectrum of vindoline (36.37; Figure 36.1; [368]), as well as analysis of the mass spectra of vindoline (36.37) and dihydrovindoline (36.41; Scheme 36.10), provided the structure shown. The analysis of the mass spectra of these compounds was, of course, greatly aided by the availability of other alkaloids with the same general system of five fused rings (see, e.g., p. 598) whose structure were

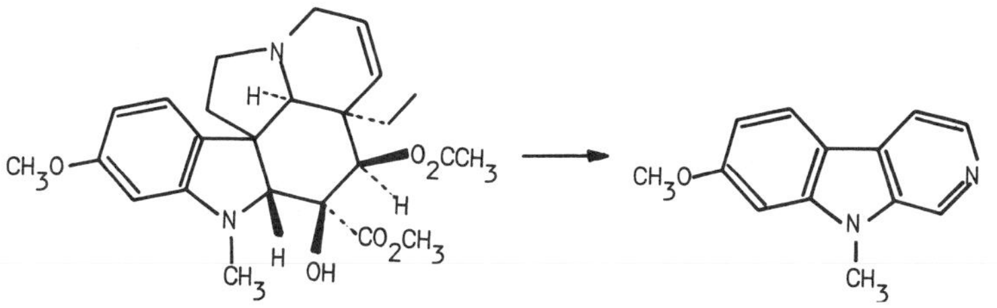

36.37 → 36.40 (Eq. 36.7)

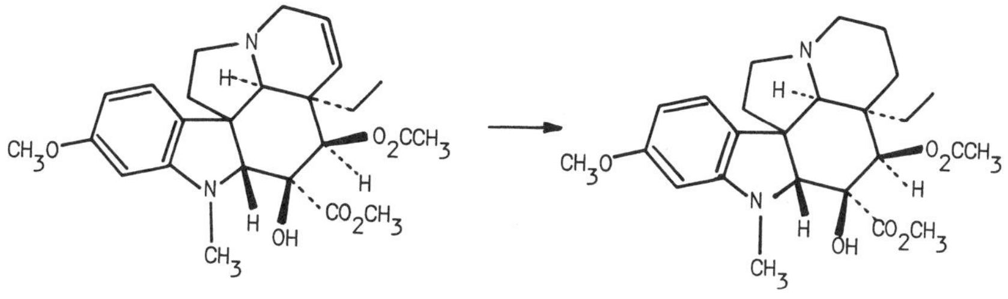

36.37 → 36.41 (Eq. 36.8)

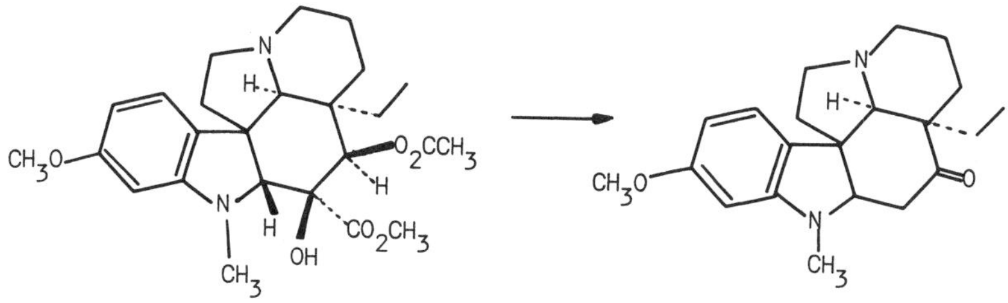

36.41 → 36.42 (Eq. 36.9)

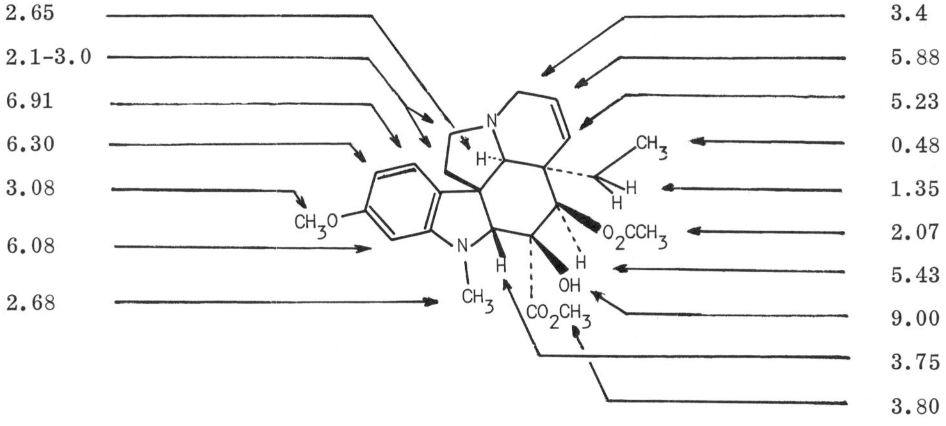

2.65, 2.1-3.0, 6.91, 6.30, 3.08, 6.08, 2.68

3.4, 5.88, 5.23, 0.48, 1.35, 2.07, 5.43, 9.00, 3.75, 3.80

FIGURE 36.1 An analysis of the proton magnetic resonance chemical shift data for vindoline. The assignments are in δ (ppm from tetramethylsilane; TMS = 0). (After Reference 368.)

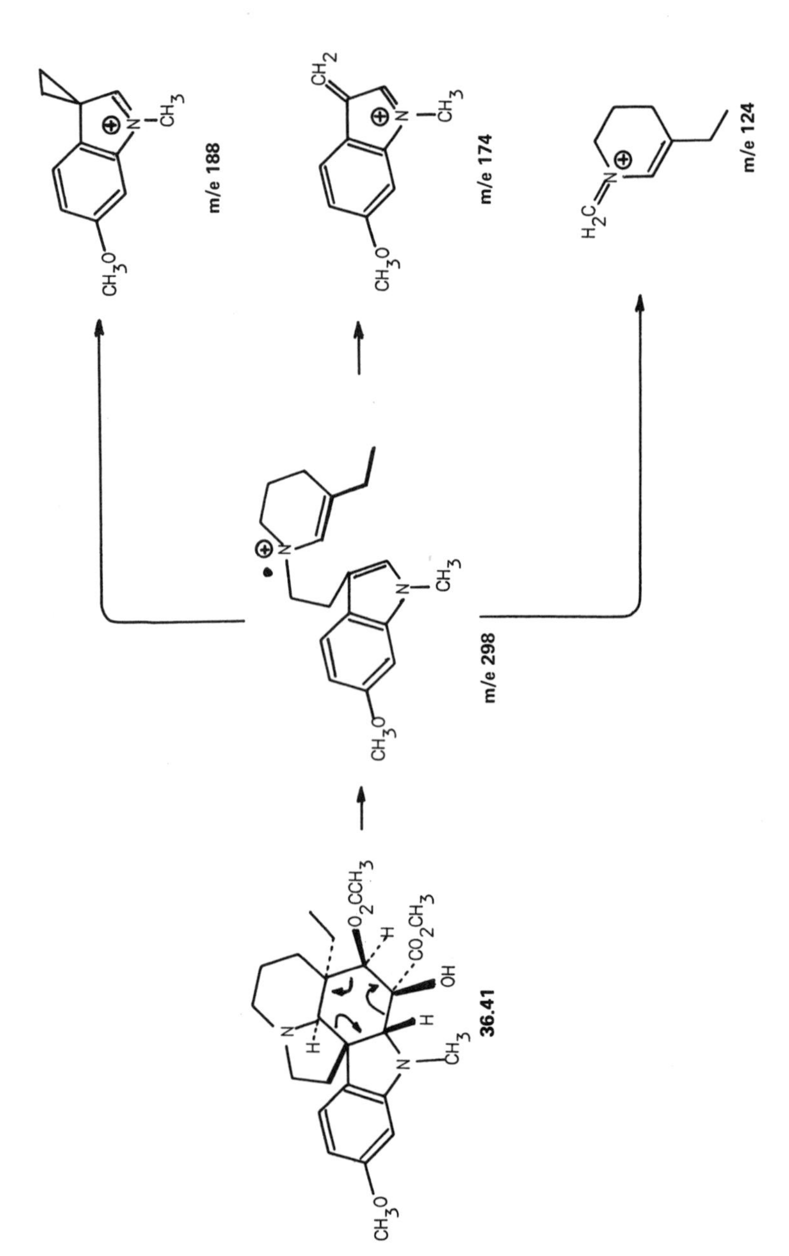

SCHEME 36.10 An analysis of the mass spectrum of dihydrovindoline. (After Reference 368.)

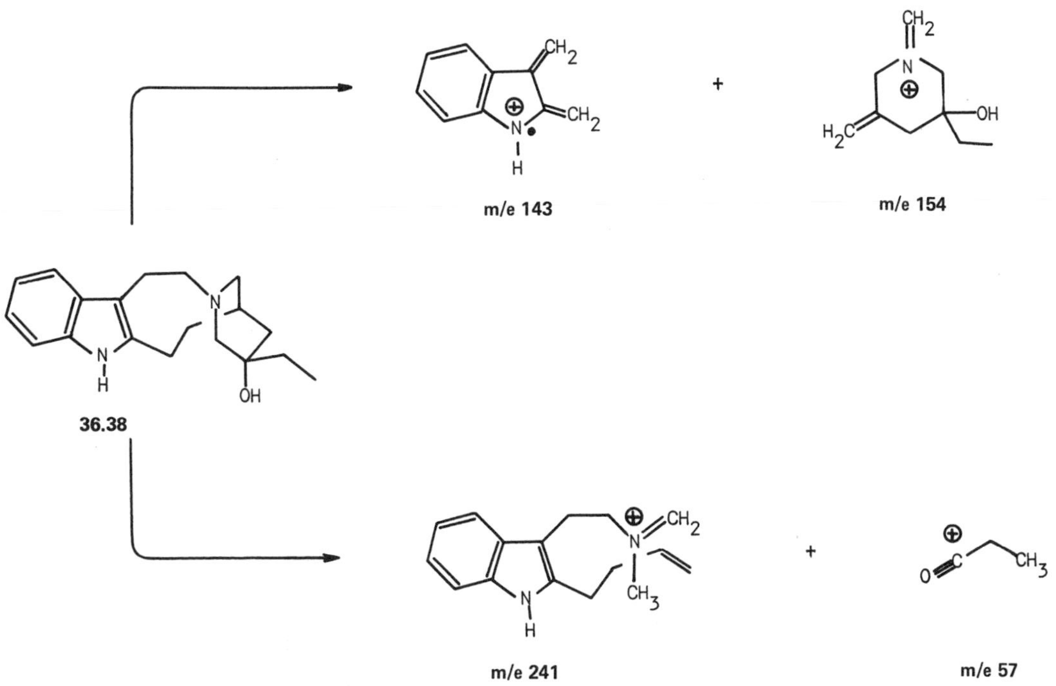

m/e 143

m/e 154

36.38

m/e 241

m/e 57

SCHEME 36.11 An analysis of the mass spectrum of velbanamine. (After Reference 365.)

known. The structure of velbanamine (36.38) was less certain, although it was obviously related to cleavamine (36.39), and the position of the hydroxyl, for example, which was recognized as tertiary, was assigned on the basis of acetylation experiments and mass spectrometry (Scheme 36.11). That the two fragments are linked as shown was derived largely on the basis of pmr experiments and the observation that cleavage of vinblastine (36.5) with 40% aqueous sulfuric acid in the absence of reducing agents leads to, among other products, an amino acid (35.42; Equation 36.10) which, on esterification followed by treatment with

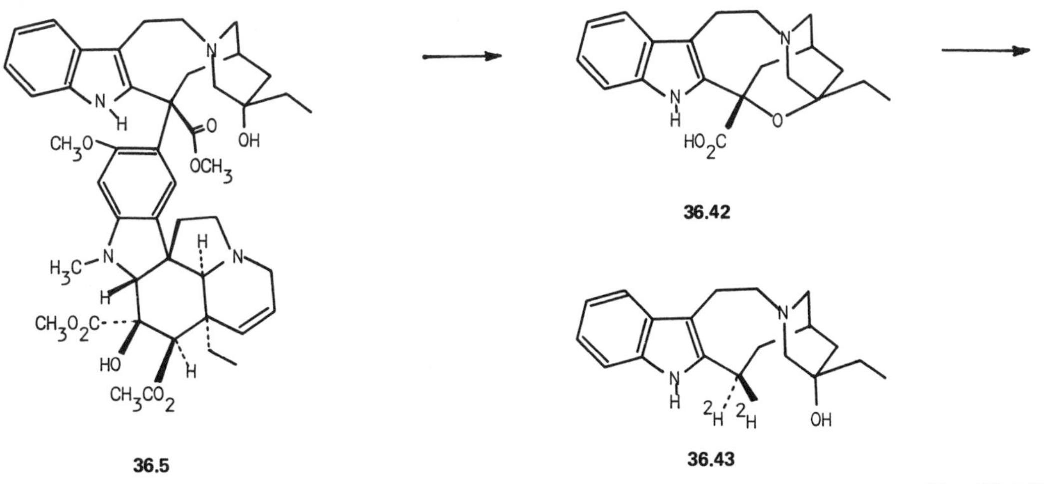

36.5

36.42

36.43

(Eq. 36.10)

deuterium chloride in deuterium oxide with stannous chloride, generates deuteriovelbanamine (36.43; [365]). Similarly, the pmr spectrum (except for the missing protons at C_4 and C_{18}) of the ester of 36.42 bears a striking resemblance to that of carbomethoxydihydrocleavamine (36.44) obtained from catharanthine (36.45; see p. 579) with zinc in acetic acid (Equation 36.11; [365]).

(Eq. 36.11)

36.45 36.44

The analysis briefly outlined above was amply confirmed by x-ray crystallography [364]. A great effort has been underway to synthesize vinblastine (36.5) [369, 370], and success has recently been reported [478].

SELECTED READING

A. R. Battersby and H. F. Hodson, in The Alkaloids, Vol. 8, R. H. F. Manske (ed.). Academic, New York, 1965, pp. 151 ff.

A. R. Battersby and H. F. Hodson, in The Alkaloids, Vol. 11, R. H. F. Manske (ed.). Academic, New York, 1968, pp. 189 ff.

R. H. F. Manske (ed.), in The Alkaloids, Vol. 8. Academic, New York, 1965, pp. 581 ff.

GENERAL READING FOR PART 6

W. I. Taylor, Indole Alkaloids. Pergamon, New York, 1966.

Part 7

ALKALOIDS DERIVED FROM INTRODUCTION OF NITROGEN INTO A TERPENOID SKELETON

37

BIOSYNTHESIS OF MONOTERPENOID, SESQUITERPENOID, DITERPENOID, AND STEROIDAL ALKALOIDS

The Dormouse turned over to shut out the sight
Of the endless chrysanthemums (yellow and white)
"How lovely," he thought, "to be back in a bed
Of delphiniums (blue) and geraniums (red)! "

A. A. Milne, When We Were Very Young (1924)

The alkaloids discussed in the earlier parts of this work share a common genesis from amino acids. Elaboration of the particular amino acid into the alkaloid may or may not have involved other fragments, e.g., a second amino acid, acetate, mevalonate (37.1), etc. In this part, representatives of families of plant bases which share a common genesis from mevalonate (37.1) alone, i.e., no amino acid is utilized for the skeletal framework (except for the ubiquitous methyl source, S-adenosylmethionine, 37.2), will be considered. Although these compounds might properly be categorized as terpenoids and steroids they are called alkaloids because they contain nitrogen.

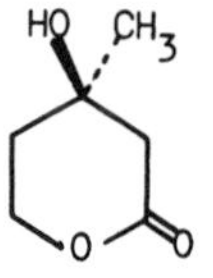

37.1

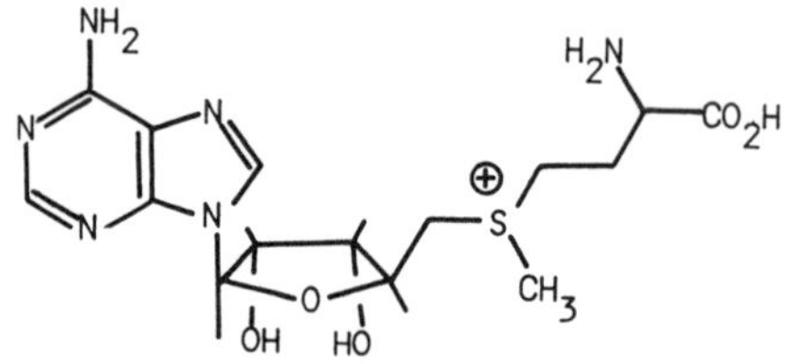

37.2

There is very little evidence currently available with regard to these alkaloids as to the timing of the introduction of nitrogen and the source of the nitrogen. Indeed, in some cases, there is little hard evidence that these bases are, in fact, derived from mevalonate (37.1; since incorporation of appropriate precursors occurs with difficulty). Thus, the origin of the "mevalonate-derived" (37.1) alkaloids is often based upon speculation resulting from the similarity of their structures to known terpenoids and steroids which fit the concepts surrounding terpenoid biogenesis.

Nevertheless, there are a few compounds where appropriate feeding experiments have been carried out and incorporation of precursors has been obtained. Thus, we will find that, as has been the case previously, methionine (37.3) serves to provide N-methyl functionality

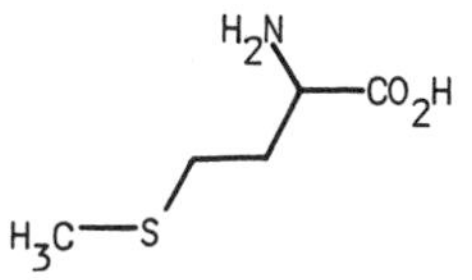

37.3

and that both mevalonate (37.1) and acetate have been found to serve as precursors of the carbon skeleton of some of the bases considered.

The alkaloids to be considered here could easily be divided into rubrics of two, three, four, and six mevalonte (37.1) units, each in a chapter of its own, were it not for two important qualifying features which cause to unite them. First, grouping them together provides insight into their overall similarity; second, there is a paucity of biosynthetic data which supports the general theory outlined within (for these compounds) and what is currently available would certainly fail to earn, for each, a chapter of its own.

The specific bases to be considered in this chapter are: the monoterpenoids (+)-β-skytanthine (37.4) and (-)-tecomanine (37.5); the sesquiterpenoid alkaloids (-)-deoxynupharidine

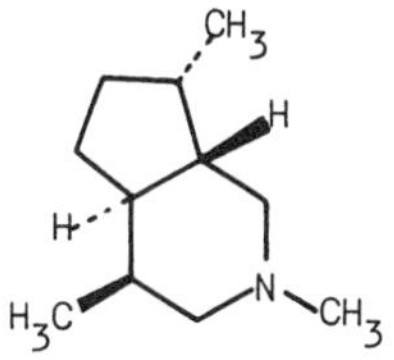

37.4

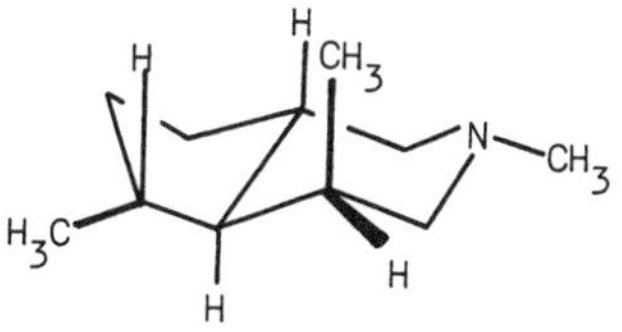

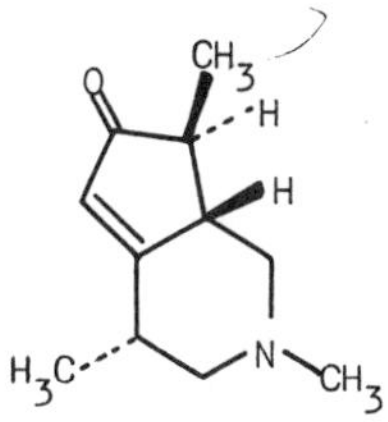

37.5

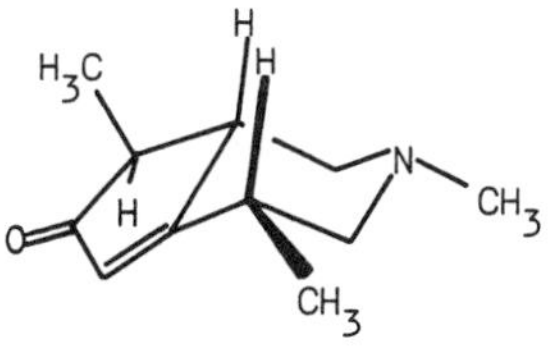

(37.6) and (-)-dendrobine (37.7); the diterpene bases atisine (37.8), veatchine (37.9), (+)-aconitine (37.10), and (+)-heteratisine (37.11); and the steroid alkaloids solanidine (37.12), (-)-conessine (37.13), and cyclobuxine-D (37.14). These bases, which are typical members of their respective classes, serve to provide a template upon which to elaborate biosynthetic speculation (mixed with some fact where it is available). As is consistent with what has gone before, only a small part of the total number of compounds which could have been considered is actually discussed.

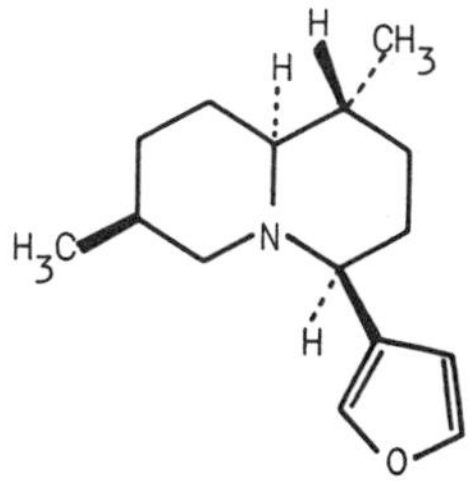

37.6

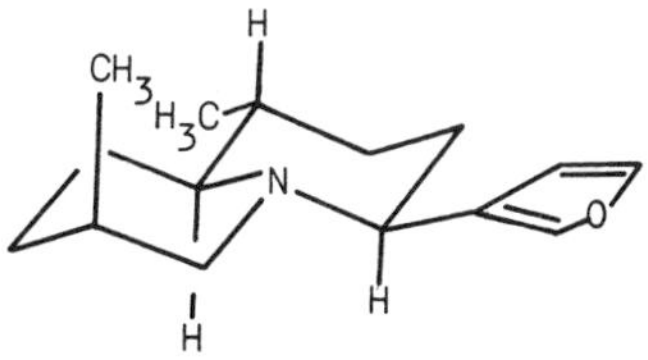

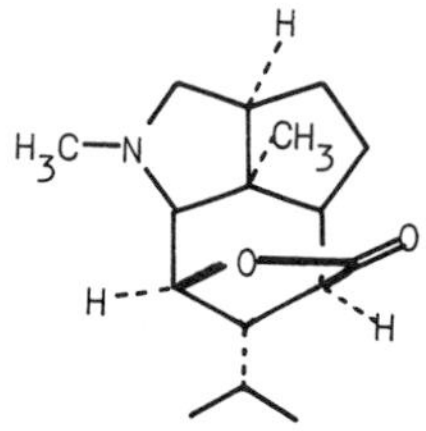

37.7

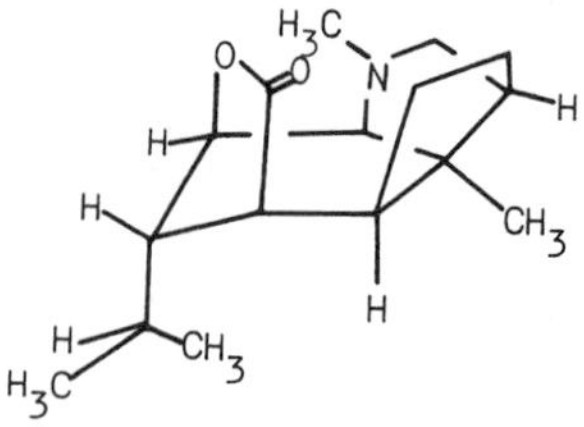

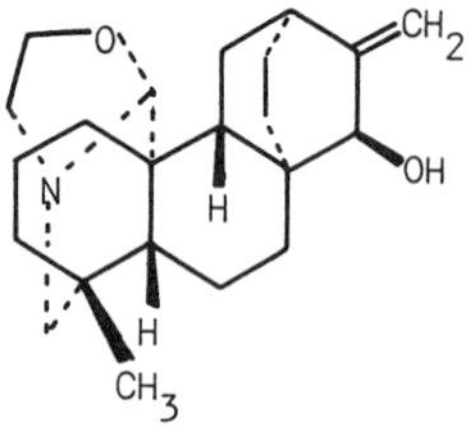

37.8

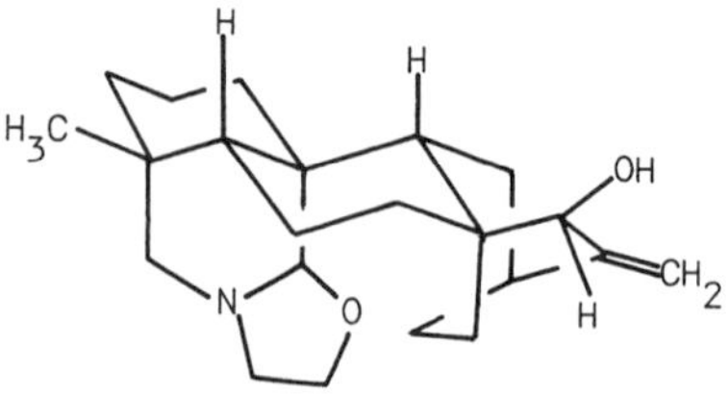

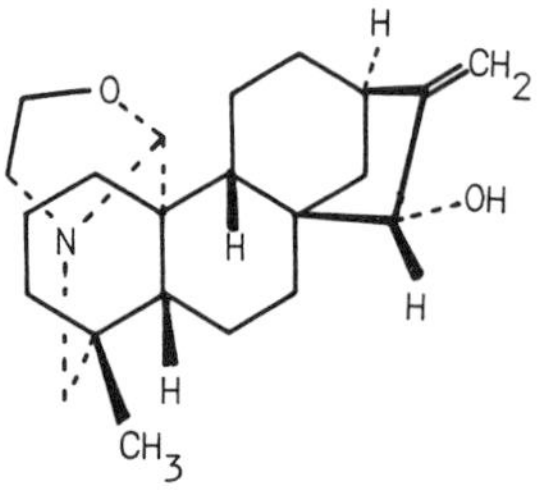

37.9

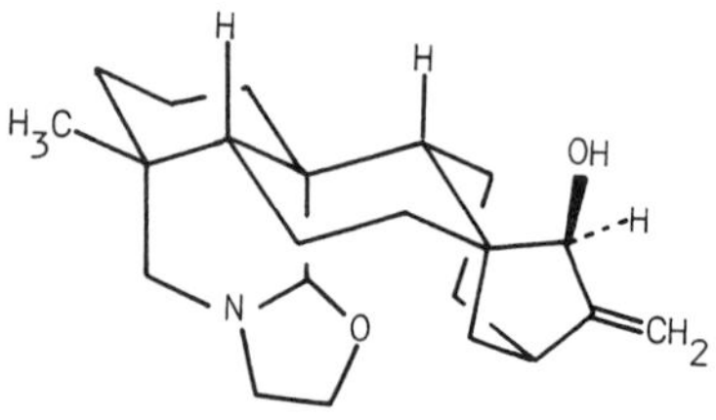

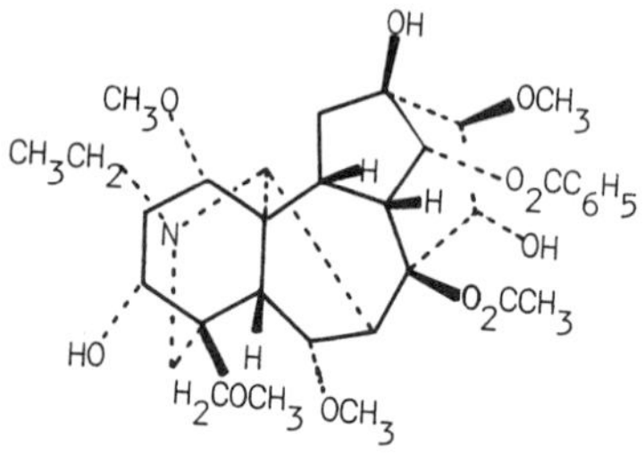

37.10

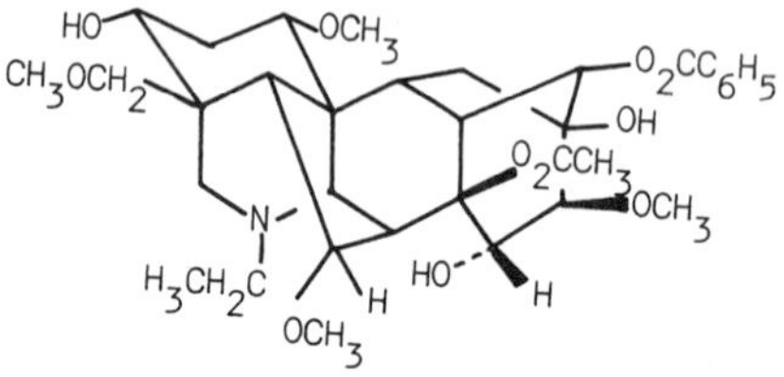

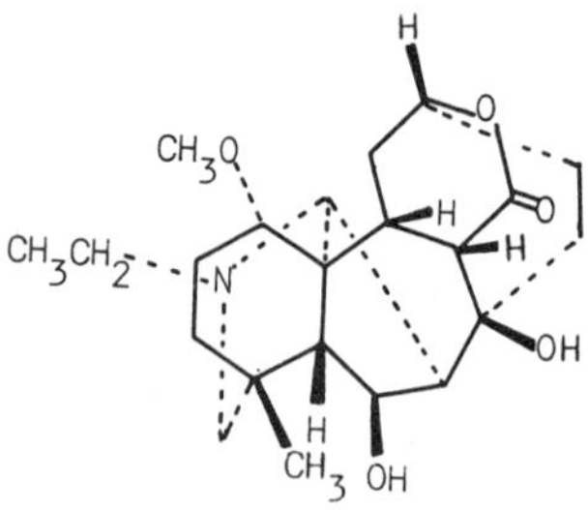

37.11

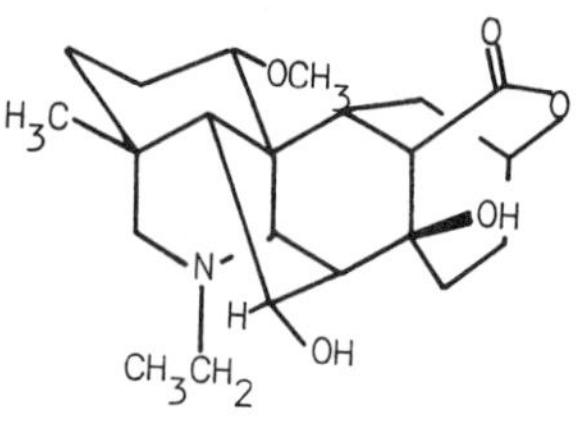

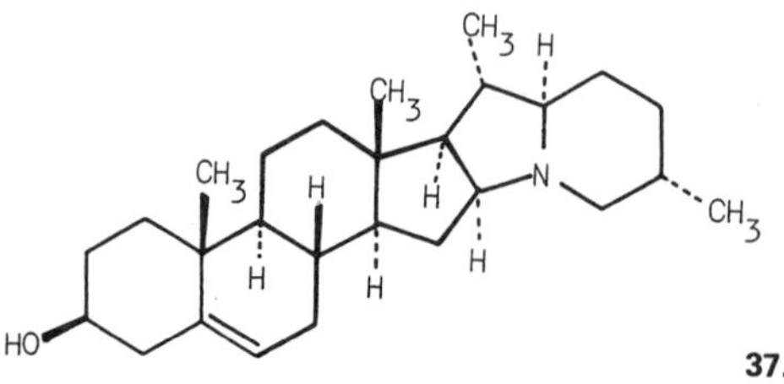

37.12

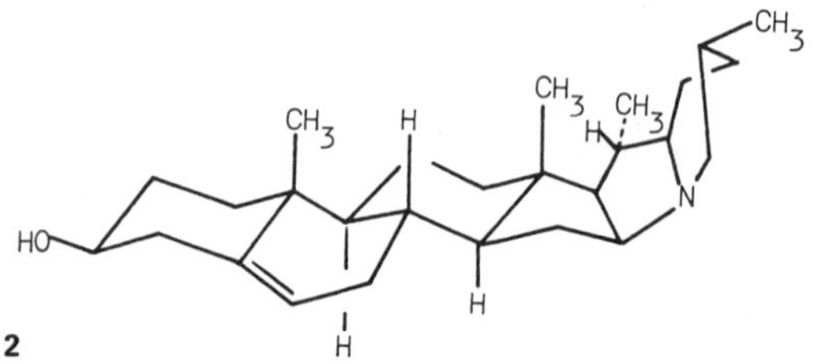

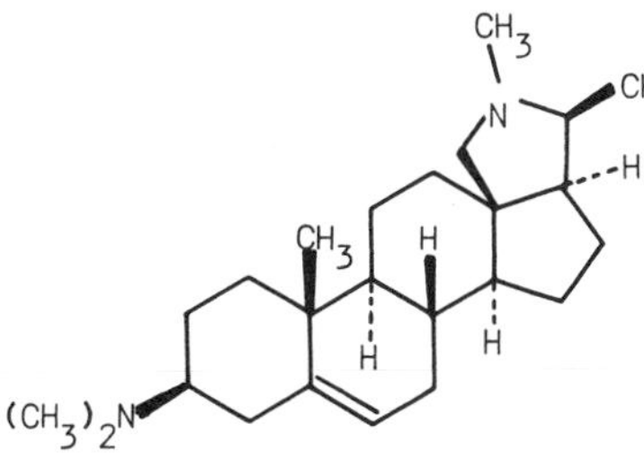

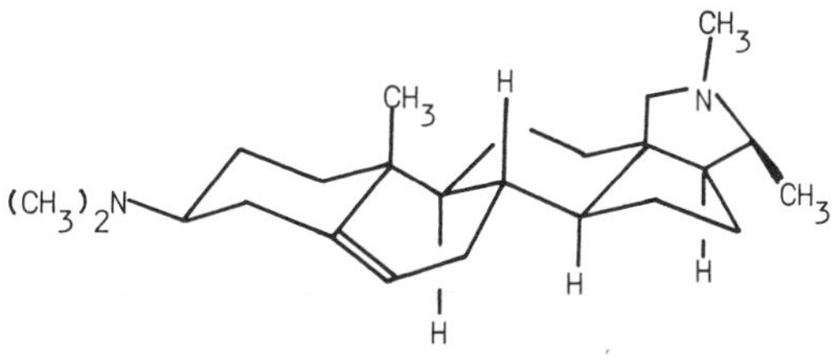

37.13

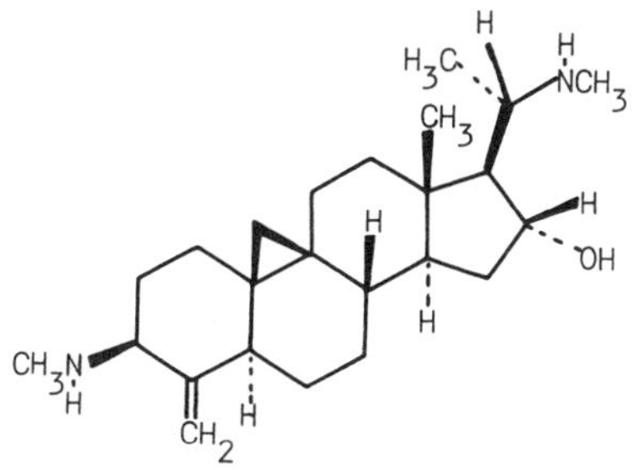

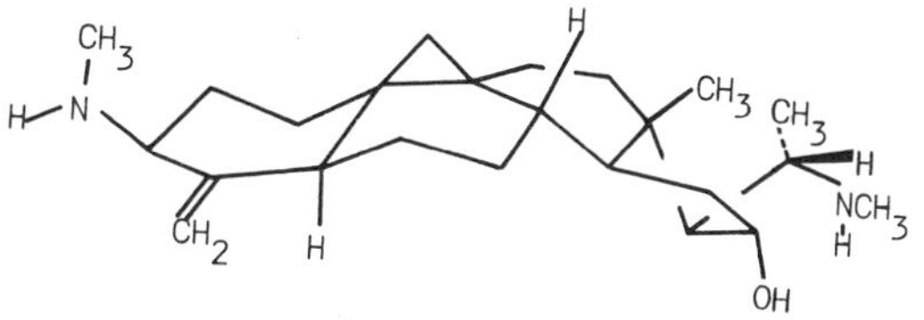

37.14

Now, although it has been dealt with previously, and, for many of the bases to be discussed, it has not been rigorously demonstrated viable, it is generally assumed that the normal genesis of the terpenoid bases* is from acetate via mevalonate (37.1) to geraniol (37.18; monoterpenoid alkaloids), and thence, presumably via farnesol (37.19; sesquiterpenoid alkaloids) to geranylgeraniol (37.20; diterpenoid alkaloids) and squalene (37.21; steroidal alkaloids). This presumed chain of events is shown diagrammatically in Scheme 37.1.

The pathway from acetate to mevalonate (37.1) and thence to isopentenyl pyrophosphate (37.22) and 3,3-dimethylallyl pyrophosphate (37.23; Scheme 37.2) has already been discussed (Chap. 2, p. 27 and Chap. 30, p. 438) and will not be elaborated upon here. Additionally, as has been demonstrated (Chap. 2, p. 27 and Chap. 30, p. 439), isopentenyl pyrophosphate (37.22) and 3,3-dimethylallyl pyrophosphate (37.23) may condense (Equation 37.1) to geranyl pyrophosphate (37.24). The detailed labeling experiments involving the C_{10} geraniol fragment in certain indole alkaloids, which, it will be recalled (Chap. 30, p. 438), were based upon earlier work on steroidal systems, have already been expounded upon (p. 443) and will

*It will be remembered (see Chap. 3) that for some time the C_5 monocarboxylic acid angelic acid (37.15) and the C_{10} dicarboxylic acid senecic acid (37.16), which occur as the esterifying acids of the necines, were thought to be mevalonate-derived (37.1). It is now known that both angelic acid (37.15) and senecic acid (37.16) are derived from isoleucine (37.17). It does not appear to have been shown that the alkaloids to be considered here cannot arise from isoleucine.

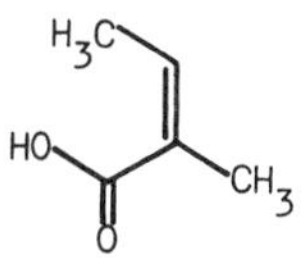

37.15 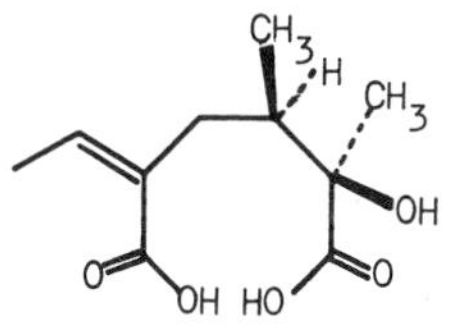 **37.16** 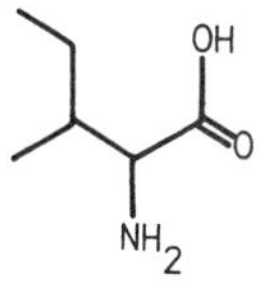 **37.17**

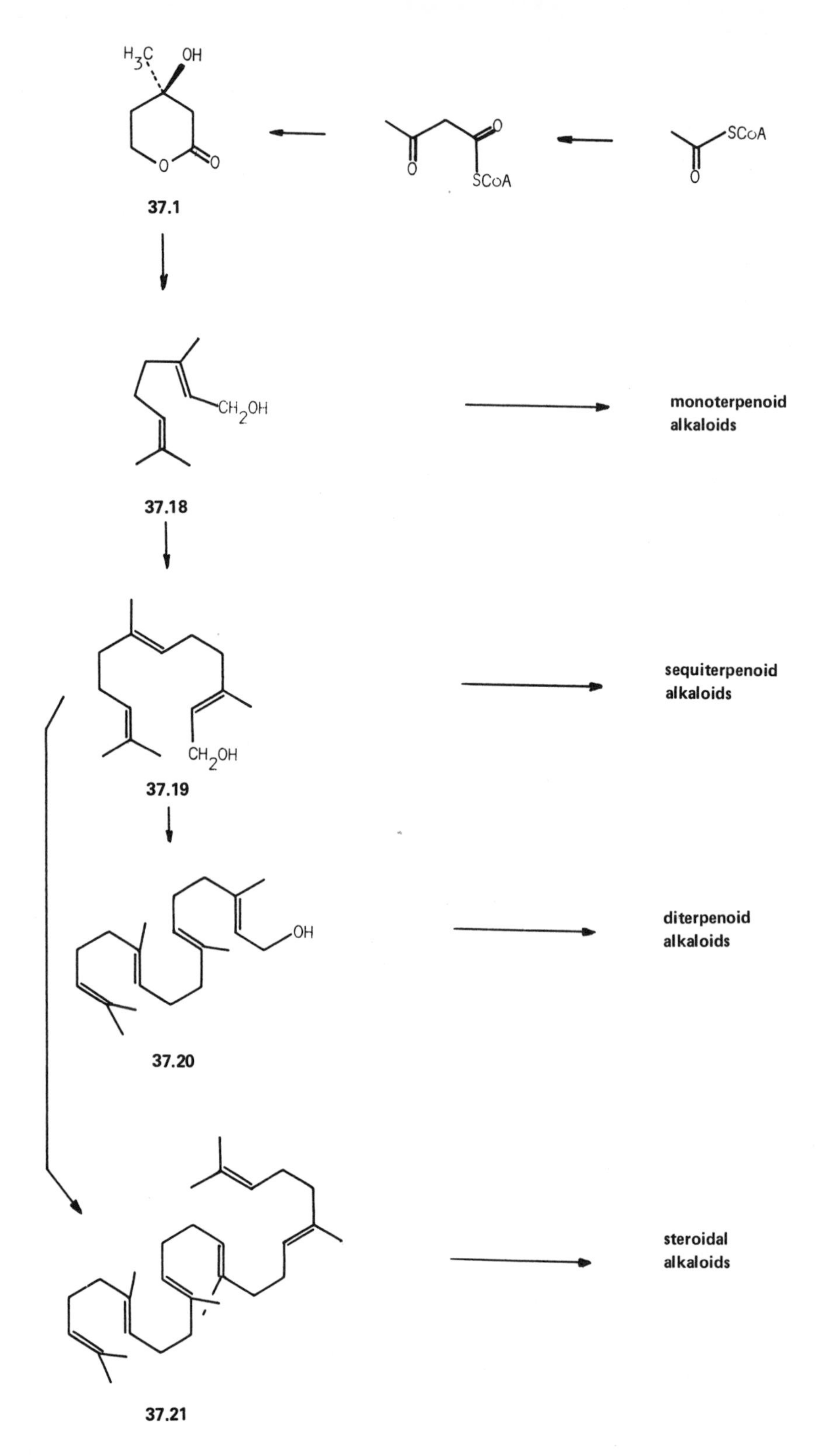

SCHEME 37.1 A presumed pathway to monoterpenoid, sesquiterpenoid, diterpenoid, and steroidal alkaloids from mevalonate.

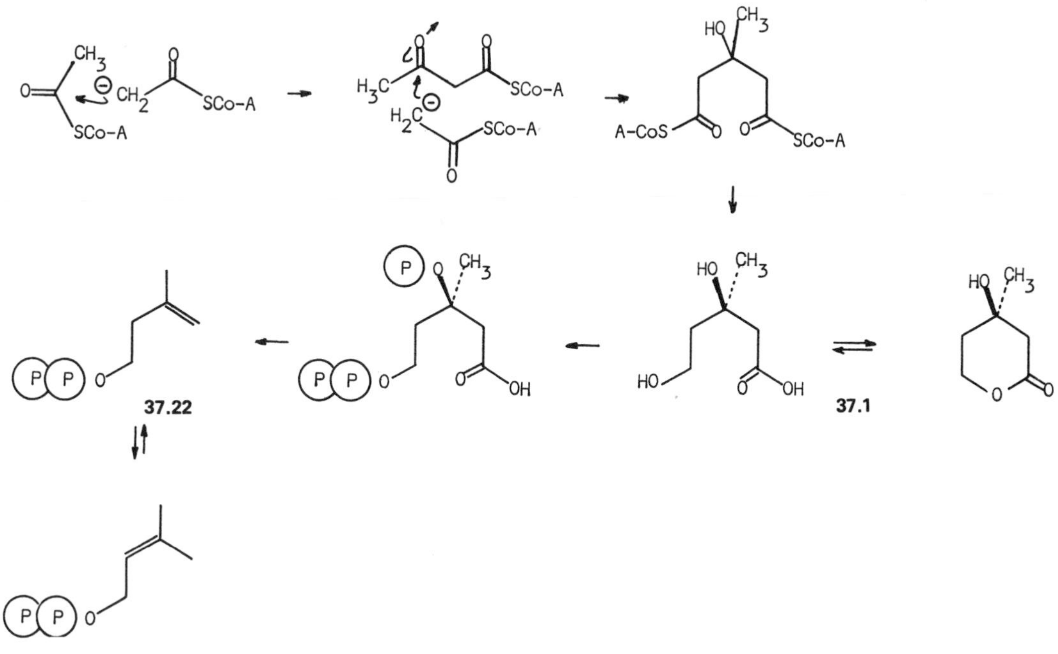

SCHEME 37.2 A pathway for the generation of isopentenyl pyrophosphate from acetyl coenzyme A via mevalonic acid.

Eq. 37.1 A pathway for the formation of geranyl pyrophosphate from isopentenyl pyrophosphate.

not be reiterated here since the tenor of work on the monoterpenoid alkaloids has not yet reached that level of sophistication.

When (±)-mevalonate-[2-^{14}C] (37.1 as the racemate) is fed to Skytanthus acutus M. [371], modest incorporation of label into (+)-β-skytanthine (37.4) is obtained. The positions in which a portion of the label occurs are determined by degradation. The degradation (Scheme 37.3) involves converting the purified (+)-β-skytanthine (37.4) into the corresponding N,N-dimethyl derivative, which on Hofmann degradation yields the alkene (37.25). Ozonolysis of the alkene (37.25) provides formaldehyde (corresponding to C_3 of β-skytanthine, 37.4) and an aminoketone (37.26). The latter undergoes a Bayer-Villiger oxidation to generate the ester 37.27, which on hydrolysis produces the alcohol 37.28 and acetic acid. The methyl group of acetic acid is presumably C_9 and the carbonyl carbon presumably C_4 of the original alkaloid. Lack of material apparently precluded further degradation but even that which had been done demonstrates incorporation of mevalonate (37.1) into the expected carbon.

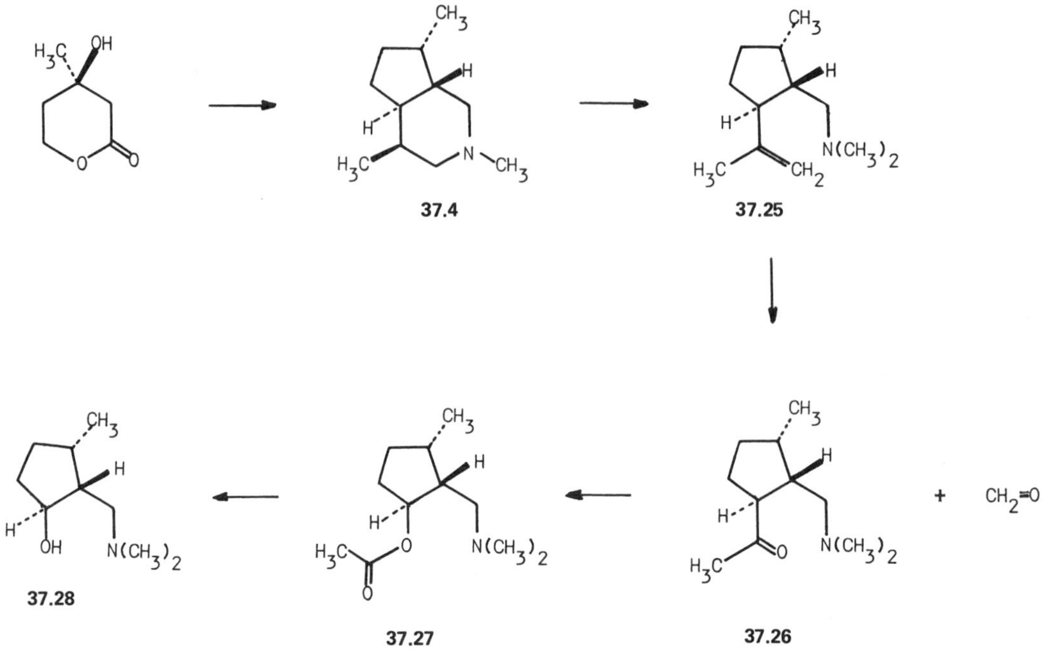

37.4 37.25

37.28 37.27 37.26

SCHEME 37.3 A pathway for the degradation of skytanthine to determine the position of label introduced on feeding mevalonate. (After Reference 371.)

Interestingly (if indeed mevalonate, 37.1, is the correct precursor), it is found that incorporation of mevalonate (37.1) into carbons C_3 and C_9 occurs with more randomization in young (1.3-year) rather than in old (3-year) plants.

The same group of workers [371], in separate experiments, demonstrated that (±)-lysine (37.29) is not incorporated into (+)-β-skytanthine (37.3) and that methionine (37.3) is the source of the N-methyl group. The Cope N-oxide method of elimination was used to implicate 37.2 as the "methylating agent."

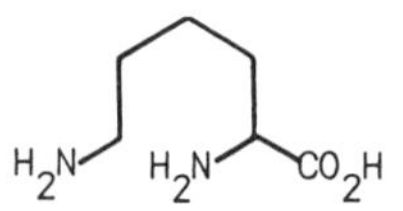

37.29

Since the early examination of β-skytanthine (37.4) from Skytanthus acutus M. there has appeared a report [372] indicating that β-skytanthine (37.4) may be an artifact which arises from use of ammonium hydroxide during its isolation. More recently [373], [2-^{14}C]-acetate, [2-^{14}C]mevalonate (37.1), and [S-^{14}C]methylmethionine (37.3) were successfully incorporated into tecomanine (37.5) in Tecoma stans Juss., suggesting, for this alkaloid too, a genesis from mevalonate (37.1; Scheme 37.4). In a further experiment [373], uniformly tritiated loganine (37.30) was not incorporated into tecomanine (37.5), suggesting that if geraniol lies on the pathway to tecomanine (37.5) it is intercepted to form the alkaloid before it goes on to loganin (37.30) or that some other iridoid is involved. δ-Skytanthine (37.31),

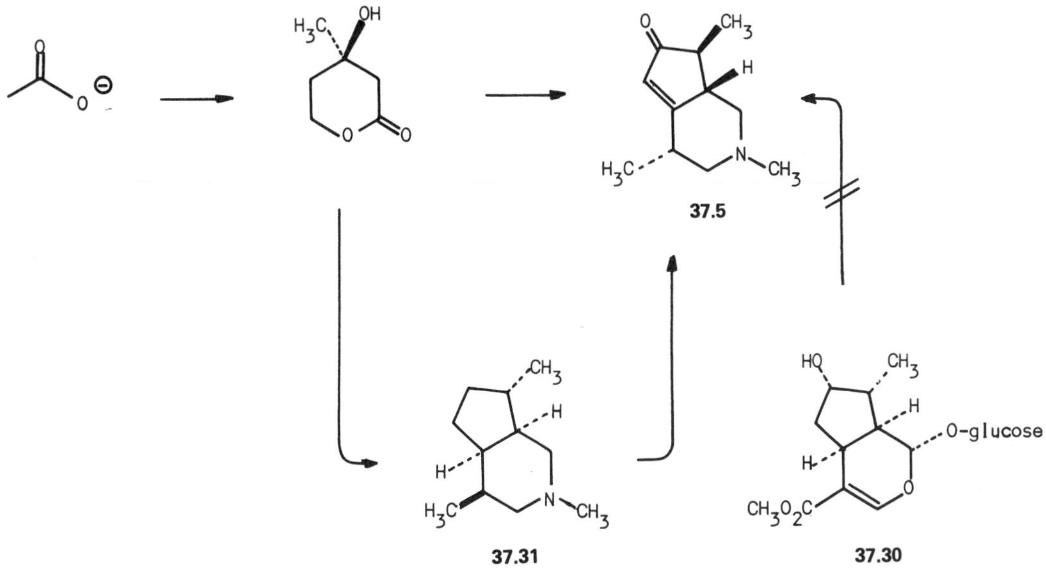

37.5

37.31

37.30

SCHEME 37.4 A pathway from acetate to tecomanine suggested by administration of labeled precursors to Tecoma stans Juss. (After Reference 373.)

also occurring in Tecoma stans Juss., was further transformed into tecomanine (37.5), suggesting that oxidation to the latter occurs at a late stage.* Utilization of this (admittedly) sketchy data permits formulation of a scheme for generation of β-skytanthine (37.4) and tecomanine (37.5; Scheme 37.5).

I. GERANIOL TO FARNESOL

The conversion of geranyl pyrophosphate (37.24) to farnesyl pyrophosphate (37.32) has been thoroughly examined, and the details of the process are well known [224]. Equation 37.2 presents a simplified version of this process. The stereochmical detail parallels that of the

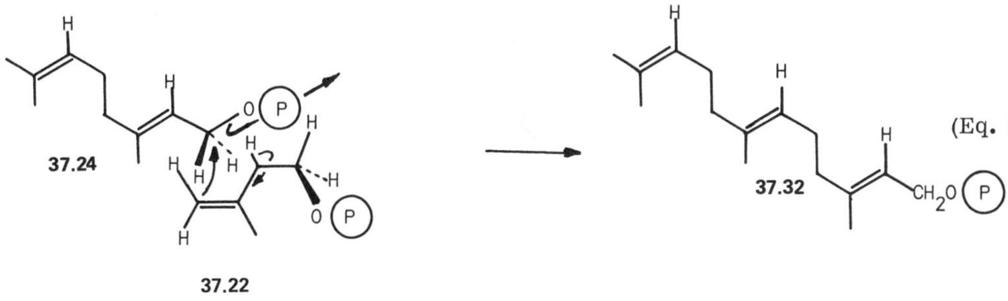

37.24

37.22

37.32

(Eq. 37.2)

*This result is very curious, given the absolute stereochemistry of these bases at the three chiral centers which remain unaffected during the potential conversion of δ-skytanthine (37.31) into tecomanine (37.5). Unfortunately, the data provided [373] do not help resolve the problem, and the rotation of neither material is given.

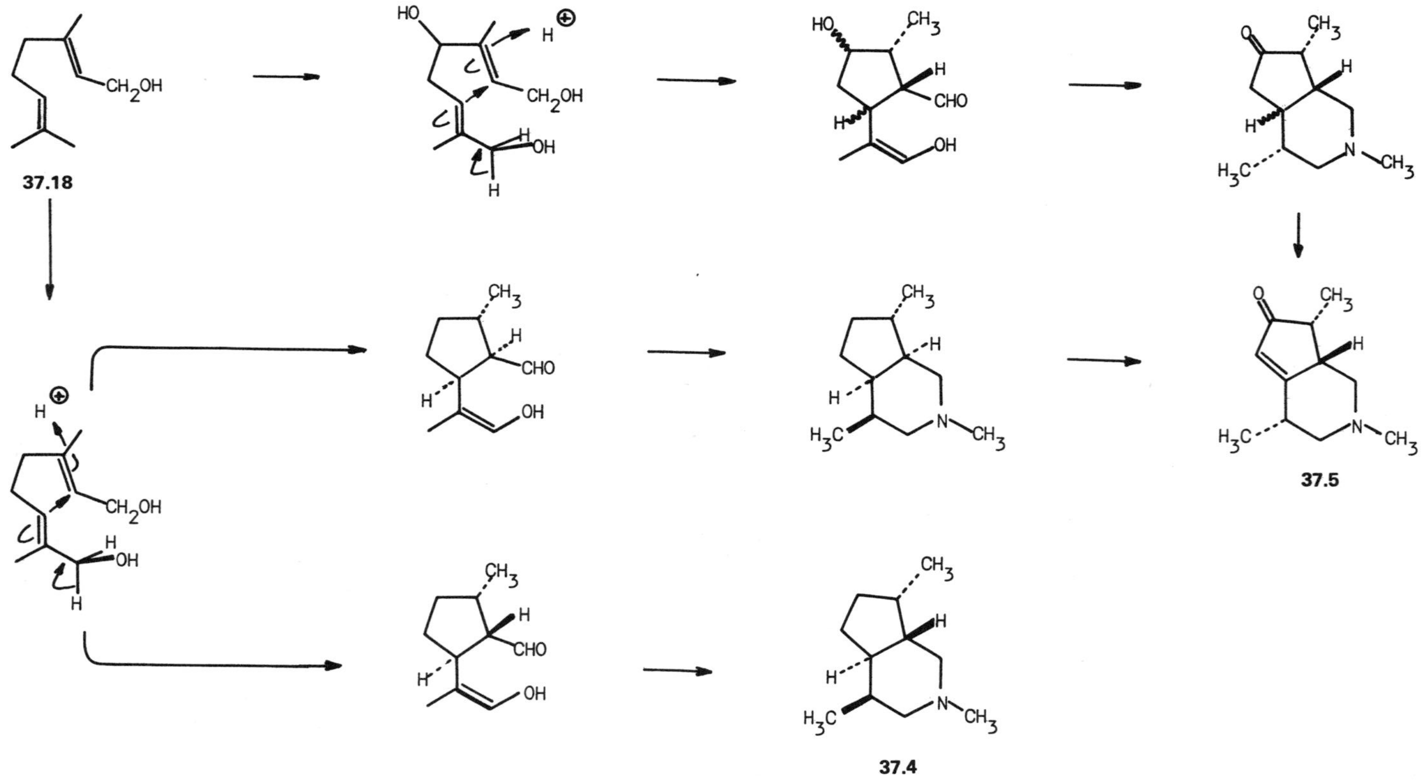

37.18

37.5

37.4

SCHEME 37.5 Potential pathways for the biosynthesis of β-skytanthine and tecomanine from geraniol.

formation of geranyl pyrophosphate (37.24) from isopentenyl pyrophosphate (37.22) and 3,3-dimethylallyl pyrophosphate (37.23), i.e., here the geranyl pyrophosphate (37.24) adds to that side of the double bond of isopentenyl pyrophosphate (37.22) on which the groups $-CH_2CH_2O$(P), $-CH_3$, $-H$, and $-H$ appear in a clockwise fashion; the 2*R*-hydrogen of the isopentenyl pyrophosphate (37.22) is lost, and the new carbon-carbon bond is formed with inversion of configuration to yield farnesyl pyrophosphate (37.32; Equation 37.2). Farnesyl pyrophosphate (37.32) may be in enzyme-catalyzed equilibrium with its double bond isomer (37.33; Equation 37.3; this should be compared to the possible geranyl pyrophosphate, 37.24, neryl pyrophosphate, 37.34, equilibrium, Equation 37.4).

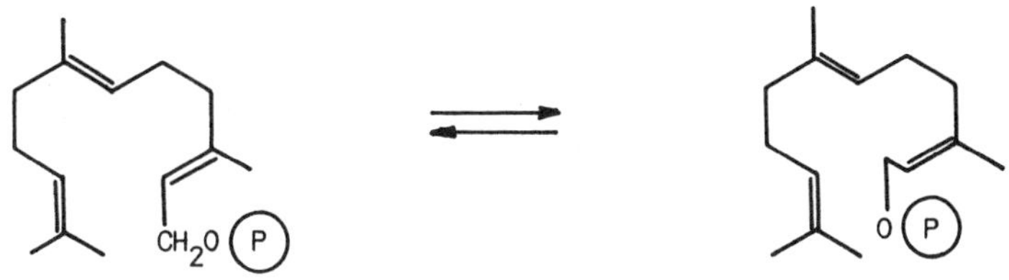

37.32 ⇌ 37.33 (Eq. 37.4)

37.24 ⇌ 37.34 (Eq. 37.5)

A potential pathway from farnesyl pyrophosphate (37.32) to (-)-deoxynupharidine (37.6) is shown in Scheme 37.6. Here, it is assumed that appropriate allylic oxidation could convert farnesyl pyrophosphate (37.32) into a polyhydroxyaldehyde such as 37.35, which via the furan derivative 37.36 could undergo aminolysis to (-)-deoxynupharidine (37.6). Alternatively (Scheme 37.7), in concert with feeding experiments [374-376], farnesyl pyrophosphate (37.32) could undergo cyclization to the trans,trans-cyclodecane cation 37.37, which after a 1,3-hydride shift (it is known that the 1-pro-R-hydrogen migrates), rearrangement to a bicyclic alkene, and oxidation, might afford the alcohol 37.88. Isomerization of 37.38 to the corresponding aldehyde and further rearrangement to the tricyclic alcohol 37.39 followed by dehydration could then produce the diene 37.40. Oxidative cleavage of one double bond and oxidation at the other would then yield 37.41. If this acid were allowed to undergo lactone formation and aminolysis, dendrobine (37.7) could form.

When [2-^{14}C]mevalonate (37.1 as the racemate) was administered [374] to *Dendrobium nobile* and the labeled dendrobine (37.7) was isolated and subjected to a Kuhn-Roth oxidative degradation, the acetic acid obtained was labeled. The extent of labeling was in concert with a genesis of the alkaloid from mevalonate (37.1) and a path such as that shown in Scheme 37.7. Support for a scheme such as the one shown (Scheme 37.7) is also derived, in part, from experiments with (1*S*)-[1-^{3}H]-2-trans,6-trans-farnesol (37.32; [375]), (5*R*)-, and (5*S*)-[5-^{3}H]mevalonic acid (37.1; [376]), and partial degradation of dendrobine (37.6) obtained from feeding experiments with [4-^{14}C]mevalonate (37.1; [377]). Thus, the feeding of the [4-^{14}C]mevalonate (37.1) should lead to label in the farnesyl pyrophosphate (37.32) as shown in Scheme 37.8. Following Scheme 37.7, this would produce the labeled dendrobine (37.7) as also shown in Scheme 37.8. That this was, at least in part, in concert with experiment, was demonstrated by reaction of the labeled dendrobine (37.7) with excess phenyl magnesium

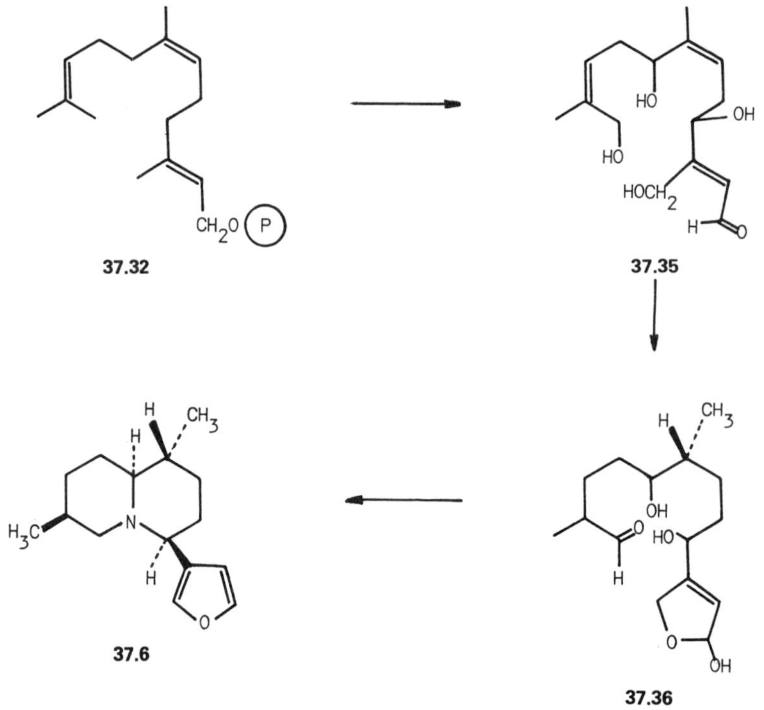

37.32

37.35

37.6

37.36

SCHEME 37.6 A potential pathway from farnesol pyrophosphate to (-)-deoxynupharidine.

bromide under forcing conditions to the corresponding tertiary alcohol, dehydration to the diphenylethene, and ozonolysis to labeled benzophenone.

II. FARNESOL TO GERANYLGERANIOL

The addition of farnesyl pyrophosphate (37.32) to the double bond of another isopentenyl pyrophosphate (37.22) creates geranylgeranyl pyrophosphate (37.42; Equation 37.5), the presumed pregenitor of diterpene alkaloids. The postulated pathway shown (Equation 37.5)

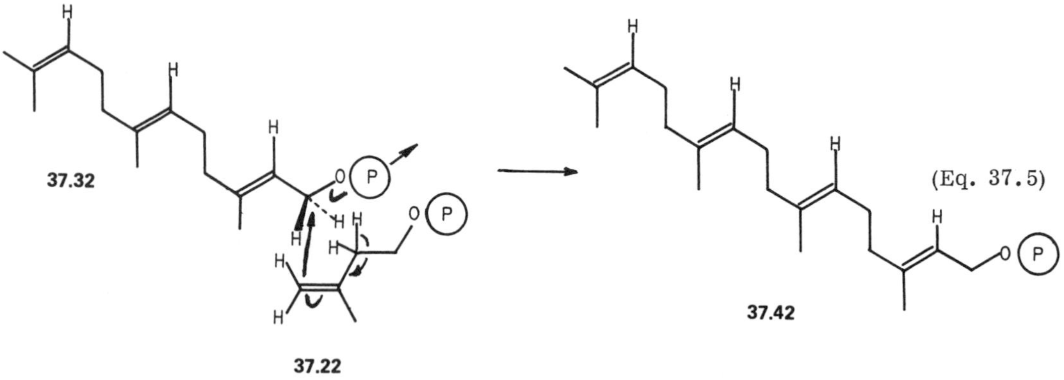

37.32 37.22 37.42 (Eq. 37.5)

bears a striking resemblance to what has been seen before, e.g., in the formation of farnesyl pyrophosphate (37.32) from geranyl pyrophosphate (37.24) and isopentenyl pyrophosphate (37.22), but has not been investigated with the same rigor.

It has been suggested [378-380], largely on the basis of structural similarities, that there is a close relationship between the diterpenes themselves and the diterpene alkaloids. Generally, however, the results obtained from administration of labeled potential precursors, e.g., acetate, mevalonate (37.1), etc., have been unsuccessful. In some cases [381], administration of labeled mevalonate (37.1) results in formation of labeled isolable plant sterols but no labeled alkaloid. However, alkaloid was presumably being synthesized, as demonstrated by assay of the total plant alkaloid content and incorporation of label from [S-^{14}C]-methylmethionine (37.3). Thus, the genesis of atisine (37.8), veatchine (37.9), (+)-aconitine (37.10), and (+)-heteratisine (37.11), as typical members of the diterpene alkaloids, is still the subject of conjecture.

Typical schemes run as follows (Scheme 37.9): Geranylgeraniol (37.43) undergoes acid-catalyzed cyclization to the carbocation 37.44 which generates a diene, and when the diene is reprotonated and further isomerized, the cation 37.45 results. When the latter (37.45) undergoes a Wagner-Meerwein rearrangement, the known diterpene (-)-kaurene (37.46) is obtained. It will, of course, be noted that (-)-kaurene (37.46) has the correct stereochemistry for the alkaloids to be considered* and distinctly resembles veatchine (37.9). It will further be noted that if the cation 37.45 undergoes the equivalent of a 1,6-endo hydride shift, a different cation (37.49), capable of a different Wagner-Meerwein rearrangement, is formed, and a hydrocarbon skeleton (37.50) distinctly resembling atisine (37.8) results. If it were then assumed that atisine (37.8; Scheme 37.10) or some species closely resembling atisine (37.8) were capable of being transformed into a more highly oxidized form, e.g., (37.51), then attack by water could potentially generate 37.52 which, further rearranging via the cyclization shown, produces 37.53, a species distinctly resembling (+)-aconitine (37.10). Last, oxidative cleavage of the same precursor (37.53) could lead to heteratisine (37.11) via an intermediate such as 37.54 (Equation 37.6).

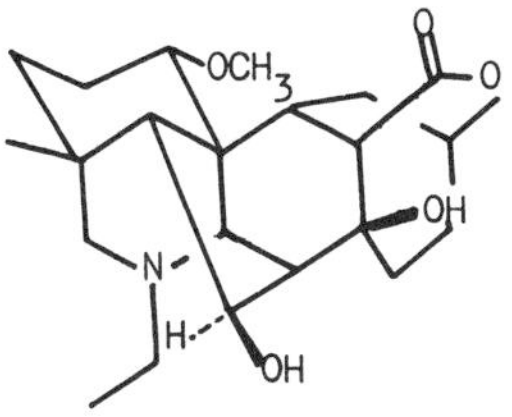

OCH_3 O O OH N H OH

37.11

*The stereochemistry shown was chosen for the cases at hand. Generally, the opposite stereochemistry, e.g., manool (37.47) and (+)-pimaric acid (37.48) is more common.

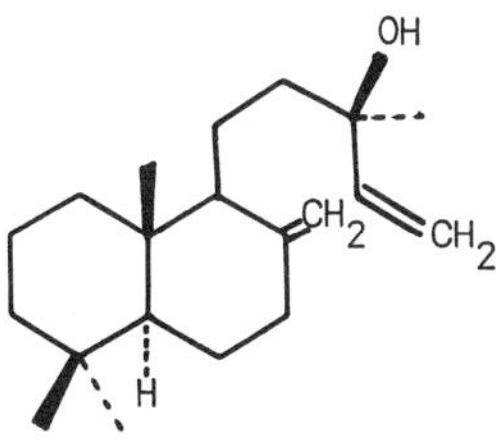

OH CH_2 CH_2 H

37.47

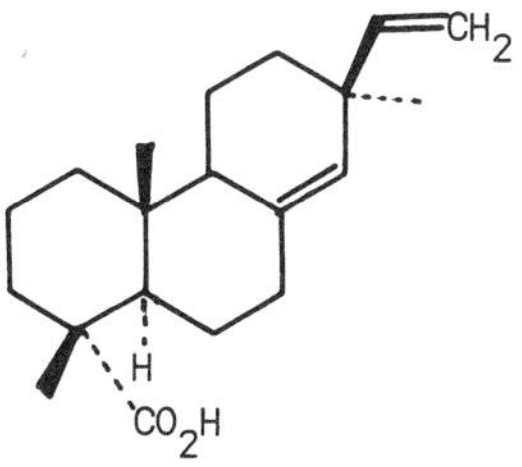

CH_2 H CO_2H

37.48

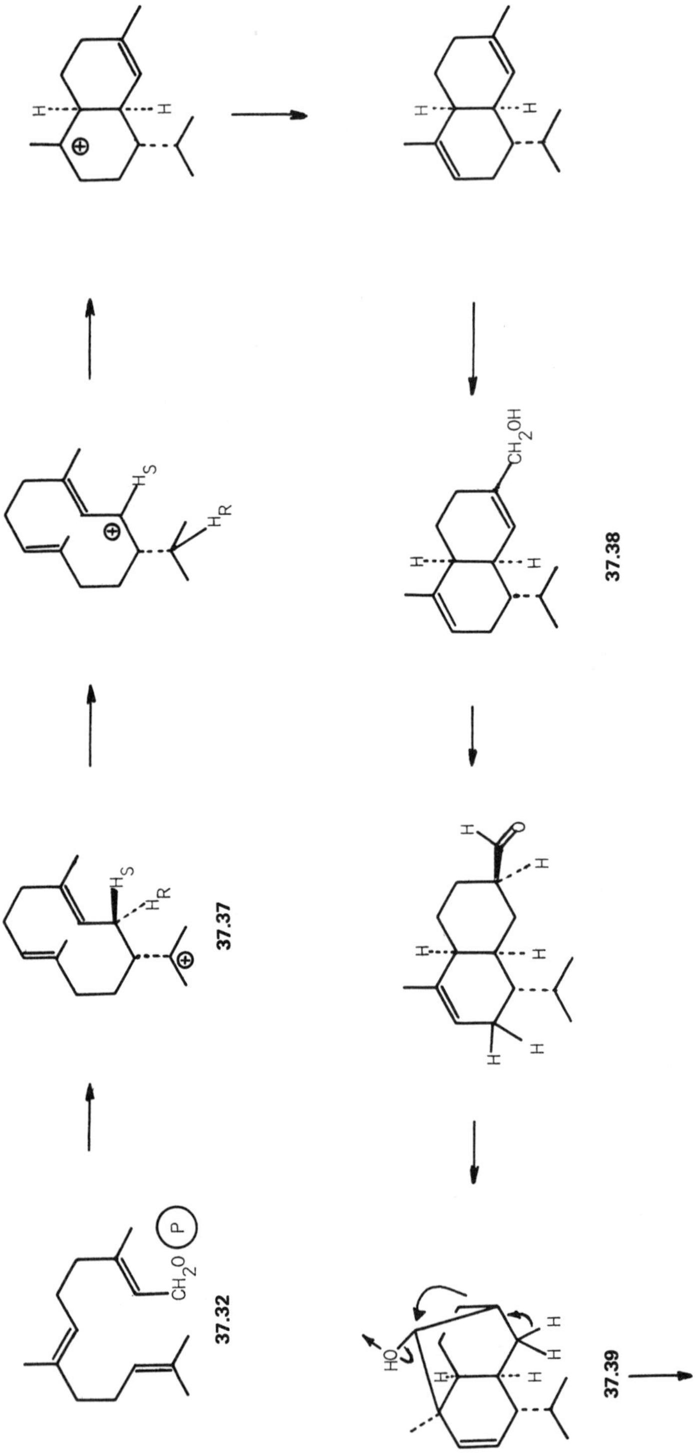

37.32

37.37

37.38

37.39

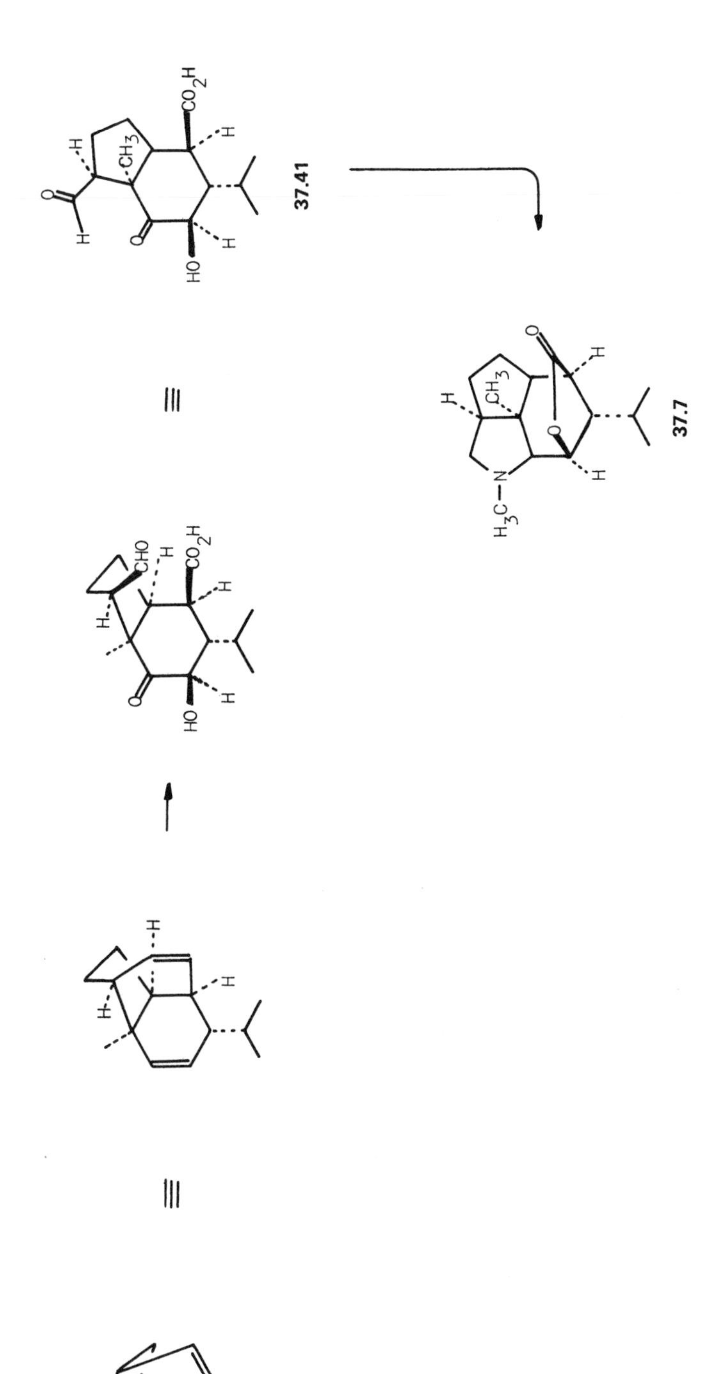

SCHEME 37.7 A possible pathway for the formation of dendrobine from farnesyl pyrophosphate.

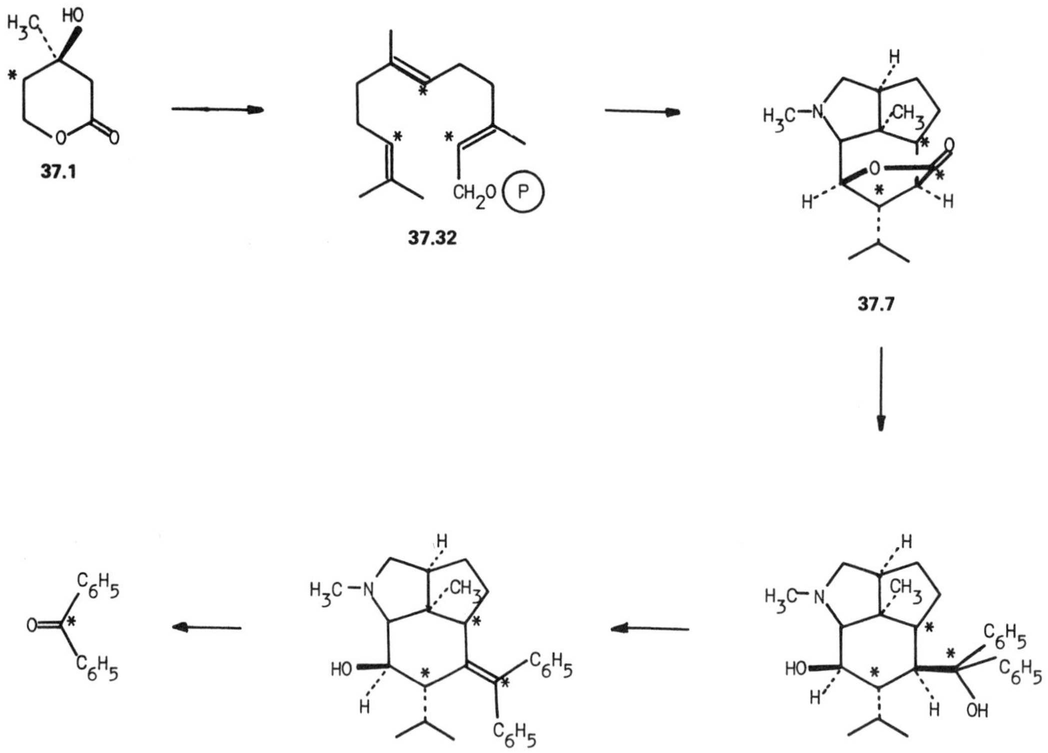

37.1

37.32

37.7

SCHEME 37.8 The partial degradation of dendrobine (*-^{14}C label from [4-^{14}C]-mevalonate). (After Reference 377.)

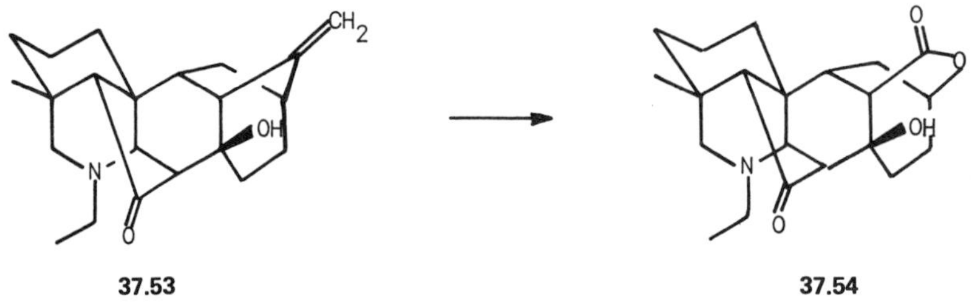

37.53

37.54

(Eq. 37.6)

It is reasonable to enquire as to when (a) the nitrogen, required for general classification of these species as alkaloids, and (b) the large number of oxygen atoms which characterize the species shown, are introduced. It is suggested here that early* oxidation occurs in allylic positions and that nitrogen is introduced late. Therefore, veatchine-like (37.9) species lead to atisine (37.8), and then, as indicated in the above schemes (Schemes 37.9 and 37.10), (+)-aconitine (37.10) and (+)-heteratisine (37.11) result.

---

*Early oxidation is suggested only for convenience, i.e., allylic positions are available for introduction of oxygen. Clearly, given the paucity of information, late stage oxidation is quite reasonable too.

III. FARNESOL TO SQUALENE

Some evidence is available (see below) that the plant sterols as well as the steroidal alkaloids discussed here are derived from squalene (37.21). However, the stereochemical work discussed below has been carried out with animal-derived enzyme preparations, and a direct correlation has not yet been established. Although the exact process by which two farnesyl pyrophosphate (37.32) molecules condense to generate squalene (37.21; Scheme 37.1; Equation 37.7) does not yet appear to be known, several reasonable proposed pathways [382], consonant with the experimental evidence [224], are extant.

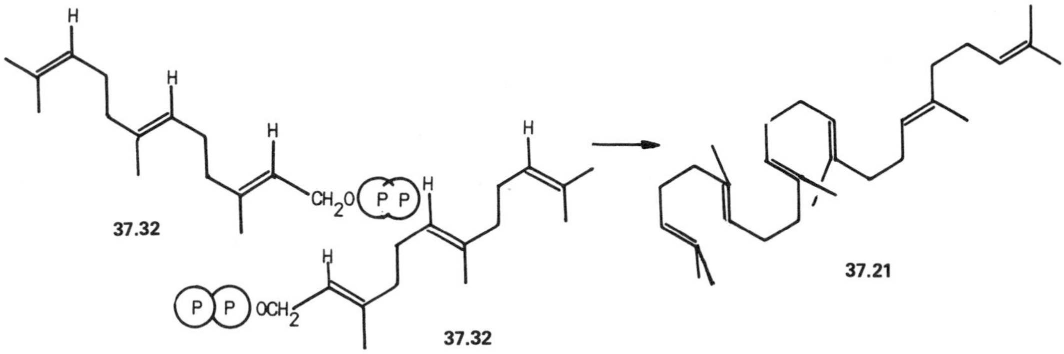

37.32 37.32 37.21

(Eq. 37.7)

The experimental evidence available for mechanistic proposals is as follows [224]: (a) Utilizing [5-^{2}H]mevalonate (37.55) and enzyme preparations from rat liver, a squalene (37.56) is obtained which has 11 rather than 12 deuterium atoms (Scheme 37.11).* Degradation of the deuterated squalene (37.56) by ozonolysis provides a (+)-trideuteriosuccinic acid (37.57) which is identified as [2S-2-^{2}H,3-$^{2}H_2$]succinic acid (37.57) by comparison to a sample of [2R-2-^{2}H]succinic acid (37.58). Therefore, the absolute configuration of the squalene

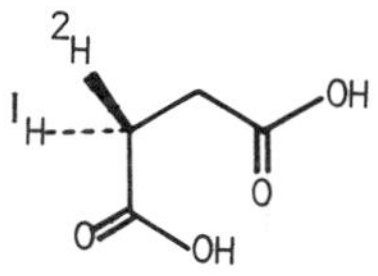

37.58

(37.56) is as shown. One hydrogen has been removed and another added during the coupling process, and the hydrogen introduced as R. (b) The hydrogen with the R configuration introduced in the coupling process of the two farnesol pyrophosphate (37.31) residues (see above) comes from reduced nicotinamide adenine diphosphate (NADH; 37.59), and it is the 4S-hydrogen of 37.59 which is transferred in this process. (c) Both the squalene (37.60) and the

---

*The reader should refer, if necessary, to Chap. 30, p. 438, and in particular to Scheme 30.3 to be reconvinced that the hydrogen at C_5 of mevalonate (37.1) are retained as shown and appear at the positions designated in geraniol (37.18) and farnesol (37.19).

37.43

37.44

37.45

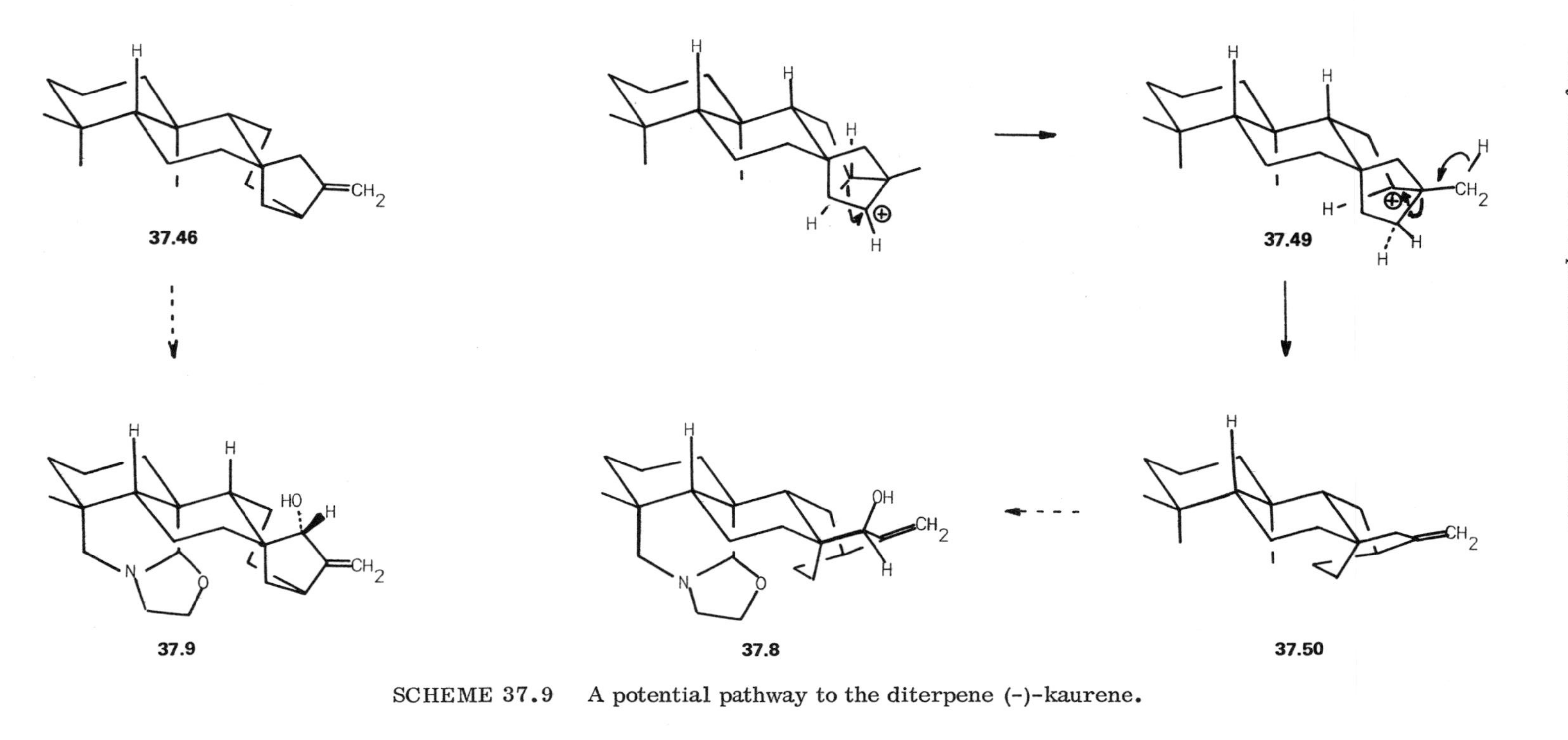

SCHEME 37.9 A potential pathway to the diterpene (-)-kaurene.

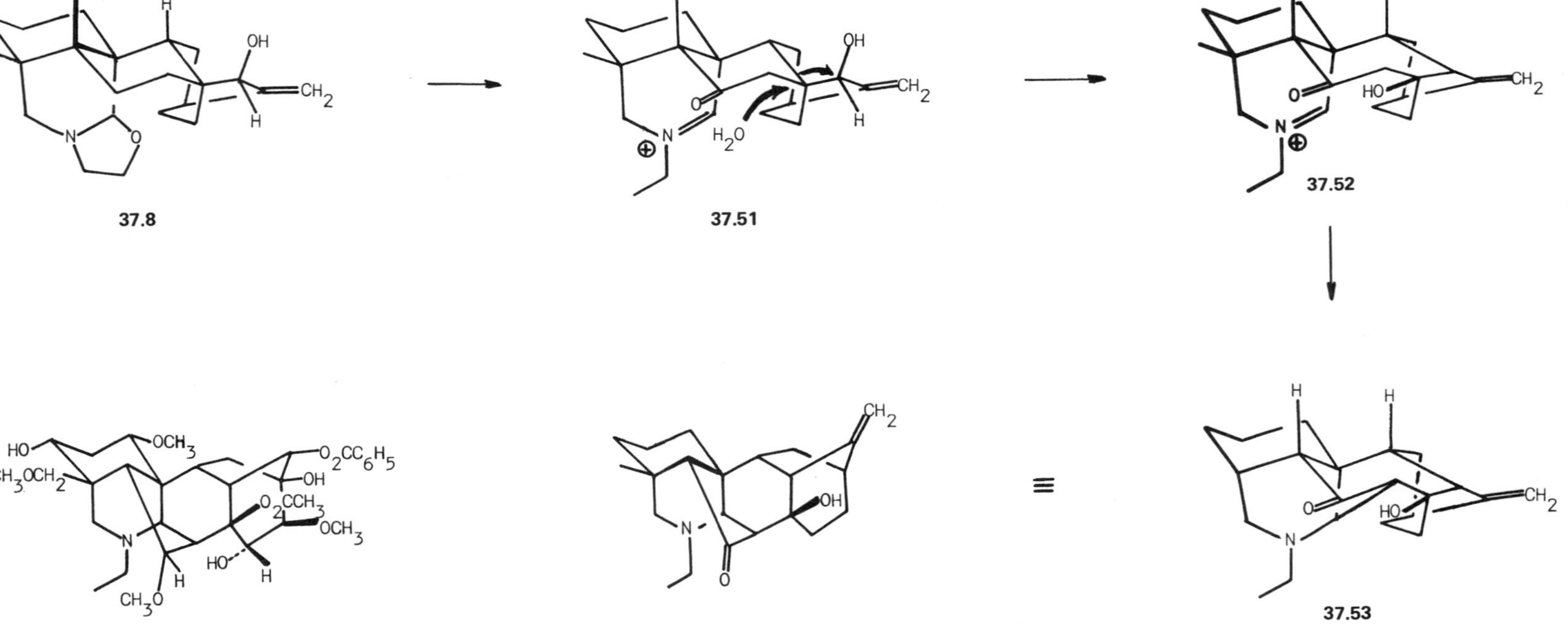

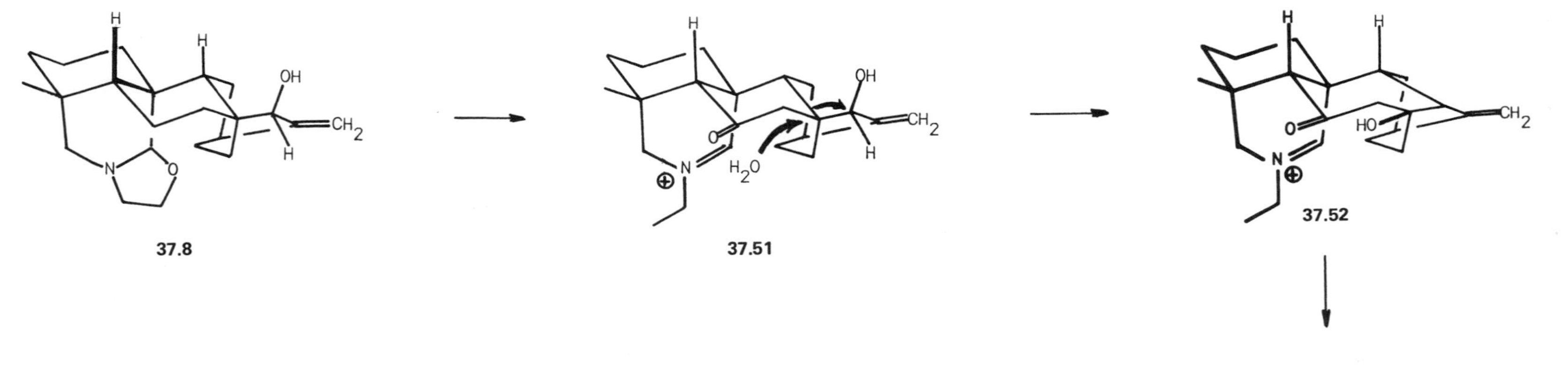

SCHEME 37.10 A possible pathway to (+)-aconitine from atisine.

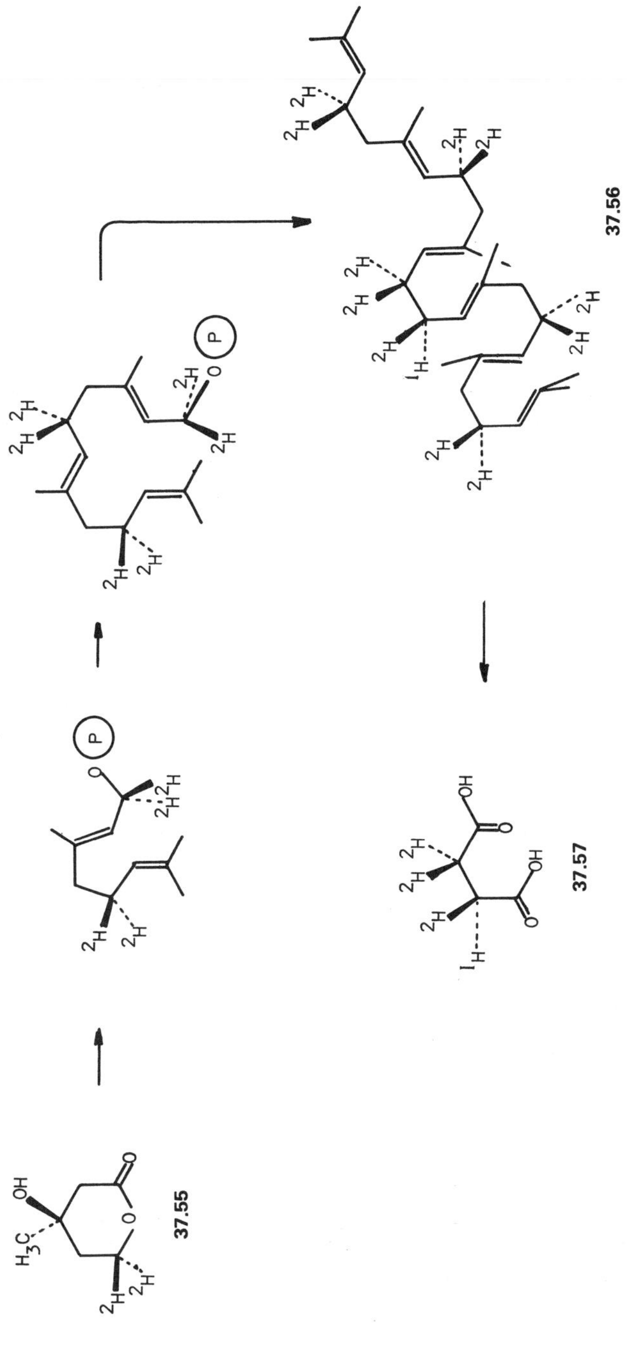

SCHEME 37.11 The experimental results on incorporation of ^{2}H-labeled mevalonate into squalene. (After Reference 224.)

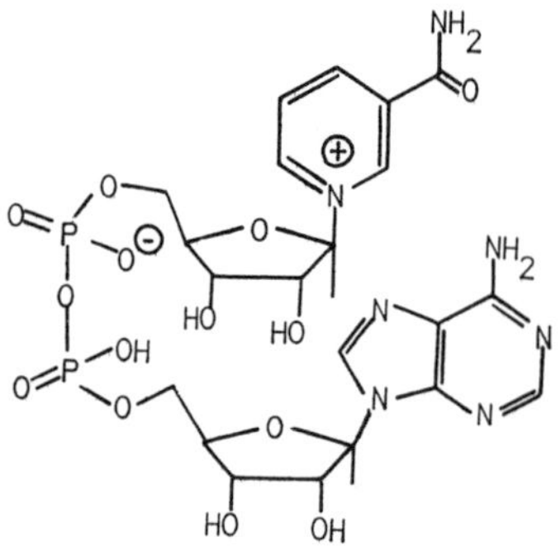

NAD

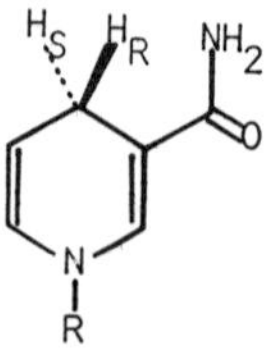

37.59
NADH

farnesol (37.61) biosynthesized from [5R-5-^{3}H,4-^{14}C]mevalonate (37.62)* have the same ^{3}H:^{14}C ratio (Scheme 37.12). Thus, the hydrogen removed from C_1 of farnesyl pyrophosphate (37.32) in the dimerization to squalene (37.60) is an S-hydrogen, and one of the two farnesyl pyrophosphates (37.32) becomes different from the other, i.e., one acts as a

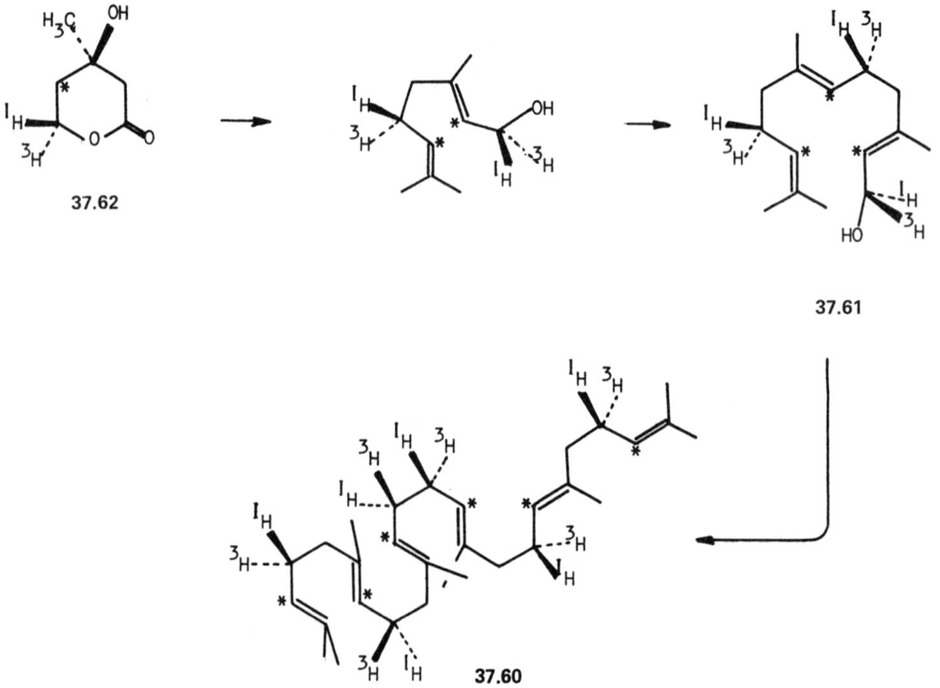

37.62

37.61

37.60

SCHEME 37.12 The biosynthesis of squalene (and farnesol) from [5R-5-^{3}H,4-^{14}C]-mevalonate, which indicates that the ^{3}H:^{14}C ratio is the same in both. (After Reference 224.)

*It is important to realize that a vanishingly small number of molecules of the mevalonate precursor (37.62) contain both ^{3}H and ^{14}C.

reactant and the other as the substrate. (d) In the condensation reaction itself between the two farnesyl pyrophosphate (37.32) residues, solvent protons are not incorporated. One and only one proton comes in, and that one is from NADH (37.59). Thus, it is unlikely that a carbon-carbon double bond is generated. (e) The farnesyl residue (37.32) which does not exchange hydrogen suffers inversion of configuration in the process of squalene (37.21) formation.

Several early schemes [382] to account for this were proposed. In one, isomerization of farnesyl pyrophosphate (37.32) to the corresponding allyl pyrophosphate occurs first and then the two fragments are allowed to dimerize. When this is followed by stereospecific elimination of phosphate with concurrent reduction by NADH (37.59), squalene (37.21) results. It is now generally recognized that, at least in systems derived from animals, a phosphorylated presqualene alcohol (37.63) is involved [383-386]. Thus, as shown in Scheme 37.13, farnesyl pyrophosphate (37.32) dimerizes, perhaps with the involvement of a phosphate ligand [382], although exogenous or enzyme-bound nucleophile cannot be excluded, to yield the phosphorylated presqualene alcohol (37.63). Then, in a process which has been written stepwise, but which may be concerted, the 4S-hydrogen of NADH (37.59) is transferred to the alcohol 37.63 and squalene (37.21) results directly (Scheme 37.14).*

IV. SQUALENE TO CHOLESTEROL AND CYCLOARTENOL

The conversion of squalene (37.21) into squalene-2,3-oxide (37.65; [387]) and thence, via lanosterol (37.66; [388]) to cholesterol (37.67; [389]) has been demonstrated, by selective feeding experiments, to obtain in nonphotosynthetic organisms. Additionally, the methyl and hydrogen migrations shown on the path from squalene (37.21) to lanosterol (37.66; Scheme 37.15) are consonant with labeling studies [390-392]. On the other hand, the details of the conversion of lanosterol (37.66) to cholesterol (37.67; Equation 37.8), which involves

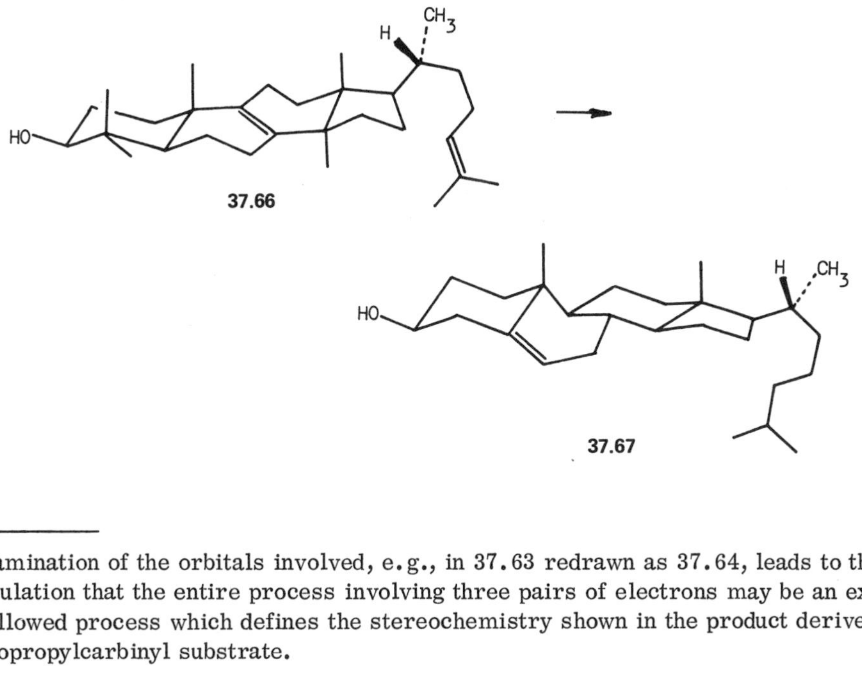

HO H CH_3

37.66

(Eq. 37.8)

HO H CH_3

37.67

---

*Examination of the orbitals involved, e.g., in 37.63 redrawn as 37.64, leads to the amusing speculation that the entire process involving three pairs of electrons may be an example of an allowed process which defines the stereochemistry shown in the product derived from the cyclopropylcarbinyl substrate.

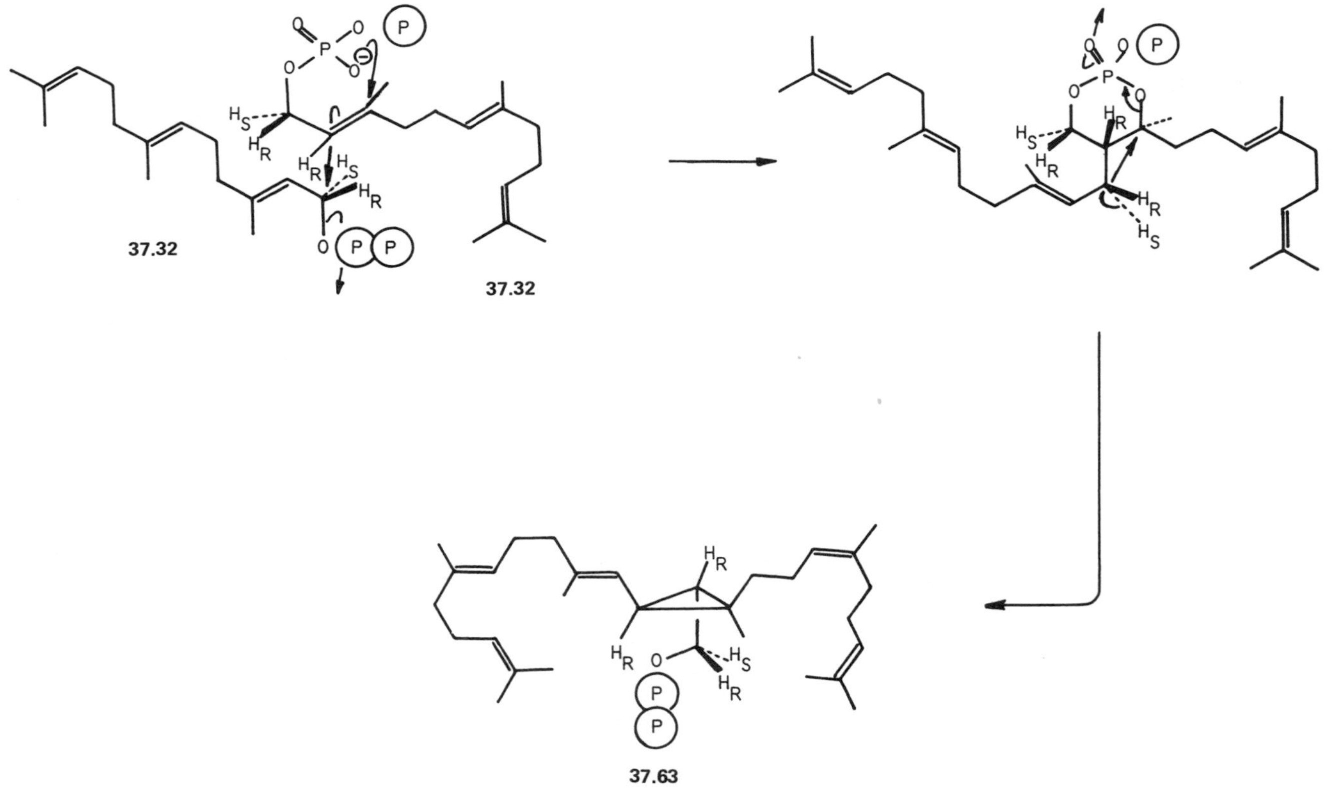

SCHEME 37.13 A possible pathway for the formation of presqualene alcohol from farnesyl pyrophosphate.

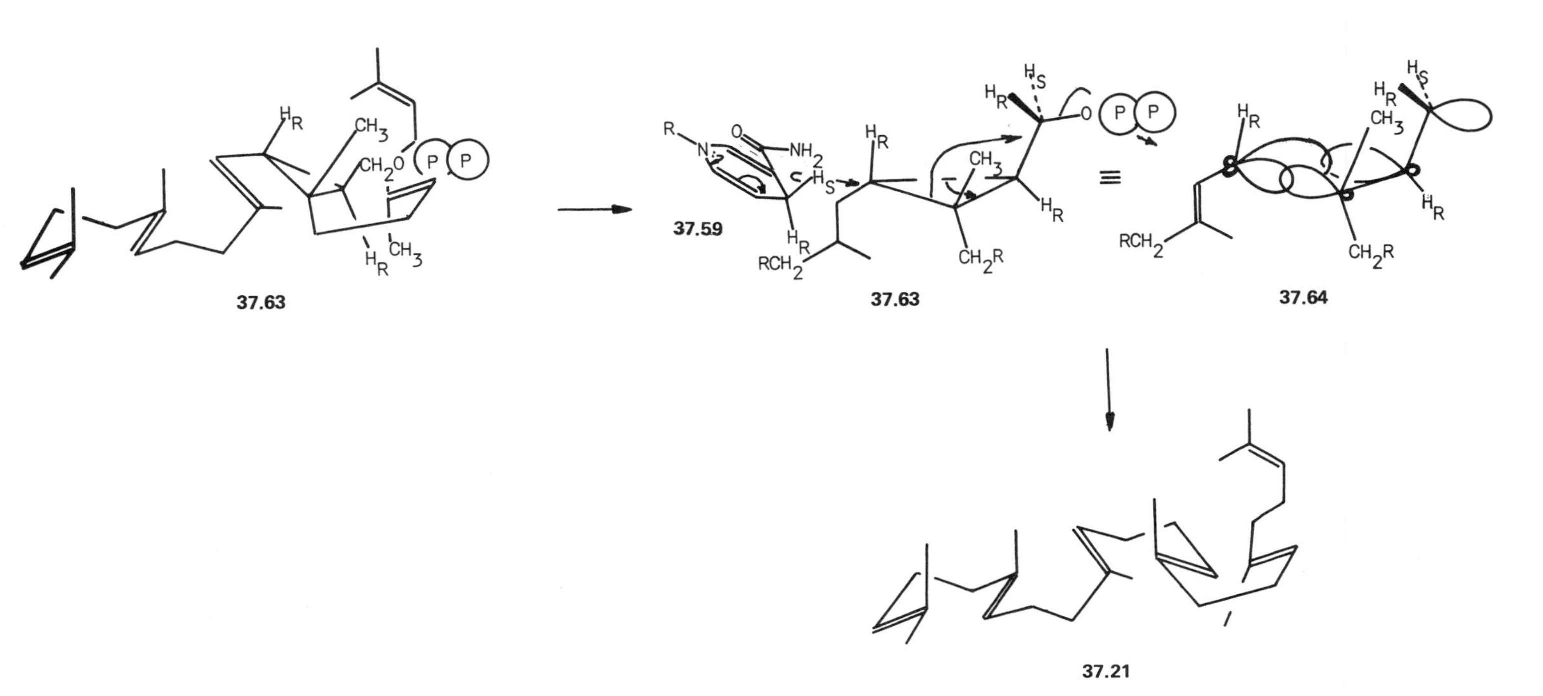

SCHEME 37.14 A possible pathway for the formation of squalene from presqualene alcohol.

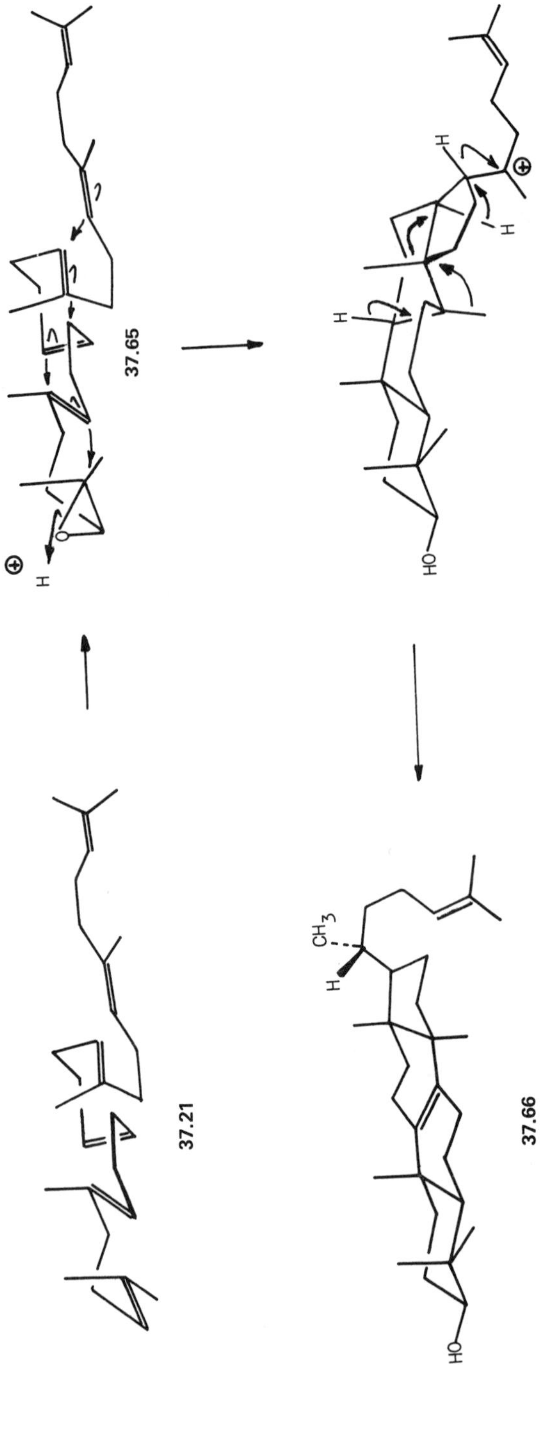
37.21
37.65
37.66
H
O
H
H
H
HO
HO
H
CH3

SCHEME 37.15 A pathway from squalene to lanosterol. (After References 387 and 388.)

(a) removal of three methyl groups; (b) reduction of the double bond (Δ^{24}) in the side chain, and (c) relocation of the double bond at $\Delta^{8(9)}$ to $\Delta^{5(6)}$, are still somewhat obscure, although there is no doubt about the overall process.

Some intermediates in the transformation of lanosterol (37.66) into cholesterol (37.67) have been suggested [393a], but even if they were to be obligatory, it need not follow that the steroidal bases, found in photosynthetic organisms, on which much work remains to be done, follow the same pattern. Indeed, it is now generally acknowledged [393b] that the biosynthesis of the plant steroidal material—not necessarily the steroid alkaloids, however—proceeds from squalene (37.21) via squalene-2,3-oxide (37.65) to cycloartenol (37.68), which bears an obvious relationship to cyclobuxin-D (37.14), and thence to the plant steroids.

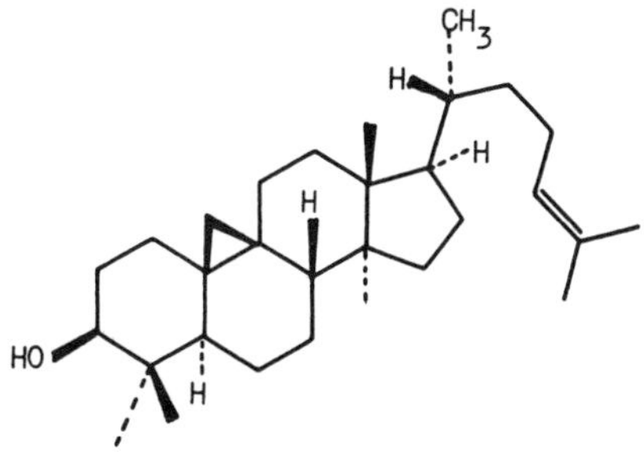

37.68

V. THE STEROID ALKALOIDS

Of the three steroid alkaloids to be discussed here, namely, solanidine (37.12), (-)-conessine (37.13), and cyclobuxine-D (37.14), there apparently exists specific feeding experiments with regard to the genesis of solanidine (37.12) alone. Feeding experiments, followed by isolation of solanidine (37.12), have tested the incorporation of [2-^{14}C]acetate and (±)-[2-^{14}C]-mevalonate (37.1; [394-396]) into this alkaloid in Solanum lacinatum, [3-^{14}C]-3-hydroxy-3-methylglutaric acid (37.69) in Solanum tuberosum [397], cholesterol (37.67; [398], and

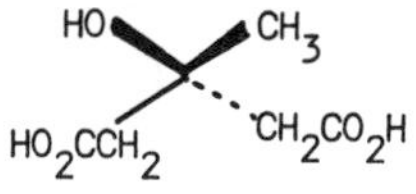

37.69

[26,27-$^{14}C_2$]cycloartenol (37.68) and [26,27-$^{14}C_2$]lanosterol (37.66) in Solanum chacoense [399]. Generally, incorporation has been quite low and there has been at least one suggestion [397] that some labeled precursors are degraded to acetate prior to alkaloid formation. On the other hand, incorporation of cholesterol (37.67), cycloartenol (37.68), and lanosterol (37.66) suggests that pathways from these already elaborated fragments are available (although they need not be obligatory). Finally, although labeled alkaloid is generated, degradative work remains to be carried out to determine the position of the label in many of the cases where incorporation does seem to have been shown.*

Scheme 37.16 outlines a possible pathway to solanidine (37.12) from cholesterol (37.67). Thus, as shown, it is suggested that the D ring and side chain undergo oxidation† to yield 37.70 which on reductive aminolysis leads directly to solanidine (37.12). If cholesterol (37.67) were to undergo oxidative cleavage of the side chain leaving only carbons C_{20} and C_{21}, oxidation at C_{18} and at C_3 would lead to a fragment such as 37.71. Here, aminolysis, cyclization, and reduction followed by methylation (Scheme 37.17) would lead to (-)-conessine (37.13). Feeding experiments with labeled precursors do not appear to have been carried out to support or refute such a suggestion. Finally, if the oxidation reactions were consummated between lanosterol (37.66) and cholesterol (37.67), but this time oxidation occurred at C_{19} instead of C_{18}, a fragment such as 37.72 could be generated from lanosterol (37.66). Here, cyclopropane ring formation with the participation by the $\Delta^{8(9)}$ double bond as well as aminolysis and methylation could be expected to afford cyclobuxine-D (37.14) (Scheme 37.18).‡ Again, labeled precursors do not appear to have been fed to test either this hypothesis of the even more likely possibility of direct formation of cyclobuxine-D (37.14) from cycloartenol (37.68).

---

*As noted earlier, in many plant systems, cycloartenol (37.68) may serve as the key intermediate lying between squalene (37.21) and the elaborated plant bases [400a] and that lanosterol (37.65) is not involved as shown here. While lanosterol (37.65) or a suitable precursor might be converted to cycloartenol (37.68) by a process such as that shown in Scheme 37.18, it has recently been shown by use of a chiral tritium labeled (at methyl) squalene-2,3-oxide (37.65) that the cyclopropane ring in cycloartenol (37.68), which arises by, formally, a hydrogen loss from the methyl group, proceeds with retention of configuration at the methyl [400b].

†Although the substitution pattern, position of the double bond, and stereochemistry of the functionalized steroid system in solanidine (37.12) are identical with that of cholesterol (37.67), oxidation at unactivated positions, which is not without precedent, is not particularly attractive. It is not impossible that oxidation on the side chain might occur prior to reduction of the Δ^{24} double bond in an intermediate lying between lanosterol (37.66) and cholesterol (37.67) and that incorporation of the latter represents a minor aberrant pathway.

‡The alternative process, i.e., hydride abstraction from C_{18} and attack on a protonated double bond, is clearly equivalent [400b].

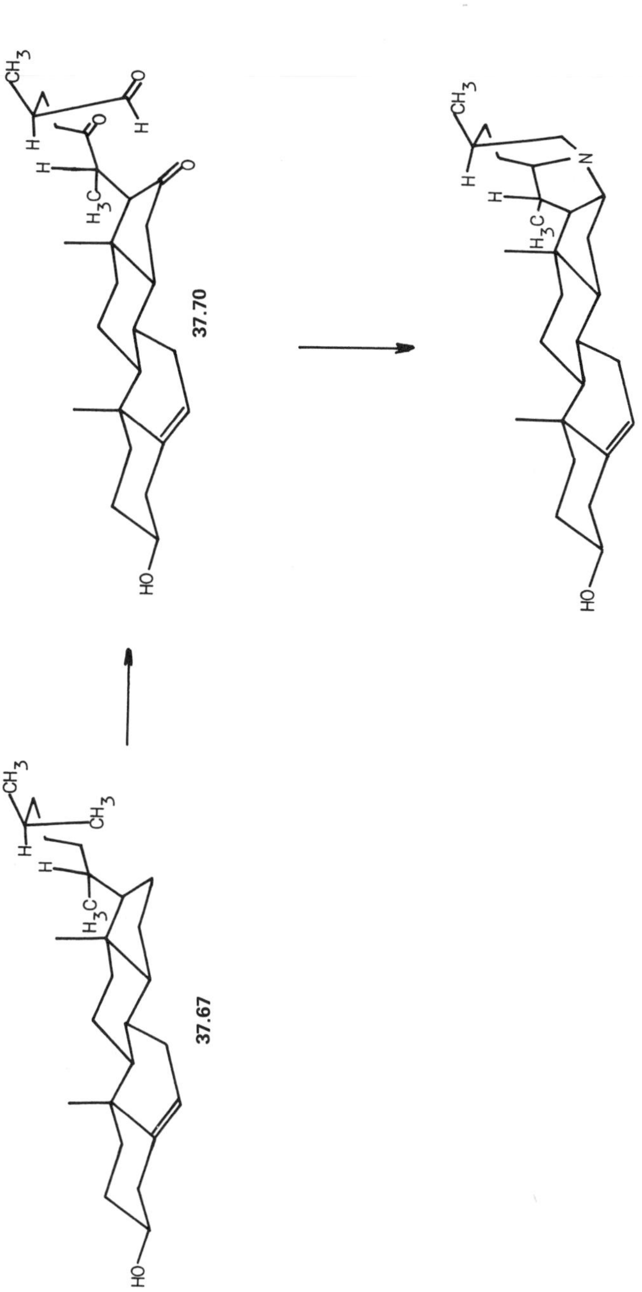

37.67 → 37.70 → 37.12

SCHEME 37.16 A potential pathway to solanidine from cholesterol.

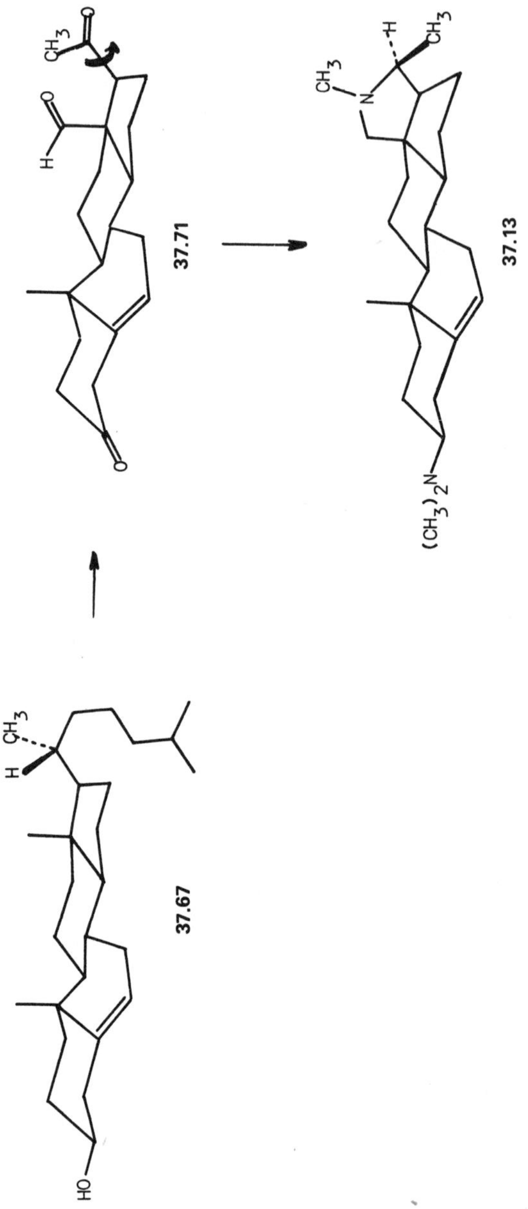

37.67

37.71

37.13

SCHEME 37.17 A potential pathway to (-)-conessine from cholesterol.

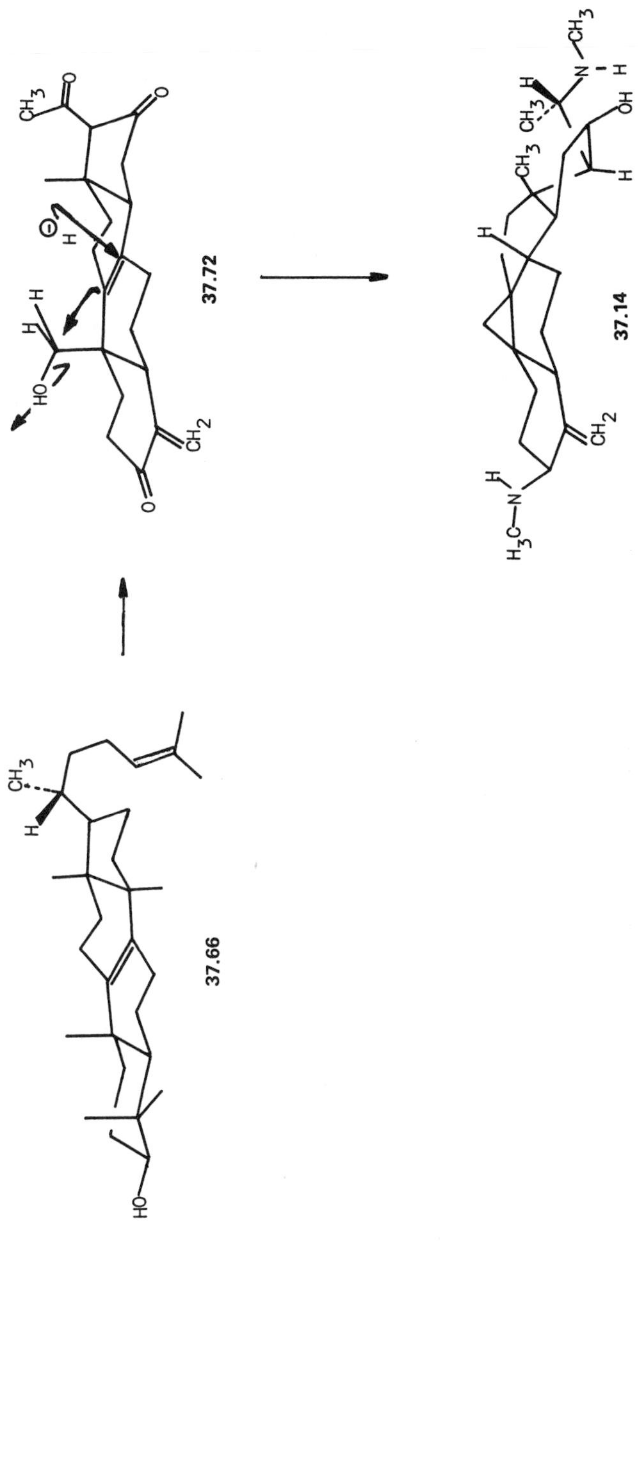

37.66

37.72

37.14

SCHEME 37.18 A potential pathway from lanosterol to cyclobuxine-D.

38

CHEMISTRY OF THE MONOTERPENOID ALKALOIDS

> Simplicity is the most deceitful mistress that ever betrayed man.
>
> H. B. Adams, The Education of Henry Adams (1907)

The two bases, namely, β-skytanthine (38.1) and tecomanine (38.2), to be considered as representatives of the modestly populous group of monoterpene alkaloids of known structure, are found in the Apocynaceae (in Skytanthus acutus Meyen) and the Bignoniaceae (in Tecoma stans Juss.), respectively. Extracts of Tecoma stans Juss. have been reported [401] to show antidiabetic properties. It is interesting to note that these monoterpenoid alkaloids come from families in which indole alkaloids derived from tryptophan (38.3) and loganin (38.4) are also found. Indeed, the absolute configuration of both β-skytanthine (38.1) and tecomanine (38.2) can be demonstrated to be that shown by relation to the absolute configuration of loganin (38.4) through related nonbasic materials, e.g., (+)-β-nepetalinic acid (38.5; [402]).

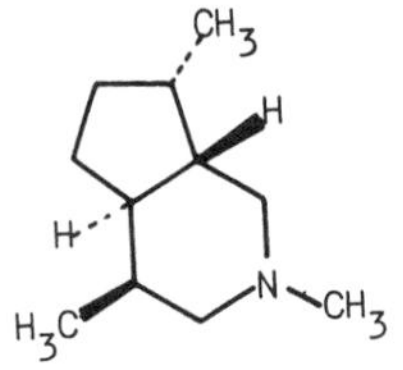

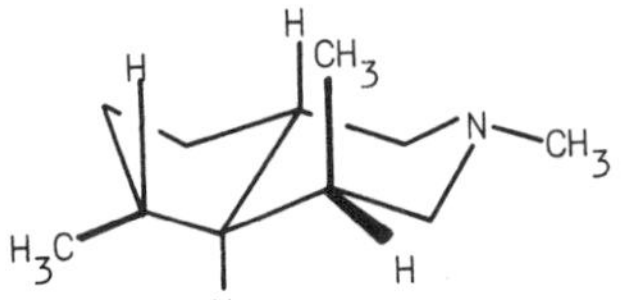

38.1

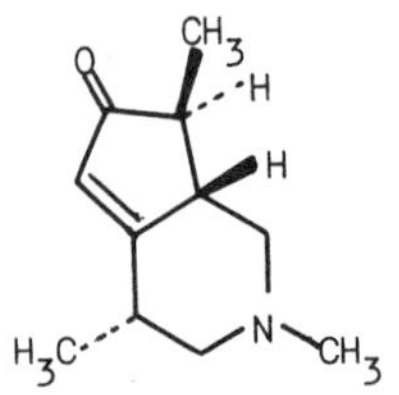

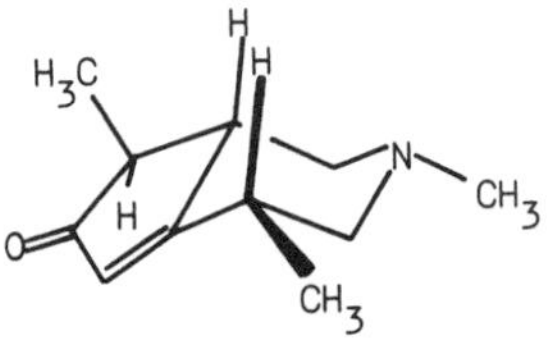

38.2

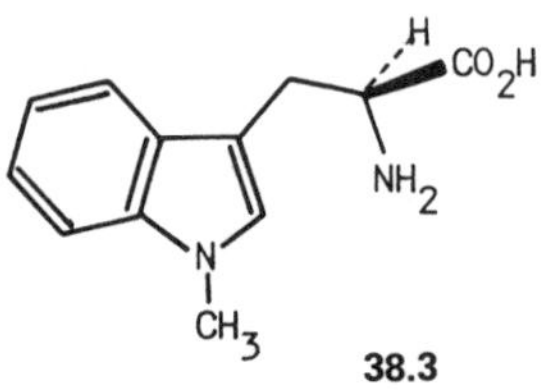

38.3

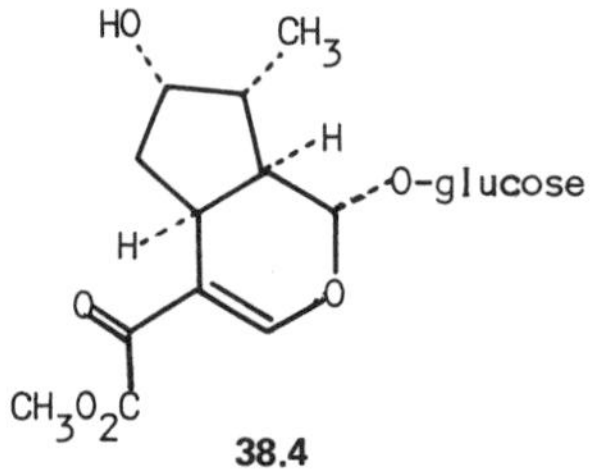

38.4

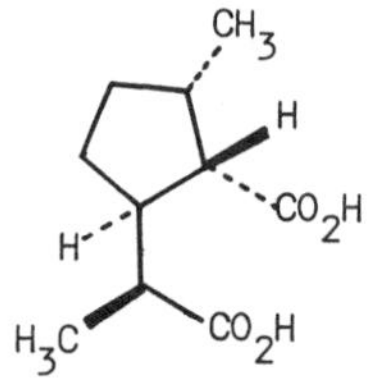

38.5

I. β-SKYTANTHINE

Chilean Skytanthus acutus Meyen yields a mixture of diasteromeric bases which are isomers of β-skytanthine (38.1), the major basic constituent of the plant [403]. β-Skytanthine (38.1), $C_{11}H_{21}N_2$, bp 54°C, 1.5 torr, $[\alpha]_D$ +42° ($CHCl_3$), readily yields a methiodide which in turn undergoes the Hofmann degradation to yield a methine (38.6). Ozonolysis of the latter generates formaldehyde and a ketone (38.7) which gives a positive iodoform test. Bayer-Villager oxidation (peroxytrifluoroacetic acid) of the ketone (38.7) yields an acetate, the hydrolysis of which produces an alcohol (38.8) capable of oxidation to a cyclopentanone derivative (C=O at 1745 cm^{-1}). These reactions are outlined in Scheme 38.1. Additionally (Equation 38.1), dehydrogenation of β-skytanthine (38.1) yields a new racemic base, $C_{10}H_{13}N$, identical

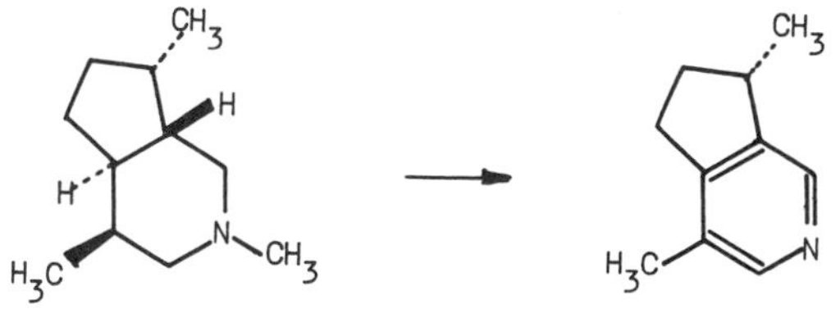

38.1 → 38.9 (Eq. 38.1)

with the known (±)-actinidine (38.9 and its mirror image). The structure of (-)-actinidine (38.9) has been proved by degradation and synthesis [404].

Thus, as shown in Equation 38.1, (-)-actinidine 38.9, bp 100-103°C, 9 torr, $[\alpha]_D$ -7.2° ($CHCl_3$) from Actinidia polygama Miq [404] yields, among other products, 5-methylpyridine-3,4-dicarboxylic acid (38.10) on oxidation. The N-oxide of actinidine (38.11) produces an

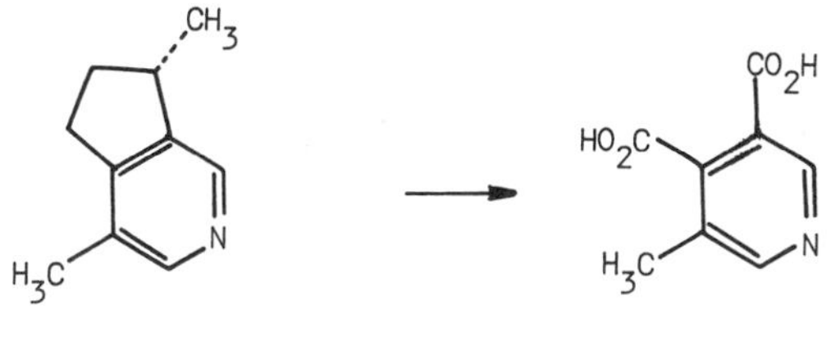

38.9 → 38.10 (Eq. 38.2)

acetate (38.12) on treatment with hot acetic anhydride, and this ester (38.12), on hydrolysis, generates an alcohol which can be oxidized to a ketone (38.13), demonstrating the presence of a methylene group gamma (γ) to the nitrogen of the pyridine ring (Scheme 38.2).

Finally, β-nepetalinic acid (38.5) can be converted into an imide (38.14) which yields the dichloride (38.15) on phosphorus pentachloride dehydration; then, palladium-catalyzed hydrogenolysis of 38.15 generates actinidine (38.9; Scheme 38.3). Alternatively [405], β-nepetalinic acid (38.5) can be reduced to the diol 38.16, the latter tosylated, and the tosylate condensed with methylamine to β-skytanthine (38.1; Scheme 38.4). In an exactly analogous fashion, the other nepetalinic acid isomers can undergo the same reaction sequence and the isomers of β-skytanthine (38.1) prepared.

The structure of β-nepetalinic acid (38.5) is given [406, 407] by generation of this acid from trans-cis-nepetalactone (38.17), a major constituent of volatile oil of catnip (Nepeta cataria L.). Thus, as shown in Scheme 38.5, hydrolysis of the nepatalactone (38.17) yields

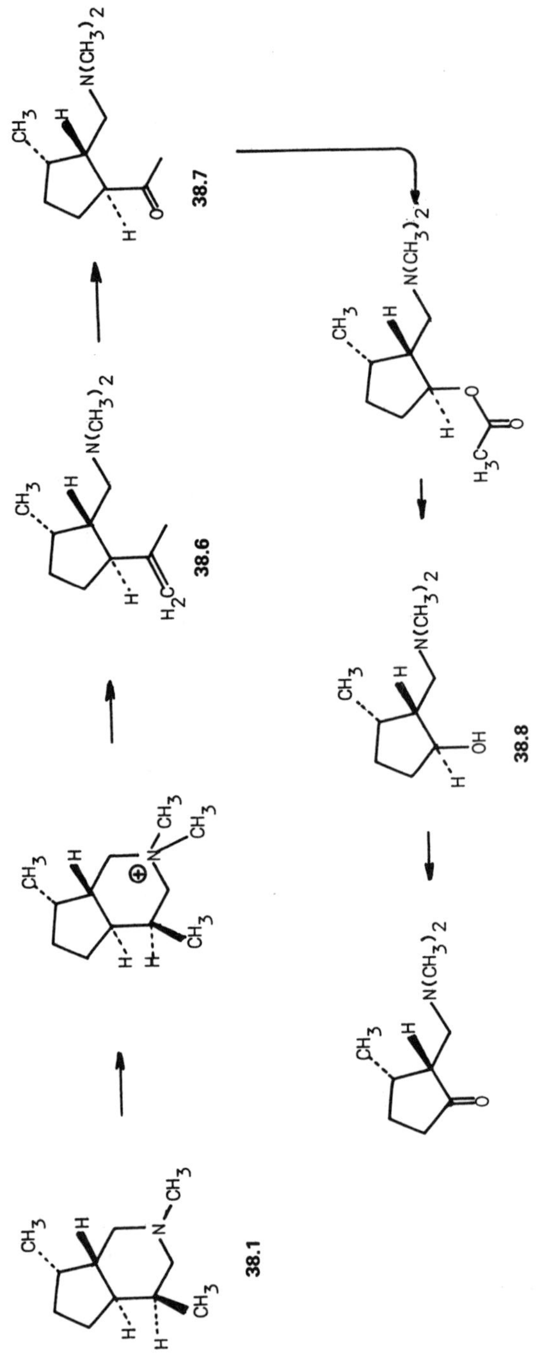

SCHEME 38.1 A partial degradation of the alkaloid β-skytanthine. (After Reference 404.)

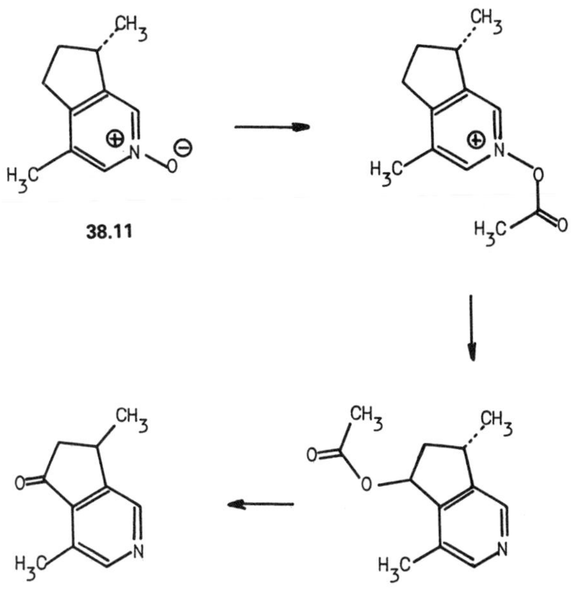

38.11

38.13 38.12

SCHEME 38.2 A partial degradation of actinidine which confirmed the position of the methylene unit γ to the nitrogen. (After Reference 404.)

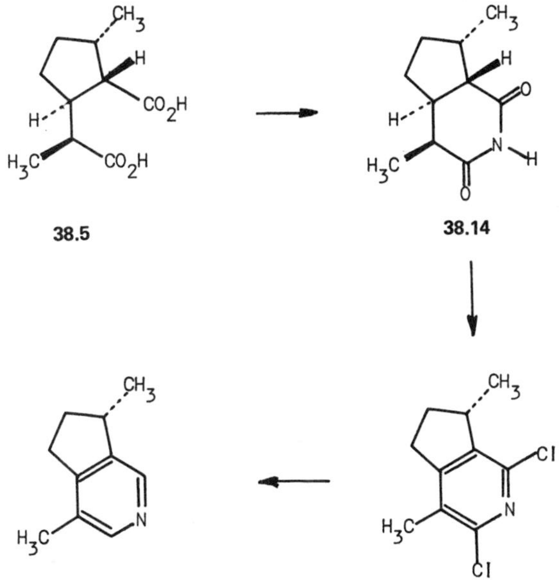

38.5 38.14

38.9 38.15

SCHEME 38.3 A synthesis of actinidine from β-nepetalinic acid. (After Reference 404.)

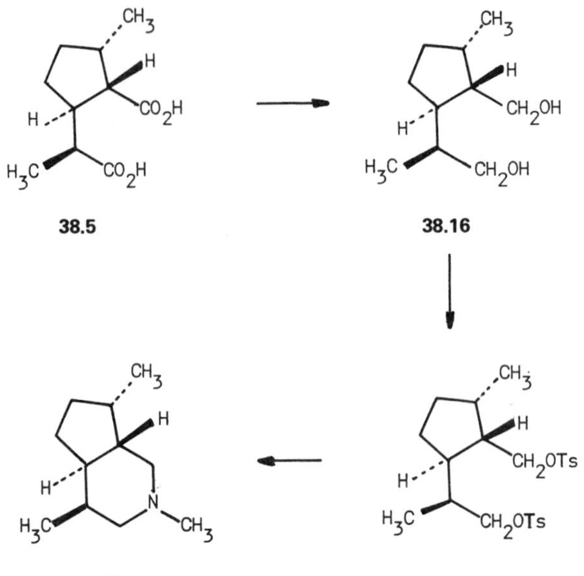

CH_3 H CO_2H H H_3C CO_2H

38.5

CH_3 H CH_2OH H H_3C CH_2OH

38.16

CH_3 H CH_2OTs H H_3C CH_2OTs

CH_3 H H H_3C N CH_3

38.1

SCHEME 38.4 A synthesis of β-skytanthine from β-nepetalinic acid. (After Reference 405.)

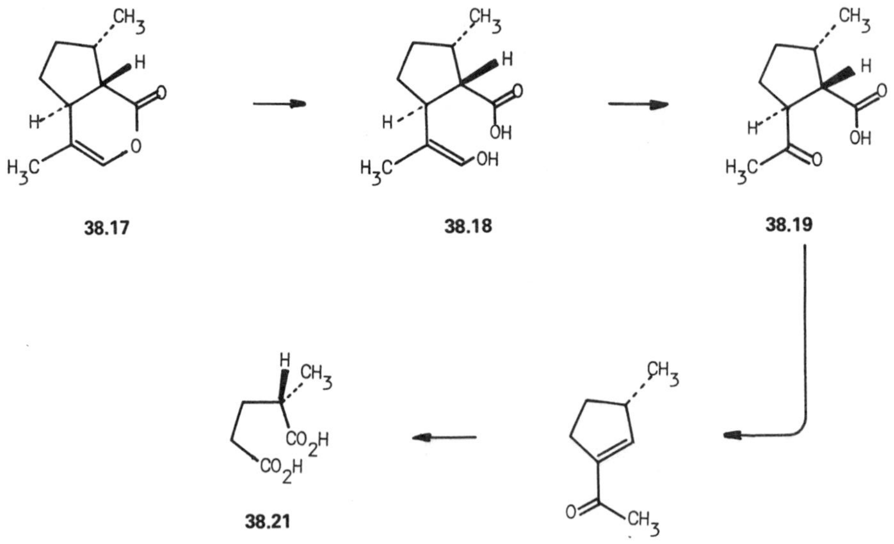

CH_3 H O H H_3C O

38.17

CH_3 H O H OH H_3C OH

38.18

CH_3 H O H OH H_3C O

38.19

H CH_3 CO_2H CO_2H

38.21

CH_3 O CH_3

38.20

SCHEME 38.5 Establishment of the absolute stereochemistry and the structure of trans-cis-nepetalactone. (After References 406 and 407.)

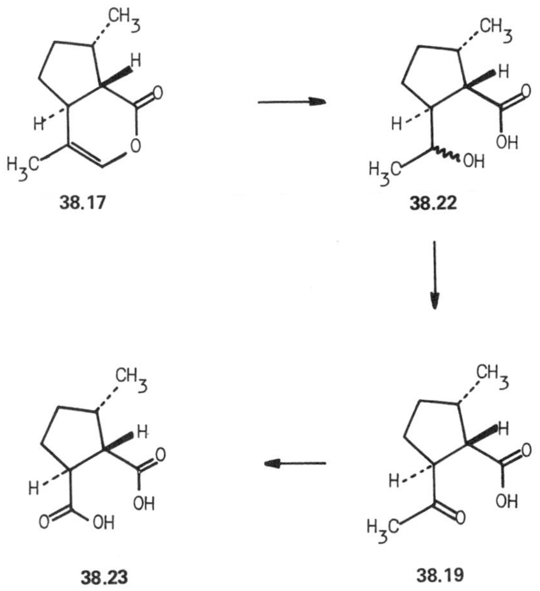

38.17 → 38.22 → 38.19 → 38.23

SCHEME 38.6 Establishment of the structure of trans-cis-nepetalactone. (After References 406 and 407.)

nepetalic acid (38.18; as a lactol in equilibrium with the corresponding aldehyde). Oxidation of the nepetalic acid (38.18) with basic hydrogen peroxide forms the nepetonic acid (38.19), which with lead dioxide undergoes decarboxylation to the corresponding unsaturated ketone (38.20). Oxidation of the ketone with potassium permanganate to (+)-α-methylglutaric acid (38.21) establishes the absolute configuration at the C-methyl carbon.

Meanwhile, ozonolysis of trans-cis-nepetalactone (38.17), followed by reduction with sodium borohydride, generates a nepetolic acid which is one of the epimeric pair corresponding to 38.22 (Scheme 38.6). Chromic acid oxidation of this alcohol yields the nepetonic acid 38.19 which undergoes cleavage via the haloform reaction to yield trans-cis-nepetic acid (38.23), the infrared spectrum of the dimethyl ester of which is superimposable upon that of (±)-trans-cis-dimethyl-3-methylcyclopentane-2,3-dicarboxylate (38.23 and its mirror image).

II. TECOMANINE

Prior to x-ray crystallography [408], which established the absolute structure of (-)-tecomanine (38.2), the gross structure of this base from Tecoma stans Juss., $C_{17}H_{17}NO$, bp 125°C, 0.1 torr, $[\alpha]_D$ -175° ($CHCl_3$), was provided by a series of degradative steps which succeeded in converting tecomanine (38.2) into (±)-actinidine (38.9). This transformation [409] is shown as Scheme 38.7 and involves reduction of the carbon-carbon double bond of tecomanine (38.2; UV, λ_{max} 226 nm, log ϵ = 4.10) to a mixture of saturated cyclopentanones (IR, 1740 cm^{-1}; dihydrotecomanine, 38.24) followed by Hwang Minlon modified Wolff-Kishner reduction of the carbonyl in 38.24 to a mixture of what is presumably isomeric skytanthine alkaloids (38.25). Finally, dehydrogenation of the mixture 38.25 generates (±)-actinidine (38.9). When the initial hydrogenation of tecomanine (38.2) is carried out in ethanol over a palladium catalyst instead of over platinum in acetic acid, a single dihydrotecomanine (one isomer of 38.24 is obtained, and Wolff-Kishner reduction yields a single isomer of skytanthine (one isomer of 38.25). This material could not be correlated with any of the known skytanthine (38.25) isomers.

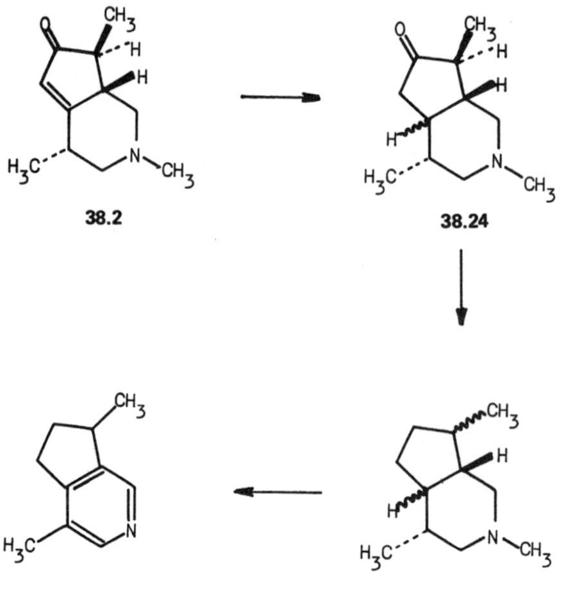

38.2

38.24

38.9

38.25

SCHEME 38.7 Establishment of the structure of tecomanine. (After Reference 409.)

SELECTED READING

W. A. Ayer and T. E. Habgood, in The Alkaloids, Vol. 11, R. H. F. Manske (ed.). Academic, New York, 1968, pp. 459 ff.

R. K. Hill, in Chemistry of the Alkaloids, S. W. Pelletier (ed.). Van Nostrand-Reinhold, New York, 1970, p. 412.

39

CHEMISTRY OF THE SESQUITERPENOID ALKALOIDS

Remember that the most beautiful things in the world are the most useless; peacocks and lilies for instance.

J. Ruskin, The Stones of Venice, Volume I, Chapter 2 (1853)

The two alkaloids considered in this chapter are derived from different families, i.e., deoxynupharidine (39.1) from Nuphar japonicum (Nymphaceae) (water lilies) and dendrobine (39.2) from the plant Dendrobium nobile (Orchidaceae), the source of the Chinese drug

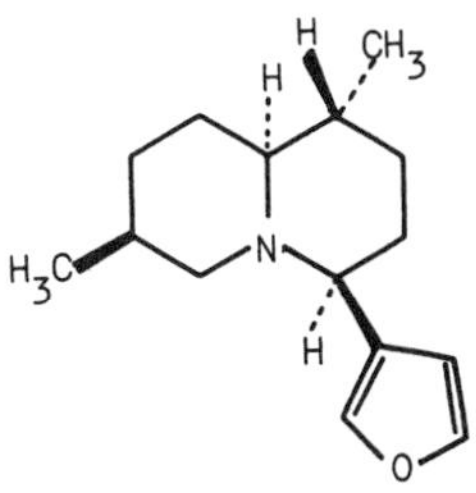

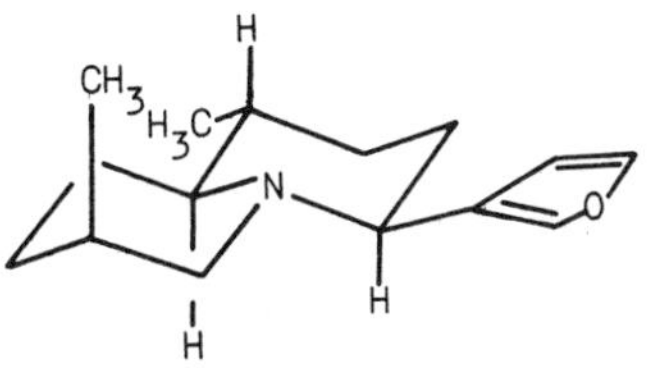

39.1

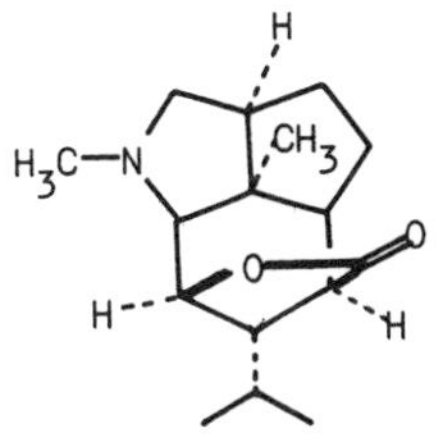

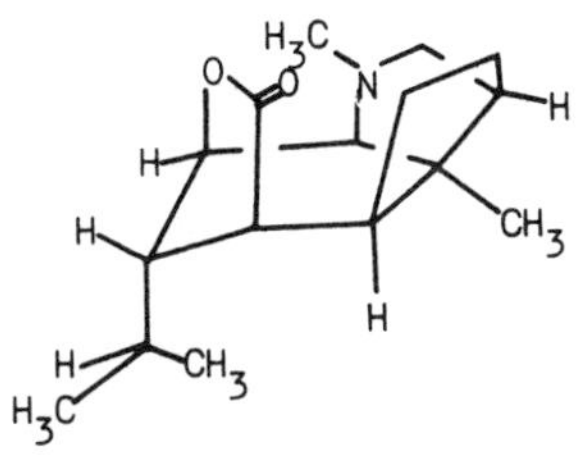

39.2

Chin-Shih-Hu, but are both presumably sesquiterpenoid in origin. These and other related sesquiterpenoid alkaloids have recently been the subject of a review [410], and while there are still relatively few such bases known, the reasonable suggestion has been made [410] that the diversity of families from which the known ones have been isolated implies many more will be found.

I. DEOXYNUPHARIDINE

Prior to the delineation of the absolute stereochemistry of (-)-deoxynupharidine (39.1), $C_{15}H_{23}NO$, mp 21°C $[\alpha]_D$ -112°, by x-ray crystallography [411], complete degradation and synthesis of the base (39.1) had determined its structure. Indeed, even the absolute stereochemistry was known since oxidation of (-)-deoxynupharidine (39.1) to (R)-(-)-2-methyladipic acid (39.3; Equation 39.1) and relation of the other functional groups to the site specified (see below) had been completed [412-414].

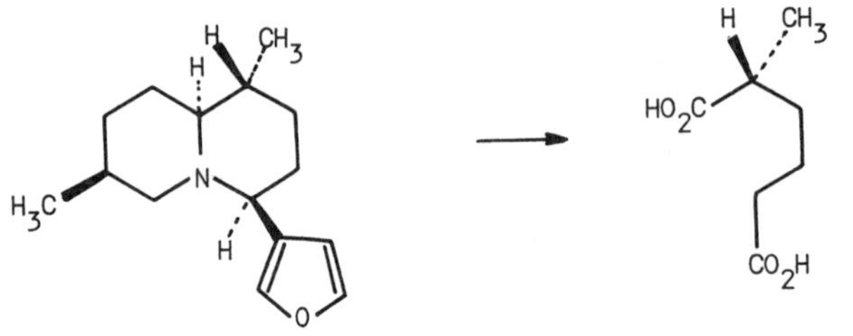

39.1 → 39.3 (Eq. 39.1)

Among the early chemistry carried out on deoxynupharidine (39.1) was the palladium-catalyzed dehydrogenation which yields a variety of products. One of the products, $C_{15}H_{19}NO$ (39.4), could be oxidized to pyridine-2,5-dicarboxylic acid (39.5; [413]), suggesting the presence of a piperidine or another pyridine-related nucleus in the parent alkaloid (39.1; Equation 39.2).

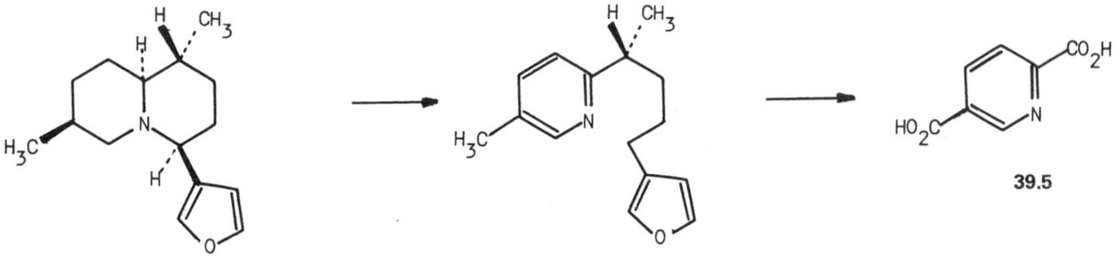

39.1 → 39.4 → 39.5 (Eq. 39.2)

Meanwhile, catalytic hydrogenation of deoxynupharidine (39.1), which had been shown to be a tertiary amine containing at least two C-methyl groups, yields both tetrahydro- and hexahydroderivatives (39.6 and 39.7, respectively; Equation 39.3). That the latter (39.7)

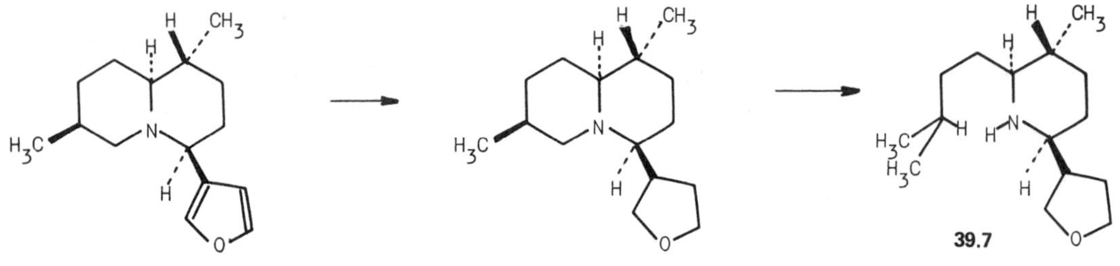

39.1 → 39.6 → 39.7 (Eq. 39.3)

was, in fact, a product of hydrogenolysis was established by showing it to be a secondary amine [415].

Utilization of the Hofmann degradation and intermediary hydrogenations as well as more careful examination of the intermediate products finally established the gross structure of deoxynupharidine (39.1). Thus (Scheme 39.1), formation of the methiodide of deoxynupharidine (39.8) followed by the Hofmann reaction yields a methine (39.9) which undergoes hydrogenation to a hexahydromethine (39.10). Additional methylation, elimination, and reduction and a final methylation and elimination produce two olefins (39.11 and 39.12) which can be reduced to a saturated nitrogen-free product, $C_{15}H_{30}O$ (anhydronupharanediol, 39.13) or utilized directly for further examination.

Now, ozonolysis of the methine base 39.9 yields 3-formylfuran (39.14), indicating the presence of a furan ring (Equation 39.4), and this was confirmed by degradation of anhydronupharanediol (39.13; Scheme 39.2).

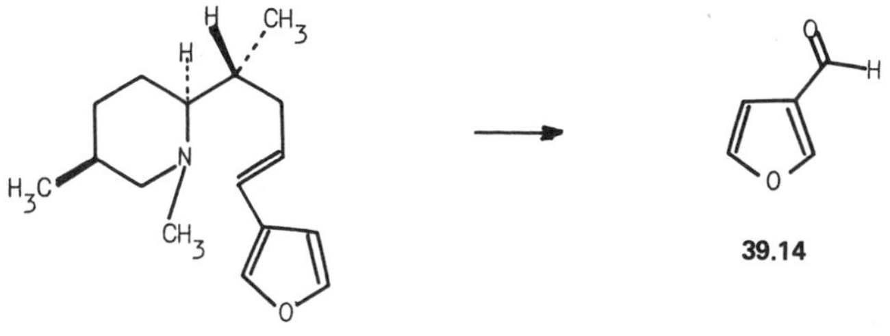

H CH3 H N H3C CH3 O

39.9

O H O

39.14

(Eq. 39.4)

Thus, treatment of anhydronupharanediol (39.13) with hydrogen iodide yields (reversibly with silver oxide) a diiodide (39.15) which can be converted, via the corresponding bis-N,N-dimethylamine derivative, to the diene 39.16 and thence into an α-ketoaldehyde (39.17). Additionally, anhydronupharanediol (39.13) generates a γ-lactone (39.18) on chromic oxide oxidation, which in turn affords a hydroxymethyl ester that can be further oxidized with chromium trioxide to nupharanedioic acid (39.19) and isocaproic acid (39.20; [416]).

Although these reactions established the position of one of the C-methyl groups and the presence of the furan ring, it was only later through examination of the alkenes 39.11 and 39.12 (see above) that the position of the second methyl group was established, and through oxidation of the α-ketoaldehyde (39.17) that the structure of anhydronupharanediol (39.13) was elucidated [415, 417].

As shown in Scheme 39.3, ozonolysis of the mixture of alkenes 39.11 and 39.12 followed by permanganate oxidation affords both acidic and neutral compounds. The acidic fraction yields isocaproic acid (39.20), isovaleric acid (39.21), and a monobasic acid $C_{10}H_{18}O_3$ (39.22). The neutral fraction affords a methyl ketone (39.23) which undergoes the iodoform reaction to yield a new acid (39.24). Indeed, since the monobasic acid $C_{10}H_{18}O_3$ (39.22) can be converted into the same ketone (39.23) by formation of the corresponding methyl ester, reaction to the tertiary alcohol (39.25) with methyl Grignard, dehydration to the corresponding alkene, and ozonolysis, the position of the second methyl group (as shown) is also established. Finally, when the α-ketoaldehyde (39.17) is oxidized with lead dioxide (Equation 39.5), 5,9-dimethyldecanoic acid (39.26) is obtained [417].

H O O

39.17

OH O

39.26

(Eq. 39.5)

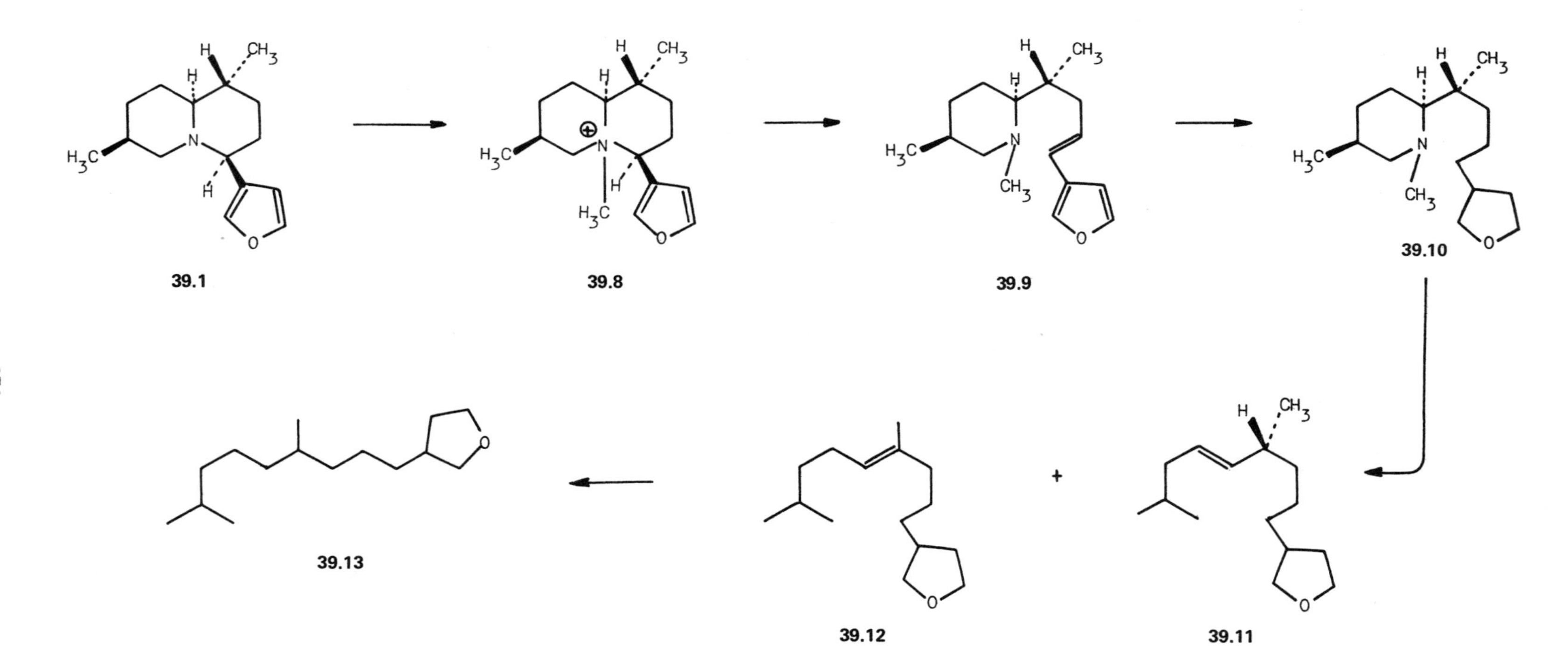

SCHEME 39.1 Degradation, via the Hofmann reaction, of deoxynupharidine to anhydronupharanediol. (After Reference 415.)

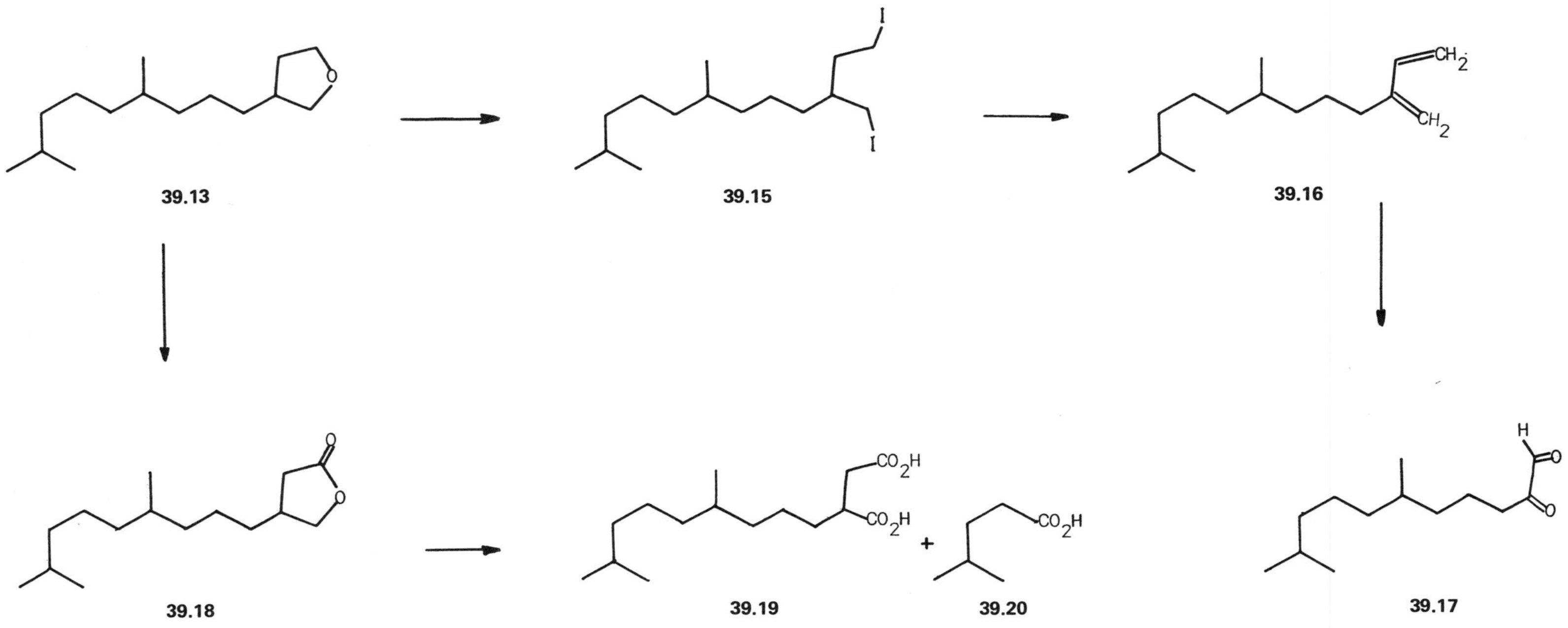

39.13

I

I

39.15

CH_2

CH_2

39.16

H

39.17

39.18

CO_2H

CO_2H

+

CO_2H

39.19

39.20

SCHEME 39.2 Degradation of anhydronupharanediol. (After Reference 416.)

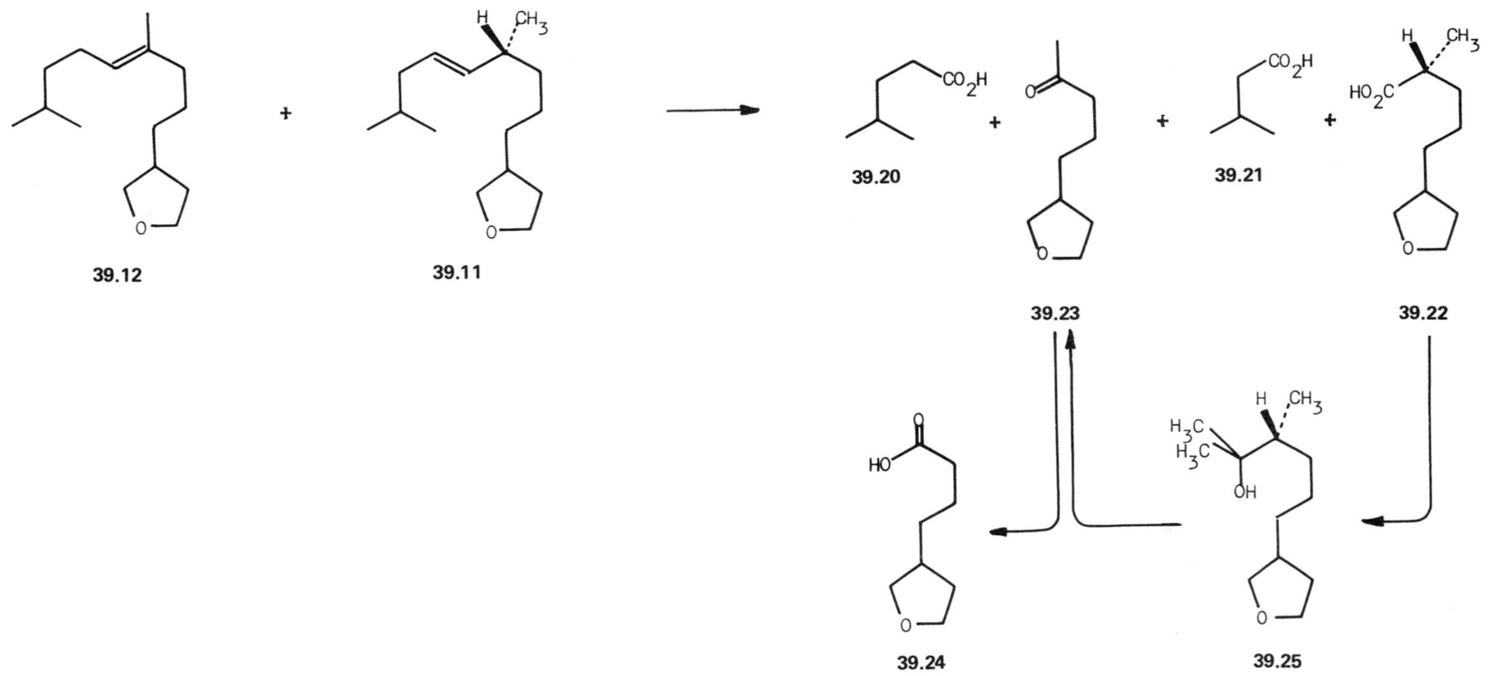

SCHEME 39.3 Degradation of the Hofmann products obtained from deoxynupharidine by ozonolysis. (After Reference 417.)

With the structure of deoxynupharidine (39.1) resolved, attention was turned to its synthesis. Several syntheses of racemic deoxynupharidine (39.1) and its diasteromers were announced before the base itself was finally prepared [418]. The successful synthesis required much information about the stereochemistry of the dimethylquinolizidine system. This information was obtained in the following fashion (Scheme 39.4). Condensation of α-picoline (39.27) with formaldehyde to yield the diol 39.28 is followed by acetic anhydride dehydration to generate the acetate 39.29. Condensation of the acetate 39.29 with malonic ester affords the diester 39.30, which on catalytic hydrogenation provides the amino ester 39.31. This β-dicarbonyl system (39.31), on acid hydrolysis, leads to the amino ester 39.32, which thermally recyclizes to the lactams 39.33 and 39.34. Separation of the lactams is accomplished chromatographically and the stereochemistry of each established by comparison of the amines 39.35 and 39.36 prepared by reduction of the lactams 39.33 and 39.34 with lithium aluminum hydride to the corresponding amines obtained by reduction of the tosylates of lupinine (39.37) and epilupinine (39.38), whose structures had been previously established (see Chap. 11, p. 123) (Equation 39.6).

In an analogous fashion, i.e., by comparison to the reduction products of the corresponding primary alcohols, only one of which can internally hydrogen bond, the relative configurations of the two bases 39.39 and 39.40 were established, and utilizing many infrared correlations, the relative configuration of the methyl groups in the isomer 39.41 was ascertained.

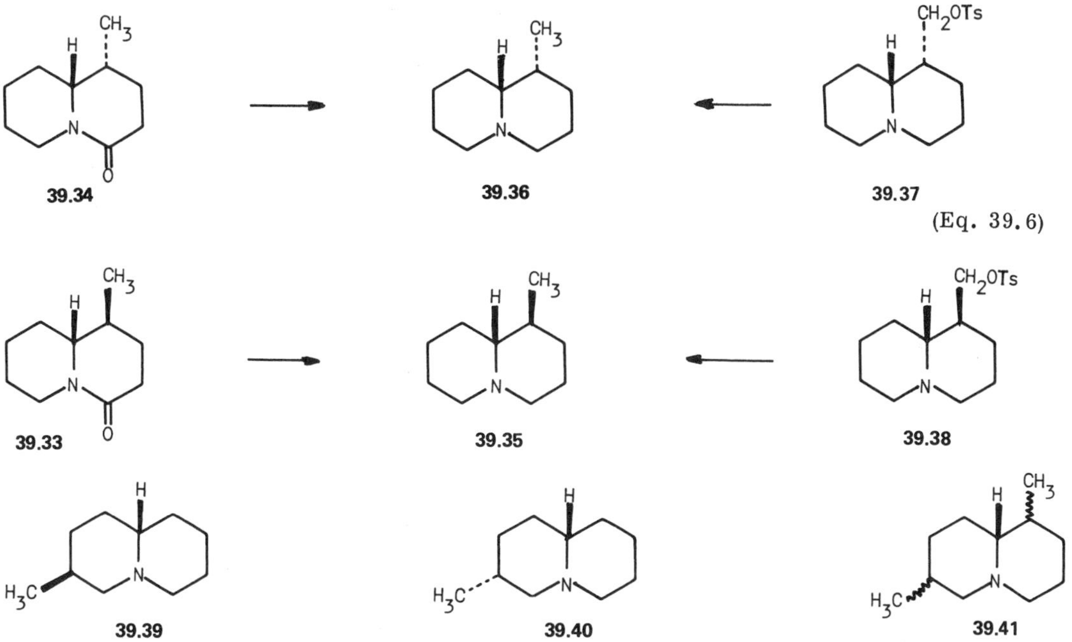

39.34 → 39.36 ← 39.37

(Eq. 39.6)

39.33 → 39.35 ← 39.38

39.39 39.40 39.41

For the synthesis of deoxynupharidine (39.1; Scheme 39.5), the lutidyl ethyl ester, 39.42, is methylated to 39.43 and the latter alkylated with ethyl β-iodopropionate to the diester 39.44. Hydrolysis, decarboxylation, and reesterification lead to the monoester 39.45, and catalytic hydrogenation and cyclization of the latter affords the lactams 39.46 which can be separated and reduced to the isomers 39.41. Once the decision as to which lactam is which had been made, the corresponding amino acid 39.47 could be converted, in alkali, with β-furoylchloride and then diazomethane, to the amido ester 39.48 and cyclization to dehydrodeoxynupharidine (39.49) effected. Catalytic reduction of 39.49 affords racemic deoxynupharidine (39.1 and its mirror image).

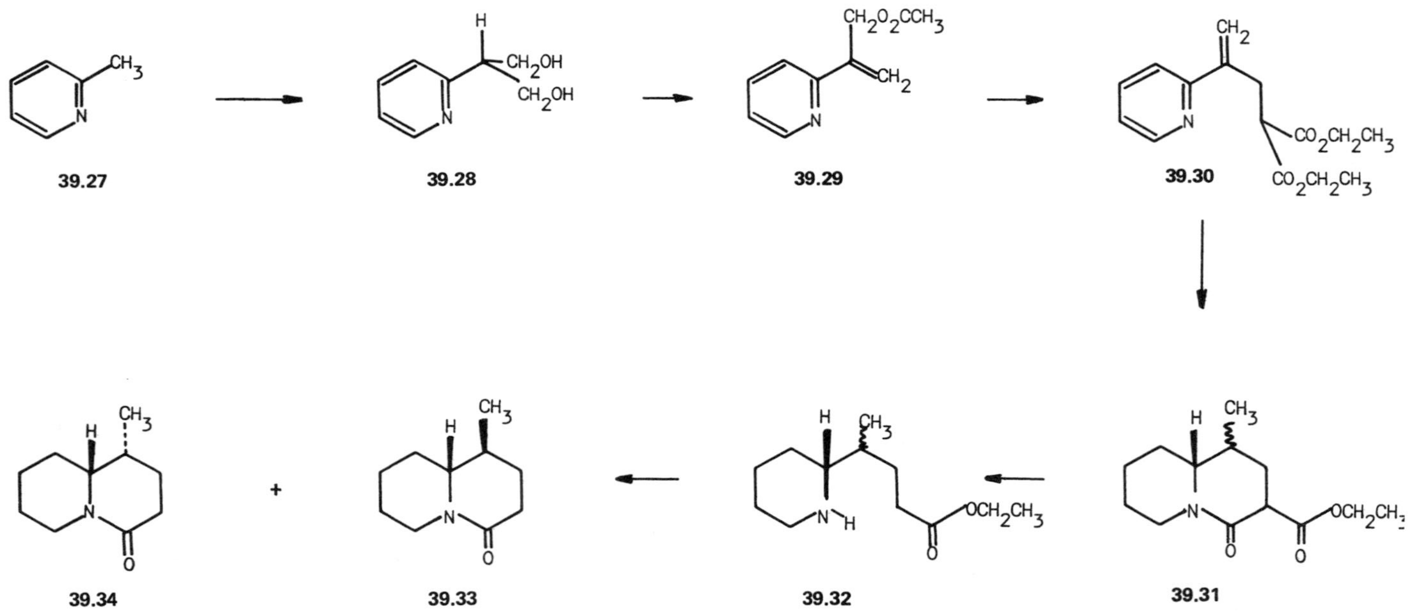

SCHEME 39.4 Synthesis of quinolizidones required to determine the stereochemistry of the dimethylquinolizidone system. (After Reference 418.)

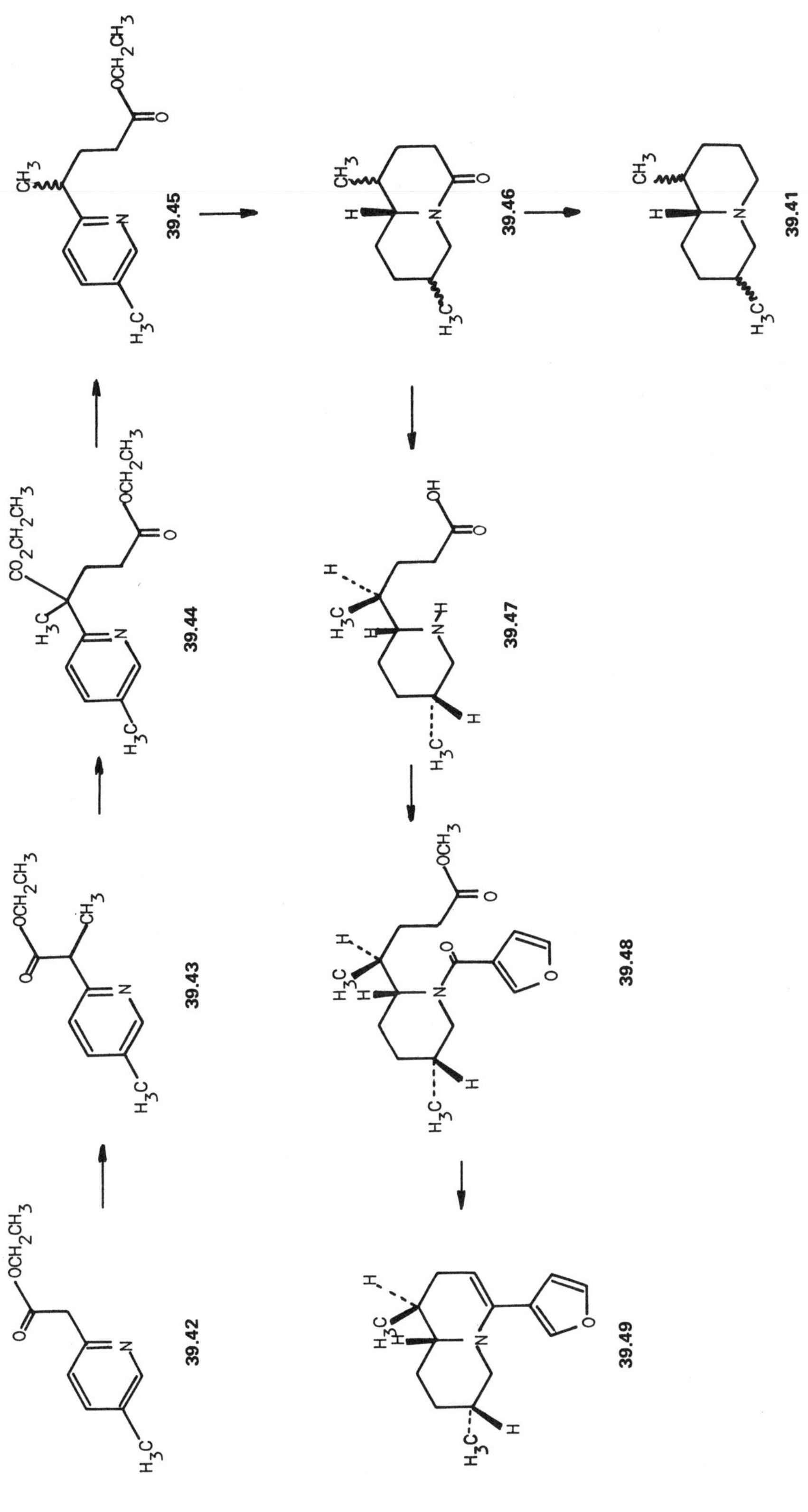

SCHEME 39.5 The total synthesis of deoxynupharidine. (After Reference 418.)

II. DENDROBINE

Dendrobine (39.2), $C_{16}H_{25}NO_2$, mp 134°C $[\alpha]^{14}$ -51.5° (ethanol) was first isolated in the early 1930s from the stem of the plant Dendrobium nobile Lindl. [419], but the work on its structure began in earnest some 30 years later. The base itself was quickly identified as containing a γ-lactone (1763 cm^{-1}) a tertiary nitrogen, one N-methyl, and at least two C-methyls. Proton magnetic resonance (pmr) provided the additional information that two C-methyls were present in an isopropyl group and that a methyl on a quaternary carbon was present too. Additionally, since there is no unsaturation other than the carbonyl group of the γ-lactone, dendrobine (39.2) must be tetracyclic.

Degradative work on dendrobine (39.2), which fully established its structure, was reported in 1964 and 1965 [420, 421]. Each new compound cited as a degradation product was fully characterized, and pmr played a critical role in determining the relative positions of the functional groups as well as their nature (Table 39.1).

TABLE 39.1

Proton Magnetic Resonance (pmr) Analysis of Dendrobine (39.2) and Derivatives[a]

Compound	Position of signal (and comments) at 60 MHz in $CDCl_3$; values in δ (ppm); TMS = 0.00	Interpretation
Dendrobine (39.2)	0.96-1.0 (6H) two doublets	$(CH_3)_2CH$
	1.33 (3H) singlet	Quaternary methyl
	2.49 (3H) singlet	N—CH_3
	4.80 (1H) quartet	—CO_2—CH
Dendrobinediol (39.50)	0.86-1.04 (6H) two doublets	$(CH_3)_2CH$
	1.22 (3H) singlet	Quaternary methyl
	2.33 (3H) singlet	N—CH_3
Cyanonordendrobine (39.52)	0.93-1.03 (6H) two doublets	$(CH_3)_2CH$—
	1.40 (3H) singlet	Quaternary methyl
	3.30 (1H) doublet (J = 3.5 Hz)	N—CH
	3.13-3.88 (2H) septet (J_{AB} = 10 Hz)	N—CH_A—CH_B—CH
	4.38 (1H) quartet	—CO_2—CH
N-Carbamoylnordendrobine (39.53)	0.87-1.14 (6H) two doublets	$(CH_3)_2CH$
	1.30 (3H) singlet	Quaternary methyl
	2.95-3.79 (2H) septet (J_{AB} = 9.5 Hz)	N—CH_A—CH_B—CH
	3.80 (1H) doublet (J = 3.5 Hz)	N—CH
	4.62 (2H) broad singlet	$CONH_2$
	4.90 (1H) quartet	CO_2—CH

TABLE 39.1 (cont.)

Compound	Position of signal (and comments) at 60 MHz in $CDCl_3$; values in δ (ppm); TMS = 0.00	Interpretation
Oxodendrobine (39.55)	0.37-1.08 (6H) two doublets	$(CH_3)_2CH$
	1.42 (3H) singlet	Quaternary methyl
	2.74 (3H) singlet	$CONCH_3$
	3.24 (1H) doublet (J = 4.0 Hz)	N—CH
	4.72 (1H) quartet	CO_2—CH
Dendrobinediol methiodide methine (39.57)	0.97-1.12 (6H) two doublets	$(CH_3)_2CH$
	1.02 (3H) singlet	Quaternary methyl
	2.18 (6H) singlet	$N(CH_3)_2$
	Broad solvent dilution shifting singlet (1H)	OH
Ketoester of oxodendrobinic acid (39.64)	0.77-1.06 (6H) two doublets	$(CH_3)_2CH$
	1.50 (3H) singlet	Quaternary methyl
	3.00 (3H) singlet	N—CH_3
	3.42 (1H) singlet	N—CH
	3.72 (3H) singlet	CO_2—CH_3
Anhydrodihydromethine monobenzoate (39.66)	1.00-1.12 (6H) two doublets	$(CH_3)_2CH$
	0.96 (3H) singlet	Quaternary methyl
	2.22 (6H) singlet	$N(CH_3)_2$
	2.91 (1H) multiplet	Allylic proton
	4.00-4.84 (2H) octet (J_{AB} = 10 Hz)	CH—CH_A—CH_B—$OCOC_6H_5$
	5.42 (1H) multiplet	Olefinic proton
	7.30-8.10 (5H) multiplet	Aromatic protons

[a]After Reference 420.

When dendrobine (39.2) is reduced with lithium aluminum hydride, the lactone carbonyl disappears, and a diol (dendrobinediol, 39.50) is formed (Equation 39.7).

Selenium dehydrogenation of dendrobine (39.2) yields 4-isopropyl-2-pyridone (39.51; Equation 39.8) and other products. Then, since dendrobinediol (39.50) does not yield any pyridone when treated under the same conditions, it can be concluded that the carbonyl of the pyridone (39.51) originated as the lactone carbonyl of dendrobine (39.2) and that, in addition, the isopropyl group is present in dendrobine (39.2) on a carbon β to the carbonyl of the γ-lactone and the nitrogen is separated from the carbonyl by four carbon atoms (see below).

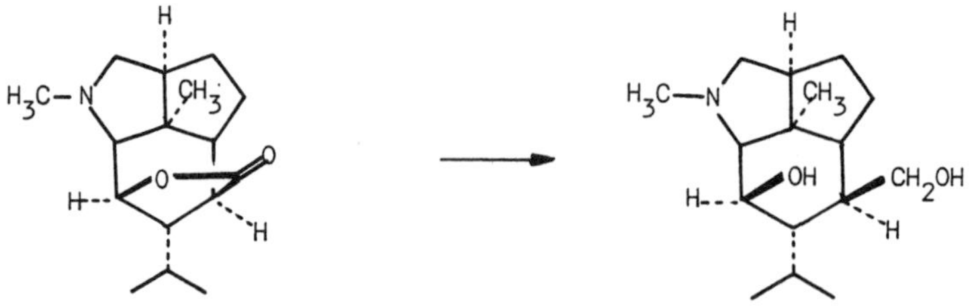

H_3C-N CH_3 O O H H

39.2

H_3C-N CH_3 OH CH_2OH H H

39.50

(Eq. 39.7)

39.2

39.51

(Eq. 39.8)

Treatment of dendrobine (39.2) with cyanogen bromide yields cyanonordendrobine (39.52) which undergoes hydrolysis to N-carbamoylnordendrobine (39.53) and then, with sodium nitrite, the latter affords nordendrobine (39.54; Equation 39.9). Nordendrobine (39.54) undergoes Eschweiler-Clarke methylation to regenerate dendrobine (39.2).

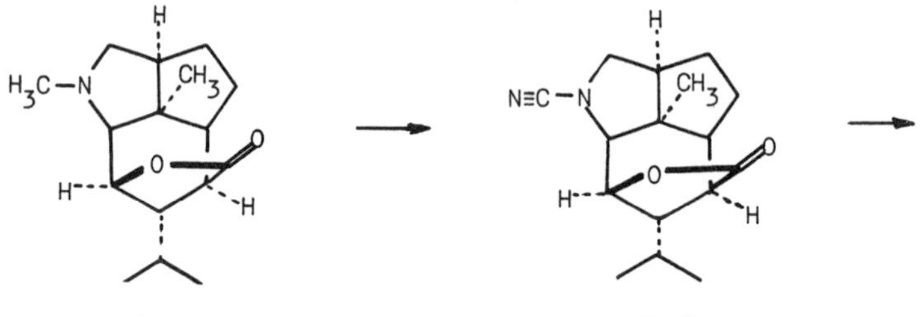

39.2

39.52

(Eq. 39.9)

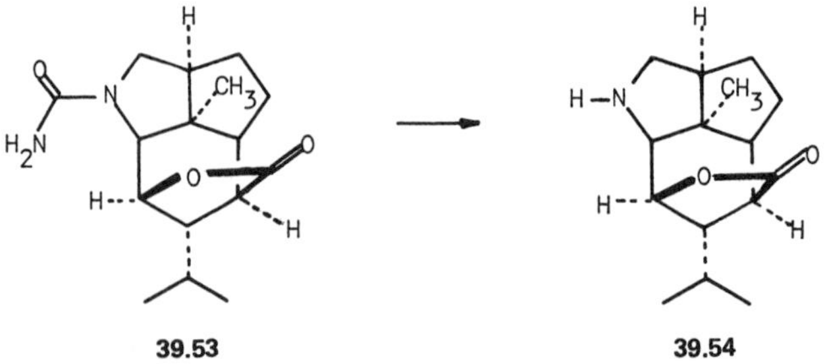

39.53

39.54

When dendrobine (39.2; Equation 39.10) is oxidized with potassium permanganate under carefully controlled conditions, oxodendrobine (39.55) is formed. Reduction of oxodendrobine (39.55) with lithium aluminum hydride generates dendrobinediol (39.50), indicating that the permanganate oxidation proceeds without skeletal rearrangement.

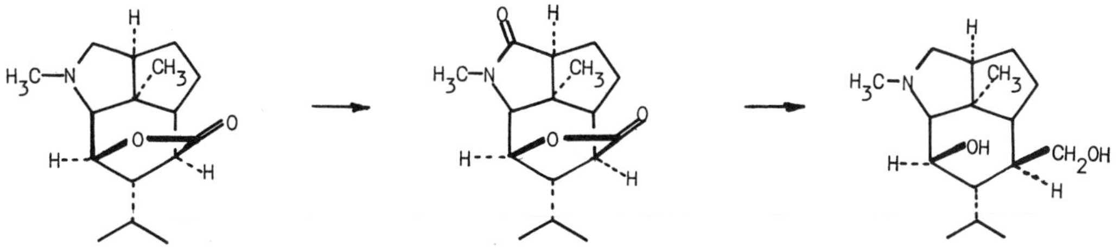

39.2 39.55 39.50 (Eq. 39.10)

Methylation of dendrobinediol (39.50) with methyl iodide yields a methiodide (39.56) which undergoes the Hofmann degradation to a "methine" (39.57; Scheme 39.6; [420]). The methine (39.57) contains a "new" carbonyl group. Although a second Hofmann reaction cannot be effected, the methine 39.57 is capable of reduction to a dihydromethine (39.58), the diacetate derivative (39.59) of which yields an N-oxide (39.60) that undergoes pyrolysis to a nitrogen-free alkene (39.61). Osmolation of the alkene (39.61) followed by lead tetraacetate oxidation then yields a ketodiacetate (39.62).

When oxodendrobine (39.55) is hydrolyzed with aqueous sodium hydroxide, oxodendrobinic acid (39.63) is obtained. This acid is capable of conversion to a methyl ester with diazomethane, and the latter yields a ketoester (39.64) on oxidation with chromic anhydride in acetic acid (Equation 39.11).

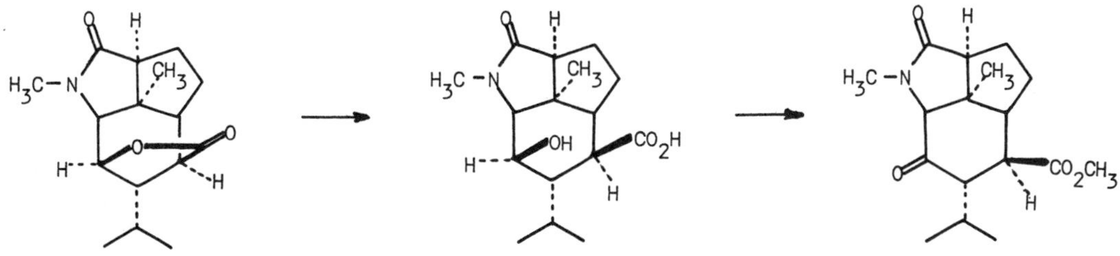

39.55 39.63 39.64

(Eq. 39.11)

The dihydromethine (39.58) is capable of being converted to a monobenzoate, and various esterification-deesterification and oxidation reactions can be utilized to establish that the dihydromethine (39.58) possesses one primary and one secondary hydroxyl function. Thus (Equation 39.12), phosphorus oxychloride/pyridine dehydration of dihydromethine monobenzoate (39.65) results in formation of anhydrodihydromethine monobenzoate (39.66).

Finally (Equation 39.13), nordendrobine (39.54) can be hydrolyzed to nordendrobinic acid (39.67) with barium hydroxide. If nordendrobinic acid (39.67) is subjected to high-vacuum distillation, nordendrobine (39.54) and nordendrobinic acid lactam (39.68) are isolated. The formation of this lactam (39.68) accounts for the generation of 4-isopropyl-2-pyridone (39.51; Equation 39.8) on selenium dehydrogenation of dendrobine (39.2).

As previously mentioned, analysis of the pmr spectra (Table 39.1) of the compounds described above strongly suggests that 39.2 accurately describes the structure of dendrobine. Furthermore, complete pmr analysis is in concert with the configuration of the isopropyl group as trans to the B,C rings (Table 39.1) which are cis fused, and easy formation of the internal quaternary salt 39.69 from dendrobinediol (39.50) on thionyl chloride treatment (Equation 39.14) establishes the configuration as shown or as the mirror image of that shown. Then, equilibration studies among the various carbonyl compounds mentioned above

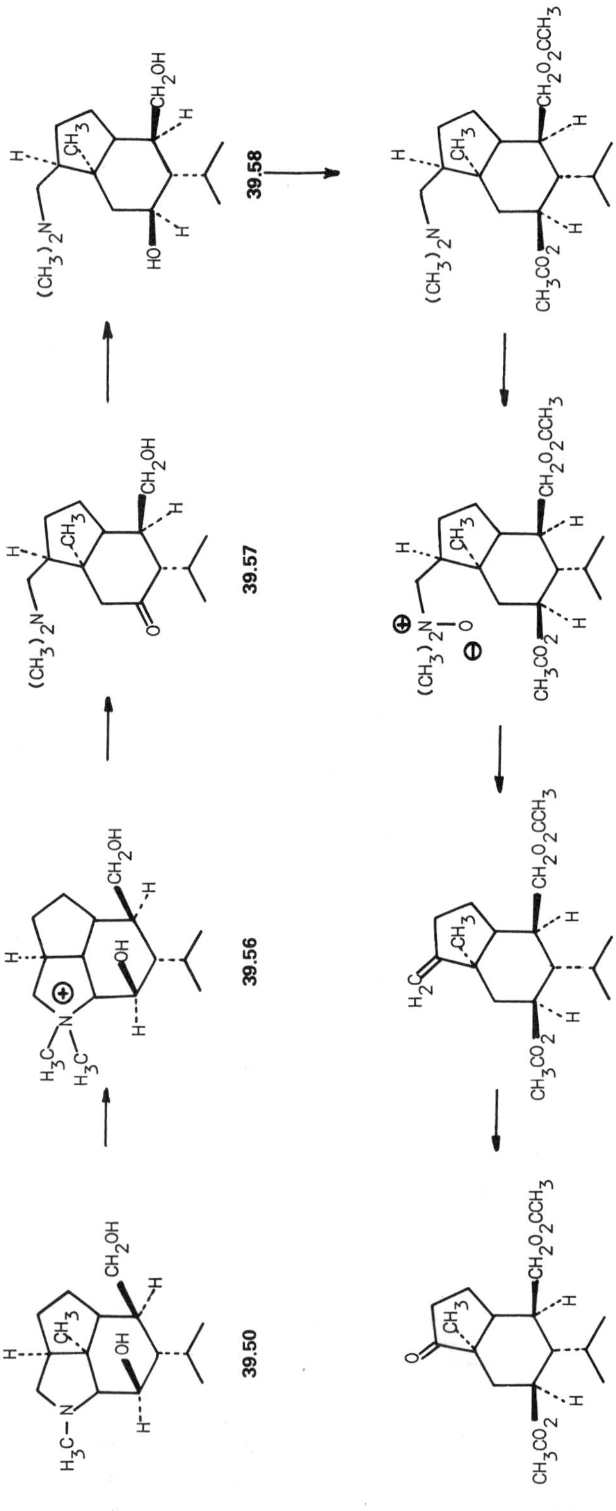

SCHEME 39.6 A degradation of dendrobinediol. (After Reference 420.)

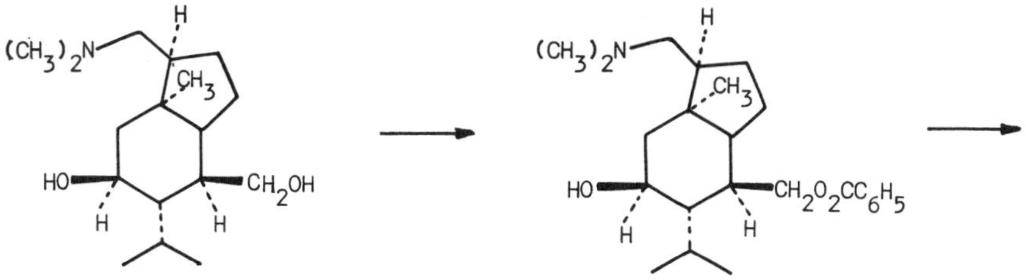

39.58 → **39.65** →

39.66

(Eq. 39.12)

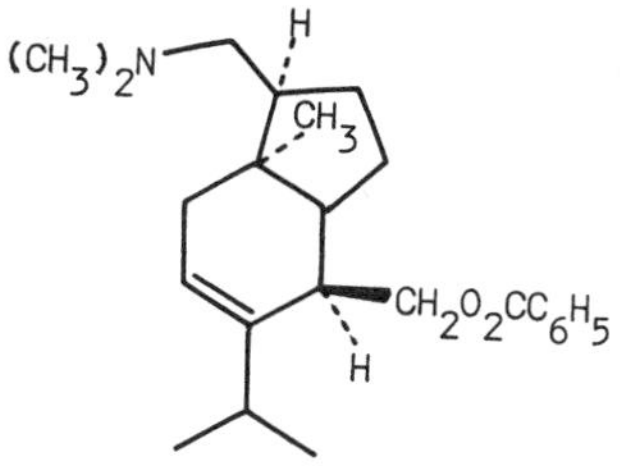

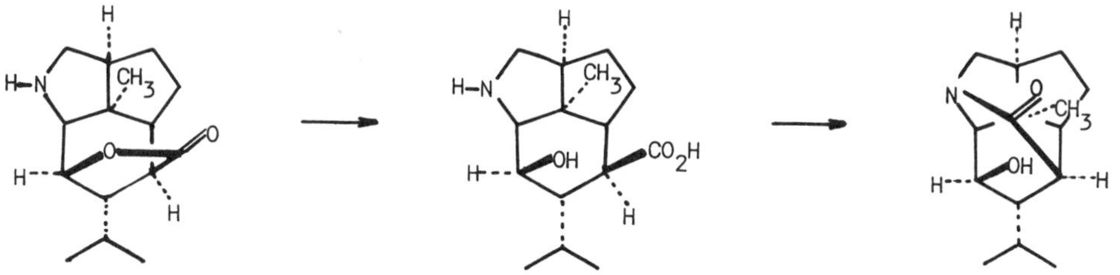

39.54 → **39.67** → **39.68**

(Eq. 39.13)

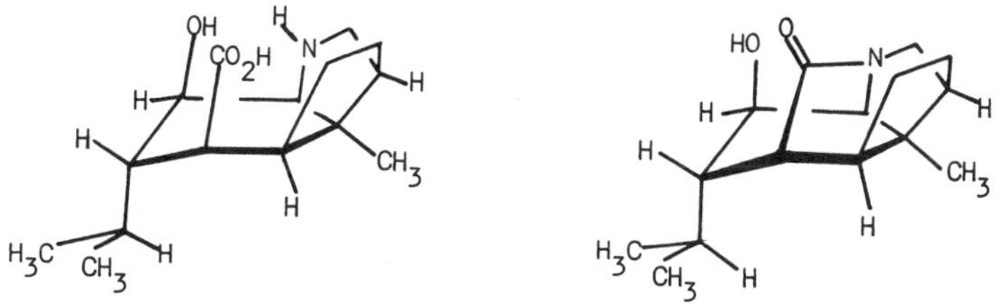

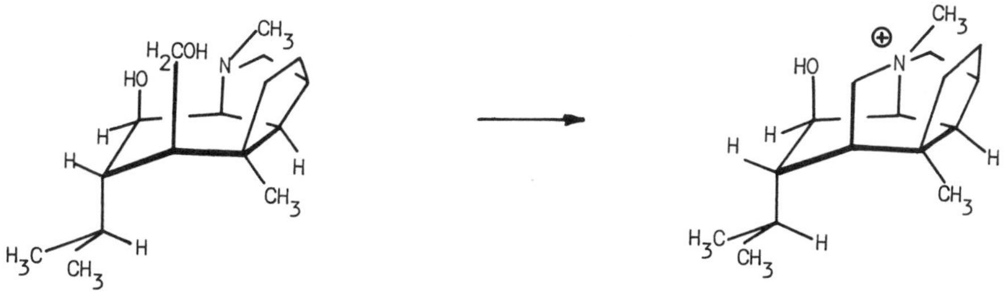

39.50 → **39.69**

(Eq. 39.14)

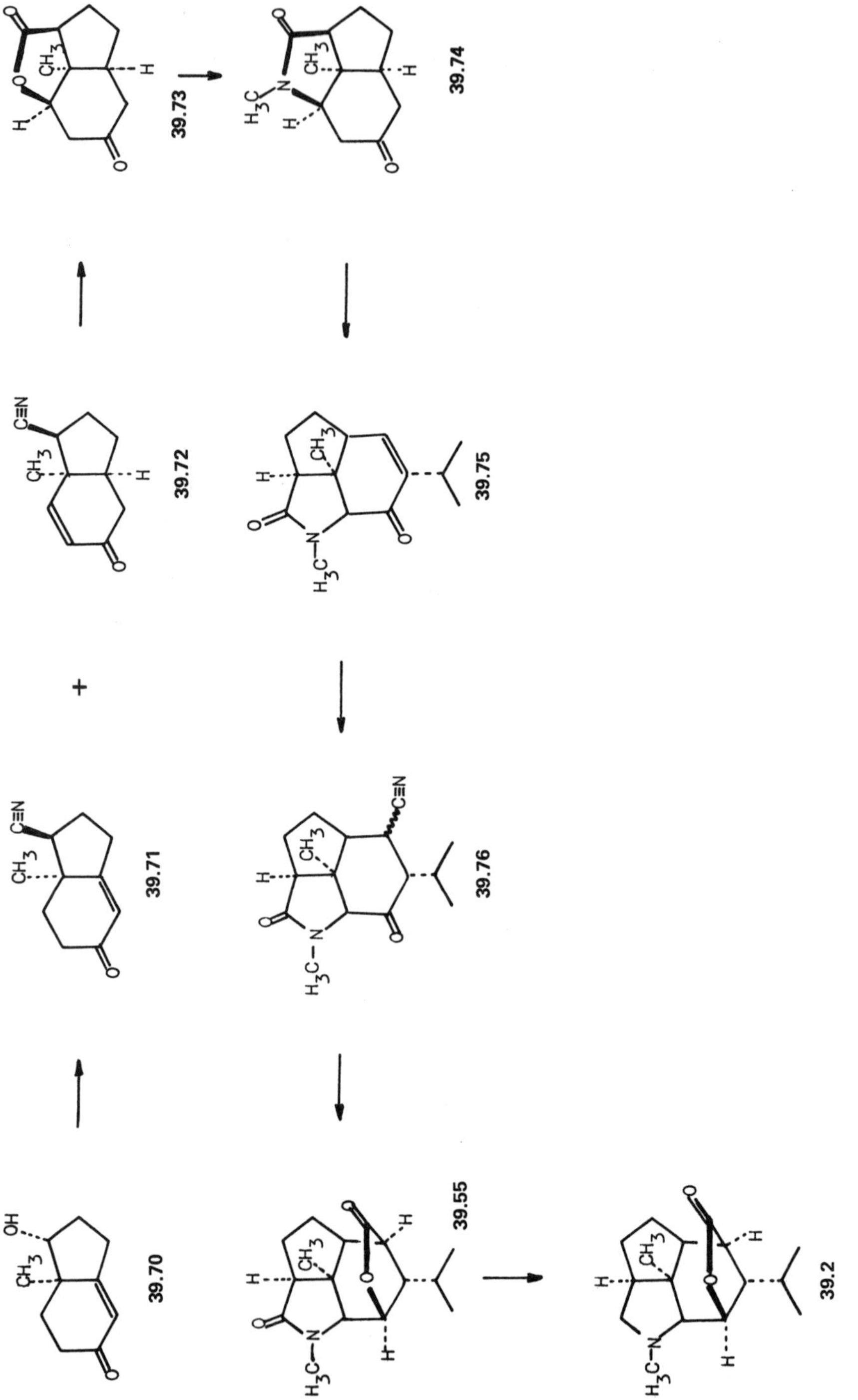

SCHEME 39.7 The total synthesis of (±)-dendrobine. (After Reference 423.)

confirm that the isopropyl and carbonyl ester groupings occupy the thermodynamically stable conformation, and use of ORD octant rules strongly suggests [422] the absolute configuration to be that shown (39.2). In confirmation of the structure proposed, (±)-dendrobine has been synthesized [423], and the synthesis is presented as Scheme 39.7.

The hydroxyketone 39.70 is converted to the nitrile 39.71, via the tosylate of 39.70, and reduction over a poisoned palladium catalyst generates the corresponding saturated keto-nitrile which undergoes bromination and dehydrohalogenation to a mixture of enones (39.71 and 39.72). Although the undesired alkene 39.71 predominates in the mixture (the ratio of 39.71:39.72 is 3:1), they can be separated. The ketone function in 39.72 is then protected, the nitrile hydrolyzed to the corresponding acid, the protecting group removed, and cyclization to the lactone 39.73 effected. On treatment of the lactone 39.73 with methylamine, the lactam 39.74 is generated. Then, in 10% overall yield, 39.74 is converted to a tertiary alcohol with isopropyl magnesium bromide, dehydration to an alkene effected, and hydroxylation, elimination, and oxidation to generate the enone 39.75 carried through. When the enone is hydrocyanated with diethylaluminum cyanide, a mixture of cyanoketones 39.76 is produced. Reduction of 39.76 with sodium borohydride followed by basic hydrolysis and acidification generates (±)-oxodendrobine (39.55) and an isomer. When (±)-oxodendrobine (39.55) is reduced by treatment of the amide with Meerwein's reagent and then sodium borohydride, (±)-dendrobine (39.2 and its mirror image) is produced.

SELECTED READING

J. T. Wrobel, in The Alkaloids, Vol. 9, R. H. F. Manske (ed.). Academic, New York, 1967, pp. 441 ff.

40

CHEMISTRY OF THE DITERPENOID ALKALOIDS

Whoever thinks a faultless piece to see,
Thinks what ne'er was, nor is, nor e'er shall be.

A. Pope, Essay on Criticism, Part 2

The diterpene alkaloids occur commonly either as amino alcohols or as esters of amino alcohols (often with acetic or benzoic acids as the esterifying acids), and these permutations and combinations thus give rise to over 100 such bases. These alkaloids occur mainly in the families Ranunculaceae and Cornaceae (Aconitum and Delphinium in the former and Garrya in the latter); of the large number of such compounds four main skeletal types are common, and the bases considered here typify those. As has often been the case, the alkaloids were first noticed because of the folklore surrounding the highly toxic plant extracts in which they were contained, and they were actually found in the course of a search for medicinally valuable palliatives. Although a good deal of chemical degradative work and ingenious deduction was applied in describing their structures and establishing their stereochemistry. Occasionally it was necessary to utilize x-ray crystallography either to remove final doubt or ultimately to establish the structure of these rather complicated materials with their large number and type of functional groups. The specific bases to be considered are veatchine (40.1), atisine (40.2), aconitine (40.3), and heteratisine (40.4).

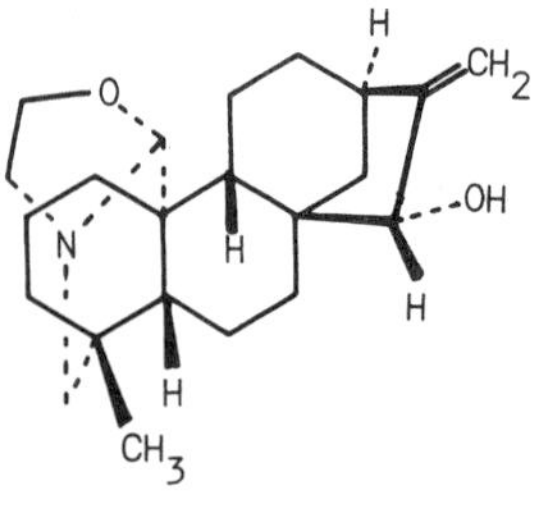

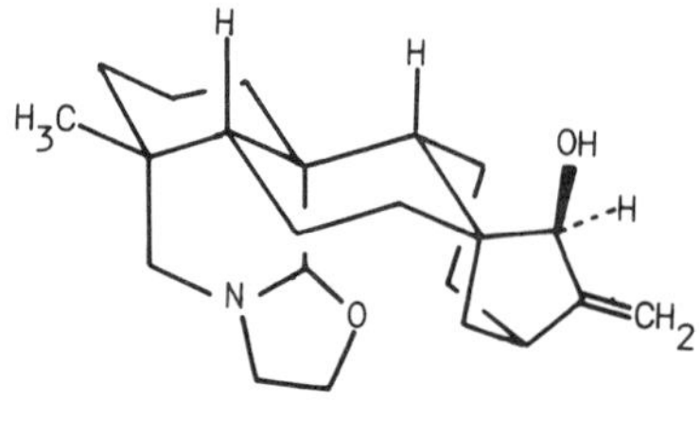

40.1

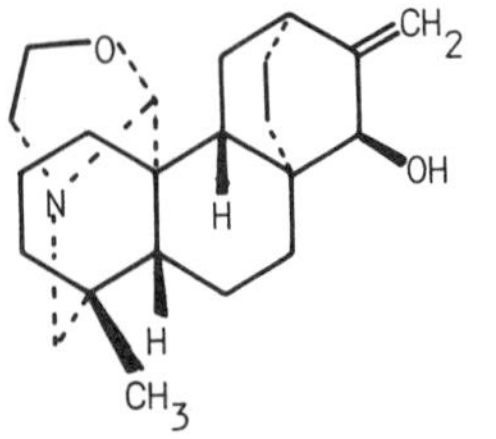

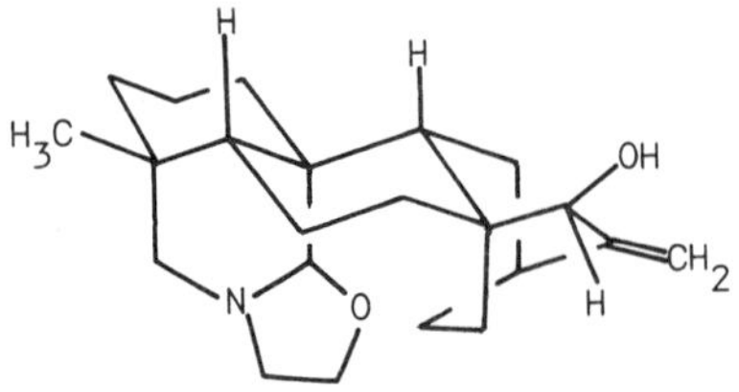

40.2

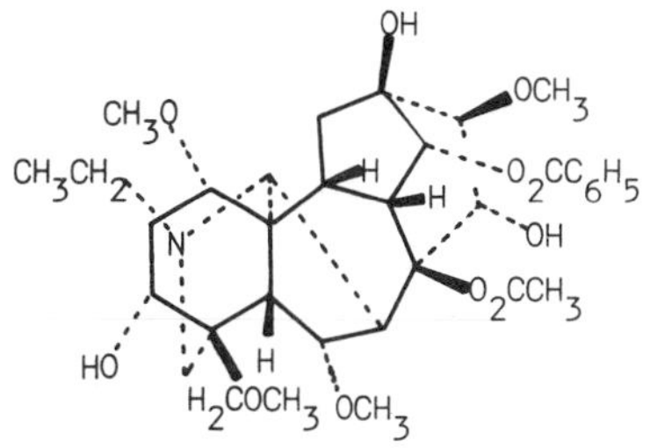

40.3

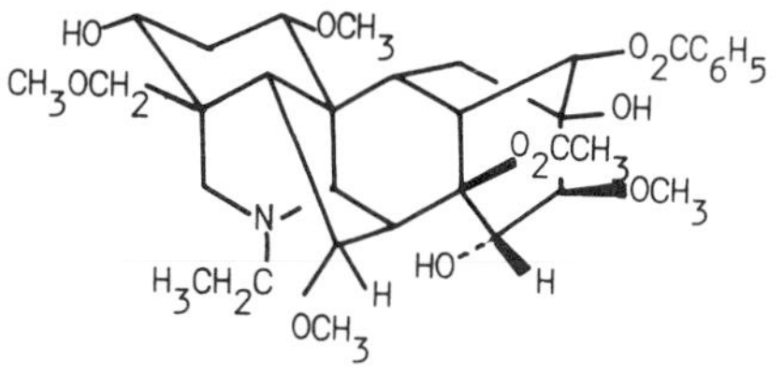

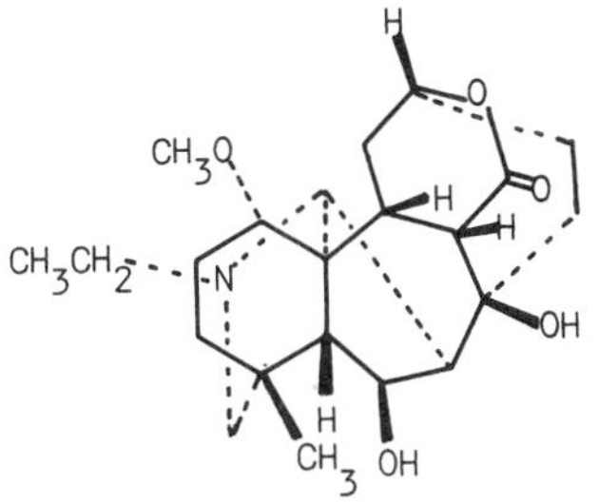

40.4

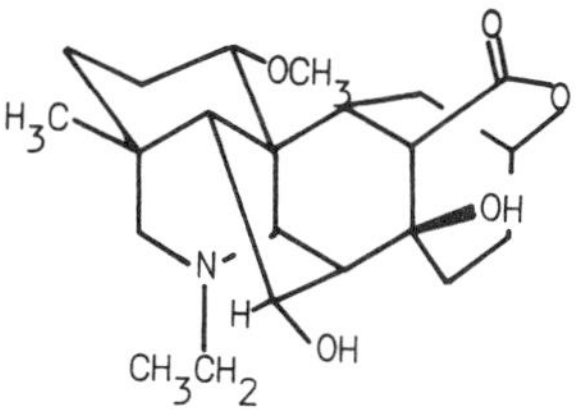

I. VEATCHINE

The salient chemistry surrounding the proof of structure of veatchine (40.1) and its cooccurring isomer garryine (40.5) has been succinctly reviewed [424]. Briefly, both veatchine (40.1) and garryine (40.5) are found in the bark of <u>Garrya veatchii</u> Kellog. and are separable from each other by countercurrent distribution. The bases both correspond to $C_{22}H_{33}O_2N$, although garryine (40.5) crystallizes (the free base is an oil) as a hydrate (mp 74-82°C) $[\alpha]_D$ -84° while veatchine (40.1) crystallizes in an anhydrous form (mp 119°C) $[\alpha]_D$ -69°. Demonstrating their inherent similarity, both alkaloids, on reduction with lithium aluminum hydride, yield dihydroveatchine (40.6) and then, on catalytic hydrogenation, tetrahydroveatchine (40.7; Equation 40.1).* Although both veatchine (40.1) and garryine (40.5) form only their respective monoacetates, dihydro- and tetrahydroveatchine (40.6 and 40.7, respectively) form diacetates, and if it is assumed that hydrogenolysis of a ring is effected on the first reduction (no carbonyl group or equivalent being noted), then with only one double bond present, both veatchine (40.1) and garryine (40.5) must be hexacyclic and dihydroveatchine (40.6) and tetrahydroveatchine (40.7) pentacyclic.

Selenium dehydrogenation of both veatchine (40.1) and garryine (40.5) yields 1-methyl-7-ethylphenanthrene (40.8) and an azaphenanthrene (presumably 40.9; Equation 40.2). On this basis alone, reasonable structures which account for five of the six rings of veatchine (40.1) and garryine (40.5) can be written. However, more gentle selenium treatment of these two bases provides further information about the structures of veatchine (40.1) and garryine (40.5). Thus, (Equation 40.3) pyrolysis of veatchine (40.1) or garryine (40.5) in the presence of selenium results in the formation of two isomeric compounds, $C_{22}H_{29}NO$, corresponding to the loss of ethylene oxide. One of these compounds (40.10) is clearly a

*Although, in tetrahydroveatchine (40.7), the methyl group is shown (E) to the one-carbon bridge of the fused bicyclo[3.2.1]octane system, it appears that the stereochemistry of the reduction is not known. The prediction that it is (E) is based upon the assumption that the exo face of the encumbered bicyclic system is, as suggested by models, more open than the endo face with respect to catalytic hydrogenation (see Reference 425).

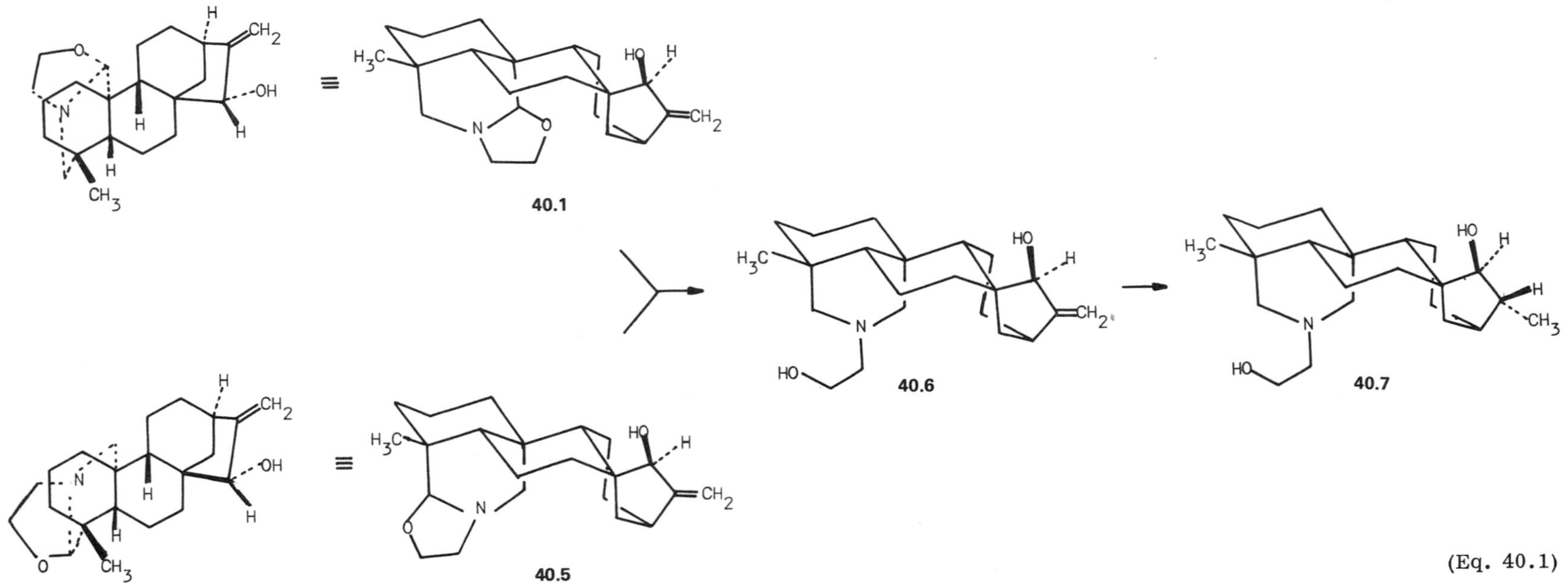

40.1

40.5

40.6

40.7

(Eq. 40.1)

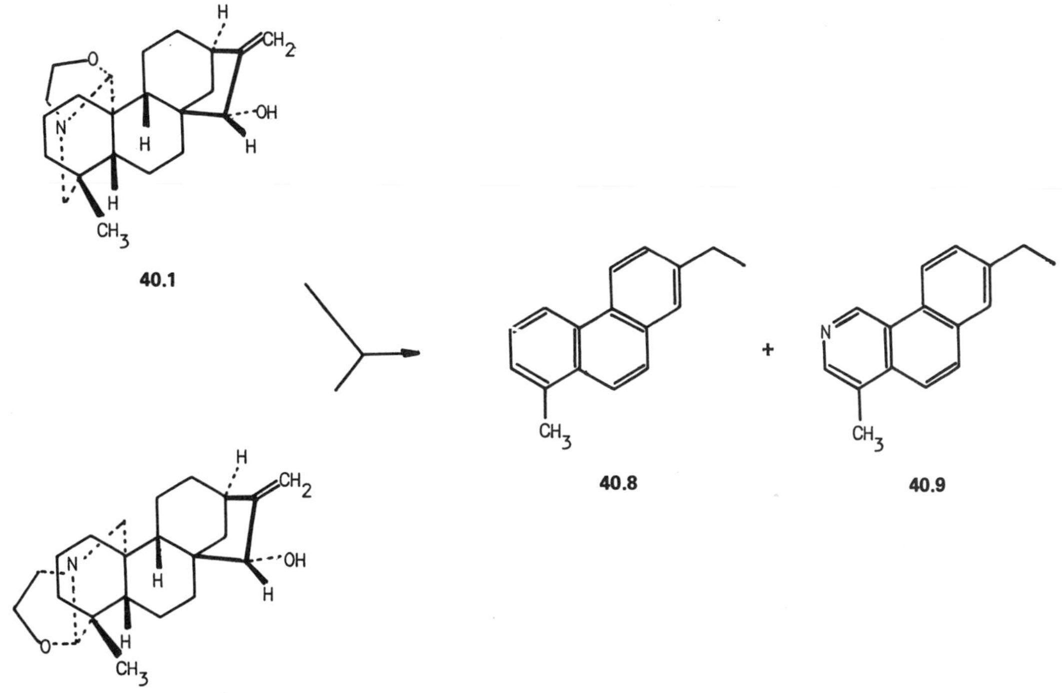

40.1

40.8

40.9

40.5

(Eq. 40.2)

40.1

40.10

40.11

40.5

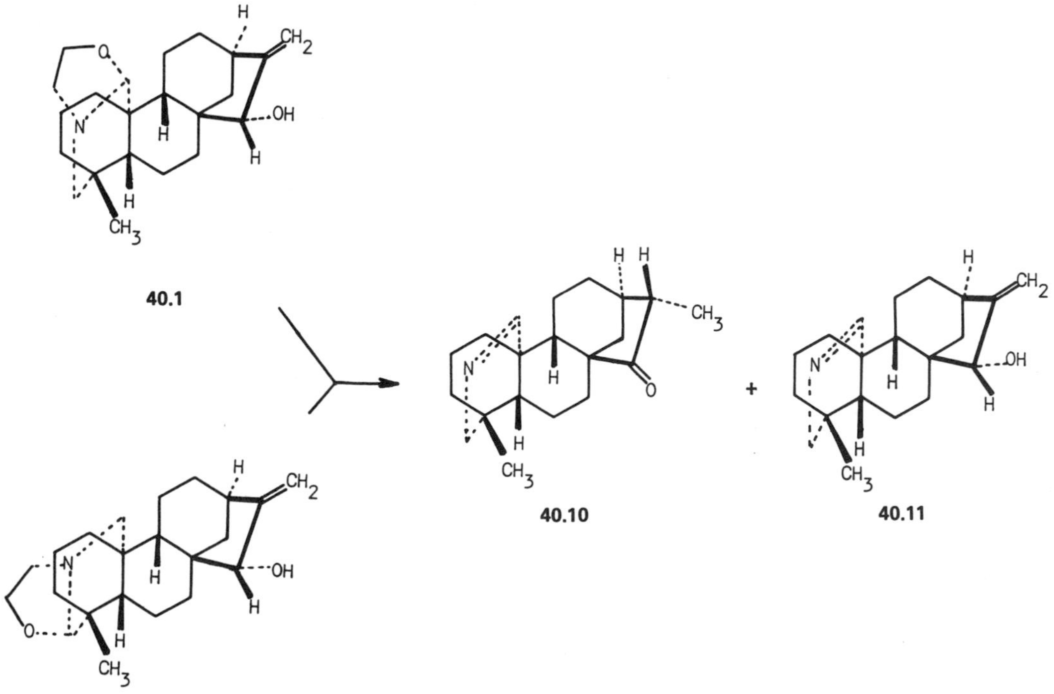

(Eq. 40.3)

cyclopentanone derivative (λ_{max} 1735 cm^{-1}) and also contains a Δ^1-piperideine moiety (λ_{max} 1650 cm^{-1}).* The second material (40.11) also has the Δ^1-piperideine functionality, but the remainder of the molecule is unchanged from that originally present in veatchine (40.1; [426, 427].†

---

*The position of the carbon-nitrogen double bond in 40.10 and 40.11 is not certain but is written to show the most hindered carbon atom as sp^2 hybridized.

†It is worth examining a potential pathway for this process. An allowed route is shown in Equation 40.4). Interestingly, garryfoline (40.12), the isomer of veatchine (40.1) epimeric at the hydroxyl-bearing carbon (C_{16}), which for steric reasons should not succeed in the process, nevertheless undergoes a facile acid-catalyzed rearrangement to the corresponding methyl ketone (cuauchichicine, 40.13) whereas veatchine (40.1) does not. Garryfoline (40.12; Equation 40.5), on initial protonation, forms a tertiary carbonium ion which can undergo, as usual, rearrangement to the ketone 40.13 by exo-2,3-hydride migration but, also as usual, not an endo-2,3-hydride shift, exactly what would be required in the case of veatchine (40.1) if the same product were formed. Thus, the geometry of the two epimeric alcohols, veatchine (40.1) and garryfoline (40.12), is established.

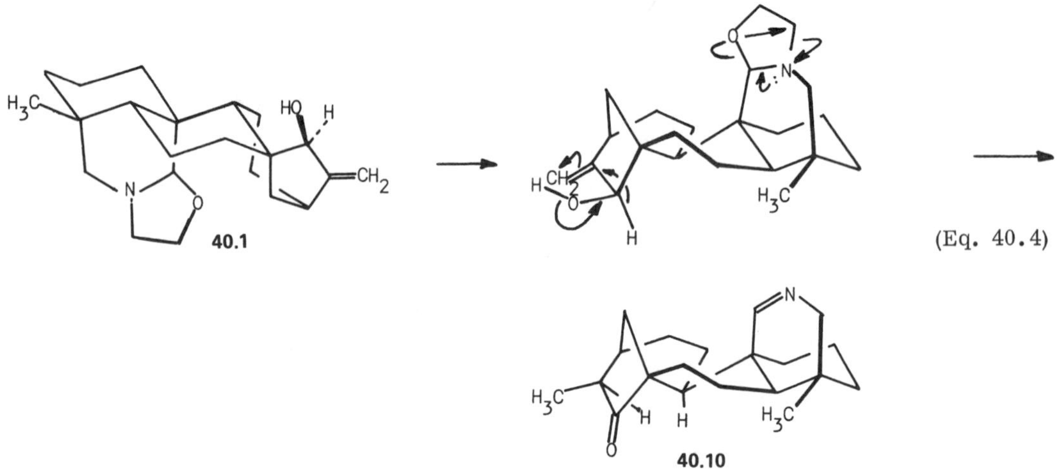

40.1

(Eq. 40.4)

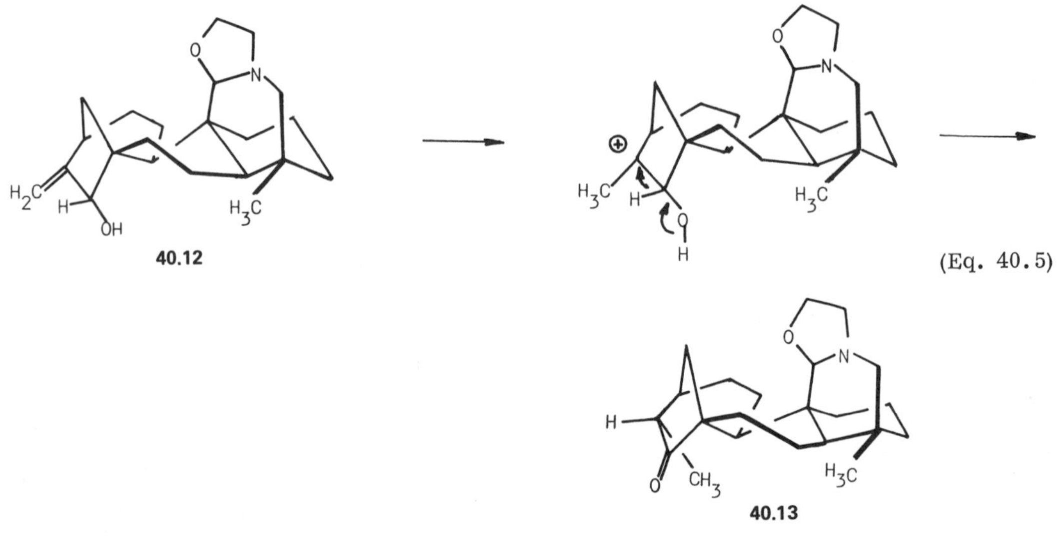

40.10

40.12

(Eq. 40.5)

40.13

When the cyclopentanone (40.10) is reduced with lithium aluminum hydride, an amino alcohol (40.14) is obtained. This alcohol (40.14) can be N-alkylated with 2-bromoethanol to provide an isomer (40.15) of tetrahydroveatchine (40.7). On the other hand, when the unrearranged selenium pyrolysis product (40.11) is reduced with lithium hydride to 40.16 and alkylated with 2-bromoethanol, dihydroveatchine (40.6) results. Careful oxidation of dihydroveatchine (40.6) with osmium tetroxide regenerates garryine (40.5; but not veatchine, 40.1).* Finally, when the alcohol 40.16 is oxidized with chromic anhydride in pyridine, a keto group, conjugated with a double bond (λ_{max} 236 nm, log ϵ 4.0), is introduced along with a Δ^1-piperideine function to form the iminoketone 40.17. Reduction of 40.17 with lithium aluminum hydride yields a secondary alcohol (40.18) different from 40.16 but which can be hydrogenated to the amino alcohol 40.14. Thus, the two products 40.10 and 40.11 are interrelated, and both are formed without deep-seated structural rearrangement. These reactions are shown in Scheme 40.1.†

Gentle oxidation of veatchine (40.1) with permanganate yields two oxoveatchines, oxoveatchine A (40.19) and oxoveatchine B (40.20; Equation 40.6). The former (40.19) contains a five-membered and the latter (40.20) on a six-membered lactam system ($_{max}$ 1700 cm^{-1}

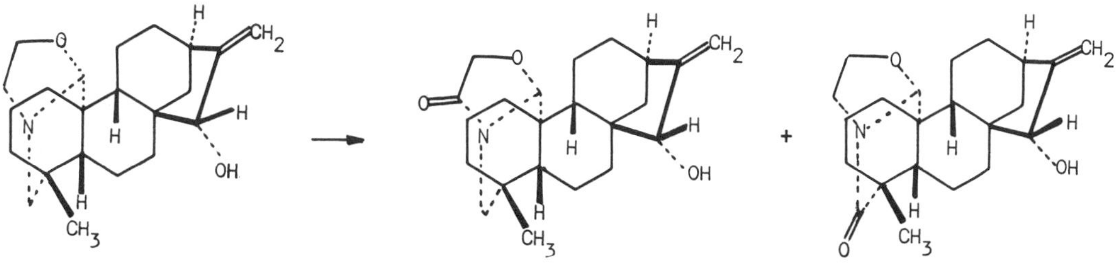

40.1 40.19 40.20

(Eq. 40.6)

and 1630 cm^{-1} respectively). On the other hand (Equation 40.7), garryine (40.5) treated under the same conditions yields a single six-membered lactam, oxogarryine (40.21). Further oxidation of either lactam 40.19 or 40.20, or vigorous oxidation of veatchine (40.1) with permanganate (Equation 40.8) yields two lactam dicarboxylic acids, one with a five-membered ring (40.22) and the other with a six-membered ring (40.23). Both acids provide dimethyl

*As noted previously, the position of the carbon-nitrogen double bond in 40.11 would lead one to the conclusion that veatchine (40.1) and not garryine (40.5) should form on alkylation. That this does not occur is attributed to a mobile equilibrium between the two possible imines, with 40.11 being the thermodynamic isomer and the reaction occurring from the kinetic isomer instead.

†The material presented in Scheme 40.1 demonstrates the type of problem associated with the assignment of the configuration of the methyl group at C_{17} in the various isomers in the system. Thus, according to Equation 40.1 and examination of models, reduction of the exomethylene double bond in dihydroveatchine (40.6) should produce the endo-methyl isomer 40.7, i.e., methyl (E) to the one-carbon bridge. On the other hand, rearrangement of veatchine (40.1) should yield the exo-methyl isomer 40.10, i.e., methyl (Z) to the one-carbon bridge. Now, catalytic reduction of 40.18 should produce, as in the case of dihydroveatchine (40.6), the endo-methyl derivative, i.e., methyl (E) to the one-carbon bridge or the C_{17} epimer of 40.14, which of course cannot have occurred if 40.14, presumably the thermodynamic isomer at C_{17}, is generated by lithium aluminum hydride reduction of 40.10!

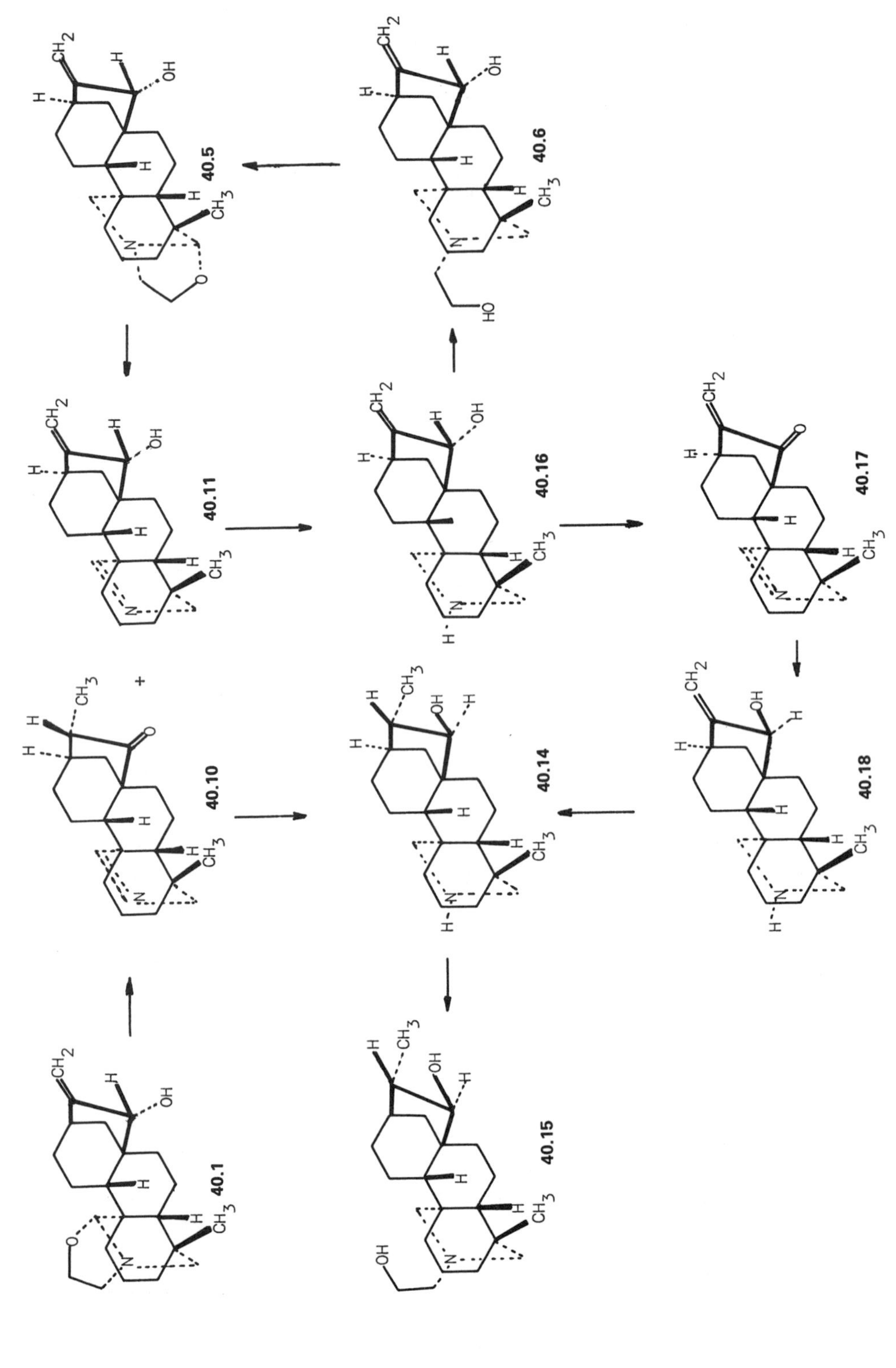

SCHEME 40.1 Correlation of the two isomeric alkaloids, veatchine and garryine.

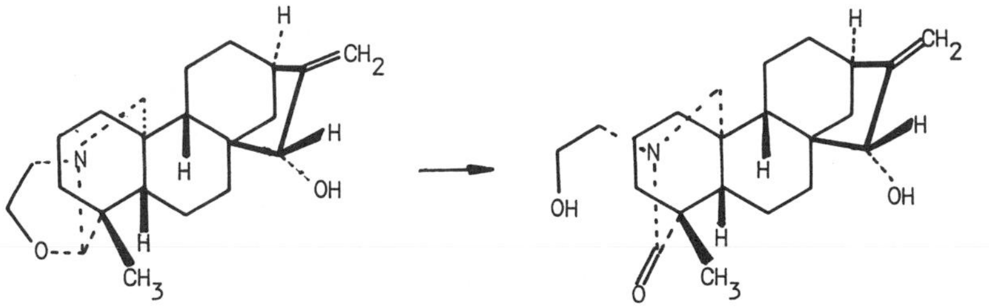

40.5 → 40.21 (Eq. 40.7)

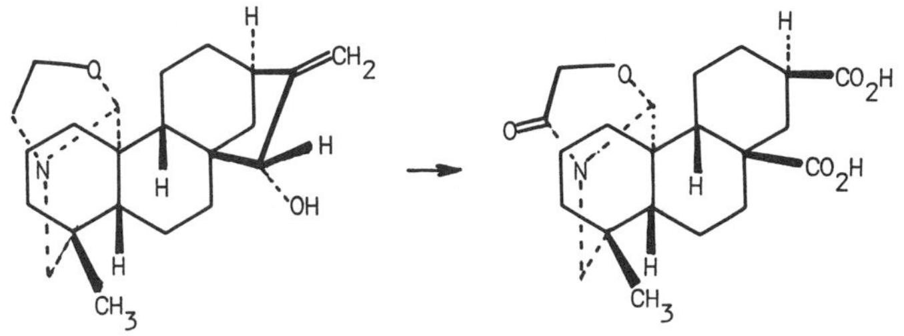

40.1 → 40.22 (Eq. 40.8)

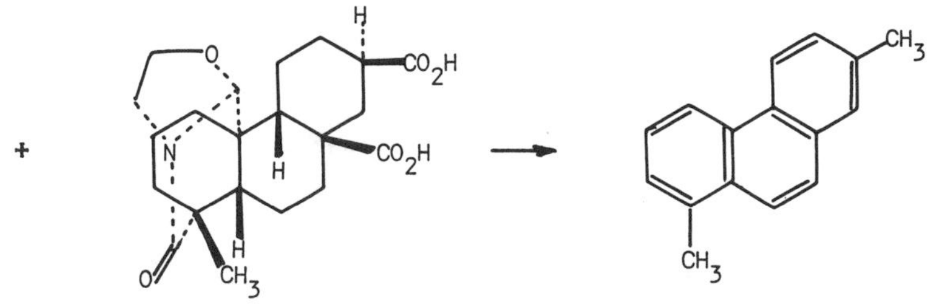

+ 40.23 → 40.24

esters, only one methyl group of which is easily removed on saponification, identifying one carboxylic acid as tertiary. Then, the anhydrides of both lactam dicarboxylic acids 40.22 and 40.23 can be individually prepared, and since they are very similar in the IR to glutaric anhydride, they are identified as cyclohexane dicarboxylic acids bearing the carboxylate functions in a cis-1,3-diaxial arrangement. Selenium dehydrogenation of the lactam dicarboxylic acid 40.23 leads to pimanthrene (40.24), from which it is concluded that aromatization is accompanied by reduction of one of the carboxylate groups (because reduction is unusual, this did lead to some question about assumptions regarding the structure of veatchine, 40.1, but the structure is beyond question now, and there is no doubt that reduction occurs [428]).

Finally, the decision as to which isomer is garryine (40.5) and which is veatchine (40.1) was reached on the basis of the observation that garryine (40.5) undergoes reaction with methyl magnesium iodide to methyldihydrogarryine (40.25;* Equation 40.9). The structure

*The structure of the methyldihydrogarryine (40.25) is apparently not known with certainty, but the methyl shown in 40.25 is placed on that side which would obtain if the alkoxy group were displaced, i.e., the product of an S_N2 process.

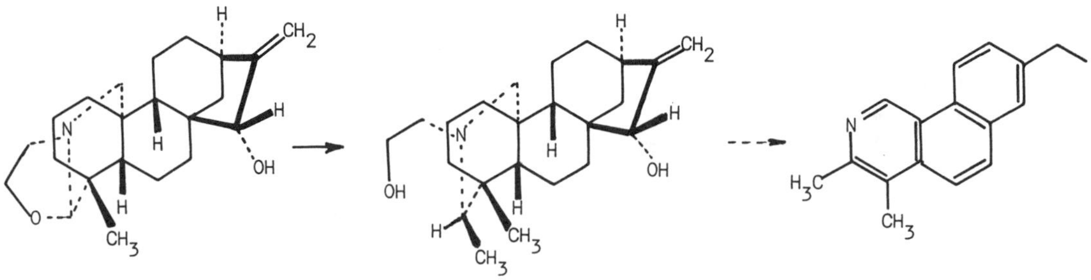

40.5 40.25 40.26

(Eq. 40.9)

of methyldihydrogarryine (40.25) is based upon the observation that the azaphenanthrene (presumably 40.26) obtained on selenium dehydrogenation of 40.25 is not identical with the known azaphenanthrene 40.27, the latter expected, but not found, on dehydrogenation of the corresponding product of the Grignard reagent with veatchine (40.1).*

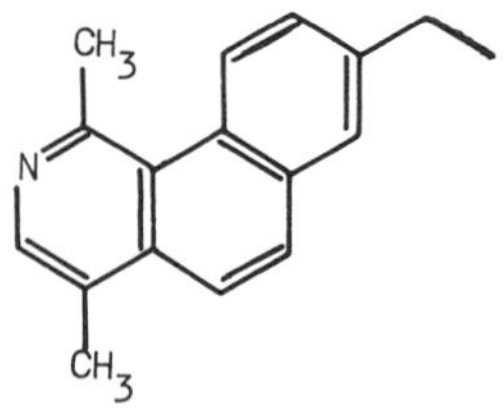

40.27

The absolute configuration of veatchine (40.1) was established [425] by degradation to diterpene derivatives whose absolute configurations could be related through the use of optical rotatory dispersion (ORD) to those of known compounds. As shown in Scheme 40.2, a crude mixture of veatchine (40.1) and garryfoline (40.12) was reduced and acetylated to yield a mixture of dihydroveatchine diacetate (40.28) and dihydrogarryfoline diacetate (40.29). Oxidative cleavage of the exocyclic methylene unit in 40.29 (which was crystalline while the epimer 40.28 was not) yields the ketoacetate 40.30, which on reduction with calcium in liquid ammonia provides 17-nor-16-oxo-15-deoxydihydrogarryfoline acetate (40.31). The positive Cotton effect and the amplitude of that effect exhibited by the ketone 40.31 is similar to that observed for the 17-nor-16-ketone (40.32) obtained from phyllocladene (40.33), whose absolute configuration is known [429].

The remaining chiral centers were established (Scheme 40.3) by treating a crude mixture of garryfoline (40.12) and veatchine (40.1) in ether with hydrogen chloride and acetylating the resulting products to yield veatchine diacetate chloride (40.34) and cuauchichicine acetate chloride (40.35). Deacetylation and elimination of ethylene oxide by treatment with base yield a separable mixture of azomethines 40.36 and 40.37. The azomethine from cuauchichicine (i.e., 40.37) was then deaminated with nitrous acid to generate the hemiacetal 40.38. Wolff-Kishner reduction of 40.38 produced the primary alcohol 40.39, in which, presumably, the basic conditions have permitted the methyl group α to the carbonyl to epimerize to the most stable (i.e., exo or (Z) to the one-carbon bridge of the bicyclo[3.2.1]octane system)

---

*This is reported to be a poor reaction. A displacement reaction such as that suggested here would, of course, be difficult for steric reasons on veatchine (40.1) as the substrate.

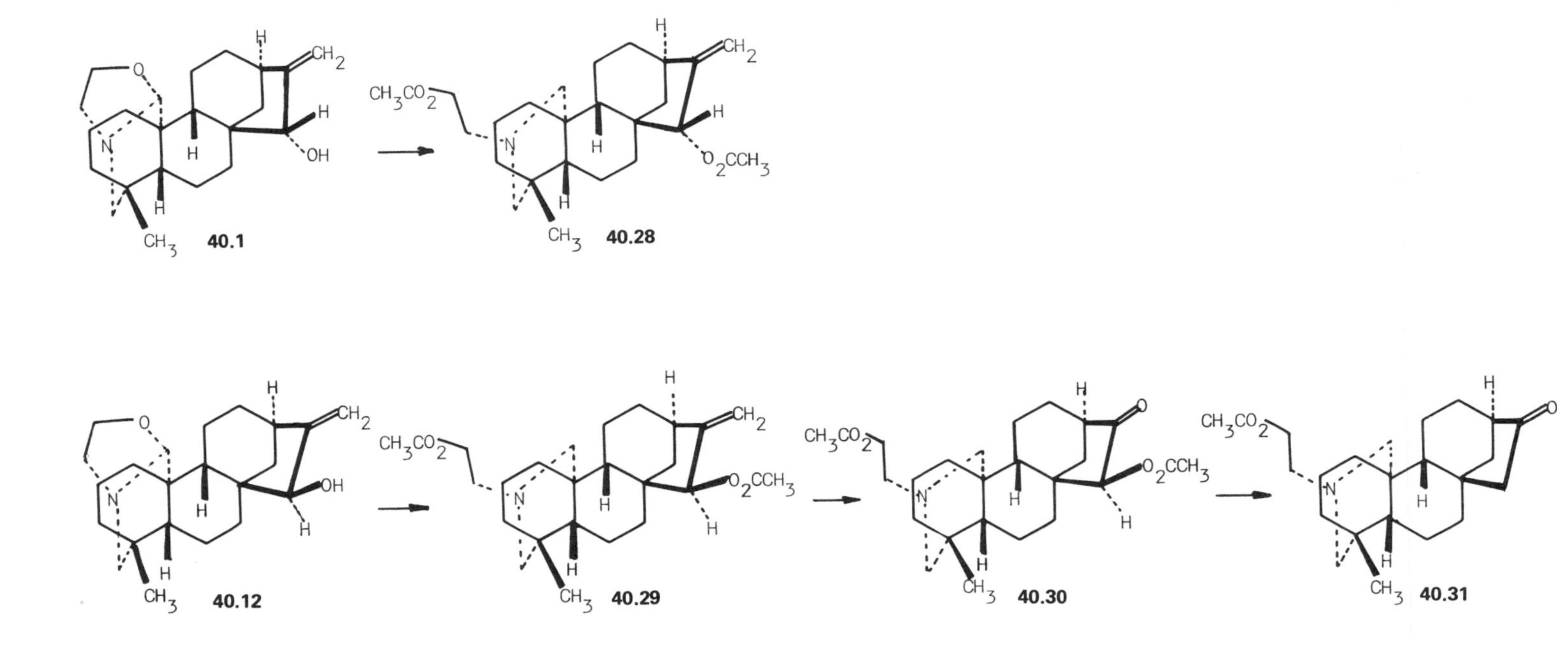

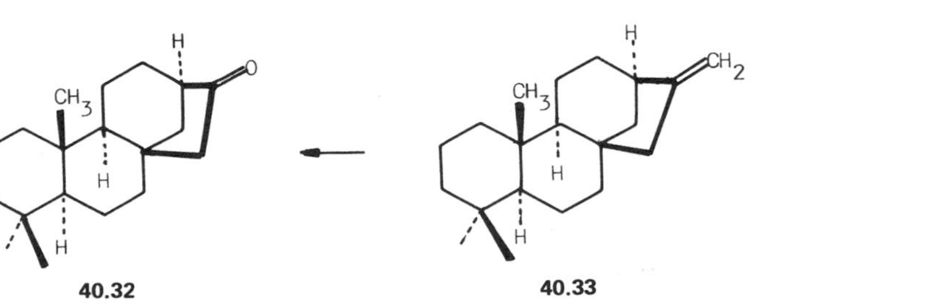

SCHEME 40.2 Establishment of the configuration at C_8 and C_{12} of veatchine and garryfoline by comparison to phyllocladene.

SCHEME 40.3 Establishment of the configuration at C_8 and C_{13} of veatchine and garryfoline by comparison to (-)-β-dihydrokaurene.

configuration. Then, oxidation of the hydroxyl to the corresponding aldehyde and reduction as above provided a crystalline hydrocarbon shown to be (-)-β-dihydrokaurene (40.40), the absolute configuration of which is known [430]. Finally, veatchine (40.1) is reported to have been synthesized [431, 432], and the synthesis is shown in Scheme 40.4.

6-Methoxy-1-tetralone (40.41) is alkylated with γ-bromocrotonate, the resulting conjugated naphthylidene crotonate partially reduced with Raney nickel, and cyclization to the oxophenanthrene 40.42 effected with acetic anhydride. Hydrocyanation of 40.42 with hydrogen cyanide and diethyl aluminum chloride yields a mixture of cyanoketones which, on recrystallization from acetone-containing hydrochloric acid, affords pure trans epimer 40.43. When the cyanoketone 40.43 is subjected to the Wittig reaction with p-tolyloxymethylenetriphenylphosphorane and the resulting products hydrolyzed, the formyl derivative 40.44 is obtained. Methylation of 40.44 with methyl iodide in the presence of potassium t-butoxide generates, stereoselectively, the 1-methyl derivative 40.45. Then, alkaline hydrolysis of 40.45 followed by ethylation to generate a mixture of epimeric ethoxylactams and reduction with lithium aluminum hydride produces the amine 40.46, and a modified Birch reduction followed by mesylation at nitrogen and acid hydrolysis leads to the conjugated ketone 40.47.

Hydrocyanation of 40.47 followed by treatment of the corresponding ketal with methyl lithium leads, on hydrolysis, to the dione 40.48 which undergoes cyclization in the presence of dilute base to provide the pentacyclic hydroxyketone 40.49. Acetylation, stereoselective reduction with sodium borohydride, and mesylation then provide 40.50 which produces the alkene 40.51 on collidine-promoted elimination. Utilizing the bulky bis-3-methyl-2-butylborane for oxidative hydroboration, the diol 40.52 can then be generated stereoselectively and the corresponding monobrosylate induced to rearrange in methanolic dioxane to the ketone 40.53. Then, the corresponding exo-methylene derivative 40.54 is generated via the Wittig reaction and the N-mesyl function replaced by an N-carboethoxy protecting group through reduction and treatment with ethyl chloroformate. Bromination with N-bromosuccinimide followed by epoxidation with perbenzoic acid and zinc-catalyzed debromination then produces a mixture of isomeric allylic alcohols 40.55 and 40.56, which are separable. When the appropriate isomer (i.e., 40.56) is refluxed with diethylene glycol and potassium hydroxide in the presence of hydrazine, the carboethoxy group is removed, and alkylation with ethylene chlorohydrin generates racemic dihydroveatchine (40.6 and its mirror image). It was then reported that oxidation converts dihydroveatchine (40.6) into veatchine (40.1); however, no experimental evidence is presented [432-434] that this last necessary step can, in fact, be consummated.

II. ATISINE

Atisine (40.2), $C_{22}H_{33}O_2N$, mp 57-60°C, is isomeric with veatchine (40.1) and garryine (40.5). In addition, an isomer of atisine (40.2), isoatisine (40.57), bears the same relationship to atisine (40.2) as garryine (40.5) bears to veatchine (40.1). Indeed, parallels in the chemistry of these pairs of bases were pointed out quite early [426], and the observation that veatchine (40.1) and garryine (40.5) provide 7-ethyl-1-methylphenanthrene (40.8; Equation 40.2) on selenium dehydrogenation whereas atisine (40.2) and isoatisine (40.57) yield 6-ethyl-1-methylphenanthrene (40.58; Equation 40.10) suggested [435] that these alkaloids differed

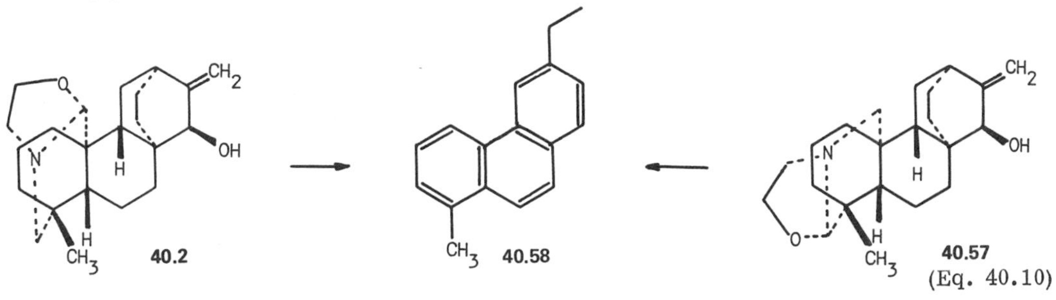

40.2 → 40.58 ← 40.57

(Eq. 40.10)

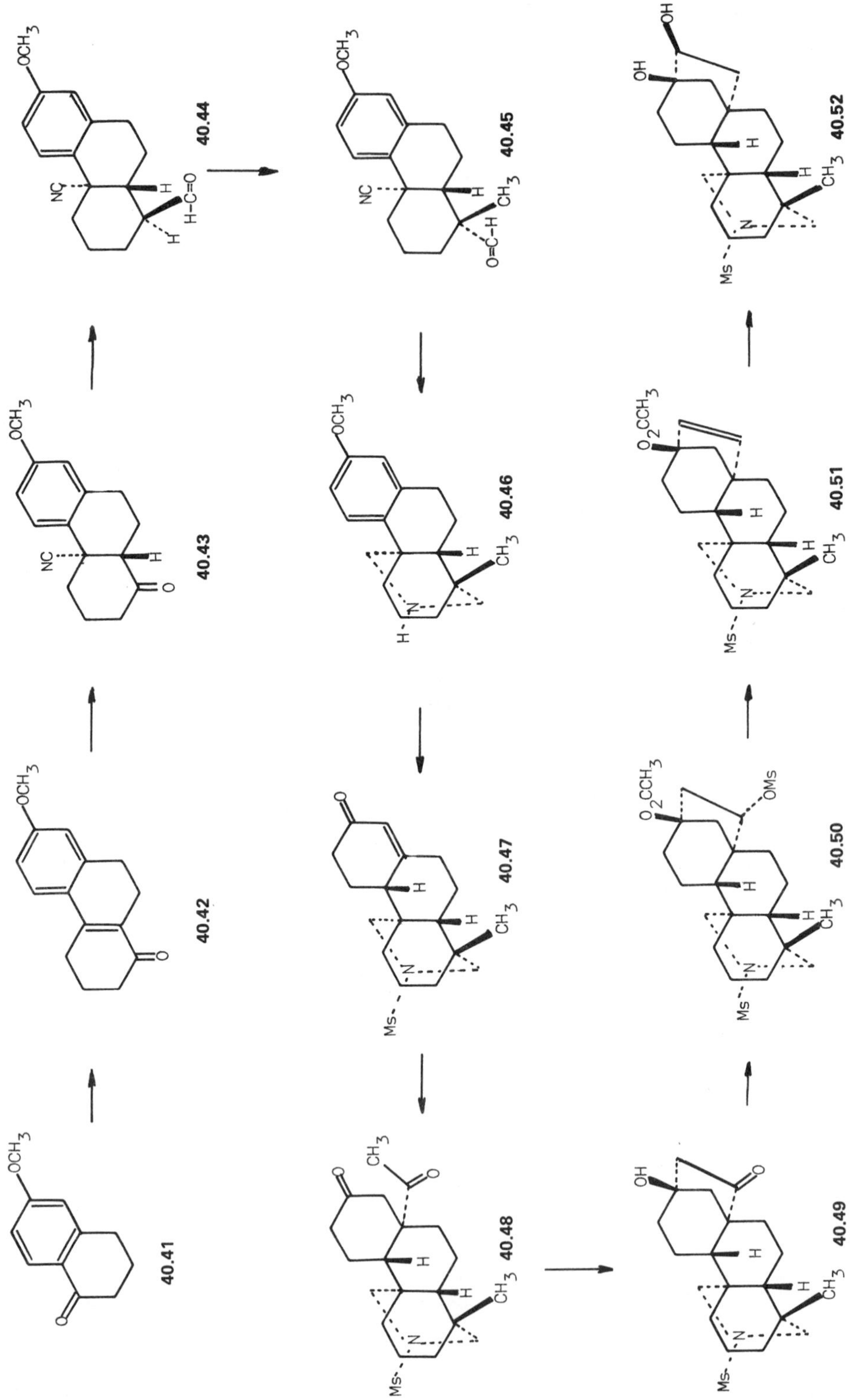
40.41
40.42
40.43
40.44
40.45
40.46
40.47
40.48
40.49
40.50
40.51
40.52

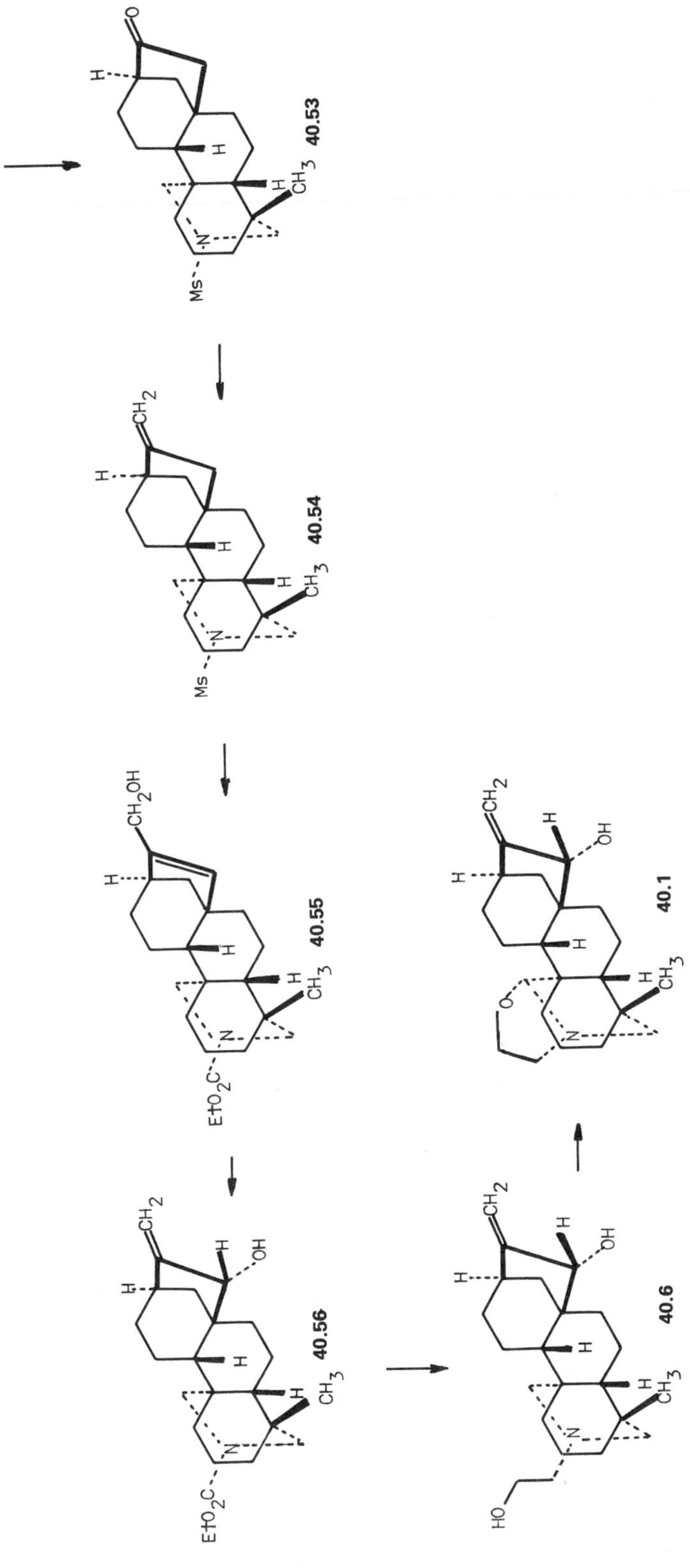

SCHEME 40.4 A synthesis of veatchine. (After References 431 and 432.)

only in the attachment of ring D to ring C. Credence was lent to this suggestion by the conversion of both atisine (40.2) and veatchine (40.1) to the same degradation product [436].

Thus, as shown in Scheme 40.5, the azomethine acetates from veatchine (40.1) and atisine (40.2; i.e., 40.59 and 40.60, respectively) which were prepared by selenium treatment of the respective alkaloids (see above) can be reduced, acetylated, and hydrolyzed to generate the corresponding acetamides, 40.61 and 40.62. Then, oxidation of each of these compounds with permanganate and periodate provides the dicarboxylic acids 40.63 and 40.64, respectively. The corresponding monomethyl esters of the acids 40.63 and 40.64 can be prepared by esterification and partial hydrolysis; the bromides (40.65 and 40.66) generated from the monoesters via the Hunsdiecker reaction; and the same monoester (40.67) isolated from both 40.65 and 40.66 by reductive dehalogenation.

The relative positioning of the methylene and hydroxyl groups suggested by the initial selenium dehydrogenation reactions was established as shown in Scheme 40.6 [437]. Thus, hydration of the azomethine 40.60, prepared as with veatchine (40.1; see above), leads to the diol 40.68, which can be reduced, acetylated, partially hydrolyzed, and oxidized to the keto acid 40.69. Bayer-Villiger oxidation of 40.69 (with trifluoroacetyl peroxide) followed by hydrolysis and further oxidation with dichromate then yields the ketoacid 40.70. Dibromination of the ketoacid (40.70) followed by dehydrohalogenation generates the phenol 40.71 and a γ-lactone [presumably 40.72; λ_{max} 1796, 1734, and 1639 cm^{-1}]. Since the formation of the γ-lactone 40.72 defines the relative positions of the carbonyl and carboxylic acid groups and the genesis of the keto group in 40.70 demands loss of two carbon atoms, the relative positions of the methylene and hydroxyl groups are established.

The relative stereochemistry of a portion of the atisine (40.2) molecule is defined by the correlation with veatchine (40.1; see above) and the remainder deduced by noting (a) that both epimeric alcohols 40.73 and 40.62 readily form acetates, a situation considered unlikely if the methylene and hydroxyl groups were on the other possible branch of the bicyclic system and (b) the hydroxyl groups in 40.73 and 40.62 should provide differential adsorption on alumina [438].

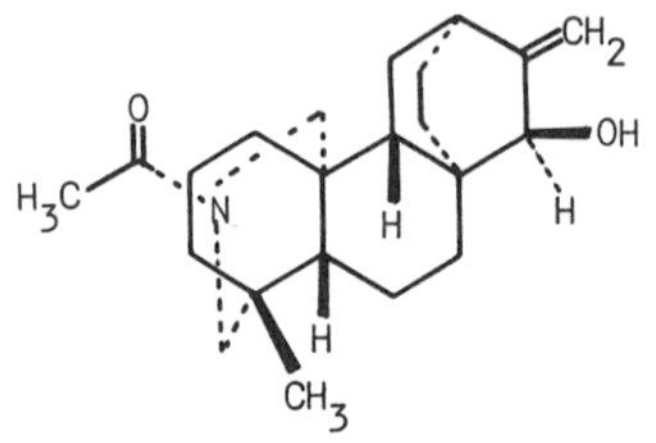

40.62 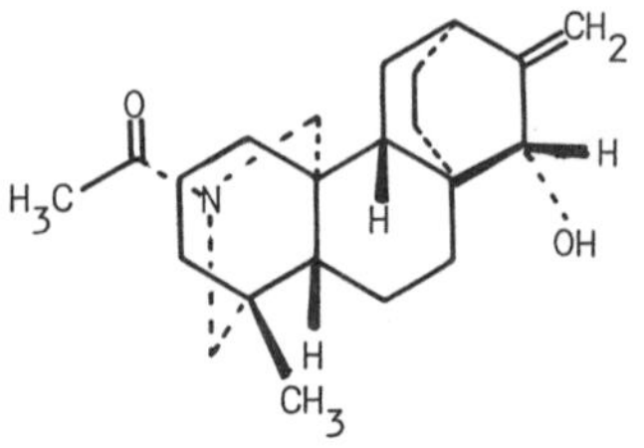 **40.73**

The absolute stereochemistry of atisine (40.2) was then determined by converting [439] the methyl ether of (+)-podocarpic acid (40.74)* into the antipode of the phenol 40.71 already generated from atisine (40.2). The transformation (Scheme 40.7) involves conversion of the methyl ether of (+)-podocarpic acid (40.74) into the corresponding azide and photochemical insertion into the angular methyl group (20% yield) to generate the lactam (40.76). Demethylation with hydrogen bromide and reduction with lithium aluminum hydride yield the aminophenol 40.77, and acetylation and hydrolysis at the phenolic center only generate the desired 40.78, the antipode of 40.71.

---

*Although already known in other ways, the absolute stereochemistry of the methyl ether of (+)-podocarpic acid (40.74) is placed on unequivocal grounds by the x-ray determination [440] of the absolute stereochemistry of a derivative, namely, methyl 6α-bromo-12-methoxy-oxo-podocarpate (40.75).

Atisine (40.2) has also been synthesized (Scheme 40.8; [431]). When the dimesylate 40.50 (Scheme 40.4) is treated with alkali, fragmentation accompanies elimination, and 40.79 results. Then, 40.79 is converted, via the corresponding alcohol, to the mesylate 40.80 and the latter induced to undergo cyclization to yield 40.81. This is followed by utilization of the Wittig reagent to provide, as earlier, the exo-methylene unit, reductive elimination of the N-mesylate, acetylation, and N-bromosuccinimide bromination to generate, as in the synthesis of veatchine (40.1), a mixture of allylic bromides. On epoxidation and reduction of the mixture, both 40.82 and 40.83, its epimer, are formed. This formally completes the synthesis since it had earlier been shown [441] that 40.83, a degradation product of atisine (40.2) could be reconverted into atisine (40.2) by a series of reactions which includes hydrazine-catalyzed deacetylation, alkylation with ethylene chlorohydrin, and oxidative oxidation.

Finally [472], despite its synthesis and the interrelationship with (+)-podocarpic acid (40.75), the structure of atisine (40.2) remains somewhat in doubt. The problem is that the ^{13}C nuclear magnetic resonance (NMR) spectrum in $CDCl_3$ solution shows <u>two</u> sets of signals for the E and F rings of the base. Careful analysis indicates that this is due to the presence of two C_{20} epimers (presumably 40.2a and 40.2b) which exist in equilibrium, and indeed,

40.2a

40.2b

the assignment of the β-configuration to the hydrogen at C_{20} was, presumably, made on the basis of a steric argument. Indeed, since both atisine (40.2) and veatchine (40.1) are isolated by acid treatment of the crude plant material and since oxazolidine ring isomerization is facile (supposedly via an intermediate as shown in Equation 40.11), the question as to which C_{20} epimer (if not both, or neither) is "naturally occurring" remains to be answered.

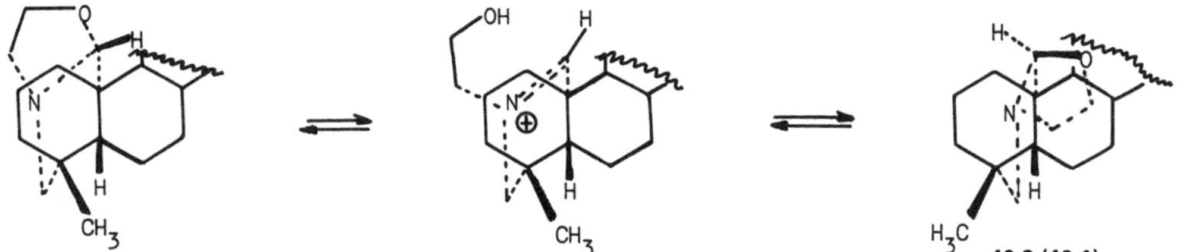

40.1 (40.2)

40.2 (40.1)

(Eq. 40.11)

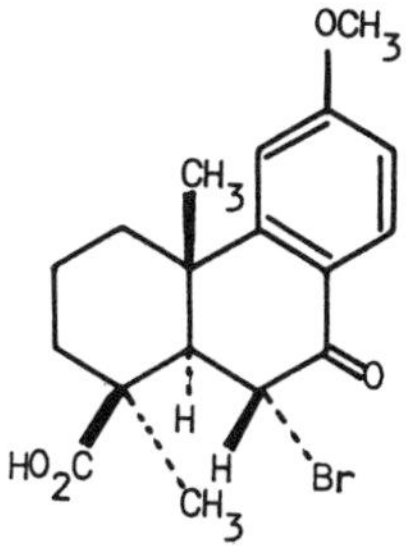

40.75

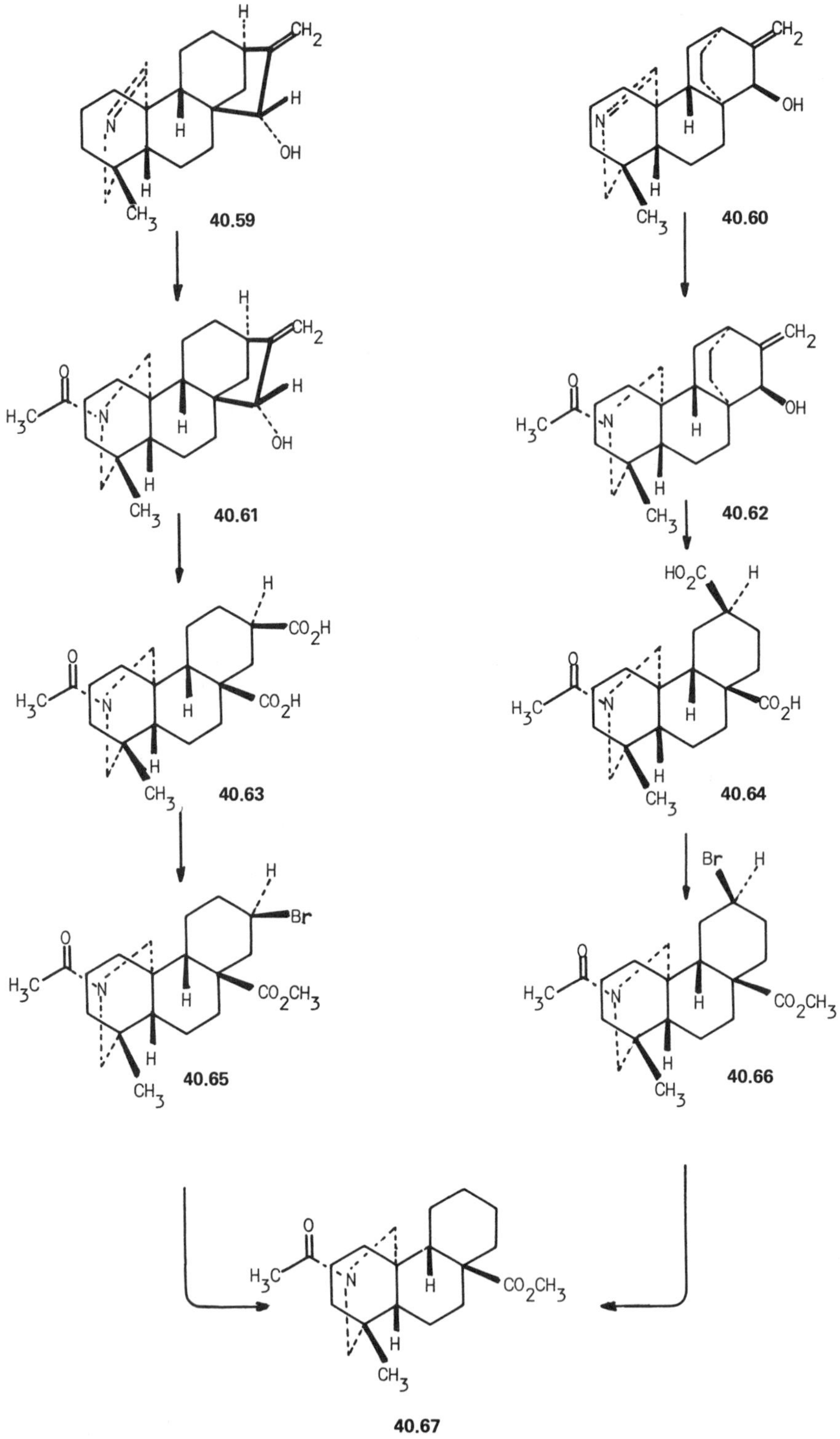

40.59

40.60

40.61

40.62

40.63

40.64

40.65

40.66

40.67

SCHEME 40.5 Degradation of veatchine and atisine to the same product. (After Reference 436.)

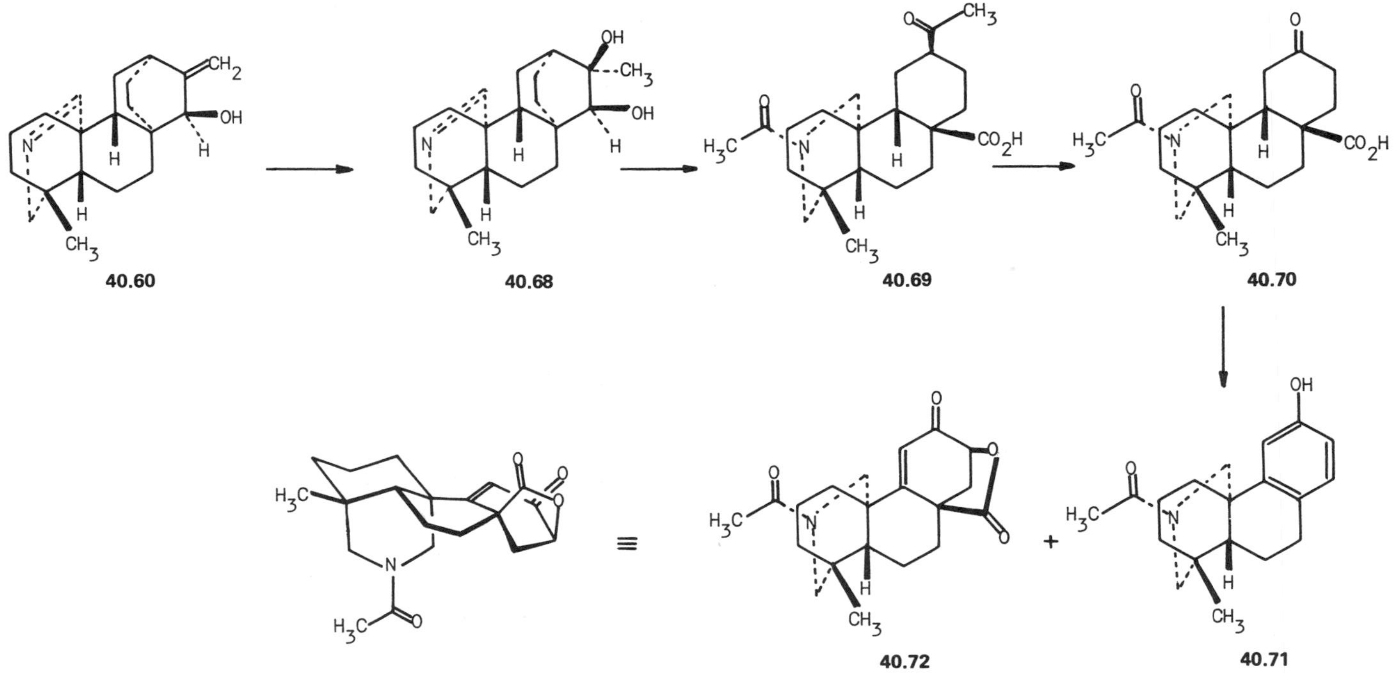

SCHEME 40.6 The establishment of the relative positions of the methylene and hydroxyl groups in atisine. (After Reference 437.)

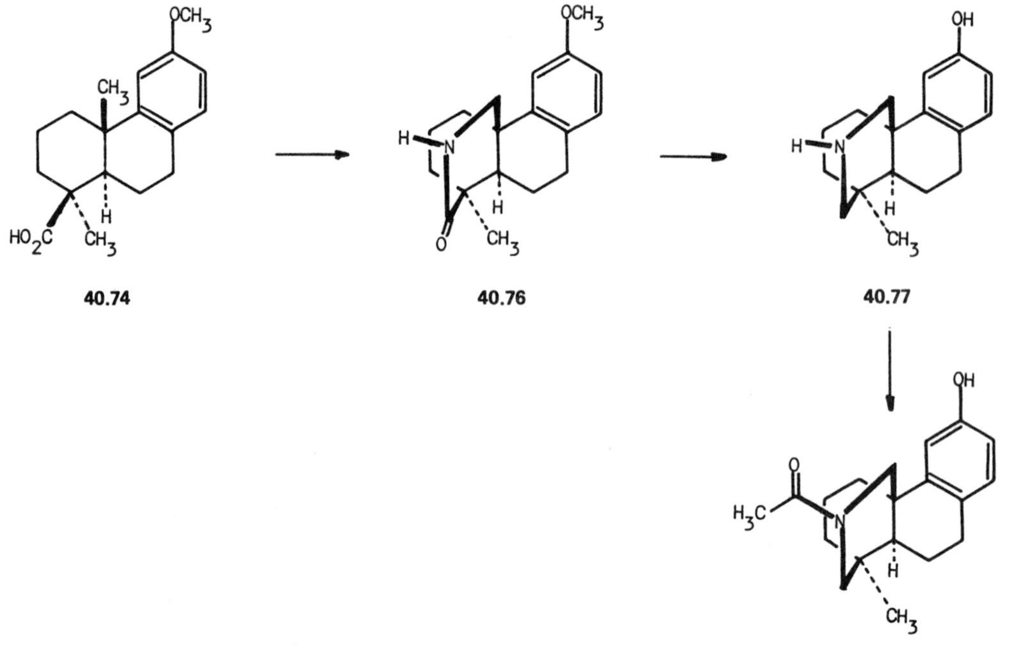

40.74

40.76

40.77

40.78

SCHEME 40.7 Determination of the absolute stereochemistry of atisine. (After Reference 439.)

III. ACONITINE

The alkaloid aconitine (40.3), $C_{34}H_{47}O_{11}N$, mp 205°C $[\alpha]_D$ +19°, and related esters, occur widely in a large number of Aconitum, and it has been reported that aconitine (40.3) is the most readily accessible of the alkaloids found in this species because it is easily separated as an insoluble perchlorate from the crude gummy mixture of bases initially obtained from the plant. Although first isolated in the early part of the nineteenth century, the reliable work done prior to about 1950 was limited to functional group analysis, since the bewildering variety of reactions observed for aconitine (40.3) left little more than a trail of confusion. Nevertheless, the structure was almost completely worked out [442] by 1959 when the x-ray crystallographic analysis on an aconitine (40.3) derivative (i.e., demethanolaconinone hydroiodide trihydrate, 40.84) became available [443]; the latter not only confirmed the deductions on the structure of aconitine (40.3) but also established the absolute configuration at 13 of the 15 asymmetric centers.

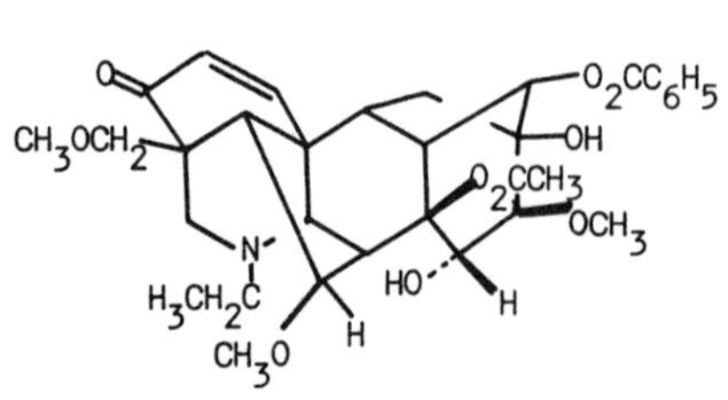

40.84

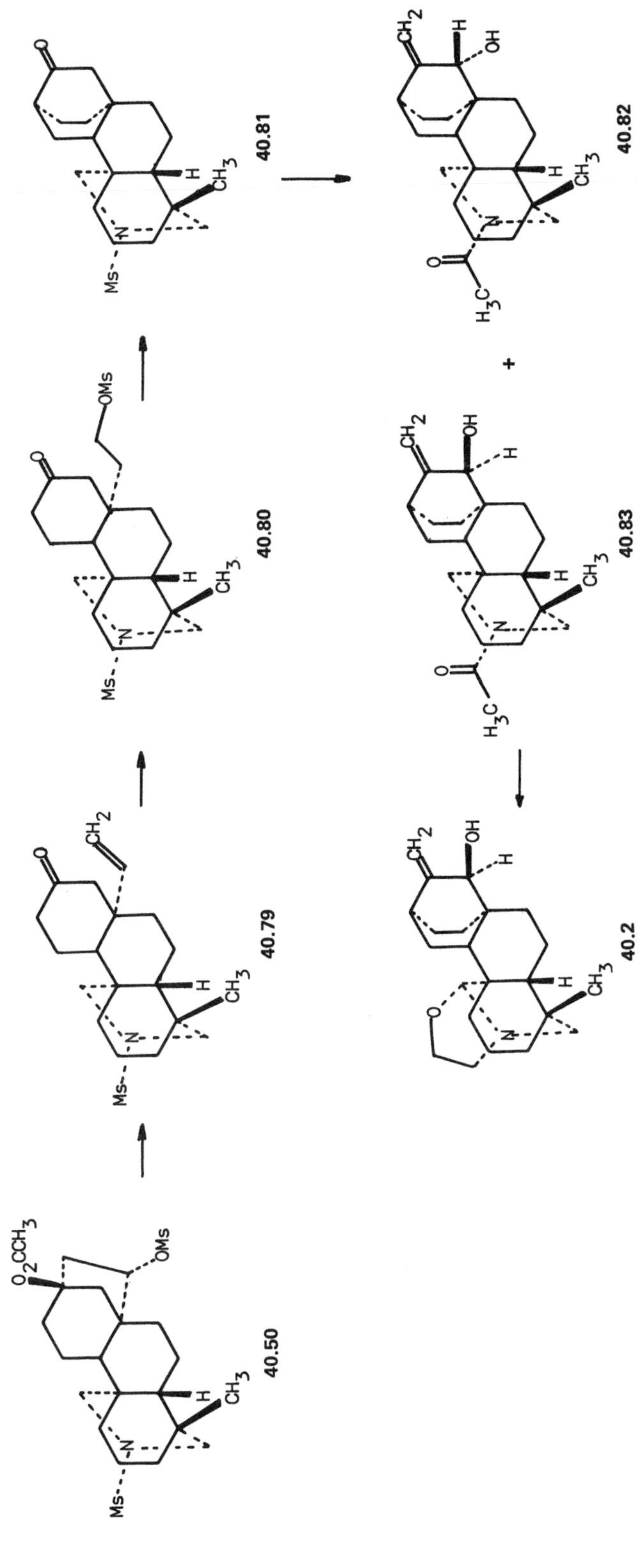

40.50 → 40.79 → 40.80 → 40.81 → 40.82 + 40.83 → 40.2

SCHEME 40.8 The total synthesis of atisine. (After Reference 431.)

The early work established that aconitine (40.3) gave an intractable mixture on selenium dehydrogenation, and attempted Hofmann degradation met with a signal lack of success. However, Zeisel determination indicated the presence of four methoxy groups, and hydrolysis with base showed the presence of one acetoxy and one benzoyloxy group. Finally, in addition to the three free hydroxyl groups, all of which could be acetylated to yield triacetylaconitine, nitrous acid treatment of aconitine (40.3), which yields an N-nitroso derivative (presumably 40.85) and results in dealkylation (Equation 40.12), indicated that an N-ethyl group was present.

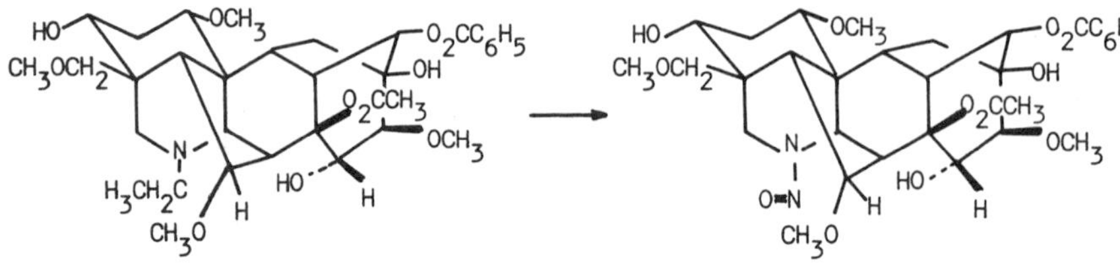

(Eq. 40.12)

40.3 40.85

On oxidation with permanganate in acetone, aconitine (40.3) yields oxonitine (40.86) and oxoaconitine (40.87; Equation 40.13), among other products. An interesting series of experiments [444] demonstrated that the N-formyl group in 40.86 is not derived from the N-ethyl

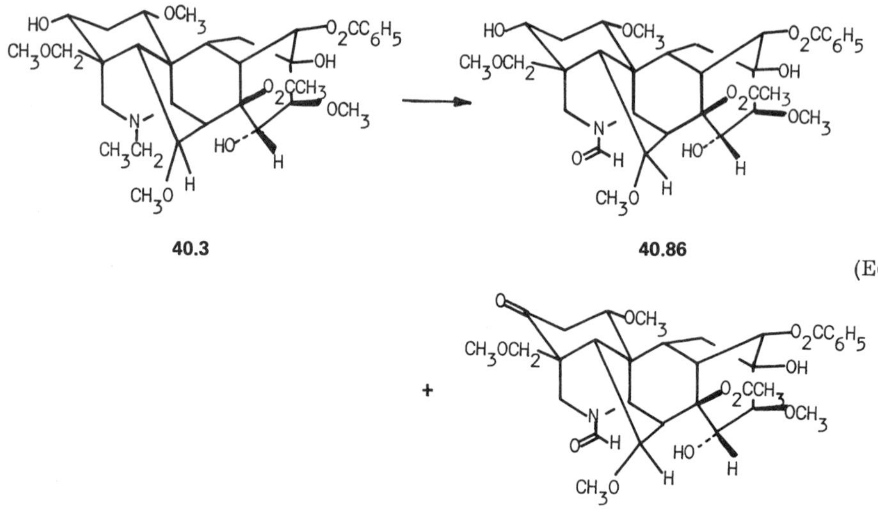

40.3 40.86

(Eq. 40.13)

40.87

group originally present in aconitine (40.3)! When oxonitine (40.86) is heated to above its melting point, acetic acid is lost and pyrooxonitine (40.88; Equation 40.14) is formed. The facility with which this loss occurs and the formation of a carbonyl group during the process suggest that the proton lost is most probably cis to the acetate and that a hydroxyl group is also present on the carbon bearing the proton (thus generating an enol capable of tautomerizing to the observed ketone).

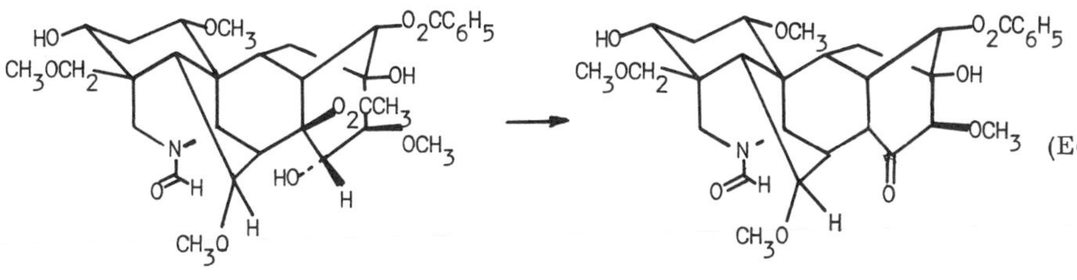

40.86 → **40.88** (Eq. 40.14)

As shown in Scheme 40.9, basic hydrolysis of oxonitine (40.86) yields oxonine (40.89), which on oxidation with chromic anhydride in pyridine provides a diketone (40.91) which is considered to have formed via an acyloin rearrangement of 40.90. Then, lead tetraacetate oxidation of 40.91 leads to the keto acid 40.92. Since oxidation of oxonitine (40.86) itself under the same conditions (chromic anhydride in pyridine) yields oxoaconitine (40.87) and not the diketone 40.91, the benzoyl group must be present on the hydroxyl being oxidized in the formation of 40.90 and vicinal to the tertiary hydroxyl involved in the rearrangement.

Oxonine (40.89) undergoes another striking rearrangement on periodate cleavage and air oxidation of a basic solution of the product of that cleavage. As shown in Scheme 40.10, oxonine (40.89) takes up one equivalent of periodate* to yield a secoketoaldehyde (40.93; [445]).† If the secoketoaldehyde 40.93 then undergoes a retroaldol cleavage to 40.95, recyclization to the aldehyde liberated by periodate oxidation and dehydration would then provide the α,β-unsaturated ketone 40.96. A second dehydration and enolization leads to the aldehyde 40.97 and then aerial oxidation in base, perhaps via the peroxide 40.98, to the phenolic ketone 40.99. Methylation of 40.99 to the dimethyl ether, which is crystalline, can then be affected, and the latter can be oxidized with chromic anhydride to the ketone 40.100. Hydrolysis of the formyl group and β-elimination of methanol is accomplished by refluxing 40.100 in methanolic hydrochloric acid, and acetylation then produces 40.101 (Scheme 40.11). Finally (Scheme 40.12), when aconitine N-oxide (40.102; [442]) is prepared from aconitine (40.3) by peracetic acid treatment and subjected to pyrolysis, ethylene and acetic acid are lost and the pentacyclic nitrone 40.103 is generated. Reduction of 40.103 with zinc in acetic acid produces an intermediate (presumably 40.104) which yields aconitine (40.3) on ethylation. Aconitine (40.3) has not yet been synthesized.

IV. HETERATISINE

The alkaloid heteratisine (40.4), $C_{22}H_{33}NO_5$, mp 267-268°C, $[\alpha]_D$ +29° ($CHCl_3$), is reported to be a minor constituent of Aconitum heterophyllum, and its structure was solved almost simultaneously by a combination of proton magnetic resonance (pmr) and degradative studies [446, 447] and x-ray crystallographic analysis [448]. The functional groups were readily

*It should be noted that there are two sets of vicinal hydroxyl groups. One set at C_{13}-C_{14} is presumably oxidized, while the second, C_8-C_{15}, remains unaffected. Similarly, although the proton at C_{15} is available for loss, the hydroxyl at C_{15} is not exposed to attacking reagent, and it is therefore not surprising that reaction is difficult at that site.

†The secoketoaldehyde (40.93) is reported to be without carbonyl absorption (aside from the N-formyl group) and may thus exist as the cyclic hydrate 40.94. An alternative cyclic hemiacetal structure (Reference 445) does not appear stereochemically viable.

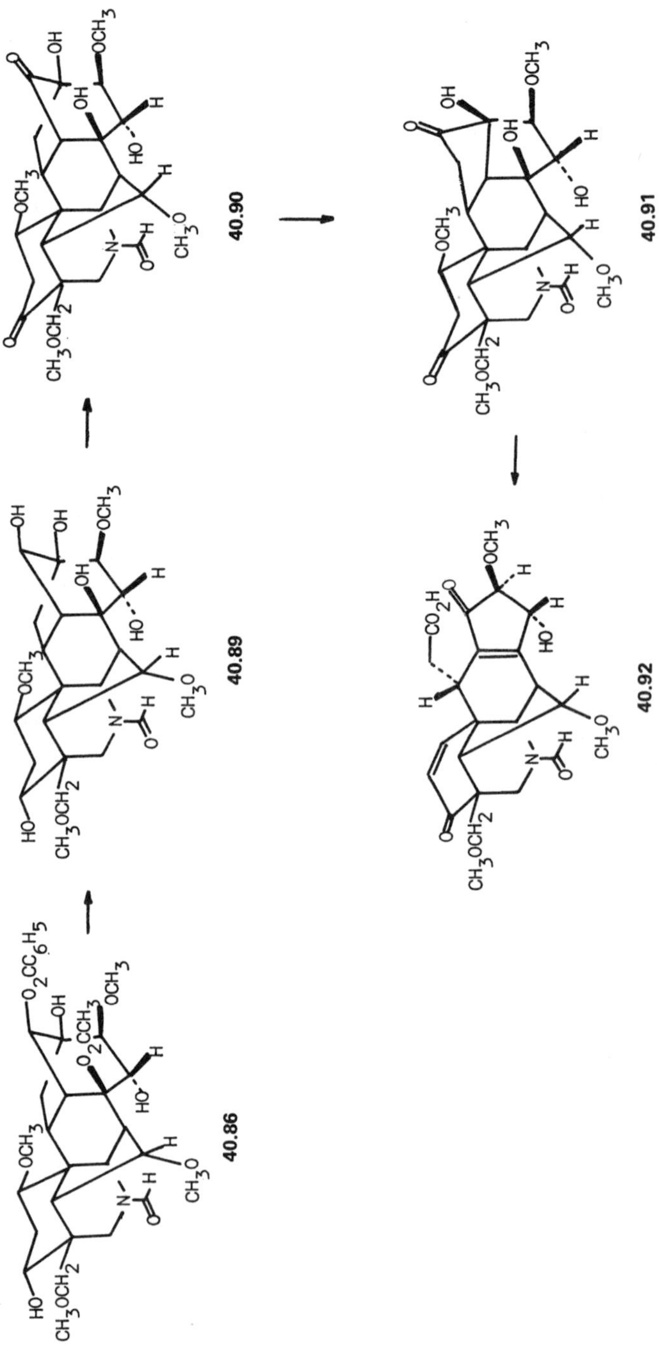

SCHEME 40.9 A partial degradation of oxonitine to oxonine and the oxidation of oxonine.

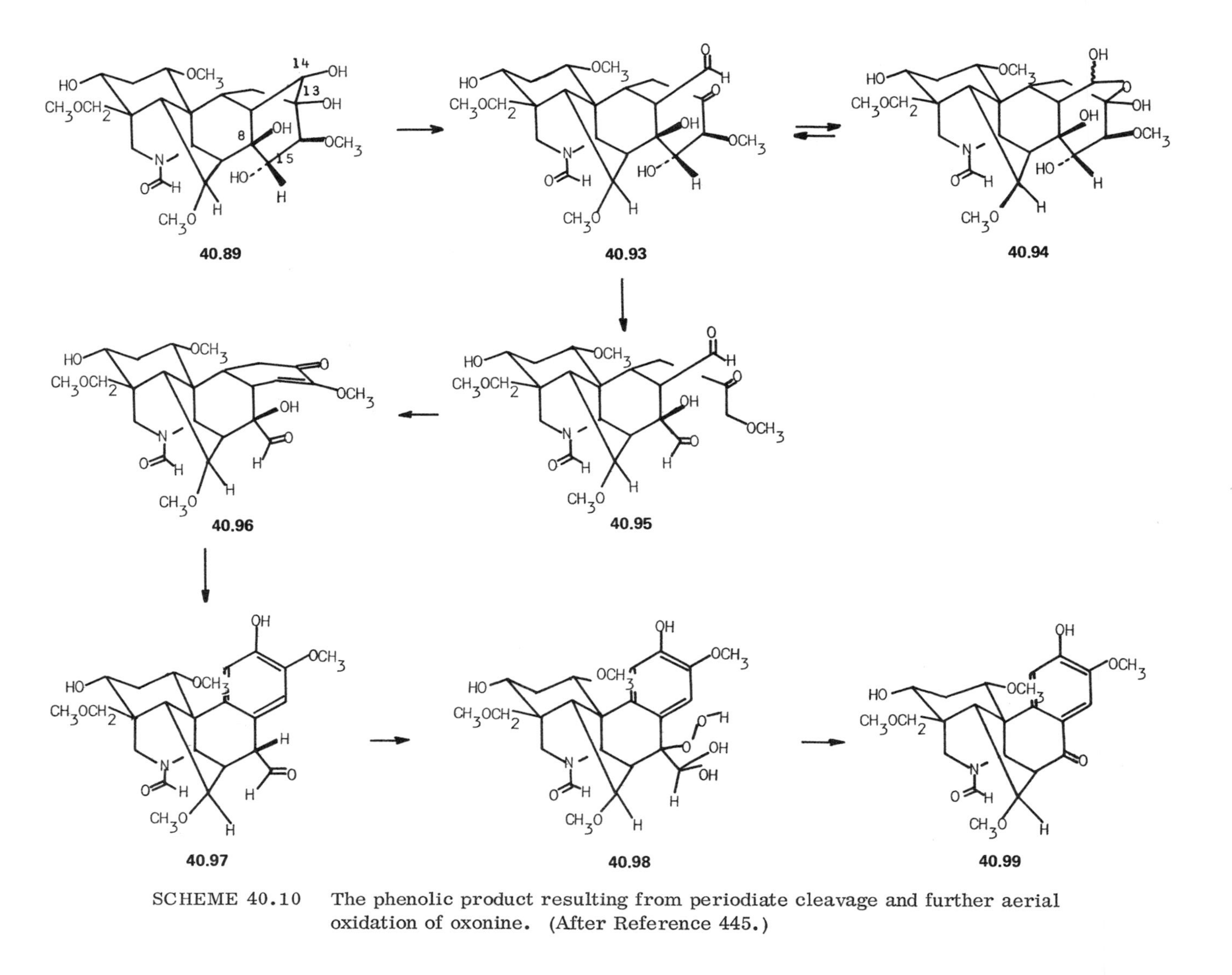

SCHEME 40.10 The phenolic product resulting from periodiate cleavage and further aerial oxidation of oxonine. (After Reference 445.)

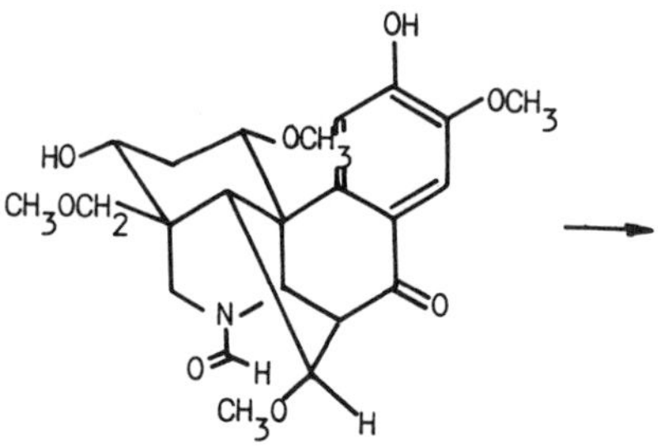

40.99

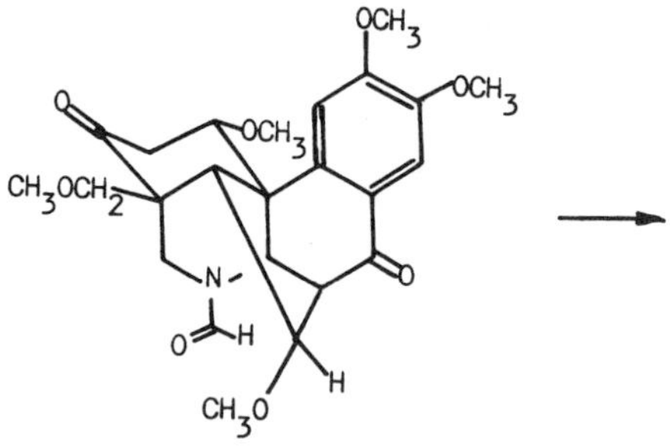

40.100

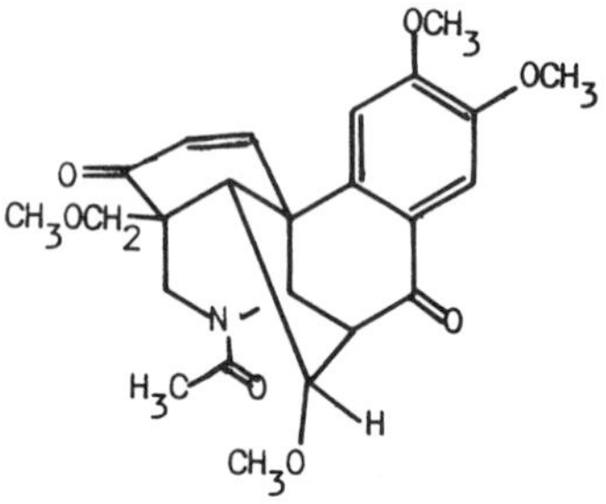

40.101

SCHEME 40.11 Further degradation of oxonine.

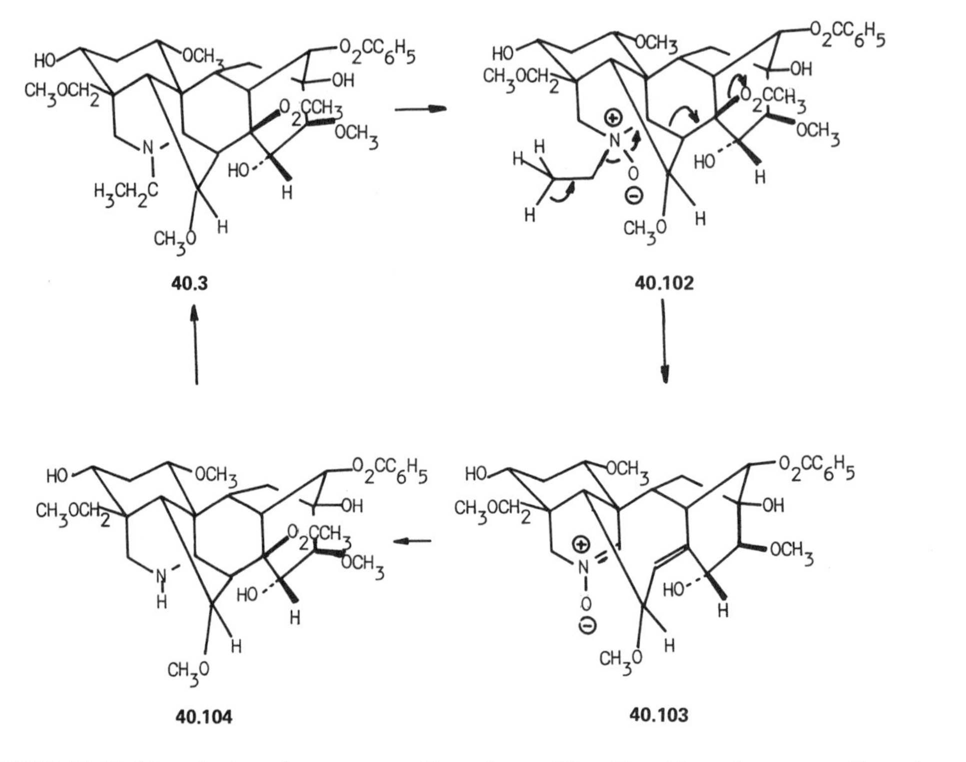

SCHEME 40.12 A ring cleavage reaction of aconitine N-oxide and regeneration of aconitine. (After Reference 442.)

identified and all of the heteroatoms accounted for by IR and PMR spectroscopy on heteratisine (40.4) and its readily formed basic monoacetate (40.105; Equation 40.15). For the parent base (40.4), hydroxyl group(s) and the δ-lactone are identified as being present by IR

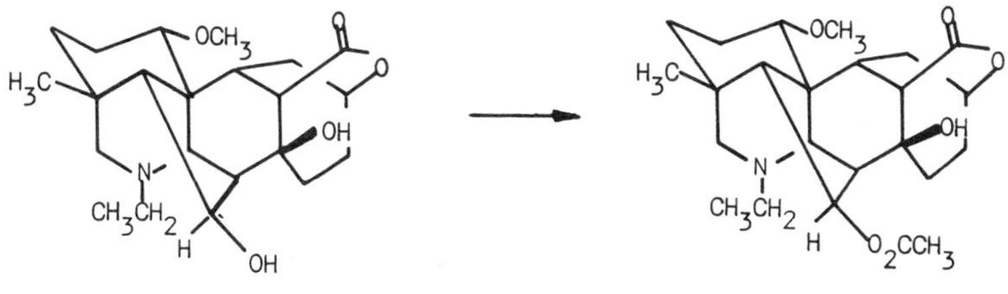

40.4 → 40.105 (Eq. 40.15)

($3472\ cm^{-1}$ and $1739\ cm^{-1}$, respectively), while the methoxyl, δ = 3.25 ppm (2H), singlet; the N-ethyl (CH_3), δ = 1.02 ppm (3H), triplet, J = 7.5 Hz; the quaternary methyl, δ = 0.97 ppm (3H), singlet; the hydrogen on the carbon of the secondary hydroxyl, δ = 4.5 ppm (1H), multiplet; and the hydrogen on the alcohol portion of the lactone, δ = 4.7 ppm (1H), multiplet are all clearly seen in the PMR spectrum. For the monoacetate (40.105), the PMR data are similar, but both the hydroxyl that remains unesterified (tentatively identified, therefore, as a tertiary alcohol; $3395\ cm^{-1}$) and the acetyl ($1714\ cm^{-1}$) can be observed along with the δ-lactone ($1746\ cm^{-1}$) in the infrared.

Oxidation of the acetate 40.105 (Scheme 40.13) with chromic anhydride-pyridine yields two major products, namely, oxoheteratisine acetate (40.106) and N-desethyldehydroheteratisine acetate (40.107). Hydrolysis of the former (40.106) yields oxoheteratisine (40.108), and further oxidation of this deacetylated derivative under the same conditions provides dehydrooxoheteratisine (40.109), which proved identical with oxoheteratisinone (40.109) obtained directly from oxidation of heteratisine (40.4) with chromic anhydride in pyridine. On the other hand, direct oxidation of heteratisine (40.4) with chromic anhydride in acetic acid yields dehydroheteratisine (heteratisinone; 40.110; Equation 40.16). For all the above derivatives, PMR and IR spectra are in accord with their respective formulations.

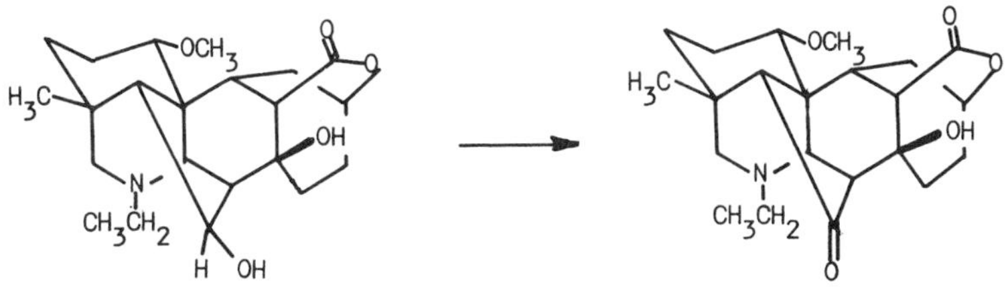

40.4 → 40.110 (Eq. 40.16)

A few interesting rearrangements were observed during the course of the study of the above reactions. Thus, while alkaline hydrolysis of heteratisine (40.4) and oxoheteratisine (40.108) simply opens the lactone ring, the corresponding ketones described above, i.e., oxoheteratisinone (40.109) and heteratisinone (40.110) undergo serious structural changes when treated under the same conditions. As shown in Scheme 40.14 for oxoheteratisinone (40.109; the exactly analogous reaction also occurs with heteratisinone, 40.110), alkaline hydrolysis results not only in lactone ring opening but also in a retroaldol reaction to yield the β-ketoacid 40.111, which then undergoes decarboxylation to the dione 40.112. Alternatively (Scheme 40.15), oxoheteratisinone (40.109), on treatment with potassium t-butoxide in

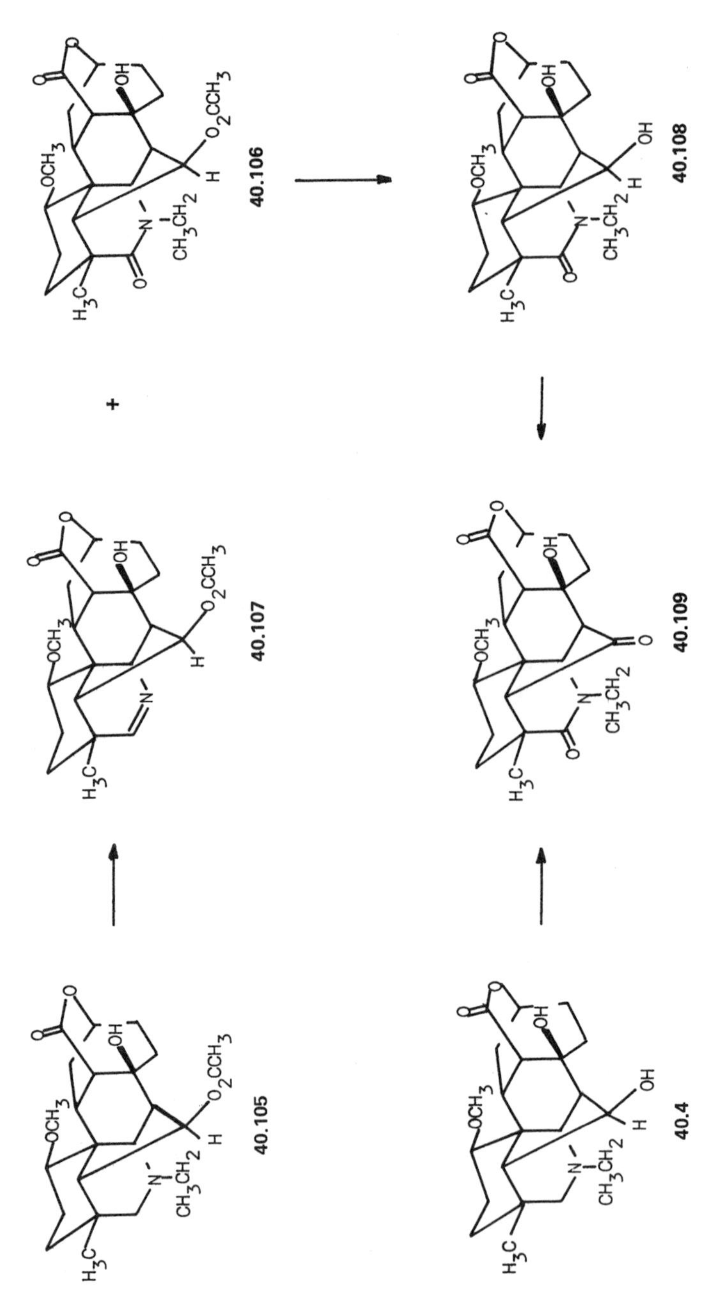

SCHEME 40.13 The products of oxidation reactions on heteratisine and the monoacetate of heteratisine.

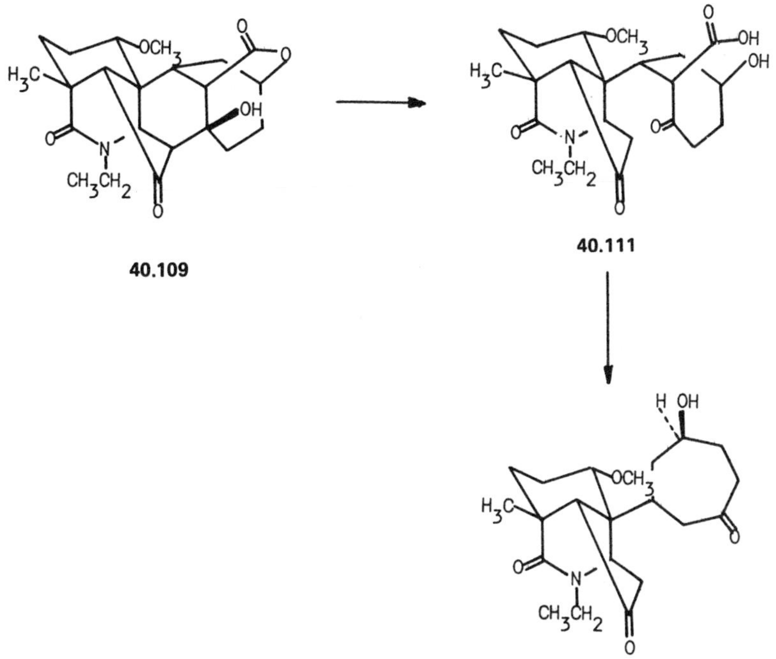

SCHEME 40.14 The alkaline hydrolysis and rearrangement of oxoheteratisinone.

t-butanol followed by esterification of the resultant product, leads to the γ-lactone carboxylic ester (40.115). Presumably, the ester 40.115 forms via the dione 40.113, which then reacts with adventitious moisture to generate 40.114, the latter undergoing cyclization to the acid which forms 40.115 on esterification.

When heteratisine monoacetate (40.105) is pyrolyzed (Scheme 40.16), acetate is lost, and (major product) pyroheteratisine (40.118) results. It has been suggested [449] that the reaction occurs by initial isomerization to the acetate of the tertiary alcohol 40.116, concerted elimination to the imine 40.117, and intramolecular hydride transfer to yield 40.118. On basic hydrolysis (Scheme 40.17), 40.118 undergoes ring opening, attack of hydroxide β to the carbonyl, and a retroaldol reaction to 40.119. Then, decarboxylation and aldol condensation with loss of water lead to 40.120.

The absolute configuration of heteratisine (40.4) has been determined by study of the optical rotatory dispersion curves of pyroheteratisine (40.118) and its hydrogenation product dihydropyroheteratisine (40.121; Equation 40.17) and comparison of the curves to diterpene alkaloids of known absolute configuration [447].

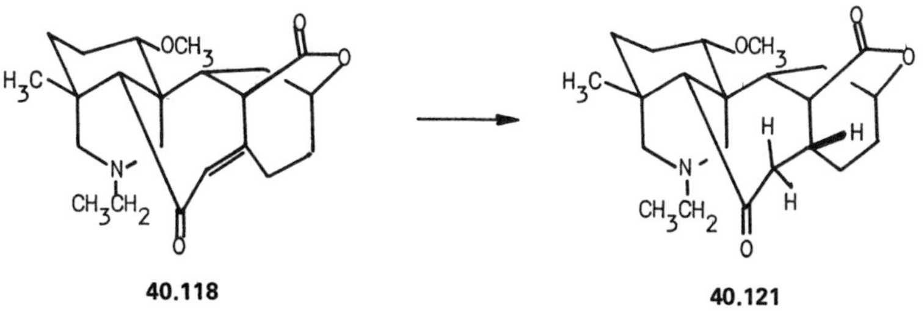

(Eq. 40.17)

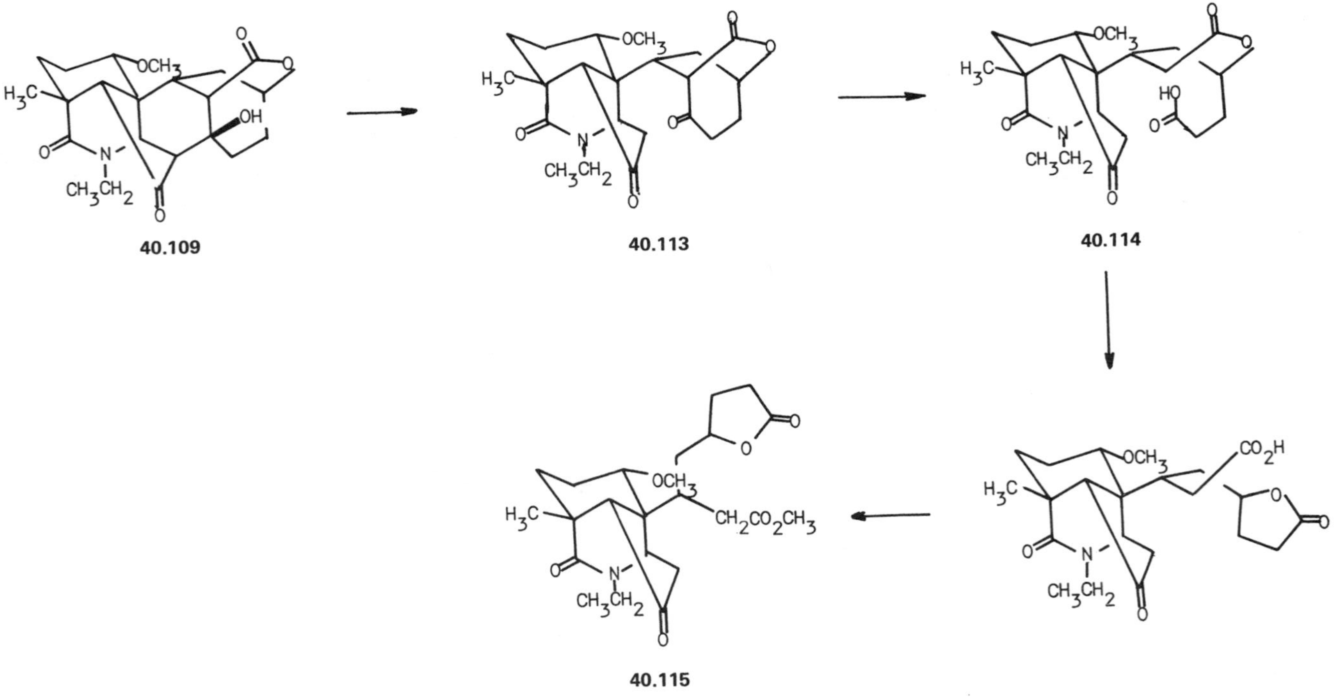

SCHEME 40.15 The t-butoxide-initiated reaction causing rearrangement in the skeleton of oxoheteratisinone.

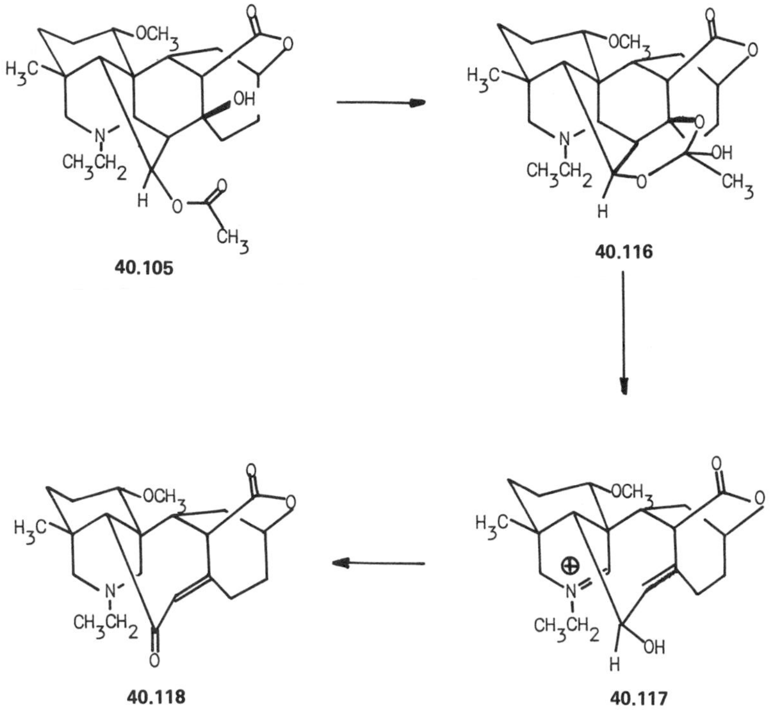

40.105

40.116

40.118

40.117

SCHEME 40.16 A possible pathway for the formation of pyroheteratisine from heteratisine monoacetate. (After Reference 449.)

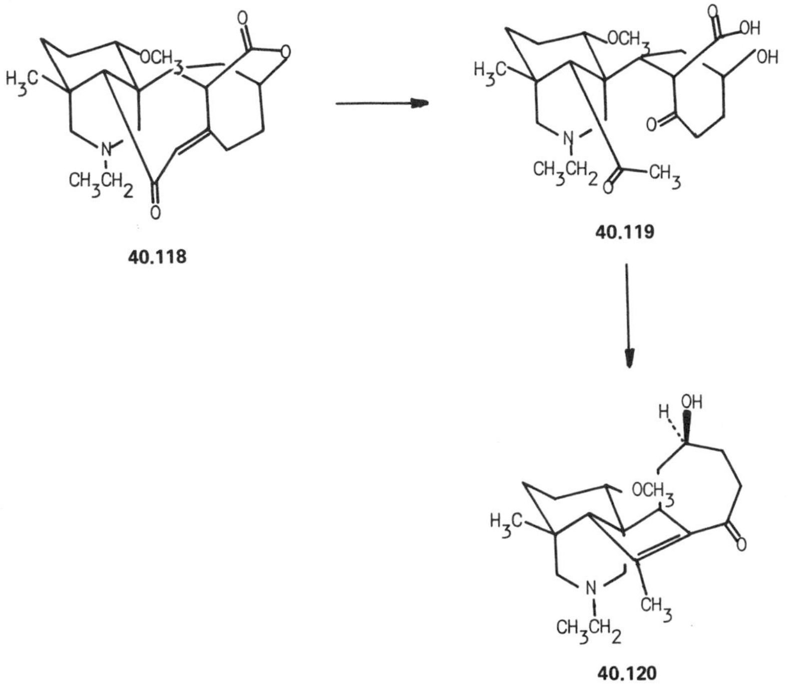

40.118

40.119

40.120

SCHEME 40.17 A possible pathway for the basic hydrolysis of pyroheteratisine.

SELECTED READING

L. H. Keith and S. W. Pelletier, in Chemistry of the Alkaloids, S. W. Pelletier (ed.). Van Nostrand-Reinhold, New York, 1970, pp. 549 ff.

S. W. Pelletier and L. H. Keith, in Chemistry of the Alkaloids, S. W. Pelletier (ed.). Van Nostrand-Reinhold, New York, 1970, p. 503.

S. W. Pelletier and L. H. Keith, in The Alkaloids, Vol. 12, R. H. F. Manske (ed.). Academic, New York, 1970, pp. 2 ff.

E. S. Stern, in The Alkaloids, Vol. 7, R. H. F. Manske (ed.). Academic, New York, 1960, 473 ff.

E. S. Stern, in The Alkaloids, Vol. 4, R. H. F. Manske (ed.). Academic, New York, 1954, pp. 275 ff.

K. Wiesner and Z. Valenta, in Progress in the Chemistry of Organic Natural Products, Vol. 16, L. Zechmeister (ed.). Springer-Verlag, Vienna, 1958, pp. 26 ff.

41

CHEMISTRY OF THE STEROID ALKALOIDS

> The Promised Land always lies on the other side of
> a wilderness.
>
> Havelock Ellis, The Dance of Life, Chapter 5

The compounds grouped here as steroid alkaloids comprise a large number of plant bases, occasionally occurring as the aglycone portion of a glycoalkaloid bonded to one or more six-carbon sugars, made up of a steroid-type backbone onto which a nitrogen atom has been appended. Therefore, the chemistry of the alkaloids is dictated by excision of the nitrogen atom somewhere along the degradative path and comparison of the remaining fragment or fragments to steroid congeners for which a vast literature exists. In many cases phenanthrene (41.1) derivatives, the typical steroid degradation products, and some nitrogen-containing fragments, e.g., pyridine (41.2) derivatives, are found on selenium dehydrogenation,

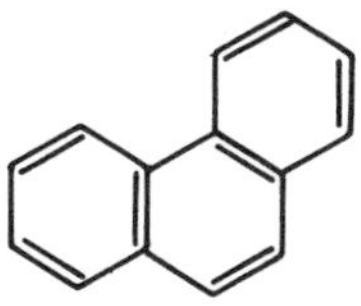

41.1 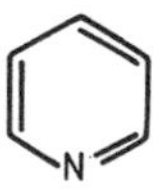 **41.2**

and this has served to help in the initial identification of the steroidal nature of the starting alkaloid. The specific compounds we shall consider are the Solanum alkamine solanidine (41.3), the Holarrhena alkaloid conessine (41.4), and the Buxus base cyclobuxine-D (41.5).

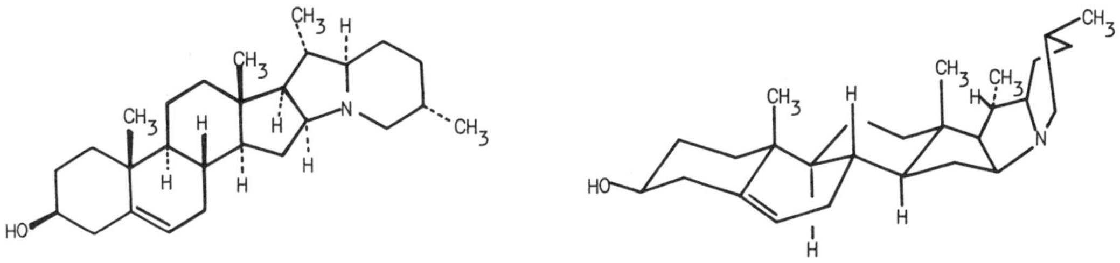

41.3

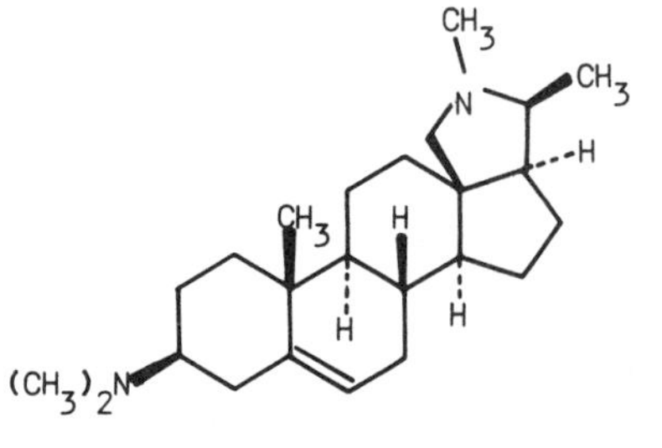

41.4

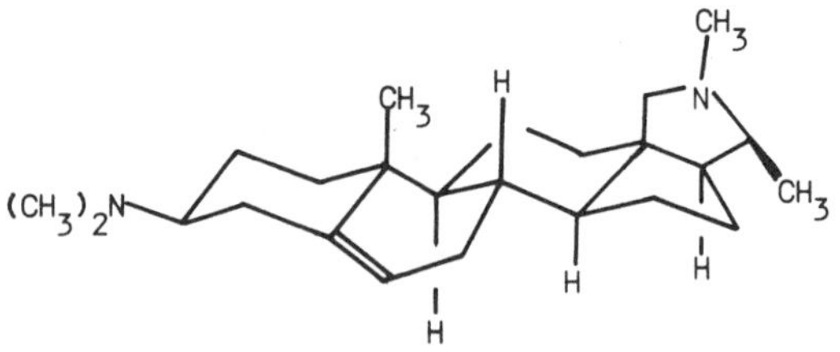

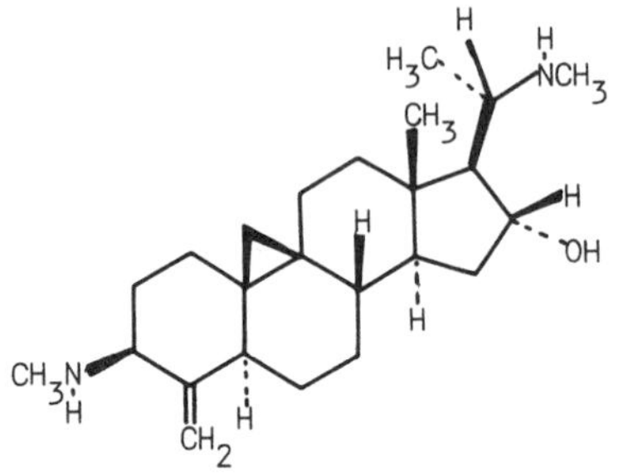

41.5

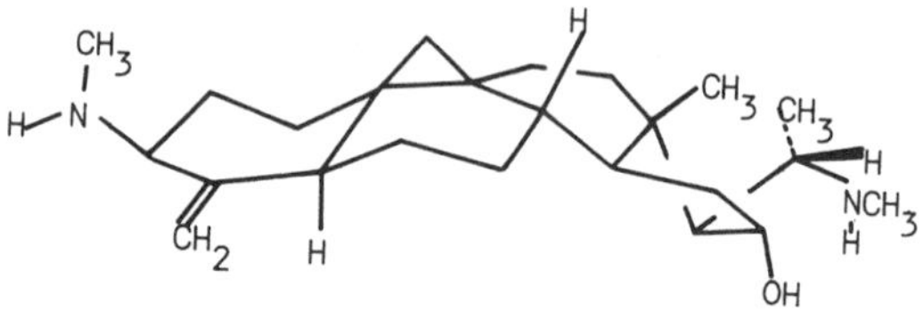

I. SOLANIDINE

The hydrolysis of the glycoalkaloid solanine (41.6), isolated from, e.g., potato sprouts, results in the formation of the aglycone solanidine (41.3) and the trisaccharide L-rhamnoside-D-galactoside-D-glucose (salanose, 41.7; Equation 41.1). The alkaloid (the aglycone) solanidine (41.3), $C_{27}H_{43}NO$, mp 219°C, $[\alpha]_D$ -27° (chloroform), represented by the aglycone portion of the glycoalkaloid, is readily identified as a tertiary amine containing one reducible double bond and possessing a secondary hydroxyl group. With seven sites of unsaturation demanded by the formula, solanidine (41.3) must therefore be hexacyclic. Selenium dehydrogenation of solanidine (41.3) under conditions suitable for steroid degradation generates γ-methylcyclopentenophenanthrene (41.8) and 2-ethyl-5-methylpyridine (41.9; Equation 41.2). These fragments immediately suggested the gross structure of solanidine (41.3). Then, since the Meerwein-Ponndorf-Oppenauer-Verley oxidation of solanidine (41.3) generates an α,β-unsaturated ketone [as indicated by ultraviolet (UV) spectroscopy; Δ^4-solaniden-3-one, 41.10; Equation 41.3], which can be reduced to a pair of epimeric alcohols neither of which is identical to solanidine (41.3), it was assumed that, as had been established for many examples in the steroid field, the double bond was originally a Δ^5-double bond in solanidine (41.3) and that it had migrated to the Δ^4-position during the oxidation.

Although relationships between solanidine (41.3) and sapogenins (e.g., kryptogenin, 41.11; see below) began to be established by chemical means [400], it was x-ray crystallographic study [450] which provided the complete details of the structure of this base and, in fact, laid the groundwork for details of the structures of the sapogenins themselves.

Solanidine (41.3) and several of its isomers have been synthesized. Thus, as shown in Scheme 41.1 [451], the sapogenin kryptogenin (41.11) can be converted to a solanidine (41.3) isomer, presumably 41.12, by zinc reduction of the C_{16}-2,4-dinitrophenylhydrazone derivative to yield the pyrroline 41.13 and then refluxing the latter in ethylene glycol in the presence of potassium hydroxide. Interestingly, kryptogenin (41.11) itself can be prepared (Scheme 41.2; [452]) from dehydropregnenolone (41.14). Thus, dehydropregnenolone (41.14) can be

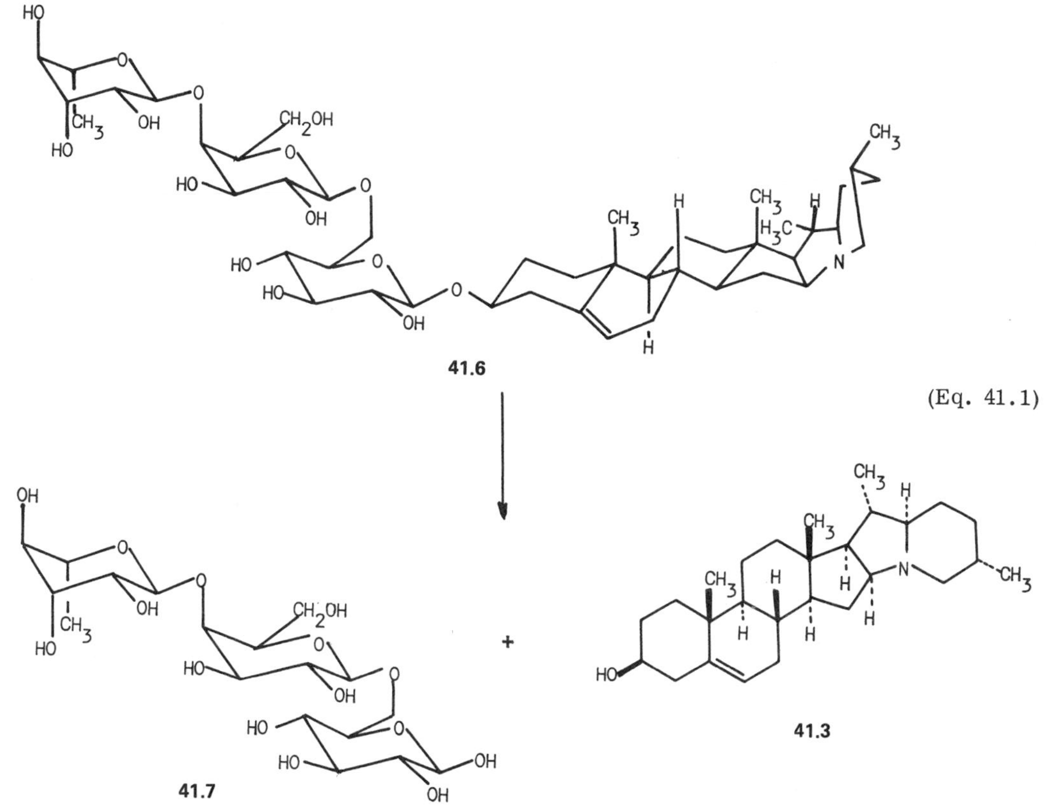

41.6

(Eq. 41.1)

41.7

41.3

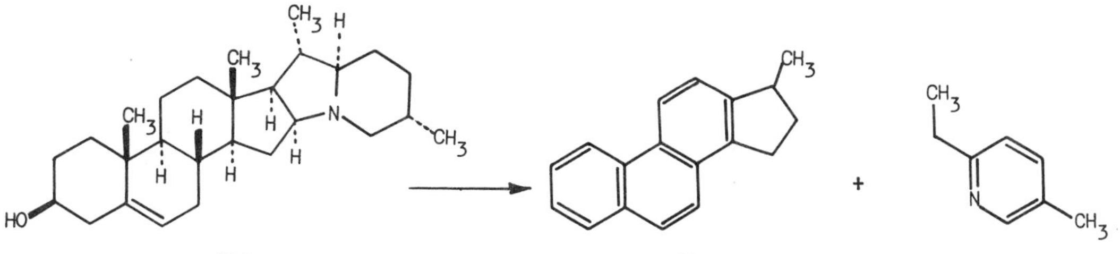

41.3

41.8

41.9

(Eq. 41.2)

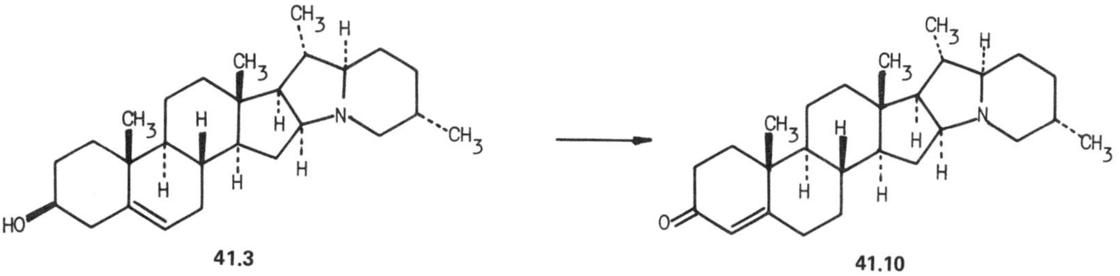

41.3

41.10

(Eq. 41.3)

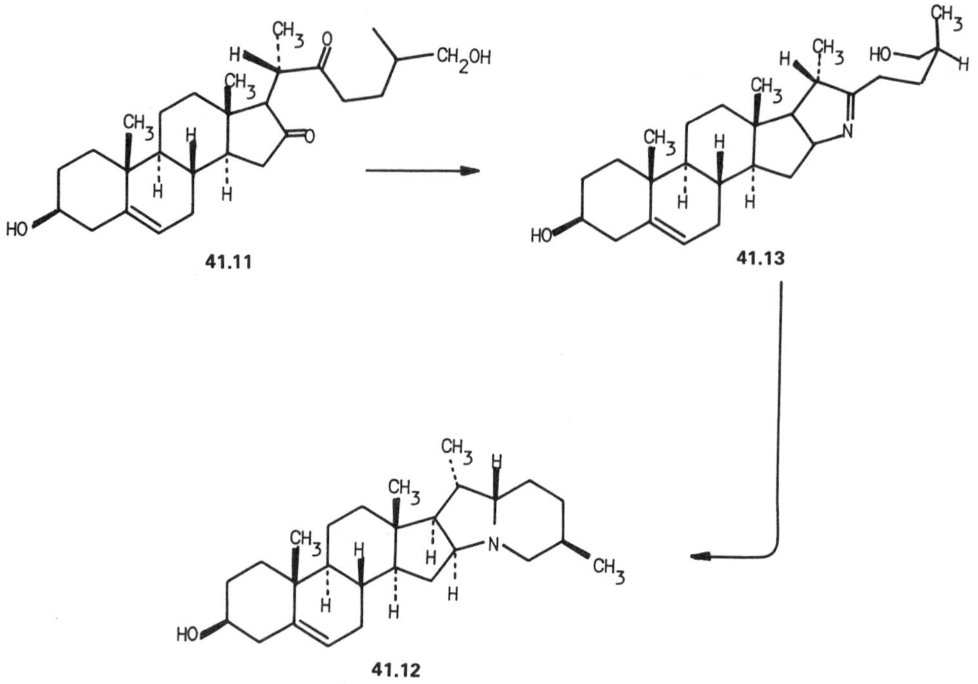

41.11

41.13

41.12

SCHEME 41.1 The synthesis of a solanidine isomer from kryptogenin. (After Reference 451.)

converted to the epoxide 41.15, which on treatment with hydrazine hydrate yields a mixture of the (Z) and (E) alcohols, 41.16 and 41.17, respectively. Oxidation of the (E) alcohol (41.17) with manganese dioxide then yields the corresponding ketone 41.18, which when allowed to undergo a Michael condensation with the optically active nitroacetate 41.19 provides the nitroketone 41.20. Under the conditions of the Nef reaction, kryptogenin (41.11) forms. Alternatively, for solanidine (41.3) itself ([453]; Scheme 41.3), if the ketone 41.18 is allowed to react with the nitroester 41.21, a mixture of adducts, from which 41.22 can be isolated, is formed. Reduction of 41.22 with zinc in acetic acid then provides the amide 41.23, which on further reduction with lithium aluminum hydride yields the desired alkaloid (41.3). Now, as shown in Scheme 41.4, dehydropregnenolone (41.14) is prepared from androstenolone (41.24; [454]) by conversion of the latter (41.24) into the corresponding cyanohydrin epimers and dehydration to the unsaturated nitrile 41.25 with phosphorus oxychloride. Then, treatment of 41.25 with methyl magnesium bromide and hydrolysis affords 41.14.

The total synthesis of androstenolone (41.24; [455, 456]) has also been described and is briefly shown in Scheme 41.5. Reduction of 1,6-dimethoxynaphthalene (41.26) with sodium in ethanol to 5-methoxytetralone-2 and methylation with methyl iodide affords the methyl ketone 41.27. Condensation of the latter (41.27) with methyl vinyl ketone and conversion to the phenol generates 41.28 which undergoes partial catalytic reduction to the cis-decalin 41.29. Acetylation and further catalytic hydrogenation then lead to a mixture of all four possible racemic monoacetates, which on oxidation, equilibration, and hydrolysis of the acetate provides the desired diasteromeric trans B/C ring-fused racemates. Separation and resolution of this mixture through the brucine succinates then result in isolation of the desired ketone 41.30. Methylation of 41.30 by condensation with ethyl formate in the presence of sodium methoxide to 41.31 followed by reaction with methyl iodide to 41.32 and cleavage of

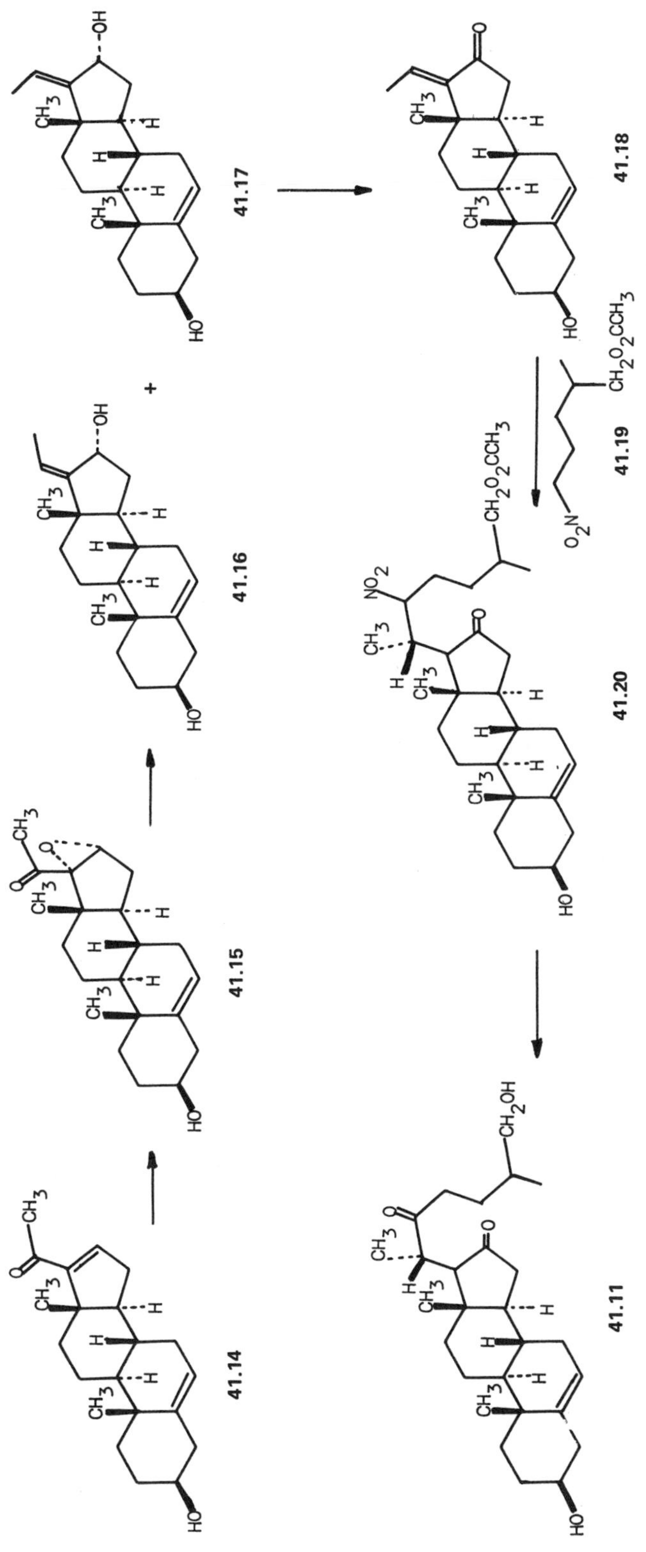

SCHEME 41.2 The synthesis of kryptogenin from dehydropregnenolone. (After Reference 452.)

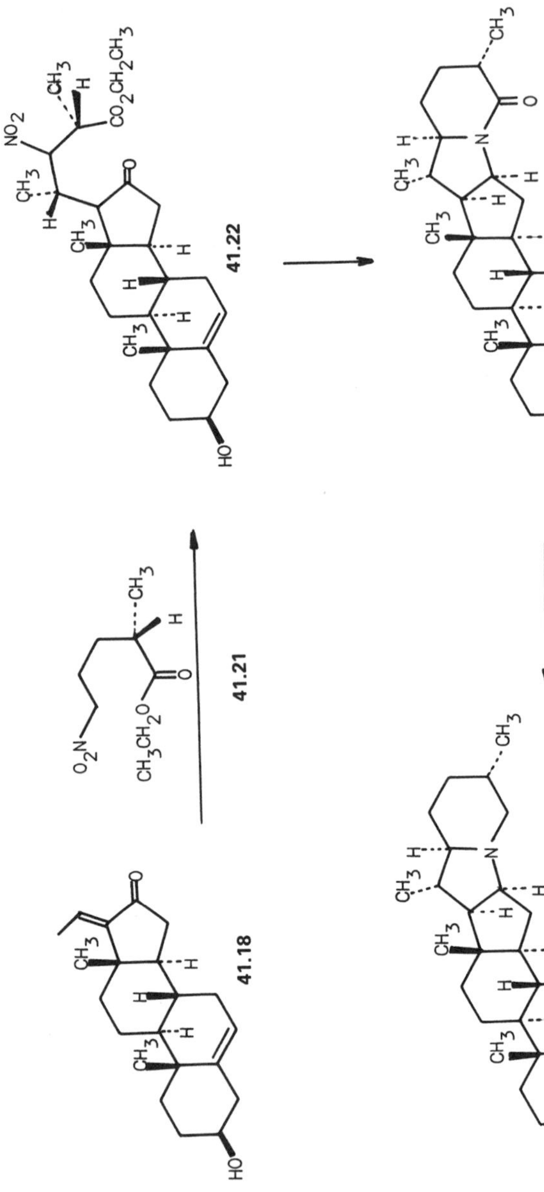

41.18

41.21

41.22

41.23

41.3

SCHEME 41.3 The synthesis of solanidine from a dehydropregnenolone derivative. (After Reference 453.)

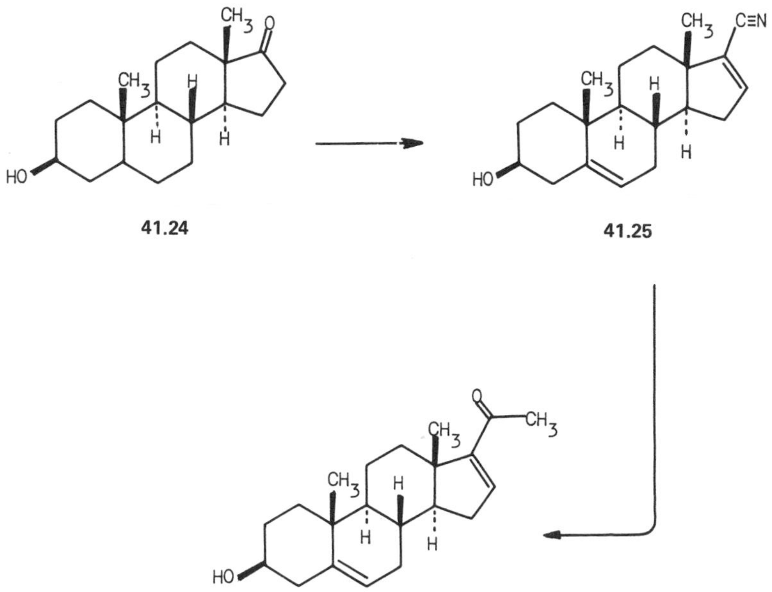

41.24

41.25

41.14

SCHEME 41.4 The conversion of androstenolone into dehydropregnenolone. (After Reference 454.)

the aldehyde with sodium carbonate affords a mixture of optically active ketones, one of which, on oxidation, yields the diketone 41.33. This diketone (41.33) is known as the Reich diketone.

When the Reich diketone (41.33) is subjected to bromination and dehydrohalogenation, the enone 41.34 is generated, and formation of the enol acetate 41.35 from this ketone and reduction with potassium amide in liquid ammonia with the addition of ammonium chloride provide the ketone 41.36 in which the remaining double bond is no longer in conjugation with the carbonyl. Then, reduction of both carbonyl groups with lithium aluminum hydride and alkylation of the less hindered hydroxyl with trityl chloride lead to 41.37, which on oxidation and hydrolytic removal of the triphenylmethyl group, affords the hydroxyketone 41.38, known as the Koster and Logemann (K-L) ketone. Now, benzoylation of the K-L ketone (41.38) and treatment with sodium triphenyl methide followed by carbon dioxide carbonation and esterification with diazomethane provides the ester 41.39. When 41.39 is allowed to undergo the Reformatsky reaction with methyl bromoacetate, dehydration, reduction, and hydrolysis followed by acetylation, the half acid ester acetate 41.40 results. The next higher homolog of 41.40, i.e., 41.41, prepared from 41.40 via the Arndt-Eistert process, when subjected to pyrolysis in the presence of acetic anhydride, results in cyclization to epiandrosterone acetate (41.42). Epiandrosterone acetate (41.42) can, by a Walden inversion, be converted into androsterone acetate (41.43) as well as into the 2-bromoketone (41.44) by bromination of the corresponding dione (androstanedione) which is obtained by hydrolysis and oxidation of 41.42 or 41.43. The bromination is carried out with one equivalent of bromine in acetic acid [457] and elimination by heating in the presence of collidine provides Δ^4-androstenedione (41.45; [458]).*

---

*It is interesting to speculate that this reaction may be occurring via a Favorskii-type intermediate.

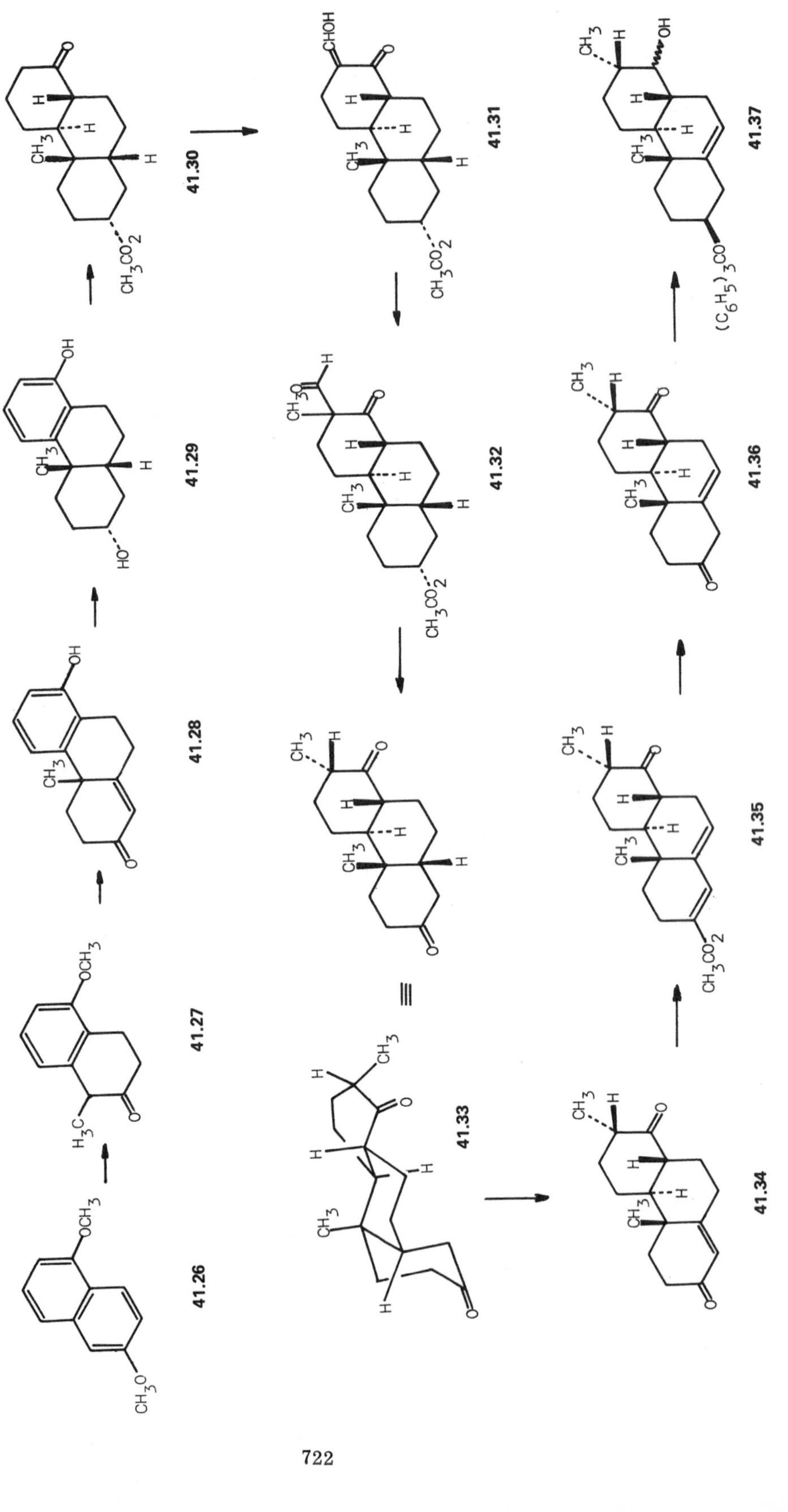

41.26

41.27

41.28

41.29

41.30

41.31

41.32

41.33

41.34

41.35

41.36

41.37

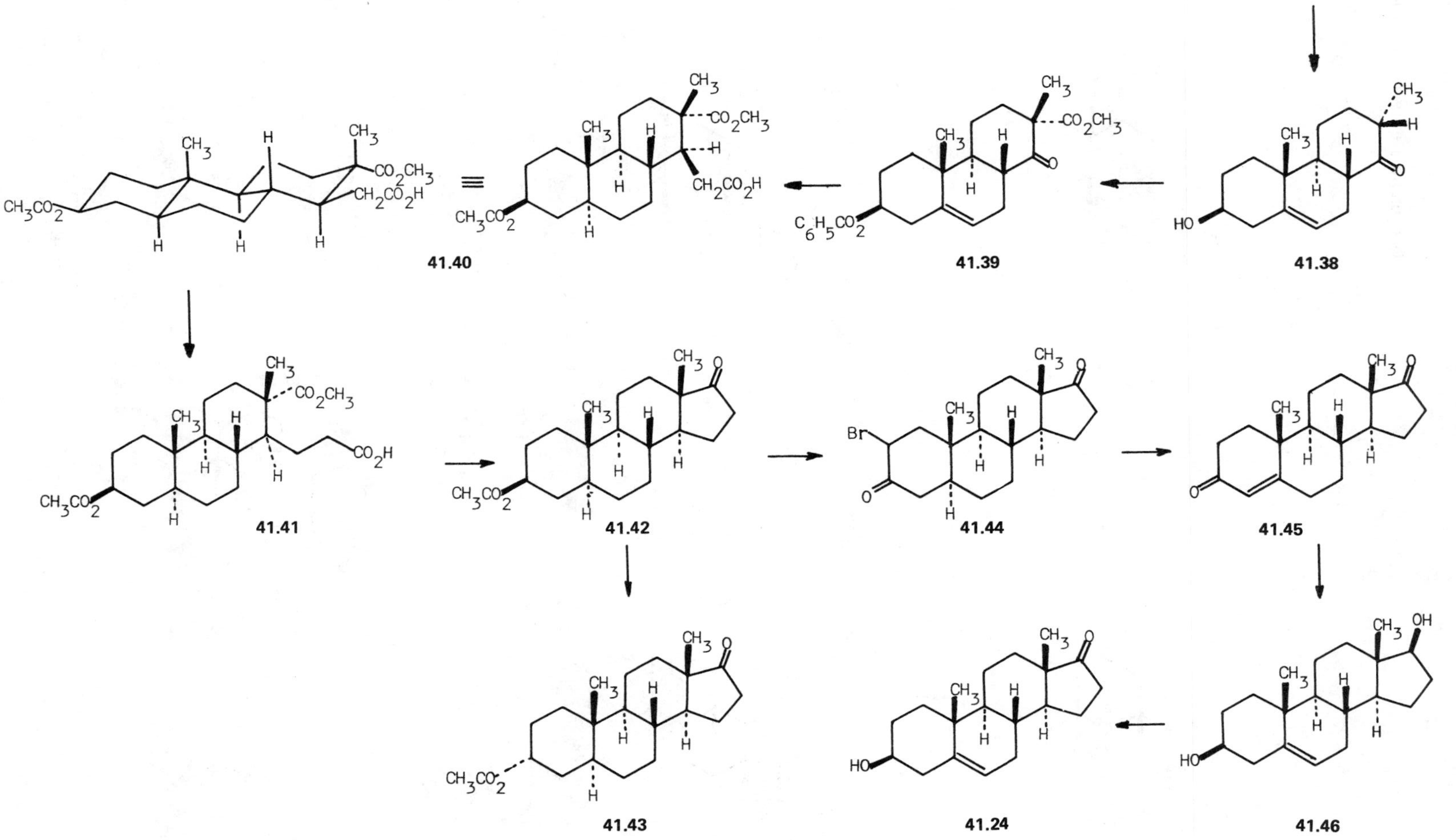

SCHEME 41.5 The total synthesis of androstenolone. (After References 455 and 460.)

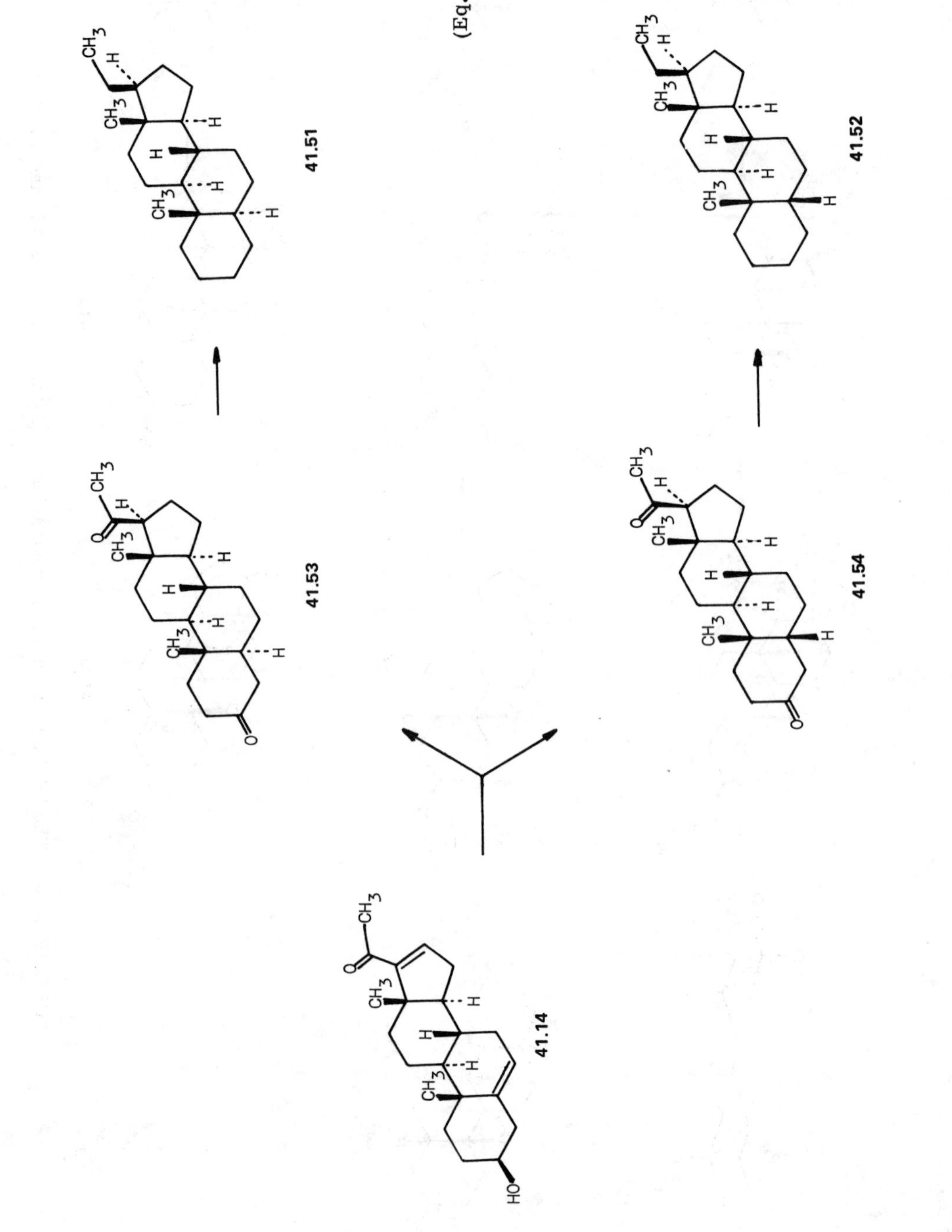

(Eq. 41.4)

41.51

41.52

41.53

41.54

41.14

As above, when the enolacetate of androstenedione (41.45) is treated with potassium amide in liquid ammonia, the solution acidified with ammonium chloride, and reduction of the resultant hydroxy ketone with lithium aluminum hydride carried out, androst-4-en-3β:17β-diol (41.46) results. This time, formation of the trityl ether of the less hindered 3β-hydroxyl, oxidation, and hydrolytic deprotection of the hydroxyl result in the formation of androstenolone (41.24).

II. CONESSINE

The oxygen-free base conessine (41.4), $C_{24}H_{40}N_2$, mp 125°C, $[\alpha]_D$ -1.9° (chloroform) was first isolated from Wrightia antidysenterica J. Grah. in 1858 (the name of the plant has since been changed to Holarrhena antidysenterica Wall.). As implied by the name of the species, extracts from this and related plants have found some use in the treatment of dysentery.

Both nitrogen atoms of conessine (41.4) were recognized as being present as tertiary amines, and (Scheme 41.6) a bisquaternary salt (41.47) can be formed which undergoes thermal decomposition to trimethylamine and a compound containing a $\Delta^{3,5}$-diene system (UV λ_{max} 235 nm, apoconessine, 41.48). Since a tertiary amine is still present in 41.48, this suggests that one of the original nitrogen atoms was present as a dimethylamino moiety. Additionally, since conessine (41.4) itself can be hydrogenated to a dihydro base, one double bond is present and the alkaloid must be pentacyclic.

Thermal selenium-catalyzed decomposition of conessine (41.4) gives rise to γ-ethylcyclopentophenanthrene (41.49), suggesting a relationship to steroids [459], and this is confirmed by Emde (sodium amalgam) degradation of apoconessine (41.48; which, of course, will not undergo another Hofmann degradation) to a triunsaturated hydrocarbon (41.50) that yields a mixture of 5α- and 5β-pregnane (41.51 and 41.52), respectively) on hydrogenation. [Although historically imperfect, it should be clear to the reader that e.g., dehydropregnenolone (41.14; Scheme 41.4) can, by catalytic hydrogenation and then oxidation of the hydroxyl function, be converted to the 5α- and 5β-pregnanediones (41.53 and 41.54, respectively), and then Clemmensen reduction provides the 5α and 5β-pregnanes (41.51 and 41.52; Equation 41.4)].

While this conversion establishes the gross structure of the conessine (41.4) backbone,* it was, at the time, still necessary to assume the original double bond was between C_5 and C_6 and that the dimethylamino group was at C_3. That these assignments were indeed correct was established as follows [460]. One of the alkaloids occurring along with conessine (41.4) is isoconessimine (41.55). This base can be generated from conessine (41.4) by von Braun degradation (cyanogen bromide demethylation) of conessine (41.4), and methylation of isoconessimine (41.55) by formaldehyde and formic acid (Clarke-Eschweiler) regenerates conessine (41.4; Equation 41.5). Now, when N-acetylisoconessimine (41.56) is subjected to

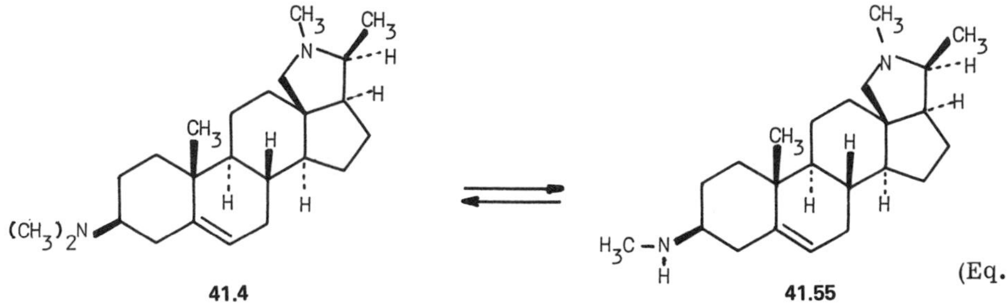

41.4 ⇌ 41.55 (Eq. 41.5)

*Indeed, at those chiral carbons unaffected by the transformations involved, even the absolute configuration is therefore known!)

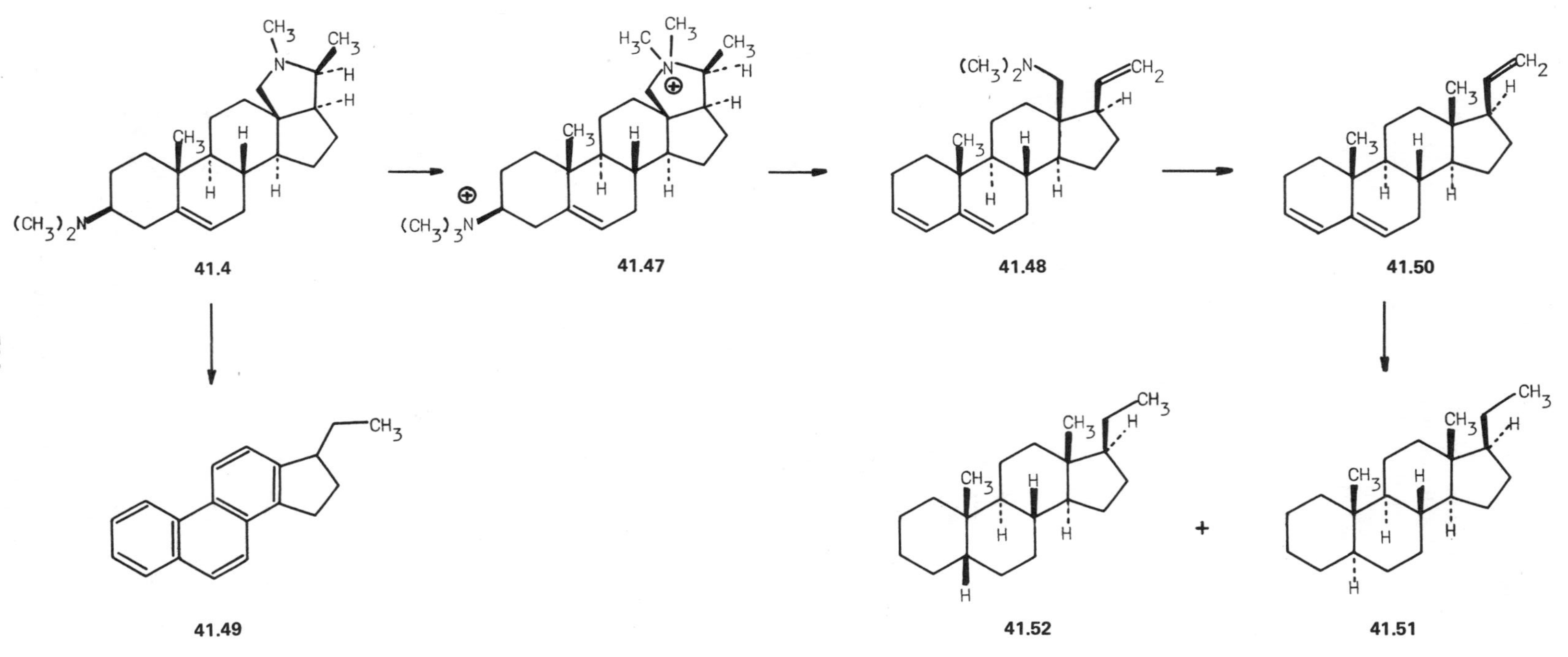

SCHEME 41.6 The degradation of conessine.

Hofmann degradation and the methochloride of the resulting aminoalkene reduced by the Emde method, $\Delta^{5,20}$-3β-N-methyl-N-acetylaminopregnadiene (41.57; Equation 41.6) is generated.

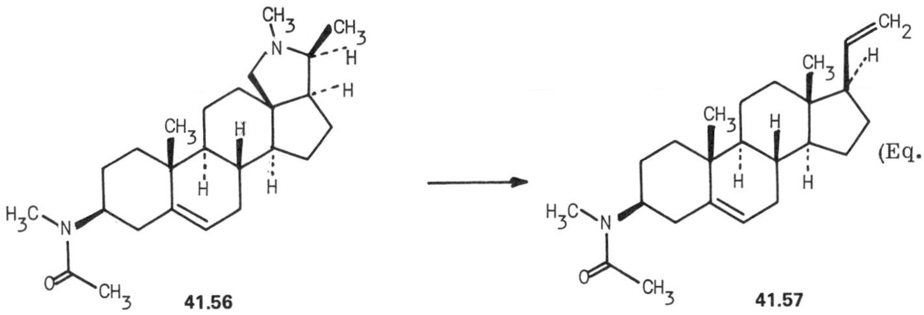

41.56 → 41.57 (Eq. 41.6)

The same material (41.57) is produced from the known 3β-hydroxypregnadiene (41.58) via displacement of the tosylate by methylamine and acetylation of the secondary amine which results (Equation 41.7). Finally, in one of the most interesting reactions (Scheme 41.7)

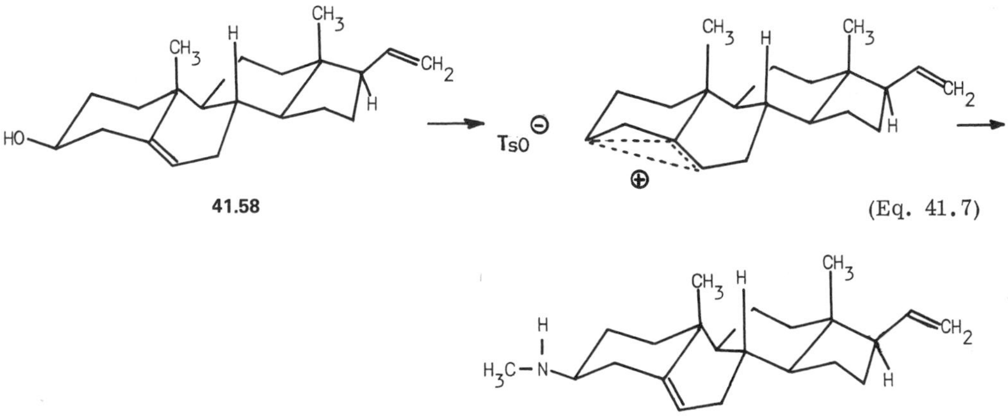

41.58 (Eq. 41.7)

observed in this family of materials [461], it is found that when conessine dimethiodide (41.47) is heated with potassium hydroxide in ethylene glycol, an isomer of conessine (41.4), i.e., heteroconessine (41.59), is formed. This presumably occurs via the alkene 41.60 which recyclizes to 41.59 and then undergoes demethylation, although, a priori, proton abstraction to the ylid 41.61 and reprotonation cannot be excluded. Along the same lines, if the Hofmann degradation of 41.47 is "carefully controlled," conessimethine (41.62; Equation 41.8) occurs, and this too can be recyclized to heteroconessine (41.59). That this process occurs establishes that the nitrogen is bonded to C_{20} rather than C_{21}, and the preferential formation of heteroconessine (41.59) is explained on the basis of a cis to trans isomerization which puts the C_{20} methyl and the C_{16} methylene further apart in 41.59 than in 41.4. Somewhat later [462] it was shown that careful oxidation (Equation 41.9) of dihydroconessine (41.63) and dihydroheteroconessine (41.64) yields the same imminum perchlorate salt (presumably 41.65) and that the pyrrolidine 41.66 results on liberation of the nitrogenous product from the perchlorate salt with base.

Several syntheses of conessine (41.4) are available. In one of them (Scheme 41.8; [463]), 5-methyl-6-methoxy-1-tetralone (41.67) is allowed to condense with dimethyl carbonate and

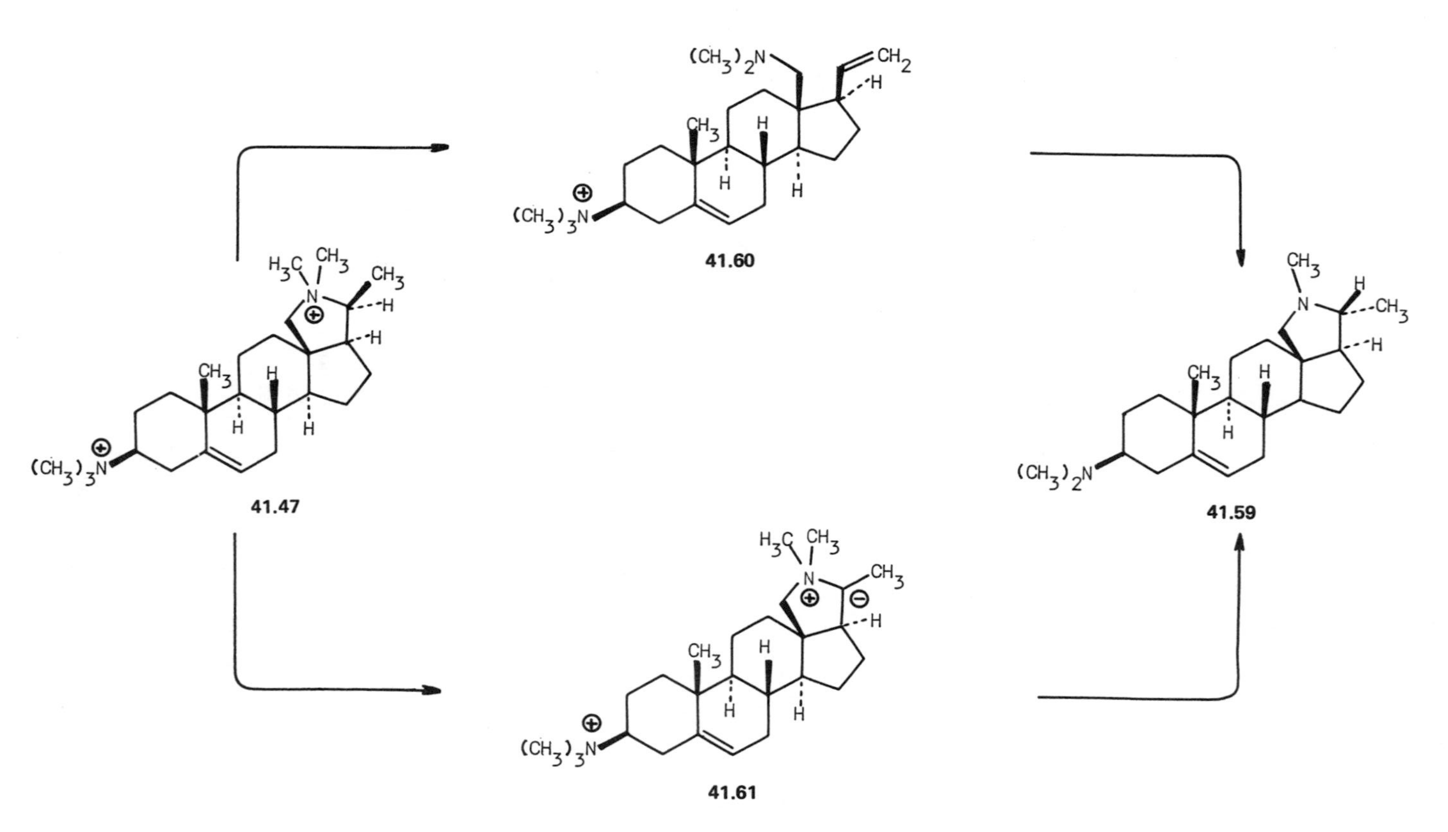

SCHEME 41.7 Possible pathways for the isomerization of the conessine system to the heteroconessine system. (After Reference 461.)

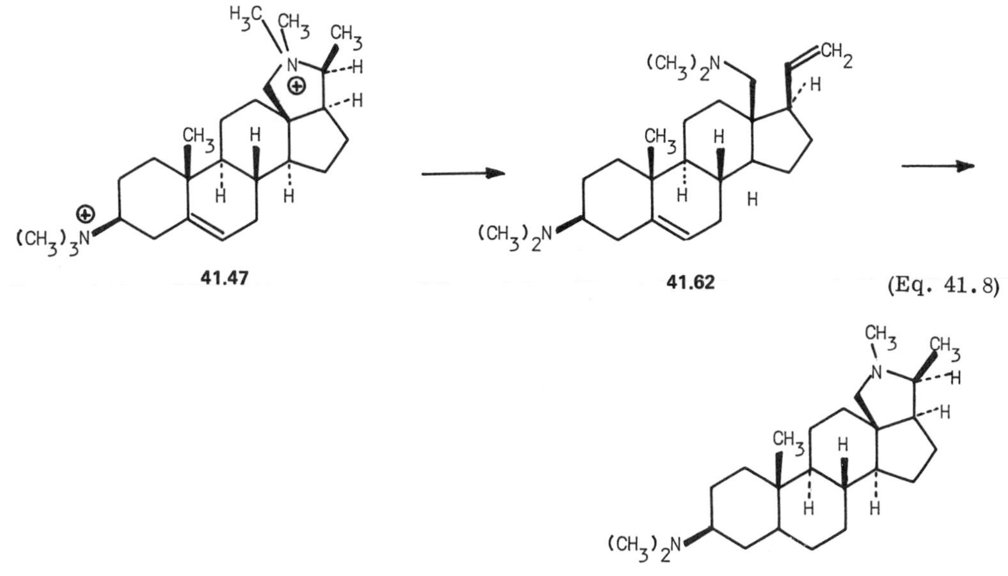

41.47

41.62

(Eq. 41.8)

41.59

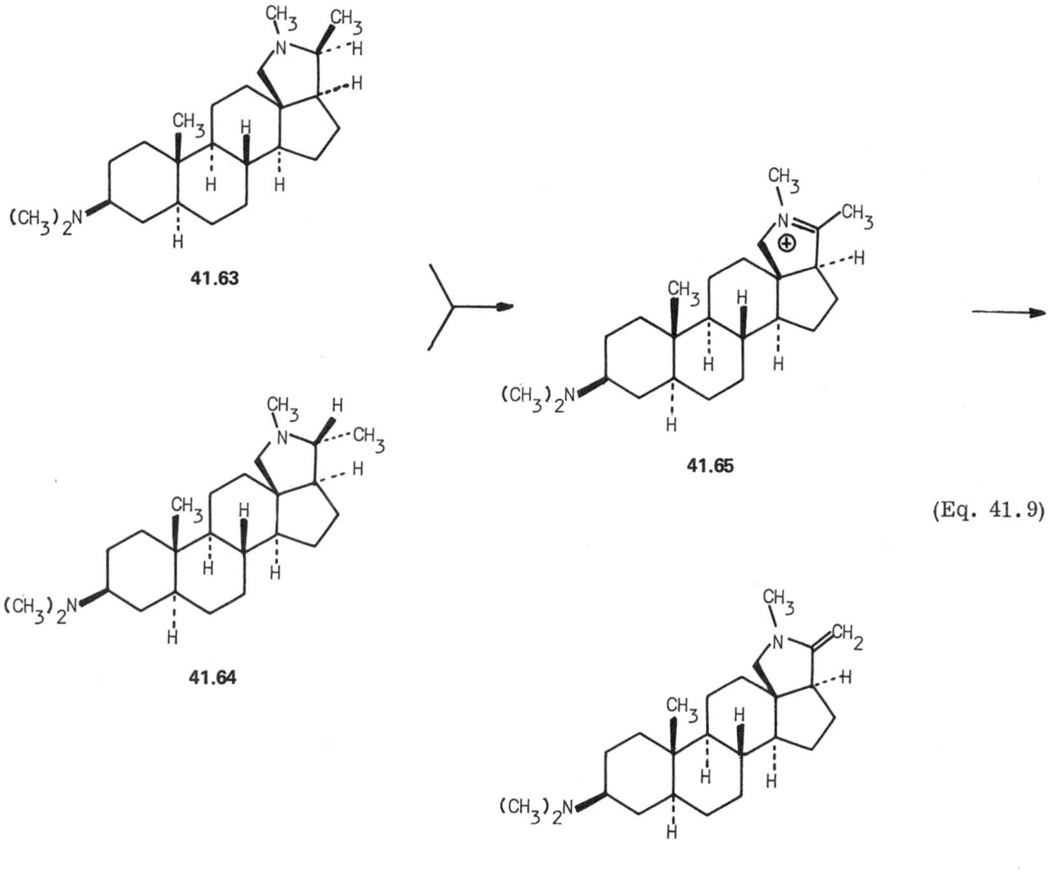

41.63

41.64

41.65

(Eq. 41.9)

41.66

41.67 → 41.68 → 41.69 → 41.70 → 41.71 → 41.72 → 41.73 → 41.74 → 41.75 → 41.76

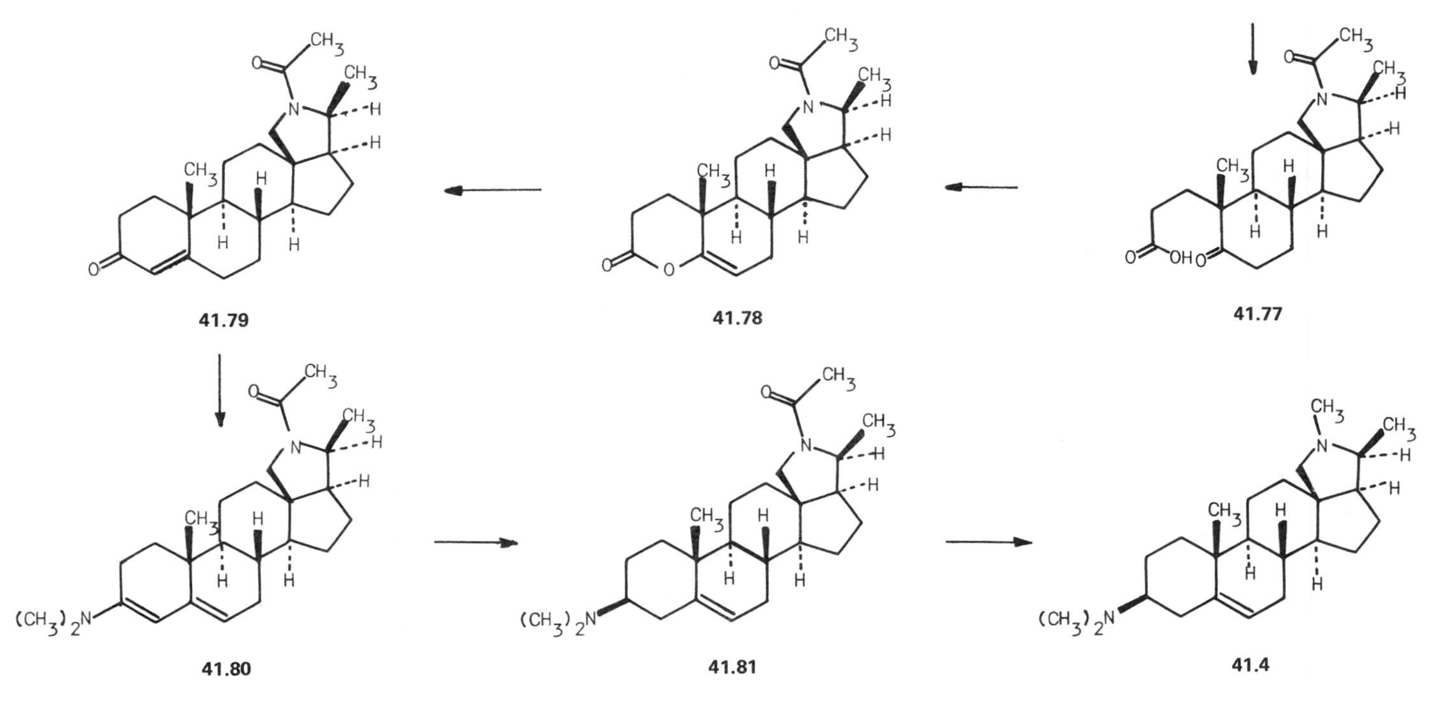

SCHEME 41.8 The total synthesis of conessine. (After Reference 463.)

and then methyl isopropenyl ketone to yield the tricyclic enone 41.68. Then, catalytic hydrogenation and treatment with phosphorus oxychloride followed by base-catalyzed elimination generate the alkene 41.69. Ozonolysis of 41.69 to the corresponding ketoaldehyde and recyclization (acetic acid-hydrogen chloride) afford 41.70, and after the ketone is protected as an ethylene ketal and the carbomethoxy group and double bond reduced, the tosylate of the resulting hydroxyketone (41.71) reacts with hydroxylamine to form the nitrone 41.72. Reduction of the nitrone (41.72) to the corresponding amine with hydrogen on a rhodium catalyst followed by treatment with hydrogen bromide to cleave the O-methyl ether, selective acetylation, hydrogenation (ruthenium oxide, 2000 psi), and oxidation provides the ketone 41.73. Then, base-catalyzed addition of acrylonitrile to 41.73 and hydrolysis generates the acid 41.74, and bromination, hydrolysis to the acyloin, and gentle oxidation provide 41.75. Methylation of 41.75 generates 41.76, and now base treatment allows inversion to the required B/C trans ring system. Then, catalytic hydrogenation followed by reduction with calcium in liquid ammonia, reacetylation on nitrogen and oxidation of the anion of the hydroxy acid with chromic anhydride in pyridine provides the keto acid 41.77. With acetic anhydride and sodium acetate, the enol lactone 41.78 is then formed, and on treatment with methyl magnesium iodide the lactone 41.78 undergoes addition and (presumably via the corresponding diketone) aldol cyclization to the α,β-unsaturated ketone 41.79. On treatment with dimethylamine in the presence of a dehydrating agent and a trace of acid, 41.79 generates the enamine 41.80, which with sodium borohydride provides the alkene 41.81. Deacetylation with calcium in liquid ammonia and Eschweiler-Clarke methylation then yield racemic conessine (41.4 and its mirror image).

III. CYCLOBUXINE-D

The leaves of Buxus sempervirens L. have apparently found use medicinally for many disorders, from skin and venereal diseases to treatment of malaria and tuberculosis. The acetone-insoluble portion of the extract providing the strong bases yields cyclobuxine-D (41.5), $C_{25}H_{42}N_2O$, mp 247°C, $[\alpha]_D^{23}$ +98° (chloroform), as the major alkaloid [464, 465].

The proton magnetic resonance (PMR) spectrum of cyclobuxine-D (41.5) is quite informative. Thus, two different terminal methylene protons are distinguished at δ 4.80 and δ 4.57 ppm (2H); a proton on a carbon bearing an hydroxyl group and flanked by two protons of a methylene and one methine proton is observed at δ 4.71 ppm (1H), J_{AX} = 4.7 Hz, J_{MX} = 9.5 Hz; two N-methyl groups at δ 3.47 (3H) and δ 3.43 ppm (3H) are seen; two tertiary C-methyls at δ 1.13 ppm (3H) and δ 0.97 ppm (3H) are found; and one secondary C-methyl at δ 2.08 ppm (3H), doublet, J = 6 Hz and a cyclopropyl methylene at δ 0.05 ppm and δ 0.28 ppm (2H), J_{AB} = 4 Hz can be detected. The pmr spectra of the N,N'-dimethyl derivative of cyclobuxine-D (41.82) as well as those of O,N,N'-triacetylcyclobuxine-D (41.83) and O-acetyl-N,N'-dimethylcyclobuxine-D are in concert with that of the parent alkaloid (41.5) as delineated

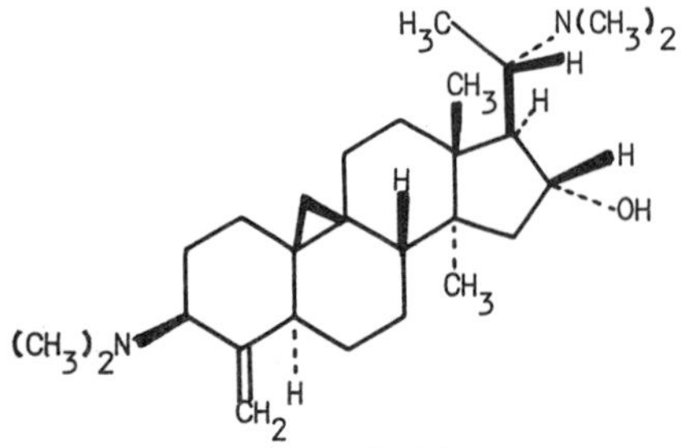

41.82

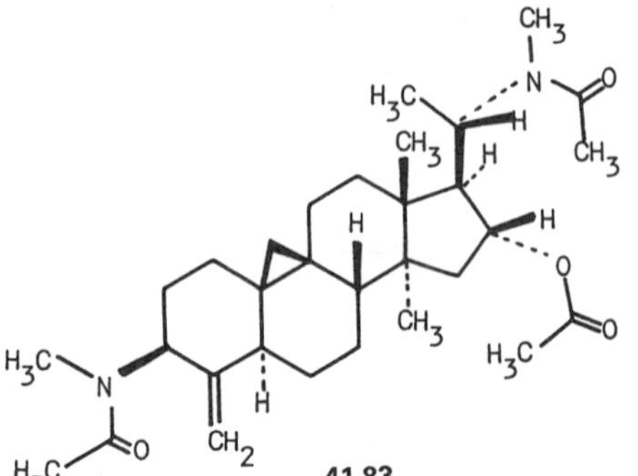

41.83

above, and the changes affected in the spectra on their formation in line with those expected. Additionally, cyclobuxine-D (41.5), on hydrogenation over a platinum catalyst in ethanolic acetic acid (Equation 41.10) provides a dihydro derivative (dihydrocyclobuxine-D, 41.84),

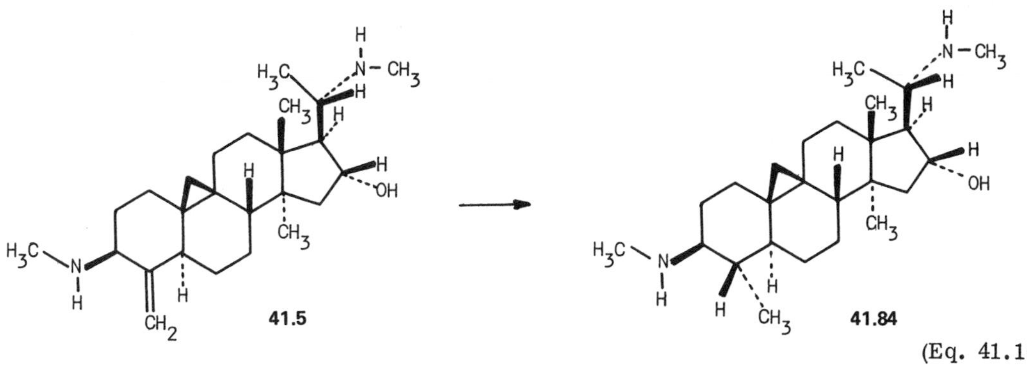

41.5 → 41.84

(Eq. 41.10)

the PMR spectrum of which no longer contains the resonances assigned to the terminal methylene functionality (see above) but has, instead, a "new" (additional) secondary C-methyl, i.e., a secondary C-methyl (3H), doublet, J = 7 Hz at δ 1.09 ppm and a second such group (3H), doublet, J = 7 Hz at δ 0.78 ppm.

On selenium-catalyzed dehydrogenation (Equation 41.11) of cyclobuxine-D (41.5), a complex mixture of anthracenes (e.g., 41.85) and phenanthrenes (e.g., 41.86 and 41.87) is obtained, and the shown structures have been suggested on the basis of spectral properties.

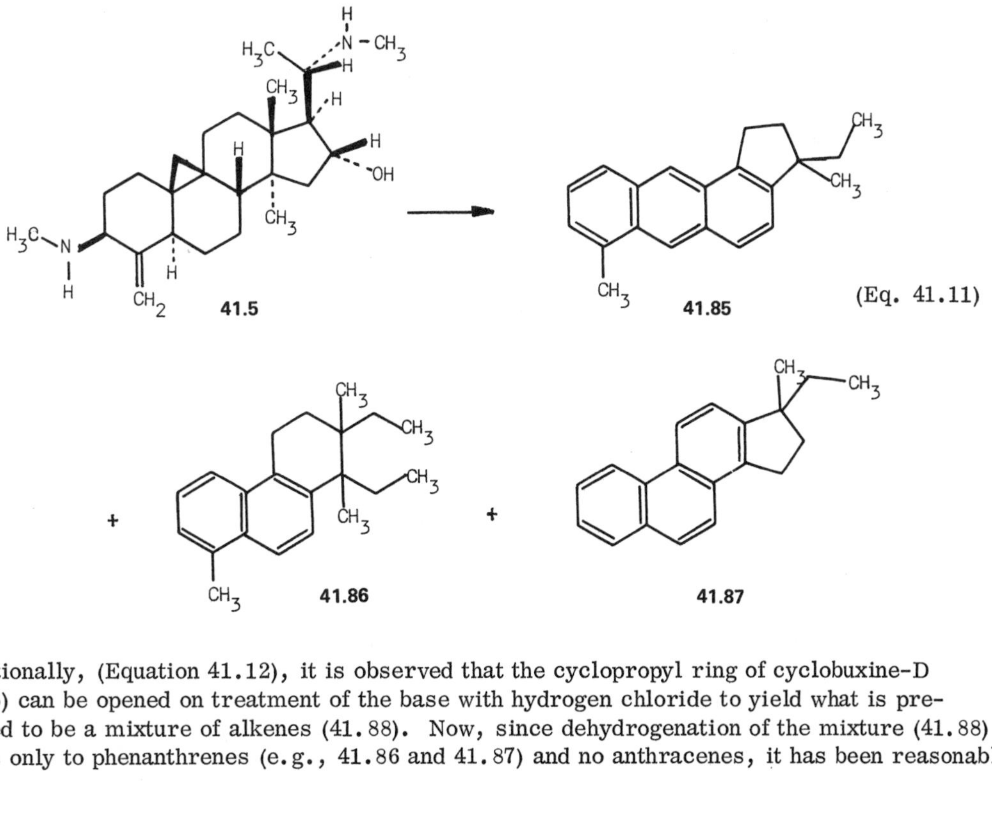

41.5 → 41.85 + 41.86 + 41.87

(Eq. 41.11)

Additionally, (Equation 41.12), it is observed that the cyclopropyl ring of cyclobuxine-D (41.5) can be opened on treatment of the base with hydrogen chloride to yield what is presumed to be a mixture of alkenes (41.88). Now, since dehydrogenation of the mixture (41.88) leads only to phenanthrenes (e.g., 41.86 and 41.87) and no anthracenes, it has been reasonably

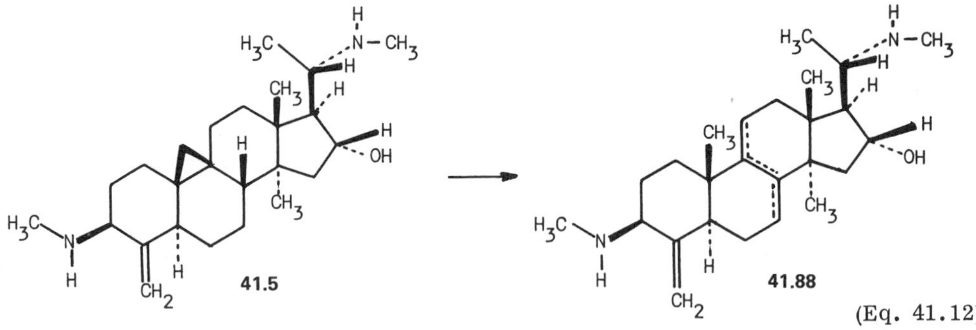

41.5 → 41.88

(Eq. 41.12)

suggested that it is the presence of the three-membered ring which provides a route to the anthracenes.

When N,N'-dimethylcyclobuxine-D (41.82) is treated with methyl iodide (Scheme 41.9) in warm chloroform, a small amount of dimethiodide (41.89) and a larger amount of monomethiodide (41.90) result. Hofmann degradation of the latter yields a conjugated alkene (41.91; λ_{max} 229 nm, log ϵ = 4.22, ethanol) providing evidence that one of the N-methyl functionalities is at C_3. Additionally (Scheme 41.10), treatment of cyclobuxine-D (41.5) with p-nitrobenzylchloroformate provides a di-p-nitrobenzylcarbamate (41.92), and ozonolysis of this derivative followed by hydrogenolysis of the resulting ketocarbamate results in formation of an unstable α-aminoketone (41.93). Oxidation and hydrolysis of the latter then provides a diosphenol (41.94) which is more highly conjugated (λ_{max} 296.5 nm, log ϵ = 3.95) than a typical steroidal diosphenol (e.g., the diosphenol from cholesterol, 41.95 has λ_{max} 278 nm), which suggests that the cyclopropane ring identified as being present in the parent base (41.5) is in conjugation with the α,β-unsaturated carbonyl moiety in the diosphenol (41.94). Support for this contention arises from the observation that the corresponding "decyclized" diosphenol (presumably 41.96) exhibits the normal spectrum (λ_{max} 278 nm).

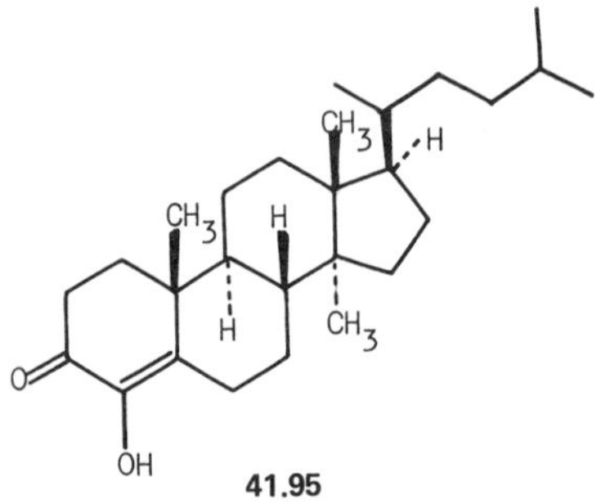

41.95 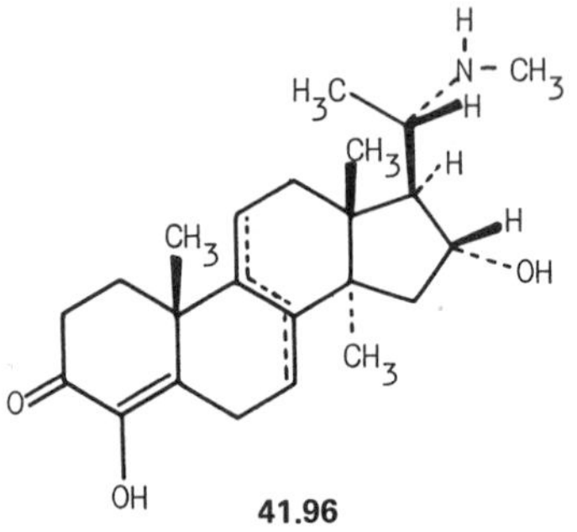 41.96

Finally, along these lines (Equation 41.13), when the di-p-nitrobenzylcarbamate of dihydrocyclobuxine-D (41.97) is oxidized with chromic anhydride in acetic acid, a ketone (in a five-membered ring, 1730 cm^{-1}) is obtained, and hydrolysis of this derivative results in the formation of a mixture of (E)- and (Z)-des-N'-16-dehydrodihydrocyclobuxines (41.98 and 41.99, respectively), presumably via methylamine elimination. Since it is presumed that this mixture of isomers can only result if the second N-methylamino function (it will be remembered that the other N-methylamino group is at C_3) is β to the carbonyl, the gross structure of cyclobuxine-D (41.5) is established.

All the above information was then confirmed and the absolute configuration of cyclobuxine-D (41.5) established by correlating the alkaloid with cycloeucalenol (41.100), whose

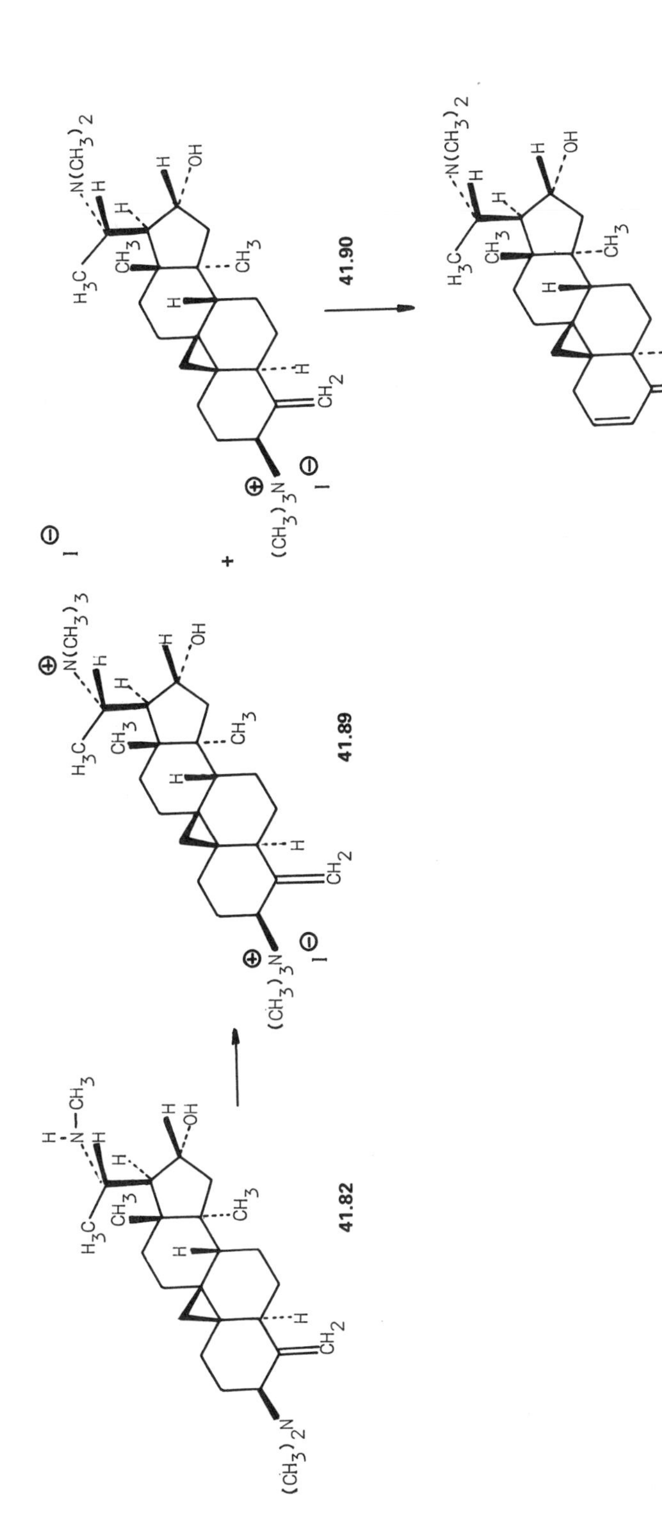

SCHEME 41.9 The formation and Hofmann degradation of the monomethiodide of N,N'-dimethylcyclobuxine-D.

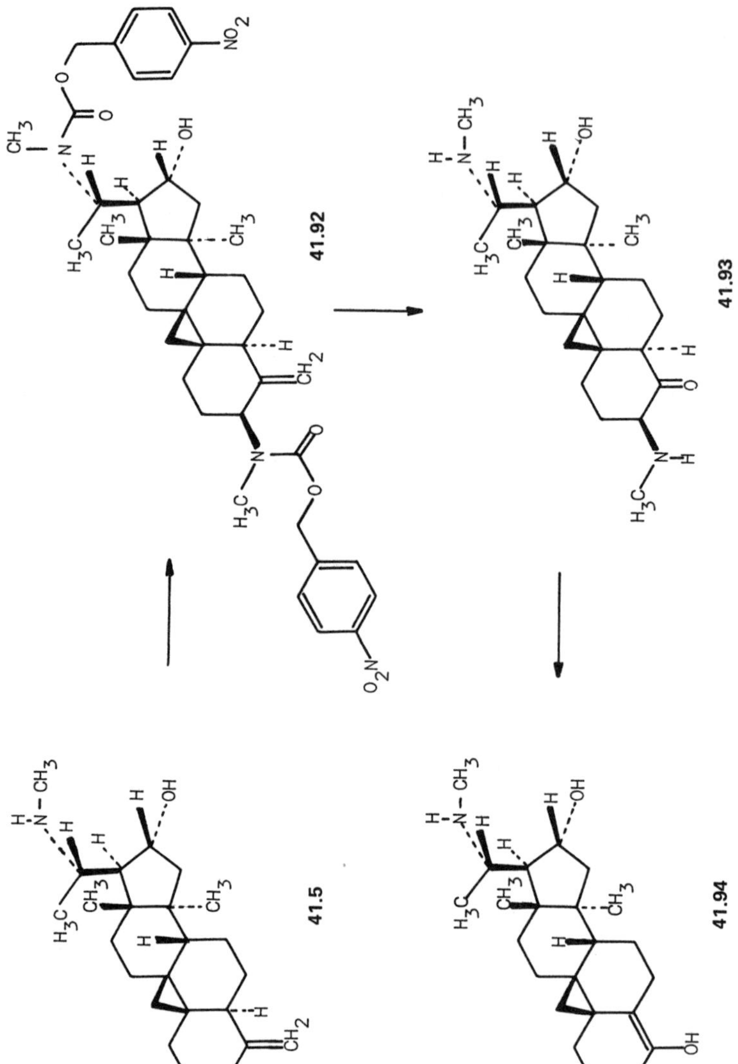

41.5

41.92

41.93

41.94

SCHEME 41.10 Degradation of cyclobuxine-D to a diosphenol.

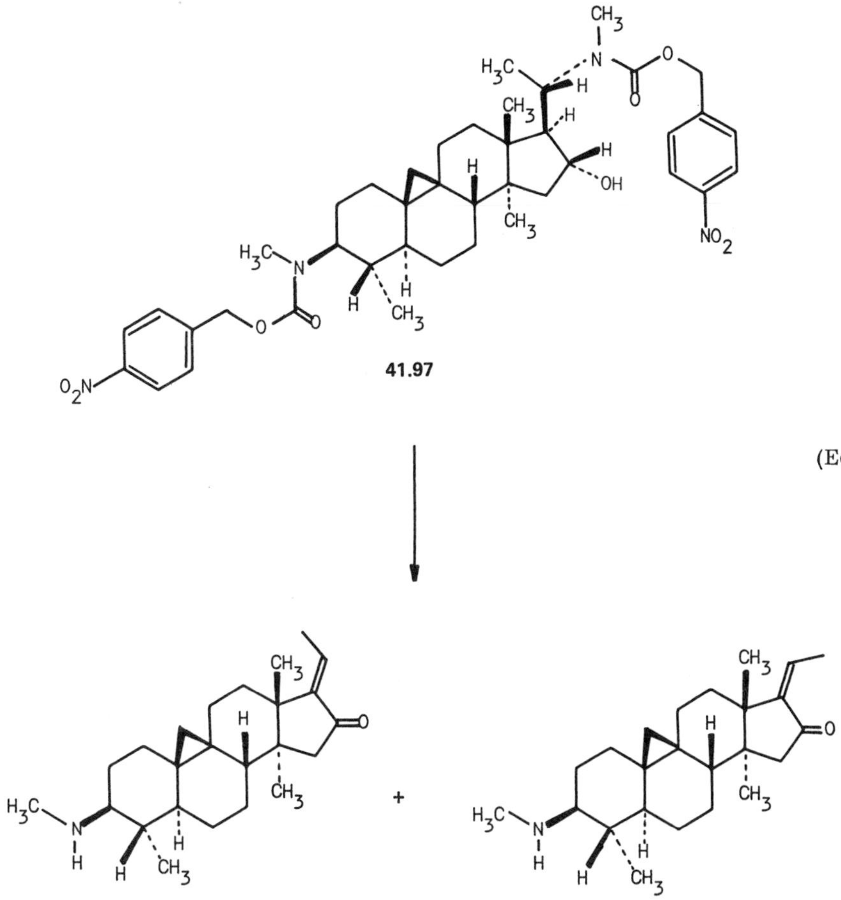

41.97

(Eq. 41.13)

41.98 + 41.99

absolute configuration had been determined through interrelationship with lanosterol (41.101; [466, 467]). First ([465]; Scheme 41.11), reaction of dihydrocyclobuxine-D (41.84) with N-chlorosuccinimide provides the corresponding dichloramine, which on treatment with sodium methoxide and hydrolysis of the resulting diimine affords the ketoenone (41.102). Hydrogenation on a palladium catalyst then yields 4α,14α-dimethyl-9β,19-cyclo-5α-pregnane-3,20-dione (41.103). The same dione (41.103) is produced from cycloeucalenol (41.100). Thus (Scheme 41.12), ozonolysis of the acetate of 41.100 yields the ketone 41.104, which on reduction to the alcohol, conversion to the chloride with phosphorus oxychloride, and pyridine-catalyzed elimination generates the alkene 41.105. Ozonolysis, permanganate oxidation of the resulting aldehyde to the corresponding acid, and esterification with diazomethane afford 41.106. Then, the alkene 41.107 is prepared by allowing 41.106 to react with phenyl magnesium bromide, acetylation of the resulting diol, and iodine-catalyzed elimination of the tertiary acetate. Allylic bromination of 41.107 with N-bromosuccinimide and dehydrohalogenation then generate the diene 41.108, which undergoes hydrolysis and oxidation to a mixture of the ketoalcohol 41.109 and the dione 41.103. The configurations of the four remaining asymmetric centers in cyclobuxine-D (41.5) were established by ORD comparisons to known compounds. Thus (Equation 41.14), the ozonolysis product 41.110 of O,N,N'-triacetylcyclobuxine-D (41.83) has an ORD curve similar to that of cholestan-4-one (41.111), which is inconsistent with an α-3-acetylmethylamino group since a large negative contribution to the

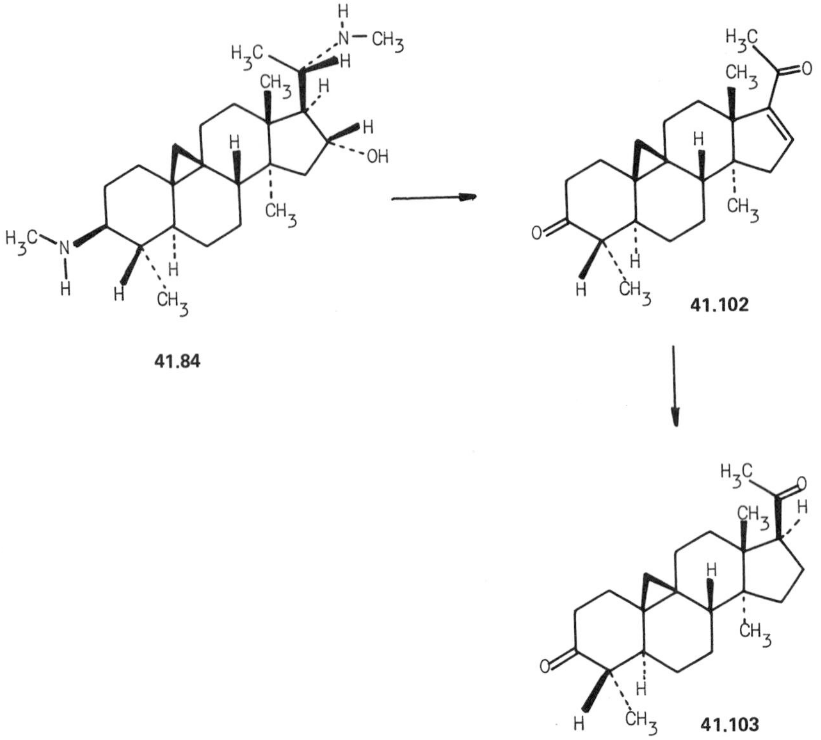

41.84

41.102

41.103

SCHEME 41.11 Degradation of dihydrocyclobuxine-D to 4α,14α-dimethyl-9β-19-5α-pregnane-3,20-dione. (After Reference 465.)

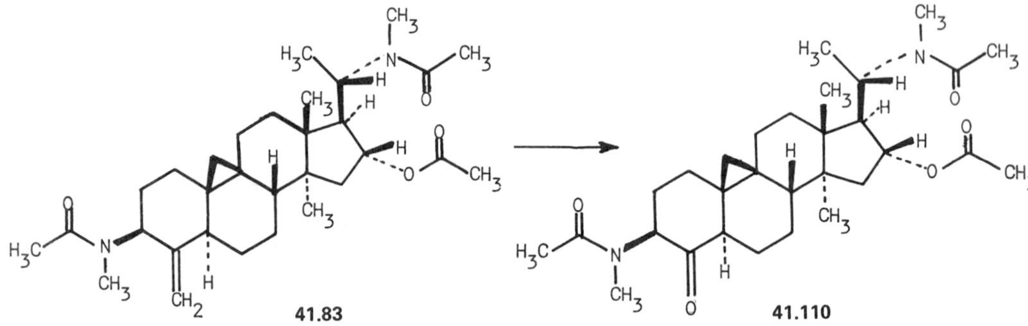

41.83

41.110

(Eq. 41.14)

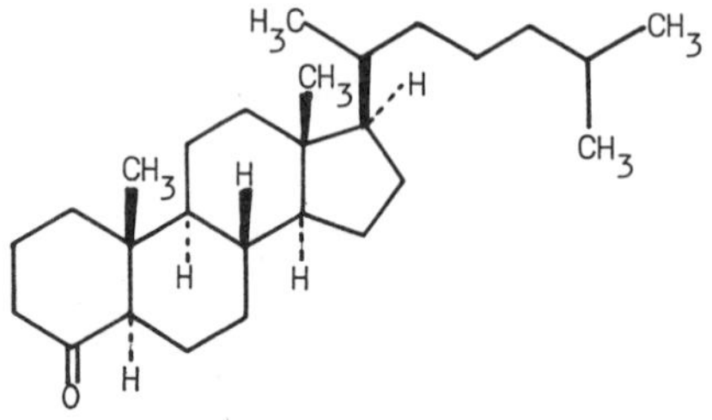

41.111

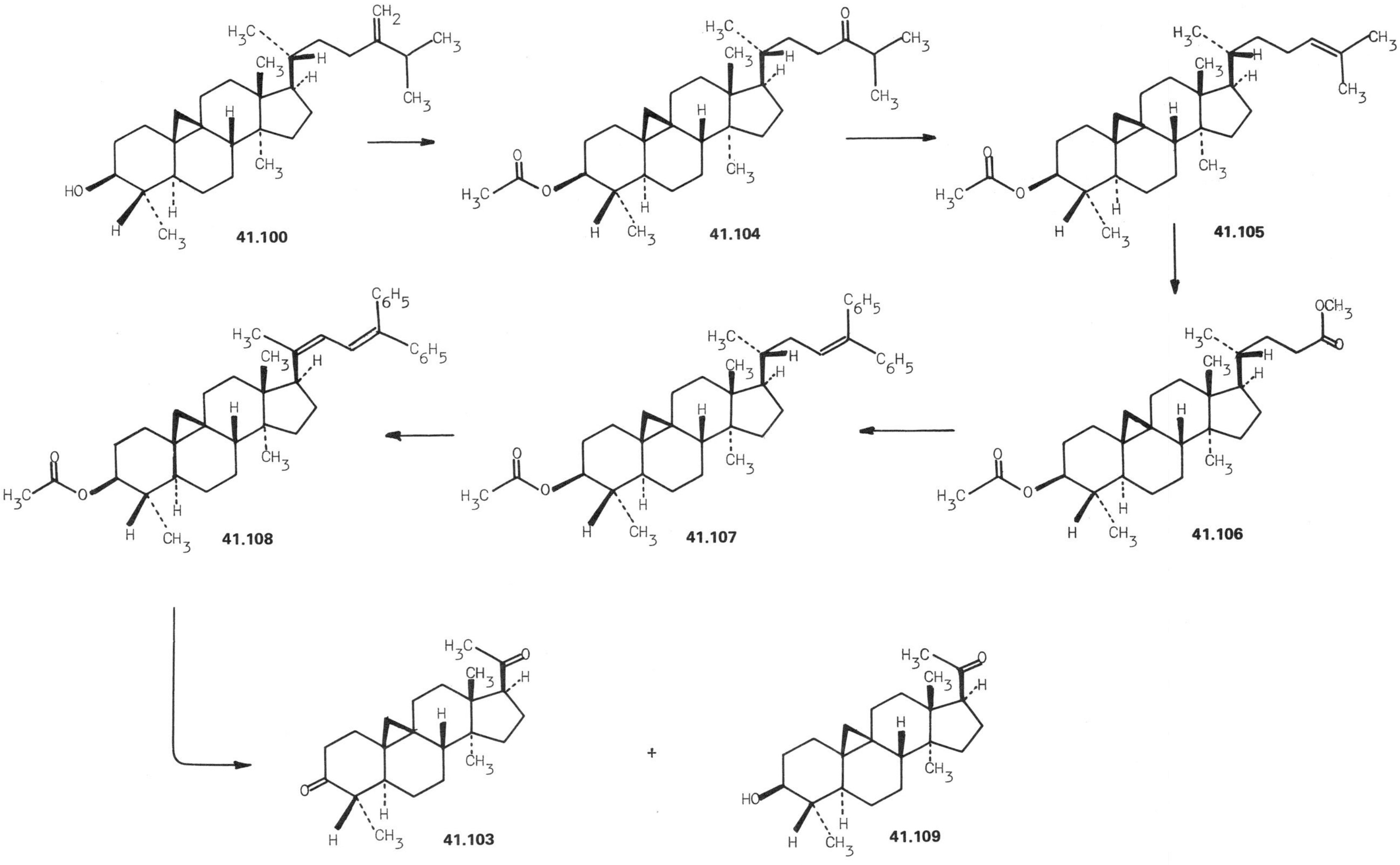

41.100

41.104

41.105

41.106

41.107

41.108

41.103

41.109

SCHEME 41.12 The conversion of cycloeucalenol into 4α-14α-dimethyl-9β,19-cyclo-5α-pregnane-3,20-dione.

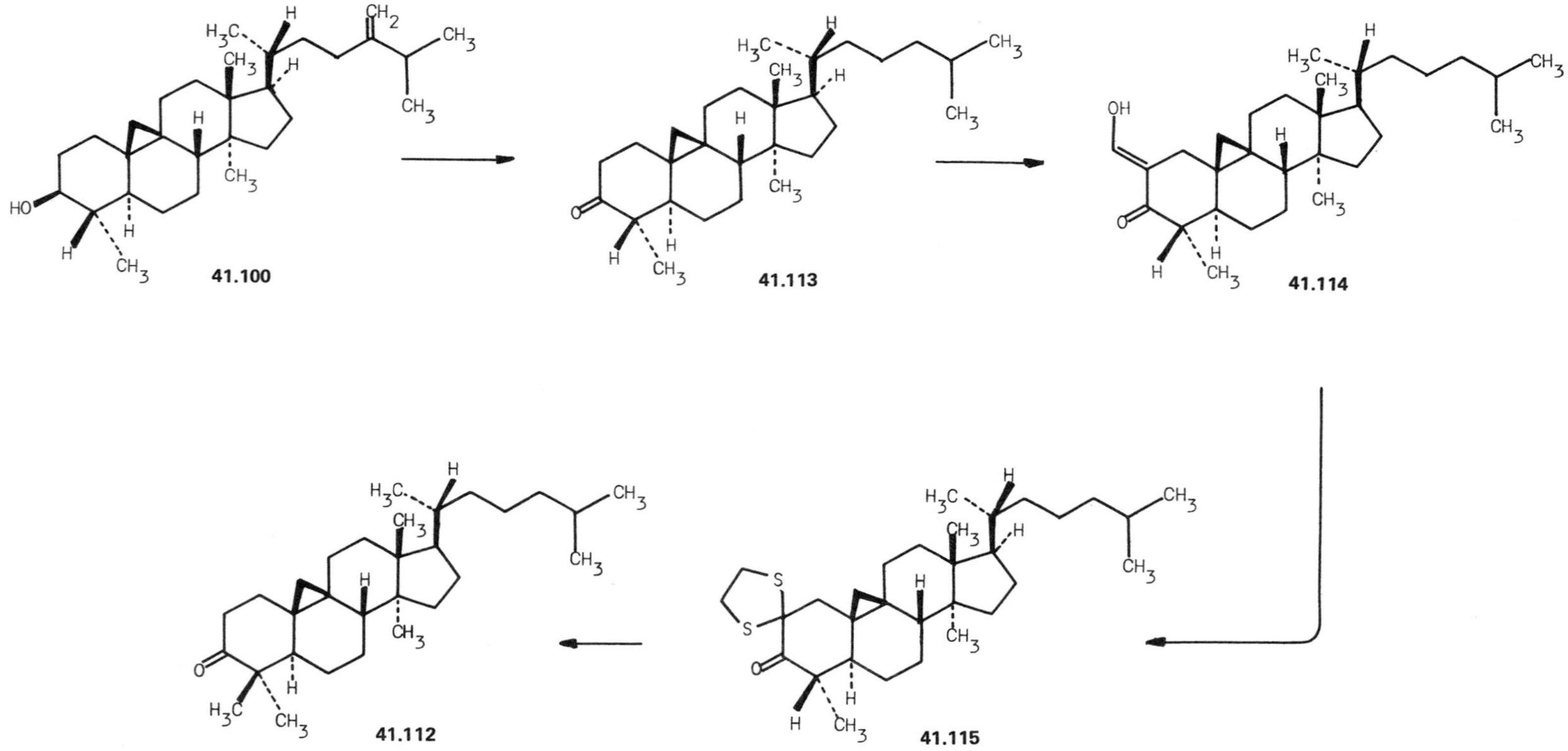

41.100

41.113

41.114

41.115

41.112

SCHEME 41.13 The interrelationship between cycloeucalenol and dihydrocycloartanone. (After Reference 467.)

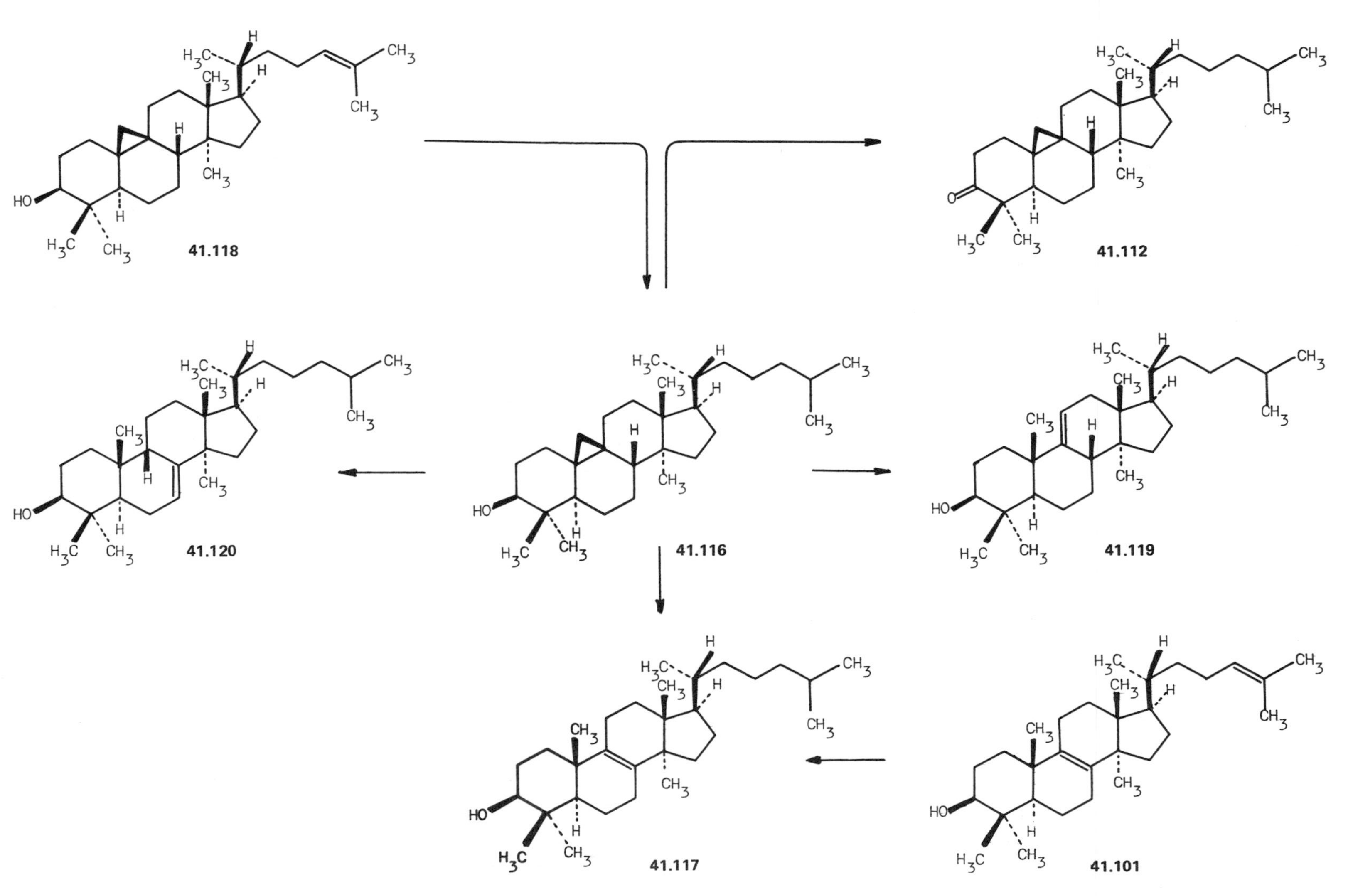

SCHEME 41.14 The interrelationship between dihydrocycloartanone and dihydrolanosterol. (After Reference 466.)

curve would be expected in that case. Hence, the methylamino function in cyclobuxine-D (41.5) is presumed β. Similar reasoning then assigned the hydrogen at C_5 as α while the α-orientation of the hydroxyl group at C_{16} is suggested by examination of N'-derivatives of the alkaloid (41.5) itself and the C_{16} epimer of 41.5 (the latter prepared by borohydride reduction of the corresponding ketone). Finally, the α-configuration is assigned to the methylamino group at C_{20} on the basis of evidence for cis interaction between derivatives of the amine and the hydroxyl group at C_{16}.

Second [467], cycloeucalenol (41.100) was related to dihydrocycloartanone (41.112; Scheme 41.13). Ozonolysis of cycloeucalenol (41.100) to the ketone, Clemmenson reduction of the carbonyl, and oxidation of the hydroxyl, lead to the ketoalkane 41.113. Then, oxidation at C_3* to the dithioketal 41.115, methylation at C_4, and reductive desulfurization provide 41.112.

Finally, dihydrocycloartanone (41.112), the oxidation product of dihydrocycloartanol (41.116), was related to Δ^8-lanostenol (dihydrolanosterol, 41.117) by observing (Scheme 41.14; [466]) that the product of hydrogenation of lanosterol (41.101), i.e., Δ^8-lanostenol (41.117) is identical to one of the products obtained on treatment of dihydrocycloartenol (41.116), the product of hydrogenation of cycloartenol (41.118), with hydrogen chloride (the other products were $\Delta^{9(11)}$-lanostenol, 41.119, and Δ^7-lanostenol, 41.120, as expected).

Cyclobuxine-D (41.5) has not yet been synthesized.

SELECTED READING

K. S. Brown, in Chemistry of the Alkaloids, S. W. Pelletier (ed.). Van Nostrand-Reinhold, New York, 1970, pp. 631 ff.

V. Cerny and F. Sorm, in The Alkaloids, Vol. 9, R. H. F. Manske (ed.). Academic, New York, 1967, pp. 305 ff.

L. F. Fieser and M. Fieser, Steroids. Reinhold, New York, 1950.

G. Habermehl, in The Alkaloids, Vol. 9, R. H. F. Manske (ed.). Academic, New York, 1967, pp. 427 ff.

O. Jeger and V. Prelog, in The Alkaloids, Vol. 7, R. H. F. Manske (ed.). Academic, New York, 1960, pp. 319 ff., 363 ff.

S. M. Kupchan and A. W. By, in The Alkaloids, Vol. 10, R. H. F. Manske (ed.). Academic, New York, 1968, pp. 193 ff.

V. Prelog and O. Jeger, in The Alkaloids, Vol. 7, R. H. F. Manske (ed.). Academic, New York, 1960, pp. 343 ff.

V. Prelog and O. Jeger, in The Alkaloids, Vol. 3, R. H. F. Manske (ed.). Academic, New York, 1953, pp. 247 ff.

Y. Sato, in Chemistry of the Alkaloids, S. W. Pelletier (ed.). Van Nostrand-Reinhold, New York, 1970, pp. 591 ff.

---

*The ketone 41.113 is condensed with ethyl formate to form the hydroxymethylene derivative 41.114, which on reaction with the ditosylate of propane-1,3-dithiol in the presence of potassium acetate yields the dithioketal 41.115.

EPILOGUE

The major objective of what has been discussed in this brief sketch of some of the superficial aspects of the alkaloids is to provide the beginning student with an appreciation of order in structure type and biogenetic reasoning. Hopefully, some chemistry was also learned. Nevertheless, this work is only the barest outline of some of what is known, and entire classes of compounds have been omitted. Additionally, as workers become more sophisticated and compounds more involved, it is found that processes not described herein are operative and surprises are still in store. Thus, the adventure is not over and before concluding, a few exceptions to what you have learned may be in order. Thus, for example, it was long believed that the alkaloid S-(+)-coniine (42.1), from the poison hemlock (Conium maculatum L.), along with a number of related cooccurring bases, were all generated from lysine (42.2), as shown for similarly constituted compounds, e.g., sedridine (42.3; see Part 3): and coniine (42.1) was forced into the Procrustean bed of its "obvious relatives." Experimental labeling studies in Conium maculatum L. [468, 469] are better fit, however, by a biosynthesis based upon the condensation of acetate units to a polyketooctanoic acid (i.e., 42.4).

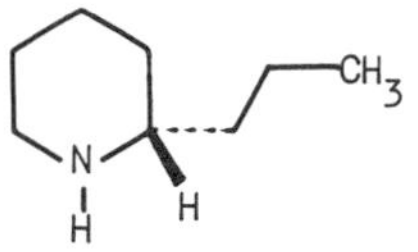

42.1 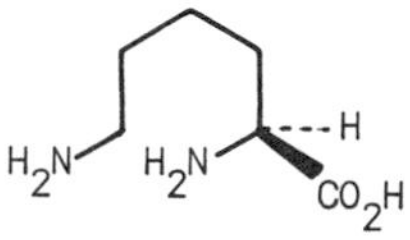 **42.2**

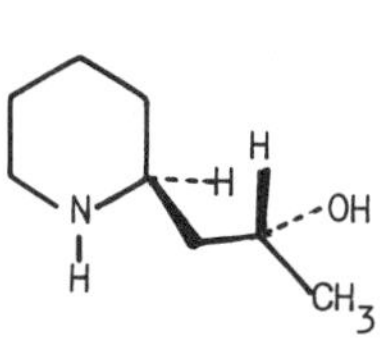

42.3 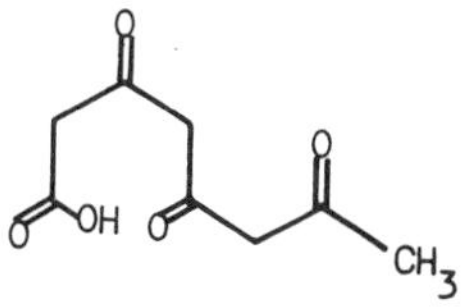 **42.4**

Two more cases warrant inclusion as examples of "surprises." First, securinine (42.5; [470], the major alkaloid of Securinega surroticosa, Rehd., is known to arise from two amino acid fragments, i.e., lysine (42.2) and tyrosine (42.6). Scheme 42.1, considerably different from all that we have seen before (in the sense of reactions at the aromatic nucleus) has been suggested and is in concert with all feeding experiments done to date. Second, for dioscorine (42.7), an alkaloid from the tropical yam, Dioscorea hispida, Dennst., it is now clear that a Δ^1-piperideine (42.8) derived from lysine (42.2) will not account for the six-membered

42.2

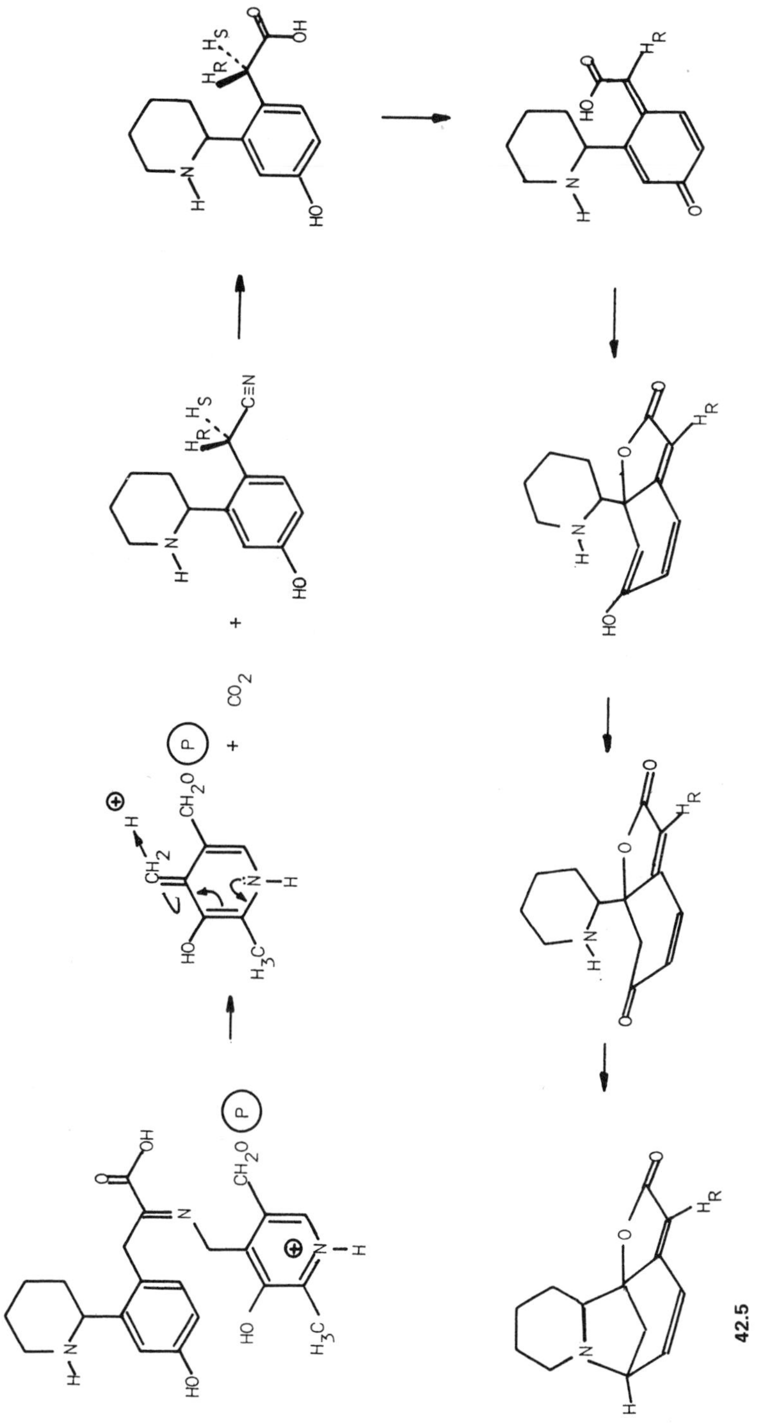

42.5

SCHEME 42.1 A possible route to securinine. (After Reference 470.)

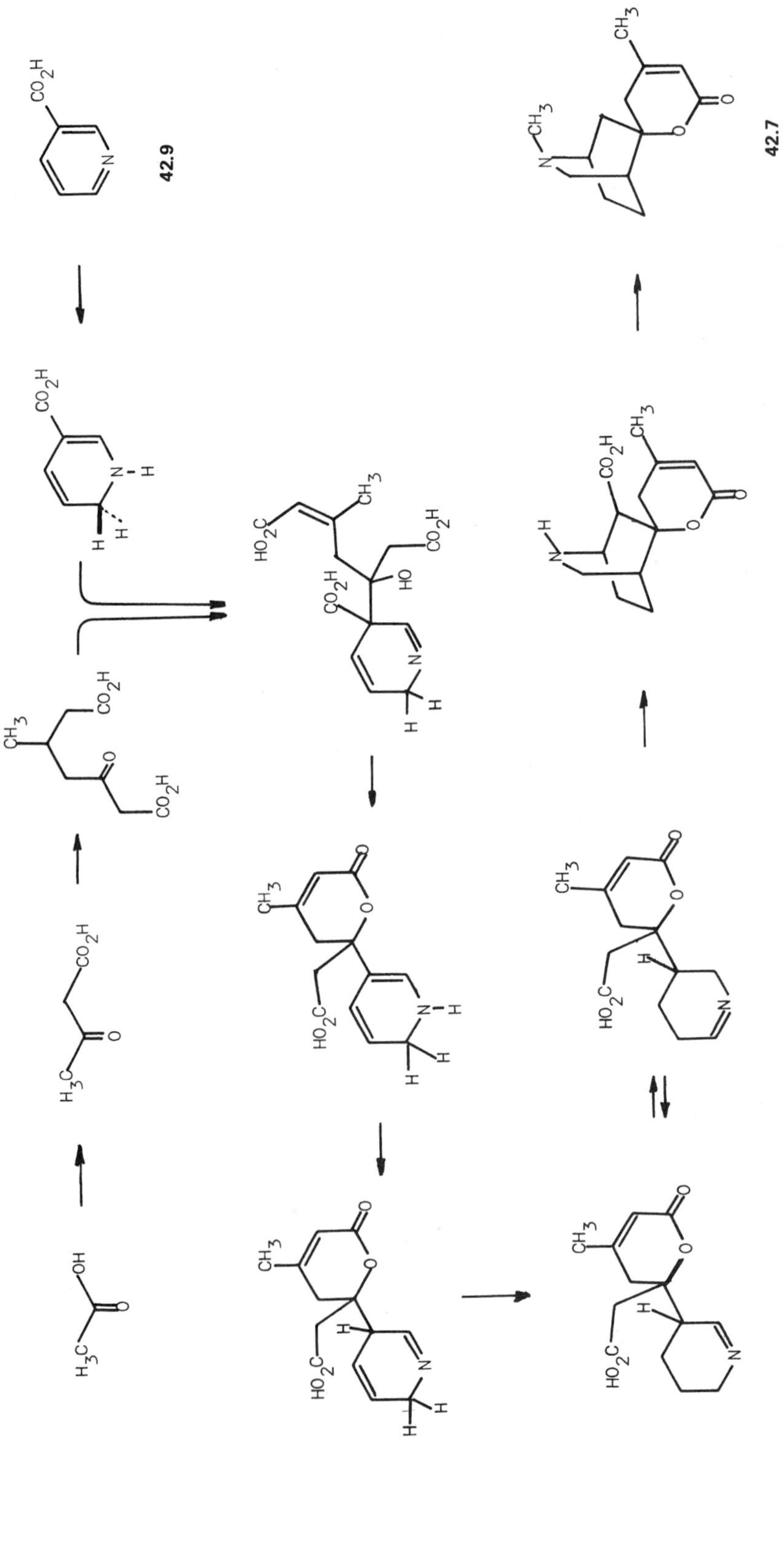

42.9

42.7

SCHEME 42.2 A possible route to dioscorine.

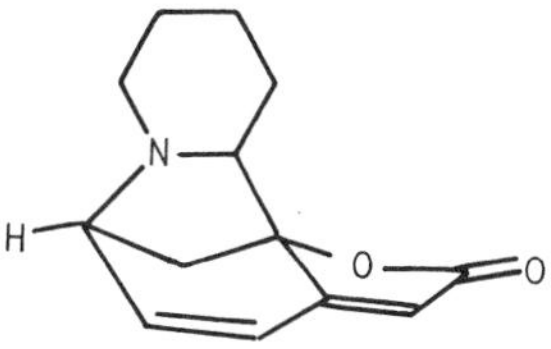

42.5

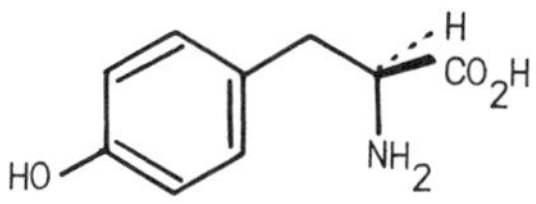

42.6

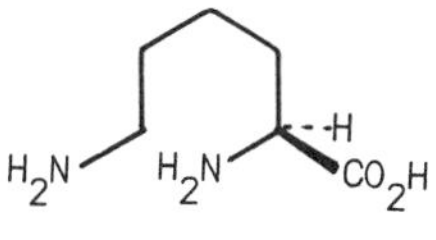

42.2

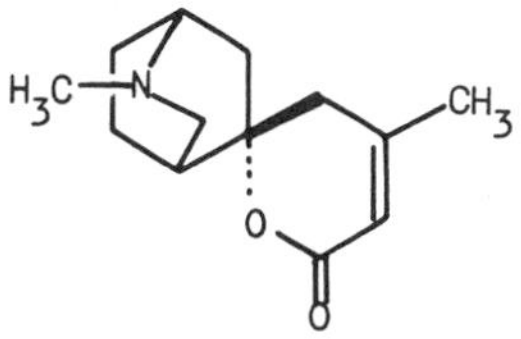

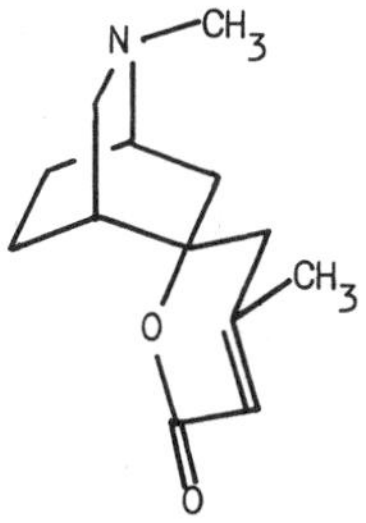

42.7

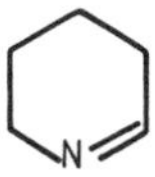

42.8

nitrogen ring. Indeed, it now appears that nicotinic acid (42.9), fed as [2-^{14}C]nicotinic acid (42.9; Scheme 42.2), is utilized for the ring while the side chain is acetate derived [471].

In conclusion, therefore, what you have learned should be treated as a slippery stepping stone to higher ground. A vantage point from where, when reached, marginal differentiation in structure is minimized and broad relationships which allow generalizations and understanding maximized. But you must remember that close inspection of the terrain of the vista before you may show stumbling blocks. The exceptions to the generalizations provided will thus be brought into relief and held forth as challenges yet to be overcome.

REFERENCES

1. G. Guroff, J. W. Daly, D. M. Herina, J. Renson, B. Witkop, and S. Udenfriend, Science, 157, 1524 (1967).
2. A. H. Soloway, J. Theor. Biol., 13, 100 (1966).
3. K. R. Hanson and I. A. Rose, Accounts Chem. Res., 8, 1 (1975).
4. L. W. Clark, in The Chemistry of Carboxylic Acids and Esters, S. Pati (ed.). Wiley, New York, 1969, pp. 598 ff.
5. H. Kohn, S. J. Benkovic, and R. A. Olofson, J. Amer. Chem. Soc., 94, 3759 (1972).
6. R. Breslow, J. Amer. Chem. Soc., 80, 3719 (1958).
7. K. R. Hanson, J. Amer. Chem. Soc., 88, 2731 (1966).
8. J. A. Bassham, in Organic Chemistry of Life, Readings from Scientific American, S. H. Freeman (ed.). San Francisco, 1973, pp. 383 ff.
9. H. R. Mahler and E. H. Cordes, Biological Chemistry. Harper & Row, New York, 1966, pp. 540 ff.
10. E. Blumenthal and J. B. M. Herbert, Trans. Faraday Soc., 41, 611 (1940).
11. I. A. Rose, in The Enzymes, Vol. 2, 3rd ed., P. O. Boyer (ed.). Academic, New York, 1970, p. 309.
12. J. R. Butler, W. L. Alworth, and M. J. Nugent, J. Amer. Chem. Soc., 96, 1617 (1974).
13. S. Escher, P. Loew, and D. Arigoni, Chem. Commun., 823 (1970).
14. A. R. Battersby, S. H. Brown, and T. G. Payne, Chem. Commun., 826 (1970).
15. A. R. Battersby, A. R. Burnett, and P. G. Parsons, Chem. Commun., 827 (1970).
16. R. Guarnaccia, L. Botta, and C. J. Coscia, J. Amer. Chem. Soc., 96, 7079 (1974).
17. R. R. Betzenberg, L. M. Hall, M. Marshall, and P. P. Cohen, J. Biol. Chem., 229, 1019 (1957).
18. E. Leistner and I. D. Spenser, J. Amer. Chem. Soc., 95, 4715 (1973).
19. I. D. Spenser, in Comprehensive Biochemistry, Vol. 20, M. Florkin and E. H. Stotz (eds.). Elsevier, New York, 1968, pp. 231 ff.
20. C. C. J. Culvenor, S. R. Jones, J. A. Lamberton, and L. W. Smith, Aust. J. Chem., 23, 1279 (1970).
21. E. Jucker and A. Lindenmann, Swiss Patent 442,318 (January, 1968).
22. A. Willstätter, Ber., 29, 943 (1896).
23. R. Adams and D. Fleš, J. Amer. Chem. Soc., 81, 4946, 5803 (1959).
24. T. A. Geissman and A. C. Weiss, J. Org. Chem., 27, 139 (1962).
25. C. C. J. Culvenor and T. A. Geissman, J. Amer. Chem. Soc., 83, 1647 (1961).
26. I. D. Spenser, Sixth Natural Products Conference. Kingston, Jamaica, January, 1976.
27. R. N. Gupta and I. D. Spenser, Can. J. Chem., 49, 384 (1971).
28. C. Schopf, H. Arm, and F. Brown, Ber., 85, 937 (1952).
29. W. M. Golebiewski and I. D. Spenser, J. Amer. Chem. Soc., 98, 6726 (1976).
30. K. Wiesner, in Progress in the Chemistry of Organic Natural Products, Vol. 20, L. Zechmeister (ed.). Springer-Verlag, Vienna, 1962, p. 271.
31. A. C. Cope, H. L. Dryden, Jr., and C. F. Howell, Org. Syn., Coll. Vol. 4, 816 (1963).

32. M. M. Robinson, W. G. Pierson, L. Dorfman, and B. F. Lambert, J. Org. Chem., 31, 3213 (1966).
33. V. Boekelheide and J. P. Lodge, Jr., J. Amer. Chem. Soc., 73, 305 (1951).
34. R. C. Cookson, Chem. Ind. (London), 337 (1953).
35. E. E. van Tamelen and R. L. Foltz, J. Amer. Chem. Soc., 82, 2400 (1960).
36. S. N. Srivastava and M. Przybylska, Tetrahedron Lett., 2697 (1968).
37. L. Mandell, K. P. Singh, J. T. Gresham, and W. J. Freeman, J. Amer. Chem. Soc., 87, 5234 (1965).
38. K. Wiesner, Z. Valenta, W. A. Ayer, L. R. Flowler, and J. E. Francis, Tetrahedron, 4, 87 (1958).
39. M. Przybylska and F. R. Ahmed, Acta Crystallogr., 11, 718 (1958).
40. K. Wiesner and L. Poon, Tetrahedron Lett., 4937 (1967).
41. G. Stork, R. A. Kretchmer, and R. H. Schlessinger, J. Amer. Chem. Soc., 90, 1647 (1968).
42. E. Leete, E. G. Gros, and T. J. Gilbertson, Tetrahedron Lett., 587 (1964).
43. E. Leete, J. Amer. Chem. Soc., 89, 7081 (1967).
44. T. J. Gilbertson and E. Leete, J. Amer. Chem. Soc., 89, 7085 (1967).
45. A. A. Liebman, B. P. Mundy, and H. Rapoport, J. Amer. Chem. Soc., 89, 664 (1967).
46. I. Zelitch, J. Biol. Chem., 240, 1869 (1965).
47. E. Leete, private communication.
48. C. R. Hutchinson, M. T. Stephen Hsia, and R. A. Carver, J. Amer. Chem. Soc., in press.
49. M. L. Rueppel, B. P. Muncy, and H. Rapoport, Phytochemistry, 13, 141 (1974).
50. E. Leete, Chem. Commun., 9 (1975).
51. E. Leete and S. A. Slattery, J. Amer. Chem. Soc., 98, 6326 (1976).
52. E. Leete, Bioorganic Chemistry, in press (1977).
53. E. Späth and G. Koller, Ber., 56, 2454 (1923).
54. A. Pinner, Ber., 26, 292 (1893).
55. A. M. Duffield, H. Budzikiewicz, and C. Djerassi, J. Amer. Chem. Soc., 87, 2926 (1965).
56. E. Späth and H. Bretschneider, Ber., 61, 327 (1928).
57. E. Leete, private communication.
58. E. Späth and L. Mamoli, Ber., 69, 1082 (1936).
59. P. M. Quan, T. K. B. Karns, and L. D. Quin, J. Org. Chem., 30, 2796 (1965).
60. K. R. Hanson and E. A. Havir, Enzyme, 7, 75 (1972).
61. E. Leete, R. M. Bowman, and M. F. Maneul, Phytochemistry, 10, 3029 (1971).
62. G. P. Basmadjian and A. G. Paul, Lloydia, 34, 91 (1971).
63. G. J. Kapadia, G. S. Ral, E. Leete, and M. G. E. Fayze, Y. N. Naishnav, and H. M. Fales, J. Amer. Chem. Soc., 92, 6943 (1970).
64. H. R. Schütte and G. Seelig, Ann., 730, 186 (1970).
65. A. R. Battersby and R. J. Parry, Chem. Commun., 901 (1971).
66. W. I. Taylor and A. R. Battersby (eds.), Oxidative Coupling of Phenols. Dekker, New York, 1967.
67. S. Agurell, I. Granelli, K. Leander, and J. Rosenblom, Acta Chem. Scand. [B] 28, 1175 (1974).
68. D. H. R. Barton and T. Cohen, Festschrift A. Stoll. Birkhäuser, Basel, 1957, p. 117.
69. R. J. Suhadolnik, A. G. Fischer, and J. Zulalian, Proc. Chem. Soc., 132 (1963).
70. C. Fuganti and M. Mazza, Chem. Commun., 1196 (1971).
71. A. I. Feinstein, Ph.D. Thesis, Iowa State University, Ames, Iowa, 1967.
72. G. W. Kirby and J. Michael, Chem. Commun., 415 (1971).
73. H. M. Fales and W. C. Wildman, J. Amer. Chem. Soc., 85, 2025 (1963).
74. W. C. Wildman and D. T. Bailey, J. Amer. Chem. Soc., 91, 150 (1969).

75. P. W. Jeffs, in MTP International Review of Science, Vol. 9, Series 1: The Alkaloids, K. Wiesner (ed.). University Park Press, Baltimore, 1974, p. 294 (and references therein).
76. W. C. Wildman, in The Alkaloids, Vol. 11, R. H. F. Manske (ed.). Academic, New York, 1968, pp. 308 ff.
77. C. Fuganti and M. Mazza, Chem. Commun., 1466 (1970).
78. A. Jindra, P. Kovacs, Z. Pittnerova, and M. Psenak, Phytochemistry, 5, 1303 (1966).
79. E. Brockmann-Hanssen, C. C. Fu, A. Y. Leung, and G. Zanati, J. Pharm. Sci., 60, 1672 (1971).
80. M. L. Wilson and J. Coscia, J. Amer. Chem. Soc., 97, 431 (1975).
81. A. R. Battersby, R. C. F. Jones, and R. Kazlauskas, Tetrahedron Lett., 1873 (1975).
82. A. R. Battersby, T. J. Brocksom, and R. Ramage, Chem. Commun., 464 (1968).
83. H. I. Parker, G. Blaochke, and K. Rapoport, J. Amer. Chem. Soc., 94, 1276 (1972).
84. R. J. Miller, C. Jolles, and H. Rapoport, Phytochemistry, 12, 597 (1973).
85. D. H. R. Barton, A. A. L. Gunatilaka, R. M. Letcher, A. M. F. T. Lobo, and D. A. Widdowson, J. Chem. Soc. [Perkin I], 874 (1973).
86. D. H. R. Barton, C. J. Potter, and D. A. Widdowson, J. Chem. Soc. [Perkin I], 346 (1974); D. H. R. Barton, R. D. Bracho, C. J. Potter, and D. A. Widdowson, J. Chem. Soc. [Perkin I], 2278 (1974).
87. D. H. R. Barton, R. B. Boar, and D. A. Widdowson, J. Chem. Soc., C, 1213 (1970).
88. F. R. Stermitz and J. N. Seiber, J. Org. Chem., 31, 2925 (1966).
89. D. W. Brown, S. F. Dyke, G. Hardy, and M. Sainsbury, Tetrahedron Lett., 1515 (1969).
90. J. R. Gear and I. D. Spenser, Can. J. Chem., 41, 783 (1963).
91. A. R. Battersby, Proc. Chem. Soc. (London), 189 (1963).
92. D. H. R. Barton, Proc. Chem. Soc. (London), 293 (1963).
93. A. R. Battersby, R. J. Francis, E. A. Reveda, and J. Staunton, Chem. Commun., 89 (1965).
94. A. R. Battersby, R. J. Francis, M. Hirst, R. Southgate, and J. Staunton, Chem. Commun., 602 (1967).
95. A. R. Battersby, J. Staunton, H. R. Wiltshire, B. J. Bircher, and C. Fuganti, J. Chem. Soc. [Perkin I], 1162 (1975).
96. H. Böhm and H. Rönsch, Z. Naturforsch., B, 23, 1552 (1968).
97. J. Hrbek, F. Šantavý, and L. Doeljš, Collect. Czech. Chem. Commun., 35, 3712 (1970).
98. M. Shamma, The Isoquinoline Alkaloids. Academic, New York, 1972, p. 462.
99. A. R. Battersby, P. Böhler, M. H. G. Munro, and R. Ramage, Chem. Commun., 1066 (1969).
100. E. Leete and P. E. Nemeth, J. Amer. Chem. Soc., 82, 6055 (1960).
101. A. R. Battersby, R. Binks, J. J. Reynolds, and D. A. Yeowell, J. Chem. Soc., 4257 (1964).
102. A. R. Battersby and R. B. Herbert, Proc. Chem. Soc. (London), 346 (1965).
103. E. Leete, Tetrahedron Lett., 333 (1965).
104. A. R. Battersby, R. B. Herbert, E. McDonald, R. Ramage, and J. H. Clements, Chem. Commun., 603 (1966).
105. A. R. Battersby, P. W. Sheldrake, and J. A. Miller, Tetrahedron Lett., 3315 (1974).
106. H. E. Miller, H. Rosler, A. Wohlpary, H. Wyler, M. E. Wilcox, H. Frohfer, T. J. Mabry, and A. S. Dreiding, Helv. Chim. Acta, 51, 1470 (1968).
107. W. La Barre, D. R. McAllister, J. S. Slotkin, O. C. Stewart, and S. Tax, Science, 114, 582 (1951).
108. E. Späth, Monatsh. Chem., 40, 129 (1919).
109. E. Späth, Monatsh. Chem., 42, 97 (1921).

110. A. Brossi, F. Schenker, and W. Leimgruber, Helv. Chim. Acta, 47, 2089 (1964).
111. E. Späth and F. Kesztler, Ber., 69, 755 (1936).
112. E. Leete, J. Amer. Chem. Soc., 88, 4218 (1966).
113. C. Djerassi, T. Nakano, and J. M. Bobbitt, Tetrahedron, 2, 58 (1958).
114. J. M. Bobbitt and T. T. Chou, J. Org. Chem., 21, 1106 (1959).
115. C. Djerassi, H. W. Brewer, C. Clarke, and L. J. Durham, J. Amer. Chem. Soc., 84, 3210 (1962).
116. C. Djerassi, S. K. Figdor, J. M. Bobbitt, and F. X. Markley, J. Amer. Chem. Soc., 79, 2203 (1957).
117. D. G. O'Donovan and H. Horan, J. Chem. Soc., C, 2791 (1968).
118. E. Späth and M. Pailer, Monatsh. Chem., 78, 348 (1948).
119. M. Pailer, Monatsh. Chem., 79, 127 (1948).
120. M. Pailer, Monatsh. Chem., 79, 331 (1948).
121. R. P. Evstigneeva and N. A. Preobrazhensky, Tetrahedron, 4, 223 (1958).
122. E. E. van Tamelen, P. W. Aldrich, and J. B. Hester, Jr., J. Amer. Chem. Soc., 81, 6214 (1959).
123. P. Böhlmann, Ber., 91, 2157 (1958).
124. A. R. Battersby, G. C. Davidson, and B. J. T. Harper, J. Chem. Soc., 1748 (1959).
125. A. R. Battersby, B. Gregory, H. Spencer, J. C. Turner, M. M. Janot, P. Potier, P. Francois, and J. Levisallas, Chem. Commun., 219 (1967).
126. O. Kennard, P. J. Roberts, N. W. Isaacs, F. H. Allen, W. D. S. Motherwell, K. H. Gibson, and A. R. Battersby, Chem. Commun., 899 (1971).
127. K. Leander and B. Luning, Tetrahedron Lett., 1393 (1968).
128. A. Brossi and S. Teitel, Helv. Chim. Acta, 54, 1564 (1971).
129. E. H. Warnhoff, Chem. Ind. (London), 1385 (1957).
130. M. Shiro, T. Sato, and H. Koyana, Chem. Ind. (London), 1229 (1966).
131. K. Torssell, Tetrahedron Lett., 623 (1974) (and references therein).
132. H. Irie, Y. Nishitani, M. Sugira, and S. Uyeo, Chem. Commun., 1313 (1970).
133. Y. Tsuda, T. Sano, J. Taga, K. Isobe, J. Toda, H. Irie, H. Tanaka, S. Takagi, M. Yamaki, and M. Murata, Chem. Commun., 933 (1975).
134. T. Kitigawa, W. I. Taylor, S. Uyeo, and H. Yajima, J. Chem. Soc., 1066 (1955).
135. S. Mizukami, Tetrahedron, 11, 89 (1960).
136. W. C. Wildman, J. Clardy, and J. A. Chan, J. Org. Chem., 37, 49 (1972).
137. S. Uyeo, T. Kitagawa, and Y. Yamamoto, Chem. Pharm. Bull. (Tokyo), 12, 408 (1964).
138. S. Kobayashi and S. Uyeo, J. Chem. Soc., 638 (1957).
139. D. J. Williams and D. Rogers, Proc. Chem. Soc., 357 (1960).
140. D. H. R. Barton and G. W. Kirby, Proc. Chem. Soc., 392 (1960); J. Chem. Soc., 806 (1962).
141. W. C. Wildman, J. Amer. Chem. Soc., 78, 4180 (1956).
142. R. J. Highet and P. F. Highet, Tetrahedron Lett., 4099 (1966).
143. H. Muxfeldt, R. S. Schneider, and J. B. Mooberry, J. Amer. Chem. Soc., 88, 3670 (1966).
144. R. V. Stevens, L. E. DuPree, Jr., and P. L. Lowenstein, J. Org. Chem., 37, 977 (1972).
145. T. Sato and H. Koyama, J. Chem. Soc., B, 1070 (1971).
146. J. Karle, J. A. Estlin, and I. L. Karle, J. Amer. Chem. Soc., 89, 6510 (1967).
147. R. W. King, C. F. Murphy, and W. C. Wildman, J. Amer. Chem. Soc., 87, 4912 (1965).
148. C. F. Murphy and W. C. Wildman, Tetrahedron Lett., 3863 (1964).
149. W. C. Wildman and D. T. Bailey, J. Amer. Chem. Soc., 89, 5514 (1967).

150. J. B. Hendrickson, T. L. Bogard, M. E. Fisch, S. Grossert, and N. Yoshimura, J. Amer. Chem. Soc., 96, 7781 (1974).
151. J. C. Clardy, F. M. Hauser, D. Dahm, R. A. Jacobson, and W. C. Wildman, J. Amer. Chem. Soc., 92, 6337 (1970).
152. H. M. Fales, D. H. S. Horn, and W. C. Wildman, Chem. Ind. (London), 1415 (1959).
153. Y. Tsuda and D. Isobe, Chem. Commun., 1555 (1971).
154. A. Brossi and S. Teitel, J. Org. Chem., 35, 3559 (1970).
155. R. K. Hill and R. M. Carlson, Tetrahedron Lett., 1157 (1964).
156. A. Pictet and A. Gams, Ber., 44, 2480 (1911).
157. F. E. King and P. L'Ecuyer, J. Chem. Soc., 427 (1937).
158. H. Corrodi and E. Hardegger, Helv. Chim. Acta, 39, 889 (1956).
159. E. Fujita, J. Pharm. Soc. Jpn., 72, 213 (1952).
160. T. Kametani, H. Iida, S. Kano, S. Tanaka, K. Fukumoto, S. Shibaya, and H. Yagi, J. Het. Chem., 4, 85 (1967).
161. A. J. Everett, L. A. Lowe, and S. Wilkinson, Chem. Commun., 1020 (1970).
162. A. R. Battersby and T. H. Brown, Proc. Chem. Soc., 85 (1964).
163. K. W. Bently and S. F. Dyke, J. Org. Chem., 22, 429 (1957).
164. V. Preininger, J. Hrbek, Jr., Z. Samek, and F. Šantavý, Arch. Pharm. (Weinheim), 302, 808 (1969).
165. J. M. Gulland and R. Robinson, J. Chem. Soc., 980 (1923).
166. D. Mackay and D. C. Hodgkins, J. Chem. Soc., 3261 (1955).
167. J. Kalvoda, P. Buchschacher, and O. Jeger, Helv. Chim. Acta, 38, 1847 (1955).
168. M. Gates and G. Tschudi, J. Amer. Chem. Soc., 78, 1380 (1956).
169. D. Elad and D. Ginsburg, J. Chem. Soc., 3052 (1954).
170. K. W. Bently and R. Robinson, J. Chem. Soc., 947 (1952).
171. M. Carmack, B. C. McKusick, and V. Prelog, Helv. Chim. Acta, 34, 1601 (1951).
172. A. Mondon and H. U. Menz, Tetrahedron, 20, 1729 (1964).
173. V. Boekelheide and G. R. Wenzinger, J. Org. Chem., 29, 1307 (1964).
174. R. K. Hill and W. R. Schearer, J. Org. Chem., 27, 921 (1962).
175. W. Nowacki and G. F. Bonsma, Z. Krist., 110, 89 (1958).
176. A. Mondon and H. J. Nestler, Angew. Chem., 76, 651 (1964).
177. M. J. Martell, Jr., T. O. Soine, and L. B. Kier, J. Amer. Chem. Soc., 85, 1022 (1963).
178. A. W. Sangster and K. L. Stuart, Chem. Rev., 65, 69 (1965).
179. A. R. Battersby and R. Binks, J. Chem. Soc., 2888 (1955).
180. A. C. Baker and A. R. Battersby, J. Chem. Soc., C, 1317 (1967).
181. J. D. Perrins, J. Chem. Soc., 15, 339 (1862).
182. W. H. Perkin, Jr., J. Chem. Soc., 57, 992 (1890).
183. W. H. Perkin, Jr., J. N. Ray, and R. Robinson, J. Chem. Soc., 127, 740 (1925).
184. R. D. Haworth and W. H. Perkin, Jr., J. Chem. Soc., 1769 (1926).
185. A. Pictet and A. Gams, Ber., 44, 2480 (1911).
186. A. R. Battersby, Proc. Chem. Soc., 189 (1963).
187. C. K. Bradsher and F. H. Day, Tetrahedron Lett., 409 (1971).
188. W. H. Perkin, Jr., J. Chem. Soc., 115, 713 (1919).
189. M. H. Benn and R. E. Mitchell, Can. J. Chem., 47, 3701 (1969).
190. H.-W. Bersch, Arch. Pharm. (Weinheim), 291, 491 (1958).
191. G. Snatzke, J. Hrbek, Jr., L. Hruban, A. Horeau, and F. Šantavý, Tetrahedron, 26, 5013 (1970).
192. W. Oppolzer and K. Keller, J. Amer. Chem. Soc., 93, 3836 (1971).
193. A. R. Battersby and H. Spencer, J. Chem. Soc., 1087 (1965).
194. W. H. Perkin, Jr., and R. Robinson, J. Chem. Soc., 99, 775 (1911).
195. A. Klásek, V. Šimánek, and F. Šantavý, Tetrahedron Lett., 4549 (1968).

196. L. Dolejš and V. Hanuš, Tetrahedron, 23, 2997 (1967).
197. M. Shamma, J. A. Weiss, S. Pfeifer, and H. Dohnert, Chem. Commun., 212 (1968).
198. W. Klötzer, S. Teitel, and A. Brossi, Helv. Chim. Acta, 54, 2057 (1971).
199. A. R. Battersby, E. McDonald, M. G. H. Munro, and R. Ramage, Chem. Commun., 934 (1967).
200. A. R. Battersby, R. B. Herbert, E. McDonald, R. Ramage, and J. H. Clements, J. Chem. Soc. (Perkin I], 1741 (1972).
201. A. R. Battersby, R. B. Bradbury, R. B. Herbert, M. H. G. Munro, and R. Ramage, J. Chem. Soc. [Perkin I], 450 (1967).
202. I. R. C. Bick, J. Harley-Mason, N. Shepard, and M. J. Vernengo, J. Chem. Soc., 1896 (1961).
203. M. V. King, J. L. DeVries, and R. Prpinshy, Acta Crystallogr., 5, 437 (1952).
204. H. Corrodi and D. Hardegger, Helv. Chim. Acta, 38, 2030 (1955).
205. E. E. van Tamelen, T. A. Spencer, Jr., D. A. Allen, Jr., and R. L. Orvis, J. Amer. Chem. Soc., 81, 6431 (1959); Tetrahedron, 14, 8 (1961).
206. T. J. Mabry, H. Wyler, I. Parikh, and A. S. Dreiding, Tetrahedron, 23, 3111 (1967).
207. T. J. Mabry, H. Wyler, G. Sassu, M. Mercier, I. Parikh, and A. S. Dreiding, Helv. Chim. Acta, 77, 640 (1962).
208. A. I. Scott, Bioorgan. Chem., 3, 398 (1974).
209. K. Bowden and L. Marion, Can. J. Chem., 29, 1037 (1951).
210. D. O'Donovan and E. Leete, J. Amer. Chem. Soc., 85, 461 (1963).
211. D. Gross, A. Nemechova, and H. R. Schütte, Z. Planzenphysiol., 57, 60 (1967); and Biochem. Physiol. Pflanz., 166, 281 (1974).
212. D. G. O'Donovan and M. F. Kenneally, J. Chem. Soc., C, 1109 (1967).
213. K. Stolle and D. Gröger, Arch. Pharm. (Weinheim), 301, 561 (1968).
214. F. Weygand and H. G. Floss, Angew. Chem., 75, 783 (1963).
215. H. G. Floss, U. Mothes, and H. Günther, Z. Naturforsch., 19B, 784 (1964).
216. H. Plieninger, H. Immel, and L. Völkl, Ann., 672, 223 (1967).
217. B. Naidoo, J. M. Cassady, G. E. Blain, and H. G. Floss, Chem. Commun., 671 (1970).
218. H. G. Floss, Tetrahedron, 32, 873 (1976).
219. I. Kompiš, M. Hesse, and H. Schmid, Lloydia, 34, 269 (1971).
220. A. I. Scott and S. L. Lee, J. Amer. Chem. Soc., 97, 6906 (1975).
221. A. I. Scott, in MTP International Review of Science, Vol. 9, D. H. Hey and K. F. Wiesner (eds.). Butterworths, London, 1973, pp. 105 ff.
222. R. Thomas, Tetrahedron Lett., 544 (1961).
223. E. Wenkert, J. Amer. Chem. Soc., 84, 98 (1962).
224. G. Popják and J. W. Cornforth, Biochem. J., 101, 553 (1966).
225. D. V. Banthorp, B. V. Charlwood, and M. J. O. Francis, Chem. Rev., 72, 115 (1972).
226. A. R. Battersby, A. R. Burnett, and P. G. Parsons, Chem. Commun., 1280 (1968).
227. A. R. Battersby, A. R. Burnett, and P. G. Parsons, Chem. Commun., 826 (1970).
228. E. Leete, Tetrahedron, 14, 35 (1961).
229. A. R. Battersby, Pure Appl. Chem., 14, 117 (1976).
230. A. R. Battersby and E. S. Hall, Chem. Commun., 793 (1969).
231. A. R. Battersby, Accounts Chem. Res., 5, 148 (1969).
232. A. I. Scott, M. B. Slayton, P. B. Reichard, and J. G. Sweeny, Bioorg. Chem., 1, 157 (1971).
233. A. Ahond, A. Cone, C. K. Fan, Y. Langlois, and P. Potier, Chem. Commun., 517 (1970).
234. A. I. Scott, Benjamin Rush Memorial Lecture, University of Pennsylvania, Philadelphia, 1975.

235. N. Kowanko and E. Leete, J. Amer. Chem. Soc., 84, 4919 (1962).
236. A. R. Battersby, R. T. Brown, R. S. Kapil, J. A. Knight, J. A. Martin, and A. O. Plunkett, Chem. Commun., 810, 888 (1966).
237. E. Leete and J. N. Wemple, J. Amer. Chem. Soc., 88, 4743 (1966).
238. S. I. Heimberger and A. I. Scott, J. Chem. Soc., 217 (1973).
239. A. R. Battersby and E. S. Hall, Chem. Commun., 793 (1960).
240. E. Wenkert and B. Wickberg, J. Amer. Chem. Soc., 87, 1580 (1965).
241. J. P. Kutney, J. F. Beck, V. R. Nelson, and R. S. Sood, J. Amer. Chem. Soc., 93, 255 (1971).
242. A. A. Gorman, M. Hesse, and H. Schmid, in Specialist Periodical Reports, The Alkaloids, 1, J. E. Saxton (Senior Reporter). The Chemical Society, London, 1971, pp. 200 ff.
243. H. R. Schütte and B. Maier, Arch. Pharm. (Weinheim), 298, 495 (1965).
244. G. W. Kirby, S. W. Shah, and E. J. Herbert, J. Chem. Soc., C, 1916 (1969).
245. E. S. Hall, F. McCapra, and A. I. Scott, Tetrahedron, 23, 4131 (1967).
246. C. R. Hutchinson, J. Amer. Chem. Soc., 96, 6806 (1974).
247. M. M. Rapport, A. A. Green, and I. H. Page, J. Biol. Chem., 108, 735 (1948).
248. M. M. Rapport, J. Biol. Chem., 108, 329 (1948).
249. K. E. Hamlin and F. E. Fischer, J. Amer. Chem. Soc., 73, 5007 (1956).
250. R. A. Abramovitch and D. Shapiro, J. Chem. Soc., 4589 (1956).
251. H. Wieland, W. Konz, and H. Mittasch, Ann., 513, 1 (1934).
252. T. Hoshino and K. Shimodaira, Ann., 520, 19 (1935).
253. A. Hofmann, R. Heim, A. Brack, and H. Kobel, Experientia, 14, 107 (1958).
254. A. Hofmann and F. Troxler, Experientia, 15, 101 (1959).
255. H. von Euler and G. Hellström, Z. Physiol. Chem., 217, 23 (1933).
256. T. Wieland and C. Y. Hsing, Ann., 526, 188 (1936).
257. H. Kuhn and O. Stein, Ber., 70, 567 (1937).
258. J. Fritzsch, Ann., 64, 360 (1848).
259. P. S. Massayetov, J. Gen. Chem. USSR, 16, 139 (1946); Chem. Abstr., 40, 6754 (1946).
260. E. Späth and E. Lederer, Ber., 63, 120 (1930).
261. A. Hofmann, Botanical Museum Leaflets, Harvard University, 20, 194 (1963).
262. A. Stoll, A. Hofmann, and F. Troxler, Helv. Chim. Acta, 32, 506 (1949).
263. P. A. Stadler and A. Hofmann, Helv. Chim. Acta, 45, 2005 (1962).
264. E. D. Kornfeld, E. J. Fornefeld, G. B. Kline, M. J. Mann, R. G. Jones, and R. B. Woodward, J. Amer. Chem. Soc., 76, 5356 (1954); 78, 3087 (1956).
265. R. J. Gould, Jr., L. C. Craig, and W. A. Jacobs, J. Biol. Chem., 145, 487 (1942).
266. P. C. Julian, W. J. Karpel, A. Magnani, and E. W. Meyer, J. Amer. Chem. Soc., 70, 180 (1948).
267. G. R. Clemo and G. A. Swan, J. Chem. Soc., 617 (1946).
268. E. Schittler and R. Speitel, Helv. Chim. Acta, 31, 1199 (1948).
269. G. A. Swan, J. Chem. Soc., 1543 (1950).
270. E. E. van Tamelen, M. Shamma, A. W. Burgstahler, J. Wolinshy, R. Tamm, and W. P. Aldrich, J. Amer. Chem. Soc., 80, 5006 (1958).
271. B. Witkop, J. Amer. Chem. Soc., 71, 2559 (1949).
272. Y. Ban and O. Yonemitsu, Chem. Ind. (London), 948 (1961).
273. C. Djerassi, R. Riniker, and B. Riniker, J. Amer. Chem. Soc., 78, 6362 (1956).
274. V. Prelog, Helv. Chim. Acta, 36, 308 (1953).
275. J. D. Morrison and H. S. Mosher, Asymmetric Organic Reactions. Prentice-Hall, Englewood Cliffs, N.J., 1971, pp. 50 ff.
276. W. Konigs, Ber., 27, 900 (1894).
277. W. von Miller and G. Rohde, Ber., 33, 3214 (1900).

278. P. Rabe, E. Ackermann, and W. Schneider, Ber., 40, 3655 (1907).
279. A. Wohl and R. Maag, Ber., 42, 627 (1909).
280. V. Prelog and E. Zalán, Helv. Chim. Acta, 27, 535 (1944).
281. L. H. Welsh, J. Amer. Chem. Soc., 71, 3500 (1949).
282. J. M. Bijvoet, A. F. Peerdeman, and A. J. van Bommel, Nature, 168, 271 (1951).
283. O. L. Carter, A. T. McPhail, and G. A. Sim, J. Chem. Soc., A, 365 (1967).
284. R. B. Woodward and W. E. von Doering, J. Amer. Chem. Soc., 67, 860 (1945).
285. M. Uskoković, J. Gutzwiller, and T. Henderson, J. Amer. Chem. Soc., 92, 203 (1970).
286. E. C. Taylor and S. F. Martin, J. Amer. Chem. Soc., 94, 4218 (1972).
287. P. Rabe and K. Kindler, Ber., 51, 466 (1918).
288. E. Wenkert and N. V. Bringi, J. Amer. Chem. Soc., 81, 1474 (1959).
289. M.-M. Janot and R. Goutarel, Bull. Soc. Chim. Fr., 588 (1951).
290. E. Ochiai and M. Ishikawa, Pharm. Bull. Jpn., 6, 208 (1958).
291. E. Ochiai and M. Ishikawa, Tetrahedron, 7, 228 (1959).
292. E. Wenkert and D. K. Roychaudhuri, J. Amer. Chem. Soc., 80, 1613 (1958).
293. E. Wenkert, B. Wickberg, and C. L. Leicht, J. Amer. Chem. Soc., 83, 5038 (1961).
294. E. E. van Tamelen and C. Placeway, J. Amer. Chem. Soc., 83, 2594 (1961).
295. J. M. Müller, E. Schlittler, and H. J. Bein, Experientia, 8, 338 (1952).
296. N. Neuss, H. E. Boaz, and J. W. Forbes, J. Amer. Chem. Soc., 76, 2463 (1954).
297. C. F. Huebner, H. B. MacPhillamy, A. F. St. André, and E. Schlittler, J. Amer. Chem. Soc., 77, 472 (1955).
298. C. F. Huebner, A. F. St. André, E. Schlittler, and A. Uffer, J. Amer. Chem. Soc., 77, 5725 (1955).
299. J. N. Shoolery and W. E. Rosen, J. Amer. Chem. Soc., 83, 2816 (1961).
300. R. B. Woodward, F. E. Bader, H. Bickel, A. J. Frey, and R. W. Kierstead, Tetrahedron, 2, 1 (1958).
301. S. Siddiqui and R. H. Siddiqui, J. Indian Chem. Soc., 8, 667 (1931).
302. D. Mukherji, R. Robinson, and E. Schlittler, Experientia, 5, 215 (1949).
303. R. B. Woodward, Angew. Chem., 68, 13 (1956).
304. J. E. Saxton, in The Alkaloids, Vol. 7, R. H. F. Manske (ed.). Academic, New York, 1960, pp. 1 ff.
305. M. F. Bartlett, R. Sklar, W. I. Taylor, E. Schlittler, R. L. S. Amai, P. Beak, N. V. Bringi, and E. Wenkert, J. Amer. Chem. Soc., 84, 622 (1962).
306. S. Masamune, S. K. Ang, C. Egli, N. Nakatsuka, S. K. Sarkar, and Y. Yasumari, J. Amer. Chem. Soc., 89, 2506 (1967).
307. A. Stoll and A. Hofmann, Helv. Chim. Acta, 36, 1143 (1953).
308. K. Bodendorg and H. Eder, Naturwissenschaften, 40, 342 (1953).
309. W. Arnold, W. von Philipsborn, H. Schmid, and P. Karrer, Helv. Chim. Acta, 40, 705 (1957).
310. S. Silvers and A. Tulinsky, Tetrahedron Lett., 339 (1962).
311. W. Klyne and J. Buckingham, Atlas of Stereochemistry. Oxford University Press, New York, 1974, p. 150.
312. E. E. van Tamelen and L. K. Oliver, J. Amer. Chem. Soc., 92, 2136 (1970).
313. O. Hesse, Ann., 203, 144 (1880).
314. J. A. Goodson and T. A. Henry, J. Chem. Soc., 127, 1640 (1925).
315. T. A. Govindachari and S. Rajappa, Proc. Chem. Soc., 134 (1959).
316. H. Manohar and S. Ramaseshan, Tetrahedron Lett., 814 (1961).
317. J. A. Hamilton, T. A. Hamor, J. M. Robertson, and G. A. Sim, J. Chem. Soc., 5061 (1962).
318. J. E. Saxton, in The Alkaloids, Vol. 8, R. H. F. Manske (ed.). Academic, New York, 1965, pp. 174 ff.

319. T. R. Govindachari and S. Rajappa, Tetrahedron, 15, 132 (1962).
320. R. Robinson, in Progress in Organic Chemistry, Vol. 1, J. W. Cook (ed.). Academic, New York, 1952, pp. 1 ff.
321. H. L. Holmes, in The Alkaloids, Vol. 1, R. H. F. Manske (ed.). Academic, New York, 1950, pp. 375 ff.
322. H. L. Holmes, in The Alkaloids, Vol. 2, R. H. F. Manske (ed.). Academic, New York, 1952, pp. 513 ff.
323. J. B. Hendrickson, in The Alkaloids, Vol. 6, R. H. F. Manske (ed.). Academic, New York, 1960, pp. 179 ff.
324. G. R. Clemo and T. P. Metcalf, J. Chem. Soc., 1518 (1937).
325. A. Hanssen, Ber., 20, 451 (1887).
326. H. Leuchs, Ber., 41, 1711 (1908).
327. H. Wieland and K. Kaziro, Ann., 506, 60 (1933).
328. A. F. Peerdeman, Acta Crystallogr., 9, 824 (1956).
329. R. B. Woodward, M. P. Cava, W. D. Ollis, A. Hunger, H. U. Daeniker, and K. Schenker, Tetrahedron, 19, 247 (1963).
330. N. Neuss and M. Gorman, Tetrahedron Lett., 206 (1961).
331. N. Camerman and J. Trotter, Acta Crystallogr., 17, 384 (1964).
332. G. H. Svoboda, I. S. Johnson, M. Gorman, and N. Neuss, J. Pharm. Sci., 51, 707 (1962).
333. M. Gorman and N. Neuss, 144th Meeting of the American Chemical Society, Los Angeles, California, 1963, p. 38M, as reported by W. I. Taylor, in The Alkaloids, Vol. 8, R. H. F. Manske (ed.). Academic, New York, 1965, p. 218.
334. M. F. Bartlett, D. F. Dickel, and W. I. Taylor, J. Amer. Chem. Soc., 80, 129 (1958).
335. G. Arai, J. Coppola, and G. A. Jeffrey, Acta Crystallogr., 13, 553 (1960).
336. G. Büchi, D. L. Coffen, K. Kocsis, P. E. Sonnet, and R. E. Siegler, J. Amer. Chem. Soc., 88, 3099 (1966).
337. G. Büchi, P. Kulsa, K. Ogasawara, and R. L. Rosati, J. Amer. Chem. Soc., 92, 999 (1970).
338. M. M. Janot, H. Pourrat, and J. LeMen, Bull. Soc. Chim. Fr., 707 (1954).
339. C. Djerassi, S. E. Flores, H. Budzikiewicz, J. M. Silson, L. J. Durham, J. LeMen, M. M. Janot, M. Plat, M. Gorman, and N. Neuss, Proc. Nat. Acad. Sci. U.S., 48, 113 (1962).
340. K. Biemann and G. Spiteller, J. Amer. Chem. Soc., 84, 4578 (1962).
341. G. Stork and J. E. Dolfini, J. Amer. Chem. Soc., 85, 2872 (1963).
342. B. M. Craven and D. E. Zacharias, Experientia, 24, 770 (1968).
343. F. E. Ziegler and G. B. Bennet, J. Amer. Chem. Soc., 93, 5930 (1971).
344. O. Clauder, K. Gesztes, and K. Szasz, Tetrahedron Lett., 1147 (1962).
345. J. Trojánek, O. Štrouf, J. Holubek, and Z. Čehan, Tetrahedron Lett., 707 (1961).
346. E. Wenkert and B. Wickberg, J. Amer. Chem. Soc., 87, 1580 (1965).
347. E. Wenkert, J. Amer. Chem. Soc., 84, 98 (1962).
348. J. Trojánek, Z. Koblicová, and D. Bláha, Chem. Ind. (London), 1261 (1965).
349. M. F. Bartlett and W. I. Taylor, J. Amer. Chem. Soc., 82, 5941 (1960).
350. M. E. Kuehne, J. Amer. Chem. Soc., 86, 2946 (1964).
351. H. M. Gordin, J. Amer. Chem. Soc., 27, 1418 (1905).
352. T. A. Hamor, J. M. Robertson, H. N. Shrinastava, and J. V. Silverton, Proc. Chem. Soc., 78 (1960).
353. E. S. Hall, F. McCapra, and A. I. Scott, Tetrahedron, 23, 4131 (1967).
354. L. Marion and R. H. F. Manske, Can. J. Res., B16, 432 (1938).
355. T. Lobayashi and R. Kikumoto, Tetrahedron, 18, 813 (1962).
356. S. R. Mason and G. W. Vane, J. Chem. Soc., B, 370 (1966).

357. E. S. Hall, F. McCapra, and A. I. Scott, Tetrahedron, 20, 565 (1964).
358. H. Wieland, K. Bähr, and B. Witkop, Ann., 547, 156 (1941).
359. H. Asmis, P. Waser, H. Schmid, and P. Karrer, Helv. Chim. Acta, 38, 1661 (1955).
360. H. Asmis, E. Bächli, H. Schmid, and P. Karrer, Helv. Chim. Acta, 37, 1993 (1954).
361. K. Bernauer, F. Berlage, W. von Philipsborn, H. Schmid, and P. Karrer, Helv. Chim. Acta, 41, 2293 (1958).
362. C. Weissmann, O. Heshmet, K. Bernauer, H. Schmid, and P. Karrer, Helv. Chim. Acta, 43, 1165 (1960).
363. H. A. Hiltebrand, Dissertation, Universität Zurich, 1964.
364. J. W. Moncreif and W. N. Lipscomb, J. Amer. Chem. Soc., 87, 4963 (1965).
365. N. Neuss, M. Gorman, W. Hargrove, N. J. Cone, K. Biemann, G. Büchi, and R. E. Manning, J. Amer. Chem. Soc., 86, 1440 (1964).
366. R. L. Nobel, D. T. Baer, and J. H. Cutts, Ann. N.Y. Acad. Sci., 76, 882 (1958).
367. R. Neuss, M. Gorman, H. E. Goaz, and N. J. Cone, J. Amer. Chem. Soc., 84, 1509 (1962).
368. M. Gorman, N. Neuss, and K. Biemann, J. Amer. Chem. Soc., 84, 1058 (1962).
369. J. Kutney, private communication.
370. N. Langlois, F. Guéritte, Y. Langlois, and P. Potier, J. Amer. Chem. Soc., 98, 7017 (1976).
371. H. Auda, H. R. Juneja, E. J. Eisenbraun, G. R. Waller, W. R. Kays, and H. H. Appel, J. Amer. Chem. Soc., 89, 2467 (1967).
372. H. H. Appel and P. M. Streeter, Rev. Lat. Amer. Quim., 1, 63 (1970); Chem. Abstr., 74, 72815v (1971).
373. D. Gross, W. Berg, and H. R. Schütte, Biochem. Physiol. Pflanz., 163, 576 (1972).
374. Y. Yamazaki, N. Matsuo, and K. Arai, Chem. Pharm. Bull. (Tokyo), 14, 1058 (1966).
375. A. Corbella, P. Gariboldi, and G. Jommi, Chem. Commun., 729 (1973); A. Corbella, P. Gariboldi, G. Jommi, and M. Sisti, Chem. Commun., 288 (1975).
376. D. Arigoni, Pure Appl. Chem., 41, 219 (1975).
377. O. E. Edwards, J. L. Douglas, and B. Mootoo, Can. J. Chem., 48, 2517 (1970).
378. E. Wenkert, Chem. Ind. (London), 282 (1955).
379. W. B. Whalley, Tetrahedron, 18, 43 (1962).
380. Z. Valenta and K. Wiesner, Chem. Ind. (London), 354 (1956).
381. E. J. Herbert and G. W. Kirby, Tetrahedron Lett., 1505 (1963).
382. G. Popják, DeW. S. Goodman, J. W. Cornforth, R. H. Cornforth, and R. Ryhage, J. Biol. Chem., 236, 1934 (1961).
383. W. W. Epstein and H. C. Rilling, J. Biol. Chem., 245, 4597 (1970).
384. L. J. Altman, L. Ash, R. C. Kowerski, W. W. Epstein, B. R. Larson, H. C. Rilling, F. Muscio, and D. E. Gregonis, J. Amer. Chem. Soc., 94, 3257 (1972).
385. G. Popják, J. Edmond, and S. M. Wong, J. Amer. Chem. Soc., 95, 2713 (1973).
386. C. D. Poulter, O. J. Muscio, and R. J. Goodfellow, Biochemistry, 13, 1530 (1974).
387. D. H. R. Barton, A. F. Gosden, G. Mellows, and D. A. Widdowson, Chem. Commun., 1067 (1968).
388. T. T. Tchen and K. Bloch, J. Amer. Chem. Soc., 77, 6085 (1955).
389a. R. B. Clayton and K. Bloch, J. Biol. Chem., 218, 319 (1956).
389b. L. J. Goad, Symp. Biochem. Soc., 29, 45 (1970).
390. R. K. Maudgal, T. T. Tchen, and K. Bloch, J. Amer. Chem. Soc., 80, 2589 (1958).
391. J. W. Cornforth, R. H. Cornforth, A. Pelter, M. G. Horning, and G. Popják, Proc. Chem. Soc., 112 (1958).
392. J. W. Cornforth, R. H. Cornforth, A. Pelter, M. G. Horning, and G. Popják, Tetrahedron, 5, 311 (1959).
393a. J. H. Richards and J. B. Hendrickson, The Biosynthesis of Steroids, Terpenes and Acetogenins. W. A. Benjamin, New York, 1964, Chap. 10.

393b. R. B. Clayton, in Aspects of Terpenoid Chemistry and Biochemistry, T. W. Goodwing (ed.). Academic, New York, 1971, p. 1 ff.
394. Y. Sato and H. G. Latham, J. Amer. Chem. Soc., 78, 3146 (1956).
395. A. R. Guseva, V. A. Paseschnichenko, and M. G. Borikhina, Biokhimiya, 26, 723 (1961); 27, 853 (1962).
396. A. R. Guseva, V. A. Paseschnichenko, and M. G. Borikhina, Prikl. Biokhim. Mikrobiol., 9, 764 (1973).
397. S. J. Jadav, D. K. Salunkhe, R. E. Wyse, and R. R. Dalvi, J. Food Sci., 38, 453 (1973).
398. T. Tschesche and H. Hulpke, Z. Naturforsch., B, 22, 791 (1967).
399. H. Ripperger, W. Moritz, and K. Schreiber, Phytochemistry, 10, 2699 (1971).
400a. J. D. Ehrhardt, L. Hirth, and G. Ourisson, Phytochemistry, 6, 815 (1967).
400b. L. J. Altman, C. Y. Han, A. Bertolino, G. Handy, D. Laungani, W. Muller, S. Schwartz, D. Shanker, W. H. de Wolf, and F. Yang, J. Amer. Chem. Soc., 100, 3235 (1978).
401. W. C. Wildman, J. LeMen, and K. Wiesner, in Cyclopentanoid Terpene Derivatives, W. I. Taylor and A. R. Battersby (eds.). Dekker, New York, 1969, pp. 239 ff.
402. P. J. Lentz and M. G. Rossman, Chem. Commun., 1269 (1969).
403. C. Djerassi, J. P. Kutney, and M. Shamma, Tetrahedron, 18, 183 (1962).
404. T. Sakan, A. Fujino, F. Murai, Y. Butsugan, and A. Suzue, Bull. Chem. Soc. Jpn., 32, 315 (1959).
405. E. J. Eisenbraun, A. Bright, and H. H. Appel, Chem. Ind. (London), 1242 (1962).
406. S. M. McElvain and E. J. Eisenbraun, J. Amer. Chem. Soc., 77, 1599 (1955).
407. R. B. Bates, E. J. Eisenbraun, and S. M. McElvain, J. Amer. Chem. Soc., 80, 3420 (1958).
408. G. Jones, G. Ferguson, and W. C. Marsh, Chem. Commun., 994 (1971).
409. G. Jones, H. M. Fales, and W. C. Wildman, Tetrahedron Lett., 397 (1963).
410. O. E. Edwards, in Cyclopentanoid Terpene Derivatives, W. I. Taylor and A. R. Battersby (eds.). Dekker, New York, 1969, pp. 357 ff.
411. K. Oda and H. Koyama, J. Chem. Soc., B, 1450 (1970).
412. M. Kotake, I. Kawasaki, T. Okamoto, S. Matsutani, S. Kusumoto, and T. Kaneko, Bull. Chem. Soc. Jpn., 35, 1335 (1962).
413. M. Kotake, S. Kusumoto, and T. Ohara, Ann., 606, 148 (1957).
414. D. C. Aldridge, J. J. Armstrong, R. N. Speake, and W. B. Turner, J. Chem. Soc., C, 1667 (1967).
415. Y. Arata and T. Ohashi, J. Pharm. Soc. Jpn., 77, 229 (1957).
416. Y. Arata, J. Pharm. Soc. Jpn., 77, 225 (1957).
417. Y. Arata, M. Koseki, and K. Sakai, J. Pharm. Soc. Jpn., 77, 232 (1957).
418. T. Bohlmann, E. Winterfeldt, P. Studt, H. Laurent, G. Boroschewski, and K. M. Klein, Ber., 94, 3151 (1961).
419. H. Suzuke, I. Keimatsu, and M. Ito, J. Pharm. Soc. Jpn., 54, 138, 146 (1934).
420. Y. Inubushi, Y. Sasaki, Y. Tsuda, B. Yasui, T. Konita, J. Matsumoto, E. Katarao, and J. Nakano, Tetrahedron, 20, 2007 (1964).
421. Y. Inubushi, Y. Sasaki, Y. Tsuda, and J. Nakano, Tetrahedron Lett., 1519 (1965).
422. Y. Inubshi, E. Katarao, Y. Tsuda, and B. Yasui, Chem. Ind. (London), 1689 (1964).
423. Y. Inubushi, T. Kiruchi, T. Ibuka, K. Tanaka, and K. Tokane, Chem. Commun., 1252 (1972).
424. K. Wiesner and Z. Valenta, in Progress in the Chemistry of Organic Natural Products, Vol. 16, L. Zechmeister (ed.). Springer-Verlag, Vienna, 1958.
425. H. Vorbruggen and C. Djerassi, J. Amer. Chem. Soc., 84, 2990 (1960).
426. K. Wiesner, S. K. Figdor, M. F. Bartlett, and D. R. Henderson, Can. J. Chem., 30, 608 (1952).

427. K. Wiesner, W. I. Taylor, S. K. Figdor, M. F. Bartlett, J. R. Armstrong, and J. A. Edwards, Ber., 86, 800 (1953).
428. K. Wiesner, J. R. Armstrong, M. F. Bartlett, and J. A. Edwards, J. Amer. Chem. Soc., 76, 6068 (1954).
429. P. K. Grant and R. Hodges, Tetrahedron, 8, 261 (1960).
430. L. H. Briggs, B. F. Cain, B. R. Davis, and J. K. Wilmhurst, Tetrahedron Lett., 8 (1959).
431. W. Nagata, T. Sugasawa, M. Narisada, T. Wakabayashi, and Y. Hayase, J. Amer. Chem. Soc., 89, 1483 (1967).
432. W. Nagata, M. Narisada, T. Wakabayashi, and T. Sugasawa, J. Amer. Chem. Soc., 89, 1499 (1967).
433. S. W. Pelletier and K. Kawazu, Chem. Ind. (London), 1879 (1963).
434. S. W. Pelletier, K. Kawazu, and K. W. Gopinath, J. Amer. Chem. Soc., 87, 5229 (1965).
435. K. Wiesner, J. R. Armstrong, M. F. Bartlett, and J. A. Edwards, Chem. Ind. (London), 132 (1954).
436. S. W. Pelletier and D. M. Locke, J. Amer. Chem. Soc., 87, 761 (1965).
437. D. Dvornik and O. E. Edwards, Can. J. Chem., 42, 137 (1964).
438. S. W. Pelletier, Tetrahedron, 14, 76 (1961).
439a. J. W. ApSimon and O. E. Edwards, Can. J. Chem., 40, 896 (1962).
439b. N. N. Girotra and L. H. Zalkow, Tetrahedron, 21, 101 (1965).
440. S. W. Pelletier and W. A. Jacobs, J. Amer. Chem. Soc., 78, 4139 (1956).
441. S. W. Pelletier and W. A. Jacobs, J. Amer. Chem. Soc., 78, 4144 (1956).
442. K. Wiesner, M. Götz, D. L. Simmons, L. R. Fowler, F. W. Bachelor, R. F. C. Brown, and G. Büchi, Tetrahedron Lett., 15 (1959).
443. M. Przybylska and L. Marion, Can. J. Chem., 37, 1117, 1843 (1959).
444. R. B. Turner, J. P. Jeschke, and M. S. Gibson, J. Amer. Chem. Soc., 32, 5182 (1960).
445. P. DeMayo, Molecular Rearrangements, Part 2. Wiley Interscience, New York, 1964, p. 931.
446. R. Aneja and S. W. Pelletier, Tetrahedron Lett., 669 (1964).
447. R. Aneja and S. W. Pelletier, Tetrahedron Lett., 215 (1965).
448. M. Przybylska, Can. J. Chem., 41, 2991 (1963).
449. O. E. Edwards, Chem. Commun., 318 (1965).
450. H. Höhne, S. Schreiber, H. Ripperger, and H. H. Worch, Tetrahedron, 22, 673 (1966).
451. F. C. Uhle and F. Sallmann, J. Amer. Chem. Soc., 83, 1190 (1960).
452. S. V. Kessar, Y. P. Gupta, and A. L. Rampal, Tetrahedron Lett., 4319 (1966).
453. S. V. Kessar, R. K. Mahajan, S. S. Bandhi, and A. L. Rampal, Tetrahedron Lett., 1547 (1968).
454. A. Butenandt and J. Schmidt-Thomé, Ber., 71, 1487 (1938); 72, 182 (1939).
455. J. W. Cornforth and R. Robinson, J. Chem. Soc., 1855 (1949).
456. H. M. E. Cardwell, J. W. Cornforth, S. R. Duff, H. Holtermann, and R. Robinson, J. Chem. Soc., 361 (1953).
457. A. Butenandt and H. Dannenberg, Ber., 69, 1158 (1936).
458. L. Ruzicka, P. A. Plattner, and A. Eschbacker, Helv. Chim. Acta, 21, 866 (1938).
459. R. D. Haworth, J. McKenna, and N. Singh, J. Chem. Soc., 831 (1949).
460. S. Siddiqui, J. Indian Chem. Soc., 11, 283 (1934).
461. H. Favre, R. D. Haworth, J. McKenna, R. G. Powell, and G. H. Whitfield, J. Chem. Soc., 1115 (1953).
462. H. Favre and B. Marinier, Can. J. Chem., 36, 429 (1958).

463. G. Stork, S. D. Darling, I. T. Harrison, and P. S. Wharton, J. Amer. Chem. Soc., 84, 2018 (1962).
464. K. S. Brown and S. M. Kupchan, J. Amer. Chem. Soc., 86, 4414 (1964).
465. K. S. Brown and S. M. Kupchan, J. Amer. Chem. Soc., 86, 4424 (1964).
466. D. H. R. Barton, R. P. Budhiraja, and J. F. McGhie, Proc. Chem. Soc., 170 (1963).
467. J. S. G. Cox, R. E. King, and T. J. King, J. Chem. Soc., 514 (1959).
468. E. Leete, J. Amer. Chem. Soc., 86, 2509 (1964).
469. E. Leete, Accounts Chem. Res., 4, 100 (1971).
470. W. M. Golebiewshi, P. Horeswood, and I. D. Spenser, Chem. Commun., 217 (1976). I am deeply grateful to Professor Spenser for providing this material prior to publication.
471. E. Leete, J. Amer. Chem. Soc., 99, 648 (1977).
472. S. J. Pelletier and N. V. Modey, J. Amer. Chem. Soc., 99, 284 (1977).
473. O. Møller, E. M. Steinberg, and K. Torssell, Acta Chem. Scand., B32, 98 (1978).
474a. K. Hermann and A. S. Dreiding, Helv. Chim. Acta, 60, 673 (1977).
474b. G. Büchi, H. Fliri, and R. Shapiro, J. Org. Chem. 43, 4765 (1978).
475. R. T. Brown, J. Leonard, and S. K. Sleigh, Chem. Commun., 636 (1977).
476. J. Stöckigt and M. H. Zenk, Chem. Commun., 646 (1977).
477. E. Leete, J. Org. Chem. 44, 165 (1979).
478. Atta-ur-Rahman, A. Basha, and M. Ghazala, Tetrahedron Lett., 2351 (1976).

AUTHOR INDEX

The numbers given below refer to the page(s) on which an author's work is cited, whether or not the author is referred to by name on that page.

SUBJECT INDEX